Lecture Notes in Computer Science 9015

Commenced Publication in 1973
Founding and Former Series Editors:
Gerhard Goos, Juris Hartmanis, and Jan van Leeuwen

Yevgeniy Dodis Jesper Buus Nielsen (Eds.)

Theory
of Cryptography

12th Theory of Cryptography Conference, TCC 2015
Warsaw, Poland, March 23-25, 2015
Proceedings, Part II

 Springer

Volume Editors

Yevgeniy Dodis
New York University, Department of Computer Science
251 Mercer Street, New York, NY 10012, USA
E-mail: dodis@cs.nyu.edu

Jesper Buus Nielsen
Aarhus University, Department of Computer Science
Åbogade 34, 8200 Aarhus N, Denmark
E-mail: jbn@cs.au.dk

ISSN 0302-9743   e-ISSN 1611-3349
ISBN 978-3-662-46496-0   e-ISBN 978-3-662-46497-7
DOI 10.1007/978-3-662-46497-7
Springer Heidelberg New York Dordrecht London

Library of Congress Control Number: 2015933013

LNCS Sublibrary: SL 4 – Security and Cryptology

Typesetting: Camera-ready by author, data conversion by Scientific Publishing Services, Chennai, India

Printed on acid-free paper

Springer is part of Springer Science+Business Media (www.springer.com)

Preface

The 2015 Theory of Cryptography Conference (TCC) was held at the Sheraton Warsaw Hotel in Warsaw, Poland, during March 23–25. TCC 2015 was sponsored by the International Association for Cryptologic Research (IACR). The general chair of the conference was Stefan Dziembowski. We would like to thank him for his hard work in organizing the conference.

The conference received 137 submissions, a record number for TCC. Each submission was reviewed by at least three Program Committee (PC) members. Because of the large number of submissions and the very high quality, the PC decided to accept 52 papers, a significant extension of capacity over previous TCCs. Still the PC had to reject many good papers. After the acceptance notification, authors of the accepted papers were given three weeks to revise their paper in response to the reviews. The revisions were not reviewed. The present proceedings consists of the revised versions of the 52 accepted papers.

The submissions were reviewed by a PC consisting of 25 leading researchers from the field. A complete list of the PC members can be found after this preface. Each PC member was allowed to submit two papers. PC-authored papers were held to a higher standard. Initially each paper was given at least three independent reviews (four for PC-authored papers), using external reviewers when appropriate. Following the individual review period there was a discussion phase. This year the review software was extended to allow PC members to interact with authors by sending them questions directly from the discussion page of a submission, and having the answer automatically appear on the discussion page. In order to minimize the chance of making decisions based on misunderstandings, the PC members were strongly encouraged to use this "chat" feature to discuss with the authors all technical issues that arose during the review and the discussion phases. As a result, the feature was used extensively where appropriate, and completely replaced (what we felt was) a much more limited and less effective "rebuttal" phase that was used in recent CRYPTO and Eurocrypt conferences. In particular, this allowed the PC members to spend their effort on the issues that were most likely to matter at the end. We believe that our experiment with the increased interaction with authors was a great success, and that it gives a better quality-to-effort ratio than a rebuttal phase. Thus, we encourage future program chairs to continue increasing the interaction with the authors. This year we also experimented with cross-reviews, letting authors of similar submissions comment on the relation between these submissions. This was less of a success. Although the chance to compare the other submissions was welcomed by some authors, the cross-reviews were found to be controversial by other authors, and it is not clear that the comparisons contributed much more than having a dedicated PC member read all the papers and form an independent opinion.

We would like to thank the PC for their dedication, high standards, and hard work. Indeed, most of the PC members truly went above and beyond. Having such a great PC made it easy to chair the conference. We would also like to thank all the external reviewers who decided to dedicate their time and effort to reviewing for TCC 2015. Your help was indispensable. A list of all external reviewers follows this preface. We apologise for any omissions. Great thanks to Shai Halevi for developing and maintaining the *websubrev* software, which was an invaluable help in running the PC. Thanks in particular for extending the system with a "chat" feature. Tuesday evening the conference had a rump session chaired by Krzysztof Pietrzak from IST Austria. We would like to thank him for his hard work in making it an enjoyable event. We also thank the Warsaw Center of Mathematics and Computer Science (WCMCS) and Google Inc. for contributing to the financing of the conference. Last but not least, thanks to everybody who submitted a paper to TCC 2015!

This year we had two invited speakers. Leonid Reyzin from Boston University talked about "Wyner's Wire-Tap Channel, Forty Years Later" and John Steinberger from Tsinghua University talked about "Block Ciphers: From Practice Back to Theory." We are grateful to both speakers for their interesting contributions to the program.

This was the first year where TCC presented the Test of Time (ToT) award to a paper that has appeared at TCC in yore and has stood the test of time. This year the award was given to Silvio Micali and Leonid Reyzin for the paper "Physically Observable Cryptography," which was presented at TCC 2004. The ToT paper was chosen by a committee selected by the TCC Steering Committee. The ToT committee has the following quotation for the ToT paper:

> *For pioneering a mathematical foundation of cryptography in the presence of information leakage in physical systems.*

The 52 papers selected for this year's TCC testify to the fact that the theory of cryptography community is a thriving and expanding community of the highest scientific quality. We are convinced that this year's TCC program contained many papers that will stand the test of time. Have fun reading these proceedings.

January 2015

Yevgeniy Dodis
Jesper Buus Nielsen

TCC 2015

12th IACR Theory of Cryptography Conference

Sheraton Warsaw Hotel, Warsaw, Poland
March 23–25

Sponsored by *International Association for Cryptologic Research (IACR)*

General Chair

Stefan Dziembowski   University of Warsaw, Poland

Program Chair

Yevgeniy Dodis   New York University, USA
Jesper Buus Nielsen   Aarhus University, Denmark

Program Committee

Joël Alwen	IST Austria, Austria
Benny Applebaum	Tel Aviv University, Israel
Nir Bitansky	Tel Aviv University, Israel
Elette Boyle	Technion, Israel
Kai-Min Chung	Academia Sinica, Taiwan
Nico Döttling	Aarhus University, Denmark
Sebastian Faust	EPFL, Switzerland
Serge Fehr	CWI, Holland
Sanjam Garg	University of California, Berkeley, USA
Shai Halevi	IBM Research, USA
Martin Hirt	ETH Zurich, Switzerland
Dennis Hofheinz	KIT, Karlsruhe, Germany
Thomas Holenstein	ETH Zurich, Switzerland
Yuval Ishai	Technion, Israel
Kaoru Kurosawa	Ibaraki University, Japan
Allison Lewko	Columbia University, USA
Mohammad Mahmoody	University of Virginia, USA
Moni Naor	Weizmann Institute of Science, Israel
Chris Peikert	Georgia Institute of Technology, USA

External Reviewers

Sponsoring and Co-Financing Institutions

TCC 2015 was co-financed by the Warsaw Center of Mathematics and Computer Science (WCMCS). The conference was also generously sponsored by Google Inc. The conference organizers are grateful for this financial support.

Wyner's Wire-Tap Channel, Forty Years Later (Invited Talk)

Leonid Reyzin

Boston University
Department of Computer Science
Boston, MA 02215, USA

Abstract. Wyner's information theory paper "The Wire-Tap Channel" turns forty this year. Its importance is underappreciated in cryptography, where its intellectual progeny includes pseudorandom generators, privacy amplification, information reconciliation, and many flavors of randomness extractors (including plain, strong, fuzzy, robust, and nonmalleable). Focusing mostly on privacy amplification and fuzzy extractors, this talk demonstrates the connection from Wyner's paper to today's research, including work on program obfuscation.

Table of Contents – Part II

Functional Encryption

Obfuscation

Table of Contents – Part I

Foundations

Symmetric Key

Multiparty Computation

Concurrent and Resettable Security

Non-malleable Codes and Tampering

Privacy Amplification

Encryption and Key Exchange

Constrained Key-Homomorphic PRFs from Standard Lattice Assumptions
(Or: How to Secretly Embed a Circuit in Your PRF)[*]

Zvika Brakerski[1],[**] and Vinod Vaikuntanthan[2],[***]

[1] Weizmann Institute of Science, Rehovot, Israel
zvika.brakerski@weizmann.ac.il
[2] Massachusetts Institute of Technology, Cambridge, MA, USA
vinodv@csail.mit.edu

Abstract. Boneh et al. (Crypto 13) and Banerjee and Peikert (Crypto 14) constructed pseudorandom functions (PRFs) from the Learning with Errors (LWE) assumption by embedding combinatorial objects, a path and a tree respectively, in instances of the LWE problem. In this work, we show how to generalize this approach to embed *circuits*, inspired by recent progress in the study of Attribute Based Encryption.

Embedding a universal circuit for some class of functions allows us to produce *constrained keys* for functions in this class, which gives us the first standard-lattice-assumption-based constrained PRF (CPRF) for general bounded-description bounded-depth functions, for arbitrary polynomial bounds on the description size and the depth. (A constrained key w.r.t a circuit C enables one to evaluate the PRF on all x for which $C(x) = 1$, but reveals nothing on the PRF values at other points.) We rely on the LWE assumption and on the one-dimensional SIS (Short Integer Solution) assumption, which are both related to the worst case hardness of general lattice problems. Previous constructions for similar function classes relied on such exotic assumptions as the existence of multilinear maps or secure program obfuscation. The main drawback of our construction is that it does not allow collusion (i.e. to provide more than a single constrained key to an adversary). Similarly to the aforementioned previous works, our PRF family is also *key homomorphic*.

Interestingly, our constrained keys are very short. Their length does not depend directly either on the size of the constraint circuit or on the input length. We are not aware of any prior construction achieving this property, even relying on strong assumptions such as indistinguishability obfuscation.

---

[*] An extended version of this manuscript can be found in [11].

[**] Supported by the Israel Science Foundation (Grant No. 468/14) and by the Alon Young Faculty Fellowship.

[***] Research supported by DARPA Grant number FA8750-11-2-0225, Alfred P. Sloan Research Fellowship, NSF CAREER Award CNS-1350619, NSF Frontier Grant CNS-1414119, Microsoft Faculty Fellowship, and a Steven and Renee Finn Career Development Chair from MIT.

Y. Dodis and J.B. Nielsen (Eds.): TCC 2015, Part II, LNCS 9015, pp. 1–30, 2015.

1   Introduction

A pseudorandom function family (PRF) [14] is a finite set of functions $\{F_s : D \to R\}_s$, indexed by a seed (or key) s, such that for a random s, F_s is efficiently computable given s, and is computationally indistinguishable from a random function from D to R, given oracle access. Since the introduction of this concept, PRFs have been one of the most fundamental building blocks in cryptography. Many variants of PRFs with additional properties have been introduced and have found a plethora of applications in cryptography. In this work, we will focus on *Constrained PRFs* and *Key-Homomorphic PRFs*.

Constrained PRFs. Constrained PRFs (CPRFs) have been introduced simultaneously by Boneh and Waters [9], Kiayias et al. [18] (as "Delegatable PRFs") and by Boyle, Goldwasser and Ivan [10] (as "Functional PRFs"). Here an adversary is allowed to ask for a constrained key which should allow it to evaluate the PRF on a subset of the inputs, while revealing nothing about the values at other inputs. It has been shown [9,18,10] how to construct CPRFs for function classes of the form $x \in [i, j]$ (where the input is interpreted as an integer) based on any one-way function. This in particular allows for the "puncturing" technique of Sahai and Waters [26] that found many uses in the obfuscation literature. Further, [9] showed how to achieve more complicated function classes such as bit fixing functions and even arbitrary circuits, but those require use of cryptographic multilinear maps. They also introduce a number of applications for such CPRFs, including broadcast encryption schemes and identity based key exchange. Hofheinz et al. [17] show how to achieve adaptively secure CPRFs from indistinguishability obfuscation using a random oracle.

The original definition of CPRFs requires resilience to arbitrary collusion. Namely, a constrained key for C_1, C_2 should give no more information than a constrained key for $C_1 \vee C_2$ and must not reveal anything about values where $C_1(x) = C_2(x) = $ 'false. Many of the applications of CPRFs (e.g. for broadcast encryption and identity based key exchange) rely on collusion resilience. Unfortunately, our construction in this work will not allow collusions, and therefore will not be useful for these applications. We hope that future works will be able to leverage our ideas into collusion resilient CPRFs.

Key-Homomorphic PRFs. In key-homomorphic PRFs, there is a group structure associated with the set of keys, and it is required that for any input x and keys s, t, $F_s(x) + F_t(x) = F_{s+t}(x)$. A construction in the random oracle model was given by Naor, Pinkas and Reingold [22], and the first construction in the standard model was given by Boneh et al. [8] based on the Learning with Errors assumption (LWE), building on a (non key homomorphic) lattice-based PRF of Banerjee, Peikert and Rosen [4]. This was followed by an improved construction by Banerjee and Peikert [3] based on quantitatively better lattice assumptions. The LWE based constructions achieved a slightly weaker notion, namely "almost" key-homomorphism, in which $\|(F_s(x) + F_t(x)) - F_{s+t}(x)\|$ is small, for an appropriately defined norm. This notion is sufficient for the known applications. Applications of key-homomorphic PRFs include distributed key-distribution, symmetric proxy re-encryption, updatable encryption and PRFs secure against related-key attacks [22,8,19].

Our Results. We view the main contribution of this work as showing how to impose *hidden semantics* into the evaluation process of LWE-based PRFs. Namely, we allow multiple computation paths for computing $F_s(x)$, such that we can selectively block some of these paths based on logic described by a circuit. This is done by extending ideas from the ABE literature, and in particular the ABE scheme of Boneh et al. [7] (see more about this connection below).

It is particularly interesting that previous constructions of PRFs [8,3] can be viewed as a special case of our framework, but ones that only allow a single computational path. Our work therefore highlights that the techniques used for constructing PRFs and for constructing ABE are special cases of the same grand schema. This could hopefully lead to new insights and constructions.

We employ our methods towards presenting a family of (single key secure) constrained key-homomorphic PRFs based on worst-case general lattice assumptions. This is a first step in solving the open problem posed in [9] of achieving (collusion resilient) CPRFs from standard assumptions.

Our construction is selectively secure in the constraint query, namely the adversary needs to decide on the constraint before seeing the public parameters, but is adaptive with regards to PRF oracle queries. We achieve the latter without "complexity leveraging", contrary to [9], and thus we do not require sub-exponential hardness assumptions as they do. This is done by employing our technique of embedding semantics into the evaluation process again. In particular, we embed the semantics of an *admissible hash function*, introduced by Boneh and Boyen [6] into the PRF, which allows us to handle adaptive queries.

Our proofs rely on two closely related hardness assumptions: The Learning with Errors (LWE) assumption, and the one-dimensional Short Integer Solution (1D-SIS) assumption. Both assumptions can be tied to the worst case hardness of general lattice problems such as GapSVP and SIVP, with similar parameters. LWE is sufficient for proving pseudorandomness in the absence of a constrained key. However, once the adversary is given a constrained key, the situation becomes more delicate. In particular, even showing *correctness* in this setting is not straightforward. (Correctness refers to the property that evaluation using the constrained key and using the actual seed result in the same output.) One can show unconditionally that the value computed using the constrained key is *close* (in norm) to the real value of the function but not that they are always equal. A similar issue comes up in the security proof (since the reduction "fabricates" oracle answers in a similar way to the constrained evaluation). Our solution is to use *computational* arguments. Namely to show that it is computationally intractable, under the 1D-SIS assumption, to come up with an input for which the constrained evaluation errs. Therefore even the correctness of our scheme relies on computational assumptions. We note that similar techniques can be used to strengthen the almost key-homomorphism property into computational key-homomorphism where it is computationally hard to find an input for which key homomorphism does not hold.

The following theorem presents the simplest application of our method, we explain how it can be extended below.

Theorem 1.1. *Let $\mathcal{C}_{\ell,d}$ be the class of size-ℓ depth-d circuits. Then for all polynomials ℓ, d, there exists a $\mathcal{C}_{\ell,d}$-constrained (almost) key-homomorphic family of PRFs without collusion, based on the (appropriately parameterized) LWE and 1D-SIS assumptions (and hence on the worst-case hardness of appropriately parameterized* GapSVP *and* SIVP *problems).*

Interestingly, we can go beyond bounded size circuits. In fact, we can support any function family with bounded length description, so long as there is a universal evaluator of depth d that takes a function description and an input, and executes the function on the input. Namely, consider a sequence of universal circuits $\{\mathcal{U}_k\}_{k\in\mathbb{N}}$, where $\mathcal{U}_k : \{0,1\}^\ell \times \{0,1\}^k \to \{0,1\}$. This sequence defines a class of functions $\{0,1\}^* \to \{0,1\}$, where each function F in the class is represented by a string $f \in \{0,1\}^\ell$, and for $x \in \{0,1\}^k$, it holds that $F(x) = \mathcal{U}_k(f,x)$. We call such a function class ℓ-uniform. We are only able to support $\mathcal{U}_k$ whose depth is bounded by some a-priori polynomial in the security parameter d, however in some cases this is sufficient to support all k's that are polynomial in the security parameter. The following theorem states our result with regards to such families.

Theorem 1.2. *Let $\mathcal{C}_{\ell,d}$ be a class of ℓ-uniform functions with depth-d evaluator. Then for all polynomials ℓ, d, there exists a $\mathcal{C}_{\ell,d}$-constrained (almost) key-homomorphic family of PRFs without collusion, based on the (appropriately parameterized) LWE and 1D-SIS assumptions (and hence on the worst case hardness of appropriately parameterized* GapSVP, SIVP*).*

Lastly, we show that the bit-length of the constrained keys in our scheme can be reduced to poly(λ) for some *fixed* polynomial. Namely, completely independent of all of the parameters of the scheme. This is done by using an ABE scheme with short secret keys as a black box. In particular we resort to the same scheme, namely the ABE scheme of Boneh et al. [7], which inspired our constrained PRF construction. This is done by encrypting all of the "components" of the constrained key, and providing them in the public parameters of the construction. Then, the actual constrained key is an ABE secret key which only allows to decrypt the relevant components. We note that this short representation for constrained keys is not homomorphic (however the scheme is still almost key homomorphic with respect to the seed). A theorem statement follows.

Theorem 1.3. *There exists a constrained PRF scheme with the same properties as in Theorem 1.2, and under the same hardness assumptions, where the constrained keys are of asymptotic bit-length* poly(λ)*, for an a-priori fixed polynomial.*

See Section 2 for an extended overview of the construction.

Relation to the ABE Construction of Boneh et al. [7]. Our techniques are greatly influenced by the aforementioned LWE-based ABE construction of Boneh et al. [7]. Recall that in ABE, messages are encrypted relative to *attributes* and decryption keys are drawn relative to *functions*. Decryption is possible only if the function f of the decryption key accepts the attribute x of the ciphertext. In order to decrypt a ciphertext, [7] first applies a public procedure that depends on f, x on the ciphertext and then applies the decryption key on the resulting value. Their construction makes sure that for any f, encryptions with regards to all accepting x's will derive a decryptable ciphertext (and all non-accepting x's cannot be decrypted).

Our constrained key for a circuit C is almost identical to an encryption of 0 with attribute C in [7]. The randomness in the encryption roughly corresponds to the seed of the PRF. An application of the PRF on the constrained key includes applying the public procedure of the ABE on the ciphertext, with respect to the function $f = \mathcal{U}$, the universal circuit for the function class to which C belongs. However, there is the question of how to represent the input: We need to be able to evaluate C on any possible input while preserving security. One of our main technical ideas is in showing that this is possible, and in fact can be achieved regardless of the input length. Combined with the framework from [7], we can guarantees that for all x, regardless which C was used to generate the "ciphertext", the output of the public procedure will only depend on x and not on C. The basic idea is therefore to use this value as the PRF value. This does not work as is (for example, it does not imply pseudorandomness for non-accepting x's) and additional ideas are required.

As mentioned above, the PRFs of [8,3] that seem to stem from different ideas and have quite different proofs than [7] can be shown to be special cases of the above paradigm, except f is taken to be an arbitrary formula (a multiplication tree). For details see Section 2.

The novelty in our approach is to show the extra power that is obtained from generalizing these two approaches. We use the universal circuit as a way to embed an undisclosed computation into an LWE instance, and show how to achieve pseudorandomness using tools such as admissible hash functions (which are also embedded into an LWE instance).

Relation with the Constrained PRF of Hofheinz et al. [17]. The work of [17] constructs adaptively secure collusion-resistant CPRFs, namely ones where the challenge x^* needs not be provided ahead of time. Their building blocks are "universal parameters" and *adaptively secure* ABE, which are used as black-box. Note that we achieve adaptive security w.r.t the challenge (but not with respect to the constraint) while relying on techniques which are only known to imply *selectively secure* ABE. Further, whereas [17] use ABE only to implement access control and therefore need to rely on strong assumptions to implement the PRF so as to interface with the ABE, we use ABE techniques to achieve both pseudorandomness and access control. On the flip side, our construction is not collusion resistant, contrary to [17].

Open Problems. The main drawback of our CPRF is its vulnerability to collusion, which severely limits its applicability as a building block. It is an open problem to achieve bounded collusion resilience, even for two constrained keys instead of one and even at the cost of increasing the parameters. Any improvement on this front should be very interesting. Another avenue for research is trying to extend the construction so that there is no restriction on the constraint circuit size, similarly to the multilinear map based construction of [9]. Finally, it would also be interesting to apply this methodology of imposing semantics on a cryptographic computation to other primitives in order to allow more fine-grained access control.

2   Overview of Our Construction

We recall that the LWE assumption asserts that for a uniform vector $\mathbf{s}$ and a matrix $\mathbf{A}$ of appropriate dimensions (over $\mathbb{Z}_q$ for an appropriate q), it holds that $(\mathbf{A}, \mathbf{s}^T \mathbf{A} + \mathbf{e}^T)$, is indistinguishable from uniform, where $\mathbf{e}$ is taken from an appropriate distribution over *low norm* vectors and referred to as the *noise vector*. In this outline we will ignore the generation of $\mathbf{e}^T$ and its evolution during computation process, and just denote it by noise (but of course care will need to be taken in the formal arguments).

The PRF of Banerjee and Peikert [3]. A high-level methodology for constructing PRFs, taken by [8,3] and also in this work, is to take $\mathbf{s}$ as the seed, and to generate for each PRF input x, an LWE matrix $\mathbf{A}_x$ such that the values $\mathbf{s}^T \mathbf{A}_x +$ noise for the different inputs x are jointly indistinguishable from uniform. Note that almost key homomorphism follows naturally for any implementation of this template, up to the accumulation of noise. The noise issue is handled by taking the PRF value to be a properly scaled down and rounded version of the above, so that the effect of the noise is minimal (and its norm can be bounded below 1). This property is also inherited by our scheme.

    As a starting point for deriving our construction, let us revisit the key-homomorphic PRF construction of [3]. Their PRF family was associated with a combinatorial object – a binary tree. Each node v of the tree was associated with an LWE matrix $\mathbf{A}_v$, where the PRF input x determined the matrices for the leaves, and matrices for internal nodes are derived as follows. Given a node v whose children are associated with $\mathbf{A}_l, \mathbf{A}_r$, they define $\mathbf{A}_v = \mathbf{A}_l \cdot \mathbf{G}^{-1}(\mathbf{A}_r)$. In this notation, $\mathbf{G}^{-1}(\cdot)$ is the binary decomposition operator, which breaks each entry in the matrix into the bit vector of length $\log(q)$ of its binary representation. Note that $\mathbf{G}^{-1}(\cdot)$ will always have small norm, and that the inverse operator $\mathbf{G}$, representing binary composition, is linear so it can be represented by a matrix. Thus for all $\mathbf{A}$ it holds that $\mathbf{G} \cdot \mathbf{G}^{-1}(\mathbf{A}) = \mathbf{A}$.

    Going back to the PRF of [3], the derivation procedure described above allows to associate a matrix with the root of the tree, which depends only on the input x (and on the topology of the tree which is fixed). We will use the root's matrix as our $\mathbf{A}_x$. The proof hinges on the invariant that LWE instances will be multiplied on the right only by low-norm matrices (of the form $\mathbf{G}^{-1}(\cdot)$), and therefore $\mathbf{s}^T \mathbf{A}_l \mathbf{G}^{-1}(\mathbf{A}_r) + \mathsf{noise} \approx (\mathbf{s}^T \mathbf{A}_l + \mathsf{noise})\mathbf{G}^{-1}(\mathbf{A}_r)$, which allows to replace $(\mathbf{s}^T \mathbf{A}_l + \mathsf{noise})$ with a new uniform vector and propagate to the right.

From Embedded Trees to Embedded Circuits. We show that the operation $\mathbf{A}_v = \mathbf{A}_l \cdot \mathbf{G}^{-1}(\mathbf{A}_r)$ is in fact a special case of a more general operation, inspired by the recent Attribute Based Encryption (ABE) construction of Boneh et al. [7]. We will associate a matrix $\mathbf{A}_v$ *as well as a binary value* x_v with each node, and pay special attention to the matrix $(\mathbf{A}_v - x_v \mathbf{G})$. In particular, considering a node v with children l, r, it holds that

$$(\mathbf{A}_l - x_l \mathbf{G}) \cdot \mathbf{G}^{-1}(\mathbf{A}_r) + (\mathbf{A}_r - x_r \mathbf{G}) \cdot x_l = \mathbf{A}_l \mathbf{G}^{-1}(\mathbf{A}_r) - x_r x_l \mathbf{G} .$$

This generalization associates the semantics of the multiplication operation with the syntactic definition $\mathbf{A}_v = \mathbf{A}_l \mathbf{G}^{-1}(\mathbf{A}_r)$, and it also maintains the invariant that the matrices $(\mathbf{A}_l - x_l \mathbf{G})$ and $(\mathbf{A}_r - x_r \mathbf{G})$ are only multiplied on the right by low norm elements, so that

$$\mathbf{s}^T \left((\mathbf{A}_l - x_l \mathbf{G}) \cdot \mathbf{G}^{-1}(\mathbf{A}_r) + (\mathbf{A}_r - x_r \mathbf{G}) \cdot x_l \right) + \mathsf{noise} \approx$$
$$\left(\mathbf{s}^T(\mathbf{A}_l - x_l \mathbf{G}) + \mathsf{noise} \right) \cdot \mathbf{G}^{-1}(\mathbf{A}_r) + \left(\mathbf{s}^T(\mathbf{A}_r - x_r \mathbf{G}) + \mathsf{noise} \right) \cdot x_l ,$$

which will play an important role in the security proof. Put explicitly, if the evaluator holds $\mathbf{s}^T(\mathbf{A}_l - x_l \mathbf{G}) + \mathsf{noise}$ and $\mathbf{s}^T(\mathbf{A}_l - x_l \mathbf{G}) + \mathsf{noise}$, then it can compute $\mathbf{s}^T(\mathbf{A}_v - x_l \cdot x_r \mathbf{G}) + \mathsf{noise}$ (and we will obviously define $x_v = x_l \cdot x_r$).

This semantic relation can be extended beyond multiplication gates, and in particular NAND gates can be supported in a fairly similar manner. Furthermore, there is no need to stick to tree structure and one can support arbitrary DAGs, which naturally correspond to circuits. Extending the above postulate, if our DAG corresponds to a circuit C, then having $\mathbf{s}^T(\mathbf{A}_i - x_i \mathbf{G}) + \mathsf{noise}$, for all leaves (= inputs), allows to compute $\mathbf{s}^T(\mathbf{A}_x - C(x)\mathbf{G}) + \mathsf{noise}$. Recalling that the value of the PRF on input x is $\mathbf{s}^T \mathbf{A}_x + \mathsf{noise}$, the aforementioned information allows us to evaluate the PRF at points where $C(x) = 0$. It can also be shown that it is computationally hard to compute the value at points where $C(x) = 1$. We note that this process is practically identical to the public part of the decryption procedure in the [7] ABE (as we explained in Section 1). We also note that since [3] were trying to minimize the complexity of evaluating their PRF, it made no sense in their construction to consider DAGs which only increase the complexity. However, as we show here, there are benefits to embedding a computational process in the PRF evaluation.

Utilizing the Universal Circuit. The tools we describe so far indeed seem to get us closer to our goal of producing constrained keys, but we are still not quite there. What we showed is that for any circuit C, we can devise a PRF with a constrained key for C. Note that we use the negated definition to the one we used before, and allow to evaluate when $C(x) = 0$ and not when $C(x) = 1$. This will be our convention throughout this overview.

In order to reverse the order of quantifiers, we take C to be the universal circuit $\mathcal{U}(F, x)$, and the constrained keys will be of the form $\mathbf{s}^T(\mathbf{A}_i - f_i \mathbf{G}) + \mathsf{noise}$, where the f_i is the ith bit of the description of the constraint F, as well as values for the x wires, which will be of the form $\mathbf{s}^T(\hat{\mathbf{A}}_b - b\mathbf{G}) + \mathsf{noise}$, for both $b \in \{0, 1\}$. These values will allow us to execute F on *any* input x. Note that we can use the same matrices $\hat{\mathbf{A}}_0, \hat{\mathbf{A}}_1$ for all input wires, hence we don't need to commit to the input size when we provide the constrained key.[1] From this description it is obvious why our construction is not collusion resistant: Given two constrained keys for two non identical functions, there exists an i such that the adversary gets both $\mathbf{s}^T \mathbf{A} + \mathsf{noise}$ and $\mathbf{s}^T(\mathbf{A}_i - \mathbf{G}) + \mathsf{noise}$. Recovering $\mathbf{s}^T$ from these values is straightforward and hence all security is lost. Note that for the input values, unlike the function description, we use two different matrices for 0 and 1: $\hat{\mathbf{A}}_0, \hat{\mathbf{A}}_1$, so a similar problem does not occur.

The Problem with Correctness, and a Computational Solution. We introduced two ways to compute the value of the PRF at x: One is to compute $\mathbf{A}_x$ and use the seed $\mathbf{s}^T$ to compute $\mathbf{s}^T \mathbf{A}_x + \mathsf{noise}$, and the other is to use the constrained key to obtain $\mathbf{s}^T(\mathbf{A}_x - F(x)\mathbf{G}) + \mathsf{noise}$, which for $F(x) = 0$ gives $\mathbf{s}^T \mathbf{A}_x + \mathsf{noise}$. The problem is that the *noise value* in these two methods could differ. It is possible to make the difference small by scaling down and rounding, but this is not going to suffice for our purposes (mostly because a similar problem comes up in the security proof). We solve this issue using the 1D-SIS assumption as follows. We first note that the evaluation using the constrained key is essentially evaluation of a linear function with small coefficients on the vectors constituting the constrained key (essentially they get multiplied by bits and by low norm matrices $\mathbf{G}^{-1}(\cdot)$). Secondly, the only way for the two computation paths to not agree is if the value $\mathbf{s}^T \mathbf{A}_x$ is very close to an integer multiple of a number p (which is part of the PRF description). Finally, we notice that by LWE, the vectors in the constrained key are indistinguishable from uniform and independent. Thus, if we encounter such x for which correctness does not work, we can also find a short linear combination of random elements whose scaled down rounded value is close to an integer. In other words, given a uniform vector $\mathbf{v}$ in $\mathbb{Z}_q$, we can find $\mathbf{z}$ such that $\lfloor \langle \mathbf{v}, \mathbf{z} \rangle / p \rfloor$ is "close" to an integer. This is similar to solving a one-dimensional instance of the SIS problem, i.e. $\langle \mathbf{v}, \mathbf{z} \rangle = 0 \pmod{p}$. Indeed, one can show that the 1D-SIS problem is as hard as standard worst-case hard lattice problems via a reduction from [24].

Pseudorandomness and Adaptive Security. Given a constrained key for F, one can compute $\mathbf{s}^T(\mathbf{A}_x - F(x)\mathbf{G}) + \mathsf{noise}$, and indeed if $F(x) = 1$ it is hard to compute $\mathsf{PRF}_\mathbf{s}(x) = \mathbf{s}^T \mathbf{A}_x + \mathsf{noise}$. However, we want to argue that this value is *pseudorandom* and furthermore that it remains pseudorandom after adaptive queries to the PRF. Namely, after the adversary sees as many values of the form $\mathsf{PRF}_\mathbf{s}(x) = \mathbf{s}^T \mathbf{A}_x + \mathsf{noise}$ as it wishes.

---

[1] Recall that in [8,3] there are only two matrices altogether. This is sufficient here for the input wires for the same reason, but we need additional matrices to encode the constraint description.

To achieve these goals, we add another feature to the PRF. We consider a new independent LWE matrix $\mathbf{D}$, and define $\mathsf{PRF_s}(x) = \mathbf{s}^T \mathbf{A}_x \cdot \mathbf{G}^{-1}(\mathbf{D}) + \mathsf{noise}$. First of all, we note that given the constrained key, we can still compute the PRF for values where $C(x) = 0$, by first computing $(\mathbf{s}^T \mathbf{A}_x + \mathsf{noise})$ as before, and then multiplying by $\mathbf{G}^{-1}(\mathbf{D})$, which has low norm. However, in general we have

$$\mathsf{PRF_s}(x) \approx \Big(\mathbf{s}^T(\mathbf{A}_x - F(x)\mathbf{G}) + \mathsf{noise}\Big) \cdot \mathbf{G}^{-1}(\mathbf{D}) + F(x)\Big(\mathbf{s}^T\mathbf{D} + \mathsf{noise}\Big) ,$$

and it can be shown that for $F(x) = 1$, the second term randomizes the expression, by the LWE assumption.

This handles pseudorandomness for a single query, but not for the case of adaptive queries (since we can only use the pseudorandomness of $(\mathbf{s}^T\mathbf{D} + \mathsf{noise})$ once). To handle adaptive queries we embed semantics into the matrix $\mathbf{D}$ itself. Namely, $\mathbf{D} = \mathbf{D}_x$ will be derived by an application of the universal circuit to the input x and *an admissible hash function* h. Admissible hash functions, introduced by Boneh and Boyen [6], allow (at a very high level) to partition the input space such that with noticeable probability all of the adaptive queries have value $h(x) = 0$, but the challenge query will have $h(x) = 1$. This means that in the proof of security, we can hold a constrained key for h, which will allow us to compute $(\mathbf{s}^T\mathbf{D}_x + \mathsf{noise})$, for all the queries of the adversary, but leave the challenge query unpredictable (to make it pseudorandom, we will multiply in the end by another final $\mathbf{D}'$). This concludes the security argument for adaptive queries.

Key-Homomorphism. As we mention above, key-homomorphism follows since we use the template $\mathsf{PRF_s}(x) = \mathbf{s}^T \mathbf{A}_x + \mathsf{noise}$. We note that the existence of noise means that homomorphism may not be accurate and with some low probability $(\mathsf{PRF_s}(x) + \mathsf{PRF_{s'}}(x))$ will only be close to $\mathsf{PRF_{s+s'}}(x)$ and not identical. However this property is sufficient for many applications.

We point out that our constrained keys are a collection elements of the form $(\mathbf{s}^T \mathbf{A}_i + \mathsf{noise})$, and therefore the scheme is also homomorphic with respect to constrained keys, i.e. constrained keys for the same F w.r.t different keys $\mathbf{s}, \mathbf{s}'$ can be added to obtain a constrained key w.r.t $\mathbf{s} + \mathbf{s}'$.

Reducing the Constrained Key Size. From the above, it follows that the constrained key contains $\ell + 2$ vectors, where ℓ is the bit length of a description of F relative to the universal circuit for the function class. Note that this does not depend directly on the input size to the function. However, indirectly the depth of the universal circuit affects the modulus q that needs to be used.

We show that we can remove the dependence on ℓ altogether using an ABE scheme with short secret keys, such as that of [7]. To do this, we notice that for each constraint function F, the adversary gets either $\mathbf{s}^T \mathbf{A}_i + \mathsf{noise}$ or $\mathbf{s}^T(\mathbf{A}_i - \mathbf{G}) + \mathsf{noise}$, according to the value of the bit f_i. We can prepare for both options by encrypting both vectors using the ABE, each with its own attribute $(i, 0)$ and $(i, 1)$ respectively. All of these encryptions, for all i, will be placed in the public parameters. Then in order to provide a constrained key, we will provide an ABE secret key for the function that takes (i, b) and returns 0 if and only if $f_i = b$. Given this key, the user can decrypt exactly those vectors that constitute its constrained key. Note that this function can be computed by a depth $O(\log(\ell)) = O(\log(\lambda))$ circuit, and thus the size of the secret key can be made asymptotically independent of all parameters except λ, e.g. by setting the parameters to support depth $\log^2(\lambda)$ circuits.

3 Preliminaries

We first recall some background. For an integer modulus q, let $\mathbb{Z}_q = \mathbb{Z}/q\mathbb{Z}$ denote the ring of integers modulo q. For an integer $p \leq q$, we define the modular "rounding" function

$$\lfloor \cdot \rceil_p : \mathbb{Z}_q \to \mathbb{Z}_p \text{ that maps } x \to \lfloor (p/q) \cdot x \rceil$$

and extend it coordinate-wise to matrices and vectors over $\mathbb{Z}_q$. We denote the elements of the standard basis by $\mathbf{u}_1, \mathbf{u}_2, \ldots$, where the dimension will be clear from the context.

We denote distributions (or random variables) that are computationally indistinguishable by $X \overset{c}{\approx} Y$. This refers to the standard notion of negligible distinguishing gap for any polynomial time distinguisher. Our reductions preserve the uniformity of the adversary so by assuming the hardness of our assumption for uniform adversary we get security for our construction against uniform adversaries, and likewise for non-uniform assumptions and adversaries.

The Gadget Matrix. Let $\ell = \lceil \log q \rceil$ and define the "gadget matrix" $\mathbf{G} = \mathbf{g} \otimes \mathbf{I}_n \in \mathbb{Z}_q^{n \times n\ell}$ where

$$\mathbf{g} = (1, 2, 4, \ldots, 2^{\ell-1}) \in \mathbb{Z}_q^\ell$$

We will also refer to this gadget matrix as the "powers-of-two" matrix. We define the inverse function $\mathbf{G}^{-1} : \mathbb{Z}_q^{n \times m} \to \{0, 1\}^{n\ell \times m}$ which expands each entry $a \in \mathbb{Z}_q$ of the input matrix into a column of size ℓ consisting of the bit decomposition of a. We have the property that for any matrix $\mathbf{A} \in \mathbb{Z}_q^{n \times m}$,

$$\mathbf{G} \cdot \mathbf{G}^{-1}(\mathbf{A}) = \mathbf{A}$$

Norms for Vectors and Matrices. We will always use the infinity norm for vectors and matrices. Namely for a vector $\mathbf{x}$, the norm $\|\mathbf{s}\|$ is the maximal absolute value of an element in $\mathbf{x}$. Similarly, for a matrix $\mathbf{A}$, $\|\mathbf{A}\|$ is the maximal absolute value of any of its entries. If $\mathbf{x}$ is n-dimensional and $\mathbf{A}$ is $n \times m$, then $\|\mathbf{x}^T \mathbf{A}\| \leq n \cdot \|\mathbf{x}\| \cdot \|\mathbf{A}\|$. We remark that L_1 or L_2 norms can also be used and even achieve somewhat tighter parameters, but the proofs become more complicated.

3.1   Constrained Pseudorandom Function: Definition

In a constrained PRF family [9,10,18], one can compute a constrained PRF key K_C corresponding to any Boolean circuit C. Given K_C, anyone can compute the PRF on inputs x such that $C(x) = 0$. Furthermore, K_C does not reveal any information about the PRF values at the other locations. Below we recall their definition, as given by [9].

Syntax   A *constrained* pseudo-random function (PRF) family is defined by a tuple of algorithms (KeyGen, Eval, Constrain, ConstrainEval) where:

- **Key Generation** KeyGen$(1^\lambda, 1^{k_{in}}, 1^{k_{out}})$ is a PPT algorithm that takes as input the security parameter λ, an input length k_{in} and an output length k_{out}, and outputs a PRF key K;
- **Evaluation** Eval(K, x) is a deterministic algorithm that takes as input a key K, a string $x \in \{0,1\}^{k_{in}}$ and outputs $y \in \{0,1\}^{k_{out}}$;
- **Constrained Key Generation** Constrain(K, C) is a PPT algorithm that takes as input a PRF key K, a circuit $C : \{0,1\}^{k_{in}} \to \{0,1\}$ and outputs a constrained key K_C;
- **Constrained Evaluation** ConstrainEval(K_C, x) is a deterministic algorithm that takes as input a constrained key K_C and a string $x \in \{0,1\}^{k_{in}}$ and outputs either a string $y \in \{0,1\}^{k_{out}}$ or $\perp$.

We define the notion of *(single key) selective-function* security for constrained PRFs.

Definition 3.1. *A family of PRFs (KeyGen, Eval, Constrain, ConstrainEval) is a single-key selective-function constrained PRF (henceforth, referred to simply as constrained PRF) if it satisfies the following properties:*

- **Functionality computationally preserved under constraining.** *For every* PPT *adversary* (A_0, A_1), *consider an experiment where we choose* $K \leftarrow$ KeyGen$(1^\lambda, 1^{k_{in}}, 1^{k_{out}})$, $(C, \sigma_0) \leftarrow A_0(1^\lambda)$, *and* $K_C \leftarrow$ Constrain(K, C). *Then:*

$$\Pr\left[x^* \leftarrow A_1^{\mathsf{Eval}(K, \cdot)}(1^\lambda, K_C, \sigma_0); \; : \begin{array}{c} C(x^*) = 0 \;\wedge \\ \mathsf{Eval}(K, x^*) \neq \mathsf{ConstrainEval}(K_C, x^*) \end{array} \right]$$

  is negligible in the security parameter, where C, K, K_C *are selected as described above.*
  In words, it is computationally hard to find an x^* *such that* $C(x^*) = 0$, *and yet the result of the constrained evaluation differs from the actual PRF evaluation.*
- **Pseudorandom at constrained points.** *For every* PPT *adversary* (A_0, A_1, A_2), *consider an experiment where* $K \leftarrow$ KeyGen$(1^\lambda, 1^{k_{in}}, 1^{k_{out}})$, $(C, \sigma_0) \leftarrow A_0(1^\lambda)$, *and* $K_C \leftarrow$ Constrain(K, C). *Then:*

$$\Pr\left[\begin{array}{l} b \leftarrow \{0,1\}; \\ (x^*, \sigma_1) \leftarrow A_1^{\mathsf{Eval}(K, \cdot)}(1^\lambda, K_C, \sigma_0); \; : \begin{array}{c} C(x^*) = 1 \;\wedge \\ A_2(1^\lambda, y^*, \sigma_1) = b \end{array} \\ \texttt{If } b = 0, \; y^* = \mathsf{Eval}(K, x^*), \\ \texttt{Else} \quad y^* \leftarrow \{0,1\}^{k_{out}} \end{array}\right] \leq \frac{1}{2} + \mathsf{negl}(\lambda)$$

The correctness and security properties could potentially be combined into one game, but we choose to present them as two distinct properties for the sake of clarity.

3.2   Learning with Errors

The Learning with Errors (LWE) problem was introduced by Regev [25] as a generalization of "learning parity with noise" [5,2]. We now define the decisional version of LWE. (Unless otherwise stated, we will treat all vectors as column vectors in this paper).

Definition 3.2 (Decisional LWE (DLWE) [25]). *Let λ be the security parameter, $n = n(\lambda)$, $m = m(\lambda)$, and $q = q(\lambda)$ be integers and $\chi = \chi(\lambda)$ be a probability distribution over $\mathbb{Z}$. The $\text{DLWE}_{n,q,\chi}$ problem states that for all $m = \text{poly}(n)$, letting $\mathbf{A} \leftarrow \mathbb{Z}_q^{n \times m}$, $\mathbf{s} \leftarrow \mathbb{Z}_q^n$, $\mathbf{e} \leftarrow \chi^m$, and $\mathbf{u} \leftarrow \mathbb{Z}_q^m$, the following distributions are computationally indistinguishable:*

$$\left(\mathbf{A}, \mathbf{s}^T \mathbf{A} + \mathbf{e}^T \right) \overset{c}{\approx} \left(\mathbf{A}, \mathbf{u}^T \right)$$

There are known quantum (Regev [25]) and classical (Peikert [23]) reductions between $\text{DLWE}_{n,q,\chi}$ and approximating short vector problems in lattices. Specifically, these reductions take χ to be a discrete Gaussian distribution $D_{\mathbb{Z},\alpha q}$ for some $\alpha < 1$. We write $\text{DLWE}_{n,q,\alpha}$ to indicate this instantiation. We now state a corollary of the results of [25,23,20,21]. These results also extend to additional forms of q (see [20,21]).

Corollary 3.3 ([25,23,20,21]). *Let $q = q(n) \in \mathbb{N}$ be either a prime power $q = p^r$, or a product of co-prime numbers $q = \prod q_i$ such that for all i, $q_i = \text{poly}(n)$, and let $\alpha \geq \sqrt{n}/q$. If there is an efficient algorithm that solves the (average-case) $\text{DLWE}_{n,q,\alpha}$ problem, then:*

- *There is an efficient quantum algorithm that solves $\mathsf{GapSVP}_{\tilde{O}(n/\alpha)}$ (and $\mathsf{SIVP}_{\tilde{O}(n/\alpha)}$) on any n-dimensional lattice.*
- *If in addition $q \geq \tilde{O}(2^{n/2})$, there is an efficient classical algorithm for $\mathsf{GapSVP}_{\tilde{O}(n/\alpha)}$ on any n-dimensional lattice.*

Recall that GapSVP_γ is the (promise) problem of distinguishing, given a basis for a lattice and a parameter d, between the case where the lattice has a vector shorter than d, and the case where the lattice doesn't have any vector shorter than $\gamma \cdot d$. SIVP is the search problem of finding a set of "short" vectors. The best known algorithms for GapSVP_γ ([27]) require at least $2^{\tilde{\Omega}(n/\log \gamma)}$ time. We refer the reader to [25,23] for more information.

In this work, we will only consider the case where $q \leq 2^n$. Furthermore, the underlying security parameter λ is assumed to be polynomially related to the dimension n.

3.3 One-Dimensional Short Integer Solution (SIS) and Variants

We present a special case of the well known Short Integer Solution (SIS) problem [1].

Definition 3.4. *The One-Dimensional Short Integer Solution problem, denoted* 1D-SIS$_{q,m,t}$*, is the following problem. Given a uniformly distributed vector* $\mathbf{v} \xleftarrow{\$} \mathbb{Z}_q^m$*, find* $\mathbf{z} \in \mathbb{Z}^m$ *such that* $\|\mathbf{z}\| \leq t$ *and also* $\langle \mathbf{v}, \mathbf{z} \rangle \in [-t, t] + q\mathbb{Z}$.

For appropriately chosen moduli q, the 1D-SIS$_{q,m,t}$ problem is as hard as worst-case lattice problems. This follows from the techniques in the classical worst-case to average-case reduction of Ajtai [1]. We state below the version due to Regev [24].

Corollary 3.5 (Section 4 in [24] and Proposition 4.7 in [13]). *Let* $n \in \mathbb{N}$ *and* $q = \prod_{i \in n} p_i$*, where all* $p_1 < p_2 < \ldots < p_n$ *are co-prime. Let* $m \geq c \cdot n \log q$ *(for some universal constant c). Assuming that* $p_1 \geq t \cdot \omega(\sqrt{mn \log n})$*, the one-dimensional SIS problem* 1D-SIS$_{q,m,t}$ *is at least as hard as* SIVP$_{t \cdot \tilde{O}(\sqrt{mn})}$ *and* GapSVP$_{t \cdot \tilde{O}(\sqrt{mn})}$.

Proof. The hardness of a closely related problem is established by combining the techniques in [24, Section 4] and [13, Proposition 4.7]: Given $\mathbf{a} \xleftarrow{\$} \mathbb{Z}_q^{m+1}$, find $\mathbf{y}$ with $\|\mathbf{y}\| \leq t$ such that $\langle \mathbf{a}, \mathbf{y} \rangle = 0 \pmod{q}$.

We now show how to convert an instance for this problem into an instance of 1D-SIS. Given an instance $\mathbf{a} \in \mathbb{Z}_q^{m+1}$, we consider the first component a_1. If this element is not a unit (i.e. invertible) in $\mathbb{Z}_q$, then the reduction aborts. Otherwise it defines $\mathbf{v} = a_1^{-1} \cdot [a_2, \ldots, a_{m+1}]$. Given a solution $\mathbf{z}$ for 1D-SIS on input $\mathbf{v}$, we define $\mathbf{y}$ by letting $\mathbf{y} = [-\langle \mathbf{v}, \mathbf{z} \rangle, x_1, \ldots, x_m]$. It is easy to verify that $\langle \mathbf{a}, \mathbf{y} \rangle = a_1 \cdot (-\langle \mathbf{v}, \mathbf{z} \rangle + \langle \mathbf{v}, \mathbf{z} \rangle) = 0 \pmod{q}$. Further, by definition, $\|\mathbf{y}\| \leq t$.

Next, we define a related problem which will be useful for our reductions.

Definition 3.6. *Let* $q = p \cdot \prod_{i \in n} p_i$*, where all* $p_1 < p_2 < \ldots < p_n$ *are all co-prime and co-prime with p as well. Further let* $m \in \mathbb{N}$*. The* 1D-SIS-R$_{q,p,t,m}$ *problem is the following: Given* $\mathbf{v} \xleftarrow{\$} \mathbb{Z}_q^m$*, find* $\mathbf{z} \in \mathbb{Z}^m$ *with* $\|\mathbf{z}\| \leq t$ *such that* $\langle \mathbf{v}, \mathbf{z} \rangle \in [-t, t] + (q/p)\mathbb{Z}$.

The following corollary establishes the hardness of 1D-SIS-R based on 1D-SIS.

Corollary 3.7. *Let* q, p, t, m *be as in Definition 3.6. Then* 1D-SIS-R$_{q,p,t,m}$ *is at least as hards as* 1D-SIS$_{q/p,t,m}$.

Proof. The reduction works in the obvious way: Given an input $\mathbf{v} \in \mathbb{Z}_{q/p}^m$ for 1D-SIS$_{q/p,t,m}$, we embed $\mathbf{v}$ in $\mathbf{v}' \in \mathbb{Z}_q^m$, using CRT representation. Namely $\mathbf{v}' = \mathbf{v}$ $\pmod{q/p}$ and $\mathbf{v}' = \mathbf{r} \pmod{p}$, where $\mathbf{r} \xleftarrow{\$} \mathbb{Z}_p^m$. Then given a solution $\mathbf{z}$ for 1D-SIS-R$_{q,p,t,m}$ with input $\mathbf{v}'$, we claim that $\mathbf{z}$ is also a solution for 1D-SIS$_{q/p,t,m}$ with input $\mathbf{v}$. This follows since by definition $\|\mathbf{z}\| \leq t$, and since $\langle \mathbf{v}, \mathbf{z} \rangle \equiv \langle \mathbf{v}', \mathbf{z} \rangle$ $\pmod{q/p}$.

3.4   Admissible Hash Functions

The concept of admissible hash functions was defined by Boneh and Boyen [6] to convert selectively secure identity based encryption (IBE) schemes into fully secure ones. In this paper, we use admissible hash functions for our PRF construction. Our definition of admissible hash functions below will follow that of Cash, Hofheinz, Kiltz and Peikert [12] with minor changes (in particular, note that we do not require that the bad set is efficiently recognizable).

Definition 3.8 ([6,12]). *Let $\mathcal{H} = \{\mathcal{H}_\lambda\}_\lambda$ be a family of hash functions such that $\mathcal{H}_\lambda \subseteq (\{0,1\}^* \to \{0,1\}^\ell)$ for some $\ell = \ell(\lambda)$. We say that $\mathcal{H}$ is a family of admissible hash functions if for every $H \in \mathcal{H}$ there exists a set bad_H of "bad string-tuples" such that the following two properties hold:*

1. For every PPT algorithm $\mathcal{A}$, there is a negligible function ν such that

$$\Pr[(x^{(0)}, \ldots, x^{(t)}) \in \mathsf{bad}_H \mid H \leftarrow \mathcal{H}_\lambda, (x^{(0)}, \ldots, x^{(t)}) \leftarrow \mathcal{A}(1^\lambda, H)] \leq \nu(\lambda)$$

where the probability is over the choice of $H \leftarrow \mathcal{H}_\lambda$ and the coins of $\mathcal{A}$.
2. Let $\mathcal{L} = \{0,1\}^{2\ell}$, and for all $L \in \mathcal{L}$ define $\Pi_L : \{0,1\}^\ell \to \{0,1\}$ to be the string comparison with wildcards function. Namely, write L as a pair of strings $(\alpha, \beta) \in \{0,1\}^\ell$, and define

$$\Pi_{L=(\alpha,\beta)}(w) = 1 \Leftrightarrow \forall i \in [\ell] \left((\alpha_i = 0) \vee (\beta_i = w_i) \right) .$$

Intuitively, Π is a string comparison function with wildcards. It compares w and β only at those points where $\alpha_i = 1$. Note that this representation is somewhat redundant but it will be useful for our application.
Then, we require that for every polynomial $t = t(\lambda)$ there exists a noticeable function $\Delta_t(\lambda)$ and an efficiently sampleable distribution $\mathcal{L}_t$ over $\mathcal{L}$ such that for every $H \in \mathcal{H}_\lambda$ and sequences $(x^{(0)}, \ldots, x^{(t)}) \notin \mathsf{bad}_H$ with $x^{(0)} \notin \{x^{(1)}, \ldots, x^{(t)}\}$, we have:

$$\Pr_{L \leftarrow \mathcal{L}_t} [\Pi_L(H(x^{(0)})) \wedge \overline{\Pi_L(H(x^{(1)}))} \wedge \cdots \wedge \overline{\Pi_L(H(x^{(t)}))}] \geq \Delta_t(\lambda)$$

It has been shown by [6] that a family of admissible hash functions can be constructed based on any collision resistant hash function. In particular one can instantiate it based on the SIS problem (for virtually any parameter setting for which the problem is hard), which is at least as hard as LWE. Therefore throughout this manuscript we assume the existence of an LWE-based family of admissible hash functions, which will not add an additional assumption to our construction.

3.5   Attribute-Based Encryption

We define (leveled) attribute-based encryption, following [16,15]. An attribute-based encryption scheme for a class of predicate circuits $\mathcal{C}$ (namely, circuits with a single bit output) consists of four algorithms ($\mathcal{ABE}$.Setup, $\mathcal{ABE}$.KeyGen, $\mathcal{ABE}$.Enc, $\mathcal{ABE}$.Dec).

$\mathcal{ABE}$.Setup($1^\lambda, 1^\ell, 1^d$) → (pp, msk) : The setup algorithm gets as input the security parameter λ, the length ℓ of the attributes and the maximum depth
of the predicate circuits d, and outputs the public parameter (pp, mpk), and
the master key msk. All the other algorithms get pp as part of their input.

$\mathcal{ABE}$.KeyGen(msk, C) → sk_C : The key generation algorithm gets as input msk
and a predicate specified by $C \in \mathcal{C}$ (of depth at most d). It outputs a secret
key (C, sk_C).

$\mathcal{ABE}$.Enc(pp, $\mathbf{x}, m$) → ct : The encryption algorithm gets as input mpk, attributes
$\mathbf{x} \in \{0, 1\}^\ell$ and a message $m \in \mathcal{M}$. It outputs a ciphertext $(\mathbf{x}, \mathsf{ct})$.

$\mathcal{ABE}$.Dec($(C, \mathsf{sk}_C), (\mathbf{x}, \mathsf{ct})$) → m : The decryption algorithm gets as input a circuit C and the associated secret key sk_C, attributes $\mathbf{x}$ and an associated
ciphertext ct, and outputs either $\perp$ or a message $m \in \mathcal{M}$.

Correctness. We require that for all ℓ, d, all $(\mathbf{x}, C)$ such that $\mathbf{x} \in \{0, 1\}^\ell$, C has
depth at most d and $C(\mathbf{x}) = 1$, for all (pp, msk) ← $\mathcal{ABE}$.Setup($1^\lambda, 1^\ell, 1^d$), all
sk_C ← $\mathcal{ABE}$.KeyGen(msk, C), all ct ← $\mathcal{ABE}$.Enc(pp, $\mathbf{x}, m$), and all $m \in \mathcal{M}$,

$$\mathsf{Dec}((C, \mathsf{sk}_C), (\mathbf{x}, \mathsf{ct})) = m) .$$

Security Definition. We define selective security of ABE, which is sufficient
for our purposes. We allow the adversary to make multiple challenge message
queries, which is equivalent to the single query case but will be easier for us to
work with.

Definition 3.9. *For a stateful adversary $\mathcal{A}$, we define the advantage function*
$\mathsf{Adv}_{\mathcal{A}}^{ABE}$ *to be*

$$\Pr \left[b = b' : \begin{array}{l} b \xleftarrow{\$} \{0, 1\}; \\ \mathbf{x}_1, \ldots, \mathbf{x}_Q \leftarrow \mathcal{A}(1^\lambda, 1^\ell, 1^d); \\ (\mathsf{pp}, \mathsf{msk}) \leftarrow \mathcal{ABE}.\mathsf{Setup}(1^\lambda, 1^\ell, 1^d); \\ \{(m_{0,i}, m_{1,i})\}_{i \in [Q]} \leftarrow \mathcal{A}^{\mathcal{ABE}.\mathsf{KeyGen}(\mathsf{msk}, \cdot)}(\mathsf{pp}), \forall i.|m_{0,i}| = |m_{1,i}|; \\ \mathsf{ct}_i \leftarrow \mathcal{ABE}.\mathsf{Enc}(\mathsf{pp}, \mathbf{x}_i, m_{b,i}); \\ b' \leftarrow \mathcal{A}^{\mathcal{ABE}.\mathsf{KeyGen}(\mathsf{msk}, \cdot)}(\mathsf{ct}_1, \ldots, \mathsf{ct}_Q) \end{array} \right] - \frac{1}{2}$$

*with the restriction that all queries C that $\mathcal{A}$ makes to $\mathcal{ABE}$.KeyGen(msk, $\cdot$) satis-
fies $C(\mathbf{x}_i) = 0$ for all i (that is, sk_C does not decrypt the ciphertext corresponding
to any of the $\mathbf{x}_i$). An attribute-based encryption scheme is* selectively secure *if
for all PPT adversaries $\mathcal{A}$, the advantage $\mathsf{Adv}_{\mathcal{A}}^{ABE}$ is a negligible function in λ.*

We will use a special type of attribute-based encryption scheme with succinct
keys, namely one where $|\mathsf{sk}_C|$ does not grow with the size of the circuit C, but
rather only its depth.

Theorem 3.10 ([7]). *Let λ be the security parameter, and $d \in \mathbb{N}$. Let $n = n(\lambda, d)$, $q = q(\lambda, d) = n^{O(d)}$, and let χ be a poly(n)-bounded error distribution.
Then, there is a selectively secure ABE scheme for the class of depth-d-bounded
circuits, based on the hardness of $\mathrm{DLWE}_{n,q,\chi}$. Furthermore, the secret key sk_C
for a circuit C has size poly(λ, n, d).*

4   Embedding Circuits into Matrices

In this section, we present the core techniques that we use in our construction. In essence, we use a method, developed in a recent work by Boneh et al. [7] to "embed" bits $x_1, \ldots, x_k$ into matrices $\mathbf{A}_1, \ldots, \mathbf{A}_k$ and compute a circuit F on these matrices. This is done through a pair of algorithms (ComputeA, ComputeC) satisfying the following properties:

1. The deterministic algorithm ComputeA takes as input a circuit $F : \{0,1\}^k \to \{0,1\}$ and k matrices $\mathbf{A}_1, \ldots, \mathbf{A}_k$, and outputs a matrix $\mathbf{A}_F$; and
2. The deterministic algorithm ComputeC takes as input a bit string $\mathbf{x} = (x_1, \ldots, x_k) \in \{0,1\}^k$, and k LWE samples $\mathbf{s}^T(\mathbf{A}_i + x_i\mathbf{G}) + \mathbf{e}_i$, and outputs an LWE sample $\mathbf{s}^T(\mathbf{A}_F + F(\mathbf{x}) \cdot \mathbf{G}) + \mathbf{e}_F$ associated to the output matrix $\mathbf{A}_F$ and the output bit $F(\mathbf{x})$.

These algorithms are closely modeled on the work of Boneh et al. [7]. We now describe how these algorithms work, and what their properties are.

The Algorithm ComputeA. Given a circuit F, input matrices $\mathbf{A}_1, \ldots, \mathbf{A}_k$ (corresponding to the k input wires) and an auxiliary matrix $\mathbf{A}_0$, the ComputeA procedure works inductively, going through the gates of the circuit F from the input to the output. Assume without loss of generality that the circuit F is composed of NOT and AND gates. For every AND gate $g = (u, v; w)$, assume inductively that we have computed matrices $\mathbf{A}_u$ and $\mathbf{A}_v$ for the input wires u and v. Define

$$\mathbf{A}_w = -\mathbf{A}_u \cdot \mathbf{G}^{-1}(\mathbf{A}_v)$$

For every NOT gate $g = (u; w)$, define

$$\mathbf{A}_w = \mathbf{A}_0 - \mathbf{A}_u$$

The Algorithm ComputeC. Given a circuit F, an input $x \in \{0,1\}^k$ and LWE samples $(\mathbf{A}_i, \mathbf{y}_i)$, the ComputeC algorithm works as follows. For each AND gate $g = (u, v; w)$, assume that we have computed LWE samples $(\mathbf{A}_u, \mathbf{y}_u)$ and $(\mathbf{A}_v, \mathbf{y}_v)$ for the input wires u and v. Define

$$\mathbf{y}_w = x_u \cdot \mathbf{y}_v - \mathbf{y}_u \cdot \mathbf{G}^{-1}(\mathbf{A}_v)$$

where x_u and x_v are the bits on wires u and v when evaluating the circuit F on input x. For every NOT gate $g = (u; w)$, define

$$\mathbf{y}_w = \mathbf{y}_0 - \mathbf{y}_u$$

We will need the following lemma about the behavior of ComputeA and ComputeC. (We remind the reader that we use $|| \cdot ||$ to denote the ℓ_∞ norm).

Lemma 4.1. *Let F be a depth-d Boolean circuit on k input bits, and let $x \in \{0,1\}^k$ be an input. Let $\mathbf{A}_0, \mathbf{A}_1, \ldots, \mathbf{A}_k \in \mathbb{Z}_q^{n \times m}$ and $\mathbf{y}_0, \ldots, \mathbf{y}_k \in \mathbb{Z}_q^m$ be such that*

$$||\mathbf{y}_i - \mathbf{s}^T(\mathbf{A}_i + x_i\mathbf{G})|| \leq B \qquad for \ i = 0, 1, \ldots, k.$$

for some $\mathbf{s} \in \mathbb{Z}_q^n$ and $B = B(\lambda)$. Let $\mathbf{A}_F \leftarrow \mathsf{ComputeA}(F, \mathbf{A}_0, \ldots, \mathbf{A}_k)$ and $\mathbf{y}_F \leftarrow \mathsf{ComputeC}(F, x, \mathbf{A}_0, \ldots, \mathbf{A}_k, \mathbf{y}_0, \ldots, \mathbf{y}_k)$. Then, $\|\mathbf{y}_F - \mathbf{s}^T(\mathbf{A}_F + F(\mathbf{x}) \cdot \mathbf{G})\| \leq m^{O(d)} \cdot B$.

Furthermore, $\mathbf{y}_F$ is a "low-norm" linear function of $\mathbf{y}_0, \ldots, \mathbf{y}_k$. That is, there are matrices $\mathbf{Z}_0, \ldots, \mathbf{Z}_k$ (which depend on the function F, the input $\mathbf{x}$, and the input matrices $\mathbf{A}_0, \ldots, \mathbf{A}_k$) such that $\mathbf{y}_F = \sum_{i=0}^k \mathbf{y}_i \mathbf{Z}_i$ and $\|\mathbf{Z}_i\| \leq m^{O(d)} \cdot B$.

Proof. We show this by induction on the levels of the circuit F, starting from the input. Consider two cases.

AND gate. Consider an AND gate $g = (u, v; w)$ where the input wires are at level L, and assume that $\mathbf{y}_u = \mathbf{s}^T(\mathbf{A}_u + x_u \mathbf{G}) + \mathbf{e}_u$ and $\mathbf{y}_v = \mathbf{s}^T(\mathbf{A}_v + x_v \mathbf{G}) + \mathbf{e}_v$, with $\|\mathbf{e}_u\|, \|\mathbf{e}_v\| \leq (m+1)^L \cdot B$. Now,

$$\mathbf{y}_w = x_u \cdot \mathbf{y}_v - \mathbf{y}_u \cdot \mathbf{G}^{-1}(\mathbf{A}_v)$$

$$= x_u \cdot \left(\mathbf{s}^T(\mathbf{A}_v + x_v \mathbf{G}) + \mathbf{e}_v\right) - \left(\mathbf{s}^T(\mathbf{A}_u + x_u \mathbf{G}) + \mathbf{e}_u\right) \cdot \mathbf{G}^{-1}(\mathbf{A}_v)$$

$$= \mathbf{s}^T\left(x_u \mathbf{A}_v + x_u x_v \mathbf{G} - \mathbf{A}_u \mathbf{G}^{-1}(\mathbf{A}_v) - x_u \mathbf{A}_v\right) + \left(-\mathbf{e}_u \mathbf{G}^{-1}(\mathbf{A}_v) + x_u \mathbf{e}_v\right)$$

$$= \mathbf{s}^T(\mathbf{A}_w + x_w \mathbf{G}) + \mathbf{e}_w$$

where $\mathbf{A}_w = -\mathbf{A}_u \cdot \mathbf{G}^{-1}(\mathbf{A}_v)$, $x_w = x_u x_v$, and

$$\|\mathbf{e}_w\| \leq m \cdot \|\mathbf{e}_u\| + \|\mathbf{e}_v\| \leq (m+1) \cdot (m+1)^L \cdot B \leq (m+1)^{L+1} \cdot B$$

NOT gate. In a similar vein, for a NOT gate $g = (u; w)$, assume that $\mathbf{y}_u = \mathbf{s}^T(\mathbf{A}_u + x_u \mathbf{G}) + \mathbf{e}_u$, with $\|\mathbf{e}_u\| \leq (m+1)^L \cdot B$. Then,

$$\mathbf{y}_w = \mathbf{y}_0 - \mathbf{y}_u = \mathbf{s}^T(\mathbf{A}_0 + \mathbf{G} - \mathbf{A}_u - x_u \mathbf{G}) + (\mathbf{e}_0 - \mathbf{e}_u)$$
$$= \mathbf{s}^T(\mathbf{A}_w + (1 - x_u)\mathbf{G}) + \mathbf{e}_w$$

where $\mathbf{A}_w = \mathbf{A}_0 - \mathbf{A}_u$, $x_w = 1 - x_u$, and

$$\|\mathbf{e}_w\| \leq \|\mathbf{e}_0\| + \|\mathbf{e}_u\| \leq B + (m+1)^L \cdot B \leq (m+1)^{L+1} \cdot B$$

Thus, $\mathbf{y}_F = \mathbf{s}^T \mathbf{A}_F + \mathbf{e}_F$ where $\|\mathbf{e}_F\| \leq m^{O(d)} \cdot B$. Furthermore, both transformations are linear functions on $\mathbf{y}_u$ and $\mathbf{y}_v$, as required.

5 Constrained PRF

5.1 Construction

A family of functions $\mathcal{F} \subseteq (\{0,1\}^* \rightarrow \{0,1\})$ is z-uniform if each function $F \in \mathcal{F}$ can be described by a string in $\{0,1\}^z$ (we associate F with its description), and there exists a uniform circuit family $\{\mathcal{U}_k\}_{k \in \mathbb{N}}$ such that $\mathcal{U}_k : \{0,1\}^z \times \{0,1\}^k \rightarrow$

$\{0, 1\}$ such that for all $x \in \{0, 1\}^k$ it holds that $\mathcal{U}_k(F, x) = F(x)$. We assume for the sake of simplicity that the depth of $\mathcal{U}_k$ grows monotonically with k and for all d we let k_d to be the maximal input size for which $\mathcal{U}_k$ has depth at most d. We define $\mathcal{F}_d$ to be such that $F \in \mathcal{F}$ is undefined for inputs of length $k > k_d$. We call such a family d-depth-bounded.

Our constrained PRF for a z-uniform d-depth-bounded family $\mathcal{F}$ works as follows.

- KeyGen($1^\lambda, 1^z, 1^d$): The key generation algorithm takes as input the maximum size z and depth d of the constraining circuits. Let $\mathcal{H}$ be a family of admissible hash functions (see Section 3.4) and let $\ell = \ell(\lambda)$ be the output length of hash functions in the family.
  Let $n = n(\lambda, d)$, $q = q(\lambda, d)$, $p = p(\lambda, d)$ be parameters chosen as described in Section 5.2 below, let $m = n \lceil \log q \rceil$.
  Generate $z + 2\ell + 3$ matrices as follows: let $\mathbf{A}_0$ and $\mathbf{A}_1$ be the "input matrices", let $\mathbf{B}_1, \mathbf{B}_2, \ldots, \mathbf{B}_z$ be the "function matrices", let $\mathbf{C}_1, \ldots, \mathbf{C}_{2\ell}$ be the "partitioning matrices", and let $\mathbf{D}$ be an "auxiliary matrix". All of these matrices are uniform in $\mathbb{Z}_q^{n \times m}$ (note that the "gadget matrix" $\mathbf{G}$ has the same dimensions). In addition sample an admissible hash function $H \xleftarrow{\$} \mathcal{H}_\lambda$. The public parameters consist of

$$\mathcal{PP} = (H, \mathbf{A}_0, \mathbf{A}_1, \mathbf{B}_1, \ldots, \mathbf{B}_z, \mathbf{C}_1, \ldots, \mathbf{C}_{2\ell}, \mathbf{D})$$

  The seed of the PRF is a uniformly random vector $\mathbf{s} \in \mathbb{Z}_q^n$.
- Eval($\mathbf{s}, \mathcal{PP}, \mathbf{x}$) takes as input the PRF seed $\mathbf{s}$, the public parameters $\mathcal{PP}$, and an input $x \in \{0, 1\}^k$ such that $k \leq k_d$ (i.e. $\mathcal{U}_k$ is of depth $\leq d$), and works as follows.
  Recall that $\mathcal{U}_k : \{0, 1\}^z \times \{0, 1\}^k \to \{0, 1\}$ is the universal circuit that takes a description of a function F and an input x and outputs $\mathcal{U}_k(F, x) = F(x)$. Let $\Pi : \{0, 1\}^{2\ell} \times \{0, 1\}^\ell \to \{0, 1\}$ denote the circuit that computes $\Pi(L, w) = \Pi_L(w)$ from Definition 3.8. Note that Π can be implemented by a binary circuit of depth $\log(\ell) + O(1)$.
  Let $(x_1, \ldots, x_k)$ denote the bits of x. Let $w = H(x)$, and let $w_1, \ldots, w_\ell$ be its bits. Compute

$$\mathbf{B}_\mathcal{U} \leftarrow \mathsf{ComputeA}(\mathcal{U}_k, \mathbf{B}_1, \ldots, \mathbf{B}_z, \mathbf{A}_{x_1}, \mathbf{A}_{x_2}, \ldots, \mathbf{A}_{x_k}) \qquad (1)$$
$$\mathbf{C}_\Pi \leftarrow \mathsf{ComputeA}(\Pi, \mathbf{C}_1, \ldots, \mathbf{C}_{2\ell}, \mathbf{A}_{w_1}, \mathbf{A}_{w_2}, \ldots, \mathbf{A}_{w_\ell}) \qquad (2)$$

  and output
$$\mathsf{PRF}_\mathbf{s}(\mathbf{x}) = \left\lfloor \mathbf{s}^T \mathbf{B}_\mathcal{U} \cdot \mathbf{G}^{-1}(\mathbf{C}_\Pi) \cdot \mathbf{G}^{-1}(\mathbf{D}) \right\rceil_p$$

- Constrain($\mathbf{s}, \mathcal{PP}, F$) takes as input the PRF key $\mathbf{s}$ and a circuit F (of size at most z) and does the following. Compute

$$\mathbf{a}_b = \mathbf{s}^T(\mathbf{A}_b + b \cdot \mathbf{G}) + \mathbf{e}_{1,b}^T \in \mathbb{Z}_q^m \text{ for } b \in \{0, 1\}$$
$$\mathbf{b}_i = \mathbf{s}^T(\mathbf{B}_i + f_i \cdot \mathbf{G}) + \mathbf{e}_{2,i}^T \in \mathbb{Z}_q^m \text{ for all } i \in [z]$$

where the vectors $\mathbf{e}$ are drawn from an error distribution χ to be specified later (in Section 5.2).

The constrained seed K_F is the tuple $(\mathbf{a}_0, \mathbf{a}_1, \mathbf{b}_1, \dots, \mathbf{b}_z) \in (\mathbb{Z}_q^m)^{z+2}$.

— ConstrainEval$(K_F, \mathcal{PP}, \mathbf{x})$ takes as input the constrained key K_F and an input $\mathbf{x}$. It computes

$$\mathbf{b}_{\mathcal{U},\mathbf{x}} \leftarrow \mathsf{ComputeC}\Big(\mathcal{U}, (\mathbf{b}_1, \dots, \mathbf{b}_z, \mathbf{a}_{x_1}, \dots, \mathbf{a}_{x_k}), (f_1, \dots, f_z, x_1, \dots, x_k) \Big)$$

and outputs $\lfloor \mathbf{b}_{\mathcal{U},\mathbf{x}} \cdot \mathbf{G}^{-1}(\mathbf{C}_\Pi) \cdot \mathbf{G}^{-1}(\mathbf{D}) \rfloor_p$, where $\mathbf{C}_\Pi$ is defined as above.

5.2   Setting the Parameters

Let us start by providing a typical parameter setting, and then explain how parameters can be modified and the effect on security.

Consider setting $n(\lambda, d) = (\lambda \cdot d)^c$, for a constant c that will be discussed shortly. We will set χ to be a discrete Gaussian distribution $D_{\mathbb{Z}, \alpha q}$ s.t. $\alpha q = \Theta(\sqrt{n})$. We define $n' = \lambda$ and let $p_1, \dots, p_{n'} = m^{O(d + \log \ell)}$ be all primes, and $p = \mathrm{poly}(\lambda)$ (in fact, there is a lot of freedom in the choice of p, and it can be as large as $m^{O(d + \log \ell)}$ under the same asymptotic hardness). Finally, let $q = p \cdot (\alpha q) \cdot \prod_{i \in [n']} p_i = m^{n' \cdot O(d + \log \ell)} = 2^{\tilde{O}(\lambda \cdot d)} = 2^{\tilde{O}(n^{1/c})}$ (recall that $\ell = \mathrm{poly}(\lambda)$).

This parameter setting translates into a PRF with $m = n \lceil \log q \rceil \cdot \Theta(\log \lambda)$ output bits per input, whose security is based (as we show in the next section) on the hardness of approximating lattice problems to within a factor of $2^{\tilde{O}(n^{1/c})}$.

Taking larger values of c will increase the hardness of the underlying lattice problem, but at the cost of considerably increasing the element sizes.

5.3   Security

Throughout this section, we let $\mathcal{F}$ be a family of z-uniform functions and let d be a depth bound (both can depend on λ). We let $n = n(\lambda, d)$, $m = m(\lambda, d)$, $q = q(\lambda, d)$, $p = p(\lambda, d)$ and the noise distributions $\chi = \chi(\lambda, d)$ be as defined in Section 5.2. We let $\mathcal{H}$ be the family of admissible hash functions as described in Section 3.4, with range $\{0, 1\}^\ell$.

Theorem 5.1. *Let $\mathcal{F}$ be a family of z-uniform functions and let d be a depth bound (both can depend on λ). Let $n = n(\lambda, d)$, $m = m(\lambda, d)$, $q = q(\lambda, d)$, $p = p(\lambda, d)$ and the noise distributions $\chi = \chi(\lambda, d)$ be as defined in Section 5.2. Further let $m' = m \cdot (z + 2\ell + 3)$, and $\gamma = \omega(\sqrt{n \log \lambda}) \cdot p \cdot m^{O(d + \log \ell)}$. Assuming the hardness of DLWE$_{n,q,\chi}$, 1D-SIS-R$_{q,p,\gamma,m'}$ and the admissible hash function family $\mathcal{H}$, the scheme $\mathcal{CPRF} = (\mathsf{KeyGen}, \mathsf{Eval}, \mathsf{Constrain}, \mathsf{ConstrainEval})$ is a single-key secure selective-function secure constrained PRF for $\mathcal{F}$.*

We note that the hardness of all three assumptions translates to the worst case hardness of approximating lattice problems such as GapSVP and SIVP to within sub-exponential factors.

Proof. Let $\mathcal{A}$ be a PPT selective-constraint adaptive-input adversary against $\mathcal{CPRF}_{z,d}$. Let $t = \mathrm{poly}(\lambda)$ be the (polynomial) number of input queries made by $\mathcal{A}$ (w.l.o.g). Let ϵ be the advantage of $\mathcal{A}$ in the constrained PRF game. We let $B = \alpha q \cdot \omega(\sqrt{\log \lambda})$. It holds that with all but negligible probabilities, all samples that we take from χ will have absolute value at most B. For the duration of the proof we assume that this is indeed the case.

The proof will proceed by a sequence of hybrids (or experiments) where the challenger samples a bit $b \in \{0, 1\}$ and interacts with $\mathcal{A}$. We let $\mathrm{Adv}_{\mathsf{H}}(\mathcal{A})$ denote the probability that $\mathcal{A}$ outputs b in hybrid H.

Hybrid H_0. This hybrid is the legitimate constrained PRF security game. The challenger generates $(\mathbf{s}, \mathcal{PP}) \leftarrow \mathsf{KeyGen}(1^\lambda, 1^z, 1^d)$. It gets $F \in \{0, 1\}^z$ from $\mathcal{A}$ and produces a constrained key $K_F \leftarrow \mathsf{Constrain}(\mathbf{s}, \mathcal{PP}, F)$. It then sends $\mathcal{PP}, K_F$ to $\mathcal{A}$. At this point $\mathcal{A}$ adaptively makes queries $x^{(i)} \in \{0, 1\}^*$, and the challenger computes $y^{(i)} \leftarrow \mathsf{Eval}(\mathbf{s}, \mathcal{PP}, x^{(i)})$ and returns it to $\mathcal{A}$. Finally, $\mathcal{A}$ outputs $x^* \in \{0, 1\}^*$. If $b = 0$ then the challenger returns $y^* \leftarrow \mathsf{Eval}(\mathbf{s}, \mathcal{PP}, x^*)$, and if $b = 1$ it returns a random y^*. Therefore, we have

$$\mathrm{Adv}_{\mathsf{H}_0}(\mathcal{A}) \geq 1/2 + \epsilon \ .$$

Hybrid H_1. This is the notorious "artificial abort" phase. Let $\Delta_t = \Delta_t(\lambda)$ be the noticeable function from Definition 3.8. This hybrid is identical to the previous one, except in the last step the challenger flips a coin and with probability $1 - \Delta_t/2$ aborts the experiment (hence giving the adversary no information on b).

The adversary's advantage thus degrades appropriately:

$$\mathrm{Adv}_{\mathsf{H}_1}(\mathcal{A}) \geq (\Delta_t/2) \cdot (1/2 + \epsilon) + (1 - \Delta_t/2) \cdot (1/2) = 1/2 + \epsilon \cdot \Delta_t/2 \ .$$

Hybrid H_2. In this hybrid, we associate some meaning with the artificial abort. Intuitively, the abort will be associated with a failure of the admissible hash function to partition the queries correctly. We are guaranteed that correct partitioning happens with probability $\geq \Delta_t$ (except for sequences that are hard to generate), but we would like to make it *(almost) exactly* $\Delta_t/2$ so as to not correlate the adversary's success probability with the string L (the loss of the 2 factor is due to probability estimation).

Specifically, in this hybrid, rather than flipping a coin at the end of the experiment, the challenger does the following. For all $\vec{x} = (x^{(1)}, \ldots, x^{(t)}, x^*)$, we define the event $\mathsf{GoodPartition}_{L,\vec{x}}$ to be the event in which $\Pi_L(H(x^{(1)})) = \cdots = \Pi_L(H(x^{(t)})) = 0$ and $\Pi_L(H(x^*)) = 1$, and define $\delta_{\vec{x}} = \mathrm{Pr}_{L \xleftarrow{\$} \mathcal{L}_t}[\mathsf{GoodPartition}_{\vec{x},L}]$. The challenger will first compute an estimate $\tilde{\delta}_{\vec{x}}$ of $\delta_{\vec{x}}$ by sampling multiple values of L from $\mathcal{L}_t$ and using Chernoff (both additive and multiplicative). Using $\mathrm{poly}(\lambda)$-many samples we can compute $\tilde{\delta}_{\vec{x}}$ such that

$$\mathrm{Pr}\left[\left|\delta_{\vec{x}} - \tilde{\delta}_{\vec{x}}\right| > \Delta_t/4\right] \leq 2^{-\lambda} \ .$$

and in addition if $\delta_{\vec{x}} \geq \Delta_t/2$ then

$$\Pr\left[\left|\frac{\delta_{\vec{x}}}{\tilde{\delta}_{\vec{x}}} - 1\right| > \epsilon/2\right] \leq 2^{-\lambda} .$$

The challenger will then perform as follows: (i) It first verifies that $\tilde{\delta}_{\vec{x}} \geq \frac{3}{4}\Delta_t$, and aborts if this is not the case. (ii) It then samples $L \xleftarrow{\$} \mathcal{L}_t$ and aborts if GoodPartition$_{\vec{x},L}$ did not occur (note that by our definitions above, this happens with probability $1 - \delta_{\vec{x}}$ over the choice of L). (iii) Then it flips a coin with probability $\frac{\tilde{\delta}_{\vec{x}} - \Delta_t/2}{\tilde{\delta}_{\vec{x}}}$ and aborts if the outcome is 1. Otherwise it carries out the experiment towards completion.

To analyze the effect on the success probability, we first notice that the probability that $\tilde{\delta}_{\vec{x}} < \frac{3}{4}\Delta_t$ (abortion is step (i)) is negligible. This is since, except with $2^{-\lambda}$ probability, this indicates that $\delta_{\vec{x}} < \Delta_t$, which implies that $\vec{x} \in \mathsf{bad}_H$. Definition 3.8 guarantees that this happens with probability at most $\nu(\lambda) = \mathsf{negl}(\lambda)$.

If the above abort did not occur, we know that $\delta_{\vec{x}} \geq \Delta_t/2$ (except with probability $2^{-\lambda}$), we first notice that the total probability of abort in steps $(ii) + (iii)$

$$1 - \delta_{\vec{x}} + \delta_{\vec{x}} \cdot \frac{\tilde{\delta}_{\vec{x}} - \Delta_t/2}{\tilde{\delta}_{\vec{x}}} = 1 - \frac{\delta_{\vec{x}}}{\tilde{\delta}_{\vec{x}}}\Delta_t/2 \in \left[(1 - \Delta_t/2) - \epsilon\Delta_t/4, \ (1 - \Delta_t/2) + \epsilon\Delta_t/4\right]$$

It therefore follows that if there was no abort in step (i), then the adversary's view in H_2 is within statistical distance $2^{-\lambda} + \epsilon\Delta_t/4$ from its view in H_1.

Putting all steps together, we get that

$$\mathsf{Adv}_{\mathsf{H}_2}(\mathcal{A}) \geq 1/2 + \epsilon \cdot \Delta_t/2 - \nu(\lambda) - O(2^{-\lambda}) - \epsilon\Delta_t/4 = 1/2 + \epsilon \cdot \Delta_t/4 - \mathsf{negl}(\lambda) .$$

Hybrid H_3. In this hybrid, the challenger first samples $L \xleftarrow{\$} \mathcal{L}_t$, and then, for each $x^{(i)}$ in turn, it checks whether $\Pi_L(H(x^{(i)})) = 0$, and immediately aborts if not. Similarly, upon receiving x^*, it checks whether $\Pi_L(H(x^*)) = 1$ and immediately aborts if not. Otherwise it continues the same as H_2.

It is rather straightforward to see that the $\mathcal{A}$'s advantage does not change. The cases in which we abort are exactly the same as the ones in the previous hybrid (since it is sufficient that a single $x^{(i)}$ does not give the required value in order to abort). Further, the sampling of L has been completely independent of all the other randomness in the experiment so it might as well happen in the beginning. We conclude that

$$\mathsf{Adv}_{\mathsf{H}_3}(\mathcal{A}) = \mathsf{Adv}_{\mathsf{H}_2}(\mathcal{A}) \geq 1/2 + \epsilon \cdot \Delta_t/4 - \mathsf{negl}(\lambda) .$$

Hybrid H_4. In this hybrid, the challenger changes the way the matrices $\mathbf{A}, \mathbf{B}, \mathbf{C}$ are generated. Recall that our security game is constraint-selective, namely $\mathcal{A}$ produces the constraint F before seeing the public parameters.

Therefore, here, the challenger waits until receiving F from $\mathcal{A}$ and only generates the public parameters at that point (note that by then L has also been specified). To generate the public parameters, the matrix $\mathbf{D}$ is produced identically to before. In addition, the challenger samples matrices $\{\hat{\mathbf{A}}_\beta\}_{\beta \in \{0,1\}}$, $\{\hat{\mathbf{B}}_i\}_{i \in [z]}$, $\{\hat{\mathbf{C}}_i\}_{i \in [2\ell]}$ It then sets

$$\mathbf{A}_\beta = \hat{\mathbf{A}}_\beta - \beta \mathbf{G}$$
$$\mathbf{B}_i = \hat{\mathbf{B}}_i - f_i \mathbf{G}$$
$$\mathbf{C}_i = \hat{\mathbf{C}}_i - L_i \mathbf{G}$$

The remainder of the experiment remains unchanged.

Since the distributions of the $\mathbf{A}, \mathbf{B}, \mathbf{C}$ matrices is identical to their original uniform distributions, it follows that

$$\mathrm{Adv}_{\mathsf{H}_4}(\mathcal{A}) = \mathrm{Adv}_{\mathsf{H}_3}(\mathcal{A}) \ .$$

Hybrid H_5. In this hybrid, the adversary changes the way it computes the outputs $y^{(i)}$. Recall that $K_F = (\mathbf{a}_0, \mathbf{a}_1, \mathbf{b}_1, \ldots, \mathbf{b}_z)$ is the constrained key given to $\mathcal{A}$. Let us denote

$$\mathbf{c}_i = \mathbf{s}^T(\mathbf{C}_i + L_i \mathbf{G}) + \mathbf{e}_{3,i}^T \quad \text{for all } i \in [z]$$
$$\mathbf{d} = \mathbf{s}^T \mathbf{D} + \mathbf{e}_4^T$$

where $\mathbf{e}_{3,i}$ are sampled coordinate-wise from χ, and $\mathbf{e}_4$ is sampled coordinate-wise from χ'.

In this hybrid, in order to answer input queries, the challenger first computes

$$\mathbf{b}_{\mathcal{U}, x^{(i)}} \leftarrow \mathsf{ComputeC}\left(\mathcal{U}, (\mathbf{b}_1, \ldots, \mathbf{b}_z, \mathbf{a}_{x_1}, \ldots, \mathbf{a}_{x_k}), (f_1, \ldots, f_z, x_1^{(i)}, \ldots, x_k^{(i)})\right)$$

and then, letting $w^{(i)} = H(x^{(i)})$

$$\mathbf{c}_{\Pi, w^{(i)}} \leftarrow \mathsf{ComputeC}\left(\Pi, (\mathbf{c}_1, \ldots, \mathbf{c}_{2\ell}, \mathbf{a}_{w_1}, \ldots, \mathbf{a}_{w_\ell}), (L_1, \ldots, L_{2\ell}, w_1^{(i)}, \ldots, w_\ell^{(i)})\right)$$

We recall that by Lemma 4.1 it holds that

$$\mathbf{b}_{\mathcal{U}, x^{(i)}}^T = \mathbf{s}^T(\mathbf{B}_{\mathcal{U}, x^{(i)}} + F(x^{(i)}) \cdot \mathbf{G}) + \mathbf{e}_{\mathcal{U}}^T$$
$$\mathbf{c}_{\Pi, w^{(i)}}^T = \mathbf{s}^T(\mathbf{C}_{\Pi, w^{(i)}} + \Pi_L(w^{(i)}) \cdot \mathbf{G}) + \mathbf{e}_\Pi^T \ ,$$

for some $\mathbf{e}_{\mathcal{U}}, \mathbf{e}_\Pi$ for which $\|\mathbf{e}_{\mathcal{U}}\| \le B \cdot m^{O(d)}$, $\|\mathbf{e}_\Pi\| \le B \cdot m^{O(\log \ell)}$.

We recall that by definition

$$
\begin{aligned}
\mathsf{PRF_s}(x^{(i)}) &= \left\lfloor \mathbf{s}^T \mathbf{B}_{\mathcal{U},x^{(i)}} \cdot \mathbf{G}^{-1}(\mathbf{C}_{\Pi,w^{(i)}}) \mathbf{G}^{-1}(\mathbf{D}) \right\rfloor_p \\
&= \Big\lfloor \mathbf{s}^T (\mathbf{B}_{\mathcal{U},x^{(i)}} + F(x^{(i)})\mathbf{G}) \cdot \mathbf{G}^{-1}(\mathbf{C}_{\Pi,w^{(i)}}) \mathbf{G}^{-1}(\mathbf{D}) \\
&\qquad\qquad - F(x^{(i)})\mathbf{s}^T \mathbf{C}_{\Pi,w^{(i)}} \mathbf{G}^{-1}(\mathbf{D}) \Big\rfloor_p \\
&= \Big\lfloor \mathbf{s}^T (\mathbf{B}_{\mathcal{U},x^{(i)}} + F(x^{(i)})\mathbf{G}) \cdot \mathbf{G}^{-1}(\mathbf{C}_{\Pi,w^{(i)}}) \mathbf{G}^{-1}(\mathbf{D}) \\
&\qquad -F(x^{(i)})\mathbf{s}^T (\mathbf{C}_{\Pi,w^{(i)}} + \Pi_L(w^{(i)})\mathbf{G}) \mathbf{G}^{-1}(\mathbf{D}) \\
&\qquad +F(x^{(i)}) \Pi_L(w^{(i)}) \mathbf{s}^T \mathbf{D} \Big\rfloor_p \\
&= \Big\lfloor \mathbf{b}_{\mathcal{U},x^{(i)}}^T \cdot \mathbf{G}^{-1}(\mathbf{C}_{\Pi,w^{(i)}}) \mathbf{G}^{-1}(\mathbf{D}) - F(x^{(i)}) \mathbf{c}_{\Pi,w^{(i)}}^T \mathbf{G}^{-1}(\mathbf{D}) \\
&\qquad +F(x^{(i)}) \Pi_L(w^{(i)}) \mathbf{d}^T + \mathbf{e}'^T \Big\rfloor_p \,,
\end{aligned}
\tag{3}
$$

where

$$
\mathbf{e}'^T = -\mathbf{e}_{\mathcal{U}}^T \mathbf{G}^{-1}(\mathbf{C}_{\Pi,w^{(i)}}) \mathbf{G}^{-1}(\mathbf{D}) + F(x^{(i)}) \mathbf{e}_{\Pi}^T \mathbf{G}^{-1}(\mathbf{D}) - F(x^{(i)}) \Pi_L(w^{(i)}) \mathbf{e}_d^T
\tag{4}
$$

which implies that $\|\mathbf{e}'\| \le E$ for some $E = (m^{O(d)} + m^{O(\log \ell)}) \cdot B$.

To analyze the distinguishing probability between these hybrids, for any input x (and $w = H(x)$) we define the event $\mathsf{Borderline}_x$ as the event where there exists $j \in [m]$ such that:

$$
\begin{aligned}
(\mathbf{b}_{\mathcal{U},x}^T \cdot \mathbf{G}^{-1}(\mathbf{C}_{\Pi,w}) \cdot \mathbf{G}^{-1}(\mathbf{D}) &- F(x) \cdot \mathbf{c}_{\Pi,w}^T \cdot \mathbf{G}^{-1}(\mathbf{D}) \\
&+ F(x) \cdot \Pi_L(w) \cdot \mathbf{d}^T) \cdot \mathbf{u}_j \in [-E, E] + (q/p)\mathbb{Z} \,,
\end{aligned}
$$

where we recall that $\mathbf{u}_j$ is the jth indicator vector. Namely, this is the probability that one of the coordinates of the vector $\mathbf{b}_{\mathcal{U},x}^T \cdot \mathbf{G}^{-1}(\mathbf{C}_{\Pi,w})\mathbf{G}^{-1}(\mathbf{D}) - F(x)\mathbf{c}_{\Pi,w}^T \mathbf{G}^{-1}(\mathbf{D}) + F(x)\Pi_L(w)\mathbf{d}^T$ is "dangerously close" to being rounded in the wrong direction.

By definition of rounding, if $\neg\mathsf{Borderline}_{x^{(i)}}$, then

$$
\begin{aligned}
\mathsf{PRF_s}(x^{(i)}) = \lfloor \mathbf{b}_{\mathcal{U},x^{(i)}}^T \cdot \mathbf{G}^{-1}(\mathbf{C}_{\Pi,w^{(i)}}) \mathbf{G}^{-1}(\mathbf{D}) &- F(x^{(i)}) \mathbf{c}_{\Pi,w^{(i)}}^T \mathbf{G}^{-1}(\mathbf{D}) \\
&+ F(x^{(i)}) \Pi_L(w^{(i)}) \mathbf{d}^T \rfloor_p \,.
\end{aligned}
$$

The challenger in this hybrid, given a query $x^{(i)}$, will first check whether $\mathsf{Borderline}_{x^{(i)}}$. If the event happens, the challenger aborts. Otherwise it returns $\mathsf{PRF_s}(x^{(i)})$ as defined above. Note that the challenger only needs to respond to queries $x^{(i)}$ for which $\Pi_L(w^{(i)}) = \Pi_L(H(x^{(i)})) = 0$, which do not depend on $\mathbf{d}$, a fact that will be important later on.

Finally, on the challenge query x^*, unless abort is needed, it holds that $F(x^*) = 1$ and $\Pi_L(w^*) = 1$ (where $w^* = H(x^*)$) and therefore, unless the event $\mathsf{Borderline}_{x^*}$ happens, it holds that

$$\mathsf{PRF_s}(x^*) = \left\lfloor \mathbf{b}_{\mathcal{U},x^*}^T \cdot \mathbf{G}^{-1}(\mathbf{C}_{\Pi,w^{(i)}})\mathbf{G}^{-1}(\mathbf{D}) - \mathbf{c}_{\Pi,w^*}^T \mathbf{G}^{-1}(\mathbf{D}) + \mathbf{d}^T \right\rceil_p .$$

The challenger will therefore abort if $\mathsf{Borderline}_{x^*}$ and return the aforementioned value otherwise (that is if the bit b is 0; if $b = 1$ then of course a uniform value is returned).

It follows that if we define $\mathsf{Borderline} = (\vee_i \mathsf{Borderline}_{x^{(i)}}) \vee \mathsf{Borderline}_{x^*}$, then

$$|\mathrm{Adv}_{\mathsf{H}_5}(\mathcal{A}) - \mathrm{Adv}_{\mathsf{H}_4}(\mathcal{A})| \leq \Pr_{\mathsf{H}_5}[\mathsf{Borderline}] .$$

We will bound $\Pr_{\mathsf{H}_5}[\mathsf{Borderline}]$ as a part of our analysis in the next hybrid.

As a final remark on this hybrid, we note that in order to execute this hybrid, the challenger does not need to access $\mathbf{s}$ itself, but rather only the $\mathbf{a}_\beta, \mathbf{b}_i, \mathbf{c}_i, \mathbf{d}$ vectors. This will be useful in the next hybrid.

Hybrid H_6. In this hybrid, all $\mathbf{a}_\beta, \mathbf{b}_i, \mathbf{c}_i, \mathbf{d}$ are sampled from the uniform distribution. Everything else remains the same. We note that by definition, in hybrid H_5:

$$\mathbf{a}_\beta^T = \mathbf{s}^T \hat{\mathbf{A}}_\beta + \mathbf{e}_{1,\beta}^T$$
$$\mathbf{b}_i^T = \mathbf{s}^T \hat{\mathbf{B}}_i + \mathbf{e}_{2,i}^T$$
$$\mathbf{c}_i^T = \mathbf{s}^T \hat{\mathbf{C}}_i + \mathbf{e}_{3,i}^T$$
$$\mathbf{d}^T = \mathbf{s}^T \mathbf{D} + \mathbf{e}_4^T ,$$

where all $\hat{\mathbf{A}}_\beta, \hat{\mathbf{B}}_i, \hat{\mathbf{C}}_i, \mathbf{D}$ are uniformly distributed, and all $\mathbf{e}_{1,\beta}^T, \mathbf{e}_{2,i}^T, \mathbf{e}_{3,i}^T, \mathbf{e}_4^T$ are sampled coordinate-wise from χ. The $\mathsf{DLWE}_{n,q,\chi}$ assumption therefore asserts that:

$$|\mathrm{Adv}_{\mathsf{H}_6}(\mathcal{A}) - \mathrm{Adv}_{\mathsf{H}_5}(\mathcal{A})| \leq \mathrm{negl}(\lambda) .$$

Furthermore, since $\mathsf{Borderline}$ is an efficiently recognizable event, it also holds that

$$\left| \Pr_{\mathsf{H}_6}[\mathsf{Borderline}] - \Pr_{\mathsf{H}_5}[\mathsf{Borderline}] \right| = \mathrm{negl}(\lambda) . \tag{5}$$

In H_6, the probability of $\mathsf{Borderline}$ can be bounded under the 1D-SIS-R assumption.

Claim. Under the 1D-SIS-R$_{q,p,\gamma,m'}$ assumption, it holds that $\Pr_{\mathsf{H}_6}[\mathsf{Borderline}] = \mathrm{negl}(\lambda)$, where $m' = m \cdot (2 + z + 2\ell + 1)$, and $\gamma = p \cdot B \cdot m^{O(d+\log \ell)}$.

Proof. Let $\mathbf{v} \in \mathbb{Z}_q^{(2+z+2\ell+1)m}$ be an input to 1D-SIS-R$_{q,p,\gamma,m'}$. Then define $\mathbf{a}_\beta, \mathbf{b}_i, \mathbf{c}_i, \mathbf{d}$ be so that their concatenation is $\mathbf{v}$.

The reduction executes H_6 as the challenger, using the vectors defined above. We claim that if Borderline occurs, then we solve 1D-SIS-R. This follows since if Borderline occurs then we found x, j such that

$$(\mathbf{b}_{\mathcal{U},x}^T \cdot \mathbf{G}^{-1}(\mathbf{C}_{\Pi,w})\mathbf{G}^{-1}(\mathbf{D}) - F(x^{(i)})\mathbf{c}_{\Pi,w}^T\mathbf{G}^{-1}(\mathbf{D}) + F(x)\Pi_L(w)\mathbf{d}^T)\mathbf{u}_j$$
$$\in [-E, E] + (q/p)\mathbb{Z} .$$

However, by Lemma 4.1, it follows that

$$\mathbf{b}_{\mathcal{U},x}^T = \sum_{\beta \in \{0,1\}} \mathbf{a}_\beta^T\mathbf{R}'_{1,\beta} + \sum_{i \in [z]} \mathbf{b}_i^T\mathbf{R}'_{2,i}$$

$$\mathbf{c}_{\Pi,x}^T = \sum_{\beta \in \{0,1\}} \mathbf{a}_\beta^T\mathbf{R}''_{1,\beta} + \sum_{i \in [2\ell]} \mathbf{c}_i^T\mathbf{R}''_{3,i}$$

where $\left\|\mathbf{R}'_{1,\beta}\right\|, \left\|\mathbf{R}'_{2,i}\right\| \leq m^{O(d)}$ and $\left\|\mathbf{R}''_{1,\beta}\right\|, \left\|\mathbf{R}''_{3,i}\right\| \leq m^{O(\log \ell)}$. It follows that there exists an (efficiently derivable) matrix $\mathbf{R}_0$ such that

$$\mathbf{b}_{\mathcal{U},x}^T \cdot \mathbf{G}^{-1}(\mathbf{C}_{\Pi,w})\mathbf{G}^{-1}(\mathbf{D}) - F(x^{(i)})\mathbf{c}_{\Pi,w}^T\mathbf{G}^{-1}(\mathbf{D}) + F(x)\Pi_L(w)\mathbf{d}^T = \mathbf{v}^T\mathbf{R}_0 ,$$

and $\|\mathbf{R}_0\| \leq m^{O(d+\log \ell)}$.

Finally,

$$\langle \mathbf{v}, \mathbf{R}_0 \cdot \mathbf{u}_j \rangle \in [-E, E] + (q/p)\mathbb{Z} ,$$

with $\|\mathbf{R}_0 \cdot \mathbf{u}_j\| \leq \|\mathbf{R}_0\| \leq m^{O(d+\log \ell)}$ and $E = B \cdot m^{O(d+\log \ell)} = m^{O(d+\log \ell)}$. Thus $\mathbf{R}_0 \cdot \mathbf{u}_j$ is a valid solution for 1D-SIS-R$_{q,p,\gamma,m'}$. The claim thus follows.

Putting together Claim 6 and Eq. (5), we get that

$$\Pr_{H_5}[\text{Borderline}] \leq \Pr_{H_6}[\text{Borderline}] + \text{negl}(\lambda) \leq \text{negl}(\lambda) .$$

and thus, finally

$$\left|\text{Adv}_{H_5}(\mathcal{A}) - \text{Adv}_{H_6}(\mathcal{A})\right| \leq \text{negl}(\lambda) .$$

Finally, we notice that the vector $\mathbf{d}$ is only used when answering the challenge query in the case of $b = 0$. This means that in the adversary's view, the answer it gets when $b = 0$ is uniform and independent of its view so far, exactly the same as the case $b = 1$ where an actual random vector is returned. It follows that

$$\text{Adv}_{H_6}(\mathcal{A}) = 1/2 .$$

On the other hand

$$\text{Adv}_{H_6}(\mathcal{A}) \geq 1/2 + \epsilon\Delta_t/4 - \text{negl}(\lambda) ,$$

and thus

$$\epsilon \leq \frac{\text{negl}(\lambda)}{\Delta_t/4} = \text{negl}(\lambda) .$$

It follows that $\mathcal{A}$ cannot achieve a noticeable advantage in the constrained PRF experiment under the DLWE$_{q,n,\chi}$ assumption.

5.4   Computational Functionality Preserving

We now prove the computational functionality preservation of our scheme, as per Definition 3.1. Throughout this section, we let $\mathcal{F}$ be a family of z-uniform functions and let d be a depth bound (both can depend on λ). We let $n = n(\lambda, d)$, $m = m(\lambda, d)$, $q = q(\lambda, d)$, $p = p(\lambda, d)$ and the noise distributions $\chi = \chi(\lambda, d)$ be as defined in Section 5.2. We let $\mathcal{H}$ be the family of admissible hash functions as described in Section 3.4, with range $\{0, 1\}^\ell$.

Theorem 5.2. *Let $\mathcal{F}$ be a family of z-uniform functions and let d be a depth bound (both can depend on λ). Let $n = n(\lambda, d)$, $m = m(\lambda, d)$, $q = q(\lambda, d)$, $p = p(\lambda, d)$ and the noise distributions $\chi = \chi(\lambda, d)$ be as defined in Section 5.2. Further let $m' = m \cdot (z + 2\ell + 3)$, and $\gamma = \omega(\sqrt{n \log \lambda}) \cdot p \cdot m^{O(d + \log \ell)}$.*

Assuming the hardness of $\mathrm{DLWE}_{n,q,\chi}$ and $1\text{D-SIS-R}_{q,p,\gamma,m'}$, the scheme $\mathcal{CPRF}$ is computationally functionality preserving.

We note that the hardness of both assumptions translates to the worst case hardness of approximating lattice problems such as GapSVP and SIVP to within sub-exponential factors.

Proof (outline).   The theorem follows from an argument practically identical to that made in Hybrids $\mathsf{H}_5, \mathsf{H}_6$ of the proof of Theorem 5.1.
   Recall that we showed that Borderline events only happen with negligible probability, and therefore with all but negligible probability, it holds that the PRF value at point $x^{(i)}$ is exactly equal to

$$\left\lfloor \mathbf{b}_{\mathcal{U},x^{(i)}}^T \cdot \mathbf{G}^{-1}(\mathbf{C}_{\Pi,w^{(i)}})\mathbf{G}^{-1}(\mathbf{D}) - F(x^{(i)})\mathbf{c}_{\Pi,w^{(i)}}^T \mathbf{G}^{-1}(\mathbf{D}) + F(x^{(i)})\Pi_L(w^{(i)})\mathbf{d}^T \right\rceil_p .$$

However, when $F(x^{(i)}) = 0$, this term simplifies to

$$\left\lfloor \mathbf{b}_{\mathcal{U},x^{(i)}}^T \cdot \mathbf{G}^{-1}(\mathbf{C}_{\Pi,w^{(i)}})\mathbf{G}^{-1}(\mathbf{D}) \right\rceil_p$$

which is exactly $\mathsf{ConstrainEval}(K_F, \mathcal{PP}, x^{(i)})$ by definition. Functionality is thus preserved with all but negligible probability.

5.5   Other Properties

We describe several other properties that our construction satisfies.

Unconditional Almost-Correctness. We have shown that our constrained PRF satisfies a computational correctness property, namely that it is hard to find an input $\mathbf{x}$ such that $\mathsf{PRF}_K(\mathbf{x}) \neq \mathsf{ConstrainEval}(K_F, \mathcal{PP}, \mathbf{x})$. We are also able to show unconditionally that the constrained evaluation and the actual PRF evaluation do not differ by much, for any input $\mathbf{x}$. Indeed, by Equation 3 and 4, we have

$$\|\mathsf{PRF}_K(\mathbf{x}) - \mathsf{ConstrainEval}(K_F, \mathcal{PP}, \mathbf{x})\|_\infty \leq m^{O(d)} \cdot B$$

Key Homomorphism. Our PRF is also "almost key homomorphic" in the sense that $\mathsf{PRF_s(x)} + \mathsf{PRF_{s'}(x)}$ is close to $\mathsf{PRF_{s+s'}(x)}$ for any keys $\mathbf{s}$ and $\mathbf{s'}$ and any input $\mathbf{x}$. Recall that our PRF is

$$\mathsf{PRF_s(x)} = \left\lfloor \mathbf{s}^T \mathbf{B}_{\mathcal{U}} \cdot \mathbf{G}^{-1}(\mathbf{C}_\Pi) \cdot \mathbf{G}^{-1}(\mathbf{D}) \right\rceil_p$$

For any keys $\mathbf{s}_i$ and input $\mathbf{x}$, denoting $\mathbf{s}_i^T \mathbf{B}_{\mathcal{U}} \cdot \mathbf{G}^{-1}(\mathbf{C}_\Pi) \cdot \mathbf{G}^{-1}(\mathbf{D})$ as $\mathbf{h}_i$, we have

$$\left\| \mathsf{PRF}_{\sum \mathbf{s}_i}(\mathbf{x}) - \sum \mathsf{PRF}_{\mathbf{s}_i}(\mathbf{x}) \right\|_\infty = \left\| \left\lfloor \sum_i \mathbf{h}_i \right\rceil_p - \sum_i \lfloor \mathbf{h}_i \rceil_p \right\|_\infty \leq k+1$$

Constrained-Key Homomorphism. Our constrained keys are "almost homomorphic" as well, in the same sense as above. That is, if K_F and K'_F are constrained versions of PRF keys K and K' for the same function F, the summation $K_F + K'_F$ is a constrained version of $K + K'$ for the function F. For any input $\mathbf{x}$, we then have that $\mathsf{ConstrainEval}(K_F + K'_F, \mathcal{PP}, \mathbf{x})$ is close to $\mathsf{PRF}_{K+K'}(\mathbf{x})$.

We remark that techniques similar to what we used in showing computational correctness can be used to strengthen the almost key-homomorphism property into computational key-homomorphism where it is computationally hard to find an input for which key homomorphism does not hold.

6   Succinct Constrained Keys

In this section we show how to reduce the size of the constrained key so that asymptotically it depends only on the security parameter and independent of the function class. The construction builds upon the scheme $\mathcal{CPRF}$ from Section 5 but reduces the key size by utilizing an attribute based encryption scheme (ABE). In particular, the constrained keys in our new system have size $\mathsf{poly}(\lambda)$, independent of the parameters of the constraining circuit (namely, its size or depth).

Our *succinct* constrained PRF $\mathcal{SCPRF}$ for a z-uniform d-depth-bounded family $\mathcal{F}$ works as follows.

- $\mathsf{KeyGen}(1^\lambda, 1^z, 1^d)$: The key generation algorithm takes as input the maximum size z and depth d of the constraining circuits. Let $t = O(\log z)$ to be specified later.
  It starts by calling $\mathcal{CPRF}.\mathsf{KeyGen}(1^\lambda, 1^z, 1^d)$ to obtain the seed $\mathbf{s}$, and public parameters $\mathcal{PP} = (H, \mathbf{A}_0, \mathbf{A}_1, \{\mathbf{B}_i\}_{i \in [z]})$.
  It then generates: $\mathbf{a}_\beta = \mathbf{s}^T(\mathbf{A}_\beta + \beta \mathbf{G}) + \mathbf{e}_{1,\beta}^T$ and $\mathbf{b}_{i,\beta} = \mathbf{s}^T(\mathbf{B}_i + \beta \mathbf{G}) + \mathbf{e}_{2,i,\beta}^T$. Note that any possible constrained key of $\mathcal{CPRF}$ consists of $\mathbf{a}_0$ and $\mathbf{a}_1$, together with a subset of $\{\mathbf{b}_{i,\beta}\}_{i \in [z], \beta \in \{0,1\}}$.
  Next it generates parameters for the ABE scheme $(\mathcal{ABE}.\mathsf{msk}, \mathcal{ABE}.\mathsf{pp}) \leftarrow \mathcal{ABE}.\mathsf{Setup}(1^\lambda, 1^t)$, and generates $\mathsf{ct}_{i,\beta} \leftarrow \mathcal{ABE}.\mathsf{Enc}(\mathcal{ABE}.\mathsf{pp}, (i, \beta), \mathbf{b}_{i,\beta})$, encryptions with (i, β) as the "attributes" and $\mathbf{b}_{i,\beta}$ as the "message".
  The public parameters consist of

  $$\mathcal{SCPRF}.\mathcal{PP} = (\mathcal{CPRF}.\mathcal{PP}, \mathcal{ABE}.\mathcal{PP}, \mathbf{a}_0, \mathbf{a}_1, \{\mathsf{ct}_{i,\beta}\}_{i,\beta})$$

The seed for $\mathcal{SCPRF}$ contains a seed for $\mathcal{CPRF}$, namely a uniformly random vector $\mathbf{s} \in \mathbb{Z}_q^n$, and in addition the ABE master secret key $\mathcal{ABE}.\mathsf{msk}$. We note that in fact $\mathbf{s}$ can be retrieved from the public parameters using $\mathcal{ABE}.\mathsf{msk}$ and therefore it is not necessary to give it explicitly. However, it is more natural to think of $\mathbf{s}$ as a part of the seed. In particular $\mathbf{s}$ will be used to evaluate $\mathcal{SCPRF}$ (see Eval below) and $\mathcal{ABE}.\mathsf{msk}$ will be used to produce constrained keys (see Constrain below).

- Eval$(\mathbf{s}, \mathcal{PP}, \mathbf{x})$ takes as input the PRF seed $\mathbf{s}$, the public parameters $\mathcal{PP}$ which contains $\mathcal{CPRF}.\mathsf{pp}$, and an input $x \in \{0,1\}^k$ such that $k \leq k_d$ (i.e. $\mathcal{U}_k$ is of depth $\leq d$), and outputs the result of the CPRF evaluation, namely $\mathcal{CPRF}.\mathsf{Eval}(\mathbf{s}, \mathcal{CPRF}.\mathsf{pp}, \mathbf{x})$.

- Constrain$(\mathcal{ABE}.\mathsf{msk}, F)$ takes as input the ABE master secret key $\mathcal{ABE}.\mathsf{msk}$ and a circuit F (represented as a string in $\{0,1\}^z$) and does the following. Consider the function:

$$\phi_F(i, \beta) = \begin{cases} 1, \text{ if } F_i = \beta \\ 0, \text{ otherwise} \end{cases}$$

Note that ϕ_F can be computed by a depth $O(\log z)$ circuit (whose depth is independent of the depth of F itself), the parameter t from above is set to be equal to this depth. We recall Section 3.5
The constrained key for F is the ABE token for ϕ_F, namely

$$K_F = \mathcal{ABE}.\mathsf{KeyGen}(\mathcal{ABE}.\mathsf{msk}, \phi_F)$$

- ConstrainEval$(K_F, \mathcal{PP}, \mathbf{x})$ takes as input the constrained key K_F, the public parameters $\mathcal{PP}$ and an input $\mathbf{x}$.
Recalling that $\mathcal{PP} = (\mathcal{CPRF}.\mathsf{pp}, \mathcal{ABE}.\mathsf{pp}, \mathbf{a}_0, \mathbf{a}_1, \{\mathsf{ct}_{i,\beta}\})$, and that K_F is the ABE decryption key for the function ϕ_F, it first decrypts to obtain $\mathbf{b}_i = \mathcal{ABE}.\mathsf{Dec}(K_F, \mathsf{ct}_{i,F_i})$, and then applies the constrained evaluation algorithm $\mathcal{CPRF}.\mathsf{ConstrainEval}((\mathbf{a}_0, \mathbf{a}_1, \{\mathbf{b}_i\}), \mathcal{CPRF}.\mathcal{PP}, \mathbf{x})$.

The correctness follows in a straightforward manner from the correctness of $\mathcal{ABE}$ and $\mathcal{CPRF}$. The constrained key size of $\mathcal{SCPRF}$ is derived from that of $\mathcal{ABE}$ and is $\mathrm{poly}(\lambda, t) = \mathrm{poly}(\lambda, \log z)$. It follows that there exists a $\mathrm{poly}(\lambda)$ asymptotic upper bound on the key sizes that applies for all polynomial values of z. Security is stated in the following theorem, the proof can be found in the extended version [11].

Theorem 6.1. *If $\mathcal{CPRF}$ is a single-key secure constrained pseudorandom function for function class $\mathcal{F}$ (Definition 3.1), which is built according to the template in Section 5, and if $\mathcal{ABE}$ is a selectively secure ABE scheme (Definition 3.9), then the scheme $\mathcal{SCPRF}$ described above is a secure single-key CPRF for $\mathcal{F}$.*

References

1. Ajtai, M.: Generating hard instances of lattice problems (extended abstract). In: Miller, G.L. (ed.) Proceedings of the Twenty-Eighth Annual ACM Symposium on the Theory of Computing, Philadelphia, Pennsylvania, USA, May 22-24, pp. 99–108. ACM (1996)

2. Alekhnovich, M.: More on average case vs approximation complexity. In: Proceedings of 44th Symposium on Foundations of Computer Science (FOCS 2003), Cambridge, MA, USA,, October 11-14, pp. 298–307. IEEE Computer Society (2003)

3. Banerjee, A., Peikert, C.: New and improved key-homomorphic pseudorandom functions. In: Garay, J.A., Gennaro, R. (eds.) CRYPTO 2014, Part I. LNCS, vol. 8616, pp. 353–370. Springer, Heidelberg (2014)

4. Banerjee, A., Peikert, C., Rosen, A.: Pseudorandom functions and lattices. In: Pointcheval, D., Johansson, T. (eds.) EUROCRYPT 2012. LNCS, vol. 7237, pp. 719–737. Springer, Heidelberg (2012)

5. Blum, A., Furst, M.L., Kearns, M.J., Lipton, R.J.: Cryptographic primitives based on hard learning problems. In: Stinson, D.R. (ed.) CRYPTO 1993. LNCS, vol. 773, pp. 278–291. Springer, Heidelberg (1994)

6. Boneh, D., Boyen, X.: Secure identity based encryption without random oracles. In: Franklin, M. (ed.) CRYPTO 2004. LNCS, vol. 3152, pp. 443–459. Springer, Heidelberg (2004)

7. Boneh, D., Gentry, C., Gorbunov, S., Halevi, S., Nikolaenko, V., Segev, G., Vaikuntanathan, V., Vinayagamurthy, D.: Fully key-homomorphic encryption, arithmetic circuit ABE and compact garbled circuits. In: Nguyen, P.Q., Oswald, E. (eds.) EUROCRYPT 2014. LNCS, vol. 8441, pp. 533–556. Springer, Heidelberg (2014)

8. Boneh, D., Lewi, K., Montgomery, H.W., Raghunathan, A.: Key homomorphic prfs and their applications. In: Canetti, R., Garay, J.A. (eds.) CRYPTO 2013, Part I. LNCS, vol. 8042, pp. 410–428. Springer, Heidelberg (2013)

9. Boneh, D., Waters, B.: Constrained pseudorandom functions and their applications. In: Sako, K., Sarkar, P. (eds.) ASIACRYPT 2013, Part II. LNCS, vol. 8270, pp. 280–300. Springer, Heidelberg (2013)

10. Boyle, E., Goldwasser, S., Ivan, I.: Functional signatures and pseudorandom functions. In: Krawczyk, H. (ed.) PKC 2014. LNCS, vol. 8383, pp. 501–519. Springer, Heidelberg (2014)

11. Brakerski, Z., Vaikuntanathan, V.: Constrained key-homomorphic prfs from standard lattice assumptions or: How to secretly embed a circuit in your prf. Cryptology ePrint Archive, Report 2015/032 (2015), http://eprint.iacr.org/

12. Cash, D., Hofheinz, D., Kiltz, E., Peikert, C.: Bonsai trees, or how to delegate a lattice basis. J. Cryptology 25(4), 601–639 (2012)

13. Gentry, C., Peikert, C., Vaikuntanathan, V.: Trapdoors for hard lattices and new cryptographic constructions. Electronic Colloquium on Computational Complexity (ECCC) 14(133) (2007)

14. Goldreich, O., Goldwasser, S., Micali, S.: How to construct random functions. J. ACM 33(4), 792–807 (1986); Extended abstract in FOCS 84

15. Gorbunov, S., Vaikuntanathan, V., Wee, H.: Attribute-based encryption for circuits. In: Boneh, D., Roughgarden, T., Feigenbaum, J. (eds.) Symposium on Theory of Computing Conference, STOC 2013, Palo Alto, CA, USA, June 1-4, pp. 545–554. ACM (2013)

16. Goyal, V., Pandey, O., Sahai, A., Waters, B.: Attribute-based encryption for fine-grained access control of encrypted data. In: Juels, A., Wright, R.N., di Vimercati, S.D.C. (eds.) Proceedings of the 13th ACM Conference on Computer and Communications Security CCS 2006, October 30 - November 3, pp. 89–98. ACM (2006)

17. Hofheinz, D., Kamath, A., Koppula, V., Waters, B.: Adaptively secure constrained pseudorandom functions. In: Cryptology ePrint Archive, Report 2014/720 (2014), http://eprint.iacr.org/

18. Kiayias, A., Papadopoulos, S., Triandopoulos, N., Zacharias, T.: Delegatable pseudorandom functions and applications. In: Sadeghi, A., Gligor, V.D., Yung, M. (eds.) 2013 ACM SIGSAC Conference on Computer and Communications Security, CCS 2013, Berlin, Germany, November 4-8, pp. 669–684. ACM (2013)
19. Lewi, K., Montgomery, H.W., Raghunathan, A.: Improved constructions of prfs secure against related-key attacks. In: Boureanu, I., Owesarski, P., Vaudenay, S. (eds.) ACNS 2014. LNCS, vol. 8479, pp. 44–61. Springer, Heidelberg (2014)
20. Micciancio, D., Mol, P.: Pseudorandom knapsacks and the sample complexity of LWE search-to-decision reductions. In: Rogaway, P. (ed.) CRYPTO 2011. LNCS, vol. 6841, pp. 465–484. Springer, Heidelberg (2011)
21. Micciancio, D., Peikert, C.: Trapdoors for lattices: Simpler, tighter, faster, smaller. In: Pointcheval, D., Johansson, T. (eds.) EUROCRYPT 2012. LNCS, vol. 7237, pp. 700–718. Springer, Heidelberg (2012)
22. Naor, M., Pinkas, B., Reingold, O.: Distributed pseudo-random functions and kdcs. In: Stern, J. (ed.) EUROCRYPT 1999. LNCS, vol. 1592, pp. 327–346. Springer, Heidelberg (1999)
23. Peikert, C.: Public-key cryptosystems from the worst-case shortest vector problem: extended abstract. In: Proceedings of the 41st Annual ACM Symposium on Theory of Computing, STOC 2009, Bethesda, MD, USA, 2009, May 31 - June 2, pp. 333–342 (2009)
24. Regev, O.: Lattices in computer science - average case hardness. Lecture Notes for Class (scribe: Elad Verbin) (2004),
http://www.cims.nyu.edu/ regev/teaching/lattices_fall_2004/ln/averagecase.pdf
25. Regev, O.: On lattices, learning with errors, random linear codes, and cryptography. In: Proceedings of the 37th Annual ACM Symposium on Theory of Computing, Baltimore, MD, USA, May 22-24, pp. 84–93 (2005)
26. Sahai, A., Waters, B.: How to use indistinguishability obfuscation: deniable encryption, and more. In: Shmoys, D.B. (ed.) Symposium on Theory of Computing, STOC 2014, New York, NY, USA, 2014, May 31 - June 03, pp. 475–484. ACM (2014)
27. Schnorr, C.: A hierarchy of polynomial time lattice basis reduction algorithms. Theor. Comput. Sci. 53, 201–224 (1987)

Key-Homomorphic
Constrained Pseudorandom Functions

Abhishek Banerjee[1,*], Georg Fuchsbauer[2,**], Chris Peikert[1,***],
Krzysztof Pietrzak[2,†], and Sophie Stevens[3,‡]

[1] School of Computer Science, College of Computing, Georgia Institute of Technology, USA
[2] Institute of Science and Technology Austria
[3] University of Bristol, UK

Abstract. A pseudorandom function (PRF) is a keyed function $F \colon \mathcal{K} \times \mathcal{X} \to \mathcal{Y}$ where, for a random key $k \in \mathcal{K}$, the function $F(k, \cdot)$ is indistinguishable from a uniformly random function, given black-box access. A *key-homomorphic* PRF has the additional feature that for any keys k, k' and any input x, we have $F(k + k', x) = F(k, x) \oplus F(k', x)$ for some group operations $+, \oplus$ on $\mathcal{K}$ and $\mathcal{Y}$, respectively. A *constrained* PRF for a family of sets $\mathcal{S} \subseteq \mathcal{P}(\mathcal{X})$ has the property that, given any key k and set $S \in \mathcal{S}$, one can efficiently compute a "constrained" key k_S that enables evaluation of $F(k, x)$ on all inputs $x \in S$, while the values $F(k, x)$ for $x \notin S$ remain pseudorandom even given k_S.

In this paper we construct PRFs that are simultaneously constrained *and* key homomorphic, where the homomorphic property holds even for constrained keys. We first show that the multilinear map-based bit-fixing and circuit-constrained PRFs of Boneh and Waters (Asiacrypt 2013) can be modified to also be *key-homomorphic*. We then show that the LWE-based key-homomorphic PRFs of Banerjee and Peikert (Crypto 2014) are essentially already *prefix-constrained* PRFs, using a (non-obvious) definition of constrained keys and associated group operation. Moreover, the constrained keys themselves are pseudorandom, and the constraining and evaluation functions can all be computed in low depth.

As an application of key-homomorphic constrained PRFs, we construct a proxy re-encryption scheme with fine-grained access control. This scheme allows storing encrypted data on an untrusted server, where each file can be encrypted relative to some attributes, so that only parties whose constrained keys match the attributes can decrypt. Moreover, the server can re-key (arbitrary subsets of) the ciphertexts without learning anything about the plaintexts, thus permitting efficient and fine-grained revocation.

* Supported by the third author's grants.
** Research supported by ERC starting grant (259668-PSPC).
*** This material is based upon work supported by the National Science Foundation under CA-REER Award CCF-1054495, by DARPA under agreement number FA8750-11-C-0096, and by the Alfred P. Sloan Foundation. Any opinions, findings, and conclusions or recommendations expressed in this material are those of the author(s) and do not necessarily reflect the views of the National Science Foundation, DARPA or the U.S. Government, or the Sloan Foundation. The U.S. Government is authorized to reproduce and distribute reprints for Governmental purposes notwithstanding any copyright notation thereon.
† Research supported by ERC starting grant (259668-PSPC).
‡ Work done while visiting the Institute of Science and Technology Austria.

Y. Dodis and J.B. Nielsen (Eds.): TCC 2015, Part II, LNCS 9015, pp. 31–60, 2015.

1   Introduction

Pseudorandom functions (PRFs), like the AES block cipher, are the workhorses of cryptography. They allow for efficient and elegant solutions to all the basic symmetric-key cryptographic tasks, including authentication and encryption. Not surprisingly, PRFs with additional properties have been intensively investigated, as those properties often allow for useful additional functionalities. We discuss two such properties below.

Key-homomorphic PRFs. A PRF [GGM86] is an efficiently computable keyed function $F \colon \mathcal{K} \times \mathcal{X} \to \mathcal{Y}$. The security property requires that no efficient adversary can distinguish $F(k, \cdot)$ instantiated with a random key $k \leftarrow \mathcal{K}$ from a uniformly random function, given oracle access.

A *key-homomorphic* PRF has the additional feature that for any keys k, k' and any input x, we have $F(k + k', x) = F(k, x) \oplus F(k', x)$ for some group operations $+$ and $\oplus$ on $\mathcal{K}$ and $\mathcal{Y}$, respectively. Naor, Pinkas and Reingold [NPR99] observed that the simple PRF $F(k, x) = H(x)^k$, where $H(\cdot)$ is a random oracle that maps into a group where the DDH problem is assumed to be hard, is a key-homomorphic PRF. The first (almost) key-homomorphic PRFs in the standard model was constructed by Boneh *et al.* [BLMR13] from lattice assumptions, and later generalized and improved by Banerjee and Peikert [BP14].

Applications of key-homomorphic PRFs include an elegant solution to one-round *distributed* PRFs for any threshold [BLMR13]. Here, for some parameters $\ell \leq n$, a user sends an input x to ℓ servers, who each return a short answer from which the user can compute $F(k, x)$. Security requires that to any subset of $\ell - 1$ servers, $F(k, \cdot)$ is pseudorandom. For $\ell = n$, one can simply share the key as $k = k_1 + k_2 + \ldots + k_n$, each server computes $F(k_i, x)$, and these can then be combined to $\sum_{i=1}^{n} F(k_i, x) = F(\sum_{i=1}^{n} k_i, x) = F(k, x)$. Boneh *et al.* [BLMR13] provide a solution for general $\ell \leq n$. Symmetric-key proxy re-encryption is another interesting application, which we will discuss in detail in Section 1.2.

Constrained PRFs. A *constrained* PRF for a family of sets $\mathcal{S} \subseteq \mathcal{P}(\mathcal{X})$ has the property that, given any key k and set $S \in \mathcal{S}$, one can efficiently compute a "constrained" key k_S that enables evaluation of $F(k, x)$ on all inputs $x \in S$, while the values $F(k, x)$ for $x \notin S$ remain pseudorandom even given k_S.

Constrained PRFs were introduced independently in [BW13, KPTZ13, BGI14]. All three papers note that the classical GGM construction [GGM86] already gives a prefix-constrained PRF, where from a key $k \in \{0, 1\}^n$, for any $v \in \{0, 1\}^{\leq n}$ one can compute a key k_v that enables the computation of $F(k, x)$ for all inputs x that start with v. Boneh and Waters [BW13] construct bit-fixing and circuit-fixing constrained PRFs from multilinear maps. In the bit-fixing construction, for every $v \in \{0, 1, ?\}$ one can compute a key k_v that enables the computation of $F(k, x)$ for any x for which $x_i = v_i$ when $v_i \neq ?$. The more general circuit-constrained construction allows generating constrained keys for any circuit C, where with k_C one can evaluate the PRF on input x if and only if $C(x) = 1$.

Prefix-constrained PRFs (or rather, "punctured" PRFs, which can be constructed from them) are a main tool in almost all the applications of indistinguishability obfuscation [GGH+13b, SW14, PST14]. The papers [BW13, BGI14, KPTZ13] discuss many more applications of constrained PRFs.

1.1 Results and Techniques

Key-Homomorphic Constrained PRFs. In this paper we construct PRFs that are simultaneously key-homomorphic and constrained. The key-homomorphic property holds not only for PRF keys, but also for constrained keys. We first show that the multilinear-map-based bit-fixing and circuit-constrained PRFs due to Boneh and Waters [BW13] can be modified to also be *key-homomorphic*. We then show that the LWE-based key-homomorphic PRFs of Banerjee and Peikert [BP14] are essentially already *prefix-constrained* PRFs, using a (non-obvious) definition of constrained keys and associated group operation. Moreover, the constrained keys themselves are pseudorandom, and the constraining and evaluation functions can all be computed in low depth. The latter feature can be important for applications of obfuscation, e.g., [GGH+13b, SW14], where the use of low-depth constrained/punctured PRFs may avoid the need for costly "bootstrapping" operations and fully homomorphic encryption.

Given the usefulness of the individual key-homomorphic and constraining properties for PRFs, we expect their combination to find even more exciting applications. We discuss one such application, symmetric-key proxy re-encryption, in Section 1.2. We next give a brief overview of our constructions, their salient features, and our proof techniques.

Bit-Fixing PRFs from MDDH. Leveled multilinear maps [GGH13a] are defined over a sequence of groups $(\mathbb{G}_1, \ldots, \mathbb{G}_\kappa)$, where $\mathbb{G}_i$ is generated by an element g_i, as bilinear maps $e_{i,j} \colon \mathbb{G}_i \times \mathbb{G}_j \to \mathbb{G}_{i+j}$; i.e., they satisfy $e_{i,j}(g_i^a, g_j^b) = (g_{i+j})^{a \cdot b}$ for all a, b. The multilinear decisional Diffie-Hellman assumption states that given random elements $g_1^{c_1}, \ldots, g_1^{c_{\kappa+1}}$, it is hard to distinguish $g_\kappa^{\Pi_{j=1}^{\kappa+1} c_j}$ from a random group element in $\mathbb{G}_\kappa$.

Using such groups, Boneh and Waters [BW13] define a bit-fixing constrained PRF for bit strings of length n as follows. A key K consists of a sequence of multilinear groups of prime order p and values $(k, \{d_{i,\beta}\}_{i \in [n], \beta \in \{0,1\}})$ from $\mathbb{Z}_p$. The PRF is defined as $F(K, x) := g_n^{k \cdot \Pi_{i \in [n]} d_{i,x_i}}$. While this construction does not appear to be key-homomorphic, in Section 3 we make it so, by observing that we can "outsource" the values $d_{i,\beta}$ as $D_{i,\beta} := g_1^{d_{i,\beta}}$ to public parameters pp, and redefine

$$F(pp, k, x) := e\left(D_{1,x_1}, e(D_{2,x_2}, e(\ldots, e(D_{n-1,x_{n-1}}, D_{n,x_n})))\right)^k = g_n^{k \cdot \Pi_{i \in [n]} d_{i,x_i}}.$$

We show that the values $D_{i,\beta}$ can be published without compromising security, that is, the function values are pseudorandom under the MDDH assumption. Because the secret key is now just k, the PRF is easily seen to be key-homomorphic.

Low-Depth Prefix-Fixing PRFs from LWE. In Section 4 we construct key-homomorphic prefix-fixing constrained PRFs from the LWE assumption, and hence from the conjectured hardness of worst-case lattice problems [Reg09, Pei09, BLP$^+$13]. In addition, natural instantiations of this construction have polylogarithmic circuit depth. To our knowledge, these are the first sublinear-depth constrained PRFs (whether key-homomorphic or not), and as such they can admit much more efficient obfuscation under existing paradigms. (Recall that the basic GGM construction, which yields a prefix-constrained PRF, is highly sequential.)

Our LWE-based construction is an extension of the recent class of key-homomorphic PRFs of Banerjee and Peikert [BP14], which generalizes and improves a previous construction of Boneh et al. [BLMR13]. We show that the BP construction can be made prefix-constrained, and that the constraining algorithm is also key-homomorphic. Notably, the approximation factors for the underlying LWE assumption are essentially the same as in [BP14], e.g., they can be as small as quasi-polynomial $\lambda^{\omega(\lambda)}$ in the security parameter.

To show all this, we start with the observation that the security proof for the BP (and BLMR) construction is very "GGM-like," i.e., it proceeds in a sequence of hybrids, one for each successive bit of the PRF input. However, the functions computed in the hybrids do not quite fit the usual GGM paradigm, because each successive output of the PRG is broken into two pieces: one piece is fed as input into the next PRG step, while the "leftover" piece is retained and then later "folded" back into the final output of all the PRG steps. A natural way to define a constrained key for a *partial* function evaluation, then, is to include all the leftover pieces in the constrained key—and this is indeed the approach we take.

The main technical challenge we face is in defining a suitable *group structure* on the leftover pieces, for key homomorphism. At first sight, this appears easy: since the leftover pieces are eventually combined with the final PRG output via a linear function, it would appear that one could simply add constrained keys by adding their corresponding leftover pieces. While this does indeed work—at least syntactically—it yields a useless construction! The problem is that essentially any application of key-homomorphic (constrained) PRFs (e.g., proxy re-encryption as described below in Section 1.2) will require a statistical "secret sharing"-like property on the (constrained) secret keys. For example, the sum of any fixed key with a uniformly random key must be uniformly random, so that the original key is completely hidden. Formally speaking, for any particular constraint we need the space of constrained keys to be a *finite* additive group (so that it supports a uniform distribution), and for the function to be key-homomorphic under this group structure.

Resolving the difficulty. Going back to the BP construction, the leftover pieces in constrained keys come from a certain finite subset $\mathcal{P} \subset \mathbb{Z}^m$, namely, a fundamental region of a special lattice $\mathcal{L}$. Obviously, the sum of two uniformly random $\mathcal{P}$-elements is *not* uniform in $\mathcal{P}$—indeed, it is typically not in $\mathcal{P}$ at all! So we cannot naïvely use the ambient group $\mathbb{Z}^m$ (which is infinite). Another idea would be to use the *finite* quotient group $\mathbb{Z}^m/\mathcal{L}$, i.e., addition modulo the lattice. This also does not work, because the function is not key-homomorphic under this form of addition.

Our solution to the above problem involves a novel method of adding modulo $\mathcal{L}$ "with carries." That is, the sum of two leftover $\mathcal{P}$-elements is mapped back to $\mathcal{P}$ by reducing modulo the lattice $\mathcal{L}$, i.e., shifting by an appropriate lattice vector $\mathbf{x} \in \mathcal{L}$. The vector $\mathbf{x}$ is then treated as a "carry" term that is "folded into" the sum of the next two $\mathcal{P}$-elements in the key, and so on. (The ultimate effect is analogous to grade-school addition, except that here the "base" in which the "numbers" are written is a high-dimensional lattice.) We show that by appropriately defining the effect of the carry terms, the PRF is indeed key-homomorphic under this form of addition.

All of the above applies to the so-called "noisy" version of the BP construction, an intermediate object that has perfect constraining, homomorphic, and pseudorandomness properties, but high circuit depth and (even worse) *exponentially* large keys. Similar to [BPR12, BP14], we show that by appropriately "rounding" this noisy construction, the keys and depth can be made small while preserving the other desirable properties (at least against computationally bounded attackers). Interestingly, this rounding transformation requires us to work with a "geometrically nice" set $\mathcal{P}$ of representatives modulo the special lattice $\mathcal{L}$ (which fortunately exists), whereas [BP14] works with any set of representatives.

1.2 Applications

Using symmetric encryption, one can store data on an untrusted server simply by first encrypting the files to be stored. Key-homomorphic and/or constrained PRFs enable symmetric encryption schemes with additional properties which are useful in this setting.

Assume there are many parties who should get access to the stored data, but that we occasionally need to revoke the access of some party. A simple solution is to re-encrypt all the data with a fresh key, and then give this key to only the parties who should continue to have access. Unfortunately, this requires either that the server knows the secret key k, or that we must download, re-encrypt, and upload the entire database. Boneh *et al.* [BLMR13] show how by using a *key-homomorphic* PRF, one can construct a so-called proxy re-encryption scheme, where the server can locally transform ciphertexts under a key k to ciphertexts under a new key k' without learning the plaintexts. We discuss their construction in Section 5.1.

The second functionality we consider is fine-grained access control, where different parties should get access to different subsets of the stored data. The trivial but inefficient solution is to encrypt each file under a separate key, and then send the appropriate keys to each party. *Constrained* PRFs (CPRF) provide a more elegant solution: every encrypted file is associated with some attribute vector x, and every party gets a constrained key k_p that allows her to evaluate the PRF on only those inputs satisfying an appropriate predicate p. The PRF then allows her to decrypt only those files whose attributes x satisfy her predicate. Of course, the expressive power of the system depends upon the predicates supported by the CPRF. A circuit CPRF allows for any efficiently computable predicate p, whereas prefix CPRFs allow for predicates that are satisfied by inputs starting with a particular prefix. Using *key-homomorphic constrained* PRFs as constructed in this paper, in Section 5 we construct a scheme for outsourced storage that supports proxy re-encryption and fine-grained access control simultaneously.

The "obvious" way to outsource storage to an untrusted server using CPRFs is to encrypt a message m for some attributes x as $c = m \oplus F(k, x)$. Now, only a party who has a constrained key k_p where $p(x) = 1$ can decrypt the ciphertext (c, x), by computing $m = c \oplus F(k, x)$. This simple solution has two problems.

First, given two ciphertexts $(c, x), (c', x)$ for the same attributes x, one can compute the XOR of the messages as $c \oplus c' = m \oplus F(k, x) \oplus m' \oplus F(k, x) = m \oplus m'$, breaking the security of the encryption scheme.

Second, a single ciphertext $c = m \oplus F(k, x)$ for a known m reveals $F(k, x) = c \oplus m$. This is a problem because the security game for CPFRs only guarantees that $F(k, x)$ is pseudorandom if the adversary was given constrained keys (for predicates $p(.)$ where $p(x) = 0$), but does not guarantee anything if she is also given outputs $F(k, x')$ for some $x' \neq x$. For the GGM based prefix CPRF there is in fact a simple attack (cf. Footnote 4).

To handle these problems, we show how to "randomize" predicates, in the sense that p^+ is a randomization of p if there exists some encoding $[\cdot, \cdot]$ such that for all (x, r) we have $p^+([x, r]) = 1$ if and only if $p(x) = 1$. Let $\mathcal{P}$ denote the predicates supported by the CPRF considered. We require $p^+ \in \mathcal{P}$ as this will assure that the set of predicates for the encryption scheme is the same as for the CPRF. We will need some other properties from the encoding which we outline below. Although we don't give a generic result showing how to randomize any set of predicates, we show very simple constructions that work for prefix, bit-fixing and circuit CPRFs (that is, all the predicates for which CPRFs have been constructed to date). For bit-fixing and and circuit CPRFs the encoding is simply concatenation $[x, r] = x \| r$. For prefix CPRFs the encoding is a simple prefix-free encoding (cf. the paragraph above Thm. 4).

To solve the problems outlined above, we make encryption probabilistic: we encrypt m as $(r, m \oplus F(k, [x, r]))$ using randomness r. A constrained key for the predicate $p(\cdot)$ for the encryption scheme is now defined as a constrained key for the predicate $p^+(.)$ for the CPRF. Note that with this key we can compute $F(k, [x, r])$ and thus decrypt if $p(x) = 1$ for any r.

Arguing security is more delicate, and will require two extra properties. First, we want $[\cdot, \cdot]$ to be injective, which will ensure that the value $F(k, [x, r])$ used in the challenge ciphertext has never been output before with high probability (i.e., unless we happened to choose the same randomness r for a previous query). Second, we want that for every $[x, r]$, there exists a predicate $p_{[x,r]} \in \mathcal{P}$ such that $p_{[x,r]}([x, r]) = 1$ but $p_{[x,r]}([x', r']) = 0$ for all $(x', r') \neq (x, r)$ (but $p_{[x,r]}(z)$ can be 1 for values z outside the range of $[\cdot, \cdot]$). In the reduction, this latter property allows us to learn the values $F(k, [x, r])$ required to answer encryption queries in the CPA game by querying our oracle (playing the CPRF security game) for the constrained key with predicate $p_{[x,r]}$. The above property ensures that every such query will exclude at most one possible candidate for our challenge ciphertext. Thus, if at some point the adversary asks for a challenge ciphertext using attributes x, we can chose our CPRF challenge as $[x, r]$ (which will be answered either by $F(k, [x, r])$ or uniform), and as we chose r uniformly at random, this query will most likely not be invalid (in the sense that it could be computed using some previously issued constrained key).

Efficient re-encryption. Using proxy re-encryption as outlined above requires the server to re-encrypt the *entire* database to ensure that a revoked party loses access. When using fine-grained access control, a party to be revoked might have access to only a small fraction of the database, so we could re-encrypt only that portion. This would make re-encryption (potentially much) more efficient, but would require some extra key-management, as now different parts of the database are encrypted under different keys.

2 Preliminaries

2.1 Key-Homomorphic Constrained Pseudorandom Functions

We now formally define key-homomorphic constrained pseudorandom functions. We model constrainability as a directed acyclic graph (DAG) on some (typically huge) set of nodes. We restrict our attention to DAGs having a unique node that has no incoming edges, called the *root node*.

Definition 1. *A constrained function family $\mathcal{C}$ is given by:*

- *a directed acyclic graph $D = (V, E)$ with unique root node $r \in V$,*
- *for each node $u \in V$, a* key space $\mathcal{K}_u$ *with an efficiently samplable probability distribution $\mathcal{D}_u$ over it,*
- *for every edge $(u, v) \in E$, a* constraining function $C_{u,v} \colon \mathcal{K}_u \to \mathcal{K}_v$ *that is efficiently computable.*

The functions $C_{u,v}$ must satisfy the following consistency *property: for any $u, v \in V$ and any two paths $P = (u = u_0, u_1, \ldots, u_k = v)$ and $P' = (u = u'_0, u'_1, \ldots, u'_\ell = v)$ from u to v, we have that*

$$C_{u_{k-1}, u_k} \circ \cdots \circ C_{u_1, u_2} \circ C_{u_0, u_1} = C_{u'_{\ell-1}, u'_\ell} \circ \cdots \circ C_{u'_1, u'_2} \circ C_{u'_0, u'_1} \ .$$

For notational convenience, we let $C_{u,v} \colon \mathcal{K}_u \to \mathcal{K}_v$ denote the above (composed) functions, and also define $C_u := C_{r,u}$ for any node $u \in V$ that is reachable from the root node r. For consistency with the typical PRF notation, we define $F_k(u) = C_u(k)$ (and to also cover constrained PRFs, if u represents a subset of inputs then $\mathsf{Constrain}_k(u) = C_u(k)$).

Lastly, a constrained function family may also have a Setup *algorithm, which samples some (public) parameters that are provided as input to all of the other algorithms.*

For the reader who may be familiar with constrained PRFs, we stress that in the above definition, the DAG nodes roughly correspond with (subsets of) *PRF inputs*, while the input k_u and output k_v of constraining function $C_{u,v}$ correspond to (constrained) *secret keys*. Despite these rough correspondences, we stress that in our model there are no distinct notions of PRF "inputs" or "outputs," only DAG nodes. This is without loss of generality: a PRF input can simply be represented as a node w with no outgoing edges, and the corresponding output is the key k_w. In fact, our model is somewhat more general because it allows for defining and proving the pseudorandomness of *constrained keys* themselves (even for nodes having outgoing edges), which can be useful in certain settings.

Definition 2. *Pseudorandomness for a constrained function family* $C = (D = (V, E),$ $\{K_u\}, \{C_{u,v}\})$ *is defined as follows. It is parameterized by a subset* $R \subseteq V$ *of what we call "challenge" nodes. We consider two closely related experiments ("games"), called "real" and "ideal," which proceed as follows:*

1. Initialize: *For the root node* $r \in V$ *we choose a value* $k = k_r \leftarrow K_r$ *according to the associated distribution* $\mathcal{D}_r$. *If the family has a* Setup *algorithm, it is run and its output is provided to the adversary.*
2. Query: *The adversary adaptively issues queries* $v \in V$, *subject to the condition that no query in* R *and any other query have a common descendant in* D. *That is, there are no distinct queries* $u \in R$, $u' \in V$ *and node* $w \in V$ *such that there exists a (possibly trivial) path from* u *to* w *and one from* u' *to* w.
   - *In the "real game," every query* v *is answered with* $k_v = F_k(v) = C_v(k)$.
   - *In the "ideal game," if* $v \in V \setminus R$ *then it is answered as in the real game, otherwise it is answered with an independent value* $k_v^* \leftarrow \mathcal{D}_v$. *(Repeated queries are answered consistently.)*

The family C *is said to be* pseudorandom *if for any polynomial-time adversary, its advantage in distinguishing the real and ideal games is negligible in the security parameter.*

In short, the definition above means that constrained keys *for the set R of challenge nodes* are pseudorandom. The condition on legal queries is necessary to prevent trivial distinguishers that work by observing the inconsistency of the ideal-game answers. In a bit more detail, given answers $k_u, k_{u'}$ for some nodes $u \in R$, $u' \in V$ (respectively) that have a common descendant $w \in V$, the distinguisher could check whether $C_{u,w}(k_u) = C_{u',w}(k_{u'})$. This always holds in the real game, but in the ideal game, where k_u is chosen independently of everything else, it would typically fail to hold.

Definition 3. *A constrained function family is* (key) homomorphic *if all the key spaces* K_u *are additive groups and if the constraining functions* $C_{u,v}$ *are additive homomorphisms, i.e., for every* $(u, v) \in E$ *and every* $k_1, k_2 \in K_u$, *we have*

$$C_{u,v}(k_1) + C_{u,v}(k_2) = C_{u,v}(k_1 + k_2) \ .$$

For key-homomorphic PRFs, all applications we know of implicitly require the key spaces K_u to be *finite* groups, and the associated distributions $\mathcal{D}_u$ to be *uniform* distributions. In short, this is because the security proofs all rely on statistical "secret sharing"-type properties, e.g., the sum of any group element and a uniformly random one is uniformly random. All our final constructions have finite key spaces with uniform distributions.

3   Bit-Fixing and Circuit-Constrained Constructions from MDDH

Boneh and Waters [BW13] constructed a "bit-fixing" constrained PRF for input space $\mathcal{X} = \{0, 1\}^n$, where one can derive constrained keys for any subset of inputs that can be described by arbitrarily fixing the values of any desired input bits. Any such subset can be described by a string $\mathbf{v} \in \{0, 1, ?\}^n$, as the set of all $x \in \{0, 1\}^n$ that match $\mathbf{v}$ at all positions where $\mathbf{v}$ is different from '?':

$$S_{\mathbf{v}} := \{x \in \{0,1\}^n \mid \forall i \in [n] : x_i = \mathbf{v}_i \lor \mathbf{v}_i = ?\} . \qquad (1)$$

Although not considered in [BW13], their construction can easily be generalized to allow computation of a constrained key for a set $S_{\mathbf{w}}$ not only from the root key, but also from any key for a set $S_{\mathbf{v}}$ for which $S_{\mathbf{w}} \subseteq S_{\mathbf{v}}$. In our DAG-based model, then, the nodes of the DAG consist of the strings $\mathbf{v} \in \{0, 1, ?\}^n$, and there is an edge $(\mathbf{v}, \mathbf{w})$ if and only if $S_{\mathbf{v}} \supseteq S_{\mathbf{w}}$ (equivalently, $\mathbf{w}_i = \mathbf{v}_i$ whenever $\mathbf{v}_i \neq ?$).

The original BW construction does not appear to be key homomorphic. However, we show how to make it so by defining public parameters for the function (which consist of elements previously contained in the secret key), and only keeping one $\mathbb{Z}_p$ element as the original secret key.

After these two modifications, we show that the PRF remains a bit-fixing constrained pseudorandom function family as defined in Definition 2. The set of challenge nodes is $R = \{0, 1\}^n$, corresponding to all "fully constrained" keys. That is, constrained keys for terminal nodes in the DAG are pseudorandom, but for nodes with outgoing edges they are not.

3.1 Preliminaries

Multilinear groups. Candidates for sequences of groups with leveled multilinear maps were first proposed by Garg, Gentry and Halevi [GGH13a]. These constructions implement *graded encodings*, which could be viewed as approximate multilinear groups. We present our results in the language of multilinear groups.

Leveled multilinear groups are generated by a group generator $\mathcal{G}$, which takes as input the security parameter 1^λ and $\kappa \in \mathbb{N}$, which determines the number of levels. $\mathcal{G}(1^\lambda, \kappa)$ outputs a sequence of groups $\mathbb{G} = (\mathbb{G}_1, \ldots, \mathbb{G}_\kappa)$ of prime order $p > 2^\lambda$. We assume that the description of each group contains a canonical generator g_i. For all $i, j \geq 1$ with $i + j \leq \kappa$, there exists a bilinear map $e_{i,j} \colon \mathbb{G}_i \times \mathbb{G}_j \to \mathbb{G}_{i+j}$, which satisfies:

$$\forall a, b \in \mathbb{Z}_p : e_{i,j}(g_i^a, g_j^b) = (g_{i+j})^{a \cdot b} .$$

We will omit the indices of e and write $e(h_1, h_2, \ldots, h_n)$ or $e(\{h_i\}_{i=1}^n)$ as a shorthand for $e(h_1, e(h_2, e(\ldots, e(h_{n-1}, h_n))))$. We make the following hardness assumption:

Assumption 1 *The κ-Multilinear Decisional Diffie-Hellman (κ-MDDH) assumption states that given $(\mathbb{G}_1, \ldots, \mathbb{G}_\kappa) \leftarrow \mathcal{G}(1^\lambda, \kappa)$ and $g = g_1, g^{c_1}, \ldots, g^{c_{\kappa+1}}$ for (uniformly) random $c_1, \ldots, c_{\kappa+1} \leftarrow \mathbb{Z}_p$, it is hard to distinguish $g_\kappa^{\prod_{j \in [\kappa+1]} c_j} \in \mathbb{G}_\kappa$ from a random group element in $\mathbb{G}_\kappa$.*

3.2 Key-Homomorphic Bit-Fixing PRF

Setup($1^\lambda, 1^n$): On input the security parameter λ and the input length n, run $\mathcal{G}(1^\lambda, n)$ to compute a sequence of groups $\mathbb{G} = (\mathbb{G}_1, \ldots, \mathbb{G}_n)$ of prime order p, with generators $g := g_1, \ldots, g_n$. Choose $(d_{1,0}, d_{1,1}), \ldots, (d_{n,0}, d_{n,1}) \leftarrow \mathbb{Z}_p^2$ uniformly at random and set $D_{i,\beta} := g^{d_{i,\beta}}$ for $i \in [n]$ and $\beta \in \{0,1\}$. Output the parameters of the scheme as

$$pp := \big(\mathbb{G} = (\mathbb{G}_1, \ldots, \mathbb{G}_n), \{D_{i,\beta}\}_{i\in[n],\,\beta\in\{0,1\}}\big) \ .$$

They define the domain as $\mathcal{X} = \{0,1\}^n$ and the range of the PRF as $\mathcal{Y} = \mathbb{G}_n$. For a key $k \in \mathbb{Z}_p$, the PRF value on input $x = (x_1, \ldots, x_n) \in \{0,1\}^n$ is defined as

$$F(pp, k, x) := e\big(\{D_{i,x_i}\}_{i\in[n]}\big)^k \ = \ g_n^{k\cdot\Pi_{i\in[n]}\,d_{i,x_i}} \ .$$

Constrain$(pp, k, \mathbf{w})$: On input pp, a key $k \in \mathbb{Z}_p \cup \bigcup_{i=1}^{n-1}\mathbb{G}_i$ and a vector $\mathbf{w} \in \{0,1,?\}^n$, which describes the constrained set as $S_{\mathbf{w}} := \{x \in \{0,1\}^n \mid \forall i \in [n] : x_i = \mathbf{w}_i \vee \mathbf{w}_i = ?\}$, let $W := \{i \in [n] : \mathbf{w}_i \neq ?\}$ be the set of indices that $\mathbf{w}$ fixes.

– If $k \in \mathbb{Z}_p$ (that is, k is a master key) then return

$$k_{\mathbf{w}} := e\big(\{D_{i,v_i}\}_{i\in W}\big)^k \ = \ (g_{|W|})^{k\cdot\Pi_{i\in W}\,d_{i,\mathbf{w}_i}} \ .$$

– Otherwise, we have $k = k_{\mathbf{v}}$ for some set $\mathbf{v} \in \{0,1,?\}^n$, for which we let V be the set of fixed indices. If $V \not\subseteq W$ or $\mathbf{v}_i \neq \mathbf{w}_i$ for some $i \in V$ then return $\perp$ (since $S_{\mathbf{v}} \not\supseteq S_{\mathbf{w}}$); else return

$$k_{\mathbf{w}} := e\big(k_{\mathbf{v}}, e(\{D_{i,v_i}\}_{i\in W\setminus V})\big)$$
$$= e\big((g_{|V|})^{k\cdot\Pi_{i\in V}\,d_{i,\mathbf{v}_i}}, (g_{|W\setminus V|})^{\Pi_{i\in W\setminus V}\,d_{i,\mathbf{w}_i}}\big) = (g_{|W|})^{k\cdot\Pi_{i\in W}\,d_{i,\mathbf{w}_i}} \ .$$

Eval(pp, k, x): – If $k \in \mathbb{Z}_p$, return $F(pp, k, x) = e(\{D_{i,x_i}\}_{i\in[n]})^k = g_n^{k\cdot\Pi_{i\in[n]}\,d_{i,x_i}}$.
– Otherwise, $k = k_{\mathbf{v}}$ for some $\mathbf{v} \in \{0,1,?\}^n$. Let $V := \{i \in [n] : \mathbf{v}_i \neq ?\}$ and $\overline{V} := \{i \in [n] : \mathbf{v}_i = ?\}$ be its complement. If $x_i \neq \mathbf{v}_i$ for some $i \in V$ then return $\perp$ (since $x \notin S_{\mathbf{v}}$); else return

$$e\big(k_{\mathbf{v}}, e(\{D_{i,x_i}\}_{i\in\overline{V}})\big) = e\big((g_{|V|})^{k\cdot\Pi_{i\in V}\,d_{i,\mathbf{v}_i}}, (g_{|\overline{V}|})^{\Pi_{i\in\overline{V}}\,d_{i,x_i}}\big)$$
$$= (g_n)^{k\cdot\Pi_{i\in[n]}\,d_{i,x_i}} = F(pp, k, x) \ .$$

3.3 Properties

Key homomorphism. The construction is key-homomorphic in the sense of [BLMR13], but it also satisfies Definition 2, which requires that Constrain is homomorphic as well. The PRF can be described in the language of Definition 1 as follows. (Note that we identify the set $S_{\mathbf{v}}$, defined in (1), with the vector $\mathbf{v}$ defining it.)

– The set of vertices of the graph D is defined as $V := \{\mathbf{v} : \mathbf{v} \in \{0,1,?\}^n\}$ and the root node is $\mathbf{r} := (?, \ldots, ?)$, representing the set $\mathcal{X} = \{0,1\}^n$. There is an edge from $\mathbf{v}$ to $\mathbf{w}$ if all bits fixed by $\mathbf{v}$ are fixed by $\mathbf{w}$ to the same value, i.e., for all $i \in [n]$: if $\mathbf{v}_i \in \{0,1\}$ then $\mathbf{w}_i = \mathbf{v}_i$.
– The additive group associated to $\mathbf{r}$ is $\mathcal{K}_{\mathbf{r}} := (\mathbb{Z}_p, +)$; for all other vertices $\mathbf{v}$ it is $\mathcal{K}_{\mathbf{v}} := (\mathbb{G}_{|V|}, \cdot)$ with $V := \{i \in [n] : \mathbf{v}_i \neq ?\}$, i.e., the positions of 0's and 1's in $\mathbf{v}$. For all $\mathbf{v}$, the distribution $\mathcal{D}_{\mathbf{v}}$ is the uniform distribution over $\mathcal{K}_{\mathbf{v}}$.

- $C_{\mathbf{v},\mathbf{w}} \colon \mathcal{K}_{\mathbf{v}} \to \mathcal{K}_{\mathbf{w}}$, for all $\mathbf{v},\mathbf{w}$ for which $(\mathbf{v},\mathbf{w})$ is an edge in D, is defined as

$$C_{\mathbf{v},\mathbf{w}}(k) := \mathsf{Constrain}(pp, k, \mathbf{w}) \ .$$

(Note that for $\mathbf{w} \in \{0,1\}$ we have $\mathsf{Constrain}(pp, k, \mathbf{w}) = \mathsf{Eval}(pp, k, \mathbf{w})$.)

By construction, running $\mathsf{Constrain}(pp, k_{\mathbf{v}}, \mathbf{w})$ on any key $k_{\mathbf{v}} \in \mathcal{K}_{\mathbf{v}}$ derived for some $\mathbf{v} \in \{0,1,?\}^n$ from some master key $k \in \mathbb{Z}_p$ always yields $(g_{|W|})^{k \cdot \prod_{i \in W} d_{i,w_i}}$ if $(\mathbf{v},\mathbf{w})$ is an edge in D. This shows consistency, which requires that for any nodes $\mathbf{v},\mathbf{w} \in \{0,1,?\}^n$ and any two paths $P = (\mathbf{v} = \mathbf{v}_0, \mathbf{v}_1, \ldots, \mathbf{v}_k = \mathbf{w})$ and $P' = (\mathbf{v} = \mathbf{v}'_0, \mathbf{v}'_1, \ldots, \mathbf{v}'_\ell = \mathbf{w})$ from $\mathbf{v}$ to $\mathbf{w}$ in D, we have

$$C_{\mathbf{v}_{k-1},\mathbf{v}_k} \circ \cdots \circ C_{\mathbf{v}_1,\mathbf{v}_2} \circ C_{\mathbf{v}_0,\mathbf{v}_1} \ = \ C_{\mathbf{v}'_{\ell-1},\mathbf{v}'_\ell} \circ \cdots \circ C_{\mathbf{v}'_1,\mathbf{v}'_2} \circ C_{\mathbf{v}'_0,\mathbf{v}'_1} \ .$$

Finally, our construction is *homomorphic*, that is, for every edge $(\mathbf{v},\mathbf{w})$ in D:

$$C_{\mathbf{v},\mathbf{w}}(k_1) \cdot C_{\mathbf{v},\mathbf{w}}(k_2) = C_{\mathbf{v},\mathbf{w}}(k_1 + k_2) \ . \tag{2}$$

To show this, let $pp = (\mathbb{G}, \{D_{i,\beta}\}_{i \in [n], \beta \in \{0,1\}}) \leftarrow \mathsf{Setup}(1^\lambda, 1^n)$. For all $k_1, k_2 \in \mathbb{Z}_p$ we then have the following:

1. $F(pp, k_1 + k_2, x) = g_n^{(k_1+k_2) \cdot \prod_{i \in [n]} d_{i,x_i}} = F(pp, k_1, x) \cdot F(pp, k_2, x)$.
2. For any $\mathbf{v} \in \{0,1,?\}^n$ we have:

$$\mathsf{Constrain}(pp, k_1 + k_2, \mathbf{v}) = (g_{|V|})^{(k_1+k_2) \cdot \prod_{i \in V} d_{i,v_i}}$$
$$= \mathsf{Constrain}(pp, k_1, \mathbf{v}) \cdot \mathsf{Constrain}(pp, k_2, \mathbf{v}) \ .$$

3. For any $\mathbf{v},\mathbf{w} \in \{0,1,?\}^n$ for which $v_i = ?$ or $v_i = w_i$ for all $i \in [n]$: if $k_{\mathbf{v}} = \mathsf{Constrain}(pp, k, \mathbf{v})$ and $k'_{\mathbf{v}} = \mathsf{Constrain}(pp, k', \mathbf{v})$ then

$$\mathsf{Constrain}(pp, k_{\mathbf{v}} \cdot k'_{\mathbf{v}}, \mathbf{w}) = e(k_{\mathbf{v}} \cdot k'_{\mathbf{v}}, D_{W \setminus V}(\mathbf{w})) = (g_{|W|})^{(k+k') \cdot \prod_{i \in W} d_{i,w_i}}$$
$$= \mathsf{Constrain}(pp, k_{\mathbf{v}}, \mathbf{w}) \cdot \mathsf{Constrain}(pp, k'_{\mathbf{v}}, \mathbf{w}) \ .$$

By 1. we have $C_{\mathbf{r},x}(k_1 + k_2) = C_{\mathbf{r},x}(k_1) \cdot C_{\mathbf{r},x}(k_2)$ for all $x \in \mathcal{X}$; by 2. we have $C_{\mathbf{r},\mathbf{v}}(k_1 + k_2) = C_{\mathbf{r},\mathbf{v}}(k_1) \cdot C_{\mathbf{r},\mathbf{v}}(k_2)$ for all $\mathbf{v}$; and by 3. we have $C_{\mathbf{v},\mathbf{w}}(k_1 + k_2) = C_{\mathbf{v},\mathbf{w}}(k_1) \cdot C_{\mathbf{v},\mathbf{w}}(k_2)$ for $\mathbf{v} \neq \mathbf{r}$; together this shows Equation (2).

Security. We show that publishing part of the secret key as parameters does not make the construction insecure. In particular, we show that our construction satisfies Definition 2, when the challenge set R is $\{0,1\}^n \subseteq V = \{0,1,?\}^n$, that is, the set of leaves of the DAG, which corresponds to the PRF domain $\mathcal{X} = \{0,1\}^n$.

We need to show that when $pp \leftarrow \mathsf{Setup}$ and $k \leftarrow \mathbb{Z}_p$ then an adversary that is given an oracle, which when queried on $\mathbf{v} \in \{0,1,?\}^n \setminus \{0,1\}^n$ returns $\mathsf{Constrain}(pp, k, \mathbf{v})$ and when queried on $x \in \{0,1\}^n$ returns either $F(pp, k, x)$ or a random element, cannot distinguish these two cases—provided it does not query a descendant $x \in R$ of some other query $u \in V$.

Theorem 1. *If there exists a PPT adversary breaking security of the above key-homo-morphic bit-fixing PRF for n-bit-inputs, with challenge set $R = \{0,1\}^n$, with advan-tage $\varepsilon(\lambda)$ and making $q(\lambda)$ queries for challenge elements, then there exists a PPT algorithm that breaks the n-Multilinear Decisional Diffie-Hellman assumption with ad-vantage $2^{-n} \cdot q(\lambda)^{-1} \cdot \varepsilon(\lambda)$.*

The proof first reduces the original game to a game where the adversary can only ask for one challenge query, which loses a factor $q(\lambda)$, by a standard hybrid argument. That game is then reduced to MDDH, following the proof from [BLMR13]; in particular, since in the simulation the reduction knows the values $\{D_{i,\beta}\}$, it can output them as public parameters (which do not exist in the original proof). See the full version for a detailed proof.

3.4   Circuit-Constrained PRF with Key-Homomorphic Evaluation

Boneh and Waters [BW13] give a second construction based on multilinear maps, which allows for constraining keys to more expressive sets, namely, all sets that can be decided by a circuit of some fixed depth. By defining public parameters, we construct a variant that is key-homomorphic as defined by Boneh *et al.* [BLMR13]. That is, we have that for all pp, k_1, k_2, x,

$$F(pp, k_1 + k_2, x) = F(pp, k_1, x) \cdot F(pp, k_2, x) \ . \tag{3}$$

However, our construction is not key-homomorphic in the sense of Definition 3, as the key-constraining function is not homomorphic.

The PRF is set up for input length n and circuit depth ℓ. The parameters are a se-quence $(\mathbb{G}_1, \ldots, \mathbb{G}_\kappa)$ of groups $\mathbb{G}_i$ of prime order p, generated by g_i, where $\kappa = n + \ell$; as well as elements $D_{i,\beta}$, uniformly chosen from $\mathbb{G}_1$, for $i \in [n]$ and $\beta \in \{0,1\}$. The secret-key space is $\mathbb{Z}_p$ and the PRF on input $x = (x_1, \ldots, x_n) \in \{0,1\}^n$ is defined as

$$F(pp, k, x) := e\big(e(\{D_{i,x_i}\}_{i \in [n]})^k, g_\ell\big) \ = \ g_\kappa^{k \cdot \prod_{i \in [n]} d_{i,x_i}} \tag{4}$$

(with $d_{i,\beta}$ such that $D_{i,\beta} = g^{d_{i,\beta}}$). It is thus defined exactly as for the bit-fixing con-struction, except that there are more groups in the sequence, and satisfies (3).

Removing the values $d_{i,\beta}$ from the secret key of the construction in [BW13] entails another syntactical change. Above, we defined the PRF value F in terms of k and the parameter values $D_{i,\beta}$, whereas in [BW13] they are defined directly as the last term of Equation (4). In [BW13], components of constrained keys (those corresponding to input wires) are defined as $K_w := g^{r_w \cdot d_{w,1}}$, which we replace by $K_w := (D_{w,1})^{r_w}$.

The values $d_{i,\beta}$ are not used anywhere else. Our construction still satisfies pseudo-randomness, since, as for the bit-fixing PRF, in the security proof the simulator knows the values $D_{i,\beta}$.

4   Prefix-Fixing Construction from LWE

In this section we prove that variants of the LWE-based key-homomorphic PRF of Baner-jee and Peikert [BP14] also support prefix constraints, and that the constraining functions are key-homomorphic as well. After recalling some standard background and notation in Section 4.1, the contents of this section have the following high-level structure:

- In Section 4.2 we define a key-homomorphic, prefix-constrained pseudorandom function family called Constrain, which we refer to as the "noisy" family. However, the functions in this family are highly sequential, with circuit depth proportional to the input length. More significantly, they have huge keys, of size *exponential* in the input length, so they cannot actually be used in reality. The purpose of defining this family is to give us a baseline object that has "perfect" consistency, homomorphic, and pseudorandomness properties (but terrible space and depth complexity), which we rely upon in the later subsections.
- In Section 4.3 we specialize the noisy Constrain family to be "errorless," i.e., all error terms are set to zero. We call the resulting family PConstrain. As a specialization of Constrain, it inherits that latter's perfect consistency and homomorphic properties. We show that the PConstrain functions (1) have small keys, (2) can be computed in low depth (e.g., logarithmic in the input length) by a slight modification to the Constrain algorithms, and (3) have outputs that are "close" to those of the noisy Constrain functions (under a mild condition on the input). However, we are still not quite done yet, because the errorless PConstrain functions are not pseudorandom.
- In Section 4.4 we combine the previous two families to obtain a family $\overline{\text{PConstrain}}$ that has essentially all the desired properties: small keys, low depth, pseudorandomness, consistency, and homomorphism. (The latter two of these properties do not hold perfectly, but only *computationally*: no efficient adversary can make any queries that reveal a violation of either property.) The $\overline{\text{PConstrain}}$ functions are defined simply as appropriately *rounded* functions from *errorless* family PConstrain. As such, they inherit the latter's small keys and low depth. In addition, they are pseudorandom because they coincide with the rounded *noisy* pseudorandom Constrain functions; this follows from the fact that the (unrounded) errorless PConstrain and noisy Constrain functions have "close" outputs, and the rounding precision is taken to be sufficiently coarse to conceal this difference. Finally, consistency and homomorphism hold for $\overline{\text{PConstrain}}$ essentially because rounding can be seen as adding a particular kind of (deterministic) error, so $\overline{\text{PConstrain}}$ may be seen as an instantiation of Constrain.

4.1 Preliminaries

We first recall some standard background from [MP12, BP14]. For an integer modulus $q \geq 1$, let $\mathbb{Z}_q = \mathbb{Z}/q\mathbb{Z}$ denote the quotient ring of integers modulo q, where for convenience we always let $q = 2^\ell$ be a power of two. For $\ell = \log q \geq 2$, define the "gadget" (row) vector

$$\mathbf{g} = (1, 2, 4, \ldots, 2^{\ell-1}) \in \mathbb{Z}_q^\ell \ ,$$

and the (deterministic) "binary decomposition" function $\mathbf{g}^{-1} \colon \mathbb{Z}_q \to \{0, 1\}^\ell$ as follows: identifying each $a \in \mathbb{Z}_q$ with its integer representative in $\{0, \ldots, q-1\}$, let $\mathbf{g}^{-1}(a) = (x_0, x_1, \ldots, x_{\ell-1}) \in \{0, 1\}^\ell$ where $a = \sum_{i=0}^{\ell-1} x_i 2^i$ is the binary representation of a.

Note that by definition, $\langle \mathbf{g}, \mathbf{g}^{-1}(a) \rangle = a$ for all $a \in \mathbb{Z}_q$, which explains our choice of notation.

Similarly, for vectors and matrices over $\mathbb{Z}_q$ we define the function $\mathbf{G}^{-1} \colon \mathbb{Z}_q^{n \times m} \to \{0, 1\}^{n\ell \times m}$ by applying $\mathbf{g}^{-1}$ entry-wise. Notice that for all $\mathbf{A} \in \mathbb{Z}_q^{n \times m}$ we have

$$\mathbf{G} \cdot \mathbf{G}^{-1}(\mathbf{A}) = \mathbf{A}, \quad \text{where} \quad \mathbf{G} = \mathbf{I}_n \otimes \mathbf{g} = \operatorname{diag}(\mathbf{g}, \ldots, \mathbf{g}) \in \mathbb{Z}_q^{n \times n\ell}$$

is the block-diagonal matrix having n copies of $\mathbf{g}$ as diagonal blocks, and zeros elsewhere. We let $\mathcal{P} \subseteq \mathbb{Z}^{n\ell}$ denote a certain set of canonical representatives of the additive quotient group $\mathbb{Z}_q^{n\ell}/(\mathbb{Z}_q^n \cdot \mathbf{G})$. Specifically, as shown in [MP12], we can use[1]

$$\mathcal{P} := \{-\tfrac{q}{4}, \ldots, \tfrac{q}{4} - 1\}^{n\ell} .$$

We define a bijection $\operatorname{Decode} \colon \mathbb{Z}_q^{n\ell} \to \mathcal{P} \times \mathbb{Z}_q^n$ as $\operatorname{Decode}(\mathbf{u}) = (\mathbf{v}, \mathbf{s})$, where

$$\mathbf{u} = \mathbf{v} + \mathbf{s} \cdot \mathbf{G} .$$

As shown in [MP12], there is an efficient algorithm for computing Decode in depth proportional to $\ell = \log q$, and clearly $\operatorname{Decode}^{-1}(\mathbf{v}, \mathbf{s}) = \mathbf{v} + \mathbf{s} \cdot \mathbf{G}$.

We recall the following easy lemma about the spectral norm, denoted $s_1(\cdot)$, of binary matrices. (See, e.g, [BP14, Lemma 3.1] for a proof.) Recall that $s_1(\mathbf{M}) = \max_{\|\mathbf{u}\|=1} \|\mathbf{u}\mathbf{M}\|$, where the maximum is taken over real unit vectors $\mathbf{u}$.

Lemma 1. *If $\mathbf{S}$ is a binary (i.e., 0-1) m-by-m matrix, then $s_1(\mathbf{S}) \leq m$.*

Binary trees. A *full* binary tree T is one in which each node is either a leaf, or has two (nonempty) children. We let $|T|$ denote the number of leaves in T, and index the leaves from 0 to $|T| - 1$ by the inorder traversal of T. If $|T| \geq 1$, we let $T.l$ and $T.r$ respectively denote its left and right subtrees, both of which are nonempty.

Given matrices $\mathbf{A}_0, \mathbf{A}_1 \in \mathbb{Z}_q^{n \times n\ell}$, we define the function $\mathbf{A}_T(x) \colon \{0, 1\}^{|T|} \to \mathbb{Z}_q^{n \times n\ell}$ as follows:

$$\mathbf{A}_T(x) := \begin{cases} \mathbf{A}_x & \text{if } |T| = 1, \\ \mathbf{A}_{T.l}(x_l) \cdot \mathbf{G}^{-1}(\mathbf{A}_{T.r}(x_r)) & \text{otherwise,} \end{cases}$$

where in the second case we parse the input $x = x_l x_r$ where $|x_l| = |T.l|$ and $|x_r| = |T.r|$.

Rounding. For a positive integer e, we define the integer rounding function $\lfloor \cdot \rceil_e \colon \mathbb{Z} \to e\mathbb{Z}$ as $\lfloor x \rceil_e := \lfloor x/e \rceil \cdot e$, and extend it component-wise to vectors and matrices. In words, $\lfloor x \rceil$ simply rounds x to the nearest integer multiple of e.[2]

---

[1] This choice of $\mathcal{P}$ is possible because we have taken q to be a power of two. It may be possible to generalize our results to other values of q using the alternative lattice bases given in [MP12], but it seems to substantially complicate the proofs.

[2] We point out that this function differs slightly from the "modular" rounding function considered in prior works, which mapped $\mathbb{Z}_q$ to $\mathbb{Z}_p$ as $\lfloor x \rceil_p = \lfloor x \cdot p/q \rceil \bmod p$. Here e corresponds with q/p, but the rounding input and output have the same "scale."

4.2 "Noisy" Function Family $\mathcal{C}$

As in previous work [BPR12, BP14], we first define and analyze a certain family $\mathcal{C}$ of "noisy" constraining functions, which have *huge* (exponential-size) keys, because each key contains many error terms. To avoid technical complications related to efficient computation on exponential-size inputs, throughout this section the error terms are always sampled "lazily," i.e., not until they are needed. In Section 4.2 we show that the constraining functions are consistent, in Section 4.2 we show that they are homomorphisms under an appropriate group operation on the key spaces, and in Section 4.2 we show that the family is pseudorandom.

The public parameters of the noisy family are two matrices $\mathbf{A}_0, \mathbf{A}_1 \in \mathbb{Z}_q^{n \times n\ell}$, chosen uniformly at random. Following Definitions 1 and 3, to describe our family we need to give a DAG with a unique root node, a key space with an additive group structure for each node in the DAG, and a constraining function for each edge in the DAG.

DAG. For a full binary tree T, our DAG corresponds to prefix-fixing constraints on $\{0,1\}^{|T|}$, i.e., the nodes are identified with the strings in $\{0,1\}^{\leq |T|}$, and there is an edge (w, wx) for every w and $x \neq \varepsilon$ such that $|wx| \leq |T|$. This DAG clearly has a unique root node, namely, the empty string ε.

Key spaces. For any full binary tree T and $0 \leq j < |T|$, we define

$$
\mathcal{R}_{T,j} := \begin{cases} \mathbb{Z}_q^n & \text{if } |T| = 1, \\ \mathcal{R}_{T.l,j} & \text{if } j \leq |T.l|, \\ \mathcal{P} \times \mathcal{R}_{T.r,j-|T.l|} & \text{otherwise.} \end{cases}
$$

For convenience in our recursive algorithms, we also define $\mathcal{R}_{T,|T|} = \mathcal{P} \times \mathbb{Z}_q^n$. In words, $\mathcal{R}_{T,j}$ has one $\mathcal{P}$-component for each *left* subtree "hanging off" the path from the root to the jth leaf. (Recall that we number the leaves starting from zero.) We also define, for $0 \leq j \leq |T|$,

$$
\mathcal{E}_{T,j} := \prod_{y \in \{0,1\}^{\leq |T|-j}} \mathbb{Z}^{n\ell} = (\mathbb{Z}^{n\ell})^{2^{|T|-j+1}-1} .
$$

In $\mathcal{E}_{T,j}$, the several $\mathbb{Z}^{n\ell}$ components (which represent the error vectors) are indexed by the binary strings of length at most $|T| - j$, which is why there are $2^{|T|-j+1} - 1$ components.

For each $w \in \{0,1\}^{\leq |T|}$, the key space $\mathcal{K}_{T,w}$ and associated distribution for w are defined as:

$$
\mathcal{K}_{T,w} := \mathcal{R}_{T,|w|} \times \mathcal{E}_{T,|w|} ,
$$
$$
\mathcal{D}_{T,w} := U(\mathcal{R}_{T,|w|}) \times (\chi^{n\ell})^{2^{|T|-|w|+1}-1} ,
$$

where χ is some error distribution over $\mathbb{Z}$ that will be used in the security proof. Note that $\mathcal{K}_{T,w}$ does not depend on the actual bits in w, only on its length $|w|$.

To make $\mathcal{K}_{T,w}$ an additive group (for $|w| > 0$), we stress that we do *not* simply treat it as a product group of its components—indeed, $\mathcal{P} \subset \mathbb{Z}^{n\ell}$ is not even closed under addition, so it is not a group. Instead, in Section 4.2 below we define a special addition operation on $\mathcal{R}_{T,|w|}$ to make it a group. Then $\mathcal{K}_{T,w}$ is simply the product group of this group with $\mathcal{E}_{T,|w|}$, with the usual addition operation on the latter.

Constraining functions. It now remains to define (consistent) constraining functions $\mathsf{Constrain}_{T,w,x} \colon \mathcal{K}_{T,w} \to \mathcal{K}_{T,wx}$ for all strings w, x such that $x \neq \varepsilon$ and $|wx| \leq |T|$; for convenience, we also define $\mathsf{Constrain}_{T,w,x}$ to be the identity function for $x = \varepsilon$. Functional pseudocode for the constraining functions is given in Algorithm 4.1. We remark that it would have been sufficient to define functions $\mathsf{Constrain}_{T,x,w}$ for $|x| = 1$ alone. Indeed, by Lemma 2 below it follows that our pseudocode is actually *equivalent* to the sequential composition of such functions, and hence has circuit depth proportional to $|x|$. We choose to present the constraining functions for general x here because in Section 4.4 we show that a slight modification yields highly parallel functions.

In summary, the constraining functions are defined recursively on the tree structure. In the base case $|T| = 1$, for key $(\mathbf{v}, (\mathbf{e}_x)_{x\in\{0,1\}^{\leq 1}}) \in \mathcal{K}_{T,w} = \mathbb{Z}_q^n \times (\mathbb{Z}^{n\ell})^3$, we simply compute and decode the "noisy" value $\mathbf{v}\mathbf{A}_x + \mathbf{e}_x \in \mathbb{Z}_q^{n\ell}$. There are three recursive cases, depending on whether we are constraining entirely within the left subtree, within the right subtree, or across the two subtrees. In the first two cases, we simply recurse on the appropriate subtree. In the third case, we recursively constrain over the remainder of the left subtree, then over the desired portion of the right subtree. Lastly, whenever we finish constraining over an *entire* (sub)tree we need to appropriately "fold" the results, which consist of some leftover value in $\mathcal{P} \subset \mathbb{Z}^{n\ell}$ from the left subtree and some value in $\mathcal{P} \times \mathbb{Z}_q^n$ from the completed right subtree, into a value in $\mathcal{P} \times \mathbb{Z}_q^n$ for the entire tree.

We remark that although our presentation is (necessarily) quite different, our constraining functions correspond to the *partial* evaluations of the noisy function family from [BP14], which the simulator computes internally when answering queries in the security proof.

Consistency. We first show consistency.

Lemma 2 (Consistency). *For any full binary tree T, parameters $\mathbf{A}_0, \mathbf{A}_1$, and strings w, x, z where $|wxz| \leq |T|$, we have that*

$$\mathsf{Constrain}_{T,wx,z} \circ \mathsf{Constrain}_{T,w,x} = \mathsf{Constrain}_{T,w,xz} \; .$$

Proof. We proceed by induction on $|T|$. The base case of $|T| = 1$ is trivial, because $\mathsf{Constrain}_{T,w,\varepsilon}$ is the identity function.

We have three inductive cases. In the first two cases, where $|wxz| \leq |T.l|$ or $|w| \geq |T.l|$, the claim follows immediately by the inductive hypothesis on $T.l$ or $T.r$, respectively. The last inductive case is $|w| < |T.l|$ and $|wxz| > |T.l|$. We analyze this in two subcases.

Algorithm 4.1. $\mathsf{Constrain}_{T,w,x}\colon \mathcal{K}_{T,w} \to \mathcal{K}_{T,wx}$ for $|wx| \leq |T|$, $x \neq \varepsilon$

Input: $(\mathbf{v}, (\mathbf{e}_y)_{|y| \leq |T|-|w|}) \in \mathcal{K}_{T,w} = \mathcal{R}_{T,|w|} \times \mathcal{E}_{T,|w|}$

1. **if** $|T| = 1$ **then** $\triangleright$ base case, so $\mathbf{v} \in \mathbb{Z}_q^n$
2.   **return** $\mathsf{Decode}(\mathbf{v} \cdot \mathbf{A}_x + \mathbf{e}_x)$
3. **else if** $|wx| \leq |T.l|$ **then** $\triangleright$ constrains entirely in left subtree...
4.   **return** $\left(\mathsf{Constrain}_{T.l,w,x}\left(\mathbf{v}, (\mathbf{e}_y)_{|y| \leq |T.l|-|w|}\right), (\mathbf{e}_{xy})_{|y| \leq |T|-|wx|}\right)$ $\triangleright$ ... so just
      recurse.
5. **else if** $|w| < |T.l|$ **then** $\triangleright$ incomplete left subtree...
6.   parse $x = x_l x_r$ where $|wx_l| = |T.l|$
7.   let $(\mathbf{v}_l, \star) = \mathsf{Constrain}_{T,w,x_l}\left(\mathbf{v}, (\mathbf{e}_y)_{|y| \leq |T.l|-|w|}\right)$ $\triangleright$ ... complete left subtree
      (self-recurse)...
8.   **return** $\mathsf{Constrain}_{T,wx_l,x_r}\left(\mathbf{v}_l, (\mathbf{e}_{x_l y})_{|y| \leq |T.r|}\right)$ $\triangleright$ ... and self-recurse to finish.
9. **else** $\triangleright$ constrains entirely in right subtree...
10.   parse $w = w_l w_r$ where $|w_l| = |T.l|$ and $\mathbf{v} = (\mathbf{v}_l, \mathbf{v}_r) \in \mathcal{P} \times \mathcal{R}_{T.r,|w_r|}$
11.   let $(\mathbf{k}_r, \star) = \mathsf{Constrain}_{T.r,w_r,x}\left(\mathbf{v}_r, (\mathbf{e}_y)_{|y| \leq |T|-|w|}\right)$ $\triangleright$ ... so recurse.
12.   **if** $|wx| = |T|$ **then** $\triangleright$ constrains over entire tree (so $\mathbf{k}_r \in \mathcal{P} \times \mathbb{Z}_q^n$)...
13.   **return** $\mathsf{Decode}(\mathbf{v}_l \cdot \mathbf{G}^{-1}(\mathbf{A}_{T.r}(w_r x))) + \mathsf{Decode}^{-1}(\mathbf{k}_r))$ $\triangleright$ ... so fold results.
14.   **else** $\triangleright$ doesn't complete the tree...
15.   **return** $(\mathbf{v}_l, \mathbf{k}_r)$ $\triangleright$ ... so append results.

The first subcase is $|wx| > |T.l|$. Here we parse $x = x_l x_r$ with $|wx_l| = |T.l|$. By definition, we have

$$\mathsf{Constrain}_{T,w,x} = \mathsf{Constrain}_{T,wx_l,x_r} \circ \mathsf{Constrain}_{T,w,x_l}$$
$$\mathsf{Constrain}_{T,w,xz} = \mathsf{Constrain}_{T,wx_l,x_rz} \circ \mathsf{Constrain}_{T,w,x_l} \ .$$

The claim then follows by the inductive hypothesis on $T.r$, by composing $\mathsf{Constrain}_{T,wx,z}$ on the left of the first equation above.

The second subcase is $|wx| \leq |T.l|$. This proceeds essentially identically to the first subcase, where we instead parse $z = z_l z_r$ with $|wxz_l| = |T.l|$.

Homomorphism. Before we can prove that the constraining functions are homomorphisms, we must make $\mathcal{K}_{T,w} = \mathcal{R}_{T,|w|} \times \mathcal{E}_{T,|w|}$ an additive group for $|w| > 0$. (Recall that $\mathcal{R}_{T,0} = \mathbb{Z}_q^n$, which is already a group.) We do so by defining a special group operation $\mathsf{Add}_{T,w}$ on the set $\mathcal{R}_{T,|w|}$—note that the operation depends on w itself, not just its length. We then let $\mathcal{K}_{T,w}$ be the product group with $\mathcal{E}_{T,|w|}$, under its usual addition operation. For convenience, we overload $\mathsf{Add}_{T,w}$ to also refer to the group operation for this product group, where the intended domain should be clear by context.

For convenience in the recursive definitions, we let $\mathsf{Add}_{T,w}$ take an auxiliary input $\mathbf{t} \in \mathbb{Z}_q^n$, which should be thought of as a kind of "carry" term that comes from reducing the sum of two $\mathcal{P}$-elements (in $\mathbb{Z}^{n\ell}$) back to $\mathcal{P}$. Initializing this carry input to zero yields the group operation. Formally, we define

$$\mathsf{Add}_{T,w}\left(\mathbf{t} \in \mathbb{Z}_q^n, \mathbf{v}, \mathbf{v}'\right) :=$$

$$\begin{cases} \mathbf{t} + \mathbf{v} + \mathbf{v}' & \text{if } |T| = 1, |w| = 0, \\ \mathsf{Add}_{T.l,w}(\mathbf{t}, \mathbf{v}, \mathbf{v}') & \text{if } |w| \leq |T.l|, \\ \left(\bar{\mathbf{v}}_l, \mathsf{Add}_{T.r,w_r}\left(\bar{\mathbf{t}}, \mathbf{v}_r, \mathbf{v}_r'\right)\right) & \text{if } |T.l| < |w| < |T|, \\ \mathsf{Decode}\left(\mathbf{t} \cdot \mathbf{A}_T(w) + \mathsf{Decode}^{-1}(\mathbf{v}) + \mathsf{Decode}^{-1}(\mathbf{v}')\right) & \text{if } |w| = |T|, \end{cases}$$

where in the third case we parse $w = w_l w_r$ for $|w_l| = |T.l|$ and $\mathbf{v} = (\mathbf{v}_l, \mathbf{v}_r), \mathbf{v}' = (\mathbf{v}_l', \mathbf{v}_r') \in \mathcal{P} \times \mathcal{R}_{T.r,|w_r|}$, and let $(\bar{\mathbf{v}}_l, \bar{\mathbf{t}}) = \mathsf{Decode}(\mathbf{t} \cdot \mathbf{A}_{T.l}(w_l) + \mathbf{v}_l + \mathbf{v}_l')$.

We now prove that the Constrain functions are homomorphisms.

Lemma 3 (Homomorphism). *For any parameters $\mathbf{A}_0, \mathbf{A}_1$ and any full binary tree T, any bit strings w, x such that $|wx| \leq |T|$, and any $\mathbf{t} \in \mathbb{Z}_q^n$ and $\mathbf{k}, \mathbf{k}' \in \mathcal{K}_{T,w}$, we have*

$$\mathsf{Constrain}_{T,w,x}(\mathsf{Add}_{T,w}(\mathbf{t}, \mathbf{k}, \mathbf{k}'))$$
$$= \mathsf{Add}_{T,wx}(\mathbf{t}, \mathsf{Constrain}_{T,w,x}(\mathbf{k}), \mathsf{Constrain}_{T,w,x}(\mathbf{k}')) . \quad (5)$$

In particular, by setting $\mathbf{t} = \mathbf{0}$ we have that $\mathsf{Constrain}_{T,w,x}$ is an additive homomorphism.

Proof. The claim is trivial for $x = \varepsilon$, so from now on we assume that $x \neq \varepsilon$. Let $i = |w|$ and $j = |wx|$, so $0 \leq i < j \leq |T|$. Parse $\mathbf{k} = (\mathbf{v}, (\mathbf{e}_y)), \mathbf{k}' = (\mathbf{v}', (\mathbf{e}_y'))$, and let $\bar{\mathbf{k}} = (\bar{\mathbf{v}}, (\bar{\mathbf{e}}_y)) = \mathsf{Add}_{T,w}(\mathbf{t}, \mathbf{k}, \mathbf{k}')$.

As usual, we proceed by induction over $|T|$. In the base case, where $w = \varepsilon$ and $|x| = |T| = 1$, we have

$$\mathsf{Constrain}_{T,\varepsilon,x}(\bar{\mathbf{v}}, (\bar{\mathbf{e}}_y)) = \mathsf{Decode}(\bar{\mathbf{v}} \cdot \mathbf{A}_x + \bar{\mathbf{e}}_x)$$
$$= \mathsf{Decode}((\mathbf{t} + \mathbf{v} + \mathbf{v}') \cdot \mathbf{A}_x + (\mathbf{e}_x + \mathbf{e}_x'))$$
$$= \mathsf{Decode}(\mathbf{t} \cdot \mathbf{A}_x + (\mathbf{v} \cdot \mathbf{A}_x + \mathbf{e}_x) + (\mathbf{v}' \cdot \mathbf{A}_x + \mathbf{e}_x'))$$
$$= \mathsf{Add}_{T,x}(\mathbf{t}, \mathsf{Constrain}(\mathbf{v}, (\mathbf{e}_y)), \mathsf{Constrain}(\mathbf{v}', (\mathbf{e}_y'))) .$$

We now consider the inductive cases. Because Constrain simply passes an appropriate subset of the input error terms (the $\mathbb{Z}^{n\ell}$ vectors in the key) to the output, for simplicity of exposition we suppress the error terms in the remainder of the proof. The final claim then follows by the product group structure of $\mathcal{K}_{T,w}$.

The first inductive case, where $i < j \leq |T.l|$, follows immediately from the inductive hypothesis on $\mathsf{Constrain}_{T.l,w,x}$. For the second inductive case, where $i < |T.l| < j$, we defer to the final paragrap of the proof. For the third inductive case, where $|T.l| \leq i$, parse $w = w_l w_r$ and $\mathbf{v} = (\mathbf{v}_l, \mathbf{v}_r), \mathbf{v}' = (\mathbf{v}'_l, \mathbf{v}'_r), \bar{\mathbf{v}} = (\bar{\mathbf{v}}_l, \bar{\mathbf{v}}_r)$, and note that by definition,

$$\bar{\mathbf{v}}_r = \mathrm{Add}_{T.r, w_r x}(\bar{\mathbf{t}}, \mathbf{v}_r, \mathbf{v}'_r) \ , \tag{6}$$

$$\bar{\mathbf{v}}_l + \bar{\mathbf{t}} \cdot \mathbf{G} = \mathbf{t} \cdot \mathbf{A}_{T.l}(w_l) + \mathbf{v}_l + \mathbf{v}'_l \tag{7}$$

for some $\bar{\mathbf{t}} \in \mathbb{Z}_q^n$. As in the code for Constrain, let $\mathbf{k}_r = \mathrm{Constrain}_{T.r, w_r, x}(\mathbf{v}_r)$ and similarly for $\mathbf{k}'_r, \bar{\mathbf{k}}_r$. Then by the inductive hypothesis on $\mathrm{Constrain}_{T.r, w_r, x}$ and Equation (6), we have

$$\bar{\mathbf{k}}_r = \mathrm{Constrain}_{T.r, w_r, x}(\bar{\mathbf{v}}_r) = \mathrm{Add}_{T.r, w_r x}(\bar{\mathbf{t}}, \mathbf{k}_r, \mathbf{k}'_r) \ . \tag{8}$$

Now if $j = |wx| < |T|$, then by definition of $\mathrm{Constrain}_{T,w,x}$ and $\mathrm{Add}_{T,wx}$ (respectively), the left- and right-hand sides of Equation (5) are respectively just $\bar{\mathbf{v}}_l$ prepended to the left- and right-hand sides of Equation (8), so they are equal. But if $j = |wx| = |T|$, then the output of $\mathrm{Constrain}_{T,w,x}(\bar{\mathbf{v}})$ is "folded," (i.e., in $\mathcal{P} \times \mathbb{Z}_q^n$), as are $\mathbf{k}_r, \mathbf{k}'_r, \bar{\mathbf{k}}_r$ as defined above. We proceed by applying the folding operation to both sides of Equation (8), namely, apply Decode^{-1}, add $\bar{\mathbf{v}}_l \cdot \mathbf{G}^{-1}(\mathbf{A}_{T.r}(w_r x))$, and apply Decode. For the left-hand side we get exactly $\mathrm{Constrain}_{T,w,x}(\bar{\mathbf{v}})$, which is the left-hand side of Equation (5). On the right-hand side, by definition of $\mathrm{Add}_{T.r, w_r x}$, by Equation (7), and by definition of $\mathbf{A}_T(\cdot)$, we get Decode applied to

$$\bar{\mathbf{v}}_l \cdot \mathbf{G}^{-1}(\mathbf{A}_{T.r}(w_r x)) + \mathrm{Decode}^{-1}(\mathrm{Add}_{T.r, w_r x}(\bar{\mathbf{t}}, \mathbf{k}_r, \mathbf{k}'_r))$$

$$= (\bar{\mathbf{v}}_l + \bar{\mathbf{t}} \cdot \mathbf{G}) \cdot \mathbf{G}^{-1}(\mathbf{A}_{T.r}(w_r x)) + \mathrm{Decode}^{-1}(\mathbf{k}_r) + \mathrm{Decode}^{-1}(\mathbf{k}'_r)$$

$$= (\mathbf{t} \cdot \mathbf{A}_{T.l}(w_l) + \mathbf{v}_l + \mathbf{v}'_l) \cdot \mathbf{G}^{-1}(\cdots) + \mathrm{Decode}^{-1}(\mathbf{k}_r) + \mathrm{Decode}^{-1}(\mathbf{k}'_r)$$

$$= \mathbf{t} \cdot \mathbf{A}_T(wx) + (\mathbf{v}_l \cdot \mathbf{G}^{-1}(\cdots) + \mathrm{Decode}^{-1}(\mathbf{k}_r))$$

$$+ (\mathbf{v}'_l \cdot \mathbf{G}^{-1}(\cdots) + \mathrm{Decode}^{-1}(\mathbf{k}'_r)) \ .$$

As desired, this is the right-hand side of Equation (5), by definition of $\mathrm{Constrain}_{T,w,x}$ and $\mathrm{Add}_{T,wx}$.

Going back to the second inductive case, where $i < |T.l| < j$, it follows by writing $\mathrm{Constrain}_{T,w,x} = \mathrm{Constrain}_{T,wx_l,x_r} \circ \mathrm{Constrain}_{T,w,x_l}$ where $x = x_l x_r$ for $|wx_l| = |T.l|$, then applying the inductive hypothesis on $T.l$ and $T.r$. This completes the proof of Lemma 3.

Pseudorandomness. We now show that the function family $\mathcal{C}$ defined above is pseudorandom according to Definition 2, with all nodes $R = \{0,1\}^{\leq |T|}$ as challenge nodes. This follows immediately from the PRG-like property demonstrated in Lemma 4 below, together with the fact (shown in prior works [BGI14, KPTZ13, BW13]) that the GGM construction [GGM86] instantiated with such a PRG family yields a prefix-constrained PRF.[3] In a bit more detail: in the following lemma we show that for any string $w \in \{0,1\}^{<|T|}$, the function $G_{T,w} \colon \mathcal{K}_{T,w} \to \mathcal{K}_{T,w0} \times \mathcal{K}_{T,w1}$ defined as

---

[3] It is easy to verify that this remains true even for our slightly stronger definition, where the adversary can query the function at inputs corresponding to internal nodes of the GGM tree.

$G(\mathbf{k}) = (\mathsf{Constrain}_{T,w,0}(\mathbf{k}), \mathsf{Constrain}_{T,w,1}(\mathbf{k}))$ is a pseudorandom generator, under the LWE assumption. Instantiating the GGM construction with these PRGs yields our constraining functions $\mathsf{Constrain}_{T,w,x}$, therefore they are pseudorandom.

Lemma 4. *Let T be any full binary tree and $w \in \{0,1\}^{<|T|}$ be any string. Then assuming the hardness of decision-$\mathsf{LWE}_{n,q,\chi}$, for $\mathbf{k} \leftarrow \mathcal{D}_{T,w}$ we have*

$$(\mathsf{Constrain}_{T,w,0}(\mathbf{k}), \mathsf{Constrain}_{T,w,1}(\mathbf{k})) \overset{c}{\approx} \mathcal{D}_{T,w0} \times \mathcal{D}_{T,w1} \ .$$

The proof of this lemma involves a simulator embedding the appropriate LWE challenge in the base case in the computation of $\mathsf{Constrain}_{T,w,b}$. The rest of the proof consists of showing the outputs corresponding to each distribution (LWE vs uniform) are distributed accordingly. We defer the details to the full version.

4.3   Parallel Errorless Constrain

In this subsection we consider the "errorless" variants of our Constrain functions, which we call PConstrain, and show that they can be computed in low depth. We also show that the output of PConstrain is typically close to that of Constrain, when the errors used in the latter are small.

Parallel Evaluation. The PConstrain functions simply correspond to the Constrain functions when all the error vectors are set to zero, that is, $\mathsf{PConstrain}_{T,w,x}(\mathbf{v}, \mathbf{s}) = \mathsf{Constrain}_{T,w,x}(\mathbf{v}, \mathbf{s}, \mathbf{0})$. In particular, this implies that the PConstrain functions are both consistent and homomorphisms, because the Constrain functions are. In addition, the errorless setting allows PConstrain to be computed with good parallelism (i.e., in low depth) by an alternative algorithm that "short circuits" the computation via a base case that constrains over an *entire (sub)tree* in just one step. More specifically, we modify the base case (Lines 1 and 2) of Algorithm 4.1 as shown in Algorithm 4.2 below. The rest of the algorithm remains unchanged, apart from the fact that PConstrain does not take or output any error terms.

In Lemma 5 we prove that the alternative algorithm is correct. Then in Section 4.3 we describe how PConstrain can be evaluated in low depth.

Algorithm 4.2. $\mathsf{PConstrain}_{T,w,x}: \mathcal{R}_{T,|w|} \to \mathcal{R}_{T,|wx|}$ for $|wx| \leq |T|$, $x \neq \varepsilon$

Input: $\mathbf{v} \in \mathcal{R}_{T,|w|}$

1. **if** $w = \varepsilon$ and $|x| = |T|$ **then**   ▷ base case
2.   **return** $\mathsf{Decode}(\mathbf{v} \cdot \mathbf{A}_T(x))$
3. *The remaining code is the same as in Algorithm 4.1, but without any error terms* $(\mathbf{e}_y)$.

For convenience, we define the function $\mathsf{Project}_{T,w}: \mathcal{K}_{T,w} \to \mathcal{R}_{T,|w|}$, which just outputs the $\mathcal{R}_{T,|w|}$-component of its input (dropping the $\mathcal{E}_{T,|w|}$-component, i.e., the errors), and $\mathsf{PrjConstrain}_{T,w,x} = \mathsf{Project}_{T,wx} \circ \mathsf{Constrain}_{T,w,x}$.

The following lemma states that what the algorithm above does is indeed correct. It is proved by a simple induction for *complete inputs* only, that is, for inputs $x = \varepsilon$ and $w \in \{0,1\}^{|T|}$. The complete proof appears in the full version.

Lemma 5. *For any fully binary tree T, any bit strings w, x with $|wx| \leq |T|$, and any $\mathbf{v} \in \mathcal{R}_{T,|w|}$,*

$$\mathsf{PConstrain}_{T,w,x}(\mathbf{v}) = \mathsf{PrjConstrain}_{T,w,x}(\mathbf{v}, \mathbf{0}) \ .$$

Parallel Evaluation of PConstrain. We now analyze the parallel complexity of the PConstrain functions according to Algorithm 4.2 (and Algorithm 4.1) above. Our main goal is to bound what we call the "nonlinear depth" of $\mathsf{PConstrain}_{T,w,x}$ in terms of the topology of T and the strings w, x. Nonlinear depth only takes into account the nonlinear Decode and $\mathbf{G}^{-1}$ operations; the remaining operations are all linear over $\mathbb{Z}_q$. For an implementation of PConstrain by an arithmetic or boolean circuit, the depth will depend on the precise circuit model used and the implementation of the linear and nonlinear operations, but in any case the final depth will be proportional to the nonlinear depth.

To state our claim we recall from [BP14] the notions of "left depth" and "right depth" of the jth leaf in a binary tree T, and of T itself. The left depth $l_T(j)$ (respectively, right depth $r_T(j)$) of the jth leaf is the number of edges from a parent to its left (resp., right) child on the path from the root to that leaf. The left and right depths $l(T), r(T)$ are respectively the maximum left and right depths over all leaves in T.

Lemma 6. *The function $\mathsf{PConstrain}_{T,w,x}(\mathbf{v})$ can be computed via (1) a preprocessing phase (independent of $\mathbf{v}$) of nonlinear depth at most $r(T)$, and (2) an online phase (dependent on $\mathbf{v}$) of nonlinear depth at most $l_T(|w|) + r_T(|x|) \leq l(T) + r(T)$.*

We remark that in [BP14], the nonlinear depth of computing the (non-constrained) PRF is just $r(T)$, so one can obtain an extremely parallel PRF using a "left spine" tree with $r(T) = 1$ and $l(T) = |T| - 1$ (this corresponds to the function from [BLMR13]). But here, evaluating the PRF from a constrained key can require nonlinear depth proportional to the sum of T's left and right depths. Therefore, to get good parallelism for all w, x we must use a shallow tree T, e.g., one with depth $O(\log|T|)$. We defer the proof of this Lemma to the full version.

Closeness. We next analyze the size of the $\mathcal{P}$-components of $\mathbf{d}$ discussed above, as they relate to the errors in the original key $\mathbf{k}_\varepsilon = (\mathbf{s}, (\mathbf{e}_y))$. Recalling that each $\mathcal{P}$-component of $\mathbf{d}$ corresponds to some left-child subtree in T, it is therefore sufficient to analyze the accumulated error in fully constrained keys over arbitrary trees. For this purpose we define a "growth factor" Φ_T associated with an arbitrary full binary tree, defined recursively as follows:

$$\Phi_T := \begin{cases} 1 & \text{if } |T| = 1, \\ \sqrt{(\Phi_{T.l} \cdot n\ell)^2 + (\Phi_{T.r})^2} & \text{otherwise.} \end{cases} \tag{9}$$

We next state a lemma that is essentially a restatement of [BP14, Lemma 3.7].

Lemma 7 (Error Bound). *For any w such that $|w| = |T|$, let*

$$(\mathbf{k}, \star) = \mathsf{Constrain}_{T,\varepsilon,w}(\mathbf{0}, (\mathbf{e}_y))$$

for $(\mathbf{e}_y) \leftarrow \mathcal{E}_{T,0}$, where the error distribution χ is subgaussian with parameter r. Then $\mathsf{Decode}^{-1}(\mathbf{k}) = \mathbf{e} \pmod q$ for some $\mathbf{e} \in \mathbb{Z}^{n\ell}$ that is subgaussian with parameter $r \cdot \Phi_T$.

More generally, let $\mathbf{d} = \mathsf{PrjConstrain}_{T,\varepsilon,w}(\mathbf{0}, (\mathbf{e}_y)) \in \mathcal{R}_{T,|w|}$ for nonempty $w \in \{0,1\}^{\leq |T|}$. Then if $q \geq 4r \cdot \Phi_T \cdot \omega(\sqrt{\log \lambda})$, the following are true with $1 - \mathrm{negl}(\lambda)$ probability over the choice of $(\mathbf{e}_y) \leftarrow \mathcal{E}_{T,0}$: (1) the $\mathbb{Z}_q^n$-component of $\mathbf{d}$ is $\mathbf{0}$, and (2) each $\mathcal{P}$-component of $\mathbf{d}$ for subtree T' is subgaussian with parameter $r \cdot \Phi_{T'}$.

4.4 "Rounded" Function Family $\overline{\mathcal{C}}$

We now define our final "rounded" family of constraining functions, denoted $\overline{\mathcal{C}}$, which we prove to be pseudorandom, as well as (computationally) key-homomorphic and consistent. In $\overline{\mathcal{C}}$ we use the same DAG on $\{0,1\}^{\leq |T|}$ as in the noisy function family, but we define somewhat different "rounded" (and errorless) key spaces, and thereby different constraining functions and group operations.

We note that in this scenario, we would only be able to achieve a *computational* version of consistency and homomorphism. That is to say that it is computationally infeasible to find inputs on which our family is not consistent (respectively, homomorphic). We discuss about these properties in more detail in the full version.

Rounding and key spaces. The family $\overline{\mathcal{C}}$ is parameterized by a "rounding factor" $e_{T'}$ for each subtree T' of T. For convenience of analysis, we choose these factors to all divide q, hence they are also powers of two. The factors are defined recursively to satisfy the inequalities

$$e_{T'} \geq \begin{cases} r \cdot \lambda^{\omega(1)} & \text{if } |T'| = 1, \\ (e_{T'.l} \cdot (n\ell) + e_{T'.r}) \cdot \lambda^{\omega(1)} & \text{otherwise.} \end{cases} \tag{10}$$

Note that by inspection of Equations (9) and (10), for all subtrees T' we have

$$e_{T'} \geq r \cdot \Phi_{T'} \cdot \lambda^{\omega(1)} \ .$$

Next, mirroring the definitions from Section 4.2, we define the "rounded" domain $\overline{\mathcal{K}}_{T,j}$ for $0 \leq j < |T|$ as follows

$$\overline{\mathcal{R}}_{T,j} := \begin{cases} \mathbb{Z}_q^n & \text{if } |T| = 1, \\ \overline{\mathcal{R}}_{T.l,j} & \text{if } j \leq |T.l|, \\ \lfloor \mathcal{P} \rceil_{e_{T.l}} \times \overline{\mathcal{R}}_{T.r,j-|T.l|} & \text{otherwise.} \end{cases}$$

As with $\mathcal{R}$, we also define $\overline{\mathcal{R}}_{T,|T|} = \lfloor \mathcal{P} \rceil_{e_T} \times \mathbb{Z}_q^n$. Note that for every subtree T' of T, we have that $\lfloor \mathcal{P} \rceil_{e_{T'}} \subseteq \mathcal{P}$ (because every $e_{T'}$ divides q), we have $\overline{\mathcal{R}}_{T,j} \subseteq \mathcal{R}_{T,j}$. The key space for $w \in \{0,1\}^{\leq |T|}$ and its associated distribution are then defined to be

$$\overline{\mathcal{K}}_{T,w} := \overline{\mathcal{R}}_{T,|w|} \ ,$$
$$\overline{\mathcal{D}}_{T,w} := U(\overline{\mathcal{K}}_{T,w}) \ .$$

Constraining functions. We first define $\mathsf{Round}_{T,j} \colon \mathcal{R}_{T,j} \to \overline{\mathcal{R}}_{T,j}$ for $0 \leq j < |T|$ as follows:

$$\mathsf{Round}_{T,j}(\mathbf{v}) := \begin{cases} \mathbf{v} & \text{if } |T| = 1 \\ \mathsf{Round}_{T.l,j}(\mathbf{v}) & \text{if } 0 < j \leq |T.l| \\ (\lfloor \mathbf{v}_l \rceil_{e_{T.l}}, \mathsf{Round}_{T.r,j-|T.l|}(\mathbf{v}_r)) & \text{otherwise,} \end{cases}$$

where we parse $(\mathbf{v}, \mathbf{s}) = (\mathbf{v}_l, \mathbf{v}_r) \in \mathcal{P} \times \mathcal{R}_{T.r, j-|T.l|}$ in the last case above. For $(\mathbf{v}, \mathbf{s}) \in \mathcal{R}_{T,|T|}$, we simply define $\mathsf{Round}_{T,|T|}(\mathbf{v}, \mathbf{s}) := (\lfloor \mathbf{v} \rceil_{e_T}, \mathbf{s})$.

With this definition in mind, the "rounded" constraining functions $\overline{\mathsf{PConstrain}}_{T,w,x} \colon \mathcal{R}_{T,|w|} \to \overline{\mathcal{K}}_{T,w}$ are simply defined as

$$\overline{\mathsf{PConstrain}}_{T,w,x} := \mathsf{Round}_{T,|wx|} \circ \mathsf{PConstrain}_{T,w,x} \ .$$

Pseudorandomness. We now show that the construction of the family $\overline{\mathcal{C}}$ from Section 4.4 is a constrained PRF, according to Definition 2. Here, we prove selctive security of the function as defined in Definition 2, and use the Security of the Constrain family of functions, as defined in Section 4.2 above. We note that this theorem is very similar to analogous ones proved in prior work [BPR12, BP14], and thus we defer the proof to the full version.

Theorem 2. *The family $\overline{\mathcal{C}}$ described above is pseudorandom for the set of challenge nodes $\{0,1\}^{\leq |T|}$, assuming that the family $\mathcal{C}$ is also pseudorandom over the same set of challenge nodes, where the χ distribution of $\mathcal{C}$ is a subgaussian distribution over $\mathbb{Z}$ with parameter r, where r is the number used to define the rounding factors in Equation (10).*

5 Proxy Re-encryption with Fine-Grained Access Control

Below we explain the symmetric proxy re-encryption scheme as defined by Boneh *et al.* [BLMR13]. Using this scheme as a starting point, we then construct our scheme which additionally allows for fine-grained access control.

5.1 Symmetric-key Proxy Re-encryption from Key Homomorphic PRFs

As an application of key homomorphic PRFs Boneh *et al.* [BLMR13] construct a *symmetric-key proxy re-encryption* scheme, a symmetric-key analogue of public-key proxy re-encryption [BBS98, CH07, ABH09, LV08]. A symmetric proxy re-encryption scheme is a symmetric encryption scheme, where given a ciphertext $c = \mathsf{Enc}(k, m)$ of some message m under key k, a proxy can translate this ciphertext to a new ciphertext $c' = \mathsf{Enc}(k', m)$ under a new key given only some re-encryption token Δ. The security definition requires roughly that the token *only* allows to translate ciphertexts in this way, but does not reveal anything about the encrypted message or the involved keys. Given a key-homomorphic PRF $F \colon \mathcal{K} \times \mathcal{X} \to \mathcal{Y}$, where $(\mathcal{K}, \circ), (\mathcal{Y}, \otimes)$ are groups such that

$$F(k \circ k', x) = F(k, x) \otimes F(k', x)$$

and any symmetric encryption scheme (enc: $\mathcal{Y} \times \mathcal{M} \to \mathcal{C}$, dec: $\mathcal{Y} \times \mathcal{C} \to \mathcal{M}$) we construct $\Pi_{proxy} = (\mathsf{Setup}, \mathsf{KeyGen}, \mathsf{Enc}, \mathsf{Dec}, \mathsf{ReKeyGen}, \mathsf{ReEnc})$ as follows. $\mathsf{Setup}(1^\lambda)$ outputs public parameters pp to be used by F. All algorithms will have pp as input, which we will not denote explicitly. The key generation algorithm KeyGen simply outputs a random key $k \leftarrow \mathcal{K}$ for F. Encryption of the proxy re-encryption scheme is defined as $\mathsf{Enc}(k, m) = (r, c_1, c_2)$ where

$$c_1 = \kappa \otimes F(k, r) \,,\; c_2 = \mathsf{enc}(\kappa, m) \text{ for random } (r, \kappa) \leftarrow \mathcal{X} \times \mathcal{Y} \qquad (11)$$

Decryption is $\mathsf{Dec}(r, c_1, c_2) = \mathsf{dec}(c_1 \otimes F(k, r)^{-1}, c_2) = m$. The re-encryption-key generation $\mathsf{ReKeyGen}$ takes two keys k, k' and outputs a re-encryption token

$$\mathsf{ReKeyGen}(k, k') = k^{-1} \circ k' \;.$$

The re-encryption procedure ReEnc takes a re-encryption token $\Delta = \mathsf{ReKeyGen}(k, k')$ and a ciphertext under key k and outputs a ciphertext of the same plaintext under the key k' as

$$\mathsf{ReEnc}(\Delta, (r, c_1, c_2)) = (r, c_1 \otimes F(\Delta, r), c_2) \;.$$

Note $(r, c_1 \otimes F(\Delta, r), c_2) = (r, \kappa \otimes F(k, r) \otimes F(\Delta, r)), c_2) = (r, \kappa \otimes F(k', r), c_2)$ is indeed an encryption of m under key k' as required. We refer the reader to [BLMR13] for a formal definition of symmetric-key proxy re-encryption and the proof of the following

Theorem 3 ([BLMR13]). *If F is a secure key-homomorphic PRF where the input space $\mathcal{X}$ is of superpolynomial size, then Π_{proxy} is a secure proxy re-encryption scheme.*

The superpolynomial domain is required in order for the probability that any two of the randomly chosen $r \in \mathcal{X}$ collide to be negligible.

5.2   Fine-Grained Access Control from Constrained PRFs

Assume the PRF F from the previous section is not only key-homomorphic, but also a constrained PRF. That is, there is a function Constrain: $\mathcal{K} \times \mathcal{P} \to \mathcal{K}_P$ which given a key k and some predicate p outputs a constrained key k_p that allows to evaluate $F(k, \cdot)$ on all inputs x where $p(x) = 1$.

Consider the proxy re-encryption scheme outlined above, but where we slightly change the encryption procedure from Eq. (11), and now instead of choosing r at random during encryption, it is given as part of the input. We call this input x the *attributes* of the ciphertext. I.e., we let $\mathsf{Enc}(k, m, x) = (x, c_1, c_2)$ with

$$c_1 = \kappa \otimes F(k, x)\,, \quad c_2 = \mathsf{enc}(\kappa, m) \ \text{ for random } \kappa \leftarrow \mathcal{Y}\,.$$

This little change now gives us an extra property: given a constrained key k_p for a predicate p, one can decrypt ciphertexts with attribute x iff $p(x) = 1$. The correctness property of Enc as a proxy re-encryption scheme is not affected by this change.

Informally, the security notion (which we will define formally later) requires that ciphertexts encrypted for some attributes x under key k hide the plaintext as long as it cannot be trivially computed from the outputs of the queries of the adversary (where we allow adversaries to make re-encryption queries and ask for constrained keys).

The security notion of constrained PRFs implies that given keys $k_{p_1}, \dots, k_{p_\ell}$ for predicates where $p_i(x) = 0$ for all $i = 1, \dots, \ell$, the output $F(k, x)$ is pseudorandom. It might therefore seem that the key κ is pseudorandom given the encapsulated key $c_1 = \kappa \otimes F(k, x)$. Unfortunately, as discussed in the introduction, this is not true, because in a CPA attack the adversary can not only ask for constrained keys, but also for ciphertexts which reveal function values. We therefore will use a carefully defined probabilistic encoding of attributes such that the functionality of the scheme is preserved, while solving the problems discussed in the introduction.

Randomizable Predicates. How to appropriately define the required encoding is best explained by an example: Consider a bit-fixing CPRF F with inputs from $\{0, 1\}^n$. Recall that given a constrained key $k_p \leftarrow \mathsf{Constrain}(k, p)$ for a predicate $p \in \{0, 1, ?\}^n$, we can compute $F(k, x)$ for any attribute x where for every $i \in [n]$ we have $(x[i] = p[i] \vee p[i] = ?)$. For any such predicate p, we denote with p^+ the predicate on $n + \lambda$ bits (where λ is a statistical security parameter) as $p^+(x\|\alpha) = p(x)$, so p^+ simply evaluates p on the first n bits.

In the encryption scheme, the predicate space is still $\{0, 1\}^n$, but F is evaluated on inputs of length $n + \lambda$ and a constrained key for $p \in \{0, 1, ?\}^n$ is computed as $k_{p^+} \leftarrow \mathsf{Constrain}(k, p\|?^\lambda)$. During encryption we now compute the encapsulated key as $c_1 = \kappa \otimes F(k, x\|\alpha)$ for some random α (α is also output as part of the ciphertext). Note that this preserves the proxy re-encryption property: given k_{p^+} one can compute $F(k, x\|\alpha)$ on any (x, α) where $p(x) = 1$. On the other hand, we'll show that the $c_1 = \kappa \otimes F(k, x\|\alpha)$ part of the challenge ciphertext hides the encapsulated key κ because $F(k, x\|\alpha)$ is pseudorandom.

Definition 4. *A randomization of a set of predicates $\mathcal{P}$ is given by an efficient injective encoding $[\cdot, \cdot] : \mathcal{P}_{in} \times \{0, 1\}^\lambda \to \mathcal{P}_{out}$ ($\mathcal{P}_{in}, \mathcal{P}_{out} \subseteq \mathcal{P}$ and λ being a statistical security parameter) and a mapping $\phi : \mathcal{P}_{in} \to \mathcal{P}_{out}$ (we'll use p^+ as shortcut for $\phi(p)$) such that $p^+([x_1, x_2]) = 1 \iff p(x_1) = 1$. For every $[x, r]$ we require that there exists a $p_{[x,r]} \in \mathcal{P}$ s.t. $p_{[x,r]}([x, r]) = 1$ but $p_{[x,r]}([x', r']) = 0$ for all $(x', r') \neq (x, r) \in \mathcal{P}_{in} \times \{0, 1\}^\lambda$.*

For a CPRF for predicates $\mathcal{P}$ that can be randomized, we define $\mathsf{Constrain}^+(k, p) \equiv \mathsf{Constrain}(k, p^+)$. Note that a key $k_{[x',r']} \leftarrow \mathsf{Constrain}(k, p_{[x,r]})$ allows to evaluate

$F(k, \cdot)$ *only on the value* $[x, r]$ *in the range of* $[\cdot, \cdot]$ *(but might allow to evaluate it on other points not in the range of the encoding).*[4]

With this definition, the encoding for bit-fixing CPRFs we discussed above can be cast as a randomized predicate with $[x_1, x_2] = x_1 \| x_2$ simply being concatenation and $p^+ = p \| ?^\lambda$ for any $p \in \{0, 1, ?\}^n$.

For prefix CPRF, we let $\tau \colon \{0, 1, 2\} \to \{0, 1\}^2$ be an encoding of a ternary to a binary alphabet (say, $0, 1, 2$ maps to $00, 01, 10$). Then we can use the encoding $[x_1, x_2] = \tau(x_1 \| 2 \| x_2)$ and set $\phi(x) = \tau(x)$ (so $\mathsf{Constrain}^+(k, x) = \mathsf{Constrain}(k, \tau(x)))$.[5]

We will prove the following theorem.

Theorem 4. *If F is a secure key-homomorphic constrained PRF, the scheme $\Pi_{fg\text{-proxy}}$ defined in Section 5.4 is a secure proxy re-encryption scheme with fine-grained access control (as defined in Section 5.3).*

5.3 Definition of Proxy Re-encryption with Fine-Grained Access Control

A *proxy re-encryption scheme with fine-grained access control* for predicates $\mathcal{P}$ over $\mathcal{X}$, where for $p \in \mathcal{P}, x \in \mathcal{X}$ we denote by $p(x) = 1$ that p holds on x, is given by algorithms

$$\Pi_{fg\text{-proxy}} = (\mathsf{Setup}, \mathsf{KeyGen}, \mathsf{Enc}, \mathsf{Dec}, \mathsf{Constrain}, \mathsf{ReKeyGen}, \mathsf{ReEnc}) \ .$$

$\mathsf{Setup}(1^\lambda)$. Setup outputs a set of public parameters pp, which are an implicit input to all other algorithms.

$\mathsf{KeyGen}(1^\lambda)$. Key generation outputs a key $k \in \mathcal{K}$.

$\mathsf{Enc} \colon \mathcal{K} \times \mathcal{X} \times \mathcal{M} \to \mathcal{X} \times \mathcal{C}$. Encryption takes a key k, attributes x and a message m and outputs a ciphertext $(x, c) \leftarrow \mathsf{Enc}(k, x, m)$.

$\mathsf{Constrain}_{\mathsf{ENC}} \colon \mathcal{K} \times \mathcal{P} \to \mathcal{K}_\mathcal{P}$. Constraining takes a key k and a predicate p and outputs a constrained key $k_p \leftarrow \mathsf{Constrain}_{\mathsf{ENC}}(k, p)$ (we use the subscript ENC to avoid confusion with the $\mathsf{Constrain}$ algorithm of the CPRF).

$\mathsf{Dec} \colon (\mathcal{K}_\mathcal{P} \cup \mathcal{K}) \times \mathcal{X} \times \mathcal{C} \to \mathcal{M}$. Decryption takes k and a ciphertext (x, c) and outputs $m \leftarrow \mathsf{Dec}(k, x, c)$; except when $k = k_p \in \mathcal{K}_\mathcal{P}$ and $p(x) = 0$, then it outputs $\bot$.

---

[4] Looking forward, this condition will allow us to replace (in the reduction) an output value $F(k, [x, r])$ with a constrained key, while only excluding one possible challenge ciphertext. We observe that without this condition simple concatenation $[x_1, x_2] = x_1 \| x_2$ would already give a randomized predicate for prefix predicates, but this would lead to a trivially insecure encryption scheme $\mathsf{Enc}(k, x, m) = (r, m \otimes F(k, [x, r]))$ if using a GGM based prefix CPRF. In such CPRFs, given some $F(k, x \| r)$ (that an adversary can learn via an encryption query) one can compute $F(k, x \| r \| r')$ for any r'. Using this fact we can break security of the encryption scheme by asking for a challenge for attribute $x' = x \| r$ which we'll be able to decrypt.

[5] The extra symbol 2 in-between the prefix x_1 and the randomness part x_2 is there so the condition from Def. 4 is satisfied. In particular, note that for any $z = [x_1, x_2] = \tau(x_1 \| 2 \| x_2)$, the constrained key $k_z = \mathsf{Constrain}(k, z)$ allows to evaluate $F(k, \cdot)$ only on inputs of the form $z \| w$, but this is in the range of the encoding $[\cdot, \cdot]$ only if w is empty (i.e., only on the unique input z in the range of $[\cdot, \cdot]$). Note that with this encoding the attack from Footnote 4 does no longer work.

ReKeyGen: $\mathcal{K} \times \mathcal{K} \to \mathcal{K}$. Re-encryption key-generation takes two keys and outpus a re-encryption key $k_\Delta \leftarrow$ ReKeyGen(k, k').

ReEnc: $\mathcal{K} \times \mathcal{C} \to \mathcal{C}$. Re-encryption takes a re-encryption key (from k to k') and a ciphertext under k, and outputs a ciphertext of the same plaintext under k'.

Correctness. For any pp output by Setup (which is an implicit input to all algorithms) and all k, x, m and p with $p(x) = 1$, let $c \leftarrow$ Enc(k, x, m). Then we require the following: Dec$(k, x, c) = m$. For all $k_p \leftarrow$ Constrain$_{\mathsf{ENC}}(k, p)$: Dec$(k_p, x, c) = m$. For any $k', k_\Delta \leftarrow$ ReKeyGen(k, k'), $c' \leftarrow$ ReEnc(k_Δ, c) we have Dec$(k', c', x) = m$.

Security. The notion of security for proxy re-encryption with fine-grained access control below is a generalization of the security notion for proxy re-encryption of [BLMR13].

We consider a game between an adversary $\mathcal{A}$ and a challenger. The challenger runs $pp \leftarrow$ Setup(1^λ) (and pp is given to $\mathcal{A}$ and to all other algorithms as input), initializes a counter ctr $:= 1$ and samples a random bit $b \in \{0, 1\}$. Then $\mathcal{A}$ can make the following queries.

Uncorrupted Key-Generation: Challenger samples $k^{\mathsf{ctr}} \leftarrow$ KeyGen(1^λ) and increases ctr (the key is not given at $\mathcal{A}$).

Corrupted Key-Generation: Challenger samples $k^{\mathsf{ctr}} \leftarrow$ KeyGen(1^λ) and increases ctr. The key is given to $\mathcal{A}$.

Re-encryption Key-Generation: On input $(i, j), i, j \leq$ ctr return ReKeyGen(k^i, k^j) to $\mathcal{A}$. We require that both keys k^i, k^j are uncorrupted.

Constrained Key Request: On input (i, p) return Constrain$_{\mathsf{ENC}}(k^i, p)$ to $\mathcal{A}$.

Encryption: On input (i, x, m) return Enc(k^i, x, m) to $\mathcal{A}$.

Re-Encryption: On input (i, j, c) return ReEnc$($ReKeyGen$(k^i, k^j), c)$ to $\mathcal{A}$. We require that k^j was generated using uncorrupted key generation.

Challenge: This oracle is queried only once in an input (i^*, x^*, m_0^*, m_1^*), we require that k^i was generated using uncorrupted key generation, and for every "Constrained Key Request" query (i, p) where k^i was generated using uncorrupted key generation, we have $p(m_0^*) = p(m_1^*) = 0$ (this also holds for queries to be made after this challenge query).
The challenger returns Enc(k^i, x^*, m_b^*) to $\mathcal{A}$.

Guess: $\mathcal{A}$ outputs a guess bit b' (the experiment stops at this point).

Definition 5. *$\Pi_{\text{fg-proxy}}$ is a secure proxy re-encryption scheme with fine-grained access control if for all polynomial-time adversaries $\mathcal{A}$, the advantage $|\Pr[b = b'] - 1/2|$ in the above game is negligible in the security parameter λ.*

A Remark on Selective Security. Note that the above notion considers *selective* security in the sense that the adversary must commit whether a key is corrupted or not during its generation (the challenge is chosen adaptively, and for this we'll have to assume adaptive security of the underlying constrained PRF). This will be useful in the security proof, where the reduction will sample corrupted keys itself, and implicitly uses the key of the challenger in the constrained PRF security game to generate uncorrupted keys. We can get selective security via "complexity leveraging", but this loses a huge exponential (in the number m of generated keys) factor in the security reduction[6] as we have to guess initially which keys will be corrupted. When the encryption scheme is actually used to outsource data to an untrusted server, we can assume that re-encryption-key generation queries are not arbitrary, but only applied to consecutive keys, i.e., we only can ask for re-encryption keys $\mathsf{ReKeyGen}(k^i, k^{i+1})$. In this case, adaptive security can be proven losing only a quadratic factor (as for the reduction it will be sufficient to only guess which key will be the first corrupted key before and after the key chosen for the challenge.)

5.4 Construction of Proxy Re-encryption with Fine-Grained Access Control from Key-Homomorphic Constrained PRFs

We now describe how to construct a scheme $\Pi_{fg\text{-}proxy}$ from a key-homomorphic constrained PRF F for predicates $\mathcal{P}$ that can be randomized (cf. Def. 4) and any symmetric encryption scheme (enc, dec).

$\mathsf{Setup}(1^\lambda)$ samples and outputs public parameters pp as used by F.
$\mathsf{KeyGen}(1^\lambda)$ outputs a random key $k \in \mathcal{K}$ for F.
$\mathsf{Enc}(k, x, m)$ picks a random $\alpha \in \{0, 1\}^\lambda$, a random key κ for enc and sets (with $[\cdot, \cdot]$ as in Def 4)

$$\mathsf{Enc}(k, m, x) = ([x, \alpha], c_1, c_2), \quad \text{where } c_1 = \kappa \otimes F(k, [x, \alpha]) \text{ and } c_2 = \mathsf{enc}(\kappa, m)$$

$\mathsf{Dec}(k_p, x, c = ([x, \alpha], c_1, c_2))$ checks if $p(x) = 1$. If so, it computes $\kappa = c_1 \otimes F(k_p, [x, \alpha])^{-1}$ and returns $\mathsf{dec}(\kappa, c_2)$.
$\mathsf{Constrain}_{\mathsf{ENC}}(k, p)$ returns $k_p \leftarrow \mathsf{Constrain}^+(k, p)$ (cf. Def. 4)
$\mathsf{ReKeyGen}(k, k')$ returns $k_\Delta = k' \circ k^{-1}$.
$\mathsf{ReEnc}(k_\Delta, c = ([x, \alpha], c_1, c_2)$ returns $c' = ([x, \alpha], F(k_\Delta, [x, \alpha]) \otimes c_1, c_2)$.

Proof of Theorem 4. We now show that the scheme constructed in Section 5.4 satisfies the security notion from Definition 5. We construct an adversary $\mathcal{B}$, who given an adversary $\mathcal{A}$ that breaks the security of the scheme, breaks the security of the underlying constrained PRF with almost the same advantage (we lose an exponentially small additive term due to the possibility of collisions in the randomness used for encryption).

At setup, adversary $\mathcal{B}$ gets the public parameters pp for F, and forwards them to $\mathcal{A}$. Now, $\mathcal{B}$ has access to an oracle $\mathsf{Constrain}(k, \cdot)$ (below $\mathsf{Constrain}^+(k, \cdot)$ is as in Def. 5). $\mathcal{B}$ will answer $\mathcal{A}$'s queries as follows.

Corrupted Key-Generation: $\mathcal{B}$ samples a key $k^{\mathsf{ctr}} \leftarrow \mathsf{KeyGen}(1^\lambda)$, increases ctr and gives the key to $\mathcal{A}$.
Uncorrupted Key-Generation: $\mathcal{B}$ samples a key k_Δ^{ctr} and implicitly sets $k^{\mathsf{ctr}} = k \circ k_\Delta^{\mathsf{ctr}}$ where k is the key used in the $\mathsf{Constrain}(k, \cdot)$ oracle of the security game against the CPRF. Note that k^{ctr} is uniform.

---

[6] That is, an attacker with advantage ε against the scheme is turned into an adversary with advantage $\varepsilon/2^m$ against the constrained PRF.

Re-encryption Key-Generation: On input (i, j) where $i, j \leq$ ctr are uncorrupted keys, $\mathcal{B}$ must return ReKeyGen(k^i, k^j) to $\mathcal{A}$. It can compute these without knowing k as

$$\mathsf{ReKeyGen}(k^i, k^j) = k \circ k_\Delta^j \circ (k \circ k^i)^{-1} = k_\Delta^j \circ (k_\Delta^i)^{-1}$$

Constrained Key Request: On input (i, p) $\mathcal{B}$ queries its oracle for the key $k_{p+} \leftarrow$ Constrain$^+(k, p)$, then computes $k_{p+}^i = k_{p+} \circ$ Constrain$^+(k_\Delta^i, p)$ and returns this key to $\mathcal{A}$.

Encryption: On input (i, m, x) compute $([x, \alpha], c_1, c_2) \leftarrow$ Enc(k, m, x) as in Sect. 5.4, note that for this we have to learn $F(k, [x, \alpha])$. For this $\mathcal{B}$ queries for the constrained key $k_{[x,\alpha]} \leftarrow$ Constrain$(k, p_{[x,\alpha]})$ (cf. Def. 4), and then computes $F(k, [x, \alpha])$ using this key.
Return $c = ([x, \alpha], c_1', c_2)$ to $\mathcal{A}$ where $c_1' = c_1 \otimes F(k_\Delta^i, [x, \alpha])$ (this step re-encrypts from k to k^i).

Re-Encryption: On input (i, j, c) return ReEnc(ReKeyGen$(k^i, k^j), c)$ to $\mathcal{A}$ (note that we already explained how to compute ReEnc(ReKeyGen$(k^i, k^j), c))$.

Challenge and Guess: If $\mathcal{A}$ outputs the challenge (i^*, x^*, m_0^*, m_1^*) (where for any predicate p where there was a constrain key-request (i, p) we have $p(x^*) = 0$).
$\mathcal{B}$ samples a random α and forwards the challenge $[x^*, \alpha]$ to its CPRF challenger (note that as α is random, with overwhelming probability $\mathcal{B}$ hasn't made the query Constrain$(k, p_{[x,\alpha]})$ in a previous encryption query, and thus this is a legal challenge.
$\mathcal{B}$ gets from his challenger a value γ which is either $F(k, [x^*, \alpha])$ or a uniformly random, depending on whether the challenger's bit b was 0 or 1.
$\mathcal{B}$ samples a random bit β, a random key κ and computes $c = ([x, \alpha], \kappa \otimes \gamma,$ enc$(\kappa, m_\beta^*))$. $\mathcal{B}$ sends c to $\mathcal{A}$, who answers with β'.
If $\beta = \beta'$ $\mathcal{B}$ outputs the guess bit $b' = 0$ (guessing γ is pseudorandom), otherwise $b' = 1$ (guessing γ is uniform).

We analyze the probability that $b = b'$. Conditioned on $b = 0$, c is correctly generated and thus $\mathcal{A}$ has some non-negligible advantage δ in guessing correctly. If $b = 1$, the $c_1 = \kappa \otimes \gamma$ part of the ciphertext is independent of κ, and thus $\mathcal{A}$'s advantage is some negligible ε_{enc} (by the security of enc).

$$\Pr[b = b'] - 1/2$$
$$1 \geq \frac{\Pr[b = b'|\beta = 0] - 1/2}{2} + \frac{\Pr[b = b'|\beta = 1] - 1/2}{2} \geq \frac{\delta}{2} - \frac{\varepsilon_{enc}}{2},$$

which is non-negligible assuming ε_{enc} is negligible but δ is not.

References

[ABH09]   Ateniese, G., Benson, K., Hohenberger, S.: Key-private proxy re-encryption. In: Fischlin, M. (ed.) CT-RSA 2009. LNCS, vol. 5473, pp. 279–294. Springer, Heidelberg (2009)

[BBS98]   Blaze, M., Bleumer, G., Strauss, M.: Divertible protocols and atomic proxy cryptography. In: Nyberg, K. (ed.) EUROCRYPT 1998. LNCS, vol. 1403, pp. 127–144. Springer, Heidelberg (1998)

[BGI14]   Boyle, E., Goldwasser, S., Ivan, I.: Functional Signatures and Pseudorandom Functions. In: Krawczyk, H. (ed.) PKC 2014. LNCS, vol. 8383, pp. 501–519. Springer, Heidelberg (2014)

[BLMR13]   Boneh, D., Lewi, K., Montgomery, H., Raghunathan, A.: Key homomorphic PRFs and their applications. In: Canetti, R., Garay, J.A. (eds.) CRYPTO 2013, Part I. LNCS, vol. 8042, pp. 410–428. Springer, Heidelberg (2013)

[BLP$^+$13]   Brakerski, Z., Langlois, A., Peikert, C., Regev, O., Stehlé, D.: Classical hardness of learning with errors. In: STOC, pp. 575–584 (2013)

[BP14]   Banerjee, A., Peikert, C.: New and improved key-homomorphic pseudorandom functions. In: Garay, J.A., Gennaro, R. (eds.) CRYPTO 2014, Part I. LNCS, vol. 8616, pp. 353–370. Springer, Heidelberg (2014)

[BPR12]   Banerjee, A., Peikert, C., Rosen, A.: Pseudorandom functions and lattices. In: Pointcheval, D., Johansson, T. (eds.) EUROCRYPT 2012. LNCS, vol. 7237, pp. 719–737. Springer, Heidelberg (2012)

[BW13]   Boneh, D., Waters, B.: Constrained pseudorandom functions and their applications. In: Sako, K., Sarkar, P. (eds.) ASIACRYPT 2013, Part II. LNCS, vol. 8270, pp. 280–300. Springer, Heidelberg (2013)

[CH07]   Canetti, R., Hohenberger, S.: Chosen-ciphertext secure proxy re-encryption. In: Ning, P., De Capitani di Vimercati, S., Syverson, P.F. (eds.) ACM CCS 2007, pp. 185–194. ACM Press (October 2007)

[GGH13a]   Garg, S., Gentry, C., Halevi, S.: Candidate Multilinear Maps from Ideal Lattices. In: Johansson, T., Nguyen, P.Q. (eds.) EUROCRYPT 2013. LNCS, vol. 7881, pp. 1–17. Springer, Heidelberg (2013)

[GGH$^+$13b] Garg, S., Gentry, C., Halevi, S., Raykova, M., Sahai, A., Waters, B.: Candidate indistinguishability obfuscation and functional encryption for all circuits. In: 54th FOCS, pp. 40–49. IEEE Computer Society Press (October 2013)

[GGM86]   Goldreich, O., Goldwasser, S., Micali, S.: How to construct random functions. Journal of the ACM 33, 792–807 (1986)

[KPTZ13]   Kiayias, A., Papadopoulos, S., Triandopoulos, N., Zacharias, T.: Delegatable pseudorandom functions and applications. In: Sadeghi, A.-R., Gligor, V.D., Yung, M. (eds.) ACM CCS 2013, pp. 669–684. ACM Press (November 2013)

[LV08]   Libert, B., Vergnaud, D.: Unidirectional Chosen-Ciphertext Secure Proxy Re-encryption. In: Cramer, R. (ed.) PKC 2008. LNCS, vol. 4939, pp. 360–379. Springer, Heidelberg (2008)

[MP12]   Micciancio, D., Peikert, C.: Trapdoors for lattices: Simpler, tighter, faster, smaller. In: Pointcheval, D., Johansson, T. (eds.) EUROCRYPT 2012. LNCS, vol. 7237, pp. 700–718. Springer, Heidelberg (2012)

[NPR99]   Naor, M., Pinkas, B., Reingold, O.: Distributed pseudo-random functions and KDCs. In: Stern, J. (ed.) EUROCRYPT 1999. LNCS, vol. 1592, pp. 327–346. Springer, Heidelberg (1999)

[Pei09]   Peikert, C.: Public-key cryptosystems from the worst-case shortest vector problem. In: STOC, pp. 333–342 (2009)

[PST14]   Pass, R., Seth, K., Telang, S.: Indistinguishability obfuscation from semantically-secure multilinear encodings. In: Garay, J.A., Gennaro, R. (eds.) CRYPTO 2014, Part I. LNCS, vol. 8616, pp. 500–517. Springer, Heidelberg (2014)

[Reg09]   Regev, O.: On lattices, learning with errors, random linear codes, and cryptography. J. ACM 56(6), 1–40 (2005)

[SW14]   Sahai, A., Waters, B.: How to use indistinguishability obfuscation: deniable encryption, and more. In: Shmoys, D.B. (ed.) 46th ACM STOC, pp. 475–484. ACM Press (May/June 2014)

Aggregate Pseudorandom Functions and Connections to Learning

Aloni Cohen[1], Shafi Goldwasser[1,2], and Vinod Vaikuntanathan[1]

[1] MIT, USA
[2] Weizmann Institute of Science, Israel

Abstract. In the first part of this work, we introduce a new type of pseudo-random function for which "aggregate queries" over exponential-sized sets can be efficiently answered. We show how to use algebraic properties of underlying classical pseudo random functions, to construct such "aggregate pseudo-random functions" for a number of classes of aggregation queries under cryptographic hardness assumptions. For example, one aggregate query we achieve is the product of all function values accepted by a polynomial-sized read-once boolean formula. On the flip side, we show that certain aggregate queries are impossible to support. Aggregate pseudo-random functions fall within the framework of the work of Goldreich, Goldwasser, and Nussboim [GGN10] on the "Implementation of Huge Random Objects," providing truthful implementations of pseudo-random functions for which aggregate queries can be answered.

In the second part of this work, we show how various extensions of pseudo-random functions considered recently in the cryptographic literature, yield impossibility results for various extensions of machine learning models, continuing a line of investigation originated by Valiant and Kearns in the 1980s. The extended pseudo-random functions we address include constrained pseudo random functions, aggregatable pseudo random functions, and pseudo random functions secure under related-key attacks.

1 Introduction

Pseudo-random functions (PRF), introduced by Goldreich, Goldwasser and Micali [GGM86], are a family of indexed functions for which there exists a polynomial-time algorithm that, given an index (which can be viewed as a secret key) for a function, can evaluate it, but no probabilistic polynomial-time algorithm without the secret key can distinguish the function from a truly random function – even if allowed oracle query access to the function. Pseudo-random functions have been shown over the years to be useful for numerous cryptographic applications. Interestingly, aside from their cryptographic applications, PRFs have also been used to show impossibility of computational learning in the membership queries model [Val84], and served as the underpinning of the proof of Razborov and Rudich [RR97] that natural proofs would not suffice for unrestricted circuit lower bounds.

Y. Dodis and J.B. Nielsen (Eds.): TCC 2015, Part II, LNCS 9015, pp. 61–89, 2015.

Since their inception in the mid eighties, various *augmented* pseudo random functions with *extra* properties have been proposed, enabling more sophisticated forms of access to PRFs and more structured forms of PRFs. This was first done in the work of Goldreich, Goldwasser, and Nussboim [GGN10] on how to efficiently construct "huge objects" (e.g. a large graph implicitly described by access to its adjacency matrix) which maintain combinatorial properties expected of a *random* "huge object." Furthermore, they show several implementations of varying quality of such objects for which complex global properties can be computed, such as computing cliques in a random graph, computing random function inverses from a point in the range, and computing the parity of a random function's values over huge sets. More recently, further augmentations of PRFs have been proposed, including: the works on constrained PRFs[1] [KPTZ13a, BGI14a, BW13a] which can release auxiliary secret keys whose knowledge enables computing the PRF in a restricted number of locations without compromising pseudo-randomness elsewhere; key-homomorphic PRFs [BLMR13] which are homomorphic with respect to the keys; and related-key secure PRFs [BC10, ABPP14]. These constructions yield fundamental objects with often surprising applications to cryptography and elsewhere. A case in point is the truly surprising use of constrained PRFs [SW14], to show that indistinguishability obfuscation can be used to resolve a long-standing problem of deniable encryption, among many others.

In the first part of this paper, we introduce a new type of augmented PRF which we call *aggregate pseudo random functions* (AGG-PRF). An AGG-PRF is a family of indexed functions each associated with a secret key, such that *given the secret key*, one can compute aggregates of the values of the function *over super-polynomially large sets* in *polynomial time*; and yet without the secret key, access to such aggregated values cannot enable a polynomial time adversary (distinguisher) to distinguish the function from random, even when the adversary can make *aggregate queries*. Note that the distinguisher can request and receive an aggregate of the function values over sets (of possibly super-polynomial size) that she can specify. Examples of aggregate queries can be the sum/product of all function values belonging to an exponential-sized interval, or more generally, the sum/product of all function values on points for which some polynomial time predicate holds. Since the sets over which our function values are aggregated are super-polynomial in size, they cannot be directly computed by simply querying the function on individual points. AGG-PRFs cast in the framework of [GGN10] are (truthful, pseudo) implementations of random functions supporting aggregates as their "complex queries." Indeed, our first example of an AGG-PRF for computing parities over exponential-sized intervals follows directly from [GGN10] under the assumption that one-way functions exist.

We show AGG-PRFs under various cryptographic hardness assumptions (one-way functions and DDH) for a number of types of aggregation operators such as sums and products and for a number of set systems including intervals, hypercubes, and (the supports of) restricted computational models such as decision

---

[1] Constrained PRFs are also known as Functional PRFs and as Delegatable PRFs.

trees and read-once Boolean formulas. We also show negative results: there are no AGG-PRFs for more expressive set systems such as (the supports of) CNF formulas. For a detailed description of our results, see Section 1.1.

In the second part of this paper, we embark on a study of the connection between the new augmented PRF constructions of recent years (constrained, related-key, aggregate) and the theory of computational learning. We recall at the outset that the fields of cryptography and machine learning share a curious historical relationship. The goals are in complete opposition and at the same time the aesthetics of the models, definitions and techniques bear a striking similarity. For example, a cryptanalyst can attack a cryptosystem using a range of powers from only seeing ciphertext examples to requesting to see decryptions of ciphertexts of her choice. Analogously, machine learning allows different powers to the learner such as random examples versus membership queries and shows that certain powers allow learners to learn concepts in polynomial time whereas others will fail. Even more directly, problems which pose challenges for machine learning such as Learning Parity with Noise (LPN) have been used as the underpinning for building secure cryptosystems, and as mentioned above [Val84] observes that the existence of PRFs in a complexity class $\mathcal{C}$ implies the existence of concept classes in $\mathcal{C}$ which can not be learned under membership queries, and [KV94] extends this direction to some public key constructions.

In the decades since the introduction of PAC learning, new computational learning models have been proposed, such as the recent "restriction access" model [DRWY12] which allows the learner to interact with the target concept by asking membership queries, but also to obtain an entire circuit that computes the concept on a random subset of the inputs. For example, in one shot, the learner can obtain a circuit that computes the concept class on all n-bit inputs that start with $n/2$ zeros. At the same time, the cryptographic research landscape has been swiftly moving in the direction of augmenting traditional PRFs and other cryptographic primitives to include higher functionalities. This brings to mind natural questions:

- *Can one leverage augmented pseudo-random function constructions to establish limits on what can and cannot be learned in augmented machine learning models?*
- *Going even further afield, can augmented cryptographic constructs suggest interesting learning models?*

We address these questions in the second part of this paper. For a detailed description of our findings, see Section 1.2.

1.1   Our Results: Aggregate Pseudo Random Functions

Aggregate Pseudo Random Functions (AGG-PRF) are indexed families of pseudo-random functions for which a distinguisher (who runs in time polynomial in the security parameter) can request and receive the value of an aggregate (for example, the sum or the product) of the function values over certain large sets and yet cannot distinguish oracle access to the function from oracle access to a truly random function. At the same time, given the function index (in other words, the secret key), one can compute such aggregates over potentially super-polynomial size sets in polynomial time. Such an efficent aggregation algorithm cannot possibly exist for random functions. Thus, this is a PRF family that is very unlike random functions (in the sense of being able to efficiently aggregate over superpolynomial size sets), and yet is computationally indistinguishable from random functions.

To make this notion precise, we need two ingredients. Let $\mathcal{F} = \{\mathcal{F}_\lambda\}_{\lambda > 0}$ where each $\mathcal{F}_\lambda = \{f_K : \mathcal{D}_\lambda \to \mathcal{R}_\lambda\}_{K \in \mathcal{K}_\lambda}$ is a collection of functions on a domain $\mathcal{D}_\lambda$ to a range $\mathcal{R}_\lambda$, computable in time $\mathsf{poly}(\lambda)$.[2] The first ingredient is a collection of sets (also called a set system) $\mathcal{S} = \{S \subseteq \mathcal{D}\}$ over which the aggregates can be efficiently computed given the index K of the function. The second ingredient is an aggregation function $\Gamma : \mathcal{R}^* \to \{0,1\}^*$ which takes as input a tuple of function values $\{f(x) : x \in S\}$ for some set $S \in \mathcal{S}$ and outputs the aggregate $\Gamma(f(x_1), \ldots, f(x_{|S|}))$.

The sets are typically super-polynomially large, but are efficiently recognizable. That is, for each set S, there is a corresponding $\mathsf{poly}(\lambda)$-size circuit C_S that takes as input an $x \in \mathcal{D}$ and outputs 1 if and only if $x \in S$.[3] Throughout this paper, we will consider relatively simple aggregate functions, namely we will treat the range of the functions as an Abelian group, and will let Γ denote the group operation on its inputs. Note that the input to Γ is super-polynomially large (in the security parameter λ), making the aggregate computation non-trivial.

This family of functions, equipped with a set system $\mathcal{S}$ and an aggregation function Γ is called an aggregate PRF family (AGG-PRF) if the following two requirements hold:

1. *Aggregatability:* There exists a polynomial (in the security parameter λ) time algorithm that given an index K to the PRF $f_K \in \mathcal{F}$ and a circuit C_S that recognizes a set $S \in \mathcal{S}$, can compute Γ over the PRF values $f_K(x)$ for all $x \in S$. That is, it can compute

$$AGG_{K,\Gamma}(S) := \Gamma_{x \in S} \, f_K(x)$$

2. *Pseudorandomness:* No polynomial-time distinguisher which can specify a set $S \in \mathcal{S}$ as a query and can receive as an answer either $AGG_{K,\Gamma}(S)$ for a random function $f_K \in \mathcal{F}$ or $AGG_{h,\Gamma}(S)$ for a truly random functions h, can distinguish between the two cases.

We show a number of constructions of AGG-PRF for various set systems under different cryptographic assumptions. We describe our constructions below, starting from the least expressive set system.

---

[2] In this informal exposition, for the sake of brevity, we will sometimes omit the security parameter and refrain from referring to ensembles.

[3] All the sets we consider are efficiently recognizable, and we use the corresponding circuit as the representation of the set. We occasionally abuse notation and use S and C_S interchangeably.

Interval Sets. We first present AGG-PRFs over interval set systems with respect to aggregation functions that compute any group operation. The construction can be based on any (standard) PRF family.

Theorem 1 (Group Summation Over Intervals, from One-Way Functions [GGN10]). [4] *Assume one-way functions exist. Then, there exists an AGG-PRF family that maps $\mathbb{Z}_p$ to a group G, with respect to a collection of sets defined by intervals $[a, b] \subseteq \mathbb{Z}_p$ and the aggregation function computing the group operation on G.*

The construction works as follows. Let $F : \{0,1\}^n \times \{0,1\}^n \to \{0,1\}$ be a (standard) pseudo-random function family based on the existence of one-way functions [GGM86, HILL99]. Construct an AGG-PRF family G supporting efficient computation of group aggregation functions. Define

$$G(k, x) = F(k, x) - F(k, x - 1)$$

To aggregate G, set

$$\sum_{x \in [a,b]} G(k, x) = F(k, b) - F(k, a - 1)$$

Given k, this can be efficiently evaluated.

Another construction from [GGN10] achieves summation over the integers for PRFs whose range is $\{0, 1\}$. We omit the details of the construction, but state the theorem for completeness.

Theorem 2 (Integer Summation Over Intervals, from One-Way Functions [GGN10]). *Assume one-way functions exist. Then, there exists an AGG-PRF family that maps $\mathbb{Z}_{2^\lambda}$ to $\{0, 1\}$, with respect to a collection of sets defined by intervals $[a, b] \subseteq \mathbb{Z}_{2^\lambda}$ and the aggregation function computing the summation over $\mathbb{Z}$.*

Hypercubes. We next construct AGG-PRFs over hypercube set systems. This partially addresses Open Problem 5.4 posed in [GGN10], whether one can efficiently implement a random function with range $\{0, 1\}$ with complex queries that compute parities over the function values on hypercubes. Under subexponential DDH hardness, Theorem 3 answers the question for products rather than parities for a function whose range is a DDH group.

Throughout this section, we take $\mathcal{D}_\lambda = \{0, 1\}^\ell$ for some polynomial $\ell = \ell(\lambda)$. A hypercube $S_{\mathbf{y}}$ is defined by a vector $\mathbf{y} \in \{0, 1, \star\}^\ell$ as

$$S_{\mathbf{y}} = \{\mathbf{x} \in \{0, 1\}^\ell : \forall i, y_i = \star \text{ or } x_i = y_i\}$$

We present a construction under the sub-exponential DDH assumption.

---

[4] Observed even earlier by Reingold and Naor and appeared in [GGI$^+$02] in the context of small space streaming algorithms.

Theorem 3 (Hypercubes from DDH). *Let* $\mathcal{HC} = \{\mathcal{HC}_{\ell(\lambda)}\}_{\lambda>0}$ *where* $\mathcal{HC}_\ell = \{0,1,\star\}^\ell$ *be the set of hypercubes on* $\{0,1\}^\ell$. *Then, there is a construction of AGG-PRF supporting the set system* $\mathcal{HC}$ *with the product aggregation function, assuming the subexponential DDH assumption.*

We sketch the construction from DDH below. Our DDH construction is the Naor-Reingold PRF [NR04]. Namely, the function is parametrized by an ℓ-tuple $\boldsymbol{k} = (k_1, \ldots, k_\ell)$ and is defined as

$$F(\boldsymbol{k}, x) = g^{\prod_{i:x_i=1} k_i}$$

Let us illustrate aggregation over the hypercube $\boldsymbol{y} = (1, 0, \star, \star, \ldots, \star)$. To aggregate the function F, observe that

$$\prod_{\{x:\ x_1=1, x_2=0\}} F(\boldsymbol{k}, x) = \prod_{\{x:\ x_1=1, x_2=0\}} g^{\prod_{i:x_i=1} k_i}$$
$$= g^{\sum_{\{x:x_1=1,x_2=0\}} \prod_{i:x_i=1} k_i}$$
$$= g^{(k_1)(1)(k_2+1)(k_3+1)\cdots(k_\ell+1)}$$

which can be efficiently computed given $\boldsymbol{k}$.

Decision Trees. A decision tree T on ℓ variables is a binary tree where each internal node is labeled by a variable x_i, the leaves are labeled by either 0 or 1, one of the two outgoing edges of an internal node is labeled 0, and the other is labeled 1. Computation of a decision tree on an input $(x_1, \ldots, x_\ell)$ starts from the root, and at each internal node n, proceeds by taking either the 0-outgoing edge or 1-outgoing edge depending on whether $x_n = 0$ or $x_n = 1$, respectively. Finally, the output of the computation is the label of the leaf reached through this process. The size of a decision tree is the number of nodes in the tree.

A decision tree T defines a set $S = S_T = \{x \in \{0,1\}^\ell : T(x) = 1\}$. We show how to compute product aggregates over sets defined by polynomial size decision trees, under the subexponential DDH assumption.

The construction is simply a result of the observation that the set $S = S_T$ can be written as a disjoint union of polynomially many hypercubes. Computing aggregates over each hypercube and multiplying the results together gives us the decision tree aggregate.

Theorem 4 (Decision Trees from DDH). *Assuming the sub-exponential hardness of the decisional Diffie-Hellman assumption, there is an AGG-PRF that supports aggregation over sets recognized by polynomial-size decision trees.*

Read-Once Boolean Formulas. Finally, we show a construction of AGG-PRF over read-once Boolean formulas, the most expressive of our set systems, under the subexponential DDH assumption. A read-once Boolean formula a Boolean circuit composed of AND, OR and NOT gates with fan-out 1, namely each input literal feeds into at most one gate, and each gate output feeds into at most one other gate. Thus, a read-once formula can be written as a binary tree where each internal node is labeled with an AND or OR gate, and each literal (variable or its negation) appears in at most one leaf.

Theorem 5 (Read-Once Boolean Formulas from DDH). *Under the subexponential decisional Diffie-Hellman assumption, there is an AGG-PRF that supports aggregation over sets recognized by read-once Boolean formulas.*

Our aggregate PRF is, once again, the Naor-Reingold PRF. The index of the PRF consists of a $(\ell + 1)$-tuple of integers in $\mathbb{Z}_p$, namely $\boldsymbol{K} = (K_0, \ldots, K_\ell) \in \mathbb{Z}_p^{\ell+1}$. The function is defined as

$$f_{\boldsymbol{K}}(x) = g^{K_0 \prod_{i \in [\ell]} K_i^{x_i}}$$

We compute aggregates by recursion on the levels of the formula. We start by noting that it is enough to compute

$$A(C, 1) := \sum_{x:C(x)=1} \prod_{i \in [1 \ldots \ell]} K_i^{x_i}$$

because once this is done, it is easy to compute

$$\prod_{x:C(x)=1} f_{\boldsymbol{k}}(x) = g^{K_0 \cdot A(C,1)}$$

For the purposes of this informal exposition, assume that ℓ is a power of two. Let C be the formula, with either $C = C_L \wedge C_R$ or $C = C_L \vee C_R$ for subformula C_L and C_R. We show how to recursively compute $A(C, 1)$ for these sub-circuits and thus for C.

Limits of Aggregation. A natural question to ask is whether one can support aggregation over sets defined by general circuits. It is however easy to see that you cannot support any class of circuits for which deciding satisfiability is hard (for example, AC^0), or even ones for which counting the number of SAT assignments is hard (DNFs, for example) as follows. Suppose C is a circuit which is either unsatisfiable or has a unique SAT assignment. Solving satisfiability for such circuits is known to be sufficient to solve SAT in general [VV86]. The algorithm for SAT simply runs the aggregator with a random PRF key K, and outputs YES if and only if the aggregator returns a non-zero value. Note that if the formula is unsatisfiable, we will always get 0 from the aggregator. Otherwise, we get $f_k(x)$, where x is the (unique) satisfying assignment. Now, this might end up being 0 accidentally, but cannot be 0 always since otherwise, we will turn it into a PRF distinguisher. The distinguisher has the satisfying assignment hardcoded into it non-uniformly, and it simply checks if $PRF_K(x)$ is 0.

Theorem 6 (Impossibility for General Set Systems). *Suppose there is an efficient algorithm which on an index for $f \in \mathcal{F}$, a set system defined by $\{x : C(x) = 1\}$ for a polynomial size Boolean circuit C, and an aggregation function Γ, outputs the $\Gamma_{x:C(x)=1} f(x)$. Then, there is efficient algorithm that takes circuits C as input and w.h.p. over its coins, decides satisfiability for C.*

Related Work to Aggregate PRFs. As described above, the work of [GGN10] studies the general question of how one can efficiently construct random, "close-to" random, and "pseudo-random" large objects, such as functions or graphs, which "truthfully" obey global combinatorial properties rather simply appearing to do so to a polynomial time observer.

Formally, using the [GGN10] terminology, a PRF is a *pseudo-implementation* of a random function, and an AGG-PRF is a pseudo-implementation of a "random function that also answers aggregate queries" (as we defined them). Furthermore, the aggregatability property of AGG-PRF implies it is a *truthful* pseudo-implementation of such a function. Whereas in this work, we restrict our attention to aggregate queries, [GGN10] considers additional "complex-queries," such as in the case of a uniformly selected N node graph, providing a clique of size $\log_2 N$ that contains the queried vertex in addition to answering adjacency queries.

Our notion of aggregate PRFs bears resemblance to the notion of "algebraic PRFs" defined in the work of Benabbas, Gennaro and Vahlis [BGV11]. There are two main differences. First, algebraic PRFs support efficient aggregation over very specific subsets, whereas our constructions of aggregate PRFs support expressive subset classes, such as subsets recognized by hypercubes, decision trees and read-once Boolean formulas. Secondly, in the security notion for aggregate PRFs, the adversary obtains access to an oracle that computes the function as well as one that computes the aggregate values over super-polynomial size sets, whereas in algebraic PRFs, the adversary is restricted to accessing the function oracle alone. Our constructions from DDH use an algebraic property of the Naor-Reingold PRF in a similar manner as in [BGV11].

1.2   Our Results: Augmented PRFs and Computational Learning

As discussed above, connections between PRFs and learning theory date back to the 80's in the pioneering work of [Val84] showing that PRF in a complexity class C implies the existence of concept classes in C which can not be learned with membership queries. In the second part of this work, we study the implications of the slew of augmented PRF constructions of recent years [BW13a, BGI14a, KPTZ13b, BC10, ABPP14] and our new aggregate PRF to computational learning.

Constrained PRFs and Limits on Restriction Access Learnability. Recently, Dvir, Rao, Wigderson, and Yehudayoff [DRWY12] introduced a new learning model where the learner is allowed non-black-box information on the computational device (such as circuits, DNF,formulas) that decides the concept; their learner receives a simplified device resulting from partial assignments to input variables (i.e. restrictions). These partial restrictions lie somewhere in between function evaluation (full restrictions) which correspond to learning with membership queries and the full description of the original device (the empty restriction). The work of [DRWY12] studies a PAC version of restriction access, called PAC_{RA}, where the learner receives the circuit restricted with respect to random partial assignments. They show that both decision trees and DNF formulas can be learned efficiently in this model. Indeed, the PAC_{RA} model seems like quite a powerful generalization, if not too unrealistic, of the traditional PAC learning model, as it returns to the learner a computational description of the simplified concept.

Yet, in this section we will show limitations of this computational model under cryptographic assumptions. We show that the *constrained pseudo-random function families* introduced recently in [BW13b, BGI14b, KPTZ13a] naturally define a concept class which is not learnable by an even stronger variant of the restriction access learning model which we define. In the stronger variant, which we name *membership queries with restriction access (MQ_{RA})* the learner can adaptively specify any restriction of the circuit from a specified class of restrictions S and receive the simplified device computing the concept on this restricted domain in return. As this setting requires substantial notation, we define this new model very informally, and defer the formal definitions and theorems to the full version.

Definition 1 (Membership Queries with Restriction Access (MQ_{RA})).
Let $C : X \to \{0,1\}$ be a concept class, and $S = \{S \subseteq X\}$ be a collection of subsets of the domain. S is the set of allowable restrictions for concepts $f \in C$. Let Simp be "simplification rule" which, for a concept f and restriction S outputs a "simplification" of f restricted to S.

An algorithm A is an $(\epsilon, \delta, \alpha)$-$MQ_{RA}$ learning algorithm for representation class C with respect to a restrictions in S and simplification rule Simp if, for every $f \in C$, $\Pr[A^{\text{Simp}(f,\cdot)} = h] \geq 1 - \delta$ where h is an ϵ-approximation to f – and furthermore, A only requests restrictions for an α-fraction of the whole domain X.

Informally, constrained PRFs are PRFs with two additional properties: 1) for any subset S of the domain in a specified collection S, a *constrained key* K_S can be computed, knowledge of which enables efficient evaluation of the PRF on S; and 2) even with knowledge of constrained keys $K_{S_1}, \ldots, K_{S_m}$ for the corresponding subsets, the function retains pseudo-randomness on all points not covered by any of these sets. Connecting this to restriction access, the constrained keys will allow for generation of restriction access examples (restricted implementations with fixed partial assignments) and the second property implies that those examples do not aid in the learning of the function.

Theorem 7 (Informal). *Suppose F is a family of constrained PRFs which can be constrained to sets in S. If F is computable in circuit complexity class C, then C is hard to MQ_{RA}-learn with restrictions in S.*

Corollary 1 (Informal). *Existing constructions of constrained PRFs [BW13a] yield the following corollaries:*

- *If one-way functions exist, then poly-sized circuits can not be learned with restrictions on sub-intervals of the input-domain; and*
- *Assuming the sub-exponential hardness of the multi-linear Diffie-Hellman problem, NC^1 cannot be learned with restriction on hypercubes.*

New Learning Models Inspired by the Study of PRFs We proceed to define two new learning models inspired by recent directions in cryptography. The first model is the *related concept* model inspired by work into related-key attacks in cryptography. While we have cryptography and lower bounds in mind, we argue that this model is in some ways natural. The second model, learning with *aggregate queries*, is directly inspired by our development of aggregate pseudo-random functions in this work; rather than being a natural model in its own right, this model further illustrates how cryptography and learning are duals in many senses.

The Related Concept Learning Model. The idea that some functions or concepts are related to one another is quite natural. For a DNF formula, for instance, related concepts may include formulas where a clause has been added or formulas where the roles of two variables are swapped. For a decision tree, we could consider removing some accepting leaves and examining the resulting behavior. For a circuit, a related circuit might alter internal gates or fix the values on some wires. A similar phenomena occurs in cryptography, where secret keys corresponding to different instances of the same cryptographic primitive or even secret keys of different cryptographic primitives are related (if, for example, they were generated by a pseudo random process on the same seed).

We propose a new computational learning model where the learner is explicitly allowed to specify membership queries not only for the concept to be learned, but also for "related" concepts, given by a class of allowed transformations on the concept. We will show both a separation from membership queries, and a general negative result in the new model. Based on recent constructions of related-key secure PRFs by Bellare and Cash [BC10] and Abdalla et al [ABPP14], we demonstrate concept classes for which access to these related concepts is of no help.

To formalize the related concept learning model, we will consider *keyed concept classes* – classes indexed by a set of keys. This will enable the study of related concepts by instead considering concepts whose keys are related in some way. Most generally, we think of a key as a succinct representation of the computational device which decides the concept. This is a general framework; for example, we may consider the bit representation of a particular log-depth circuit as a key for a concept in the concept class NC^1. For a concept f_k in concept class $\mathcal{C}$, we allow the learner to query a membership oracle for f_k and also for 'related' concepts $f_{\phi(k)} \in \mathcal{C}_K$ for ϕ in a specified class of allowable functions Φ. For example: let $K = \{0,1\}^\lambda$ and let $\Phi^\oplus = \{\phi_\Delta : k \mapsto k \oplus \Delta\}_{\Delta \in \{0,1\}^\lambda}$. Informally:

Definition 2 (Φ-Related-Concept Learning Model (Φ-RC)). *For $\mathcal{C}_K$ a keyed concept class, let $\Phi = \{\phi : K \to K\}$ be a set of functions on K that contains the identity function* id. *A related-concept oracle RC_k, on query (ϕ, x), responds with $f_{\phi(k)}(x)$, for all $\phi \in \Phi$ and $x \in X$.*

An algorithm A is an (ϵ, δ)-Φ-RK learning algorithm for a $\mathcal{C}_k$ if, for every $k \in K$, when given access to the oracle $RK_k(\cdot)$, the algorithm A outputs with probability at least $1 - \delta$ a function $h : \{0,1\}^n \to \{0,1\}$ that ϵ-approximates f_k.

Yet again, we are able to demonstrate the limitations of this model using the power of a strong type of pseudo-random function. We show that *related-key secure PRF families (RKA-PRF)* defined and instantiated in [BC10] and [ABPP14] give a natural concept class which is not learnable with related key queries. RKA-PRFs are defined with respect to a set Φ of functions on the set of PRF keys. Informally, the security notion guarantees that for a randomly selected key k, no efficient adversary can distinguish oracle access to f_k and $f_{\phi(k)}$ (for many adaptively chosen functions $\phi \in \Phi$) from an oracle that returns completely random values. We leverage this strong pseudo-randomness property to show hard-to-learn concepts in the related concept model.

Theorem 8 (Informal). *Suppose $\mathcal{F}$ is a family of RKA-PRFs with respect to related-key functions Φ. If $\mathcal{F}$ is computable in circuit complexity class $\mathcal{C}$, then $\mathcal{C}$ is hard to learn in the Φ'-RC model for some Φ'.*

Existing constructions of RKA-PRFs [ABPP14] yield the following corollary:

Corollary 2 (Informal). *Assuming the hardness of the DDH problem, and collision-resistant hash functions, NC^1 is hard to Φ-RC-learn for an class of affine functions Φ.*

The Aggregate Learning Model. The other learning model we propose is inspired by our aggregate PRFs. Here, we consider a new extension to the power of the learning algorithm. Whereas membership queries are of the form "What is the label of an example x?", we grant the learner the power to request the evaluation of simple functions on tuples of examples $(x_1, ..., x_n)$ such as "How many of $x_1, ..., x_n$ are in C?" or "Compute the product of the labels of $x_1, ..., x_n$?". Clearly, if n is polynomial then this will result only a polynomial gain in the query complexity of a learning algorithm in the best case. Instead, we propose to study cases when n may be super-polynomial, but the description of the tuples is succinct. For example, the learning algorithm might query the number of x's in a large interval that are positive examples in the concept.

As with the restriction access and related concept models – and the aggregate PRFs we define in this work – the Aggregate Queries (AQ) learning model will be considered with restrictions to both the types of aggregate functions Γ the learner can query, and the sets $\mathcal{S}$ over which the learner may request these functions to be evaluated on. We now present the AQ learning model informally:

Definition 3 (($\Gamma, \mathcal{S}$)-Aggregate Queries (AQ) Learning). *Let $\mathcal{C} : X \to \{0, 1\}$ be a concept class, and let $\mathcal{S}$ be a collection of subsets of X. Let $\Gamma : \{0, 1\}^* \to V$ be an aggregation function. For $f \in \mathcal{C}$, let AGG_f be an "aggregation" oracle, which for $S \in \mathcal{S}$, returns $\Gamma_{x \in S} f(x)$. Let MEM_f be the membership oracle, which for input x returns $f(x)$.*

An algorithm $\mathcal{A}$ is an (ϵ, δ)-($\Gamma, \mathcal{S}$)-AQ learning algorithm for $\mathcal{C}$ if for every $f \in \mathcal{C}$,

$$\Pr[\mathcal{A}^{MEM_f(\cdot), \mathsf{AGG}_f(\cdot)} = h] \geq 1 - \delta$$

where h is an ϵ-approximation to f.

Initially, AQ learning is reminiscent of learning with statistical queries (SQ). In fact, this apparent connection inspired this portion of our work. But the AQ setting is in fact incomparable to SQ learning, or even the weaker "statistical queries that are independent of the target" as defined in [BF02]. On the one hand, AQ queries provide a sort of noiseless variant of SQ, giving more power to the AQ learner; on the other hand, the AQ learner is restricted to aggregating over sets in $\mathcal{S}$, whereas the SQ learner is not restricted in this way, thereby limiting the power of the AQ learner. The AQ setting where $\mathcal{S}$ contains every subset of the domain is indeed a noiseless version of "statistical queries independent of the target," but even this model is a restricted version of SQ. This does raise the natural question of a noiseless version of SQ and its variants; hardness results in such models would be interesting in that they would suggest that the hardness comes not from the noise but from an inherent loss of information in statistics/aggregates.

We will show both a simple separation from learning with membership queries (in the full version), and under cryptographic assumptions, a general lower bound on the power of learning with aggregate queries. The negative examples will use the results in Section 1.1.

Theorem 9. *Let $\mathcal{F}$ be a boolean-valued aggregate PRF with respect to set system $\mathcal{S}$ and aggregation function Γ. If $\mathcal{F}$ is computable in complexity class $\mathcal{C}$, then $\mathcal{C}$ is hard to $(\Gamma, \mathcal{S})$-AQ learn.*

Corollary 3. *Using the results from Section 3, we get the following corollaries:*

- *The existence of one way functions implies that $P/poly$ is hard to $(\sum, \mathcal{S}_{[a,b]})$-AQ learn, with $\mathcal{S}_{[a,b]}$ the set of sub-intervals of the domain as defined in Section 3.*
- *The DDH assumption implies that NC^1 is hard to $(\sum, \mathcal{S}_{[a,b]})$-AQ learn, with $\mathcal{S}_{[a,b]}$ being the set of sub-intervals of the domain as defined in Section 3.*
- *The subexponential DDH Assumption implies that NC^1 is hard to $(\prod, \mathcal{R})$-AQ learn, with $\mathcal{R}$ the set of read-once boolean formulas defined in Section 3.*

Open Questions. As discussed in the introduction, augmented pseudo-random functions often have powerful and surprising applications, perhaps the most recent example being constrained PRFs [BW13a, KPTZ13a, BGI14a]. Perhaps the most obvious open question that emerges from this work is to find applications for aggregate PRFs. We remark that a primitive similar to aggregate PRFs was used in [BGV11] to construct delegation protocols.

Perhaps a more immediate concern is that all our aggregate PRF constructions (except for intervals) requires sub-exponential hardness assumptions. We view it as an important open question to base these constructions on polynomial assumptions.

In this work we restricted our attention to particular types of aggregation functions and subsets over which the aggregation takes place, although our definition captures more general scenarios. We looked at aggregation functions that compute group operations over Abelian groups. Can we support more general aggregation functions that are not restricted to group operations, for example the majority aggregation function, or even non-symmetric aggregation functions? We show positive results for intervals, hypercubes, and sets recognized by read-once formulas and decision trees. On the other hand, we show that it is unlikely that we can support general sets, for example sets recognized by CNF formulas. This almost closes the gap between what is possible and what is hard. A concrete open question in this direction is to construct an aggregate PRF computing summation over an Abelian group for sets recognized by DNFs, or provide evidence that this cannot be done.

Organization. This paper is organized into two parts that can be read essentially independently of each other. In the first part (Sections 2 and 3), we present the definition and constructions of aggregate pseudo-random functions. In the second part (Section 4), we show connections between various notions of augmented PRFs and their applications to augmented learning models.

2   Aggregate PRF

We will let λ denote the security parameter throughout this paper.

Let $\mathcal{F} = \{\mathcal{F}_\lambda\}_{\lambda>0}$ be a function family where each function $f \in \mathcal{F}_\lambda$ maps a domain $\mathcal{D}_\lambda$ to a range $\mathcal{R}_\lambda$. An *aggregate function* family is associated with two objects:

1. an ensemble of sets $\mathcal{S} = \{\mathcal{S}_\lambda\}_{\lambda>0}$ where each $\mathcal{S}_\lambda$ is a collection of subsets of the domain $S \subseteq \mathcal{D}_\lambda$; and
2. an "aggregation function" $\Gamma_\lambda : (\mathcal{R}_\lambda)^* \to \mathcal{V}_\lambda$ that takes a tuple of values from the range $\mathcal{R}_\lambda$ of the function family and "aggregates" them to produce a value in an output set $\mathcal{V}_\lambda$.

Let us now make this notion formal. To do so, we will impose restrictions on the set ensembles and the aggregation function. First, we require set ensemble $\mathcal{S}_\lambda$ to be *efficiently recognizable*. That is, there is a polynomial-size Boolean circuit family $\mathcal{C} = \{\mathcal{C}_\lambda\}_{\lambda>0}$ such that for any set $S \in \mathcal{S}_\lambda$ there is a circuit $C = C_S \in \mathcal{C}_\lambda$ such that $x \in S$ if and only if $C(x) = 1$. Second, we require our aggregation functions Γ to be efficient in the length of its inputs, and symmetric; namely the output of the function does not depend on the order in which the inputs are fed into it. Summation over an Abelian group is an example of a possible aggregation function. Third and finally, elements in our sets $\mathcal{D}_\lambda$, $\mathcal{R}_\lambda$, and $\mathcal{V}_\lambda$ are all representable in $\mathsf{poly}(\lambda)$ bits, and the functions $f \in \mathcal{F}_\lambda$ are computable in $\mathsf{poly}(\lambda)$ time.

Define the aggregate function $\mathsf{AGG} = \mathsf{AGG}^\lambda_{f,\mathcal{S}_\lambda,\Gamma_\lambda}$ that is indexed by a function $f \in \mathcal{F}_\lambda$, takes as input a set $S \in \mathcal{S}_\lambda$ and "aggregates" the values of $f(x)$ for all $x \in \mathcal{S}_\lambda$. That is, $\mathsf{AGG}(S)$ outputs

$$\Gamma\big(f(x_1), f(x_2), \dots, f(x_{|S|})\big)$$

where $S = \{x_1, \ldots, x_{|S|}\}$. More precisely, we have

$$AGG^\lambda_{f,\mathcal{S}_\lambda,\Gamma_\lambda} : \mathcal{S}_\lambda \to \mathcal{V}_\lambda$$
$$S \mapsto \Gamma_{x_i \in S}\big(f(x_1), \ldots, f(x_{|S|})\big)$$

We will furthermore require that the AGG can be computed in $\mathsf{poly}(\lambda)$ time. We require this in spite of the fact that the sets over which the aggregation is done can be exponentially large! Clearly, such a thing is impossible for a random function f but yet, we will show how to construct *pseudo-random* function families that support efficient aggregate evaluation. We will call such a pseudo-random function (PRF) family an *aggregate PRF* family. In other words, our objective is two fold:

1. Allow anyone who knows the (polynomial size) function description to efficiently compute the aggregate function values over exponentially large sets; but at the same time,
2. Ensure that the function family is indistinguishable from a truly random function, even given an oracle that computes aggregate values.

A simple example of aggregates is that of computing the summation of function values over sub-intervals of the domain. That is, let domain and range be $\mathbb{Z}_p$ for some $p = p(\lambda)$, let the family of subsets be $\mathcal{S}_\lambda = \{[a,b] \subseteq \mathbb{Z}_p : a, b \in \mathbb{Z}_p; a \leq b\}$, and the aggregation function be $\Gamma_\lambda(y_1, \ldots, y_k) = \sum_{i=1}^k y_i \pmod{p}$. In this case, we are interested in computing

$$AGG^\lambda_{f,\mathcal{S}_\lambda,\mathsf{sum}}([a,b]) = \sum_{a \leq x \leq b} f(x)$$

We will, in due course, show both constructions and impossibility results for aggregate PRFs, but first let us start with the formal definition.

Definition 4 (Aggregate PRF). *Let $\mathcal{F} = \{\mathcal{F}_\lambda\}_{\lambda > 0}$ be a function family where each function $f \in \mathcal{F}_\lambda$ maps a domain $\mathcal{D}_\lambda$ to a range $\mathcal{R}_\lambda$, $\mathcal{S}$ be an efficiently recognizable ensemble of sets $\{\mathcal{S}_\lambda\}_{\lambda > 0}$, and $\Gamma_\lambda : (\mathcal{R}_\lambda)^* \to \mathcal{V}_\lambda$ be an aggregation function. We say that $\mathcal{F}$ is an $(\mathcal{S}, \Gamma)$-aggregate pseudorandom function family (also denoted $(\mathcal{S}, \Gamma)$-AGG-PRF) if there exists an efficient algorithm $\mathsf{Aggregate}_{k,\mathcal{S},\Gamma}(S)$: On input a subset $S \in \mathcal{S}$ of the domain, outputs $v \in \mathcal{V}$, such that*

– **Efficient aggregation:** *For every $S \in \mathcal{S}$, $\mathsf{Aggregate}_{k,\mathcal{S},\Gamma}(S) = AGG_{k,\mathcal{S},\Gamma}(S)$ where $AGG_{k,\mathcal{S},\Gamma}(S) := \Gamma_{x \in S}\ F_k(x)$.[56]*

---

[5] We omit subscripts on AGG and Aggregate when clear from context.

[6] AGG is defined to be the correct aggregate value, while Aggregate is the algorithm by which we compute the value AGG. We make this distinction because while a random function cannot be efficiently aggregated, the aggregate value is still well-defined.

- **Pseudorandomness:** *For all probabilistic polynomial-time (in security parameter λ) algorithms A, and for randomly selected key $k \in K$:*

$$| \Pr_{f \leftarrow \mathcal{F}_\lambda} [A^{f_k, AGG_{f_k}, S, \Gamma}(1^\lambda)] - \Pr_{h \leftarrow \mathcal{H}_\lambda} [A^{h, AGG_h, S, \Gamma}(1^\lambda)]| \leq \mathsf{negl}(\lambda)$$

where $\mathcal{H}_\lambda$ is the set of all functions $D_\lambda \to R_\lambda$.

Remark 1. In this work, we restrict our attention to aggregation functions that treat the range $\mathcal{V}_\lambda = \mathcal{R}_\lambda$ as an Abelian group and compute the group sum (or product) of its inputs. We denote this setting by $\Gamma = \sum$ (or $\prod$, respectively). Supporting other types of aggregation functions (ex: max, a hash) is a direction for future work.

2.1 A General Security Theorem for Aggregate PRFs

How does the security of a function family in the AGG-PRF game relate to security in the normal PRF game (in which A uses only the oracle f and not AGG_f)?

In this section, we show a general security theorem for aggregate pseudo-random functions. Namely, we show that any "sufficiently secure" PRF is also aggregation-secure (for any collection of efficiently recognizable sets and any group-aggregation operation), in the sense of Definition 4, by way of an *inefficient* reduction (with overhead polynomial in the size of the domain). In Section 3, we will use this to construct AGG-PRFs from a subexponential-time hardness assumption on the DDH problem. We also show that no such *general* reduction can be efficient, by demonstrating a PRF family that is not aggregation-secure. As a general security theorem cannot be shown without the use of complexity leveraging, this suggests a natural direction for future study: to devise constructions for similarly expressive aggregate PRFs from polynomial assumptions.

Lemma 1. *Let $\mathcal{F} = \{\mathcal{F}_\lambda\}_{\lambda > 0}$ be a pseudo-random function family where each function $f \in \mathcal{F}_\lambda$ maps a domain D_λ to a range $\mathcal{R}_\lambda$. Suppose there is an adversary A that runs in time $t_A = t_A(\lambda)$ and achieves an advantage of $\epsilon_A = \epsilon_A(\lambda)$ in the aggregate PRF security game for the family $\mathcal{F}$ with an efficiently recognizable set system $\mathcal{S}_\lambda$ and an aggregation function Γ_λ that is computable in time polynomial in its input length. Then, there is an adversary B that runs in time $t_B = t_A + \mathsf{poly}(\lambda, |D_\lambda|)$ and achieves an advantage of $\epsilon_B = \epsilon_A$ in the standard PRF game for the family $\mathcal{F}$.*

Proof. Let $f_K \leftarrow \mathcal{F}_\lambda$ be a random function from the family $\mathcal{F}_\lambda$. We construct the adversary B which is given access to an oracle $\mathcal{O}$ which is either f_K or a uniformly random function $h : D_\lambda \to \mathcal{R}_\lambda$.

B works as follows: It queries the PRF on all inputs $x \in \mathcal{D}_\lambda$, builds the function table T_K of f_K and runs the adversary A, responding to its queries as follows:

1. Respond to its PRF query $x \in \mathcal{D}_\lambda$ by returning $T_K[x]$; and
2. Respond to its aggregate query (Γ, S) by (a) going through the table to look up all x such that $x \in S$; and (b) applying the aggregation function honestly to these values.

Finally, when A halts and returns a bit b, B outputs the bit b and halts.

B takes $O(|\mathcal{D}_\lambda|)$ time to build the truth table of the oracle. For each aggregate query (Γ, S), B first checks for each $x \in \mathcal{D}_\lambda$ whether $x \in S$. This takes $|\mathcal{D}_\lambda| \cdot \mathsf{poly}(\lambda)$ time, since S is efficiently recognizable. It then computes the aggregation function Γ over $f(x)$ such that $x \in S$, taking $\mathsf{poly}(|\mathcal{D}_\lambda|)$ time, since Γ is computable in time polynomial in its input length. The total time, therefore, is

$$t_B = t_A + \mathsf{poly}(\lambda, |\mathcal{D}_\lambda|)$$

Clearly, when $\mathcal{O}$ is the pseudo-random function f_K, B simulates an aggregatable PRF oracle to A, and when $\mathcal{O}$ is a random function, B simulates an aggregate random oracle to A. Thus, B has the same advantage in the PRF game as A does in the aggregate PRF game.

The above gives an inefficient reduction from the PRF security of a function family $\mathcal{F}$ to the AGG-PRF security of the same family running in time polynomial in the size of the domain. Can this reduction be made efficient; that is, can we replace $t_B = t_A + \mathsf{poly}(\lambda)$ into the Lemma 1?

This is not possible. Such a reduction would imply that every PRF family that supports efficient aggregate functionality AGG is AGG-PRF secure; this is clearly false. Take for example a pseudorandom function family $\mathcal{F}_0 = \{f : \mathbb{Z}_{2p} \to \mathbb{Z}_p\}$ such that for all f, there is no x with $f(x) = 0$. It is possible to construct such a pseudorandom function family $\mathcal{F}_0$ (under the standard definition). While 0 is not in the image of any $f \in \mathcal{F}_0$, a random function with the same domain and range will, with high probability, have 0 in the image. For an aggregation oracle AGG_f computing *products* over $\mathbb{Z}_p$: $\mathsf{AGG}_f(\mathbb{Z}_{2p}) \neq 0$ if $f \in \mathcal{F}_0$, while $\mathsf{AGG}_f(\mathbb{Z}_{2p}) = 0$ with high probability for random f.

Thus, access to aggregates for products over $\mathbb{Z}_p{}^7$ would allow an adversary to trivially distinguish $f \in \mathcal{F}_0$ from a truly random map.

2.2   Impossibility of Aggregate PRF for General Sets

It is natural to ask whether whether an aggregate PRF might be constructed for more general sets than we present in Section 3. There we constructed aggregate PRF for the sets of all satisfying assignments for read-once boolean formula and

---

[7] Taken with respect to a set ensemble $\mathcal{S}$ containing, as an element, the whole domain $\mathbb{Z}_{2p}$. While this is not necessary (a sufficiently large subset would suffice), it is the case for the ensembles $\mathcal{S}$ we consider in this work.

decision trees. As we show in the following, it is impossible to extend this to support the set of satisfying assignmnets for more general circuits.

Theorem 10. *Suppose there is an algorithm that has a PRF description K, a circuit C, and a fixed aggregation rule (sum over a finite field, say), and outputs the aggregate value*

$$\sum_{x:C(x)=1} f_K(x)$$

Then, there is an algorithm that takes circuits C as input and w.h.p. over it coins, decides the satisfiability of C.

Proof. The algorithm for SAT simply runs the aggregator with a randomly chosen K, and outputs YES if and only if the aggregator returns 1. The rationale is that if the formula is unsatisfiable, you will always get 0 from the aggregator.[8] Otherwise, you will get $f_K(x)$, where x is the satisfying assignment. (More generally, $\sum_{x:C(x)=1} f_K(x)$). Now, this might end up being 0 accidentally, but cannot be 0 always since otherwise, you will get a PRF distinguisher. The distinguisher has the satisfying assignment hardcoded into it non-uniformly,[9] and it simply checks if $f_K(x) = 0$.

This impossibility result can be generalized for efficient aggregation of functions that are not pseudo-random. For instance, if $f(x) \equiv 1$ was the constant function 1, the same computing the aggregate over f satisfying inputs to C would not only reveal the satisfiability of C, but even the number of satisfying assignments! In the PRF setting though, it seems that aggregates only reveal the (un)satisfiability of a circuit C, but not the number of satisfying assignments. Further studying the relationship between the (not necessarily pseudo-random) function f, the circuit representation of C, and the tractability of computing aggregates is an interesting direction. A negative result for a class for which satisfiability (or even counting assignments) is tractable would be very interesting.

3 Constructions of Aggregate PRF

In this section, we show several constructions of aggregate PRFs. In Section 3.1, we show as a warm-up a generic construction of aggregate PRFs for intervals (where the aggregation is any group operation). This construction is black-box: given any PRF with the appropriate domain and range, we construct a related

---

[8] This proof may be extended to the case when the algorithm's output is not restricted to be 0 when the input circuit C is unsatisfiable, and even arbitrary outputs for sufficiently expressive classes of circuits.

[9] As pointed out by one reviewer, for sufficiently expressive classes of circuits C, this argument can be made uniform. Specifically, we use distinguish the challenge y from a pseudo-random generator from random by choosing $C := C_y$ that is satisfiable if and only if y is in the PRG image, and modify the remainder of the argument accordingly.

family of aggregate PRFs and with no loss in security. In Section 3.2, we show a construction of aggregate PRFs for products over bit-fixing sets (hypercubes), from a strong decisional Diffie-Hellman assumption. We then generalize the DDH construction: in Section 3.3, to the class of sets recognized by polynomial-size decision trees; and in Section 3.4, to sets recognized by read-once Boolean formulas. In these last three constructions, we make use of Lemma 1 to argue security.

3.1 Generic Construction for Interval Sets

Our first construction is from [GGN10][10]. The construction is entirely black-box: from any appropriate PRF family $\mathcal{G}$, we construct a related AGG-PRF family $\mathcal{F}$. Unlike the proofs in the sequel, this reduction exactly preserves the security of the starting PRF.

Let $\mathcal{G}_\lambda = \{g_K : \mathbb{Z}_{n(\lambda)} \to R_\lambda\}_{K \in \mathcal{K}_\lambda}$ be a PRF family, with $R = R_\lambda$ being a group where the group operation is denoted by $\oplus$[11]. We construct an aggregatable PRF $\mathcal{F}_\lambda = \{f_K\}_{K \in \mathcal{K}_\lambda}$ for which we can efficiently compute summation of $f_K(x)$ for all x in an interval $[a, b]$, for any $a \le b \in \mathbb{Z}_n$. Let $\mathcal{S}_{[a,b]} = \{[a, b] \subseteq \mathbb{Z}_n : a, b \in \mathbb{Z}_n; a \le b\}$ be the set of all interval subsets of $\mathbb{Z}_n$, $[a, b] = \{x \in \mathbb{Z}_n : a \le x \le b\}$. Define $\mathcal{F} = \{f_K : \mathbb{Z}_n \to R\}_{K \in \mathcal{K}}$ as follows:

$$f_K(x) = \begin{cases} g_K(0) & : x = 0 \\ g_K(x) \ominus g_K(x-1) & : x \ne 0 \end{cases}$$

Lemma 2. *Assuming that $\mathcal{G}$ is a pseudo-random function family, $\mathcal{F}$ is a $(\mathcal{S}_{[a,b]}, \oplus)$-aggregate pseudo-random function family.*

Proof. It follows immediately from the definition of f_K that one can compute the summation of $f_K(x)$ over any interval $[a, b]$. Indeed, rearranging the definition yields

$$\sum_{x \in [0,b]} f_K(x) = g_K(b) \quad \text{and} \quad \sum_{x \in [a,b]} f_K(x) = g_K(b) \oplus -g_K(a-1)$$

We reduce the pseudo-randomness of $\mathcal{F}$ to that of $\mathcal{G}$. The key observation is that each query to the f_K oracle as well as the aggregation oracle for f_K can be answered using at most two black-box calls to the underlying function g_K. By assumption on $\mathcal{G}$, replacing the oracle for g_K with a uniformly random function $h : \mathbb{Z}_n \to R$ is computationally indistinguishable. Furthermore, the function f defined by replacing g by h, namely

$$f'(x) = \begin{cases} h(0) & : x = 0 \\ h(x) \ominus h(x-1) & : x \ne 0 \end{cases}$$

is a truly random function. Thus, the simulated oracle with g_K replaced by h implements a uniformly random function that supports aggregate queries. Security according to Definition 4 follows immediately.

---

[10] See Example 3.1 and Footnote 18.

[11] The only structure of $\mathbb{Z}_n$ we us is the *total order*. Our construction directly applies to any finite, totally-ordered domain D by first mapping D to $\mathbb{Z}_n$, preserving order.

Another construction from the same work achieves summation over the integers for PRFs whose range is $\{0,1\}$. We omit the details of the construction, but state the theorem for completeness.

Theorem 11 (Integer Summation Over Intervals, from One-Way Functions [GGN10]). *Assume one-way functions exist. Then, there exists an* $(\mathcal{S}_{[a,b]}, \sum)$-*AGG-PRF family that maps* $\mathbb{Z}_{2^\lambda}$ *to* $\{0,1\}$, *where* $\sum$ *denotes summation over* $\mathbb{Z}$.

3.2 Bit-Fixing Aggregate PRF from DDH

We now construct an aggregate PRF computing products for bit-fixing sets. Informally, our PRF will have domain $\{0,1\}^{poly(\lambda)}$, and support aggregation over sets like $\{x : x_1 = 0 \wedge x_2 = 1 \wedge x_7 = 0\}$. We will naturally represent such sets by a string in $\{0,1,\star\}^{poly(\lambda)}$ with 0 and 1 indicating a fixed bit location, and $\star$ indicating a free bit location. We call each such set a 'hypercube.' The PRF will have a multiplicative group $\mathcal{G}$ as its range, and the aggregate functionality will compute group products.

Our PRF is exactly the Naor-Reingold PRF [NR04], for which we demonstrate efficient aggregation and security. We begin by stating the decisional Diffie-Hellman assumption.

Let $\mathcal{G} = \{\mathcal{G}_\lambda\}_{\lambda > 0}$ be a family of groups of order $p = p(\lambda)$. The decisional Diffie-Hellman assumption for $\mathcal{G}$ says that the following two ensembles are computationally indistinguishable:

$$\{(\mathcal{G}_\lambda, g, g^a, g^b, g^{ab}) : G \leftarrow \mathcal{G}_\lambda; \; g \leftarrow G; \; a, b \leftarrow \mathbb{Z}_p\}_{\lambda > 0}$$
$$\approx_c \{(G, g, g^a, g^b, g^c) : G \leftarrow \mathcal{G}_\lambda; \; g \leftarrow G; \; a, b, c \leftarrow \mathbb{Z}_p\}_{\lambda > 0}$$

We say that the $(t(\lambda), \epsilon(\lambda))$-DDH assumption holds if for every adversary running in time $t(\lambda)$, the advantage in distinguishing between the two distributions above is at most $\epsilon(\lambda)$.

Construction. Let $\mathcal{G} = \{\mathcal{G}_\lambda\}_{\lambda > 0}$ be a family of groups of order $p = p(\lambda)$, each with a canonical generator g, for which the decisional Diffie Hellman (DDH) problem is hard. Let $\ell = \ell(\lambda)$ be a polynomial function. We will construct a PRF family $\mathcal{F}_\ell = \{\mathcal{F}_{\ell,\lambda}\}_{\lambda > 0}$ where each function $f \in \mathcal{F}_{\ell,\lambda}$ maps $\{0,1\}^{\ell(\lambda)}$ to $\mathcal{G}_\lambda$. Our PRF family is exactly the Naor-Reingold PRF [NR04]. Namely, each function f is parametrized by $\ell + 1$ numbers $\boldsymbol{K} := (K_0, K_1, \ldots, K_\ell)$, where each $K_i \in \mathbb{Z}_p$.

$$f_{\boldsymbol{K}}(x_1, \ldots, x_\ell) = g^{K_0 \prod_{i=1}^{\ell} K_i^{x_i}} = g^{K_0 \prod_{i:x_i=1} K_i} \qquad \in \mathcal{G}_\lambda$$

The aggregation algorithm $\mathsf{Aggregate}$ for bit-fixing functions gets as input the PRF key $\boldsymbol{K}$ and a bit-fixing string $y \in \{0, 1, \star\}^\ell$ and does the following:

- Define the strings K'_i as follows:

$$K'_i = \begin{cases} 1 & \text{if } y_i = 0 \\ K_i & \text{if } y_i = 1 \\ 1 + K_i & \text{otherwise} \end{cases}$$

- Output $g^{K_0 \prod_{i=1}^\ell K'_i}$ as the answer to the aggregate query.

Letting $\mathcal{HC} = \{\mathcal{HC}_{\ell(\lambda)}\}_{\lambda > 0}$ where $\mathcal{HC}_\ell = \{0, 1, \star\}^\ell$ is the set of hypercubes on $\{0, 1\}^\ell$, we now prove the following:

Theorem 12. *Let $\epsilon > 0$ be a constant, choose the security parameter $\lambda = \Omega(\ell^{1/\epsilon})$, and assume the $(2^{\lambda^\epsilon}, 2^{-\lambda^\epsilon})$-hardness of DDH over the group $\mathcal{G}$. Then, the collection of functions $\mathcal{F}$ defined above is a secure aggregate PRF with respect to the subsets $\mathcal{HC}$ and the product aggregation function over $\mathcal{G}$.*

Correctness. We show that the answer we computed for an aggregate query $y \in \{0, 1, \star\}^\lambda$ is correct. Define the sets

$$\mathsf{Match}(y) := \{x \in \{0, 1\}^\lambda : \forall i, y_i = \star \text{ or } x_i = y_i\} \text{ and } \mathsf{Fixed}(y) := \{i \in [\lambda] : y_i \in \{0, 1\}\}$$

Thus, $\mathsf{Match}(y)$ is the set of all 0-1 strings x that match all the fixed locations of y, but can take any value on the wildcard locations of y. $\mathsf{Fixed}(y)$ is the set of all locations i where the bit y_i is fixed. Note that:

$$
\begin{aligned}
\mathsf{AGG}(\boldsymbol{K}, y) &= \textstyle\prod_{x \in \mathsf{Match}(y)} f_{\boldsymbol{K}}(x) && \text{(by definition of AGG)} \\
&= \textstyle\prod_{x \in \mathsf{Match}(y)} g^{K_0 \prod_{i=1}^\ell K_i^{x_i}} && \text{(by definition of } f_{\boldsymbol{K}}) \\
&= g^{K_0 \sum_{x \in \mathsf{Match}(y)} \prod_{i=1}^\ell K_i^{x_i}} \\
&= g^{K_0 \left(\prod_{i \in \mathsf{Fixed}(y)} K_i^{y_i} \right) \cdot \left(\prod_{i \in [\ell] \setminus \mathsf{Fixed}(y)} (1 + K_i) \right)} && \text{(inverting sums and products)} \\
&= g^{K_0 \prod_{i=1}^\ell K'_i} && \text{(by definition of } K'_i) \\
&= \mathsf{Aggregate}(\boldsymbol{K}, y) && \text{(by definition of } \mathsf{Aggregate})
\end{aligned}
$$

Security. We will rely on the following theorem from [NR04].

Theorem 13 (Theorem 4.1, [NR04]). *Suppose there is an adversary A that runs in time $t(\lambda)$ and has an advantage of $\gamma(\lambda)$ in the (regular) PRF game. Then, there is an adversary B that runs in time $\mathsf{poly}(\lambda) \cdot t(\lambda)$ and breaks the DDH assumption with advantage $\gamma(\lambda)/\lambda$.*

The aggregate PRF security proof proceeds as follows. First, we choose the security parameter $\lambda = \Omega(\ell^{1/\epsilon})$ as in the theorem statement. We use Lemma 1 to conclude that if there is an adversary distinguisher D breaking the aggregate PRF security of $\mathcal{F}$ in $\mathsf{poly}(\lambda)$ time with $1/\mathsf{poly}(\lambda)$ advantage, then there is an adversary A that breaks the regular PRF security of $\mathcal{F}$ in $\mathsf{poly}(\lambda) \cdot 2^{O(\ell)} = \mathsf{poly}(\lambda) \cdot 2^{\lambda^\epsilon} = 2^{O(\lambda^\epsilon)}$ time with $1/\mathsf{poly}(\lambda)$ advantage. Using Theorem 13 now tells us that there is an adversary B that wins the DDH distinguishing game in $2^{O(\lambda^\epsilon)}$ time with $1/\text{poly}(\lambda)$ advantage, breaking the subexponential DDH assumption. This establishes the aggregate security of the PRF and thus Theorem 12.

Obtaining a security proof based on polynomial assumptions is an interesting open question.

3.3 Decision Trees

We generalize the previous construction from DDH to support sets specified by polynomial-sized decision trees by observing that such decision trees can be written as disjoint unions of hypercubes.

A decision tree family $\mathcal{T}_\lambda$ of size $p(\lambda)$ over $\ell(\lambda)$ variables consists of binary trees with at most $p(\lambda)$ nodes, where each internal node is labeled with a variable x_i for $i \in [\ell]$, the two outgoing edges of an internal node are labeled 0 and 1, and the leaves are labeled with 0 or 1. On input an $x \in \{0,1\}^\ell$, the computation of the decision tree starts from the root, and upon reaching an internal node n labeled by a variable x_i, takes either the 0-outgoing edge or the 1-outgoing edge out of the node n, depending on whether x_i is 0 or 1, respectively.

We now show how to construct a PRF family $\mathcal{F}_\ell = \{\mathcal{F}_{\ell,\lambda}\}_{\lambda>0}$ where each $\mathcal{F}_{\ell,\lambda}$ consists of functions that map $\mathcal{D}_\lambda := \{0,1\}^\ell$ to a group $\mathcal{G}_\lambda$, that supports aggregation over sets recognized by decision trees. That is, let $\mathcal{S}_\lambda = \{S \subseteq \{0,1\}^\ell : \exists$ a decision tree $T_S \in \mathcal{T}_\lambda$ that recognizes $S\}$.

Our construction uses a hypercube-aggregate PRF family $\mathcal{F}'_\ell$ as a sub-routine. First, we need the following simple lemma.

Lemma 3 (Decision Trees as Disjoint Unions of Hypercubes). *Let $S \subseteq \{0,1\}^\ell$ be recognized by a decision tree T_S of size $p = p(\lambda)$. Then, S is a disjoint union of at most p hybercubes $H_{y_1}, \ldots, H_{y_p}$, where each $y_i \in \{0,1,\star\}^\ell$ and $H_{y_i} = \text{Match}(y_i)$. Furthermore, given T_S, one can in polynomial time compute these hypercubes.*

Given the lemma, Aggregate is simple: on input a set S represented by a decision tree T_S, compute the disjoint hypercubes $H_{y_1}, \ldots, H_{y_p}$. Run the hypercube aggregation algorithm to compute

$$g_i \leftarrow \text{Aggregate}_{\mathcal{F}}(K, y_i)$$

and outputs $g := \prod_{i=1}^p g_i$.

Basing the construction on the hypercube-aggregate PRF scheme from Section 3.2, we get a decision tree-aggregate PRF based on the sub-exponential DDH assumption. The security of this PRF follows from Lemma 1 by an argument identical to the one in Section 3.2.

3.4 Read-Once Formulas

Read-once boolean formula provide a different generalization of hypercubes and they too admit an efficient aggregation algorithm for the Naor-Reingold PRF, with a similar security guarantee.

A boolean formula on ℓ variables is a circuit on $x = (x_1, \ldots, x_\ell) \in \{0,1\}^\ell$ composed of only AND, OR, and NOT gates. A *read-once boolean formula* is a boolean formula with fan-out 1, namely each input literal feeds into at most one gate, and each gate output feeds into at most one other gate.[12] Let R_λ be the family of all read-once boolean formulas over $\ell(\lambda)$ variables. Without loss of generality, we restrict these circuits to be in a standard form: namely, composed of fan-in 2 and fan-out 1 AND and OR gates, and any NOT gates occurring at the inputs.

In this form, the circuit for any read-once boolean formula can be identified with a labelled binary tree; we identify a formula by the label of its root C_ϕ. Nodes with zero children are variables or their negation, labelled by x_i or $\bar{x}_i$, while all other nodes have 2 children and represent gates with fan-in 2. For such a node with label C, its children have labels C_L and C_R. Note that each child is itself a read-once boolean formula on fewer inputs, and their inputs are disjoint Let the gate type of a node C be $\mathsf{type}(C) \in \{AND, OR\}$.

We describe a recursive aggregation algorithm for computing products of PRF values over all accepting inputs for a given read-once boolean formula C_ϕ. Looking forward, we require the formula to be read-once in order for the recursion to be correct. The algorithm described reduces to that of Section 3.2 in the case where ϕ describes a hypercube.

Construction. The aggregation algorithm for read-once Boolean formulas takes as input the PRF key $\boldsymbol{K} = (K_0, \ldots, K_\ell)$ and a formula $C_\phi \in R_\lambda$ where C_ϕ only reads the variables $x_1, \ldots, x_m$ for some $m \leq \ell$. We abuse notation and interpret C_ϕ to be a formula on both $\{0,1\}^\ell$ and $\{0,1\}^m$ in the natural way.

$$\mathsf{AGG}_{k,\prod}(C_\phi) = \prod_{x:C_\phi(x)=1} g^{K_0 \prod_{i \in [\ell]} K_i^{x_i}} \tag{1}$$

$$= g^{K_0 \sum_{x:C_\phi(x)=1} \prod_{i \in [\ell]} K_i^{x_i}} \tag{2}$$

$$= g^{K_0 \cdot A(C_\phi, 1) \cdot \prod_{m < j \leq \ell}(1 + K_j)} \tag{3}$$

where we define $A(C,1) := \sum_{\{x \in \{0,1\}^m : C(x)=1\}} \prod_{i \in [m]} K_i^{x_i}$. If $A(C,1)$ is efficiently computable, then **Aggregate** will simply compute it and return (3). To this end, we provide a recursive procedure for computing $A(C,1)$.

Generalizing the definition for any sub-formula C with variables named x_1 to x_m, define the values $A(C,0)$ and $A(C,1)$:

$$A(C,b) := \sum_{\{x \in \{0,1\}^m : \ C(x)=b\}} \prod_{i \in [m]} K_i^{x_i}.$$

---

[12] We allow a formula to ignore some inputs variables; this enables the model to express hypercubes directly.

Recursively compute $A(C, b)$ as follows:

- If C is a literal for variable x_i, then by definition:

$$A(C, b) = \begin{cases} K_i & \text{if } C = x_i \\ 1 & \text{if } C = \bar{x}_i \end{cases}$$

- Else, if type$(C) = AND$: Let C_L and C_R be the children of C. By hypothesis, we can recursively compute $A(C_L, b)$ and $A(C_R, b)$ for $b \in \{0, 1\}$. Compute $A(C, b)$ as:

$$A(C, 1) = A(C_L, 1) \cdot A(C_R, 1)$$
$$A(C, 0) = A(C_L, 0) \cdot A(C_R, 0) + A(C_L, 1) \cdot A(C_R, 0) + A(C_L, 0) \cdot A(C_R, 1)$$

- Else, type$(C) = OR$: Let C_L and C_R be the children of C. By hypothesis, we can recursively compute $A(C_L, b)$ and $A(C_R, b)$ for $b \in \{0, 1\}$. Compute $A(C, b)$ as:

$$A(C, 1) = A(C_L, 1) \cdot A(C_R, 1) + A(C_L, 1) \cdot A(C_R, 0) + A(C_L, 0) \cdot A(C_R, 1)$$
$$A(C, 0) = A(C_L, 0) \cdot A(C_R, 0))$$

Lemma 4. $A(C, b)$ *as computed above is equal to* $\sum_{\{x \in \{0,1\}^m : C(x) = b\}} \prod_{i \in [m]} K_i^{x_i}$

Proof. For C a literal, the correctness is immediate. We must check the recursion for each type$(C) \in \{AND, OR\}$ and $b \in \{0, 1\}$. We only show the case for $b = 1$ when C is an OR gate; the other three cases can be shown similarly.

Let $S_{b_L, b_R} = \{x = (x_L, x_R) : (C_L(x_L), C_R(x_R)) = (b_L, b_R)\}$ be the set of inputs (x_L, x_R) to C such that $C_L(x_L) = b_L$ and $C_R(x_R) = b_R$. The set $\{x : C(x) = 1\}$ can be decomposed into the disjoint union $S_{0,1} \sqcup S_{1,0} \sqcup S_{1,1}$. Furthermore,

$$A(C, 1) = \sum_{x \in S_{0,1}} \prod_{i \in [m]} K_i^{x_i} + \sum_{x \in S_{1,0}} \prod_{i \in [m]} K_i^{x_i} + \sum_{x \in S_{1,1}} \prod_{i \in [m]} K_i^{x_i}$$

Because C is read-once, the sets of inputs on which C_L and C_R depend are disjoint; this implies that $A(C_L, b_L) \cdot A(C_R, b_R) = \sum_{x \in S_{b_L, b_R}} \prod_{i \in [m]} K_i^{x_i}$, yielding the desired recursion.

Theorem 14. *Let $\epsilon > 0$ be a constant, choose the security parameter $\lambda = \Omega(\ell^{1/\epsilon})$, and assume $(2^{\lambda^\epsilon}, 2^{-\lambda^\epsilon})$-hardness of the DDH assumption. Then, the collection of functions $\mathcal{F}_\lambda$ defined above is a secure aggregate PRF with respect to the subsets R_λ and the product aggregation function over the group $\mathcal{G}$.*

Proof. Correctness is immediate from Lemma 4, and Equation (3). Security follows from the decisional Diffie-Hellman assumption in much the same way it did in the case of bit-fixing functions.

4 Connection to Learning

We now turn a discussion of the connection between augmented PRFs and computational learning. After the preliminaries, we present the learning with aggregate queries model and the corresponding hardness results implied by our constructions of AGG-PRFs. We defer the discussion of learning with restriction queries and access to related concepts to the full version of this paper.

4.1 Preliminaries

Notation: For a probability distribution D over a set X, we denote by $x \leftarrow D$ to mean that x is sampled according to D, and $x \leftarrow X$ to denote uniform sampling form X. For an algorithm A and a function $\mathcal{O}$, we denote that A has oracle access to $\mathcal{O}$ by $A^{\mathcal{O}(\cdot)}$.

We recall the definition of a "concept class". In this section, we will often need to explicitly reason about the representations of the concept classes discussed. Therefore we make use of the notion of a "representation class" as defined by [KV94] alongside that of concept classes. This unified formalization enables us to discuss both these traditional learning models (namely, PAC and learning with membership queries) as well as the new models we present below. Our definitions are parametrized by $\lambda \in \mathbb{N}$.[13]

Definition 5 (Representation class [KV94]). *Let $K = \{K_\lambda\}_{\lambda \in \mathbb{N}}$ be a family of sets, where each $k \in K_\lambda$ has description in $\{0,1\}^{s_k(\lambda)}$ for some polynomial $s_k(\cdot)$. Let $X = \{X_\lambda\}_{\lambda \in \mathbb{N}}$ be a set, where each X_λ is called a* domain *and each $x \in X_\lambda$ has description in $\{0,1\}^{s_x(\lambda)}$ for some polynomial $s_x(\cdot)$. With each λ and each $k \in K_\lambda$, we associate a Boolean function $f_k : X_\lambda \to \{0,1\}$.[14] We call each such function f_k a* concept, *and k its* index *or its* description. *For each λ, we define the* concept class $C_\lambda = \{f_k : k \in K_\lambda\}$ *to be the set of all concepts with index in K_λ. We define the* representation class $C = \{C_\lambda\}$ *to be the union of all concept classes C_λ.*

This formalization allows us to easily associate complexity classes with concepts in learning theory. For example, to capture the set of all DNF formulas on λ inputs with size at most $p(\lambda)$ for a polynomial p, we will let $X_\lambda = \{0,1\}^\lambda$, and $K_\lambda^{p(\lambda)}$ be the set of descriptions of all DNF formulas on λ variables with size at most $p(\lambda)$ under some reasonable representation. Then a concept $f_k(x)$ evaluates the formula k on input x. Finally, $\mathsf{DNF}_\lambda^{p(\lambda)} = \{f_k : k \in K_\lambda^{p(\lambda)}\}$ is the concept class, and $\mathsf{DNF}^{p(\lambda)} = \{DNF_\lambda^{p(\lambda)}\}_{\lambda \in \mathbb{N}}$. $DNF^{p(\lambda)}$ is the representation class that computes all DNF formulas on λ variables with description of size at most $p(\lambda)$ in the given representation.

As a final observation, note that a Boolean-valued PRF family $\mathcal{F} = \{\mathcal{F}_\lambda\}$ where $\mathcal{F}_\lambda = \{f_k : X_\lambda \to \{0,1\}\}$ with keyspace $K = \{K_\lambda\}$ and domain $X = \{X_\lambda\}$

---

[13] When clear from the context, we will omit the subscript λ.

[14] This association is an efficient procedure for evaluating f_k. Concretely, we might consider that there is a universal circuit F_λ such that for each λ, $f_k(\cdot) = F_\lambda(k, \cdot)$.

satisfies the syntax of a representation class as defined above. This formalization is useful precisely because it captures both PRF families and complexity classes, enabling lower bounds in various learning models.

In proving lower bounds for learning representation classes, it will be convenient to have a notion of containment for two representation classes.

Definition 6 ($\subseteq$). *For two representation classes $\mathcal{F} = \{\mathcal{F}_\lambda\}$ and $\mathcal{G} = \{\mathcal{G}_\lambda\}$ on the same domain $X = \{X_\lambda\}$, and with indexing sets $I = \{I_\lambda\}$ and $K = \{K_\lambda\}$ respectively, we say $\mathcal{F} \subseteq \mathcal{G}$ if for all sufficiently large λ, for all $i \in I_\lambda$, there exists $k \in K_\lambda$ such that $g_k \equiv f_i$.*

Informally, if a representation class contains a PRF family, then this class is hard to MQ-learn (as in [Val84]). We apply similar reasoning to more powerful learning models. For example, if $\mathcal{G}$ is the representation class $DNF^{p(\lambda)}$ as defined above, then $\mathcal{F} \subseteq DNF^{p(\lambda)}$ is equivalent to saying that for all sufficiently large λ, the concept class $\mathcal{F}_\lambda$ can be decided by a DNF on λ inputs of $p(\lambda)$ size.

We now recall some standard definitions.

Definition 7 (ϵ-approximation). *Let $f, h : X \to \{0, 1\}$ be arbitrary functions. We say h ϵ-approximates f if $\Pr_{x \leftarrow X}[h(x) \neq f(x)] \leq \epsilon$.*

In general, ϵ-approximation is considered under a general distribution on X, but we will consider only the uniform distribution in this work.

Definition 8 (*PAC Learning*). *For a concept $f : X_\lambda \to \{0, 1\}$, and a probability distribution D_λ over X_λ, the example oracle $EX(f, D_\lambda)$ takes no input and returns $(x, f(x))$ for $x \leftarrow D_\lambda$. An algorithm $\mathcal{A}$ is an (ϵ, δ)-PAC learning algorithm for representation class $\mathcal{C}$ if for all sufficiently large λ, $\epsilon = \epsilon(\lambda) > 0$, $\delta = \delta(\lambda) > 0$ and $f \in \mathcal{C}_\lambda$,*

$$\Pr[\mathcal{A}^{EX(f,D_\lambda)} = h : \ h \text{ is an } \epsilon\text{-approximation to } f] \geq 1 - \delta$$

Definition 9 (*MQ Learning*). *For a concept $f : X_\lambda \to \{0, 1\}$, the membership oracle $MEM(f)$ takes as input a point $x \in X_\lambda$ and returns $f(x)$. An algorithm $\mathcal{A}$ is an (ϵ, δ)-MQ learning algorithm for representation class $\mathcal{C}$ if for all sufficiently large λ, $\epsilon = \epsilon(\lambda) > 0, \delta = \delta(\lambda) > 0$, and $f \in \mathcal{C}_\lambda$,*

$$\Pr[\mathcal{A}^{MEM(f)} = h : \ h \text{ is an } \epsilon\text{-approximation to } f] \geq 1 - \delta$$

We consider only PAC learning *with uniform examples*, where D_λ is the uniform distribution over X_λ. In this case, MQ is strictly stronger than PAC: everything that is PAC learnable is MQ learnable.

Observe that for any $f : X_\lambda \to \{0, 1\}$, either $h(x) = 0$ or $h(x) = 1$ will $\frac{1}{2}$-approximate f. Furthermore, if $\mathcal{A}$ is inefficient, f may be learned exactly. For a learning algorithm to be non-trivial, we require that it is *efficient* in λ, and that it at least *weakly* learns $\mathcal{C}$.

Definition 10 (Efficient- and Weak- Learning).

- $\mathcal{A}$ *is said to be* efficient *if the time complexity of $\mathcal{A}$ and h are polynomial in $1/\epsilon, 1/\delta$, and λ.*
- $\mathcal{A}$ *is said to* weakly *learn $\mathcal{C}$ if there exist some polynomials $p_\epsilon(\lambda), p_\delta(\lambda)$ for which $\epsilon \leq \frac{1}{2} - \frac{1}{p_\epsilon(\lambda)}$ and $\delta \leq 1 - \frac{1}{p_\delta(\lambda)}$.*
- *We say a representation class is* learnable *if it is both efficiently and weakly learnable. Otherwise, it is* hard to learn.

Lastly, we recall the efficiently recognizable ensembles of sets as defined in Section 2. We occasionally call such ensembles indexed, or succinct. Throughout this section, we require this property of our set ensembles $\mathcal{S}$. Both the MQ_{RA} and AQ learning models that we present are defined with respect to $\mathcal{S} = \{\mathcal{S}_\lambda\}$, an efficiently recognizable ensemble of subsets of the domain X_λ.

4.2   Learning with Aggregate Queries

This computational learning model is inspired by our aggregate PRFs. Rather than being a natural model in its own right, this model further illustrates how cryptography and learning are in some senses duals. Here, we consider a new extension to the power of the learning algorithm. Whereas membership queries are of the form "What is the label of an example x?", we grant the learner the power to request the evaluation of simple functions on tuples of examples $(x_1, ..., x_k)$ such as "How many of $(x_1...x_k)$ are in C?" or "Compute the product of the labels of $(x_1, ..., x_k)$?". Clearly, if k is polynomial then this will result only a polynomial gain in the query complexity of a learning algorithm in the best case. Instead, we propose to study cases when k may be super polynomial, but the description of the tuples is succinct. For example, the learning algorithm might query the number of x's in a large interval that are positive examples in the concept.

As with the restriction access and related concept models – and the aggregate PRFs we define in this work – the Aggregate Queries (AQ) learning model will be considered with restrictions to both the types of aggregate functions Γ the learner can query, and the sets $\mathcal{S}$ over which the learner may request these functions to be evaluated on. We now present the AQ learning model informally:

Definition 11 (($\Gamma, \mathcal{S}$)-Aggregate Queries (AQ) Learning). *Let $\mathcal{C}$ be a representation class with domains $X = \{X_\lambda\}$, and $\mathcal{S} = \{\mathcal{S}_\lambda\}$ where each $\mathcal{S}_\lambda$ is a collection of efficiently recognizeable subsets of the X_λ. $\Gamma : \{0,1\}^* \to V_\lambda$ be an aggregation function [as in def.]. Let $\mathsf{AGG}_k^\lambda \triangleq \mathsf{AGG}_{f_k, \mathcal{S}_\lambda, \Gamma_\lambda}^\lambda$ be the aggregation oracle for $f_k \in \mathcal{C}_\lambda$, for $S \in \mathcal{S}_\lambda$ and Γ_λ.*

An algorithm $\mathcal{A}$ is an (ϵ, δ)-($\Gamma, \mathcal{S}$)-AQ learning algorithm for $\mathcal{C}$ if, for all sufficiently large λ, for every $f_k \in \mathcal{C}_\lambda$, $\Pr[\mathcal{A}^{MEM_{f_k}(\cdot), \mathsf{AGG}_k^\lambda(\cdot)} = h] \geq 1 - \delta$ where h is an ϵ-approximation to f_k.

Hardness of Aggregate Query Learning

Theorem 15. *Let $\mathcal{F}$ be a boolean-valued aggregate PRF with respect to set system $\mathcal{S} = \{S_\lambda\}$ and accumulation function $\Gamma = \{\Gamma_\lambda\}$. For a representation class $\mathcal{C}$, if $\mathcal{F} \subseteq \mathcal{C}$, then $\mathcal{C}$ is hard to $(\Gamma, \mathcal{S})$-AQ learn.*

Looking back to our constructions of aggregate pseudorandom function families from the prequel, we have the following corollaries.

Corollary 4. *The existence of one-way functions implies that $P/poly$ is hard to $(\sum, \mathcal{S}_{[a,b]})$-AQ learn, with $\mathcal{S}_{[a,b]}$ the set of sub-intervals of the domain as defined in Section 3.*

Corollary 5. *The DDH Assumption implies that NC^1 is hard to $(\sum, \mathcal{S}_{[a,b]})$-AQ learn, with $\mathcal{S}_{[a,b]}$ the set of sub-intervals of the domain as defined in Section 3.*

Corollary 6. *The subexponential DDH Assumption implies that NC^1 is hard to $(\prod, \mathcal{R})$-AQ learn, with $\mathcal{R}$ the set of read-once boolean formulas defined in Section 3.*

Proof (Proof of Theorem 15). Interpreting $\mathcal{F}$ itself as a concept class, we will show an efficient reduction from violating the pseudorandomness property of $\mathcal{F}$ to weakly $(\Gamma, \mathcal{S})$-AQ learning $\mathcal{F}$. By assumption, $\mathcal{F} \subseteq \mathcal{C}$, implying that $\mathcal{C}$ is hard to learn as well.

Reduction: Suppose for contradiction that there exists an efficient weak learning algorithm $\mathcal{A}$ for $\mathcal{F}$. We define algorithm $\mathcal{B}$ violating the aggregate PRF security of $\mathcal{F}$. In the PRF security game, $\mathcal{B}$ is presented with two oracles: $F(\cdot)$ and AGG_F^λ for a function F chosen according to the secret bit $b \in \{0,1\}$. In EXP(0), $F = f_k$ for random $k \in K_\lambda$; by assumption $f_k \in \mathcal{C}_\lambda$. In EXP(1), F is a uniformly random function from X to $\{0,1\}$. The learning algorithm $\mathcal{A}$ is presented with precisely the same oracles. $\mathcal{B}$ runs $\mathcal{A}$, simulating its oracles by passing queries and responses to its own oracles. $X_\mathcal{A} = \{x \in X_\lambda : \mathcal{A} \text{ queried } (\psi, x) \text{ for some } \psi\}$. Once $\mathcal{A}$ terminates, it outputs hypothesis h.

After receiving hypothesis h, $\mathcal{B}$ estimates the probability

$$p = \Pr_{x \leftarrow X \backslash X_\mathcal{A}}[h(x) = F(x)]$$

(using polynomial in $\lambda, p_\epsilon(\lambda)$ samples). In EXP(0), this probability is at least $1 - \epsilon$ with probability at least $1 - \delta$; in EXP(1), it is exactly $1/2$. To sample uniform $x \in X \backslash X_\mathcal{A}$, we simply take a uniform $x \in X$: with high probability $x \in X \backslash X_\mathcal{A}$. If the estimate is close to ϵ, guess EXP(0); otherwise, flip an fair coin $b' \in \{0,1\}$ and guess EXP(b'). The advantage ADV_λ^{APRF} of $\mathcal{B}$ in the PRF security game is at least $\frac{1}{3p_\delta(n)}$ for all sufficiently large λ (as shown below), directly violating the security of $\mathcal{F}$.

Let

$$p_b \triangleq \Pr_{x \in X \backslash X_\mathcal{A}}[h(x) \neq F(x)|EXP(b)]$$

be the probability taken with respect to experiment EXP(b). In EXP(1), F is a uniformly random function. Thus, $p_1 = \frac{1}{2}$. With high probability, $\mathcal{B}$ will output a random bit $b' \in \{0, 1\}$, guessing correctly with probability $1/2$.

In EXP(0), h is an ϵ-approximation to F with probability at least $1 - \delta$. In this case, $p_0 \geq 1 - \epsilon \geq \frac{1}{2} + \frac{1}{p_\epsilon(\lambda)}$. By a Hoeffding bound, $\mathcal{B}$ will guess $b' = 0$ with high probability by estimating p using only polynomial in λ, $p_\epsilon(\lambda)$ samples. On the other hand, if h is not an ϵ-approximation, $\mathcal{B}$ will $b' = 0$ with probability at least $1/2$.

Let $\mathrm{negl}(\lambda)$ be the error probability from the Hoeffding bound, which can be made exponentially small in λ. The success probability is:

$$\Pr[b = b' | b = 0] \geq (1 - \delta)(1 - \mathrm{negl}(\lambda)) + \frac{\delta}{2}$$

which, for $1 - \delta \geq \frac{1}{p_\delta(\lambda)}$ is at least $\frac{1}{3p_\delta(\lambda)} + \frac{1}{2}$ for sufficiently large λ. Thus $\mathcal{B}$ a non-negligible advantage of $1/3p_{\delta(\lambda)}$ in the $(\Gamma, \mathcal{S})$-aggregate-PRF security game.

Acknowledgements. Aloni Cohen's research was supported in part by NSF Graduate Student Fellowship and NSF grants CNS1347364, CNS1413920 and FA875011-20225.

Shafi Goldwasser's research was supported in part by NSF Grants CNS1347364, CNS1413920 and FA875011-20225.

Vinod Vaikuntanathan's research was supported by DARPA Grant number FA8750-11-2-0225, an Alfred P. Sloan Research Fellowship, an NSF CAREER Award CNS-1350619, NSF Frontier Grant CNS-1414119, a Microsoft Faculty Fellowship, and a Steven and Renee Finn Career Development Chair from MIT.

References

[ABPP14] Abdalla, M., Benhamouda, F., Passelègue, A., Paterson, K.G.: Related-key security for pseudorandom functions beyond the linear barrier. In: Garay, J.A., Gennaro, R. (eds.) CRYPTO 2014, Part I. LNCS, vol. 8616, pp. 77–94. Springer, Heidelberg (2014)

[BC10] Bellare, M., Cash, D.: Pseudorandom functions and permutations provably secure against related-key attacks. In: Rabin, T. (ed.) CRYPTO 2010. LNCS, vol. 6223, pp. 666–684. Springer, Heidelberg (2010)

[BF02] Bshouty, N.H., Feldman, V.: On using extended statistical queries to avoid membership queries. The Journal of Machine Learning Research 2, 359–395 (2002)

[BGI14a] Boyle, E., Goldwasser, S., Ivan, I.: Functional signatures and pseudorandom functions. In: Krawczyk [Kra14], pp. 501–519

[BGI14b] Boyle, E., Goldwasser, S., Ivan, I.: Functional signatures and pseudorandom functions. In: Krawczyk [Kra14], pp. 501–519

[BGV11] Benabbas, S., Gennaro, R., Vahlis, Y.: Verifiable delegation of computation over large datasets. In: Rogaway, P. (ed.) CRYPTO 2011. LNCS, vol. 6841, pp. 111–131. Springer, Heidelberg (2011)

[BLMR13]   Boneh, D., Lewi, K., Montgomery, H., Raghunathan, A.: Key homomorphic PRFs and their applications. In: Canetti, R., Garay, J.A. (eds.) CRYPTO 2013, Part I. LNCS, vol. 8042, pp. 410–428. Springer, Heidelberg (2013)

[BW13a]   Boneh, D., Waters, B.: Constrained pseudorandom functions and their applications. In: Sako and Sarkar [SS13], pp. 280–300

[BW13b]   Boneh, D., Waters, B.: Constrained pseudorandom functions and their applications. In: Sako and Sarkar [SS13], pp. 280–300

[DRWY12]   Dvir, Z., Rao, A., Wigderson, A., Yehudayoff, A.: Restriction access. In: Goldwasser, S. (ed.) Innovations in Theoretical Computer Science 2012, Cambridge, MA, USA, January 8-10, pp. 19–33. ACM (2012)

[GGI⁺02]   Gilbert, A.C., Guha, S., Indyk, P., Kotidis, Y., Muthukrishnan, S., Strauss, M.: Fast, small-space algorithms for approximate histogram maintenance. In: Proceedings on 34th Annual ACM Symposium on Theory of Computing, Montréal, Québec, Canada, May 19-21, pp. 389–398 (2002)

[GGM86]   Goldreich, O., Goldwasser, S., Micali, S.: How to construct random functions. J. ACM, 33(4):792–807 (1986); Extended abstract in FOCS 84

[GGN10]   Goldreich, O., Goldwasser, S., Nussboim, A.: On the implementation of huge random objects. SIAM J. Comput. 39(7), 2761–2822 (2010)

[HILL99]   Håstad, J., Impagliazzo, R., Levin, L.A., Luby, M.: A pseudorandom generator from any one-way function. SIAM J. Comput. 28(4), 1364–1396 (1999)

[KPTZ13a] Kiayias, A., Papadopoulos, S., Triandopoulos, N., Zacharias, T.: Delegatable pseudorandom functions and applications. In: Sadeghi et al. [SGY13], pp. 669–684

[KPTZ13b] Kiayias, A., Papadopoulos, S., Triandopoulos, N., Zacharias, T.: Delegatable pseudorandom functions and applications. In: Sadeghi et al. [SGY13], pp. 669–684

[Kra14]   Krawczyk, H. (ed.): PKC 2014. LNCS, vol. 8383. Springer, Heidelberg (2014)

[KV94]   Kearns, M.J., Valiant, L.G.: Cryptographic limitations on learning boolean formulae and finite automata. J. ACM 41(1), 67–95 (1994)

[NR04]   Naor, M., Reingold, O.: Number-theoretic constructions of efficient pseudo-random functions. J. ACM 51(2), 231–262 (2004)

[RR97]   Razborov, A.A., Rudich, S.: Natural proofs. J. Comput. Syst. Sci. 55(1), 24–35 (1997)

[SGY13]   Sadeghi, A.-R., Gligor, V.D., Yung, M. (eds.): 2013 ACM SIGSAC Conference on Computer and Communications Security, CCS 2013, Berlin, Germany, November 4-8. ACM (2013)

[SS13]   Sako, K., Sarkar, P. (eds.): ASIACRYPT 2013, Part II. LNCS, vol. 8270, pp. 2013–2019. Springer, Heidelberg (2013)

[SW14]   Sahai, A., Waters, B.: How to use indistinguishability obfuscation: deniable encryption, and more. In: Shmoys, D.B. (ed.) Symposium on Theory of Computing, STOC 2014, May 31-June 03, pp. 475–484. ACM, New York (2014)

[Val84]   Leslie, G.: Valiant. A theory of the learnable. Communications of the ACM 27(11), 1134–1142 (1984)

[VV86]   Valiant, L.G., Vazirani, V.V.: NP is as easy as detecting unique solutions. Theor. Comput. Sci. 47(3), 85–93 (1986)

Oblivious Polynomial Evaluation and Secure Set-Intersection from Algebraic PRFs

Carmit Hazay*

Faculty of Engineering, Bar-Ilan University, Israel
carmit.hazay@biu.ac.il

Abstract. In this paper we study the two fundamental functionalities *oblivious polynomial evaluation in the exponent* and *set-intersection*, and introduce a new technique for designing efficient secure protocols for these problems (and others). Our starting point is the [6] technique (CRYPTO 2011) for verifiable delegation of polynomial evaluations, using *algebraic PRFs*. We use this tool, that is useful to achieve *verifiability* in the *outsourced setting*, in order to achieve *privacy* in the *standard two-party* setting. Our results imply new simple and efficient oblivious polynomial evaluation (OPE) protocols. We further show that our OPE protocols are readily used for secure set-intersection, implying much simpler protocols in the plain model. As a side result, we demonstrate the usefulness of algebraic PRFs for various search functionalities, such as keyword search and oblivious transfer with adaptive queries. Our protocols are secure under full simulation-based definitions in the presence of malicious adversaries.

Keywords: Efficient Secure Computation, Oblivious Polynomial Evaluation, Secure Set-Intersection, Committed Oblivious PRF.

1 Introduction

Efficient secure two-party computation. Secure two-party computation enables two parties to mutually run a protocol that computes some function f on their private inputs, while preserving a number of security properties. Two of the most important properties are privacy and correctness. The former implies data confidentiality, namely, nothing leaks by the protocol execution but the computed output. The latter requirement implies that the protocol enforces the integrity of the computations made by the parties, namely, honest parties learn the correct output. Feasibility results are well established [49,23,39,5], proving that any efficient functionality can be securely computed under full simulation-based definitions (following the ideal/real paradigm). Security is typically proven with respect to two adversarial models: the semi-honest model (where the adversary follows the instructions of the protocol but tries to learn more than it should from the protocol transcript), and the malicious model (where the adversary follows an arbitrary polynomial-time strategy), and feasibility holds in the presence of both types of attacks.

---

* Research partially supported by a grant from the Israel Ministry of Science and Technology (grant No. 3-10883).

Y. Dodis and J.B. Nielsen (Eds.): TCC 2015, Part II, LNCS 9015, pp. 90–120, 2015.

Following these works, many constructions focused on improving the *efficiency* of the computational and communication costs. Conceptually, this line of works can be split into two sub-lines: **(1)** Improved generic protocols that compute any boolean or arithmetic circuit; see [47,30,44,36,38,7,16,43] for just a few examples. **(2)** Protocols for concrete functionalities. In the latter approach attention is given to constructing efficient protocols for specific functions while exploiting their internal structure. This approach has been proven useful for many different functions in both the semi-honest and malicious settings. Notable examples are calculating the kth ranked element [1], pattern matching and related search problems [29,48], set-intersection [31,28] and oblivious pseudorandom function (PRF) evaluation [20].

In this paper we study the two fundamental functionalities *oblivious polynomial evaluation in the exponent* and *set-intersection* and introduce a new technique for designing efficient secure protocols for these problems in the presence of semi-honest and malicious attacks with simulation-based security proofs. We further demonstrate that our technique is useful for various search functionalities.

Algebraic PRFs. Informally, an algebraic pseudorandom function (PRF) is a PRF with a range that forms an Abelian group such that group operations are efficiently computable. In addition, certain algebraic operations on these outputs can be computed significantly more efficiently if one possesses the key of the pseudorandom function that was used to generate them. This property is denoted by *closed form efficiency* and allows to compute a batch of l PRF values much more efficiently than by computing the l values separately and then combing them. Algebraic PRFs were exploited in [6] to achieve faster verifiable polynomial evaluations (in the exponent). Specifically, in their setting, a client outsources a d-degree polynomial to an untrusted server together with some authenticating information, while the client stores a short secret key. Next, when the client provides an input for this polynomial the server computes the result and an authentication message that allows the client to verify this computation in sub-linear time in d.

More concretely, let $Q(\cdot) = (q_0, \ldots, q_d)$ be the polynomial stored on the server in the clear. Then the client additionally stores a vector of group elements $\{g^{aq_i+r_i}\}_{i=0}^d$ where $a \leftarrow \mathbb{Z}_p$ and p is a prime, and r_i is the ith coefficient of a polynomial $R(\cdot)$ of the same degree as $Q(\cdot)$. Then for every client's input t the server returns $y = Q(t)$ and $u = g^{aQ(t)+R(t)}$ and the client accepts u if and only if $u = g^{ay+R(t)}$. Interestingly, in case $g^{r_i} = \mathsf{PRF}_K(i)$, where PRF is an algebraic PRF, the closed form efficiency property enables the client to compute the value $g^{R(t)}$ in sub-linear time in d. Stated differently, verifiability is achieved by viewing $g^{aq_i+r_i}$ as a (one-time) message authentication code (MAC) for g^{q_i} where batch verification of multiple MACs can be computed more efficiently than verifying each MAC separately.

In this work we demonstrate the usefulness of algebraic PRFs for various two-party problems by designing secure protocols based on this primitive. In particular, we modify the way [6] use algebraic PRFs so that instead of achieving *verifiability* in the *outsourced setting*, we achieve *privacy* in the *standard two-party* setting. It is worth noting that although the main focus of [6] is correctness, they do discuss how to achieve one-sided privacy by encrypting the coefficients of the polynomial (since the polynomial must be specified explicitly). Nevertheless, it is not clear how to maintain the privacy of the input to the polynomial in their protocol. In this work, we use algebraic PRFs to mask the polynomial in a different way that does not allow the verifiability of the polynomial evaluation but allows the extractability of the polynomial more easily, and demonstrate an alternative way to achieve correctness. We focus our attention on the *plain model* where no trusted setup is required.

Oblivious polynomial evaluation. The oblivious polynomial evaluation (OPE) functionality is an important functionality in the field of secure two-party computation. It considers a setting where party P_0 holds a polynomial $Q(\cdot)$ and party P_1 holds an element t, and the goal is that P_1 obtains $Q(t)$ and nothing else while P_0 learns nothing. OPE has proven to be a useful building block and can be used to solve numerous cryptographic problems; e.g., secure equality of strings, set-intersection, approximation of a Taylor series, RSA key generation, oblivious keyword search, set membership, data entanglement and more [22,37,21,20,41,3].

Despite its broad applicability the study of OPE was demonstrated using only few concrete secure protocols, initiated in [40] and further continued in [9,50,24]. In particular, the only protocol with a complete simulation-based proof in the presence of malicious attacks is the protocol in [24]. This protocol evaluates a d-degree polynomial over a composite order group $\mathbb{Z}_N$ with $O(sd)$ modular exponentiations, where N is an RSA composite and s is a statistical security parameter.

The general (and currently the most practical) approach of [16,15] for arithmetic circuits follows the preprocessing model: in an offline phase some shared randomness is generated independently of the function and the inputs; in an online phase the actual secure computation is performed. One of the main advantages of these protocols is that the basic operations are almost as cheap as those used in the passively secure protocols. To get good performance, these protocols use the somewhat-homomorphic SIMD approach that handles many values in parallel in a single ciphertext, and thus more applicable for large degree polynomials. Similarly, protocols for Boolean circuits apply the cut-and-choose technique which requires to repeat the computation s times in order to prevent cheating except with probability 2^{-s} [35].

In some applications such as password-based authenticate key exchange protocols or when sampling an element from a d-wise independence space, the polynomial degree is typically small and even a constant. In these cases, our protocols have clear benefits since they are much simpler, efficient and easily implementable.

Secure set-intersection. In the set-intersection problem parties P_0, P_1, holding input sets X, Y of sizes m_X and m_Y, respectively, wish to compute $X \cap Y$. This problem has been intensively studied by researchers in the last few years mainly due to its potential applications for dating services, datamining, recommendation systems, law enforcement and more; see [21,34,13,31,32,25,28] for a few examples. For instance, consider two security agencies that wish to compare their lists of suspects without revealing their contents, or an airline company that would like to check its list of passengers against the list of people that are not allowed to go abroad.

Two common approaches are known to solve this problem securely in the plain model: (**1**) oblivious polynomial evaluation and (**2**) committed oblivious PRF evaluation. In the former approach party P_0 computes a polynomial $Q(\cdot)$ such that $Q(x) = 0$

for all $x \in X$. This polynomial is then encrypted using homomorphic encryption and sent to P_1, that computes the encryption of $r_y \cdot Q(y) + y$ for all $y \in Y$, and using fresh randomness r_y. This approach (or a variant of it) was taken in [21,34,13,28].

The second approach uses a secure implementation of oblivious pseudorandom function evaluation. Namely, P_0 chooses a PRF key K and computes the set $\mathsf{PRF}_X = \{\mathsf{PRF}_K(x)\}_{x \in X}$. The parties then execute an oblivious PRF protocol where P_0 inputs K, whereas P_1 inputs the set Y and learns the set $\mathsf{PRF}_Y = \{\mathsf{PRF}_K(y)\}_{y \in Y}$. Finally, P_0 sends the set PRF_X to P_1 that computes $\mathsf{PRF}_X \cap \mathsf{PRF}_Y$ and extracts the actual intersection. This idea was introduced in [20] and further used in [25,31,32]. Other solutions in the random oracle model such as [12,11,2] take a different approach by applying the random oracle on (one of) the sets members, or apply oblivious transfer extension [18].

In a recent result [45], the authors overview exiting solutions for set-intersection in the semi-honest setting and compare their efficiency. One of their conclusions is that OPE-based approaches are inferior to oblivious-transfer extension based approaches. It is an interesting question to test whether this conclusion also for the case for the malicious setting as well.

To the best of our knowledge, the most efficient protocol in the malicious plain model that does not require a trusted setup or rely on non-standard assumptions is the protocol of [28] that incurs computation of $O(m_X + m_Y \log(m_X + m_Y))$ modular exponentiations. A more efficient protocol with $O(m_X + m_Y)$ communication and computational costs was introduced by [31] in the common reference string (CRS) model (where the CRS includes a safe RSA composite that determines the group order and implies high overhead when mutually produced). Another drawback of this protocol is that its security proof runs an exhaustive search on the input domain of the PRF in order to extract P_0's input. This implies that the proof works for small domain PRFs and that the complexity of the simulator grows linearly with the size of the PRF's input domain.

Committed oblivious PRF evaluation. The oblivious PRF evaluation functionality $\mathcal{F}_{\mathrm{PRF}}$ that *obliviously* evaluates a PRF is defined by $(K, x) \mapsto (-, \mathsf{PRF}_K(x))$. This functionality is very important in the context of secure computation since it essentially implements a random oracle. That is, the party with the PRF key, say P_0, mimics the random oracle role via interaction. Therefore, if the protocol that realizes $\mathcal{F}_{\mathrm{PRF}}$ is simulation-based secure then both desirable properties of a random oracle, *programmability* and *observability*, can be achieved by this protocol. First, since the simulator can force any output for a corrupted P_1, it essentially programs the function's output. In addition, it can also observe (via extraction) the input to the functionality. Nevertheless, the usefulness of oblivious PRF evaluation is reflected via an additional property of committed key that implies that the *same key* is used for multiple PRF evaluations.

Committed oblivious PRF (CPRF) evaluation has been used to compute secure set-intersection [31,25], oblivious transfer with adaptive queries [20], keyword search [20], pattern matching [25,19] and more. It is therefore highly important to design efficient protocols for this functionality. Current implementations of the [42] algebraic PRF, discussed in this paper, employ an oblivious transfer protocol for each input bit [20,25] and are only secure for a single PRF evaluation. Consequently, the protocol of [25] does not achieve full security against malicious adversaries. In addition, the protocol from [31] (that implements a variant of the [17] PRF) requires a trusted setup of a safe RSA composite and suffers from the drawbacks specified above.

1.1 Our Results

In this paper we use algebraic PRFs to design alternative simple and efficient protocols for polynomial evaluation, set-intersection, committed oblivious PRF evaluation and search problems. Below, we demonstrate the broad usefulness of our technique.

Oblivious polynomial evaluation (Section 3). We present secure protocols in the plain model for OPE *in the exponent* with simulation-based security against semi-honest and malicious attacks. We stress that evaluating a polynomial in the exponent has strong applicability in the context of set membership where the goal is to privately verify membership in some secret set, as well as achieving d-wise independence. We use algebraic PRFs to build simple two-phases OPE protocols as follows. In the first phase party P_0, holding the polynomial $g^{Q(\cdot)}$, publishes its masked polynomial $g^{Q(\cdot)+R(\cdot)}$ where the set $g^{R(\cdot)}$ is determined by an algebraic PRF. Next, P_1 locally computes $g^{Q(t)+R(t)}$ and the parties run an unmasking secure computation for obliviously evaluating $g^{R(t)}$ for P_1.

The efficiency of the latter phase is dominated by the overhead of the closed form efficiency property of the specific PRF. In this work, we consider two PRF implementations used by [6]: (**1**) a PRF with security under the strong-DDH assumption. (**2**) The Naor-Reingold PRF [42] with security under the DDH assumption. More concretely, the efficiency of our protocols is only $d + 1$ modular exponentiations for the first phase of sending the masked polynomial, and $d+1+O(1)$ (resp. $O(\log d)$) modular exponentiations for the second phase of obliviously evaluating the pseudorandom polynomial under the strong-DDH (resp. DDH) assumption. For simplicity, we only consider univariate polynomials. Our technique can be applied for multivariate polynomials as well (with total degree d or of degree d in each variable); see [6] for further details. To the best of our knowledge, our protocols are the first to obliviously evaluate both univariate and multivariate polynomials that efficiently.

Secure set-intersection (Section 4). In this work we demonstrate that algebraic PRFs are useful for both approaches of OPE and committed oblivious PRF that enable to design set-intersection protocols. We first show that our protocols for OPE readily induce secure protocols for set-intersection. That is, first P_0 encodes the set X by a polynomial $g^{Q(\cdot)}$ as specified above, and masks it. Next, for each $y \in Y$ party P_1 verifies whether the masked polynomial evaluation of y equals the evaluation $g^{R(y)}$, and concludes whether the element is in the intersection. We stress that this naive approach requires a multiplicative overhead (in the sets sizes) since for each element in its input Y, P_1 needs to evaluate a polynomial of degree m_X. To reduce the computational overhead, Freedman et al. [21] introduced a balanced allocation scheme [4] into their protocol that splits the elements into $\mathcal{B} = \frac{m_X}{\log \log m_X}$ bins, with maximum number of $\mathcal{M} = O(m_X/\mathcal{B} + \log \log \mathcal{B}) = O(\log \log m_X)$ elements in each bin. In that case, the elements mapped by P_0 to a certain bin must only be compared to those mapped by P_1 to the same bin. Therefore, P_1 should only evaluate an $\mathcal{M}$-degree polynomial for each $y \in Y$, rather than a polynomial of degree m_X. Nevertheless, their solution with hash functions is only applicable in the semi-honest setting. Following that, Hazay and Nissim [28] introduced a maliciously secure protocol which implies the computation of $O(m_X + m_Y \log(m_X + m_Y))$ modular exponentiations. Their construction is fairly complicated and combines both approaches of OPE and oblivious PRF evaluation.

We introduce the hashing technique into our constructions and provide a generic description that can be instantiated with different hash functions. Our protocols are far less complicated and maintain a modular description. Specifically, we devise an alternative zero-knowledge proof for verifying the correctness of the hashed polynomials while exploiting the algebraic properties of the PRF. Under the strong-DDH assumption our protocol matches the communication overhead of the protocol from [31] (that also relies on a dynamic hardness assumption) and implies the computation of $O(m_X + m_Y \log \log m_X)$ exponentiations, with the benefits that it operates over prime order groups, it does not require a trusted setup and the proof complexity does not depend on the PRF's input domain size. Under the DDH assumption our protocol, using hash functions, implies the computation of $O(m_X + m_Y \log m_X)$ exponentiations which improves the overhead of the [28] protocol. Next we show that algebraic PRFs are useful for applications that rely on committed oblivious PRF evaluation. Our results for set-intersection are summarized in Table 1.

Committed oblivious PRF evaluation (Section 5). Observing that the batch computation for l PRF values $\text{PRF}'_K(x) = \prod_{i=0}^{l} [\text{PRF}_K(i)]^{x^i}$ is a PRF as well (by fixing l properly), we derive new PRF constructions in prime order groups and more interestingly, simple committed oblivious PRF evaluation protocols. Our strong-DDH based PRF requires constant overhead, and our DDH-based protocol is the *first* committed oblivious PRF implementation for the [42] function. Our protocols using committed oblivious PRF imply set-intersection protocols with $O(m_X + m_Y)$ costs under the strong-DDH assumption and $((m_X + m_Y) \log(m_X + m_Y))$ communication and computation costs under the DDH assumption, where the former analysis matches the overhead from [31]. In particular, plugging-in our protocols for committed oblivious PRF evaluation in the protocols cited above implies malicious security fairly immediately. Finally, we note that committed oblivious PRF evaluation is also useful for search functionalities that support database search and data retrievals, such as in keyword search and oblivious transfer with adaptive queries.

2 Preliminaries

2.1 Basic Notations

We denote the security parameter by n. We say that a function $\mu : \mathbb{N} \to \mathbb{N}$ is *negligible* if for every positive polynomial $p(\cdot)$ and all sufficiently large n it holds that $\mu(n) < \frac{1}{p(n)}$. We use the abbreviation PPT to denote probabilistic polynomial-time. We further denote by $a \leftarrow A$ the random sampling of a from a distribution A, by $[d]$ the set of elements $(1, \ldots, d)$ and by $[0, d]$ the set of elements $(0, \ldots, d)$.

We define a d-degree polynomial $Q(\cdot)$ by its set of coefficients $(q_0, \ldots, q_d)$, or simply write $Q(x) = q_0 + q_1 x + \ldots q_d x^d$. Typically, these coefficients will be picked

Table 1. Comparisons with secure set-intersection constructions. We highlight the constructions with the best performance under each assumption.

Reference	Modeling	Hardness Assumption	Overhead (Number of Exp.)
[31]	CRS of a safe prime	Decisional d-DHI	$O(m_X + m_Y)$
[28]	plain model	DDH	$O(m_X + m_Y \log(m_X + m_Y))$
[18]	random oracle	random oracle	$O(n)$, where n is sec. parameter
This Work – OPE	plain model	d-strong DDH	$O(m_X + m_Y \log \log m_X)$
This Work – OPE	**plain model**	**DDH**	$\mathbf{O(m_X + m_Y \log m_X)}$
This Work – CPRF	**plain model**	**d-strong DDH**	$\mathbf{O(m_X + m_Y)}$
This Work – CPRF	plain model	DDH	$O((m_X + m_Y) \log(m_X + m_Y))$

from $\mathbb{Z}_p$ for a prime p. We further write $g^{Q(\cdot)}$ to denote the coefficients of $Q(\cdot)$ in the exponent of a generator g of a multiplicative group $\mathbb{G}$ of prime order p.

2.2   Zero-Knowledge Proofs

To prevent malicious behavior, the parties must demonstrate that they are well-behaved. To achieve this, our protocols utilize zero-knowledge (ZK) proofs of knowledge. Our proofs are Σ-protocols with a constant overhead. A generic efficient technique that enables to transform any Σ-protocol into a zero-knowledge proof of knowledge can be found in [26]. This transformation requires additional 6 exponentiations.

1. π_{DL}, for demonstrating the knowledge of a solution x to a discrete logarithm problem [46].
$$\mathcal{R}_{\mathrm{DL}} = \{((\mathbb{G}, g, h), x) \mid h = g^x\}.$$

2. π_{DDH}, for demonstrating that an El Gamal ciphertext is an encryption of zero [10].
$$\mathcal{R}_{\mathrm{DDH}} = \{((\mathbb{G}, g, h, g_1, h_1), x) \mid g_1 = g^x \wedge h_1 = h^x\}.$$

3. π_{MULT}, for proving that a ciphertext c_2 encrypts a product of two plaintexts values. Namely,
$$\mathcal{R}_{\mathrm{MULT}} = \left\{ ((\mathbb{G}, \mathrm{PK}, c_0, c_1, c_2), (a_0, a_1, r_0, r_1, r_2,)) \left| \begin{array}{l} c_i = \mathsf{Enc}_{\mathrm{PK}}(a_i; r_i) \\ \text{for } i \in \{0, 1\} \\ \wedge\, c_2 = \mathsf{Enc}_{\mathrm{PK}}(a_0 \cdot a_1; r_2) \end{array} \right. \right\}$$

where multiplication is performed in the corresponding plaintext group. A zero-knowledge proof for the El Gamal PKE, that is based on the Damgård and Jurik technique [14], can be found in [28].

4. π_{Eq}, for demonstrating equality of two exponentiations. Namely,
$$\mathcal{R}_{\mathrm{Eq}} = \left\{ ((\mathrm{PK}, c_1, c_1', c_2, c_2'), (m, r_1, r_2)) \left| \begin{array}{l} c_1' = c_1^m \cdot \mathsf{Enc}_{\mathrm{PK}}(0; r_1) \\ \wedge c_2' = c_2^m \cdot \mathsf{Enc}_{\mathrm{PK}}(0; r_2) \end{array} \right. \right\}$$

where exponentiation, as well as multiplication with an encryption of zero, are computed componentwise. A variant of this zero-knowledge proof was presented and discussed in [27] for Paillier encryption scheme and can be easily extended for this relation as well. We leave the details of this proof to the full version.

3   Protocols for Oblivious Polynomial Evaluation

In this section we introduce our new constructions for oblivious polynomial evaluation (OPE) in the exponent, implementing functionality $\mathcal{F}_{\text{OPE}} : (g^{Q(\cdot)}, t) \mapsto (-, g^{Q(t)})$ for $Q(\cdot) = (q_0, \ldots, q_d)$. In particular, we assume common knowledge of the public parameters: a multiplicative group $\mathbb{G}$ of order p and a generator g for $\mathbb{G}$, and that the polynomial coefficients are in $\mathbb{Z}_p$. In our solution, party P_0 generates these parameters and publishes its masked polynomial $g^{Q(\cdot)+R(\cdot)}$, where the set of values $g^{R(\cdot)}$ is determined by an algebraic PRF that has a closed form efficient computation for univariate polynomials (see Section 3.1). Next, P_1 computes $g^{Q(t)+R(t)}$ and the parties run an unmasking secure computation for obliviously evaluating $g^{R(t)}$ for P_1. Importantly, the closed form efficiency property of the PRF allows the parties to mutually compute $g^{R(t)}$ in *sub-linear* time in d. Before presenting our OPE constructions we formally define algebraic pseudorandom functions.

3.1   Algebraic Pseudorandom Functions [6]

Algebraic PRFs are PRFs with two additional algebraic properties. First, they map their inputs into some Abelian group, where certain algebraic operations on these outputs can be computed signicantly faster if one possesses the PRF key. These properties were exploited in [6] to achieve faster polynomial evaluations (in the exponent), where the coefficients of these polynomials lie in the PRF range. Several constructions, implying different overheads, were introduced in [6]; we focus our attention on their constructions for univariate polynomials. Our protocols can be applied for multivariate polynomials as well (with total degree d or of degree d in each variable). We begin with the formal definition of algebraic PRFs.

Definition 3.1 (Algebraic PRFs). *We say that* $\mathcal{PRF} = (\text{KeyGen}, \text{PRF}, \text{CFEval})$, *is an algebraic PRF if* KeyGen, PRF *are polynomial-time algorithms specified as follows:*

- KeyGen, *given a security parameter* 1^n, *and a parameter* $m \in \mathbb{N}$ *that determines the domain size of the PRF, outputs a pair* $(K, param) \leftarrow \mathcal{K}_n$, *where* $\mathcal{K}_n$ *is the key space for a security parameter* n. K *is the secret key of the PRF, and* $param$ *encodes the public parameters.*
- PRF, *given a key* K, *public parameters* $param$, *and an input* $x \in \{0,1\}^m$, *outputs a value* $y \in Y$, *where* Y *is some set determined by* $param$.
- *In addition, the following properties hold:*
  Pseudorandomness. *We say that* $\mathcal{PRF}$ *is pseudorandom if for every PPT adversary* $\mathcal{A}$, *and every polynomial* $m = m(n)$, *there exists a negligible function* negl *such that*

$$|\Pr[\mathcal{A}^{\text{PRF}_K(\cdot)}(1^n, param) = 1] - \Pr[\mathcal{A}^{f_n(\cdot)}(1^n, param) = 1]| \leq \text{negl}(n),$$

where $(K, param) \leftarrow \mathsf{KeyGen}(1^n, m)$ and $f_n : \{0,1\}^m \mapsto Y$ is a random function.

Algebraic. We say that $\mathcal{PRF}$ is algebraic if the range Y of $\mathsf{PRF}_K(\cdot)$ for every $n \in \mathbb{N}$ and $(K, param) \leftarrow \mathcal{K}_n$ forms an Abelian multiplicative group. We require that the group operation on Y be efficiently computable given $param$.

Closed form efficiency. Let N be the order of the range sets of PRF for security parameter n. Let $z = (z_1, \ldots, z_l) \in (\{0,1\}^m)^l$, $k \in \mathbb{N}$, and efficiently computable $h : \mathbb{Z}_N^k \mapsto \mathbb{Z}_N^l$ with $h(x) = \langle h_1(x), \ldots, h_l(x) \rangle$. We say that (h, z) is closed form efficient for PRF if there exists an algorithm $\mathsf{CFEval}_{h,z}$ such that for every $x \in \mathbb{Z}_N^k$,

$$\mathsf{CFEval}_{h,z}(x, K) = \prod_{i=1}^{l} [\mathsf{PRF}_K(z_i)]^{h_i(x)}$$

and the running time of CFEval is polynomial in n, m, k but sublinear in l.

The last property is very important for our purposes since it allows to run certain computations very fast when the secret key is known. We next describe two implementations for algebraic PRFs introduced in [6].

Algebraic PRFs from Strong DDH. Let $\mathcal{G}$ be a computational group scheme. The following construction $\mathcal{PRF}_1$ is an algebraic PRF with polynomial sized domains.

$\mathsf{KeyGen}(1^n, m)$: Generate a group description $(\mathbb{G}, p, g) \leftarrow \mathcal{G}(1^n)$. Choose $k_0, k_1 \leftarrow \mathbb{Z}_p$. Output $param = (m, p, g, \mathbb{G})$, $K = (k_0, k_1)$.

$\mathsf{PRF}_K(x)$: Interpret x as an integer in $\{0, \ldots, D = 2^m\}$ where D is polynomial in n. Compute and output $g^{k_0 k_1^x}$.

Closed form efficiency for polynomials of degree d. We now show an efficient closed form for $\mathcal{PRF}_1$ for polynomials of the form (where evaluation is computed in the exponent)

$$Q(x) = \mathsf{PRF}_K(0) \cdot \mathsf{PRF}_K(1)^x \cdot \ldots \cdot \mathsf{PRF}_K(d)^{x^d} = \prod_{i=0}^{d} \mathsf{PRF}_K(i)^{x^i}$$

where $d \leq D$. Let $h : \mathbb{Z}_p \mapsto \mathbb{Z}_p^{d+1}$, be defined as $h(x) \overset{\text{def}}{=} (1, x, \ldots, x^d)$ and $(z_0, \ldots, z_d) = (0, \ldots, d)$. Then, we can define

$$\mathsf{CFEval}_h(x, K) \overset{\text{def}}{=} g^{\frac{k_0(k_1^{d+1} x^{d+1} - 1)}{k_1 x - 1}}.$$

Specifically, we write

$$\prod_{1=0}^{d} [\mathsf{PRF}_K(z_i)]^{h_i(x)} = \prod_{i=0}^{d} [g^{k_0 k_1^i}]^{x^i} = g^{k_0 \sum_{i=0}^{d} k_1^i x^i}.$$

Correctness of CFEval follows by the identity $\sum_{i=0}^{d} k_0 k_1^i x^i = \frac{k_0((k_1 x)^{d+1} - 1)}{k_1 x - 1}$.

Theorem 3.2 ([6]). *Suppose that the D-Strong DDH assumption holds. Then, $\mathcal{PRF}_1$ is a pseudorandom function.*

Algebraic PRFs From DDH. Let $\mathcal{G}$ be a computational group scheme. Define $\mathcal{PRF}_2$ as follows.

KeyGen($1^n, m$): Generate a group description $(p, g, \mathbb{G}) \leftarrow \mathcal{G}(1^n)$. Choose $k_0, k_1, \ldots, k_m \leftarrow \mathbb{Z}_p$. Output $param = (m, p, g, \mathbb{G})$, $K = (k_0, k_1, \ldots k_m)$.
PRF$_K(x)$: Interpret $x = (x_1, \ldots, x_m)$ as an m-bit string. Compute and output $g^{k_0 \prod_{i=1}^m k_i^{x_i}}$.

This function is known by the Naor-Reingold function [42].

Closed form for polynomials of degree d. We describe an efficient closed form for $\mathcal{PRF}_2$ for computing polynomials of the same form as above. That is,

$$Q(x) = \mathsf{PRF}_K(0) \cdot \mathsf{PRF}_K(1)^x \cdot \ldots \cdot \mathsf{PRF}_K(d)^{x^d} = \prod_{i=0}^d \mathsf{PRF}_K(i)^{x^i}.$$

Let $h : \mathbb{Z}_p \mapsto \mathbb{Z}_p^{d+1}$, defined as $h(x) = (1, x, \ldots, x^d)$ and let $z = (z_1, \ldots, z_l) = (0, \ldots, d)$ then

$$\mathsf{CFEval}_{h,z}(x, K) \stackrel{\text{def}}{=} g^{k_0(1+k_1 x)(1+k_2 x^2)\ldots(1+k_m x^{2^m})}$$

with $m = \lceil \log d \rceil$ (clearly, d must be a power of 2).

Theorem 3.3 ([42]). *Suppose that the DDH assumption holds. Then, $\mathcal{PRF}_2$ is a pseudorandom function.*

To this end, we only consider $z = (0, \ldots, d)$ and omit z from the subscript, writing $\mathsf{CFEval}_h(x, K)$ instead.

3.2 Our OPE Constructions

We describe our protocol for oblivious polynomial evaluation in the $\mathcal{F}_{\mathrm{MaskPoly}}$-hybrid setting, where the parties have access to a trusted party that computes functionality $\mathcal{F}_{\mathrm{MaskPoly}} : (K, t) \mapsto (-, g^{R(t)})$ relative to some prime order group $\mathbb{G}$ and generator g that are picked by P_0, for $g^{R(\cdot)} = (g^{r_0}, \ldots, g^{r_d})$ and $g^{r_i} = \mathsf{PRF}_K(i)$ for all i. For simplicity, we first describe a semi-honest variant of our protocol and then show how to enhance its security into the malicious setting. Formally, let $\mathcal{PRF} = \langle \mathsf{KeyGen}, \mathsf{PRF}, \mathsf{CFEval} \rangle$ denote an algebraic PRF with a range group $\mathbb{G}$ (cf. Definition 3.1), then our semi-honest protocol follows.

Protocol 1 (Protocol π_{OPE} with Semi-Honest Security.)

- **Input:** *Party P_0 is given a d-degree polynomial $g^{Q(\cdot)} = (g^{q_0}, \ldots, g^{q_d})$ with coefficients q_i's from $\mathbb{Z}_p$ with respect to prime order group $\mathbb{G}$ and generator g. Party P_1 is given an element t from $\mathbb{Z}_p$. Both parties are given a security parameter 1^n, group description $\mathbb{G}, p$ and g.*

- **The protocol:**

    1. **Masking the Polynomial.** P_0 invokes $(K, param) \leftarrow \mathsf{KeyGen}(1^n, \lceil \log d \rceil)$ where param includes a group description $\mathbb{G}$ of prime order p and a generator g. It next defines a sequence of d elements $\widetilde{R}(\cdot) = (\tilde{r}_0, \ldots, \tilde{r}_d)$ over $\mathbb{G}$ where $\tilde{r}_i = \mathsf{PRF}_K(i)$ for all i.

       P_0 sends P_1 param and the masked polynomial $C(\cdot) = \left(g^{q_0}\tilde{r}_0, \ldots, g^{q_d}\tilde{r}_d\right)$, where multiplication is implemented (componentwise) in $\mathbb{G}$.

    2. **Unmasking the Result.** Upon receiving the masked polynomial $C(\cdot) = (c_0, \ldots, c_d)$, party P_1 computes the polynomial evaluation $C(t) = \prod_{i=0}^{d}(c_i)^{t^i}$. I.e., $C(\cdot)$ is evaluated in the exponent. Next, the parties invoke an ideal execution of $\mathcal{F}_{\mathrm{MaskPoly}}$ where the input of P_0 is K and the input of P_1 is t. Let Z denote the output of P_1 from this ideal call, then P_1's output is $C(t)/Z$ where division in implemented in $\mathbb{G}$.

Note that correctness holds since party P_1 computes in Step 2 the polynomial evaluation

$$C(t) = \prod_{i=0}^{d}(c_i)^{t^i} = \prod_{i=0}^{d}(g^{q_i}\tilde{r}_i)^{t^i} = \prod_{i=0}^{d}(g^{q_i}g^{r'_i})^{t^i} = g^{Q(t)+R(t)}$$

and then "fixes" its computation by dividing out $Z = g^{R(t)}$. In addition, privacy holds due to the pseudorandomness of $\mathcal{PRF}$ that hides the coefficients of $Q(\cdot)$. Next, we prove the following theorem. The proof is straightforward and is left for the full version.

Theorem 3.4. Assume $\mathcal{PRF} = \langle \mathsf{KeyGen}, \mathsf{PRF}, \mathsf{CFEval} \rangle$ is an algebraic PRF, then Protocol 1 securely realizes functionality $\mathcal{F}_{\mathrm{OPE}}$ in the presence of semi-honest adversaries in the $\mathcal{F}_{\mathrm{MaskPoly}}$-hybrid model.

Efficiency. In the first phase P_0 computes $d + 1$ modular exponentiations as it can first compute the PRF evaluations in $\mathbb{Z}_p$ (using the PRF key K) and then raise the outcomes to the power of g. Next, P_0 multiplies each PRF evaluation $\mathsf{PRF}_K(i)$ with g^{q_i} (where these computations can be combined into a single exponentiation per index i). Efficiency of the second phase is dominated by the degree of $Q(\cdot)$ and the implementation of functionality $\mathcal{F}_{\mathrm{MaskPoly}}$. In Section 3.3 we discuss several ways to realize $\mathcal{F}_{\mathrm{MaskPoly}}$. (1) Assuming the strong-DDH assumption, our protocol requires a constant number of modular exponentiations. (2) Assuming the DDH assumption our protocol requires $O(\log d)$ modular exponentiations. Therefore, the overall cost is $2(d+1) + O(1)$ (resp. $O(\log d)$ exponentiations.

Security in the Presence of Malicious Adversaries. We next prove the security of Protocol 1 in the presence of malicious attacks. We observe that if the protocol that implements $\mathcal{F}_{\mathrm{MaskPoly}}$ is secure in the presence of malicious corruptions then the entire protocol is secure against malicious attacks as well. Intuitively, security against a corrupted P_1 is immediately implied since a corrupted P_1 does not learn anything beyond $g^{R(t')}$, where t' is P_1's input to $\mathcal{F}_{\mathrm{MaskPoly}}$. More concretely, in the security proof the simulator publishes a random polynomial $\widetilde{S}(\cdot) = g^{S(\cdot)}$ first, and then extracts P_1's input t' to π_{MaskPoly}. Finally, the simulator forces P_1's output within π_{MaskPoly} to be

$g^{S(t')}/g^{Q(t')}$. In case P_0 is corrupted we need to demonstrate how to extract the coefficients of $g^{Q(\cdot)}$. This is achieved by the fact that P_0 is committed to the PRF key K within $\mathcal{F}_{\text{MaskPoly}}$.

To conclude, in order to obtain malicious security the only modification we need to consider with respect to π_{OPE} is to employ a maliciously secure implementation of functionality $\mathcal{F}_{\text{MaskPoly}}$. In the hybrid setting this does not make a difference for the protocol description. In Section 3.3 we discuss secure implementations of functionality $\mathcal{F}_{\text{MaskPoly}}$. The proof for the following theorem is simple and left for the full version.

Theorem 3.5. *Assume* $\mathcal{PRF} = \langle \text{KeyGen}, \text{PRF}, \text{CFEval} \rangle$ *is an algebraic PRF, then Protocol 1 securely realizes functionality* $\mathcal{F}_{\text{OPE}}$ *in the presence of malicious adversaries in the* $\mathcal{F}_{\text{MaskPoly}}$*-hybrid model.*

3.3   Secure Protocols for π_{MaskPoly}

In this section we describe a concrete protocol that implements functionality $\mathcal{F}_{\text{MaskPoly}}$: $(K, t) \mapsto (-, g^{R(t)})$, used as a subprotocol within our main protocol π_{OPE} for oblivious polynomial evaluation from Section 3.2. This computation corresponds to the polynomial evaluation $\widetilde{R}(x) = \text{PRF}_K(0) \cdot \text{PRF}_K(1)^x \cdot \ldots \cdot \text{PRF}_K(d)^{x^d}$ with respect to function PRF. In what follows, we discuss a detailed secure implementation for $\mathcal{PRF}_1$ that is described in Section 3.1 and then briefly discuss how to implement function $\mathcal{PRF}_2$, formally described in Section 3.1, using similar ideas.

We recall that when implementing functionality $\mathcal{F}_{\text{MaskPoly}}$ relative to $\mathcal{PRF}_1$ the parties compute the value $g^{\sum_{i=0}^{d} k_0 k_1^i x^i} = g^{\frac{k_0((k_1 x)^{d+1} - 1)}{k_1 x - 1}}$, so that P_0 enters a PRF key $K = (k_0, k_1)$ and learns nothing and P_1 enters $x = t$ and learns this outcome. This is a simple computation that requires a constant number of exponentiations and can be easily implemented securely. Achieving malicious security requires to ensure correctness of computations which we obtain using simple zero-knowledge proofs of knowledge. Loosely speaking, the parties first generate a joint public key for the additive El Gamal PKE such that no party knows the secret key (we omit the details here). Next, each party commits to its input and the parties jointly compute $k_1 t$. A slight complication arises since the parties need to compute the inverse of $k_1 t - 1$. Relying on the fact that $(k_1 t - 1)^{-1} = (k_1 t - 1)^{p-2} \bmod p$ and that

$$\frac{k_0((k_1 t)^{d+1} - 1)}{k_1 t - 1} = \frac{k_0(k_1 t)^{d+1} - k_0}{k_1 t - 1} = \frac{k_0 k_1^{d+1} t^{d+1}}{k_1 t - 1} - \frac{k_0}{k_1 t - 1},$$

we let the parties compute the inverse of $k_1 t - 1$ first and then complete the computation by multiplying the result with $k_0(k_1 t)^{d+1}$ and k_0. Formally, our protocol uses the following tools:

1. Distributed additive El Gamal. We denote this scheme by $\Pi = (\pi_{\text{KeyGen}}, \text{Enc}, \pi_{\text{DEC}})$.
2. Zero-knowledge proofs of knowledge: π_{DL} for proving a discrete logarithm and π_{Eq} for proving consistency of exponents, which are formally stated in Section 2.2.

Finally, we implicity assume that a party rerandomizes its homomorphic computations on the ciphertexts. Such that rerandomization is carried out by multiplying the outcome with a random encryption of zero. We now describe our protocol is details.

Protocol 2 (Protocol π_{MaskPoly} with Malicious Security.)

- **Input:** *Party P_0 is given a PRF key $K = (k_0, k_1)$. Party P_1 is given an element t. Both parties are given a security parameter 1^n, a polynomial degree d and $(\mathbb{G}, p, g)$ for a group description $\mathbb{G}$ of prime order p and a generator g.*
- **Convention:** *Homomorphic operations on ciphertexts are computed componentwise.*
- **The protocol:**
    1. **Distributed key generation.** *P_0 and P_1 run protocols $\pi_{\text{KeyGen}}(1^n, 1^n)$ in order to generate additive El Gamal public key $\text{PK} = \langle \mathbb{G}, p, g, h \rangle$ for which the corresponding shares of the secret key SK are $(\text{SK}_0, \text{SK}_1)$. P_0 then sends P_1 encryptions of k_0 and k_1, denoted by c_{k_0} and c_{k_1}, and proves their knowledge using π_{DL}.*
    2. **Computing encryption of $k_1 t$.** *Upon receiving ciphertexts c_{k_0} and c_{k_1}, P_1 sends P_0 an encryption of its input t, denoted by c_t. It further computes the encryption of $k_1 t$, denoted by $c_{k_1 t}$, and proves consistency relative to c_t and $c_{k_1 t}$ using the zero-knowledge proof π_{Eq}.*
    3. **Computing encryptions of k_1^{d+1} and t^{d+1}.** *P_0 computes the encryption of k_1^{d+1}, denoted by $c_{k_1^{d+1}}$, and proves consistency between g^{d+1} and $c_{k_1^{d+1}}$ using π_{Eq}. Similarly, P_0 computes the encryption of t^{d+1}, denoted by $c_{t^{d+1}}$, and proves correctness.*
    4. **Computing encryption of $(k_1 t - 1)^{-1}$.** *The parties compute the inverse of $(k_1 t - 1)$, by first computing the encryption of $k_1 t - 1$ given ciphertext $c_{k_1 t}$ from above, and then raising the result to the power of $p - 2$. Let c_{inv} denote the outcome.*
    5. **Computing encryptions of $k_0 (k_1 t - 1)^{-1}$ and $k_0 k_1^{d+1} (k_1 t - 1)^{-1}$.** *Given ciphertexts c_{inv}, $c_{k_1^{d+1}}$ and c_{k_0}, P_0 computes the encryptions of $k_0 (k_1 t - 1)^{-1}$ and $k_0 k_1^{d+1} (k_1 t - 1)^{-1}$ and proves consistency relative to c_{inv}, $c_{k_1^{d+1}}$ and c_{k_0} using π_{Eq} (where the proof of the later computation involves running π_{Eq} twice). Let c_0 and c_0' denote the respective outcomes.*
    6. **Computing encryption of $k_0 k_1^{d+1} t^{d+1} (k_1 t - 1)^{-1}$.** *Given ciphertexts $c_{t^{d+1}}$ and c_0', P_1 computes the encryption of $k_0 k_1^{d+1} t^{d+1} (k_1 t - 1)^{-1}$ and proves consistency using π_{Eq}. Let c_1 denote the respective outcome.*
    7. **Outcome.** *Finally, the parties decrypt c_1 / c_0 for P_1 that outputs the result.*

Theorem 3.6. *Assume $\Pi = (\pi_{\text{KeyGen}}, \text{Enc}, \pi_{\text{DEC}})$, π_{DL} and π_{Eq} are as above, then Protocol 2 securely realizes functionality $\mathcal{F}_{\text{MaskPoly}}$ with respect to $\mathcal{PRF}_1$ in the presence of malicious adversaries.*

We leave the proof to the full version. Next, we note that the implementation of the other PRF $\mathcal{PRF}_2$ follows similarly. Namely, recall that the parties compute the value $g^{k_0(1+k_1 x)(1+k_2 x^2) \dots (1+k_m x^{2^m})}$ which can be carried out in $O(m)$ time as follows. First, P_0 commits to its key $(k_0, k_1, \dots, k_m)$, whereas P_1 commits to the elements $(x, x^2, \dots, x^{2^m})$ together with a ZK proof of consistency. Next, given the product $\tilde{g} = g^{k_0(1+k_1 x)(1+k_2 x^2) \dots (1+k_{m'}' x^{2^{m'}})}$ for some integer $m' < m$, the parties compute

$$\tilde{g} \cdot \tilde{g}^{k_{m'+1} x^{2^{(m'+1)}}} = \tilde{g}^{(1+k_{m'+1} x^{2^{(m'+1)}})} = g^{k_0(1+k_1 x)(1+k_2 x^2) \dots (1+k_{m'+1} x^{2^{(m'+1)}})}$$

where $\hat{g} = \tilde{g}^{k_{m'+1}}$ is carried out by P_0 and proven correct with respect the commitment of $g^{k_{m'+1}}$. This computation is followed by P_1 computing $\hat{g}^{x^{2^{(m'+1)}}}$ which is also verified against the commitment of $g^{x^{2^{(m'+1)}}}$ where the commitment is realized using El Gamal. See the ZK proof π_{Eq} for more details.

4 Secure Set-Intersection

One important application that benefits from our OPE construction is the set-intersection functionality which is defined by having each party's input consists of a *set* of elements from domain $\{0,1\}^t$. Formally:

Definition 4.1. *Let X and Y be subsets of a predetermined arbitrary domain $\{0,1\}^t$ and m_X and m_Y the respective upper bounds on the sizes of X and Y.[1] Then functionality $\mathcal{F}_\cap$ is defined by:*

$$(X, Y) \mapsto (m_Y, (X \cap Y, m_X)).$$

To achieve a secure set-intersection protocol, we modify protocol π_{OPE} from Section 3.2 as follows. First, P_0 prepares a polynomial $Q(\cdot)$ with coefficients in $\mathbb{Z}_p$ and the set of roots X. It then masks $Q(\cdot)$ as in Protocol 1 using the sequence of pseudorandom elements $\widetilde{R}(\cdot)$. The parties then interact with a trusted party that computes functionality $\mathcal{F}_{\text{EqMask}}$, which is a slight variation of functionality $\mathcal{F}_{\text{MaskPoly}}$. Namely, instead of implementing $\mathcal{F}_{\text{MaskPoly}}$ the functionality checks for equality with respect to P_1's polynomial evaluations of $g^{Q(\cdot)}\widetilde{R}(\cdot)$ and $\widetilde{R}(\cdot)$ on the set Y. This modification in the functionality's description is required due to the fact that we cannot let P_1 learn $Q(y)$ for arbitrary $y \in Y$ (even if P_1 is honest), since that would leak information about X. More specifically, $\mathcal{F}_{\text{EqMask}}$ is defined by $(K, \{(y_i, T_i)\}_{y_i \in Y}) \mapsto (-, \{b_i\}_i)$, where $b_i = 1$ only if $T_i = g^{R(y_i)}$ and 0 otherwise, $g^{R(\cdot)} = (g^{r_0}, \ldots, g^{r_m X})$ and $g^{r_i} = \text{PRF}_K(i)$ for all i. Stated differently, $b_i = 1$ if and only if $Q(y_i) = 0$ (or $y_i \in X \cap Y$) with overwhelming probability. Finally, P_1 outputs the set of elements $Z \subseteq Y$ for which $b_i = 1$.

Our implementation for $\mathcal{F}_{\text{MaskPoly}}$ from Section 3.3 easily supports this functionality, since P_0 can run its zero-knowledge proofs with respect to a single set of ciphertexts encrypting its PRF key. In addition, in order to enable the extraction of the set X by the simulator we add zero-knowledge proofs of knowledge for the relation $\mathcal{R}_{\text{DL}}$, formally defined in Section 2.2. This technicality arises because P_0 sends elements in $\mathbb{G}$ yet the polynomial $Q(\cdot) + R(\cdot)$ is evaluated *in the exponent*, implying that X and Y must be sampled from $\mathbb{Z}_p$ as well. Note that P_0 may fix X and its masked polynomial in $\mathbb{G}$. Nevertheless, P_1 needs to know the discrete logarithms of Y with respect to some group generator g in order to evaluate the masked polynomial.

Formally, let $d = m_X - 1$, then define our set-intersection protocol as follows,

Protocol 3 (Protocol $\pi_\cap$ with malicious security.)

- **Input:** *Party P_0 is given a set X of size m_X. Party P_1 is given a set Y of size m_Y. Both parties are given a security parameter 1^n.*
- **The protocol:**
    1. **Masking the input polynomial.** P_0 *defines an d-degree polynomial $Q(\cdot) = (q_0, \ldots, q_d)$ with coefficients in $\mathbb{Z}_p$ and the set of roots X, for $d = m_X - 1$. It then*

---

[1] In order to deal with a proof technicality, where a corrupted party inputs less elements than its set size, prior constructions assume a super polynomial lower bound on the input domain sizes. Since we do not wish to restrict the input domains, we assume that the set sizes are not strict and may denote some upper bound on the actual numbers of elements.

invokes $(K, param) \leftarrow$ KeyGen$(1^n, d)$ where $param$ includes a group description $\mathbb{G}$ of prime order p and a generator g, and defines a new d-degree polynomial $\tilde{R}(\cdot) = (\tilde{r}_0, \ldots, \tilde{r}_d)$ over $\mathbb{G}$, where r_i is defined by PRF$_K(i)$ for all i.
P_0 sends P_1 $param$ and the masked polynomial $C(\cdot) = (g^{q_0}\tilde{r}_0, \ldots, \ldots, g^{q_d}\tilde{r}_d)$, where multiplication is implemented in $\mathbb{G}$. P_0 further proves the knowledge of the discrete logarithm of $c_i = g^{q_i}\tilde{r}_i$ for all i with respect to a generator g, by invoking an ideal execution of $\mathcal{F}_{\mathrm{DL}}$ on input $\{((g, c_i), \log_g c_i)\}_{i \in [0,d]}$.[2] The input of P_1 for $\mathcal{F}_{\mathrm{DL}}$ is $\{(g, c_i)\}_{i \in [0,d]}$.

2. **Unmasking the result.** Upon receiving the masked polynomial $C(\cdot) = (c_0, \ldots, c_d)$ and upon receiving from $\mathcal{F}_{\mathrm{DL}}$ the value 1, denoting "accept" for all i, party P_1 computes the polynomial evaluation $C(y) = \prod_{i=0}^{d}(c_i)^{y^i}$ for all $y \in Y$ (picked in a random order). I.e., $C(\cdot)$ is evaluated in the exponent.
   Next, the parties invoke an ideal execution of $\mathcal{F}_{\mathrm{EqMask}}$, where the input of P_0 is K and the input of P_1 is the set $\{(y, C(y))\}_{y \in Y}$. P_1 outputs y if and only if the output from $\mathcal{F}_{\mathrm{EqMask}}$ on $(y, C(y))$ is 1.

Correctness follows easily since P_1 outputs only elements in Y that zeros polynomial $Q(\cdot)$, whom its roots are the set X. Next, we prove the following theorem.

Theorem 4.2. *Assume* $\mathcal{PRF} = \langle$KeyGen$, F,$ CFEval$\rangle$ *is an algebraic PRF, then Protocol 3 securely realizes functionality* $\mathcal{F}_{\cap}$ *in the presence of malicious adversaries in the* $\{\mathcal{F}_{\mathrm{DL}}, \mathcal{F}_{\mathrm{EqMask}}\}$*-hybrid model.*

Proof: We prove security for each corruption case separately.

P_0 *is corrupted.* Let $\mathcal{A}$ be a PPT adversary corrupting party P_0, we design a PPT simulator $\mathcal{SIM}$ that simulates the view $\mathcal{A}$, playing the role of the honest P_1 while extracting $\mathcal{A}$'s input set X, details follow.

1. Given input $(1^n, X, z)$, $\mathcal{SIM}$ invokes $\mathcal{A}$ on this input and receives $\mathcal{A}$'s first message, $(\mathbb{G}, p, g)$ and a d-degree polynomial $C(\cdot) = (c_0, \ldots, c_d)$.
2. $\mathcal{SIM}$ emulates the ideal calls of $\mathcal{F}_{\mathrm{DL}}$ by playing the role of the trusted party that receives from $\mathcal{A}$ tuples $\{((g, c_i), c_i')\}_{i \in [0,d]}$ and records these values. $\mathcal{SIM}$ verifies whether $c_i = g^{c_i'}$ for all i and records 1 only if these conditions are met, and 0 otherwise. In case $\mathcal{SIM}$ records 0 it aborts and outputs whatever $\mathcal{A}$ does.
3. $\mathcal{SIM}$ defines the input set X' as follows. For every i let $\tilde{r}_i = $ PRF$_K(i)$ and $r_i = \log_g \tilde{r}_i$ and let $q_i' = c_i' - r_i$.[3] $\mathcal{SIM}$ fixes polynomial $Q'(\cdot) = (q_0', \ldots, q_d')$ and defines X' to be the set of roots of $Q'(\cdot)$. $\mathcal{SIM}$ computes X' by factoring $Q'(\cdot)$ over $\mathbb{Z}_p$ and sends the set X' to the trusted party, receiving back m_Y.
4. $\mathcal{SIM}$ emulates the ideal call of $\mathcal{F}_{\mathrm{MaskPoly}}$ by playing the role of the trusted party that receives from $\mathcal{A}$ a PRF key K.
5. $\mathcal{SIM}$ outputs whatever $\mathcal{A}$ does.

Note that the adversary's view is identical to its view in the hybrid execution since it does not get any output from the internal ideal calls as well as from $\mathcal{F}_{\cap}$. We now claim

---

[2] We implicitly assume that P_0 knows the discrete logarithms of the r_i's by its knowledge of K. This is the case for all PRF implementations presented in [6].

[3] See Footnote 2.

that P_1's output is identical with overwhelming probability in both executions due to the following. In the hybrid execution the correctness of the ideal call for $\mathcal{F}_{\mathrm{EqMask}}$ ensures that P_1 obtains the correct equality bit for every $y \in Y$. Namely, if $C(y) \neq \widetilde{R}(y)$ then the honest P_1 obtains 0 from $\mathcal{F}_{\mathrm{EqMask}}$ and does not output y. On the other hand, if $C(y) = \widetilde{R}(y)$ then P_0 receives 1 and returns y. Stating differently, P_1 returns $y \in Y$ only if $C(y)/\widetilde{R}(y) = 1$ where division is computed component-wise. Next, in the simulation $\mathcal{SIM}$ defines the input set X' of the adversary as the set of roots with respect to the unmasked polynomial $C(\cdot)/\widetilde{R}(\cdot)$ (computed component-wise), where the masking is defined by the PRF key K input by the adversary to $\mathcal{F}_{\mathrm{EqMask}}$. Therefore the intersection is computed with respect to the same set X'.

P_1 *is corrupted.* Let $\mathcal{A}$ be a PPT adversary corrupting party P_1, we design a PPT simulator $\mathcal{SIM}$ that generates the view of $\mathcal{A}$ as follows. $\mathcal{SIM}$ first sends a random polynomial $\widetilde{S}(\cdot)$. Next, upon receiving the adversary's set of elements Y' to $\mathcal{F}_{\mathrm{MaskPoly}}$, $\mathcal{SIM}$ forwards it to the trusted party for $\mathcal{F}_{\cap}$. Let Z' denotes the output returned by the trusted party, then $\mathcal{SIM}$ completes the simulation by forcing the output of $\mathcal{A}$ within $\mathcal{F}_{\mathrm{EqMask}}$ to be consistent with the set Z. More formally,

1. Given input $(1^n, Y, z)$, $\mathcal{SIM}$ invokes $\mathcal{A}$ on this input and sends it $(\mathbb{G}, p, g)$.
2. $\mathcal{SIM}$ picks a random d-degree polynomial $\widetilde{S}(\cdot) = (\tilde{s}_0, \ldots, \tilde{s}_d) = (g^{s_0}, \ldots, g^{s_d})$ with coefficients in $\mathbb{G}$ and sends it to $\mathcal{A}$. (We assume that the simulator knows m_X as part of its auxiliary information. This can also be assured by modifying the definition of the functionality, given m_X to P_1 as part of its input).
3. $\mathcal{SIM}$ emulates the ideal calls of $\mathcal{F}_{\mathrm{DL}}$ by playing the role of the trusted party that receives from $\mathcal{A}$ tuples $\{(g, \tilde{s}_i)\}_{i \in [0,d]}$ and sends $\mathcal{A}$ the value 1 for all i (denoting accept calls).
4. $\mathcal{SIM}$ then emulates the ideal call of $\mathcal{F}_{\mathrm{EqMask}}$ by playing the role of the trusted party that receives from $\mathcal{A}$ the set $\{(y'_j, T_{y'_j})\}_{j \in [m_Y]}$. $\mathcal{SIM}$ sends the set $Y' = \{y'_j\}_{j \in [m_Y]}$ to the trusted party, receiving back the intersection $Z = X \cap Y'$. For all $y'_j \in Z$, $\mathcal{SIM}$ emulates the ideal response of $\mathcal{F}_{\mathrm{EqMask}}$ as follows. If $T_{y'_j} = g^{S(y'_j)}$ then $\mathcal{SIM}$ sends $\mathcal{A}$ the value 1. Otherwise it sends 0. For all $y'_j \notin Z$, $\mathcal{SIM}$ always replies with 0.
5. $\mathcal{SIM}$ outputs whatever $\mathcal{A}$ does.

Note that the protocol never verifies that $\mathcal{A}$'s inputs to $\mathcal{F}_{\mathrm{EqMask}}$ are consistent pairs $\{(y'_j, T_{y'_j})\}_j$ of which $T_{y'_j} = g^{S(y'_j)}$ for all $j \in [m_Y]$. We prove that this is not required. Specifically, the differences between the hybrid and simulated executions are as follows. First, $\mathcal{SIM}$ sends in the simulation a random polynomial instead of a real masked polynomial. In addition, $\mathcal{SIM}$ fixes the output of $\mathcal{F}_{\mathrm{EqMask}}$ based on the correctness of $\mathcal{A}$'s computations which deviates from the way this functionality is defined. Consider a hybrid game Hyb where the simulator $\mathcal{SIM}_{\mathrm{Hyb}}$ uses the real input X of P_0 to define polynomial $Q(\cdot)$, but decides on the output of $\mathcal{F}_{\mathrm{EqMask}}$ according to the strategy specified in the simulation. Namely for every pair $(y'_j, T_{y'_j})$, $\mathcal{SIM}_{\mathrm{Hyb}}$ verifies first whether $T_{y'_j} = C(y'_j)$ and returns 1 if equality holds. Clearly, the views induced in Hyb and in the simulation are computationally indistinguishable due to the pseudorandomness of F. This argument is similar to the argument presented in the proof of Protocol 1.

Next, we claim that the distributions induced by the views of the hybrid execution and game Hyb are statistically close.

Formally, for every y'_j consider two cases. (i) $y'_j \notin X$ which implies that y'_j is not in the intersection and that $b_j = 0$ in the simulation of Hyb. Next, define a Bad event in which $\mathcal{A}$ receives $b_j = 1$ from the trusted party for $\mathcal{F}_{\text{EqMask}}$ in the hybrid execution. Clearly, this event holds only if $T_{y'_j} = \text{CFEval}(y'_j, K) = g^{R(y'_j)}$ for K the PRF key entered by the honest P_0, which implies that $\mathcal{A}$ must correctly guess $\text{CFEval}(y'_j, K)$. We claim that the probability this event occurs is negligible due to the pseudorandomness of F and CFEval (in Section 5 we discuss the pseusorandomness of CFEval). Specifically, any successful guess with a non-negligible probability implies an attack on the PRF. Thus, the probability that Bad occurs is negligible. It therefore holds that the adversary's views are statistically close condition on the event that y'_j is not in the intersection. (ii) $y'_j \in X$ which implies that y'_j is in the intersection. Nevertheless, here there is no analogue bad event. This is because $b_j = 1$ only when $T_{y'_j} = C(y'_j) = \text{CFEval}(y'_j, K)$, which implies that $b_j = 1$ in both executions due to correctness of $\mathcal{F}_{\text{EqMask}}$. ■

Efficiency. As in Protocol 1, the efficiency of Protocol 3 is dominated by the implementation of functionality $\mathcal{F}_{\text{EqMask}}$. Our protocols from Section 3.3 can be easily modified to support this functionality without significantly effecting their overhead, since the parties can first compute the encryption of the closed form efficiency of the PRF and then compare it with the input of P_1. Therefore, the overall communication complexity is $O(m_X)$ group elements for sending the first message and $O(m_Y)$ (resp. $O(m_Y \log m_Y)$) group elements for the second phase of implementing $\mathcal{F}_{\text{EqMask}}$ for each $y \in Y$, depending on the underlying PRF. In particular, the number of modular exponentiations implies multiplicative costs in the sets sizes since P_1 evaluates its masked polynomial for each element in Y. Next, we demonstrate how to reduce this cost.

4.1   Improved Constructions Using Hash Functions

We now show how to reduce the computational overhead using hash functions by splitting the set elements into smaller bins. Our protocol is applicable for different hash functions such as: simple hashing, balanced allocations [4] and Cuckoo hashing [33]. For simplicity, we first describe our protocol for the simple hashing case; see Section 4.1 for a discussion about extensions to the other two hashing. Informally, the parties first agree on a hash function that is picked from a family of hash functions and induces a set of bins with some upper bound on the number of elements in each bin. Next, P_0 maps its elements into these bins and generates a polynomial for each such bin, which is computed as in Protocol 3 but with a smaller degree. Finally, P_0 masks all the polynomials and sends them to P_1. Upon receiving the masked polynomials, P_1 maps its elements into the same set of bins and evaluates the masked polynomials for these mapped bins. In the last step, the parties unmask these evaluations. To be precise, we need to specify how the masking procedure works and ensure that the parties do not deviate from the instructions of the protocol.

We fix some notations first. We denote by h the hash function picked by the parties, by $\mathcal{B}$ the number of bins and by $\mathcal{M}$ the maximum number of elements allocated to any single bin (where $\mathcal{B}$ and $\mathcal{M}$ are parameters specified by the concrete hash function in use and further depend on m_X). Note that the potential number of allocated elements is bounded by $\mathcal{B}\mathcal{M}$ which may be higher than the exact number m_X. This implies that the protocol must ensure that P_0 does not take advantage of that and introduce more set elements into the protocol execution. In addition, it must be ensured that a corrupted P_0 does not mask the zero polynomial, which would imply that P_1 accepts any value it substitutes in the masked polynomial. On the other hand, the protocol must ensure that a corrupted P_1 does not gain any information by entering incorrect values. Verifying that a polynomial is not all zeros can be easily done by substitution a random element in it and checking that the result is different than zero. In Section 4.1 we demonstrate how to enforce P_0's correct behaviour by designing a new proof that exploits the algebraic properties of the underlying PRF. The verification procedure for P_1 is even simpler as demonstrated below.

Next, we explain how the masking procedure is computed. Denote by $Q_j(\cdot)$ the polynomial associated with the jth bin. If the degree of $Q_j(\cdot)$ is smaller than $\mathcal{M} - 1$ then P_0 fixes the values of the $\mathcal{M}_1 - \deg(Q_j(\cdot))$ leading coefficients to be zeros. It then masks the ith coefficient of $Q_j(\cdot)$ by multiplying it with $\mathrm{PRF}_K((j-1) \cdot \mathcal{M} + i)$ for $i \in [0, \mathcal{M} - 1]$. Furthermore, unmasking is computed by comparing the evaluation of the jth polynomial to the following computation

$$\prod_{i=0}^{j\mathcal{M}-1} \mathrm{PRF}_K(i)^{x^i} \Big/ \prod_{i=0}^{(j-1)\mathcal{M}-1} \mathrm{PRF}_K(i)^{x^i}$$

$$= \mathrm{PRF}_K((j-1)\mathcal{M})^{x^{(j-1)\mathcal{M}}} \cdot \ldots \cdot \mathrm{PRF}_K(j\mathcal{M}-1)^{x^{j\mathcal{M}-1}},$$

which is exactly the set of PRF values that mask polynomial $Q_j(\cdot)$.

More formally, our protocol uses two functionalities in order to ensure correctness. First, the parties call functionality $\mathcal{F}_{\mathrm{Bins}}$ for proving that the masked polynomials sent by P_0 are correctly defined. Namely, $\mathcal{F}_{\mathrm{Bins}} : (K, \{C_j(\cdot) = (c_0^j, \ldots, c_{\mathcal{M}-1}^j)\}_{j \in [\mathcal{B}]}) \mapsto (-, b)$ and $b = 1$ only if none of the unmasked polynomials $\{Q_j(\cdot)\}_j$ is the zero polynomial and the overall degrees of these polynomials $\{Q_j(\cdot)\}_j$ is bounded by m_X. In addition, the parties call functionality $\mathcal{F}_{\mathrm{EqMaskHash}}$ in order to correctly unmask polynomial evaluations $\{C_{h(y)}(y)\}_{y \in Y}$ for P_1. We continue with the detailed description of our set-intersection protocol in the hybrid model. In Sections 4.1 and 4.1 we discuss how to securely implement these functionalities.

Protocol 4 (Protocol $\pi_\cap$ with Malicious Security and Hash Functions.)

- **Input:** *Party P_0 is given a set X of size m_X. Party P_1 is given a set Y of size m_Y. Both parties are given a security parameter 1^n.*
- **The protocol:**
    1. **Fixing the parameters of the hash function.** *The parties fix the parameters $\mathcal{B}$ and $\mathcal{M}$ of the hash function and picks a hash function $h : \{0,1\}^t \mapsto [\mathcal{B}]$. P_0 invokes $(K, param) \leftarrow \mathsf{KeyGen}(1^n, \mathcal{M} - 1)$ where $param$ includes a group description $\mathbb{G}$ of prime order p and a generator g.*

2. **Masking the input polynomial.** *For every* $x \in X$, P_0 *maps* x *into bin* $h(x)$. *Let* $\mathcal{B}_j$ *denote the set of elements mapped into bin* j. *Next,* P_0 *constructs a polynomial* $Q_j(\cdot) = (q_0^j, \ldots, q_d^j)$ *with coefficients in* $\mathbb{Z}_p$ *and the set of roots* $\mathcal{B}_j$. *If* $|\mathcal{B}_j| < \mathcal{M}$, P_0 *fixes the leading* $\mathcal{M} - |\mathcal{B}_j| - 1$ *coefficients to zero.*

   For each $j \in [\mathcal{B}]$, P_0 *defines a new* $(\mathcal{M}-1)$-*degree polynomial* $\widetilde{R}_j(\cdot) = (\tilde{r}_0^j, \ldots, \tilde{r}_{\mathcal{M}-1}^j)$ *over* $\mathbb{G}$, *where* $\tilde{r}_i^j$ *is defined by* $\mathsf{PRF}_K((j-1)\mathcal{M} + i)$ *for all* $i \in [0, \mathcal{M}-1]$. P_0 *sends* P_1 param *and the masked polynomials* $\{C_j(\cdot)\}_j = \{g^{q_0^j}\tilde{r}_0^j, \ldots, \ldots, g^{q_{\mathcal{M}-1}^j}\tilde{r}_{\mathcal{M}-1}^j\}_j$, *where multiplication is implemented in* $\mathbb{G}$. P_0 *further proves the knowledge of the discrete logarithm of* $c_i^j = g^{q_i^j}\tilde{r}_i^j$ *for all* i *and* j *with respect to a generator* g, *by invoking an ideal execution of* $\mathcal{F}_{\mathrm{DL}}$ *on input* $\{((g, c_i^j), \log_g c_i^j)\}_{i \in [0, \mathcal{M}-1], j \in [\mathcal{B}]}$.[4] *The input of* P_1 *for* $\mathcal{F}_{\mathrm{DL}}$ *is* $\{(g, c_i^j)\}_{i \in [0, \mathcal{M}-1], j \in [\mathcal{B}]}$.

   Finally, P_0 *proves correctness using* $\mathcal{F}_{\mathrm{Bins}}$ *where* P_0 *enters* K *and* P_1 *enters the masked polynomials.*

3. **Unmasking the result.** *Upon receiving the polynomials* $\{C_j(\cdot) = (c_0^j, \ldots, c_{\mathcal{M}-1}^j)\}_{j \in [\mathcal{B}]}$ *and upon receiving accepting messages from* $\mathcal{F}_{\mathrm{DL}}, \mathcal{F}_{\mathrm{Bins}}$, *party* P_1 *computes the following for every* $y \in Y$ *(picked in a random order). It first maps* y *into bin* $h(y)$ *and then computes the polynomial evaluation* $C_{h(y)}(y) = \prod_{i=(h(y)-1)\mathcal{M}}^{h(y)\mathcal{M}-1}(c_i^{h(y)})^{y^i}$. *I.e.,* $C_{h(y)}(\cdot)$ *is evaluated in the exponent.*

   Next, the parties invoke an ideal execution of $\mathcal{F}_{\mathrm{EqMaskHash}}$, *where the input of* P_0 *is* K *and the input of* P_1 *is the set* $\{(y, h(y), C_{h(y)}(y))\}_{y \in Y}$.

   P_1 *outputs* y *only if the output from* $\mathcal{F}_{\mathrm{EqMaskHash}}$ *on* $(y, h(y), C_{h(y)}(y))$ *is* 1.

Theorem 4.3. *Protocol 4 securely realizes functionality* $\mathcal{F}_{\cap}$ *in the presence of malicious adversaries in the* $\{\mathcal{F}_{\mathrm{DL}}, \mathcal{F}_{\mathrm{Bins}}, \mathcal{F}_{\mathrm{EqMaskHash}}\}$-*hybrid model.*

Security follows easily from the secure implementations of $\mathcal{F}_{\mathrm{Bins}}$ and $\mathcal{F}_{\mathrm{EqMaskHash}}$ and the proof of Protocol 3. We discuss these protocols next. We stress that P_1 needs to ensure in Protocol 4 that P_0 indeed uses the same PRF key for both sub-protocols (for instance by ensuring that P_0 enters the same commitment of K).

A Secure Protocol for $\mathcal{F}_{\mathrm{Bins}}$. In this section we design a protocol π_{Bins} for securely implementing functionality $\mathcal{F}_{\mathrm{Bins}} : (K, \{C_j(\cdot)\}_{j \in [\mathcal{B}]}) \mapsto (-, b)$ for which $b = 1$ only if none of the unmasked polynomials $\{Q_j(\cdot)\}_j$ is the zero polynomial and the overall degrees of all polynomials $\{Q_j(\cdot)\}_j$ is bounded by m_X. To prove that none of the polynomials is the all zeros polynomial we evaluate each masked polynomial on a random element and then verify that the result is different than zero. In particular, for each j the parties first agree on a random element z_j and then compute the polynomial evaluation $C_j(z_j)$. Next, the parties verify whether $C_j(z_j) = \widetilde{R}_j(z_j)$ where $\widetilde{R}_j(\cdot)$ is the masking polynomial of $C_j(\cdot)$. Note that if $Q_j(\cdot)$ is not the all zeros polynomial then $C_j(z_j) \neq \widetilde{R}_j(z_j)$ with overwhelming probability over the choice of z_j. This is because there exists a coefficient $q_{i,j} \neq 0$ which implies that for $C_j(z_j) = Q_j(z_j) \cdot \widetilde{R}_j(z_j)$. Now since $Q_j(z_j) \neq 0$ it holds that $C_j(z_j) \neq \widetilde{R}_j(z_j)$. On the other hand, in case $Q_j(\cdot)$ is the zero polynomial then it holds that $C_j(z_j) = \widetilde{R}_j(z_j)$ for all z_j. This is because $Q_j(z_j) = 0$ as all its coefficients equal zero.

---

[4] See Footnote 2.

The more challenging part is to prove that the overall degrees of all polynomials $\{Q_j(\cdot)\}_j$ is bounded by $m_X + \mathcal{B}$.[5] Our proof ensures that as follows. First, P_0 picks a PRF key K and forwards P_1 a commitment of K together with encryptions of $f = (f_0 = \mathsf{PRF}_K(0), \ldots, f_{\mathcal{B}\mathcal{M}-1} = \mathsf{PRF}_K(\mathcal{B}\mathcal{M} - 1))$ (that are encrypted using the El Gamal encryption scheme). Next, P_0 proves that it computed the sequence f correctly. This can be achieved by exploiting the closed form efficiency property of the PRF. Namely, the parties mutually compute the encryption of $\prod_{i=0}^{\mathcal{B}\mathcal{M}-1} \mathsf{PRF}_K(i)^{z^i}$ for some random z, and then compare it with the encryption of $\prod_{i=0}^{\mathcal{B}\mathcal{M}-1} f_i^{z^i}$. In particular, the latter computation is carried out on the ciphertexts that encrypt the corresponding values from f by utilizing the homomorphic property of El Gamal. Then, equality is verified such that P_0 proves that the two ciphertexts encrypt the same value. Finally, the parties divide the vector of ciphertexts f with the polynomials coefficients $\{C_j(\cdot)\}_{j \in [\mathcal{B}]}$ component-wise (note that both vectors have the same length). P_0 then proves that the overall degrees of the polynomials is as required using a sequence of zero-knowledge proofs. The last part of our proof borrows ideas from [28]. We continue with the formal description of our protocol.

Protocol 5 (Protocol π_{Bins} with Malicious Security.)

- **Input:** *Party P_0 is given a PRF key K for function PRF. Both parties are given a security parameter 1^n, masked polynomials $\{C_j(\cdot) = (c_0^j, \ldots, c_{\mathcal{M}-1}^j)\}_{j \in [\mathcal{B}]}$, $(\mathbb{G}, p, g)$ for a group description $\mathbb{G}$ of prime order p and a generator g, and an integer m_X.*
- **The protocol:**
  1. **Setup.** *P_0 generates $(\mathsf{PK}, \mathsf{SK}) \leftarrow \mathsf{Gen}(1^n)$ for the El Gamal encryption scheme for group $\mathbb{G}$. It then computes the set $f = (f_0 = \mathsf{PRF}_K(0), \ldots, f_{\mathcal{B}\mathcal{M}-1} = \mathsf{PRF}_K(\mathcal{B}\mathcal{M} - 1))$ and sends to P_1 their encryptions under PK, denoted by $(e_0, \ldots, e_{\mathcal{B}\mathcal{M}-1})$, as well as PK.*
  2. **Proving the correctness of f.** *The parties pick $z \leftarrow \mathbb{Z}_p$ at random and compute $e_f = \prod_{i=0}^{\mathcal{M}\mathcal{B}-1} e_i^{z^i}$. Next, the parties compute the encryption of the product $\prod_{i=0}^{\mathcal{B}\mathcal{M}-1} \mathsf{PRF}_K(i)^{z^i}$, denoted by e_{PRF}, which corresponds to the closed form efficiency function of PRF. Finally, P_0 proves that the two ciphertexts encrypt the same plaintext by proving that e_f / e_{PRF} is a Diffie-Hellman tuple using π_{DL} (see Section 2.2).*
  3. **Proving a bound m_X on the overall degrees.** *If π_{DL} is verified correctly, the parties compute the differences with respect to the masked polynomials $\{C_j(\cdot)\}_j$ and plaintexts f, component-wise. Namely, for all $j \in [\mathcal{B}]$ and $i \in [0, \mathcal{M} - 1]$ the parties compute the encryption of $c_i^j / f_{(j-1)\mathcal{M}+i}$. We denote the result vector of ciphertexts by c_{Diff}. P_0 then sets $Z_{i,j} = 1$ for $0 \le i \le \deg(Q_j(\cdot))$, and otherwise $Z_{i,j} = 0$. P_0 computes $z_{i,j} = \mathsf{Enc}_{\mathsf{PK}}(Z_{i,j})$ and sends $\{z_{i,j}\}_{i,j}$ to P_1. P_0 proves that $Z_{0,j}, Z_{1,j}, \ldots, Z_{M-1,j}$ is monotonically non-increasing. For that, P_0 and P_1 compute encryptions of $Z_{i,j} - Z_{i+1,j}$ and $Z_{i,j} - Z_{i+1,j} - 1$, and P_0 proves that $Z_{i,j} - Z_{i+1,j} \in \{0, 1\}$ by showing that one of these encryptions denotes a Diffie-Hellman tuple using π_{DDH}.*

     P_0 completes the proof that the values $Z_{i,j}$ were constructed correctly by proving for all i, j that one of the encryptions $\{e_{(j-1)\mathcal{M}+i}, z'_{i,j}\}$ is an encryption of zero, where $z'_{i,j}$ is an encryption of $1 - Z_{i,j}$.[6]

---

[5] For technical reasons, we require that in case of an empty bin, P_0 fixes the polynomial that is associated with this bin to be 1.

[6] We wish to avoid the case where $e_{(j-1)\mathcal{M}+i}$ is an encryption of a non-zero value while $z'_{i,j}$ encrypts zero.

Finally, to prove that the sum of degrees of the polynomials $\{Q_j(\cdot)\}$ equals m_X, both parties compute an encryption τ of $T = \sum_{i,j} Z_{i,j} - B - m_X$. Then P proves that $(\text{PK}, \text{Enc}_{\text{PK}}(T))$ is a Diffie-Hellman tuple using π_{DDH}.

4. **Checking zero polynomials.** *If all the proofs are verified correctly, then for any $j \in [\mathcal{B}]$ the parties compute $C_j(z_j)$ where $z_j \leftarrow \mathbb{Z}_p$. The parties call $\mathcal{F}_{\text{EqMaskHash}}$ with inputs $(K, \{z_j, j, C_j(z_j)\}_{j \in [\mathcal{B}]})$. Let $\{b_j\}_{j \in [\mathcal{B}]}$ be P_1's output from this ideal call.[7]*

5. *P_1 outputs $b = 1$ only if $b_j = 0$ for all j.*

Theorem 4.4. *Assume that El Gamal is IND-CPA, then Protocol 5 securely realizes functionality $\mathcal{F}_{\text{Bins}}$ in the presence of malicious adversaries in the $\{\mathcal{F}_{\text{DL}}, \mathcal{F}_{\text{DDH}}, \mathcal{F}_{\text{EqMaskHash}}\}$-hybrid model.*

The details of the proof are omitted here. Next, we note that the efficiency of our protocol is dominated by Steps 2 and 4, where in the former step the parties compute the closed form efficiency relative to the set f in time $O(\mathcal{B}\mathcal{M}) = O(m_X)$ and in the latter step the parties substitute a random element in every polynomial C_j. Overall, the overhead of this step relative to PRF $\mathcal{PRF}_1$ implies $O(\mathcal{B}) = O(m_X)$ group elements and modular exponentiations. For PRF $\mathcal{PRF}_2$ this step implies $O(\mathcal{B} \log m_X) = O((m_X \setminus \log \log m_X) \cdot \log m_X)$ cost; see a discussion below.

A Secure Protocol for $\mathcal{F}_{\text{EqMaskHash}}$. The next protocol is designed in order to compare the result of P_1's polynomial evaluations on the set Y with the masking polynomials. Basically, for every $y \in Y$, P_1 computes first $C_{h(y)}(y)$. The parties then run a protocol for comparing $\{C_{h(y)}(y)\}_{y \in Y}$ with $\{\tilde{R}_{h(y)}(y)\}_{y \in Y}$. To do so, P_1 must also input the value $h(y)$ which determines the bin's name. Nevertheless, we do not require from the parties to mutually compute $h(y)$ since that would imply a far less efficient protocol. Instead, we demonstrate that P_1 cannot learn additional information by entering an inconsistent bin number. Finally, for every j, P_1 outputs 1 only if equality holds.

Formally, we define $\mathcal{F}_{\text{EqMaskHash}}$ by $(K, \{y, h(y), C_{h(y)}(y)\}_{y \in Y}) \mapsto (-, \{b_j\}_j)$, where $b_j = 1$ if $C_{h(y)}(y) = \prod_{i=0}^{h(y)\mathcal{M}-1} \text{PRF}_K(i)^{y^i} \big/ \prod_{i=0}^{(h(y)-1)\mathcal{M}-1} \text{PRF}_K(i)^{y^i}$. The actual implementation of this functionality depends on the underlying PRF. We consider two different implementations here. First, considering our protocol from Section 3.3 designed for $\mathcal{PRF}_1$, an analogue protocol for our purposes can be similarly designed with the modification that the parties now compare $C_{h(y)}(y)$ against the result of the following formula evaluation,

$$g^{\frac{k_0\left((k_1 x)^{(h(y)+1)\mathcal{M}-1} - (k_1 x)^{h(y)\mathcal{M}-1}\right)}{k_1 x - 1}}$$

where $h(y)$ is only known to P_1. Note that our protocol from Section 3.3 does not need to rely on the fact that both parties know the polynomial degree d for computing this formula. It is sufficient to prove that the computation of some ciphertext c to the power of $h(y)$ is consistent with a ciphertext encrypting $g^{h(y)}$, where such a ciphertext can be provided by P_1. See this protocol from Section 3.3 and the ZK proof π_{Eq} for more details. The overall overhead of the modified protocol is also constant.

---

[7] Note that z_j may be an element that is not mapped to the jth bin.

Next, considering the unmasking protocol for $\mathcal{PRF}_2$, the parties compute the following formula that corresponds to the masking of the polynomial that is associated with bin $h(y)$,

$$g^{k_0\left(1+k_1 x\right)\left(1+k_2 x^2\right)\ldots\left(1+k_m x^{2^{\log(h(y)\mathcal{M}-1)}}\right)}$$
$$\Big/ \; g^{k_0\left(1+k_1 x\right)\left(1+k_2 x^2\right)\ldots\left(1+k_m x^{2^{2^{\log((h(y)-1)\mathcal{M}-1)}}}\right)}.$$

Note that computing this formula requires $O(\log m_X)$ exponentiations on the worst case if the bin number implies a high value so that $h(y)\mathcal{M}$, which determines the polynomial degree, is $O(m_X)$.

Security is stated as follows. If P_0 is corrupted then security follows similarly to the security proof of the protocols implementing $\mathcal{F}_{\mathrm{MaskPoly}}$ (Section 3.3) since P_0 enters the same input for both functionalities and runs the same computations with respect to its PRF key. The interesting and less trivial corruption case is of P_1. We consider two bad events here: (1) A corrupted P_1 enters y, h' for which $h' \neq h(y)$. This implies that the parties will not compute the correct unmasking. (2) A corrupted P_1 enters consistent $y, h(y)$, but an incorrect value $C_{h(y)}(y)$. Note that upon extracting P_1's input to the protocol execution, the simulator can always tell whether this input corresponds to the first or the second case, or neither.

Specifically, in the first case the parties compute the unmasking on y for which element y in not allocated to the specified bin h'. This implies that P_1 would always obtain 0 from the protocol execution unless it correctly guesses $\widetilde{R}_{h'}(y)$, which only occurs with a negligible probability due to the security of the PRF. Therefore we can successfully simulate this case by always returning zero. We further note that the security argument of the later case boils down to the security presented in the proof for a single polynomial shown in the proof of Theorem 4.2, since in this case P_1 enters $h(y)$ that is consistent with y so the parties compute the correct masking for y.

Using More Than One Hash Function. In some cases, such as for balanced allocation hash function [4], better performance are obtained by using a pair of hash functions h_1, h_2, which allocate elements into two distinct bins. That is, the input to the functionality are defined by $(K, \{y, h_1(y), h_2(y), C_{h(y)}(y)\}_{y \in Y}) \mapsto (-, \{b_j\}_j)$. This poses a problem in our setting since a corrupted P_1 may deviate from the protocol by substituting a different element with respect to each hash function, and learn some information about P_0's input. Specifically, if P_1 learns that some element $y \in X$ was not allocated to $h_1(y)$ it can conclude that P_0 has $\mathcal{M}$ additional elements that are already mapped into bin $h_1(y)$. Note that this leaked information cannot be simulated since it depends on the real input X. In this case we need to verify that P_1 indeed maps the same element into both bins correctly. A simple observation shows that if this is not the case then the simulation fails only for elements that are in the intersection. Meaning, there exists a bin for which the membership result is positive (since otherwise the adversary anyway learns 0, and it cannot distinct the cases of non-membership and incorrect behaviour). We thus define the polynomials slightly different, forcing correct behaviour.

Specifically, the polynomial $Q_j(\cdot)$ that is associated with the set of elements $\mathcal{B}_j$ (namely, the elements that are mapped to the jth bin) is defined as follows. For each $x \in$

$\mathcal{B}_j$, set $Q_j(x) = g^{h_1(x)+h_2(x)}$ where $h_1(x)$ and $h_2(x)$ are viewed as elements in $\mathbb{Z}_p$. Next, in the unmasking phase, for any tuple (y, h_1, h_2, C_y) entered by P_1, the parties compare C_y with both $\left(\prod_{i=0}^{h_1 \mathcal{M}-1} \mathsf{PRF}_K(i)^{y^i} \Big/ \prod_{i=0}^{(h_1-1)\mathcal{M}-1} \mathsf{PRF}_K(i)^{y^i} \right) \cdot g^{h_1 \cdot h_2}$ and $\left(\prod_{i=0}^{h_2 \mathcal{M}-1} \mathsf{PRF}_K(i)^{y^i} \Big/ \prod_{i=0}^{(h_2-1)\mathcal{M}-1} \mathsf{PRF}_K(i)^{y^i} \right) \cdot g^{h_1 \cdot h_2}$ such that the functionality returns 1 to P_1 if equality holds with respect to one of the comparisons. Therefore, P_1 will learn that an element $y \in X$ only if it entered h_1 and h_2 such that $h_1 + h_2 = h_1(y) + h_2(y)$. Note that this implies that if one of the h_1, h_2 values is inconsistent with $h_1(y), h_2(y)$ yet equality holds, then the other value is also inconsistent with high probability. In this case, P_1's output will always be 0 since the incorrect polynomials will be unmasked.

We further need to prove that for any $y \notin X$ the protocol returns 0 with overwhelming probability. Specifically, we need to prove that the probability that either $Q_{h_1(y)}(y) = g^{h_1(y)+h_2(y)}$ or $Q_{h_2(y)}(y) = g^{h_1(y)+h_2(y)}$, is negligible. In order to simplify our proof, we modify our construction and fix $Q_j(x) = \mathsf{PRF}_K(g^{h_1(x)+h_2(x)})$ for any $x \in \mathcal{B}_j$ using a PRF K key that is mutually picked by both parties. In this case, we can easily claim that the probability that the protocol returns 1 for $y \notin X$ is negligible since that implies that either $Q_{h_1(y)}(y)$ or $Q_{h_2(y)}(y)$ equal the pseudorandom value $\mathsf{PRF}_K(g^{h_1(y)+h_2(y)})$ for $y \notin X$. We stress that the PRF key for this purpose can be publicly known since pseudorandomness is still maintained as long as the algorithm for generating the bin polynomials does not use this key. We further note that both algebraic PRFs that we consider in this paper can be easily evaluated over an encrypted plaintext given the PRF key since it only require linear operations.

Finally, a similar solution can be easily adapted for Cuckoo hashing with a stash [33] (by treating the stash as a third polynomial). Nevertheless, Cuckoo hashing using a stash suffers from the following drawback. It has been proven in [33] that for any constant s, using a stash of size s implies an overflow with probability $O(n^s)$ (taken over the choice of the hash functions). Specifically, if the algorithm aborts whenever the original choice of hash functions results in more than s items being moved to the stash, then this means that the algorithm aborts with probability of at most $O(n^s)$. Consequently, P_1 can identify with that probability whether a specific potential input of P_0 does not agree with the hash functions h_1 and h_2. This probability is low but not negligible. On the other hand, Broder and Mitzenmacher [8] have shown for balanced allocations hash function that asymptotically, when mapping n items into n bins, the number of bins with i or more items falls approximately like $2^{2.6i}$. This means that if $\mathcal{M} = \omega(\log \log n)$ then except with negligible probability no bin will be of size greater than $\mathcal{M}$. Nevertheless, (ignoring the abort probability), Cuckoo hashing performs better than balanced allocation hash functions, and by tuning the parameters accordingly this abort probability can be ignored for most practical applications.

Efficiency. The efficiency here depends on the parameters $\mathcal{B} = O(m_X / \log \log m_X)$ and $\mathcal{M} = O(\log \log m_X)$ that are specified by the underlying hash function, as well as the PRF implementation that induce the overhead of the implementations of $\mathcal{F}_{\mathrm{Bins}}$ and $\mathcal{F}_{\mathrm{EqMaskHash}}$. Concretely, when implementing the algebraic PRF with $\mathcal{PRF}_1$ the number of exponentiations computed by the parties is $O(\mathcal{BM} + m_Y \mathcal{M}) =$

Functionality $\mathcal{F}_{\text{CPRF}}$

Functionality $\mathcal{F}_{\text{CPRF}}$ communicates with with parties P_0 and P_1, and adversary $\mathcal{SIM}$.

1. Upon receiving a message (key, K) from P_0, send message key to $\mathcal{SIM}$ and record K.
2. Upon receiving (input, x) from P_1, send message input to adversary $\mathcal{SIM}$. Upon receiving an approve message, send $\text{PRF}_K(x)$ to P_1. Otherwise, send $\perp$ to P_1 and abort.

Fig. 1. The committed oblivious PRF evaluation functionality.

$O(m_X + m_Y \log \log m_X)$, whereas the number of transmitted group elements is $O(\mathcal{B}\mathcal{M} + m_Y) = O(m_X + m_Y)$. Moreover, implementing the algebraic PRF using $\mathcal{PRF}_2$ implies the overhead of $O(m_X + m_Y \log m_X)$ exponentiations and the communication is as above.

5   Committed Oblivious PRF Evaluation

The oblivious PRF evaluation functionality $\mathcal{F}_{\text{PRF}}$ is an important functionality that is defined by $(K, x) \mapsto (-, \text{PRF}_K(x))$. One known example for a protocol that implements $\mathcal{F}_{\text{PRF}}$ is the instantiation based on the Naor-Reingold pseudorandom function [42] (specified in Section 3.1), that is implemented by the protocol presented in [20] (and proven secure in the malicious setting in [25]). This protocol involves executing an oblivious transfer for every bit of the input x. Nevertheless, it has major drawback since it does not enforce the usage of *the same* key for multiple evaluations, which is often required. In this section, we observe first that the algebraic closed form efficiency of PRFs $\mathcal{PRF}_1$ and $\mathcal{PRF}_2$, specified in Section 3.1, are PRFs as well. Moreover, the protocols for securely evaluating these functions induce efficient implementations for the *committed* oblivious PRF evaluation functionality with respect to these new PRFs in the presence of adaptive inputs. This is because the PRF evaluations protocols are implemented with respect to the same set of key commitments. We formally define this functionality in Figure 1.

More formally, let PRF be an algebraic PRF from a domain $\{0, 1\}^m$ into a group $\mathbb{G}$. Then, define the new function $\text{PRF}' : \mathbb{Z}_p \mapsto \mathbb{G}$ by $\text{PRF}'_K(x) = \prod_{i=0}^{l}[\text{PRF}_K(i)]^{x^i}$. Note that the domain size of PRF' is bounded by $l + 1$, since upon observing $l + 1$ evaluations of PRF' it is possible to interpolate the coefficients of the polynomial $\{\text{PRF}_K(i)\}_i$ (in the exponent). On the other hand, it is easy to verify that if $l + 1 \leq 2^m$ then PRF' is a PRF. The proof is straight forward and thus omitted.

Theorem 5.1. *Assume $F : \{0, 1\}^m \mapsto \mathbb{G}$ is a PRF, then PRF' is a PRF for $(l + 1) \leq 2^m$.*

We implement PRF' using the two PRFs from Section 3.1 and obtain two new PRF constructions under the strong-DDH and DDH assumptions. Let $K = (k_0, k_1)$ be the key for the PRF $\mathcal{PRF}_1$ with the strong-DDH based security, and recall that the closed form efficiency for this function is defined by

$$\mathsf{PRF}'_K(x) = \mathsf{CFEval}_h(x, K) = g^{\frac{k_0(k_1^{d+1}x^{d+1}-1)}{k_1x-1}}.$$

This implies that PRF$'$ only requires a constant number of modular exponentiations. See Section 3.3 for secure implementations of obliviously evaluating PRF$'$. Next, let $K = (k_0, \ldots, k_m)$ be the key for the Naor-Reingold PRF, and recall that the closed form efficiency of this function is defined by

$$\mathsf{PRF}'_K(x) = \mathsf{CFEval}_{h,z}(x, K) = g^{k_0(1+k_1,x)(1+k_2x^2)\ldots(1+k_mx^{2^m})}$$

which requires $O(\log l) = O(m)$ operations, namely, a logarithmic number of operations in the domain size where x is an m-bits string. This is the same order of overhead induced by the [20] implementation that requires an OT for each input bit. Nevertheless, our construction has the advantage that it also achieves easily the property of a committed key since multiple evaluations can be computed with respect to the same PRF key. Plugging-in our protocol inside the protocols for keyword search, OT with adaptive queries [20] and set-intersection [25] implies security against malicious adversaries fairly immediately. It is further useful for search functionalities as demonstrated below.

5.1   The Set-Intersection Protocol

We continue with describing our set-intersection protocol. Informally, P_0 generates a PRF key for PRF and evaluates this function on its set X. It then sends the evaluation results to P_1 and the parties engage in a committed oblivious PRF protocol that evaluates PRF on the set Y. P_1 then concludes the intersection. In order to handle a technicality in the security proof, we require that P_0 must generate its PRF key *independently* of its input X, since otherwise it may maliciously pick a secret key that implies collisions on elements from X and Y, causing the simulation to fail. We ensure key independence by asking the parties to mutually generate the PRF key after P_0 has committed to its input. Then upon choosing the PRF key, the parties invoke two variations of functionality $\mathcal{F}_{\mathrm{CPRF}}$, denoted by $\mathcal{F}^0_{\mathrm{CPRF}}$ and $\mathcal{F}^1_{\mathrm{CPRF}}$. Formally, we define $\mathcal{F}^0_{\mathrm{CPRF}}$ as follows: $((K, (x_1, \ldots, x_{m_X}), R), (c_{\mathrm{KEY}}, (c_1, \ldots, c_{m_X}), \mathrm{PK})) \mapsto (-, (\mathsf{PRF}_K(x_1), \ldots, \mathsf{PRF}_K(x_{m_X})))$ only if c_i encrypts x_i for all i and c_{KEY} is a commitment of K where verification is carried out using randomness R. In the final step, the parties call functionality $\mathcal{F}^1_{\mathrm{CPRF}}$ in order to evaluate the PRF on the set Y and is defined by $((K, R), (c_{\mathrm{KEY}}, (y_1, \ldots, y_{m_Y}))) \mapsto (-, (\mathsf{PRF}_K(y_1), \ldots, \mathsf{PRF}_K(y_{m_Y})))$ only if c_{KEY} is a commitment of K where verification is carried out using randomness R. In both executions the output is given to P_1 that computes the intersection of the results.

Implementing $\mathcal{F}^0_{\mathrm{CPRF}}$ and $\mathcal{F}^1_{\mathrm{CPRF}}$. Implantation-wise, there is not much of a difference between the protocols for the two functionalities, which mainly differ due to the identity of the party that enters the input values to the PRF (where the same committed key is used for both protocol executions). We note that the realization of $\mathcal{F}^0_{\mathrm{CPRF}}$ and $\mathcal{F}^1_{\mathrm{CPRF}}$ can be carried out securely based on the implementations of the closed form efficiency functions shown in Section 3.3, since our committed PRFs are based on these functions. More concretely, the difference with respect to functionality $\mathcal{F}_{\mathrm{CPRF}}^0$ is that now when P_0 is corrupted the simulator needs to extract the randomness used for committing to the PRF key and the x_i's elements which can be achieved using the proof of knowledge π_{DL} since the parties use the El Gamal PKE. Specifically, P_0 proves the knowledge of the discrete logarithm of $(c_1, \ldots c_{m_X})$ with respect to a generator g, by invoking an ideal execution of $\mathcal{F}_{\mathrm{DL}}$ on input $\{((g, c_i), \log_g c_i)\}_{i \in [m_X]}$.[8] The input of P_1 for $\mathcal{F}_{\mathrm{DL}}$ is $\{(g, c_i)\}_{i \in [m_X]}$. In case P_1 does not receive an "accept" message from $\mathcal{F}_{\mathrm{DL}}$ it aborts. Next, the parties continue with the PRF evaluations where the ZK proofs are carried out with respect to the same key commitment. We note that extracting the PRF key and the set $(x_1, \ldots, x_{m_X})$ is already implied by the protocols from Section 3.3 due to the ZK proofs of knowledge. Finally, the implementation of $\mathcal{F}_{\mathrm{CPRF}}^1$ follows similarly but without the additional proof we added above for $\mathcal{F}_{\mathrm{CPRF}}^0$ in order to extract the randomness of the committed input.

Next, we describe our set-intersection protocol using committed oblivious PRF.

Protocol 6 (Protocol $\pi_\cap$ with malicious security from committed oblivious PRF.)

- **Input:** *Party P_0 is given a set X of size m_X. Party P_1 is given a set Y of size m_Y. Both parties are given a security parameter 1^n.*
- **The protocol:**
  1. **Distributed key generation.** *P_0 and P_1 run protocol $\pi_{\mathrm{KeyGen}}(1^n, 1^n)$ in order to generate additive El Gamal public key $\mathrm{PK} = \langle \mathbb{G}, p, g, h \rangle$ where the corresponding shares of the secret key SK are $(\mathrm{SK}_0, \mathrm{SK}_1)$.*
  2. **Input commitment and PRF key generation.** *P_0 sends encryptions of its input X under PK; denote this set of ciphertexts by $C = (c_1, \ldots c_{m_X})$.*
     P_0 invokes $(K, param) \leftarrow \mathsf{KeyGen}(1^n, d = \log(m_X + m_Y))$ where param includes a group description $\mathbb{G}$ of prime order p and a generator g, and sends P_1 param and a ciphertext $\mathsf{Enc}_{\mathrm{PK}}(K; R)$.
     P_1 picks a new key $(K', param) \leftarrow \mathsf{KeyGen}(1^n, d = \log(m_X + m_Y))$ and sends it to P_0. The parties then compute the encryption c_{KEY} of $\widetilde{K} = KK'$, relying on the homomorphic property of El Gamal.
  3. **PRF evaluations on X.** *The parties call functionality $\mathcal{F}_{\mathrm{CPRF}}^0$ where P_0 enters the set X, key $\widetilde{K}$ and randomness R and P_1 enters C, c_{KEY} and PK. Denote by $\mathrm{PRF}_X = \{\mathrm{PRF}'_{\widetilde{K}}(x)\}_{x \in X}$ the output of P_1 from this ideal call only if C is a vector of ciphertexts that encrypts X and c_{KEY} is a commitment of $\widetilde{K}$, where verification is computed using randomness R.*
  4. **Oblivious PRF evaluations on Y.** *The parties call functionality $\mathcal{F}_{\mathrm{CPRF}}^1$ where P_0 enters the key $\widetilde{K}$ and randomness R and P_1 enters the commitment c_{KEY} and the set Y. Denote by $\mathrm{PRF}_Y = \{f_y\}_{y \in Y}$ the output of P_1 from this ideal call only if c_{KEY} is a commitment of $\widetilde{K}$ where verification is computed using randomness R.*
     P_1 outputs all $y \in Y$ for which $f_y \in \mathrm{PRF}_X$.

Theorem 5.2. *Assume $\mathrm{PRF}'_K(\cdot)$ is a PRF defined as above and that El Gamal is IND-CPA, then Protocol 6 securely realizes functionality $\mathcal{F}_\cap$ in the presence of malicious adversaries in the $\{\mathcal{F}_{\mathrm{DL}}, \mathcal{F}_{\mathrm{CPRF}}^0, \mathcal{F}_{\mathrm{CPRF}}^1\}$-hybrid model.*

---

[8] We abuse notation and write $\log c$ to denote the discrete logarithm of the two group elements in ciphertext c.

Proof: We prove security for each corruption case separately. We assume that the simulator is given m_X and m_Y as part of its auxiliary input.

P_0 *is corrupted.* Let $\mathcal{A}$ be a PPT adversary corrupting party P_0, we design a PPT simulator $\mathcal{SIM}$ that generates the view of $\mathcal{A}$ as follows.

1. Given $(1^n, X, z)$, $\mathcal{SIM}$ engages in an execution of $\pi_{\mathrm{KeyGen}}(1^n, 1^n)$ with $\mathcal{A}$. Denote the outcome by PK.
2. Upon receiving from $\mathcal{A}$ its commitment for the PRF key $K \leftarrow \mathsf{KeyGen}(1^n, d = \log(m_X + m_Y))$, $\mathcal{SIM}$ picks a new key share K' and sends it to $\mathcal{A}$ using PK. Denote the combined key by $\widetilde{K} = KK'$.
3. $\mathcal{SIM}$ extracts the adversary's input X' from the input to the ideal call $\mathcal{F}^0_{\mathrm{CPRF}}$. It then sends X' to the trusted party and completes the execution as would the honest P_1 do on an arbitrary set.

In the hybrid setting, computational indistinguishability between the hybrid and simulated executions is trivially claimed since the adversary does not receive any message from P_1 that depends on Y. An important observation here is that the probability of the event for which there exists $y \in Y$ such that $y \notin X'$ and yet $\mathrm{PRF}_{\widetilde{K}}(y) \in \mathrm{PRF}_{X'}$ is negligible, since the key $\widetilde{K}$ is picked independently of the set X'. This argument follows from similarly to the proof in [25] and implies that P_1's output in both executions is identical condition that the above event does not occur.

P_1 *is corrupted.* Let $\mathcal{A}$ be a PPT adversary corrupting party P_1, we design a PPT simulator $\mathcal{SIM}$ that generates the view of $\mathcal{A}$ as follows.

1. Given $(1^n, Y, z)$, $\mathcal{SIM}$ engages in an execution of $\pi_{\mathrm{KeyGen}}(1^n, 1^n)$ with $\mathcal{A}$. Denote the outcome by PK.
2. $\mathcal{SIM}$ picks a PRF key share $K \leftarrow \mathsf{KeyGen}(1^n, d = \log(m_X + m_Y))$ and sends its encryption to $\mathcal{A}$ using PK. Upon receiving $\mathcal{A}$'s key share K' the simulator sets the combined key by $\widetilde{K} = KK'$.
3. $\mathcal{SIM}$ picks a set of m_X arbitrary elements $X_{\mathcal{SIM}}$ from $\mathbb{Z}_p$. It then emulates the ideal call $\mathcal{F}^0_{\mathrm{CPRF}}$ and hands the adversary a random set U of size m_X and proper length.
4. Finally, the simulator extracts the adversary's input Y' to the ideal call $\mathcal{F}^1_{\mathrm{CPRF}}$ and sends this set to the trusted party, receiving back $Z = X \cap Y'$. The simulator completes the execution as follows. For each element $y' \in Y' \cap Z$ it programs the ideal answer of $\mathcal{F}^1_{\mathrm{CPRF}}$ to be $r \in U$ where r is picked from the remaining elements from the set U that were not picked thus far. Otherwise, the simulator returns a fresh random element from $\mathbb{Z}_p$.

Security here follows from the IND-CPA security of the El Gamal PKE and the security of the PRF. That is, the simulated view is different from the hybrid view relative to the encrypted input of P_0 and the fact that the simulator uses a random function to evaluate the sets in $X'_{\mathcal{SIM}}$ and Y'. Therefore, the proof can be shown by defining a hybrid game where in the first game the simulator encrypts P_0's real input X but completes the simulation as in the original simulation. Indistinguishability between the simulation and the hybrid game follows easily by a reduction to the IND-CPA security of El Gamal since the simulator never uses the secret key of the encryption scheme. Indistinguishability between the hybrid game and the hybrid execution follows by a reduction to the pseudorandomness of the PRF. ∎

Efficiency. The overhead of protocol 6 depends on the implementations of $\mathcal{F}^0_{\text{CPRF}}$ and $\mathcal{F}^1_{\text{CPRF}}$ discussed above. Our protocol obtains $O(m_X + m_Y)$ communication and computation overheads under the strong-DDH assumption and $O((m_X + m_Y)\log(m_X + m_Y))$ overheads under the DDH assumption, where the former analysis matches the [31] analysis (such that both constructions rely on dynamic assumptions).

5.2 Search Functionalities

In search functionalities a receiver searches in a sender's database, retrieving the appropriate record(s) according to some search query. The database for search functionalities can be described by pairs of queries/records $\{(q_i, T_i)\}_i$ such that the answer to a query q_i is a record T_i.[9] In a private setting we need to ensure that nothing beyond these records leaks to the receiver, while the sender does not learn anything about the receiver's search queries. Committed oblivious PRF evaluation is a useful tool for securely implementing various search functionalities [20]. First, in the setup phase the database is encoded and handed to the receiver. That is, for each query q_i the sender defines the pair $(\text{PRF}_K(q_i\|1), \text{PRF}_K(q_i\|2) \oplus T_i)$. Next, in the query phase the parties run a committed oblivious PRF evaluation protocol twice such that the sender inputs K and the receiver inputs a query q. The receiver's output are the values $\text{PRF}_K(q\|1)$ and $\text{PRF}_K(q\|2)$, where the first outcome is used to find the encrypted record while the second outcome is used to extract the record. (Alternative implementations involve a single invocation of PRF by splitting $\text{PRF}_K(q)$ into two parts). Examples for such functionalities are keyword search, oblivious transfer with adaptive queries and pattern matching (and all its variants). The functionality of committed oblivious PRF is important in this context since the sender must be enforced to use the same PRF key.

References

1. Aggarwal, G., Mishra, N., Pinkas, B.: Secure computation of the k th-ranked element. In: Cachin, C., Camenisch, J.L. (eds.) EUROCRYPT 2004. LNCS, vol. 3027, pp. 40–55. Springer, Heidelberg (2004)
2. Ateniese, G., De Cristofaro, E., Tsudik, G.: (if) size matters: Size-hiding private set intersection. IACR Cryptology ePrint Archive, 2010:220 (2010)
3. Ateniese, G.: Dagdelen, I. Damgård, D. Venturi. Entangled cloud storage. IACR Cryptology ePrint Archive, 2012:511 (2012)
4. Azar, Y., Broder, A.Z., Karlin, A.R., Upfal, E.: Balanced allocations. SIAM J. Comput. 29(1), 180–200 (1999)

---

[9] This may not be the most concise description of the database but it is the simplest. In particular, it will do for our purposes.

5. Beaver, D.: Foundations of secure interactive computing. In: Feigenbaum, J. (ed.) CRYPTO 1991. LNCS, vol. 576, pp. 377–391. Springer, Heidelberg (1992)
6. Benabbas, S., Gennaro, R., Vahlis, Y.: Verifiable delegation of computation over large datasets. In: Rogaway, P. (ed.) CRYPTO 2011. LNCS, vol. 6841, pp. 111–131. Springer, Heidelberg (2011)
7. Bendlin, R., Damgård, I., Orlandi, C., Zakarias, S.: Semi-homomorphic encryption and multiparty computation. In: Paterson, K.G. (ed.) EUROCRYPT 2011. LNCS, vol. 6632, pp. 169–188. Springer, Heidelberg (2011)
8. Broder, A.Z., Mitzenmacher, M.: Using multiple hash functions to improve IP lookups. In: INFOCOM, pp. 1454–1463 (2001)
9. Chang, Y.-C., Lu, C.-J.: Oblivious polynomial evaluation and oblivious neural learning. Theor. Comput. Sci. 341(1-3), 39–54 (2005)
10. Chaum, D., Pedersen, T.P.: Wallet databases with observers. In: Brickell, E.F. (ed.) CRYPTO 1992. LNCS, vol. 740, pp. 89–105. Springer, Heidelberg (1993)
11. De Cristofaro, E., Kim, J., Tsudik, G.: Linear-complexity private set intersection protocols secure in malicious model. In: Abe, M. (ed.) ASIACRYPT 2010. LNCS, vol. 6477, pp. 213–231. Springer, Heidelberg (2010)
12. De Cristofaro, E., Tsudik, G.: Practical private set intersection protocols with linear complexity. In: Sion, R. (ed.) FC 2010. LNCS, vol. 6052, pp. 143–159. Springer, Heidelberg (2010)
13. Dachman-Soled, D., Malkin, T., Raykova, M., Yung, M.: Efficient robust private set intersection. In: Abdalla, M., Pointcheval, D., Fouque, P.-A., Vergnaud, D. (eds.) ACNS 2009. LNCS, vol. 5536, pp. 125–142. Springer, Heidelberg (2009)
14. Damgård, I., Jurik, M., Nielsen, J.B.: A generalization of paillier's public-key system with applications to electronic voting. Int. J. Inf. Sec. 9(6), 371–385 (2010)
15. Damgård, I., Keller, M., Larraia, E., Pastro, V., Scholl, P., Smart, N.P.: Practical covertly secure MPC for dishonest majority – or: Breaking the SPDZ limits. In: Crampton, J., Jajodia, S., Mayes, K. (eds.) ESORICS 2013. LNCS, vol. 8134, pp. 1–18. Springer, Heidelberg (2013)
16. Damgård, I., Pastro, V., Smart, N., Zakarias, S.: Multiparty computation from somewhat homomorphic encryption. In: Safavi-Naini, R., Canetti, R. (eds.) CRYPTO 2012. LNCS, vol. 7417, pp. 643–662. Springer, Heidelberg (2012)
17. Dodis, Y., Yampolskiy, A.: A verifiable random function with short proofs and keys. In: Vaudenay, S. (ed.) PKC 2005. LNCS, vol. 3386, pp. 416–431. Springer, Heidelberg (2005)
18. Dong, C., Chen, L., Wen, Z.: When private set intersection meets big data: An efficient and scalable protocol. IACR Cryptology ePrint Archive, 2013:515 (2013)
19. Faust, S., Hazay, C., Venturi, D.: Outsourced pattern matching. In: Fomin, F.V., Freivalds, R., Kwiatkowska, M., Peleg, D. (eds.) ICALP 2013, Part II. LNCS, vol. 7966, pp. 545–556. Springer, Heidelberg (2013)
20. Freedman, M.J., Ishai, Y., Pinkas, B., Reingold, O.: Keyword search and oblivious pseudorandom functions. In: Kilian, J. (ed.) TCC 2005. LNCS, vol. 3378, pp. 303–324. Springer, Heidelberg (2005)
21. Freedman, M.J., Nissim, K., Pinkas, B.: Efficient private matching and set intersection. In: Cachin, C., Camenisch, J.L. (eds.) EUROCRYPT 2004. LNCS, vol. 3027, pp. 1–19. Springer, Heidelberg (2004)
22. Gilboa, N.: Two party RSA key generation (Extended abstract). In: Wiener, M. (ed.) CRYPTO 1999. LNCS, vol. 1666, pp. 116–129. Springer, Heidelberg (1999)
23. Goldreich, O., Micali, S., Wigderson, A.: How to play any mental game or a completeness theorem for protocols with honest majority. In: STOC, pp. 218–229 (1987)
24. Hazay, C., Lindell, Y.: Efficient oblivious polynomial evaluation with simulation-based security. IACR Cryptology ePrint Archive, 2009:459 (2009)

25. Hazay, C., Lindell, Y.: Efficient protocols for set intersection and pattern matching with security against malicious and covert adversaries. J. Cryptology 23(3), 422–456 (2010)

26. Hazay, C., Lindell, Y.: Efficient Secure Two-Party Protocols – Techniques and Constructions. Springer (2010)

27. Hazay, C., Mikkelsen, G.L., Rabin, T., Toft, T.: Efficient RSA key generation and threshold paillier in the two-party setting. In: Dunkelman, O. (ed.) CT-RSA 2012. LNCS, vol. 7178, pp. 313–331. Springer, Heidelberg (2012)

28. Hazay, C., Nissim, K.: Efficient set operations in the presence of malicious adversaries. J. Cryptology 25(3), 383–433 (2012)

29. Hazay, C., Toft, T.: Computationally secure pattern matching in the presence of malicious adversaries. In: Abe, M. (ed.) ASIACRYPT 2010. LNCS, vol. 6477, pp. 195–212. Springer, Heidelberg (2010)

30. Ishai, Y., Prabhakaran, M., Sahai, A.: Founding cryptography on oblivious transfer – efficiently. In: Wagner, D. (ed.) CRYPTO 2008. LNCS, vol. 5157, pp. 572–591. Springer, Heidelberg (2008)

31. Jarecki, S., Liu, X.: Efficient oblivious pseudorandom function with applications to adaptive OT and secure computation of set intersection. In: Reingold, O. (ed.) TCC 2009. LNCS, vol. 5444, pp. 577–594. Springer, Heidelberg (2009)

32. Jarecki, S., Liu, X.: Fast secure computation of set intersection. In: Garay, J.A., De Prisco, R. (eds.) SCN 2010. LNCS, vol. 6280, pp. 418–435. Springer, Heidelberg (2010)

33. Kirsch, A., Mitzenmacher, M., Wieder, U.: More robust hashing: Cuckoo hashing with a stash. In: Halperin, D., Mehlhorn, K. (eds.) ESA 2008. LNCS, vol. 5193, pp. 611–622. Springer, Heidelberg (2008)

34. Kissner, L., Song, D.: Privacy-preserving set operations. In: Shoup, V. (ed.) CRYPTO 2005. LNCS, vol. 3621, pp. 241–257. Springer, Heidelberg (2005)

35. Lindell, Y.: Fast cut-and-choose based protocols for malicious and covert adversaries. In: Canetti, R., Garay, J.A. (eds.) CRYPTO 2013, Part II. LNCS, vol. 8043, pp. 1–17. Springer, Heidelberg (2013)

36. Lindell, Y., Oxman, E., Pinkas, B.: The IPS compiler: Optimizations, variants and concrete efficiency. In: Rogaway, P. (ed.) CRYPTO 2011. LNCS, vol. 6841, pp. 259–276. Springer, Heidelberg (2011)

37. Lindell, Y., Pinkas, B.: Privacy preserving data mining. J. Cryptology 15(3), 177–206 (2002)

38. Lindell, Y., Pinkas, B.: Secure two-party computation via cut-and-choose oblivious transfer. In: Ishai, Y. (ed.) TCC 2011. LNCS, vol. 6597, pp. 329–346. Springer, Heidelberg (2011)

39. Micali, S., Rogaway, P.: Secure computation (abstract). In: Feigenbaum, J. (ed.) CRYPTO 1991. LNCS, vol. 576, pp. 392–404. Springer, Heidelberg (1992)

40. Naor, M., Pinkas, B.: Oblivious transfer and polynomial evaluation. In: STOC, pp. 245–254 (1999)

41. Naor, M., Pinkas, B.: Oblivious polynomial evaluation. SIAM J. Comput. 35(5), 1254–1281 (2006)

42. Naor, M., Reingold, O.: Number-theoretic constructions of efficient pseudo-random functions. In: FOCS, pp. 458–467 (1997)

43. Nielsen, J.B., Nordholt, P.S., Orlandi, C., Burra, S.S.: A new approach to practical active-secure two-party computation. In: Safavi-Naini, R., Canetti, R. (eds.) CRYPTO 2012. LNCS, vol. 7417, pp. 681–700. Springer, Heidelberg (2012)

44. Nielsen, J.B., Orlandi, C.: LEGO for two-party secure computation. In: Reingold, O. (ed.) TCC 2009. LNCS, vol. 5444, pp. 368–386. Springer, Heidelberg (2009)

45. Pinkas, B., Schneider, T., Zohner, M.: Faster private set intersection based on OT extension. In: Proceedings of the 23rd USENIX Security Symposium, San Diego, CA, USA, August 20-22, pp. 797–812 (2014)

46. Schnorr, C.-P.: Efficient identification and signatures for smart cards. In: Brassard, G. (ed.) CRYPTO 1989. LNCS, vol. 435, pp. 239–252. Springer, Heidelberg (1990)
47. Schoenmakers, B., Tuyls, P.: Practical two-party computation based on the conditional gate. In: Lee, P.J. (ed.) ASIACRYPT 2004. LNCS, vol. 3329, pp. 119–136. Springer, Heidelberg (2004)
48. Vergnaud, D.: Efficient and secure generalized pattern matching via fast fourier transform. In: Nitaj, A., Pointcheval, D. (eds.) AFRICACRYPT 2011. LNCS, vol. 6737, pp. 41–58. Springer, Heidelberg (2011)
49. Yao, A.C.-C.: How to generate and exchange secrets (extended abstract). In: FOCS, pp. 162–167 (1986)
50. Zhu, H., Bao, F.: Augmented oblivious polynomial evaluation protocol and its applications. In: de Capitani di Vimercati, S., Syverson, P.F., Gollmann, D. (eds.) ESORICS 2005. LNCS, vol. 3679, pp. 222–230. Springer, Heidelberg (2005)

Verifiable Random Functions
from Weaker Assumptions

Tibor Jager

Horst Görtz Institute for IT Security,
Ruhr-University Bochum, Germany
tibor.jager@rub.de

Abstract. The construction of a verifiable random function (VRF) with large input space and full adaptive security from a static, non-interactive complexity assumption, like decisional Diffie-Hellman, has proven to be a challenging task. To
date it is not even clear that such a VRF exists. Most known constructions either
allow only a small input space of polynomially-bounded size, or do not achieve
full adaptive security under a static, non-interactive complexity assumption.

The only known constructions without these restrictions are based on non-
static, so-called "q-type" assumptions, which are parametrized by an integer q.
Since q-type assumptions get stronger with larger q, it is desirable to have q as
small as possible. In current constructions, q is either a polynomial (e.g., Hohenberger and Waters, Eurocrypt 2010) or at least linear (e.g., Boneh *et al.*, CCS
2010) in the security parameter.

We show that it is possible to construct relatively simple and efficient verifiable random functions with full adaptive security and large input space from
non-interactive q-type assumptions, where q is only *logarithmic* in the security
parameter. Interestingly, our VRF is essentially identical to the verifiable *unpredictable* function (VUF) by Lysyanskaya (Crypto 2002), but very different from
Lysyanskaya's VRF from the same paper. Thus, our result can also be viewed as
a new, direct VRF-security proof for Lysyanskaya's VUF. As a technical tool, we
introduce and construct *balanced* admissible hash functions.

1 Introduction

Verifiable random functions. Verifiable random functions (VRFs) can be seen as the
public-key equivalent of pseudorandom functions. Each function V_{sk} is associated with
a secret key sk and a corresponding public verification key vk. Given sk, an element
X from the domain of V_{sk}, and $Y = V_{sk}(X)$, it is possible to create a *non-interactive,
publicly verifiable* proof π that Y was computed correctly. For security, *unique provability* is required. This means that for each X only one unique value Y such that the
statement "$Y = V_{sk}(X)$" can be proven may exist. Note that unique provability is a
very strong requirement: not even the party that creates sk (possibly maliciously) may
be able to create fake proofs. These additional features should not affect the pseudo-
randomness of the function on other inputs. Verifiable random functions are strongly
related to verifiable *unpredictable* functions (VUFs), where the weaker notion of *unpredictability* instead of pseudorandomness is required.

Y. Dodis and J.B. Nielsen (Eds.): TCC 2015, Part II, LNCS 9015, pp. 121–143, 2015.

Their strong properties make VRFs useful for applications like resettable zero-knowledge proofs [30], lottery systems [31], transaction escrow schemes [26], updatable zero-knowledge databases [27], or e-cash [3,4]. VRFs can also be seen as verifiably unique digital signatures (called *invariant signatures* in [23]), their uniqueness makes them *strongly unforgeable* [10,35].

The difficulty of constructing VRFs. In particular the *unique provability* requirement makes it very difficult to construct verifiable random functions. For instance, the natural attempt of combining a pseudorandom function with a non-interactive zero-knowledge proof system fails, since zero-knowledge proofs are inherently simulatable, which contradicts uniqueness. More generally, any reduction which attempts to prove pseudorandomness of a candidate construction faces the following problem.

 - On the one hand, the reduction *must* be able to compute the unique function value $Y := V_{sk}(X)$ for preimages X selected by the attacker, along with a proof of correctness π. Due to the unique provability, there exists only one unique value Y such that the statement "$Y = V_{sk}(X)$" can be proven, thus the reduction is not able to "lie" by outputting false values $\tilde{Y}$.
   Note that this stands in contrast to typical reductions for pseudorandom functions, like the Naor-Reingold construction [33] for instance, where due to the absence of proofs the reduction is be able to output incorrect values.
 - On the other hand, the reduction *must not* be able to compute $Y^* = V_{sk}(X^*)$ for a particular X^*, as it must be able to use an attacker that distinguishes Y^* from random to break a complexity assumption.

Most previous works [29,28,16,17,1] constructed VRFs with only small input spaces of polynomially-bounded size.[1] The only two exceptions are due to Hohenberger and Waters [25] and Boneh *et al.* [9], who constructed verifiable random functions with full adaptive security that allow an input space of exponential size.

VRFs with large input spaces from non-interactive assumptions. Hohenberger and Waters [25] provided the first fully-secure VRF with exponential-size input space whose security is based on a non-interactive complexity assumption. The security proof relies on a q-type assumption, where an algorithm receives as input a list of group elements

$$(g, h, g^x, \dots, g^{x^{q-1}}, g^{x^{q+1}}, \dots, g^{x^{2q}}, T) \in \mathbb{G}^{2q+1} \times \mathbb{G}_T$$

where $e : \mathbb{G} \times \mathbb{G} \to \mathbb{G}_T$ is a bilinear map. The assumption is that no efficient algorithm is able to distinguish $T = e(g, h)^{x^q}$ from a random group element with probability significantly better than $1/2$. The proof given in [25] requires that $q = \Theta(Q \cdot k)$, where k is the security parameter and Q is the number of function evaluations $V_{sk}(X)$ queried by the attacker in the security experiment. Note that in particular Q can be very large, as it is only bounded by a polynomial in the security parameter.

---

[1] Or, alternatively but usually equivalently, based on interactive complexity assumptions or with only weaker *selective* security.

The construction of Boneh *et al.* [9] is based on the assumption where the algorithm receives as input a list of group elements

$$(g, h, g^x, \ldots, g^{x^q}, T) \in \mathbb{G}^{q+2} \times \mathbb{G}_T$$

and the algorithm has to distinguish $T = e(g, h)^{1/x}$ from random. The proof in [9] requires $q = \Theta(k)$. *Is it possible to construct VRFs with large input and full adaptive security from weaker q-type assumptions?*

Our contribution. We construct verifiable random functions with exponential-size input space, full adaptive security, and based on a q-type assumption with *very small q* . More precisely, $q = O(\log k)$ depends only *logarithmically* on the security parameter. The VRF construction essentially corresponds to the verifiable *unpredictable* function of Lysyanskaya [28], which inspired many very similar VRF constructions with either weaker security or based on stronger assumptions [25,1,16].

As a technical tool, we introduce the notion of *balanced* admissible hash functions (balanced AHFs), which are standard admissible hash functions [8] with an extra property (cf. the explanations below and in Section 4), and may be useful for applications beyond VRFs. We show how to construct balanced AHFs from codes with suitable minimal distance.

VRF construction. Let $\mathbb{G}, \mathbb{G}_T$ be groups with bilinear map $e : \mathbb{G} \times \mathbb{G} \to \mathbb{G}_T$, and let $C : \{0,1\}^k \to \{0,1\}^n$ be a hash function. We construct a VRF with domain $\{0,1\}^k$ and range $\mathbb{G}_T$. The verification key of our VRF consists of C along with $2n+2$ random elements of $\mathbb{G}$

$$vk = \big(g, h, (g_{i,j})_{(i,j)\in[n]\times\{0,1\}}\big)$$

The secret key consists of the discrete logarithms $\alpha_{i,j}$ such that $g^{\alpha_{i,j}} = g_{i,j}$ for $(i,j) \in [n] \times \{0,1\}$.

The function is evaluated on input $X \in \{0,1\}^k$ by first computing

$$(C_1, \ldots, C_n) := C(X) \qquad \text{and} \qquad \alpha_X := \prod_{i=1}^{n} \alpha_{i,C_i}$$

and finally

$$V_{sk}(X) := e(g, h)^{\alpha_X}$$

A proof that $V_{sk}(X) = e(g, h)^{\alpha_X}$ consists of group elements $(\pi_1, \ldots, \pi_n)$ where

$$\pi_i := \pi_{i-1}^{\alpha_{i,C_i}}$$

for $i \in [n]$ and with $\pi_0 := g$. Correctness of proofs is verified with the bilinear map.

Similarity to Lysyanskaya's VUF. We note that our VRF construction is nearly identical to a VUF (resp. unique signature) construction of Lysyanskaya [28], but very different from the VRF construction of [28]. To explain this in more detail, recall that Lysyanskaya [28] followed a much more complex approach:

1. Construct a VUF based on a "computational" complexity assumption (in contrast to a "decisional" complexity assumption)
2. Turn this VUF into a VRF with single-bit output, by using a Goldreich-Levin hard-core predicate [22]. This step is not as simple as it may appear, because Micali *et al.* [29] show in their initial VRF paper that this only yields a VRF with *polynomially-bounded* input space (due to the fact that the randomness of the Goldreich-Levin hard-core predicate must be public to allow verifiability, which in turn leads to the problems discussed in [34]).
3. Turn this single-bit-VRF into a VRF with many-bit output (still with poly-bounded input space), by applying a generic construction from [29]. Note that this generic construction requires many evaluations of the underlying single-bit VRF.
4. In order to extend the VRF to a larger input space, apply another generic tree-based construction of [29]. Note that again this requires many evaluations of the underlying VRF.

In contrast, our direct VRF security proof of (essentially) the VUF-construction of Lysyanskaya yields directly a – in comparison much more simple and efficient – VRF with exponential-sized input space, adaptive security, and many-bit output. We rely on the new notion of *balanced* admissible hash functions in our security analysis.

Our security analysis and the need for balanced AHFs. We prove security under the qDDH-assumption, which states that given

$$(g, h, g^x, \ldots, g^{x^q}, T)$$

it is hard to distinguish $T = e(g, h)^{x^{q+1}}$ from random.

A qDDH-challenge is embedded into the view of the attacker by setting

$$g_{i,j} := g^{x + \alpha_{i,j}}$$

where $\alpha_{i,j} \overset{\$}{\leftarrow} \mathbb{Z}_{|G|}$ is a random blinding term, but *only for $O(\log k)$ carefully selected* indices (i, j). This careful embedding essentially *partitions* the domain $\{0, 1\}^k$ of the VRF into two sets $\mathcal{X}_0, \mathcal{X}_1$, such that

– For all values $X \in \mathcal{X}_1$ we have

$$V_{sk}(X) = e(g^{\prod_{i=0}^q \gamma_i x^i}, h) \quad \text{and} \quad \pi_j = g^{\prod_{i=0}^q \gamma_{j,i} x^i} \quad \forall 1 \leq j \leq n \qquad (1)$$

where the γ_i and $\gamma_{j,i}$ are integers in $\mathbb{Z}_{|G|}$ which are known to the reduction. Note that the polynomials in the exponent of Equations (1) have degree at most q, thus $V_{sk}(X)$ and $\pi_1, \ldots, \pi_n$ can be computed, given the values $(g, g^x, \ldots, g^{x^q})$ from the qDDH challenge and the integers $\gamma_i, \gamma_{j,i}$.
– For all values $X^* \in \mathcal{X}_0$ the reduction is able to compute integers γ_i such that

$$Y^* = e(g^{\prod_{i=0}^q \gamma_i x^i}, h) \cdot T^{\gamma_{q+1}}$$

such that if $T = e(g, h)^{x^{q+1}}$ then it holds that $Y^* = V_{sk}(X^*)$. Note that if T is random, then so is Y^*.

Let $\{X^{(1)}, \ldots, X^{(Q)}\}$ denote the set of inputs on which the VRF-attacker queries the evaluation of the VRF with corresponding proof, and let X^* denote the element such that the attacker attempts to distinguish $V_{sk}(X^*)$ from random. The reduction will succeed, if it holds that $\{X^{(1)}, \ldots, X^{(Q)}\} \subseteq \mathcal{X}_1$ and $X^* \in \mathcal{X}_0$.

Instantiating C with an admissible hash function ensures that with non-negligible probability it simultaneously holds that $\{X^{(1)}, \ldots, X^{(Q)}\} \subseteq \mathcal{X}_1$ and $X^* \in \mathcal{X}_0$. However, unfortunately this is not yet sufficient to make the analysis of the success probability of our reduction go through, due to the incompatibility of partitioning proofs with "decisional" complexity assumptions, like qDDH. Intuitively, the problem stems from the fact that two different sequences of queries made by the attacker may cause the simulator to abort with different probabilities. This issue was explained in great detail in [37,5,14].

Therefore we introduce the stronger notion of *balanced* AHFs. Essentially, a balanced AHF ensures that the upper bound γ_{max} and the lower bound γ_{min} on the probability in

$$\gamma_{max} \geq \Pr[\{X^{(1)}, \ldots, X^{(Q)}\} \subseteq \mathcal{X}_1 \wedge X^* \in \mathcal{X}_0] \geq \gamma_{min}$$

are reasonably close. This is a typical requirement for partitioning proofs based on decisional complexity assumptions, it occurs both in reductions with and without the "artificial abort" [37,5]. This suggests that the notion of balanced AHFs may find applications beyond the construction of VRFs.

We stress that we achieve a reduction from a q-type assumption with $q = O(\log k)$ only if we instantiate the VRF construction with a *specific* AHF, essentially the code-based AHF of [19,28]. The reason is that this is the only construction we are aware of which allows us to embed the given qDDH-challenge into at most $O(\log k)$ carefully selected public-key elements $g_{i,j}$ in the way described above. We still have to prove that their AHF is also a *balanced* AHF.

More related work. VRFs were introduced by Micali, Rabin, and Vadhan [29], along with verifiable *unpredictable* functions (VUFs), a generic conversion from VUFs to VRFs based on Goldreich-Levin hard-core predicates [22], and a VUF-construction (with small input space) based on the RSA assumption. Specific, number-theoretic constructions of VRFs can be found in [29,28,16,17,1,25,9]. Note that most of these constructions either do not achieve full adaptive security for large input spaces, or are based on much stronger, *interactive* complexity assumptions. In particular, the VRF construction of Dodis [16] with outer error-correcting code is based on a q-type assumption (there called the sf-DDH *assumption of order q*) with $q = O(\log k)$, but this assumption is *interactive*. We wish to avoid interactive assumptions to prevent circular arguments, as explained by Naor [32].

Abdalla *et al.* gave generic constructions of VRFs from so-called *VRF-suitable identity-based KEMs* [1,2]. While the conference version of this paper [1] considered only selective security, the full version [2] contains proofs that the construction from [1] achieves full security, under either under the complexity assumption from [25] with polynomially-bounded q, or, alternatively, under a q-type assumption with $q = O(k)$ when combined with an admissible hash function.

Brakerski *et al.* [11] introduced the relaxed notion of *weak* VRFs, along with simple and efficient constructions, and proofs that neither VRFs, nor weak VRFs can be constructed (in a black-box way) from one-way permutations. Fiore and Schröder [18] proved that verifiable random functions are not even implied (in a black-box sense) by trapdoor permutations. Several works introduced related primitives, like simulatable VRFs [12] and constrained VRFs [21].

At Eurocrypt 2006, Cheon [15] described an algorithm, which computes the discrete logarithm x on input $(g, g^x, \ldots, g^{x^q})$. This algorithm is faster by a factor of $\sqrt{q}$ than generic algorithms for the standard discrete logarithm problem where only (g, g^x) is given. This shows that q-type assumptions are particularly problematic when q is large. The security loss must be compensated with larger group parameters, at the cost of efficiency. We stress that Cheon's algorithm is only much faster than generic algorithms for the standard discrete logarithm problem if q is very large (say, $q = 2^{40}$). However, Cheon's algorithm gives no apparent reason to criticise q-type assumptions for small q, like $q \leq 40$.

On avoiding q-Type assumptions altogether. Chase and Meiklejohn [13] present a conversion that allows to replace q-type assumption in certain applications with a *static* (that is, not q-type) subgroup hiding assumption, by leveraging the *dual-systems* techniques of Waters [36]. It is natural to ask whether these techniques can be used to construct verifiable random functions from static assumptions. Unfortunately, the conversion of [13] requires to add *randomization*. Thus, when applying it to known VRF constructions like [17], then this contradicts the unique provability requirement. Accordingly, Chase and Meiklejohn were able to prove that the VRF of Dodis and Yampolski [17] forms a secure pseudorandom function under a static assumption, but not that it is a secure verifiable random function.

We leave the construction of a verifiable random function with large input space and full adaptive security from a static assumption, like Decisional Diffie-Hellman, as an open problem.

2 Preliminaries

For a vector $K \in \{0, 1\}^n$ we write K_i to denote the i-th component of K. If A is a finite set, then $a \xleftarrow{\$} A$ denotes the action of sampling a uniformly random from A. If A is a probabilistic algorithm, then we write $a \xleftarrow{\$} A$ to denote the action of computing a by running A with uniformly random coins. We define $[n] := \{1, \ldots, n\} \subset \mathbb{N}$ as the set of all positive integers up to n.

2.1 Verifiable Unpredictable/Random Functions

Let $(\mathsf{Gen}, \mathsf{Eval}, \mathsf{Vfy})$ be the following algorithms.

- Algorithm $(vk, sk) \xleftarrow{\$} \mathsf{Gen}(1^k)$ takes as input a security parameter k and outputs a key pair (vk, sk). We say that sk is the *secret key* and vk is the *verification key*.
- Algorithm $(Y, \pi) \xleftarrow{\$} \mathsf{Eval}(sk, X)$ takes as input secret key sk and $X \in \{0, 1\}^k$, and outputs a function value $Y \in \mathcal{Y}$, where $\mathcal{Y}$ is a finite set, and a proof π. We write $V_{sk}(X)$ to denote the function value Y computed by Eval on input (sk, X).

– Algorithm $\mathsf{Vfy}(vk, X, Y, \pi) \in \{0, 1\}$ takes as input verification key vk, $X \in \{0, 1\}^k$, $Y \in \mathcal{Y}$, and proof π, and outputs a bit.

Initialize :	**Evaluate**(X) :	**Challenge**(X^*) :
$b \xleftarrow{\$} \{0, 1\}$	$(Y, \pi) \xleftarrow{\$} \mathsf{Eval}(sk, X)$	$(Y_0, \pi) \xleftarrow{\$} \mathsf{Eval}(sk, X^*)$
$(vk, sk) \xleftarrow{\$} \mathsf{Gen}(1^k)$	**Return** (Y, π)	$Y_1 \xleftarrow{\$} \mathcal{Y}$
Return vk		**Return** Y_b

Finalize$^{\mathsf{VUF}}(X^*, Y^*)$:	**Finalize**$^{\mathsf{VRF}}(b')$:
$(Y, \pi) \xleftarrow{\$} \mathsf{Eval}(sk, X^*)$	**If** $b' = b$ **then**
If $Y^* = Y$ **then**	**Return** 1
Return 1	**Else Return** 0
Else Return 0	

Fig. 1. Procedures defining the security experiments for VUFs and VRFs

Definition 1. *We say that* $(\mathsf{Gen}, \mathsf{Eval}, \mathsf{Vfy})$ *is a* verifiable random function *(VRF) if all the following properties hold.*

Correctness. *Algorithms* $\mathsf{Gen}, \mathsf{Eval}, \mathsf{Vfy}$ *are polynomial-time algorithms, and for all* $(vk, sk) \xleftarrow{\$} \mathsf{Gen}(1^k)$ *and all* $X \in \{0, 1\}^k$ *holds: if* $(Y, \pi) \xleftarrow{\$} \mathsf{Eval}(sk, X)$, *then* $\mathsf{Vfy}(vk, X, Y, \pi) = 1$.

Unique Provability. *For all* $(vk, sk) \xleftarrow{\$} \mathsf{Gen}(1^k)$ *and all* $X \in \{0, 1\}^k$, *there does not exist any tuple* (Y_0, π_0, Y_1, π_1) *such that* $Y_0 \neq Y_1$ *and* $\mathsf{Vfy}(vk, X, Y_0, \pi_0) = \mathsf{Vfy}(vk, X, Y_1, \pi_1) = 1$.

Pseudorandomness. *Consider an attacker* $\mathcal{A}$ *with access (via oracle queries) to the procedures defined in Figure 1. Let* $G^{\mathcal{A}}_{\mathsf{VRF}}$ *denote the game where* $\mathcal{A}$ *first queries* **Initialize**, *then* **Challenge**, *then* **Finalize**^{VRF}, *where the output of* **Finalize**^{VRF} *is the output of the game. Moreover,* $\mathcal{A}$ *may arbitrarily issue* **Evaluate***-queries, but only after querying* **Initialize** *and before querying* **Finalize**^{VRF}. *We say that* $\mathcal{A}$ *is* legitimate, *if* $\mathcal{A}$ *never queries* **Evaluate**(X) *and* **Challenge**(X^*) *with* $X = X^*$ *throughout the game.*
We define the advantage of $\mathcal{A}$ *in breaking the pseudorandomness as*

$$\mathsf{Adv}^{\mathsf{VRF}}_{\mathcal{A}}(k) := 2 \cdot \Pr[G^{\mathcal{A}}_{\mathsf{VUF}} = 1] - 1$$

Definition 2. *We say that* $(\mathsf{Gen}, \mathsf{Eval}, \mathsf{Vfy})$ *is a* verifiable unpredictable function *(VUF) if the* correctness *and* unique provability *properties from Definition 1 hold, and we have:*

Unpredictability. *Consider an attacker* $\mathcal{A}$ *with access (via oracle queries) to the procedures defined in Figure 1. Let* $G^{\mathcal{A}}_{\mathsf{VUF}}$ *denote the game where* $\mathcal{A}$ *first queries* **Initialize**, *then an arbitrary number of* **Evaluate***-queries, then* **Finalize**^{VUF}, *and the output of* **Finalize**^{VUF} *is the output of the game. We say that* $\mathcal{A}$ *is legitimate, if* $\mathcal{A}$ *never queries* **Evaluate**(X) *and* **Challenge**(X^*) *with* $X = X^*$ *throughout the game.*

We define the advantage of $\mathcal{A}$ *in breaking the unpredictability as*

$$\mathsf{Adv}_{\mathcal{A}}^{\mathsf{VUF}}(k) := \Pr[G_{\mathsf{VUF}}^{\mathcal{A}} = 1]$$

2.2 q-Diffie-Hellman Assumptions

In the sequel let $\mathbb{G}, \mathbb{G}_T$ begroups of prime order, with bilinear map $e : \mathbb{G} \times \mathbb{G} \to \mathbb{G}_T$.

Initialize$^{q\mathsf{CDH}}$:
_____
$g, h \xleftarrow{\$} \mathbb{G}; x \xleftarrow{\$} \mathbb{Z}_{|G|}$
Return $(g, g^x, \ldots, g^{x^q}, h)$

Finalize$^{q\mathsf{CDH}}(T)$:
_____
If $T = e(g^{x^{q+1}}, h)$ **then Return** 1
Else Return 0

Initialize$^{q\mathsf{DDH}}$:
_____
$g, h \xleftarrow{\$} \mathbb{G}; x \xleftarrow{\$} \mathbb{Z}_{|G|}; b \xleftarrow{\$} \{0, 1\}$
$T_0 := e(g, h)^{x^{q+1}}, T_1 \xleftarrow{\$} \mathbb{G}_T$
Return $(g, g^x, \ldots, g^{x^q}, h, T_b)$

Finalize$^{q\mathsf{DDH}}(b')$:
_____
If $b' = b$ **then Return** 1
Else Return 0

Fig. 2. Procedures defining the q-Diffie Hellman assumptions

Definition 3. *Let* $G_{\mathcal{B}}^{q\mathsf{DDH}}$ *be the game with* $\mathcal{B}$ *and the procedures defined in Figure 2, where* $\mathcal{B}$ *calls* **Initialize**$^{q\mathsf{DDH}}$*, then* **Finalize**$^{q\mathsf{DDH}}$*, and the output of* **Finalize**$^{q\mathsf{DDH}}$ *is the output of the game. We denote with*

$$\mathsf{Adv}_{\mathcal{B}}^{q\mathsf{DDH}}(k) := 2 \cdot \Pr\left[G_{\mathcal{B}}^{q\mathsf{DDH}} = 1\right] - 1$$

the advantage of $\mathcal{A}$ *in breaking the qDDH-assumption in* $(\mathbb{G}, \mathbb{G}_T)$.

Definition 4. *Let* $G_{\mathcal{B}}^{q\mathsf{CDH}}$ *be the game with* $\mathcal{B}$ *and the procedures defined in Figure 2, where* $\mathcal{B}$ *calls* **Initialize**$^{q\mathsf{CDH}}$*, then* **Finalize**$^{q\mathsf{CDH}}$*, and the output of* **Finalize**$^{q\mathsf{CDH}}$ *is the output of the game. We denote with*

$$\mathsf{Adv}_{\mathcal{B}}^{q\mathsf{CDH}}(k) := \Pr\left[G_{\mathcal{B}}^{q\mathsf{CDH}} = 1\right]$$

the advantage of $\mathcal{A}$ *in breaking the qCDH-assumption in* $(\mathbb{G}, \mathbb{G}_T)$.

3 Main Construction

Let $\mathbb{G}, \mathbb{G}_T$ be groups of prime order with bilinear map $e : \mathbb{G} \times \mathbb{G} \to \mathbb{G}_T$, such that each group element has a unique representation, and that group membership can be tested efficiently.

Let $\mathcal{VF} = (\text{Gen}, \text{Eval}, \text{Vfy})$ be the following construction.

Generation. Algorithm $\text{Gen}(1^k)$ chooses an admissible hash function $C : \{0,1\}^k \rightarrow \{0,1\}^n$ and two random generators $g, h \xleftarrow{\$} \mathbb{G}$. Then it computes $g_{i,j} := g^{\alpha_{i,j}}$, where $\alpha_{i,j} \xleftarrow{\$} \mathbb{Z}_{|\mathbb{G}|}$ and for $(i,j) \in [n] \times \{0,1\}$. The keys are defined as

$$vk := \left(C, g, h, (g_{i,j})_{(i,j)\in[n]\times\{0,1\}} \right) \quad \text{and} \quad sk := (\alpha_{i,j})_{(i,j)\in[n]\times\{0,1\}}$$

Evaluation. On input $X \in \{0,1\}^k$, algorithm $\text{Eval}(sk, X)$ first computes $C(X)$. For $i \in [n]$ let $C(X)_i$ denote the i-th bit of $C(X) \in \{0,1\}^n$. Then the algorithm determines the function value by computing $a_X := \prod_{i=1}^{n} \alpha_{i,C(X)_i}$ and setting

$$Y := e(g, h)^{a_X}.$$

The corresponding proof $\pi = (\pi_1, \ldots, \pi_n)$ is computed recursively by first defining $\pi_0 := g$ and then setting

$$\pi_i := \pi_{i-1}^{\alpha_{i,C(X)_i}} \quad \text{for all} \quad i \in [n]$$

The algorithm outputs (Y, π).

Verification. Algorithm $\text{Vfy}(vk, X, Y, \pi)$ checks the consistency of π using the bilinear map. It first tests if X and π contain only valid group elements. Then it computes $C(X) = (C(X)_1, \ldots, C(X)_n) \in \{0,1\}^n$, defines $\pi_0 := g$, and outputs 1 if and only if all the following equations are satisfied.

$$e(\pi_i, g) = e(\pi_{i-1}, g_{i,C(X)_i}) \quad \text{for all} \quad i \in [n]$$
$$Y = e(\pi_n, h)$$

It is straightforward to verify that the above construction is *correct* in the sense of Definitions 1 and 2. Furthermore, the *unique provability* follows from the group structure and the fact that even an unbounded attacker is not able to devise a proof π for a different group element. It remains to prove *pseudorandomness*.

4 Balanced Admissible Hash Functions

Standard admissible hash functions (AHFs) were introduced by Boneh and Boyen [8], a simplified definition was given by Freire et al. [19]. For our application, we will need AHFs with stronger properties, therefore we have to extend the notion of AHFs to *balanced* AHFs. The essential difference between balanced AHFs and the standard definition (e.g. [20, Definition 3]) is that previous works required only a reasonable *lower* bound on the probability in Equation (3) below. In contrast, the security analysis of our VRF construction will essentially require reasonable *upper and lower bounds*, and that these bounds are sufficiently close.

Definition 5. *Let $k \in \mathbb{N}$ and $n = n(k)$ be a polynomial, and let $C : \{0,1\}^k \rightarrow \{0,1\}^{n(k)}$ be an efficiently computable function. Let $F_K : \{0,1\}^k \rightarrow \{0,1\}$ be defined as*

$$F_K(X) := \begin{cases} 0, & \text{if } \forall i : C(X)_i = K_i \quad \vee \quad K_i = \perp \\ 1, & \text{else.} \end{cases} \tag{2}$$

We say that C is a balanced admissible hash function *(balanced AHF), if there exists an efficient algorithm* AdmSmp$(1^k, Q, \delta)$, *which takes as input (Q, δ) where $Q = Q(k) \in \mathbb{N}$ is polynomially bounded and $\delta = \delta(k) \in (0, 1]$ is non-negligible, and outputs $K \in \{0, 1, \perp\}^n$ such that for all $X^{(1)}, \ldots, X^{(Q)}, X^* \in \{0, 1\}^k$ with $X^* \notin \{X^{(1)}, \ldots, X^{(Q)}\}$ holds that*

$$\gamma_{\max}(k) \geq \Pr[F_K(X^{(1)}) = \cdots = F_K(X^{(Q)}) = 1 \wedge F_K(X^*) = 0] \geq \gamma_{\min}(k) \quad (3)$$

where $\gamma_{\max}(k)$ and $\gamma_{\min}(k)$ satisfy that the function $\tau(k)$ defined as

$$\tau(k) := 2 \cdot \gamma_{\min}(k) \cdot \delta(k) - \gamma_{\max}(k) + \gamma_{\min}(k) \quad (4)$$

is non-negligible. The probability is taken over the choice of K.

Remark 1. The definition of τ essentially condenses two requirements, namely (1) that $\gamma_{\min}$ is non-negligible, and (2) that the difference $\gamma_{\max} - \gamma_{\min}$ is "reasonably" small, where "reasonably" depends on $\gamma_{\min}$ and δ. The definition of function τ may appear very specific, however, such a term appears typically in security analyses that follow the approach of Bellare and Ristenpart [5]. Therefore we think this is exactly what is needed for typical applications of balanced AHFs. See Lemma 1, for instance.

Instantiating balanced admissible hash functions. Efficient *standard* admissible hash functions are known to exist [28,8,19]. For instance, there is a simple construction from codes with suitable minimal distance [28,19]. In this section we will show that such codes also yield a *balanced* AHF. In contrast to [28,19], we have to show both upper and lower bounds, and choose certain parameters more carefully to ensure that (4) is a non-negligible function.

Theorem 1. *Let $(C_k)_{k \in \mathbb{N}}$ with $C_k : \{0, 1\}^k \to \{0, 1\}^n$ be a family of codes with minimal distance nc for a constant c. Then $(C_k)_{k \in \mathbb{N}}$ is a family of balanced admissible hash functions. Moreover,* AdmSmp$(1^k, Q, \delta)$ *outputs $K \in \{0, 1, \perp\}^n$ with exactly $d = \left\lfloor \frac{\ln(2Q+Q/\delta)}{-\ln((1-c))} \right\rfloor$ components not equal to $\perp$.*

Proof. Consider the algorithm AdmSmp which sets

$$d := \left\lfloor \frac{\ln(2Q + Q/\delta)}{-\ln((1 - c))} \right\rfloor$$

and chooses K uniformly random from $(\{0, 1\} \cup \{\perp\})^n$ with exactly d components not equal to $\perp$.[2]

Fix $X^{(1)}, \ldots, X^{(Q)}, X^* \in \{0, 1\}^k$ with $X^* \notin \{X^{(1)}, \ldots, X^{(Q)}\}$ for the analysis of this algorithm.

Upper bound. Note that we have $\Pr[F_K(X^*) = 0] = 2^{-d}$, and thus

$$\gamma_{\max} := 2^{-d} = \Pr[F_K(X^*) = 0]$$
$$\geq \Pr[F_K(X^*) = 0] \cdot \Pr[F_K(X^{(1)}) = \cdots = F_K(X^{(Q)}) = 1 \mid F_K(X^*)]$$
$$= \Pr[F_K(X^*) = 0 \wedge F_K(X^{(1)}) = \cdots = F_K(X^{(Q)}) = 1].$$

---

[2] Note that this algorithm is identical to the algorithm from [20, Theorem 2], except that we have chosen d slightly differently.

Lower bound. We first observe that for any two strings $X, X^* \in \{0,1\}^k$ with $X \neq X^*$ holds that

$$\Pr[F_K(X) = 0 \mid F_K(X^*) = 0] \leq (1 - c)^d.$$

To see this, consider an experiment where two code words $C(X)$ and $C(X^*)$ are given, with $X, X^* \in \{0,1\}^k$ and $X \neq X^*$, and we sample d pairwise distinct positions $i_1, \ldots, i_d \xleftarrow{\$} [n]$. Since $C(X)$ and $C(X^*)$ differ in at least nc positions, the probability that $C(X)_{i_1} = C(X^*)_{i_1}$ is at most $(n - nc)/n = 1 - c$. The probability that $C(X)_{i_j} = C(X^*)_{i_j}$ for all $j \in [d]$ is thus at most $(1 - c)^d$.

A union bound yields that

$$\Pr[F_K(X^{(1)}) = 0 \vee \cdots \vee F_K(X^{(Q)}) = 0 \mid F_K(X^*) = 0] \leq Q(1 - c)^d$$

which implies

$$\Pr[F_K(X^{(1)}) = 1 \wedge \cdots \wedge F_K(X^{(Q)}) = 1 \mid F_K(X^*) = 0] \geq 1 - Q(1 - c)^d$$

This yields the lower bound

$$\gamma_{\min} := (1 - Q(1 - c)^d) \cdot 2^{-d}$$
$$\leq \Pr[F_K(X^{(1)}) = 1 \wedge \cdots \wedge F_K(X^{(Q)}) = 1 \mid F_K(X^*) = 0] \cdot \Pr[F_K(X^*) = 0]$$
$$= \Pr[F_K(X^{(1)}) = \cdots = F_K(X^{(Q)}) = 1 \wedge F_K(X^*) = 0]$$

Balancedness of bounds. Finally, it remains to show that for polynomial Q and non-negligible δ the function τ from (4) is non-negligible. We first compute (omitting the parameter k from functions to simplify notation):

$$\tau := 2 \cdot \delta \cdot \gamma_{\min} - \gamma_{\max} + \gamma_{\min}$$
$$= 2 \cdot \delta \cdot (1 - Q(1 - c)^d) \cdot 2^{-d} - 2^{-d} + (1 - Q(1 - c)^d) \cdot 2^{-d}$$
$$= 2^{-d} \cdot (2\delta - (2\delta + 1) \cdot Q(1 - c)^d)$$

Now we will show that if d is chosen as above, then both 2^{-d} and $2\delta - (2\delta + 1) \cdot Q(1 - c)^d$ are non-negligible. Thus, their product is non-negligible as well.

We have

$$2^{-d} = 2^{-\left\lfloor \frac{\ln(2Q + Q/\delta)}{-\ln((1-c))} \right\rfloor} \geq 2^{\frac{\ln(2Q + Q/\delta)}{\ln((1-c))}}$$

and

$$2\delta - (2\delta + 1) \cdot Q(1 - c)^d = 2\delta - (2\delta + 1) \cdot Q(1 - c)^{\left\lfloor \frac{\ln(2Q + Q/\delta)}{-\ln((1-c))} \right\rfloor}$$
$$\geq 2\delta - (2\delta + 1) \cdot Q(2Q + Q/\delta)^{-1}$$
$$= 2\delta - (2\delta Q + Q)(2Q + Q/\delta)^{-1}$$
$$= 2\delta - \delta(2\delta Q + Q)(2\delta Q + Q)^{-1} = \delta$$

which both are non-negligible since c is a constant, $Q \in \mathbb{N}$, and $\delta \in (0, 1]$ is non-negligible.

5 $\mathcal{VF}$ is a Verifiable Random Function

Theorem 2. *If $\mathcal{VF}$ is instantiated with the balanced admissible hash function from Theorem 1, then for any legitimate attacker $\mathcal{A}$ that breaks the* pseudorandomness *of $\mathcal{VF}$ in time $t_{\mathcal{A}}$ with advantage $\mathsf{Adv}_{\mathcal{A}}^{\mathsf{VRF}}$ by making at most Q* Eval-*queries, there exists an algorithm $\mathcal{B}$ that breaks the q-DDH assumption with $q = \left\lfloor \frac{\ln(2Q+Q/\delta)}{-\ln((1-c))} \right\rfloor - 1$ in time $t_{\mathcal{B}} \approx t_{\mathcal{A}}$ and with advantage*

$$\mathsf{Adv}_{\mathcal{B}}^{q\mathsf{DDH}}(k) \geq \tau(k)$$

where $2 \cdot \delta$ is a non-negligible lower bound on $\mathsf{Adv}_{\mathcal{A}}^{\mathsf{VRF}}(k)$, and $\tau(k)$ is a non-negligible function.

Initialize :

bad $:= 0$

$K \xleftarrow{\$} \mathsf{AdmSmp}(1^k, Q, \delta)$

For $(i,j) \in [n] \times \{0,1\}$ **do**

$\quad \alpha_{i,j} \xleftarrow{\$} \mathbb{Z}_{|\mathbb{G}|}$

$\quad$ **If** $K_i = j$ **then** $g_{i,j} := g^{x+\alpha_{i,j}}$

$\quad$ **Else** $g_{i,j} := g^{\alpha_{i,j}}$

$vk := \big(C, g, h, (g_{i,j})_{(i,j)}\big)$

Return vk

Evaluate(X) **:**

$(Y, \pi) := \bot$

If $F_K(X) \neq 1$ **then**

$\quad$ bad $:= 1;$

Else

$\quad Y := e(g^{P_{K,n,X(x)}}, h)$

$\quad$ **For** $j \in [n]$ **do**

$\quad\quad \pi_j := g^{P_{K,j,X(x)}}$

$\quad \pi := (\pi_1, \ldots, \pi_n)$

Return (Y, π)

Challenge(X^*) **:**

$Y^* := \bot$

If $F_K(X) = 1$ **then**

$\quad$ bad $:= 1$

Else

$\quad$ Compute $\gamma_0, \ldots, \gamma_{q+1}$ s.t.

$\quad\quad P_{K,n,X^*}(x) = \sum_{i=0}^{q+1} \gamma_i x^i$

$\quad Y^* := T^{\gamma_{q+1}} \cdot \prod_{i=1}^{q} e((g^{x^i})^{\gamma_i}, h)$

Return Y^*

Finalize$^{\mathsf{VRF}}(b')$ **:**

If bad $= 1$ **then** $c' \xleftarrow{\$} \{0,1\}$

Else $c' := b'$

Return c'

Fig. 3. Procedures for the simulation of the VRF pseudorandomness experiment by $\mathcal{B}$

Proof. Algorithm $\mathcal{B}$ receives as input $(g, g^x, \ldots, g^{x^q}, h, T)$ and runs algorithm $\mathcal{A}$ as a subroutine. Whenever $\mathcal{A}$ queries **Initialize, Evaluate, Challenge,** or **Finalize,** $\mathcal{B}$ executes the corresponding procedure from Figure 3. Let us give some remarks on these procedures.

Initialization. The values (g, h, g^x) in **Initialize** are from the qDDH-challenge. Recall that $2 \cdot \delta$ is a non-negligible lower bound on $\mathsf{Adv}_{\mathcal{A}}^{\mathsf{VRF}}(k)$, and Q is the upper bound on the number of **Evaluate**-queries.

Note that $\mathcal{B}$ computes the $g_{i,j}$-values exactly as in the original Gen-algorithm, by choosing $\alpha_{i,j} \xleftarrow{\$} \mathbb{Z}_{|G|}$ and setting $g_{i,j} := g^{\alpha_{i,j}}$, but with the exception that

$$g_{i,K_i} := g^{x + \alpha_{i,K_i}}.$$

for all $(i, j) \in [n] \times \{0, 1\}$ with $K_i = j$. Due to our choice of an admissible hash function according to Theorem 1, there are *exactly* $q + 1$ components K_i of K which are not equal to $\bot$.

Finally, note that all g_{i,K_i}-values are distributed correctly, and that this set-up defines the secret key *implicitly* as $sk := (\log_g g_{i,j})_{(i,j) \in [n] \times \{0,1\}}$. Thus, the function $V_{sk}(X)$ is well-defined for all X (but $\mathcal{B}$ will not be able to evaluate V_{sk} on all inputs X, as explained below).

Helping definitions. In order to explain how $\mathcal{B}$ responds to **Evaluate** and **Challenge** queries made by $\mathcal{A}$, let us define two sets $I_{K,w,X}$ and $J_{K,w,X}$, which depend on an AHF key K, a VRF input $X \in \{0,1\}^k$, and integer $w \in \mathbb{N}$ with $1 \leq w \leq n$, as

$$I_{K,w,X} := \{i \in [w] : K_i = C(X)_i\} \quad \text{and} \quad J_{K,w,X} := [w] \setminus I_{K,w,X}$$

Note that $I_{K,w,X}$ denotes the set of all indices $i \in [w] \subseteq [n]$ such that $K_i = C(X)_i$, and $J_{K,w,X}$ denotes the set of all indices in $[w]$ which are not contained in $I_{K,w,X}$. Based on these sets, we define polynomials $P_{K,w,X}(x)$

$$P_{K,w,X}(x) = \prod_{i \in I_{K,w,X}} (x + \alpha_{i,K_i}) \cdot \prod_{i \in J_{K,w,X}} \alpha_{i,K_i} \in \mathbb{Z}_{|G|}[x]$$

Now we can make the following observations:

1. For all X with $F_K(X) = 1$, the set $I_{K,w,X}$ contains at most q elements, and thus the polynomial $P_{K,w,X}(x)$ has degree at most q.
   This implies that if $F_K(X) = 1$, then $\mathcal{B}$ can efficiently compute $g^{P_{K,w,X}(x)}$ for all $w \in [n]$. To this end, $\mathcal{B}$ first computes the coefficients $\gamma_0, \ldots, \gamma_q$ of the polynomial $P_{K,w,X}(x) = \sum_{i=0}^{q} \gamma_i x^i$ with degree at most q, and then

   $$g^{P_{K,w,X}(x)} := g^{\sum_{i=0}^q \gamma_i x^i} = \prod_{i=0}^{q} (g^{x^i})^{\gamma_i}$$

   using the terms $(g, g^x, \ldots, g^{x^q})$ from the q-DDH challenge.
2. If $F_K(X) = 0$, then $P_{K,n,X}(x)$ has degree $q + 1$. We do not know how $\mathcal{B}$ can efficiently compute $g^{P_{K,n,X}(x)}$ in this case.

*Responding to **Evaluate**-queries.* If $F_K(X) = 1$, then procedure **Evaluate** computes the group elements $g^{P_{K,w,X}(x)}$ as explained above. Note that in this case the response to the **Evaluate**(X)-query of $\mathcal{A}$ is correct. However, if $F_K(X) = 0$, then the response of $\mathcal{B}$ is incorrect.

Responding to the **Challenge***-query.* If $F_K(X^*) = 0$, then procedure **Challenge** computes

$$Y^* := T^{\gamma_{q+1}} \cdot \prod_{i=1}^{q} e((g^{x^i})^{\gamma_i}, h) = T^{\gamma_{q+1}} \cdot e(g^{\sum_{i=1}^{q} \gamma_i x^i}, h)$$

where $\gamma_0, \ldots, \gamma_{q+1}$ are the coefficients of the degree-$(q+1)$-polynomial $P_{K,n,X^*}(x) = \sum_{i=0}^{q+1} \gamma_i x^i$. Note that if $T = e(g,h)^{x^{q+1}}$, then it holds that $Y^* = V_{sk}(X^*)$. Moreover, if T is uniformly random, then so is Y^*.

Analysis of $\mathcal{B}$'s running time. The running time $t_{\mathcal{B}}$ of $\mathcal{B}$ consists essentially of the running time $t_{\mathcal{A}}$ of $\mathcal{A}$ plus a minor number of additional operations, thus we have $t_{\mathcal{B}} \approx t_{\mathcal{A}}$.

Analysis of $\mathcal{B}$'s success probability. The simulation of the challenger by $\mathcal{B}$ is perfect, unless bad := 1 is set. This happens only if $\mathcal{A}$ queries **Evaluate**(X) with $F_K(X) \neq 1$, or **Challenge**(X^*) with $F_K(X^*) = 1$. Since the AHF key K is information-theoretically hidden in vk, the terms γ_{max} and γ_{min} from Equation (3) are upper and lower bounds on the probability that bad := 1 is never set throughout the experiment.

Lemma 1.

$$\mathsf{Adv}_{\mathcal{B}}^{q\mathsf{CDH}}(k) \geq 2 \cdot \gamma_{min} \cdot \delta - \gamma_{max} + \gamma_{min}$$

The proof of Lemma 1 follows the approach of Bellare and Ristenpart [5] very closely, therefore it is deferred to Appendix A. This approach allows us to provide an analysis without the "artificial abort" of Waters [37]. The latter has also been used to analyze the VRF of Hohenberger and Waters [24], but leads to a less tight reduction.

Remark 2. Note that the lower bound on $\mathsf{Adv}_{\mathcal{B}}^{q\mathsf{CDH}}(k)$ in Lemma 1 is only useful, if δ and γ_{min} are non-negligible and γ_{max} and γ_{min} are sufficiently close. This is where we need the balancedness of admissible hash function C.

Observe that since we instantiate C with a balanced AHF and δ is a non-negligible lower bound on $\mathsf{Adv}_{\mathcal{A}}^{\mathsf{VRF}}(k)/2$, the function

$$\tau(k) := 2 \cdot \gamma_{min} \cdot \delta - \gamma_{max} + \gamma_{min}$$

is non-negligible. This concludes the proof of Theorem 2.

6 $\mathcal{VF}$ is a Verifiable Unpredictable Function

In this section we prove that construction $\mathcal{VF}$ also is a secure VUF. Note that this construction is essentially identical to the VUF of Lysyanskaya [28], only the proof is based on a different complexity assumption.

The main purpose of this section is to show that for the VUF-security proof of $\mathcal{VF}$ an even weaker (but still $O(\log k)$) q-type assumption is sufficient. We can base security on a $q\mathsf{CDH}$ assumption that is weaker in two ways. First, it is the *computational* version of the $q\mathsf{DDH}$ assumption. Second, we need only $q = \lfloor (\ln 2Q)/c \rfloor - 1$. Thus, in contrast to the VRF-security proof, q is independent of the advantage of the attacker.

6.1   Admissible Hash Functions

In order to prove that $\mathcal{VF}$ is a VUF, it will suffice to instantiate $\mathcal{VF}$ with a standard (that is, not necessarily balanced) admissible hash function C. We recall the standard definition of admissible hash functions (AHFs) from Freire et al. [19].

Definition 6 ([19]). *Let $k \in \mathbb{N}$ and $n = n(k)$ be a polynomial, and let $C : \{0,1\}^k \to \{0,1\}^{n(k)}$ be an efficiently computable function. Let $F_K : \{0,1\}^k \to \{0,1\}$ be defined as in Equation (2). We say that C is an* admissible hash function *(AHF), if there exists an efficient algorithm* $\mathsf{AdmSmp}(1^k, Q)$, *which takes as input polynomial $Q = Q(k) \in \mathbb{N}$, and computes $K \in (\{0,1\} \cup \{\bot\})^n$ such that for all $X^{(1)}, \ldots, X^{(Q)}, X^* \in \{0,1\}^k$ with $X^* \notin \{X^{(1)}, \ldots, X^{(Q)}\}$ holds that*

$$\Pr[F_K(X^{(1)}) = \cdots = F_K(X^{(Q)}) = 1 \ \wedge \ F_K(X^*) = 0] \geq \gamma_{\min}(k) \qquad (5)$$

such that $\gamma_{\min}(k)$ non-negligible. The probability is taken over the choice of K.

Instantiating Admissible Hash Functions. A simple and efficient construction of AHFs can be found in [19] (based on [28]), we capture their existence in the following lemma.

Lemma 2 ([28,19]). *Let S be a set and $(C_k)_{k\in\mathbb{N}}$ with $C_k : \{0,1\}^k \to S^n$ be a family of codes, with minimal distance nc for a constant c and such that $|S|$ is bounded by a polynomial in k. Then $(C_k)_{k\in\mathbb{N}}$ is an admissible hash function, where $\mathsf{AdmSmp}(Q)$ outputs $K \in S \cup \{\bot\}^n$ with exactly $d := \lfloor (\ln 2Q)/c \rfloor$ components not equal to $\bot$ and $\gamma_{\min} \geq (1 - Q(1-c)^d) \cdot 2^{-d}$.*

Remark 3. Note that even though the last two statements of the above theorem were not made explicit in previous works, they are implicitly contained in the proof of [20, Theorem 2].

6.2   Security Analysis

Theorem 3. *If $\mathcal{VF}$ is instantiated with the admissible hash function from Lemma 2, then for any legitimate attacker $\mathcal{A}$ that breaks the* unpredictability *of $\mathcal{VF}$ in time $t_{\mathcal{A}}$ with advantage $\mathsf{Adv}_{\mathcal{A}}^{\mathsf{VUF}}$ by making at most Q Eval-queries, there exists an algorithm $\mathcal{B}$ that breaks the qCDH assumption with $q = \lfloor (\ln 2Q)/c \rfloor - 1$ in time $t_{\mathcal{B}} \approx t_{\mathcal{A}}$ and with advantage*

$$\mathsf{Adv}_{\mathcal{B}}^{\mathsf{qCDH}}(k) \geq \mathsf{Adv}_{\mathcal{A}}^{\mathsf{VUF}}(k) \cdot (1 - Q(1-c)^d) \cdot 2^{-d}$$

where $d := \lfloor (\ln 2Q)/c \rfloor = q + 1$.

The proof of this theorem is nearly identical to the proof of Theorem 2, but the analysis of the success probability of $\mathcal{B}$ is much simpler, because we consider *unpredictability* instead of *pseudorandomness*. Therefore we only sketch the proof.

Proof. Algorithm $\mathcal{B}$ receives as input $(g, g^x, \ldots, g^{x^q}, h, T)$ and runs algorithm $\mathcal{A}$ as a subroutine. Whenever $\mathcal{A}$ issues a query (**Initialize, Evaluate, Finalize**), then $\mathcal{B}$ executes the corresponding procedure from Figure 4.

Initialize(X) :

bad $:= 0$

$K \overset{\$}{\leftarrow} \mathrm{AdmSmp}(1^k, Q, \delta)$

For $(i,j) \in [n] \times \{0,1\}$ **do**

$\quad \alpha_{i,j} \overset{\$}{\leftarrow} \mathbb{Z}_{|G|}$

$\quad$ **If** $K_i = j$ **then** $h_{i,j} := g^{x + \alpha_{i,j}}$

$\quad$ **Else** $h_{i,j} := g^{\alpha_{i,j}}$

$vk := \left(C, g, h, (h_{i,j})_{(i,j)} \right)$

Return vk

Evaluate(X) :

$(Y, \pi) := \perp$

If $F_K(X) \neq 1$ **then**

$\quad$ bad $:= 1$;

Else

$\quad Y := e(g^{P_{K,n,X}(x)}, h)$

$\quad$ **For** $j \in [n]$ **do**

$\quad\quad \pi_j := g^{P_{K,j,X}(x)}$

$\quad \pi := (\pi_1, \ldots, \pi_n)$

Return (Y, π)

Finalize$^{\mathsf{VUF}}(X^*, Y^*)$:

If $F_K(X^*) = 0$ **then**

$\quad$ bad $:= 1$

If bad $= 1$ **then Return** $\perp$

Compute $\gamma_0, \ldots, \gamma_{q+1}$

$\quad$ s.t. $P_{K,n,X^*}(x) = \sum_{i=0}^{q+1} \gamma_i x^i$

$T := \left(Y^* / e(g^{\sum_{i=1}^q \gamma_i x^i}, h) \right)^{1/\gamma_{q+1}}$

Return T

Fig. 4. Procedures for the simulation of the VUF unpredictability experiment by $\mathcal{B}$

The running time $t_{\mathcal{B}}$ of $\mathcal{B}$ consists essentially of the running time $t_{\mathcal{A}}$ of $\mathcal{A}$ plus a minor number of additional operations, thus we have $t_{\mathcal{B}} \approx t_{\mathcal{A}}$. Note that $\mathcal{B}$ simulates the original VUF security experiment perfectly, if bad $= 0$ throughout the game. Note also that

$$Y^* = e(g,h)^{\sum_{i=0}^{q+1} \gamma_i x^i} \implies T = e(g,h)^{x^{q+1}}$$

The choice of K is information-theoretically hidden in vk. Thus,

$$\mathsf{Adv}_{\mathcal{B}}^{q\mathsf{CDH}}(k) \geq \mathsf{Adv}_{\mathcal{A}}^{\mathsf{VUF}}(k) \cdot \Pr[\mathsf{bad} = 0]$$

$$\geq \mathsf{Adv}_{\mathcal{A}}^{\mathsf{VUF}}(k) \cdot \gamma_{\min}(k) = \mathsf{Adv}_{\mathcal{A}}^{\mathsf{VUF}}(k) \cdot (1 - Q(1-c)^d) \cdot 2^{-d}$$

Acknowledgements. We thank the anonymous reviewers of TCC 2015 for their helpful comments.

References

1. Abdalla, M., Catalano, D., Fiore, D.: Verifiable random functions from identity-based key encapsulation. In: Joux, A. (ed.) EUROCRYPT 2009. LNCS, vol. 5479, pp. 554–571. Springer, Heidelberg (2009)
2. Abdalla, M., Catalano, D., Fiore, D.: Verifiable random functions: Relations to identity-based key encapsulation and new constructions. Journal of Cryptology 27(3), 544–593 (2014)

3. Au, M.H., Susilo, W., Mu, Y.: Practical compact e-cash. In: Pieprzyk, J., Ghodosi, H., Dawson, E. (eds.) ACISP 2007. LNCS, vol. 4586, pp. 431–445. Springer, Heidelberg (2007)
4. Belenkiy, M., Chase, M., Kohlweiss, M., Lysyanskaya, A.: Compact e-cash and simulatable VRFs revisited. In: Shacham, H., Waters, B. (eds.) Pairing 2009. LNCS, vol. 5671, pp. 114–131. Springer, Heidelberg (2009)
5. Bellare, M., Ristenpart, T.: Simulation without the artificial abort: Simplified proof and improved concrete security for Waters' IBE scheme. In: Joux, A. (ed.) EUROCRYPT 2009. LNCS, vol. 5479, pp. 407–424. Springer, Heidelberg (2009)
6. Bellare, M., Ristenpart, T.: Simulation without the artificial abort: Simplified proof and improved concrete security for Waters' IBE scheme. Cryptology ePrint Archive, Report 2009/084 (2009), http://eprint.iacr.org/
7. Bellare, M., Rogaway, P.: The security of triple encryption and a framework for code-based game-playing proofs. In: Vaudenay, S. (ed.) EUROCRYPT 2006. LNCS, vol. 4004, pp. 409–426. Springer, Heidelberg (2006)
8. Boneh, D., Boyen, X.: Secure identity based encryption without random oracles. In: Franklin, M. (ed.) CRYPTO 2004. LNCS, vol. 3152, pp. 443–459. Springer, Heidelberg (2004)
9. Boneh, D., Montgomery, H.W., Raghunathan, A.: Algebraic pseudorandom functions with improved efficiency from the augmented cascade. In: Al-Shaer, E., Keromytis, A.D., Shmatikov, V. (eds.) ACM CCS 2010, Chicago, Illinois, USA, October 4–8, pp. 131–140. ACM Press (2010)
10. Boneh, D., Shen, E., Waters, B.: Strongly unforgeable signatures based on computational Diffie-Hellman. In: Yung, M., Dodis, Y., Kiayias, A., Malkin, T. (eds.) PKC 2006. LNCS, vol. 3958, pp. 229–240. Springer, Heidelberg (2006)
11. Brakerski, Z., Goldwasser, S., Rothblum, G.N., Vaikuntanathan, V.: Weak verifiable random functions. In: Reingold, O. (ed.) TCC 2009. LNCS, vol. 5444, pp. 558–576. Springer, Heidelberg (2009)
12. Chase, M., Lysyanskaya, A.: Simulatable VRFs with applications to multi-theorem NIZK. In: Menezes, A. (ed.) CRYPTO 2007. LNCS, vol. 4622, pp. 303–322. Springer, Heidelberg (2007)
13. Chase, M., Meiklejohn, S.: Déjà Q: Using dual systems to revisit q-type assumptions. In: Nguyen, P.Q., Oswald, E. (eds.) EUROCRYPT 2014. LNCS, vol. 8441, pp. 622–639. Springer, Heidelberg (2014)
14. Chatterjee, S., Sarkar, P.: HIBE with short public parameters without random oracle. In: Lai, X., Chen, K. (eds.) ASIACRYPT 2006. LNCS, vol. 4284, pp. 145–160. Springer, Heidelberg (2006)
15. Cheon, J.H.: Security analysis of the strong Diffie-Hellman problem. In: Vaudenay, S. (ed.) EUROCRYPT 2006. LNCS, vol. 4004, pp. 1–11. Springer, Heidelberg (2006)
16. Dodis, Y.: Efficient construction of (distributed) verifiable random functions. In: Desmedt, Y.G. (ed.) PKC 2003. LNCS, vol. 2567, pp. 1–17. Springer, Heidelberg (2002)
17. Dodis, Y., Yampolskiy, A.: A verifiable random function with short proofs and keys. In: Vaudenay, S. (ed.) PKC 2005. LNCS, vol. 3386, pp. 416–431. Springer, Heidelberg (2005)
18. Fiore, D., Schröder, D.: Uniqueness Is a Different Story: Impossibility of Verifiable Random Functions from Trapdoor Permutations. In: Cramer, R. (ed.) TCC 2012. LNCS, vol. 7194, pp. 636–653. Springer, Heidelberg (2012)
19. Freire, E.S.V., Hofheinz, D., Paterson, K.G., Striecks, C.: Programmable hash functions in the multilinear setting. In: Canetti, R., Garay, J.A. (eds.) CRYPTO 2013, Part I. LNCS, vol. 8042, pp. 513–530. Springer, Heidelberg (2013)
20. Freire, E.S.V., Hofheinz, D., Paterson, K.G., Striecks, C.: Programmable hash functions in the multilinear setting. Cryptology ePrint Archive, Report 2013/354 (2013), http://eprint.iacr.org/

21. Fuchsbauer, G.: Constrained Verifiable Random Functions. In: Abdalla, M., De Prisco, R. (eds.) SCN 2014. LNCS, vol. 8642, pp. 95–114. Springer, Heidelberg (2014)

22. Goldreich, O., Levin, L.A.: A hard-core predicate for all one-way functions. In: 21st ACM STOC, Seattle, Washington, USA, May 15–17, pp. 25–32. ACM Press (1989)

23. Goldwasser, S., Ostrovsky, R.: Invariant signatures and non-interactive zero-knowledge proofs are equivalent (extended abstract). In: Brickell, E.F. (ed.) CRYPTO 1992. LNCS, vol. 740, pp. 228–245. Springer, Heidelberg (1993)

24. Hohenberger, S., Waters, B.: Realizing Hash-and-Sign Signatures under Standard Assumptions. In: Joux, A. (ed.) EUROCRYPT 2009. LNCS, vol. 5479, pp. 333–350. Springer, Heidelberg (2009)

25. Hohenberger, S., Waters, B.: Constructing verifiable random functions with large input spaces. In: Gilbert, H. (ed.) EUROCRYPT 2010. LNCS, vol. 6110, pp. 656–672. Springer, Heidelberg (2010)

26. Jarecki, S., Shmatikov, V.: Handcuffing big brother: an abuse-resilient transaction escrow scheme. In: Cachin, C., Camenisch, J.L. (eds.) EUROCRYPT 2004. LNCS, vol. 3027, pp. 590–608. Springer, Heidelberg (2004)

27. Liskov, M.: Updatable zero-knowledge databases. In: Roy, B. (ed.) ASIACRYPT 2005. LNCS, vol. 3788, pp. 174–198. Springer, Heidelberg (2005)

28. Lysyanskaya, A.: Unique signatures and verifiable random functions from the DH-DDH separation. In: Yung, M. (ed.) CRYPTO 2002. LNCS, vol. 2442, pp. 597–612. Springer, Heidelberg (2002)

29. Micali, S., Rabin, M.O., Vadhan, S.P.: Verifiable random functions. In: 40th FOCS, October 17–19, pp. 120–130. IEEE Computer Society Press, New York (1999)

30. Micali, S., Reyzin, L.: Soundness in the public-key model. In: Kilian, J. (ed.) CRYPTO 2001. LNCS, vol. 2139, pp. 542–565. Springer, Heidelberg (2001)

31. Micali, S., Rivest, R.L.: Micropayments revisited. In: Preneel, B. (ed.) CT-RSA 2002. LNCS, vol. 2271, pp. 149–163. Springer, Heidelberg (2002)

32. Naor, M.: On cryptographic assumptions and challenges (invited talk). In: Boneh, D. (ed.) CRYPTO 2003. LNCS, vol. 2729, pp. 96–109. Springer, Heidelberg (2003)

33. Naor, M., Reingold, O.: Number-theoretic constructions of efficient pseudo-random functions. In: 38th FOCS, Miami Beach, Florida, October 19–22, pp. 458–467. IEEE Computer Society Press (1997)

34. Naor, M., Reingold, O.: From unpredictability to indistinguishability: A simple construction of pseudo-random functions from MACs (extended abstract). In: Krawczyk, H. (ed.) CRYPTO 1998. LNCS, vol. 1462, pp. 267–282. Springer, Heidelberg (1998)

35. Steinfeld, R., Pieprzyk, J., Wang, H.: How to Strengthen Any Weakly Unforgeable Signature into a Strongly Unforgeable Signature. In: Abe, M. (ed.) CT-RSA 2007. LNCS, vol. 4377, pp. 357–371. Springer, Heidelberg (2006)

36. Waters, B.: Dual system encryption: Realizing fully secure IBE and HIBE under simple assumptions. In: Halevi, S. (ed.) CRYPTO 2009. LNCS, vol. 5677, pp. 619–636. Springer, Heidelberg (2009)

37. Waters, B.: Efficient identity-based encryption without random oracles. In: Cramer, R. (ed.) EUROCRYPT 2005. LNCS, vol. 3494, pp. 114–127. Springer, Heidelberg (2005)

A   Proof of Lemma 1

Let $G_{\mathcal{B}(\mathcal{A})}^{q\mathsf{DDH}}$ denote the $q\mathsf{DDH}$ security experiment with $\mathcal{B}$ running $\mathcal{A}$ as a subroutine as described above. Let good denote the event that variable bad is never set to 1. Then, since $\mathcal{B}$ outputs a random bit if bad $:= 1$ is set, it holds that

$$\Pr[G_{\mathcal{B}(\mathcal{A})}^{q\mathsf{DDH}} = 1] = \Pr[G_{\mathcal{B}(\mathcal{A})}^{q\mathsf{DDH}} = 1 \wedge \mathsf{good}] + \Pr[\neg\mathsf{good}] \cdot \Pr[G_{\mathcal{B}(\mathcal{A})}^{q\mathsf{DDH}} = 1 \mid \neg\mathsf{good}]$$

$$= \Pr[G_{\mathcal{B}(\mathcal{A})}^{q\mathsf{DDH}} = 1 \wedge \mathsf{good}] + \Pr[\neg\mathsf{good}] \cdot 1/2$$

and therefore

$$\mathsf{Adv}_{\mathcal{B}}^{q\mathsf{DDH}}(k) = 2 \cdot \Pr[G_{\mathcal{B}(\mathcal{A})}^{q\mathsf{DDH}} = 1] - 1$$

$$= 2 \cdot \Pr[G_{\mathcal{B}(\mathcal{A})}^{q\mathsf{DDH}} = 1 \wedge \mathsf{good}] - \Pr[\mathsf{good}] \tag{6}$$

Thus, it remains to derive suitable bounds on $\Pr[G_{\mathcal{B}(\mathcal{A})}^{q\mathsf{DDH}} = 1 \wedge \mathsf{good}]$ and $\Pr[\mathsf{good}]$. We will need the following lemma from [5,7].

Lemma 3 ([5,7]). *Let G_i and G_j be two games which proceed identical until* bad $= 1$. *Then*

- $\Pr[G_i \text{ sets } \mathsf{bad} = 1] = \Pr[G_j \text{ sets } \mathsf{bad} = 1]$
- $\Pr[G_i = b \wedge G_i \text{ does not set } \mathsf{bad} = 1] = \Pr[G_j = b \wedge G_j \text{ does not set } \mathsf{bad} = 1]$
  for any b.

A simpler-to-analyze game. Following Bellare and Ristenpart [5], we now gradually make changes to game $G_{\mathcal{B}(\mathcal{A})}^{q\mathsf{DDH}}$, until we reach game G_3, which will be easier to analyze. In the sequel let good_i denote the event that bad is never set to bad $= 1$ in Game i.

Game 0. We define $G_0 := G_{\mathcal{B}(\mathcal{A})}^{q\mathsf{DDH}}$, which implies

$$\Pr[G_{\mathcal{B}(\mathcal{A})}^{q\mathsf{DDH}} = 1 \wedge \mathsf{good}] = \Pr[G_0 = 1 \wedge \mathsf{good}_0] \quad \text{and} \quad \Pr[\mathsf{good}] = \Pr[\mathsf{good}_0]$$

Game 1. In this game the procedures **Initialize**$_1$, **Evaluate**$_1$, **Challenge**$_1$, and **Finalize**$_1$ described in Figure 5 are used. Note that **Initialize**$_1$ generates a normal VRF key pair (vk, sk), and **Evaluate**$_1$ and **Challenge**$_1$ use the secret key sk to evaluate the VRF and to create the challenge.

However, note that sk is only used in **Evaluate**$_1(X)$-queries with $F_K(X) = 1$, and **Challenge**$_1(X^*)$-queries with $F_K(X^*) = 0$. This mimics the simulation of $\mathcal{B}$ perfecty, in particular all outputs computed by these procedures are distributed *exactly* like in Game 0. This implies that

$$\Pr[G_1 = 1 \wedge \mathsf{good}_1] = \Pr[G_0 = 1 \wedge \mathsf{good}_0] \quad \text{and} \quad \Pr[\mathsf{good}_1] = \Pr[\mathsf{good}_0]$$

Procedures for Game G_1:

Evaluate$_1(X)$:

$(Y, \pi) := \perp$
If $F_K(X) \neq 1$ then
  bad $:= 1$
Else
  $(Y, \pi) \xleftarrow{\$} \mathsf{Eval}(sk, X)$
Return (Y, π)

Challenge$_1(X^*)$:

$Y^* := \perp$
If $F_K(X) = 1$ then
  bad $:= 1$
Else
  If $b = 1$ then
    $(Y^*, \pi) \xleftarrow{\$} \mathsf{Eval}(sk, X)$
  Else $Y^* \xleftarrow{\$} \mathbb{G}_T$
  Return Y^*

Finalize$_1(b')$:

If bad $= 1$ then $c' \xleftarrow{\$} \{0, 1\}$
Else $c' := b'$
If $c' = b$ then Return 1
Else Return 0

Initialize$_1(X)$:

bad $:= 0$
$(vk, sk) \xleftarrow{\$} \mathsf{Gen}_C(1^k)$
$b \xleftarrow{\$} \{0, 1\}$
$K \xleftarrow{\$} \mathsf{AdmSmp}(1^k, Q, \delta)$
Return vk

Procedures for Game G_2 (new instructions are highlighted in boxes):

Evaluate$_2(X)$:

$(Y, \pi) := \perp$
If $F_K(X) \neq 1$ then
  bad $:= 1$
  $\boxed{(Y, \pi) \xleftarrow{\$} \mathsf{Eval}(sk, X)}$
Else
  $(Y, \pi) \xleftarrow{\$} \mathsf{Eval}(sk, X)$
Return (Y, π)

Challenge$_2(X^*)$:

$Y^* := \perp$
If $F_K(X) = 1$ then
  bad $:= 1$
  $\boxed{\text{If } b = 1 \text{ then}}$
    $\boxed{(Y^*, \pi) \xleftarrow{\$} \mathsf{Eval}(sk, X)}$
  $\boxed{\text{Else } Y^* \xleftarrow{\$} \mathbb{G}_T}$
Else
  If $b = 1$ then
    $(Y^*, \pi) \xleftarrow{\$} \mathsf{Eval}(sk, X)$
  Else $Y^* \xleftarrow{\$} \mathbb{G}_T$
  Return Y^*

Finalize$_2(b')$:

If bad $= 1$ then $\boxed{c' := b'}$
Else $c' := b'$
If $c' = b$ then Return 1
Else Return 0

Procedures for Game G_3 (new instructions are highlighted in boxes):

Evaluate$_3(X)$:

$\boxed{\mathbf{X} := \mathbf{X} \cup \{X\}}$
$(Y, \pi) \xleftarrow{\$} \mathsf{Eval}(sk, X)$
Return (Y, π)

Challenge$_3(X^*)$:

If $b = 1$ then
  $(Y^*, \pi) \xleftarrow{\$} \mathsf{Eval}(sk, X)$
Else $Y^* \xleftarrow{\$} \mathbb{G}_T$
Return Y^*

Initialize$_3(X)$:

bad $:= 0$
$(vk, sk) \xleftarrow{\$} \mathsf{Gen}_C(1^k)$
$b \xleftarrow{\$} \{0, 1\}$
$\boxed{\mathbf{X} := \emptyset}$
Return vk

Finalize$_3(b')$:

$\boxed{K \xleftarrow{\$} \mathsf{AdmSmp}(1^k, Q, \delta)}$
$\boxed{\text{For } X \in \mathbf{X} \text{ do}}$
  $\boxed{\text{If } F_K(X) \neq 1 \text{ then bad} := 1}$
$\boxed{\text{If } F_K(X^*) = 1 \text{ then bad} := 1}$
$c' := b'$
If $c' = b$ then Return 1
Else Return 0

Fig. 5. Procedures defining the sequence of games in the proof of Lemma 1

Game 2. In this game we set $\mathbf{Initialize}_2 := \mathbf{Initialize}_1$, and define $\mathbf{Finalize}_2$, $\mathbf{Evaluate}_2$, and $\mathbf{Challenge}_2$ as depicted in Figure 5. Note that Games G_2 and G_1 proceed identical until bad is set, thus by Lemma 3 we have

$$\Pr[G_2 = 1 \wedge \mathsf{good}_2] = \Pr[G_1 = 1 \wedge \mathsf{good}_1] \quad \text{and} \quad \Pr[\mathsf{good}_2] = \Pr[\mathsf{good}_1]$$

Game 3. Note that the outputs of procedures $\mathbf{Evaluate}_2$ and $\mathbf{Challenge}_2$ are independent of K, only $\mathbf{Finalize}_2$ depends on K. Therefore we can simplify our description of the game, by choosing K only at the end of the game, and checking only then if bad needs to be set to bad $:= 1$.

Formally, in Game G_3 the procedures $\mathbf{Initialize}_3$, $\mathbf{Evaluate}_3$, $\mathbf{Challenge}_3$, and $\mathbf{Finalize}_3$ described in Figure 5 are used. All changes are purely conceptual, thus we have

$$\Pr[G_3 = 1 \wedge \mathsf{good}_3] = \Pr[G_2 = 1 \wedge \mathsf{good}_2] \quad \text{and} \quad \Pr[\mathsf{good}_3] = \Pr[\mathsf{good}_2]$$

Note also that now K is chosen only *after* $\mathcal{A}$ asks $\mathbf{Finalize}_3$.

Analysis of Game G_3. It remains to derive bounds on $\Pr[G_3 = 1 \wedge \mathsf{good}_3]$ and $\Pr[\mathsf{good}_3]$. Let $\mathcal{X}$ denote the set

$$\mathcal{X} := \{(X^{(1)}, \ldots, X^{(Q)}, X^*) : X^* \neq X^{(i)}, 1 \leq i \leq Q\}$$

of all sequences of queries a legitimate attacker $\mathcal{A}$ may ask, and let $\mathbf{X}^* \in \mathcal{X}$. Let $\gamma(\mathbf{X}^*)$ denote the probability of good_3 (over the choice of K), if the particular sequence $\mathbf{X}^*$ of queries is asked. Note that $\gamma(\mathbf{X}^*)$ equals the probability in Equation (3), so that $\gamma_{\min}$ is a lower bound on the smallest value of $\gamma(\mathbf{X}^*)$ over all $\mathbf{X}^* \in \mathcal{X}$, and $\gamma_{\max}$ is an upper bound on the largest value of $\gamma(\mathbf{X}^*)$ over all $\mathbf{X}^* \in \mathcal{X}$. Let $Q(\mathbf{X}^*)$ denote the event that the execution of Game G_3 results in the particular sequence $\mathbf{X}^*$. Then we can state the following lemma (which corresponds to [6, Lemma 3.4]).

Lemma 4. *For any $\mathbf{X}^*$ as defined above holds that*

$$\Pr[G_3 = 1 \wedge \mathsf{good}_3 \wedge Q(\mathbf{X}^*)] = \gamma(\mathbf{X}^*) \cdot \Pr[G_3 = 1 \wedge Q(\mathbf{X}^*)]$$
$$\Pr[\mathsf{good}_3 \wedge Q(\mathbf{X}^*)] = \gamma(\mathbf{X}^*) \cdot \Pr[Q(\mathbf{X}^*)]$$

The proof of Lemma 4 is nearly identical to the proof of [6, Lemma 3.4], and therefore deferred to Appendix B.

Now we can compute

$$\mathsf{Adv}_{\mathcal{B}}^{q\mathsf{DDH}}(k) = 2 \cdot \Pr[G_{\mathcal{B}(\mathcal{A})}^{q\mathsf{DDH}} = 1 \wedge \mathsf{good}] - \Pr[\mathsf{good}] \tag{7}$$

$$= 2 \cdot \Pr[G_3 = 1 \wedge \mathsf{good}_3] - \Pr[\mathsf{good}_3] \tag{8}$$

$$= 2 \cdot \sum_{\mathbf{X}^* \in \mathcal{X}} \Pr[G_3 = 1 \wedge \mathsf{good}_3 \wedge \mathsf{Q}(\mathbf{X}^*)] - \sum_{\mathbf{X}^* \in \mathcal{X}} \Pr[\mathsf{good}_3 \wedge \mathsf{Q}(\mathbf{X}^*)] \tag{9}$$

$$= 2 \cdot \sum_{\mathbf{X}^* \in \mathcal{X}} \gamma(\mathbf{X}^*) \cdot \Pr[G_3 = 1 \wedge \mathsf{Q}(\mathbf{X}^*)] - \sum_{\mathbf{X}^* \in \mathcal{X}} \gamma(\mathbf{X}^*) \cdot \Pr[\mathsf{Q}(\mathbf{X}^*)] \tag{10}$$

$$\geq 2 \cdot \gamma_{\mathsf{min}} \cdot \sum_{\mathbf{X}^* \in \mathcal{X}} \Pr[G_3 = 1 \wedge \mathsf{Q}(\mathbf{X}^*)] - \gamma_{\mathsf{max}} \cdot \sum_{\mathbf{X}^* \in \mathcal{X}} \Pr[\mathsf{Q}(\mathbf{X}^*)]$$

$$= 2 \cdot \gamma_{\mathsf{min}} \cdot \Pr[G_3 = 1] - \gamma_{\mathsf{max}} \tag{11}$$

$$= 2 \cdot \gamma_{\mathsf{min}} \cdot (\mathsf{Adv}_{\mathcal{A}}^{\mathcal{VF}}(k) + 1)/2 - \gamma_{\mathsf{max}}$$

$$= \gamma_{\mathsf{min}} \cdot \mathsf{Adv}_{\mathcal{A}}^{\mathcal{VF}}(k) - \gamma_{\mathsf{max}} + \gamma_{\mathsf{min}}$$

$$\geq 2 \cdot \gamma_{\mathsf{min}} \cdot \delta - \gamma_{\mathsf{max}} + \gamma_{\mathsf{min}} \tag{12}$$

Here, (7) is due to Equation (6), (8) follows from the sequence of games described above, (9) and (11) follow from the fact that we sum over mutually exclusive events $\mathsf{Q}(\mathbf{X}^*)$ with $\sum_{\mathbf{X}^* \in \mathcal{X}} \Pr[\mathsf{Q}(\mathbf{X}^*)] = 1$, (10) is by Lemma 4, and (12) by the definition of $\delta \leq \mathsf{Adv}_{\mathcal{A}}^{\mathcal{VF}}(k)/2$.

B   Proof of Lemma 4

The execution of AdmSmp in Game 3 uses random coins which are independent of the rest of the game. Therefore, the set of random coins underlying Game 3 can be seen as a cross product $\Omega = \Omega' \times R_K$, where each member is a pair $(\omega', r_K) \in \Omega$ such that r_K denotes the random coins used by algorithm AdmSmp, and ω' denotes all other coins of the experiment and the attacker.

Note that that any particular choice $\mathbf{X}^*$ of a sequence of queries made by $\mathcal{A}$ depends only on ω', because in Game 3 algorithm AdmSmp is executed in the $\mathbf{Finalize}_3$-procedure, when the sequence of queries $\mathbf{X}^*$ issued by the attacker is already fixed. Thus, for all $\mathbf{X}^* \in \mathcal{X}$ let $\Omega'(\mathbf{X}^*)$ denote the set of all $\omega' \in \Omega'$ that produce the particular sequence of queries $\mathbf{X}^*$. Similarly, note that the probability that Game 3 outputs 1 depends only on Ω'.

Let $\Omega_1' \subseteq \Omega'$ denote the set of all $\omega' \in \Omega'$ such that the experiment outputs 1. Let $R_{\mathsf{good}}(\mathbf{X}^*) \subseteq R_K$ denote the set of all coins leading to an AHF key K such that for $\mathbf{X}^* = (X^{(1)}, \ldots, X^{(Q)}, X^*)$ holds that

$$F_K(X^{(1)}) = \cdots = F_K(X^{(Q)}) = 1 \quad \wedge \quad F_K(X^*) = 0$$

Then the set of coins such that $G_3 = 1$ is $\Omega_1' \times R_K$, and the set of coins leading to $\mathsf{good}_3 \wedge \mathsf{Q}(\mathbf{X}^*)$ is $\Omega'(\mathbf{X}^*) \times R_{\mathsf{good}}(\mathbf{X}^*)$. Now we can compute

$$\begin{aligned}
\Pr[G_3 = 1 \wedge \mathsf{good}_3 \wedge \mathsf{Q}(\mathbf{X}^*)] &= \frac{|(\Omega_1' \times R_K) \cap (\Omega'(\mathbf{X}^*) \times R_{\mathsf{good}}(\mathbf{X}^*))|}{|\Omega' \times R_K|} \\
&= \frac{|(\Omega_1' \cap \Omega'(\mathbf{X}^*)) \times R_{\mathsf{good}}(\mathbf{X}^*)|}{|\Omega' \times R_K|} \\
&= \frac{|\Omega_1' \cap \Omega'(\mathbf{X}^*)| \cdot |R_{\mathsf{good}}(\mathbf{X}^*)|}{|\Omega'| \cdot |R_K|} \\
&= \frac{|\Omega_1' \cap \Omega'(\mathbf{X}^*)| \cdot |R_K|}{|\Omega'| \cdot |R_K|} \cdot \frac{|R_{\mathsf{good}}(\mathbf{X}^*)|}{|R_K|} \\
&= \frac{|(\Omega_1' \cap \Omega'(\mathbf{X}^*)) \times R_K|}{|\Omega' \times R_K|} \cdot \frac{|R_{\mathsf{good}}(\mathbf{X}^*)|}{|R_K|} \\
&= \Pr[G_3 = 1 \wedge \mathsf{Q}(\mathbf{X}^*)] \cdot \gamma(\mathbf{X}^*)
\end{aligned}$$

and

$$\begin{aligned}
\Pr[\mathsf{good}_3 \wedge \mathsf{Q}(\mathbf{X}^*)] &= \frac{|\Omega'(\mathbf{X}^*) \times R_{\mathsf{good}}(\mathbf{X}^*)|}{|\Omega' \times R_K|} \\
&= \frac{|\Omega'(\mathbf{X}^*)| \cdot |R_{\mathsf{good}}(\mathbf{X}^*)|}{|\Omega'| \cdot |R_K|} \\
&= \frac{|\Omega'(\mathbf{X}^*)| \cdot |R_K|}{|\Omega'| \cdot |R_K|} \cdot \frac{|R_{\mathsf{good}}(\mathbf{X}^*)|}{|R_K|} \\
&= \frac{|\Omega'(\mathbf{X}^*) \times R_K|}{|\Omega' \times R_K|} \cdot \frac{|R_{\mathsf{good}}(\mathbf{X}^*)|}{|R_K|} \\
&= \Pr[\mathsf{Q}(\mathbf{X}^*)] \cdot \gamma(\mathbf{X}^*)
\end{aligned}$$

Multi-Client Verifiable Computation with Stronger Security Guarantees

S. Dov Gordon[1], Jonathan Katz[2], Feng-Hao Liu[2],
Elaine Shi[2], and Hong-Sheng Zhou[3]

[1] Applied Communication Sciences, USA
sgordon@appcomsci.com
[2] University of Maryland, USA
{jkatz,fenghao,elaine}@cs.umd.edu
[3] Virginia Commonwealth University, USA
hszhou@vcu.edu

Abstract. At TCC 2013, Choi et al. introduced the notion of *multi-client* verifiable computation (MVC) in which a set of clients outsource to an untrusted server the computation of a function f over their collective inputs in a sequence of time periods. In that work, the authors defined and realized multi-client verifiable computation satisfying soundness against a malicious server and privacy against the semi-honest corruption of a single client. Very recently, Goldwasser et al. (Eurocrypt 2014) provided an alternative solution relying on multi-input functional encryption.

Here we conduct a systematic study of MVC, with the goal of satisfying stronger security requirements. We begin by introducing a simulation-based notion of security that provides a unified way of defining soundness and privacy, and automatically captures several attacks not addressed in previous work. We then explore the feasibility of achieving this notion of security. Assuming no collusion between the server and the clients, we demonstrate a protocol for multi-client verifiable computation that achieves stronger security than the protocol of Choi et al. in several respects. When server-client collusion is possible, we show (somewhat surprisingly) that simulation-based security cannot be achieved, even assuming only semi-honest behavior.

1 Introduction

Protocols for *verifiable computation* (or *secure outsourcing*) allow computationally weak clients to delegate to a more powerful server the computation of a function f on a series of dynamically chosen inputs $x^{(1)}, x^{(2)}, \ldots$. The main desideratum is that, following a pre-processing stage whose complexity depends on f, the work of the client per function evaluation should be significantly lower than the cost of computing the function itself [20]. The initial proposal and construction of non-interactive verifiable computation [20] led to a long line of follow-up work [1, 3, 7–10, 16–18, 21, 26, 27, 33–36].

Y. Dodis and J.B. Nielsen (Eds.): TCC 2015, Part II, LNCS 9015, pp. 144–168, 2015.

We are interested here in the *multi-client* setting introduced by Choi et al. [15]. Imagine that n clients wish to compute some function f over their joint inputs $\{(x_1^{(\mathsf{ssid})}, \dots, x_n^{(\mathsf{ssid})})\}_{\mathsf{ssid}}$ for a series of subsessions identified by ssid. (One can view the ssid as encoding a current time period, though there are other possibilities as well.) As in earlier work, we assume no client-client communication, and focus on *non-interactive* solutions in which each evaluation of the function requires only a single round of communication between each client and the server.

In earlier works on multi-client verifiable computation [15, 24], the primary goal is to achieve security (soundness and privacy) against a *malicious server*, assuming that *clients behave honestly*. Soundness means that a malicious server should not be able to fool a client into accepting a wrong result; privacy means that clients' inputs should remain hidden from the server. (Choi et al. also considered privacy against clients, but while still assuming semi-honest client behavior.)

1.1   Our Contributions

In this paper, we conduct a systematic study of multi-client verifiable computation with stronger security guarantees. The primary question we address is security when *clients* may be malicious. These malicious clients may potentially be colluding with each other, or with the server.

Formal Security Modeling. We begin by introducing a simulation-based notion of security in the universal composability framework, which provides a unified way of defining soundness and privacy. As a technical advantage, it means that protocols satisfying the definition achieve a strong, simulation-based notion of security not considered in previous work. Our definition also automatically captures *adaptive soundness* as well as *selective-failure* attacks, which were not handled in prior work on the multi-client setting[1].

Impossibility When the Server and Clients Collude. Ideally, one would like to achieve a strong notion of security where a subset of the clients may be

---

[1] Intuitively, a scheme suffers from selective-failure attacks if the server can learn some secret information from the "decision" of the clients, upon receiving output from the server. In the single-client setting, previous schemes in the work [16, 20] can be completely broken by the attacks, unless the clients are willing to redo the expensive preprocessing upon any server failure. In the multi-client setting, the same attacks also apply to the scheme by Choi et al. [15], which is basically an extension of [20]. We note that there is no simple fix to the approaches taken in [15, 16, 20] using known techniques. Previous schemes that are not vulnerable to such attacks (such as the scheme in [36]) used completely different approaches.

Adaptive soundness is a technical issue pointed out by Bellare et al. [5] – if a Yao's garbled circuit is published first and later the adversary can choose inputs based on the garbled circuit, then it is not known how to prove security other than just assuming the garbling scheme itself is secure. Previously, the schemes of [15, 20] used Yao's garbled circuits in this way, so the schemes suffer from such drawback. See the work of Choi et al. [15] for further discussions about adaptive soundness and selective failure attacks.

corrupted, and may be colluding with the server. Unfortunately, we show that simulation-secure MVC is impossible to realize (for general functions) when the server colludes with clients. This impossibility result holds even in the standalone setting, even when the server colludes with only a single, semi-honest client, and even in the presence of trusted setup assumptions such as PKI, common reference strings (CRS), shared secret randomness, etc. Intuitively, our lower bound result is due to a connection we establish between MVC and virtual black-box (VBB) obfuscation, whose impossibility is known [4]. More details can be found in Section 5.

Feasibility Result: When Server and Clients do not Collude. In contrast to the above, we show positive results for the case when client-server collusion is assumed not to occur. We show a construction that achieves security (i.e., soundness and privacy) against either a malicious server, or an arbitrary set of malicious, colluding clients. Our construction achieves both *adaptive soundness* and security against *selective abort*.

Our construction relies only on *falsifiable* assumptions. While it is alternatively possible to construct MVC schemes using a new notion, *multi-input functional encryption*, by Goldwasser et al. [24], this notion inherently requires (indistinguishable) obfuscation, which requires non-falsifiable assumptions or exponential assumptions [22]. Moreover, current constructions of obfuscation have prohibitively large overhead.

1.2   Techniques and New Primitives

Techniques used for achieving our upper bound results can be of independent interest. When server-client collusion is not allowed, we take a two-step approach to achieve simulation-security of MVC. As a stepping stone, we identify a new building block named multi-sender attribute-based encryption (mABE).

Our Two-Step Approach for MVC. We start with a protocol which achieves the simulation-based security against either (i) a malicious server or (ii) any coalition of semi-honest clients. Although this is also achieved by the protocol of Choi et al. [15]—even if not claimed explicitly there—our construction has the advantages of offering adaptive soundness based on standard assumptions as well as resilience to selective-failure attacks.

We then present a generic compiler that upgrades our intermediate solution (as well as the one by Choi et al. [15]) to handle an arbitrary subset of malicious clients. While we could rely on standard techniques, distributing commitments to random tapes during setup, and asking each party to prove in zero knowledge that they have acted honestly, we instead offer a compiler that does not require committed randomness, allowing us to reduce our setup assumptions to a simple common reference string. We demonstrate that as long as our semi-honest protocol offers a sufficiently strong notion of privacy, our compiler provides security against malicious corruption. This gives us a non-interactive multi-client verifiable computation protocol secure against a malicious adversary under all possible cases of non-client-server collusions, in the standard model under *falsifiable* assumptions.

A New Building Block: mABE. We identify a new building block, *multi-sender attribute-based encryption*, which can be of independent interest.

Recall that in the single sender setting, Parno et al. [36] showed that an attribute-based encryption (ABE) (that supports functions and their complements) implies publicly verifiable computation (without input privacy). Later, Goldwasser et al. [26] showed (i) how to compile an ABE scheme to a private-index functional encryption scheme using fully homomorphic encryption (FHE), and (ii) that private-index functional encryption implies input-private publicly verifiable computation.

We conduct a parallel study in the multi-sender setting. The multi-sender counterpart, multi-sender ABE (mABE) is defined as follows. Each sender $P_i \in \{P_1, \ldots, P_n\}$ has an attribute value x_i, as well as two input messages $(m_0^{(i)}, m_1^{(i)})$. A single receiver (or server) can use a decryption key for function f_i to learn $m_b^{(i)}$ if and only if $b = f_i(x_1, \ldots, x_n)$. We show how to construct an mABE scheme secure against a malicious server or semi-honest senders. To construct this mABE scheme, we first observe a special "local encoding" property of the LWE-based ABE scheme by Gorbunov, Vaikuntanathan, and Wee [30]. We then combine this observation with a proxy-OT protocol proposed by Choi et al. [15].

After obtaining the mABE construction, we then apply Goldwasser et al's compiler techniques [26] to transform it into an *attribute-hiding* mABE scheme (which can also be thought of as a multi-sender, private-index functional encryption scheme). Finally, just as single-sender private-index functional encryption implies input-private verifiable computation, we show that attribute-hiding mABE implies multi-client verifiable computation with input privacy, secure against a malicious server or an arbitrary subset of semi-honest clients. We can then use the compiler described previously to obtain security in the face of malicious corruptions, so long as there is no client-server collusion.

Sacrificing Input Privacy to Allow Server-Client Collusion. Since attribute hiding mABE implies multi-client verifiable computation, it follows that attribute hiding mABE is also impossible for general functions under server-client collusion. However, it is still interesting to consider settings without input privacy/attribute hiding. In this work, we show that any mABE (without attribute hiding) construction that is secure under some arbitrary corruption pattern implies public input MVC under the same corruption pattern; in particular, this gives a method of handling server-client collusion. We also show that an mABE scheme secure under server-client collusion, even in the standalone setting, implies extractable witness encryption (equivalently, point-filter obfuscation) [25]. So building MVC protocols without input privacy via this method would inherently require non-falsifiable assumptions. We note that it is also possible to construct MVC protocols without input privacy against an arbitrary corruption based on other non-falsifiable assumptions, such as SNARKs. It is an interesting question – whether we can construct a secure MVC without input privacy against an arbitrary corruption based on falsifiable assumptions, yet one should keep in mind that any possible solution should avoid the route using mABE, as implied by the result above.

1.3 Related Work

Non-interactive verifiable computation was first proposed by Gennaro, Gentry, and Parno [20]. Since then, various improvements have been proposed [1, 16, 17, 21, 27, 36], and constructions for specific functionalities [9, 18, 34, 35].

Various works have considered server-aided secure computation with the goal of eliminating client-to-client interaction. Most of these existing works do not achieve complete non-interactivity, in the sense that they still require multiple rounds of server-client interaction. Kamara et al. [31, 32] consider server-aided multi-party computation, but their approach is not non-interactive.

Multi-input functional encryption [24] is also related to non-interactive multi-party outsourcing. The earlier work by Shi et al. [37] shares similar goals, but for specific functionalities such as summation and variance. Shi et al. [37] also describe various application domains such as secure sensor network aggregation. In the multi-input functional encryption model or that of Shi et al. [37], the server learns the final outcome of the computation, and verifiability (i.e., soundness) is not an inherent part of the problem formulation. Interestingly, Goldwasser et al. [24], observe that multi-input functional encryption can, in fact, be leveraged to construct multi-client verifiable computation. This solution uses the technique developed in [36] which solves the selective-failure issue. However, Goldwasser et al. [24] do not consider malicious clients or client-server collusion. In addition, another major drawback is that known multi-input functional encryption schemes rely on non-falsifiable assumptions related to obfuscation, and this is somewhat necessary as pointed out by [24].

2   Multi-Client Verifiable Computation

2.1 Definitions

We start by introducing the notion of non-interactive multi-client verifiable computation (MVC) that has the following structure: let κ be the security parameter, n be the number of clients $\mathsf{P}_1, \ldots, \mathsf{P}_n$ who are delegating some computation on some n-ary function $f : \mathcal{X}^n \to \mathcal{Y}^n$ to a distinguished server Serv and would like to verify the correctness of their answers. Here we assume each client's input message space is $\mathcal{X}$, and output message space $\mathcal{Y}$, for some polynomial-length (in the security parameter) $|\mathcal{X}|$ and $|\mathcal{Y}|$.

Intuitively, MVC protocols have the following properties: (1) All participants are allowed to access to a certain initial setup $\mathcal{G}$ (e.g., PKI, CRS). (2) Then an offline stage follows; in the offline stage, each client sends a single message to the server Serv. (3) In the online stage, in a single time period (subsession), each client is only allowed to send an outgoing message to the server and then receive an incoming message from the server. In the whole paper, we assume that the clients cannot communicate with each other directly, and can only send a single round of message to the server per time period (subsession). Next, we give more details.

Definition 1 (Non-interactive Multi-client Verifiable Computation).
Let κ be the security parameter, n be the number of clients and f be an n-ary function being computed. A non-interactive multi-client verifiable computation consists of n clients $\mathsf{P}_1 \ldots \mathsf{P}_n$ and a server Serv with the following structure:

Setup stage: *All parties P_i's, $i \in [n]$ and Serv have access to a setup $\mathcal{G}$, where party P_i obtains $(\mathsf{pub}, \mathsf{sk}_i)$ upon queries for some secret and public information.*

Offline stage: *Each client P_i sends a single message to the server. The server stores these as $\hat{f}$, an encoded version of f.*

Online stage: *This step is a query-response move: at each sub-session (or time period) ssid, upon receiving an input (ssid, x_i) for $i \in [n]$, the client $\mathsf{P}_i(\mathsf{pub}, \mathsf{sk}_i, x_i)$ computes some message $(\hat{x}_i, \tau_i)$. Then he sends $\hat{x}_i$ to the server and stores τ_i as a secret.*

The server Serv carries out the computation on the messages received, and sends each client P_i for $i \in [n]$ an encoded output $(\mathsf{ssid}, \hat{y}_i)$.

Each client computes and some output $y_i \cup \{\bot\}$ based on $(\mathsf{pub}, \mathsf{sk}_i, \hat{y}_i, \tau_i)$, where $\bot$ means that he is not convinced with the outcome.

Remark 1. For the setup $\mathcal{G}$, we do not specify whether it is trusted in our definition. For our positive results, we want to minimize the requirements, and we showed that a self-registered PKI is enough for semi-honest client or malicious server corruptions. For the case of malicious clients corruptions, we further need an additional CRS. On the other hand, for our lower bound results, we rule out a large class of instantiations of $\mathcal{G}$, including the trusted PKI, CRS, shared secret randomness, and their combinations.

Note that the trusted PKI is a setup where a trusted party generates public- and secret-key pairs for each user, and publishes the public keys to all users. The self-registered PKI is a weaker setup where each user generates their own key pairs, and registers the public keys with the setup so that the setup can publish the public keys to all users.

2.2 Security Definition

The security definition for non-interactive multi-client verifiable computation, MVC, turns out to be subtle. An MVC protocol cannot achieve the standard multi-party computation security, which requires that malicious clients have only one chance to provide their inputs, and cannot switch inputs later. In the non-interactive setting, if the server and some clients are simultaneously corrupted, then after gathering the transcripts of the honest clients, by definition the malicious clients can now select different inputs for themselves and learn the corresponding outputs. For example, consider $n = 2$. If client P_1 and the server are corrupted, then they effectively have access to oracle $f_1(*, x_2)$ where f_1 is the output of the first party, and x_2 is the honest input of P_2. The notation $*$ means that client P_1 can choose arbitrary inputs for itself and query this oracle a polynomial number of times. So our security definition would allow the adversary to learn $f_1(*, x_2)$ in the ideal world, and guarantees that this is the most that he can learn.

Multi-Client Private Verifiable Computation

The functionality is parameterized with an n-ary function $f : \mathcal{X}^n \to \mathcal{Y}^n$. The functionality interacts with n clients P_i for $i \in [n]$, a distinguished server Serv, and the simulator $\mathcal{S}im$.

Initialization:
Upon receiving (Init) from client P_i, send (Init, P_i) to notify the simulator $\mathcal{S}im$. Later, when $\mathcal{S}im$ returns (Init, P_i), send a notification (Init, P_i) to the server Serv.
Upon receiving (Init) from the server Serv, send (Init, Serv) to notify the simulator $\mathcal{S}im$.

Computation:
Upon receiving (Input, ssid, x_i) from client P_i, send (ssid, P_i) to notify $\mathcal{S}im$. Later, when $\mathcal{S}im$ returns (ssid, P_i), store (ssid, x_i), and send a notification (Input, ssid, P_i) to server Serv.
Upon receiving (Input, ssid, 1) from server Serv, retrieve (ssid, x_i) for all $i \in [n]$. If some (ssid, x_i) has not been stored yet, send (Output, ssid, fail) to the server and all clients.

- **Server is not corrupted:** Compute $(y_1, \ldots, y_n) \leftarrow f(x_1, \ldots, x_n)$. Later when $\mathcal{S}im$ returns (ssid, P_i, ϕ), if $\phi = $ ok, send (Output, ssid, y_i) to client P_i; if $\phi = $ fail, send (Output, ssid, fail) to client P_i.
- **Server is corrupted:** Let $\mathcal{I} \subseteq [n]$ denote the set of indices corresponding to corrupted clients. Let $\overline{\mathcal{I}} := [n] \setminus \mathcal{I}$. Let $\mathbf{x}_{\mathcal{I}}^*$ denote the corrupted clients' inputs, $\mathbf{x}_{\overline{\mathcal{I}}}$ denote the remaining clients' inputs. Without loss of generality, we can renumber the clients such that $\mathcal{I} := \{1, 2, \ldots, |\mathcal{I}|\}$.
  The functionality provides to $\mathcal{S}im$ blackbox oracle access to the following oracle $O_{f,\mathcal{I}}$ where $\mathcal{S}im$ can choose inputs $\mathbf{x}_{\mathcal{I}}^*$ for corrupted clients to query:

  Oracle $O_{f,\mathcal{I}}(\mathbf{x}_{\mathcal{I}}^*)$:
  Compute $(y_1, \ldots, y_n) \leftarrow f(\mathbf{x}_{\mathcal{I}}^*, \mathbf{x}_{\overline{\mathcal{I}}})$.
  Output $\{y_i\}_{i \in \mathcal{I}}$ to $\mathcal{S}im$, and internally remember the last seen $\{y_i\}$ for $i \in \overline{\mathcal{I}}$.

  At any time (not necessarily simultaneously for all i), on receiving (ssid, P_i, ϕ) from $\mathcal{S}im$ for some $i \in \overline{\mathcal{I}}$, the functionality[a] sends to P_i (Output, ssid, y_i) corresponding to the last seen y_i if $\phi = $ ok, otherwise it sends (Output, ssid, fail) to P_i.

---

[a] Restricting to sending the last seen outputs does not lose generality, since the simulator can always repeat a previous query to the oracle $O_{f,\mathcal{I}}$.

Fig. 1. Functionality $\mathcal{F}_{\mathsf{pVC}}$

On the other hand if interaction is allowed, it is well-understood that this issue can be avoided by standard techniques.

Based on this observation, we formally define the ideal functionality for *private* MVC in Figure 1 that captures the above issues, and soundness and privacy. The security of the protocol above follows the standard real/ideal paradigm [28, 29]. Here we only include the universal composability (UC) definition by Canetti [13, 14]. The standalone security definition can be found in [12, 23].

Definition 2 (UC Security [14]). *We say a protocol Π securely realizes $\mathcal{F}$ if for any PPT adversary $\mathcal{A}$ in the real world, there exists a PPT simulator Sim in the ideal world, so that no PPT environment $\mathcal{Z}$ is able to tell the real world execution from the ideal world execution, i.e., $\mathsf{EXEC}_{\mathcal{A},\Pi,\mathcal{Z}} \approx \mathsf{EXEC}_{Sim,\mathcal{F},\mathcal{Z}}$.*

We can also define a notion of verifiable computation without input privacy. This is essentially the same definition, except that the server learns all the inputs of the clients. We present a formal description and provide a construction of this relaxed notion in the full version of this paper. In the following remarks we highlight and clarify a few properties of the stronger definition above:

Soundness Against Selective Failure Attacks: Our ideal-functionality models a reactive functionality that has multiple sub-sessions after a common pre-processing (i.e., Initialization) phase. Our definition implies this soundness, where learning the decision bit of the clients does not help the server to fool the clients. In particular, following the convention of simulation-based definition, our security definition requires the clients to report the outputs (and acceptance decisions) to the environment.

Communication Model. We assume that the adversary controls the communication medium between all parties. Our protocol later relies on PKI setup, and we can implement a secure channel with PKI. Therefore, while not explicitly stated, all our protocols are described assuming the secure channel ideal world.

Semi-Honest v.s. Malicious Corruption. Semi-honestly corrupted participants follow the protocol faithfully, but the adversary sees the internal states of all semi-honestly corrupted parties.

As mentioned above, due to the non-interactive nature, if the server and at least one client are simultaneously corrupted either in the malicious or semi-honest model, then our ideal functionality $\mathcal{F}_{\mathsf{pVC}}$ implements a blackbox-access oracle which the simulator can query multiple times by specifying inputs for the malicious clients. For malicious corruption, the simulator can ask the ideal functionality to send outputs to different clients corresponding to different corrupted clients' inputs. For example, suppose P_1 and the server are maliciously corrupted, the simulator can ask the functionality to send $f_2(x_1, x_2, x_3)$ to P_2, and send $f_3(x_1', x_2, x_3)$ to P_3. For semi-honest corruption, the outputs sent back to the clients always correspond to inputs chosen by the environment.

Static Corruption. We assume a static corruption model in this paper, where some protocol participants are corrupted at the beginning of protocol execution.

UC and Stand-Alone Security. In the paper we use both the UC definition and standalone security definition. In the standalone security, the environment machine $\mathcal{Z}$ (i.e., the distinguisher) provides inputs to all protocol participants and the adversary at the beginning of protocol execution, and it receives outputs from these entities when the execution is complete. The environment and the adversary are not allowed to communicate during the protocol execution. Protocols secure in the standalone security model can be composed sequentially. On the other hand, in the UC framework, the environment and the adversary are always allowed to communicate. Protocols secure in the UC framework can be composted with arbitrary protocols. It is obvious that UC security implies stand-alone security.

Efficiency. An important feature of MVC is the *online* efficiency of the clients. Usually, we require the clients' computation time be much less than the complexity of the function f, so that over many online computations, the total cost of the clients will have low amortized cost. However, for private MVC, in some cases it is also interesting if the clients' computation time is similar to f, e.g. when the function f is simple. For example, it client P_1 and P_2 want to do a secure comparison over their inputs. The privacy requirement makes it interesting regardless of whether the clients' online computation time is smaller than the function being delegated. We do not specify a definition of efficiency but discuss it for each scheme individually.

3 Malicious Server or Semi-honest Client Corruptions

In this section and the following section, we will demonstrate constructions that achieve security against malicious adversaries, as long as there is no simultaneous server-client corruption.

Roadmap. As described in Section 1.2, our plan of action is: 1) define and obtain an mABE scheme; 2) use Goldwasser et al's compiler techniques [26] to achieve *attribute-hiding* mABE; and 3) show that attribute-hiding mABE implies private MVC.

All of the above primitives are proven secure under a malicious server or *semi-honestly* corrupted clients in this section. Then, in the following Section 4, we show a generic *compiler* based on non-interactive zero-knowledge proofs, such that any protocol secure against semi-honest corruption of an arbitrary subset of clients, and additionally offering clients *perfect privacy* from one another, can be transformed into a protocol that is secure against either a malicious server or an arbitrary subset of malicious clients.

For convenience, in the remainder of the section, we focus on the case when only the first client P_1 learns output, and the remaining clients learn nothing. Based on this, we can obtain a protocol where every party learns outputs through simple parallel repetition.

3.1   Multi-sender ABE

We define a multi-sender, two-outcome ABE scheme. Intuitively, the mABE functionality implements the following: consider n senders and a server. The first sender P_1 chooses two messages m_0 and m_1, and each P_i for $i \in [n]$ has an attribute x_i. The goal is for the server to m_b where $b = f(x_1, x_2, \ldots, x_n)$ while keeping m_{1-b} secret. We require the mABE scheme to be *non-interactive*, i.e., after an initial preprocessing phase in which the server learns an encoding of the function f, in each online phase, each sender sends a single message to the server, and the server can learn m_b.

We note that our mABE formulation can also be regarded as a generalization of the proxy oblivious transfer (POT) primitive proposed by Choi et al. [15]. We present the definition of POT in Appendix A. In other words, sender P_1 obliviously transfers one of m_0 and m_1 to the server, where which message is transferred is determined by a policy function f over all senders' attributes.

Figure 2 formally describes the mABE ideal functionality. We define mABE for the single-key setting, since our verifiable computation application is inherently single-key.

---

mABE Functionality

Functionality $\mathcal{F}_{\mathsf{mABE}}^f$ interacts with multiple senders $P_1, \ldots, P_n$, a server Serv, as well as a simulator $\mathcal{S}im$. The functionality is parameterized by a function $f : (\{0,1\}^\ell)^n \to \{0,1\}$.

- Upon receiving $(\mathsf{ssid}, m_0, m_1, x_1)$ from the sender P_1, notify $\mathcal{S}im$ with (ssid, P_1). Later, if $\mathcal{S}im$ replies with (ssid, P_1), store $(\mathsf{ssid}, m_0, m_1, x_1)$, and notify Serv with $(\mathsf{ssid}, P_1, x_1)$.
- Similarly, upon receiving (ssid, x_i) from other senders P_i for $i \in \{2, .., n\}$, notify $\mathcal{S}im$ with (ssid, P_i). Later when $\mathcal{S}im$ replies with (ssid, P_i), if no (ssid, x_i) recorded yet, store it, and notify Serv with $(\mathsf{ssid}, P_i, x_i)$.
- Upon receiving $(\mathsf{ssid}, 1)$ from Serv, if all $(\mathsf{ssid}, m_0, m_1, x_1)$, and (ssid, x_i) for $i \in \{2, .., n\}$ are recorded, return $(\mathsf{ssid}, m_{f(x_1, \ldots, x_n)})$ to Serv. Otherwise, if some tuple for ssid has not been recorded, return fail to Serv.

Fig. 2. Functionality $\mathcal{F}_{\mathsf{mABE}}^f$

We now present our (non-interactive) protocol that realizes $\mathcal{F}_{\mathsf{mABE}}^f$ for any efficiently computable f. We use as building blocks a non-interactive POT protocol, and any two-outcome attribute-based encryption (ABE) scheme with a special structure where the attributes of ciphertexts can be encoded bit-by-bit. We formalize this local encoding property in the following. and observe that the ABE construction by Gorbunov, Vaikuntanathan, and Wee [30] satisfies this special property. Also, we remark that one can build a two-outcome ABE from a standard one, as shown by Goldwasser et al. [26]. Here we use ABE to denote the two-outcome ABE for simplicity.

Definition 3 (Two-outcome ABE with Local Encoding). *A two-outcome attribute-based encryption scheme* ABE *for a class of boolean functions* $\mathcal{F} = \{\mathcal{F}_\ell\}_{\ell \in \mathbb{N}}$ *from* $\{0,1\}^k \rightarrow \{0,1\}$, *is a tuple of polynomial time algorithms:* ABE.{Setup, KeyGen, Enc, Dec} *as follows:*

- ABE.Setup(1^k) *outputs a master public key* mpk_{ABE} *and a master secret key* msk_{ABE}.
- ABE.KeyGen($\text{msk}_{\text{ABE}}, f$) *On inputs* msk_{ABE} *and a function* $f \in \mathcal{F}$, *output a function key* sk_f.
- ABE.Enc($\text{mpk}_{\text{ABE}}, x, m_0, m_1$) *takes as input the master public key* mpk_{ABE}, *an attribute* $x \in \{0,1\}^\ell$ *for some* ℓ, *and two messages* m_0, m_1, *outputs a ciphertext* c.
- ABE.Dec(sk_f, c) *takes as input a key* sk_f *and a ciphertext and outputs a message* m^*.

Local encoding. We say that a two-outcome ABE scheme satisfies *local encoding* if the encryption algorithm ABE.Enc can be equivalently expressed as the following, where enc is a sub-algorithm:

1. select common randomness R;
2. for all $i \in [k]$, compute $\hat{x}[i] = \text{enc}(\text{mpk}_{\text{ABE}}, x[i]; R)$;
3. $\hat{m} = \text{enc}(\text{mpk}_{\text{ABE}}, m_0, m_1; R)$.

Finally, the ciphertext c can be written as $c := (\hat{x}[1], \hat{x}[2], \ldots, \hat{x}[k], \hat{m})$.

The correctness property guarantees that the decryptor can learn one of the messages m_b for $b = f(x)$, and the security guarantees that this is the only thing he can learn. We present the formal definitions in the appendix and also refer the readers to the work by Goldwasser et al. [26].

We present our construction of mABE in the $\mathcal{G}^{\text{ABE}}$ setup model, where $\mathcal{G}^{\text{ABE}}$ serves as a self-registered PKI which allows the sender to generate (mpk_{ABE}, msk_{ABE}) $\leftarrow$ ABE.KeyGen(1^k), and register mpk_{ABE}. When queried by players other than the sender, it returns mpk_{ABE}.

Construction of mABE. Let $f : (\{0,1\}^\ell)^n \rightarrow \{0,1\}$ be a policy function, and without loss of generality, we let Serv denote the server, and let $P_1, \ldots, P_n$ denote the senders. We make use of $(n-1) \cdot \ell$ instances of the functionality $\mathcal{F}_{\text{POT}}$ indexed by (i,j) such that for $i \in \{2, \ldots, n\}$, all $j \in [\ell]$, in the (i,j)-th instance, P_1 plays the sender, P_i plays the chooser, and Serv plays the server. In the protocol below, we assume the existence of private channels; i.e. we assume that all parties encrypt their messages before sending them. This step is left implicit.[2] The parties act as follows:

---

[2] Recall that our protocol for realizing $\mathcal{F}_{\text{POT}}$ relies on a setup phase for establishing a PKI, so we could rely on this PKI for encrypting messages. If we instead were to use a protocol for $\mathcal{F}_{\text{POT}}$ that did not rely on a PKI, we could simply add the establishment of a PKI to the setup phase of this protocol. Finally, we note that the assumption of private channels is not necessary: we could instead choose to leak P_{n+1}'s output to an eavesdropper. This would suffice for our purposes, but makes the resulting ideal functionality and the security proof a bit more involved.

- **Offline Stage**: Every party receives a function f as input. P_1 calls the setup $\mathcal{G}^{ABE}$ to receive (mpk_{ABE}, msk_{ABE}), and computes some $sk_f = ABE.KeyGen(msk_{ABE}, f)$. He sends sk_f to the server. All the other clients runs an empty step.
- **Online Stage**:
  - On input (sid, m_0, m_1, x_1), the sender P_1 does the following *in parallel*:
    1. Sample a random string R. Compute $C = enc(mpk_{ABE}, m_0, m_1; R)$, $\hat{x}_1 = enc(mpk_{ABE}, x_1, R)$ (bit-by-bit) and sends them to the receiver Serv.
    2. For $i \in \{2, \ldots, n\}, j \in [\ell]$, P_1 computes $\hat{c}_{i,j,0} = enc(mpk_{ABE}, 0; R)$, and $\hat{c}_{i,j,1} = enc(mpk_{ABE}, 1; R)$, and then sends $(\hat{c}_{i,j,0}, \hat{c}_{i,j,1})$ to the (i,j)-th instance of $\mathcal{F}_{POT}$.
  - For $i \in \{2, \ldots, n\}$, upon receiving (sid, x_i), the party P_i sends, in parallel, $x_i[j]$ to the (i,j)-th instance of $\mathcal{F}_{POT}$ for all $j \in [\ell]$. Here $x_i[j]$ denotes the j-th bit of x_i.
  - Party Serv receives $enc(mpk_{ABE}, m_0, m_1)$, $enc(mpk_{ABE}, x_1)$ (bit-by-bit) from the sender P_1, and $enc(mpk_{ABE}, x_2), \ldots, enc(mpk_{ABE}, x_n)$ (bit-by-bit) via the instances of the functionality $\mathcal{F}_{POT}$. He outputs m' by running the ABE decryption algorithm on the received ciphertexts using decryption key sk_f.

Then we are able to achieve the following theorem. We present the proof in the full version of this paper.

Theorem 1. *Assuming the existence of two-outcome ABE for a function* $f : (\{0,1\}^{\ell})^n \to \{0,1\}$ *with the additional encoding property as above, then the protocol above securely realizes the ideal functionality* $\mathcal{F}^f_{mABE}$ *in the* $(\mathcal{F}_{POT}, \mathcal{G}^{ABE})$*- hybrid model, against either (1) malicious server corruption, or (2) any semi-honest (static) corruption among any fixed set of clients.*

Using mABE as a building block, we can easily achieve verifiable computation without privacy. In the full version of this paper, we present the formal definition of MVC without privacy, and the protocol that achieves this notion using mABE. We note that the construction is very similar to the one in the next section (see Theorem 3).

3.2 Achieving Attribute Hiding

In Figure 3, we define an attribute-hiding version of mABE, where the sender attributes are not leaked to the receiver. The attribute-hiding mABE functionality, denoted $\mathcal{F}_{ah\text{-}mABE}$, is defined in almost the same way as $\mathcal{F}_{mABE}$, except that when the functionality notifies the server, it only notifies $(ssid, P_i)$, without leaking the attributes x_i's.

We present our protocol that realizes $\mathcal{F}_{ah\text{-}mABE}$ in the $\mathcal{G}^{FHE}$ setup plus $\mathcal{F}_{mABE}$ hybrid model, where $\mathcal{G}^{FHE}$ serves as a self-registered PKI which allows the sender to generate $(pk_{FHE}, sk_{FHE}) \leftarrow FHE.KeyGen(1^k)$, and register pk_{FHE}. When queried by parties other than the sender, it returns pk_{FHE}. Our construction can be viewed as a distributed version of that of Goldwasser et al. [26], who constructed attribute-hiding ABE (or functional encryption) from a non-hiding one. Briefly speaking,

---

ah-mABE **Functionality**

Functionality $\mathcal{F}^f_{\text{ah-mABE}}$ interacts with multiple senders $P_1, \ldots, P_n$, a server Serv, as well as a simulator $\mathcal{S}im$. The functionality is parameterized by a function $f : (\{0,1\}^\ell)^n \to \{0,1\}$.

- Upon receiving $(\text{ssid}, m_0, m_1, x_1)$ from the sender P_1, notify $\mathcal{S}im$ with (ssid, P_1). Later, if $\mathcal{S}im$ replies with (ssid, P_1), store $(\text{ssid}, m_0, m_1, x_1)$, and notify Serv with (ssid, P_1).
- Similarly, upon receiving (ssid, x_i) from other senders P_i for $i \in \{2, .., n\}$, notify $\mathcal{S}im$ with (ssid, P_i). Later when $\mathcal{S}im$ replies with (ssid, P_i), if no (ssid, x_i) recorded yet, store it, and notify Serv with (ssid, P_i).
- Upon receiving $(\text{ssid}, 1)$ from Serv, if all $(\text{ssid}, m_0, m_1, x_1)$, and (ssid, x_i) for $i \in \{2, .., n\}$ are recorded, return $(\text{ssid}, m_{f(x_1, \ldots, x_n)})$ to Serv. Otherwise, if some tuple for ssid has not been recorded, return fail to Serv.

Fig. 3. Functionality $\mathcal{F}^f_{\text{ah-mABE}}$

the first party P_1 generates a garbled circuit of the FHE decryption circuit, and then all parties input ciphertexts of their attributes to $\mathcal{F}_{\text{mABE}}$, to allow the server to learn *only* a set of labels to the garbled circuit. Then the server can learn only the outcome by evaluating the garbled circuit. Intuitively, since the attributes are encrypted, and the server can learn *only* a set of labels of the garbled circuit, the server can only learn the outcome but not the attributes of the parties.

Construction of ah-mABE. Let $f : (\{0,1\}^\ell)^n \to \{0,1\}$ be a policy function, let $P_1, \ldots, P_n$ be the senders, and let Serv be the receiver. Denote $g := \text{Eval}_{\text{FHE}}(\text{pk}_{\text{FHE}}, f', (c, c', c_1), \ldots, c_n)$ where pk_{FHE} is an FHE public key, $c, c', c_1 \ldots, c_n$ are ciphertexts and f' is an n-nary function that on input $((m_0, m_1, x_1), \ldots, x_n)$ outputs $m_{f(x_1, \ldots, x_n)}$. Assume the function g has an λ-bit output, and denote g_i as the function that outputs the i-bit of g. Then the parties do as follows:

- Upon receiving input $(\text{ssid}, m_0, m_1, x_1)$, P_1 does the following:
  - Obtain $(\text{pk}_{\text{FHE}}, \text{sk}_{\text{FHE}})$, and compute $(\Gamma, \{L_i^0, L_i^1\}_{i \in [\lambda]}) \leftarrow \text{Gb.Garble}(1^k, \text{Dec}_{\text{FHE}}(\text{sk}_{\text{FHE}}, \cdot))$ where $\text{Dec}_{\text{FHE}}(\text{sk}_{\text{FHE}}, \cdot)$ is a circuit that takes a λ-bit ciphertext as input and outputs a single bit message.
  - Send (ssid, Γ) to the receiver Serv, and in parallel,
  - Compute $\hat{m}_0 \leftarrow \text{Enc}_{\text{FHE}}(\text{pk}_{\text{FHE}}, m_0)$, $\hat{m}_1 \leftarrow \text{Enc}_{\text{FHE}}(\text{pk}_{\text{FHE}}, m_1)$, $\hat{x}_1 \leftarrow \text{Enc}_{\text{FHE}}(\text{pk}_{\text{FHE}}, x_1)$, and send $(\text{ssid}, L_j^0, L_j^1, (\hat{m}_0, \hat{m}_1, \hat{x}_1))$ to the functionality $\mathcal{F}^{g_j}_{\text{mABE}}$ for all $j \in [\lambda]$.
- For $i \in [n] \setminus \{1\}$, upon receiving input (ssid, x_i), P_i first calls $\mathcal{G}^{\text{FHE}}$ to obtain pk_{FHE}. Then he computes $\hat{x}_i \leftarrow \text{Enc}_{\text{FHE}}(\text{pk}_{\text{FHE}}, x_i)$ and sends $(\text{ssid}, \hat{x}_i)$ to the functionality $\mathcal{F}^{g_j}_{\text{mABE}}$ for all $j \in [\lambda]$.

– Upon receiving input $(\text{ssid}, \hat{x}_1, \ldots, \hat{x}_n, \{L^{d_i}\}_{i \in [\lambda]}, \Gamma)$ from the ideal functionalities and P_1, the receiver Serv computes $\mathsf{Gb.Eval}(\Gamma, \{L^{d_i}\}_{i \in [\lambda]})$, and outputs the result of the evaluation.

Then we are able to achieve the following theorem. We present the proof in the full version of this paper.

Theorem 2. *Assuming the existence of a fully homomorphic encryption scheme and a garbling scheme, the protocol above securely realizes the ideal functionality $\mathcal{F}_{\mathsf{ah-mABE}}^f$ for any efficiently computable f in the $(\mathcal{F}_{\mathsf{mABE}}, \mathcal{G}^{\mathsf{FHE}})$-hybrid model, against either (1) malicious server corruption, or (2) semi-honest (static) corruption among any fixed set of senders.*

Using the functionality $\mathcal{F}_{\mathsf{ah-mABE}}$, we are able to build an MVC scheme that also achieves input and output privacy, in a similar fashion that (single-sender) private-index functional encryption implies private verifiable computation [26]. As before, we assume f outputs only one bit and only the first party receives the output. The construction is in the $\mathcal{F}_{\mathsf{ah-mABE}}^f$ hybrid model. More formally, let $f : (\{0,1\}^\ell)^n \to \{0,1\}$ be a function to be delegated, let $\mathsf{P}_1, \ldots, \mathsf{P}_n$ be the clients and Serv be the server. The the parties do as the following:

– Upon receiving input (ssid, x_1), P_1 samples two random inputs $m_0, m_1 \leftarrow \{0,1\}^\ell$ and sends $(\text{ssid}, m_0, m_1, x_1)$ to the functionality $\mathcal{F}_{\mathsf{ah-mABE}}^f$. Locally, he stores m_0, m_1.
– For $i \in [n] \setminus \{1\}$, upon receiving message (ssid, x_i), P_i sends (ssid, x_i) to the functionality $\mathcal{F}_{\mathsf{ah-mABE}}^f$.
– Upon receiving (ssid, m) from $\mathcal{F}_{\mathsf{ah-mABE}}^f$, the server sends P_1 the message (ssid, m).
– Upon receiving (ssid, m) from the server, P_1 checks whether $m = m_b$ for some $b \in \{0,1\}$. If so, he outputs b, and otherwise he outputs $\perp$.

In particular we show the following theorem. We present the proof in the full version of this paper.

Theorem 3. *The protocol above securely realizes $\mathcal{F}_{\mathsf{pVC}}$ in the $\mathcal{F}_{\mathsf{ah-mABE}}^f$ hybrid world, against either (1) malicious server corruption, or (2) semi-honest (static) corruption of any fixed set of clients.*

Remark 2. Actually the above theorem can be more general – we can show that the protocol is secure against *any* (static) pattern of corruption in the $\mathcal{F}_{\mathsf{ah-mABE}}^f$ hybrid world (we will include this in the proof). However, in the previous Theorems 1 and 2, we only know how to realize $\mathcal{F}_{\mathsf{ah-mABE}}^f$ against either (1) malicious server corruption, or (2) semi-honest (static) corruption of any fixed set of clients. Therefore, by putting things together we can obtain an input-private verifiable computation (pVC) protocol against such patterns of corruption. In Section 5, we will show that the corruption pattern cannot be extended – it is impossible to construct general pVC protocols against arbitrary server-client collusions. This in particular implies that it is impossible to construct a protocol for $\mathcal{F}_{\mathsf{ah-mABE}}^f$ against arbitrary server-client collusions.

Efficiency of our construction. We outline the efficiency of a scheme where every client receives 1 bit of output — this can be achieved by a parallel repetition of our basic construction where only P_1 receives output. For such a private MVC scheme, the server runs in $\text{poly}(\kappa) \cdot O(|f| \cdot n)$. If we instantiate using the ABE construction of Gorbunov et al. [30], the run-time and the communication cost for each client is $O(d \cdot n\ell\kappa)$, where d is the depth of the function f being delegated, ℓ is the input length, and κ is the security parameter. In Appendix B we also offer more detailed discussion. We note that if some non-falsifiable assumption is used, it is possible to remove the dependence on the circuit depth. As mentioned, the focus of this paper is on using falsifiable assumptions.

Also we note that efficiency of Choi et al.'s construction [15] does not depend on circuit depth — however they security is weaker in many respects. An interesting direction for future research is to construct a scheme (or prove impossibility) where the client online computation and communication does not depend on the number of parties n and the circuit depth d, by only using standard assumptions.

4   From Semi-honest to Malicious Clients Corruptions

In the previous section, we considered the case where the clients can be corrupted in the semi-honest way. In this section, we present a simple compiler that upgrades the previous protocol to one that is secure against any maliciously corrupted clients, and remains non-interactive. That is, the resulting protocol is secure against either malicious server, or against a set of malicious clients. Our construction only needs an additional setup $\mathcal{F}_{\mathsf{CRS}}$.

We note that if we allow more rounds of communication, it is already known how to achieve security against arbitrary malicious corruptions (i.e. of clients and/or the server) [2]. However in the non-interactive multi-client verifiable computation MVC, there are no known constructions. We have already demonstrated security against a malicious server, and we will consider arbitrary corruptions of both the server and the clients in Section 5. Here we address the case where multiple clients are corrupted, and demonstrate that if an MVC protocol offers security against the semi-honest corruption of an arbitrary subset of the clients, and, additionally it offers the clients *perfect privacy* from one another (as defined in Definition 4), then there exists a simple compiler for guaranteeing security against the malicious corruption of clients. Of course, if we are allowed for a trusted PKI during the setup phase, we could include honestly generated, committed randomness for each party, and then use a NIZK to prove that all messages were honestly generated. However, we are interested in avoiding the use of trusted PKI, instead allowing each party to register the key of their choice; see Remark 1 for a discussion about trusted PKI and self-registered PKI.

Definition 4. *An* MVC *protocol* Π *has* perfect client privacy *if for all inputs* $x_1, \ldots, x_n$, *for an adversary* $\mathcal{A}$ *that semi-honestly corrupts some subset of the parties* $\{P_i\}_{i \in \mathcal{I}}$ *where* $\mathcal{I} \subset [n]$, *and for every random tape* $r_{\mathcal{A}}$ *belonging to* $\mathcal{A}$, *there exists a simulator* $\mathcal{S}im$ *such that the following distributions are identical*

$$\left\{\mathsf{View}_{\Pi(x_1, \ldots, x_n), \mathcal{A}}\right\} \equiv \left\{\mathcal{S}im(\{x_i, y_i\}_{i \in \mathcal{I}}, r_{\mathcal{A}})\right\}$$

where $(y_1, \ldots, y_n) \leftarrow f(x_1, \ldots, x_n)$, and $\mathsf{View}_{\Pi(x_1,\ldots,x_n),\mathcal{A}}$ *is the view of the adversary when the inputs to the clients are* $(x_1, \ldots, x_n)$. *In particular, the view contains random string* $r_{\mathcal{A}}$, *inputs* $\{x_i\}_{i\in\mathcal{I}}$, *and the message received from the server, and the messages generated by honest clients.*

Note that what distinguishes this from a standard requirement for semi-honest corruption is that we require indistinguishability to hold for every random tape of the adversary, rather than only on average. Intuitively, if a protocol meets this requirement, we can simplify the standard compilation techniques, since the adversary is free to use the random tape of his choice. To achieve security in the presence of malicious adversaries, it suffices to have the clients prove (using a NIZK) that their messages are consistent with *some* random string. Formally, we are able to achieve the following theorem. We present the proof in the full version of this paper.

Theorem 4. *Suppose there exists an* MVC *protocol* Π *in self-registered PKI setup hybrid model that is secure against semi-honest client corruptions, and that* Π *has perfect client privacy. Then there exists an* MVC *protocol* Π' *in the ZK and the self-registered PKI setup hybrid model, which is secure against malicious client corruptions.*

In order to apply the compiler results to our protocol, we need to show that our constructions have the desired property. We show this by the following claim:

Claim. If the underlying ABE and FHE and the garbling schemes is perfectly correct, then the private MVC protocol from Section 3 has perfect client privacy.

Proof. Since $\mathsf{P}_2, \ldots, \mathsf{P}_n$ do not receive messages or output, their views can be simulated easily. Now, we give a simulation of P_1's view. In the honest protocol, P_1 samples random strings m_0, m_1 and some r for generating ciphertexts of the ABE, FHE and garbling schemes. Suppose these schemes have perfect correctness. Then for every r the honest P_1 will receive either m_0 or m_1 from the server, depending on $b := f(x_1, \ldots, x_n)$. Therefore, given the result of the computation, b, and the random tape of P_1, $R = (m_0, m_1, r)$, the simulator can simply output m_b as the message from the server, producing an identical view. This completes the simulation of his view.

As a consequence of this theorem and Theorems 1, 2, 3, and the fact that noninteractive ZK can be implemented in the CRS model, we are able to construct a private MVC protocol using CRS and self-registered PKI. We summarize this by the following theorem:

Theorem 5. *Assume the existence of a fully homomorphic encryption scheme, a garbling scheme, and an ABE that has local encoding property. Assume the primitives have perfect correctness. Then for any efficiently computable* f, *there exists an* MVC *protocol that securely realizes the ideal functionality* $\mathcal{F}_{\mathsf{pVC}}^f$ *in the CRS and self-registered PKI hybrid model, against (1) any malicious server corruption, or (2) any malicious (static) corruption among any fixed set of clients.*

5   When the Server and Some Clients Are Corrupted

In this section, we consider the remaining, more complicated case where the server and clients can be corrupted at the same time. We show that even for a seemingly simple case where only one client and the server are corrupted together, it is impossible to construct private MVC for general functions, under a large class of instantiations of $\mathcal{G}$ setup including trusted PKI (which is stronger than self-registered PKI, see Remark 1), CRS, shared secret randomness, etc. The lower bound holds even in the standalone setting, and for semi-honest corruptions.

In particular, we consider the case with two clients and one server, where the function being delegated is a universal circuit $U(\cdot, \cdot)$, the first client's input is a circuit C, the second client's is a string x. The server returns $U(C, x) = C(x)$ to both parties. If there exists a private MVC protocol with respect to such U, i.e. if there exists a protocol that realizes $\mathcal{F}_{\mathsf{pVC}}$, then, even if it is only secure against semi-honest corruption and only in the standalone setting, we can construct an obfuscator for any circuit. (We refer the reader to the remark following Definition 2 for a definition of the standalone setting.) By previous lower bounds for obfuscation [4], this leads to an impossibility result. We present the formal statement below.

We note that there is a similar lower bound argument in the server-aided MPC setting in the work [2]. Our lower bound further shows that even a natural relaxation of security (where the ideal functionality can be called multiple times if there is server-client corruption) is not achievable for all functionalities.

Theorem 6. *Suppose there exist an instantiation of $\mathcal{G}$ setup and a private MVC protocol Π (i.e., one that realizes $\mathcal{F}_{\mathsf{pVC}}$) for all efficiently computable functions in the $\mathcal{G}$ setup hybrid world, against semi-honest corruptions for arbitrary parties in the standalone model, then there exists an obfuscator for any circuit C secure under the virtual black-box simulation.*

Proof. Consider the case where two clients want to delegate the computation of the universal circuit $U(\cdot, \cdot)$ to the server; the first client provides a circuit C, and the second provides an input x. Then the honest server returns $C(x)$ to both parties. Suppose there exists an instantiation of setup $\mathcal{G}$ and a secure protocol Π that achieve this goal, then we construct an obfuscator O that on input C does the following:

- O simulates the setup $\mathcal{G}$ and the role of each client in the offline stage to obtain $\mathsf{pub}, \mathsf{sk}_1, \mathsf{sk}_2, \hat{f}$.
- O simulates the first client's procedure on input C in the online stage. Let $\hat{C}$ be the message that P_1 sends to the server.
- O outputs $(\hat{C}, \mathsf{pub}, \mathsf{sk}_2, \hat{f})$ as an obfuscation of C, i.e. $O(C)$.

To evaluate $O(C)$ on input x, the evaluator simulates P_2's online phase to create an encoding of x using $\mathsf{pub}, \mathsf{sk}_2$, and then simulates the (corrupted) server to evaluate $\hat{C}, \hat{x}$ with the encoded version of the function $\hat{f}$.

The correctness follows immediately from the correctness of the protocol Π, and the efficiency of the obfuscator follows directly from the efficiency of the parties in the protocol Π. In the rest of the proof, we are going to show the virtual black-box (VBB) simulation property. In particular we will turn the protocol simulator into a VBB obfuscation simulator.

Now we analyze the construction. In particular, we want to show given an adversary $\mathcal{A}$ attacking the security of the obfuscation, we are going to construct a simulator Sim such that the probability $\mathcal{A}(O(C)) = 1$ is close to that of $Sim^C(1^k) = 1$ up to a negligible factor for all polynomial-sized circuits C. We do this by defining a particular adversary $\mathcal{A}^*$ that attacks $\Pi^{\mathcal{G}}$, and using the protocol simulator that is guaranteed to exist for this adversary by the security of the MVC protocol.

Given $\mathcal{A}$ and any poly-sized circuit C, we define the following experiment in the $\mathcal{G}$-hybrid world. Let $\mathcal{Z}^*$ be an environment and $\mathcal{A}^*$ be an adversary attacking protocol $\Pi^{\mathcal{G}}$. $\mathcal{A}^*$ corrupts the server and the second party at the beginning. He queries the ideal functionality $\mathcal{G}$ and stores the reply $(\mathsf{pub}, \mathsf{sk}_2)$. Upon receiving a message from P_1 during the offline stage (on behalf of the server), he uses this message, along with the offline message that an honest P_2 would send, to construct $\hat{f}$ as the server would do. When P_1 sends a message $\hat{C}$ to the server during the online phase, $\mathcal{A}^*$ interprets $(\hat{C}, \mathsf{pub}, \mathsf{sk}_2, \hat{f})$ as an obfuscation of $O(C)$. Then $\mathcal{A}^*$ runs $\mathcal{A}$ on the interpreted $O(C)$ and passes $\mathcal{A}$'s output to $\mathcal{Z}^*$. $\mathcal{Z}^*$ outputs this as the output of the experiment $\mathsf{EXEC}_{\mathcal{A}^*, \Pi, \mathcal{Z}^*}$.

Now we are ready to construct the simulator Sim. By the premise that $\Pi^{\mathcal{G}}$ realizes $\mathcal{F}_{\mathsf{pVC}}$, for this $\mathcal{A}^*$, there exists Sim^* such that for this particular $\mathcal{Z}^*$, we have $\mathsf{EXEC}^{\mathcal{G}}_{\mathcal{A}^*, \Pi, \mathcal{Z}^*} \approx \mathsf{EXEC}_{Sim^*, \mathcal{F}_{\mathsf{pVC}}, \mathcal{Z}^*}$. Given such Sim^*, we define a simulator Sim for the VBB obfuscation as follows:

- Sim basically simulates the execution of $\mathsf{EXEC}_{Sim^*, \mathcal{F}_{\mathsf{pVC}}, \mathcal{Z}^*}$.
- Whenever the protocol simulator Sim^* queries the oracle $O_{U,\{2\}}(\cdot)$ in the ideal functionality with some modified P_2's input x_2', Sim simulates it using a black-box query to C with input x_2' and returns $C(x_2')$ to Sim^*.
- Then Sim outputs whatever the output of the experiment $\mathsf{EXEC}_{Sim^*, \mathcal{F}_{\mathsf{pVC}}, \mathcal{Z}^*}$.

Then we are going to prove that Sim is a good VBB simulator by the following lemma:

Lemma 1. *For the simulator Sim described above, there exists a negligible function $\nu(\cdot)$ such that $|\Pr[\mathcal{A}(O(C)) = 1] - \Pr[Sim^C(1^k) = 1]| < \nu(k)$.*

Proof. Assume there is a non-negligible function ε with $\Pr[\mathcal{A}(O(C)) = 1] - \Pr[Sim^C(1^k) = 1] > \varepsilon$. We show that the real and simulation worlds in the protocol are distinguishable.

According to the description of $\mathsf{EXEC}^{\mathcal{G}}_{\mathcal{A}^*, \Pi, \mathcal{Z}^*}$, the output of such experiment is identical to that of $\mathcal{A}(O(C))$. On the other hand, the output of $\mathsf{EXEC}_{Sim^*, \mathcal{F}_{\mathsf{pVC}}, \mathcal{Z}^*}$ is exactly the same as that of $Sim^C(1^k)$. So this means the executions of the protocol are distinguishable by ε, which reaches a contradiction. Thus we complete the proof.

Acknowledgments. This research was sponsored in part by the U.S. Army Research Laboratory and the U.K. Ministry of Defence and was accomplished under Agreement Number W911NF-06-3-0001. The views and conclusions contained in this document are those of the authors and should not be interpreted as representing the official policies, either expressed or implied, of the U.S. Army Research Laboratory, the U.S. Government, the U.K. Ministry of Defence, or the U.K. Government. The U.S. and U.K. Governments are authorized to reproduce and distribute reprints for Government purposes notwithstanding any copyright notation hereon.

References

1. Applebaum, B., Ishai, Y., Kushilevitz, E.: From secrecy to soundness: Efficient verification via secure computation. In: Abramsky, S., Gavoille, C., Kirchner, C., Meyer auf der Heide, F., Spirakis, P.G. (eds.) ICALP 2010. LNCS, vol. 6198, pp. 152–163. Springer, Heidelberg (2010)
2. Asharov, G., Jain, A., López-Alt, A., Tromer, E., Vaikuntanathan, V., Wichs, D.: Multiparty computation with low communication, computation and interaction via threshold FHE. In: Pointcheval, D., Johansson, T. (eds.) EUROCRYPT 2012. LNCS, vol. 7237, pp. 483–501. Springer, Heidelberg (2012)
3. Backes, M., Fiore, D., Reischuk, R.M.: Verifiable delegation of computation on outsourced data. In: Sadeghi, A.-R., Gligor, V.D., Yung, M. (eds.) ACM CCS 2013, pp. 863–874. ACM Press (November 2013)
4. Barak, B., Goldreich, O., Impagliazzo, R., Rudich, S., Sahai, A., Vadhan, S.P., Yang, K.: On the (im)possibility of obfuscating programs. In: Kilian, J. (ed.) CRYPTO 2001. LNCS, vol. 2139, pp. 1–18. Springer, Heidelberg (2001)
5. Bellare, M., Hoang, V.T., Rogaway, P.: Adaptively secure garbling with applications to one-time programs and secure outsourcing. In: Wang, X., Sako, K. (eds.) ASIACRYPT 2012. LNCS, vol. 7658, pp. 134–153. Springer, Heidelberg (2012)
6. Bellare, M., Hoang, V.T., Rogaway, P.: Foundations of garbled circuits. In: Yu, T., Danezis, G., Gligor, V.D. (eds.) ACM CCS 2012, pp. 784–796. ACM Press (October 2012)
7. Ben-Sasson, E., Chiesa, A., Genkin, D., Tromer, E., Virza, M.: SNARKs for C: Verifying program executions succinctly and in zero knowledge. In: Canetti, R., Garay, J.A. (eds.) CRYPTO 2013, Part II. LNCS, vol. 8043, pp. 90–108. Springer, Heidelberg (2013)
8. Ben-Sasson, E., Chiesa, A., Tromer, E., Virza, M.: Succinct non-interactive zero knowledge for a von neumann architecture. In: Usenix Security Symposium (2014)
9. Benabbas, S., Gennaro, R., Vahlis, Y.: Verifiable delegation of computation over large datasets. In: Rogaway, P. (ed.) CRYPTO 2011. LNCS, vol. 6841, pp. 111–131. Springer, Heidelberg (2011)
10. Bitansky, N., Canetti, R., Chiesa, A., Tromer, E.: Recursive composition and bootstrapping for SNARKS and proof-carrying data. In: Boneh, D., Roughgarden, T., Feigenbaum, J. (eds.) 45th ACM STOC, pp. 111–120. ACM Press (June 2013)
11. Brakerski, Z.: Fully homomorphic encryption without modulus switching from classical GapSVP. In: Safavi-Naini, R., Canetti, R. (eds.) CRYPTO 2012. LNCS, vol. 7417, pp. 868–886. Springer, Heidelberg (2012)
12. Canetti, R.: Security and composition of multiparty cryptographic protocols. Journal of Cryptology 13(1), 143–202 (2000)

13. Canetti, R.: Universally composable security: A new paradigm for cryptographic protocols. Cryptology ePrint Archive, Report 2000/067 (2000), http://eprint.iacr.org/2000/067

14. Canetti, R.: Universally composable security: A new paradigm for cryptographic protocols. In: 42nd FOCS, pp. 136–145. IEEE Computer Society Press (October 2001)

15. Choi, S.G., Katz, J., Kumaresan, R., Cid, C.: Multi-client non-interactive verifiable computation. In: Sahai, A. (ed.) TCC 2013. LNCS, vol. 7785, pp. 499–518. Springer, Heidelberg (2013)

16. Chung, K.-M., Kalai, Y., Vadhan, S.P.: Improved delegation of computation using fully homomorphic encryption. In: Rabin, T. (ed.) CRYPTO 2010. LNCS, vol. 6223, pp. 483–501. Springer, Heidelberg (2010)

17. Chung, K.-M., Kalai, Y.T., Liu, F.-H., Raz, R.: Memory delegation. In: Rogaway, P. (ed.) CRYPTO 2011. LNCS, vol. 6841, pp. 151–168. Springer, Heidelberg (2011)

18. Fiore, D., Gennaro, R.: Publicly verifiable delegation of large polynomials and matrix computations, with applications. In: Yu, T., Danezis, G., Gligor, V.D. (eds.) ACM CCS 2012, pp. 501–512. ACM Press (October 2012)

19. Garg, S., Gentry, C., Sahai, A., Waters, B.: Witness encryption and its applications. In: Boneh, D., Roughgarden, T., Feigenbaum, J. (eds.) 45th ACM STOC, pp. 467–476. ACM Press (June 2013)

20. Gennaro, R., Gentry, C., Parno, B.: Non-interactive verifiable computing: Outsourcing computation to untrusted workers. In: Rabin, T. (ed.) CRYPTO 2010. LNCS, vol. 6223, pp. 465–482. Springer, Heidelberg (2010)

21. Gennaro, R., Gentry, C., Parno, B., Raykova, M.: Quadratic span programs and succinct nIZKs without pCPs. In: Johansson, T., Nguyen, P.Q. (eds.) EUROCRYPT 2013. LNCS, vol. 7881, pp. 626–645. Springer, Heidelberg (2013)

22. Gentry, C., Lewko, A.B., Sahai, A., Waters, B.: Indistinguishability obfuscation from the multilinear subgroup elimination assumption. IACR Cryptology ePrint Archive, 2014:309 (2014)

23. Goldreich, O.: Foundations of Cryptography: Basic Applications, vol. 2. Cambridge University Press, Cambridge (2004)

24. Goldwasser, S., Gordon, S.D., Goyal, V., Jain, A., Katz, J., Liu, F.-H., Sahai, A., Shi, E., Zhou, H.-S.: Multi-input functional encryption. In: Nguyen, P.Q., Oswald, E. (eds.) EUROCRYPT 2014. LNCS, vol. 8441, pp. 578–602. Springer, Heidelberg (2014)

25. Goldwasser, S., Kalai, Y.T., Popa, R.A., Vaikuntanathan, V., Zeldovich, N.: How to run turing machines on encrypted data. In: Canetti, R., Garay, J.A. (eds.) CRYPTO 2013, Part II. LNCS, vol. 8043, pp. 536–553. Springer, Heidelberg (2013)

26. Goldwasser, S., Kalai, Y.T., Popa, R.A., Vaikuntanathan, V., Zeldovich, N.: Reusable garbled circuits and succinct functional encryption. In: Boneh, D., Roughgarden, T., Feigenbaum, J. (eds.) 45th ACM STOC, pp. 555–564. ACM Press (June 2013)

27. Goldwasser, S., Lin, H., Rubinstein, A.: Delegation of computation without rejection problem from designated verifier CS-proofs. IACR Cryptology ePrint Archive, 2011:456 (2011)

28. Goldwasser, S., Micali, S.: Probabilistic encryption. J. Comput. Syst. Sci. 28(2), 270–299 (1984)

29. Goldwasser, S., Micali, S., Rackoff, C.: The knowledge complexity of interactive proof systems. SIAM J. Comput. 18(1), 186–208 (1989)

30. Gorbunov, S., Vaikuntanathan, V., Wee, H.: Attribute-based encryption for circuits. In: Boneh, D., Roughgarden, T., Feigenbaum, J. (eds.) 45th ACM STOC, pp. 545–554. ACM Press (June 2013)

31. Kamara, S., Mohassel, P., Raykova, M.: Outsourcing multi-party computation. Cryptology ePrint Archive, Report 2011/272 (2011), http://eprint.iacr.org/

32. Kamara, S., Mohassel, P., Riva, B.: Salus: a system for server-aided secure function evaluation. In: Yu, T., Danezis, G., Gligor, V.D. (eds.) ACM CCS 2012, pp. 797–808. ACM Press (October 2012)

33. Kosba, A.E., Papadopoulos, D., Papamanthou, C., Sayed, M.F., Shi, E., Triandopoulos, N.: Trueset: Nearly practical verifiable set computations. In: Usenix Security Symposium (2014)

34. Papamanthou, C., Shi, E., Tamassia, R.: Signatures of correct computation. In: Sahai, A. (ed.) TCC 2013. LNCS, vol. 7785, pp. 222–242. Springer, Heidelberg (2013)

35. Papamanthou, C., Tamassia, R., Triandopoulos, N.: Optimal verification of operations on dynamic sets. In: Rogaway, P. (ed.) CRYPTO 2011. LNCS, vol. 6841, pp. 91–110. Springer, Heidelberg (2011)

36. Parno, B., Raykova, M., Vaikuntanathan, V.: How to delegate and verify in public: Verifiable computation from attribute-based encryption. In: Cramer, R. (ed.) TCC 2012. LNCS, vol. 7194, pp. 422–439. Springer, Heidelberg (2012)

37. Shi, E., Chan, T.-H.H., Rieffel, E.G., Chow, R., Song, D.: Privacy-preserving aggregation of time-series data. In: NDSS 2011. The Internet Society (February 2011)

A   Preliminaries

Here we present the definitions we use in this paper.

A.1   Two-outcome ABE

Definition 5 (Correctness of Two-outcome ABE [26]). *For any polynomial* $n(\cdot)$, *for every sufficiently large security parameter* κ, *if* $n = n(\kappa)$, *for all boolean functions* $f \in \mathcal{F}_n$, *attributes* $x \in \{0,1\}^n$, *messages* $M_0, M_1 \in \mathcal{M}$, *there exists some negligible* $\nu(\cdot)$ *such that*

$$\Pr\left[\begin{array}{l} (\mathsf{mpk}_{\mathsf{ABE}}, \mathsf{msk}_{\mathsf{ABE}}) \leftarrow \mathsf{ABE.Setup}(1^\kappa); \\ \mathsf{sk}_f \leftarrow \mathsf{ABE.KeyGen}(\mathsf{msk}_{\mathsf{ABE}}, f); \\ c \leftarrow \mathsf{ABE.Enc}(\mathsf{mpk}_{\mathsf{ABE}}, x, m_0, m_1); \\ m = \mathsf{ABE.Dec}(\mathsf{sk}_f, c): \\ m = m_{f(x)} \end{array}\right] = 1 - \nu(\kappa).$$

If $\nu = 0$, *then the scheme has perfect correctness.*

Then we define the security for single-key two-outcome ABE.

Definition 6 (Security of Two-outcome ABE [26]). *Let* ABE *be a two-outcome ABE scheme for the class of boolean functions* $\mathcal{F} = \{\mathcal{F}_n\}_{n \in \mathbb{N}}$ *and associated message space* $\mathcal{M}$ *and let* $\mathcal{A} = (\mathcal{A}_1, \mathcal{A}_2, \mathcal{A}_3)$ *be a triple of PPT adversaries. Consider the following experiment.*

- $(\mathsf{mpk}_{\mathsf{ABE}}, \mathsf{msk}_{\mathsf{ABE}}) \leftarrow \mathsf{ABE.Setup}(1^\kappa)$
- $(f, \mathsf{st}_1) \leftarrow \mathcal{A}_1(\mathsf{mpk}_{\mathsf{ABE}})$
- $\mathsf{sk}_f \leftarrow \mathsf{ABE.KeyGen}(\mathsf{msk}_{\mathsf{ABE}}, f)$
- $(m, m_0, m_1, x, \mathsf{st}_2) \leftarrow \mathcal{A}_2(\mathsf{st}_1, \mathsf{sk}_f)$
- choose a bit b at random. Then let

$$c = \begin{cases} \mathsf{ABE.Enc}(\mathsf{mpk}_{\mathsf{ABE}}, x, m, m_b), & \text{if } f(x) = 0, \\ \mathsf{ABE.Enc}(\mathsf{mpk}_{\mathsf{ABE}}, x, m_b, m), & \text{otherwise.} \end{cases}$$

- $b' \leftarrow \mathcal{A}_3(\mathsf{st}_2, c)$. If $b = b'$, and there exists n such that, for all $f \in \mathcal{F}_n$, messages $m.m_0, m_1 \in \mathcal{M}$, $|m_0| = |m_1|$, $x \in \{0,1\}^n$, output 1. Else output 0.

We say the scheme is a full-secure single-key two-outcome ABE if for all PPT adversaries $\mathcal{A}$, and for all sufficiently large κ, the probability that the experiment outputs 1 is bounded by $1/2 + \nu(k)$ for some negligible function ν.

A.2   Garbling Schemes

Definition 7 (Garbling Schemes [6]). A garbling scheme for a family of circuits $C = \{C_n\}_{n \in \mathbb{N}}$ with C_n a set of boolean circuits taking as input n bits, is a tuple of PPT algorithms $\mathsf{Gb} = \mathsf{Gb}.\{\mathsf{Garble, Enc, Eval}\}$ such that

- $\mathsf{Gb.Garble}(1^\kappa, C)$ takes as input the security parameter κ and a circuit $C \in C_n$ for some n and outputs the garbled circuit Γ and a secret key sk.
- $\mathsf{Gb.Enc}(\mathsf{sk}, x)$ takes as input x and outputs an encoding c,
- $\mathsf{Gb.Eval}(\Gamma, c)$ takes as input a garbled circuit Γ and an encoding c, and outputs a value y which should be $C(x)$.

The correctness and efficiency properties are straight-forward. Next we consider a special property of the encoding of the Yao's garbled scheme, which we will use in this paper. The secret key has the form $\mathsf{sk} = \{L_i^0, L_i^1\}_{i \in [n]}$, and the encoding of an input x of n bits is of the form $c = (L^{x_1}, L^{x_2}, \ldots, L^{x_n})$, where x_i is the i-th bit of x.

Then we are going to define the security of garbling schemes.

Definition 8 (Input and Circuit Privacy). A garbling scheme Gb for a family of circuits $\{C_n\}_{n \in \mathbb{N}}$ is input and circuit private if there exists a PPT simulator $\mathcal{S}im$ such that for every adversaries $\mathcal{A}$ and D, for all sufficiently large κ,

$$\left| \Pr \begin{bmatrix} (x, C, \alpha) \leftarrow A(1^k); \\ (\Gamma, \mathsf{sk}) \leftarrow \mathsf{Gb.Garble}(1^\kappa, C); \\ c \leftarrow \mathsf{Gb.Enc}(\mathsf{sk}, x): \\ D(\alpha, x, C, \Gamma, c) = 1 \end{bmatrix} - \Pr \begin{bmatrix} (x, C, \alpha) \leftarrow A(1^k); \\ (\tilde{\Gamma}, \tilde{c}) \leftarrow \mathcal{S}im(1^\kappa, C(x), 1^{|C|}, 1^{|x|}): \\ D(\alpha, x, C, \tilde{\Gamma}, \tilde{c}) = 1 \end{bmatrix} \right| = \nu(k)$$

for some negligible $\nu(\cdot)$, where we consider only $\mathcal{A}$ such that for some n, $x \in \{0,1\}^n$ and $C \in C_n$.

A.3   Extractable Witness Encryption

Definition 9 (Witness Encryption [19]). *A witness encryption for a language* $L \in NP$ *with corresponding witness relation* R_L *consists of two polynomial-time algorithms* WE.$\{$Enc, Dec$\}$ *such that*

- *Encryption* WE.Enc($1^\kappa, x, b$): *takes as input a security parameter* κ, *a statement* $x \in \{0, 1\}^*$, *a bit* b *and outputs a ciphertext* c.
- *Decryption* WE.Dec(w, c): *takes as input a witness* $w \in \{0, 1\}^*$ *and a ciphertext* c *and outputs a bit* b *or* $\perp$.

Correctness: For all $(x, w) \in R_L$, *for all bits* b *for every sufficiently large security parameter* κ, *we have*

$$\Pr[c \leftarrow \text{WE.Enc}(1^\kappa, x, b) : \text{WE.Dec}(w, c) = b] = 1 - \nu(\kappa),$$

for some negligible ν.

Definition 10 (Extractable Security [25]). *A witness encryption scheme for a language* $L \in NP$ *is secure if for all PPT adversaries* $\mathcal{A}$, *and all poly* q, *there exists a PPT extractor* E *and a poly* p *such that for all auxiliary input* z *and all* $x \in \{0, 1\}^*$, *the following holds:*

$$\Pr[b \leftarrow \{0, 1\}; c \leftarrow \text{WE.Enc}(1^\kappa, x, b) : \mathcal{A}(x, c, b) = b] \geq 1/2 + 1/q(|x|)$$
$$\Rightarrow \Pr[E(x, z) = w : (x, w) \in R_L] \geq 1/p(|x|).$$

A.4   Obfuscations

Definition 11 (Circuit Obfuscator [4]). *A probabilistic algorithm* O *is a (circuit) obfuscator for the collection* $\mathcal{F}$ *of circuits if the following holds:*

- *(functionality) For every circuit* $C \in \mathcal{F}$, *the string* $O(C)$ *describes a circuit that computes the same function as* C.
- *(polynomial slowdown) There is a polynomial* p *such that for every circuit* $C \in \mathcal{F}$, *we have* $|O(C)| \leq p(|C|)$.
- *("virtual black box" (VBB) property) For any PPT* $\mathcal{A}$, *there is a PPT* $\mathcal{S}im$ *and a negligible function* ν *such that for all circuits* $C \in \mathcal{F}$, *it holds that*

$$\left| \Pr\left[\mathcal{A}(O(C)) = 1\right] - \Pr\left[\mathcal{S}im^C(1^{|C|}) = 1\right] \right| \leq \nu(|C|).$$

We say that O *is efficient if it runs in polynomial time. If we omit specifying the collection* $\mathcal{F}$, *then it is assumed to be the collection of all circuits.*

A.5   Proxy Oblivious Transfer

Choi et al. [15] recently defined and constructed proxy oblivious transfer. Instead of taking the game based security definitions from the paper by Choi et al., here we define the security of POT in the real/ideal paradigm, which provides a stronger security guarantee. In the ideal functionality below, we omit the session id for notational simplicity. We remark that in each session, the functionality could accept multiple new inputs; we assign a sub-session id, i.e., ssid, for each new input.

Theorem 7 ([15]). *There is a* non-interactive *protocol which realizes* $\mathcal{F}_{\mathsf{POT}}$ *in the self-registered PKI setup* $\mathcal{G}^{\mathsf{Diffie-Hellman}}$*-hybrid model, against (1) any malicious server corruption, or (2) any semi-honest (static) corruption among any fixed set of clients.*

Proxy Oblivious Transfer

Functionality $\mathcal{F}_{\mathsf{POT}}$ interacts with a sender P_S, a chooser P_C, a receiver P_R, and the simulator $\mathcal{S}im$.

- Upon receiving $(\mathsf{ssid}, m_0, m_1)$ from the sender P_S, notify $\mathcal{S}im$ with $(\mathsf{ssid}, \mathsf{P}_S)$. Later, when $\mathcal{S}im$ replies with $(\mathsf{ssid}, \mathsf{P}_S)$, if no value $(\mathsf{ssid}, m_0', m_1')$ has been recorded yet, store it and notify P_R with $(\mathsf{ssid}, \mathsf{P}_S)$.
- Similarly, upon receiving (ssid, b) from the chooser P_C, notify $\mathcal{S}im$ with $(\mathsf{ssid}, \mathsf{P}_C)$. Later, when $\mathcal{S}im$ replies with $(\mathsf{ssid}, \mathsf{P}_C)$, if no value (ssid, b') has been recorded yet, store it and notify P_R with $(\mathsf{ssid}, \mathsf{P}_C)$.
- Upon receiving $(\mathsf{ssid}, 1)$ from P_R, if both $(\mathsf{ssid}, m_0, m_1)$ and (ssid, b) are recorded, send (ssid, m_b) to the receiver P_R; else send fail.

Fig. 4. Functionality $\mathcal{F}_{\mathsf{POT}}$

Choi et al. [15] constructed a *non-interactive* protocol in the offline/online model that realizes the ideal functionality $\mathcal{F}_{\mathsf{POT}}$. In this model, the two clients run some protocol in the offline stage, prior to learning their inputs, and then complete the protocol in the online stage, after receiving their inputs. In their construction, the clients do not need to interact in the offline stage, and in the online stage both the sender and chooser send a single message to the server. The construction relies on the existence of non-interactive key agreement schemes (e.g., the Diffie-Hellman key exchange scheme).

B   Instantiations and Efficiency

In this section, we discuss the instantiations of our building blocks. We need a two-outcome attribute encryption scheme with the local encoding property, a fully homomorphic encryption scheme, and a garbling scheme. In particular we can use any instantiation of FHE schemes, e.g. one by Brakerski [11], and any instantiation of Yao's garbling scheme.

The attribute based encryption (ABE) constructed by Gorbunov, Vaikun-tanathan, Wee [30] actually achieves the requirements of regular ABE with the local encoding property. Goldwasser et al. [26] showed a generic way to achieve two-outcome ABE from a regular one. So by plugging the GSW ABE scheme and using the generic technique, we achieve the two-outcome ABE as required by Definition 3.

For our private MVC scheme, the server clearly runs in $poly(\kappa, f)$. For the clients, $P_2, \ldots, P_n$ runs in time $O(\ell\kappa)$, where ℓ is the input length; P_1 generates $O(n\ell\kappa)$ ABE ciphertexts plus a garble circuit of size $O(\kappa)$, where n is the number of parties, ℓ is the input length, and κ is the security parameter. However, the ciphertexts' length for the currently best known ABE construction of Gorbunov, Vaikuntanathan, Wee [30] depends on the circuit depth (independent of the size). Therefore, P_1's running time (the communication complexity as well) depends on $O(d \cdot n\ell\kappa)$, where d is the depth of the function being delegated. The construction of Choi et al. [15] has better online efficiency for clients that is independent of the function complexity, but has the issues of adaptive soundness and is vulnerable to selective failure attacks. The construction using multi-input functional encryption [24] can achieve better efficiency but their solution inherently requires the existence of indistinguishable obfuscation, which is a stronger assumption and has large overhead.

Public Verification of Private Effort*

Giulia Alberini[1], Tal Moran[2,**], and Alon Rosen[2,***]

[1] School of Computer Science, McGill University, Canada
giulia.alberini@mail.mcgill.ca
[2] Efi Arazi School of Computer Science, IDC Herzliya, Israel
{talm,alon.rosen}@idc.ac.il

Abstract. We introduce a new framework for polling responses from a large population. Our framework allows gathering information without violating the responders' anonymity and at the same time enables public verification of the poll's result. In contrast to prior approaches to the problem, we do not require trusting the pollster for faithfully announcing the poll's results, nor do we rely on strong identity verification.

We propose an "effort based" polling protocol whose results can be publicly verified by constructing a "responder certification graph" whose nodes are labeled by responders' replies to the poll, and whose edges cross-certify that adjacent nodes correspond to honest participants. Cross-certification is achieved using a newly introduced (privately verifiable) "Private Proof of Effort" (PPE). In effect, our protocol gives a general method for converting privately-verifiable proofs into a publicly-verifiable protocol. The soundness of the transformation relies on expansion properties of the certification graph.

Our results are applicable to a variety of settings in which crowd-sourced information gathering is required. This includes crypto-currencies, political polling, elections, recommendation systems, viewer voting in TV shows, and prediction markets.

Keywords: Polling, anonymity, protocols, random graphs, public verifiability, proof of work, CAPTCHA.

1 Introduction

The Internet enables reciprocal communication on a massive scale. Thus, it has the potential to allow new forms of information gathering and "crowd-sourced" decision making. Some examples (already in widespread use) are political polling, elections (which are a mechanism for achieving consensus among voters about which candidate to put in office), recommendation systems (e.g., based on users'

---

* This version is an extended abstract. A complete version of the paper appears on the Cryptology ePrint Archive [1].
** Supported by ISF grant no. 1790/13 and by the European Union Seventh Framework Programme (FP7/2007-2013) under grant agreement no. 293843.
*** Supported by ISF grant no. 1255/12 and by the ERC under the EUs Seventh Framework Programme (FP/2007-2013) ERC Grant Agreement no. 307952.

Y. Dodis and J.B. Nielsen (Eds.): TCC 2015, Part II, LNCS 9015, pp. 169–198, 2015.

opinions about products and services), prediction markets (leveraging the "wisdom of the crowds" to predict future events) and "crypto-currencies" (such as Bitcoin [16]).

We can think of all these cases as a generalized "opinion poll":the outcome is the result of aggregating the opinions of a large population of Internet users. The "protocols" that implement the poll (and the methods of computing the results) are different in each case, but in all of them we can categorize the participants into three types (some parties may belong to multiple categories):

1. *pollsters* are responsible for collecting the information and publishing the result.
2. *responders* are the parties who provide inputs to the poll.
3. *verifiers* are interested in (should agree on) the result, but may not be active participants.

Although at first glance the examples mentioned above may not necessarily appear to be a *distributed* protocol problem (e.g., in elections there is a central election authority that can broadcast results to everyone), it is natural to consider the case when the central authorities are *untrusted*, and can potentially act maliciously. Viewed this way, verifiable polling is a generalization of the fundamental problem of achieving consensus between mutually-distrustful parties. While in the general polling setting, inputs of various parties could differ and are aggregated into the poll's "tally", the basic consensus problem focuses on the special case in which parties only have to agree on a specific output if all of their inputs match. Correctness of the consensus is guaranteed by the verifiability property of the polling protocol.

In their general form, verifiable opinion polls are also useful as building blocks in more complex protocols. For example, the main technical innovation of Bitcoin, a recently popular "crypto-currency", is in achieving a distributed, decentralized consensus about the currency's public transaction ledger (the record of all Bitcoin transactions) [16].

In the "traditional" setting for the verifiable polling problem (and its variants), the number and identities of the parties are known in advance. Using standard cryptographic techniques, solutions are known to many of them. Techniques for verifiable voting, for example, provide solutions that hide the individual responses of the participants, even after revealing the tally (see subsection 1.5 for references).

Unfortunately, adapting the traditional solutions to work in a decentralized internet environment is non-trivial. One of the major problems encountered in this setting is the lack of identity verification. Strong identity verification on a large scale is expensive, and in many cases completely impractical (e.g., when participants are spread across national boundaries, there might not be a single entity trusted by all of them to certify identities). The mechanisms for identity verification become even more complex when anonymity (or pseudonymity) of the participants is required. In the absence of identity verification, it is impossible to distinguish a fake identity from a real one; this opens the door to "Sybil attacks" (attacks based on creating multiple fake identities).

There are various methods used to mitigate Sybil attacks without requiring identity verification. A recurring idea is to force participants to prove they expended some valuable resource: for example, spending money or performing a computational task. This serves to limit the number of fake identities an adversary can create. In this paradigm, we have no choice but to relax our requirements from the poll: rather than requiring "one vote per participant", we now allow "one vote per effort" (where an "effort unit" corresponds to expending some resource). We call this *effort-based polling.*

The Bitcoin protocol is an excellent example of this type: in effect, consensus is achieved by having parties constantly "vote" on which version of the transaction ledger they accept, where for each "vote" the party must also generate a "proof-of-work" to prove that the required amount of computational effort was expended.[1]

Proofs of work are one of the very few examples of proofs of effort that are *publicly* verifiable. However, they suffer from significant drawbacks. First of all, they are inherently wasteful in that the computation "does nothing" except prove work (indeed, this is one of the strongest arguments against the Bitcoin currency [12]). Secondly, and perhaps more importantly, a party with access to more computing power than most honest responders may gain a hugely disproportionate influence on the results (not to mention the wide disparities between the responders themselves).

1.1   Privately Verifiable Proofs of Effort

An alternative to publicly-verifiable proofs of work, and one that may be potentially easier to achieve, is that of *privately*-verifiable proofs of resource expenditure. One well known example is that of enforcing human involvement in each response. In voting for the "American Idol" TV show, for example, online viewers must solve a CAPTCHA [19] for each vote, but the total number of votes is effectively unlimited. (What makes the CAPTCHA solution privately verifiable is the fact that all currently known CAPTCHAs are *private coin*: in effect, every CAPTCHA is generated together with its solution.)

Beyond being easier to achieve (and not being "wasteful") The "human effort" requirement may be useful when there is a "resource gap" between honest and malicious parties. For example, show producers have significantly more money and access to more computing power than most honest viewers (and there are wide disparities between the viewers themselves)—using proof-of-work in this context could give them a hugely disproportionate influence on the results.

In effect, what CAPTCHAs enable us to achieve is what we call a *privately-verifiable proof of effort* (PPE). Informally, this is an interactive protocol between two parties: if both parties are honest the test returns "true" to both, otherwise the test returns "false" to the honest participant.

---

[1] The outcome of a Bitcoin "poll" is not a majority-vote, but a randomized selection in which the probability for selecting a "candidate" is proportional to the total effort expended for that candidate. However, this still fits in our generalized polling framework.

Definition 1 (PPE, Informal). *A two-party protocol is a PPE if it satisfies:*

1. **Effort** *If both parties honestly follow the protocol, they expend one "effort unit".*
2. σ-**Completeness.** *If both parties honestly followed the protocol, they will both output "true" at the end of the protocol with probability at least $1 - \sigma$.*
3. ε-**Soundness.** *If one party is malicious (invests less than the required effort) and the other honestly follows the protocol, the honest party will output "false" with probability at least $1 - \varepsilon$.*

We note that this definition is necessarily informal, since the term "effort unit" is itself not well defined. In our analysis, we sidestep the problem by reversing the definition: instead of defining a PPE as a proof of effort, we define a "proof of effort" as successful completion of PPE with at least one honest participant (formally, we follow Canetti et al.'s framework for defining CAPTCHAs [4] and define effort in terms of oracle calls; see section 2 for details).

The peer-to-peer nature of PPEs makes them potentially easier to realize than their publicly-verifiable counterparts (which require costly distributed coordination). In section A, we list several potential mechanisms for PPEs, most of which do not require human involvement (making them fully automatizable, and hence scalable). These include proofs of storage, human interaction (including symmetric CAPTCHAs) and leveraging social networks.

1.2   Our Results

While PPEs seem easier to realize, it is not at all clear how to utilize them in order to deal with the problem of a cheating pollster. For instance, in the American Idol example, a malicious CAPTCHA generator can use the solutions to the CAPTCHAs without expending any human effort. Thus, existing CAPTCHAs cannot be publicly verified (hence cannot be used to achieve a consensus about the result of the poll when the generator is untrusted).

Our main result is a new protocol for *publicly*-verifiable effort-based polling, based on any *privately*-verifiable proof-of-effort (PPE). The protocol uses PPEs to generate a "responder certification graph": each responder is a node in the graph while an edge between two responders corresponds to a PPE execution. Loosely speaking, we guarantee that, as long as enough honest users participate in the protocol, a large number of cheating nodes will be publicly detected (note that, unlike most standard definitions of an "honest party", in our case every party controlled by the adversary is considered "cheating", even if it follows the honest protocol exactly).

If each node in the graph is published together with their response to the poll, the poll results cannot be skewed significantly by the pollster without being detected.

In its simplest variant, our protocol assumes that the responder certification graph is sampled at random. This sampling can be performed in a publicly-verifiable way, say by applying a "random-looking" function (e.g., SHA-1) to two nodes' indices to determine if there is an edge between them in the graph. Since our protocol's analysis relies only on expansion properties of such randomly chosen graphs, the construction can potentially be derandomized—using an explicit graph with the appropriate expansion properties, we could remove our assumption about SHA-1 and improve the protocol parameters, at the cost of making the protocol more complex.

We note that while the structure of certification graph is fixed (it depends only on the number of nodes), we allow the adversary to specify the number of nodes (within bounds) and to *arbitrarily* control the assignment of honest nodes to vertices in the graph. We prove that security holds *for every assignment*.

1.3   Main Theorems

The total number of nodes in the certification graph is denoted m and corresponds to the total number of responders (some of whom may be controlled by the adversary). The number of honest responders is denoted by n. We denote by d the average degree of the responder certification graph: this is the number of PPE executions each responder is expected to participate in.

We model our assumption that the pollsters have bounded resources by specifying that a cheating pollster cannot participate in too many successful PPEs with honest responders. In terms of the certification graph, this assumption implies a bound a on the number of "attack edges"—PPE executions in which the cheating pollster participates as one party and convinces an honest responder to accept. The ratio a/d gives a lower bound on the number of "cheating" nodes; an attacker can always create this many cheating nodes without detection by following the protocol honestly. Thus, our security guarantees make sense only when $a/d \ll n$ (we can think of a/dn as a small constant).

We denote by κ the security parameter. Our main theorems guarantee the soundness (a malicious pollster can't cheat undetectably) and completeness (an honest execution will be accepted) of our protocol. For simplicity, we will consider PPEs for which ε (the soundness error of a PPE) is negligible in the security parameter and omit it. For our completeness proof, we require an additional independence property: that for a given node, the probability of failure in each PPE execution is independent (the probability can depend on the node, however).

Theorem 1 (Soundness—Informal). *Let A be an adversarial pollster that cannot succeed in more than a PPEs with honest responders. If there are at least $n \geq \alpha m$, $\alpha \in (0, 1)$, honest responders to the poll and A controls more than $\Omega(\frac{a}{d})$ of the responses in the poll outcome, then verification will fail with overwhelming probability (in κ).*

See section 4 for the full theorem and proof. Note that our proof holds in the random oracle model, but under a very reasonable assumption about the cryptographic hash function (that the generated graph has good expansion parameters) it holds in the standard model as well.

Theorem 2 (Completeness—Informal). *If the pollster is honest, and malicious responders are bounded by $O(m)$ successful PPEs, the probability that verification fails is negligible in κ.*

See section 5 for the full theorem and proof. The bound on successful PPEs by malicious responders is required to guarantee robustness of the protocol— when the pollster is honest the verification should succeed even if some of the responders are malicious.

1.4   Comparison to Verifiable Voting

At a high level, our polling protocol has the same form as most universally verifiable voting protocols (involving an "election authority", "voters", "receipts" and "verification procedure"):

1. The pollster sets up the poll and publishes public parameters on a bulletin-board (modeled as a broadcast channel). This corresponds to the role of the "election authority".
2. Honest responders (corresponding to the "voters") send their responses to the pollster and engage in an interactive proof protocol to ensure that they are expending the correct amount of "effort" for each response. This protocol includes interaction with the pollster and also, unlike most voting protocols, interaction with a subset of other responders.
   The pollster signs the transcript of each communication with a responder and sends this signature to the responder (think of this as the "receipt" in the voting protocol).
3. The pollster publishes the empirical distribution of responses, together with a proof of correctness.
4. The verification procedure consists of both a *local* verification step performed by the responders (which in a voting protocol corresponds to verifying that the voter's receipt appears on the bulletin board) and a *global* verification step performed by the verifiers (which corresponds to the "universal verification" step in voting protocols). Note that responders can also act as verifiers if they wish.

A significant difference between effort-based polling and verifiable voting is the issue of voter identity. In our polling protocol, parties are identified only by self-chosen pseudonyms (for our purposes, a pseudonym is a verification key of a public-key signature scheme). We do not limit the number of pseudonyms a party may generate, or require parties to link their pseudonyms to their real identities.

In contrast, most voting protocols assume each party in the protocol has been identified by a trusted authority, in order to ensure that each voter gets only a single vote. By relaxing this requirement to "one vote per effort expended", we can dispense with the complexity, expense and privacy implications of securely identifying responders.

In particular, our protocol is compatible with completely anonymous polling (if responders communicate with the pollster over *anonymous channels*)—in addition to hiding the link between their real identities and their responses, use of anonymous channels can hide the fact of participation in the poll, with the degree of anonymization depending only on the anonymous channel (in contrast, cryptographic voting protocols that support hiding the voters' participation require a separate non-anonymous registration step, and anonymity depends on the election trustees in addition to the anonymous channel).

1.5   Related Work

Sybil Defense. In a "Sybil attack", an adversary creates multiple "fake" identities in order to manipulate a protocol. The problem of establishing trustworthy virtual identities has plagued the Internet from its inception [10]. It is particularly acute in distributed systems with no central authority—without additional assumptions, vulnerability to some forms of Sybil attacks is unavoidable in this case [9]. The paper by [10] deals with the problem of establishing identities. One of the first discussions of trust metrics based on social graphs appears in [13]. The term "Sybil Attack" (attributed to Brian Zill from MSR) was introduced in [9], where it is shown that in the absence of a central certifying authority, some attacks are always possible.

A reputation system for p2p with similar ideas to pagerank (doesn't handle sybils) is developed in [11], and the possibility of using "Turing tests" to limit Sybil nodes is mentioned in [3]. In [7] it is shown that there exists no symmetric sybil-proof reputation mechanism. Since the existing sybil-defense protocols all care about reputations (e.g., determining which nodes are "real" and which are sybils), they all strongly rely on breaking symmetry: having at least one trusted node. Our protocol is symmetric, however we can sidestep the impossibility proof because we don't care about individual nodes' reputations—only about the aggregate opinion of all the nodes.

The technique of random walks on a social networks to bound the effect of Sybil attack is introduced in [23] (see [22] for an expanded version with full proofs). A 2006 survey of sybil attack literature can be found in [14]. An improved version of [23] (slightly different protocol, same goal but better parameters) appears in [21], and a newer protocol to identify sybil nodes in a social graph is presented in [8]. The protocol makes very similar assumptions about the social graph, and Bayesian methods to compute the probabilities that nodes are sybils. Finally, [18] uses the social network graph to aggregate votes for online content. The "vote collector" is assumed to be honest, and votes are collected using max-flow in the social graph.

Most of these techniques implicitly or explicitly use assumptions about expansion properties of social-network graphs. We also make use of the idea that if "adding edges is hard" in an expander graph the adversary is limited in the effect bad nodes can have, but in our case the graph is artificially generated, so we can prove (in the random-oracle model, at least) that our graph has the required properties. On the other hand, the labeling of the graph is adversarial;

despite this, we get results that are—in some sense—stronger than the results on social networks: we can bound the total number of "bad" nodes (rather than just their influence).

Verifiable and Private Polling. A widely used technique for privacy-preserving polling is called "randomized response" and was introduced in [20]. The first suggestion for cryptographic verifiability in voting, which also gives a mechanism for establishing anonymous channels (mix-nets) was made in [5]. More recently, the works of [6,17] propose taking into account *human* voters in End-to-End verifiability, and introduce the notion of separate verification steps for the voter and external observers. Another incarnation of this idea is verifiable (for the pollster) privacy-preserving polling using scratch-off cards [15].

2   Model and Definitions

We now introduce a formal model for capturing the notion of verifiable effort-based polling. The definition will have to address both the syntax of a polling protocol and the issue of the "effort" involved in the protocol execution. To model the effort expended by each one of the protocol participants, we give parties access to an *effort oracle*. The effort spent by each party is measured as the number of calls that party makes to the oracle. To justify this measure, we propose to use "peer-to-peer" protocols that presumably require the expenditure of one call to an effort oracle per successful execution. One well known example for such a protocol is a CAPTCHA, automatically generated challenges that should be solvable only if given a call to an effort oracle (and moreover accommodate automatic verification of the solution). Other options, (some of which may be more practical) are described in section A.

Before delving into the details, we note that for the convenience of the reader, the ePrint version of this paper contains a table of the parameters and notation used in the paper [1].

2.1   Verifiable Effort-Based Polling

An m-responder polling scheme is a multi-party protocol between a pollster, denoted P and m responders, denoted $R_1, \ldots, R_m$. The ith responder holds an input $x_i \in D \cup \{\bot\}$, where D is the domain from which the responses are taken and $\bot$ denotes lack of participation in the poll. In practice m will be an upper bound on the number of responders; We denote by $n < m$ the actual number of (honest) participants. The number of honest responders is known only to the adversary. Thus, the adversary can create "fake" responders by replacing some of the $\bot$ inputs with adversarially chosen values. As the adversary knows all the inputs and controls all the outputs in our protocols, we do not need to consider corrupted responders—the adversary can just replace an honest responder's input with a different one to simulate a corrupted responder.

We give parties access to an oracle denoted E, and let R_i^E (resp. P^E) denote the execution of R_i (resp. P) with access to the oracle E. Let e_i denote the total number of oracle calls made by R_i to E. Let $\langle P^E, R_1^E(x_1), \ldots, R_m^E(x_m)\rangle$ be a random variable describing the output of a protocol execution, where the probabilities are taken over the parties' coin tosses. The output of the protocol takes the form $(\overline{Y}, \overline{z})$, where $\overline{Y} = (\overline{y}, \overline{w})$ denotes the output of the pollster ($\overline{y} = (y_1, \ldots, y_m)$ indicates the outputs of the responders as announced by the pollster, and $\overline{w}$ contains a proof of correctness of the result) and $\overline{z} = (z_1, \ldots, z_m)$ denotes the local outputs of the responders, where z_i corresponds to the local output of R_i following the protocol execution. The role of the local outputs z_i is to enable local verification by the parties.

To make the polling scheme publicly-verifiable we will additionally require the existence of a verifier V that takes $\overline{Y}$ and $\overline{z}$ as inputs (the verification procedure can use the output of the local verification; e.g., global verification could fail if too many responders complain).

Definition 2 (Verifiable Effort-Based Polling). *Let $\kappa, m, a \in \mathbb{N}$ and let $\alpha, \theta \in [0, 1]$ and $B : \mathbb{N} \times \mathbb{N} \mapsto \mathbb{N}$. An m-responder effort-based polling scheme is said to be (α, B)-sound and θ-robust if there exists a probabilistic polynomial-time algorithm V such that for any $x_1, \ldots, x_m \in D \cup \{\perp\}$ with $n = \#\{i \in [m]|x_i \neq \perp\}$, the following properties are satisfied:*

Soundness: *For every PPT (Probabilistic Polynomial Time) P^*, if $n \geq \alpha m$ and $\Delta(\overline{x}, \overline{y}) \geq B(a, m)$ then*

$$\Pr\left[V(\overline{Y}, \overline{z}) = \texttt{accept}\right] < 2^{-\kappa},$$

where the probability is taken over $(\overline{Y}, \overline{z}) \leftarrow (P^{*E}, R_1^E(x_1), \ldots, R_m^E(x_m))$, a is the total number of oracle calls made by P^* to E, and $\Delta(\overline{x}, \overline{y})$ is the minimum Hamming distance between $\overline{x}$ and some permutation of $\overline{y}$ (i.e., this corresponds to the number of responses changed/added by the adversary).

Completeness: *For every subset $\{i_1, \ldots, i_t\} \subseteq [m]$ of responders (correspond-ing to malicious responders), if $e_{i_1} + \ldots + e_{i_t} < \theta m$ then*

$$\Pr\left[V(\overline{Y}, \overline{z}) = \texttt{accept}\right] > 1 - 2^{-\kappa},$$

where the probability is taken over $(\overline{Y}, \overline{z}) \leftarrow (P^E, R_1^E(x_1), \ldots, R_m^E(x_m))$.

Informally, we can interpret (α, B)-soundness as a guarantee that if at least an α-fraction of responders are honest, then the adversary cannot change too many responses without getting caught. The influence of the adversary is captured by the function B. Generally, we would expect $B(a, m)$ to be proportional to the number of responses an honest user could add using a calls to the effort oracle. Thus, an intuitive measure of the protocol's soundness is a bound on the multiplicative advantage of the adversary:

$$C(a) = B(a, m)\frac{d}{a}$$

If the multiplicative advantage is bounded by C, then any adversary who can change $C \cdot \ell$ responses using an optimal cheating strategy could have altered ℓ

responses (in expectation) by honestly following the protocol and expending the same amount of effort.

The θ-robustness of the protocol guarantees that if the total effort available to malicious responders is less than θm, then they cannot cause the verification procedure to fail except with negligible probability.

2.2   Formally Defining Proofs of Effort

In the "effort-oracle" model we can fully formalize definition 1. Note that while we define PPE to be a two-party protocol, we require soundness to hold even in a concurrent setting, in which a malicious party A^* participates concurrently in multiple executions of the protocol with other parties. To achieve this, we assume each protocol execution has a unique identifier id (e.g., in practice this could be a concatenation of the identities of the participating parties and the current time).

Definition 3 (One-Sided PPE). *A protocol $\Pi^E(P, V)$ between a prover P and a verifier V is a one-sided PPE if it satisfies the following properties:*

1. **Efficiency** *An honest execution of $\Pi^E(P, V)$ requires P and V to make at most one oracle call to E (each).*
2. **σ-Completeness** *If P and V execute an instance of $\Pi^E(P, V)$ and both honestly follow the protocol, then with probability at least $1 - \sigma$ V will output "true" at the end of the protocol.*
3. **ε-Soundness** *For every PPT (Probabilistic Polynomial time) P^* that executes an instance of $\Pi^E(P^*, V)$ using identifier id, if V honestly follows the protocol but P^* does not make at least one oracle call to E with input id, then the probability that V outputs "true" is at most ε.*

Definition 4 (Two-Sided PPE). *A protocol $\Pi^E(A, B)$ between two parties A and B is a two-sided (symmetric) PPE if it simultaneously satisfies the one-sided PPE definition for A as a prover and B as verifier, and vice versa.*

3   The Protocol

The main technical innovation in this paper is the construction of the Pollster's proof for the correctness of the published results. To do this, we borrow ideas from the literature on defense against Sybil attacks using pre-existing trust relations.

To account for the possibility that an honest responder can fail a PPE execution independently of his honesty, we denote by η_E the fraction of failing PPEs that the protocol tolerates before discarding someone's vote. On the other hand, we indicate by η_V the upper bound on the fraction of responders whose vote can be discarded by the pollster (if the number of discarded votes is greater than η_V, the overall verification will fail). Moreover, in order to avoid denial-of-service attacks caused by malicious responders that intentionally fail all their PPEs, our protocol will require to register for the poll by solving a single-sided PPE (i.e., a PPE that requires effort only from the voters side in order to be successful). With high probability this kind of attack will then be unsuccessful whenever the cheating responders are limited in the amount of effort they can expend. Following, is a high-level description of our protocol. The full formal protocol description can be found in the ePrint version of the paper [1].

1. **Parameter Announcement.** This phase consists of a single broadcast by the pollster, consisting of the public parameters for the poll. The pollster generates a unique, random identifier id for the poll and public key parameters for a digital signature scheme. We denote by SK, VK the secret and public key respectively (note that these are required only for completeness— responders will have their own signature and verification keys).
   The public parameters are the tuple $(id, questions, p, VK)$, where $questions$ is the set of poll questions. p is a probability that determines the expected degree of the certification graph.

2. **Registration.** Each responder R_i samples a private key SK_i and a public key VK_i for their signature scheme, and sends $(addr_i, VK_i)$ to the pollster (where $addr_i$ is the responder's network address). Each responder then solves a single-sided PPE (verified by the pollster). If verification was successful, the pollster adds $(addr_i, VK_i)$ to its list of registered responders.
   When the registration phase is over, the pollster broadcasts the list of registered public keys. Note that the network addresses are not required to appear in the broadcast list. The order of public keys in the list maps each registered responder to a unique index (i.e., the ith key in the list is mapped to index i).
   For each index i, we define N_i to be "the neighborhood of i" in the certification graph. N_i is computed from i and m (the total number of parties) using a cryptographic hash H: $j \in N_i$ iff $H(i, j) \le p$, where the output of H is treated as a binary fraction in $[0, 1]$ (e.g., H could be SHA-1). Since all of the parameters are public, every party can compute the list of its neighbors in the graph.
   However, while R_i may know the verification key of every neighbor, it does not necessarily know their network addresses. The parties can communicate via the pollster, or alternatively, the pollster can send each party i the network addresses of all its neighbors in the graph.

3. **Responder Certification (PPE execution).** As just described, every pair of responders is paired in a PPE instance with probability p. Now, for each $R_j \in N_i$, responder R_i engages in a PPE with R_j. The actual execution is peer-to-peer, however the communication may be facilitated by the pollster (e.g., the pollster's website can be used as a conduit for a VOIP chat). If the PPE execution succeeded (both parties received "true"), the parties sign each other's public keys (concatenated with a unique "poll identifier", to prevent the signatures being reused in other polls) and send the signed values to each other.

4. **Poll Response.** Every responder R_i sends to the pollster the results of the certification phase (a signature on VK_i from each neighbor with which it successfully completed a PPE) and x_i, the actual response to the poll questions.

5. **Results and Proof.** We can think of the responders as nodes of a graph G_c in which they are connected by edges if and only if they were supposed to interact through a PPE. Let $V = \{1, \ldots, m\}$ denote the set of responders and $E := \{(i,j)|i,j \in V, H(i,j) < p\}$ the set of edges. We call $G_c = (V, E)$ the "certification graph". Note that anyone can compute G_c given the serial numbers associated to the responders and p. Then, as a "proof of correctness" the pollster publishes the graph consisting of the following[2]:

   Node labels: For each responder R_i the pollster publishes
   $$(x_i, sig_{SK_i}(x_i), VK_i, id_i).$$

   Edge signature: For each successful PPE the pollster publishes
   $(sig_j(VK_i), sig_i(VK_j))$, where VK_i, VK_j are the public keys of the responders involved in it.

   List of deleted nodes: The list of all nodes whose response will not count in the result because they failed more than a η_E fraction of the PPEs.

   The empirical distribution of the responses can be computed by counting the votes associated to the non-deleted nodes. Note that the graph published by the pollster, call it G_p, is composed of the same nodes as G_c, but it's missing all the edges associated to unsuccessful PPEs. So, $G_p = (V, E')$ is a subgraph of $G_c = (V, E)$ where (i,j) is in E' if and only if R_i and R_j *successfully* interacted through a PPE.

6. **Verification.** The procedure is divided in two steps:

   Local verification (performed by each responder) consists of verifying that the corresponding node was published correctly, as were the edge signatures in which he was involved (no adjacent edge is missing, and all the adjacent edges in the graph were verified with a successful PPE). If any of these verifications fail, the responders sends a "complaint".

   Global verification (can be performed by anyone) consists of checking that all the nodes, and no others, that failed more then $\eta_E d$ edges are indeed marked as deleted. To verify if a node i is marked correctly, the verifier needs to find its neighbors in the graph (by computing the hash function $H(i,j)$ for every $j \neq i$) and checking how many of the signed edges appear in the published graph. Then, the verifiers need to check that no more than a η_V fraction of the nodes were deleted and that not "too many" valid complaints were sent.

An adversarial pollster can attempt to manipulate results either by changing the responses associated with honest nodes or by "controlling" many nodes (they will be nodes that do not correspond to any honest participant, but appear in G_p and their "behaviour" is dictated by the pollster), such that the overall

---

[2] the information as described is redundant (e.g., the list of deleted nodes can be computed from the list of edge signatures and node labels), but we describe it in this way to make the description of the verification process simpler.

empirical distribution differs from the empirical distribution over the honest nodes. In the former case, the local verification will detect the adversary and many valid complaints will be sent. In the latter case, we use an expansion property of the graph to prove that any large enough set of "bad" nodes (nodes that are controlled by the adversary) must have many edges to its complement in the graph. Thus, an adversary who wants to control a big enough set of nodes must succeed in many PPE executions with honest nodes; since the adversary is bounded in the number of successful PPE executions, it will be caught with high probability.

The protocol also provides a measure of robustness against malicious responders. Cheating responders cannot undetectably *modify* the results for the same reason that a cheating pollster cannot. However, they can attempt to launch a denial-of-service attack by causing verification to fail. As explained above, the single-sided PPE in step 4 will prevent this form of attack, as long as the cheating responders are limited enough in the amount of effort they can expend.

4   Soundness

To prove the soundness of our protocol we need to show that the number of votes that the adversary can control is at most proportional to the amount of effort that he is willing to invest. That is, whenever the adversary is able to control a "meaningful" amount of votes that is significantly greater that the number of votes that she could have controlled by honestly following the protocol (with the same effort investment), our verification procedure will fail with overwhelming probability. The proof of such a result will rely both on the security of the signature schemes and on an expansion property of the graph G_p published by the pollster as proof of correctness.

The use of the signatures is entirely straightforward: they prevent the adversary from changing honest users' votes and from claiming a failed PPE with an honest user was successful (to do this, the adversary would have to forge the honest node's signature). Similarly, the pollster's signature on the honest user's registration information and the signatures of its neighboring nodes allows the honest user to verifiably complain about being omitted from the count despite successfully completing the requisite number of PPEs. The "meat" of the security proof is in the analysis of the certification graph, and that will be the focus of this section.

As described in the previous sections, in a m-responders polling scheme, each responder holds an input $x_i \in D \cup \{\bot\}$, where D is the domain from which the responses are taken and $\bot$ denotes lack of participation in the poll. We call a node *honest* if its corresponding party participated in the poll (its input was not $\bot$). We call a node *bad* if it is not honest but its response in the output is not $\bot$. Finally, we say a node is *deleted* if it failed more than $\eta_E d$ of the PPEs it was assigned (note that both honest and bad nodes may be deleted), where d is the number of PPE executions each responder is expected to participate in. Note that for soundness to hold, we need that at least a certain portion, say α, of the responders are actually honest. That is, we need at least αm responders to participate to the poll by sending an input. The adversary could in theory be the one controlling the remaining $(1 - \alpha)m$ votes by replacing $\perp$ as an actual vote in the output and by spending the effort he has available. We prove that if the number of controlled nodes is significantly greater than the number of votes he would have controlled by acting honestly, then he will be detected with high probability.

In order to prove soundness, we bound separately the number of bad nodes (corresponding to "fake" parties generated by the adversary) and the number of changes the adversary can make to the input of honest nodes (that is, responder R_i voted x_i and the pollster output $y_i \in D \setminus \{x_i\}$ or $y_i = \perp$ instead). To prove the first bound, we rely on an expansion property of the graph G_p output by the pollster. In the following subsection we give a general definition of such a property and we prove some lemmas that will be useful for our proof.

4.1   Large-Set Expanding Property

The LSE property is similar to the "jumbled" graphs of Thomason, but is weaker since we don't care if small sets do not expand. This lets us get better LSE parameters for random graphs than are possible for the standard jumbled graphs (formally, we use the $G(n, p)$ model for random graphs; a graph is distributed according to $G(n, p)$ if it has n vertices and for each pair of vertices the corresponding edge exists with probability p).

Definition 5 (Large-Set Expanding (LSE)). A graph $G = (V, E)$, with $m = |V|$, is said to be (K, ρ, q)-LSE if for every pair of disjoint sets $A, B \subset V$ such that $K \leq |A| \leq m/2$, $|B| \geq m - |A| - \rho$ it holds that the set of edges between A and B, denoted by $e(A, B)$, has cardinality greater than $|A||B|q$.

In our analysis K will denote a bound on both the maximum number of bad nodes that we will allow and on the minimum number of good nodes that we require, ρ will be the maximum number of deleted nodes and q a function of the probability that two voters have to run a PPE.

Lemma 1. Let $G(m, p) = (V, E)$ be a random graph with $p = d/m$, For every $\rho \geq 1$, $\rho \in \mathbb{N}$ and every $b > 1$, if

$$d > \frac{4b^2 m}{m - 2\rho}(\ln m + 1)$$

then G is $(K, \rho, \frac{b-1}{b}p)$-LSE with probability at least $1 - 2^{-\kappa}$ for $K = \kappa + (\rho + 2)\ln m + \rho$ (where the probability is over the choice of graph).

Proof. Consider an arbitrary pair of sets $A, B \subset V$ such that $K \leq |A| \leq m/2$, $|B| = m - |A| - r$ with $1 \leq r \leq \rho$. Define the random variable $X_{i,j}$ to be the indicator variable for the event $(i, j) \in E$.

Since G is a random (m, p)-graph, the $X_{i,j}$'s are independent and $Pr[X_{i,j} = 1] = p$. Then

$$|e(A, B)| = \sum_{i \in A} \sum_{j \in B} X_{i,j}$$

$$E[|e(A, B)|] = |A||B|p = \mu$$

For $A, B \subset V$ such that $K \le |A| \le \frac{m}{2}$ and $m - |A| - \rho \le |B| \le m - |A|$, let $Bad(A, B)$ be the event that

$$|e(A, B)| < \frac{b-1}{b} \mu$$

(For A, B not satisfying the size restrictions, we define $Bad(A, B)$ to be the null event.)

To prove the lemma, we must bound the probability that $Pr\,[\exists A, B \subset V : Bad(A, B)]$. First, since the $X_{i,j}$'s are independent, by the Chernoff bound we have for any disjoint sets A and B:

$$Pr[|e(A, B)| < \frac{b-1}{b}\mu]$$

$$\le \exp\{-\frac{\mu}{2b^2}\} = \exp\{-\frac{|A||B|p}{2b^2}\}$$

$$= \exp\{-\frac{|A|(m - |A| - r)p}{2b^2}\} \le \exp\{-\frac{|A|(m/2 - r)p}{2b^2}\}$$

Next, we bound the probability that there exist two sets A and B of fixed sizes $|A| = x$, $|B| = m - x - r$ such that $Bad(A, B)$ occurs. Denote

$$\varepsilon = Pr\left[\bigcup_{\substack{A, B \subset V, A \cap B = \emptyset \\ |A| = x, |B| = m - x - r}} Bad(A, B)\right]$$

By the union bound, this probability is bounded by

$$\varepsilon \le \sum_{\substack{A, B \subset V, A \cap B = \emptyset \\ |A| = x, |B| = m - x - r}} Pr\left[|e(A, B)| < \frac{b-1}{b}\mu\right]$$

$$\le \sum_{\substack{A, B \subset V, A \cap B = \emptyset \\ |A| = x, |B| = m - x - r}} \exp\{-\frac{|A||B|p}{2b^2}\}$$

$$= \sum_{\substack{A, B \subset V, A \cap B = \emptyset \\ |A| = x, |B| = m - x - r}} \exp\{-\frac{x(m - x - r)p}{2b^2}\}$$

$$= \binom{m}{x}\binom{m-x}{r}\exp\{-\frac{x(m-x-r)p}{2b^2}\}$$

$$\leq \binom{m}{x}\binom{m}{r}\exp\{-\frac{x(m-x-r)p}{2b^2}\}$$

$$\leq \left(\frac{me}{x}\right)^x\left(\frac{me}{r}\right)^r\exp\{-\frac{x(m-x-r)p}{2b^2}\}$$

Since $|A| = x \leq \frac{m}{2}$,

$$\exp\{-\frac{x(m-x-r)p}{2b^2}\} \leq \exp\{-\frac{x(m/2-r)p}{2b^2}\}$$

Hence

$$\varepsilon \leq \exp\left\{x(\ln m + 1 - \ln x) + r(\ln m + 1 - \ln r) - x\left(\frac{m}{2} - r\right)\frac{p}{2b^2}\right\}$$

$$\leq \exp\left\{-x\left(\frac{d}{4b^2} - \frac{dr}{2b^2 m} - \ln m - 1 + \ln x\right) + r(\ln m + 1 - \ln r)\right\}$$

$$\leq \exp\left\{-x\left(\frac{d}{4b^2} - \frac{dr}{2b^2 m} - \ln m\right) + r(\ln m + 1)\right\}$$

$$\leq \exp\left\{-x + r(\ln m + 1)\right\}$$

Where the last two inequalities hold as long as $\ln x > 1$ (which is always true assuming $K > 3$), $\ln r \geq 0$ (which is always true for $r \geq 1$) and $d > \frac{4b^2 m}{m-2r}(\ln m + 1)$.

Applying the union bound again, we get

$$\Pr\left[\exists A, B \subset V : Bad(A, B)\right]$$

$$\leq \sum_{x=K}^{m/2}\sum_{r=1}^{\rho}\Pr\left[\bigcup_{\substack{A\subset V\\|A|=x}}\bigcup_{\substack{B\subset V\\|B|=m-x-r}}\{|e(A,B)| < \frac{b-1}{b}\mu\}\right]$$

$$\leq \frac{m}{2}\rho e^{-K+\rho(\ln m+1)}$$

$$\leq 2^{-\kappa}$$

since $K > \kappa\ln 2 + (\rho + 1)\ln m + \ln\rho + \rho$.

In our analysis, we will use this lemma to prove that the certification graph G_c is indeed expanding with specific parameters K, ρ, and q. We will then need to use the following lemma, in order to prove that our protocol is sound:

Lemma 2. *Consider a graph $G = (V, E)$ with $m = |V|$ nodes. Let $G' = (V, E')$ be the graph obtained from G by deleting at most s edges per node. If G is (K, ρ, q)-LSE, then G' is $(K, \rho, q - \frac{2s}{m-2\rho})$-LSE.*

Proof. For simplicity let $q' = q - \frac{2s}{m-2\rho}$. Consider $A, B \subset V$ such that $K \leq |A| \leq m/2$ and $m - |A| - \rho \leq |B| \leq m - |A|$. We want to prove that $|e_{G'}(A, B)| > |A||B|q'$, where $e_G(\cdot, \cdot)$ indicates the set of edges between A and B in the graph G.

First, by assumption the maximum number of edges that can be missing in G' from v are exactly s. Therefore, the maximum number of edges that can be missing in G' from the set of all edges with at least one node in A is $|A|s$. In the worst case, for us, all the missing edges were part of $e(A, B)$ in G. Thus,

$$|e_{G'}(A, B)| \geq |e_G(A, B)| - |A|s$$

Now we can use the fact that G is (K, ρ, q)-LSE to obtain the following:

$$|e_{G'}(A, B)| \geq |e_G(A, B)| - s|A| > |A||B|q - s|A| .$$

It remains to show that $|A||B|q' \leq |A||B|q - s|A|$. From $q' = q - \frac{2s}{m-2\rho}$ and $|A| \leq m/2$ we get

$$|A||B|q' = |A||B| \left(q - \frac{2s}{m - 2\rho} \right)$$
$$\leq |A||B| \left(q - \frac{s}{m - |A| - \rho} \right)$$
$$= |A||B|q - |A|s \left(\frac{|B|}{m - |A| - \rho} \right)$$
$$\leq |A||B|q - |A|s ,$$

from which we can conclude $|e_{G'}(A, B)| > |A||B|q'$ as required.

4.2   Main Theorem and Proofs

We can now apply the results obtained in the previous subsection specifically to our protocol. Let a denote the maximum number of effort oracle calls that the adversary is willing to make and let $K = \kappa + (\eta_V m + 2) \ln m + \eta_V m$.[3] Formally, we prove

Theorem 3 (Soundness). *Let*

$$b = \sqrt{\frac{d(\frac{1}{2} - \eta_V)}{2(\ln m - 1)}} .$$

If $b > (\frac{1}{2} - \eta_V)/(\frac{1}{2} - \eta_V - \eta_E) > 1$, then the protocol of section 3 is an (α, B)-sound verifiable polling protocol for

$$\alpha = K/m + \eta_V$$

---

[3] Recall that η_V is a parameter denoting the max fraction of nodes that can be deleted before verification fails.

and

$$B(a,m) = \max\left\{ K, \left(\frac{b}{(b-1)(\frac{1}{2} - \eta_V) - b\eta_E} \right) \frac{a}{d} \right\} + \theta m \ .$$

When a is sufficiently large (so we can ignore the K "free" responses), this implies the multiplicative advantage of the adversary is bounded by

$$C(a) = \left(\frac{b}{(b-1)(\frac{1}{2} - \eta_V) - b\eta_E} \right) + \frac{\theta m d}{a} \ .$$

One way to interpret this is that an adversary gets resources equivalent to θm honest users "for free", but any more powerful adversary has multiplicative advantage bounded by

$$C^* = \left(\frac{b}{(b-1)(\frac{1}{2} - \eta_V) - b\eta_E} \right) + 1$$

(recall that an honest user must solve, in expectation, d PPEs during the protocol execution, so an adversary more powerful than that must have $a > \theta m d$).

Proof. As we discussed at the beginning of the section, there are two ways for the pollster to affect the vote count:

1. By possibly controlling some of the nodes.
2. By replacing or deleting the votes of honest participants.

For the latter, the bound relies on the security of the signature scheme and on the local verification of honest parties. In fact, the signature scheme ensures that the adversary cannot modify responses (with a $y_i \neq \perp$) (since that would require forging a signature compatible with the node's verification key). Thus, the local verification of honest nodes will catch the adversary deleting or completely replacing nodes; Global verification fails whenever more then θm nodes complain—thus, the number of deleted/replaced nodes in a successful protocol execution can be at most θm.

It is left to show that if the number of controlled nodes is higher than B, then global verification will fail. The proof proceeds as follow:

– Using Lemma 1 and Lemma 2 we prove that G_p is LSE with high probability.
– We will then have a lower bound on the number crossing edges between a possible set of bad nodes and the set of honest nodes.
– We conclude by noticing that the pollster, in order to control a set of nodes larger than B, would have had to succeeded in more than a PPEs involving honest participants.

Let F denote the nodes in G_p corresponding to voters that have failed more than $\eta_E d$ PPEs. It must be that $|F| \leq \eta_V m$, otherwise the verification procedure would fail. Let B and H denote the set of bad and honest nodes, respectively, that have not been labeled as "deleted". Thus B, H and F are disjoint sets whose union is V. That is, since we have a total of m nodes, if $|B| = x$ then $|H| = m - x - |F|$. Recall that a successful PPE corresponds to an edge in G_p. Thus, a lower bound on the number of edges in G_p between the sets B and H translates to a lower bound on the number of PPEs in which the adversary must have succeeded, and hence on the number of oracle calls made by the adversary.

Note that from Lemma 1, we know that G_c is $(K, \eta_V m, \frac{b-1}{b} p)$-LSE with probability at least $1 - 2^\kappa$. Thus, from Lemma 2, we can conclude that, with probability at least $1 - 2^\kappa$, G_p is $(K, \eta_V m, \frac{b-1}{b} p - \frac{2\eta_E d}{m - 2\eta_V m})$-LSE. Wlog assume

$$|B| < m/2 \quad \text{and} \quad |B| \geq \max\left\{ K, \left(\frac{b}{(b-1)(\frac{1}{2} - \eta_V) - b\eta_E} \right) \frac{a}{d} \right\}$$

(the case $|H| < m/2$ is analogous, using $|H| \geq (\alpha - \eta_V)m \geq K$). Then, we get:

$$|e(B, H)| > |B||H| \left(\frac{b-1}{b} p - \frac{2\eta_E d}{m - 2\eta_V m} \right)$$

$$\geq |B| (m - |B| - \eta_V m) \left(\frac{(b-1)d(1 - 2\eta_V) - 2bd\eta_E}{mb(1 - 2\eta_V)} \right)$$

$$\geq \left(\frac{2b}{(b-1)(1 - 2\eta_V) - 2b\eta_E} \right) \frac{a}{d} \left(\frac{m}{2} - \eta_V m \right) \left(\frac{(b-1)d(1 - 2\eta_V) - 2bd\eta_E}{mb(1 - 2\eta_V)} \right)$$

$$= \left(\frac{2b}{(b-1)(1 - 2\eta_V) - 2b\eta_E} \right) \frac{a}{d} \left(\frac{m(1 - 2\eta_V)}{2} \right) \left(\frac{d[(b-1)(1 - 2\eta_V) - 2b\eta_E]}{mb(1 - 2\eta_V)} \right)$$

$$= a$$

Thus, with probability at least $1 - 2^\kappa$, $|e(B, H)| > a$ which contradicts the assumption of the adversary being limited to a successful PPEs.

Therefore, the number of votes controlled by a pollster that invests a effort oracle calls must be lower than $\max\{K, \left(\frac{b}{(b-1)(\frac{1}{2} - \eta_V) - b\eta_E} \right) \frac{a}{d}\} + \theta m$, as wanted.

5   Completeness

It is now left to show that in the case of an honest pollster, the verification procedure will succeed with overwhelming probability. Even when dealing with an honest pollster, we still need to take into account the possibility that malicious voters might try to force the verification to fail. This can be done by registering for the poll but aborting in all the PPE executions. Such a strategy will force the verification procedure to label the node as *deleted* and all its edges as *failing*. It will thus increase the number of *deleted* nodes which, for the verification to output accept, needs to be smaller than $\eta_V m$.

To make sure that such an attack would require the adversary to expend actual effort, we require each responder to solve a single-sided PPE (where the effort is required only from the responders) in order to be allowed to participate to the poll. We think of the number of maliciously controlled nodes as bounded by θm, where θ will depend on the "effort" invested by the malicious voters. Theorem

4 gives a bound on η_V as a function of θ and κ that will enable the verification procedure, in case of an honest pollster, to output accept with probability at least $(1 - 2^\kappa)$.

To prove the main theorem of this section, we will require the following lemma (whose proof is below):

Lemma 3. *Assuming static corruption, the probability that malicious responders with a θm bound on effort (in total) can control $\max\{3\theta md, 3\kappa\}$ edges in the certification graph is bounded by $e^{-\kappa}$.*

Theorem 4 (Completeness). *Let θm denote the maximum number of effort oracle calls that can be made by malicious responders and*

$$\eta_V^{min} = \theta + \frac{3 \cdot \max\left\{\frac{\kappa}{md}, \theta\right\}}{\eta_E} + \frac{2\sigma}{\eta_E}\left(1 + \max\left\{2, \frac{2\kappa}{md\sigma}\right\}\right)$$

If the pollster follows the protocol honestly, $\eta_E > 0$ and $\eta_V \geq \eta_V^{min}$ then the probability that the verification procedure outputs accept is at least $1 - 2^{-\kappa}$.

We note that for non-trivial soundness, the values of η_E and η_V are further constrained. See the ePrint version of this paper for a discussion on choosing the parameters [1].

Proof. Let $G(m, p) = (V, E)$, with $p = d/m$, be the random graph generated by the pollster. Recall that an edge (i, j) is labeled as *failing* whenever the PPE between i and j fails. We denote by σ the probability that such an event occurs between honest voters. Moreover, η_E is the highest fraction of PPEs that can fail before a node/voter gets labeled as *deleted*, and η_V is the maximal fraction of deleted nodes accepted by the verification procedure.

Let $X_{i,j}$ denote the indicator random variable for the event "$(i, j) \in E$ is a failing edge". Note that, if i and j are both honest, the $X_{i,j}$'s are independent and $Pr[X_{i,j} = 1] = \sigma p$. Let $X = \sum_{i\in V}\sum_{j\in V} X_{i,j}$. Then, $E[X] = m^2\sigma p = md\sigma$ and X denotes the number of failing edges in the graph. Since each edge affect 2 nodes, $2X$ is actually the cardinality of the set containing (with repetitions) all the nodes affected by failing X edges. Note that for a node to be labeled as deleted, such a node needs to be connected to at least $d\eta_E$ failing edges, which means that such a node has been counted at least $d\eta_E$ times in $2X$. Thus, the expected number of deleted nodes in case of honest responders is bounded by $2X/d\eta_E$. Now, in our analysis, we need to take into account that, in the worst case scenario, there will be θm nodes maliciously controlled who will intentionally fail all their PPE's. Therefore, we will have to account for the following:

1. The malicious nodes (which are θm) will be failing nodes;
2. Enough bad edges will cause an honest node to be marked deleted. However, by Lemma 3, with high probability the malicious responders cannot affect more than $3 \cdot \max\{\kappa, \theta md\}$ honest edges. Which means that at most another $3 \cdot \max\{\kappa/d, \theta m\}/\eta_E$ nodes can be "forced" to be labeled as deleted.

To conclude, we want to prove that the probability that $\frac{2X}{d\eta_E} + \theta m + \frac{3\theta m}{\eta_E}$ is greater than $\eta_V m$ is negligible. Let $\eta_V = \theta + \frac{3\theta}{\eta_E} + \frac{2\sigma}{\eta_E}(1 + \delta)$. (we will set δ below.) Then,

$$\Pr\left[\frac{2X}{d\eta_E} + \theta m + \frac{3\theta m}{\eta_E} > \eta_V m\right] = \Pr\left[\frac{2X}{d\eta_E} > \frac{2\sigma}{\eta_E}(1 + \delta)m\right]$$
$$= \Pr\left[X > (1 + \delta)md\sigma\right]$$

By the Chernoff Bound,

$$\Pr\left[X > (1 + \delta)md\sigma\right] \leq \exp\left\{-\frac{\delta^2}{2 + \delta}md\sigma\right\}$$

Setting $\delta = \max\left\{2, \frac{2\kappa}{md\sigma}\right\}$, we ensure that $\Pr\left[X > (1 + \delta)md\sigma\right] \leq e^{-\kappa}$.

We conclude by proving Lemma 3, as a corollary of the following claim:

Claim. Let $S \subset V$ be an arbitrary set of vertices and denote $\delta = \max\{2, 2\kappa/(|S|d)\}$. Then

$$\Pr\left[|\{(i, j) \in E | i \in S\}| > (1 + \delta)d|S|\right] < e^{-\kappa}$$

(i.e., the probability S has more than $(1 + \delta)d|S|$ edges is bounded by $e^{-\kappa}$).

Proof. For every pair of vertices $i, j \in V$, let $X_{i,j}$ be the indicator variable for the event $(i, j) \in E$. By definition, $E[X_{i,j}] = p$. Denote $X = \sum_{\substack{i \in S \\ j \in V}} X_{i,j}$ the number of edges adjacent to S. Then $E[X] = mp|S| = d|S|$. By Chernoff,

$$\Pr\left[X_i > (1 + \delta)d|S|\right] \leq \exp\left\{-\frac{\delta}{2/\delta + 1}d|S|\right\} \leq \exp\left\{-\frac{\delta}{2}d|S|\right\} = \exp\{-\kappa\}$$

Proof (Proof of Lemma 3). Since the pollster randomly shuffles the nodes in the certification path during the registration phase, any set of responders is assigned a random set of nodes in the certification graph. By symmetry, we can consider the probability for any specific set of size θm. The result follows by setting $|S| = \theta m$ in Claim 5.

6 Discussion and Open Questions

General Verifiable Computation Among Anonymous Participants. While we state our main results in terms of polling, the security guarantee we give is that the final published graph does not contain too many "bad" nodes. It may be possible to leverage this technique for doing more general computations, where the edges in the graph correspond to a private computation between two parties, and the final goal is a joint, publicly-verifiable computation (in this case, the "responses" might be some intermediate public values of the computation).

Parallel and Distributed Verification. The verification procedure in our protocol is highly parallelizable: each responder must verify three properties, each of which can be done by reading only a small part of the graph:

- that her own node was correctly published on the bulletin-board (requires $O(m)$ evaluations of the hash function, but only $O(d)$ communication),
- that the total number of deleted nodes was small (requires reading a small list of nodes),
- and that no edges were missed (this is a local property of each potential edge that can be computed from the node labels and the size of the graph).

The only non-local part in the verification is the aggregation of the results from all the nodes. However, by publishing a small amount of additional information, this computation can be distributed as well. Given an aggregation tree, where each node aggregates the results from its children, a verifier can check a single local neighborhood and a path from that neighborhood to the root in the tree. Thus, if we can assume that enough honest responders will participate in verification, the total amount communication for each responder can be made logarithmic in the size of the graph.

Practicality of the Protocol. The parameters achieved by our protocol are not quite good enough to be practical for interaction-based PPEs (the degree of the graph would be about 180 for reasonable parameters). However, this may already be good enough for PPEs that can be automated (for example, the social-network based PPE). Moreover, we believe further research can significantly improve the efficiency.

Improving Efficiency by Using Hypergraphs. Our bound on the degree of the graph may be slightly high for some uses of the protocol. However, we can extend the PPE definition to a multi-party setting, in which several parties certify each other simultaneously (e.g., using a multi-person chat, such as "Google Hangout" or "Skype"). This has the potential of significantly lowering the degree. Extending our protocol in this way may be an interesting direction for future work.

Improving Efficiency by Using Explicit Graphs. Our bound on the degree of the graph is for a randomly chosen graph. In particular, our soundness analysis includes the event that the chosen graph is not a good expander as a failure mode. Thus, we require the properties to hold for random graphs *with overwhelming probability*. However, it is fairly easy to prove that graphs with better parameters (e.g., lower degree for the same expansion rate) *exist*: if we have an explicit representation of such a graph, soundness will hold unconditionally.

References

1. Alberini, G., Moran, T., Rosen, A.: Public verification of private effort. Cryptology ePrint Archive, Report 2014/983 (2014), http://eprint.iacr.org/2014/983
2. Ateniese, G., Kamara, S., Katz, J.: Proofs of storage from homomorphic identification protocols. In: Matsui, M. (ed.) ASIACRYPT 2009. LNCS, vol. 5912, pp. 319–333. Springer, Heidelberg (2009)
3. Awerbuch, B., Scheideler, C.: Group spreading: A protocol for provably secure distributed name service. In: Díaz, J., Karhumäki, J., Lepistö, A., Sannella, D. (eds.) ICALP 2004. LNCS, vol. 3142, pp. 183–195. Springer, Heidelberg (2004)
4. Canetti, R., Halevi, S., Steiner, M.: Mitigating dictionary attacks on password-protected local storage. In: Dwork, C. (ed.) CRYPTO 2006. LNCS, vol. 4117, pp. 160–179. Springer, Heidelberg (2006)
5. Chaum, D.: Untraceable electronic mail, return addresses, and digital pseudonyms. Communications of the ACM 24(2), 84–88 (1981)
6. Chaum, D.: E-voting: Secret-ballot receipts: True voter-verifiable elections. IEEE Security & Privacy 2(1), 38–47 (2004)
7. Cheng, A., Friedman, E.: Sybilproof reputation mechanisms. In: ACM SIGCOMM Workshop on Economics of Peer-to-Peer Systems, P2PECON 2005, pp. 128–132. ACM, New York (2005)
8. Danezis, G., Mittal, P.: Sybilinfer: Detecting sybil nodes using social networks. In: NDSS. The Internet Society (2009)
9. Douceur, J.R.: The sybil attack. In: Druschel, P., Kaashoek, M.F., Rowstron, A. (eds.) IPTPS 2002. LNCS, vol. 2429, pp. 251–260. Springer, Heidelberg (2002)
10. Ellison, C.M.: Establishing identity without certification authorities. In: USENIX SSYM 1996, p. 7. USENIX Association, Berkeley (1996)
11. Kamvar, S.D., Schlosser, M.T., Garcia-Molina, H.: The eigentrust algorithm for reputation management in p2p networks. In: WWW, pp. 640–651 (2003)
12. Krugman, P.: Bits and barbarism. New York Times (December 2013), http://www.nytimes.com/2013/12/23/opinion/krugman-bits-and-barbarism.html
13. Levien, R., Aiken, A.: Attack-resistant trust metrics for public key certification. In: USENIX SSYM 1998, p. 18. USENIX Association, Berkeley (1998)
14. Levine, B.N., Shields, C., Margolin, N.B.: A survey of solutions to the sybil attack. Tech. Report 2006-052, University of Massachusetts Amherst, Amherst, MA (October 2006)
15. Moran, T., Naor, M.: Polling with physical envelopes: A rigorous analysis of a human-centric protocol. In: Vaudenay, S. (ed.) EUROCRYPT 2006. LNCS, vol. 4004, pp. 88–108. Springer, Heidelberg (2006)
16. Nakamoto, S.: Bitcoin: A peer-to-peer electronic cash system (May 2009)
17. Neff, C.A.: Practical high certainty intent verification for encrypted votes (October 2004), http://www.votehere.net/vhti/documentation/vsv-2.0.3638.pdf
18. Tran, D.N., Min, B., Li, J., Subramanian, L.: Sybil-resilient online content voting. In: Rexford, J., Sirer, E.G. (eds.) NSDI, pp. 15–28. USENIX Association (2009)
19. Von Ahn, L., Blum, M., Hopper, N.J., Langford, J.: Captcha: Using hard ai problems for security. In: Biham, E. (ed.) EUROCRYPT 2003. LNCS, vol. 2656, pp. 294–311. Springer, Heidelberg (2003)
20. Warner, S.: Randomized response: a survey technique for eliminating evasive answer bias. Journal of the American Statistical Association, 63–69 (1965)
21. Yu, H., Gibbons, P.B., Kaminsky, M., Xiao, F.: Sybillimit: A near-optimal social network defense against sybil attacks. IEEE/ACM Trans. Netw. 18(3), 885–898 (2010)

22. Yu, H., Kaminsky, M., Gibbons, P.B., Flaxman, A.D.: Sybilguard: defending against sybil attacks via social networks. Expanded Technical Report IRP-TR-06-01, Intel Research, Pittsburgh, Pittsburgh, PA (June 2006), http://www.pittsburgh.intel-research.net/people/gibbons/papers/sybilguard-tr.pdf
23. Yu, H., Kaminsky, M., Gibbons, P.B., Flaxman, A.D.: Sybilguard: defending against sybil attacks via social networks. IEEE/ACM Trans. Netw. 16(3), 576–589 (2008)

A   Implementing PPEs, Extensions and Selective Polling

The peer-to-peer nature of PPEs seems to facilitate implementation with relatively simple mechanisms. Below we give several examples.

Bitcoin and Proofs of Storage. The original motivation for proofs of storage (PoS) is to allow clients to outsource data storage "to the cloud". In this setting a storage provider stores a large file on behalf of a client. Roughly, a PoS protocol allows the provider to prove to the client that it is still storing the file (can reconstruct the entire file), using a small amount of communication.

Since storage is a valuable resource, it is tempting to use proofs-of-storage as the "effort unit" in an effort-based polling scheme (e.g., one unit of effort could be storing 1GB of data for 1 day).

Moreover, publicly-verifiable proofs of storage have been constructed [2]—given a "public-key" generated for a specific file, anyone can verify that the PoS that the storage provider publishes for that file.

The problem here is that "backup" is a peer-to-peer concept. In particular, any solution must prevent malicious parties from sending each other "fake" data to store: e.g., they store a short seed instead of a large pseudorandom file generated by that seed.

However, the existing PoS protocols can be trivially used to construct a PPE: an honest user will send good (incompressible) files to its peers (e.g., by encrypting the file), and it can verify using the PoS that the files were stored as required. The soundness and completeness properties of this PPE are inherited directly from the PoS protocol, hence we can hope for almost perfect completeness and negligible soundness-error.

This implementation of PPEs may be most interesting in the context of Bitcoin. One of the strong arguments against the currency is the inherent waste of the Bitcoin protocol [12]; this is a direct consequence of using proof-of-work as the basis of its effort-based polling scheme. If we could replace proof-of-work with, for example, "proof-of-backup", instead of generating heat as a side-effect, the Bitcoin network would function as a distributed backup system in addition to a currency.

Human Interaction. The simplest type of PPE consists of human interaction: participants certify each other's effort by simply talking with each other (e.g., using VOIP, video, or even textual chat). This is at least as hard to pass than a "real" Turing test (which consists solely of textual interaction), so its soundness properties seem to be very robust.

To prevent a proxying attack (in which Eve convinces Alice that she has expended effort by relaying Alice's challenges to Bob and vice versa), the protocol can include Bob reading aloud his partner's identity. Thus, to act as a person-in-the-middle, Eve would have to translate Bob saying Eve's public key to Bob saying Alice's public key, which seems like it would require some actual effort.

Symmetric CAPTCHAs. In this version of the PPE, each party generates a "real" CAPTCHA to be solved by the other party while simultaneously solving the CAPTCHA she received.

To prevent a proxying attack, we bind the CAPTCHA to the parties' identities using a combination of cryptographic commitments and a Message Authentication Code (MAC).

Define the CAPTCHA as a problem-generator $G(r)$ that given a random input r generates a CAPTCHA C along with its solution V.

1. When Bob generates a CAPTCHA for Alice, he chooses a secret MAC key and sets as the random input to G the MAC of the pair of public keys (Alice,Bob). He sends $G(r)$ to Alice.
2. Alice solves C, and sends a commitment to her solution to Bob.
3. Bob then sends his secret MAC key to Alice.
4. Alice verifies that the challenge she received is correctly generated (i.e., bound to Alice and Bob's public keys). If not, she aborts
5. Otherwise, Alice opens her commitment
6. Bob verifies that Alice correctly solved the challenge.

This protocol ensures that Eve can't use Alice to solve Bob's challenge to her, since Alice would refuse to open her commitment if she sees the CAPTCHA wasn't meant for her. Note that in terms of the effort required, this is not harder than just solving a CAPTCHA—the rest of the protocol can be completely automated.

Leveraging Existing Social Networks. Instead of an online effort, a possible PPE implementation can use an existing social network (basing the "effort" on the assumption that becoming "well connected" in a social network is difficult). For example, two parties can verify that they have several short, vertex-disjoint paths between them in the social network (or use some other measure of distance for which the effort assumption seems reasonable).

In this version of the protocol, parties are not guaranteed anonymity (since they must reveal their identities in order to verify their distance in the social network), but the public transcript of the protocol does not reveal anything about their identities or their social-network neighborhood.

The main problem here is preventing an adversary from using the same social-network identity in multiple different PPE invocations. The fact that the PPE is a private-coin primitive makes this problem easy to solve, assuming the social network allows users to publish information linked to their real identity (e.g., a "home page"). Party i chooses a random nonce r_i and publish a commitment to r_i on their homepage. When executing the PPE with party j, i will publish r_i and privately open the commitment to r_i towards party j; Party j can verify by looking at i's homepage that the nonce is the correct one. Assuming the homepage provides a consistent view to all honest users, i cannot use a different nonce in different PPE invocations. However, the public transcript cannot be linked to i's social-network identity due to the hiding property of the commitment.

Other PPE Extensions. Our basic definition of PPE only guarantees that "effort" is expended by the parties. This can be easily extended to capture more complex conditions that are hard to verify publicly but may be easy to verify in a peer-to-peer manner. For example, limiting a poll to a small geographic area. While certifying location in a publicly-verifiable way is difficult, verifying that someone else is physically nearby can be much easier (e.g., using speed of response or shared environmental cues, such as noise or micro-local weather conditions). By extending the PPE to verify physical proximity, we can guarantee the vast majority of participants must be local (assuming a large enough fraction is).

Another example is polling groups whose membership is secret (e.g., a poll of the "Anonymous" organization). If members of the group can recognize each other (e.g., they have a "secret handshake"), we can use the same technique to guarantee that our poll is targeting the group.

Limiting a poll to specific communities in an existing social network can be done similarly. Thus we can conduct verifiable polls on a social-network graph while keeping the graph itself secret—this can be important, since the structure of the social network often reveals a large amount of information about the identity of its nodes.

B   Choosing Parameters

Below is a table containing a list of the most common parameters used throughout the paper. We partition the parameters into *fixed* parameters (in Table 1)—those that depend on assumptions about adversarial behavior and the effectiveness of the PPEs, and *tunable* parameters (in Table 2)—those that can be set by the poll designer (subject to certain constraints) and *computed* parameters (in Table 3)—these are functions of the previous parameters.

Table 1. Fixed Parameters

Symbol	Description
m	Total number of responders to the poll / Number of nodes in the graph.
n	Number of honest responders.
a	Upper bound on the number of oracle calls that the adversary can successfully perform / Upper bound on the number of attack edges.
θ	Upper bound on the fraction of malicious responders: the total number of oracle calls made by malicious responders is at most θm.
σ	Probability of a PPE failing when both parties honestly follow the protocol.
ϵ	Probability of a PPE succeeding when one party does not make at least one oracle call.

Table 2. Tunable Parameters

Symbol	Description
κ	Security parameter.
d	Expected degree of the graph (expected number of PPE executions per responder). This can be tuned by changing p ($p = d/m$).
α	Minimum fraction of honest responders required to guarantee soundness.

Table 3. Computed Parameters

Symbol	Description
p	Edge probability. Every pair of responders will be required to engage in a PPE with probability p.
η_E	Upper bound on the fraction of PPE's that a responder can fail without getting deleted.
η_V	Upper bound on the fraction of nodes that can be deleted without causing the verification procedure to fail.
K	Number of nodes in the graph that the adversary can control "for free".
C^*	Upper bound on the multiplicative advantage of the adversary (an adversary has no more influence than an honest user that can invest C^* times the effort).

B.1   Constraints on Parameters

First, from Theorem 3 we have:

$$\sqrt{\frac{d(\frac{1}{2} - \eta_V)}{2(\ln m - 1)}} > (\frac{1}{2} - \eta_V)/(\frac{1}{2} - \eta_V - \eta_E)$$

which implies that

$$d > \frac{\frac{1}{2} - \eta_V}{(\frac{1}{2} - \eta_V - \eta_E)^2}(2\ln m - 2) . \tag{1}$$

By the definitions of α and K in Theorem 3, we get

$$\alpha \geq K/m + \eta_V = \frac{\kappa + 2\ln m}{m} + \eta_V(lnm + 2) \geq \frac{\kappa + 2\ln m}{m} \tag{2}$$

Isolating η_V instead of α, we have:

$$\eta_V \leq \eta_V^{\max} = \frac{\alpha - \frac{\kappa + 2\ln m}{m}}{2 + \ln m} \tag{3}$$

Combining this with the bound on η_V from Theorem 4, we get

$$\eta_V^{\min} = \theta + \frac{3 \cdot \max\left\{\frac{\kappa}{md}, \theta\right\}}{\eta_E} + \frac{2\sigma}{\eta_E}\left(1 + \max\left\{2, \frac{2\kappa}{md\sigma}\right\}\right) \leq \eta_V \leq \eta_V^{\max}$$

Which implies the following bound on η_E:

$$\eta_E \geq \eta_E^{\min} = \eta_E \cdot \frac{\eta_V^{\min} - \theta}{\eta_V^{\max} - \theta} = \frac{3 \cdot \max\left\{\frac{\kappa}{md}, \theta\right\} + 2\sigma\left(1 + \max\left\{2, \frac{2\kappa}{md\sigma}\right\}\right)}{\eta_V^{\max} - \theta} \tag{4}$$

Finally, note that we must have $\theta < \eta_V \leq \eta_V^{\max}$, but this is not sufficient. Since we need $\frac{1}{2} - \eta_V - \eta_E > 0$:

$$\frac{1}{2} < \eta_V + \eta_E \geq \eta_V^{\min} + \eta_E^{\min}$$

$$\geq \theta + \frac{3\theta + 2\sigma\left(1 + \max\left\{2, \frac{2\kappa}{md\sigma}\right\}\right)}{\eta_V^{\max} - \theta}$$

Assuming $\theta < \frac{1}{2}\eta_V^{\max}$, this implies

$$\frac{1}{2} > \theta + \frac{6\theta + 4\sigma\left(1 + \max\left\{2, \frac{2\kappa}{md\sigma}\right\}\right)}{\eta_V^{\max}}$$

$$= \theta\left(1 + \frac{6}{\eta_V^{\max}}\right) + \frac{4\sigma}{\eta_V^{\max}}\left(1 + \max\left\{2, \frac{2\kappa}{md\sigma}\right\}\right)$$

This gives us the following bound on θ:

$$\theta < \frac{\frac{1}{2} - \frac{4\sigma}{\eta_V^{\max}}\left(1 + \max\left\{2, \frac{2\kappa}{md\sigma}\right\}\right)}{1 + \frac{6}{\eta_V^{\max}}} \tag{5}$$

Since the θ must be non-negative, we also have a bound on σ:

$$\frac{12\sigma}{\eta_V^{\max}} \leq \frac{4\sigma}{\eta_V^{\max}}\left(1 + \max\left\{2, \frac{2\kappa}{md\sigma}\right\}\right) \leq \frac{1}{2}$$

hence

$$\sigma \leq \frac{\eta_V^{\max}}{24} \tag{6}$$

B.2   Examples of Parameter Settings

For simplicity we will consider PPEs for which the soundness error ϵ is negligible and we will omit it. Moreover, depending on the context in which we would like to use our protocol and the level of security we would like to achieve, different type of PPEs might be more suitable. As presented in section A, there are multiple ways we could think of implementing PPEs and, naturally, each implementation comes with its own advantages/disadvantages. For instance, opting for a proof-of-storage based implementation can provide us with PPEs with almost perfect completeness ($\sigma = 0$), but requires a lot of communication. On the other hand, other implementations which would give us a worse completeness error (e.g., based on CAPTCHAs), might have higher error but require fewer (or different) resources.

In Table 4 the reader can find example parameter settings for two parameter regimes: in Scenarios 1 and 2, there are 5000 responders and PPEs are error-free, while Scenarios 3 and 4 have 100000 responders with PPEs that have a non-negligible (albeit small) error rate. The first scenario in each pair has degree close to the minimum possible for those parameters, while the second demonstrates the soundness advantage of increasing the degree (we note that the values are based on our worst-case bounds—in practice it may be possible to achieve better parameters).

Table 4. Possible Parameters

Symbol	Scenario 1	Scenario 2	Scenario 3	Scenario 4
κ	40	40	40	40
m	5,000	5,000	100,000	100,000
θ	1/1000	1/1000	1/10000	1/10000
σ	0	0	1/1000	1/1000
η_E	1/8	1/8	0.23	0.23
η_V	0.025	0.025	0.028	0.028
α	0.28	0.28	0.38	0.38
K	1246	1246	35,192	35,192
d	60	120	165	240
C^*	200	10	670	23

As to be expected, higher the degree of the graph (that is the number of PPEs each responder is required to carry out) lower is the advantage the adversary gets.

Primary-Secondary-Resolver Membership Proof Systems

Moni Naor* and Asaf Ziv**

Weizmann Institute of Science,
Department of Computer Science and Applied Mathematics, Israel
{moni.naor,asaf.ziv}@weizmann.ac.il

Abstract. We consider *Primary-Secondary-Resolver Membership Proof Systems* (PSR for short) and show different constructions of that primitive. A PSR system is a 3-party protocol, where we have a *primary*, which is a trusted party which commits to a set of members and their values, then generates public and secret keys in order for *secondaries* (provers with knowledge of both keys) and *resolvers* (verifiers who only know the public key) to engage in interactive proof sessions regarding elements in the universe and their values. The motivation for such systems is for constructing a secure Domain Name System (DNSSEC) that does not reveal any unnecessary information to its clients.

We require our systems to be complete, so honest executions will result in correct conclusions by the *resolvers*, sound, so malicious *secondaries* cannot cheat *resolvers*, and zero-knowledge, so *resolvers* will not learn additional information about elements they did not query explicitly. Providing proofs of membership is easy, as the *primary* can simply precompute signatures over all the members of the set. Providing proofs of non-membership, i.e. a denial-of-existence mechanism, is trickier and is the main issue in constructing PSR systems.

The construction we present in this paper uses a set of cryptographic keys for all elements of the universe which are not members, which we implement using hierarchical identity based encryption. In the full version of this paper we present a full analysis for two additional strategies to construct a denial of existence mechanism. One which uses cuckoo hashing with a stash, where in order to prove non-membership, a secondary must prove that a search for an element will fail. Another strategy uses a verifiable "random looking" function and proves non-membership by proving an element's value is between two consecutive values of members.

For all three constructions we suggest fairly efficient implementations, of order comparable to other public-key operations such as signatures and encryption. The first approach offers perfect ZK and does not reveal the size of the set in question, the second can be implemented based on very solid cryptographic assumptions and uses the unique structure of cuckoo hashing, while the last technique has the potential to be highly efficient, if one could construct an efficient and secure VRF/VUF or if one is willing to live in the random oracle model.

---

* Incumbent of the Judith Kleeman Professorial Chair.
** Research supported in part by grants from the Israel Science Foundation, BSF and Israeli Ministry of Science and Technology and from the I-CORE Program of the Planning and Budgeting Committee and the Israel Science Foundation.

Y. Dodis and J.B. Nielsen (Eds.): TCC 2015, Part II, LNCS 9015, pp. 199–228, 2015.

1   Introduction

We consider the cryptographic primitive called *Primary-Secondary-Resolver Membership Proof Systems* (PSR for short) and show efficient constructions of that primitive. The motivation for this type of systems comes from trying to improve DNSSEC which is a security extension of DNS (Domain Name System) (plain DNS communication doesn't guarantee security (confidentiality and authenticity) for the users). The basic problem is as follows, we have a trustworthy source, called the *primary*, which maps all valid names (e.g. URLs) in its domain to their corresponding values (e.g. IP addresses). This *primary* doesn't communicate directly with users (*resolvers*) who wish to make DNS queries for names; it has the *secondaries* for that, which are DNS servers that receive some initial information from the *primary* and are in charge of responding to *resolvers'* queries. As there may be many such *secondary* servers, we cannot be sure they are all honest and we do not wish to give them the ability to fool *resolvers* with a false response to a DNS query. We would like to give them enough information so as to give correct responses to DNS queries and a short proof of some sort to help convince the *resolver* of the authenticity of the data they received. On the other hand, we do not wish the *resolvers* get more information about the domain than a simple answer to their query, i.e. whether the answer is positive or negative is all a *resolver* should be able to deduce (the issue of releasing too much information about the domain has been an obstacle in getting the current DNSSEC adapted [4]).

A PSR system consists of a setup algorithm, used by the *primary* which receives a privileged subset R from a universe U of names (e.g. the list of hosts in its domain) and a set of corresponding values V, mapping each element $x_i \in R$ to its value $v_i \in V$ (e.g. mapping all URLs in a domain to their IPs). The *primary* generates a public key PK (one may think of it as a signature key), which should be available to all parties of the protocol. It also generates a secret key SK which provides *secondaries* the ability to answer queries honestly. We will be interested only in *efficient* constructions where the public key size and the amount of communication between the *secondaries* and the *resolvers* are independent of the cardinality of the set R.

1.1   Our Contributions

In a companion paper to this work [19] the notion of PSR systems was introduced (albeit it was defined as a one-round proof protocol), as well as an efficient construction named NSEC5 was suggested. NSEC5 is based on RSA and analyzed in the random oracle model. The main application of PSR systems is for a secure Domain Name Server that does not reveal information about the underlying set. That paper also gave a lower bound that shows that in order to preserve soundness and prevent an adversarial *resolver* from learning additional information about elements they didn't query, the *secondary* must perform some non-trivial computation: it must do the computational work needed in a a public key identification scheme, for which the best known implementations are signatures (in the random oracle model these two are equivalent). (This showed that none of the prior approaches to DNSSEC such as NSEC3 yield a solution that is secure against zone enumeration, i.e. listing of the set R).

We consider PSR Systems that are more general than those of [19] and define PSR systems with interactive proofs as well as systems that are *perfect* zero-knowledge.

In this paper we investigate in depth PSR systems. Our main interest is efficiency, where we are interested in the computational and communication load on all three parties, but in particular in the secondary-resolver part that is performed online. Our main goal in this work is to provide PSR systems that are efficient and based on reasonable and well studied assumptions. We aim for efficiency that is of the order of other public-key primitives such as encryption and signatures.

We provide three general techniques to constructing PSR systems and present efficient implementations to each of them. We use signatures and various different cryptographic primitives in our constructions such as: hierarchical identity based encryption schemes, one-time signatures, cuckoo hashing (with a stash) with commitments and fixed-set non-membership proofs, verifiable random/unpredictable functions and pseudorandom functions with interactive zero-knowledge proofs. Our constructions are based on solid cryptographic assumptions: the discrete logarithm assumption and factoring, the existence of universal one way hash functions and various Diffie-Hellman assumptions. Some of our constructions even achieve perfect zero-knowledge.

It is quite clear that the more challenging case in constructing PSR systems is dealing with the *non*-members of the set. For the members of the set a precomputed signature by the *primary* solves the problem. We suggest three approaches for constructing PSR systems. All constructions use (regular) signatures to handle proofs of membership, as we precompute a signature over every $x_i \in R$ and its value v_i. Thus, the difference between the constructions is how they handle proofs of non-membership, i.e. we offer different denial of existence mechanisms.

In our first approach the *primary* matches encryption keys to elements of the universe U. A *secondary* with knowledge of such a key can use it to generate a proof of non-membership for the corresponding element. The *primary* precomputes a set of secret keys K, from which it can derive the keys corresponding *only* to the set of elements $U \backslash R$ and sends it to the *secondaries* as part of their secret key. As long as we make sure the *secondaries* cannot produce any key for an element in R, we can construct a denial of existence mechanism in a number of ways. *Resolvers* can encrypt a random challenge, which can be decrypted only with the secret key corresponding to the queried element $x \in U$, thus non-membership can be proven only for elements outside of R. One can also just send that secret key to *resolvers* when queried, making them verify the correctness of the key by encrypting and decrypting random challenges by themselves. The *secondaries* can also generate signatures for the queried element under a secret key corresponding to that element and verified with a corresponding public key.

In order to implement those constructions efficiently we use *Hierarchical Identity Based Encryption* (or HIBE for short). One can think of a set of identities as nodes in a full binary tree, where with the secret key for an identity, one can produce the key to any of its descendants. We think of the leaves as elements in the universe, so by making sure the set of keys K doesn't contain any secret key to an element in R or any of its ancestors, but contains at least an ancestor key to the rest of the elements in U, we get an efficient denial of existence mechanism. Lastly we consider a construction that uses a chain of signatures from the root of the tree to the leaf, where each signature signs the public key needed to verify the next signature in the chain. All those constructions manage to achieve *perfect* zero-knowledge.

The idea of the second approach is to imitate an oblivious search for the element, where by oblivious we mean that the locations examined are determined by the element searched and some hash functions. The point is to show that the searched element is in none of the probed locations. For the data structure we use cuckoo hashing [36] where (unless we are unlucky) each element resides in one of two locations. That is, as a denial of existence mechanism, we need to prove non equality just twice. To handle the unlucky case we use a cuckoo hashing scheme with a stash [26] to store some extra elements. We need to prove non equality to these elements as well, however we have the advantage that these elements are fixed for all possible searches. To handle the "normal" case the *primary* places Pedersen commitments [37] for the relevant elements in the cells of the cuckoo hash tables (including "dummies" in the empty cells) and signs these commitments. The *secondary* is provided with the signed commitments and proves the committed values are not equal to the queried element. For the stash non equality we use a generalization of the Feige-Fiat-Shamir identification protocol [15]. Both proofs are zero knowledge and are rather efficient as the computation needed in order to execute these two interactive zero-knowledge protocols is dominated by only a constant number of exponentiations. As Pedersen commitments rely on the discrete logarithm assumption and the Feige-Fiat-Shamir protocol relies on the factoring assumption, the result is a PSR system which reveals the size of the set R but is very efficient and is based on conservative and well studied cryptographic assumptions.

Our third approach to constructing PSR systems applies a "random looking" function F, for which we can prove the value $F(x)$ in a zero knowledge fashion, without revealing information about the value of the function at other locations. The *primary* precomputes the values of F over the set R, sorts them lexicographically and signs them in pairs, $\{Sign(y_i, y_{i+1})\}_{i=0}^{r}$. In order to prove non-membership for an element $x \notin R$ one simply provides a proof that $F(x) = y$ and the signature $Sign(y_i, y_{i+1})$ for which $y_i < y < y_{i+1}$ (we choose F to have negligible probability for collisions). This construction reveals the size of the set R during multiple executions of the protocol as a *resolver* which issues enough random queries will eventually witness all signatures $Sign(y_i, y_{i+1})$ and learn the size of R, but in some applications such as DNSSEC, revealing the size of the set is acceptable. In order to construct the function F we use variants of *Verifiable*

Random Functions (VRF) and *Verifiable Unpredictable Functions* (VUF) [30], the Naor-Reingold PRF [32] with zero knowledge interactive proofs, the GHR signature scheme [16] and a random oracle construction which uses the famous BLS signature scheme [7]. The scheme NSEC5 presented in [19] (which resides in the random oracle model) falls into this category as well.

For all three constructions we suggest fairly efficient implementations. The first approach offers perfect ZK and does not reveal the size of the set in question, the second can be implemented based on very solid cryptographic assumptions and uses the unique structure of cuckoo hashing, while the last technique has the potential to be highly efficient, if one could construct an efficient and secure VRF/VUF or one is willing to live in the random oracle model.

Structural Issues: We analyze and prove that PSR systems with one-round proofs are secure even in a concurrent setting. This means that in the case of one-round proofs, even a coordinated attack of *resolvers* trying to learn information about elements in the universe which they did not query explicitly will fail with overwhelming probability. In the case of many-rounds proofs we show that providing each *secondary* with an independent set of keys also results in a concurrently secure PSR system. We prove that PSR systems exist if and only if one way functions exist, which in turn helps us get a black box separation from *zero knowledge sets* [29], which is a more restrictive membership proving system (see details in Section 1.3), thus showing that the two primitives are indeed inherently different.

1.2   A Guide for Reading the Paper

In Section 2 we present our model, the definition of PSR systems, our requirements of completeness, soundness and zero-knowledge and in Section 3 we show cases where the system is secure in a concurrent setting. In Section 4 we show a HIBE based construction which achieves perfect ZK. In Section 5 we give a short description and intuition regrading our cuckoo hashing with a stash based PSR and the one based on "random looking" functions (the full version [34] gives a more detailed description). We also present a signature based PSR system and use it to prove that the existence of one way functions is equivalent to the existence of PSR systems, which leads us to a black box separation between PSR systems and ZKS [29]. In Section 6 we present concluding remarks.

1.3   Related Work

There are several types of cryptographic primitives that are related to PSR systems. Consider *zero-knowledge sets*, introduced by Micali, Rabin and Kilian [29] (ZKS for short) and its generalization zero-knowledge elementary databases. The latter is a primitive, defined in the common reference string model or the trusted parameters model, where a user (prover) can commit to a database and later open and prove its values to a verifier in a zero knowledge fashion. The existence of ZKS implies the existence of a PSR system, as a zero-knowledge elementary database construction implements a PSR System (the other direction is not true as implied by Corollary 2). However, the problem is that even the best known constructions of ZKS are inefficient. The point is that in a ZKS the requirements are too stringent: even the *primary* cannot cheat. This is not something of interest in our setting, since the *primary* is a trustworthy party that commits to a set of its choosing and it does not make sense for it to cheat. We are only interested in preventing the *secondaries* from cheating. Hence we introduced a more complex setting with three parties, at the benefit of gaining efficiency.

Chase et al. [12] introduce the notion of trapdoor mercurial commitments (TMC for short) and construct ZKS based on TMCs. They show a few implementations of their new primitive where their most efficient implementation is a constant factor improvement on the original MRK construction, while both rely on the discrete logarithm assumption. Catalano et al. [11] extend their notion of TMC to trapdoor q-mercurial commitments (q-TMC for short) and by that further improve the efficiency of ZKS implementation by shortening the non-memberships proofs by a constant factor, at the expense of slowing down the verification process. Their construction of q-TMC relies on the q-strong Diffie-Hellman assumption. Later, Libert and Yung [28] introduced a new construction for q-TMCs, based on the q-Diffie Hellman exponent assumption, and managed to shorten the memberships proofs by a constant factor as well. All those ZKS constructions have the same basic structure: a tree (either binary as in [29,12] or with arity q as in [11,28]), where the leaves represent the elements in the universe and a proof of membership or non-membership is a path of commitments from the root to the leaf. All four ZKS constructions use proofs made up of $O(\log |U|)$ group elements and require $O(\log |U|)$ modular exponentiations for verification.

Prabhakaran and Xue introduced *statistically hiding sets* [38] (SHS for short), which are a slight variation on ZKS. Their definition of statistical hiding is formulated with computationally unbounded simulation, which means it is a relaxation of the security requirement of ZKS as they do not require efficient simulation. Their construction uses accumulators, first presented in [5], in order to accumulate a set of values into one value, where there is a short proof for every value in the set. Although it is more efficient than ZKS and can be extended to statistical hiding databases, their underlying assumptions are rather new and strong. They use the strong RSA assumption and an assumption they call the knowledge of exponent assumption. They require the use of a hash function which maps elements to large prime numbers and a trapdoor DDH group.

Ostrovsky, Rackoff and Smith [35] generalized ZKS by defining Consistent Query Protocols, which allow more general queries than membership queries. They also suggested a relaxation for ZK proofs, allowing the server to leak an upper bound T on the size of the database (called size-T-Privacy). Our privacy requirement, f-ZK, is a generalization of this size-T-Privacy requirement.

Another related line of investigation is that of data structures that come with a guarantee of correctness. That is when the data structure, like a dictionary, returns an answer it also provides a proof that the answer is correct in the sense that it is consistent with some external information. One motivation for these investigations comes from data structure for managing CRLs (certificate revocation lists). The difference with the current work is that no additional information than the result of the query should leak.

A recent paper by Ghosh, Ohrimenko and Tamassia [18] introduces two new primitives which are related notions to PSR systems: a 2-party and a 3-party protocols for proving values of elements in a database and their order (lists). The 2-party protocol they define is Zero Knowledge Lists (ZKL for short), where their construction of the primitive is too inefficient for our needs, as it builds upon ZKS (which, as we mentioned, does not have an efficient implementation yet). The 3-party protocol is Privacy Preserving Authenticated Lists (PPAL) which unlike ZKL is closer in spirit to our PSR systems but it cannot answer non-membership queries (their construction only handles queries for elements in the list and returns their order in the list combined with a proof). Besides that, their constructions are also analyzed in the random oracle model, where we strive to find constructions in the standard model.

2   Model and Security Definitions

We model Primary-Secondary-Resolver Membership Proof systems as a 3-party protocol where the *primary*, a trusted party, commits to a set R, a subset of the universe U, where each element $x_i \in R$ is coupled with a value $v_i \in V$. The *primary* generates two keys for the committed set, the secret key SK given only to *secondaries* in the system and the public key PK given to all parties of the protocol, i.e. *secondaries* and *resolvers*. The *resolvers* in the system engage in an interactive protocol with the *secondaries* in order to learn whether a given $x \in U$ is in R or not and if yes then they obtain its value v_x. The *secondaries* use their secret key to generate proofs (possibly interactive) for the correct statement regarding the queried element, while *resolvers* verify the correctness of the proofs they get. We require that the *secondaries* won't be able to cheat the *resolvers* and if the *secondaries* are following the protocols then the *resolvers* should be able to verify the correctness of the responses with overwhelming success probability. Another important requirement we would like from such a system is zero-knowledge, i.e. for *resolvers* to learn as little as possible about elements they didn't query explicitly. See Figure 1 for an illustration of the 3-parties' engagement in the protocol.

Remark 1. Note that we chose to focus on the static version of this problem, i.e. when the sets R and V are determined at the beginning of the process and do not change throughout the process. The dynamic case for this problem is out of the scope of this paper, though we discuss the issues of defining requirements for the dynamic case, as well as give guidelines on how to transform our constructions into ones which can handle dynamic changes in the full version of the paper [34].

2.1   PSR Systems

The system consists of three algorithms: the *Setup* algorithm is used by the *primary* to generate the public key PK which it publishes to all parties in the

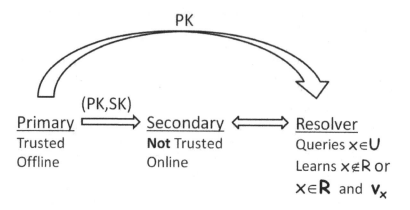

Fig. 1. Illustration of a PSR system

protocol and the secret key SK, delivered to the *secondaries*. The *resolvers* use the *Verify* algorithm in order to initiate an interactive proof session with the *secondaries* who use the *Prove* algorithm to prove interactively the correct membership statement about the element, queried by the *resolver*.

Definition 1. *Let U be a universe of elements. A Primary-Secondary-Resolver system (PSR for short) is specified by three probabilistic polynomial-time algorithms (Setup, Prove, Verify):*

$Setup(R, V, 1^k)$: *On input k the security parameter, a privileged set $R \subseteq U$ and its values V, where $|R| = |V| = r$ (for every $x_i \in R$ the corresponding value is $v_i \in V$), this algorithm outputs two strings: (PK, SK) which are the public and secret keys for the system.*

$Verify(x, PK)$: *The algorithm gets as input $x \in U$ and the public key PK. It initiates an interactive proof protocol over the element $x \in U$ with a secondary of its choice and verifies the correctness of the proof given by the secondary. It outputs 1 when it accepts the interactive proof and 0 otherwise.*

$Prove(x, PK, SK)$: *On input $x \in U$ and both the public and secret keys (PK, SK) this algorithm proves interactively to a resolver either the statement $x \in R$ and its value is v_x or $x \notin R$.*

We require the above three algorithms to satisfy three properties: completeness, soundness and zero knowledge.

2.2   Completeness and Soundness

The *completeness* requirement means that when the parties at hand are honest and follow the protocol, then the system works properly. The *resolvers* will learn successfully whether the element $x \in U$, which they queried, is in R (and its value) or not. We do allow a *negligible* probability of failure.

Definition 2. *Completeness: For all $R \subseteq U$, for all V and $\forall x \in U$,*

$$\Pr \left[\begin{array}{l} (PK, SK) \xleftarrow{R} Setup(R, V, 1^k); \\ Verify(x, PK) \xleftarrow{R} Prove(x, PK, SK); \\ Verify(x, PK) = 1 \end{array} \right] \geq 1 - \mu(k)$$

For a negligible function $\mu(k)$.

As for *soundness*, we want that even a malicious *secondary* A, would not be able to convince an honest *resolver* of a false statement with more than a negligible probability. We require this to hold even when the adversary gets to choose R and V, then gets the keys (PK, SK) and then chooses $x \in U$ on which it wishes to cheat. At the end of the protocol A outputs either 0 if it tries to convince the *resolver* that $x \notin R$ or $(1, v)$ if it tries to convince him that $x \in R$ and its value is v.

Definition 3. *Soundness: for all probabilistic polynomial time stateful adversaries A we have*

$$\Pr \left[\begin{array}{l} (R, V) \xleftarrow{R} A(1^k); \\ (PK, SK) \xleftarrow{R} Setup(R, V, 1^k); \\ x \xleftarrow{R} A(PK, SK); \\ Verify(x, PK) \xleftarrow{R} A(x, PK, SK); \\ Verify(x, PK) = 1 \wedge \\ ((A(x, PK, SK) = 0 \wedge x \in R) \vee \\ (A(x, PK, SK) = (1, v) \wedge (x \notin R \vee (x = x_i \wedge v \neq v_i)))) \end{array} \right] \leq \mu(k)$$

For a negligible function $\mu(k)$.

Note that our definitions are strong because they ensure (up to negligible probability) that an adversary cannot find any $x \in U$ violating either completeness or soundness, even after getting its relevant keys, i.e. (PK, SK) for a *secondary* in the soundness condition and PK for a *resolver* in the completeness condition.

2.3   Zero-Knowledge

We want to restrict the amount of information learned about the set R by *resolvers* during the interactive proofs. Besides the answer to the question being asked by the *resolver* we would like him to learn as little as possible about the set R. In some cases we let some information about the set R leak during the protocol (or many executions of the protocol on different elements), which is why we choose to define zero-knowledge with respect to a function f acting on R. We show two constructions of PSR systems which don't leak any information about the set R (see Sections 4 and 5.1), while the rest of the constructions leak the size of the set R (see Section 5) We define this property as f-**Zero-Knowledge**

(f-ZK for short), where $f(R)$ is some information about the set which we can tolerate leaking to *resolvers*.

We require that the *resolver* cannot distinguish between: (1) a real system which provides the original proofs, and (2) a simulator that can only obtain the answer to each query asked by the *resolver* online, but must still be able to "forge" a satisfactory proof for that response. This allows us to deduce that the *resolver* has not learned much about R from the proofs, for if it had, it would be able to distinguish between an interaction with the simulator and one with the real *secondary* (at least after it gets R explicitly).

The PSR Simulator: Let SIM be an interactive polynomial time algorithm with restricted oracle access to the set R, which means it can query the oracle only on elements which the adversary communicating with him queried explicitly. On its first step SIM receives $f(R)$ and outputs a fake public key PK^*, a fake secret key SK_{SIM} and $f(R)$. On its following steps an adversary interacts with the simulator and queries different elements in the universe. Following every such query x_i the simulator queries its oracle for x_i and either learns $x_i \notin R$ or $x_i \in R$ and its value is v_i. SIM proves interactively the statement on x_i to the adversary. The simulator is successful if its output, i.e. its random tape, public key and transcripts of the interactive protocols, is indistinguishable from that of a real PSR system.

The first step of the interactive protocol for the PSR system[1] is:

$$(PK, SK, f(R)) \xleftarrow{R} Setup(R, V, 1^k)$$

and for the simulator the first step is:

$$(PK^*, SK_{SIM}, f(R)) \xleftarrow{R} SIM^R(f(R), 1^k)$$

The rest is a series of interactive proofs of membership between the adversary and either a PSR system or a simulator, where the simulator uses the fake public key PK^* and the fake secret key SK_{SIM} to respond to queries and the system uses the real keys (PK, SK).

Definition 4. *Let $f()$ be some function from 2^U to some domain and let algorithms $(Setup, Prove, Verify)$ be a PSR system. We say that it is f-zero knowledge (f-ZK for short) if it satisfies the following property for a negligible function $\mu(k)$:*

There exists a simulator SIM such that for every probabilistic polynomial time algorithms A (adversary) and D (distinguisher), a set $R \subseteq U$ and V, the distinguisher D cannot distinguish (See Remark 2 below) between the following two views (interactions of A with a PSR system or a PSR simulator) with an advantage greater than $\mu(k)$, even for D that knows R:

$$view^{real} = \{r_{real}, PK, f(R), (x_1, \pi_1), (x_2, \pi_2), \ldots\}$$

---

[1] Note that the Setup algorithm is not defined to output $f(R)$, but it is obviously a simple modification, as it gets R and can compute $f(R)$ easily. We add this output in order to generate comparable views.

and

$$view^{SIM} = \{r_{SIM}, PK^*, f(R), (x_1, \pi_1^*), (x_2, \pi_2^*), \ldots\}$$

where the two views are generated by the protocols described above, π_i and π_i^* are the transcripts for the interactive protocols over the element x_i and r_{SIM} and r_{real} are the random tapes of the simulator and secondaries respectively.

Remark 2. We have three notions of Zero-knowledge for PSR systems: computational ZK, which means that the distinguisher cannot *computationally* distinguish between the two views, *statistical* ZK, where the distributions of the two views are statistically close and *perfect* ZK where the two distributions are identical. Note that the perfect and statistical ZK have the added advantage of being secure in an information theoretic sense, which guarantees *everlasting privacy.* As both these ZK properties are information theoretic, they require their underlying assumptions to hold *only during the execution of the protocol,* while for computational ZK, we require the assumptions to hold 'forever' in order to prevent an adversary from breaking the privacy of the scheme at a later point in time. Our HIBE and signature based constructions (Section 4 and 5.1 respectively) achieve perfect ZK, the cuckoo hashing construction (Section 5.2) achieves statistical ZK, while the last construction (Section 5.3) achieves computational ZK.

In our companion paper [19], we prove two very important facts about non-interactive PSR systems. The first is that f-ZK, where $f(R)$ is the cardinality of the set R, implies prevention of zone enumeration, i.e. if a PSR is f-ZK, then a *resolver* cannot learn any information about an element it didn't query explicitly. All of the constructions in this paper are at least f-ZK for this f (the HIBE and signature based constrictions are even perfect ZK), which means they all prevent zone enumeration. The second important result is that PSR systems require a heavy computational task from the *secondaries*, such as public key cryptography or public key authentication, in order to maintain both soundness and f-ZK. This fact is crucial to understanding why the *secondaries* work hard in our constructions. Note that both these proofs were for the single-round PSR and in the random oracle model, but the proofs generalize to our (possibly interactive) setting as well. The prevention of zone enumeration holds as is in the standard model for interactive proofs, while the reduction to public key authentication for interactive PSRs in the standard model is only selectively secure, as opposed to existentially secure in the random oracle model. We state the resulting theorem:

Theorem 1. *Given an f-ZK PSR system (where $f(R) = |R|$ or $f(R) = null$), one can construct a public-key identification or a selectively secure public key authentication protocol from the PSR system where the prover's complexity is similar to the* secondary's. *The construction is black box.[2].*

---

[2] See the original paper for the proof and definitions for public key authentication.

3   Concurrent Zero Knowledge

In this section we prove that in some cases PSR systems are not only f-ZK as defined earlier, but also concurrent zero knowledge with respect to that same function f. Concurrent ZK was introduced by Dwork, Naor and Sahai [14] as an extension to zero knowledge. In order for an interactive proof system to be concurrent ZK we require that if we have up to a polynomial number of provers and verifiers, where the verifiers are controlled by a malicious adversary and work concurrently (one could start an interactive proof with a prover, put it on hold and finish an earlier interaction), then still no information is leaked to the adversary controlling the verifiers.

We use similar definitions to the ones defined by Rosen [40] and adapt them to our setting. For an interactive proof system $\langle P, V \rangle$, we define a nonuniform probabilistic polynomial time concurrent adversary A. A gets some input I (for PSR systems $I = PK$), controls a polynomial number of verifiers and has access to an unbounded number of copies of the prover P. A can use verifiers to interact with the provers and controls the scheduling of all the messages in the system, meaning that A controls when any verifiers output a message and when every prover outputs a message. We denote by $view_A^P(I)$ the view of the adversary, which is a random variable which contains the random tape of A and all the concurrent interaction of A with the provers (copies of P).

Roughly speaking, a protocol is concurrent ZK if for every such adversary A there is a probabilistic polynomial time simulator S_A such that the two ensembles $\{view_A^P(I)\}$ and $\{S_A(I)\}$ are computationally indistinguishable, where I is some $x \in L$ and $S_A(I)$ is the output of a simulator which uses the adversary A as an oracle. But PSR systems, as we defined them, consist of multiple executions of membership/non-membership interactive proofs using the keys (PK, SK). Thus it is more natural for us to define $I = PK$ and compare between the view of an adversary communicating with *secondaries* (provers) on the public key PK and the view of an adversary communicating with the simulator on the fake public key PK^*.

Thus we define a **concurrent PSR simulator** as a probabilistic polynomial time algorithm SIM, with restricted oracle access to the set R, such that on its first step of the computation, SIM gets $f(R)$ and outputs a fake public key PK^*, a fake secret key SK_{SIM} and $f(R)$. SIM is not allowed to query its oracle on $x \in U$ if it was not explicitly queried by a *resolver* (verifier) on it. When an adversary interacts with a simulator, the copies of the prover are replaced with the simulator itself which acts as a prover (i.e. it emulates all the provers), uses the fake cryptographic keys it generated and can query its oracle for the element queried by the *resolvers*.

We consider two different concurrent settings: where all the *secondaries* get the exact same pair of keys and when each *secondary* and *resolver* get a pair of keys generated independently for them. We prove, that in the case we use independent keys, every PSR system which is f-ZK in the sequential (regular) setting is also f-CZK, thus by making the *primary* work $k \cdot m$ times harder, one can get a concurrently secure PSR system with k *secondaries* and m *resolvers*, from a sequentially secure PSR system. When all *secondaries* get the exact same pair of keys we prove that non-interactive PSRs remain concurrently secure as well.

We denote by $\{view_A^{SIM}(f(R))\}$ the view which contains $f(R), PK^*$, the random tape of A and the concurrent interaction between SIM and A. We denote by $\{view_A^{real}(R)\}$ the view which contains $f(R), PK$ the random tape of A and the concurrent interaction between the real PSR system and A, where the keys are generated by the setup algorithm of the PSR and the provers are honest *secondaries* in a real PSR system.

Definition 5. *A PSR system is f-Concurrent Zero Knowledge (f-CZK) if for every nonuniform probabilistic polynomial time concurrent adversary A and every $R \subseteq U$ there exists a concurrent PSR simulator SIM, such that the two views: $\{view_A^{SIM}(f(R))\}$ and $\{view_A^{real}(R)\}$ are indistinguishable, even for a distinguisher which knows R.*

Note that the way we defined the f-ZK simulator in Section 2.3 the simulation occurs online, meaning there is no rewinding. Rewinding usually raises an obstacle in going from regular ZK to concurrent ZK, so this is a good property to have for the simulator. We prove that a non interactive PSR system (one-round proofs) is always an f-CZK PSR system. On the other hand, we show that for many-round PSR systems this is not necessarily the case: we provide a counter example with more than one round proofs which is not concurrent zero knowledge.

Theorem 2. *If (Setup, Prove, Verify) constitute an f-ZK PSR system with one round proofs then it is also f-Concurrent Zero Knowledge.*

Proof. Assume towards contradiction that there exists a concurrent adversary A such that there exists a distinguisher D, that can distinguish between an interaction of A with a real PSR system and an interaction of A with a concurrent PSR simulator. We describe an adversary B which uses A as a subroutine in order to generate two views (one of B interacting with the system and one interacting with the f-ZK simulator) which can be distinguished with a non-negligible advantage. B simply acts as a mediator between the concurrent adversary A and the prover (system/simulator). Every time A issues a new query to some prover, B simply sends the first message of the interaction to the prover and records the response. Notice that although A might be asking for different provers, B only uses the one prover it has access to and as this is only a two message protocol, B simply records the response to the query. When A asks for the response of that interaction, B sends back the recorded response. When A wishes to terminate the interaction, B terminates the interaction with the prover. At the end of the interaction the view generated by the adversary B isn't in the concurrent setting as in practice B executed the interactions with the provers sequentially. This is not a problem as we can describe a distinguisher D' which uses D to distinguish between interactions with an f-ZK PSR simulator and ones with a real PSR system.

When D' gets a view of the interaction between B and the prover it also gets B's random tape, so D' can run it again with the same random bits (the random tape is included in the view) and rearrange the view it got to look like a concurrent view (i.e. rearrange the order of the messages). Now D' runs D on the newly generated view and outputs its output. If D succeeds in distinguishing between the provers with non-negligible advantage ε, then so does D' as the view of the adversary A interacting with the provers is identical to the view of B interacting with the same provers after D' completed its transformation of the view to look concurrent. Thus we reach a contradiction, which means that non-interactive f-ZK PSR systems are also f-concurrent zero knowledge in the same sense: computational, statistical or perfect ZK.   □

Counter Example for a Many-Round PSR: We show that Theorem 2 does not hold when we try to generalize it to many-rounds PSRs. Suppose that we have a one-round proof f-ZK PSR. We modify it by adding two more rounds to its proof. During the setup algorithm the *primary* selects some pseudorandom function F, such that for an adversary (who doesn't know the secret key), the probability of guessing $F(x)$ for a randomly chosen x will be negligible in the security parameter for the PSR. The first round of the interaction will be the *resolver* asking to learn the value $F(x_1)$ for x_1 of its choice (under honest behavior it should be uniformly random). The second round will be the *secondary* sending an element x_2, chosen uniformly at random, to the *resolver* and if the *resolver* returns the correct value $F(x_2)$ then the *secondary* returns a description of R. Otherwise it continues to the original one round proof of the PSR. One can see this is still an f-ZK PSR, as guessing $F(x_2)$ for a randomly chosen x_2 is successful with only negligible probability, even after seeing several values of F. Thus the *resolver* will learn more than it should about R only with negligible probability, making the new PSR secure if the original one was secure.

On the other hand, in a concurrent setting, a malicious *resolver* can simply interact with a *secondary* and when it gets its challenge x_2, stop the interaction and start a new one with a new *secondary*. In the first round, the *resolver* will set $x_1' = x_2$, i.e. it asks the new *secondary* what is the value of $F(x_2)$; it will then return the answer to the first *secondary*, which should accept it as the correct answer and then it will "spill the beans" and reveal the entire set R, thus violating the f-CZK property (no concurrent simulator can do it for a random set R).

Concurrent Zero-Knowledge with Independent Keys: The reason the above counter example was successful is that the provers were confined by the common key of the PRF they all shared. We claim that in case we have a concurrent execution of the PSR system but where each prover (*secondary*) - verifier (*resolver*) couple receives different and independently chosen keys (that is for each *secondary-resolver* the *primary* executes the setup algorithm independently), then the resulting PSR systems are f-CZK[3].

---

[3] Note that it is critical to use different keys for every couple (*secondary-resolver*) running concurrently, otherwise in the scenario described in the counter example, either a malicious *resolver* can communicate with two *secondaries* using the same keys and break the f-CZK property, or two malicious *resolvers* can collide and interact with one *secondary* using the same keys to break the f-CZK property.

Proof Sketch: the way the concurrent simulator will work is by running the (regular) simulator for each *secondary* independently. We now use a hybrid argument to show that if we are in the described setting and we have an adversary A that can generate two *distinguishable* views for the concurrent setting, then we can construct an adversary B that can generate *distinguishable* views for the sequential setting. If there is a distinguisher D that can distinguish with non-negligible advantage between the two views (generated by A) then it can also distinguish between at least two adjacent hybrids with non-negligible advantage, due to the hybrid argument. This means that there is some index i for which we can construct the adversary B as follows: the first $i-1$ provers will be simulated by B to be a real PSR system *secondaries* (this is done by running the setup algorithm $i-1$ times), the ith prover will be the prover interacting with B (either a simulator or a real *secondary*) and the rest of the provers will be simulated by B using the strategy employed by the (regular) simulator. The two possible views resulting from interacting with this adversary B will be distinguishable with a non-negligible advantage due to the hybrid argument, thus contradicting the assumption that the PSR system is f-ZK.   □

Remark 3. In the full version of the paper [34] we claim and give a proof sketch to show that in the Universally Composable security (UC security) framework, introduced by Canetti [10], PSR systems which have non-interactive proofs or use independent keys are also secure in the UC framework.

4   HIBE Based Construction of PSR Systems

In this section we introduce a PSR system based on *Hierarchical Identity Based Encryption* (or HIBE for short). We think of the universe of elements U, as the leaves of a full binary tree. The *primary* can generate an encryption key for any node in the tree, where this encryption key holds the power to prove non-membership for every element in the universe which is a descendant of that node. A proof of non-membership for an element $x \in U$ uses the encryption key of the leaf that corresponds to x, while an encryption key for an internal node can generate the keys of its descendants. Thus if the *primary* generates the encryption key for the root node, it can then generate a set of keys K which contains keys *only* to the elements in $U \backslash R$. In order to do that the *primary* removes the entire path of keys from the root to a leaf $x \in R$ and generates keys to the siblings of each node along that path. One might notice the similarity to revocation schemes, as we "revoke" all keys for the elements in R and as shown by Naor et al. [31], this process results in a forest of $O(|R| \cdot \log \frac{|U|}{|R|})$ full binary trees (See Figure 2 for an example).

In order to generate this set of keys K we will use a HIBE scheme, which is an identity based encryption scheme (i.e. an element's encoding is its identity) with the special property we need: that every key can generate keys to its descendants in the hierarchy tree. For high efficiency we use the HIBE construction of Boneh et al. [6], which we describe in more details in Section 4.4. Agrawal, Boneh and

Boyen also offer two HIBE constructions [3,2] based on lattices, which give us also two lattice based assumptions from which we can construct a PSR system. The HIBE construction is *perfect ZK*, in the sense that it doesn't reveal any information about the set R to any adversarial *resolver*, not even its cardinality, while providing perfect simulation.

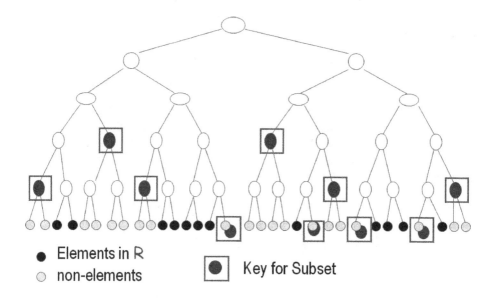

Fig. 2. A full binary tree that represents a set R and its set of keys K

4.1   HIBE Definition

An IBE (Identity Based Encryption) is a scheme where one can encrypt messages to users using their names/IDs or any other unique identifiers one chooses to use. A trusted party generates a master public key (also called system parameters sometimes) and a master secret key, where the first is used by users to encrypt messages under any identity they wish, while the latter is used to generate secret keys for all identities in the scheme, which are then distributed to the users (each user gets its own secret key). A user can then use its secret key to decrypt messages intended for him. A HIBE is an hierarchical IBE, which means that identities in the scheme are defined by up to ℓ coordinates and anyone who has a secret key for its identity x, can generate secret keys to any of its descendants, i.e. to any identity with x as its prefix.

We use the following definition for HIBE which is similar to that of Gentry and Silverberg [17]. An **ID-tuple** is a description of a user in the system defined by $(I_1, \ldots, I_t)$ where $t \leq \ell$ and ℓ is the maximum depth of the hierarchy of identities, i.e. the maximal number of coordinates in an identity. In our construction we use binary vectors as the identities.

Definition 6. *A HIBE is defined by five algorithms: Setup, MKeyGen, KeyGen, Encrypt and Decrypt.*

Setup. *Gets a security parameter k and the depth of the hierarchy ℓ and generates the master public key MK_P, which should be distributed to all the users in the system and a master secret key MK_S given only to the root user, both corresponding to the HIBE of depth ℓ.*

MKeyGen. *Gets the master key MK_S and a target identity $ID = (I_1, \ldots, I_t)$ and generates a private key (from a distribution of valid keys) denoted as SK_{ID}, which user ID can use to decrypt messages intended for him and also to generate properly distributed private keys (i.e. with same distribution, as if it was generated using MK_S) to any of its descendants (any user who has the identity ID as a prefix to its own identity).*

KeyGen. *Gets a private key SK_{ID} for identity $ID = (I_1, \ldots, I_t)$ and some descendant of that identity of any level, $ID^* = (I_1, \ldots, I_t, I_{t+1}, \ldots, I_m)$ and generates a private key SK_{ID^*} from its proper distribution. It is critical that for every identity, two different ancestors produce the same distribution on the generation of its private key. Sometimes this algorithm is described only for one level deeper than that of ID, but this can be extended by invoking the algorithm recursively.*

Encrypt. *Gets the master public key, a message m and a target identity ID and outputs a ciphertext CT which is an encryption of m intended for ID.*

Decrypt. *Gets a private key for identity ID and a ciphertext CT intended for that identity and decrypts it to retrieve the original message m.*

We include the description of the HIBE by Boneh et al. [6] in Section 4.4, which is the most efficient HIBE implementation we could find for our purposes. It uses only a constant number of pairing computations and exponentiations and a logarithmic number (in the size of the universe U) of multiplications in a group, for the algorithms used by the *secondaries* and *resolvers*: Encrypt, Decrypt and KeyGen for leaves in the tree. Not all algorithms are as efficient as those three, but we may allow the *primary* setup to take longer time as it commits to the set R only once.

4.2   HIBE Security

There are four types of security notions for HIBE. We have indistinguishability under chosen plaintext attack and under chosen ciphertext attack, where in the first an adversary can issue queries to different secret keys in the HIBE and in the second it can also issue decryption queries where it can ask to decrypt ciphertexts. For the needs of our construction the weaker notion of security will suffice, i.e. indistinguishability under chosen plaintext attack. We can also talk about the difference between selective and existential security, where in the first an adversary selects a priori the target identity it wishes to be tested on and in the second it can choose the target identity after it issues some queries. Again we only need the weaker notion of security for our construction, i.e. selective security. We use the definitions of security as defined by Boneh et al. [6].

Definition 7. *Indistinguishability under selective identity chosen plaintext attack (IND-sID-CPA). We say that a HIBE system is (t, q, ε) IND-sID-CPA if any t-time adversary A that uses q queries wins the following game with an advantage of at most ε. This is a communication game between an adversary A and a challenger which controls the HIBE system at hand.*

step 1: *A sends a target identity ID^* to the challenger and two equal length messages m_0, m_1 on which it wishes to be tested.*

step 2: *The challenger runs the HIBE's setup algorithm, sends the master public key to the adversary and keeps the master secret key to himself.*

step 3: *A adaptively issues up to q key queries to the challenger, where it asks to know the private key of an identity ID. The challenger responds with the correct keys to all queries. The only restriction is that A didn't issue a key query on identity ID^* or a prefix of it.*

step 4: *The challenger draws a bit at random $b \in \{0, 1\}$, computes $CT = Encrypt(MK_P, ID^*, m_b)$ and sends CT to A.*

step 5: *A issues more queries (where the total number of queries is at most q) where again A cannot issue key queries to prefixes of the identity ID^* or to ID^* itself. When A finishes with the queries it issues a guess $b' \in \{0, 1\}$ and wins the game if $b' = b$.*

Notation. If we have a HIBE which is (t, q, ε) IND-sID-CPA secure, t, q are polynomials and ε is negligible in the scheme's security parameter, then we simply say it is IND-sID-CPA secure.

Remark 4. In a recent paper, Lewko and Waters [27] examine the difficulty in proving full (existential) security for HIBEs. They show that proving full security for a large class of HIBEs results in an exponential degradation (in the depth of the hierarchy) in security. Luckily for us we only need selective chosen plaintext security (the weakest security notion for HIBEs), which most if not all HIBEs achieve, without the exponential degradation.

4.3 PSR from HIBE

Suppose that all possible queries that *resolvers* issue are in the domain $\{0, 1\}^\ell$. We can assume that, as we may use a collision resistant hash function h in order to map our domain of queries into a domain with the appropriate ℓ. We will use a HIBE of depth ℓ. As we do in all constructions, for $x \in R$ we will use *consistent* signatures on the element and its value, i.e. a signing algorithm that produces the same signature on the same message. We will use the HIBE scheme to deal with non-membership proofs. In order to prove non-membership in R, the *secondaries* will get as part of the secret key SK, a set of secret HIBE keys K, from which they can generate a secret key corresponding to any $x \notin R$ (the secret key is $SK_{h(x)}$) and prove its possession by decrypting random challenges encrypted by the *resolvers* under the queried element's identity $h(x)$ (alternatively the key may be given to the *resolvers* who should verify its correctness).

We do not want the *secondary* to be able to prove the non membership of an actual member $x \in R$, so we make sure it cannot obtain the secret keys to any element in R. Thus *secondaries* will not be able to prove false statements with overwhelming probability, as in order to prove false statements the *secondary* will have to either forge signatures or decrypt a message it doesn't have the private key for.

In order to give *secondaries* the correct set of private keys, consider the full binary tree of depth ℓ. The *primary* removes all nodes which are in R or are ancestors/prefixes of elements in R. All the remaining nodes in the tree (both internal and leaves) comprise a forest of full binary trees of different depths. The *primary* then generates the secret key to all the roots of the binary trees in the forest and distributes it to the *secondaries*. Now, the union of all those keys, denoted as K, can generate all keys corresponding to leaves that are not members of R. As mentioned before, the number of trees in the forest can be shown to be $O(r \log \frac{|U|}{r})$ [31].

We now describe the PSR construction that uses a HIBE which is required to be *only* IND-sID-CPA secure (see Definition 7 for details) and an existentially unforgeable signature scheme.

Setup$(R, V, 1^k)$: Use the setup algorithm for the signature scheme in order to obtain the keys (PK_{sig}, SK_{sig}, h) where h is a collision resistant hash function that maps elements from U to $\{0,1\}^{\ell}$. Use the setup algorithm for the HIBE scheme and obtain the master public key MK_P and the master secret key MK_S for a HIBE of depth ℓ. Set the public key to be $PK = (PK_{sig}, MK_P, h)$.

Now generate the forest of full binary trees, as specified above, by removing all the nodes in the full binary tree of depth ℓ, which are prefixes of $h(x_i)$ for every $x_i \in R$. For every root t_j in that forest, generate its corresponding secret key k_j (using the MKeyGen algorithm) and set $K = \{(t_j, k_j)\}$. Now sign every element $x_i \in R$ with its value: $s_i = (Sign_{SK_{sig}}(x_i, v_i), (x_i, v_i))$ and set the secret key to be $SK = (K, \{s_i\}_{i=1}^r)$.

Verify(x, PK): Gets an element $x \in U$ and the public key and initiates an interactive protocol with a *secondary*. It draws uniformly at random a message m from the message domain of the HIBE scheme and encrypts it under the public key of $h(x)$: $CT = Encrypt(m, h(x), MK_P)$. It send (CT, x) to a *secondary*. If it gets in return back m, it returns 1 and "$x \notin R$"; if it gets in return a signature s and a pair (x, v) where it verifies correctly that s is a valid signature on (x, v) then it accepts that $x \in R$ and its value is v and returns 1. Otherwise it returns 0.

Prove(x, PK, SK): Gets the public and private keys and also (CT, x) from a *resolver*. If there exists a signature s_i for which $x_i = x$, then it returns s_i. Otherwise the secret key SK contains, in its HIBE set of keys K, a key for a prefix of $h(x)$. The *secondary* generates the secret key for $h(x)$ (using the HIBE KeyGen algorithm), decrypts CT under that secret key and returns m to the verifier.

Theorem 3. *The three algorithms described above constitute a (perfect) ZK PSR (i.e. f is the null function and the simulation is perfect).*

Proof. In order to prove the above scheme constitutes a PSR system we need to prove it fulfills the three properties required from a PSR system: completeness, soundness and zero-knowledge.

Perfect Completeness. For all $R \subseteq U$, for all V and for all $x \in U$ we need to show that after obtaining the keys (PK, SK) from the setup algorithm, it always holds that an honest *secondary* manages to convince an honest *resolver* of the true statement regarding the queried element x. For every element $x_i \in R$ the *primary* precomputed $s_i = (Sign_{SK_{sig}}(x_i, v_i), (x_i, v_i))$ which is part of the secret key and thus the *secondary* will always succeed in proving membership statements. As for statements of the type $x \notin R$, using the set of HIBE keys K given to the *secondaries*, they can derive a secret key for every $x \in U \backslash R$ (actually a key for every such $h(x)$). Using that key $SK_{h(x)}$, *secondaries* can always decrypt a random challenge issued by *resolvers* and thus will always manage to prove statements of non-membership.

Soundness. Assume for contradiction that there exits some polynomial time adversary that using (PK, SK) can provide for some $x \notin R$ a proof that $x \in R$ with non-negligible probability. This means it can forge a signature with non-negligible probability for that x and some value v, violating the unforgeability assumption on the underlying signature scheme. The same holds if an adversary is trying to prove for some $x \in R$ with value v a different value $v' \neq v$, i.e. due to the existential unforgeability of the signature scheme proving a false value for $x \in R$ is infeasible as well.

If we assume to have such an adversary A that can provide for some $x \in R$ a proof that $x \notin R$ with non-negligible probability ε, then we can use A to construct an adversary B that wins the IND-sID-CPA security game (Definition 7) with a non-negligible advantage $\frac{\varepsilon}{2}$. If A can cheat with probability ε for the set $R \subseteq U$ and some $x \in R$ then the adversary B (trying to win the IND-sID-CPA security game) will first select $h(x)$ as its target identity (h will be chosen by him as well), choose two random messages as the challenge messages $\{m_0, m_1\}$ and get the HIBE master public key, MK_P. Then B runs the setup algorithm for the PSR over U and R while using MK_P as its master public key for the HIBE in the PSR and will use the key queries in the security game to generate the set of HIBE keys K. Note that as $x \in R$ all the key queries will be for non-prefixes of $h(x)$ as K doesn't contain any ancestors of $h(R) = \{h(x_i)|x_i \in R\}$.

Thus B will generate a valid pair of keys (PK, SK) for a PSR and hand them to the adversarial *secondary* A. B will now send the random challenge it got form the IND-sID-CPA security game (an encryption under $h(x)$ of m_0 or m_1) to A which will try to decrypt the ciphertext. A succeeds in decrypting the challenge with probability ε and if the decryption A offers matches one of the two original challenge messages (m_0, m_1) then B chooses this message and else it guesses uniformly at random. Thus B wins the IND-sID-CPA security game with an advantage of about $\frac{\varepsilon}{2}$ [4]. Thus violating the security assumption made on the HIBE scheme being used.

We also note that it is infeasible for an adversary to find an element on which it can provide a false proof. As the adversary gets both keys we can assume it knows R. The adversary cannot find an element $x \notin R$ and provide a false proof with non-negligible probability as this again violates the unforgeability of the signature scheme. Regarding $x \in R$ as we know that the HIBE is *selectively* secure then we know that if the target identity is chosen in advance, then any polynomial time adversary has at most a negligible advantage ε in distinguishing between the two target messages, which makes its probability of decrypting the target ciphertext at most 2ε (by the reduction shown above). So as this time there are $|R| = r$ target identities, any adversary has at most a probability of $2\varepsilon \cdot r$ (still negligible as r is polynomial) to decrypt a random challenge under one of the identities of $h(R)$, thus it is also infeasible to find $x \in R$ for which a *secondary* can cheat on.

Perfect ZK. In order to show that this PSR is indeed zero knowledge we need to show a suitable simulator SIM which can fool any adversary into believing it is a real PSR system. SIM simply chooses the function h as the *primary* does, runs the setup algorithm for the HIBE to obtain (MK_P, MK_S) and the setup algorithm for the signature scheme to obtain (Pk_{sig}, SK_{sig}). SIM then sets the fake public key to be $PK^* = (MK_P, PK_{sig}, h)$ and the fake secret key to be $SK_{SIM} = (SK_{sig}, MK_S)$. Note that the fake public key is generated the exact same way the original public key is generated and the fake secret key has the master secret key for the HIBE instead of the subset of the keys (K) and the secret key for the signature scheme instead of the signatures on the elements of R and their values $(\{s_i\}_{i=1}^r)$. When SIM is queried on an element $x \in U$, it queries its oracle to R on x. If $x \in R$ and its value is v_x it returns $s = (Sign_{SK_{sig}}(x, v_x), (x, v_x))$. If $x \notin R$ then SIM gets (CT, x) and it can generate the secret key for $h(x)$ using the master secret key MK_S, decrypt the challenge and return it to the adversary.

We claim that the two views generated by the simulator and a real PSR system are not only indistinguishable but identically distributed, thus making this construction **perfect** zero-knowledge. The public keys are generated by the same algorithm. The signatures (proofs regarding $x \in R$) are generated online instead of during the setup algorithm as the *primary* does, but yield the same distribution over the signatures, due their consistency. Proofs for elements $x \notin R$ are also identical as both the simulator and a PSR system decrypt successfully the random challenges on elements outside of R with probability 1 and simply return it. This concludes the proof that this PSR system is perfect ZK. □

Remark 5. Note that we can also use two variants of HIBEs, one where *secondaries* deliver the queried element's decryption key to the *resolver* (and by that

---

[4] There is a probability that A decrypts CT to a wrong message that happens to be m_{1-b} while m_b was chosen as the challenge. But, as $\{m_0, m_1\}$ are chosen uniformly at random and are not known at all to A this probability is negligible.

make it verify the key's correctness by itself) and one where we use signatures instead of encryption, i.e. *secondaries* produce signatures over the queried element with its corresponding secret key.

4.4 HIBE Construction by Boneh, Boyen and Goh

We describe the construction by Boneh et al. [6] as it is the most efficient HIBE implementation for our needs. Its greatest virtue, with respect to our construction, is the fact that generating secret keys for nodes get more efficient the deeper the node is in the hierarchy. Thus generating keys for leaves is very efficient, which is critical for us, since this is done online by the *secondaries* generating non-membership proofs. Let $\mathbb{G}$ be a bilinear group of prime order p and let $e : \mathbb{G} \times \mathbb{G} \to \mathbb{G}_1$ be an admissible bilinear map (i.e. its bilinear-$\forall g_1, g_2 \in \mathbb{G}$ it holds that $e(g_1^x, g_2^y) = e(g_1, g_2)^{xy}$, non-degenerate - $e(g, g) \neq 1$ and efficiently computable). We choose arbitrarily how to map J_0, J_1 to $\mathbb{Z}_p^*$, since the original HIBE can handle identities of the type $ID \in (\mathbb{Z}_p^*)^\ell$ (or shorter), while we only require binary identities of length at most ℓ. This means that for some node in level k of the tree, $u = x_1 \ldots x_k$ where $x_i \in \{0, 1\}$ has identity $I_u = (J_{x_1}, \ldots, J_{x_k}) = (I_1, \ldots, I_k)$, which will be also its public key. We also assume that the messages to be encrypted are elements in $\mathbb{G}_1$. We choose ℓ, the depth of the hierarchy, to be $\lceil \log |U| \rceil$, in order for the leaves of the full binary tree of depth ℓ to represent the elements in the universe.

The HIBE system works as follows:

- *Setup*($1^k, 1^\ell$): Gets k the security parameter and ℓ the depth of the hierarchy. To generate the public master key for the HIBE of maximum depth ℓ, draw uniformly at random: $g \in \mathbb{G}$, $\alpha \in \mathbb{Z}_p^*$, set $g_1 = g^\alpha$ and pick some more random elements $g_2, g_3, h_1, \ldots, h_\ell \in \mathbb{G}$. Next compute $Aux = (h_1^{J_0}, h_1^{J_1}, \ldots, h_\ell^{J_0}, h_\ell^{J_1})$ and define the master secret key to be $MK_S = g_2^\alpha$ and the public master key to be: $MK_P = (g, g_1, g_2, g_3, h_1, \ldots, h_\ell, Aux)$.
- *MKeyGen*(MK_S, ID): To generate a private key for $ID = (I_1, \ldots, I_k) \in (\mathbb{Z}_p^*)^k$ pick uniformly at random $r \in \mathbb{Z}_p$ and output:

$$SK_{ID} = (g_2^\alpha \cdot (h_1^{I_1} \cdots h_k^{I_k} \cdot g_3)^r, g^r, h_{k+1}^r, \ldots, h_\ell^r) \in \mathbb{G}^{\ell-k+2}$$

Note that the deeper the node the smaller the private key.
- *KeyGen*(SK_{ID}, ID^*): For $ID^* = (I_1, \ldots, I_m) \in (\mathbb{Z}_p^*)^m$ and a private key of its ancestor $ID = (I_1, \ldots, I_k)$ $(m > k)$ do the following in order to generate a properly distributed key:
  If $SK_{ID} = (g_2^\alpha \cdot (h_1^{I_1} \cdots h_k^{I_k} \cdot g_3)^{r'}, g^{r'}, h_{k+1}^{r'}, \ldots, h_\ell^{r'}) = (a_0, a_1, b_{k+1}, \ldots, b_\ell)$
  then choose uniformly at random $t \in \mathbb{Z}_p$ and output: $SK_{ID^*} =$

$$(a_0 \cdot b_{k+1}^{I_{k+1}} \cdots b_m^{I_m} (h_1^{I_1} \cdots h_m^{I_m} \cdot g_3)^t, a_1 \cdot g^t, b_{m+1} \cdot h_{m+1}^t, \ldots, b_\ell \cdot h_\ell^t) \in \mathbb{G}^{\ell-m+2}.$$

This can be computed using $4 + (\ell - m)$ exponentiations and $O(\ell)$ multiplications by utilizing Aux. This private key is a properly distributed key for

$ID^* = (I_1, \dots, I_m)$ with $r = r' + t \in \mathbb{Z}_p$. Note that the deeper the node – the shorter the key, thus computing a secret key for a leaf is very efficient. If ID^* is a leaf ($m = \ell$) we get:

$$SK_{ID^*} = (a_0 \cdot b_{k+1}^{I_{k+1}} \cdots b_\ell^{I_\ell} (h_1^{I_1} \cdots h_\ell^{I_\ell} \cdot g_3)^t, a_1 \cdot g^t) \in \mathbb{G}^2.$$

Computing secret keys for the leaves takes only 4 exponentiations and $O(\ell)$ multiplications, since by utilizing Aux, the *secondary* multiplies all the b_i's where $I_i = J_1$ and then raises them to the power of J_1 and similarly for J_0; exponentiations of $h_i^{J_j}$ are already calculated and included in Aux.

- $Encrypt(MK_P, ID, m)$: To encrypt a message $m \in \mathbb{G}_1$ under the public key $ID = (I_1, \dots, I_k)$ draw uniformly at random $s \in \mathbb{Z}_p$ and output:

$$CT = (e(g_1, g_2)^s \cdot m, g^s, (h_1^{I_1} \cdots h_k^{I_k} \cdot g_3)^s) \in \mathbb{G}_1 \times \mathbb{G}^2$$

Which takes 1 pairing computation, 3 exponentiations and $O(\ell)$ multiplications (we can also add $e(g_1, g_2)$ to MK_P in order to avoid computing pairings in the encryption).

- $Decrypt(SK_{ID}, CT)$: Consider a ciphertext $CT = (A, B, C)$ encrypted for $ID = (I_1, \dots, I_k)$ where the private key is $SK_{ID} = (a_0, a_1, b_{k+1}, \dots, b_\ell)$. Output:

$$A \cdot \frac{e(a_1, C)}{e(B, a_0)} = e(g_1, g_2)^s \cdot m \cdot \frac{e(g^r, (h_1^{I_1} \cdots h_k^{I_k} \cdot g_3)^s)}{e(g^s, g_2^\alpha \cdot (h_1^{I_1} \cdots h_k^{I_k} \cdot g_3)^r)} =$$

$$= e(g_1, g_2)^s \cdot m \cdot \frac{1}{e(g, g_2)^{s\alpha}} = m$$

Which takes only two pairing computations and one multiplication.

This HIBE achieves selective-ID security for both chosen plaintext and chosen ciphertext attacks (IND-sID-CPA and IND-sID-CCA respectively) under the ℓ-weak decisional Bilinear Diffie-Hellman Inversion assumption (ℓ-wBDHI, see definition in [6]) in the standard model and is fully secure in the random oracle model, where ℓ is the number of levels of the hierarchy.

Performance. As for the performance of the resulting PSR, the setup algorithm's running time is dominated by the generation of the set of private keys K which is of size $O(r \log \frac{|U|}{r})$. In order to provide proofs of non-membership, the *secondaries* have to decrypt a message intended for an identity of depth ℓ, for which they have to first generate a proper key. This takes 4 exponentiations and $O(\ell)$ multiplications. The *secondaries* then decrypt the message, which takes 2 pairing computations and one multiplication. For a *resolver* to issue a query for an element it has to encrypt one message which takes 3 exponentiations and $O(\ell)$ multiplications (we avoid the pairing computation in the encryption by adding $e(g_1, g_2)$ to MK_P).

So in total a *secondary* has to do at most 2 pairing computations, 4 exponentiations and $O(\ell)$ multiplications, while a *resolver* has to do only 3 exponentiations and $O(\ell)$ multiplications. As mentioned before, we can also have a variant of the protocol where the *resolvers* receive the secret key itself (and have them encrypt and decrypt random challenges by themselves). This moves the computational load of 2 pairing computations to the *resolvers*. The *primary* has to work harder as the setup algorithm is more costly, but that is understandable as the *primary* has to set up the system only once.

5   PSR System Constructions

In the full version of this paper [34] we present two additional strategies for constructing PSR systems and another construction which follows the lines of the HIBE construction but uses one-time signatures. We describe them here informally, where the full version contains a more comprehensive and formal treatment of these constructions.

5.1   Using One-Time Signatures

One-time signatures are signatures with a very weak security/unforgeability requirement, where an adversary who witnesses at most one signature of its choice cannot forge a signature, which will be verified successfully. We utilize the same strategy we used for the HIBE construction (Section 4), with the difference of using one-time signatures to produce a chain of signatures from the root of a binary tree to the leaf corresponding to the queried element (again where a *secondary* cannot generate this proof for elements in the set R). The chain of signatures consists of public keys corresponding to the nodes along the path, signed using the secret key of their parents. In the full version of this paper [34] we prove this construction is a non-interactive PSR system and has perfect ZK.

Now as we can construct both types of signatures (one-time and regular) from universally one way hash functions (UOWHF) [33], we can conclude that the existence of UOWHFs implies the existence of PSR systems with perfect ZK. UOWHFs in turn can be constructed from one-way functions [39]. PSR systems imply identification schemes, as shown in our companion paper [19], which in turn imply the existence of one-way functions, as shown by Impagliazzo and Luby [24] (see also [23]).

Thus the point of this construction is not efficiency, but to use this black box constructions to prove the following corollary:

Corollary 1. *Single round PSR systems exist if and only if one-way functions exist. If many rounds PSR systems exist then a single round PSR system exists.*

This also gives us a separation result from ZKS [29], since Chase et al. [12] proved that interactive ZKS and collision resistant hash functions (CRH) are existentially equivalent and Simon [43] showed a separation result, which states that no CRH can be constructed from one-way functions (or even permutations) in a black box manner. Thus we get the following corollary:

Corollary 2. *One cannot construct ZKS (and even interactive ZKS) in a black box manner from PSR systems (interactive or not).*

5.2   Using Cuckoo Hashing with a Stash

We now discuss an instantiation of the second approach for constructing PSRs mentioned in the introduction, imitating an oblivious search for the element, where the locations examined are determined by the element searched and some hash functions. The point is that the *secondary* needs to show that the searched element is in none of the probed locations.

Cuckoo Hashing is a scheme first introduced by Pagh and Rodler [36]. If we have a set $|R| = r$ for which we want to prove (non) membership, we use two tables T_1 and T_2 of size $(1 + \varepsilon)r$ (where ε is constant) and two hash functions (F_1, F_2), which map elements in the universe into those two tables. Every element $x \in R$ is placed in either location $F_1(x)$ in table T_1 or location $F_2(x)$ in table T_2. This off course may fail for the choice of some functions (F_1, F_2) (with probability $O(\frac{1}{r})$), thus we also use a stash, to store elements we could not place in the cuckoo hash tables due to collisions. Kirsch, Mitzenmacher and Wieder [26] show that the probability that the stash is larger than s is bounded by $O(r^{-s})$. This helps us bound the amount of information that leaks on the set R, by the choice of the functions (F_1, F_2). In order to prove $x \notin R$ we need to show that x was not placed in the stash and not in either of the two possible locations in the cuckoo hash tables.

In order to prove non-membership in the tables we use commitment schemes (see [20] for definitions) with inequality proofs, where we require the commitments to be: *hiding*, so that commitments to two different values are identically distributed, and *binding*, so that even the commiter cannot open a commitment to a value, different than the committed value. We also want the proofs of inequality to be complete (honest execution results in correct conclusions), sound (commiter can't cheat) and have indistinguishability between two proofs of inequality, i.e. proving the inequality of x to two commitments to elements different than x is indistinguishable. In order to prove an element was not placed in the stash we use a scheme for proving non-membership in a fixed set, from which we require the exact same conditions as we require from commitments, with the difference that we need to commit to a set of elements instead of a single element.

We chose to use Pedersen commitments [37] with ZK proofs of inequality. The inequality proofs use ZK proofs of equality for Pedersen commitments (based on the adaptation of Schnorr's identification protocol [41]) and the ZK proofs of inequality for discrete logarithms, suggested by Camenisch and Shoup [9]. In order to construct a scheme to prove non-membership in a fixed set (our stash S), we use a generalization of the Feige-Fiat-Shamir identification protocol [15] combined with the set lower bound technique of Goldwasser and Sipser [21], to allow *secondaries* to prove they know a *large* fraction of the secrets (corresponding to the queried element) as opposed to knowing none of them (when the queried element is placed in the stash). Every element in the universe is mapped to n challenges and the *primary* distributes the corresponding secret to every challenge that doesn't correspond to an element in the stash S. This way we get that for every $x \in S$ the *secondaries* know none of the secrets, but they know a large fraction of the secrets for every element $x \notin S$.

All and all we get an interactive PSR system which leaks the cardinality of the set R and is quite efficient. The denial-of-existence mechanism we described requires a constant number of exponentiations for both parties (9 for the *secondary* and 8 for the *resolver*) in order to prove inequalities for Pedersen's commitments and at most $n = \log |U|$ modular multiplications and a Gaussian elimination process (for a matrix of size $\frac{n}{4} \times \frac{n}{3}$), for the fixed set non-membership proof system, suggested to implement the stash. Its great advantage is that it uses very conservative and well studied assumptions: factoring (for the Feige-Fiat-Shamir protocol) and discrete logarithm (for the Pedersen commitments).

5.3   Using Verifiable Random Looking Functions

In the full version of this paper [34] we show a few constructions for PSR systems based on variants of *Verifiable Random/Unpredictable Functions* [30] (VRF and VUF for short), a construction that uses *Pseudorandom Functions with interactive ZK proofs* and discuss constructions in the random oracle model. All these constructions employ the same strategy which uses functions that map elements in the universe to some large domain $\{0, 1\}^m$, where a *secondary*, holding a secret key, can prove to a *resolver*, holding a public key, the value of $F(x)$. Another important property we require from our functions is to appear random, in the sense that an adversary (without knowledge of the secret key) who knows the set R, cannot distinguish between the set of values $F(R) = \{F(x) | x \in R\}$ and a set of random values $\{r_i | i \in [r] : r_i \overset{R}{\in} \{0, 1\}^m\}$, even after a series of queries to the function (which do not include queries to elements in R)[5].

The PSR system itself does the following: the *primary* computes the values of the function over the set R and arranges them in lexicographical order $y_1, \ldots, y_r$. Next it signs all the couples of adjacent values and gives the signatures to the *secondaries*: $Sign(y_i, y_{i+1})$ (adding an opening and a closing value 0^m and 1^m). Now in order to prove non-membership for an element $x \notin R$ the *secondary* simply computes $F(x)$ and finds and index i for which $y_i < F(x) < y_{i+1}$, proves to the *resolver* that it computed $F(x)$ honestly and sends the signature $Sign(y_i, y_{i+1})$. The *resolver* is convinced after verifying that $F(x)$ was computed correctly and that its value is truly in between two values of the set R, precomputed by the *primary* (i.e. it verifies that $y_i < F(x) < y_{i+1}$).

This construction leaks the size of the set R and can be instantiated using different implementations for the function F, thus resulting in different efficiency and PSR systems which are based on different cryptographic assumptions. The VRF and VUF constructions are non-interactive and can be implemented using the constructions of [22,8,1,25] (VRF) and [16,25] (VUF) in the standard model. As a PRF with interactive ZK proofs we suggest the Naor-Reongold PRF [32]. In the random oracle model we can get very efficient implementations by using functions comprised of the BLS signature scheme [7] and random oracles or the NSEC5 construction suggested in our companion paper [19].

---

[5] Both VRFs and PRFs achieve this property naturally by their definitions, while we need to modify VUFs a bit in order to get this property.

6   Conclusions and Future Directions

We introduced PSR systems and presented three general strategies for constructing them, with different implementations for the underlying primitives. Our focus in this paper was on trying to find efficient constructions, based on solid cryptographic assumptions. A construction can be measured by a few standards: efficiency, the underlying cryptographic assumptions and the ZK requirement (for which f does the f-ZK requirement hold and whether it is computational, statistical or perfect ZK). There is no clear overall winner that dominates in all criteria.

If the (null f) ZK property is critical (e.g. in case the *primary* does not want to reveal the size of the set), then the HIBE construction (Section 4) and the signature based PSR (Section 5.1) both achieve perfect f-ZK, where f is the null function. Both schemes are one-round PSRs and hence they are also secure in a concurrent setting (as proved in Theorem 2). The rest of the constructions reveal the size of the set R and do not achieve perfect ZK. The HIBE construction by Boneh et al. is efficient (Section 4.4), as *secondaries* and *resolvers* use only $O(\log|U|)$ group multiplications and a constant number of pairing computations and modular exponentiations for their computations. It is based on the $O(\log|U|)$-weak decisional Bilinear Diffie-Hellman Inversion assumption[6] (ℓ-wBDHI, see definition in [6]). The downside for this scheme is the computational load on the *primary*, which has to compute keys for $O(|R|\log\frac{|U|}{|R|})$ nodes, which may result in superlinear time for generating the scheme's keys, but at least it is only executed once.

Our cuckoo hash based PSR construction (Section 5.2) offers both an appealing technique and an efficient implementation, based on very solid and well studied cryptographic assumptions: factoring and the discrete logarithm (defined in the original papers [15] and [37] respectively). If the security of the PSR is the most important thing for its users (e.g. a database containing top secret information), it makes sense to use the cuckoo hashing construction as it is based on two very well studied assumptions and has the statistical ZK property, which gives us everlasting privacy (see Remark 2). This technique's efficiency depends on the implementations of the commitment scheme and the fixed set non-membership, which using the implementations we suggest (see full version of the paper for exact details [34]) results in the *resolvers* and *secondaries* doing a constant number of modular exponentiations and $O(\log|U|)$ modular multiplications, which is about as efficient as the HIBE construction asymptotically.

Our PSR based on random looking functions (Section 5.3) reveals the size of the set R, but has the potential of being very efficient if we can construct a VRF/VUF which is both efficient and secure. We would like to use such a function which is secure for large domains but can be evaluated and verified with, say, a constant number of operations ([13] is that efficient but lacks security), as *secondaries*

---

[6] Boneh et al. prove this assumption holds in the generic group model [42]. This means that using generic algorithms (ones that don't exploit any special properties of the group elements' encodings), one cannot construct a polynomial time algorithm to break the assumption, which is an encouraging result towards using this assumption.

only have to evaluate the function on the queried element and generate its proof, while the *resolvers* verify the value and one signature. We note that the four secure VRFs [22,8,1,25] are not a lot less efficient than the HIBE construction. If we can implement an efficient division intractable hashing family then we can use the GHR signature scheme [16] to implement the random looking function, which is highly efficient, as computing and verifying each value requires one hash computation and one modular exponentiation. These constructions also have the added advantage of being non-interactive, which also makes them concurrently secure.

If one is willing to live with random oracles, then this technique yields very efficient PSR systems. Both the BLS signature scheme [7] based PSR and the NSEC5 construction described in our companion paper [19] are very efficient, while the first relies on a gap Diffie-Hellman group (see definition in the original paper [7]) and the latter on the RSA hardness assumption (see definition in [19]).

The implementations proposed are fairly efficient, but undoubtedly it is possible to optimize them or come up with other ones. In terms of readiness to deployment, i.e. whether the implementations are mature, then probably HIBE is the best bet unless one is willing to trust random oracles in which case both the BLS and NSEC5 schemes are good.

Acknowledgments. We thank our co-authors from [19], Sharon Goldberg, Dimitrios Papadopoulos, Leonid Reyzin and Sachin Vasant for many helpful discussions and Yevgeniy Dodis for suggesting the question of whether single-round PSRs can be based on one-way functions. We thank Pavel Hubáček for carefully reading the paper.

References

1. Abdalla, M., Catalano, D., Fiore, D.: Verifiable random functions: Relations to identity-based key encapsulation and new constructions. J. Cryptology 27(3), 544–593 (2014)
2. Agrawal, S., Boneh, D., Boyen, X.: Efficient lattice (H)IBE in the standard model. In: Gilbert, H. (ed.) EUROCRYPT 2010. LNCS, vol. 6110, pp. 553–572. Springer, Heidelberg (2010)
3. Agrawal, S., Boneh, D., Boyen, X.: Lattice basis delegation in fixed dimension and shorter-ciphertext hierarchical IBE. In: Rabin, T. (ed.) CRYPTO 2010. LNCS, vol. 6223, pp. 98–115. Springer, Heidelberg (2010)
4. Bau, J., Mitchell, J.C.: A security evaluation of DNSSEC with NSEC3. In: NDSS 2010, The Internet Society (2010)
5. Benaloh, J.C., de Mare, M.: One-way accumulators: A decentralized alternative to digital signatures. In: Helleseth, T. (ed.) EUROCRYPT 1993. LNCS, vol. 765, pp. 274–285. Springer, Heidelberg (1994)
6. Boneh, D., Boyen, X., Goh, E.-J.: Hierarchical identity based encryption with constant size ciphertext. In: Cramer, R. (ed.) EUROCRYPT 2005. LNCS, vol. 3494, pp. 440–456. Springer, Heidelberg (2005)
7. Boneh, D., Lynn, B., Shacham, H.: Short signatures from the weil pairing. J. Cryptology 17(4), 297–319 (2004)
8. Boneh, D., Montgomery, H.W., Raghunathan, A.: Algebraic pseudorandom functions with improved efficiency from the augmented cascade. In: ACM CCS 2010, pp. 131–140. ACM (2010)

9. Camenisch, J., Shoup, V.: Practical verifiable encryption and decryption of discrete logarithms. In: Boneh, D. (ed.) CRYPTO 2003. LNCS, vol. 2729, pp. 126–144. Springer, Heidelberg (2003)
10. Canetti, R.: Universally composable security: A new paradigm for cryptographic protocols. IACR Cryptology ePrint Archive 2000, 67 (2000)
11. Catalano, D., Fiore, D., Messina, M.: Zero-knowledge sets with short proofs. In: Smart, N.P. (ed.) EUROCRYPT 2008. LNCS, vol. 4965, pp. 433–450. Springer, Heidelberg (2008)
12. Chase, M., Healy, A., Lysyanskaya, A., Malkin, T., Reyzin, L.: Mercurial commitments with applications to zero-knowledge sets. In: Cramer, R. (ed.) EUROCRYPT 2005. LNCS, vol. 3494, pp. 422–439. Springer, Heidelberg (2005)
13. Dodis, Y., Yampolskiy, A.: A verifiable random function with short proofs and keys. In: Vaudenay, S. (ed.) PKC 2005. LNCS, vol. 3386, pp. 416–431. Springer, Heidelberg (2005)
14. Dwork, C., Naor, M., Sahai, A.: Concurrent zero-knowledge. In: ACM 1998, pp. 409–418. ACM (1998)
15. Feige, U., Fiat, A., Shamir, A.: Zero-knowledge proofs of identity. J. Cryptology 1(2), 77–94 (1988)
16. Gennaro, R., Halevi, S., Rabin, T.: Secure hash-and-sign signatures without the random oracle. In: Stern, J. (ed.) EUROCRYPT 1999. LNCS, vol. 1592, pp. 123–139. Springer, Heidelberg (1999)
17. Gentry, C., Silverberg, A.: Hierarchical id-based cryptography. In: Zheng, Y. (ed.) ASIACRYPT 2002. LNCS, vol. 2501, pp. 548–566. Springer, Heidelberg (2002)
18. Ghosh, E., Ohrimenko, O., Tamassia, R.: Verifiable member and order queries on a list in zero-knowledge. IACR Cryptology ePrint Archive 2014, 632 (2014)
19. Goldberg, S., Naor, M., Papadopoulos, D., Reyzin, L., Vasant, S., Ziv, A.: NSEC5: provably preventing DNSSEC zone enumeration. IACR Cryptology ePrint Archive 2014, 582 (2014)
20. Goldreich, O.: The Foundations of Cryptography - Volume 1, Basic Techniques. Cambridge University Press (2001)
21. Goldwasser, S., Sipser, M.: Private coins versus public coins in interactive proof systems. In: ACM 1986, pp. 59–68. ACM (1986)
22. Hohenberger, S., Waters, B.: Constructing verifiable random functions with large input spaces. In: Gilbert, H. (ed.) EUROCRYPT 2010. LNCS, vol. 6110, pp. 656–672. Springer, Heidelberg (2010)
23. Impagliazzo, R.: Pseudo-random generators for cryptography and for randomized algorithms. Ph.D. thesis, University of California, Berkeley (1990)
24. Impagliazzo, R., Luby, M.: One-way functions are essential for complexity based cryptography (extended abstract). In: FOCS 1989, pp. 230–235. IEEE Computer Society (1989)
25. Jager, T.: Verifiable random functions from weaker assumptions. IACR Cryptology ePrint Archive 2014, 799 (2014)
26. Kirsch, A., Mitzenmacher, M., Wieder, U.: More robust hashing: Cuckoo hashing with a stash. SIAM J. Comput. 39(4), 1543–1561 (2009)
27. Lewko, A., Waters, B.: Why proving HIBE systems secure is difficult. In: Nguyen, P.Q., Oswald, E. (eds.) EUROCRYPT 2014. LNCS, vol. 8441, pp. 58–76. Springer, Heidelberg (2014)
28. Libert, B., Yung, M.: Concise mercurial vector commitments and independent zero-knowledge sets with short proofs. In: Micciancio, D. (ed.) TCC 2010. LNCS, vol. 5978, pp. 499–517. Springer, Heidelberg (2010)
29. Micali, S., Rabin, M.O., Kilian, J.: Zero-knowledge sets. In: FOCS 2003, pp. 80–91. IEEE Computer Society (2003)

30. Micali, S., Rabin, M.O., Vadhan, S.P.: Verifiable random functions. In: FOCS 1999, pp. 120–130. IEEE Computer Society (1999)

31. Naor, D., Naor, M., Lotspiech, J.: Revocation and tracing schemes for stateless receivers. In: Kilian, J. (ed.) CRYPTO 2001. LNCS, vol. 2139, pp. 41–62. Springer, Heidelberg (2001)

32. Naor, M., Reingold, O.: Number-theoretic constructions of efficient pseudo-random functions. In: FOCS 1997, pp. 458–467. IEEE Computer Society (1997)

33. Naor, M., Yung, M.: Universal one-way hash functions and their cryptographic applications. In: ACM 1989, pp. 33–43. ACM (1989)

34. Naor, M., Ziv, A.: Primary-secondary-resolver membership proof systems. IACR Cryptology ePrint Archive 2014, 905 (2014)

35. Ostrovsky, R., Rackoff, C., Smith, A.: Efficient consistency proofs for generalized queries on a committed database. IACR Cryptology ePrint Archive 2004, 170 (2004)

36. Pagh, R., Rodler, F.F.: Cuckoo hashing. In: Meyer auf der Heide, F. (ed.) ESA 2001. LNCS, vol. 2161, pp. 121–133. Springer, Heidelberg (2001)

37. Pedersen, T.P.: Non-interactive and information-theoretic secure verifiable secret sharing. In: Feigenbaum, J. (ed.) CRYPTO 1991. LNCS, vol. 576, pp. 129–140. Springer, Heidelberg (1992)

38. Prabhakaran, M., Xue, R.: Statistically hiding sets. In: Fischlin, M. (ed.) CT-RSA 2009. LNCS, vol. 5473, pp. 100–116. Springer, Heidelberg (2009)

39. Rompel, J.: One-way functions are necessary and sufficient for secure signatures. In: ACM 1990, pp. 387–394. ACM (1990)

40. Rosen, A.: Concurrent Zero-Knowledge - With Additional Background by Oded Goldreich. In: Information Security and Cryptography. Springer (2006)

41. Schnorr, C.-P.: Efficient identification and signatures for smart cards. In: Brassard, G. (ed.) CRYPTO 1989. LNCS, vol. 435, pp. 239–252. Springer, Heidelberg (1990)

42. Shoup, V.: Lower bounds for discrete logarithms and related problems. In: Fumy, W. (ed.) EUROCRYPT 1997. LNCS, vol. 1233, pp. 256–266. Springer, Heidelberg (1997)

43. Simon, D.R.: Findings collisions on a one-way street: Can secure hash functions be based on general assumptions? In: Nyberg, K. (ed.) EUROCRYPT 1998. LNCS, vol. 1403, pp. 334–345. Springer, Heidelberg (1998)

Tight Parallel Repetition Theorems
for Public-Coin Arguments Using KL-Divergence

Kai-Min Chung[1],[*] and Rafael Pass[2],[**]

[1] Academia Sinica, Taiwan
kmchung@iis.sinica.edu.tw
[2] Cornell University, USA
rafael@cs.cornell.edu

Abstract. We present a new and conceptually simpler proof of a tight parallel-repetition theorem for public-coin arguments [Pass-Venkitasubramaniam, STOC'07], [Håstad et al, TCC'10], [Chung-Liu, TCC'10]. We follow the same proof framework as the previous non-tight parallel-repetition theorem of Håstad et al—which relied on *statistical distance* to measure the distance between experiments—and show that it can be made tight (and further simplified) if instead relying on *KL-divergence* as the distance between the experiments.

We then use this new proof to present the first tight "Chernoff-type" parallel repetition theorem for arbitrary public-coin arguments, demonstrating that parallel-repetition can be used to simultaneously decrease both the soundness and completeness error of *any* public-coin argument at a rate matching the standard Chernoff bound.

1 Introduction

Ideally, we would like the soundness error of an interactive proof [GMR89, BM88] or argument systems [BCC88] to be negligible. But, in many settings, our starting point is a protocol with somewhat large soundness error. For example, to design an interactive argument for a language L, it may be easier to first design a protocol with soundness error $1/2$. This leads to the question of *soundness amplification*: How can we to decrease the soundness error of a given protocol? (Ideally, we would like to simultaneously decrease both the soundness and completeness error; we return to this question shortly.) A natural approach to performing such soundness amplification is through a *direct product theorem*: Roughly speaking, a direct product theorem for a class of problems states that if an adversary can

[*] Chung is supported in part by NSF Award CNS-1217821, NSF Award CCF-1214844, Pass' Sloan Fellowship, and Ministry of Science and Technology MOST 103-2221-E-001-022-MY3; part of this work was done while being at Cornell University.
[**] Pass is supported in part by an Alfred P. Sloan Fellowship, Microsoft New Faculty Fellowship, NSF CAREER Award CCF-0746990, NSF Award CCF-1214844, NSF Award CNS-1217821, AFOSR YIP Award FA9550-10-1-0093, and DARPA and AFRL under contract FA8750-11-2-0211.

Y. Dodis and J.B. Nielsen (Eds.): TCC 2015, Part II, LNCS 9015, pp. 229–246, 2015.

solve an instance of a problem with probability at most δ, then his chance of solving multiple independent instances should decrease exponentially, ideally to δ^k, if we have k independent instances. For the case of interactive proofs/arguments, the two most natural ways of running several instances are *sequential repetition* and *parallel repetition*. In the case of sequential repetitions, we run k instances of some underlying protocol sequentially, one after the other, and the verifier finally accepts if all instances were accepted. It is well-known that sequential repetition decreases the soundness error of both interactive proofs and arguments at an essentially optimal rate; see [BM88, Gol01, DP98]. However, sequential repetition increases the number of communication rounds of the original protocol. In the case of parallel repetition, we instead run the k protocols in parallel. It is known that parallel repetition decrease soundness error at an optimal rate for the case of interactive proofs (i.e., for the case of statistical soundness). For arguments (i.e., computational soundness), however, surprising things start happening: The seminal work of Bellare, Impagliazzo, and Naor [BIN97] demonstrate protocols for which parallel repetition fails to amplify soundness *at all*! These counter examples, however, are for *private-coin* protocols.

On the other hand, for the case of *public-coin* protocols, parallel repetition theorems have been established: Pass and Venkitasubramaniam [PV07] first showed a tight parallel repetition theorem for constant-round protocols. Håstad, Pass, Wikström and Pietrzak [HPWP10] next extended it to arbitrary (i.e., not necessarily constant-round) protocols; the rate at which the soundness decreases, however, was no longer optimal—roughly speaking, when $\delta = (1 - \mu)$, k repetitions decreases the error to $e^{-\Omega(\mu^2 k)}$ as opposed to $\delta^k = e^{-\Omega(\mu k)}$. Finally, Chung and Liu presented an optimal parallel repetition theorem—where k repetitions sufficed to decrease the error to δ^k. A more comprehensive survey of known parallel repetition theorems can be found in Section 1.3.

1.1   Our Results

A New Proof of Tight Parallel Repetition for Public-coin Protocols In this work, we revisit the result of Chung and Liu. Our central contribution is a new proof of their tight parallel repetition theorem. Our proof follows the same framework as the "simple" proof of the *non-tight* parallel-repetition theorem of Håstad et al—which relied on statistical distance to measure the distance between experiments—and shows that it can be made tight (and further simplified) if instead relying on *KL-divergence*[1] as the distance between the experiments. (KL-divergence was previously instrumental for proving tight parallel theorems for two-prover games [BRR+09] in an *information-theoretic* setting. Our new proof demonstrates that also in the computational setting, KL-divergence appears to be the right measure of distance when analyzing reductions through hybrid experiments.) As such, our proof significantly simplifies and "demystifies" the proof of [CL10], which directly analyzed the success probability through an intricate

---

[1] Recall that $\mathbf{KL}(X||Y) = \sum_{x \in \mathrm{supp}(X)} \Pr[X = x] \cdot \log \frac{\Pr[X=x]}{\Pr[Y=x]}$. For convenience, here we define KL-divergence with log base e. The choice is inconsequential.

inductive argument relying on Holder's inequality, providing little intuition for "why" the reduction works.

Additionally, as we now turn to discussing, our new proof enables considering more general scenarios (whereas the analysis in [CL10] is explicitly set up to analyze the particular direct-product case), and we believe this technique may be useful more broadly.

Tight Chernoff-type Parallel-repetition Theorem for Any Public-coin Protocols. So far we have only discussed *direct-product* parallel repetition theorems, where the parallel verifier accepts iff all parallel sessions are accepting. If the starting protocol also has a large completeness error, then parallel repetition with a "direct product" verifier also increases the completeness error. Ideally, we would like to have a way to simultaneously decrease both the completeness and the soundness error: just as for error reduction of the class **BPP**, the idea is to consider a *threshold verifier*, who accept whenever the fraction of accepting sessions is greater than a certain threshold γ (that is greater than the soundness error δ, or else there is no hope to reduce the soundness error). For error reduction of **BPP**, it follows by a standard Chernoff bound that such an approach works. For interactive arguments, such "Chernoff-type" parallel repetition theorems where first studied by Impagliazzo, Jaiswal, and Kabanets [IJK09] for the case of three-message protocols. Håstad et al. [HPWP10] extend the results of [IJK09] also to public-coin protocols and Chung and Liu [CL10] improved the error decrease rate obtaining "tight" Chernoff-type parallel repetition theorems (matching the parameters of the standard Chernoff bound)[2] in two setting:

- For the case of constant-round protocols (by relying on the direct product parallel repetition theorem of [PV07]).
- When the gap between the threshold γ and the soundness error δ is a constant (i.e., $\gamma - \delta = \Omega(1)$); this is done by relying on a generic reduction to the direct product case, which is only efficient when the gap is a constant.

In particular, for non-constant round protocols, previous result only enabled simultaneously decreasing the completeness and soundness error at a rate matching the standard Chernoff bound whenever the gap between the soundness error and "1-the completeness error" is a constant (as opposed to it being an inverse polynomial). We show that using our new proof technique, the analysis for the direct product case directly extends also to the case of threshold verifiers, yielding a Chernoff-type parallel repetition theorem for *any* public-coin protocol, and *any* threshold $\gamma > \delta$.

Specifically, we demonstrate the following theorem, which matches a "**KL**-version" Chernoff bound, and directly implies both tight direct product theorems and tight Chernoff-type theorems.

Theorem 1 (informal). *For a public-coin interactive argument with soundness error $\delta \in (0,1)$, k-fold parallel repetition with threshold $\gamma > \delta \in (0,1)$ decreases*

---

[2] Also the standard Chernoff bound is not "optimal" so we content ourselves with matching the parameters of the standard Chernoff bound.

soundness error to $e^{-k \cdot \mathbf{KL}(\gamma||\delta)} +$ ngl, *where* ngl *is a negligible function in the security parameter.*[3] *In particular,*[4]

- *For threshold* $\gamma = 1$ *(the direct product setting), the soundness error is decreased to* $\delta^k +$ ngl.
- *For threshold* $\gamma = (1 + \mu)\delta$ *(the "multiplicative" Chernoff-bound setting), the soundness error is decreased to* $e^{-\Omega(\mu^2 \delta k)} +$ ngl *for* $\mu \in (0, 1)$ *and* $e^{-\Omega(\mu \delta k)} +$ ngl *for* $\mu > 1$.

1.2   Proof Overview

We now explain our new proofs of a tight parallel repetition theorem for public-coin protocols. For simplicity of exposition, we start by focusing on the direct product case, and next extend the analysis to deal with threshold verifiers.

We first set-up some notation. Let us consider a public-coin protocol (P, V) with m rounds, where at each round $j \in [m]$, the verifier V sends a uniformly random message x_j to P, receives back a second-message y_j, and at the end *deterministically* decides to accept or reject based on the transcript $(x_1, y_1, \ldots, x_m, y_m)$. We denote by (P^k, V^k) the k-fold parallel repetition of (P, K); here V^k sends a message $\boldsymbol{x} = (x_{j,1}, \ldots, x_{j,k})$, receives back a message $\boldsymbol{y} = (y_{j,1}, \ldots, y_{j,k})$ at each round $j \in [m]$, and accepts iff $(x_{1,i}, y_{1,i}, \ldots, x_{m,i}, y_{m,i})$ is accepting for every instance $i \in [k]$. We refer to the different parallel executions of the protocol (P, V) inside (P^k, V^k) as the parallel *sessions*.

To prove that parallel repetition reduces the soundness error, we show how to transform any *parallel prover* P^{k*} that convinces V^k with probability $\epsilon \geq 1.1\delta^k$ to a *single-instance* prover P^* that convinces V with probability at least δ. This implies that parallel repetition reduces the soundness error at an essentially optimal rate (from δ to $1.1\delta^k$). We may without loss of generality assume that P^{k*} is deterministic—its optimal random coins can always be fixed non-uniformly.[5]

More precisely, P^* will internally emulate an execution of P^{k*} and use this execution in order to convince an external verifier V. On a high-level, the general strategy is quite straight forward. P^* picks one of the k sessions, i; this session will be externally forwarded (between P^{k*} and V), and all the other sessions, $-i$, will be appropriately emulated internally. In other words, the external verifier V is "embedded" in some session i of V^k, and P^* internally emulates P^{k*} and the remaining $k - 1$ sessions $-i$ of V^k while forwarding P^{k*}'s messages $y_{j,i}$'s for session i to V; see Figure 1 for an illustration for the case of a one-round protocol.

Recall that since we have assumed that P^{k*} is deterministic, the interaction between P^{k*} and V^k is determined solely by V^k's message $\boldsymbol{x}_1, \ldots, \boldsymbol{x}_m$, where

---

[3] As shown by [DJMW12], under some cryptographic assumptions, the additive negligible term is necessary.

[4] For the direct product setting, it follows by the fact that $\mathbf{KL}(1||\delta) = \log(1/\delta)$. For the Chernoff-bound setting, it follows by the fact that $\mathbf{KL}((1 + \mu)\delta||\delta) = \Theta(\mu^2 \delta)$ for $\mu \in (0, 1)$, and $\mathbf{KL}((1 + \mu)\delta||\delta) = \Theta(\mu \delta)$ for $\mu > 1$.

[5] Alternatively, "close to optimal" coins can be uniformly fixed by sampling.

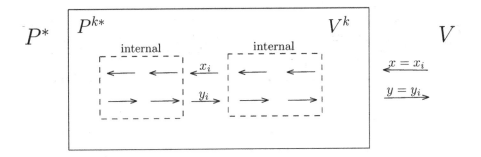

Fig. 1. Interaction between P^* and V for a one-round protocol: P^* embeds the external verifier V in session i of V^k and internally emulates P^{k*} and the remaining $k-1$ sessions $-i$ of V^k while forwarding P^{k*}'s message y for session i to V

each $\boldsymbol{x}_j = (x_{j,1}, \dots, x_{j,k})$. Thus, P^* needs to decide *the session i* to embed V at beginning, and then at each round j, given an external message $x_{j,i}$, to *choose the remaining $k - 1$ messages* $\boldsymbol{x}_{j,-i}$. We now recall the *rejection sampling* strategy of [HPWP10].

The Rejection Sampling Strategy. We consider a rejection sampling prover, P^*_{rej}, that selects the session $i \in [k]$ uniformly at random, and then at each round j, upon receiving the external verifier V's message $x_{j,i}$, P^*_{rej} selects $\boldsymbol{x}_{j,-i}$ using rejection sampling as follows: P^*_{rej} repeatedly samples a random continuation of (P^{k*}, V^k) (i.e., samples uniformly random $\boldsymbol{x}_{j,-i}, \boldsymbol{x}_{j+1} \dots, \boldsymbol{x}_m)^6$ until it finds an *accepting continuation*, i.e., V^k accepts at the end of interaction (or a certain a-prior bound M on the number of samples is reached, in which case P^* aborts and fails). Then, P^* selects the corresponding messages in the accepting continuation as the messages of V_{-i} at round j.

To analyze the success probability of P^*_{rej}, let us first allow P^*_{rej} to make an unbounded number of samples (i.e., set $M = \infty$). Note that in this case, at each round j, P^*_{rej} simply selects $\boldsymbol{x}_{j,-i}$ conditioned on P^{k*} convincing V^k. See Figure 2 for an illustration. As we shall see, if P^{k*} convinces V^k with probability ϵ, then P^* convinces V with probability $\geq \epsilon^{1/k} > \delta$. We then deal with the bounded-sample case at the end (looking forward, as long as we make poly$(1/\epsilon)$ queries, having such a cut-off only slightly affects the success probability of P^*).

The main idea for analyzing (the unbounded sample version of) P^* is to consider an Ideal experiment, where P^* succeeds with probability 1 and next show that the actual execution of (P^*, V), referred to as the Real experiment, and the Ideal experiment are close (using an appropriate choice of a distance measure), from which we can conclude that P^* succeeds with high probability in the Real experiment. Let us start by formalizing the Real experiment.

---
^{6} Note that here we use the fact that the protocol is public coin so that sampling a random continuation is simply sampling uniformly random $\boldsymbol{x}_{j,-i}, \boldsymbol{x}_{j+1} \dots, \boldsymbol{x}_m$.

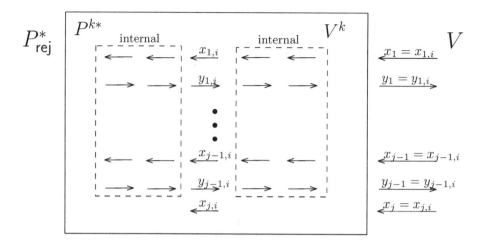

Fig. 2. Interaction between P^*_{rej} and V

The Real *Experiment.* Consider an execution of (P^*_{rej}, V) as follows. At beginning, P^*_{rej} selects a random coordinate $i \in [k]$. Then at each round $j \in [m]$, V selects a uniformly random $x_{j,i}$, and P^*_{rej} selects a random $\boldsymbol{x}_{j,-i}$ conditioned on W using rejection sampling (namely, repeatedly samples a random continuation of (P^{k*}, V^k) until it finds an *accepting continuation*, i.e., V^k accepts at the end of interaction, and selects the corresponding $\boldsymbol{x}_{j,-i}$). If no such $\boldsymbol{x}_{j,-i}$ exists, then P^*_{rej} *fails.* P^*_{rej} *succeeds* if it does not fail. The output of the experiment is defined to be $(i, \boldsymbol{x}_1, \ldots, \boldsymbol{x}_m)$.

First, note that to prove that parallel repetition works (at an optimal rate) we need to show that P^* convinces V in the Real experiment with probability at least $\epsilon^{1/k}$. Secondly, observe that an equivalent way of defining the output $(i, \boldsymbol{x}_1, \ldots, \boldsymbol{x}_m)$ of the experiment is as follows: uniformly sample $i \in [k]$, then for each $j \in [m]$, uniformly sample $x_{j,i} \in \{0,1\}^n$, and uniformly sample $\boldsymbol{x}_{-i} \in \{0,1\}^{(k-1)n}$ *conditioned* on P^{k*} convincing V^k.

The Ideal *Experiment.* Let us turn to defining the Ideal experiment. The experiment is defined identically to the Real experiment, except that now we additionally select $x_{j,i}$ conditioned on P^{k*} convincing V^k; that is, uniformly sample $i \in [k]$, then for each $j \in [m]$, uniformly sample $x_{j,i} \in \{0,1\}^n$ *conditioned* on P^{k*} convincing V^k, and uniformly sample $\boldsymbol{x}_{j,-i}$ *conditioned* on $P^{k*}(\boldsymbol{x})$ convincing V^k; again, the output of the experiment is defined to be $(i, \boldsymbol{x})$.

Note that an equivalent way of defining the Ideal experiment is to uniformly sampling $i \in [k]$, and then directly uniformly sample $(\boldsymbol{x}_1, \ldots, \boldsymbol{x}_m)$ conditioned on P^{k*} convincing V^k. Since P^{k*} convinces V^k with positive probability, it thus follows that in the Ideal experiment P^*_{rej} convinces V with probability 1.

Going from Ideal *to* Real. Observe that the only difference between the Real and the Ideal experiments is that at each round $j \in [m]$, in Real $x_{j,i}$'s are sampled uniformly at random, and in Ideal they are sampled at random conditioned on P^{k*} convincing V^k. The following natural approach is taken in [HPWP10].

Consider a set of m "hybrid" experiments, where in H_j, the messages in the first j rounds are selected just as in Ideal (i.e., both $x_{j',i}$ and $\boldsymbol{x}_{j',-i}$ for $j' \leq j$ are sampled conditioned on P^{k*} convincing V^k), and the remaining $m - j$ rounds are selected just as in Real (i.e., for $j' > j$, only $\boldsymbol{x}_{j',-i}$ is sampled conditioned on P^{k*} convincing V^k, but $x_{j',i}$ is uniformly sampled without any conditioning). Clearly $H_0 = $ Real and $H_m = $ Ideal. Furthermore, the only difference between two consecutive hybrids $j-1$ and j is whether $x_{j,i}$ is sampled conditioned on P^{k*} convincing V^k or not, where i is uniformly chosen. To bound the distance between the hybrids, [HPWP10] uses the following version of Raz' Lemma [Raz98].

Lemma 1 (Raz' Lemma [Raz98]). *Let $(H, \boldsymbol{X}) = (H, X_1, \ldots, X_k)$ be independent random variables and W be an event. Then,*

$$\frac{1}{k} \sum_{i=1}^{k} \mathbf{SD}((H, X_i)|_W, (H|_W, X_i)) \leq \sqrt{\frac{\log(1/\Pr[W])}{k}}.$$

Let W be the event that P^{k*} convinces V^k. The lemma directly implies that the statistical distance between any two consecutive hybrids H_{j-1} and H_j is at most $\sqrt{(\log(1/\Pr[W]))/k}$. Thus, by the triangle-inequality, the statistical distance between the Real and the Ideal experiments is at most $m \cdot \sqrt{(\log(1/\Pr[W]))/k}$, which yields a lower bound on the success probability of P^*_{rej} in the Real experiment that suffices to demonstrate that parallel repetition reduces the soundness error at an exponential rate.

However, the bound is not tight for two reasons. First, due to the "hybrid argument" we incur a linear loss in the number of rounds m (thus, to make the soundness error small we need the number of parallel repetitions to grow polynomially with the number of rounds in the protocol). [HPWP10] shows how to avoid the loss of m by proving a "multi-round" version of Raz' Lemma, which avoids the round-by-round hybrid argument. But the bound still does is not tight due to the use of statistical distance to measure the distance between Real and Ideal

KL Divergence as a Distance Measure. The crux of our new proof is to instead use Kullback-Leibler divergence (KL divergence, for short) as a distance measure between the Real and Ideal experiments. In fact, the proof of Raz' Lemma (and also the multi-round version in [HPWP10]) first provides a bound on the KL divergence between the random variables, and then arrives a bound on their statistical distance by relying on Pinsker inequality. The "translation" between KL divergence and statistical distance, however, incurs a quadratic loss. By directly working with KL divergence, we can avoid it.[7] By a calculation very similar to the proof of Raz' lemma (and essentially implicit in [HPWP10]), we directly show the following lemma:

---

[7] A similar phenomena occurred already in the context of parallel repetition for "free" two-prover games; see [BRR+09].

Lemma 2. KL(Ideal||Real) $\leq \frac{\log(1/\Pr[W])}{k}$.

Let us now show how to get a tight lower bound on the success probability of P^* in the Real experiment. Let $\mathsf{Suc_{Real}}$ and $\mathsf{Suc_{Ideal}}$ be indicator variables that indicate, respectively, whether P^* convinces V in the Real and the Ideal experiments.

$$\frac{\log(1/\Pr[W])}{k} \geq \mathbf{KL}(\text{Ideal}||\text{Real}) \geq \mathbf{KL}(\mathsf{Suc_{Ideal}}||\mathsf{Suc_{Real}}) = 1 \cdot \log\frac{1}{\Pr[\mathsf{Suc_{Real}}=1]},$$
(1)

which implies that $\Pr[\mathsf{Suc_{Real}}=1] \geq \epsilon^{1/k}$ since $\Pr[W] = \epsilon$. The second inequality follows since applying the same function to two distributions can only decrease their KL divergence, whereas the last equality follows by the definition of KL divergence and the fact that $\Pr[\mathsf{Suc_{Ideal}}=1] = 1$. This concludes that P^* convinces V with probability at least $\epsilon^{1/k}$ in the Real experiment.

Dealing with Threshold Verifiers. Our analysis directly extends also to yield tight "Chernoff-type" parallel-repetition theorems where we consider a threshold verifier $V^{k,\gamma}$ that accepts iff more that $\gamma \cdot k$ sessions are accepting. Let us consider the same rejection sampling strategy P^*_{rej} that selects a uniform $i \in [k]$, and then at each round $j \in [m]$, samples $\boldsymbol{x}_{j,-i}$ conditioned on P^{k*} convinces $V^{k,\gamma}$ (note that we do not require $V^{k,\gamma}$ accepts the i-th session). Let us also consider the same Real and Ideal experiments as above. For the same reason, we have $\mathbf{KL}(\text{Ideal}||\text{Real}) \leq \frac{\log(1/\Pr[W])}{k}$. The only difference is that in the Ideal experiment, we no longer have that the success probability is 1. However, since $V^{k,\gamma}$ accepts only when $\geq \gamma \cdot k$ sessions accepts, and $i \leftarrow [k]$ is uniform and independent of the transcript $(\boldsymbol{x}_1, \ldots, \boldsymbol{x}_m)$, P^*_{rej} convinces V with probability at least γ in Ideal. An analogous calculation shows that if P^{k*} convinces $V^{k,\gamma}$ with probability $\geq e^{-k \cdot \mathbf{KL}(\gamma||\delta)}$, then P^*_{rej} convinces V with probability at least δ in the Real experiment.

Handling The Bounded-Sample Case. In our analysis so far we have assumed that P^* can make an unbounded number of samples. Let us now show that its success probability is still high even if we impose a polynomial bound M on the number of samples it can make (and thus P^* becomes efficient). Let us first consider the Ideal experiment. The main observation is that, in the Ideal experiment, *in expectation*, P^* only needs to make $1/\epsilon$ samples to pick $\boldsymbol{x}_{j,-i}$ conditioned on $P^{k*}(\boldsymbol{x})$ convincing V^k for every $j \in [m]$ (since the prefix $(\boldsymbol{x}_1, \ldots, \boldsymbol{x}_{j-1}, x_{j,i})$ is also picked conditioned on P^{k*} convincing V^k, and P^{k*} convinces V^k with probability ϵ). Thus, if the allowed number of samples M is sufficiently larger than $1/\epsilon$, then by the Markov inequality, P^* can successfully convince V^k with probability "almost" 1, even if we restrict P^* to use at most M samples.[8] Since the

---

[8] Since $M > 1/\epsilon$, we only get an efficient reduction as long as ϵ is an inverse polynomial. As a consequence, parallel repetition of arguments cannot decrease the soundness error beyond being "negligible". As shown by [DJMW12], under some cryptographic assumptions, this is inherent.

Ideal and the Real experiments are statistically close, this directly yields a lower bound on the success probability of P^* in the Real experiment. But as we saw, working with statistical distance does not give the tight bound. To obtain a tight bound, we again work with KL divergence. Here, the only difference is that $\Pr[\mathsf{Suc}_{\mathsf{Ideal}} = 1]$ is no longer 1, but can be made arbitrarily (inverse polynomially) close to 1 by increasing M. This is sufficient to conclude that $\Pr[\mathsf{Suc}_{\mathsf{Real}} = 1]$ can be made arbitrarily (inverse polynomially) close to $\epsilon^{1/k}$ as well (since the KL divergence of two binary random variables is a "smooth" function of the probabilities of both random variables).

1.3   Related Works: When Parallel Repetition Works

Let us briefly summarize the class of argument systems for which parallel repetitions is known to decrease the soundness error.

- The seminal work of Yao [Yao82] on hardness amplification of one-way functions can be viewed as showing that parallel repetition reduces the soundness error at an optimal rate for all two-message argument systems for which the verifier's decision to accept or reject is a public function of the transcript; that is, the verifier is not employing any secret randomness to decide whether to accept or reject—we refer to such protocols as being *publicly verifiable*. An important special case of publicly-verifiable protocol are *public-coin* protocols (a.k.a. *Arthur-Merlin* protocols [BM88]) where in each round the verifier simply tosses some random coins and directly sends them to the prover (that is, the verifier doesn't employ any secret randomness).
- The seminal work of Bellare, Impagliazzo and Naor [BIN97] was the first one to explicitly study parallel repetition of argument systems and demonstrated that parallel repetition reduces the soundness error for all three-message protocols (not just publicly-verifiable ones). The results of [BIN97] demonstrated that parallel repetition reduces the soundness error of such protocols at an exponential rate, but did not establish an optimal rate (i.e., reducing the soundness error from ϵ to ϵ^k). Nevertheless, the more recent work by Canetti, Halevi and Steiner [CHS05] shows that parallel repetition indeed reduces the soundness error at an optimal rate for this class of protocols.
- [BIN97, PW07] demonstrate 4-message protocols for which parallel repetition fails; in these protocols, the verifier uses "secret randomness". Pass and Venkitasubramaniam [PV12] turn to considering public-coin protocols, and demonstrate that parallel repetition decreases the soundness error for *constant-round* public-coin protocols at an optimal rate.
- Håstad, Pass, Wikström and Pierzak [HPWP08] show that parallel repetition, in fact, works for *all* (not necessarily constant-round) public-coin protocols, and decreases the soundness error at an exponential rate. Chung and Liu [CL10] demonstrate that it in fact decreases at an optimal rate.
- [HPWP08] consider a generalization of both public-coin and three-message protocol, called *simulatable* protocols—where roughly speaking the verifier's

messages can be predicted without knowing its randomness—and demonstrate that parallel repetition reduces the error at an exponential rate; an improved "nearly" optimal rate (reducing the soundness error from ϵ to $\epsilon^{k/2}$) is obtained by [CL10].

- [HPWP08], and more explicitly [CL10], also consider protocols satisfying a "computational" simulatability property and demonstrate that parallel repetition reduces the soundness error at a nearly optimal rate also for such protocols.
- The elegant work of Haitner [Hai09] considers a certain class of protocols with "random-terminating" verifiers and demonstrates that parallel repetition reduces the soundness error at an exponential rate for such protocols; random-terminating protocols are important since *any* argument can be turned into a random-terminating one, while only slightly increasing the soundness error. [HPWP10] provide a generalization, called δ-simulatable protocols—where, very roughly speaking, we only need to predict a δ-fraction of the verifier's messages—that encompasses both simulatable and random-terminating protocols, and demonstrate that parallel repetition decreases the soundness error at an exponential rate. Optimal, or even "nearly" optimal, parallel repetition theorems for δ-simulatable (or even random-terminating) protocols are not known.

2   Preliminaries

Throughout the paper, all log are base e.

2.1   Interactive Arguments

Definition 1 (Interactive Proofs/Arguments). *A pair of interactive algorithms (P, V) is an* **interactive proof** *for a* **NP** *language L with* **completeness error** *c and* **soundness error** *s if it satisfies the following properties:*

- *Completeness: For all $x \in L$ with* **NP** *witness w,*

$$\Pr[\langle P(w), V \rangle(x) = 1] = 1 - c(|x|).$$

- *Soundness: For all adversarial provers P^*, and for every all $x \notin L$,*

$$\Pr[\langle P^*, V \rangle(x) = 1] \leq s(|x|).$$

where $\langle P, V \rangle(x)$ denotes the output of V after communicating with P if both players get x as a common input. (P, V) is an **interactive argument** *for L if P runs in polynomial time and the soundness property holds only against all non-uniform polynomial-time adversarial provers P^*. (P, V) is public-coin if verifier's messages are uniformly random strings; otherwise, (P, V) is private-coin.*

Definition 2 (Parallel Repetition with Threshold Verifiers). *Let* (P, V) *be an interactive protocol. Let* $k \in \mathbb{N}$ *and* $\gamma \in (0, 1)$. *We use* $(P^k, V^{k,\gamma})$ *to denote* k-fold parallel repetition of (P, V) with threshold γ, where P^k and $V^{k,\gamma}$ interact by executing k copies of (P, V) in parallel and at the end of execution, $V^{k,\gamma}$ accepts iff at least $\gamma \cdot k$ copies accept. For the direct product case with $\gamma = 1$, we use V^k to denote $V^{k,1}$.

2.2 Kullback-Leibler Divergence

Here we review the definition and basic properties of Kullback-Leibler divergence.

Definition 3. *Let* X *and* Y *be discrete random variables over a finite support* $[N]$. *The Kullback-Leibler divergence (KL divergence for short) of* X *from* Y *is defined to be*

$$\mathbf{KL}(X || Y) = \sum_{x \in [N]} \Pr[X = x] \log \frac{\Pr[X = x]}{\Pr[Y = x]}.$$

For $p, q \in (0, 1)$, *we use* $\mathbf{KL}(p || q)$ *to denote the KL divergence* $\mathbf{KL}(X || Y)$ *of two binary random variables* X *and* Y *with* $\Pr[X = 1] = p$ *and* $\Pr[Y = 1] = q$.

The following properties of KL divergence can be found in any Information Theory textbook (e.g., [CT06]). We first recall the chain rule for KL divergence.

Lemma 3 (chain rule). *Let* (X_1, X_2) *and* (Y_1, Y_2) *be random variables. We have*

$$\mathbf{KL}((X_1, X_2) || (Y_1, Y_2)) = \mathbf{KL}(X_1 || Y_1) + \mathop{\mathrm{E}}_{x \leftarrow X_1} [\mathbf{KL}(X_2 |_{X_1 = x} || Y_2 |_{Y_1 = x})].$$

The following lemma says that applying a deterministic function can only decrease the KL divergence.

Lemma 4. *Let* X *and* Y *be random variables and* f *a deterministic function. We have*

$$\mathbf{KL}(f(X) || f(Y)) \leq \mathbf{KL}(X || Y).$$

The following lemma allows us to decompose the KL divergence of two joint distributions by the sum of the KL divergence of their marginals.

Lemma 5 ([Hol09], Lemma 4.2). *Let* $\mathbf{X} = (X_1, \ldots, X_k)$ *be independent random variables, and* $\mathbf{Y} = (Y_1, \ldots, Y_k)$ *be random variables.*

$$\sum_{i=1}^{k} \mathbf{KL}(Y_i || X_i) \leq \mathbf{KL}(\mathbf{Y} || \mathbf{X})$$

The following lemma bounds how much conditioning can creates the KL divergence.

Lemma 6. *Let X be a random variable and W a (probabilistic) event.*

$$\mathbf{KL}(X|_W \| X) \le \log \frac{1}{\Pr[W]}.$$

The following simple lemma bounds the sensitivity of KL divergence of two binary random variables with respect to the first coordinate, which will be useful for us. The lemma follows by the fact that KL divergence is a smooth function, and is proved by a straightforward calculation. For the sake of completeness, we provide a proof in the appendix.

Lemma 7. *For every $p, q, \eta \in (0,1)$ such that $\eta \le \min\{p/2, q\}$, we have*

$$\mathbf{KL}(p\|q) - \mathbf{KL}(p - \eta\|q) \le \Theta\left(\eta \cdot \log(1/\eta)\right)$$

2.3 A Lemma on Sampling

The following simple lemma is taken from Håstad et al. [HPWP10]; for self-containment, we recall the proof.

Lemma 8 ([HPWP10]). *Let (X, Y) be a joint distribution over some finite domain. Let W be a deterministic event on (X, Y). Consider the following experiment:*

- *Sample $x \leftarrow X|_W$.*
- *Sample $y \leftarrow Y|_{W \wedge X=x}$ using rejection sampling; i.e., sample i.i.d. $y_1, y_2, \ldots \leftarrow Y|_{X=x}$ and outputs the first y_t such that $(x, y_t) \in W$.*

Let T be the number of sample used in the rejection sampling. We have $\mathrm{E}[T] = \frac{1}{\Pr[W]}$.

Proof. The lemma follows by the following calculation.

$$\mathrm{E}[T] = \sum_x \Pr[X = x|W] \cdot \mathrm{E}[T|X = x]$$

$$= \sum_x \Pr[X = x|W] \cdot \frac{1}{\Pr[W|X = x]}$$

$$= \sum_x \frac{\Pr[X = x \wedge W]}{\Pr[W]} \cdot \frac{\Pr[X = x]}{\Pr[W \wedge X = x]}$$

$$= \sum_x \frac{\Pr[X = x]}{\Pr[W]} = \frac{1}{\Pr[W]}.$$

3 Proof of the Parallel Repetition Theorem

In this section, we present the formal of our tight Chernoff-type parallel repetition theorem for public-coin protocols.

Theorem 2. *Let (P, V) be a public-coin interactive argument for a language L. There exists an oracle adversarial prover $P^{(\cdot)*}$ such that for every $k \in \mathbb{N}$, input $z \in \{0, 1\}^*$, every $\gamma, \delta, \xi \in (0, 1)$ with $\gamma > \delta$, and every deterministic parallel adversarial prover P^{k*}, if*

$$\Pr[\langle P^{k*}, V^{k,\gamma}\rangle(z) = 1] \geq \epsilon \overset{\text{def}}{=} (1 + \xi) \cdot e^{-k \cdot \mathbf{KL}(\gamma \| \delta)},$$

then

$$\Pr[\langle P^{(P^{k*})*}(k, \gamma, \delta, \xi), V\rangle(z) = 1] \geq \delta.$$

Furthermore, $P^{(\cdot)*}$ runs in time $\mathrm{poly}(|z|, k, \epsilon^{-1}, \xi^{-1}, (\gamma - \delta)^{-1})$ given oracle access to P^{k*}.

Note that in the direct product setting with $\gamma = 1$, $\mathbf{KL}(1 \| \delta) = \log(1/\delta)$ so $e^{-k \cdot \mathbf{KL}(\gamma \| \delta)} = \delta^k$. Thus, the above theorem implies a tight direct product theorem as a corollary. For the "multiplicative" Chernoff-bound setting with $\gamma = (1 + \mu)\delta$, we have $\mathbf{KL}((1 + \mu)\delta \| \delta) = \Theta(\mu^2 \delta)$ for $\mu \in (0, 1)$, and $\mathbf{KL}((1 + \mu)\delta \| \delta) = \Theta(\mu \delta)$ for $\mu > 1$, which implies bounds $e^{-\Omega(\mu^2 \delta k)}$ and $e^{-\Omega(\mu \delta k)}$, respectively. This matches the usual multiplicative Chernoff bounds.

Proof. Let m denote the round complexity of (P, V). Let us consider a $P_{\mathrm{rej}}^{(\cdot)*}$ that interacts with V by the aforementioned *rejection sampling* with $M = \Theta(\frac{k \cdot m}{\epsilon \cdot \xi \cdot (\gamma - \delta)} \cdot \log \frac{k}{\delta \cdot \xi})$. Specifically, P_{rej}^*, selects the session $i \in [k]$ uniformly at random, and then at each round j, upon receiving the external verifier V's message $x_{j,i}$, P_{rej}^* selects $x_{j,-i}$ using rejection sampling as follows: P_{rej}^* repeatedly samples a random continuation of $(P^{k*}, V^{k,\gamma})$ until it finds an *accepting continuation*, i.e., $V^{k,\gamma}$ accepts at the end of interaction (note that we do not require $V^{k,\gamma}$ accepts the i-th coordinate), or $M = \Theta(\frac{k \cdot m}{\epsilon \cdot \xi \cdot (\gamma - \delta)} \cdot \log \frac{k}{\delta \cdot \xi})$ samples is reached, in which case P_{rej}^* aborts and fails. Then, P_{rej}^* selects the corresponding messages in the accepting continuation as the messages of V_{-i} at round j.

By inspection, $P_{\mathrm{rej}}^{(\cdot)*}$ runs in time $\mathrm{poly}(|z|, k, \epsilon^{-1}, \xi^{-1}, (\gamma - \delta)^{-1})$ on input $z, k, \gamma, \delta, \xi$. It remains to show that if P^{k*} convinces $V^{k,\gamma}$ with probability at least ϵ, then $P_{\mathrm{rej}}^{(\cdot)*}$ convinces V with probability at least δ. Let W denote the event that P^{k*} convinces $V^{k,\gamma}$ in the execution of $\langle P^{k*}, V^{k,\gamma}\rangle(z)$. We consider the following Real experiment, which is the same as the execution of $\langle P_{\mathrm{rej}}^{(P^{k*})*}(k, \gamma, \delta, \xi), V\rangle(z)$ except that P_{rej}^* takes an unbounded number of samples (i.e., set $M = \infty$).

The Real Experiment. Consider an execution of (P_{rej}^*, V) as follows. At beginning, P_{rej}^* selects a random coordinate $i \in [k]$. Then at each round $j \in [m]$, V selects a uniformly random $x_{j,i}$, and P_{rej}^* selects a random $x_{j,-i}$ conditioned on W using rejection sampling (namely, repeatedly samples a random continuation of $(P^{k*}, V^{k,\gamma})$ until it finds an *accepting continuation*, i.e., $V^{k,\gamma}$ accepts at the end of interaction, and selects the corresponding $x_{j,-i}$). Let T_j denotes the number of samples P_{rej}^* takes. If no such $x_{j,-i}$ exists, then P_{rej}^* *fails*, and we set $T_j = \infty$ and all remaining $x_{j,-i}, x_{j+1}, \ldots, x_m = \bot$. P_{rej}^* *succeeds* if it does not fail. The output of the experiment is defined to be $(i, x_1, \ldots, x_m)$.

Note that the event that $P^{(\cdot)*}$ convinces V in $\langle P^{(P^{k*})*}(k, \gamma, \delta, \xi), V \rangle(z)$ corresponds to the event that in the Real experiment, P^* succeeds *and* $T_j \leq M$ for every $j \in [m]$. Let $\mathsf{Suc}_{\mathsf{Real}}$ be the indicator random variable of this event. Our goal is to lower bound

$$\Pr[\langle P^{(P^{k*})*}(k, \gamma, \delta, \xi), V \rangle(z) = 1] = \Pr[\mathsf{Suc}_{\mathsf{Real}} = 1].$$

We next compare it with an Ideal experiment, which is identical to the Real experiment, except that the messages $x_{1,i}, \ldots, x_{m,i}$ are also selected *conditioned* on W.

The Ideal Experiment. At beginning, P^*_{rej} selects a random coordinate $i \in [k]$. Then at each round $j \in [m]$, V selects a random $x_{j,i}$ conditioned on W, and P^*_{rej} selects a random $x_{j,-i}$ conditioned on W using rejection sampling. Let T_j denotes the number of samples P^*_{rej} takes. The output of the experiment is defined to be $(i, x_1, \ldots, x_m)$.

Note that sampling random $x_{1,i}, x_{1,-i}, \ldots, x_{m,i}, x_{m,-i}$ conditioned on W step by step is equivalent to sampling the whole $x_1, \ldots, x_m$ conditioned on W. Thus, the output distribution of the Ideal experiment is simply a uniformly random coordinate $i \in [k]$ and a uniformly random *accepting* transcript $(x_1, \ldots, x_m)$. Let $\mathsf{Suc}_{\mathsf{Ideal}}$ be the corresponding indicator random variable of $\mathsf{Suc}_{\mathsf{Real}}$ in the Ideal experiment; that is, $\mathsf{Suc}_{\mathsf{Ideal}}$ is the indicator random variable of the event that P^*_{rej} convinces V and $T_j \leq M$ for every $j \in [m]$.

In what follows, we will show that (i) $\Pr[\mathsf{Suc}_{\mathsf{Ideal}} = 1] \geq \gamma - (m/M\epsilon)$ and (ii) $\mathbf{KL}(\mathsf{Ideal}||\mathsf{Real}) \leq (\log(1/\Pr[W]))/k$, and derive the desired lower bound on $\Pr[\mathsf{Suc}_{\mathsf{Real}} = 1]$ from them.

Lemma 9. $\Pr[\mathsf{Suc}_{\mathsf{Ideal}} = 1] \geq \gamma - (m/M\epsilon)$.

Proof. Note that in the Ideal experiment, for every $i \in [k]$ and $j \in [m]$, the prefix $(x_1, \ldots, x_{j-1}, x_{j,i})$ is chosen randomly conditioned on W and then P^*_{rej} selects a random $x_{j,-i}$ conditioned on W using rejection sampling. Applying Lemma 8 with $X = (X_1, \ldots, X_{j-1}, X_{j,i})$, $Y = X_{j,-i}$ and event W implies that $E[T_j] = 1/\Pr[W] \leq 1/\epsilon$ for every $j \in [m]$. By the Markov inequality, we have $\Pr[T_j \leq M] \geq 1 - 1/(M\epsilon)$ for every $j \in [m]$. Also note that i is uniformly random and independent of x and T_j's so the probability that a random coordinate i is accepting is at least γ. Thus, it follows by an union bound that $\Pr[\mathsf{Suc}_{\mathsf{Ideal}} = 1] \geq \gamma - (m/M\epsilon)$.

Lemma 10. $\mathbf{KL}(\mathsf{Ideal}||\mathsf{Real}) \leq (\log(1/\Pr[W]))/k$.

Proof. It is instructive to first prove the one-round case (i.e., $m = 1$), which is equivalent to the KL-version of Raz' Lemma. In this case by definition, $\mathsf{Ideal} = (I, X_1|_W)$ and $\mathsf{Real} = (I, X_{1,I}, X_{1,-I}|_{W, X_{1,I}})$. By applying the chain rule (Lemma 3), we have

$$\mathbf{KL}(\mathsf{Ideal}||\mathsf{Real}) = \mathbf{KL}(I||I) + \mathop{E}_{I}\left[\mathbf{KL}\left(X_1|_W||(X_{1,I}, X_{1,-I}|_{W, X_{1,I}})\right)\right]$$

$$= \frac{1}{k}\sum_{i=1}^{k}\mathbf{KL}\left(X_1|_W||(X_{1,i}, X_{1,-i}|_{W, X_{1,i}})\right).$$

For each term $\mathbf{KL}(X_1|_W||(X_{1,i}, X_{1,-i}|_{W,X_{1,i}}))$, by applying the chain rule again, we have

$$
\begin{aligned}
\mathbf{KL}\left(X_1|_W||(X_{1,i}, X_{1,-i}|_{W,X_{1,i}})\right) \\
= \mathbf{KL}(X_{1,i}|_W||X_{1,i}) + \underset{X_{1,i}|_W}{\mathrm{E}}\left[\mathbf{KL}(X_{1,-i}|_{W,X_{1,i}}||X_{1,-i}|_{W,X_{1,i}})\right] \\
= \mathbf{KL}(X_{1,i}|_W||X_{1,i}).
\end{aligned}
$$

Applying Lemma 5,

$$
\begin{aligned}
\frac{1}{k}\sum_{i=1}^{k}\mathbf{KL}\left(X_1|_W||(X_{1,i}, X_{1,-i}|_{W,X_{1,i}})\right) = \frac{1}{k}\sum_{i=1}^{k}\mathbf{KL}(X_{1,i}|_W||X_{1,i}) \\
\leq \frac{1}{k}\mathbf{KL}(X_1|_W||X_1).
\end{aligned}
$$

Therefore, by Lemma 6,

$$
\mathbf{KL}(\mathsf{Ideal}||\mathsf{Real}) \leq \frac{1}{k}\mathbf{KL}(X_1|_W||X_1) \leq \frac{\log(1/\Pr[W])}{k}.
$$

We proceed to consider the general case, which is proved by the same calculation, except that we first apply an additional chain rule to break up terms corresponding to each round.

$$
\mathbf{KL}(\mathsf{Ideal}||\mathsf{Real}) = \sum_{j=1}^{m}\underset{I,X_{<j}|_W}{\mathrm{E}}\left[\mathbf{KL}(X_j|_{W,X_{<j}}||(X_{j,I}|_{W,X_{<j}}, X_{j,-I}|_{W,X_{<j},X_{j,I}}))\right].
$$

Now, for each term, the same calculation as before using the chain rule and Lemma 5 shows that

$$
\begin{aligned}
\underset{I,X_{<j}|_W}{\mathrm{E}}\left[\mathbf{KL}(X_j|_{W,X_{<j}}||(X_{j,I}|_{W,X_{<j}}, X_{j,-I}|_{W,X_{<j},X_{j,I}}))\right] \\
\leq \frac{1}{k}\underset{X_{<j}|_W}{\mathrm{E}}\left[\mathbf{KL}(X_j|_{W,X_{<j}}||X_j|_{X_{<j}})\right].
\end{aligned}
$$

Applying another chain rule and Lemma 6 gives,

$$
\begin{aligned}
\mathbf{KL}(\mathsf{Ideal}||\mathsf{Real}) \leq \frac{1}{k}\underset{X_{<j}|_W}{\mathrm{E}}\left[\mathbf{KL}(X_j|_{W,X_{<j}}||X_j|_{W,X_{<j}})\right] \\
= \frac{1}{k}\mathbf{KL}(X_{\leq m}|_W||X_{\leq m}) \\
\leq \frac{\log(1/\Pr[W])}{k}
\end{aligned}
$$

We now derive the desired lower bound on the probability $\Pr[\mathsf{Suc_{Real}} = 1]$ using Lemma 9 and 10. Let $q = \Pr[\mathsf{Suc_{Real}} = 1]$ and $\eta = m/M\epsilon$. Since our goal is to lower bound q by δ and $\gamma - \eta \geq \delta$, we can assume w.l.o.g., that $q \leq \gamma - \eta$. Lemma 10 implies that

$$
\mathbf{KL}(\gamma - \eta||q) \leq \mathbf{KL}(\mathsf{Suc_{Ideal}}||\mathsf{Suc_{Real}}) \leq \mathbf{KL}(\mathsf{Ideal}||\mathsf{Real}) \leq (\log(1/\Pr[W]))/k,
$$

244 K.-M. Chung and R. Pass

where the second inequality follows since applying the same function to two distributions can only decrease their KL divergence. Now, note that the fact that $\Pr[W] \geq (1 + \xi)e^{-k \cdot \mathbf{KL}(\gamma||\delta)}$ and Lemma 7 implies that

$$\Pr[W] \geq (1 + \xi) \cdot e^{-k \cdot \mathbf{KL}(\gamma||\delta)} \geq e^{-k \cdot \mathbf{KL}(\gamma||\delta) + \xi/2}$$
$$\geq e^{-k \cdot (\mathbf{KL}(\gamma - \eta||\delta) + \Theta(\eta \cdot \log(1/\eta))) + \xi/2} \geq e^{-k \cdot \mathbf{KL}(\gamma - \eta||\delta)},$$

where the last inequality follows by the fact that $k \cdot \Theta(\eta \cdot \log(1/\eta)) \leq \xi/2$ (which follows by the choice of M). Combining the above inequalities, we have $\mathbf{KL}(\gamma - \eta||q) \leq \mathbf{KL}(\gamma - \eta||\delta)$, which implies $q \geq \delta$.

References

[BCC88] Brassard, G., Chaum, D., Crépeau, C.: Minimum disclosure proofs of knowledge. Journal of Computer and System Sciences 37(2), 156–189 (1988)

[BIN97] Bellare, M., Impagliazzo, R., Naor, M.: Does parallel repetition lower the error in computationally sound protocols? In: FOCS, pp. 374–383 (1997)

[BM88] Babai, L., Moran, S.: Arthur-Merlin games: A randomized proof system, and a hierarchy of complexity classes. J. Comput. Syst. Sci. 36(2), 254–276 (1988)

[BRR+09] Barak, B., Rao, A., Raz, R., Rosen, R., Shaltiel, R.: Strong parallel repetition theorem for free projection games. In: Dinur, I., Jansen, K., Naor, J., Rolim, J. (eds.) APPROX and RANDOM 2009. LNCS, vol. 5687, pp. 352–365. Springer, Heidelberg (2009)

[CHS05] Canetti, R., Halevi, S., Steiner, M.: Hardness amplification of weakly verifiable puzzles. In: Kilian, J. (ed.) TCC 2005. LNCS, vol. 3378, pp. 17–33. Springer, Heidelberg (2005)

[CL10] Chung, K.-M., Liu, F.-H.: Parallel repetition theorems for interactive arguments. In: Micciancio, D. (ed.) TCC 2010. LNCS, vol. 5978, pp. 19–36. Springer, Heidelberg (2010)

[CT06] Cover, T.M., Thomas, J.A.: Elements of Information Theory (Wiley Series in Telecommunications and Signal Processing). Wiley-Interscience (2006)

[DJMW12] Dodis, Y., Jain, A., Moran, T., Wichs, D.: Counterexamples to hardness amplification beyond negligible. In: Cramer, R. (ed.) TCC 2012. LNCS, vol. 7194, pp. 476–493. Springer, Heidelberg (2012)

[DP98] Damgård, I., Pfitzmann, B.: Sequential iteration of interactive arguments and an efficient Zero-Knowledge argument for NP. In: Larsen, K.G., Skyum, S., Winskel, G. (eds.) ICALP 1998. LNCS, vol. 1443, pp. 772–783. Springer, Heidelberg (1998)

[GMR89] Goldwasser, S., Micali, S., Rackoff, C.: The knowledge complexity of interactive proof-systems. SIAM Journal on Computing 18(1), 186–208 (1989)

[Gol01] Goldreich, O.: Foundations of Cryptography. Basic tools. Cambridge University Press (2001)

[Hai09] Haitner, I.: A parallel repetition theorem for any interactive argument. In: FOCS (2009)

[Hol09] Holenstein, T.: Parallel repetition: Simplification and the no-signaling case. Theory of Computing 5(1), 141–172 (2009)

[HPWP08]   Håstad, J., Pass, R., Wikström, D., Pietrzak, K.: An efficient parallel repetition theorem (2008) (unpublished manuscript)

[HPWP10]   Håstad, J., Pass, R., Wikström, D., Pietrzak, K.: An efficient parallel repetition theorem. In: Micciancio, D. (ed.) TCC 2010. LNCS, vol. 5978, pp. 1–18. Springer, Heidelberg (2010)

[IJK09]   Impagliazzo, R., Jaiswal, R., Kabanets, V.: Chernoff-type direct product theorems. J. Cryptology 22(1), 75–92 (2009)

[PV07]   Pass, R., Venkitasubramaniam, M.: An efficient parallel repetition theorem for arthur-merlin games. In: STOC, pp. 420–429 (2007)

[PV12]   Pass, R., Venkitasubramaniam, M.: A parallel repetition theorem for constant-round arthur-merlin proofs. Transactions on Computation Theory 4(4), 10 (2012)

[PW07]   Pietrzak, K., Wikström, D.: Parallel repetition of computationally sound protocols revisited. In: Vadhan, S.P. (ed.) TCC 2007. LNCS, vol. 4392, pp. 86–102. Springer, Heidelberg (2007)

[Raz98]   Raz, R.: A parallel repetition theorem. SIAM J. Comput. 27(3), 763–803 (1998)

[Yao82]   Yao, A.C.-C.: Theory and applications of trapdoor functions (extended abstract). In: FOCS, pp. 80–91 (1982)

Proof of Lemma 7

Lemma 11 (Lemma 7 Restated). *For every $p, q, \eta \in (0,1)$ such that $\eta \le \min\{p/2, q\}$, we have*

$$\mathbf{KL}(p\|q) - \mathbf{KL}(p - \eta\|q) \le \Theta\left(\eta \cdot \log(1/\eta)\right)$$

Proof. By definition,

$$\mathbf{KL}(p\|q) = p \log \frac{p}{q} + (1-p) \log \frac{1-p}{1-q}$$

$$= p \log p + p \log \frac{1}{q} + (1-p) \log(1-p) + (1-p) \log \frac{1}{1-q}$$

$$\mathbf{KL}(p - \eta\|q) = (p-\eta) \log \frac{p-\eta}{q} + (1-p+\eta) \log \frac{1-p+\eta}{1-q}$$

$$= (p-\eta) \log(p-\eta) + (p-\eta) \log \frac{1}{q} + (1-p+\eta) \log(1-p+\eta)$$

$$+ (1-p+\eta) \log \frac{1}{1-q}$$

By further expanding, we have

$$(p-\eta) \log(p-\eta) = p \log(p-\eta) - \eta \log(p-\eta)$$

$$= p \log p + p \log\left(1 - \frac{\eta}{p}\right) - \eta \log(p-\eta)$$

$$(p-\eta) \log \frac{1}{q} = p \log \frac{1}{q} - \eta \log \frac{1}{q}$$

$$(1 - p + \eta) \log(1 - p + \eta) = (1 - p) \log(1 - p + \eta) + \eta \log(1 - p + \eta)$$
$$= (1 - p) \log(1 - p) + (1 - p) \log(1 + \frac{\eta}{1 - p})$$
$$+ \eta \log(1 - p + \eta)$$
$$(1 - p + \eta) \log \frac{1}{1 - q} = (1 - p) \log \frac{1}{1 - q} + \eta \log \frac{1}{1 - q}$$

Therefore,

$$\mathbf{KL}(p\|q) - \mathbf{KL}(p - \eta\|q)$$
$$= -p \log(1 - \frac{\eta}{p}) + \eta \log(p - \eta) + \eta \log \frac{1}{q} - (1 - p) \log(1 + \frac{\eta}{1 - p})$$
$$- \eta \log(1 - p + \eta) - \eta \log \frac{1}{1 - q}$$
$$\leq -p \log(1 - \frac{\eta}{p}) + \eta \log \frac{1}{q} - \eta \log(1 - p + \eta)$$
$$\leq 2\eta + \eta \log \frac{1}{q} - \eta \log \eta \leq \Theta(\eta \log(1/\eta)),$$

where the first inequality follows by dropping negative terms, the second inequality follows by the monotonicity of logarithm and using Taylor expansion, and the last inequality uses $\eta < q$.

Stretching Groth-Sahai:
NIZK Proofs of Partial Satisfiability

Carla Ràfols

Horst-Görtz Institut für IT Sicherheit, Ruhr-Universität Bochum, Germany
`carla.rafols@rub.de`

Abstract. Groth, Ostrovsky and Sahai constructed a non-interactive Zap for NP-languages by observing that the common reference string of their proof system for circuit satisfiability admits what they call *correlated key generation*. The latter means that it is possible to create from scratch two common reference strings in such a way that it can be publicly verified that at least one of them guarantees perfect soundness while it is computationally infeasible to tell which one. Their technique also implies that it is possible to have NIWI Groth-Sahai proofs for certain types of equations over bilinear groups in the plain model. We extend the result of Groth, Ostrovsky and Sahai in several directions. Given as input some predicate P computable by some monotone span program over a finite field, we show how to generate a set of common reference strings in such a way that it can be publicly verified that the subset of them which guarantees perfect soundness is accepted by the span program. We give several different flavors of the technique suitable for different applications scenarios and different equation types. We use this to stretch the expressivity of Groth-Sahai proofs and construct NIZK proofs of partial satisfiability of sets of equations in a bilinear group and more efficient Groth-Sahai NIWI proofs without common reference string for a larger class of equation types. Finally, we apply our results to significantly reduce the size of the signatures of the ring signature scheme of Chandran, Groth and Sahai or to have a more efficient proof in the standard model that a commitment opens to an element of a public list.

1 Introduction

Zero-knowledge proofs have played a significant role both in the theory and the practice of cryptographic protocols. Although non-interactive zero-knowledge proofs are in principle more useful for practical purposes, for roughly twenty years after their invention [4] their prohibitive costs made their interactive counterparts the only real alternative. For example, although the connection between NIZKs and signatures was early recognized [2], in practice a widely used technique was to build schemes in the random oracle based on a special kind of interactive proof of knowledge, a Σ-protocol ([24,13]). This approach turned out to be quite fruitful and it was used to build a number of schemes even with complex functionalities, like for instance the numerous kinds of distributed signature schemes based on the work of Cramer *et al.* [9] on interactive proofs of

Y. Dodis and J.B. Nielsen (Eds.): TCC 2015, Part II, LNCS 9015, pp. 247–276, 2015.

partial knowledge. In such a proof, the prover convinces the verifier that it knows a subset of the witnesses of a set of statements. Again, although De Santis *et al.* [10] had achieved similar results before for the non-interactive case, they were considered of theoretical interest only and went unnoticed by protocol designers.

The situation changed radically when in 2008, after some promising advances towards making non-interactive proofs practical, ([5,14]), Groth and Sahai ([16]) gave efficient non-interactive proofs of membership in the language of satisfiable quadratic equations in bilinear groups. Compared to previous work, their proposal had the advantage of considering a language which is both very natural and very general. The great number of protocols which use GS proofs ([19,17,6], just to name a few) shows the strength and flexibility of their framework.

MORE EXPRESSIVE IS MORE EFFICIENT. The fact that practical instantiations of GS proofs are in bilinear groups imposes some limitations on the type of equations for which satisfiability can be proven. For instance, each equation should be at most quadratic, and further, in asymmetric bilinear groups, it should have degree at most one in each variable. This can be circumvented at the cost of adding additional variables and equations but this might significantly increase the proof size. Although GS proofs can be considered practical, they remain expensive and even proofs of simple statements might require several dozens of group elements. Therefore, the design goal of obtaining proofs for a more expressive language goes hand in hand with obtaining efficiency improvements.

For instance, suppose one wants to prove a statement of the type which is informally expressed as:

$$\text{"}\hat{\mathbf{c}} \text{ is a commitment to some value } X \in \{1, \dots, L\}\text{,"} \tag{1}$$

for some $L \in \mathbb{N}$. This statement is naturally encoded as "$\hat{\mathbf{c}}$ opens to some value X which satisfies one of the equations $\{X - 1 = 0, \dots, X - L = 0\}$". However, as GS proofs do not support this kind of statements, following [6,14], the strategy is to add auxiliary variables $b_1, \dots, b_L$, prove that $\sum_{i=1}^{L} b_i = 1$ and that, for all $i \in \{1, \dots, L\}$, $b_i \in \{0, 1\}$ and $(X - i)b_i = 0$. Further, in the instantiation of GS proofs in asymmetric bilinear groups, to prove each one of the statements $b_i \in \{0, 1\}$ we must add a new additional variable $\bar{b}_i$, and prove satisfiability of the equations $\{b_i - \bar{b}_i = 0, b_i(\bar{b}_i - 1) = 0\}$. The question remains if we can encode these statements in some alternative, more efficient way by *stretching* GS proofs so that they support this sort of statements directly.

PARTIAL PROOFS. We would like a result close in spirit to [10,9] for GS proofs. In all these works, the main idea is to use as a building block a proof system $\mathcal{PS}_1$ that allows to prove a certain atomic statement x, and modify it to construct a proof system $\mathcal{PS}_2$ for statements of the kind "Given the atomic statements $x_1, \dots, x_L$, there exists a subset of indexes $A \subset \{1, \dots, L\}$ in some family of sets $\Omega \subset \mathcal{P}([L])$ such that all the atomic statements x_i, $i \in A$ are true".

The prover in $\mathcal{PS}_2$ must construct a proof given only the witnesses for the statements x_i, $i \in A$ and the proof must leak no information about the actual set A, other than it is in Ω. The common strategy of these works is to construct the prover of $\mathcal{PS}_2$ using as building block both the prover of $\mathcal{PS}_1$ — for the statements x_i, $i \in A$ — and the simulator — for the statements x_i, $i \notin A$. Since real proofs are (computationally) indistinguishable from simulated ones, the final proof output by the prover of $\mathcal{PS}_2$, which consists of proofs for all the statements $x_1, \ldots, x_L$, will reveal nothing about A. The main challenge is then to ensure that the soundness condition is met — guaranteeing that the prover cannot simulate all the proofs—, while making sure that the simulator gets a properly distributed input.

In all these works — including ours — this is done by means of secret sharing techniques, however the challenges that arise are specific to each proof system. For instance, in the work of Cramer et al. [9], both $\mathcal{PS}_1$ and $\mathcal{PS}_2$ are Σ-protocols, which are 3 round interactive protocols. In this case, the key difference between a prover and a simulator which outputs an accepting transcript is that the latter creates the transcript by altering the order in which the real protocol is executed and letting the information sent in the first round depend on the challenge, which is the information sent in the second round. In [9], the prover of $\mathcal{PS}_2$ receives a challenge c, from which it creates L challenges $c_1, \ldots, c_L$ and uses c_i as a challenge to prove the atomic statement x_i with $\mathcal{PS}_1$. The secret sharing techniques ensure that the prover of $\mathcal{PS}_2$ has the right amount of freedom in choosing these challenges: namely, they guarantee that there exists some $A \in \Omega$ such that the prover can choose all c_i, $i \in A^c$ (i.e. it can simulate the proofs of x_i, $i \in A^c$), while it is unable to guess the value of c_i, $i \in A$ (i.e. soundness must hold for x_i, $i \in A$).

THE GS PROOF SYSTEM. Clearly, the techniques of Cramer et al. are specific to Σ-Protocols. On the other hand, the techniques of de Santis et al. [10] for the non-interactive case are specific to statements related to quadratic residuosity. Neither of them does apply in any obvious way to GS proofs as the conditions which guarantee soundness or allow to simulate proofs are quite different there.

Indeed, let us recall some basics about GS proofs. They allow to prove satisfiability of several equation types over a bilinear group $gk = (q, \hat{\mathbb{G}}, \check{\mathbb{H}}, \mathbb{T}, e, \hat{g}, \check{h})$, where $\hat{\mathbb{G}}, \check{\mathbb{H}}, \mathbb{T}$ are groups of prime order q in additive notation, the elements $\hat{g}, \check{h}$ are generators of $\hat{\mathbb{G}}, \check{\mathbb{H}}$ respectively, and $e : \hat{\mathbb{G}} \times \check{\mathbb{H}} \to \mathbb{T}$ is an efficiently computable, non-degenerate bilinear map. The witness of satisfiability is a solution to the equation which consists of several elements in $\mathbb{Z}_q$, $\hat{\mathbb{G}}$ or $\check{\mathbb{H}}$. The proof is constructed in a two-step process: first, the prover commits to each element of the witness, then it shows that the committed values satisfy the equation. The common reference string (essentially) consists of some commitment keys. These keys can be generated in one of two indistinguishable modes: in the soundness mode these keys define perfectly binding commitments, and even an unbounded prover cannot convince a verifier of a false statement, and in the witness indistinguishable mode they define perfectly hiding commitments. Further, in the latter case, there exists some trapdoor which allows to simulate proofs which are identically distributed to real proofs (computed as in the WI mode). Additionally, a key is binding or hiding depending on whether or not it satisfies certain linear relations. For instance, the commitment key in the group $\hat{\mathbb{G}}$ consists of three vectors $\hat{\mathbf{u}}, \hat{\mathbf{v}}, \hat{\mathbf{w}} \in \hat{\mathbb{G}}^2$. The commitment to a scalar $x \in \mathbb{Z}_q$ is $\hat{\mathbf{c}} = x\hat{\mathbf{w}} + r\hat{\mathbf{v}}$, for some random $r \in \mathbb{Z}_q$, and to a group element $\hat{\mathbf{x}} \in \hat{\mathbb{G}}$ is $\hat{\mathbf{c}} = (\hat{\mathbf{0}}, \hat{\mathbf{x}})^{\top} + r\hat{\mathbf{v}} + s\hat{\mathbf{u}}$, for some random $r, s \in \mathbb{Z}_q$. For scalars the commitment is perfectly binding if $\hat{\mathbf{v}}, \hat{\mathbf{w}} \in \hat{\mathbb{G}}^2$ are linearly independent and perfectly hiding otherwise. For group elements the opposite is true of $\hat{\mathbf{v}}, \hat{\mathbf{u}}$: the commitment is perfectly hiding if $\hat{\mathbf{v}}, \hat{\mathbf{u}}$ are linearly independent and perfectly binding otherwise. To construct partial proofs for the GS proof system, the ideas of [10,9] must be adapted to the inner workings of GS proofs we have just described.

THE NON-INTERACTIVE ZAP OF GROTH, OSTROVSKY AND SAHAI (GOS). On the other hand, the authors of [15], observe that the common reference string of GS proofs admits what we call Or-Verifiable Correlated Key Generation: namely, that a prover can create *from scratch*, without common reference string, a pair of two keys in such a way that it can be publicly verified that at least one of the two keys is binding for quadratic equations.

More specifically, the observation of GOS is that given any vector $\hat{\mathbf{v}} \in \hat{\mathbb{G}}^2$ such that $\{\hat{\mathbf{v}}, (\hat{0}, \hat{g})^{\top}\}$ are linearly independent vectors (this condition can be obviously publicly verified by checking if the first component of $\hat{\mathbf{v}}$ is $\hat{0}$), and two vectors $\hat{\mathbf{z}}_1, \hat{\mathbf{z}}_2 \in \hat{\mathbb{G}}^2$ such that $\hat{\mathbf{z}}_1 + \hat{\mathbf{z}}_2 = (\hat{0}, \hat{g})^{\top}$, it holds that at least one of $\hat{\mathbf{z}}_1$, $\hat{\mathbf{z}}_2$ is independent of $\hat{\mathbf{v}}$ (otherwise their sum could not be independent of $\hat{\mathbf{v}}$). This means that given only some bilinear group gk (and no common reference string) and a tuple $(\hat{\mathbf{v}}, \hat{\mathbf{z}}_1, \hat{\mathbf{z}}_2)$ with the above constraints, it can be publicly verified that at least one of the pairs of correlated keys $\{\hat{\mathbf{v}}, \hat{\mathbf{z}}_1\}$, $\{\hat{\mathbf{v}}, \hat{\mathbf{z}}_2\}$ is a binding GS key for committing to scalars.

Since GS proofs have perfect soundness, if one of the keys is binding the prover cannot cheat even if it knows additional information about the common reference string (e.g. the discrete logarithm). Thus, if the prover chooses $(\hat{\mathbf{v}}, \hat{\mathbf{z}}_1, \hat{\mathbf{z}}_2)$ and then it proves a statement x with both pairs of correlated keys, the statement must be true. Further, the prover is free to choose one of the keys to be hiding for commitments to scalars and this can be done in such a way that it is computationally infeasible to tell which key is the hiding one. This implies that the proof reveals no information about the witness.

Thus, more technically if we prove the same statement x with both pairs of keys we have given a non-interactive Zap (NI Zap) for x, i.e. a witness indistinguishable proof in *the plain model*. On the other hand, if we take two different statements x_1, x_2 and we give a real proof for one of them with the binding key and we simulate the other one with the hiding key, we have given a NI Zap that at least one of x_1, x_2 is true. This seems to be exactly the right starting point for adapting the techniques for partial proofs to the GS setting. In this work we want to fully develop the approach of GOS, and we address several of its limitations:

1. The technique of GOS needs to be adapted to prove not only witness indistinguishability but also zero-knowledge. Indeed, in the construction above, one of the keys is always binding and to prove "x_1 or x_2", we always need the witness of at least one of the statements.

2. One of the pairs $\{\hat{\mathbf{v}}, \hat{\mathbf{z}}_1\}$, $\{\hat{\mathbf{v}}, \hat{\mathbf{z}}_2\}$ is a binding key to commit to scalars, but not to group elements. For which equation types does this approach allow to gain efficiency? Can we find a similar technique for commitments to group elements?

3. Can we extend the techniques to other predicates other than the OR of two equations?

In summary, the question is if we can extend the notion of Verifiable Correlated Key Generation to incorporate all these aspects and in such a way that it is useful to construct more efficient NIZK proofs of partial satisfiability for a large class of predicates.

1.1 Our Results

LABELED COMMIT-AND-PROVE SCHEMES. Recently, Escala and Groth [11] gave a complete formulation of the GS proof system as a labeled Commit-and-Prove (CaP) scheme. The labels are meant to deal with different variable and equation types. This formulation is really convenient for our purposes, as it allows to define in a precise way which equation types admit verifiable correlated key generation and which do not. One of our contributions (section 3) is to slightly modify the definition of labeled CaP of [11] so that it can accommodate both the GS proof system and our new proof system for partial satisfiability.

EXTENDING THE DEFINITION OF VERIFIABLE CORRELATED KEY GENERATION. In section 5, we extend the ideas of GOS in several directions, to use them as a building block to construct proofs of partial satisfiability.

- First, we give a new definition of verifiable correlated key generation — VCKG for short — to adapt it to more general predicates, namely to any predicate P computable by a monotone span program $\mathcal{SP}$. We also modify the definition to explicitly take as input a set of labels so that it fits with the GS CaP formulation of [11]. The motivation for doing so is that the same common reference string might be binding for some equation types and hiding for others. For instance, it is unclear how to extend the result of GOS to prove that an OR of two pairing product equations is satisfied without trusted setup. By this we do not mean that one could not use other to results to prove this (at worst we could prove this by reduction to circuit satisfiability). Our point is rather that introducing labels allows to clearly identify that the construction of GOS *only* ensures that one of the two commitment keys is binding for scalars, but not for group elements.

- Second, we define Simulatable VCKG (SVCKG), which is essentially the same as VCKG except for the fact that the generation algorithm takes as input a common reference string instead of creating the keys from scratch. The keys can be generated in two indistinguishable ways: in such a way that the indexes of the binding keys are a valid assignment of the predicate P (as in VCKG) or in such a way that they are hiding for every index. This definition is introduced with the aim of constructing NIZK proofs of partial satisfiability, and not only NIWI proofs.

In section 6 we show how to combine SVCKG (resp. VCKG) for a predicate P and vector of labels $\mathbf{T} = (\mathsf{t}_1, \ldots, \mathsf{t}_L)$ with the GS proof system to obtain a NIZK proof (resp. NI Zap) that some sets of quadratic equations $\mathcal{S}_1, \ldots, \mathcal{S}_L$ (compatible with $\mathbf{T}$) are partially satisfiable for the predicate P.

CONSTRUCTIONS. In section 7 we give several explicit constructions of (S)VCKG for different equation types and predicates P. Essentially, all we require from P is that it should be computable by a monotone span program and the equation types should all admit what we call left-simulation (or all admit right simulation). The construction of GOS (and our extension for other P) guarantees that some keys are binding for committing to scalars and this limits the equation types for which one can prove partial satisfiability. Therefore, in appendix C, we also show how to do correlated key generation for commitment keys to group elements, at some efficiency cost. These explicit constructions of SVCKG (resp. VCKG) together with the GS CaP give a quite expressive realization of NIZK proofs (resp. NI Zap) for partial satisfiability of equations in bilinear groups.

EFFICIENCY. In section 7.4 we discuss what is the size of our proofs of partial satisfiability. We then compare it with the size of the proof that 1-out-of-L sets of equations is satisfiable which results from the approach suggested by Groth ([6,14]).

APPLICATIONS. In section 9 we discuss some applications. For instance, we show how to save $O(\sqrt{N})$ group elements — where N is the size of the ring — in the signature size of the ring signature scheme of Chandran et al. [8], which is the most efficient ring signature in the standard model.

2   Preliminaries

Given some $n \in \mathbb{N}$, $\mathbf{v} \in \mathbb{Z}_q^n$ denotes a column vector unless specifically stated. Given a set of vectors $\{\mathbf{v}_1, \ldots, \mathbf{v}_r\} \subset \mathbb{Z}_q^n$, $\langle \{\mathbf{v}_1, \ldots, \mathbf{v}_r\} \rangle$ denotes their linear span. Matrices are denoted in boldface and $\mathbf{0}_{m \times n}$ denotes the all-zero $m \times n$ matrix. Given some set $\mathcal{S}$ its cardinal is written as $|\mathcal{S}|$ and $s \leftarrow \mathcal{S}$ denotes the process of sampling an element uniformly at random. For an algorithm D, we write $z \leftarrow \mathsf{D}(x, y, \ldots)$ to indicate that D is a (probabilistic) algorithm that outputs z on input $(x, y, \ldots)$. Given a positive integer L, we denote by $[L]$ the set $\{1, \ldots, L\}$.

We identify a set $A \subset [L]$ in the natural way with a vector $\mathbf{v}_A \in \{0,1\}^L$ and we denote its complementary as $A^c := [L] \backslash A$. Given a family of sets $\Omega \subset \mathcal{P}([L])$, we denote $P_\Omega : \{0,1\}^L \to \{0,1\}$ the predicate such $P_\Omega(\mathbf{v}_A) = 1$ if and only if $A \in \Omega$.

2.1   Bilinear Groups

Let $\mathcal{G}$ be some probabilistic polynomial time algorithm which on input 1^λ, where λ is the security parameter, returns the description of a bilinear group $gk =$

$(q, \hat{\mathbb{G}}, \check{\mathbb{H}}, \mathbb{T}, e, \hat{g}, \check{h})$, where $\hat{\mathbb{G}}, \check{\mathbb{H}}$ and $\mathbb{T}$ are groups of prime order q, the elements $\hat{g}, \check{h}$ are generators of $\hat{\mathbb{G}}, \check{\mathbb{H}}$ respectively, and $e : \hat{\mathbb{G}} \times \check{\mathbb{H}} \to \mathbb{T}$ is an efficiently computable, non-degenerate bilinear map.

Essentially, we take up the notation of Escala and Groth [11] for elements and operations in the bilinear group. Namely, $\hat{\mathbb{G}}, \check{\mathbb{H}}, \mathbb{T}$ are written additively, elements $\hat{x} \in \hat{\mathbb{G}}$ are written with a hat and elements in $\check{y} \in \check{\mathbb{H}}$ with an inverted hat and $\hat{0}, \check{0}$ and $0_{\mathbb{T}}$ denote the neutral elements in the respective groups. For any $\hat{x} \in \hat{\mathbb{G}}, \check{y} \in \check{\mathbb{H}}$, multiplication refers to the pairing operation, $i.e$, $\hat{x}\check{y} := e(\hat{x}, \check{y})$. Matrix/vector or matrix/matrix multiplication of elements in $\hat{\mathbb{G}}$ and $\check{\mathbb{H}}$ are done in the natural way via the pairing operation, for example, given $\hat{\mathbf{X}} \in \hat{\mathbb{G}}^{\ell \times m}$ and $\check{\mathbf{Y}} \in \check{\mathbb{H}}^{m \times n}$, $\hat{\mathbf{X}}\check{\mathbf{Y}} \in \mathbb{T}^{\ell \times n}$.

2.2   SXDH Assumption

Let $(q, \hat{\mathbb{G}}, \check{\mathbb{H}}, \mathbb{T}, e, \hat{g}, \check{h}) \leftarrow \mathcal{G}(1^\lambda)$ be a bilinear group. The Decision Diffie-Hellman Assumption in $\hat{\mathbb{G}}$ states that the two distributions $(\hat{g}, \xi\hat{g}, \rho\hat{g}, \xi\rho\hat{g})$ and $(\hat{g}, \xi\hat{g}, \rho\hat{g}, \kappa\hat{g})$, where $\xi, \rho, \kappa \leftarrow \mathbb{Z}_q$, are computationally indistinguishable. The DDH Assumption in $\check{\mathbb{H}}$ is defined in a similar way.

Definition 1. *(SXDH Assumption) The Symmetric eXternal Diffie-Hellman Assumption holds relative to the group generator algorithm $\mathcal{G}$ if the DDH Assumption holds in both $\hat{\mathbb{G}}$ and $\check{\mathbb{H}}$ for $(q, \hat{\mathbb{G}}, \check{\mathbb{H}}, \mathbb{T}, e, \hat{g}, \check{h}) \leftarrow \mathcal{G}(1^\lambda)$.*

2.3   Monotone Span Programs

Definition 2. *[18] A monotone span program (MSP) over a field $\mathbb{Z}_q$ consists of a tuple $\mathcal{SP} = (\mathbf{M}, \rho)$, where $\mathbf{M} \in \mathbb{Z}_q^{(m+1) \times d}$ is a matrix with row vectors $\{\mathbf{r}_0, \mathbf{r}_1, \ldots, \mathbf{r}_m\}$, $\rho : [m] \to [L]$ is a labeling function and $\mathbf{r}_0$ is called the target vector. $\mathcal{SP}$ is said to compute a predicate $P : \{0, 1\}^L \to \{0, 1\}$ if for any $\mathbf{v}_A \in \{0, 1\}^L$, $P(\mathbf{v}_A) = 1$ if and only if $\mathbf{r}_0 \in \langle\{\mathbf{r}_j : j \in \rho^{-1}(A)\}\rangle$.*

Without loss of generality we assume that $m > d$ and that $\mathbf{M}$ has full rank. Specially for MSPs, sometimes it is more intuitive to talk about sets: in this case we say that $\mathcal{SP}$ accepts $A \subset [L]$ if and only if $P(\mathbf{v}_A) = 1$ and that $\mathcal{SP}$ computes $\Omega \subset \mathcal{P}([L])$ if it computes P_Ω.

It is well known that there is a connection between MSPs, secret sharing schemes (sss, [25,3]) and linear codes. Borrowing some terminology from sss, we refer to the family of sets Ω computed by $\mathcal{SP}$ as an *access structure*. Also if $A \in \Omega$, A is said to be *an authorized subset*. If no proper subset of A is authorized, then we say that A is a *minimal authorized subset*. The dual access structure $\Omega^\star$ is defined as $\Omega^\star := \{[L]\backslash A : A \notin \Omega\}$. The latter notion has found applicability and various scenarios, including the proofs of partial satisfiability of [10,9].

A classical example of a (monotone) span program is the threshold one.

Example 1. $\Omega^{(k,L)} := \{A \subset [L] : |A| \geq k\}$ is called a (k,L)-threshold access structure. A span program $\mathcal{SP}_{(k,L)}$ computing $\Omega_{(k,L)}$ can be defined by letting ρ be the identity function, $\mathbf{r}_i^\top = (1,i,\ldots,i^{k-1})$ and $\mathbf{r}_0^\top = (1,0,\ldots,0)$. The dual access structure is $\Omega^*_{(k,L)} := \{A \subset [L] : |A| \geq L - k + 1\}$.

The key ingredient for our results is the following technical lemma, most specially part 1), which states some well-known or easy facts about (monotone) span programs:

Lemma 1. *Let* $\mathcal{SP} = (\mathbf{M}, \rho)$ *be a monotone span program which computes* Ω.

1) *If* $\boldsymbol{\zeta} = (\zeta_0, \zeta_1, \ldots, \zeta_m)^\top \in \mathbb{Z}_q^{m+1}$ *is such that* $\boldsymbol{\zeta}^\top \mathbf{M} = \mathbf{0}_d$, $\zeta_0 \neq 0$, *then* $\rho(\{j : \zeta_j \neq 0\}) \in \Omega$.
2) *If* $A \in \Omega$, *it is possible to sample* $\boldsymbol{\zeta} \in \mathrm{Im}(\mathbf{M}^*)$ *uniformly conditioned on a)* $\zeta_0 = 1$, *and b)* $\zeta_j = 0$ *for all* $j \in \rho^{-1}(A^c)$.
3) *Let* $\{j_1, \ldots, j_\ell\} = \rho^{-1}(A^c)$. *For any* $(a_{j_1}, \ldots, a_{j_\ell}) \in \mathbb{Z}_q^\ell$, *the probability that* $(\tau_{j_1}, \ldots, \tau_{j_\ell}) = (a_{j_1}, \ldots, a_{j_\ell})$ *is the same if a)* $\boldsymbol{\tau} \leftarrow \mathrm{Im}(\mathbf{M}^*)$ *or b)* $\boldsymbol{\tau} \leftarrow \mathrm{Im}(\mathbf{M}^*)$ *conditioned on* $\tau_0 = 0$.

Proof. 1) Observe that, since $\mathbf{M}$ is the transposed of the parity check matrix of $\mathbf{M}^*$, $\boldsymbol{\zeta}^\top \mathbf{M} = \mathbf{0}_d$ if and only if $\boldsymbol{\zeta} \in \mathrm{Im}(\mathbf{M}^*)$, that is $\boldsymbol{\zeta} = \mathbf{M}^* \boldsymbol{\omega}$ for some $\boldsymbol{\omega} \in \mathbb{Z}_q^{m+1-d}$ and in particular $\zeta_j = (\mathbf{r}_j^*)^\top \boldsymbol{\omega}$. Suppose that $B := \rho(\{j : \zeta_j \neq 0\}) \notin \Omega$. But then, by definition of Ω^*, $B^c := [L] \backslash B \in \Omega^*$, so $\mathbf{r}_0^* = \sum_{j \in \rho^{-1}(B^c)} a_j \mathbf{r}_j^*$ for some coefficients a_j. Multiplying on both sides by $\boldsymbol{\omega}$, $\zeta_0 = \sum_{j \in \rho^{-1}(B^c)} a_j \zeta_j = 0$, which contradicts the assumption $\zeta_0 \neq 0$. The rest of the proof is given in appendix A.

3   Commit-and-Prove Scheme

GS Proofs consist of a two step process: given some set of equations and a solution — which is a witness of satisfiability — a prover first commits to the solution and then proves that the committed values satisfy the set of equations. This corresponds to the notion of commit-and-prove (CaP) scheme ([20,7]), although the reformulation in these terms is not straightforward, since GS Proofs allow for a flexibility (different commitment/ equation types) which is not easily captured by the standard definition of a CaP scheme. To address this issue, Escala and Groth [11] write the GS proof system as a CaP scheme with labels. The labels are meant to specify the different commitment/equation types. For example, one might commit to the pair $(\mathsf{t}, m) = (\mathsf{sca}_{\hat{\mathbb{G}}}, m)$, which indicates that $m \in \mathbb{Z}_q$ is a variable whose commitment is in the group $\hat{\mathbb{G}}$.

Let $\mathcal{R}_\mathcal{L}$ be an efficiently verifiable ternary relation, which is described by tuples $(gk, x, W) \in \mathcal{R}_\mathcal{L}$ consisting of a group key, the statement x and the witness W. Define $\mathcal{L}_{gk}$ the language of all statements x for which there is a witness W such that $(gk, x, W) \in \mathcal{R}_\mathcal{L}$. This witness W is a set of pairs $(\mathsf{t}_i, m_i) \in \mathcal{T}_{gk} \times \tilde{\mathcal{M}}_{gk} \subset \mathcal{M}_{gk}$, where $\mathcal{T}_{gk}$ is the label space and $\mathcal{M}_{gk}$ is the labeled message space. For instance, $(\mathsf{sca}_{\hat{\mathbb{G}}}, m) \in \mathcal{M}_{gk}$, $\mathsf{sca}_{\hat{\mathbb{G}}} \in \mathcal{T}_{gk}$.

We assume that the statement x unambiguously defines some vector T_x of elements of $\mathcal{T}_{gk}$. One should think of T_x as describing the labels which define the correct format of a witness of x.

Definition 3. *(Labeled CaP scheme (modified from [11])) A commit-and-prove scheme* $\mathsf{CaP} = (\mathsf{G}, \mathsf{LabGen}, \mathsf{Com}, \mathsf{P}, \mathsf{V})$ *for* $\mathcal{L}_{gk}$ *consists of five PPT algorithms.*

- $\mathsf{G}(1^\lambda)$*: This algorithm runs in two steps. On input the security parameter* λ, $\mathsf{G}_0(1^\lambda)$ *outputs a group key* gk *which includes the description of a group, a space* $\mathcal{K}_{gk}$ *of valid commitment keys, a message space* $\mathcal{M}_{gk}$, *a randomness space* $\mathcal{R}_{gk}$ *and a commitment space* $\mathcal{C}_{gk}$. *Algorithm* $\mathsf{G}_1(gk)$ *outputs the pair* (gk, ck) *where* $ck \in \mathcal{K}_{gk}$ *is a commitment key.*
- $\mathsf{LabGen}(gk, ck, x, W)$*: This algorithm, on input* gk, $ck \in \mathcal{K}_{gk}$, *a pair* (x, W) *such that* $(gk, x, W) \in \mathcal{R}_\mathcal{L}$, *outputs a public label,* $\mathsf{k}^p = (\mathsf{t}, \tilde{\mathsf{t}})$, *and a secret label,* k^s, *for each* $\mathsf{t} \in \mathsf{T}_x$.
- $\mathsf{Com}(gk, ck, (\mathsf{k}^p, \mathsf{k}^s, m))$*: On input* gk, $ck \in \mathcal{K}_{gk}$, *a public label* $\mathsf{k}^p = (\mathsf{t}, \tilde{\mathsf{t}})$ *such that* $(\mathsf{t}, m) \in \mathcal{M}_{gk}$ *and a secret label* k^s, *this algorithm picks randomness* $(\mathsf{t}, r) \in \mathcal{R}_{gk}$ *and returns a commitment* c *with label* k^p *such that* $(\mathsf{k}^p, c) \in \mathcal{C}_{gk}$.
- $\mathsf{P}(gk, ck, x, Op, C)$*: On input* gk, $ck \in \mathcal{K}_{gk}$, $x \in \mathcal{L}_{gk}$ *and sets* $Op = \{(\mathsf{k}_i^p = (\mathsf{t}_i, \tilde{\mathsf{t}}_i), \mathsf{k}_i^s, m_i, r_i) : i \in \mathcal{I}\}$, $C = \{(\mathsf{k}_i^p, c_i) : i \in \mathcal{I}\}$ *such that for each* $i \in \mathcal{I}$, $(\mathsf{k}_i^p, \mathsf{k}_i^s, m_i, r_i)$ *is a valid opening of* (k_i^p, c_i), *and such that* $(gk, x, \{(\mathsf{t}_i, m_i) : i \in \mathcal{I}\}) \in \mathcal{R}_\mathcal{L}$, *this algorithm outputs a proof* π.
- $\mathsf{V}(gk, ck, x, C, \pi)$*: Given the group key* gk, *a commitment key* ck, *a statement* x, *a proof* π *and commitments* $(\tilde{\mathsf{t}}_i, c_i) \in \mathcal{C}_{gk}$, *algorithm* V *returns 1 if the proof is accepted and 0 otherwise.*

Compared to [11], in our definition commitments admit an extra label pair $(\mathsf{k}^p, \mathsf{k}^s)$ and an algorithm which generates this label LabGen. For the Groth-Sahai CaP scheme CaP_{GS}, we ignore these additional labels. This extra label will be necessary when we construct a CaP scheme CaP_{par} for partial satisfiability of quadratic equations based on CaP_{GS}, as we implicitly use different GS commitment keys ck_j to prove a single statement with CaP_{par}. The keys used for each commitment can be seen as public label of the commitment (but not as part of the witness (t_i, m_i)) and the secret label k^s as the trapdoor (when it exists, else it is some special symbol). That is why we also say in the definition of P "$(\mathsf{k}_i^p, \mathsf{k}_i^s, m_i, r_i)$ is a valid opening of (k_i^p, c_i)", as the opening might depend on k_i^p. Although the simulation trapdoor is not necessary to compute the commitments, we assume that the algorithm $\mathsf{Com}(gk, ck, (\mathsf{k}^p, \mathsf{k}^s, m))$ takes also as input k^s, because k^s might not only contain the simulation trapdoor but also additional information which allows to speed up the computation, like the discrete logarithms of the commitment keys given in k_i^p.[1]

Another difference with [11] is that, we explicitly define a keyspace $\mathcal{K}_{gk}$. We assume that membership in $\mathcal{K}_{gk}$ is efficiently decidable for all gk. Further, we

---

[1] Escala and Groth observed that if the prover knows the discrete logarithm of ck, computation is sped up significantly, as then most exponentiations can be replaced by operations in $\mathbb{Z}_q$.

distinguish between the group key and the commitment keys. The language for which we define the proof system depends on gk but not of ck, i.e. it is a group dependent language[2] This is done with the purpose of precisely defining the sets of hiding and binding keys to define verifiable correlated key generation for the GS CaP. For the following definition, we restrict ourselves to some CaP scheme in which algorithm LabGen is trivial (so that we can omit (k^p, k^s)).

Definition 4. *Given some* CaP *scheme, we define* $\mathcal{K}^t_{gk,\text{bind}}$ *(resp.* $\mathcal{K}^t_{gk,\text{hid}}$*) as the set of* $ck \in \mathcal{K}_{gk}$ *such that* $\text{Com}(gk, ck, t, m)$ *is perfectly binding (resp. perfectly hiding) for all* $(t, m) \in \mathcal{M}_{gk}$.

We defer the definitions of perfect completeness and perfect soundness to appendix B.1. Roughly speaking, completeness guarantees that correctly generated proofs are accepted, perfect soundness that a proof of a false statement is never accepted. Both the GS proof system and our scheme satisfy a strong notion of security, namely *composable zero-knowledge*.

Definition 5. *The commit-and-prove system* CaP *is (computationally) composable zero-knowledge if there exist PPT algorithms* SimGen, SimCom, SimProve, SimLabGen *such that for all non-uniform polynomial time stateful interactive adversaries* $\mathcal{A}$,

$$\Pr\left[(gk, ck) \leftarrow \mathsf{G}(1^\lambda) : \mathcal{A}(gk, ck) = 1\right]$$
$$\approx \Pr\left[(gk, ck, tk) \leftarrow \mathsf{SimGen}(1^\lambda) : \mathcal{A}(gk, ck) = 1\right] \text{ and}$$

$$\Pr\left[(gk, ck, tk) \leftarrow \mathsf{SimGen}(1^\lambda); (x, \mathcal{I}) \leftarrow \mathcal{A}^{\mathsf{Com}(gk,ck,\cdot),\widetilde{\mathsf{LabGen}}(gk,ck,\cdot,\cdot)}(gk, ck, tk);\right.$$
$$\left. \pi \leftarrow \mathsf{P}(gk, ck, x, \{(k^p_i, k^s_i, m_i, r_i) : i \in \mathcal{I}\}, \{(k^p_i, c_i) : i \in \mathcal{I}\}) : \mathcal{A}(\pi) = 1\right] =$$
$$\Pr\left[(gk, ck, tk) \leftarrow \mathsf{SimGen}(1^\lambda); (x, \mathcal{I}) \leftarrow \mathcal{A}^{\mathsf{SimCom}(gk,ck,\cdot),\widetilde{\mathsf{SimLabGen}}(gk,ck,\cdot)}(gk, ck, tk);\right.$$
$$\left. \pi \leftarrow \mathsf{SimProve}(gk, ck, tk, x, \{(k^p_i, k^s, s_i) : i \in \mathcal{I}\}, \{(k^p_i, c_i) : i \in \mathcal{I}\}) : \mathcal{A}(\pi) = 1\right],$$

where a) $\mathsf{SimLabGen}(gk, ck, tk, x)$ *returns* $(k^p_i = (t_i, \tilde{t}_i), k^s_i)$ *for each* $t_i \in \mathsf{T}_x$, *b)* $\mathsf{SimLabGen}(gk, ck, tk, x)$ *(resp.* $\mathsf{LabGen}(gk, ck, x)$*) returns* $k^p_i = (t_i, \tilde{t}_i)$ *for each* $t_i \in \mathsf{T}_x$ *with the distribution induced by running* SimLabGen *(resp. by* LabGen*) with the same input, c)* $\mathsf{SimCom}(gk, ck, \cdot)$ *on* (k^p_i, k^s_i) *outputs* $(k^p_i, c_i) \in \mathcal{C}_{gk}$, *d)* $\mathcal{A}$ *picks* $(x, \mathcal{I})$ *such that* $(gk, x, \{(t_i, m_i) : i \in \mathcal{I}\}) \in \mathcal{R}_{\mathcal{L}}$.

Finally a non-interactive Zap (NI Zap) is a (computationally) witness indistinguishable proof in the plain model, i.e. without common reference string. Informally, computational WI just requires that two proofs generated by the prover on a statement x with different witnesses W_0, W_1 should be computationally indistinguishable even for an adversary who chooses (x, W_0, W_1).

---

[2] As in the original definition of GS Proofs, [16].

Definition 6. *We say that the commit-and-prove system* CaP *is a NI Zap if the algorithm* $G_1(gk)$ *is trivial (i.e.* $(gk, ck = \bot) \leftarrow G_1(gk)$*) and the CaP has perfect completeness, perfect soundness and computational witness indistinguishability.*

4 Groth-Sahai NIZK Proofs

In this section we give a high-level description of the GS proof system in terms of a commit-and-prove scheme as in [11]. We concentrate on the key generation and commit phase, which are the ones necessary to understand our construction, for the full description we refer the reader to the original paper.

The GS proof system allows to prove that x is satisfiable, where x encodes some set of quadratic equations in a bilinear group of the following form:

$$\sum_{j=1}^{n} f(\alpha_j, y_j) + \sum_{i=1}^{m} f(x_i, \beta_i) + \sum_{i=1}^{m}\sum_{j=1}^{n} f(x_i, \gamma_{ij} y_j) = t, \tag{2}$$

where A_1, A_2, A_T are $\mathbb{Z}_q$-vector spaces equipped with some bilinear map $f : A_1 \times A_2 \to A_T$, $\boldsymbol{\alpha} \in A_1^n$, $\boldsymbol{\beta} \in A_2^m$, $\boldsymbol{\Gamma} = (\gamma_{ij}) \in \mathbb{Z}_q^{m \times n}$, $t \in A_T$. The modules and the map f can be defined in different ways as: (a) in pairing-product equations (PPEs), $A_1 = \hat{\mathbb{G}}$, $A_2 = \check{\mathbb{H}}$, $A_T = \mathbb{T}$, $f(\hat{x}, \check{y}) = \hat{x}\check{y} \in \mathbb{T}$, (b1) in multi-scalar multiplication equations in $\hat{\mathbb{G}}$ (MMEs), $A_1 = \hat{\mathbb{G}}$, $A_2 = \mathbb{Z}_q$, $A_T = \hat{\mathbb{G}}$, $f(\hat{x}, y) = y\hat{x} \in \hat{\mathbb{G}}$, b2) MMEs in $\check{\mathbb{H}}$ (MMEs), $A_1 = \mathbb{Z}_q$, $A_2 = \check{\mathbb{H}}$, $A_T = \check{\mathbb{H}}$, $f(x, \check{y}) = x\check{y} \in \check{\mathbb{H}}$, and (c) in quadratic equations in $\mathbb{Z}_q$ (QEs), $A_1 = A_2 = A_T = \mathbb{Z}_q$, $f(x, y) = xy \in \mathbb{Z}_q$. Each element describing an equation receives a label t_i and each equation a label L_{eq}, for instance $\mathsf{L}_{eq} = \mathsf{QE}$ is a quadratic equation, or $\mathsf{L}_{eq} = \mathsf{MLin}_{\hat{\mathbb{G}}}$ is a linear multi-scalar multiplication equation with variables in $\hat{\mathbb{G}}$. The classification of Escala and Groth of equation types (see [11], figure 6) is very fine grained with

$G(1^\lambda)$	$\mathsf{SimGen}(1^\lambda)$
$gk \leftarrow (q, \hat{\mathbb{G}}, \check{\mathbb{H}}, \mathbb{T}, e, \hat{g}, \check{h}) \leftarrow \mathcal{G}(1^\lambda)$	$gk \leftarrow (q, \hat{\mathbb{G}}, \check{\mathbb{H}}, \mathbb{T}, e, \hat{g}, \check{h}) \leftarrow \mathcal{G}(1^\lambda)$
$\omega, \sigma, \xi, \psi \leftarrow \mathbb{Z}_q^*$	$\omega, \sigma, \xi, \psi, \leftarrow \mathbb{Z}_q^*$
$\hat{\mathbf{v}} \leftarrow (\xi\hat{g}, \hat{g})^\top$, $\check{\mathbf{v}} \leftarrow (\psi\check{h}, \check{h})^\top$	$\hat{\mathbf{v}} \leftarrow (\xi\hat{g}, \hat{g})^\top$, $\check{\mathbf{v}} \leftarrow (\psi\check{h}, \check{h})^\top$
$\hat{\mathbf{u}} \leftarrow \omega\hat{\mathbf{v}}$, $\check{\mathbf{u}} \leftarrow \sigma\check{\mathbf{v}}$	$\hat{\mathbf{u}} \leftarrow \omega\hat{\mathbf{v}} + (\hat{0}, \hat{g})^\top$, $\check{\mathbf{u}} \leftarrow \sigma\check{\mathbf{v}} + (\check{0}, \check{h})^\top$
$\hat{\mathbf{w}} \leftarrow \hat{\mathbf{u}} - (\hat{0}, \hat{g})^\top$, $\check{\mathbf{w}} \leftarrow \check{\mathbf{u}} - (\check{0}, \check{h})^\top$	$\hat{\mathbf{w}} \leftarrow \hat{\mathbf{u}} - (\hat{0}, \hat{g})^\top$, $\check{\mathbf{w}} \leftarrow \check{\mathbf{u}} - (\check{0}, \check{h})^\top$
$ck \leftarrow (\hat{\mathbf{u}}, \hat{\mathbf{v}}, \hat{\mathbf{w}}, \check{\mathbf{u}}, \check{\mathbf{v}}, \check{\mathbf{w}})$	$ck \leftarrow (gk, \hat{\mathbf{u}}, \hat{\mathbf{v}}, \hat{\mathbf{w}}, \check{\mathbf{u}}, \check{\mathbf{v}}, \check{\mathbf{w}})$
Return (gk, ck)	$tk \leftarrow (ck, \omega, \sigma)$
	Return (gk, ck, tk)

Label t	Message	Randomness	Commitment	$\mathcal{K}_{gk,\mathrm{bind}}^{\mathrm{t}}$	$\mathcal{K}_{gk,\mathrm{hid}}^{\mathrm{t}}$
$\mathsf{sca}_{\hat{\mathbb{G}}}$	$(\mathsf{sca}_{\hat{\mathbb{G}}}, x)$	$(\mathsf{sca}_{\hat{\mathbb{G}}}, r)$	$\hat{\mathbf{c}} \leftarrow \hat{\mathbf{w}}x + \hat{\mathbf{v}}r$	$ck : \hat{\mathbf{w}} \notin \langle\hat{\mathbf{v}}\rangle$	$ck : \hat{\mathbf{w}} \in \langle\hat{\mathbf{v}}\rangle$
$\mathsf{com}_{\hat{\mathbb{G}}}$	$(\mathsf{com}_{\hat{\mathbb{G}}}, \hat{x})$	$(\mathsf{com}_{\hat{\mathbb{G}}}, r, s)$	$\hat{\mathbf{c}} \leftarrow \mathbf{e}_2\hat{x} + \hat{\mathbf{v}}r + \hat{\mathbf{u}}s$	$ck : \hat{\mathbf{u}} \in \langle\hat{\mathbf{v}}\rangle$	$ck : \hat{\mathbf{u}} \notin \langle\hat{\mathbf{v}}\rangle$

Fig. 1. Generator algorithms of the CaP scheme of [11] and table describing most important commitment types

the objective of augmenting the class of equations which admit zero-knowledge proofs (softening the requirement $t = 0_\mathbb{T}$ for PPEs given in [16]) and also of describing efficiency improvements which only apply to a specific equation type.

Given some equation of the form (2), the first step of the prover is to commit to all elements describing the equation according to their label, where "commit" is used in a wide sense as an equivalent to embed the elements in the right space. That is, for instance, the equation $\hat{x}_1 \check{b}_1 + \hat{x}_2 \check{b}_2 = 0_\mathbb{T}$ is described by $\hat{x}_1, \hat{x}_2, \check{b}_1, \check{b}_2$. As $\hat{x}_1, \hat{x}_2$ are variables in $\hat{\mathbb{G}}$, they have the label $\text{com}_{\hat{\mathbb{G}}}$, while the constants $\check{b}_1, \check{b}_2$ have the labels $(\text{pub}_{\check{\mathbb{H}}}, \check{b}_1)$ and $(\text{pub}_{\check{\mathbb{H}}}, \check{b}_2)$. The commitment to an element with label $\text{com}_{\hat{\mathbb{G}}}$ is described in figure 1, and the one to $\check{b}_i$ is simply $(\check{0}, \check{b}_i)^\top \in \check{\mathbb{H}}^2$. The latter deviates from the usual definition of commitment in the sense that it is not computationally hiding.

The vector of labels T_x associated to some statement x, to which we referred in the syntactic definition of Labeled CaP, is the specification (in some fixed order) of the label type of all the elements describing the equation, for instance, in the example above $\mathsf{T}_x = (\text{com}_{\hat{\mathbb{G}}}, \text{com}_{\hat{\mathbb{G}}}, (\text{pub}_{\check{\mathbb{H}}}, \check{b}_1), (\text{pub}_{\check{\mathbb{H}}}, \check{b}_2))$. Of course this vector of labels must be consistent with the equation type L_{eq}.

Recall that GS CaP uses the *parameter switching technique* of [15]. This means that the common reference string can be generated in two different, computationally indistinguishable ways: in the soundness setting, not even a computationally unbounded adversary can convince a verifier of a false statement, while in the witness indistinguishability (WI) setting, the keys are generated with a trapdoor which allows to construct simulated proofs.

In the CaP scheme for partial satisfiability we will let the prover choose some keys ck_j and some trapdoors tk_j, $j = 1, \ldots, m$. Each key ck_j will be used to prove/simulate a different atomic statement x_i (i.e. satisfiability of some equation set $\mathcal{S}_i$). It is fundamental to define precisely for which type of equations a key ck_j defines perfectly sound proofs or when it allows to simulate them, so that we can prove meaningful statements about partial satisfiability.

For this, although we do not describe all possible equation types or all possible labels in $\mathcal{T}_{gk}$ (or their corresponding commitments), we must specify how to commit to variables. The four possible label types for variables are $\text{sca}_{\hat{\mathbb{G}}}, \text{sca}_{\check{\mathbb{H}}}, \text{com}_{\hat{\mathbb{G}}}, \text{com}_{\check{\mathbb{H}}}$, which correspond, respectively, to elements 1) in $A_1 = \mathbb{Z}_q$, 2) in $A_2 = \mathbb{Z}_q$, 3) in $A_1 = \hat{\mathbb{G}}$ or 4) in $A_2 = \check{\mathbb{H}}$. The interesting thing about these commitments (see figure 1) is that they are binding or hiding depending on the way we generate the keys. Very roughly, simulation in GS Proofs works by opening "commitments" to more than one element. Thus, a necessary condition to simulate proofs for some equation of the form (2) with a certain ck is that ck defines a hiding commitment for the variables in one of the modules A_i. In summary, it is essential to discuss which keys ck are binding/hiding for each variable type.

Key space and commitments. The space of keys $\mathcal{K}_{gk}$ consists of all tuples $(gk, \hat{\mathbf{u}}, \hat{\mathbf{v}}, \hat{\mathbf{w}}, \check{\mathbf{u}}, \check{\mathbf{v}}, \check{\mathbf{w}})$ such that gk is a valid description of an asymmetric bilinear group, $\hat{\mathbf{u}}, \hat{\mathbf{v}} \in \hat{\mathbb{G}}^2$, $\check{\mathbf{u}}, \check{\mathbf{v}} \in \check{\mathbb{H}}^2$ and $\hat{\mathbf{w}} = \hat{\mathbf{u}} - (\hat{0}, \hat{g})^\top$, $\check{\mathbf{w}} = \check{\mathbf{u}} - (\check{0}, \check{h})^\top$. We define

$e_2 := (0,1)^\top$ and $\hat{e}_2 := (\hat{0}, \hat{g})^\top$. Commitments to scalars and group elements are described in the table below (for the group $\check{\mathbb{H}}$ they are defined analogously). Note that in the soundness setting the algorithm $G(1^\lambda)$ outputs keys $ck \in \mathcal{K}_{gk,\text{bind}}^{\text{sca}_{\hat{\mathbb{G}}}} \cap \mathcal{K}_{gk,\text{bind}}^{\text{com}_{\hat{\mathbb{G}}}}$ (see Definition 4) and in the WI setting, SimGen outputs $ck \in \mathcal{K}_{gk,\text{hid}}^{\text{sca}_{\hat{\mathbb{G}}}} \cap \mathcal{K}_{gk,\text{hid}}^{\text{com}_{\hat{\mathbb{G}}}}$, but in general a key might be binding/hiding only for one label t, this is why the CaP formulation of GS Proofs is really convenient to define correlated key generation.

Right vs Left-Simulatable. The simulation trapdoor $tk = (ck, \omega, \sigma)$ allows to double-open some commitments. This trapdoor allows to simulate all the considered equations, but it will be convenient to be more precise. We say that an equation is left-simulatable if there exists an efficient algorithm SimProve which takes as input $tk = \omega$ (right-simulatable if the same holds for σ). Roughly speaking, an equation x of the form (2) is left (resp. right) simulatable if it is possible to equivocate enough commitments to elements of A_1 (resp. to A_2) to $\hat{0}$ (resp. $\check{0}$) so that the equation admits the trivial solution. In any case, for our purposes it is enough to know that there are equations which can only be simulated on one side and that this can be made precise.

Admissible simulation labels. Further, we say an equation type L_{eq} admits label $\mathsf{t}_{sim} = \text{com}_{\hat{\mathbb{G}}}$ if $A_1 = \hat{\mathbb{G}}$ and it is left-simulatable or label $\mathsf{t}_{sim} = \text{sca}_{\hat{\mathbb{G}}}$ if $A_1 = \mathbb{Z}_q$ and it is left-simulatable (the same w.r.t. $\check{\mathbb{H}}$ and right-simulatable). For instance, QEs are both left and right-simulatable and admit both labels $\{\text{sca}_{\hat{\mathbb{G}}}, \text{sca}_{\check{\mathbb{H}}}\}$, while linear MMEs in $\hat{\mathbb{G}}$ admit the label $\text{sca}_{\hat{\mathbb{G}}}$ if the variables are in $\mathbb{Z}_q$ but only admit the label $\text{com}_{\hat{\mathbb{G}}}$ if the variables are in $\hat{\mathbb{G}}$ and the equation is homogeneous.

5   (Simulatable) Verifiable Correlated Key Generation: Definitions

Let $\mathsf{CaP} = (\mathsf{G} = (\mathsf{G}_0, \mathsf{G}_1), \mathsf{Com}, \mathsf{P}, \mathsf{V})$ be a commit-and-prove scheme with perfect soundness, perfect completeness and composable zero-knowledge and let $\mathsf{SimGen}, \mathsf{SimCom}, \mathsf{SimProve}$ be the corresponding simulation algorithms.

The definitions in this section are meant to capture the necessary properties that a CaP scheme must satisfy so that we can extend it to give proofs for partial relations. Given a monotone span program $\mathcal{SP} = (\mathbf{M}, \rho)$, we will require the existence of an algorithm K_{corr} (or $\mathsf{K}_{\text{scorr}}$ for the simulatable case) which outputs a set of correlated keys $\Sigma = \{ck_1, \ldots, ck_m\} \subset \mathcal{K}_{gk}$. These keys should be such that it can be publicly verified that the (unknown) subset of binding keys corresponds to some satisfying assignment of the predicate computed by $\mathcal{SP}$. Further, K_{corr} ($\mathsf{K}_{\text{scorr}}$) should also output a trapdoor for the non-binding keys.

When P is the predicate OR of two variables, the first definition (of VCKG) matches the original one of GOS. In this definition, we require the keys to be created from scratch, given only the group key gk.

On the other hand, for the second definition (SVCKG), our algorithms take as input a common reference string ck. In this case, we require the existence of an algorithm $\mathsf{K}_{\mathsf{scorr}}$ which outputs some set of keys with the same properties as in VCKG when ck is binding. We also require the existence of another algorithm which outputs only hiding keys with their simulation trapdoor. When ck is hiding, both should have identically distributed output.

To construct NIZK proofs of partial satisfiability, we will use as a building block a SVCKG scheme, while a construction of VCKG combined with the GS CaP will allow us to derive NIWI proofs of partial satisfiability in the plain model.

In both definitions, the vector $\mathbf{T}$ specifies a vector of labels. With these labels we can define precisely what we mean by "hiding key" or "binding key", as this depends on the label type. Later, when we use (S)VCKG to prove partial satisfiability, we will use the key ck_j to prove statement $x_{\rho(j)}$. These labels will guarantee that the key ck_j matches the equation type of $x_{\rho(j)}$.

Definition 7. CaP *admits P-Verifiable Correlated Key Generation for the predicate* $P_\Omega : \{0,1\}^L \to \{0,1\}$ *computed by a span program* $\mathcal{SP} = (\mathbf{M}, \rho)$ *and some label vector* $\mathbf{T} = (\mathsf{t}_1, \mathsf{t}_2, \ldots, \mathsf{t}_L)$, *if there exist two probabilistic polynomial time algorithms* $(\mathsf{K}_{\mathsf{corr}}, \mathsf{V}_{\mathsf{corr}})$, *with the following properties:*

a) *Given any* $\mathbf{v}_A \in \{0,1\}^L$ *such that* $P(\mathbf{v}_A) = 1$, *and* $gk \leftarrow \mathsf{G}_0(1^\lambda)$, *algorithm* $\mathsf{K}_{\mathsf{corr}}(gk, \mathcal{SP}, \mathbf{v}_A, \mathbf{T})$ *outputs* $(\Sigma, \mathsf{TK}_{A^c})$, *where* $\Sigma = \{ck_1, \ldots, ck_m\}$ *is such that for all* $j \in \rho^{-1}(A^c)$, $ck_j \in \mathcal{K}^{\mathsf{t}_{\rho(j)}}_{gk, \mathsf{hid}}$, *and* $\mathsf{TK}_{A^c} := \{tk_j : j \in \rho^{-1}(A^c)\}$ *is the set of the corresponding (valid) trapdoors.*

b) *For all PPT adversaries* $\mathsf{D} = (\mathsf{D}_0, \mathsf{D}_1)$ *and if* $\mathbf{v}_A, \mathbf{v}_B$ *are such that* $P(\mathbf{v}_A) = P(\mathbf{v}_B) = 1$,

$$\Pr\Big[gk \leftarrow \mathsf{G}_0(1^\lambda); (\Sigma, \mathsf{TK}_{A^c}) \leftarrow \mathsf{K}_{\mathsf{corr}}(gk, \mathcal{SP}, \mathbf{v}_A, \mathbf{T});$$

$$(\mathbf{v}_A, \mathbf{v}_B, st) \leftarrow \mathsf{D}_0(gk, \mathcal{SP}, \mathbf{T}) : \mathsf{D}_1(\Sigma, \mathsf{TK}_{A^c \cap B^c}, st) = 1\Big]$$

$$\approx \Pr\Big[gk \leftarrow \mathsf{G}_0(1^\lambda); (\Sigma, \mathsf{TK}_{B^c}) \leftarrow \mathsf{K}_{\mathsf{corr}}(gk, \mathcal{SP}, \mathbf{v}_B, \mathbf{T});$$

$$(\mathbf{v}_A, \mathbf{v}_B, st) \leftarrow \mathsf{D}_0(gk, \mathcal{SP}, \mathbf{T}) : \mathsf{D}_1(\Sigma, \mathsf{TK}_{A^c \cap B^c}, st) = 1\Big]$$

c) *Given as input* $(gk, \mathcal{SP}, \Sigma, \mathbf{T})$, $\mathsf{V}_{\mathsf{corr}}$ *outputs a bit* b *such that, for all PPT adversaries* D:

$$\Pr\Big[(\Sigma, \mathbf{T}) \leftarrow \mathsf{D}(gk, \mathcal{SP}, \mathbf{T}); \mathsf{V}_{\mathsf{corr}}(gk, \mathcal{SP}, \Sigma, \mathbf{T}) = 1 :$$

$$\rho(\{j : ck_j \in \mathcal{K}^{\mathsf{t}_{\rho(j)}}_{gk, \mathsf{bind}}\}) \notin \Omega\Big] = 0.$$

Definition 8. CaP *admits P-Simulatable Verifiable Correlated Key Generation for the predicate* $P_\Omega : \{0,1\}^L \to \{0,1\}$ *computed by a span program* $\mathcal{SP} = (\mathbf{M}, \rho)$ *and some label vector* $\mathbf{T} = (\mathsf{t}_1, \ldots, \mathsf{t}_L)$, *if there exist three probabilistic polynomial time algorithms* $(\mathsf{K}_{\mathsf{scorr}}, \mathsf{V}_{\mathsf{scorr}}, \mathsf{SimCorr})$, *with the following properties:*

a) *as in point a) of definition 7 except that* $\mathsf{K}_{\mathsf{scorr}}$ *receives* $(gk, ck) \leftarrow \mathsf{G}(1^\lambda)$ *as part of the input.*

b) *Given some* $(gk, ck, tk) \leftarrow \mathsf{SimGen}(1^\lambda)$, *algorithm* $\mathsf{SimCorr}(gk, ck, tk, \mathcal{SP}, \mathbf{T})$ *outputs* $(ck, \Sigma, \mathsf{TK})$ *such that* $\Sigma = \{ck_1, \dots, ck_m\}$ *is a set of commitment keys with* $ck_j \in \mathcal{K}^{\mathsf{t}_{\rho(j)}}_{gk,\mathsf{hid}}$ *for all* $j \in [m]$ *and a set* $\mathsf{TK} := \{tk_j : j \in [m]\}$ *such that* (ck_j, tk_j), $j \in [m]$ *is a valid pair of commitment key and trapdoor. Further,*

$$\Pr\Big[(gk, ck, tk) \leftarrow \mathsf{SimGen}(1^\lambda); (\Sigma, \mathsf{TK}_{A^c}) \leftarrow \mathsf{K}_{\mathsf{scorr}}(gk, ck, \mathcal{SP}, \mathbf{v}_A, \mathbf{T}) :$$

$$\mathsf{D}(gk, ck, \mathcal{SP}, \Sigma, \mathbf{T}) = 1\Big]$$

$$= \Pr\Big[(gk, ck, tk) \leftarrow \mathsf{SimGen}(1^\lambda); (\Sigma, \mathsf{TK}) \leftarrow \mathsf{SimCorr}(gk, ck, tk, \mathcal{SP}, \mathbf{T}) :$$

$$\mathsf{D}(gk, ck, \mathcal{SP}, \Sigma, \mathbf{T}) = 1\Big]$$

c) *Given* $(gk, ck, \mathcal{SP}, \Sigma, \mathbf{T})$, $\mathsf{V}_{\mathsf{scorr}}$ *outputs a bit* b *such that, for all PPT adversaries* D:

$$\Pr\Big[(gk, ck) \leftarrow \mathsf{G}(1^\lambda); (\Sigma, \mathbf{T}) \leftarrow \mathsf{D}(gk, ck, \mathcal{SP}, \mathbf{T}); \mathsf{V}_{\mathsf{scorr}}(gk, ck, \mathcal{SP}, \Sigma, \mathbf{T}) = 1 :$$

$$\rho(\{j : ck_j \in \mathcal{K}^{\mathsf{t}_{\rho(j)}}_{gk,\mathsf{bind}}\}) \notin \Omega\Big] = 0.$$

6   NIZK Proofs and NI Zap of Partial Satisfiability

In this section we formally put together all the pieces of the puzzle: the GS CaP, the new definition of labeled CaP and the notion of SVCKG (resp. VCKG) to construct NIZK proofs (resp. a NI Zap) for partial satisfiability.

More specifically, starting from the GS CaP described in section 4, $\mathsf{CaP}_{GS} = (\mathsf{G}, \mathsf{Com}, \mathsf{P}, \mathsf{V})$ for the language defined by relation $\mathcal{R}_\mathcal{L}$, and any construction of (S)VCKG, we build a CaP scheme $\mathsf{CaP}_{par} = (\mathsf{G}_p, \mathsf{LabGen}_p, \mathsf{Com}_p, \mathsf{P}_p, \mathsf{V}_p)$ and a NI Zap for the relation $\mathcal{R}_{par}$.

Relations of partial satisfiability. Formally, $\mathcal{R}_{par}$ consists of the tuples (gk, X, W) such that:

a) X consists of $\{\{x_i : i \in [L]\}, \mathbf{T}, \mathcal{SP}\}$, where x_i is a quadratic equation in the group described by gk and $\mathcal{SP}$ is a monotone span program which computes some predicate $P : \{0,1\}^L \to \{0,1\}$ such that CaP admits simulatable verifiable correlated key generation for P and the vector of labels $\mathbf{T}$,

b) Each statement x_i admits the simulation label t_i,

c) W is of the form $\{\tilde{W}_i : i \in [L]\}$, where for all i, $\tilde{W}_i = \{(t_{i\ell}, \tilde{m}_{i\ell}) : \ell \in [n_i], \tilde{m}_{i\ell} = (b_i, m_{i\ell})\}$ is such that a) if $b_i = 0$, $m_{i\ell} = 0$ for all $\ell \in [n_i]$ and b) if $b_i = 1$, $W_i := \{(t_{i\ell}, m_{i\ell}) : \ell \in [n_i]\}$ is such that $(gk, x_i, W_i) \in \mathcal{R}_\mathcal{L}$,

d) If $A := \{i \in [L] : b_i = 1\}$, then $P(\mathbf{v}_A) = 1$.

6.1   NIZK Proofs of Partial Satisfiability

The main idea of the construction is the following: the prover of the proof system for partial satisfiability runs $\mathsf{K}_{\mathsf{scorr}}$ on input the span program $\mathcal{SP}$ (of size m) encoded in X and some set A accepted by $\mathcal{SP}$. Algorithm $\mathsf{K}_{\mathsf{scorr}}$ returns a set of commitment keys $ck_1, \ldots, ck_m$ and the trapdoors for all ck_j, $j \in \rho^{-1}(A^c)$. The set A should correspond to the statements for which the prover has a real witness. The proof of the rest of the statements can be simulated using the trapdoors output by $\mathsf{K}_{\mathsf{scorr}}$. For zero-knowledge, the simulator will run $\mathsf{SimCorr}$ to generate only hiding keys with their respective trapdoors. The trapdoor tk_j will be used to simulate the proof of the statement $x_{\rho(j)}$, which is possible because the statement $x_{\rho(j)}$ admits the simulation label $\mathsf{t}_{\rho(j)}$.

- $\mathsf{G}_p(1^\lambda)$: Runs $(gk, ck) \leftarrow \mathsf{G}(1^\lambda)$.
- $\mathsf{LabGen}_p(gk, ck, X, W)$: Runs $(ck, \Sigma, \mathsf{TK}_{A^c}) \leftarrow \mathsf{K}_{\mathsf{scorr}}(gk, ck, \mathcal{SP}, \mathbf{v_A}, \mathbf{T})$, it parses Σ as $\{ck_1, \ldots, ck_m\}$, and for each pair (i, ℓ), it outputs $(\mathsf{k}_{i\ell}^p, \mathsf{k}_{i\ell}^s)$, where $\mathsf{k}_{i\ell}^p = (\mathsf{t}_{i\ell}, \tilde{\mathsf{t}}_{i\ell})$ and

$$(\tilde{\mathsf{t}}_{i\ell}, \mathsf{k}_{i\ell}^s) = \begin{cases} \tilde{\mathsf{t}}_{i\ell} = \{ck_j : j \in \rho^{-1}(i)\}, \mathsf{k}_{i\ell}^s = \{tk_j : j \in \rho^{-1}(i)\} & \text{if } i \in A^c, \\ \tilde{\mathsf{t}}_{i\ell} = \{ck_j : j \in \rho^{-1}(i)\}, \mathsf{k}_{i\ell}^s = 0 & \text{if } i \in A. \end{cases}$$

- $\mathsf{Com}_p(gk, ck, (\mathsf{k}_{i\ell}^p, \mathsf{k}_{i\ell}^s, m_{i\ell}))$: Parse $\mathsf{k}_{i\ell}^s$ as $(\mathsf{t}_{i\ell}, \{ck_j : j \in \rho^{-1}(i)\})$ and for each $i \in [L]$ and each $j \in \rho^{-1}(i)$, it defines

$$c_{i\ell j} := \begin{cases} c_{i\ell j} \leftarrow \mathsf{SimCom}(gk, ck_j, tk_j, \mathsf{t}_{i\ell}) & \text{if } i \in A^c, \\ c_{i\ell j} \leftarrow \mathsf{Com}(gk, ck_j, \mathsf{t}_{i\ell}, m_{i\ell}) & \text{if } i \in A. \end{cases}$$

It outputs $(\mathsf{k}_{i\ell}^p, \bigcup_{j \in \rho^{-1}(i)} c_{i\ell j})$.
- $\mathsf{P}_p(gk, ck, X, Op, C)$: Receives as input the statement X, some set $C = \bigcup_{i \in [L]} \bigcup_{j \in \rho^{-1}(i)} C_{ij}$ which is the union of sets of commitments $C_{ij} = \{c_{i\ell j} : \ell \in [n_i]\}$, and a set $Op = \bigcup_{i \in [L]} \bigcup_{j \in \rho^{-1}(i)} Op_{ij}$ which is the union of the sets $Op_{ij} := \{(\mathsf{t}_{i\ell}, ck_j, tk_j, m_{i\ell}, r_{i\ell}) : \ell \in [n_i]\}$, where each Op_{ij} is a valid opening of C_{ij} (we assume that for simulated commitments $m_{i\ell}$ is just set to 0). For each $i \in [L]$ and for each $j \in \rho^{-1}(i)$,

$$\pi_j := \begin{cases} \pi_j \leftarrow \mathsf{SimProve}(gk, ck_j, tk_j, x_i, Op_{ij}) & \text{if } i \in A^c, \\ \pi_j \leftarrow \mathsf{P}(gk, ck_j, x_i, Op_{ij}) & \text{if } i \in A. \end{cases}$$

Let $\Pi_i := \{\pi_j : j \in \rho^{-1}(i)\}$ and output $\Pi = \bigcup_{i \in [L]} \Pi_i$.
- $\mathsf{V}_p(gk, ck, X, C, \Pi)$: Given the group key gk, a commitment key ck, a statement X (which includes a description of $\mathbf{T}$), a proof Π and a set of commitments C, algorithm V_p proceeds as follows:
  - From the public types of the commitments in C, it derives a list of commitment keys $\Sigma = \{ck_1, \ldots, ck_m\}$ (or outputs failure if this is not possible). This is done by checking that for each $i \in [L]$ and each $\ell \in [n_i]$, the public types $\mathsf{k}_{i\ell}^p = (\mathsf{t}_{i\ell}, \tilde{\mathsf{t}}_{i\ell})$ are consistently assigned. That is, for each $i \in [L]$, $\tilde{\mathsf{t}}_{i\ell}$ should encode the same set of cardinal $|\rho^{-1}(i)|$ of commitment keys $\{ck_j : j \in \rho^{-1}(i)\} \subset \mathcal{K}_{gk}$, regardless of ℓ.

- It runs $b \leftarrow \mathsf{V}_{\mathsf{scorr}}(gk, ck, \Sigma, \mathbf{T})$ (the set of labels $\mathbf{T}$ is encoded in X). If $b = 0$, halts and outputs 0, else it proceeds.
- For each $i \in [L]$, and each $j \in \rho^{-1}(i)$, it verifies that each of the proofs π_j of statement x_i is satisfied individually by running $\mathsf{V}(gk, ck_j, x_i, C_{ij}, \pi_j)$.
- It outputs 0 if any of these checks fails, else it outputs 1.

- $\mathsf{SimGen}_p(1^\lambda)$: Runs $(gk, ck, tk) \leftarrow \mathsf{SimGen}(1^\lambda)$.
- $\mathsf{SimLabGen}_p(gk, ck, tk, X)$: Runs $(ck, \Sigma, \mathsf{TK}) \leftarrow \mathsf{SimCorr}(gk, ck, tk, \mathbf{T})$ and for every $\mathsf{t}_{i\ell}$, it returns $\mathsf{k}_{i\ell}^p := (\mathsf{t}_{i\ell}, \{ck_j : j \in \rho^{-1}(i)\})$ and $\mathsf{k}_{i\ell}^s := \{tk_j : j \in \rho^{-1}(i)\}$.
- $\mathsf{SimCom}_p(gk, ck, (\mathsf{k}_{i\ell}^p, \mathsf{k}_{i\ell}^s))$: This algorithm parses $\mathsf{k}_{i\ell}^p$ as $(\mathsf{t}_{i\ell}, \{ck_j : j \in \rho^{-1}(i)\})$ and $\mathsf{k}_{i\ell}^s$ as $\mathsf{k}_{i\ell}^s = \{tk_j : j \in \rho^{-1}(i)\}$. It outputs $\{(\mathsf{k}_{i\ell}^p, \bigcup_{j \in \rho^{-1}(i)} c_{i\ell j})\}$, where $c_{i\ell j} \leftarrow \mathsf{SimCom}(gk, ck_j, tk_j, \mathsf{t}_{i\ell})$.
- $\mathsf{SimProve}_p(gk, ck, tk, X, Op)$: For all $i \in [L]$, and all $j \in \rho^{-1}(i)$, and a set of commitment openings $Op = \bigcup_{i \in [L]} \bigcup_{j \in \rho^{-1}((i)} Op_{ij}$, this algorithm computes:

$$\pi_j \leftarrow \mathsf{SimProve}(gk, ck_j, tk_j, x_i, Op_{ij}).$$

It outputs $\Pi = \bigcup_{i \in [L]} \Pi_i$, where $\Pi_i := \{\pi_j : j \in \rho^{-1}(i)\}$.

Theorem 1. CaP_{par} *is a CaP scheme with perfect completeness, perfect soundness and composable zero-knowledge for $\mathcal{L}_{par}$.*

Proof. Perfect completeness follows from the completeness of the GS CaP and the fact, for all $i \in [L]$, x_i admits the simulation label t_i (else the prover could fail to compute a simulated proof for x_i). Perfect soundness follows from the perfect soundness of the GS CaP and the properties of SVCKG. Indeed, by property c) of SVCKG, if the verifier accepts the keys $\Sigma = \{ck_1, \ldots, ck_m\}$ (after running $\mathsf{V}_{\mathsf{scorr}}$), then $A := \rho(\{j : ck_j \in \mathcal{K}_{gk,\mathsf{bind}}^{\mathsf{t}_{\rho(j)}}\}) \in \Omega$. That is, there is some set $A \in \Omega$, such that for every $i \in A$ at least one $j \in \rho^{-1}(i)$ is a binding key for the label t_i. Therefore, for every $i \in A$, at least one of the proofs in the set Π_i (there are $|\rho^{-1}(i)|$ proofs of x_i) is generated with a binding key. By the perfect soundness of GS Proofs, this means $(gk, x_i, W_i) \in \mathcal{R}_{\mathcal{L}}$. Composable zero-knowledge holds because, by property b) of SVCKG, the keys output by $\mathsf{SimCorr}$ and by $\mathsf{K}_{\mathsf{scorr}}$ on a simulated key ck are identically distributed. The composable zero-knowledge property of GS Proofs guarantees that if x_i is a satisfiable quadratic equation (that is, if x_i is in the language accepted by GS Proofs), then a real proof computed with a simulated key has the same distribution as a simulated proof. On the other hand, if x_i is not satisfiable, then both in a real proof (computed with the output of $\mathsf{K}_{\mathsf{scorr}}$) or in a fake proof, the proof is simulated, so in both cases it follows the same distribution.

6.2 Non-Interactive Zap for Partial Satisfiability

The NI Zap for Satisfiability is constructed in a very similar way as the NIZK proofs of partial satisfiability, except that one uses as a building block the algorithms for VCKG (instead of SVCKG) and of course, the fact that $ck = \perp$.

It is obvious that the resulting construction is complete and soundness follows from the same arguments as before, namely from property c) of the definition of VCKG. We next sketch the proof for computational WI.

Suppose an adversary B against the WI of the Zap outputs $(X, W_0, W_1) \in \mathcal{R}_{par}$. Each W_b encodes a set $A_b \in \Omega$ which specifies for which sets of equations W_b contains a real witness. If $A_0 = A_1$, then the adversary will not be able to distinguish a proof computed with W_0 or W_1 unless it breaks the composable zero-knowledge property of GS proofs, which implies that real proofs with different witnesses are computationally indistinguishable and that simulated proofs are independent of the witness. Therefore, we can assume that $A_0 \neq A_1$. But in this case, we can use B to construct an adversary D that breaks property b) of the VCKG scheme. Indeed, D gives to its challenger $(\mathbf{v}_{A_0}, \mathbf{v}_{A_1})$ and receives $(\Sigma, \mathsf{TK}_{A_0^c \cap A_1^c})$, with Σ generated from A_b, for $b \leftarrow \{0, 1\}$. Even if b is unknown to D, it can compute a proof of the statement X. Indeed, for all $i \in [L]$, and all $j \in \rho^{-1}(i)$, D can compute a proof π_j of the statement x_i, as follows:

- if $i \in A_0 \cup A_1$, π_j is a real proof as it can extract a witness for x_i from W_0 or W_1,
- if $i \in (A_0 \cup A_1)^c = A_0^c \cap A_1^c$, π_j is a simulated proof computed with $\mathsf{TK}_{A_0^c \cap A_1^c}$
.

Finally, D gives the keys Σ and the proof of X to B, who outputs a bit b', and D forwards this bit to its challenger.

Since in the GS CaP, real proofs with a simulated key have the same distribution as simulated proofs, the proof given to B follows the same distribution as a proof generated with W_b. Thus, $|\Pr[b' = b] - 1/2|$ is non-negligible, so D is successful with non-negligible probability.

7   (Simulatable) Verifiable Correlated Key Generation: Constructions

We give some constructions of verifiable correlated key generation in different flavors. Let $\mathcal{SP}$ be a monotone span program computing $\Omega \subset \mathcal{P}([L])$. From the construction of proofs of partial satisfiability of last section, we know that to prove that there is a set of indexes $A \in \Omega$ such that all $\mathcal{S}_i$, for $i \in A$ are satisfiable, the GS CaP must admit (S)VCKG for a vector $\mathbf{T} = (\mathsf{t}_1, \ldots, \mathsf{t}_L)$ such that each equation $\mathcal{S}_i$ admits t_i as a simulation label. Therefore, we are interested in constructing P-verifiable correlated key generation for as many possible types of vectors $\mathbf{T}$ and general predicates P, since this means that our CaP for partial satisfiability will admit a wider class of languages.

7.1   SVCKG for sca$_{\hat{\mathbb{G}}}$ and MSPs

This construction of SVCK works for any $P : \{0,1\}^L \to \{0,1\}$ computed by a a monotone span program $\mathcal{SP}$, but only for the vector of labels $\mathbf{T} = (\mathrm{sca}_{\hat{\mathbb{G}}}, \ldots, \mathrm{sca}_{\hat{\mathbb{G}}})$ (the case $\mathbf{T} = (\mathrm{sca}_{\hat{\mathbb{H}}}, \ldots, \mathrm{sca}_{\hat{\mathbb{H}}})$ is defined in a similar way in $\hat{\mathbb{H}}$).

- $\mathsf{K_{scorr}}(gk, ck, \mathcal{SP}, A, \mathbf{T})$: The algorithm receives as input gk, $ck = (\hat{\mathbf{u}}, \hat{\mathbf{v}}, \hat{\mathbf{w}}, \check{\mathbf{u}}, \check{\mathbf{v}}, \check{\mathbf{w}})$, the description of a MSP $\mathcal{SP}$, a set of indexes $A \subset [L]$ such that $\mathcal{SP}$ accepts A and a vector of labels $\mathbf{T}$. It proceeds as follows:
  1 It samples $\boldsymbol{\tau}, \boldsymbol{\zeta} \in \mathbf{Im}(\mathbf{M}^*)$ uniformly at random conditioned on a) $\zeta_0 = 1$, $\zeta_j = 0$ for all $j \in \rho^{-1}(A^c)$ and b) $\tau_0 = 0$ (as in lemma 1.)
  2 It defines $\hat{\mathbf{z}}_j := \tau_j \hat{\mathbf{v}} + \zeta_j \hat{\mathbf{w}}$, $j \in [m]$ and outputs $(ck, \Sigma, \mathsf{TK}_{A^c})$, where $\Sigma = \{ck_1, \dots, ck_m\}$, $ck_j := (\hat{\mathbf{u}}_j, \hat{\mathbf{v}}, \hat{\mathbf{z}}_j, \check{\mathbf{u}}, \check{\mathbf{v}}, \check{\mathbf{w}})$, $\hat{\mathbf{u}}_j := \hat{\mathbf{z}}_j + (\hat{0}, \hat{g})^\top$ and $\mathsf{TK}_{A^c} := \{\tau_j : j \in \rho^{-1}(A^c)\}$.
- $\mathsf{V_{scorr}}(gk, ck, \mathcal{SP}, \Sigma, \mathbf{T})$: Parse each key as $ck_j = (\hat{\mathbf{u}}_j, \hat{\mathbf{v}}_j, \hat{\mathbf{w}}_j, \check{\mathbf{u}}_j, \check{\mathbf{v}}_j, \check{\mathbf{w}}_j)$, and reject if, for some $j \in [m]$, $ck_j \notin \mathcal{K}_{gk}$, $\hat{\mathbf{v}}_j \neq \hat{\mathbf{v}}$, $\check{\mathbf{u}}_j \neq \check{\mathbf{u}}$ or $\check{\mathbf{v}}_j \neq \check{\mathbf{v}}$. Else, define $\hat{\mathbf{z}}_0 := \hat{\mathbf{w}}$ and $\hat{\mathbf{Z}} := (\hat{\mathbf{z}}_0 || \hat{\mathbf{z}}_1 || \dots || \hat{\mathbf{z}}_m)$. Output 1 if $\hat{\mathbf{Z}}\mathbf{M} = \hat{\mathbf{0}}_{2 \times d}$ holds, else output 0.
- $\mathsf{SimCorr}(gk, ck, tk, \mathcal{SP}, \mathbf{T})$: The algorithm receives $(gk, ck, tk) \leftarrow \mathsf{SimGen}(1^\lambda)$, with $tk = (\omega, \sigma)$. It samples a uniform vector in $\boldsymbol{\mu}^\top = (\mu_0, \dots, \mu_m) \in \mathbf{Im}(\mathbf{M}^*)$ subject to the the restriction $\mu_0 = \omega$. For all $j \in [m]$, it defines $\hat{\mathbf{z}}_j := \mu_j \hat{\mathbf{v}}$, $\hat{\mathbf{u}}_j := \hat{\mathbf{z}}_j + (\hat{0}, \hat{g})^\top$ and $ck_j := (\hat{\mathbf{u}}_j, \hat{\mathbf{v}}, \hat{\mathbf{z}}_j, \check{\mathbf{u}}, \check{\mathbf{v}}, \check{\mathbf{w}})$. It outputs $(ck, \Sigma, \mathsf{TK})$, where $\Sigma = \{ck_1, \dots, ck_m\}$ and $\mathsf{TK} := \{\mu_j : j \in [m]\}$.

Lemma 2. *The GS CaP scheme described in section 4 admits simulatable verifiable correlated key generation for labels* $\mathbf{T} = (sca_{\hat{\mathbb{G}}}, \dots, sca_{\hat{\mathbb{G}}})$.

Proof. We prove that the algorithms described above satisfy points a), b), c) of definition 8.

By definition of $\boldsymbol{\zeta}$, $\zeta_j = 0$ for all $j \in \rho^{-1}(i)$, $i \in A^c$. Therefore, for all $i \in A^c$, $\hat{\mathbf{z}}_j = \tau_j \hat{\mathbf{v}}$, so according to figure (1), $ck_j \in \in \mathcal{K}^{sca_{\hat{\mathbb{G}}}}_{gk,hid}$ and the corresponding trapdoor is $tk_j = \tau_j$.

To see b), note that all the keys output by SimCorr are obviously in $\mathcal{K}^{sca_{\hat{\mathbb{G}}}}_{gk,hid}$ and the trapdoor is valid. We just have to argue that the output of the algorithm SimCorr has the same distribution as the output of $\mathsf{K_{scorr}}$ when $(gk, ck, tk) \leftarrow \mathsf{SimGen}(1^\lambda)$. In that case, $\hat{\mathbf{w}} = \omega \hat{\mathbf{v}}$ and $\mathsf{K_{scorr}}$ outputs $\hat{\mathbf{z}}_j = \tau_j \hat{\mathbf{v}} + \zeta_j \hat{\mathbf{w}} = (\tau_j + \omega \zeta_j)\hat{\mathbf{v}}$, for all $j \in [m] \cup \{0\}$. Let $\boldsymbol{\nu} := \boldsymbol{\tau} + \omega \boldsymbol{\zeta}$. The constraints imposed on $\boldsymbol{\tau}, \boldsymbol{\zeta}$ imply that $\boldsymbol{\nu}$ is uniform conditioned on a) $\boldsymbol{\nu} \in \mathbf{Im}(\mathbf{M}^*)$, b) $\nu_0 = \omega$ and c) $\nu_j = \tau_j$ for all $j \in \rho^{-1}(A^c)$. Because of part 3) of lemma 1, if $\{j_1, \dots, j_\ell\} = \rho^{-1}(A^c)$, the distribution of $(\tau_{j_1}, \dots, \tau_{j_\ell})$, is the uniform one conditioned on $\boldsymbol{\tau} \leftarrow \mathbf{Im}(\mathbf{M}^*)$. We conclude that $\boldsymbol{\nu}$ is a uniformly random vector in $\mathbf{Im}(\mathbf{M}^*)$ conditioned to $\nu_0 = \omega$, so the outputs of SimCorr and $\mathsf{K_{scorr}}$ are identically distributed, which proves b).

Finally, for c), let $\Sigma = \{ck_1, \dots, ck_m\}$ be some set of keys accepted by the verifier, i.e. some set such that $\Sigma \subset \mathcal{K}_{gk}$ and $\hat{\mathbf{Z}}\mathbf{M} = \hat{\mathbf{0}}_{2 \times d}$. Since if $(gk, ck) \leftarrow \mathsf{G}(1^\lambda)$, the vectors $\hat{\mathbf{v}}, \hat{\mathbf{w}}$ are a basis of $\hat{\mathbb{G}}^2$, we can write each column of $\mathbf{Z}$ (numbered from 0 to m) as $\hat{\mathbf{z}}_j = \hat{\mathbf{v}}\tau_j + \hat{\mathbf{w}}\zeta_j$, for some arbitrary values τ_j, ζ_j. In this notation, $\hat{\mathbf{Z}} = \hat{\mathbf{v}}\boldsymbol{\tau}^\top + \hat{\mathbf{w}}\boldsymbol{\zeta}^\top$. Replacing in the verification equation, we have that $(\hat{\mathbf{v}}\boldsymbol{\tau}^\top + \hat{\mathbf{w}}\boldsymbol{\zeta}^\top)\mathbf{M} = \hat{\mathbf{0}}_{2 \times d}$. But since $\hat{\mathbf{v}}, \hat{\mathbf{w}}$ are linearly independent, the equation can only hold if $\boldsymbol{\zeta}^\top \mathbf{M} = \mathbf{0}_{1 \times d}$. We can now apply lemma 1, part 1), to conclude that $A := \rho(\{j : \zeta_j \neq 0\}) \in \Omega$. But if $\zeta_j \neq 0$, then ck_j is a binding key for the label $sca_{\hat{\mathbb{G}}}$, which proves c).

7.2   VCKG for $sca_{\hat{\mathbb{G}}}$ and MSPs

This construction is almost identical to the previous one except that for the non-simulatable case. For the OR predicate of two variables, it matches exactly the GOS construction.

- $\mathsf{K}_{\mathsf{corr}}(gk, \mathcal{SP}, A, \mathbf{T})$: The algorithm first runs the key generation algorithm of the CaP scheme obtaining $gk \leftarrow \mathsf{G}_0(1^\lambda)$. Then it proceeds as in the $\mathsf{K}_{\mathsf{scorr}}$ algorithm, except that the vectors $\hat{\mathbf{z}}_j$ are now defined by $\hat{\mathbf{z}}_j = \tau_j \hat{\mathbf{v}} + \zeta_j (\hat{0}, \hat{g})^\top$.
- $\mathsf{V}_{\mathsf{corr}}(gk, \Sigma, \mathbf{T})$: The algorithm proceeds as algorithm $\mathsf{K}_{\mathsf{scorr}}$ but with $\hat{\mathbf{z}}_0 := (\hat{0}, \hat{g})^\top$.

The proof follows the same lines as the previous one, the only relevant difference is that to prove point b) of definition 7, we use the DDH Assumption in $\hat{\mathbb{G}}$. The argument is very similar to [15] and is omitted.

7.3   Other Labels

We could not construct (S)VCKG for $\mathbf{T} = (\mathsf{com}_{\hat{\mathbb{G}}}, \ldots, \mathsf{com}_{\hat{\mathbb{G}}})$ for the original GS CaP based on SXDH. The core of our construction is to use secret sharing techniques to guarantee that at least a certain subset of the vectors $\hat{\mathbf{z}}_1, \ldots, \hat{\mathbf{z}}_m$ is linearly independent of $\hat{\mathbf{v}}$. The problem is that for group elements, the soundness condition is exactly the opposite, namely it requires linear dependency of $\hat{\mathbf{v}}, \hat{\mathbf{u}}$ (see Fig. 1). In appendix C.1 we extend the GS CaP based on SXDH to admit new labels, which allow to commit to group elements and to scalars in a different way, with respective labels $\mathsf{com}_{\hat{\mathbb{G}}}^{par}$ and $\mathsf{sca}_{\hat{\mathbb{G}}}^{par}$ ($\mathsf{sca}_{\hat{\mathbb{H}}}^{par}, \mathsf{com}_{\hat{\mathbb{H}}}^{par}$). This new instantiation of GS proofs $\tilde{\mathsf{CaP}}_{GS}$ admits (S)VCKG for these new label types, i.e. for any vector $\mathbf{T} = (\mathsf{t}_1, \ldots, \mathsf{t}_L)$, for $\mathsf{t}_i \in \{\mathsf{com}_{\hat{\mathbb{G}}}^{par}, \mathsf{sca}_{\hat{\mathbb{G}}}^{par}\}$. This means that we can apply our approach to many more sets of equations $\mathcal{S}_1, \ldots, \mathcal{S}_L$, but at some efficiency cost, because $\tilde{\mathsf{CaP}}_{GS}$ is less efficient that the original GS CaP.

7.4   Efficiency Discussion

Proof size. The proof that some sets of equations $\mathcal{S}_i, i \in [L]$ are partially satisfiable requires the prover to send the keys Σ and then a proof (real or simulated) of satisfiability of each $\mathcal{S}_i$. The size of the proofs depends thus on the equations in $\mathcal{S}_i$. Therefore, to understand the performance of our proof system for partial satisfiability, the best thing is to analyze its *overhead*, which is the difference between the size of our proof and the sum of the sizes of a simulated proof of $\mathcal{S}_i$, for all $i \in [L]$. The number of elements necessary to commit to all the variables in $\mathcal{S}_i$ is also counted as part of the proof of $\mathcal{S}_i$. That is, the overhead is the difference between proving partial satisfiability and proving that all of $\mathcal{S}_i, i \in [L]$ hold, using independent variables for each of the $\mathcal{S}_i$.

Efficient encoding of Σ. It is quite important for the efficiency comparison to note that the set Σ of keys output by the correlated key generation algorithm admit a more efficient encoding. In all our constructions, the description of Σ

requires to give m vectors $\hat{\mathbf{z}}_1, \ldots, \hat{\mathbf{z}}_m$. Instead of letting the prover choose Σ and then verifying if the keys are valid with the verification algorithm $\mathsf{V}_{\mathsf{corr}}$ (or $\mathsf{V}_{\mathsf{scorr}}$) by checking whether $\hat{\mathbf{Z}}\mathbf{M} = \mathbf{0}$ (where the last m columns of $\hat{\mathbf{Z}}$ are $\hat{\mathbf{z}}_1, \ldots, \hat{\mathbf{z}}_m$ and the first is $\hat{\mathbf{z}}_0 := \hat{\mathbf{w}}$), it is enough to let the prover output only $\hat{\mathbf{z}}_1, \ldots, \hat{\mathbf{z}}_{m-d}$. Indeed, let $\mathbf{M}_0^*$ be the $(m + 1 - d) \times (m + 1 - d)$ minor formed by the first $m + 1 - d$ rows of $\mathbf{M}^*$ and $\mathbf{M}_1^*$ the minor formed by the rest of the rows. Reordering if necessary, we can assume that $\mathbf{M}_0^*$ is invertible. Then, the following holds:

Lemma 3. *Let* $\hat{\mathbf{Z}}$ *be an arbitrary matrix in* $\hat{\mathbb{G}}^{2 \times (m+1)}$, *with columns* $\hat{\mathbf{z}}_0, \hat{\mathbf{z}}_1, \ldots, \hat{\mathbf{z}}_m$, *then:*

$$\hat{\mathbf{Z}}\mathbf{M} = \mathbf{0} \iff (\hat{\mathbf{z}}_{m+1-d}|| \cdots ||\hat{\mathbf{z}}_m) = (\hat{\mathbf{z}}_0||\hat{\mathbf{z}}_1|| \cdots ||\hat{\mathbf{z}}_{m-d}) \left(\mathbf{M}_1^* (\mathbf{M}_0^*)^{-1}\right)^\top. \quad (3)$$

Proof. Denote as $\mathbf{f}_1, \mathbf{f}_2$ the rows of $\mathbf{Z}$. Note that $\hat{\mathbf{Z}}\mathbf{M} = \mathbf{0}$ if and only if $\mathbf{f}_1, \mathbf{f}_2 \in \mathrm{Im}(\mathbf{M}^*)$, since the columns of $\mathbf{M}^*$ are a basis of all vectors $\mathbf{f}$ such that $\mathbf{f}^\top \mathbf{M} = \mathbf{0}$ (by definition of the parity check matrix). On the other hand, a vector $\mathbf{f} \in \mathrm{Im}(\mathbf{M}^*)$ if and only if there exists some $\mathbf{w} \in \mathbb{Z}_q^2$ such that $\mathbf{f}^\top = (\mathbf{M}_0^*\mathbf{w}||\mathbf{M}_1^*\mathbf{w})^\top$. Since $\mathbf{M}_0^*$ has full rank this is equivalent to $\mathbf{f} \in \mathrm{Im}(\mathbf{M}^*)$ if and only if $(f_{m+1-d}, \ldots, f_m)^\top = \mathbf{M}_1^*(\mathbf{M}_0^*)^{-1}(f_0, f_1, \ldots, f_{m-d})^\top$. The statement follows from applying this reasoning to $\mathbf{f}_1, \mathbf{f}_2$.

The lemma implies that we can eliminate the test of algorithms $\mathsf{V}_{\mathsf{corr}}, \mathsf{V}_{\mathsf{scorr}}$, and instead let the verifier compute the last d columns of $\hat{\mathbf{Z}}$ on its own using $\hat{\mathbf{z}}_1, \ldots, \hat{\mathbf{z}}_{m-d}$. This means that sending Σ requires only $2(m - d)$ group elements.

Comparison with previous work. We compare our results with the approach of Groth [14] (simplified by Camenisch et al. [6]) for the statement "1-out-of-L sets of equations $\mathcal{S}_1, \ldots, \mathcal{S}_L$ are satisfiable". They construct a "compiler" which takes some sets of satisfiable equations $\mathcal{S}_1, \ldots, \mathcal{S}_L$ and turns them into a single set of equations which is only satisfiable if one of the $\mathcal{S}_i$'s is. The compiler works by (renaming if necessary) assuming the $\mathcal{S}_1, \ldots, \mathcal{S}_L$ have independent variables and adding variables $b_1, \ldots, b_{L-1}$, $b_i \in \{0, 1\}$ and defining $b_L := 1 - b_1 - \ldots - b_{L-1}$. For each $i \in [L]$, b_i modifies the equations in $\mathcal{S}_i$ so that they admit the trivial solution if $b_i = 0$ and that they remain unchanged if $b_i = 1$. The overhead is the cost of proving $b_i \in \{0, 1\}$ for all $i \in [L - 1]$, which is $(L - 1)(6|\hat{\mathbb{G}}| + 6|\hat{\mathbb{H}}|)$. Our solution is notably more efficient (only $2|L - 1||\hat{\mathbb{G}}|$) when the vector of admissible simulation labels of $\mathcal{S}_1, \ldots, \mathcal{S}_L$ is $(\mathrm{sca}_{\hat{\mathbb{G}}}, \ldots, \mathrm{sca}_{\hat{\mathbb{G}}})$, although it is not clear what happens for other $\mathbf{T}$ (see the efficiency discussion in appendix C.1).

8   Examples

In this section we give two examples of P-Simulatable Verifiable Correlated Key Generation. Throughout this section, given a set $S \subset \mathbb{Z}_q$ and some $i \in S$, $\lambda_i^S(X) := \prod_{j \in S \setminus \{i\}} \frac{X - j}{i - j}$.

Example 2. (Or of two equations).

$$\mathbf{M} = \begin{pmatrix} 1 \\ 1 \\ 1 \end{pmatrix} \qquad \mathbf{M}^* = \begin{pmatrix} 1 & 0 \\ 1 & 1 \\ -2 & -1 \end{pmatrix}$$

$$\mathbf{M}_0^* = \begin{pmatrix} 1 & 0 \\ 1 & 1 \end{pmatrix} \qquad \mathbf{M}_1^* = \begin{pmatrix} -2 & -1 \end{pmatrix} \qquad \mathbf{T}_{\mathbf{M}^*} = \mathbf{M}_1^*(\mathbf{M}_0^*)^{-1} = \begin{pmatrix} -1 & -1 \end{pmatrix}.$$

$(\mathbf{M}, \rho)$ is a monotone span program, where $\rho(1) = 1$, $\rho(2) = 2$ which computes the predicate Or of two variables, $(\mathbf{M}^*, \rho)$ computes the dual predicate. The algorithm $\mathsf{K}_{\mathsf{scorr}}$ (or $\mathsf{K}_{\mathsf{corr}}$), receives a vector $\mathbf{v}_A \in \{0,1\}^2$ such that $P(\mathbf{v}_A) = 1$. Since P is monotone, alternatively we can say that it receives a set $A \subset \{1,2\}$ such that $|A| \geq 1$. It then generates $\boldsymbol{\zeta}, \boldsymbol{\tau}$ according to one of these two possibilities:

1. If $A = \{1\}$, it sets $\boldsymbol{\zeta} = (1, -1, 0) = \mathbf{M}^* \begin{pmatrix} 1 \\ -2 \end{pmatrix}$, $\boldsymbol{\tau} = (0, r, -r) = \mathbf{M}^* \begin{pmatrix} 0 \\ -r \end{pmatrix}$, for $r \leftarrow \mathbb{Z}_q$, that is, $\boldsymbol{\zeta}, \boldsymbol{\tau} \in \mathbf{Im}(\mathbf{M}^*)$ are uniform conditioned on $\zeta_0 = 1$, $\zeta_2 = 0$ and $\tau_0 = 0$.
2. If $A = \{2\}$, $\boldsymbol{\zeta} = (1, 0, -1) = \mathbf{M}^* \begin{pmatrix} 1 \\ -1 \end{pmatrix}$, $\boldsymbol{\tau} = (0, r, -r) = \mathbf{M}^* \begin{pmatrix} 0 \\ -r \end{pmatrix}$, for $r \leftarrow \mathbb{Z}_q$, that is, $\boldsymbol{\zeta}, \boldsymbol{\tau} \in \mathbf{Im}(\mathbf{M}^*)$ are uniform conditioned on $\zeta_0 = 1$, $\zeta_1 = 0$ and $\tau_0 = 0$.

If $A = \{1, 2\}$, it can choose one of the previous alternatives arbitrarily, so we can assume w.l.o.g. that $|A| = 1$. Then, according to the type of labels which it receives, it proceeds as follows:

- If $\mathbf{T} = (\mathsf{sca}_{\hat{\mathbb{G}}}, \mathsf{sca}_{\hat{\mathbb{G}}})$, it sets $\hat{\mathbf{z}}_1 = \zeta_1 \hat{\mathbf{w}} + \tau_1 \hat{\mathbf{v}}$, $\hat{\mathbf{z}}_2 = \zeta_2 \hat{\mathbf{w}} + \tau_2 \hat{\mathbf{v}}$, the trapdoor for the key indexed by A^c is $\pm r$. That is, we have:
  1. If $A = \{1\}$, $\hat{\mathbf{z}}_1 = -\hat{\mathbf{w}} + r\hat{\mathbf{v}}$, $\hat{\mathbf{z}}_2 = -r\hat{\mathbf{v}}$, $tk_2 = -r$.
  2. If $A = \{2\}$, $\hat{\mathbf{z}}_1 = r\hat{\mathbf{v}}$, $\hat{\mathbf{z}}_2 = -\hat{\mathbf{w}} - r\hat{\mathbf{v}}$, $tk_1 = r$.

As we explained in section 6, instead of letting the prover of CaP_{par} run $\mathsf{K}_{\mathsf{scorr}}$ and output $(\hat{\mathbf{z}}_0 := \hat{\mathbf{w}}, \hat{\mathbf{z}}_1, \hat{\mathbf{z}}_2)$ and then let the verifier run $\mathsf{V}_{\mathsf{corr}}$ to see if the keys are properly generated, one can gain some efficiency by letting the prover only send $\hat{\mathbf{z}}_1$, and recover $\hat{\mathbf{z}}_2$ as:

$$\hat{\mathbf{z}}_2 = \hat{\mathbf{Z}}_0 \mathbf{T}_{\mathbf{M}^*}^\top,$$

where $\hat{\mathbf{Z}}_0 = (\hat{\mathbf{z}}_0 || \hat{\mathbf{z}}_1)$.

Example 3. k-out-of-L equations.

$$\mathbf{M} = \begin{pmatrix} 1 & 0 & \dots & 0 \\ 1 & 1 & \dots & 1 \\ \vdots & \vdots & & \vdots \\ 1 & L & \dots & L^{k-1} \end{pmatrix} \qquad \mathbf{M}^* = \begin{pmatrix} 1 & 0 & \dots & 0 \\ -\lambda_1^{[L]}(0) & -\lambda_1^{L}(0) & \dots & -\lambda_1^{[L]}(0) \\ -\lambda_2^{[L]}(0) & -2\lambda_2^{[L]}(0) & \dots & -2^{L-k}\lambda_2^{[L]}(0) \\ \vdots & \vdots & & \vdots \\ -\lambda_L^{[L]}(0) & -L\lambda_L^{[L]}(0) & \dots & -L^{L-k}\lambda_L^{[L]}(0) \end{pmatrix}.$$

This is the definition of $\mathbf{M}$, $\mathbf{M}^*$ which is consistent with the definition given in section 2.3 but as in the more efficient version of our protocol, the matrix

M does not play any role, we can choose a more efficient encoding of M^* (this allows to save in computation for the prover), namely:

$$M^* = \begin{pmatrix} 1 & 0 & \cdots & 0 \\ 1 & 1 & \cdots & 1 \\ \vdots & \vdots & & \vdots \\ 1 & L & \cdots & L^{L-k} \end{pmatrix}.$$

It is obvious that with both definitions M^* computes the same span program as in one case we just have replaced each row by some scalar multiple of itself, and this does not change the linear dependencies among the rows. It can be easily verified that:

$$T_{M^*} = \begin{pmatrix} \lambda_0^S(L-k+1) & \lambda_1^S(L-k+1) & \cdots & \lambda_{L-k}^S(L-k+1) \\ \lambda_1^S(L-k+2) & \lambda_0^S(L-k+2) & \cdots & \lambda_{L-k}^S(L-k+2) \\ \vdots & \vdots & & \vdots \\ \lambda_0^S(L) & \lambda_1^S(L) & \cdots & \lambda_{L-k}^S(L) \end{pmatrix},$$

where $S = \{0, 1, \ldots, L-k\}$. That is, if $\hat{z}_i^\top = (\hat{z}_{i1}, \hat{z}_{i2})$, then both $(\hat{z}_{01}, \hat{z}_{11}, \ldots, \hat{z}_{L1})$ and $(\hat{z}_{02}, \hat{z}_{12}, \ldots, \hat{z}_{L2})$ are evaluations of some univariate polynomial of degree at most $L - k$ in the points $0, 1, \ldots, L$ and for any $j = 1, 2$, the transformation matrix T_{M^*} which allows to compute $(\hat{z}_{L-k+1j}, \ldots, \hat{z}_{Lj})$ from $(\hat{z}_{0j}, \ldots, \hat{z}_{L-kj})$ is simply a polynomial interpolation matrix. Further, given some $A \in \Omega_{(k,L)}$, the vectors ζ, τ can be defined as the evaluation in $0, 1, \ldots, L$ of two uniformly random polynomials $\zeta(x), \tau(x)$ of degree at most $L - k$ conditioned on 1) $\zeta(0) = 1$ and $\zeta(i) = 0$ for all $i \in A^c$ (ζ always exists since $|A^c| \leq L - k$ and is unique if $|A| = k$) and 2) $\tau(0) = 0$.

Another paradigmatic example of access structure realizable by a monotone span program is the threshold hierarchical one, see Tassa [26]. Although we did not include any example, recall that our construction is also for non-ideal sss, that is, the monotone span program might have more than one row with the same label (this is important since there are not that many known instances of ideal sss for interesting access structures).

9   Applications

Next we discuss some applications of our results, but we expect that many more can be found, for instance, in the design of signature schemes with complex functionalities in the standard model in bilinear groups like attribute-based signatures. Another interesting direction to explore is the application to anonymous credentials.

Proving that some commitments open to $b \in \{0,1\}^L$, *and* $\mathbf{wt}(b) = 1$. Given some group key gk and some commitment keys $\hat{w}, \hat{v}$, our results allow to give more efficient proofs that each of the commitments in $\{\hat{c}_i : i \in [L]\} \subset \hat{\mathbb{G}}^2$ opens to a bit $b_i \in \{0,1\}$, and that $\sum_{i \in [L]} b_i = 1$.

Alternatively, if we let $\hat{\mathbf{c}}_L := \hat{\mathbf{w}} - \sum_{i \in [L-1]} \hat{\mathbf{c}}_i$, it is enough to prove that each of the commitments in $\{\hat{\mathbf{c}}_i : i \in [L-1]\}$ opens to a bit $b_i \in \{0,1\}$. In the asymmetric instantiation of bilinear groups of GS proofs, this requires $(L-1)(4|\hat{\mathbb{G}}| + 6|\check{\mathbb{H}}|)$ elements for the proofs and $2(L-1)|\hat{\mathbb{G}}|$ for the description of $\hat{\mathbf{c}}_1, \ldots, \hat{\mathbf{c}}_{L-1}$. On the other hand, we can encode the statement as a partial satisfiability statement as:

$$\text{"}(L-1)\text{-out-of-}L \text{ of } (\{\exists r_1 \in \mathbb{Z}_q : \hat{\mathbf{c}}_1 = r_1 \hat{\mathbf{v}}\}, \ldots, \{\exists r_L \in \mathbb{Z}_q : \hat{\mathbf{c}}_L = r_L \hat{\mathbf{v}}\}) \text{ hold."} \tag{4}$$

Each statement $x_i = \{\exists r_i \in \mathbb{Z}_q : \hat{\mathbf{c}}_i = r_i \hat{\mathbf{v}}\}$ can be encoded as two linear equations (with equation label $\mathsf{MLin}_{\hat{\mathbb{H}}}$), and they both admit the simulation label $\mathsf{sca}_{\hat{\mathbb{H}}}$. The size of the proof is thus $2|\check{\mathbb{H}}|$ for the description of the correlated keys Σ (see section 8), $L(2|\check{\mathbb{H}}| + 2|\hat{\mathbb{G}}|)$ for the proof (real or simulated) of x_i and $2(L-1)|\hat{\mathbb{G}}|$ for the description of $\hat{\mathbf{c}}_1, \ldots, \hat{\mathbf{c}}_{L-1}$. In conclusion, our approach saves $O(L)$ elements in the proof size.

We note that if $\hat{\mathbf{w}}, \hat{\mathbf{v}}$ are part of some common reference string generated by a trusted party, we can prove (4) in zero-knowledge, but we can also take $\hat{\mathbf{w}} = (\hat{0}, \hat{g})^\top$ and let the prover generate $\hat{\mathbf{v}}$ so that no other party knows its discrete logarithm. Then using the NI Zap for partial satisfiability, the prover can create a NIWI proof of (4) without a trusted setup.

Proving membership in a list. Chandran, Groth and Sahai [8] showed how to prove that a committed value is in some public list $\{\lambda_1, \ldots, \lambda_N\} \subset \check{\mathbb{H}}$ with proof size $O(\sqrt{N})$. The main idea is to write the list elements in a matrix $\mathbf{R}$ of size $L \times L$, $L := \sqrt{N}$ and then give two sets of commitments $\{\hat{\mathbf{c}}_i : i \in [L]\}$, $\{\hat{\mathbf{d}}_i : i \in [L]\} \subset \hat{\mathbb{G}}^2$, each opening to a different bit string of weight 1. Without going into details, using some homomorphic properties of the commitments, the prover uses one of the bit strings to (privately) select a row i of the matrix $\mathbf{R}$ and the other to select a column j. With some additional checks, this convinces the verifier that a commitment $\hat{\mathbf{c}}$ opens to some (secret) position i, j of the matrix $\mathbf{R}$. In summary, for the proof of membership in a list of size N we need to prove twice a statement of the type 4, so our results allow to save $O(\sqrt{N})$ group elements.

Ring signatures. Ring signatures [22] allow a signer to sign on behalf of an ad-hoc group to which it belongs, anonymously. The proof of membership in a list of size $O(\sqrt{N})$ was designed by Chandran, Groth and Sahai [8] with the objective of designing more efficient ring signatures. Their scheme (with a signature size of $O(\sqrt{N})$ when the ring size is N) has the shortest signature size of all the schemes known in the standard model. Our savings for the proof of membership translate directly into savings for this construction.

Simulation-sound NIZKs. Simulation-sound NIZKs [21,23] are non-interactive zero-knowledge proofs with a stronger soundness requirement. More specifically, no prover should be able to construct a false proof which is accepted by the verifier even after seeing several simulated proofs of false statements. This notion is useful to construct IND-CCA2 encryption schemes following the Naor-Yung paradigm [21].

One technique to build simulation-sound proofs of satisfiability of some set of equations $\mathcal{S}$ over a bilinear group suggested by Groth [14] and subsequently explored by several papers with small variations [6,17], is to give a GS proof of the statement: "$\mathcal{S}$ is satisfiable" or "$\hat{c}$ is a commitment to some signature". Real proofs will use a witness for $\mathcal{S}$, while simulated proofs will prove the other branch of the statement, using as simulation trapdoor the secret key of the signature.

In general, the technique of Groth for constructing simulation-sound proofs might not be the most efficient for all equation types $\mathcal{S}$, (for instance if $\mathcal{S}$ encodes membership in a linear space of $\hat{\mathbb{G}}^n$, see [1]) but when it is, one should check if our improvements apply. For instance, in the simulation sound proof of Camenisch *et al.* [6], one has to prove the OR of two equations which admit the label $(\text{sca}_{\hat{\mathbb{G}}}, \text{sca}_{\hat{\mathbb{G}}})$, and this has an overhead of only $2|\hat{\mathbb{G}}|$, as opposed to the $6|\hat{\mathbb{G}}| + 6|\hat{\mathbb{H}}|$ elements originally computed in [6]. On the other hand, our techniques do not seem to help for the simulation sound proof of [17].

References

1. Abdalla, M., Benhamouda, F., Pointcheval, D.: Disjunctions for hash proof systems: New constructions and applications. Cryptology ePrint Archive, Report 2014/483 (2014), http://eprint.iacr.org/2014/483

2. Bellare, M., Goldwasser, S.: New paradigms for digital signatures and message authentication based on non-interactive zero knowledge proofs. In: Brassard, G. (ed.) CRYPTO 1989. LNCS, vol. 435, pp. 194–211. Springer, Heidelberg (1990)

3. Blakley, G.: Safeguarding cryptographic keys. In: Proceedings of the National Computer Conference, American Federation of Information, Processing Societies Proceedings, vol. (48), pp. 313–317 (1979)

4. Blum, M., Feldman, P., Micali, S.: Non-interactive zero-knowledge and its applications. In: STOC 1988, pp. 103–112 (1988)

5. Boyen, X., Waters, B.: Compact group signatures without random oracles. In: Vaudenay, S. (ed.) EUROCRYPT 2006. LNCS, vol. 4004, pp. 427–444. Springer, Heidelberg (2006)

6. Camenisch, J., Chandran, N., Shoup, V.: A public key encryption scheme secure against key dependent chosen plaintext and adaptive chosen ciphertext attacks. In: Joux, A. (ed.) EUROCRYPT 2009. LNCS, vol. 5479, pp. 351–368. Springer, Heidelberg (2009)

7. Canetti, R., Lindell, Y., Ostrovsky, R., Sahai, A.: Universally composable two-party and multi-party secure computation. In: 34th ACM STOC, Montréal, Québec, Canada, May 19–21, pp. 494–503. ACM Press (2002)

8. Chandran, N., Groth, J., Sahai, A.: Ring signatures of sub-linear size without random oracles. In: Arge, L., Cachin, C., Jurdziński, T., Tarlecki, A. (eds.) ICALP 2007. LNCS, vol. 4596, pp. 423–434. Springer, Heidelberg (2007)

9. Cramer, R., Damgård, I.B., Schoenmakers, B.: Proof of partial knowledge and simplified design of witness hiding protocols. In: Desmedt, Y.G. (ed.) CRYPTO 1994. LNCS, vol. 839, pp. 174–187. Springer, Heidelberg (1994)

10. De Santis, A., Di Crescenzo, G., Persiano, G.: Secret sharing and perfect zero-knowledge. In: Stinson, D.R. (ed.) CRYPTO 1993. LNCS, vol. 773, pp. 73–84. Springer, Heidelberg (1994)

11. Escala, A., Groth, J.: Fine-tuning groth-sahai proofs. In: Krawczyk, H. (ed.) PKC 2014. LNCS, vol. 8383, pp. 630–649. Springer, Heidelberg (2014)

12. Escala, A., Herold, G., Kiltz, E., Ràfols, C., Villar, J.: An algebraic framework for diffie-hellman assumptions. In: Canetti, R., Garay, J.A. (eds.) CRYPTO 2013, Part II. LNCS, vol. 8043, pp. 129–147. Springer, Heidelberg (2013)

13. Fiat, A., Shamir, A.: How to prove yourself: Practical solutions to identification and signature problems. In: Odlyzko, A.M. (ed.) CRYPTO 1986. LNCS, vol. 263, pp. 186–194. Springer, Heidelberg (1987)

14. Groth, J.: Simulation-sound NIZK proofs for a practical language and constant size group signatures. In: Lai, X., Chen, K. (eds.) ASIACRYPT 2006. LNCS, vol. 4284, pp. 444–459. Springer, Heidelberg (2006)

15. Groth, J., Ostrovsky, R., Sahai, A.: New techniques for noninteractive zero-knowledge. J. ACM 59(3), 11 (2012)

16. Groth, J., Sahai, A.: Efficient noninteractive proof systems for bilinear groups. SIAM J. Comput. 41(5), 1193–1232 (2012)

17. Hofheinz, D., Jager, T.: Tightly secure signatures and public-key encryption. In: Safavi-Naini, R., Canetti, R. (eds.) CRYPTO 2012. LNCS, vol. 7417, pp. 590–607. Springer, Heidelberg (2012)

18. Karchmer, M., Wigderson, A.: On span programs. In: Structure in Complexity Theory Conference, pp. 102–111 (1993)

19. Katz, J., Vaikuntanathan, V.: Round-optimal password-based authenticated key exchange. In: Ishai, Y. (ed.) TCC 2011. LNCS, vol. 6597, pp. 293–310. Springer, Heidelberg (2011)

20. Kilian, J.: Use of randomness on algorithms and protocols. MIT Press (1990)

21. Naor, M., Yung, M.: Public-key cryptosystems provably secure against chosen ciphertext attacks. In: 22nd ACM STOC, Baltimore, Maryland, USA, May 14–16, pp. 427–437. ACM Press (1990)

22. Rivest, R.L., Shamir, A., Tauman, Y.: How to leak a secret. In: Boyd, C. (ed.) ASIACRYPT 2001. LNCS, vol. 2248, pp. 552–565. Springer, Heidelberg (2001)

23. Sahai, A.: Non-malleable non-interactive zero knowledge and adaptive chosen-ciphertext security. In: 40th FOCS, October 17–19, pp. 543–553. IEEE Computer Society Press, New York (1999)

24. Schnorr, C.-P.: Efficient identification and signatures for smart cards. In: Brassard, G. (ed.) CRYPTO 1989. LNCS, vol. 435, pp. 239–252. Springer, Heidelberg (1990)

25. Shamir, A.: How to share a secret. Commun. ACM 22(11), 612–613 (1979)

26. Tassa, T.: Hierarchical threshold secret sharing. In: Naor, M. (ed.) TCC 2004. LNCS, vol. 2951, pp. 473–490. Springer, Heidelberg (2004)

A   Proof of Lemma 1

We give the proof of parts 2),3) of lemma 1.

Proof. (Lemma 1) Since $A \in \Omega$, $A^c \notin \Omega^*$. Thus, if B is a basis of the vectors $\{\mathbf{r}_j^* : j \in \rho^{-1}(A^c)\}$, the set $\{\mathbf{r}_j^* : j \in B \cup \{0\}\}$ is a set of linearly independent vectors. Find a set of indexes $C \subset [m]$ such that $B \cup \{0\} \subset C$ and the vectors $\{\mathbf{r}_j^* : j \in C\}$ are a basis the space spanned by the rows of $\mathbf{M}^*$. Note that $\boldsymbol{\zeta} = \mathbf{M}^* \boldsymbol{\omega}_1$ and $\boldsymbol{\tau} = \mathbf{M}^* \boldsymbol{\omega}_2$, if and only if $\zeta_j = (\mathbf{r}_j^*)^\top \boldsymbol{\omega}_1$, $\tau_j = (\mathbf{r}_j^*)^\top \boldsymbol{\omega}_2$. Because the rows indexed by C are a basis of the rows of $\mathbf{M}^*$, $\zeta_j, \tau_j, j \in C$, uniquely define ζ, τ and further, if we sample a vector $\nu \leftarrow \mathbf{Im}(\mathbf{M}^*)$, $\{\nu_j : j \in C\}$ is a uniform set of values in $\mathbb{Z}_q$. This shows that there is always one and only one vector ν which is compatible with some fixed set of values $\{\nu_j : j \in C\}$. This proves part 2), as it implies that if we set $\zeta_0 = 1$ and $\zeta_j = 0$ for all $j \in B$, and $\zeta_j \leftarrow \mathbb{Z}_q$ for all $j \in C\backslash(B \cup \{0\})$, this defines a unique vector ζ which is uniform conditioned on satisfying the constraints specified in 2). To see 3), just note that, regardless of whether we sample $\tau \leftarrow \mathbf{Im}(\mathbf{M}^*)$, or $\tau \leftarrow \mathbf{Im}(\mathbf{M}^*)$ conditioned on $\tau_0 = 0$, the same arguments used so far guarantee that $\{\tau_j : j \in C\backslash\{0\}\}$ is a uniform set of values in $\mathbb{Z}_q$, and by construction of B and C, $\{\tau_j : j \in \rho^{-1}(A^c)\}$ are completely determined by a subset of this set, namely by $\{\tau_j : j \in B\}$ and therefore independent of τ_0.

B   Security Definitions

B.1   Commit-and-Prove Schemes

Below we give the remaining security definitions for commit-and-prove schemes as taken from [11] and adapted to our modifications.

Definition 9 (Perfect Completeness). *The commit-and-prove system* CaP *is (perfectly) correct if for all adversaries* $\mathcal{A}$

$$\Pr\Big[(gk, ck) \leftarrow \mathsf{G}(1^\lambda); (x, W = \{(t_i, m_i) : i \in \mathcal{I}\}) \leftarrow \mathcal{A}(gk, ck);$$

$$\{(\mathsf{k}_i^p, \mathsf{k}_i^s) : i \in \mathcal{I}\} \leftarrow \mathsf{LabGen}(gk, ck, x, W);$$

$$C = \{(\mathsf{k}_i^p, c_i) \leftarrow \mathsf{Com}(gk, ck, (\mathsf{k}_i^p, \mathsf{k}_i^s, m_i)) : i \in \mathcal{I}\};$$

$$\pi \leftarrow \mathsf{P}(gk, ck, x, Op, C) : \mathsf{V}(gk, ck, x, C, \pi) = 1\Big] = 1,$$

where $\mathcal{A}$ *outputs* (x, W) *such that* $(gk, x, W) \in \mathcal{R}_\mathcal{L}$ *and* Op *is a valid set of openings of* C.

A commit-and-prove scheme is sound if it is impossible to prove a false statement.

Definition 10 (Perfect Soundness). *The commit-and-prove system* CaP *is (perfectly) sound if there exists a deterministic (unbounded) opening algorithm* Open *such that for all adversaries* $\mathcal{A}$

$$\Pr\Big[(gk, ck) \leftarrow \mathsf{G}(1^\lambda); (x, \{(\mathsf{k}_i^p, c_i) : i \in \mathcal{I}\}, \pi) \leftarrow \mathcal{A}(gk, ck);$$

$$\{(t_i, m_i) : (t_i, m_i) \leftarrow \mathsf{Open}(gk, ck, (\mathsf{k}_i^p, c_i))\} :$$

$$\mathsf{V}(gk, ck, x, \{(\mathsf{k}_i^p, c_i) : i \in \mathcal{I}\}, \pi) = 0 \ \vee \ (gk, x, \{(t_i, m_i) : i \in \mathcal{I}\}) \in \mathcal{R}_\mathcal{L}\Big] = 1.$$

C   Verifiable Correlated Key Generation For Other Equation Types

EXTENDING THE GROTH-SAHAI CAP BASED ON SXDH. We describe an alternative instantiation of the GS CaP scheme based on SXDH. Recall that the original one does not admit correlated key generation for labels $\text{com}_{\hat{\mathbb{G}}}$ (or mixed labels $\text{com}_{\hat{\mathbb{G}}}, \text{sca}_{\hat{\mathbb{G}}}$), while this one does. We just specify how to to generate the real commitment keys and the simulated keys with the simulation trapdoor, the rest of the algorithms of the CaP are easy to derive from the original paper of Groth and Sahai [16] or from the specification of GS proofs for any matrix assumption [12].

Essentially, the new instantiation introduces new label types so that one can commit to group elements and to scalars in two different ways. The labels $\text{sca}_{\hat{\mathbb{G}}}, \text{sca}_{\check{\mathbb{H}}}, \text{com}_{\hat{\mathbb{G}}}, \text{com}_{\check{\mathbb{H}}}$ indicate that one should commit to a group element as in the original instantiation of [11]. With these labels we can do what we described before, namely, we can prove that some equation admitted by the GS proof system is satisfiable, or that a set of equations with admissible simulation labels $(\text{sca}_{\hat{\mathbb{G}}}, \ldots, \text{sca}_{\hat{\mathbb{G}}})$ is partially satisfiable.

The new labels are $\text{sca}_{\hat{\mathbb{G}}}^{par}, \text{com}_{\hat{\mathbb{G}}}^{par}$ ($\text{sca}_{\check{\mathbb{H}}}^{par}, \text{com}_{\check{\mathbb{H}}}^{par}$ in $\check{\mathbb{H}}$). The new instantiation of GS proofs we give below admits verifiable correlated key generation for $\mathbf{T} = (\mathsf{t}_1, \ldots, \mathsf{t}_L)$ where, for all $i \in [L]$, $\mathsf{t}_i \in \{\text{com}_{\hat{\mathbb{G}}}^{par}, \text{sca}_{\hat{\mathbb{G}}}^{par}\}$ (or for all $i \in [L]$, $\mathsf{t}_i \in \{\text{sca}_{\check{\mathbb{H}}}^{par}, \text{com}_{\check{\mathbb{H}}}^{par}\}$). The table in figure 2 describes how to commit with these new label types, where $\mathbf{e}_3 = (0, 0, 1)^\top$ and $\hat{\mathbf{e}}_3 = (\hat{0}, \hat{0}, \hat{g})^\top$.

For a complete description of the new CaP, we would need to specify new equation types L_{eq} and define which types of commitments are compatible with each equation type, since we are dealing with vectors of potentially different sizes. Given a quadratic equation written in the form given in equation (2) (section 4), it is enough that the commitments to all the elements involved in the equation in the same module A_i are in the same space. For instance we can define $\mathsf{L}_{eq} = \text{MLin}_{\hat{\mathbb{G}}, par}$ as a linear multi-scalar multiplication equation in which the variables in $A_1 = \hat{\mathbb{G}}$ are committed with label $\text{com}_{\hat{\mathbb{G}}}^{par}$ and the constants in $A_2 = \mathbb{Z}_q$ are in the usual space $\check{\mathbb{H}}^2$. This is cumbersome to specify but straightforward, and we omit any further details.

A commitment to an element with any of these labels is a vector of dimension 3. Further, with these new labels, we essentially commit to scalars and group elements in the same way (so that a key can be binding/hiding for scalars and group elements at the same time). Therefore, there is no longer an efficiency advantage in the proof size for equations involving scalars over equations involving group elements[3]) This has an impact on efficiency, that is why one should only use these labels to prove statements which are too expensive to prove with the

---

[3] In the original GS proof instantiation, equations involving $\text{sca}_{\hat{\mathbb{G}}}, \text{sca}_{\check{\mathbb{H}}}$ are more efficient than (similar) equations with $\text{com}_{\hat{\mathbb{G}}}, \text{com}_{\check{\mathbb{H}}}$. For instance, the equation $\hat{x}a = \hat{t}$ requires one extra proof element compared to $\hat{a}x = \hat{t}$, but this is no longer true for the new labels.

$G(1^\lambda)$	$\mathsf{SimGen}(1^\lambda)$
$gk \leftarrow (q, \hat{\mathbb{G}}, \check{\mathbb{H}}, \mathbb{T}, e, \hat{g}, \check{h}) \leftarrow \mathcal{G}(1^\lambda)$	$gk \leftarrow (q, \hat{\mathbb{G}}, \check{\mathbb{H}}, \mathbb{T}, e, \hat{g}, \check{h}) \leftarrow \mathcal{G}(1^\lambda)$
$\omega, \sigma, \xi, \chi, \psi, \phi \leftarrow \mathbb{Z}_q^*$	$\rho, \omega, \xi, \chi, \psi, \phi \leftarrow \mathbb{Z}_q^*$
$\hat{\mathbf{v}} \leftarrow (\xi\hat{g}, \hat{g})^\top$, $\check{\mathbf{v}} \leftarrow (\psi\check{h}, \check{h})^\top$	$\hat{\mathbf{v}} \leftarrow (\xi\hat{g}, \hat{g})^\top$, $\check{\mathbf{v}} \leftarrow (\psi\check{h}, \check{h})^\top$
$\hat{\mathbf{u}} \leftarrow \omega\hat{\mathbf{v}}$, $\check{\mathbf{u}} \leftarrow \sigma\check{\mathbf{v}}$	$\hat{\mathbf{u}} \leftarrow \omega\hat{\mathbf{v}} + (\hat{0}, \hat{g})^\top$, $\check{\mathbf{u}} \leftarrow \sigma\check{\mathbf{v}} + (\check{0}, \check{h})^\top$
$\hat{\mathbf{w}} \leftarrow \hat{\mathbf{u}} - (\hat{0}, \hat{g})^\top$, $\check{\mathbf{w}} \leftarrow \check{\mathbf{u}} - (\check{0}, \check{h})^\top$	$\hat{\mathbf{w}} \leftarrow \hat{\mathbf{u}} - (\hat{0}, \hat{g})^\top$, $\check{\mathbf{w}} \leftarrow \check{\mathbf{u}} - (\check{0}, \check{h})^\top$
$\hat{\mathbf{a}} \leftarrow (\chi\hat{g}, \xi\hat{g}, \hat{g})^\top$, $\check{\mathbf{a}} \leftarrow (\phi\check{h}, \psi\check{h}, \check{h})^\top$	$\hat{\mathbf{a}} \leftarrow (\chi\hat{g}, \xi\hat{g}, \hat{g})^\top$, $\check{\mathbf{a}} \leftarrow (\phi\check{h}, \psi\check{h}, \check{h})^\top$
$\hat{\mathbf{b}} \leftarrow \omega\hat{\mathbf{a}} + (\hat{g}, \hat{0}, \hat{0})^\top$, $\check{\mathbf{b}} \leftarrow \sigma\check{\mathbf{a}} + (\check{h}, \check{0}, \check{0})^\top$	$\hat{\mathbf{b}} \leftarrow \omega\hat{\mathbf{a}} + (\hat{0}, \hat{0}, \hat{g})^\top$, $\check{\mathbf{b}} \leftarrow \sigma\check{\mathbf{a}} + (\check{0}, \check{0}, \check{h})^\top$
$ck \leftarrow (\hat{\mathbf{u}}, \hat{\mathbf{v}}, \hat{\mathbf{w}}, \check{\mathbf{u}}, \check{\mathbf{v}}, \check{\mathbf{w}}, \hat{\mathbf{a}}, \hat{\mathbf{b}}, \check{\mathbf{a}}, \check{\mathbf{b}})$	$ck \leftarrow (\hat{\mathbf{u}}, \hat{\mathbf{v}}, \hat{\mathbf{w}}, \check{\mathbf{u}}, \check{\mathbf{v}}, \check{\mathbf{w}}, \hat{\mathbf{a}}, \hat{\mathbf{b}}, \check{\mathbf{a}}, \check{\mathbf{b}})$
Return (gk, ck)	$tk \leftarrow (ck, \sigma, \omega)$
	Return (gk, ck, tk)

Label t	Message	Randomness	Commitment	$\mathcal{K}_{gk,\text{bind}}^{\text{t}}$	$\mathcal{K}_{gk,\text{hid}}^{\text{t}}$
$\text{sca}_{\hat{\mathbb{G}}}$	$(\text{sca}_{\hat{\mathbb{G}}}, x)$	$(\text{sca}_{\hat{\mathbb{G}}}, r)$	$\hat{\mathbf{c}} \leftarrow \hat{\mathbf{w}}x + \hat{\mathbf{v}}r$	$ck : \hat{\mathbf{w}} \notin \langle \hat{\mathbf{v}} \rangle$	$ck : \hat{\mathbf{w}} \in \langle \hat{\mathbf{v}} \rangle$
$\text{com}_{\hat{\mathbb{G}}}$	$(\text{com}_{\hat{\mathbb{G}}}, \hat{x})$	$(\text{com}_{\hat{\mathbb{G}}}, r, s)$	$\hat{\mathbf{c}} \leftarrow \mathbf{e}_2\hat{x} + \hat{\mathbf{v}}r + \hat{\mathbf{u}}s$	$ck : \hat{\mathbf{e}}_2 \notin \langle \hat{\mathbf{u}}, \hat{\mathbf{v}} \rangle$	$ck : \hat{\mathbf{e}}_2 \in \langle \hat{\mathbf{u}}, \hat{\mathbf{v}} \rangle$
$\text{sca}_{\hat{\mathbb{G}}}^{par}$	$(\text{sca}_{\hat{\mathbb{G}}}^{par}, x)$	$(\text{sca}_{\hat{\mathbb{G}}}^{par}, r, s)$	$\hat{\mathbf{c}} \leftarrow \hat{\mathbf{e}}_3 x + \hat{\mathbf{a}}r + \hat{\mathbf{b}}s$	$ck : \hat{\mathbf{e}}_3 \notin \langle \hat{\mathbf{a}}, \hat{\mathbf{b}} \rangle$	$ck : \hat{\mathbf{e}}_3 \in \langle \hat{\mathbf{a}}, \hat{\mathbf{b}} \rangle$
$\text{com}_{\hat{\mathbb{G}}}^{par}$	$(\text{com}_{\hat{\mathbb{G}}}^{par}, \hat{x})$	$(\text{com}_{\hat{\mathbb{G}}}^{par}, r, s)$	$\hat{\mathbf{c}} \leftarrow \mathbf{e}_3\hat{x} + \hat{\mathbf{a}}r + \hat{\mathbf{b}}s$	$ck : \hat{\mathbf{e}}_3 \notin \langle \hat{\mathbf{a}}, \hat{\mathbf{b}} \rangle$	$ck : \hat{\mathbf{e}}_3 \in \langle \hat{\mathbf{a}}, \hat{\mathbf{b}} \rangle$

Fig. 2. Generator algorithms. The two last coordinates of $\hat{\mathbf{a}}$ (resp. of $\hat{\mathbf{b}}, \check{\mathbf{a}}, \check{\mathbf{b}}$) correspond to the vector $\hat{\mathbf{v}}$ (resp. to $\hat{\mathbf{u}}, \check{\mathbf{v}}, \check{\mathbf{u}}$). Table describing the most important commitment types.

normal instantiation. For instance, if one just wants to prove satisfiability of one PPE, one should use a standard commitment to group elements.

C.1 VCKG for Group Elements

To prove soundness in the constructions of section 7, we used in a fundamental way that $\hat{\mathbf{w}}$ (or $(\hat{0}, \hat{g})^\top$) and $\hat{\mathbf{v}}$ are linearly independent. By giving this new instantiation with an additional dimension, the same arguments follow in a relative straightforward way. Indeed, the main reason why this new scheme admits verifiable correlated key generation for these new label types is that *both* for binding and hiding keys, $\hat{\mathbf{b}} \notin \langle \hat{\mathbf{a}} \rangle$. Intuitively, the point is that the secret sharing techniques we are using "work well" with linear independence relations and they "fail" with linear dependence relations. When we tried to construct correlated key generation for the label types $(\text{com}_{\hat{\mathbb{G}}}, \ldots, \text{com}_{\hat{\mathbb{G}}})$, we did not know how to force the prover to choose binding keys for group elements, i.e. keys such that $\hat{\mathbf{u}} \in \langle \hat{\mathbf{v}} \rangle$.

We sketch the construction of section 7.2 for the vector of labels $(\text{com}_{\hat{\mathbb{G}}}^{par}, \ldots, \text{com}_{\hat{\mathbb{G}}}^{par})$. Algorithm $\mathsf{K}_{\text{scorr}}$ samples two vectors $\boldsymbol{\tau} \in \mathbb{Z}_q^{m+1}$, $\boldsymbol{\kappa} \in \mathbb{Z}_q^{m+1}$ uniformly at random conditioned on $\tau_0 = 0, \kappa_0 = 0$. It also samples $\boldsymbol{\zeta}$ as usual, namely, as a uniform vector conditioned on $\zeta_0 = 1$ and $\zeta_j = 0$ for all $j \in \rho^{-1}(i)$, $i \in A^c$. The vectors $\hat{\mathbf{z}}_j$ are defined as $\hat{\mathbf{z}}_j = \tau_j \hat{\mathbf{a}} + \zeta_j \hat{\mathbf{b}} + \kappa_j(\hat{0}, \hat{0}, \hat{g})^\top$ and $\Sigma = \{ck_1, \ldots, ck_m\}$, where $ck_j = (\hat{\mathbf{u}}_j, \hat{\mathbf{v}}, \hat{\mathbf{w}}_j, \check{\mathbf{u}}, \check{\mathbf{v}}, \check{\mathbf{w}}, \hat{\mathbf{a}}, \hat{\mathbf{z}}_j, \check{\mathbf{a}}, \check{\mathbf{b}})$ (and $\hat{\mathbf{u}}_j, \hat{\mathbf{w}}_j$ are changed according to

$\hat{\mathbf{z}}_j$ to guarantee that $ck_j \in \mathcal{K}_{gk}$ (that is, $\hat{\mathbf{u}}_j$ should match the last two coordinates of $\hat{\mathbf{z}}_j$, $\hat{\mathbf{w}}_j = \hat{\mathbf{u}}_j - (\hat{0}, \hat{g})^\top$). The construction also works for $\mathbf{T} = (t_1, \ldots, t_L)$, $t_i \in \{\text{com}_{\hat{\mathbb{G}}}^{par}, \text{sca}_{\hat{\mathbb{G}}}^{par}\}$, since the set of binding/hiding keys for $\text{com}_{\hat{\mathbb{G}}}^{par}$ and $\text{sca}_{\hat{\mathbb{G}}}^{par}$ is the same.

The proof is identical to the one of lemma 2. Indeed, the key observation is that if $\zeta_j = 0$, then $(\hat{0}, \hat{0}, \hat{g})^\top \in \langle \hat{\mathbf{a}}, \hat{\mathbf{z}}_j \rangle$ (unless $\kappa_j = 0$, which occurs only with negligible probability). This means that the key ck_j is hiding, and further the simulation trapdoor is (τ_j, κ_j). On the other hand, if $\zeta_j \neq 0$, since $(\hat{0}, \hat{0}, \hat{g})^\top \notin \langle \hat{\mathbf{a}}, \hat{\mathbf{z}}_j \rangle$ the key is binding. Therefore, we are in the same situation as in lemma 2. The rest of the algorithms/ proof are also straightforward. Now the matrix $\hat{\mathbf{Z}} \in \hat{\mathbb{G}}^{3 \times (m+1)}$ and the verifier checks if $\mathbf{ZM} = \hat{\mathbf{0}}_{3 \times d}$, where $\mathbf{M}$ is the matrix associated to the span program.

Example. We retake the example of the OR of two equations as defined in section 8 but for the labels $\mathbf{T} = (\text{com}_{\hat{\mathbb{G}}}^{par}, \text{com}_{\hat{\mathbb{G}}}^{par})$. The vectors $\boldsymbol{\zeta}, \boldsymbol{\tau}$ are defined as explained in section 8. Additionally, one chooses another vector $\boldsymbol{\kappa} \in \mathbf{Im}(\mathbf{M}^*)$ uniformly conditioned on $\kappa_0 = 0$, i.e. $\boldsymbol{\kappa} = (0, s, -s)$, $s \leftarrow \mathbb{Z}_q$. Let's see why the approach works. For instance, assume that $A = \{1\}$ was the set used to compute the keys. In this case, $\hat{\mathbf{z}}_1 = r\hat{\mathbf{a}} - \hat{\mathbf{b}} + s(\hat{0}, \hat{0}, \hat{g})^\top$, $\hat{\mathbf{z}}_2 = -r\hat{\mathbf{a}} - s(\hat{0}, \hat{0}, \hat{g})^\top$, $tk_2 = (r, s)$. The reason why the key $(\hat{0}, \hat{0}, \hat{g})^\top \notin \langle \hat{\mathbf{a}}, \hat{\mathbf{z}}_1 \rangle$ is because $\hat{\mathbf{b}}$ is linearly independent of $\hat{\mathbf{a}}$ in the soundness setting.

Efficiency. For the same span program $\mathcal{SP}$, the description of Σ for $\mathbf{T} = (t_1, \ldots, t_L) = (\text{sca}_{\hat{\mathbb{G}}}, \ldots, \text{sca}_{\hat{\mathbb{G}}})$ is more efficient than when $t_i \in \{\text{com}_{\hat{\mathbb{G}}}^{par}, \text{sca}_{\hat{\mathbb{G}}}^{par}\}$, because for the latter we need to send $3(m-d)|\hat{\mathbb{G}}|$ elements. Additionally, for this construction there is an overhead that depends on the number of variables and the equation type. This is because for each variable that we commit to using one of the labels $t_i \in \{\text{com}_{\hat{\mathbb{G}}}^{par}, \text{sca}_{\hat{\mathbb{G}}}^{par}\}$ we need $3|\hat{\mathbb{G}}|$, as opposed to $2|\hat{\mathbb{G}}|$ in the normal instantiation of GS proofs. For quadratic equations (but not for linear ones), this also results in a larger proofs. For each equation type, one should evaluate if the approach is competitive, but for simple statements it looks like an interesting alternative (for instance, if one wants to prove OR of two linear equations in $\hat{\mathbb{G}}$, each with one variable in $\hat{\mathbb{G}}$).

Outlier Privacy*

Edward Lui and Rafael Pass**

Cornell University, USA
{luied,rafael}@cs.cornell.edu

Abstract. We introduce a generalization of differential privacy called
tailored differential privacy, where an individual's privacy parameter is
"tailored" for the individual based on the individual's data and the data
set. In this paper, we focus on a natural instance of tailored differential
privacy, which we call *outlier privacy*: an individual's privacy parameter
is determined by how much of an *"outlier"* the individual is. We provide a new definition of an outlier and use it to introduce our notion
of outlier privacy. Roughly speaking, $\epsilon(\cdot)$-*outlier privacy* requires that
each individual in the data set is guaranteed "$\epsilon(k)$-differential privacy
protection", where k is a number quantifying the "outlierness" of the
individual. We demonstrate how to release accurate histograms that sat-
isfy $\epsilon(\cdot)$-outlier privacy for various natural choices of $\epsilon(\cdot)$. Additionally,
we show that $\epsilon(\cdot)$-outlier privacy with our weakest choice of $\epsilon(\cdot)$—which
offers no explicit privacy protection for "non-outliers"—already implies
a "distributional" notion of differential privacy w.r.t. a large and natural
class of distributions.

1 Introduction

Enormous amounts of data are collected by hospitals, social networking systems,
government agencies, and other organizations. There are huge social benefits in
analyzing this data, but we must protect the *privacy* of the individuals in the
data. The current standard definition of privacy for data analysis is *differential
privacy* [7,5], which requires that the output distribution of the data analysis
algorithm changes very little when a single individual's data is added or removed
from the data set. Accurate differentially private algorithms for a wide variety
of tasks have been developed, allowing for useful and private data analysis (e.g.,
see [6,4]).

Currently, the standard notion of differential privacy guarantees the *same*
level of privacy protection for all individuals. More precisely, in ϵ-differential
privacy, every individual has the same "ϵ-differential privacy protection", which
guarantees that the algorithm's output distribution changes by at most ϵ when

---

* A full version of this paper is available at https://eprint.iacr.org/2014/982
** Pass is supported in part by an Alfred P. Sloan Fellowship, Microsoft New Fac-
ulty Fellowship, NSF CAREER Award CCF-0746990, NSF Award CCF-1214844,
NSF Award CNS-1217821, AFOSR YIP Award FA9550-10-1-0093, and DARPA and
AFRL under contract FA8750-11-2-0211.

Y. Dodis and J.B. Nielsen (Eds.): TCC 2015, Part II, LNCS 9015, pp. 277–305, 2015.

adding or removing the individual's data from the data set. While this is a strong privacy guarantee if ϵ is very small (we elaborate more on this below), it clearly also does result in a non-trivial privacy loss for moderate values of ϵ. Additionally, it has also been established that to achieve non-trivial utility, ϵ cannot be too small—in particular, $\epsilon \gg 1/n$ where n is the number of individuals in the data set. Furthermore, to answer a counting query with ϵ-differential privacy and with error at most α, we must have $\epsilon \geq \Omega(1/\alpha)$.

An alternative idea is to provide *different levels of privacy* protection to different individuals—intuitively, some individuals require more privacy than others, and the algorithm should accommodate this. This general idea, which first appeared in the work of Ghosh and Roth [11], has been partly investigated in a mechanism design setting (e.g., see [11,8,12,16,15]), where individuals are requested to not only submit their data, but also their "privacy valuation". The mechanism then tries to accommodate each individual's privacy valuation, while at the same time releasing data that is useful. Unfortunately, however, in the most realistic setting—where an individual's privacy valuation may be correlated with her data and thus also needs to be protected—the literature is plagued by strong impossibility results.

Tailored Differential Privacy: Protecting Outliers. In this paper, we consider a different approach to deal with the issue that different individuals may have different privacy needs. Instead of having the individuals specify their own privacy valuation/parameter, an individual's privacy parameter will be determined based on the individual's data and the data set. In other words, an individual's privacy parameter will be *tailored* for the individual based on the data set—we refer to such a notion as *tailored differential privacy*. In this paper, we focus on a natural instance of tailored differential privacy: an individual's privacy parameter will be determined by how much of an *"outlier"* the individual is (w.r.t. the data set). Roughly speaking, "outliers"—intuitively, individuals that are "far away", or "vastly different" from most other individuals—will be granted higher privacy protection than individuals that "mix" with lots of other individuals. One reason for providing higher privacy protection to outliers is that we may want to limit the amount of information leaked about a group of outliers. Let us present an example to illustrate what we mean.

Example 1 (Salaries of a Company's Employees). Consider the standard ϵ-differentially private algorithm for releasing a histogram, which simply adds (Laplace) $Lap(1/\epsilon)$ noise[1] to each bin independently. Suppose such an algorithm is used to release a histogram of the salaries of a large company's employees, where the range of possible salaries is partitioned into intervals, which correspond to the bins of the histogram. Assume there exists a (small, but non-trivial) group of, say, 100 managers, and all these managers have similar salaries that belong to the same bin; assume further that the other employees in the company have much lower salaries. Since the group of managers is relatively small, we consider

---

[1] $Lap(\lambda)$ is the Laplace distribution with mean 0 and scale λ, whose associated pdf is $f_\lambda(x) = \frac{1}{2\lambda}\exp(-\frac{|x|}{\lambda})$.

them to be outliers and would like to prevent their (approximate) salary from being revealed. But, if ϵ is not small enough, by choosing the highest-salary bin with a noisy count of at least 50, the bin containing the managers can be predicted with "high" probability (roughly $1 - \exp(-50\epsilon)$).

Leaking the salary information of a small group of managers may perhaps not be considered a serious "breach" of their privacy. However, the same argument still holds if we further partition each salary bin into two sub-bins corresponding to HIV positive and HIV negative individuals. If the fraction of HIV positive managers is significantly higher than what is usual, this fact will released by the ϵ-differentially private algorithm (assuming ϵ is not too small).

In contrast, if we could provide sufficiently higher privacy protection (i.e., a sufficiently smaller privacy parameter) to each of the managers, then the amount of information leaked about the group of managers would be significantly less, and thus the managers' salary, or information about their HIV status, will not be (significantly) revealed.

In the above example, the managers are considered "outliers"—the group of outliers is "small" and other individuals in the data set are "far" from them; thus we consider it a violation of their privacy that sensitive information about them is leaked. In contrast, if the group of managers was "huge", we would no longer consider them outliers, and releasing aggregate information about a huge group of people should not be considered a violation of privacy. Indeed, note that in the above example, the sensitive information that is leaked is not about a *single* individual, it is about the *group* of managers; this clarifies why traditional differential privacy (which is only meant to mask a single individual's information) does not suffice to protect this information.

The notion of (k, ϵ)-*group differential privacy* (which in particular is implied by ϵ/k-differential privacy), on the other hand, could be used to protect information about the group of managers (if we let $k = 100$). But using such a strong notion of privacy would require adding noise proportional to $100/\epsilon$ to all the bins in the above example, and would render the released data useless. On the other hand, if we *tailor* the level of privacy required by an individual to whether the individual is an outlier or not (which, looking forward, will be enabled by our notion of *outlier privacy*), we could make sure to guarantee $(\epsilon/100)$-differential privacy for *only* the managers (and thus any information about the group of managers is protected), and only ϵ-differential privacy for everyone else.

Let us now turn to formalizing our notion of *outlier privacy*. Towards doing this, we first need to provide a mathematical definition of what if means for an individual to be an outlier.

A New Mathematical Definition of "Outliers". As mentioned above, intuitively, outliers are data points or records that are "far away" or "vastly different" from the rest of the data. There are many existing methods of identifying outliers (see [2] for a survey); for example, for a set of data points, an outlier can be defined as a data point that is not within a certain distance of any other data point. However, such methods are often problematic for high-dimensional data (which is quite common), since the data points tend to be sparsely spaced and thus every data point may be an outlier (e.g., see [13]). As far as we know, all of the existing methods for identifying an outlier only look at the data itself and do not explicitly consider the *algorithm* that will be run on the data. In contrast, similar to the notion of differential privacy, we provide a definition of an outlier that depends on the algorithm that operates on the data set. (Additionally, existing methods of identifying outliers are also designed for some specific type of data (e.g., data points in $\mathbb{R}^d$); in contrast, we seek a method that works for any type of data.)

We aim to capture the intuition that a data record t in a data set is an outlier if, "from the perspective of the algorithm", the data record is not "equivalent" to sufficiently many data records in the data set. More formally, we say that a data record t is *equivalent* to another data record t' w.r.t. an algorithm A if A can never distinguish t and t'—that is, for every data set D containing t, the output distribution of the algorithm A does not change if we replace t by t' in D. (For instance, for computing a histogram, two individuals t and t' are equivalent if they correspond to the same bin in the histogram.) We now call a data record t a k-*outlier* w.r.t. the data set D and the algorithm A if t is equivalent (w.r.t. A) to at most k records in the data set. The parameter k quantifies to what extent the data record is an outlier.

Defining Outlier Privacy. We now turn to (informally) defining our notion of outlier privacy. Roughly speaking, $\epsilon(\cdot)$-*outlier privacy* requires that for every data set D, every $k > 0$, and every k-outlier t in the data set D, t is guaranteed "$\epsilon(k)$-differential privacy protection"—that is, if we remove t from the data set, the output distribution of the algorithm changes by at most $\epsilon(k)$, where the metric used is the same as that in differential privacy.

To address the privacy issues illustrated in Example 1, let us first consider $\epsilon(\cdot)$-outlier privacy for a specific "threshold" function $\epsilon(\cdot)$, which is specified by two parameters k and ϵ; we refer to the resulting notion as (k, ϵ)-*simple outlier privacy*. Roughly speaking, (k, ϵ)-simple outlier privacy requires ϵ/k-differential privacy for k-outliers, but does not have any privacy requirements for the other individuals. By requiring ϵ/k-differential privacy for k-outliers, (k, ϵ)-simple outlier privacy provides "(k, ϵ)-group differential privacy protection" for each *group* of k-outliers where the group size is at most k—that is, if we *simultaneously* remove k or fewer k-outliers from the data set, the output distribution of the algorithm changes by at most ϵ. (This fact follows from the observation that we can remove the k-outliers in the group one at a time, each time causing the output distribution to change by at most ϵ/k; since the group size is bounded by k, the total change in the output distribution is at most ϵ.)

Note that $(100, \epsilon)$-simple outlier privacy suffices to protect the privacy of the managers in Example 1. However, it does not protect the privacy of any of the other individuals. A minimal privacy guarantee would be to require that the managers' privacy is guaranteed (as a group) and everyone else gets the "individual" differential privacy guarantee; that is, we seek an algorithm that satisfies both $(100, \epsilon)$-simple outlier privacy, and ϵ-differential privacy. Again, this can be viewed as an instance of $\epsilon(\cdot)$-outlier privacy for a slightly different threshold function $\epsilon(\cdot)$. More precisely, our notion of (k, ϵ)-*simple outlier differential privacy* requires ϵ/k-differential privacy for k-outliers and ϵ-differential privacy for the other individuals.

(k, ϵ)-simple outlier differential privacy provides just *two* separate levels of privacy protection. We may also consider a more general instance of $\epsilon(\cdot)$-outlier privacy, which we refer to as *staircase outlier privacy*. In staircase outlier privacy, there are ℓ thresholds $k_1 > \ldots > k_\ell$, and $\ell + 1$ privacy parameters $\epsilon_0 > \ldots > \epsilon_\ell$, and we require that for every $1 \leq i \leq \ell$, every k_i-outlier is protected by ϵ_i-differential privacy; also, it is required that all the individuals are protected by ϵ_0-differential privacy by default.

1.1 Our Results

Our central results consist of demonstrating efficient algorithms for releasing accurate histograms that satisfy $\epsilon(\cdot)$-outlier privacy for various natural choices of $\epsilon(\cdot)$—in particular, we consider, simple outlier privacy, simple outlier differential privacy, staircase outlier privacy, and finally $\epsilon(\cdot)$-outlier privacy for a relatively general choice of $\epsilon(\cdot)$, and provide various (different) algorithms for releasing histograms that achieve these notions. Additionally, we show that the weakest notion of just simple outlier privacy (recall that this notion only protects outliers, and requires no privacy protection for the other individuals)—which we demonstrate can be achieved using particularly simple algorithms—actually already implies a "distributional" notion of differential privacy, and thus also a distributional notion of simple outlier differential privacy. Roughly speaking, the distributional notion of differential privacy only requires the differential privacy property to hold if the data set is drawn from some class of distributions. The class of distributions can represent a set of possible distributions that contains the supposed "true distribution", or the class can represent a set of possible beliefs an adversary may have about the data set. In our result, we consider a large and natural class of distributions obtained by sampling from any population. Our class of distributions includes quite general distributions/beliefs based on biased and imperfect sampling from a population, in a setting where the adversary may even know whether certain individuals were sampled or not.

Algorithms for Simple, Simple Differentially Private, and Staircase Outlier Privacy. Let us start by giving an example of a (k, ϵ)-simple outlier private algorithm for releasing a histogram (recall that (k, ϵ)-simple outlier privacy requires ϵ/k-differential privacy for all k-outliers, and no privacy for everyone else). Consider an algorithm that computes a histogram but suppresses the counts for all bins that have a count $\leq k$. A data record t is a k-outlier if and only if its bin has a count $\leq k$, so by suppressing the counts of those bins to 0, we ensure that output of the algorithm does not change if t is removed from the database. Simple outlier privacy may seem like a weak privacy guarantee—after all, the privacy of non-outliers is not explicitly protected. However, we will show that simple outlier privacy in fact implies a certain distributional notion of differential privacy, which might provide sufficient privacy protection in many settings. Thus, simple outlier privacy already implies a distributional notion of simple outlier differential privacy.

Let us now turn to directly designing simple outlier differentially private algorithms. We are able to design a histogram algorithm that achieves (k, ϵ)-simple outlier differential privacy. Roughly speaking, the algorithm first adds sufficient noise to each bin to achieve ϵ-differential privacy; then, the algorithm goes through each bin of the histogram, and if the bin has a noisy count that is less than k, the algorithm adds sufficient noise to the bin to achieve ϵ/k-differential privacy. The algorithm then outputs the resulting noisy histogram.

Finally, by generalizing the above approach, we can design a histogram algorithm that achieves staircase outlier privacy. Roughly speaking, the algorithm first adds sufficient noise to each bin to achieve ϵ_0-differential privacy; then, the algorithm goes through each of the "levels (i.e., steps) of the staircase" starting from the top, and if a bin currently has a noisy count that is at most the threshold for the current level i, the algorithm adds sufficient noise to the bin to achieve ϵ_i-differential privacy. The algorithm then outputs the resulting noisy histogram.

Outlier Private Algorithms for General $\epsilon(\cdot)$. We also provide histogram algorithms that satisfy $\epsilon(\cdot)$-outlier privacy for a relatively general $\epsilon(\cdot)$. Let us provide some intuition for how the outlier private histogram algorithms work. The standard ϵ-differentially private algorithm for releasing a histogram simply adds (Laplace) $Lap(1/\epsilon)$ noise to each bin count independently. By adding $Lap(1/\epsilon)$ noise to each bin, when a data record t is removed from the data set, the output distribution over noisy histograms only changes by at most ϵ (w.r.t. the metric used in differential privacy). To achieve $\epsilon(\cdot)$-outlier privacy, the output distribution can only change by at most $\epsilon(k)$, where k is the count of t's bin (t is the data record that is removed). Thus, one may try adding $Lap(1/\epsilon(k))$ noise to each bin, where k is the count of the bin. However, this does not work, since the amount of noise added depends on the count k in a way that is too sensitive. In particular, when we remove t from the data set and the count of t's bin decreases from k to $k-1$, the magnitude of the noise changes from $1/\epsilon(k)$ to $1/\epsilon(k-1)$, which changes the output distribution by more than $\epsilon(k)$.

One way to fix this problem is to add noise to the $\epsilon(\cdot)$ function, so that the $1/\epsilon(k)$ and the $1/\epsilon(k-1)$ become noisy and would be "ϵ'-close" for some $\epsilon' > 0$. To allow for a variety of solutions, we will consider using any algorithm $\mathcal{A}$ that approximates $\epsilon(\cdot)$ in a "differentially private" way—that is, $\mathcal{A}(k) \approx \mathcal{A}(k-1)$ for every $k > 0$. Then, we will add $\approx Lap(1/\mathcal{A}(k_b))$ noise to each bin b, where k_b is the count for bin b. This works as long as the noise magnitude $1/\mathcal{A}(k_b)$ is large enough; the noise magnitude $1/\epsilon(k_b)$ is large enough, but since $\mathcal{A}(k_b)$ only approximates $\epsilon(k_b)$, $\mathcal{A}(k_b)$ might be too large. Thus, we will also require that $\mathcal{A}(k_b)$ is at most $\epsilon(k_b)$ with very high probability.

Comparison to Related Work. There are some similarities between simple outlier privacy and the notion of crowd-blending privacy in [9]. Crowd-blending privacy uses a notion of "ϵ-blend", where $\epsilon > 0$, whereas in our definition of an outlier, we use a notion of equivalence w.r.t. the algorithm, which corresponds to ϵ-blend with $\epsilon = 0$. Also, in (k, ϵ)-simple outlier privacy, when removing a k-outlier, the output distribution is only allowed to change by at most ϵ/k, whereas in (k, ϵ)-crowd-blending privacy, the output distribution is allowed to change by at most ϵ. Our result that simple outlier privacy implies distributional differential privacy is somewhat similar to the result in [9] that states that if one combines a crowd-blending private algorithm with a natural pre-sampling step, the combined algorithm is zero-knowledge private (which implies differential privacy; see [10]) if we view the population as the input data set to the combined algorithm. In contrast, our result achieves a distributional notion of differential privacy on the data set as opposed to the population, which is a different model and definition.

Our result that simple outlier privacy implies distributional differential privacy also has some similarities to a result in [1], where it is shown that a histogram algorithm that suppresses small counts achieves a notion of distributional differential privacy (slightly weaker than ours, since their definition permits choosing a simulator, but in our definition, the simulator has to be the algorithm itself), but for a class of distributions incomparable to the class we consider (the classes are somewhat similar, but neither is a subset of the other). However, our class of distributions includes distributions/beliefs based on biased and imperfect sampling (from a population) in a setting where the adversary may even know whether certain individuals were sampled or not; the class of distributions considered in [1] does not consider such an adversarial setting. Also, we consider the class of simple outlier private algorithms, which includes but is more general than just histogram algorithms that suppress small counts.

Some Remarks on Outlier Privacy. Our notion of $\epsilon(\cdot)$-outlier privacy usually does not satisfy composition; that is, if an algorithm A is $\epsilon_A(\cdot)$-outlier private and an algorithm B is $\epsilon_B(\cdot)$-outlier private, the composition of A and B is usually not $(\epsilon_A + \epsilon_B)(\cdot)$-outlier private. This is due to the fact that a k-outlier w.r.t. the composition of A and B might not be a k-outlier w.r.t. A or B.

In our definition of $\epsilon(\cdot)$-outlier privacy, a k-outlier t is guaranteed "$\epsilon(k)$-differential privacy protection"—that is, if we *remove* t from the data set, the output distribution of the algorithm only changes by at most $\epsilon(k)$. Note, however, that this does not mean that if we replace t with *any* other individual t', the output distribution of the algorithm only changes by at most $\epsilon(k)$. In particular, if we replace t with a "non-outlier" t', then the output distribution may change more significantly. More precisely, the only thing we can say about the change in the output distribution is that it is bounded by $\epsilon(k) + \epsilon(k')$ if t is an k-outlier and t' is an k'-outlier—this follows since removing t changes the output distribution by at most $\epsilon(k)$, and adding t' changes the output distribution by at most $\epsilon(k')$.

Possible Future Directions and Additional Applications. Our results in this paper have focused mostly on histograms. To some extent, this is because our notion of an outlier is very liberal, due to the fact that our notion of equivalence between individuals is very strict (and thus it is "easier" to be classified as an outlier). One can consider generalizing our definition of a k-outlier to a (k, ϵ')-outlier, where the definition is the same except that (k, ϵ')-outlier uses ϵ'-blending (as in [9]) to define equivalence between individuals. If we are using a notion of outlier privacy that guarantees at least ϵ_0-differential privacy for every individual, then every individual would $2\epsilon_0$-blend with every other individual (by "transitivity"), so we should choose the blending parameter ϵ' to be smaller than $2\epsilon_0$. Using the definition of a (k, ϵ')-outlier in our various notions of outlier privacy, one can perhaps construct useful algorithms that satisfy these new notions of outlier privacy. For example, the algorithm in [9] for releasing synthetic data points would satisfy our generalized notion of (k, ϵ, ϵ')-simple outlier privacy where the notion of a (k, ϵ')-outlier is used. We leave the exploration of these generalized notions of outlier privacy for future work.

In the area of robust statistics, one of the main goals is to design statistical methods and estimators that are not significantly affected by outliers. A simple approach would be to first remove the outliers from the data set, and then apply non-robust statistical methods to the remaining data set. In order to use this approach, one needs a method of identifying outliers. Our mathematical definition of an outlier, or a variant of it, can be used to remove outliers before running non-robust statistical methods or algorithms on the data. Also, our notions of outlier privacy can be adapted to define a notion of "outlier robustness" for statistical computations. We leave the exploration of such ideas for future work.

2   Outlier Privacy

A *data set* is a finite *multiset* of *data records*, where a data record is simply an element of some fixed set X, which we refer to as the *data universe*. Let $\mathcal{D}$ be the set of all data sets. Given a data set D and data records t and t', let $D_{-t} = D \setminus \{t\}$ and $(D, t') = D \uplus \{t'\}$. Given $\epsilon, \delta \geq 0$ and two random variables (or distributions) Z and Z', we shall write $Z \approx_{\epsilon, \delta} Z'$ to mean that for every $Y \subseteq \mathrm{Supp}(Z) \cup \mathrm{Supp}(Z')$, we have

$$\Pr[Z \in Y] \leq e^\epsilon \Pr[Z' \in Y] + \delta$$

and

$$\Pr[Z' \in Y] \leq e^\epsilon \Pr[Z \in Y] + \delta.$$

We shall also write $Z \approx_\epsilon Z'$ to mean $Z \approx_{\epsilon, 0} Z'$. Differential privacy ([7,5]) can now be defined in the following manner:

Definition 1 ((ϵ, δ)-Differential Privacy [7,5]). *An algorithm $\mathcal{M}$ is said to be (ϵ, δ)-differentially private if for every pair of data sets D and D' differing in only one data record, we have $\mathcal{M}(D) \approx_{\epsilon, \delta} \mathcal{M}(D')$.*

Intuitively, differential privacy protects the privacy of each individual by requiring the output distribution of the algorithm to not change much when an individual's data is added or removed from the data set. Achieving differential privacy often involves adding noise drawn from some distribution, usually the Laplace distribution. We will use $Lap(\lambda)$ to denote the Laplace distribution with mean 0 and scale λ, whose associated pdf is $f_\lambda(x) = \frac{1}{2\lambda} \exp(-\frac{|x|}{\lambda})$. For convenience, we will sometimes abuse notation and use $Lap(\lambda)$ to denote a random variable that has the Laplace distribution $Lap(\lambda)$.

We now define our notion of *tailored differential privacy* as described in the introduction. Roughly speaking, $(\epsilon(\cdot), \delta(\cdot))$-tailored differential privacy requires that each individual t in the data set D is protected by $(\epsilon(t, D), \delta(t, D))$-differential privacy, where $\epsilon(\cdot)$ and $\delta(\cdot)$ are *functions* that, on input a data record t and a data set D, outputs privacy parameters $\epsilon(t, D)$ and $\delta(t, D)$ for t. Recall that X is the set of possible data records, and $\mathcal{D}$ is the set of all data sets.

Definition 2 (Tailored Differential Privacy). *Let* $\epsilon(\cdot), \delta(\cdot) : X \times \mathcal{D} \to \mathbb{R}_{\geq 0} \cup \{\infty\}$. *An algorithm* $\mathcal{M}$ *is said to be* $(\epsilon(\cdot), \delta(\cdot))$-*tailored differentially private if for every data set* D *and every data record* $t \in D$, *we have* $\mathcal{M}(D) \approx_{\epsilon(t,D),\delta(t,D)} \mathcal{M}(D \setminus \{t\})$.

In this paper, we focus on a specific instance of tailored differential privacy, which we call *outlier privacy*. Outlier privacy tailors an individual's privacy parameter to the "outlierness" of the individual. Let us first describe our definition of an *outlier*. In the definitions below, let $\mathcal{M}$ be any algorithm that takes a data set as input. Roughly speaking, we say that a pair of data records $t, t' \in X$ are *equivalent w.r.t.* $\mathcal{M}$ (or $\mathcal{M}$-*equivalent*), denoted $t \equiv_\mathcal{M} t'$, if the algorithm $\mathcal{M}$ can never distinguish the two data records, regardless of the input data set.

Definition 3 (Equivalent w.r.t. $\mathcal{M}$, or $\mathcal{M}$-Equivalent). *Given a pair of data records* $t, t' \in X$, *we say that* t *is equivalent to* t' *w.r.t.* $\mathcal{M}$, *or* t *is* $\mathcal{M}$-*equivalent to* t', *denoted* $t \equiv_\mathcal{M} t'$, *if for every data set* D' *containing* t, *we have* $\mathcal{M}(D') = \mathcal{M}(D'_{-t}, t')$ *(in distribution).*

Using the definition of a pair of data records being equivalent w.r.t. an algorithm $\mathcal{M}$, we now define the notion of a *k-outlier*. Roughly speaking, a k-outlier is a data record that is $\mathcal{M}$-equivalent to at most k data records in the data set (including itself).

Definition 4 (k-Outlier). *Given a data set* D, *a data record* $t \in D$ *is said to be a* k-*outlier in* D *w.r.t.* $\mathcal{M}$ *if there are at most* k *data records in* D *that are equivalent to* t *w.r.t.* $\mathcal{M}$.

As the parameter k increases, the property of being a k-outlier becomes weaker (i.e., easier to satisfy), and the set of k-outliers becomes larger. Using the definition of a k-outlier, we now define our new notion of privacy called $(\epsilon(\cdot), \delta(\cdot))$-*outlier privacy*. Roughly speaking, $(\epsilon(\cdot), \delta(\cdot))$-outlier privacy requires that for every $k > 0$ and every k-outlier t in the data set, t is protected by $(\epsilon(k), \delta(k))$-differential privacy—that is, if we remove t from the data set, the output distribution of the algorithm changes by at most $(\epsilon(k), \delta(k))$, where the metric used is the same as that in (ϵ, δ)-differential privacy.

Definition 5 $((\epsilon(\cdot), \delta(\cdot))$-Outlier Privacy). *Let $\epsilon(\cdot), \delta(\cdot) : \mathbb{N} \to \mathbb{R}_{\geq 0} \cup \{\infty\}$. An algorithm $\mathcal{M}$ is said to be $(\epsilon(\cdot), \delta(\cdot))$-outlier private if for every data set D, every $k > 0$, and every k-outlier t in D, we have $\mathcal{M}(D) \approx_{\epsilon(k), \delta(k)} \mathcal{M}(D \setminus \{t\})$.*

We will often write $\epsilon(\cdot)$-*outlier private* to mean $(\epsilon(\cdot), \delta(\cdot))$-outlier private with $\delta(k) = 0$ for every k. $(\epsilon(\cdot), \delta(\cdot))$-outlier privacy generalizes differential privacy by allowing one to specify different levels of privacy protection for different individuals based on how much of an outlier the individuals are. Intuitively, one may want to provide greater privacy protection to outliers, since their privacy may be more at risk. By setting $\epsilon(\cdot)$ and $\delta(\cdot)$ to be constants ϵ and δ respectively, one recovers the definition of (ϵ, δ)-differential privacy.

2.1   Simple Outlier Privacy

Let us first consider $\epsilon(\cdot)$-outlier privacy with a specific $\epsilon(\cdot)$ function, together which we call (k, ϵ)-*simple outlier privacy*. Roughly speaking, (k, ϵ)-simple outlier privacy requires ϵ/k-differential privacy for k-outliers, but does not have any privacy requirements for the other individuals.

Definition 6 $((k, \epsilon)$-Simple Outlier Privacy). *Let $k, \epsilon > 0$. An algorithm $\mathcal{M}$ is said to be (k, ϵ)-simple outlier private if for every data set D and every k-outlier t in D, we have $\mathcal{M}(D) \approx_{\epsilon/k} \mathcal{M}(D \setminus \{t\})$.*

(k, ϵ)-simple outlier privacy is equivalent to $\epsilon(\cdot)$-outlier privacy with the function $\epsilon(\cdot)$ defined by $\epsilon(k') = \epsilon/k$ if $k' \leq k$, and $\epsilon(k') = \infty$ otherwise. By requiring ϵ/k-differential privacy for k-outliers, (k, ϵ)-simple outlier privacy provides "(k, ϵ)-group differential privacy protection" for each *group* of k-outliers where the group size is at most k—that is, if we *simultaneously* remove k or fewer k-outliers from the data set, the output distribution of the algorithm changes by at most ϵ. (This fact follows from the observation that we can remove the k-outliers in the group one at a time, each time causing the output distribution to change by at most ϵ/k; since the group size is bounded by k, the total change in the output distribution is at most ϵ.) This privacy protection for groups of k-outliers can be particularly useful when one needs to protect the privacy of a group of outliers. In some cases, in order to protect the privacy of a single outlier, one needs to protect the privacy of an entire group of outliers simultaneously. In such cases, ordinary differential privacy may not be sufficient, like in Example 1 in the introduction. For completeness, let us now formalize what we mean when we say that (k, ϵ)-simple outlier privacy provides "(k, ϵ)-group differential privacy protection" for each group of k-outliers where the group size is at most k.

Proposition 1. *Let $\mathcal{M}$ be any algorithm that is (k,ϵ)-simple outlier private. Then, for every data set D and every $A \subseteq D$ of size at most k and consisting of only k-outliers in D, we have $\mathcal{M}(D) \approx_\epsilon \mathcal{M}(D \setminus A)$.*

Proof. Let D be any data set, and let $A \subseteq D$ be of size at most k and consisting of only k-outliers in D. Let $A = \{t_1, \ldots, t_r\}$, where $r \leq k$. Now, for $i = 0, \ldots, r$, let $D^{(i)} = D \setminus \{t_1, \ldots, t_i\}$. We note that $D^{(0)} = D$ and $D^{(r)} = D \setminus A$. Since $\mathcal{M}$ is (k,ϵ)-simple outlier private and A only consists of k-outliers in D, and since k-outliers in D remain as k-outliers after removing data records from D, we have $\mathcal{M}(D^{(i)}) \approx_{\epsilon/k} \mathcal{M}(D^{(i+1)})$ for every $0 \leq i \leq r - 1$. Thus, we have $\mathcal{M}(D) \approx_\epsilon \mathcal{M}(D \setminus A)$, as required. $\qquad\square$

Let us now give some examples of simple outlier private algorithms. Our first example is an algorithm that computes a histogram but suppresses the small counts to 0. Intuitively, data records in the same bin are equivalent w.r.t. $\mathcal{M}$, while a pair of data records belonging to separate bins are not equivalent w.r.t. $\mathcal{M}$. Thus, a data record is a k-outlier if and only if its bin has a count $\leq k$, so to achieve $(k,0)$-simple outlier privacy, the algorithm "suppresses" the counts $\leq k$ to 0.

Example 2 (Simple Outlier Private Histogram with Suppression of Small Counts). Let $k > 0$. Let $\mathcal{M}$ be an algorithm that, on input a data set D, computes a histogram from D, and then for every bin count that is $\leq k$, $\mathcal{M}$ "suppresses" (i.e., changes) the bin count to 0. $\mathcal{M}$ then outputs the modified histogram.

Theorem 1. *The above algorithm $\mathcal{M}$ is $(k,0)$-simple outlier private.*

Proof. Let D be any data set, and let t be any k-outlier in D. We note that t is $\mathcal{M}$-equivalent to precisely those records that belong in the same bin as t. Since t is a k-outlier, there are at most k records in t's bin. Thus, $\mathcal{M}$ will suppress t's bin count to 0. We observe that removing t from the data set (and thus from t's bin) will still result in $\mathcal{M}$ suppressing t's bin count to 0. Thus, $\mathcal{M}$ is $(k,0)$-simple outlier private. $\qquad\square$

Instead of suppressing small counts to 0, one can add noise to the small counts to achieve (k,ϵ)-simple outlier privacy.

Example 3 (Simple Outlier Private Histogram with Noise Added to Small Counts). Let $k > 0$. Let $\mathcal{M}$ be an algorithm that, on input a data set D, computes a histogram from D, and then for each bin count that is $\leq k$, $\mathcal{M}$ adds $Lap(k/\epsilon)$ noise to the bin count independently. $\mathcal{M}$ then outputs the modified histogram.

Theorem 2. *The above algorithm $\mathcal{M}$ is (k,ϵ)-simple outlier private.*

Proof. Let D be any data set, and let t be any k-outlier in D. We note that t is $\mathcal{M}$-equivalent to precisely those records that belong in the same bin as t. Since t is a k-outlier, there are at most k records in t's bin. Thus, $\mathcal{M}$ will add

$Lap(k/\epsilon)$ noise to t's bin count. We observe that removing t from the data set (and thus from t's bin) will still result in $\mathcal{M}$ adding $Lap(k/\epsilon)$ noise to t's bin count; using the pdf of $Lap(k/\epsilon)$ and performing some standard calculations for proving differential privacy (e.g., see [7]), one can easily show that the noisy count of t's bin after removing t is ϵ/k-close (i.e., $\approx_{\epsilon/k}$) to the noisy count of t's bin before removing t. Thus, $\mathcal{M}$ is (k, ϵ)-simple outlier private.   □

The simple outlier private algorithms above also satisfy a distributional notion of differential privacy for a large and natural class of distributions, since simple outlier privacy implies such a distributional notion of differential privacy, which we show in Section 3.

Relationship of Simple Outlier Privacy to Other Privacy Definitions.
Since (k, ϵ)-simple outlier privacy requires ϵ/k-differential privacy for k-outliers (and no privacy guarantee for the other individuals), we see that ϵ/k-differential privacy implies (k, ϵ)-simple outlier privacy.

Proposition 2. *Let* $k, \epsilon > 0$. *If an algorithm* $\mathcal{M}$ *is* ϵ/k-*differentially private, then it is* (k, ϵ)-*simple outlier private.*

Proof. This follows immediately from the definition of ϵ/k-differential privacy and (k, ϵ)-simple outlier privacy.   □

Although (k, ϵ)-simple outlier privacy can be obtained by achieving ϵ/k-differential privacy, achieving ϵ/k-differential privacy normally requires substantially more "noise" to be added. As demonstrated in the above examples, one can achieve better accuracy/utility with (k, ϵ)-simple outlier privacy because only the k-outliers require ϵ/k-differential privacy.

In [9], a notion of a pair of data records "ϵ-blending with each other" is used (in their notion of crowd-blending privacy), where it is required that the algorithm cannot distinguish the two records by more than ϵ. More precisely, a data record t ϵ-*blends with* t' w.r.t. $\mathcal{M}$ if for every data set D' containing t, we have $\mathcal{M}(D') \approx_{\epsilon} \mathcal{M}(D'_{-t}, t')$. In this paper, in our definition of equivalence w.r.t. $\mathcal{M}$ and in our definition of a k-outlier, we require the "blending" to be perfect (i.e., $\epsilon = 0$), since for an $(\epsilon/2)$-differentially private algorithm, every record ϵ-blends with every other record, and thus there would be no outliers. Furthermore, by setting $\epsilon = 0$, the "blends with" relation is an equivalence relation on the set of all possible data records. For an algorithm releasing histograms, the equivalence classes are precisely the bins of the histogram. In other words, a pair of data records blend with one another if and only if they belong to the same bin. There are also some similarities between simple outlier privacy and the notion of crowd-blending privacy in [9], which we now recall.

Definition 7 (Crowd-blending privacy [9]). *An algorithm $\mathcal{M}$ is (k, ϵ)-crowd-blending private if for every data set D and every data record $t \in D$, at least one of the following conditions hold:*

- *There are at least k data records in D that ϵ-blend with t.*
- $\mathcal{M}(D) \approx_\epsilon \mathcal{M}(D \setminus \{t\})$

The first condition in crowd-blending privacy is roughly saying that t is not a $(k-1)$-outlier, except that in the definition of $(k-1)$-outlier, the weaker notion of ϵ-blending is used instead of 0-blend. In the second condition, when t is removed from D, the output distribution of $\mathcal{M}$ changes by at most ϵ, but in (k, ϵ)-simple outlier privacy, the output distribution of $\mathcal{M}$ is only allowed to change by at most ϵ/k (for reasons we have explained above). We now formally show that simple outlier privacy implies crowd-blending privacy.

Proposition 3. *If an algorithm $\mathcal{M}$ is (k, ϵ)-simple outlier private, then it is $(k + 1, \epsilon/k)$-crowd-blending private.*

Proof. Suppose an algorithm $\mathcal{M}$ is (k, ϵ)-simple outlier private. We will show that $\mathcal{M}$ is also $(k + 1, \epsilon/k)$-crowd-blending private. Let D be any data set, let $t \in D$, and let A be the multiset of all data records t' in D such that $t' \equiv_\mathcal{M} t$. If A is of size at least $k + 1$, then the first property in $(k + 1, \epsilon)$-crowd-blending privacy holds. Otherwise, t is a k-outlier in D, so by the definition of (k, ϵ)-simple outlier privacy, we have $\mathcal{M}(D) \approx_{\epsilon/k} \mathcal{M}(D \setminus \{t\})$, which is the second property in $(k + 1, \epsilon/k)$-crowd-blending privacy. □

2.2 Simultaneously Achieving Simple Outlier Privacy and Differential Privacy

Although (k, ϵ)-simple outlier privacy protects the privacy of k-outliers, there is no privacy guarantee for the other individuals. Thus, we now consider a stronger notion of outlier privacy that provides ϵ/k-differential privacy for k-outliers and ϵ-differential privacy for everyone else. In other words, the stronger notion of outlier privacy provides both (k, ϵ)-simple outlier privacy and ϵ-differential privacy. We call this notion of outlier privacy *simple outlier differential privacy*. We first generalize (k, ϵ)-simple outlier privacy to (k, ϵ, δ)-simple outlier privacy so that we can define (k, ϵ, δ)-simple outlier differential privacy.

Definition 8 ((k, ϵ, δ)-Simple Outlier Privacy). *Let $k, \epsilon > 0$. An algorithm $\mathcal{M}$ is said to be (k, ϵ, δ)-simple outlier private if for every data set D and every k-outlier t in D, we have $\mathcal{M}(D) \approx_{\epsilon/k, \delta} \mathcal{M}(D \setminus \{t\})$.*

We now define (k, ϵ, δ)-simple outlier differential privacy.

Definition 9 ((k, ϵ, δ)-Simple Outlier Differential Privacy). *Let $k, \epsilon > 0$. An algorithm $\mathcal{M}$ is said to be (k, ϵ, δ)-simple outlier differentially private if $\mathcal{M}$ is (k, ϵ, δ)-simple outlier private and (ϵ, δ)-differentially private.*

We will write (k, ϵ)-*simple outlier differentially private* to mean (k, ϵ, δ)-simple outlier differentially private with $\delta = 0$. In the definition of (k, ϵ, δ)-simple outlier differential privacy, the same parameters ϵ and δ are used for both the simple outlier privacy requirement and the differential privacy requirement; however, one can easily consider a more general definition where separate parameters are used for the two requirements. (k, ϵ)-simple outlier differential privacy is equivalent to $\epsilon(\cdot)$-outlier privacy with the function $\epsilon(\cdot)$ defined by $\epsilon(k') = \epsilon/k$ if $k' \leq k$, and $\epsilon(k') = \epsilon$ otherwise. We now describe an algorithm for releasing histograms that achieves simple outlier differential privacy.

Example 4 (Simple Outlier Differentially Private Histogram with Suppression of Small Counts). Let $k, \alpha, \epsilon > 0$. Let $\mathcal{M}$ be an algorithm that, on input a data set D, computes a histogram from D, and then adds $Lap(1/\epsilon)$ noise to each bin count independently. Then, for every new (noisy) bin count that is $\leq k + \alpha/\epsilon$, $\mathcal{M}$ "suppresses" the bin count to 0. $\mathcal{M}$ then outputs the modified histogram.

Theorem 3. *The above algorithm $\mathcal{M}$ is $(k, \epsilon, e^{-\alpha}/2)$-simple outlier differentially private.*

Proof. We first show that $\mathcal{M}$ is ϵ-differentially private. We note that $\mathcal{M}$ first computes a noisy histogram using the standard ϵ-differentially private algorithm for releasing a noisy histogram. After that, $\mathcal{M}$ does not look at the input data set anymore, so the output of $\mathcal{M}$ is simply a post-processing of the output of an ϵ-differentially private algorithm. Thus, $\mathcal{M}$ itself is ϵ-differentially private.

We now show that $\mathcal{M}$ is $(k, 0, e^{-\alpha}/2)$-simple outlier private. Let D be any data set, and let t be any k-outlier in D. We need to show that $\mathcal{M}(D) \approx_{0, e^{-\alpha}/2} \mathcal{M}(D \setminus \{t\})$. It suffices to show that regardless of whether the data set is D or $D \setminus \{t\}$, we have that with probability at least $1 - e^{-\alpha}/2$, $\mathcal{M}$ will suppress t's bin count to 0. This event occurs precisely when the new (noisy) count for t's bin is $\leq k + \alpha/\epsilon$. Since t is a k-outlier, there are at most k records in t's bin (before any noise is added), so the probability of this event is at least the probability that $Lap(1/\epsilon) \leq \alpha/\epsilon$. One can easily verify that this latter event occurs with probability at least $1 - e^{-\alpha}/2$, as required. $\qquad\square$

In the above example, instead of suppressing the noisy bin count to 0, the algorithm $\mathcal{M}$ can add $Lap(k/\epsilon)$ noise to the noisy bin count. Let us now describe such an algorithm more formally.

Example 5 (Simple Outlier Differentially Private Histogram with Noise Added to Small Counts). Let $k, \alpha, \epsilon > 0$. Let $\mathcal{M}$ be an algorithm that, on input a data set D, computes a histogram from D, and then adds $Lap(1/\epsilon)$ noise to each bin count independently. Then, for every new (noisy) bin count that is $\leq k + \alpha/\epsilon$, $\mathcal{M}$ adds $Lap(k/\epsilon)$ noise to the noisy bin count. $\mathcal{M}$ then outputs the modified histogram.

Theorem 4. *The above algorithm $\mathcal{M}$ is $(k, \epsilon, e^{-\alpha})$-simple outlier differentially private.*

Proof. We first show that $\mathcal{M}$ is ϵ-differentially private. We note that $\mathcal{M}$ first computes a noisy histogram using the standard ϵ-differentially private algorithm for releasing a noisy histogram. After that, $\mathcal{M}$ does not look at the input data set anymore, so the output of $\mathcal{M}$ is simply a post-processing of the output of an ϵ-differentially private algorithm. Thus, $\mathcal{M}$ itself is ϵ-differentially private.

We now show that $\mathcal{M}$ is $(k, \epsilon, e^{-\alpha})$-simple outlier private. Let D be any data set, and let t be any k-outlier in D. We need to show that $\mathcal{M}(D) \approx_{\epsilon/k, e^{-\alpha}} \mathcal{M}(D \setminus \{t\})$. We first show that regardless of whether the data set is D or $D \setminus \{t\}$, we have that with probability at least $1 - e^{-\alpha}/2$, the first noisy count for t's bin is $\leq k + \alpha/\epsilon$ (this is the condition that determines whether $Lap(k/\epsilon)$ noise will be further added to the noisy bin count). Since t is a k-outlier, there are at most k records in t's bin (before any noise is added), so the probability of this event is at least the probability that $Lap(1/\epsilon) \leq \alpha/\epsilon$. One can easily verify that this latter event occurs with probability at least $1 - e^{-\alpha}/2$, as required.

Now, let $\mathcal{M}'$ be the same as $\mathcal{M}$ except that for t's bin, instead of checking the condition that the first noisy count for t's bin is $\leq k + \alpha/\epsilon$, $\mathcal{M}'$ simply pretends that the condition is true. Then, we have $\mathcal{M}(D) \approx_{0, e^{-\alpha}/2} \mathcal{M}'(D)$ and $\mathcal{M}(D \setminus \{t\}) \approx_{0, e^{-\alpha}/2} \mathcal{M}'(D \setminus \{t\})$. Thus, to show that $\mathcal{M}(D) \approx_{\epsilon/k, e^{-\alpha}} \mathcal{M}(D \setminus \{t\})$, it suffices to show that $\mathcal{M}'(D) \approx_{\epsilon/k} \mathcal{M}'(D \setminus \{t\})$. Since $\mathcal{M}'$ adds $Lap(k/\epsilon)$ noise to t's bin count, it is easy to show using standard calculations that $\mathcal{M}'(D) \approx_{\epsilon/k} \mathcal{M}'(D \setminus \{t\})$, as required. $\square$

Revisiting the "Salaries of a Company's Employees" Example. The

above simple outlier differentially private histogram algorithms can be used to protect the privacy of the managers and the other employees in the example described in the introduction. As mentioned previously, one can also protect the privacy of the managers by using a group differentially private algorithm for releasing a histogram. For comparison, let us now describe the standard group differentially private algorithm for releasing a histogram.

Example 6 (The Standard Group Differentially Private Histogram). Let $k, \epsilon > 0$. Let $\mathcal{M}$ be an algorithm that, on input a data set D, computes a histogram from D, and then adds $Lap(k/\epsilon)$ noise to each bin count independently. $\mathcal{M}$ then outputs the modified histogram.

It is known that the algorithm $\mathcal{M}$ is (k, ϵ)-group differentially private (e.g., see [7]).

As we can see, the standard group differentially private histogram algorithm adds $Lap(k/\epsilon)$ noise to *all* the bins, including the bins with many individuals in them. Our simple outlier differentially private algorithms suppress or add $\approx Lap(k/\epsilon)$ noise (depending on which variant we are using) to only the bins that contain outliers, and for the other bins, our algorithms only add $Lap(1/\epsilon)$ noise, which is substantially less than $Lap(k/\epsilon)$ noise. Thus, in the "Salaries of a Company's Employees" example, our algorithms have much better accuracy.

2.3   Staircase Outlier Privacy

In simple outlier differential privacy, there are only *two* separate levels of privacy protection: ϵ/k-differential privacy for k-outliers, and ϵ-differential privacy for everyone else. We can generalize this notion of outlier privacy to have more than two levels of privacy protection. We call this generalized notion *staircase outlier privacy*. In staircase outlier privacy, there are ℓ thresholds $k_1 > \ldots > k_\ell$, and $\ell + 1$ privacy parameters $\epsilon_0 > \ldots > \epsilon_\ell$, and we require that for every $1 \leq i \leq \ell$, every k_i-outlier is protected by (ϵ_i, δ)-differential privacy; also, it is required that all the individuals are protected by (ϵ_0, δ)-differential privacy by default.

Definition 10 (Staircase Outlier Privacy). *Let $\ell > 0$, let $k_1 > \ldots > k_\ell > 0$, let $\infty \geq \epsilon_0 > \epsilon_1 > \ldots > \epsilon_\ell \geq 0$, and let $\delta \geq 0$. An algorithm $\mathcal{M}$ is said to be $((k_1, \ldots, k_\ell), (\epsilon_0, \ldots, \epsilon_\ell), \delta)$-staircase outlier private if $\mathcal{M}$ is (ϵ_0, δ)-differentially private, and for every data set D, every $1 \leq i \leq \ell$, and every k_i-outlier t in D, we have $\mathcal{M}(D) \approx_{\epsilon_i, \delta} \mathcal{M}(D \setminus \{t\})$.*

We will write $((k_1, \ldots, k_\ell), (\epsilon_0, \ldots, \epsilon_\ell))$-*staircase outlier private* to mean $((k_1, \ldots, k_\ell), (\epsilon_0, \ldots, \epsilon_\ell), \delta)$-staircase outlier private with $\delta = 0$. In the above definition, a single δ parameter is used, but one can easily generalize the above definition to allow for $\ell + 1$ different levels of δ: $\delta_0 > \delta_1 > \ldots > \delta_\ell$. Staircase outlier privacy generalizes simple outlier privacy and simple outlier differential privacy: (k, ϵ)-simple outlier privacy is equivalent to $(k, (\infty, \epsilon/k))$-staircase outlier privacy, and (k, ϵ, δ)-simple outlier differential privacy is equivalent to $(k, (\epsilon, \epsilon/k), \delta)$-staircase outlier privacy. $((k_1, \ldots, k_\ell), (\epsilon_0, \ldots, \epsilon_\ell), \delta)$-staircase outlier privacy is equivalent to $(\epsilon(\cdot), \delta)$-outlier privacy with a "staircase" $\epsilon(\cdot) : \mathbb{N} \rightarrow \mathbb{R}_{\geq 0} \cup \{\infty\}$ function, where $\epsilon(k) = \epsilon_0$ if $k > k_1$, $\epsilon(k) = \epsilon_1$ if $k_2 < k \leq k_1$, $\epsilon(k) = \epsilon_2$ if $k_3 < k \leq k_2$, and so forth. More formally, $\epsilon(\cdot)$ is defined by $\epsilon(k) = \epsilon_j$, where j is the smallest integer such that $k \leq k_j$, and $j = 0$ if no such integer exists.

For convenience and simplicity, we will define $x/0 = \infty$ and $x/\infty = 0$ for any real $x > 0$. Also, "adding $Lap(\infty)$ noise" to some value means suppressing (i.e., changing) the value to 0, and "adding $Lap(0)$ noise" to some value means adding no noise at all to the value, i.e., the value is left unmodified. Let us now describe a histogram algorithm that achieves staircase outlier privacy. Roughly speaking, the algorithm first adds noise to each bin to achieve ϵ_0-differential privacy; then, the algorithm goes through each of the "levels of the staircase" starting from the top, and if a bin currently has a noisy count that is at most the threshold for that level, the algorithm adds sufficient noise to the bin to achieve ϵ_i-differential privacy. The algorithm then outputs the resulting noisy histogram.

Example 7 (Staircase Outlier Private Algorithm for Releasing a Histogram). Let $\ell > 0$, let $k_1 > \ldots > k_\ell > 0$, and let $\infty \geq \epsilon_0 > \epsilon_1 > \ldots > \epsilon_\ell \geq 0$. Let $\alpha > 0$, and let $\mathcal{M}$ be an algorithm that, on input a data set D, computes a histogram from D, and then adds $Lap(1/\epsilon_0)$ noise to each bin count independently. Then, for $i = 1, \ldots, \ell$, $\mathcal{M}$ does the following: For every current noisy bin count that is $\leq k_i + (\alpha/\epsilon_0 + \cdots + \alpha/\epsilon_{i-1})$, $\mathcal{M}$ adds $Lap(1/\epsilon_i)$ noise to the current noisy bin count. $\mathcal{M}$ then outputs the modified histogram.

Theorem 5. *The above algorithm $\mathcal{M}$ is $((k_1, \ldots, k_\ell), (\epsilon_0, \ldots, \epsilon_\ell), \ell e^{-\alpha})$-staircase outlier private.*

Proof. We first show that $\mathcal{M}$ is ϵ_0-differentially private. We note that $\mathcal{M}$ first computes a noisy histogram using the standard ϵ_0-differentially private algorithm for releasing a noisy histogram. After that, $\mathcal{M}$ does not look at the input data set anymore, so the output of $\mathcal{M}$ is simply a post-processing of the output of an ϵ_0-differentially private algorithm. Thus, $\mathcal{M}$ itself is ϵ_0-differentially private.

We now show that for every data set D, every $1 \le i \le \ell$, and every k_i-outlier t in D, we have $\mathcal{M}(D) \approx_{\epsilon_i, \ell e^{-\alpha}} \mathcal{M}(D \setminus \{t\})$. Let D be any data set, let $1 \le i \le \ell$, and let t be any k_i-outlier in D. We need to show that $\mathcal{M}(D) \approx_{\epsilon_i, \ell e^{-\alpha}} \mathcal{M}(D \setminus \{t\})$. We first show that regardless of whether the data set is D or $D \setminus \{t\}$, we have that with probability at least $1 - \ell e^{-\alpha}/2$, it holds that at every iteration $i' \le i$ in the algorithm $\mathcal{M}$, the condition that the current noisy count for t's bin is $\le k_{i'} + (\alpha/\epsilon_0 + \cdots + \alpha/\epsilon_{i'-1})$ is true. We note that this holds if for $i' = 0, \ldots, i-1$, the noise $Lap(1/\epsilon_{i'})$ added by $\mathcal{M}$ is $\le \alpha/\epsilon_{i'}$ (note that the original true count of t's bin is $\le k_{i'}$, since t is a k_i-outlier and $k_i \le k_{i'}$). One can easily verify that each of these latter events occurs with probability at least $1 - e^{-\alpha}/2$. Thus, by the union bound, with probability at least $1 - \ell e^{-\alpha}/2$, it holds that at every iteration $i' \le i$ in the algorithm $\mathcal{M}$, the condition that the noisy count for t's bin is $\le k_{i'} + (\alpha/\epsilon_0 + \cdots + \alpha/\epsilon_{i'-1})$ is true.

Let $\mathcal{M}'$ be the same as $\mathcal{M}$ except that for every iteration $i' \le i$, instead of checking the condition that the current noisy bin count for t's bin is $\le k_{i'} + (\alpha/\epsilon_0 + \cdots + \alpha/\epsilon_{i'-1})$, $\mathcal{M}'$ simply pretends that the condition is true. Then, we have $\mathcal{M}(D) \approx_{0, \ell e^{-\alpha}/2} \mathcal{M}'(D)$ and $\mathcal{M}(D \setminus \{t\}) \approx_{0, \ell e^{-\alpha}/2} \mathcal{M}'(D \setminus \{t\})$. Thus, to show that $\mathcal{M}(D) \approx_{\epsilon_i, \ell e^{-\alpha}} \mathcal{M}(D \setminus \{t\})$, it suffices to show that $\mathcal{M}'(D) \approx_{\epsilon_i} \mathcal{M}'(D \setminus \{t\})$. Since $\mathcal{M}'$ adds $Lap(1/\epsilon_i)$ noise to t's bin during iteration i, and since all the computation afterwards can be viewed as post-processing, it is easy to show using standard calculations that $\mathcal{M}'(D) \approx_{\epsilon_i} \mathcal{M}'(D \setminus \{t\})$, as required. $\square$

In the above example, the algorithm $\mathcal{M}$ can be modified to output bits for each bin b indicating at which iterations i noise was added to bin b. The privacy guarantee (Theorem 5) and its proof would still be exactly the same, but by outputting such information, a data analyst would know exactly what noise distributions were added to the true count of each bin.

Analyzing the Accuracy/Utility of the Above Algorithm $\mathcal{M}$. Let us now investigate the utility/accuracy of the above algorithm $\mathcal{M}$. We note that $\mathcal{M}$ processes each bin separately and independently, so we can simply analyze the accuracy of a single bin b. Suppose the count of a bin b is exactly k. Let j be the smallest integer such that $k \le k_j$, and $j = 0$ if no such integer exists. From the proof of Theorem 5, it is not hard to see that with probability at least $1 - \ell e^{-\alpha}$, it holds that at every iteration $i = 1, \ldots, j$, the algorithm $\mathcal{M}$ adds $Lap(1/\epsilon_i)$ noise to bin b. This means that with probability at least $1 - \ell e^{-\alpha}$, $\mathcal{M}$ will add at least $\sum_{i=0}^{j} Lap(1/\epsilon_i)$ noise to bin b.

Let us now try to derive a probabilistic upper bound on the noise added to bin b. Let us investigate whether noise will be added to bin b on a particular iteration i'. We note that for iteration $i = 1, \ldots, i' - 1$, $\mathcal{M}$ adds either $Lap(1/\epsilon_i)$ noise or no noise to bin b, and with probability at least $1 - e^{-\alpha}$, this noise will not decrease the current noisy count by more than α/ϵ_i. Thus, by the union bound, with probability at least $1 - \ell e^{-\alpha}$, the noisy count at iteration i' will be at least $k - (\alpha/\epsilon_0 + \cdots + \alpha/\epsilon_{i'-1})$, and if this number is $> k_{i'} + (\alpha/\epsilon_0 + \cdots + \alpha/\epsilon_{i'-1})$, $\mathcal{M}$ will not add any noise to bin b at iteration i'. Let I be the set of $i' \in \{1, \ldots, \ell\}$ such that this inequality does not hold, i.e., $k - (\alpha/\epsilon_0 + \cdots + \alpha/\epsilon_{i'-1}) \leq k_{i'} + (\alpha/\epsilon_0 + \cdots + \alpha/\epsilon_{i'-1})$, which is equivalent to $k \leq k_{i'} + 2(\alpha/\epsilon_0 + \cdots + \alpha/\epsilon_{i'-1})$. Then, with probability at least $1 - \ell e^{-\alpha}$, the noise distributions added to bin b is a subset of $\{i \in I : Lap(1/\epsilon_i)\} \cup \{Lap(1/\epsilon_0)\}$ (recall that $Lap(1/\epsilon_0)$ noise is added to bin b at the beginning by default).

Suppose $j < \ell$. If the k_i's are "well-spaced" and the ϵ_i's are not "too small", then we can show that with probability at least $1 - \ell e^{-\alpha}$, $\mathcal{M}$ will add at most $\sum_{i=0}^{j+1} Lap(1/\epsilon_i)$ noise to bin b. More formally, suppose that for every $1 \leq i \leq \ell - 1$, we have $k_i > k_{i+1} + 2(\alpha/\epsilon_0 + \cdots + \alpha/\epsilon_i)$. Then, by the definition of j above, we have $k > k_i$ for $i = j+1, \ldots, \ell$, so $k > k_{i+1} + 2(\alpha/\epsilon_0 + \cdots + \alpha/\epsilon_i)$ for $i = j+1, \ldots, \ell - 1$, which is equivalent to $k > k_i + 2(\alpha/\epsilon_0 + \cdots + \alpha/\epsilon_{i-1})$ for $i = j+2, \ldots, \ell$. This means that for every $j + 2 \leq i \leq \ell$, we have $i \notin I$, so with probability at least $1 - \ell e^{-\alpha}$, $\mathcal{M}$ will add at most $\sum_{i=0}^{j+1} Lap(1/\epsilon_i)$ noise to bin b, as required. We note that $\sum_{i=0}^{j+1} Lap(1/\epsilon_i)$ noise can be substantially lower than the $Lap(1/\epsilon_\ell)$ noise added by the standard ϵ_ℓ-differentially private algorithm for releasing a histogram.

2.4   Examples of Outlier Private Histogram Algorithms for General $\epsilon(\cdot), \delta(\cdot)$

In this section, we provide some examples of outlier private histogram algorithms for general $\epsilon(\cdot)$ and $\delta(\cdot)$ functions. Let us first provide some intuition for how the outlier private histogram algorithms work. The standard ϵ-differentially private algorithm for releasing a histogram simply adds $Lap(1/\epsilon)$ noise to each bin count independently. By adding $Lap(1/\epsilon)$ noise to each bin, when a data record t is removed from the data set, the output distribution over noisy histograms only changes by at most ϵ (w.r.t. the metric used in differential privacy). To achieve $\epsilon(\cdot)$-outlier privacy, the output distribution over noisy histograms can only change by at most $\epsilon(k)$, where k is the count of t's bin (t is the data record that is removed). Thus, one may try adding $Lap(1/\epsilon(k))$ noise to each bin, where k is the count of the bin. However, this does not work, since the amount of noise added depends on the count k in a way that is too sensitive. In particular, when we remove t from the data set and the count of t's bin decreases from k to $k - 1$, the magnitude of the noise changes from $1/\epsilon(k)$ to $1/\epsilon(k-1)$, which changes the output distribution over noisy histograms by more than $\epsilon(k)$.

One way to fix this problem is to add noise to the $\epsilon(\cdot)$ function, so that the $1/\epsilon(k)$ and the $1/\epsilon(k-1)$ become noisy and would be "ϵ'-close" for some $\epsilon' > 0$. To allow for a variety of solutions, we will consider using any algorithm $\mathcal{A}$ that approximates $\epsilon(\cdot)$ in a "differentially private" way—that is, $\mathcal{A}(k) \approx \mathcal{A}(k-1)$ for every $k > 0$. Then, we will add $\approx Lap(1/\mathcal{A}(k_b))$ noise to each bin b, where k_b is the count for bin b. This works as long as the noise magnitude $1/\mathcal{A}(k_b)$ is large enough; the noise magnitude $1/\epsilon(k_b)$ is large enough, but since $\mathcal{A}(k_b)$ only approximates $\epsilon(k_b)$, $\mathcal{A}(k_b)$ might be too large. Thus, we will also require that $\mathcal{A}(k)$ is at most $\epsilon(k)$ with very high probability. Below, instead of adding Laplace noise to each bin, we consider a general algorithm $\mathcal{B}$ that outputs a noisy count, and satisfies $\mathcal{B}(k, \epsilon') \approx_{\epsilon'} \mathcal{B}(k-1, \epsilon')$ for every $k > 0$ and $\epsilon' \geq 0$, which is the property we need; adding Laplace noise satisfies this property. For generality, we also add a $\delta(\cdot)$ parameter and consider $(\epsilon(\cdot), \delta(\cdot))$-outlier privacy. Let us now describe the required properties for $\mathcal{A}$.

Definition 11 (Differentially Private Lower Bound for $(\epsilon(\cdot), \delta(\cdot))$). *Let* $\epsilon(\cdot), \delta(\cdot) : \mathbb{N} \to \mathbb{R}_{\geq 0} \cup \{\infty\}$ *be functions. An algorithm $\mathcal{A}$ is said to be an* $(\epsilon_\mathcal{A}, \delta_\mathcal{A}, \delta'_\mathcal{A})$-*differentially private lower bound for* $(\epsilon(\cdot), \delta(\cdot))$ *if $\mathcal{A}$ takes an integer $k \geq 0$ as input and satisfies the following properties:*

- $\mathcal{A}(k) \approx_{\epsilon_\mathcal{A}, \delta_\mathcal{A}} \mathcal{A}(k-1)$ *for every integer $k > 0$.*
- *For every $k \in \mathbb{N}$, with probability at least $1 - \delta'_\mathcal{A}$, $\mathcal{A}(k)$ outputs an $(\epsilon_{total}, \delta_{total})$ satisfying $\epsilon_\mathcal{A} \leq \epsilon_{total} \leq \epsilon(k)$ and $\delta_\mathcal{A} + \delta'_\mathcal{A} \leq \delta_{total} \leq \delta(k)$.*

We now describe our outlier private histogram algorithm for general $\epsilon(\cdot)$ and $\delta(\cdot)$ functions.

Example 8 (Outlier Private Histogram Algorithm for General $\epsilon(\cdot), \delta(\cdot)$). Let $\epsilon(\cdot), \delta(\cdot) : \mathbb{N} \to \mathbb{R}_{\geq 0} \cup \{\infty\}$ be monotone functions. Let $\mathcal{A}$ be any $(\epsilon_\mathcal{A}, \delta_\mathcal{A}, \delta'_\mathcal{A})$-differentially private lower bound for $(\epsilon(\cdot), \delta(\cdot))$, and suppose that $\epsilon(\cdot)$ and $\delta(\cdot)$ are bounded from below by $\epsilon_\mathcal{A}$ and $\delta_\mathcal{A} + \delta'_\mathcal{A}$ respectively, i.e., $\epsilon(k) \geq \epsilon_\mathcal{A}$ and $\delta(k) \geq \delta_\mathcal{A} + \delta'_\mathcal{A}$ for every $k \in \mathbb{N}$. Let $\mathcal{B}$ be any algorithm that satisfies $\mathcal{B}(k, \epsilon', \delta') \approx_{\epsilon', \delta'} \mathcal{B}(k-1, \epsilon', \delta')$ for every integer $k > 0$, every $\epsilon', \delta' \geq 0$.

Let $\mathcal{M}$ be an algorithm that, on input a data set D, computes a histogram from D, and then does the following for each bin b independently: Let k_b be the count for bin b. $\mathcal{M}$ runs $\mathcal{A}(k_b)$ to get its output $(\epsilon_{total}, \delta_{total})$, and then runs $\mathcal{B}(k_b, \epsilon_{total} - \epsilon_\mathcal{A}, \delta_{total} - \delta_\mathcal{A} - \delta'_\mathcal{A})$ and uses its output to replace the count k_b for bin b. After going through all the bins, $\mathcal{M}$ outputs the modified histogram (and the output $(\epsilon_{total}, \delta_{total})$ of $\mathcal{A}(k_b)$ for each bin b, if this is desired).

Theorem 6 (Outlier Private Histogram Algorithm for General $\epsilon(\cdot)$, $\delta(\cdot)$). *The above algorithm $\mathcal{M}$ is $(\epsilon(\cdot), \delta(\cdot))$-outlier private.*

Proof. Let D be any data set, let $k > 0$, and let t be any k-outlier in D. We need to show that $\mathcal{M}(D) \approx_{\epsilon(k), \delta(k)} \mathcal{M}(D \setminus \{t\})$. We note that t is equivalent to (w.r.t. $\mathcal{M}$) with precisely those records that belong to the same bin as t, so k is an upper bound on the count for t's bin. Since $\epsilon(\cdot)$ and $\delta(\cdot)$ are monotone, we can assume without loss of generality that k is equal to the count for t's bin.

Now, consider removing t from the data set D; the count for t's bin decreases by 1, but the counts of the other bins remain the same. Since $\mathcal{M}$ processes each bin separately and independently, it suffices to show that

$$\mathcal{B}(k, \epsilon_{total,k} - \epsilon_{\mathcal{A}}, \delta_{total,k} - \delta_{\mathcal{A}} - \delta'_{\mathcal{A}}) \approx_{\epsilon(k), \delta(k)} \mathcal{B}(k-1, \epsilon_{total,k-1} - \epsilon_{\mathcal{A}}, \delta_{total,k-1} - \delta_{\mathcal{A}} - \delta'_{\mathcal{A}}),$$
(1)

where $(\epsilon_{total,k}, \delta_{total,k}) \sim \mathcal{A}(k)$ and $(\epsilon_{total,k-1}, \delta_{total,k-1}) \sim \mathcal{A}(k-1)$. By definition of $\mathcal{A}$, we have $\mathcal{A}(k) \approx_{\epsilon_{\mathcal{A}}, \delta_{\mathcal{A}}} \mathcal{A}(k-1)$, so $(\epsilon_{total,k}, \delta_{total,k}) \approx_{\epsilon_{\mathcal{A}}, \delta_{\mathcal{A}}} (\epsilon_{total,k-1}, \delta_{total,k-1})$, so

$$\mathcal{B}(k, \epsilon_{total,k} - \epsilon_{\mathcal{A}}, \delta_{total,k} - \delta_{\mathcal{A}} - \delta'_{\mathcal{A}}) \approx_{\epsilon_{\mathcal{A}}, \delta_{\mathcal{A}}} \mathcal{B}(k, \epsilon_{total,k-1} - \epsilon_{\mathcal{A}}, \delta_{total,k-1} - \delta_{\mathcal{A}} - \delta'_{\mathcal{A}}).$$
(2)

By definition of $\mathcal{B}$, we have $\mathcal{B}(k, \epsilon', \delta') \approx_{\epsilon', \delta'} \mathcal{B}(k-1, \epsilon', \delta')$ for every $\epsilon', \delta' \geq 0$, and by definition of $\mathcal{A}$, with probability at least $1 - \delta'_{\mathcal{A}}$, $\mathcal{A}(k-1)$ outputs an $(\epsilon_{total,k-1}, \delta_{total,k-1})$ satisfying $\epsilon_{\mathcal{A}} \leq \epsilon_{total,k-1} \leq \epsilon(k-1)$ and $\delta_{\mathcal{A}} + \delta'_{\mathcal{A}} \leq \delta_{total,k-1} \leq \delta(k-1)$, so

$$\mathcal{B}(k, \epsilon_{total,k-1} - \epsilon_{\mathcal{A}}, \delta_{total,k-1} - \delta_{\mathcal{A}} - \delta'_{\mathcal{A}})$$
$$\approx_{\epsilon(k-1) - \epsilon_{\mathcal{A}}, \delta(k-1) - \delta_{\mathcal{A}}} \mathcal{B}(k-1, \epsilon_{total,k-1} - \epsilon_{\mathcal{A}}, \delta_{total,k-1} - \delta_{\mathcal{A}} - \delta'_{\mathcal{A}}).$$
(3)

Now, combining (2) and (3) and noting that $\epsilon(k-1) \leq \epsilon(k)$ and $\delta(k-1) \leq \delta(k)$ (since $\epsilon(\cdot)$ and $\delta(\cdot)$ are monotone), we get (1), as required. □

A typical choice for the algorithm $\mathcal{B}$ in the above example is the algorithm that adds Laplace noise: The algorithm $\mathcal{B}$, on input $k \geq 0$ and $\epsilon', \delta' \geq 0$, adds $Lap(1/\epsilon')$ noise to k and then outputs the modified (noisy) k. Let us now give some examples of the algorithm $\mathcal{A}$:

- Adding noise to k and then computing $\epsilon(\cdot)$ on the noisy k: Let $\epsilon_{\mathcal{A}}, \alpha > 0$, and suppose that $\epsilon(\cdot)$ and $\delta(\cdot)$ are bounded from below by $\epsilon_{\mathcal{A}}$ and $e^{-\alpha}/2$, respectively. Let $\mathcal{A}$ be an algorithm that, on input $k \geq 0$, samples $\lambda \sim Lap(1/\epsilon_{\mathcal{A}})$, lets $k' = \max\{\lfloor k + \lambda - \alpha/\epsilon_{\mathcal{A}} \rfloor, 0\}$, and then outputs $(\epsilon(k'), e^{-\alpha}/2)$. Then, $\mathcal{A}$ is an $(\epsilon_{\mathcal{A}}, 0, e^{-\alpha}/2)$-differentially private lower bound for $(\epsilon(\cdot), \delta(\cdot))$.
- Adding noise to $\epsilon(k)$ calibrated to global sensitivity of $\epsilon(\cdot)$: Let $\epsilon_{\mathcal{A}}, \alpha > 0$, and suppose that $\epsilon(\cdot)$ and $\delta(\cdot)$ are bounded from below by $\epsilon_{\mathcal{A}}$ and $e^{-\alpha}/2$, respectively. Let $\Delta(\epsilon) = \sup_{k' \in \mathbb{Z}_{>0}} |\epsilon(k') - \epsilon(k'-1)|$, and suppose that $\Delta(\epsilon) < \infty$. Let $\mathcal{A}$ be an algorithm that, on input $k \geq 0$, samples $\lambda \sim Lap(\Delta(\epsilon)/\epsilon_{\mathcal{A}})$, and then outputs $(\max\{\epsilon(k) + \lambda - \alpha\Delta(\epsilon)/\epsilon_{\mathcal{A}}, \epsilon_{\mathcal{A}}\}, e^{-\alpha}/2)$. Then, $\mathcal{A}$ is an $(\epsilon_{\mathcal{A}}, 0, e^{-\alpha}/2)$-differentially private lower bound for $(\epsilon(\cdot), \delta(\cdot))$.
- Adding noise to $\epsilon(k)$ calibrated to smooth sensitivity of $\epsilon(\cdot)$: Let $\epsilon_{\mathcal{A}}, \alpha > 0$, and suppose that $\epsilon(\cdot)$ and $\delta(\cdot)$ are bounded from below by $\epsilon_{\mathcal{A}}$ and $\delta_{\mathcal{A}} + e^{-\alpha}/2$, respectively. Let $\delta_{\mathcal{A}} \in (0, 1)$, and let $0 \leq \beta \leq \frac{\epsilon_{\mathcal{A}}}{2 \ln(2/\delta_{\mathcal{A}})}$. Let $S^*_{\epsilon, \beta}(k) = \sup_{k' \in \mathbb{Z}_{>0}}(|\epsilon(k) - \epsilon(k')| \cdot e^{-\beta|k-k'|})$, and suppose that $S^*_{\epsilon, \beta}(k) < \infty$ for every k. Let $\mathcal{A}$ be an algorithm that, on input $k \geq 0$, samples $\lambda \sim Lap(2S^*(k)/\epsilon_{\mathcal{A}})$, and then outputs $(\max\{\epsilon(k) + \lambda - 2\alpha S^*_{\epsilon, \beta}(k)/\epsilon_{\mathcal{A}}, \epsilon_{\mathcal{A}}\}, \delta_{\mathcal{A}} + e^{-\alpha}/2)$. Then, $\mathcal{A}$ is an $(\epsilon_{\mathcal{A}}, \delta_{\mathcal{A}}, e^{-\alpha}/2)$-differentially private lower bound for $(\epsilon(\cdot), \delta(\cdot))$ (see [14]).

- Adding noise to the "noise magnitude function" $1/\epsilon(\cdot)$, calibrated to global sensitivity of $1/\epsilon(\cdot)$: Let $\epsilon_\mathcal{A}, \alpha > 0$, and suppose that $\epsilon(\cdot)$ and $\delta(\cdot)$ are bounded from below by $\epsilon_\mathcal{A}$ and $e^{-\alpha}/2$, respectively. Let $\Delta(1/\epsilon) = \sup_{k' \in \mathbb{Z}_{>0}} |1/\epsilon(k') - 1/\epsilon(k' - 1)|$, and suppose that $\Delta(1/\epsilon) < \infty$. Let $\mathcal{A}$ be an algorithm that, on input $k \geq 0$, samples $\lambda \sim Lap(\Delta(1/\epsilon)/\epsilon_\mathcal{A})$, and then outputs
  $\left(\max\left\{ \frac{1}{\max\{1/\epsilon(k) + \lambda - \alpha\Delta(1/\epsilon)/\epsilon_\mathcal{A}, 0\}}, \epsilon_\mathcal{A} \right\}, e^{-\alpha}/2 \right)$. Then, $\mathcal{A}$ is an $(\epsilon_\mathcal{A}, 0, e^{-\alpha}/2)$-differentially private lower bound for $(\epsilon(\cdot), \delta(\cdot))$.
- Adding noise to the "noise magnitude function" $1/\epsilon(\cdot)$, calibrated to smooth sensitivity of $1/\epsilon(\cdot)$: Let $\epsilon_\mathcal{A}, \alpha > 0$, and suppose that $\epsilon(\cdot)$ and $\delta(\cdot)$ are bounded from below by $\epsilon_\mathcal{A}$ and $\delta_\mathcal{A} + e^{-\alpha}/2$, respectively. Let $\delta_\mathcal{A} \in (0, 1)$, and let $0 \leq \beta \leq \frac{\epsilon_\mathcal{A}}{2\ln(2/\delta_\mathcal{A})}$. Let $S^*_{1/\epsilon,\beta}(k) = \sup_{k' \in \mathbb{Z}_{>0}}(|1/\epsilon(k) - 1/\epsilon(k')| \cdot e^{-\beta|k-k'|})$, and suppose that $S^*_{1/\epsilon,\beta}(k) < \infty$ for every k. Let $\mathcal{A}$ be an algorithm that, on input $k \geq 0$, samples $\lambda \sim Lap(2S^*(k)/\epsilon_\mathcal{A})$, and then outputs $\left(\max\left\{ \frac{1}{\max\{1/\epsilon(k) + \lambda - 2\alpha S^*(k)/\epsilon_\mathcal{A}, 0\}}, \epsilon_\mathcal{A} \right\}, \delta_\mathcal{A} + e^{-\alpha}/2 \right)$. Then, $\mathcal{A}$ is an $(\epsilon_\mathcal{A}, \delta_\mathcal{A}, e^{-\alpha}/2)$-differentially private lower bound for $(\epsilon(\cdot), \delta(\cdot))$ (see [14]).

In the above example, the algorithm $\mathcal{M}$ can also release the output $(\epsilon_{total}, \delta_{total})$ of $\mathcal{A}(k_b)$ for each bin b. By releasing this extra information, a data analyst would know exactly what noise distribution was added to the true count of each bin.

Analyzing the Accuracy/Utility of the Above Algorithm $\mathcal{M}$. Let us now investigate the utility/accuracy of the above algorithm $\mathcal{M}$. We note that $\mathcal{M}$ processes each bin separately and independently, so we can simply analyze the accuracy of a single bin b. Suppose the count of a bin b is exactly k. For simplicity, we will assume that $\mathcal{B}$ is the algorithm described above that adds Laplace noise. Let us now consider the various algorithms for $\mathcal{A}$ described above. All of the algorithms involve adding Laplace noise to some value that is used in determining the ϵ_{total} outputted by $\mathcal{A}$. By using the cdf of the Laplace distribution, one can obtain a probabilistic upper bound on the amount of noise added, which gives a probabilistic lower bound on ϵ_{total}. Since the algorithm $\mathcal{B}$ adds $Lap(\frac{1}{\epsilon_{total} - \epsilon_\mathcal{A}})$ to bin b, we can obtain a probabilistic upper bound on the amount of noise added to bin b. If we apply this analysis to each of the above algorithms for $\mathcal{A}$, we get the following results:

- Adding noise to k and then computing $\epsilon(\cdot)$ on the noisy k: With probability at least $1 - e^{-\alpha}$, the amount of noise added to bin b is at most $Lap(1/\epsilon')$, where $\epsilon' = \epsilon(\max\{\lfloor k - 2\alpha/\epsilon_\mathcal{A} \rfloor, 0\}) - \epsilon_\mathcal{A}$.
- Adding noise to $\epsilon(k)$ calibrated to global sensitivity of $\epsilon(\cdot)$: With probability at least $1 - e^{-\alpha}$, the amount of noise added to bin b is at most $Lap(1/\epsilon')$, where $\epsilon' = \max\{\epsilon(k) - 2\alpha\Delta(\epsilon)/\epsilon_\mathcal{A} - \epsilon_\mathcal{A}, 0\}$.
- Adding noise to $\epsilon(k)$ calibrated to smooth sensitivity of $\epsilon(\cdot)$: With probability at least $1 - e^{-\alpha}$, the amount of noise added to bin b is at most $Lap(1/\epsilon')$, where $\epsilon' = \max\{\epsilon(k) - 4\alpha S^*_{\epsilon,\beta}(k)/\epsilon_\mathcal{A} - \epsilon_\mathcal{A}, 0\}$.

– Adding noise to the "noise magnitude function" $1/\epsilon(\cdot)$, calibrated to global sensitivity of $1/\epsilon(\cdot)$: With probability at least $1 - e^{-\alpha}$, the amount of noise added to bin b is at most $Lap(1/\epsilon')$, where $\epsilon' = \max\left\{\frac{1}{\max\{1/\epsilon(k) - 2\alpha\Delta(1/\epsilon)/\epsilon_\mathcal{A}, 0\}} - \epsilon_\mathcal{A}, 0\right\}$.

– Adding noise to the "noise magnitude function" $1/\epsilon(\cdot)$, calibrated to smooth sensitivity of $1/\epsilon(\cdot)$: With probability at least $1 - e^{-\alpha}$, the amount of noise added to bin b is at most $Lap(1/\epsilon')$, where $\epsilon' = \max\left\{\frac{1}{\max\{1/\epsilon(k) - 4\alpha S^*_{\epsilon,\beta}(k)/\epsilon_\mathcal{A}, 0\}} - \epsilon_\mathcal{A}, 0\right\}$.

We note that the amount of noise added in the above algorithms can be substantially lower than the $Lap(1/\epsilon(1))$ noise added by the standard $\epsilon(1)$-differentially private algorithm for releasing a histogram.

2.5 Comparing the Staircase Algorithm and the Algorithms for General $\epsilon(\cdot), \delta(\cdot)$

Suppose we want to release a histogram while satisfying $(\epsilon(\cdot), \delta)$-outlier privacy for some monotone function $\epsilon(\cdot)$ and some small $\delta > 0$. If $\epsilon(\cdot)$ only takes on a small number of possible values, then $\epsilon(\cdot)$ is a "staircase" (i.e., piecewise constant) function, so we may want to use the staircase outlier private algorithm for releasing a histogram. If $\epsilon(\cdot)$ takes on infinitely many possible values, then the staircase algorithm cannot even be used. If $\epsilon(\cdot)$ takes on a large but finite number of possible values, the staircase algorithm can still be used, but the amount of noise added to each bin may be too large. This is because the staircase algorithm goes through all the "levels of the staircase" starting from the top, each time adding noise if the current noisy count is less than the top boundary of the level. For bins with a low true count, a lot of noise is added.

For $\epsilon(\cdot)$ functions that take on infinitely many or a large number of possible values, one would want to use our outlier private algorithm for a general $\epsilon(\cdot)$. For example, consider the function $\epsilon(k) = k\epsilon_0$ for some small constant $\epsilon_0 > 0$. Such a function has global sensitivity $\Delta(\epsilon(\cdot)) := \sup_{k' \in \mathbb{Z}_{>0}} |\epsilon(k') - \epsilon(k'-1)| = \epsilon_0$, which is small. Thus, we can use our general outlier private histogram algorithm and choose $\mathcal{A}$ to be the algorithm described above that adds noise to $\epsilon(k)$ calibrated to the global sensitivity of $\epsilon(\cdot)$. If $\epsilon(\cdot)$ has high global sensitivity but low "local sensitivity" for most input values, then one can choose $\mathcal{A}$ to be the algorithm described above that adds noise to $\epsilon(k)$ calibrated to the smooth sensitivity (see [14]) of $\epsilon(\cdot)$. Recall that we allow $\epsilon(\cdot)$ to take on the value ∞ (usually for sufficiently high inputs k), meaning that there is no privacy requirement. If $\epsilon(\cdot)$ does take on the value ∞, then both the global sensitivity and the smooth sensitivity of $\epsilon(\cdot)$ would be ∞, which is not allowed. In such cases, we may want to choose $\mathcal{A}$ to be one of the algorithms described above that add noise to the "noise magnitude function" $1/\epsilon(\cdot)$ instead of $\epsilon(\cdot)$. (Recall that we define $1/\infty$ to be equal to 0.) Alternatively, we can choose $\mathcal{A}$ to be the algorithm that adds noise to k and then computes $\epsilon(\cdot)$ on the noisy k.

We note that for our outlier private algorithm for general $\epsilon(\cdot)$, the function $\epsilon(\cdot)$ needs to be bounded from below by some constant $\epsilon_{\mathcal{A}} > 0$. This is because running the algorithm $\mathcal{A}$ results in "$\epsilon_{\mathcal{A}}$-privacy loss". Our staircase algorithm does not have this restriction; the staircase algorithm works even if the lowest level has an ϵ requirement of 0, in which case the staircase algorithm suppresses counts in the lowest level to 0 with very high probability.

3 Simultaneously Achieving Simple Outlier Privacy and Distributional Differential Privacy

In this section, we show that simple outlier privacy implies a certain notion of *distributional differential privacy*, very similar to the one in [1]. Let us first state the definition of distributional differential privacy w.r.t. a set of distributions over data sets. Let Φ be any set of distributions over data sets.

Definition 12 (Distributional Differential Privacy w.r.t. Φ). *An algorithm $\mathcal{M}$ is said to be (ϵ, δ)-differentially private w.r.t. Φ if for every distribution $\phi \in \Phi$ and every $t \in \bigcup Supp(\phi)$, if we let $\mathcal{D} \sim \phi$, then*

$$\mathcal{M}(\mathcal{D})|_{t \in \mathcal{D}} \approx_{\epsilon, \delta} \mathcal{M}(\mathcal{D} \setminus \{t\})|_{t \in \mathcal{D}}.$$

The definition in [1] is slightly weaker than ours, since their definition permits choosing a simulator that is used instead of $\mathcal{M}$ on the right hand side of the $\approx_{\epsilon, \delta}$, but in our definition, the simulator has to be the algorithm $\mathcal{M}$ itself. The set of distributions Φ can represent a set of possible distributions that contains the supposed "true distribution", or Φ can represent a set of possible beliefs an adversary may have about the data set (see [1] for more information). We will consider a very large and natural class of distributions that even includes relatively "adversarial" beliefs. Let us now describe our class of distributions.

We begin with some necessary terminology and notation. A *population* is a collection of individuals each holding a data record. For simplicity and convenience, we will not distinguish between an individual and the data record the individual holds; thus, an individual is simply a data record, and a population is simply a multiset of data records. Given a population $\mathcal{P}$ and a function $\pi : \mathcal{P} \to [0, 1]$, let $Sam(\mathcal{P}, \pi)$ be the distribution over data sets obtained by sampling each individual t in the population $\mathcal{P}$ with probability $\pi(t)$ independently. We note that for $Sam(\mathcal{P}, \pi)$, two individuals in $\mathcal{P}$ with the same data record will have the same probability of being sampled. However, we can easily modify the data universe X to include personal/unique identifiers so that we can represent an individual by a unique data record in X.

Let $RS(p, p', \ell)$ be the convex hull of the set of all distributions $Sam(\mathcal{P}, \pi)$, where $\mathcal{P}$ is any population, and $\pi : \mathcal{P} \to [0, 1]$ is any function such that $|\{t \in \mathcal{P} : \pi(t) \notin [p, p'] \cup \{0\}\}| \leq \ell$, i.e., for every individual t in $\mathcal{P}$ except for at most ℓ individuals, π assigns to t some probability in $[p, p'] \cup \{0\}$. Such distributions $Sam(\mathcal{P}, \pi)$ represent sampling from the population $\mathcal{P}$ in a very natural way, where most/all individuals are sampled with probability in between p

and p' (inclusive) or with probability 0. We allow at most ℓ individuals to be sampled with probability outside this range, to model the fact that an adversary may know whether certain individuals were sampled or not. The set $RS(p, p', \ell)$ includes all such natural ways of sampling from a population, and also captures a large class of possible beliefs an adversary may have about the data set. (In fact, $RS(p, p', \ell)$ is the convex hull of such a large set of distributions.)

Let us now state our theorem that says that simple outlier privacy implies distributional differential privacy w.r.t. $RS(p, p', \ell)$.

Theorem 7. *Let M be any (k, ϵ)-simple outlier private algorithm with $k \geq 2$, let $0 < p \leq p' < 1$, and let $0 \leq \ell < k - 1$. Then, for every $0 < \epsilon_{Sam} \leq \ln 2$, M is also $(k, \epsilon_{DP}, \delta_{DP})$-distributional differentially private w.r.t. $RS(p, p', \ell)$, where*

$$\epsilon_{DP} = \max\left\{\frac{\epsilon}{k}, \ln\left(\frac{p'}{p}\frac{1-p}{1-p'}\right) + \epsilon_{Sam}\right\} \quad and$$

$$\delta_{DP} = \max\left\{\frac{1}{p}, \frac{1}{1-p'}\right\} e^{-\Omega((k-\ell)\cdot(1-p')^2 \cdot \epsilon_{Sam}^2)}.$$

Remark 1. In Theorem 7, it suffices for M to be (k, ϵ, ϵ')-*simple outlier private*, which is the same as (k, ϵ)-simple outlier private except that the notion of equivalence is replaced by the notion of ϵ'-blends. The proof would be almost exactly the same, but the ϵ_{DP} parameter we achieve would be $\epsilon_{DP} = \max\left\{\frac{\epsilon}{k}, \ln\left(\frac{p'}{p}\frac{1-p}{1-p'}\right) + \epsilon_{Sam} + \epsilon'\right\}$ instead (the δ_{DP} parameter remains the same). The reason we start off with a (k, ϵ)-simple outlier private algorithm is that, as motivated in the introduction, we want an algorithm that satisfies both (k, ϵ)-simple outlier privacy and some notion of (distributional) differential privacy.

Before we prove Theorem 7, let us make some remarks. Our result (Theorem 7) is somewhat similar to the result in [9] that states that if one combines a crowd-blending private algorithm with a natural pre-sampling step, the combined algorithm is zero-knowledge private (which implies differential privacy) if we view the population as the input data set to the combined algorithm. In contrast, our result achieves a distributional notion of differential privacy on the data set as opposed to the population, which is a different model and definition. For example, one difference is that in distributional differential privacy, the individual t whose privacy we need to protect is guaranteed to be sampled, but in the model of [9], the individual t in the population might not even be sampled at all, in which case t's privacy is already protected. This leads to differences in the privacy parameters we can achieve.

Our result also has some similarities to a result in [1], where it is shown that a histogram algorithm that suppresses small counts achieves a notion of distributional differential privacy (described above), but for a class of distributions incomparable to the class we consider (the classes are somewhat similar, but neither is a subset of the other). However, our class of distributions includes distributions/beliefs based on biased and imperfect sampling in a setting where the adversary may even know whether certain individuals were sampled or not; the class of distributions considered in [1] does not consider such an adversarial setting. Also, we consider the class of simple outlier private algorithms, which includes but is more general than just histogram algorithms that suppress small counts.

Let us now prove Theorem 7. We begin by stating a lemma about the smoothness of the Poisson binomial distribution[2] near its expectation, which has appeared in [9], and will be used later in the proof of Lemma 2.

Lemma 1 (Smoothness of the Poisson binomial distribution near its expectation). *Let $\mathcal{P}$ be any population, $0 < p \le p' < 1$, $\pi : \mathcal{P} \to [0,1]$ be any function, and $\epsilon_{Sam} > 0$. Let A be any non-empty (multi)subset of $\mathcal{P}$ such that $\pi(a) \in [p, p']$ for every $a \in A$. Let $\tilde{D} = Sam(\mathcal{P}, \pi)$, $\tilde{m} = |\tilde{D} \cap A|$, $n = |A|$, and $\bar{p} = \frac{1}{n} \sum_{a \in A} \pi(a)$. Then, for every integer $m \in \{0, \ldots, n-1\}$, we have the following:*

- *If $m + 1 \le (n+1)\bar{p} \cdot \frac{e^{\epsilon_{Sam}}}{\bar{p}e^{\epsilon_{Sam}} + (1-\bar{p})}$, then $\Pr[\tilde{m} = m] \le \frac{p'}{p} \frac{1-p}{1-p'} e^{\epsilon_{Sam}} \Pr[\tilde{m} = m+1]$.*
- *If $m + 1 \ge (n+1)\bar{p} \cdot \frac{1}{\bar{p} + (1-\bar{p})e^{\epsilon_{Sam}}}$, then $\Pr[\tilde{m} = m] \ge \frac{p}{p'} \frac{1-p'}{1-p} e^{-\epsilon_{Sam}} \Pr[\tilde{m} = m+1]$.*

The proof of Lemma 1 can be found in the full version of [9]. We now prove a lemma that roughly says that if an individual is $\mathcal{M}$-equivalent to many people in the population, then the individual's privacy is protected.

Lemma 2. *Let $\mathcal{M}$ be any algorithm, $\mathcal{P}$ be any population, $0 < p \le p' < 1$, and $\pi : \mathcal{P} \to [0,1]$ be any function. Let $t \in \mathcal{P}$, and let $A \subseteq \mathcal{P} \setminus \{t\}$ such that $A \ne \emptyset$ and for every $t' \in A$, $t' \equiv_{\mathcal{M}} t$ and $\pi(t') \in [p, p']$. Let $n = |A|$ and $\bar{p} = \frac{1}{n} \sum_{t' \in A} \pi(t')$. Then, for every $0 < \epsilon_{Sam} \le \ln 2$, we have*

$$\mathcal{M}(Sam(\mathcal{P} \setminus \{t\}, \pi) \uplus \{t\}) \approx_{\epsilon_{total}, \delta_{total}} \mathcal{M}(Sam(\mathcal{P} \setminus \{t\}, \pi)),$$

where $\epsilon_{total} = \ln\left(\frac{p'}{p} \frac{1-p}{1-p'}\right) + \epsilon_{Sam}$ and $\delta_{total} = \max\left\{\frac{1}{\bar{p}}, \frac{1}{1-\bar{p}}\right\} \cdot e^{-\Omega\left((n+1)\bar{p} \cdot (1-\bar{p})^2 \cdot \epsilon_{Sam}^2\right)}$.

Proof. Let $0 < \epsilon_{Sam} \le \ln 2$, $\tilde{D} = Sam(\mathcal{P} \setminus \{t\}, \pi)$, $\tilde{m} = |\tilde{D} \cap A|$, and $Y \subseteq Range(\mathcal{M})$. We first show that for every $m \in \{0, \ldots, n-1\}$, we have

$$\mathcal{M}(\tilde{D} \uplus \{t\})|_{\tilde{m}=m} = \mathcal{M}(\tilde{D})|_{\tilde{m}=m+1}. \tag{1}$$

It is known that there exists a "draw-by-draw" selection procedure for drawing samples from A (one at a time) such that right after drawing the jth sample, the samples chosen so far has the same distribution as $Sam(A, \pi)|_{|Sam(A,\pi)|=j}$ (e.g.,

---

[2] The Poisson binomial distribution is the distribution of the sum of independent Bernoulli random variables, where the success probabilities in the Bernoulli random variables are not necessarily the same.

see Section 3 in [3]). More formally, there exists a vector of random variables $(X_1, \ldots, X_n)$ jointly distributed over A^n such that for every $j \in [n]$, $\{X_1, \ldots, X_j\}$ has the same distribution as $Sam(A, \pi)|_{|Sam(A,\pi)|=j}$. Now, fix $m \in \{0, \ldots, n-1\}$. Then, we have $(\widetilde{D} \uplus \{t\})|_{\widetilde{m}=m} = Sam(\mathcal{P} \setminus (A \uplus \{t\}), \pi) \uplus \{X_1, \ldots, X_m\} \uplus \{t\}$ and $\widetilde{D}|_{\widetilde{m}=m+1} = Sam(\mathcal{P} \setminus (A \uplus \{t\}), \pi) \uplus \{X_1, \ldots, X_m\} \uplus \{X_{m+1}\}$. The condition (1) then follows from the fact that $t \equiv_{\mathcal{M}} t'$ for every individual $t' \in A$, and $Supp(X_{m+1}) \subseteq A$. Thus, we have shown (1).

Let $\alpha = \frac{e^{\epsilon_{Sam}}}{\bar{p} e^{\epsilon_{Sam}} + (1-\bar{p})}$ and $\beta = \frac{1}{\bar{p} + (1-\bar{p}) e^{\epsilon_{Sam}}}$. Let $\epsilon_{total} = \ln(\frac{p'}{p} \frac{1-p}{1-p'}) + \epsilon_{Sam}$, and let $\delta_{total} = \max\{\Pr[\widetilde{m} + 1 > (n+1)\bar{p} \cdot \alpha], \Pr[\widetilde{m} < (n+1)\bar{p} \cdot \beta]\}$. By Lemma 1 and (1) (and the fact that $m = n$ does not satisfy $m + 1 \leq (n+1)\bar{p} \cdot \alpha$), we have

$$\Pr[\mathcal{M}(\widetilde{D} \uplus \{t\}) \in Y]$$

$$\leq \sum_{\substack{m \in \{0, \ldots, n\} \\ m+1 \leq (n+1)\bar{p} \cdot \alpha}} \Pr[\widetilde{m} = m] \cdot \Pr[\mathcal{M}(\widetilde{D} \uplus \{t\}) \in Y \mid \widetilde{m} = m] + \Pr[\widetilde{m} + 1 > (n+1)\bar{p} \cdot \alpha]$$

$$\leq \sum_{\substack{m \in \{0, \ldots, n\} \\ m+1 \leq (n+1)\bar{p} \cdot \alpha}} \frac{p'}{p} \frac{1-p}{1-p'} e^{\epsilon_{Sam}} \Pr[\widetilde{m} = m+1] \cdot \Pr[\mathcal{M}(\widetilde{D}) \in Y \mid \widetilde{m} = m+1] + \delta_{total}$$

$$\leq e^{\epsilon_{total}} \Pr[\mathcal{M}(\widetilde{D}) \in Y] + \delta_{total} \tag{3}$$

and

$$\Pr[\mathcal{M}(\widetilde{D} \uplus \{t\}) \in Y]$$

$$\geq \sum_{\substack{m \in \{0, \ldots, n-1\} \\ m+1 \geq (n+1)\bar{p} \cdot \beta}} \Pr[\widetilde{m} = m] \cdot \Pr[\mathcal{M}(\widetilde{D} \uplus \{t\}) \in Y \mid \widetilde{m} = m]$$

$$\geq \sum_{\substack{m \in \{0, \ldots, n-1\} \\ m+1 \geq (n+1)\bar{p} \cdot \beta}} \frac{p}{p'} \frac{1-p'}{1-p} e^{-\epsilon_{Sam}} \Pr[\widetilde{m} = m+1] \cdot \Pr[\mathcal{M}(\widetilde{D}) \in Y \mid \widetilde{m} = m+1]$$

$$\geq \frac{p}{p'} \frac{1-p'}{1-p} e^{-\epsilon_{Sam}} \cdot (\Pr[\mathcal{M}(\widetilde{D}) \in Y] - \Pr[\widetilde{m} < (n+1)\bar{p} \cdot \beta])$$

$$\geq e^{-\epsilon_{total}} \cdot \Pr[\mathcal{M}(\widetilde{D}) \in Y] - \delta_{total}. \tag{4}$$

Thus, we have $\mathcal{M}(\widetilde{D} \uplus \{t\}) \approx_{\epsilon_{total}, \delta_{total}} \mathcal{M}(\widetilde{D})$. Now, we observe that

$$\delta_{total}$$

$$= \max\{\Pr[\widetilde{m} + 1 > (n+1)\bar{p} \cdot \alpha], \Pr[\widetilde{m} < (n+1)\bar{p} \cdot \beta]\}$$

$$\leq \max\left\{\frac{1}{\bar{p}} \Pr[\widetilde{m} + Bin(1, \bar{p}) > (n+1)\bar{p} \cdot \alpha], \frac{1}{1-\bar{p}} \Pr[\widetilde{m} + Bin(1, \bar{p}) < (n+1)\bar{p} \cdot \beta]\right\}$$

$$\leq \max\left\{\frac{1}{\bar{p}} \exp\left(-\Omega\left((n+1)\bar{p} \cdot (\alpha-1)^2\right)\right), \frac{1}{1-\bar{p}} \exp\left(-\Omega\left((n+1)\bar{p} \cdot (1-\beta)^2\right)\right)\right\}$$

$$\leq \max\left\{\frac{1}{\bar{p}}, \frac{1}{1-\bar{p}}\right\} \cdot \exp\left(-\Omega\left((n+1)\bar{p} \cdot (1-\bar{p})^2 \epsilon_{Sam}^2\right)\right),$$

where $Bin(1, \bar{p})$ is a binomial random variable with 1 trial and success probability $\bar{p}$, and the second last inequality follows from multiplicative Chernoff bounds (and the fact that $\alpha \leq 2$, since $\epsilon_{Sam} \leq \ln 2$). □

We now prove a lemma that roughly says that even if an individual is $\mathcal{M}$-equivalent to only a few people in the population, the individual's privacy is still protected.

Lemma 3. *Let $\mathcal{M}$ be any (k, ϵ)-simple outlier private algorithm with $k \geq 2$, let $\mathcal{P}$ be any population, and let $\pi : \mathcal{P} \to [0, 1]$ be any function. Let $t \in \mathcal{P}$, and let $A \subseteq \mathcal{P} \setminus \{t\}$ such that $t' \equiv_{\mathcal{M}} t$ for every $t' \in A$. Let $n = |A|$, $s = |\{t' \in \mathcal{P} \setminus \{t\} : t' \equiv_{\mathcal{M}} t$ and $t' \notin A\}|$, and $\bar{p} = \frac{1}{n} \sum_{t' \in A} \pi(t')$. Then, if $s < k - 1$, $\bar{p} > 0$, and $n\bar{p} \leq \frac{k-s-1}{2}$, then we have*

$$\mathcal{M}(Sam(\mathcal{P} \setminus \{t\}, \pi) \uplus \{t\}) \approx_{\epsilon/k, \delta_{total}} \mathcal{M}(Sam(\mathcal{P} \setminus \{t\}, \pi)),$$

where $\delta_{total} = e^{-\Omega(k-s)}$.

Proof. Suppose $s < k - 1$, $\bar{p} > 0$, and $n\bar{p} \leq \frac{k-s-1}{2}$. Let $\widetilde{D} = Sam(\mathcal{P} \setminus \{t\}, \pi)$ and $\widetilde{m} = |\widetilde{D} \cap A|$. We note that if $\widetilde{m} < k - s - 1$, then t is $\mathcal{M}$-equivalent to fewer than k people in $\widetilde{D} \uplus \{t\}$, and since $\mathcal{M}$ is (k, ϵ)-simple outlier private, we have

$$\mathcal{M}(\widetilde{D} \uplus \{t\})|_{\widetilde{m}<k-s-1} \approx_{\epsilon} \mathcal{M}(\widetilde{D})|_{\widetilde{m}<k-s-1}$$

Let $\delta' = \Pr[\widetilde{m} \geq k - s - 1]$. Then, we have

$$\mathcal{M}(\widetilde{D} \uplus \{t\}) \approx_{\epsilon, \delta'} \mathcal{M}(\widetilde{D}). \tag{1}$$

Let $\tau = \frac{k-s-1}{2\bar{p}}$. Then, we have $n \leq \tau$. The lemma now follows from (1) and the inequality

$$
\begin{aligned}
\delta' &= \Pr[\widetilde{m} \geq 2\tau\bar{p}] \\
&\leq \Pr[\widetilde{m} + Bin(\lfloor \tau \rfloor - n, \bar{p}) + Bin(1, (\tau - \lfloor \tau \rfloor)\bar{p}) \geq 2\tau\bar{p}] \\
&\leq e^{-\Omega(\tau\bar{p})} \\
&\leq e^{-\Omega(k-s)},
\end{aligned}
$$

where $Bin(j, q)$ denotes a binomial random variable with j trials and success probability q, and the second inequality follows from a multiplicative Chernoff bound (note that the expectation of $\widetilde{m} + B(\lfloor \tau \rfloor - n, \bar{p}) + B(1, (\tau - \lfloor \tau \rfloor)\bar{p})$ is $\tau\bar{p}$). □

We will now use the above lemmas to prove Theorem 7.

Proof (of Theorem 7). Recall that $RS(p, p', \ell)$ is the convex hull of a set of distributions, which we denote by Φ'. From the definition of distributional differential privacy w.r.t. $RS(p, p', \ell)$, it is easy to see that it suffices to show differential privacy w.r.t. Φ' instead. Let $\phi = Sam(\mathcal{P}, \pi) \in \Phi'$, where $\mathcal{P}$ is the population associated with ϕ, and $\pi : \mathcal{P} \to [0, 1]$ is the sampling probability function associated with ϕ. It is easy to see that without loss of generality, we can assume that $\pi(t') > 0$ for every $t' \in \mathcal{P}$. Let t be any individual in $\mathcal{P}$, and let $\mathcal{D} \sim Sam(\mathcal{P}, \pi)$. We need to show that

$$\mathcal{M}(\mathcal{D})|_{t \in \mathcal{D}} \approx_{\epsilon_{DP}, \delta_{DP}} \mathcal{M}(\mathcal{D} \setminus \{t\})|_{t \in \mathcal{D}}.$$

We note that $\mathcal{M}(\mathcal{D})|_{t \in \mathcal{D}} = \mathcal{M}(Sam(\mathcal{P} \setminus \{t\}, \pi) \uplus \{t\})$ and $\mathcal{M}(\mathcal{D} \setminus \{t\})|_{t \in \mathcal{D}} = \mathcal{M}(Sam(\mathcal{P} \setminus \{t\}, \pi))$. Thus, it suffices to show

$$\mathcal{M}(Sam(\mathcal{P} \setminus \{t\}, \pi) \uplus \{t\}) \approx_{\epsilon_{DP}, \delta_{DP}} \mathcal{M}(Sam(\mathcal{P} \setminus \{t\}, \pi)). \tag{1}$$

To this end, let $A = \{t' \in \mathcal{P} \setminus \{t\} : t' \equiv_{\mathcal{M}} t \text{ and } \pi(t') \in [p, p']\}$, $n = |A|$, $\bar{p} = \frac{1}{n} \sum_{t' \in A} \pi(t')$, and $s = |\{t' \in \mathcal{P} \setminus \{t\} : t' \equiv_{\mathcal{M}} t \text{ and } t' \notin A\}|$. We note that $s \le l$, which we use later in some of the inequalities below. Let $\tau = \frac{k-s-1}{2\bar{p}}$. We will consider two cases: $n > \tau$ and $n \le \tau$.

Suppose $n > \tau$. By Lemma 2, we have

$$\mathcal{M}(Sam(\mathcal{P} \setminus \{t\}, \pi) \uplus \{t\}) \approx_{\epsilon_{DP}, \delta_1} \mathcal{M}(Sam(\mathcal{P} \setminus \{t\}, \pi)),$$

where

$$\delta_1 = \max\left\{\frac{1}{\bar{p}}, \frac{1}{1-\bar{p}}\right\} \cdot e^{-\Omega\left((n+1)\bar{p} \cdot (1-\bar{p})^2 \cdot \epsilon_{Sam}^2\right)}$$

$$\le \max\left\{\frac{1}{p}, \frac{1}{1-p'}\right\} \cdot e^{-\Omega\left((k-s-1) \cdot (1-p')^2 \cdot \epsilon_{Sam}^2\right)}$$

$$\le \delta_{DP}.$$

Now, suppose $n \le \tau$. By Lemma 3, we have

$$\mathcal{M}(Sam(\mathcal{P} \setminus \{t\}, \pi) \uplus \{t\}) \approx_{\epsilon_2, \delta_2} \mathcal{M}(Sam(\mathcal{P} \setminus \{t\}, \pi)),$$

where $\epsilon_2 = \epsilon/k \le \epsilon_{DP}$ and $\delta_2 = e^{-\Omega(k-s)}$, so $\delta_2 \le \delta_{DP}$.
Thus, we have shown (1), as required.   □

References

1. Bassily, R., Groce, A., Katz, J., Smith, A.: Coupled-worlds privacy: Exploiting adversarial uncertainty in statistical data privacy. In: FOCS. pp. 439–448 (2013)
2. Chandola, V., Banerjee, A., Kumar, V.: Anomaly detection: A survey. ACM Comput. Surv. 41(3), 15:1–15:58 (2009)
3. Chen, X.H., Dempster, A.P., Liu, J.S.: Weighted finite population sampling to maximize entropy. Biometrika 81(3), 457–469 (1994)
4. Dwork, C.: The differential privacy frontier (Extended abstract). In: Reingold, O. (ed.) TCC 2009. LNCS, vol. 5444, pp. 496–502. Springer, Heidelberg (2009)
5. Dwork, C.: Differential privacy. In: Bugliesi, M., Preneel, B., Sassone, V., Wegener, I. (eds.) ICALP 2006. LNCS, vol. 4052, pp. 1–12. Springer, Heidelberg (2006)

6. Dwork, C.: Differential privacy: A survey of results. In: Agrawal, M., Du, D.-Z., Duan, Z., Li, A. (eds.) TAMC 2008. LNCS, vol. 4978, pp. 1–19. Springer, Heidelberg (2008)

7. Dwork, C., McSherry, F., Nissim, K., Smith, A.: Calibrating noise to sensitivity in private data analysis. In: Halevi, S., Rabin, T. (eds.) TCC 2006. LNCS, vol. 3876, pp. 265–284. Springer, Heidelberg (2006)

8. Fleischer, L.K., Lyu, Y.H.: Approximately optimal auctions for selling privacy when costs are correlated with data. In: Proceedings of the 13th ACM Conference on Electronic Commerce, EC 2012, pp. 568–585. ACM (2012)

9. Gehrke, J., Hay, M., Lui, E., Pass, R.: Crowd-blending privacy. In: Safavi-Naini, R., Canetti, R. (eds.) CRYPTO 2012. LNCS, vol. 7417, pp. 479–496. Springer, Heidelberg (2012)

10. Gehrke, J., Lui, E., Pass, R.: Towards privacy for social networks: A zero-knowledge based definition of privacy. In: Ishai, Y. (ed.) TCC 2011. LNCS, vol. 6597, pp. 432–449. Springer, Heidelberg (2011)

11. Ghosh, A., Roth, A.: Selling privacy at auction. In: Proceedings of the 12th ACM Conference on Electronic Commerce, EC 2011, pp. 199–208. ACM (2011)

12. Ligett, K., Roth, A.: Take it or leave it: Running a survey when privacy comes at a cost. In: Goldberg, P.W. (ed.) WINE 2012. LNCS, vol. 7695, pp. 378–391. Springer, Heidelberg (2012)

13. Narayanan, A., Shmatikov, V.: Robust de-anonymization of large sparse datasets. In: Proceedings of the 2008 IEEE Symposium on Security and Privacy, SP 2008, pp. 111–125. IEEE Computer Society (2008)

14. Nissim, K., Raskhodnikova, S., Smith, A.: Smooth sensitivity and sampling in private data analysis. In: STOC 2007, pp. 75–84 (2007)

15. Nissim, K., Vadhan, S., Xiao, D.: Redrawing the boundaries on purchasing data from privacy-sensitive individuals. In: Proceedings of the 5th Conference on Innovations in Theoretical Computer Science, ITCS 2014, pp. 411–422. ACM (2014)

16. Roth, A., Schoenebeck, G.: Conducting truthful surveys, cheaply. In: Proceedings of the 13th ACM Conference on Electronic Commerce, EC 2012, pp. 826–843. ACM (2012)

Function-Private Functional Encryption in the Private-Key Setting*

Zvika Brakerski[1,**] and Gil Segev[2,***]

[1] Department of Computer Science and Applied Mathematics,
Weizmann Institute of Science, Rehovot 76100, Israel
`zvika.brakerski@weizmann.ac.il`
[2] School of Computer Science and Engineering,
Hebrew University of Jerusalem, Jerusalem 91904, Israel
`segev@cs.huji.ac.il`

Abstract. Functional encryption supports restricted decryption keys that allow users to learn specific functions of the encrypted messages. Although the vast majority of research on functional encryption has so far focused on the privacy of the encrypted *messages*, in many realistic scenarios it is crucial to offer privacy also for the *functions* for which decryption keys are provided.

Whereas function privacy is inherently limited in the public-key setting, in the private-key setting it has a tremendous potential. Specifically, one can hope to construct schemes where encryptions of messages $m_1, \ldots, m_T$ together with decryption keys corresponding to functions $f_1, \ldots, f_T$, reveal essentially no information other than the values $\{f_i(m_j)\}_{i,j \in [T]}$. Despite its great potential, the known function-private private-key schemes either support rather limited families of functions (such as inner products), or offer somewhat weak notions of function privacy.

We present a generic transformation that yields a *function-private* functional encryption scheme, starting with any *non-function-private* scheme for a sufficiently rich function class. Our transformation preserves the message privacy of the underlying scheme, and can be instantiated using a variety of existing schemes. Plugging in known constructions of functional encryption schemes, we obtain function-private schemes based either on the Learning with Errors assumption, on obfuscation assumptions, on simple multilinear-maps assumptions, and even on the existence of any one-way function (offering various trade-offs between security and efficiency).

---

* The full version of this paper is available as [12].
** Supported by the Israel Science Foundation (Grant No. 468/14) and by the Alon Young Faculty Fellowship.
*** Supported by the European Union's Seventh Framework Programme (FP7) via a Marie Curie Career Integration Grant, by the Israel Science Foundation (Grant No. 483/13), and by the Israeli Centers of Research Excellence (I-CORE) Program (Center No. 4/11).

Y. Dodis and J.B. Nielsen (Eds.): TCC 2015, Part II, LNCS 9015, pp. 306–324, 2015.

1   Introduction

The most classical cryptographic scenario, dating back thousands of years, consists of two parties who wish to secretly communicate in the presence of an eavesdropper. This classical scenario has traditionally led the cryptographic community to view the security provided by encryption schemes as an *all-or-nothing* guarantee: The encrypted data can be fully recovered using the decryption key, but it is completely useless without it. In a wide variety of modern scenarios, however, a more fine-grained approach is urgently needed. Starting with the seminal notion of identity-based encryption [29,6,13], this need has recently motivated the cryptographic community to develop a vision of *functional encryption* [27,10,26], allowing tremendous flexibility when accessing encrypted data.

Functional encryption supports restricted decryption keys that allow users to learn specific functions of the encrypted data and nothing else. More specifically, in a functional encryption scheme, a trusted authority holds a master secret key known only to the authority. When the authority is given the description of some function f as input, it uses its master secret key to generate a functional key sk_f associated with the function f. Now, anyone holding the functional key sk_f and an encryption of some message m, can compute $f(\mathsf{m})$ but cannot learn any additional information about the message m. Extensive research has recently been devoted to studying the security of functional encryption schemes and to constructing such schemes (see, for example, [27,10,26,21,2,5,11,14,15,19] and the references therein).

Function Privacy in Functional Encryption. The vast majority of research on functional encryption to date has focused on the privacy of encrypted *messages*. In various scenarios, however, one should consider not only privacy for the encrypted messages but also privacy for the *functions* for which functional keys are provided. Consider, for example, a user who subscribes to an on-line storage service for storing her files. For protecting her privacy, the user locally encrypts her files using a functional encryption scheme prior to uploading them to the service. The user can then remotely query her data by providing the service with a functional key sk_f corresponding to any query f. Without any additional privacy guarantees, the functional key sk_f may entirely reveal the user's query f to the service, which is clearly undesirable whenever the query itself contains sensitive information.

Scenarios of such flavor have motivated the research of *function privacy* in functional encryption, first in the private-key setting by Shen, Shi and Waters [30], and very recently in the public-key setting by Boneh, Raghunathan and Segev [8,9] followed by Agrawal et al. [1]. Intuitively, function privacy requires that functional keys reveal no unnecessary information on their functionality.

The extent to which function privacy can be satisfied differs dramatically between the settings of private-key and public-key encryption. Specifically, in the public-key setting, where anyone can encrypt messages, only a limited form of function privacy can be satisfied. This is because given a functional key sk_f and the public key pk of the scheme, a malicious user can learn information about the function f by evaluating it on any point m of his choosing (by first encrypting m and then using sk_f to decrypt $f(\mathsf{m})$). This process reveals non-trivial information about f and in some cases may entirely leak the function's description (unless additional restrictions are imposed, see [8,9,1] for more details). As a result, function-private functional encryption schemes in the public-key setting are quite restricted, and furthermore such have only been presented respective to limited function families (e.g., point functions and inner products).

Our Work: Function Privacy in the Private-Key Setting. In this work, we focus on function privacy in the private-key setting. In this setting, function privacy has significantly more potential than in the public-key setting, both as a stand-alone feature and as a very useful building block (see Section 1.2 for subsequent work). Specifically, one can hope to achieve the following notion of privacy (stated informally): Any user that obtains encryptions of messages $\mathsf{m}_1, \ldots, \mathsf{m}_T$, and functional keys corresponding to functions $f_1, \ldots, f_T$, learns essentially no information other than the values $\{f_i(\mathsf{m}_j)\}_{i,j \in [T]}$. This is a strong notion of privacy which has great (realistic) potential for a wide variety of applications.

Despite its great potential, the known function-private private-key schemes either support only the inner-product functionality for attribute-based encryption [30,1], or offer only somewhat weak notions of function privacy [19,17]. We refer the reader to Section 1.3 for a detailed discussion of the known function private schemes. This state of affairs has motivated us to explore the following fundamental question:

> Can we construct private-key functional encryption schemes
> that support *rich and expressive* families of functions
> while offering *strong notions* of function privacy?

1.1   Our Contributions

Our work provides a positive answer to the above fundamental question. We present private-key functional encryption schemes that support rich and expressive families of functions, while offering strong notions of function privacy.

Specifically, we put forward a generic transformation that yields a *function-private* private-key functional encryption scheme based on any (possibly *non-function-private*) private-key functional encryption scheme that supports all functions that are computable by bounded-size circuits. In particular, our transformation can be instantiated by the recently developed functional encryption scheme of Goldwasser et al. [19] that is based on the LWE assumption, by the schemes of Garg et al. [15], Boyle et al. [11], Ananth et al. [3], and Waters [31] that are based on obfuscation assumptions, by the scheme of Garg et al. [16] that is based on simple assumptions on multilinear maps, and even by the scheme of Gorbunov et al. [21] which is somewhat less efficient but can be based on any one-way function (we refer the reader to Section 2.2 for more details). Although most of these constructions are in fact public-key schemes, they are in particular private-key ones (i.e., in some sense, these schemes currently seem significantly more powerful than what is required for our transformation).

The notions of function privacy that are satisfied by our transformation are the strongest notions that have been proposed so far (we refer the reader to Section 3 for a detailed discussion of these notions). In addition, the resulting scheme inherits the message privacy of the underlying scheme (i.e., full vs. selective security, and one-key vs. many-keys security), and supports all functions that are computable by circuits whose size is slightly smaller than those supported by the underlying scheme. Finally, we note that our transformation is in fact oblivious to the computational model that is supported by the underlying scheme and to its representation (e.g., circuits vs. Turing machines), as long as the scheme supports a universal function for the model and a few additional basic operations (see Section 1.4 below).

1.2 Subsequent Work

Our generic construction and proof techniques have already been proved fruitful by Komargodski et al. [24] and by Ananth et al. [4] beyond the context of function-private functional encryption as its own primitive. Komargodski et al. presented a construction of a private-key functional encryption scheme for *randomized* functions based on *any* private-key functional encryption scheme for *deterministic* functions that is sufficiently expressive. Their work follows-up on the work of Goyal et al. [23] who put forward the notion of functional encryption for randomized functionalities, and constructed a public-key functional encryption scheme for randomized functionalities based on the (seemingly significantly stronger) assumption of indistinguishability obfuscation. Ananth et al. presented a construction of a *fully-secure* functional encryption scheme from any *selectively-secure* functional encryption scheme that is sufficiently expressive (their transformation applies in both the private-key setting and the public-key setting). Previous constructions of fully-secure schemes were based on assumptions that seem significantly stronger, such as obfuscation and multilinear maps assumptions [11,3,31,16].

One of the key insights underlying both of these works is that in the private-key setting, where encryption is performed honestly by the owner of the master secret key, the power of obfuscation may not be needed. Instead, they observed that in some cases one can rely on the weaker notion of function privacy. More specifically, both Komargodski et al. and Ananth et al. showed that any sufficiently expressive functional encryption scheme may be appropriately utilized via our function-privacy techniques for implementing some of the proof techniques that were so far implemented based on obfuscation (including, for example, a variant of the punctured programming approach of Sahai and Waters [28]).

1.3 Additional Related Work

Function Privacy. As mentioned above, Shen, Shi and Waters [30] initiated the research on predicate privacy in attribute-based encryption in the private-key setting. They constructed a predicate-private inner-product encryption scheme in the private-key setting. Boneh, Raghunathan and Segev [8,9] initiated the research on function privacy in the public key setting. They constructed function-private public-key functional encryption schemes for point functions (equivalently, anonymous IBE) and for subspace membership. Since their work is in the public-key setting, their framework assumes that the functions come from a distribution of sufficient entropy, as otherwise it seems that no realistic notion of function privacy can be satisfied.

Agrawal et al. [1] then presented a general framework for function-private functional encryption both in the private-key setting and in the public-key setting, and explore their plausibility. Most relevant to our work, they presented the *full security* notion for function-private functional encryption in the private-key setting and presented improved constructions for the inner-product functionality in this model. We note that we refer to their notion of full security as *full function privacy* (see Definition 3.2).

Reusable Garbled Circuits. The related notion of *reusable garbled circuits* (ruGC) is defined as follows. Given a secret key, two procedures can be carried out: Garbling a circuit C (which corresponds to generating a function key) and encoding of an input x (which corresponds to an encryption of a message). Given a garbled C and an encoded input x, it is possible to publicly compute $C(x)$. The security requirement is that an adversary that chooses C to be garbled and then a sequence of inputs $x_1, \ldots, x_t$ to be encoded cannot learn more than $C(x_1), \ldots, C(x_t)$. Security is formalized in a simulation-based model: The simulator is required to simulate the garbled circuit without knowing C, and then it is fed with $C(x_i)$ in turn and is required to simulate the encoded inputs. Goldwasser et al. [19,20] constructed a simulation-secure functional encryption scheme (without function privacy) and showed how ruGC follows from that primitive[1]. The similarity to function-private functional encryption is apparent, but there are some significant differences. It follows from the result of [2] that ruGC, much like simulation secure functional encryption, cannot securely support an a-priori unbounded number of circuits, whereas we are able to guarantee function privacy for any polynomial (a-priori unknown) number of function keys. A very similar argument shows that the situation where C is chosen after the inputs x_i is also impossible in the context of ruGC (at least under a natural definition of the simulation process), whereas we would like the inputs and functions to be adaptively introduced in arbitrary order. On the flip side, ruGC provides simulation based security which seems to be a stronger notion than indistinguishability-based security achieved by our construction.

Multi-Input Functional Encryption. Goldwasser et al. [17] have recently introduced the notion of *multi-input functional encryption* (MIFE) schemes. As the name suggests, MIFE allows functional keys to correspond to multi-input functions which can be evaluated on tuples of ciphertexts. Slightly simplified, the dream

---

[1] Additional constructions were presented by Boneh et al. [7] who were able to reduce the garbling overhead from multiplicative to additive in either the size of the circuit or the size of the encoded input.

notion of security (specifically, indistinguishability-based security in the private-key setting, which is most relevant to this work) is that of an adversary that is allowed to make functional key queries and also message queries containing pairs of messages (m_0, m_1), and in response it gets an encryption of m_b, where b is a secret bit. We would like the adversary to not be able to guess b unless it obtained a key to a function that behaves differently on the m_0's and on the m_1's. This dream version of security, even just for two inputs, implies function-private private-key functional encryption: We will use the first input coordinate to encrypt the description of the function and the second to encrypt the input, and provide a function key for the universal 2-input function. However, Goldwasser et al. [17] fall short of achieving this dream version, since their adversary is not allowed to make message queries adaptively. Furthermore, their construction relies on strong notions of obfuscation (indistinguishability obfuscation and differing input obfuscation), whereas the construction in this paper only relies on private-key function encryption (which is currently known to be implied by obfuscation, but no reverse derivation is known and it is quite possible that they can be constructed under milder assumptions – see Section 2.2).

1.4 Overview of Our Approach

In this section we provide a high-level overview of our approach and techniques. We begin with a brief description of the notions of function privacy that we consider in this work, and then describe the main ideas underlying our construction.

Function Privacy. Our notion of function privacy is that of Agrawal et al. [1, Def. 2.7] (generalizing Shen, Shi and Waters [30]), which considers the privacy of functional keys and the privacy of encrypted messages in a completely symmetric manner. Specifically, we consider adversaries that issue both key-generation queries of the form (f_0, f_1) and encryption queries of the form (m_0, m_1). These queries are answered by providing a functional key for f_b and an encryption of m_b, where all queries are answered using the same bit $b \in \{0, 1\}$. We allow adversaries to adaptively issue any polynomial number of such queries (this number does not have to be bounded in advance), and their goal is to distinguish the experiment where $b = 0$ and the experiment where $b = 1$. Our only requirement from such adversaries is that for all key-generation queries (f_0, f_1) and for all encryption queries (m_0, m_1) it holds that $f_0(m_0) = f_1(m_1)$. In addition to this notion, we also consider two "selective" relaxations, and we refer the reader to Section 3 for more details on our notions of function privacy.

A Failed Attempt. Our starting point is any given private-key functional encryption scheme without function privacy. A natural approach towards achieving function privacy is to modify its key-generation algorithm so that it provides functional keys containing only *encrypted* descriptions of the associated functions. Namely, for generating a functional key for a function f, we will first encrypt the description of f using a symmetric encryption scheme $\mathcal{SKE} = (\text{SKE.KG}, \text{SKE.Enc}, \text{SKE.Dec})$ to obtain a ciphertext $c_f \leftarrow \text{SKE.Enc}(\text{SKE.k}, f)$, where SKE.k is a key for the scheme $\mathcal{SKE}$ (which does not change throughout the lifespan of the scheme). Then, the key-generation algorithm would provide a functional key for the function $U_{c_f}(\mathsf{m}, k) \overset{\text{def}}{=} (\mathsf{SKE.Dec}(k, c_f))(\mathsf{m})$ (that is, the function that first decrypts c_f using the candidate key k, and then applies the resulting function on m). The semantic security of $\mathcal{SKE}$ guarantees that the function U_{c_f} hides the description of f, as long as the key $\mathsf{SKE.k}$ is not known. In order to maintain the functionality, the message encryption algorithm must also change: Rather than encrypting the message m alone using the underlying functional encryption scheme, we will now encrypt the pair $(m, \mathsf{SKE.k})$. One can verify that the functionality of the scheme still holds since clearly $U_{c_f}(\mathsf{m}, \mathsf{SKE.k}) = f(\mathsf{m})$.

One could hope to prove that this construction is function private. Indeed, Goldwasser et al. [19] used this exact scheme to construct reusable garbled circuits. This approach by itself, however, is insufficient for our purposes. On one hand, during the proof of security we would like to rely on the semantic security of $\mathcal{SKE}$ for arguing that the function U_{c_f} hides the description of f. This implies that the key $\mathsf{SKE.k}$ should be kept completely secret. On the other hand, the functionality of the scheme must be preserved even during the proof of security. Thus, in order to allow adversaries to use the functional key for the function U_{c_f}, the key $\mathsf{SKE.k}$ must be used while encrypting messages as above. This conflict is the main challenge that our construction overcomes.

We note that also in the construction of reusable garbled circuits of Goldwasser et al. [19] this conflict arises. However, they consider only a *selective single-function* security notion asking adversaries to specify a challenge function f prior to receiving any encryptions. Within such a selective framework, the conflict is easily resolved: During the proof of security one can preserve the functionality by modifying the encryption algorithm to encrypt $f(\mathsf{m})$ instead of encrypting m itself. Thus, the description of the function f is in fact not needed, but only the value $f(\mathsf{m})$ is needed for each encrypted message m (note that $f(\mathsf{m})$ is anyway known to the adversary). This approach, however, seems inherently limited to a selective framework, whereas we would like to allow adversaries to adaptively query the key-generation and encryption oracles, at any point in time, and for any polynomial number of queries[2].

Our Scheme. To get around the aforementioned obstacle, we show that the Naor-Yung "double encryption" methodology [25] can be adapted to our setting. Instead of encrypting the description of f only once, we encrypt it twice using two independent symmetric keys. For preserving the functionality of the system, only one out of the two keys will be explicitly needed, and this allows us to attack the other key. Combined with the message privacy of the underlying functional encryption scheme, this approach enables us to prove the security of our scheme.

---

[2] The approach of Goldwasser et al. can be extended to deal with any a-priori bounded number of functions, as long as they are specified in advance (this is done using [21]). In this case, the length of ciphertexts in their scheme would be linear in the number of functions. This is in fact inherent to their setting, as they consider a simulation-based notion of security [2]. We consider indistinguishability-based notions of security, and would like to inherit the (either full or selective) security of the underlying functional encryption scheme.

More specifically, the master secret key of our scheme consists of a master secret key msk for the underlying functional encryption scheme, and two keys, SKE.k and SKE.k', for a symmetric-key CPA-secure scheme. In order to generate a functional key for a function f, we first generate two symmetric encryptions $c \leftarrow \mathsf{SKE.Enc}(\mathsf{SKE.k}, f)$ and $c' \leftarrow \mathsf{SKE.Enc}(\mathsf{SKE.k}', f)$ of the description of f. Then, we issue a functional key for the function $U_{c,c'}$ which is defined as follows on inputs of the form $(\mathsf{m}, \mathsf{m}', k, k')$: If $k \neq \perp$ then decrypt c using k for obtaining a function f, and output $f(\mathsf{m})$. Otherwise, if $k = \perp$, then decrypt c' using k' for obtaining a function f', and output $f'(\mathsf{m}')$. In order to encrypt a message m, we will use the encryption scheme of the underlying functional encryption scheme to encrypt $(\mathsf{m}, \perp, \mathsf{SKE.k}, \perp)$ using its master secret key msk. Note that this scheme works quite similarly to the aforementioned intuitive idea, only it has "placeholders" for elements that will be used in the proof.

Towards illustrating some of the ideas underlying the proof of security, consider an adversary that makes just one encryption query $(\mathsf{m}_0, \mathsf{m}_1)$, and just one key-generation query (f_0, f_1) in some *arbitrary* order (recall that we require $f_0(\mathsf{m}_0) = f_1(\mathsf{m}_1)$). The view of this adversary consists of an encryption of m_b and a functional key for f_b for a uniformly chosen bit $b \in \{0, 1\}$. The proof starts by modifying the functional key: Instead of computing $c' \leftarrow \mathsf{SKE.Enc}(\mathsf{SKE.k}', f_b)$ we compute $c' \leftarrow \mathsf{SKE.Enc}(\mathsf{SKE.k}', f_1)$. Note that since the key SKE.k' is in fact not being used, and since the functionality of the functional key is not hurt (c' is anyway not used for decryption), the CPA-security of the symmetric scheme implies that this goes unnoticed. Next, we modify the encryption algorithm to encrypt $(\perp, \mathsf{m}_1, \perp, \mathsf{SKE.k}')$ instead of $(\mathsf{m}_b, \perp, \mathsf{SKE.k}, \perp)$. This time the adversary will not notice the change due to the message-privacy of the underlying functional encryption scheme, since the new and old ciphertexts will decrypt to the same value $f_b(\mathsf{m}_b) = f_1(\mathsf{m}_1)$ under the modified functional key. Finally, we modify the functional key once again: Instead of computing $c \leftarrow \mathsf{SKE.Enc}(\mathsf{SKE.k}, f_b)$ we compute $c \leftarrow \mathsf{SKE.Enc}(\mathsf{SKE.k}, f_1)$. As before, since the key SKE.k is in fact not being used, and since the functionality of the functional key is not hurt, then the CPA-security of the symmetric scheme implies that this goes unnoticed. At this point, we observe that the view of the adversary is in fact completely independent of the choice of the bit b, and the result follows. We refer the reader to Section 4 for the formal description and proof of our scheme.

1.5   Open Problems

Our work raises various open problems on the feasibility and the design of functional encryption schemes. Some of these are outlined below.

Private-Key vs. Public-Key Functional Encryption. Our construction relies on any private-key functional encryption schemes, but in fact, most of the existing constructions of such schemes are secure even in the public-key setting (see Section 2.2). Clearly, any functional encryption scheme that is secure in the public-key setting is also secure in the private-key one. However, the existing constructions either apply in a restricted setting or rely on somewhat strong assumptions that are related to program obfuscation. Our work provides additional motivation for studying private-key FE in hope for achieving constructions with better efficiency or under improved assumptions. Alternatively, perhaps it is possible to construct a public-key functional encryption scheme based on any private-key scheme.

Simulation-Based Function Privacy. Following Shen, Shi and Waters [30] and Boneh, Raghunathan and Segev [8,9], we consider indistinguishability-based notions of function privacy. As already observed [10,26], in some cases such notions do not provide realistic security guarantees for functional encryption schemes. A (somewhat relaxed) simulation-based notion of function privacy was recently formalized by Agrawal et al. [1], and an interesting open problem is to further explore its relation to our notions and to our construction.

Relying on Restricted Function Families. Our construction relies on any private-key functional encryption scheme that supports a sufficiently rich function class. Although, as discussed above, various such schemes are known to exist, an interesting open problem is to construct a function-private scheme based on any scheme that supports more restricted function classes (e.g., inner products or subspace membership).

1.6   Paper Organization

The remainder of this paper is organized as follows. In Section 2 we introduce the basic notation and tools underlying our construction. In Section 3 we introduce the notions of function privacy that are considered in this work. In Section 4 we present our generic construction of a function-private scheme.

2   Preliminaries

In this section we present the notation and basic definitions that are used in this work. For a distribution X we denote by $x \leftarrow X$ the process of sampling a value x from the distribution X. Similarly, for a set $\mathcal{X}$ we denote by $x \leftarrow \mathcal{X}$ the process of sampling a value x from the uniform distribution over $\mathcal{X}$. For an integer $n \in \mathbb{N}$ we denote by $[n]$ the set $\{1, \ldots, n\}$. A real function over the natural numbers is *negligible* if it vanishes faster than the inverse of any polynomial.

We rely on the following standard notion of a *left-or-right oracle* when formalizing the security of encryption schemes:

Definition 2.1 (Left-or-Right Oracle). *Let $\mathcal{O}$ be a (possibly probabilistic) two-input functionality. For each $b \in \{0,1\}$ we denote by $\mathcal{O}_b$ the three-input functionality $\mathcal{O}_b(k, x_0, x_1) \overset{\mathsf{def}}{=} \mathcal{O}(k, x_b)$.*

2.1   Private-Key Encryption

A private-key encryption scheme over a message space $\mathcal{M}$ is a triplet (KG, Enc, Dec) of probabilistic polynomial-time algorithms. The key-generation algorithm

KG takes as input the unary representation 1^λ of the security parameter $\lambda \in \mathbb{N}$ and outputs a secret key k. The encryption algorithm Enc takes as input a secret key k and a message $m \in \mathcal{M}$, and outputs a ciphertext c. The decryption algorithm Dec takes as input a secret key k and a ciphertext c, and outputs a message m or the dedicated symbol $\bot$. In terms of correctness we require that for any key k that is produced by $KG(1^\lambda)$ and for every message $m \in \mathcal{M}$ it holds that $Dec(k, Enc(k, m)) = m$ with probability 1 over the internal randomness of the algorithms Enc and Dec.

In terms of security, we rely on the standard notion of a CPA-secure private-key encryption scheme that is formulated using a left-or-right encryption oracle. Recall (Definition 2.1) that for an encryption scheme $\Pi = (KG, Enc, Dec)$ and for any $b \in \{0, 1\}$ we denote by Enc_b the left-or-right encryption oracle $Enc_b(k, m_0, m_1) \overset{\text{def}}{=} Enc(k, m_b)$.

Definition 2.2 (CPA Security). *A private-key encryption scheme $\Pi = (KG, Enc, Dec)$ is CPA-secure if for any probabilistic polynomial-time adversary $\mathcal{A}$, there exists a negligible function $\nu(\lambda)$ such that*

$$\mathsf{Adv}_{\Pi, \mathcal{A}}^{\mathsf{CPA}}(\lambda) \overset{\text{def}}{=} \left| \Pr\left[\mathcal{A}^{Enc_0(k, \cdot, \cdot)}(\lambda) = 1 \right] - \Pr\left[\mathcal{A}^{Enc_1(k, \cdot, \cdot)}(\lambda) = 1 \right] \right| \leq \nu(\lambda),$$

where the probability is taken over the choice of $k \leftarrow KG(1^\lambda)$ and over the randomness of $\mathcal{A}$.

2.2 Private-Key Functional Encryption

We rely on the standard indistinguishability-based notions of full security and selective security for functional encryption schemes (see, for example, [10,26,5]), by adapting them to the private-key setting. In this paper, as we consider security notions for both messages and keys, we refer to the standard notions of security and selective security as *message privacy* and *selective-message privacy*.

A private-key functional encryption scheme over a message space $\mathcal{M}$ and a function space $\mathcal{F}$ is a quadruple (Setup, KG, Enc, Dec) of probabilistic polynomial-time algorithms. The setup algorithm Setup takes as input the unary representation 1^λ of the security parameter $\lambda \in \mathbb{N}$ and outputs a master-secret key msk. The key-generation algorithm KG takes as input a master-secret key msk and a function $f \in \mathcal{F}$, and outputs a functional key sk_f. The encryption algorithm Enc takes as input a master-secret key msk and a message $m \in \mathcal{M}$, and outputs a ciphertext c. In terms of correctness we require that for every function $f \in \mathcal{F}$ and message $m \in \mathcal{M}$ it holds that $Dec(KG(msk, f), Enc(msk, m)) = f(m)$ with all but a negligible probability over the internal randomness of the algorithms Setup, KG, Enc, and Dec.

In terms of message privacy we require that encryptions of any two messages, m_0 and m_1, are computationally indistinguishable for any adversary that may adaptive obtain functional keys for any function $f \in \mathcal{F}$ as long as $f(m_0) = f(m_1)$. This is formalized via the following definitions. Recall (Definition 2.1) that for a private-key functional encryption scheme $\Pi = (Setup, KG, Enc, Dec)$

and for any $b \in \{0, 1\}$ we denote by Enc_b the left-or-right encryption oracle $\mathsf{Enc}_b(\mathsf{msk}, \mathsf{m}_0, \mathsf{m}_1) \overset{\text{def}}{=} \mathsf{Enc}(\mathsf{msk}, \mathsf{m}_b)$.

Definition 2.3 (Valid message-privacy adversary). *A probabilistic polynomial-time algorithm $\mathcal{A}$ is a* valid message-privacy adversary *if for all private-key functional encryption schemes $\Pi = (\mathsf{Setup}, \mathsf{KG}, \mathsf{Enc}, \mathsf{Dec})$, for all $\lambda \in \mathbb{N}$ and $b \in \{0, 1\}$, and for all f and $(\mathsf{m}_0, \mathsf{m}_1)$ with which $\mathcal{A}$ queries the oracles KG and Enc_b, respectively, it holds that $f(\mathsf{m}_0) = f(\mathsf{m}_1)$.*

Definition 2.4 (Full Message Privacy). *A private-key functional encryption scheme $\Pi = (\mathsf{Setup}, \mathsf{KG}, \mathsf{Enc}, \mathsf{Dec})$ over a message space $\mathcal{M} = \{\mathcal{M}_\lambda\}_{\lambda \in \mathbb{N}}$ and a function space $\mathcal{F} = \{\mathcal{F}_\lambda\}_{\lambda \in \mathbb{N}}$ is* fully message private *if for any valid message-privacy adversary $\mathcal{A}$, there exists a negligible function $\nu(\lambda)$ such that*

$$\mathsf{Adv}^{\mathsf{MP}}_{\Pi, \mathcal{A}}(\lambda) \overset{\text{def}}{=} \left| \Pr\left[\mathcal{A}^{\mathsf{KG}(\mathsf{msk}, \cdot), \mathsf{Enc}_0(\mathsf{msk}, \cdot, \cdot)}(\lambda) = 1 \right] \right.$$
$$\left. - \Pr\left[\mathcal{A}^{\mathsf{KG}(\mathsf{msk}, \cdot), \mathsf{Enc}_1(\mathsf{msk}, \cdot, \cdot)}(\lambda) = 1 \right] \right| \leq \nu(\lambda),$$

where the probability is taken over the choice of $\mathsf{msk} \leftarrow \mathsf{Setup}(1^\lambda)$ and over the randomness of $\mathcal{A}$.

Definition 2.5 (Selective-Message Message Privacy). *A private-key functional encryption scheme $\Pi = (\mathsf{Setup}, \mathsf{KG}, \mathsf{Enc}, \mathsf{Dec})$ over a message space $\mathcal{M} = \{\mathcal{M}_\lambda\}_{\lambda \in \mathbb{N}}$ and a function space $\mathcal{F} = \{\mathcal{F}_\lambda\}_{\lambda \in \mathbb{N}}$ is* T-selective-message message private, *where $T = T(\lambda)$, if for any probabilistic polynomial-time adversary $\mathcal{A}$ there exists a negligible function $\nu(\lambda)$ such that*

$$\mathsf{Adv}^{\mathsf{sMP}}_{\Pi, \mathcal{A}, T}(\lambda) \overset{\text{def}}{=} \left| \Pr\left[\mathsf{Expt}^{\mathsf{sMP}}_{\Pi, \mathcal{A}, T}(\lambda, 0) = 1 \right] - \Pr\left[\mathsf{Expt}^{\mathsf{sMP}}_{\Pi, \mathcal{A}, T}(\lambda, 1) = 1 \right] \right| \leq \nu(\lambda),$$

where for each $b \in \{0, 1\}$ and $\lambda \in \mathbb{N}$ the experiment $\mathsf{Expt}^{\mathsf{sMP}}_{\Pi, \mathcal{A}, T}(\lambda, b)$ is defined as follows:

1. $\mathsf{msk} \leftarrow \mathsf{Setup}(1^\lambda)$.
2. $((\mathsf{m}_{0,1}, \ldots, \mathsf{m}_{0,T}), (\mathsf{m}_{1,1}, \ldots, \mathsf{m}_{1,T}), \mathsf{state}) \leftarrow \mathcal{A}(1^\lambda)$, *where $\mathsf{m}_{0,i}, \mathsf{m}_{1,i} \in \mathcal{M}_\lambda$ for all $i \in [T]$.*
3. $\mathsf{c}^*_i \leftarrow \mathsf{Enc}(\mathsf{msk}, \mathsf{m}_{b,i})$ *for all $i \in [T]$.*
4. $b' \leftarrow \mathcal{A}^{\mathsf{KG}(\mathsf{msk}, \cdot)}(\mathsf{c}^*_1, \ldots, \mathsf{c}^*_T, \mathsf{state})$, *where for each of $\mathcal{A}$'s queries f to $\mathsf{KG}(\mathsf{msk}, \cdot)$ it holds that $f(\mathsf{m}_{0,i}) = f(\mathsf{m}_{1,i})$ for all $i \in [T]$.*
5. *Output b'.*

Such a scheme Π is selective-message message private *if it is T-selective-message message private for all polynomials $T = T(\lambda)$.*

Definition 2.6 (Selective-Function Message Privacy). *A private-key functional encryption scheme $\Pi = (\mathsf{Setup}, \mathsf{KG}, \mathsf{Enc}, \mathsf{Dec})$ over a message space $\mathcal{M} = \{\mathcal{M}_\lambda\}_{\lambda \in \mathbb{N}}$ and a function space $\mathcal{F} = \{\mathcal{F}_\lambda\}_{\lambda \in \mathbb{N}}$ is* T-selective-function message* private, *where* $T = T(\lambda)$, *if for any probabilistic polynomial-time adversary* $\mathcal{A}$ *there exists a negligible function* $\nu(\lambda)$ *such that*

$$\mathsf{Adv}^{\mathsf{sfMP}}_{\Pi,\mathcal{A},T}(\lambda) \stackrel{\text{def}}{=} \left| \Pr\left[\mathsf{Expt}^{\mathsf{sfMP}}_{\Pi,\mathcal{A},T}(\lambda, 0) = 1 \right] - \Pr\left[\mathsf{Expt}^{\mathsf{sfMP}}_{\Pi,\mathcal{A},T}(\lambda, 1) = 1 \right] \right| \leq \nu(\lambda),$$

where for each $b \in \{0,1\}$ *and* $\lambda \in \mathbb{N}$ *the experiment* $\mathsf{Expt}^{\mathsf{sfMP}}_{\Pi,\mathcal{A},T}(\lambda, b)$ *is defined as follows:*

1. $\mathsf{msk} \leftarrow \mathsf{Setup}(1^\lambda)$.
2. $(f_1, \ldots, f_T, \mathsf{state}) \leftarrow \mathcal{A}(1^\lambda)$, *where* $f_i \in \mathcal{F}_\lambda$ *for all* $i \in [T]$.
3. $\mathsf{sk}_{f_i} \leftarrow \mathsf{KG}(\mathsf{msk}, f_i)$ *for all* $i \in [T]$.
4. $b' \leftarrow \mathcal{A}^{\mathsf{Enc}_b(\mathsf{msk}, \cdot, \cdot)}(\mathsf{sk}_{f_1}, \ldots, \mathsf{sk}_{f_T}, \mathsf{state})$, *where for each of* $\mathcal{A}$'s *queries* $(\mathsf{m}_0, \mathsf{m}_1)$ *to* $\mathsf{Enc}_b(\mathsf{msk}, \cdot, \cdot)$ *it holds that* $f_i(\mathsf{m}_0) = f_i(\mathsf{m}_1)$ *for all* $i \in [T]$.
5. *Output* b'.

Such a scheme Π *is* selective-function message private *if it is* T-selective-function *message private for all polynomials* $T = T(\lambda)$.

Known Instantiations. Private-key functional encryption schemes that satisfy the notions presented in Definitions 2.4–2.6 (and support circuits of any a-priori bounded polynomial size) are known to exist based on various assumptions. Most of the known schemes are in fact *public-key* schemes, which are in particular private-key ones[3]. Each of these scheme can be used to instantiate our generic transformation.

Specifically, a scheme that satisfies our notion of selective-message message privacy was constructed by Garg et al. [15] based on indistinguishability obfuscation. Schemes that satisfy the stronger notion of full message privacy (Definition 2.4) were constructed by Boyle et al. [11] and by Ananth et al. [3] based on differing-input obfuscation, by Waters [31] based on indistinguishability obfuscation, and by Garg et al. [16] based on multilinear maps. Moreover, a generic transformation from selective-message message privacy to full message privacy was recently showed by Ananth et al. [4].

A scheme that satisfies the notion of 1-selective-function message privacy was constructed by Gorbunov, Vaikuntanathan and Wee [21] under the sole assumption that public-key encryption exists. In the private-key setting, their transformation can in fact rely on any private-key encryption scheme (and thus on any one-way function). By assuming, in addition, the existence of a pseudorandom generator computable by small-depth circuits (which is known to be implied by most concrete intractability assumptions), they construct a scheme that satisfies the notion of T-selective-function message privacy for any predetermined polynomial $T = T(\lambda)$. However, the length of the ciphertexts in their scheme grows

---

[3] For indistinguishability-based message privacy in the public-key setting, considering one challenge is equivalent to considering a left-or-right encryption oracle [21]. Therefore, as public-key schemes are also private-key ones, in our indistinguishability-based definitions we directly consider left-or-right encryption oracles.

linearly with T and with an upper bound on the circuit size of the functions that the scheme allows (which also has to be known ahead of time). Goldwasser et al. [19] showed that based on the Learning with Errors (LWE) assumption, T-selective-function message privacy can be achieved where the ciphertext size grows with T and with a bound on the depth of allowed functions.

3   Modeling Function Privacy in the Private-Key Setting

In this section we introduce the notions of function privacy that are considered in this work. We consider three notions: *full function privacy, selective-message function privacy*, and *selective-function function privacy*. These are indistinguishability-based notions whose goal is to guarantee that functional keys reveal no unnecessary information on their functionality. Specifically, these notions ask that any efficient adversary that obtains encryptions of messages $m_1, \ldots, m_T$, and functional keys corresponding to functions $f_1, \ldots, f_T$, learns essentially no information other than the values $\{f_i(m_j)\}_{i,j\in[T]}$. Our notions generalize the standard message-privacy notions for functional encryption (see Section 2.2) by taking into account function privacy *in addition* to message privacy.

Full Function Privacy. The strongest notion that we consider, which we refer to as *full function privacy*, was recently put forward by Agrawal et al. [1] who generalized the notion of Shen, Shi and Waters [30] for predicate-privacy in attribute-based encryption. This notion considers both privacy of functional keys and privacy of encrypted messages in a completely symmetric manner. Specifically, we consider adversaries that interact with a left-or-right key-generation oracle $\mathsf{KG}_b(\mathsf{msk}, \cdot, \cdot)$, and with a left-or-right encryption oracle $\mathsf{Enc}_b(\mathsf{msk}, \cdot, \cdot)$ (where both oracles operate using the same bit b)[4]. We allow adversaries to adaptively interact with these oracles for any polynomial number of queries (which does not have to be bounded in advance), and the adversaries' goal is to distinguish the cases $b = 0$ and $b = 1$. Our only requirement from such adversaries is that for all (f_0, f_1) and (m_0, m_1) with which they query the oracles KG_b and Enc_b, respectively, it holds that $f_0(m_0) = f_1(m_1)$. We note that this is clearly an inherent requirement.

Definition 3.1 (Valid Function-Privacy Adversary). *A probabilistic polynomial-time algorithm $\mathcal{A}$ is a valid function-privacy adversary if for all private-key functional encryption schemes $\Pi = (\mathsf{Setup}, \mathsf{KG}, \mathsf{Enc}, \mathsf{Dec})$, for all $\lambda \in \mathbb{N}$ and $b \in \{0,1\}$, and for all (f_0, f_1) and (m_0, m_1) with which $\mathcal{A}$ queries the oracles KG_b and Enc_b, respectively, the following three conditions hold:*

---

[4] Recall (Definition 2.1) that for a probabilistic two-input functionality $\mathcal{O}$ and for $b \in \{0,1\}$, we denote by $\mathcal{O}_b$ the probabilistic three-input functionality $\mathcal{O}_b(k, x_0, x_1) \overset{\text{def}}{=} \mathcal{O}(k, x_b)$.

1. $f_0(m_0) = f_1(m_1)$.
2. *The messages m_0 and m_1 have the same length.*
3. *The descriptions of the functions f_0 and f_1 have the same length.*

Definition 3.2 (Full function privacy). *A private-key functional encryption scheme $\Pi = (\mathsf{Setup}, \mathsf{KG}, \mathsf{Enc}, \mathsf{Dec})$ over a message space $\mathcal{M} = \{\mathcal{M}_\lambda\}_{\lambda \in \mathbb{N}}$ and a function space $\mathcal{F} = \{\mathcal{F}_\lambda\}_{\lambda \in \mathbb{N}}$ is fully function private if for any valid function-privacy adversary $\mathcal{A}$, there exists a negligible function $\nu(\lambda)$ such that*

$$\mathsf{Adv}_{\Pi,\mathcal{A}}^{\mathsf{FP}}(\lambda) \overset{\text{def}}{=} \left| \Pr\left[\mathcal{A}^{\mathsf{KG}_0(\mathsf{msk},\cdot,\cdot),\mathsf{Enc}_0(\mathsf{msk},\cdot,\cdot)}(\lambda) = 1 \right] \right.$$
$$\left. - \Pr\left[\mathcal{A}^{\mathsf{KG}_1(\mathsf{msk},\cdot,\cdot),\mathsf{Enc}_1(\mathsf{msk},\cdot,\cdot)}(\lambda) = 1 \right] \right| \leq \nu(\lambda),$$

where the probability is taken over the choice of $\mathsf{msk} \leftarrow \mathsf{Setup}(1^\lambda)$ and over the randomness of $\mathcal{A}$.

Selective Notions. We consider two relaxations of our notion of full function privacy from Definition 3.2. The first, which we refer to as *selective-message function privacy* restricts the access that adversaries have to the left-or-right encryption oracle. Specifically, this notion asks that adversaries choose in advance their set of encryption queries. We note that adversaries are still given oracle access to the left-or-right key-generation oracle, with which they can interact in an adaptive manner for any polynomial number of queries. The second, which we refer to as *selective-function function privacy* restricts the access that adversaries have to the left-or-right key-generation oracle. Specifically, this notion asks that adversaries choose in advance their set of key-generation queries. We note that adversaries are still given oracle access to the left-or-right encryption oracle, with which they can interact in an adaptive manner for any polynomial number of queries. In addition, we note that our definition of a valid function-privacy adversary (Definition 3.1) naturally extends to the selective setting.

Definition 3.3 (Selective-Message Function Privacy). *A private-key functional encryption scheme $\Pi = (\mathsf{Setup}, \mathsf{KG}, \mathsf{Enc}, \mathsf{Dec})$ over a message space $\mathcal{M} = \{\mathcal{M}_\lambda\}_{\lambda \in \mathbb{N}}$ and a function space $\mathcal{F} = \{\mathcal{F}_\lambda\}_{\lambda \in \mathbb{N}}$ is T-selective-message function private, where $T = T(\lambda)$, if for any probabilistic polynomial-time adversary $\mathcal{A}$ there exists a negligible function $\nu(\lambda)$ such that*

$$\mathsf{Adv}_{\Pi,\mathcal{A},T}^{\mathsf{smFP}}(\lambda) \overset{\text{def}}{=} \left| \Pr\left[\mathsf{Expt}_{\Pi,\mathcal{A},T}^{\mathsf{smFP}}(\lambda, 0) = 1 \right] - \Pr\left[\mathsf{Expt}_{\Pi,\mathcal{A},T}^{\mathsf{smFP}}(\lambda, 1) = 1 \right] \right| \leq \nu(\lambda),$$

where for each $b \in \{0,1\}$ and $\lambda \in \mathbb{N}$ the experiment $\mathsf{Expt}_{\Pi,\mathcal{A},T}^{\mathsf{sMP}}(\lambda, b)$ is defined as follows:

1. $\mathsf{msk} \leftarrow \mathsf{Setup}(1^\lambda)$.
2. $((m_{0,1}, \ldots, m_{0,T}), (m_{1,1}, \ldots, m_{1,T}), \mathsf{state}) \leftarrow \mathcal{A}(1^\lambda)$, *where $m_{0,i}, m_{1,i} \in \mathcal{M}_\lambda$ for all $i \in [T]$.*
3. $c_i^* \leftarrow \mathsf{Enc}(\mathsf{msk}, m_{b,i})$ *for all $i \in [T]$.*

4. $b' \leftarrow \mathcal{A}^{\mathsf{KG}_b(\mathsf{msk},\cdot,\cdot)}(\mathsf{c}_1^*, \ldots, \mathsf{c}_T^*, \mathsf{state})$, *where for each of $\mathcal{A}$'s queries (f_0, f_1) to* $\mathsf{KG}_b(\mathsf{msk}, \cdot, \cdot)$ *it holds that $f_0(\mathsf{m}_{0,i}) = f_1(\mathsf{m}_{1,i})$ for all $i \in [T]$.*

5. *Output b'.*

Such a scheme Π is selective-message function private *if it is T-selective-message function private for all polynomials $T = T(\lambda)$.*

Definition 3.4 (Selective-function function privacy). *A private-key functional encryption scheme $\Pi = (\mathsf{Setup}, \mathsf{KG}, \mathsf{Enc}, \mathsf{Dec})$ over a message space $\mathcal{M} = \{\mathcal{M}_\lambda\}_{\lambda \in \mathbb{N}}$ and a function space $\mathcal{F} = \{\mathcal{F}_\lambda\}_{\lambda \in \mathbb{N}}$ is T-selective-function function private, where $T = T(\lambda)$, if for any probabilistic polynomial-time adversary $\mathcal{A}$ there exists a negligible function $\nu(\lambda)$ such that*

$$\mathsf{Adv}_{\Pi,\mathcal{A},T}^{\mathsf{sfFP}}(\lambda) \stackrel{\mathsf{def}}{=} \left| \Pr\left[\mathsf{Expt}_{\Pi,\mathcal{A},T}^{\mathsf{sfFP}}(\lambda, 0) = 1\right] - \Pr\left[\mathsf{Expt}_{\Pi,\mathcal{A},T}^{\mathsf{sfFP}}(\lambda, 1) = 1\right] \right| \le \nu(\lambda),$$

where for each $b \in \{0,1\}$ and $\lambda \in \mathbb{N}$ the experiment $\mathsf{Expt}_{\Pi,\mathcal{A},T}^{\mathsf{sMP}}(\lambda, b)$ is defined as follows:

1. $\mathsf{msk} \leftarrow \mathsf{Setup}(1^\lambda)$.

2. $((f_{0,1}, \ldots, f_{0,T}), (f_{1,1}, \ldots, f_{1,T}), \mathsf{state}) \leftarrow \mathcal{A}(1^\lambda)$, *where $f_{0,i}, f_{1,i} \in \mathcal{F}_\lambda$ for all $i \in [T]$.*

3. $\mathsf{sk}_i^* \leftarrow \mathsf{KG}(\mathsf{msk}, f_{b,i})$ *for all $i \in [T]$.*

4. $b' \leftarrow \mathcal{A}^{\mathsf{Enc}_b(\mathsf{msk},\cdot,\cdot)}(\mathsf{sk}_1^*, \ldots, \mathsf{sk}_T^*, \mathsf{state})$, *where for each of $\mathcal{A}$'s queries $(\mathsf{m}_0, \mathsf{m}_1)$ to $\mathsf{Enc}_b(\mathsf{msk}, \cdot, \cdot)$ it holds that $f_{0,i}(\mathsf{m}_0) = f_{1,i}(\mathsf{m}_1)$ for all $i \in [T]$.*

5. *Output b'.*

Such a scheme Π is selective-function function private *if it is T-selective-function function private for all polynomials $T = T(\lambda)$.*

Finally, we observe that due to the symmetry between the roles of the encryption oracle and the key-generation oracle in these two selective notions, they are in fact equivalent when switching between the encryption algorithm and key-generation algorithm of any given scheme. That is, a private-key functional encryption scheme $(\mathsf{Setup}, \mathsf{KG}, \mathsf{Enc}, \mathsf{Dec})$ is selective-message function private if and only if the scheme $(\mathsf{Setup}, \mathsf{Enc}, \mathsf{KG}, \mathsf{Dec})$ is selective-function function private. To be a little more accurate, replacing the roles of functions f and message m may require some standard "type casting" to represent a message as function and function as message. This is done using universal machines: To cast a function f as a message, we consider its description as the message to be encrypted. This means that if Enc only takes bounded length messages, then the new scheme will only support functions with bounded description lengths. To cast a message m as a function, we consider a universal function that accepts a description of a function f and outputs $f(\mathsf{m})$. Again, depending on the computational model under consideration, this may impose some restrictions. For example, if working over circuits then the universal circuit imposes an upper bound on the size of the functions supported by the new scheme, whereas that may not have been required in the original scheme before the switch (however, in this example, if the function size was a-priori unbounded in the original scheme, then the message space after the switch will become a-priori length unbounded).

4   Our Function-Private Scheme

In this section we present our generic construction of a function-private private-key functional encryption scheme. Our construction relies on the following two building blocks:

- A private-key functional encryption scheme $\mathcal{FE} = $ (FE.Setup, FE.KG, FE.Enc, FE.Dec).
- A private-key encryption scheme $\mathcal{SKE} = $ (SKE.KG, SKE.Enc, SKE.Dec).[5]

Our new functional encryption scheme $\mathcal{FPE} = $ (Setup, KG, Enc, Dec) is defined as follows.

The Setup Algorithm. On input the security parameter 1^λ the setup algorithm Setup samples FE.msk $\leftarrow$ FE.Setup(1^λ), SKE.k $\leftarrow$ SKE.KG(1^λ), and SKE.k$'$ $\leftarrow$ SKE.KG(1^λ). Then, it outputs msk $= $ (FE.msk, SKE.k, SKE.k$'$).

The Key-Generation Algorithm. On input the master secret key msk and a function f, the key-generation algorithm KG first computes $c \leftarrow$ SKE.Enc(SKE.k, f) and $c' \leftarrow$ SKE.Enc(SKE.k$'$, f). Then, it computes FE.SK$_{U_{c,c'}} \leftarrow$ FE.KG(FE.msk, $U_{c,c'}$), where the function $U_{c,c'}$ is described in Figure 1. Finally, it outputs sk$_f = $ FE.SK$_{U_{c,c'}}$.

The Encryption Algorithm. On input the master secret key msk and a message m, the encryption algorithm Enc outputs c $\leftarrow$ FE.Enc(FE.msk, (m, $\bot$, SKE.k, $\bot$)).

The Decryption Algorithm. On input a functional key sk$_f$ and a ciphertext c, the decryption algorithm Dec outputs FE.Dec(sk$_f$, c).

$$U_{c,c'}(\mathsf{m}, \mathsf{m}', k, k')$$

1. If $k \neq \bot$, compute $f \leftarrow$ SKE.Dec(k, c) and output $f(\mathsf{m})$.
2. Else, if $k' \neq \bot$, compute $f' \leftarrow$ SKE.Dec(k', c') and output $f'(\mathsf{m}')$.
3. Else, output $\bot$.

Fig. 1. The function $U_{c,c'}$

Note that if the underlying scheme $\mathcal{FE}$ supports functions that are computable by circuits of size at most s, for some sufficiently large polynomial $s = s(n)$, then our new scheme $\mathcal{FPE}$ supports functions that are computable by circuits of size $\Omega(s)$. Specifically, a functional key sk$_f$ for a function f in the new scheme consists of a functional key for the function $U_{c,c'}$ in the underlying scheme. The function $U_{c,c'}$ is computable in a straightforward manner by a circuit that contains two copies of a circuit for computing f, and two copies of $\mathcal{SKE}$'s decryption circuit. The security of our construction is captured by the following theorem:

---

[5] To be absolutely formal, this building block is implied by the former in an obvious way.

Theorem 4.1. *Assuming that the scheme $\mathcal{SKE}$ is CPA-secure the following hold:*

1. *If the scheme $\mathcal{FE}$ is fully message private then the scheme $\mathcal{FPE}$ is fully function private.*
2. *If the scheme $\mathcal{FE}$ is selective-message message private (resp. T-selective-message message private) then the scheme $\mathcal{FPE}$ is selective-message function private (resp. T-selective-message function private).*
3. *If the scheme $\mathcal{FE}$ is selective-function message private (resp. T-selective-function message private) then the scheme $\mathcal{FPE}$ is selective-function function private (resp. T-selective-function function private).*

As discussed in Section 2.2, Theorem 4.1 can be instantiated based on a variety of known functional encryption schemes (e.g., [21,3,15,19,11]) that offer full message privacy, selective-message message privacy, and selective-function message privacy.

In the full version of this paper [12] we prove Theorem 4.1 by showing that for any valid function-privacy adversary $\mathcal{A}$ for the scheme $\mathcal{FPE}$ that there exist a probabilistic-polynomial time adversary $\mathcal{B}_1$ attacking the CPA security of $\mathcal{SKE}$, and a probabilistic polynomial-time adversary $\mathcal{B}_2$ attacking the message privacy of $\mathcal{FE}$, such that

$$\mathsf{Adv}^{\mathsf{FP}}_{\mathcal{FPE},\mathcal{A}}(\lambda) \leq 2 \cdot \left(\mathsf{Adv}^{\mathsf{CPA}}_{\mathcal{SKE},\mathcal{B}_1}(\lambda) + \mathsf{Adv}^{\mathsf{MP}}_{\mathcal{FE},\mathcal{B}_2}(\lambda) \right).$$

Table 1. The differences between the experiments $\mathcal{H}^{(0)}, \ldots, \mathcal{H}^{(4)}$. Adjacent experiments that differ on the generation of c or c' are proven indistinguishable using the CPA security of $\mathcal{SKE}$. Adjacent experiments that differ on the input to FE.Enc are proven indistinguishable using the message privacy of $\mathcal{FE}$.

Experiment	Encryption oracle	Key-generation oracle
$\mathcal{H}^{(0)}$	FE.Enc(FE.msk, $(m_0, \perp, \mathsf{SKE.k}, \perp)$)	$c \leftarrow$ SKE.Enc(SKE.k, f_0) $c' \leftarrow$ SKE.Enc(SKE.k', f_0)
$\mathcal{H}^{(1)}$	FE.Enc(FE.msk, $(m_0, \perp, \mathsf{SKE.k}, \perp)$)	$c \leftarrow$ SKE.Enc(SKE.k, f_0) $c' \leftarrow$ SKE.Enc(SKE.k', f_1)
$\mathcal{H}^{(2)}$	FE.Enc(FE.msk, $(\perp, m_1, \perp, \mathsf{SKE.k'})$)	$c \leftarrow$ SKE.Enc(SKE.k, f_0) $c' \leftarrow$ SKE.Enc(SKE.k', f_1)
$\mathcal{H}^{(3)}$	FE.Enc(FE.msk, $(\perp, m_1, \perp, \mathsf{SKE.k'})$)	$c \leftarrow$ SKE.Enc(SKE.k, f_1) $c' \leftarrow$ SKE.Enc(SKE.k', f_1)
$\mathcal{H}^{(4)}$	FE.Enc(FE.msk, $(m_1, \perp, \mathsf{SKE.k}, \perp)$)	$c \leftarrow$ SKE.Enc(SKE.k, f_1) $c' \leftarrow$ SKE.Enc(SKE.k', f_1)

The proof consists of a sequence of five hybrid experiments, denoted $\mathcal{H}^{(0)}, \ldots,$ $\mathcal{H}^{(4)}$, where each two consecutive experiments are computationally indistinguishable from $\mathcal{A}$'s point of view. Each such experiment $\mathcal{H}^{(i)}$ is completely characterized by its key-generation oracle and its encryption oracle, where the differences between these experiments are summarized in Table 1.

Acknowledgments. We thank Shweta Agrawal and Vinod Vaikuntanthan for insightful discussions.

References

1. Agrawal, S., Agrawal, S., Badrinarayanan, S., Kumarasubramanian, A., Prabhakaran, M., Sahai, A.: Function private functional encryption and property preserving encryption: New definitions and positive results. Cryptology ePrint Archive, Report 2013/744 (2013)
2. Agrawal, S., Gorbunov, S., Vaikuntanathan, V., Wee, H.: Functional encryption: New perspectives and lower bounds. In: Canetti, R., Garay, J.A. (eds.) CRYPTO 2013, Part II. LNCS, vol. 8043, pp. 500–518. Springer, Heidelberg (2013)
3. Ananth, P., Boneh, D., Garg, S., Sahai, A., Zhandry, M.: Differing-inputs obfuscation and applications. Cryptology ePrint Archive, Report 2013/689 (2013)
4. Ananth, P., Brakerski, Z., Segev, G., Vaikuntanathan, V.: The trojan method in functional encryption: From selective to adaptive security, generically. Cryptology ePrint Archive, Report 2014/917 (2014)
5. Bellare, M., O'Neill, A.: Semantically-secure functional encryption: Possibility results, impossibility results and the quest for a general definition. In: Abdalla, M., Nita-Rotaru, C., Dahab, R. (eds.) CANS 2013. LNCS, vol. 8257, pp. 218–234. Springer, Heidelberg (2013)
6. Boneh, D., Franklin, M.K.: Identity-based encryption from the Weil pairing. SIAM Journal on Computing 32(3), 586–615 (2003), Preliminary version in Kilian, J. (ed.) CRYPTO 2001. LNCS, vol. 2139, pp. 213–229. Springer, Heidelberg (2001)
7. Boneh, D., Gentry, C., Gorbunov, S., Halevi, S., Nikolaenko, V., Segev, G., Vaikuntanathan, V., Vinayagamurthy, D.: Fully key-homomorphic encryption, arithmetic circuit ABE and compact garbled circuits. In: Nguyen, P.Q., Oswald, E. (eds.) EUROCRYPT 2014. LNCS, vol. 8441, pp. 533–556. Springer, Heidelberg (2014)
8. Boneh, D., Raghunathan, A., Segev, G.: Function-private identity-based encryption: Hiding the function in functional encryption. In: Canetti, R., Garay, J.A. (eds.) CRYPTO 2013, Part II. LNCS, vol. 8043, pp. 461–478. Springer, Heidelberg (2013)
9. Boneh, D., Raghunathan, A., Segev, G.: Function-private subspace-membership encryption and its applications. In: Sako, K., Sarkar, P. (eds.) ASIACRYPT 2013, Part I. LNCS, vol. 8269, pp. 255–275. Springer, Heidelberg (2013)
10. Boneh, D., Sahai, A., Waters, B.: Functional encryption: Definitions and challenges. In: Ishai, Y. (ed.) TCC 2011. LNCS, vol. 6597, pp. 253–273. Springer, Heidelberg (2011)
11. Boyle, E., Chung, K.-M., Pass, R.: On extractability obfuscation. In: Lindell, Y. (ed.) TCC 2014. LNCS, vol. 8349, pp. 52–73. Springer, Heidelberg (2014)
12. Brakerski, Z., Segev, G.: Function-private functional encryption in the private-key setting. Cryptology ePrint Archive, Report 2014/550 (2014)
13. Cocks, C.: An identity based encryption scheme based on quadratic residues. In: Honary, B. (ed.) Cryptography and Coding 2001. LNCS, vol. 2260, pp. 360–363. Springer, Heidelberg (2001)

14. De Caro, A., Iovino, V., Jain, A., O'Neill, A., Paneth, O., Persiano, G.: On the achievability of simulation-based security for functional encryption. In: Canetti, R., Garay, J.A. (eds.) CRYPTO 2013, Part II. LNCS, vol. 8043, pp. 519–535. Springer, Heidelberg (2013)

15. Garg, S., Gentry, C., Halevi, S., Raykova, M., Sahai, A., Waters, B.: Candidate indistinguishability obfuscation and functional encryption for all circuits. In: Proceedings of the 54th Annual IEEE Symposium on Foundations of Computer Science, pp. 40–49 (2013)

16. Garg, S., Gentry, C., Halevi, S., Zhandry, M.: Fully secure functional encryption without obfuscation. Cryptology ePrint Archive, Report 2014/666 (2014)

17. Goldwasser, S., Gordon, S.D., Goyal, V., Jain, A., Katz, J., Liu, F.-H., Sahai, A., Shi, E., Zhou, H.-S.: Multi-input functional encryption. In: Nguyen, P.Q., Oswald, E. (eds.) EUROCRYPT 2014. LNCS, vol. 8441, pp. 578–602. Springer, Heidelberg (2014) Merge of [18] [22]

18. Goldwasser, S., Goyal, V., Jain, A., Sahai, A.: Multi-input functional encryption. Cryptology ePrint Archive, Report 2013/727 (2013)

19. Goldwasser, S., Kalai, Y., Popa, R.A., Vaikuntanathan, V., Zeldovich, N.: Reusable garbled circuits and succinct functional encryption. In: Proceedings of the 45th Annual ACM Symposium on Theory of Computing, pp. 555–564 (2013)

20. Goldwasser, S., Kalai, Y.T., Popa, R.A., Vaikuntanathan, V., Zeldovich, N.: How to run turing machines on encrypted data. In: Canetti, R., Garay, J.A. (eds.) CRYPTO 2013, Part II. LNCS, vol. 8043, pp. 536–553. Springer, Heidelberg (2013)

21. Gorbunov, S., Vaikuntanathan, V., Wee, H.: Functional encryption with bounded collusions via multi-party computation. In: Safavi-Naini, R., Canetti, R. (eds.) CRYPTO 2012. LNCS, vol. 7417, pp. 162–179. Springer, Heidelberg (2012)

22. Gordon, S.D., Katz, J., Liu, F.-H., Shi, E., Zhou, H.-S.: Multi-input functional encryption. Cryptology ePrint Archive, Report 2013/774 (2013)

23. Goyal, V., Jain, A., Koppula, V., Sahai, A.: Functional encryption for randomized functionalities. Cryptology ePrint Archive, Report 2013/729 (2013)

24. Komargodski, I., Segev, G., Yogev, E.: Functional encryption for randomized functionalities in the private-key setting from minimal assumptions. To appear in Proceedings of the 12th Theory of Cryptography Conference (2015)

25. Naor, M., Yung, M.: Public-key cryptosystems provably secure against chosen ciphertext attacks. In: Proceedings of the 22nd Annual ACM Symposium on Theory of Computing, pp. 427–437 (1990)

26. O'Neill, A.: Definitional issues in functional encryption. Cryptology ePrint Archive, Report 2010/556 (2010)

27. Sahai, A., Waters, B.: Slides on functional encryption (2008),
http://www.cs.utexas.edu/~bwaters/presentations/files/functional.ppt

28. Sahai, A., Waters, B.: How to use indistinguishability obfuscation: deniable encryption, and more. In: Proceedings of the 46th Annual ACM Symposium on Theory of Computing, pp. 475–484 (2014)

29. Shamir, A.: Identity-based cryptosystems and signature schemes. In: Blakely, G.R., Chaum, D. (eds.) CRYPTO 1984. LNCS, vol. 196, pp. 47–53. Springer, Heidelberg (1985)

30. Shen, E., Shi, E., Waters, B.: Predicate privacy in encryption systems. In: Reingold, O. (ed.) TCC 2009. LNCS, vol. 5444, pp. 457–473. Springer, Heidelberg (2009)

31. Waters, B.: A punctured programming approach to adaptively secure functional encryption. Cryptology ePrint Archive, Report 2014/588 (2014)

Functional Encryption
for Randomized Functionalities

Vipul Goyal[1], Abhishek Jain[2,*], Venkata Koppula[3,**], and Amit Sahai[4,***]

[1] Microsoft Research, India
vipul@microsoft.com
[2] Johns Hopkins University, USA
abhishek@cs.jhu.edu
[3] University of Texas, Austin, USA
kvenkata@cs.utexas.edu
[4] UCLA and the Center for Encrypted Functionalities, USA
sahai@cs.ucla.edu

Abstract. In this work, we present the first definitions and constructions for functional encryption supporting *randomized functionalities*. The setting of randomized functionalities require us to revisit functional encryption definitions by, for the first time, explicitly adding security requirements for *dishonest encryptors*, to ensure that they cannot improperly tamper with the randomness that will be used for computing outputs. Our constructions are built using indistinguishability obfuscation.

1 Introduction

Originally, encryption was thought of as a way to encrypt "point to point" communication. However, in the contemporary world with cloud computing and complex networks, it has become clear that we need encryption to offer more functionality. To address this issue, the notion of functional encryption (FE) has been developed [25,18,5,19,4,21]. In a functional encryption for a family $\mathcal{F}$, it is possible to derive secret keys K_f for any function $f \in \mathcal{F}$ from a master secret key. Given an encryption of some input x, that user can use its secret key K_f to obtain $f(x)$, and should learn nothing else about x beyond $f(x)$.

* The author is partly funded by NSF CNS-1414023. Part of the research was conducted while visiting Microsoft Research, India.
** Part of this research was conducted during internship at Microsoft Research, India.
*** Research supported in part from a DARPA/ONR PROCEED award, NSF Frontier Award 1413955, NSF grants 1228984, 1136174, 1118096, and 1065276, a Xerox Faculty Research Award, a Google Faculty Research Award, an equipment grant from Intel, and an Okawa Foundation Research Grant. This material is based upon work supported by the Defense Advanced Research Projects Agency through the U.S. Office of Naval Research under Contract N00014-11- 1-0389. The views expressed are those of the author and do not reflect the official policy or position of the Department of Defense, the National Science Foundation, or the U.S. Government.

Y. Dodis and J.B. Nielsen (Eds.): TCC 2015, Part II, LNCS 9015, pp. 325–351, 2015.

A driving force behind functional encryption research has been to understand what class of functions can be supported by functional encryption. This remarkable line of research has progressed to now encompass all functions describable by deterministic polynomial-size circuits [24,16,15,8,11]. We continue this line of research to move even beyond deterministic polynomial-size circuits: specifically, we consider the case of *randomized* functionalities. Indeed, not only are randomized functionalities strongly motivated by real-world scenarios, but randomized functionalities present new challenges for functional encryption. Techniques developed in the context of functional encryption for deterministic circuit do not directly translate into techniques for randomized circuits. To understand the basic technical problem, below we give an illustrative example.

Let us illustrate the desiderata for functional encryption for randomized functions by considering an example of performing an audit on an encrypted database through random sampling. Suppose there is a bank that maintains large secure databases of the transactions in each of its branches. There is an auditor Alice who would like to gain access to a random sample of database entries from each branch in order to manually audit these records and check for improper transactions. We note that random sampling of transactions for manual analysis is quite common during audits. There are two primary concerns:

- The auditor wants to ensure that cheating in a branch is caught with reasonable probability.
- The organization wants to ensure that a malicious auditor cannot learn undesirable information (e.g., too much about a particular customer) from the encrypted databases. In particular, it wants to ensure that a malicious auditor cannot gain access to arbitrarily chosen parts of the database, but rather is limited to seeing only a randomly selected sample for each branch.

If we try to solve this problem naively using functional encryption, by giving the auditor a secret key SK_f that lets it obtain a random subset of an encrypted database CT, we are faced with the question: where does the randomness come from? Clearly, the randomness cannot be specified in the ciphertext alone since then a cheating encrypter (bank branch) could influence it. It cannot be specified in the decryption key alone as well: then auditor would get the same (or correlated) sample from the databases of different branches. (We also stress that since functional encryption does not guarantee function privacy, randomness present in the function f, even if chosen by a trusted party, would be known to Alice.)

Even if the randomness was chosen by an XOR of coins built into the decryption key and the ciphertext, this would allow malicious encryptors, over time, to ensure correlations among the random coins used by the auditor when inspecting different databases (or the same database after updates to it). Such correlations could potentially be used to eventually learn completely the coins embedded in the decryption key (based on the auditor's actions in response to planted improprieties in databases). Another option is to use a pseudorandom function (PRF) whose key is inbuilt in the decryption key. However again, since functional encryption does not guarantee function privacy, the PRF key could be completely leaked to a malicious auditor. As a result, the sample would not be "random" anymore in the auditor's view (since he knows the PRF key).

This scenario also illustrates the importance of dealing with *dishonest encryptors* in the context of functional encryption for randomized functionalities, because of the influence they can have on the choice of coins used in computing the output. The issue of dishonest encryptors is, in fact, also relevant to the case of deterministc functionalities.[1] However, to the best of our knowledge, this issue was never considered explicitly in previous work on functional encryption. This is perhaps because in the context of deterministic functionalities, the issue of dishonest encryptors seems very related to simple correctness, which is not the case in the current work.

Defining functional encryption for randomized functionalities. To avoid the problems sketched in the examples above, we define functional encryption for randomized functionalities using the simulation paradigm: We want that an adversary, given SK_f and an honestly generated encryption of x, be simulatable given only $f(x; r)$ where r is true randomness that is completely unknown to the adversary. At the same time, consider an adversary that can generate dishonest ciphertexts $\hat{CT}$ and learn from outside the output of decrypting $\hat{CT}$ using a secret key SK_g (that is unknown to the adversary). We want such an adversary to be simulatable given only $g(\hat{x}; r)$, where $\hat{x}$ is an input that is information-theoretically fixed by $\hat{CT}$ and r is again true randomness that is unknown to the adversary. Note that a crucial feature of our definition is that if a party uses a secret key SK_f on a particular ciphertext CT, it will always get back $f(x; r)$ for the same randomness r. In other words, the user cannot repeatedly sample the functionality to obtain multiple outputs for different random coins. This allows users of our definition to more tightly control how much information an adversary or user learns. However, given two distinct ciphertexts CT_1 and CT_2 both encrypting x, a malicious user possessing SK_f should obtain exactly two independent samples of the output of the function: $f(x; r_1)$ and $f(x; r_2)$.

Application to differentially private data release. A natural application of functional encryption would be to provide non-interactive differentially private data release with high levels of accuracy. Consider a scenario where a government would like to allow researchers to carry out research studies on different hospital patient record databases, but only if the algorithm that analyzes the patient data achieves a sufficient level of differential privacy. Without using cryptography, methods for allowing the hospitals to publish differentially private data that would allow for meaningful and diverse research studies must incur very high accuracy loss [10]. An alternative would be to have a government agency review a

---

[1] For example, the FE schemes in [16,15] are not secure against a dishonest encryptor who uses the simulator algorithm to create ciphertexts. Indeed, such an adversary can force arbitrary outputs on an honest receiver. However, a straightforward compilation of these schemes with simulation-sound NIZK proofs of knowledge yields security against dishonest encryptors.

specific research algorithm f, and if the algorithm guarantees sufficient privacy, to issue a secret key SK_f that the researcher could use to obtain the output of her algorithm on any hospital's encrypted patient records. Note that in such a setting, the hospital patient record could be encrypted and stored *without* any noise addition. The noise could be added by the algorithm f after computing the correct output. Such a setting would ensure very high accuracy (essentially the same as the interactive setting where the hospitals store data in clear and answer the researcher queries after adding noise in an online fashion).

Note however, to achieve differential privacy, such an algorithm f must be randomized. Furthermore, typical differentially private algorithms require that the randomness used to compute the output must be correctly and freshly sampled each time and be kept secret (or else the differential privacy could be completely compromised). By realizing functional encryption that would allow such randomized function evaluation, we would simultaneously remove the need for the hospital to participate in any study beyond simply releasing an encrypted database, and remove the need for the researcher to share his hypothesis and algorithm with any entity beyond the government regulatory body that issues secret keys.

1.1 Our Results

We show how to formalize the definition sketched above, generalizing the simulation-based security definitions given in [4,21]. We then construct a functional encryption scheme supporting arbitrary randomized polynomial-size circuits assuming indistinguishability obfuscation for circuits and one-way functions. We prove security in the selective model that can be amplified to full security using standard complexity leveraging.

While our focus is on simulation-based security, we note that it cannot be realized for an unbounded number of messages [4,3]. Towards that end, in Sect. 2.1, we also provide indistinguishability based security definitions for randomized functions, generalizing the case of deterministic functions [4,21]. We prove security in the selective model for an unbounded number of messages (again, this can be amplified to full security using standard complexity leveraging[2]).

The starting point for our construction is the functional encryption scheme of [11] for polynomial-size deterministic circuits. In that scheme, in essence the secret key SK_f is built upon obfuscating the function f using an indistinguishability obfuscator [2]. We show how to modify this construction to achieve our notion of functional encryption for randomized functionalities by building upon the recently introduced idea of punctured programming [26]. In particular, we embed a psuedo-random function (PRF) key into the obfuscated program, which is executed on the ciphertext, to obtain the randomness used to derive the output. We adapt ideas from [9,23] to ensure that valid ciphertexts are unique.

---

[2] Subsequent to our work, Waters [27] gave a construction of fully secure functional encryption (for deterministic functions) from indistinguishability obfuscation, without complexity leveraging. We leave the problem of adapting our techniques to the scheme of [27] for future work.

The core of our argument of security is to show that indistinguishability obfuscation guarantees the secrecy of the random coins derived by this method.

Our results immediately imply the application to differential privacy: Consider two "neighboring" databases x_0 and x_1. Differential privacy guarantees that the statistical distance between the distributions of outputs of the mechanism f for these two databases is at most e^ϵ, a small (but non-negligible) quantity. Now consider an adversary's view given an encryption of x_0. By our simulation-based notion of security, the adversary's view can be simulated given only $f(x_0; r)$ where r is true (secret) randomness. This view is e^ϵ close to the view that would be generated given only $f(x_1; r)$, by differential privacy of f. Finally we apply our definition to show that this view is negligibly close to the real adversary's view given an encryption of x_1. Thus, our functional encryption scheme when applied to f yields a computationally differentially private mechanism.

1.2 Other Applications

Subsequent to our work, Garg et al. [12] use functional encryption for randomized functions in NC^1 as a crucial tool to construct fully secure functional encryption for all circuits from multilinear maps. We refer the reader to their paper for more details.

1.3 Related Work

In an independent and concurrent work, Alwen et al. [1] also study functional encryption for randomized functions.[3] The main difference between their work and ours is that they do not consider security against malicious encryptors. In particular, they provide a construction of FE for randomized functions from FE for deterministic functions by encrypting a PRF key along with every message. This PRF key is evaluated over the identifier associated with a function key to sample randomness on the fly, which is then used to compute the function output. Interestingly, they show that a 2-ary version of randomized FE can be used to construct fully homomorphic encryption (see [1] for details). However, they do not provide a construction of such an FE scheme.

We note that while the security definition of [1] suffices for their target application, in this work, we model randomized functionalities following the standard approach in secure computation where in the ideal world, no single party has full control over the randomness used in the function evaluation and instead the randomness is chosen by the trusted party. In particular, we require that the randomness used for the computation is chosen uniformly even if either of the parties is malicious. Indeed, as discussed earlier, this is the main source of non-triviality in our results.

---

[3] See [17] for the eprint version of our work.

1.4   Organization

The rest of this paper is organized as follows. We start by presenting the formal definitions for functional encryption for randomized functionalities (Sect. 2). Next, we recall the definitions for various cryptographic primitives used in our construction (Sect. 3). We then present our construction of functional encryption for randomized functionalities (Sect. 4) and prove its security in the selective model (Sect. 5).

2   Functional Encryption for Randomized Functions

In this section, we present definitions for functional encryption for randomized functions (or rand-FE for short). We start by presenting the syntax for rand-FE and then proceed to give the security definitions for the same.

Syntax. Throughout the paper, we denote the security parameter by 1^κ. Let $\mathcal{X} = \{\mathcal{X}_\kappa\}_{\kappa \in \mathbb{N}}$, $\mathcal{R} = \{\mathcal{R}_\kappa\}_{\kappa \in \mathbb{N}}$ and $\mathcal{Y} = \{\mathcal{Y}_\kappa\}_{\kappa \in \mathbb{N}}$ be ensembles where each $\mathcal{X}_\kappa$, $\mathcal{R}_\kappa$ and $\mathcal{Y}_\kappa$ is a finite set. Let $\mathcal{F} = \{\mathcal{F}_\kappa\}_{\kappa \in \mathbb{N}}$ be an ensemble where each $\mathcal{F}_\kappa$ is a finite collection of randomized functions. Each function $f \in \mathcal{F}_\kappa$ takes as input a string $x \in \mathcal{X}_\kappa$ and randomness $r \in \mathcal{R}_\kappa$ and outputs $f(x; r) \in \mathcal{Y}_\kappa$.

A functional encryption scheme $\mathcal{FE}$ for randomized functions $\mathcal{F}$ consists of four algorithms (rFE.Setup, rFE.Enc, rFE.Keygen, rFE.Dec):

- **Setup** rFE.Setup(1^κ) is a PPT algorithm that takes as input the security parameter κ and outputs the public key MPK and the master secret key MSK.
- **Encryption** rFE.Enc(x, MPK) is a PPT algorithm that takes as input a message x and the public key MPK and outputs a ciphertext CT.
- **Key Generation** rFE.Keygen(f, MSK) is a PPT algorithm that takes as input a function $f \in \mathcal{F}$ and the master secret key MSK and outputs a secret key SK$_f$.
- **Decryption** rFE.Dec$(\mathsf{CT}, \mathsf{SK}_f)$ is a deterministic algorithm that takes as input a ciphertext CT, the public key MPK and a secret key SK$_f$ and outputs a string $y \in \mathcal{Y}_\kappa$.

Definition 1 (Correctness). *A functional encryption scheme $\mathcal{FE}$ for randomized function family $\mathcal{F}$ is correct if for every polynomial $n = n(\kappa)$, every $\boldsymbol{f} \in \mathcal{F}_\kappa^n$ and every $\boldsymbol{x} \in \mathcal{X}_\kappa^n$, the following two distributions are computationally indistinguishable:*

1. **Real:** $\left\{ \mathsf{rFE.Dec}\left(\mathsf{CT}_i, \mathsf{SK}_{f_j}\right) \right\}_{i=1,j=1}^{n,n}$, *where:*
   - $(\mathsf{MPK}, \mathsf{MSK}) \leftarrow \mathsf{rFE.Setup}(1^\kappa)$
   - $\mathsf{CT}_i \leftarrow \mathsf{rFE.Enc}(x_i, \mathsf{MPK})$ *for* $i \in [n]$
   - $\mathsf{SK}_{f_j} \leftarrow \mathsf{rFE.Keygen}(f_j, \mathsf{MSK})$ *for* $j \in [n]$
2. **Ideal:** $\{f_j\left(x_i; r_{i,j}\right)\}_{i=1,j=1}^{n,n}$ *where* $r_{i,j} \leftarrow \mathcal{R}_\kappa$

Remark 1. We note that unlike the case of deterministic functions where it suffices to define correctness for a single ciphertext and a single key, in the case of randomized functions, it is essential to define correctness for *multiple* ciphertexts and functions. To see this, consider the scenario where a secret key SK_f corresponding to a function f is implemented in such a way that it has some "fixed" randomness r hardwired in it. Now, upon decrypting any ciphertext $CT \leftarrow rFE.Enc(x, MPK)$ with SK_f, one would obtain the output $f(x; r)$ w.r.t. the *same* randomness r. Note that this clearly incorrect implementation of SK_f would satisfy the correctness definition for a single ciphertext and a single key, but will fail to satisfy our definition given above.

2.1 Security for Functional Encryption

We now present our security definitions for rand-FE. We first observe that existing security definitions for functional encryption only consider the *malicious receiver* setting, in that they intuitively guarantee that an adversary who owns a secret key SK_f corresponding to a function f cannot learn anymore than $f(x)$ from an encryption of x. In this work, we are also interested in achieving security against *malicious senders*. In particular, we would like to guarantee that an adversarial encryptor cannot force "bad" outputs on an honest receiver. As discussed earlier, this is particularly important when modeling randomized functions.

We consider a a unified adversarial model that captures both malicious receivers and malicious senders. We present both simulation-based and indistinguishability based security definitions. For simplicity, we present our security definitions for the *selective model*, where the adversary must decide the challenge messages up front, before the system parameters are chosen.

Simulation Based Security. We now present a simulation-based security definition (or, SIM-security) for rand-FE. If we only consider malicious receivers, then our definition looks essentially identical to the standard (selective) SIM-security definition for FE (for deterministic functions) [4,21]. In order to provide security against adversarial senders, we extend the existing definition. To understand the main idea behind our definition, let us consider an honest receiver who owns a secret key SK_f corresponding to a function f. Then, in order to formalize the intuition that an adversarial sender cannot force "incorrect" outputs on this honest receiver, we allow the adversary to make *decryption queries* for arbitrary ciphertexts[4] w.r.t. the secret key SK_f. In the ideal world, the simulator must be able to "extract" the plaintext x from each decryption query and compute as output $f(x; r)$ for some true randomness r. We then require that the decryption query in the real world yields an indistinguishable output.

We now proceed to give our formal definition. For simplicity, below we define security w.r.t. black-box simulators, although we note that our definition can be easily extended to allow for non-black-box simulation following [3,8]. Our definition is parameterized by q that denotes the number of challenge messages.

---

[4] This is similar in spirit to the standard chosen-ciphertext security notion for public-key encryption.

Definition 2 (SIM-security for rand-FE). *A functional encryption scheme* $\mathcal{FE}$ *for the randomized function family* $\mathcal{F}$ *is said to be* q-SIM-*secure if there exists a simulator* $S = (S_1, S_2, S_3)$ *such that for every PPT adversary* $A = (A_1, A_2, A_3)$, *the outputs of the following two experiments are computationally indistinguishable:*

Experiment $\mathsf{REAL}_A^{\mathcal{FE}}(1^\kappa)$:
  $(\boldsymbol{x}, st_1) \leftarrow A_1(1^\kappa)$ where $\boldsymbol{x} \in \mathcal{X}_\kappa^q$
  $(\mathsf{MPK}, \mathsf{MSK}) \leftarrow \mathsf{rFE.Setup}(1^\kappa)$
  $st_2 \leftarrow A_2^{\mathcal{O}_1(\mathsf{MSK}, \cdot),\ \mathcal{O}_2(\mathsf{MSK}, \cdot, \cdot)}(\mathsf{MPK}, st_1)$
  $\mathsf{CT}_i^* \leftarrow \mathsf{rFE.Enc}(x_i, \mathsf{MPK})$ for $i \in [q]$
  $\alpha \leftarrow A_3^{\mathcal{O}_1(\mathsf{MSK}, \cdot),\ \mathcal{O}_2(\mathsf{MSK}, \cdot, \cdot)}(\mathbf{CT}^*, st_2)$
  Output $(\boldsymbol{x}, \{f\}, \{g\}, \{y\}, \alpha)$

Experiment $\mathsf{IDEAL}_A^{\mathcal{FE}}(1^\kappa)$:
  $(\boldsymbol{x}, st_1) \leftarrow A_1(1^\kappa)$ where $\boldsymbol{x} \in \mathcal{X}_\kappa^q$
  $(\mathsf{MPK}, \mathbf{CT}^*, st') \leftarrow S_1(1^\kappa)$
  $st_2 \leftarrow A_2^{\mathcal{O}_1'(\cdot),\ \mathcal{O}_2'(\cdot, \cdot)}(\mathsf{MPK}, st_1)$
  $\alpha \leftarrow A_3^{\mathcal{O}_1'(\cdot),\ \mathcal{O}_2'(\cdot, \cdot)}(\mathbf{CT}^*, st_2)$
  Output $(\boldsymbol{x}, \{f'\}, \{g'\}, \{y'\}, \alpha)$

---

where,

1. **Real experiment:** *In this experiment,* $\mathcal{O}_1(\mathsf{MSK}, \cdot)$ *denotes the key generation oracle* $\mathsf{rFE.Keygen}(\cdot, \mathsf{MSK})$. *The set* $\{f\}$ *denotes the key queries made by* A_2 *and* A_3.
   $\mathcal{O}_2(\mathsf{MSK}, \cdot, \cdot)$ *denotes a decryption oracle that takes inputs of the form* (CT, g) *where* $g \in \mathcal{F}$. *If the query is from* A_3, *then we require that* $\mathsf{CT} \neq \mathsf{CT}_i^*$. $\mathcal{O}_2$ *computes* $\mathsf{SK}_g \leftarrow \mathsf{rFE.Keygen}(g, \mathsf{MSK})$ *and returns* $\mathsf{rFE.Dec}(\mathsf{CT}, \mathsf{SK}_g)$. *The set* $\{g\}$ *denotes the functions that appear in the decryption queries of* A_2 *and* A_3 *and* $\{y\}$ *denotes the responses of* $\mathcal{O}_2$.
2. **Ideal experiment:** $\mathcal{O}_1'(\cdot)$ *denotes the simulator algorithm* $S_2(st', \cdot)$ *that has oracle access to the ideal functionality* $\mathsf{KeyIdeal}(\boldsymbol{x}, \cdot)$. *The functionality* $\mathsf{KeyIdeal}$ *accepts key queries* f' *and returns* $f'(x_i, r_i)$ *for every* $x_i \in \boldsymbol{x}$ *and randomly chosen* $r_i \in \mathcal{R}_\kappa$. *The set* $\{f'\}$ *denotes the key queries made by* S_2 *to* $\mathsf{KeyIdeal}$.
   $\mathcal{O}_2'(\cdot, \cdot)$ *denotes the simulator algorithm* $S_3(st', \cdot, \cdot)$ *that has oracle access to ideal functionality* $\mathsf{DecryptIdeal}(\cdot, \cdot)$. *The functionality* $\mathsf{DecryptIdeal}$ *accepts input queries* (x, g') *and returns* $y' = g'(x; r)$ *for randomly chosen* $r \in \mathcal{R}_\kappa$. *The set* $\{g'\}$ *denotes the functions that appear in the queries of* S_3 *and* $\{y'\}$ *denotes the responses of* $\mathsf{DecryptIdeal}$.

We note that in the above selective security definition, pre-ciphertext key queries are essentially redundant since an adversary can defer all such queries to the post-ciphertext key query phase. Nevertheless, we present our definition in the above form to remain syntactically consistent with the full security definition that consists of two distinct key query phases.

Indistinguishability Based Security. Here we present indistinguishability-based security definitions for rand-FE. We give two (incomparable) definitions: the first definition, referred to as $\mathsf{IND}_{\mathsf{pre}}$-security allows for adversaries that make key queries *before* obtaining the public key. The second definition, referred to as

IND_{post}-security, allows for key queries *after* the adversary receives the public key, but puts additional constraints on the distribution of these queries. In both cases, we strengthen the adversary by allowing decryption queries in a similar manner as the SIM-security definition.

Security against key queries before public key. We first give a security definition for the case where the adversary is restricted to making key queries before obtaining the public key. Similar to the FE definition for deterministic functions [4,21], we consider two worlds: a left world where the adversary requests ciphertexts for challenge message x_0, and a right world where the challenge message is x_1. Our definition differs from standard definition for (deterministic) FE in two ways. First, instead of requiring the outputs corresponding to x_0 and x_1 to be equal (for every key query f), we now require them to be *computationally indistinguishable*[5] (given the auxiliary input of the adversary). Second, we strengthen the adversary by allowing her to make decryption queries in the same manner as the SIM-security definition.

Definition 3 (IND_{pre}-secure rand-FE). *A functional encryption scheme $\mathcal{FE}$ is IND_{pre}-secure if for every non-uniform PPT adversary $A = (A_1, A_2, A_3)$, every $z \in \{0,1\}^*$, the distributions $\mathsf{Exp}^0_{\mathcal{FE},A}(1^\kappa, z)$ and $\mathsf{Exp}^1_{\mathcal{FE},A}(1^\kappa, z)$ are computationally indistinguishable, where $\mathsf{Exp}^b_{\mathcal{FE},A}(1^\kappa, z)$ is defined as follows :*

Experiment $\mathsf{Exp}^b_{\mathcal{FE},A}(1^\kappa, z)$:
  $(\mathsf{MPK}, \mathsf{MSK}) \leftarrow \mathsf{rFE.Setup}(1^\kappa)$
  $(x_0, x_1, st_1) \leftarrow A_1^{\mathsf{rFE.Keygen}(\cdot, \mathsf{MSK})}(1^\kappa, z)$ where $x_0, x_1 \in \mathcal{X}_\kappa$
  $st_2 \leftarrow A_2^{\mathcal{O}(\mathsf{MSK}, \cdot, \cdot)}(\mathsf{MPK}, st_1)$
  $\mathsf{CT}^* \leftarrow \mathsf{rFE.Enc}(x_b, \mathsf{MPK})$
  Output $A_3(\mathsf{CT}^*, st_2)$

In the above experiment:

1. *Let $\{f\}$ denote the list of key queries made by A_1 to the key generation oracle. Then, the distributions $(z, \{f(x_0)\})$ and $(z, \{f(x_1)\})$ are computationally indistinguishable.*
2. *$\mathcal{O}(\mathsf{MSK}, \cdot, \cdot)$ denotes a decryption oracle that takes inputs of the form (CT, g) where $g \in \mathcal{F}$. It then computes $\mathsf{SK}_g \leftarrow \mathsf{rFE.Keygen}(g, \mathsf{MSK})$ and returns $\mathsf{rFE.Dec}(\mathsf{CT}, \mathsf{SK}_g)$.*

Remark 2 (Unbounded IND_{pre} security). Definition 3 can be naturally extended to allow for multiple challenge messages. The constraint on the key queries $\{f\}$ made by A_2 will now be that given the challenge message vectors $(\boldsymbol{x_0}, \boldsymbol{x_1})$, for every i, the distributions $(z, \{f(x_0[i])\})$ and $(z, \{f(x_1[i])\})$ are computationally indistinguishable. We call this *unbounded* IND_{pre} security.

Note that by a standard hybrid argument, IND_{pre} security (for one message) implies unbounded IND_{pre} security.

---

[5] We note that this condition cannot be verified efficiently.

Security against key queries after public-key. Next we give a security definition for the case where the adversary is allowed to make key queries after obtaining the public key. The crucial difference from the previous definition is that we now require that the output distributions in the left and right world should be *statistically indistinguishable.*

Definition 4 (IND$_{post}$-secure rand-FE). *A functional encryption scheme $\mathcal{FE}$ is* IND$_{post}$*-secure for the randomized function family $\mathcal{F}$ if for every non-uniform PPT adversary $A = (A_1, A_2)$, every $z \in \{0,1\}^*$, the distributions $\mathsf{Exp}^0_{\mathcal{FE},A}(1^\kappa, z)$ and $\mathsf{Exp}^1_{\mathcal{FE},A}(1^\kappa, z)$ are computationally indistinguishable, where $\mathsf{Exp}^b_{\mathcal{FE},A}(1^\kappa, z)$ is defined as follows :*

Experiment $\mathsf{Exp}^b_{\mathcal{FE},A}(1^\kappa, z)$:
  $(\mathsf{MPK}, \mathsf{MSK}) \leftarrow \mathsf{rFE.Setup}(1^\kappa)$
  $(x_0, x_1, st_1) \leftarrow A_1(1^\kappa, z)$ where $x_0, x_1 \in \mathcal{X}_\kappa$
  $\mathsf{CT}^* \leftarrow \mathsf{rFE.Enc}(x_b, \mathsf{MPK})$
  Output $A_2^{\mathsf{rFE.Keygen}(\cdot, \mathsf{MSK}), \mathcal{O}(\mathsf{MSK}, \cdot, \cdot)}(\mathsf{MPK}, \mathsf{CT}^*, st_1)$

In the above experiment:

1. *Let $\{f\}$ denote the list of key queries made by A_2 to the key generation oracle. Then the distributions $(\mathsf{MPK}, z, \{f(x_0)\})$ and $(\mathsf{MPK}, z, \{f(x_1)\})$ are statistically indistinguishable.*
2. *$\mathcal{O}(\mathsf{MSK}, \cdot, \cdot)$ denotes a decryption oracle that takes inputs of the form (CT, g) where $\mathsf{CT} \neq \mathsf{CT}^*$ and $g \in \mathcal{F}$. It computes $\mathsf{SK}_g \leftarrow \mathsf{rFE.Keygen}(g, \mathsf{MSK})$ and returns $\mathsf{rFE.Dec}(\mathsf{CT}, \mathsf{SK}_g)$.*

Remark 3 (Unbounded IND$_{post}$ *security).* Similar to Definition 3, the above definition can also be naturally extended to capture security for multiple challenge messages. We call this *unbounded* IND$_{post}$ security. Note that one-message IND$_{post}$ security implies unbounded IND$_{post}$ security.

Remark 4 (Statistical vs Computational Indistinguishability). Note that if we modify Definition 4 by requiring the output distributions to be computationally indistinguishable (as in Definition 3, then it may result in a circularity. Consider a key query f from A_2 that simply re-encrypts the plaintext underlying the challenge ciphertext CT^*_b.[6] In this case, the requirement on the output distributions is the same as our desired security guarantee for the challenge ciphertexts, which results in a vaccuous definition. By requiring the output distributions to be statistically indistinguishable, we are able to break such circularity.

SIM implies IND. It is easy to see that SIM-security implies both IND$_{pre}$ and IND$_{post}$ security. Furthermore, since IND$_{pre}$ (resp., IND$_{post}$) security for one message implies unbounded IND$_{pre}$ (resp., IND$_{post}$) security, we have that 1-SIM security implies unbounded IND$_{pre}$ and IND$_{post}$ security. We state it below:

---

[6] Note that in Definition 3, such a query is not possible since the adversary is required to make all the key queries *before* receiving the public key.

Lemma 1. *Let $\mathcal{FE}$ be a 1-SIM-secure FE scheme for randomized function family $\mathcal{F}$. Then $\mathcal{FE}$ is also unbounded $\mathsf{IND}_{\mathsf{pre}}$-secure and unbounded $\mathsf{IND}_{\mathsf{post}}$-secure for $\mathcal{F}$.*

The proof follows in the same manner as the case of deterministic functions [4]. We provide a sketch in Appendix B for the case of one message. Combining this with remarks 2 and 3 yields the proof of lemma 1 for unbounded messages.

3   Preliminaries

In this section, we present definitions for various cryptographic primitives that we shall use in our construction of functional encryption for randomized functions. We assume familiarity with standard semantically secure public-key encryption and strongly unforgeable signature schemes and omit their formal definition from this text. Below, we recall the notions of indistinguishability obfuscation, puncturable pseudorandom functions, non-interactive witness indistinguishable proof systems and perfectly binding commitment schemes.

3.1   Indistinguishability Obfuscation

Here we recall the notion of indistinguishability obfuscation that was defined by Barak et al. [2]. Intuitively speaking, we require that for any two circuits C_1 and C_2 that are "functionally equivalent" (i.e., for all inputs x in the domain, $C_1(x) = C_2(x)$), the obfuscation of C_1 must be computationally indistinguishable from the obfuscation of C_2. Below we present the formal definition following the syntax of [11].

Definition 5. *(Indistinguishability Obfuscation) A uniform PPT machine $i\mathcal{O}$ is called an indistinguishability obfuscator for a circuit class $\{\mathcal{C}_\kappa\}$ if the following holds:*

- **Correctness:** *For every $\kappa \in \mathbb{N}$, every $C \in \mathcal{C}_\kappa$, every input x in the domain of C, we have that*

$$Pr[C'(x) = C(x) : C' \leftarrow i\mathcal{O}(C)] = 1$$

- **Indistinguishability:** *For every $\kappa \in \mathbb{N}$, for all pairs of circuits $C_0, C_1 \in \mathcal{C}_\kappa$, if $C_0(x) = C_1(x)$ for all inputs x, then for all PPT adversaries $\mathcal{A}$, we have:*

$$|Pr[\mathcal{A}(i\mathcal{O}(C_0)) = 1] - Pr[\mathcal{A}(i\mathcal{O}(C_1)) = 1]| \leq \mathsf{negl}(\kappa)$$

Recently, Garg et al. [11] gave the first candidate construction for an indistinguishability obfuscator $i\mathcal{O}$ for the circuit class $P/poly$. Subsequent to their work, Pass et al [22] construct an indistinguishability obfuscator based on an "uber" assumption on multilinear encodings. More recently, Gentry et al [13] construct an indistinguishability obfuscator based on the multilinear subgroup elimination assumption.

3.2 Puncturable Pseudorandom Functions

Puncturable family of PRFs are a special case of constrained PRFs [6,7,20], where the PRF is defined on all input strings except for a set of size polynomial in the security parameter. Below we recall their definition, as given by [26].

Syntax A *puncturable* family of PRFs is defined by a tuple of algorithms (Key, Eval, Puncture) and a pair of polynomials $n(\cdot)$ and $m(\cdot)$:

- **Key Generation** $\mathsf{Key}(1^\kappa)$ is a PPT algorithm that takes as input the security parameter κ and outputs a PRF key K
- **Punctured Key Generation** $\mathsf{Puncture}(K, S)$ is a PPT algorithm that takes as input a PRF key K, a set $S \subset \{0,1\}^{n(\kappa)}$ and outputs a punctured key K_S
- **Evaluation** $\mathsf{Eval}(K, x)$ is a deterministic algorithm that takes as input a key K (punctured key or PRF key), a string $x \in \{0,1\}^{n(\kappa)}$ and outputs $y \in \{0,1\}^{m(\kappa)}$

Definition 6. *A family of PRFs* Key, Eval, Puncture *is* puncturable *if it satisfies the following properties :*

- **Functionality preserved under puncturing.** *Let* $K \leftarrow \mathsf{Key}(1^\kappa)$, $K_S \leftarrow \mathsf{Puncture}(K, S)$. *Then, for all* $x \notin S$, $\mathsf{Eval}(K, x) = \mathsf{Eval}(K_S, x)$.
- **Pseudorandom at punctured points.** *For every PPT adversary* (A_1, A_2) *such that* $A_1(1^\kappa)$ *outputs a set* $S \subset \{0,1\}^{n(\kappa)}$, $x \in S$ *and state* st, *consider an experiment where* $K \leftarrow \mathsf{Key}(1^\kappa)$ *and* $K_S \leftarrow \mathsf{Puncture}(K, S)$. *Then*

$$\left| Pr[A_2(K_S, x, \mathsf{Eval}(K, x), \mathsf{st}) = 1] - Pr[A_2(K_S, x, U_{m(\kappa)}, \mathsf{st}) = 1] \right| \leq \mathsf{negl}(\kappa)$$

where U_ℓ *denotes the uniform distribution over* ℓ *bits.*

As observed by [20,6,7], the [14] construction of PRFs from one-way functions easily yield puncturable PRFs.

Theorem 1 ([14,20,6,7]). *If one-way functions exist, then for all polynomials* $n(\kappa)$ *and* $m(\kappa)$, *there exists a puncturable PRF family that maps* $n(\kappa)$ *bits to* $m(\kappa)$ *bits.*

We note that in the above construction, the size of the punctured key K_S grows linearly with the size of the punctured set S.

3.3 Non-Interactive Witness Indistinguishable Proofs

In this section, we present the definition for non-interactive witness-indistinguishable (NIWI) proofs. We emphasize that we are interested in *proof* systems, i.e., where the soundness guarantee holds against computationally unbounded cheating provers.

Syntax. Let R be an efficiently computable relation that consists of pairs (x, w), where x is called the statement and w is the witness. Let L denote the language consisting of statements in R. A non-interactive proof system for a language L consists of a setup algorithm NIWI.Setup, a prover algorithm NIWI.Prove and a verifier algorithm NIWI.Verify, defined as follows:

- **Setup** NIWI.Setup(1^κ) is a PPT algorithm that takes as input the security parameter 1^κ and outputs a common reference string crs.
- **Prover** NIWI.Prove(crs, x, w) is a PPT algorithm that takes as input the common reference string crs, a statement x along with a witness w. $(x, w) \in R$; if so, it produces a proof string π, else it outputs fail.
- **Verifier** NIWI.Verify(crs, x, π) is a PPT algorithm that takes as input the common reference string crs and a statement x with a corresponding proof π. It outputs 1 if the proof is valid, and 0 otherwise.

Definition 7 (NIWI). *A non-interactive witness-indistinguishable proof system for a language L with a PPT relation R is a tuple of algorithms (*NIWI.Setup, NIWI.Prove, NIWI.Verify*) such that the following properties hold:*

- **Perfect Completeness:** *For every $(x, w) \in R$, it holds that*

$$\Pr[\text{NIWI.Verify}(\text{crs}, x, \text{NIWI.Prove}(\text{crs}, x, w)) = 1] = 1$$

  where crs $\leftarrow$ NIWI.Setup(1^κ), *and the probability is taken over the coins of* NIWI.Setup, NIWI.Prove *and* NIWI.Verify.
- **Statistical Soundness:** *For every adversary $\mathcal{A}$, it holds that*

$$\Pr[\text{NIWI.Verify}(\text{crs}, x, \pi) = 1 \land x \notin L \mid \text{crs} \leftarrow \text{NIWI.Setup}(1^\kappa); (x, \pi) \leftarrow \mathcal{A}(\text{crs})] = \text{negl}(1^\kappa)$$

- **Witness Indistinguishability:** *For any triplet (x, w_0, w_1) such that $(x, w_0) \in R$ and $(x, w_1) \in R$, the distributions $\{$crs, NIWI.Prove(crs, x, w_0)$\}$ and $\{$crs, NIWI.Prove(crs, x, w_1)$\}$ are computationally indistinguishable, where* crs $\leftarrow$ NIWI.Setup(1^κ).

Recently, it was shown by Sahai and Waters [26] that NIWI proofs can be constructed from indistinguishability obfuscation and one-way functions.

3.4 Commitment Schemes

A commitment scheme Com is a PPT algorithm that takes as input a string x and outputs $C \leftarrow \text{Com}(x)$. A perfectly binding commitment scheme must satisfy the *perfect binding* and *computational hiding* properties :

- **Perfectly Binding:** This property states that two different strings cannot have the same commitment. More formally, $\forall x_1 \neq x_2, s_1, s_2$ Com($x_1; s_1$) $\neq$ Com($x_2; s_2$)
- **Computational Hiding:** For all strings x_0 and x_1 (of the same length), for all PPT adversaries $\mathcal{A}$, we have that :

$$|Pr[\mathcal{A}(\text{Com}(x_0)) = 1] - Pr[\mathcal{A}(\text{Com}(x_1) = 1)]| \leq \text{negl}(\kappa)$$

For simplicity of exposition, we present our FE scheme in Sect. 4 using a non-interactive perfectly binding scheme. We stress, however, that it is actually sufficient to use a standard 2-round statistically binding scheme in our construction. Such schemes can be based on one way functions.

4 Our Construction

Let $\mathcal{F}$ denote the family of all PPT functions. We now present a functional encryption scheme $\mathcal{FE}$ for $\mathcal{F}$. For any a priori bounded $q = \mathsf{poly}(\kappa)$, we prove that $\mathcal{FE}$ is q-SIM-secure. Note that from Lemma 1, it follows that $\mathcal{FE}$ is also unbounded $\mathsf{IND}_{\mathsf{pre}}$ and $\mathsf{IND}_{\mathsf{post}}$ secure.

Note that in the case of SIM-security, the size of the secret keys in $\mathcal{FE}$ grows linearly with q. It follows from [4,3,8] that such a dependence on q is necessary.

Notation. Let (NIWI.Setup, NIWI.Prove, NIWI.Verify) be a NIWI proof system. Let Com be a perfectly binding commitment scheme. Let $i\mathcal{O}$ be an indistinguishability obfuscator for all efficiently computable circuits. Let (Key, Puncture, Eval) be a puncturable family of PRF. Let (Gen, Sign, Verify) be a strongly unforgeable one-time signature scheme. Finally, let (PKE.Setup, PKE.Enc, PKE.Dec) be a semantically secure public-key encryption scheme.

Let c-len $=$ c-len(1^κ) denote the length of ciphertexts in (PKE.Setup, PKE.Enc, PKE.Dec) . Let v-len $=$ v-len(1^κ) denote the length of verification keys in (Gen, Sign, Verify). We shall use a parameter len $= 2 \cdot$ c-len $+$ v-len in the description of our scheme. We now proceed to describe our scheme $\mathcal{FE} = ($rFE.Setup, rFE.Enc, rFE.Keygen, rFE.Dec$)$.

Setup rFE.Setup(1^κ): The setup algorithm first computes a CRS crs $\leftarrow$ NIWI.Setup for the NIWI proof system. Next, it computes two key pairs – $(PK_1, SK_1) \leftarrow$ PKE.Setup(1^κ), $(PK_2, SK_2) \leftarrow$ PKE.Setup(1^κ) – of the public-key encryption scheme. Finally, it computes a commitment $C \leftarrow$ Com(0^{len}).

The public key MPK $= ($crs$, PK_1, PK_2, C)$ and the master secret key MSK $= SK_1$. The algorithm outputs (MPK, MSK).

Encryption rFE.Enc(x, MPK): To encrypt a message x, the encryption algorithm first generates a key pair $(sk, vk) \leftarrow$ Gen(1^κ) of the one-time signature scheme. It then computes ciphertexts $c_1 \leftarrow$ PKE.Enc$(x, PK_1; r_1)$ and $c_2 \leftarrow$ PKE.Enc$(x, PK_2; r_2)$. Next, it computes a NIWI proof $\pi \leftarrow$ NIWI.Prove(crs$, z, w)$ for the NP statement $z = (z_1 \vee z_2)$ where z_1 and z_2 are defined as follows:

$$z_1 := (\exists x, s_1, s_2 \text{ such that } c_1 = \mathsf{PKE.Enc}(x, PK_1; s_1) \wedge c_2 = \mathsf{PKE.Enc}(x, PK_2; s_2)) \tag{1}$$

$$z_2 := (\exists s \text{ such that } C = \mathsf{Com}(c_1 \| c_2 \| vk, s)) \tag{2}$$

A witness $w_{\mathsf{real}} = (x, s_1, s_2)$ for z_1 is referred to as the *real* witness, while a witness $w_{\mathsf{trap}} = s$ for z_2 is referred to as the *trapdoor* witness.

The honest encryption algorithm uses the real witness w_{real} to compute π. Finally, it computes a signature $\sigma \leftarrow \mathsf{Sign}(c_1\|c_2\|\pi, sk)$ on the string $c_1\|c_2\|\pi$ using sk. The output of the algorithm is the ciphertext $\mathsf{CT} = (c_1, c_2, \pi, vk, \sigma)$.

Key Generation $\mathsf{rFE.Keygen}(f, \mathsf{MSK})$: On input f ,the key generation algorithm first chooses a fresh PRF key $K \leftarrow \mathsf{Key}(1^\kappa)$. It then computes the secret key $\mathsf{SK}_f \leftarrow i\mathcal{O}(\mathcal{G}_f)$ where the function $\mathcal{G}_f$ is described in Fig. 1. Note that $\mathcal{G}_f$ has the public key MPK, the secret key SK_1 and the PRF key K hardwired in it.

Input: Ciphertext CT
Constants: MPK, SK_1, K, f

1. Parse $\mathsf{CT} = (c_1, c_2, \pi, vk, \sigma)$.
2. If $\mathsf{Verify}(\sigma, c_1\|c_2\|\pi, vk) = 0$, then output $\perp$ and stop, otherwise continue to the next step.
3. If $\mathsf{NIWI.Verify}(\mathsf{crs}, z, \pi) = 0$, then output $\perp$ and stop, otherwise continue to the next step. Here $z = (c_1, c_2, vk, PK_1, PK_2, C)$ is the statement corresponding to π.
4. Compute $x \leftarrow \mathsf{PKE.Dec}(c_1, SK_1)$.
5. Compute $r \leftarrow \mathsf{Eval}(K, c_1\|c_2\|vk)$.
6. Output $f(x; r)$.

Fig. 1. Functionality $\mathcal{G}_f$

The algorithm outputs SK_f as the secret key corresponding to f.

Size of Function $\mathcal{G}_f$. In order to prove that $\mathcal{FE}$ is q-SIM-secure, we require the function $\mathcal{G}_f$ to be padded with zeros such that $|\mathcal{G}_f| = |\mathsf{Sim.}\mathcal{G}_f|$, where the "simulated" functionality $\mathsf{Sim.}\mathcal{G}_f$ is described later in Fig. 2. In this case, the size of SK_f grows linearly with q.

Decryption $\mathsf{rFE.Dec}(\mathsf{CT}, \mathsf{SK}_f)$: On input CT, the decryption algorithm computes and outputs $\mathsf{SK}_f(\mathsf{CT})$.

This completes the description of $\mathcal{FE}$. We prove the correctness of $\mathcal{FE}$ in Appendix A.

Theorem 2. *Assuming indistinguishability obfuscation for all polynomial-time computable circuits and one-way functions, the proposed scheme $\mathcal{FE}$ is 1-SIM-secure.*

5   Proof of Theorem 2

We now prove that the proposed scheme $\mathcal{FE}$ is 1-SIM-secure. Our proof can be naturally extended to q-SIM-security, for any a priori fixed $q = \mathsf{poly}(\kappa)$.

We first construct an ideal world adversary aka simulator S in Sect. 5.1. Next, in Sect. 5.2, we prove indistinguishability of the outputs of the real and ideal world experiments via a hybrid argument.

5.1 Description of Simulator

We describe a simulator $S = (S_1, S_2, S_3)$ that makes black-box use of a real world adversary $A = (A_1, A_2, A_3)$.

Algorithm S_1. S_1 first performs a simulated setup procedure. Namely, it first computes a CRS crs $\leftarrow$ NIWI.Setup(1^κ) for the NIWI proof system and two key pairs – $(PK_1, SK_1) \leftarrow$ PKE.Setup(1^κ) and $(PK_2, SK_2) \leftarrow$ PKE.Setup(1^κ) – for the public-key encryption scheme. Next, it chooses a key pair for the signature scheme - $(sk^*, vk^*) \leftarrow$ Gen(1^κ). Then, it computes the commitment C in the following manner: (a) First compute $c_1^* \leftarrow$ PKE.Enc($\mathbf{0}, PK_1$) and $c_2^* \leftarrow$ PKE.Enc($\mathbf{0}, PK_2$). (b) Next, compute $C \leftarrow$ Com($c_1^* \| c_2^* \| vk^*$). Let s denote the randomness used to compute C .

S_1 constructs a proof π^* by using the *trapdoor* witness s, that is, $\pi^* \leftarrow$ NIWI.Prove(crs, y, s), where the statement $y = (c_1^*, c_2^*, vk^*, PK_1, PK_2, C)$. Finally, it computes $\sigma^* \leftarrow$ Sign($c_1^* \| c_2^* \| \pi^*, sk^*$). It sets MPK $= (\text{crs}, PK_1, PK_2, C)$ and challenge ciphertext CT$^* = (c_1^*, c_2^*, \pi^*, vk^*, \sigma^*)$.

Algorithm S_2. S_2 simulates the key generation oracle. Whenever A_2 or A_3 makes a key query for a function f, S_2 performs the following sequence of steps:

1. Query the ideal functionality KeyIdeal on input f. Let y^* be the output of KeyIdeal .
2. Compute a PRF key $K \leftarrow$ Key(1^κ) and a punctured key $K' \leftarrow$ Puncture(K, $c_1^* \| c_2^* \| vk^*$).
3. Compute the secret key SK$_f \leftarrow i\mathcal{O}(\text{Sim}.\mathcal{G}_f)$ where the functionality Sim.$\mathcal{G}_f$ is described in Fig. 2. Sim.$\mathcal{G}_f$ has the public key MPK, secret key SK_1, the punctured key K', the challenge ciphertext CT* and the output value y^* hardwired in it.
4. Return SK$_f$.

Algorithm S_3. S_3 simulates the decryption oracle. Whenever A_2 or A_3 makes a decryption query (CT, g) where CT $= (c_1, c_2, \pi, vk, \sigma)$, S_3 performs the following sequence of steps:

1. If Verify($\sigma, c_1 \| c_2 \| \pi, vk$), then output $\perp$ and stop, otherwise continue to the next step.
2. If NIWI.Verify(crs, $z, \pi) = 0$, then output $\perp$ and stop, otherwise continue to the next step. Here $z = (c_1, c_2, vk, PK_1, PK_2, C)$ is the statement corresponding to π.
3. Compute $x \leftarrow$ PKE.Dec(c_1, SK_1).
4. Return DecryptIdeal(x, g).

Input: Ciphertext CT

Constants: MPK, SK_1, K', f, $\mathsf{CT}^* = (c_1^*, c_2^*, \pi^*, vk^*, \sigma^*)$, y^*

1. Parse $\mathsf{CT} = (c_1, c_2, \pi, vk, \sigma)$.
2. If $\mathsf{Verify}(\sigma, c_1\|c_2\|\pi, vk)$, then output $\bot$ and stop, otherwise continue to the next step.
3. If $\mathsf{NIWI.Verify}(\mathsf{crs}, z, \pi) = 0$, then output $\bot$ and stop, otherwise continue to the next step. Here $z = (c_1, c_2, vk, PK_1, PK_2, C)$ is the statement corresponding to π.
4. If $(c_1\|c_2\|vk = c_1^*\|c_2^*\|vk^*)$ output y and stop.
5. Compute $x \leftarrow \mathsf{PKE.Dec}(c_1, SK_1)$.
6. Compute $r \leftarrow \mathsf{Eval}(K', c_1\|c_2\|vk)$.
7. Output $f(x; r)$.

Fig. 2. Functionality $\mathsf{Sim}.\mathcal{G}_f$

5.2 Indistinguishability of the Outputs

We now describe a series of hybrid experiments $\mathsf{H}_0, \ldots, \mathsf{H}_{11}$, where H_0 corresponds to the real world and H_{11} corresponds to the ideal world experiment.

Hybrid H_0: This is the real experiment. Here, each decryption query (CT, g) is answered using a decryption key $sk_g \leftarrow i\mathcal{O}(\mathcal{G}_g)$ where $\mathcal{G}_g$ is defined in the same manner as $\mathcal{G}_f$, except that it has function g hardwired in it.

Hybrid H_1: This experiment is the same as H_0 except in the manner in which the key queries of the adversary are answered. Let $\mathsf{CT}^* = (c_1^*, c_2^*, \pi^*, vk^*, \sigma^*)$ denote the challenge ciphertext. Whenever the adversary A_2 or A_3 makes a key query f, we perform the following steps:

1. Compute a PRF key $K \leftarrow \mathsf{Key}(1^\kappa)$ and then compute a punctured key $K' \leftarrow \mathsf{Puncture}(K, c_1^*\|c_2^*\|vk^*)$.
2. Compute $r \leftarrow \mathsf{Eval}(K, c_1^*\|c_2^*\|vk^*)$ and $y^* = f(x; r)$.
3. Compute the secret key $\mathsf{SK}_f \leftarrow i\mathcal{O}(\mathsf{Sim}.\mathcal{G}_f)$ where the functionality $\mathsf{Sim}.\mathcal{G}_f$ is described in Fig. 2. Note that $\mathsf{Sim}.\mathcal{G}_f$ has the public key MPK, master secret key MSK, the punctured key K', the challenge ciphertext components ct^* and the output value y^* (as computed above) hardwired in it.
4. Return SK_f.

Hybrid H_2: This experiment is the same as H_1, except that we now answer the key queries of A_2 and A_3 in the same manner as the simulator S_2.

Hybrid H_3: This experiment is the same as H_2, except that the setup algorithm computes the commitment C in the following manner: let $\mathsf{CT}^* = (c_1^*, c_2^*, \pi^*, vk^*, \sigma^*)$ denote the challenge ciphertext. Then, $C \leftarrow \mathsf{Com}(c_1^*\|c_2^*\|vk^*)$.

Hybrid H_4: This experiment is the same as H_3, except that we modify the challenge ciphertext $\mathsf{CT}^* = (c_1^*, c_2^*, \pi^*, vk^*, \sigma^*)$: the proof string π^* is now computed using the *trapdoor* witness s where s is the randomness used to compute the commitment C.

Hybrid H_5: This experiment is the same as H_4, except that in the challenge ciphertext $\mathsf{CT}^* = (c_1^*, c_2^*, \pi^*, vk^*, \sigma^*)$, the *second* ciphertext c_2^* is an encryption of zeros, i.e., $c_2^* \leftarrow \mathsf{PKE.Enc}(\mathbf{0}, PK_2)$.

Hybrid H_6: This experiment is the same as H_5, except that for every key query f, the secret key SK_f is computed as $\mathsf{SK}_f \leftarrow i\mathcal{O}(\mathsf{Sim}.\mathcal{G}'_f)$ where $\mathsf{Sim}.\mathcal{G}'_f$ is the same as function $\mathsf{Sim}.\mathcal{G}_f$ except that:

1. It has secret key SK_2 hardwired instead of SK_1.
2. It decrypts the *second* component of each input ciphertext using SK_2. More concretely, in Step 5 of $\mathsf{Sim}.\mathcal{G}'_f$, plaintext x is computed as $x \leftarrow \mathsf{PKE.Dec}(c_2, SK_2)$.

Hybrid H_7: This experiment is the same as H_6, except that we modify the manner in which the decryption queries of A_2 and A_3 are answered: each query (CT, g) is answered using a decryption key $sk_g \leftarrow i\mathcal{O}(\mathcal{G}'_f)$ where $\mathcal{G}'_f$ is the same as function $\mathcal{G}_g$ except that:

1. It has secret key SK_2 hardwired instead of SK_1.
2. It decrypts the *second* component of each input ciphertext using SK_2. More concretely, in Step 4 of $\mathcal{G}_g$, plaintext x is computed as $x \leftarrow \mathsf{PKE.Dec}(c_2, SK_2)$.

Hybrid H_8: This experiment is the same as H_7, except that in the challenge ciphertext $\mathsf{CT}^* = (c_1^*, c_2^*, \pi^*, vk^*, \sigma^*)$, the *first* ciphertext c_1^* is an encryption of zeros, i.e., $c_1^* \leftarrow \mathsf{PKE.Enc}(\mathbf{0}, PK_1)$.

Hybrid H_9: This experiment is the same as H_8, except that we modify the manner in which the decryption queries of A_2 and A_3 are answered: each query (CT, g) is answered using a decryption key $sk_g \leftarrow i\mathcal{O}(\mathcal{G}_f)$.

Hybrid H_{10}: This experiment is the same as H_9, except that we change the manner in which the key queries are answered. For every key query f, the secret key SK_f is computed as $\mathsf{SK}_f \leftarrow i\mathcal{O}(\mathsf{Sim}.\mathcal{G}_f)$.

Hybrid H_{11}: This experiment is the same as H_{10}, except that we now answer the decryption queries of A_2 and A_3 in the same manner as the simulator algorithm S_3. Note that this is the ideal experiment.

This completes the description of the hybrid experiments. Next, we prove that for every i, the outputs of experiments H_i and H_{i+1} are computationally indistinguishable.

Lemma 2. *Assuming that $i\mathcal{O}$ is an indistinguishability obfuscator, hybrid experiments H_0 and H_1 are computationally indistinguishable.*

Proof. Note that the only difference in H_0 and H_1 is that in the former experiment, we output $i\mathcal{O}(\mathcal{G}_f)$ as the key corresponding to any key query f, while in the latter experiment, we output $i\mathcal{O}(\mathsf{Sim}.\mathcal{G}_f)$. In order to prove that these two hybrids are computationally indistinguishable, we show that for every key query f, $\mathcal{G}_f$ and $\mathsf{Sim}.\mathcal{G}_f$ have identical input-output behavior. Then, by security of indistinguishability obfuscation, we would have that $i\mathcal{O}(\mathcal{G}_f)$ and $i\mathcal{O}(\mathsf{Sim}.\mathcal{G}_f)$ are computationally indistinguishable, which in turn would imply H_0 and H_1 are computationally indistinguishable.

Observation 1. *For any input* $\mathsf{CT} = (c_1, c_2, \pi, vk, \sigma)$, $\mathcal{G}_f$ *outputs* $\perp$ *if and only if* $\mathsf{Sim}.\mathcal{G}_f$ *outputs* $\perp$.

Note that both $\mathcal{G}_f$ and $\mathsf{Sim}.\mathcal{G}_f$ output $\perp$ if and only if either the signature σ does not verify or the proof π does not verify; that is, either $\mathsf{Verify}(\sigma, c_1 \| c_2 \| \pi, vk) = 0$ or $\mathsf{NIWI}.\mathsf{Verify}(crs, y, \pi) = 0$ where $y = (c_1, c_2, vk, PK_1, PK_2, C)$. Let us call an input $\mathsf{CT} = (c_1, c_2, \pi, vk, \sigma)$ *valid* if both the signature σ and proof π verify. Next, we prove that both $\mathcal{G}_f$ in H_0 and $\mathsf{Sim}.\mathcal{G}_f$ in H_1 have the same functionality for all valid inputs.

Claim 1. *For any valid input* $\mathsf{CT} = (c_1, c_2, \pi, vk, \sigma)$, $\mathcal{G}_f(\mathsf{CT}) = \mathsf{Sim}.\mathcal{G}_f(\mathsf{CT})$.

Proof. We consider two cases : $c_1 \| c_2 \| vk \neq c_1^* \| c_2^* \| vk^*$ and $c_1 \| c_2 \| vk = c_1^* \| c_2^* \| vk^*$. For the first case, note that by the first property of constrained PRF, it follows that $\mathsf{Eval}(K, c_1 \| c_2 \| vk) = \mathsf{Eval}(K', c_1 \| c_2 \| vk) = r$. Both $\mathcal{G}_f$ in H_0 and $\mathsf{Sim}.\mathcal{G}_f$ in H_1 decrypt c_1 using SK_1 to compute x, and then output $f(x, r)$.

In the second case, $\mathcal{G}_f$ computes $r \leftarrow \mathsf{Eval}(K, c_1^* \| c_2^* \| vk^*)$, and then computes x by decrypting c_1 and outputs $y = f(x; r)$. On the other hand, $\mathsf{Sim}.\mathcal{G}_f$ simply outputs the hard-wired value y^* when $c_1 \| c_2 \| vk = c_1^* \| c_2^* \| vk^*$. However, note that $y^* = y$, thereby ensuring that $\mathcal{G}_f(\mathsf{CT}^*) = \mathsf{Sim}.\mathcal{G}_f(\mathsf{CT}^*)$.

Using the above claims, we can now describe our reduction. Assume A_2 and A_3 together make a total of ℓ key queries. We define hybrids $H_{0,i}$, $0 \le i \le \ell$, as follows: in $H_{0,i}$, we respond to the first $\ell - i$ queries using $\mathsf{FE}.\mathsf{Keygen}$ as in H_0, and respond to the last i queries as in H_1.

Claim 2. *If $\exists$ a PPT distinguisher A that can distinguish the outputs of $H_{0,i}$ and $H_{0,i+1}$ with non negligible advantage, then there exists a PPT adversary B that can break the security of $i\mathcal{O}$ with non-negligible advantage.*

Let C be the challenger for obfuscation. Adversary B works as follows:

1. It first honestly computes $(\mathsf{MPK}, st', \mathsf{CT}^*)$.
2. For the first $(\ell - i - 1)$ key queries f, B computes the key for f using $\mathsf{rFE}.\mathsf{Keygen}(\cdot, \mathsf{MSK})$. For the last i key queries f, B computes the key for f as in H_1.
3. For the $(\ell-i)$'th key query for function f, B chooses a PRF key K, computes $K' \leftarrow \mathsf{Puncture}(K, c_1^* \| c_2^* \| vk^*)$ and $y = f(x; \mathsf{Eval}(K, c_1^* \| c_2^* \| vk^*))$. It then defines programs $\mathcal{G}_f$, $\mathsf{Sim}.\mathcal{G}_f$ and sends them to C, and receives an obfuscation SK_f, which it passes on to the adversary.

4. $\mathcal{B}$ runs the rest of the experiment in the same manner as in H_0 and H_1.
5. Finally, $\mathcal{B}$ sends the output of the experiment to $\mathcal{A}$ and returns its output to $\mathcal{C}$.

Now, if $\mathcal{C}$ returns obfuscation of $\mathcal{G}_f$, then $\mathcal{B}$ perfectly simulates experiment $H_{0,i}$, else it simulates experiment $H_{0,i+1}$. Thus, if $\mathcal{A}$ distinguishes the outputs with non negligible advantage, then clearly $\mathcal{B}$ breaks the security of indistinguishability obfuscation with non negligible advantage.

Lemma 3. *Assuming* (Key, Eval, Puncture) *is a puncturable family of PRFs, hybrid experiments* H_1 *and* H_2 *are computationally indistinguishable.*

Proof. Assume A_2 and A_3 make a total of ℓ key queries. We consider ℓ intermediate hybrids $H_{1,i}$ for $0 \leq i \leq \ell$ where in $H_{1,i}$, we respond to the first $\ell - i$ key queries as in H_1, and the remaining i key queries as in H_2. We show that if there exists a PPT distinguisher $\mathcal{A}$ that can distinguish the outputs of $H_{1,i}$ and $H_{1,i+1}$ with non-negligible advantage, then there exists a PPT adversary $\mathcal{B}$ that can break the security of puncturable PRFs with non-negligible advantage. The construction of $\mathcal{B}$ is as follows :

1. $\mathcal{B}$ first computes $\mathsf{MPK}, \mathsf{MSK}, \mathsf{CT}^*$ honestly.
2. For the first $(\ell - i - 1)$ key queries from A_3, $\mathcal{B}$ responds in the same manner as in H_1. For the last i key queries, $\mathcal{B}$ responds as in H_2.
3. For the $(\ell - i)$'th key query f, $\mathcal{B}$ first sends $(c_1^*\|c_2^*\|vk^*)$ to the challenger $\mathcal{C}$ and receives (K', r), where $K' = \mathsf{Puncture}(K, c_1^*\|c_2^*\|vk^*)$ for some PRF key K and r is either $\mathsf{Eval}(K, c_1^*\|c_2^*\|vk^*)$ or a uniformly random string in $\mathcal{R}_\kappa$. It then defines the function $\mathsf{Sim}.\mathcal{G}_f$ as before. $\mathcal{B}$ sends $i\mathcal{O}(\mathsf{Sim}.\mathcal{G}_f)$ as the key for function f.
4. $\mathcal{B}$ runs the rest of the experiment in the same manner as in H_1 and H_2.
5. Finally, $\mathcal{B}$ sends the output of the experiment to $\mathcal{A}$ and returns its output to $\mathcal{C}$.

Note that if r was computed as $\mathsf{Eval}(K, c_1^*\|c_2^*\|vk^*)$, then $\mathcal{B}$ perfectly simulates experiment $H_{1,i}$, else it simulates $H_{1,i+1}$. Thus, if $\mathcal{A}$ can distinguish the outputs of $H_{1,i}$ and $H_{1,i+1}$ with non-negligible advantage, then $\mathcal{B}$ can break security of puncturable PRFs with non-negligible advantage.

Lemma 4. *Assuming* Com *is a computationally hiding commitment scheme, hybrid experiments* H_2 *and* H_3 *are computationally indistinguishable.*

Proof. Note that the only difference between experiments H_2 and H_3 is that C is computed as a commitment to 0^{len} in the former case and $(c_1^*\|c_2^*\|vk^*)$ in the latter. Then, assume that $\exists$ PPT distinguisher $\mathcal{A}$ that can distinguish the outputs of H_2 and H_3 with non-negligible advantage. Using $\mathcal{A}$, we can construct a PPT algorithm $\mathcal{B}$ that breaks the computational hiding property of Com as follows:

1. $\mathcal{B}$ first runs A_1 to obtain x. It then computes $(PK_1, SK_1) \leftarrow \mathsf{PKE.Setup}(1^\kappa)$, $(PK_2, SK_2) \leftarrow \mathsf{PKE.Setup}(1^\kappa)$, crs $\leftarrow \mathsf{NIWI.Setup}$ and $(sk^*, vk^*) \leftarrow \mathsf{Gen}(1^\kappa)$.

2. Next, it computes $c_1^* \leftarrow \mathsf{PKE.Enc}(x, PK_1)$, $c_2^* \leftarrow \mathsf{PKE.Enc}(x, PK_2)$ and constructs a valid proof π^* using the real witness. Then it signs $c_1^* \| c_2^* \| \pi^*$ using sk^* to compute σ^*. It sets $\mathsf{CT}^* = (c_1^*, c_2^*, \pi^*, vk^*, \sigma^*)$
3. $\mathcal{B}$ sends 0^{len} and $(c_1^* \| c_2^* \| vk^*)$ to $\mathcal{C}$, and receives C, which is either a commitment to 0^{len} or $(c_1^* \| c_2^* \| vk^*)$.
4. $\mathcal{B}$ simulates the rest of the experiment as in H_2 and H_3.
5. Finally, $\mathcal{B}$ sends the output of the experiment to $\mathcal{A}$ and returns its output to $\mathcal{C}$.

Now, if C is a commitment to 0^{len}, then $\mathcal{B}$ perfectly simulates H_2, else it simulates H_3. Thus, if $\mathcal{A}$ can distinguish the outputs of H_4 and H_5 with non-negligible advantage, then $\mathcal{B}$ breaks the hiding of Com.

Lemma 5. *Assuming witness indistinguishability of* NIWI, *hybrid experiments* H_3 *and* H_4 *are computationally indistinguishable.*

Proof. In H_3, we use the *real witness* for proving that c_1^* and c_2^* are encryptions of the same message, while in H_4, we use the *trapdoor witness* for proving that C is a commitment to $(c_1^* \| c_2^* \| vk^*)$. Since NIWI is witness indistinguishable, the two hybrids are computationally indistinguishable.

Lemma 6. *Assuming* $(\mathsf{PKE.Setup}, \mathsf{PKE.Enc}, \mathsf{PKE.Dec})$ *is* $\mathsf{IND\text{-}CPA}$ *secure, hybrid experiments* H_4 *and* H_5 *are computationally indistinguishable.*

Proof. We show that if there exists an efficient distinguisher $\mathcal{A}$ that can distinguish between H_4 and H_5, then there exists an efficient adversary $\mathcal{B}$ that breaks $\mathsf{IND\text{-}CPA}$ security. $\mathcal{B}$ is defined as follows:

1. $\mathcal{B}$ first receives a public key pk from $\mathsf{IND\text{-}CPA}$ challenger $\mathcal{C}$.
2. $\mathcal{B}$ computes $(PK_1, SK_1) \leftarrow \mathsf{PKE.Setup}(1^\kappa)$, $\mathsf{crs} \leftarrow \mathsf{NIWI.Setup}$, $(sk^*, vk^*) \leftarrow \mathsf{Gen}(1^\kappa)$ and sets $PK_2 = pk$. Next, it encrypts the challenge message x using PK_1 to compute ciphertext c_1^*
3. $\mathcal{B}$ sends $(\mathbf{0}, x)$ as its challenge messages to $\mathcal{C}$, and receives a ciphertext c. It sets $c_2^* = c$. Next, it computes the commitment $C = \mathsf{Com}(c_1^* \| c_2^* \| vk^*)$.
4. $\mathcal{B}$ runs the rest of the experiment in the same manner as in H_4 and H_5.
5. Finally, $\mathcal{B}$ sends the output of the experiment to $\mathcal{A}$.
6. If $\mathcal{A}$ outputs H_4, then $\mathcal{B}$ outputs that c is an encryption of x. Else it outputs c is an encryption of $\mathbf{0}$.

Now, if c is an encryption of x, then $\mathcal{B}$ perfectly simulates experiment H_4, else it simulates H_5. Then, clearly, if $\mathcal{A}$'s output is correct, then so is $\mathcal{B}$'s output. Hence, if $\mathcal{A}$ can distinguish the outputs of the two experiments with non negligible advantage, then $\mathcal{B}$ can win the $\mathsf{IND\text{-}CPA}$ game with the same advantage.

Lemma 7. *Assuming* NIWI *is statistically sound,* $i\mathcal{O}$ *is an indistinguishability obfuscator and* Com *is perfectly binding, hybrid experiments* H_5 *and* H_6 *are computationally indistinguishable.*

Proof. As in the proof of Lemma 2, we first argue that both $\mathsf{Sim}.\mathcal{G}_f$ and $\mathsf{Sim}.\mathcal{G}_f'$ have identical input-output behavior.

Observation 2. *For all inputs* $\mathsf{CT} = (c_1, c_2, \pi, vk, \sigma)$, $\mathsf{Sim}.\mathcal{G}_f(\mathsf{CT}) = \bot$ *if and only if* $\mathsf{Sim}.\mathcal{G}'_f(\mathsf{CT}) = \bot$.

Both $\mathsf{Sim}.\mathcal{G}_f$ and $\mathsf{Sim}.\mathcal{G}'_f$ output $\bot$ if and only if either $\mathsf{Verify}(\sigma, c_1\|c_2\|\pi, vk) = 0$ or $\mathsf{NIWI}.\mathsf{Verify}(\mathsf{crs}, y, \pi) = 0$ where $y = (c_1, c_2, vk, PK_1, PK_2, C)$. Therefore, we only need to consider valid inputs. Next, we show that any valid input must satisfy one of the two properties listed below.

Claim 3. *Any valid ciphertext* $\mathsf{CT} = (c_1, c_2, \pi, vk, \sigma)$ *should satisfy one of the following properties :*
- c_1 *and* c_2 *are encryptions of the same message*
- $c_1\|c_2\|vk = c_1^*\|c_2^*\|vk^*$.

Proof. Suppose, on the contrary, there exists a valid input such that it satisfies neither of the properties. Since NIWI is statistically sound, if the input is valid, then the statement $y = (c_1, c_2, vk, PK_1, PK_2, C)$ must have either a real witness or a trapdoor witness. Since c_1 and c_2 are encryptions of different messages, a real witness does not exist. Therefore, for the input to be valid, there must exist a trapdoor witness; that is, there exists an s such that $C = \mathsf{Com}(c_1\|c_2\|vk; s)$. However, since $C = \mathsf{Com}(c_1^*\|c_2^*\|vk^*)$ and Com is perfectly binding, it follows that $(c_1\|c_2\|vk) = (c_1^*\|c_2^*\|vk^*)$. Thus, we have a contradiction.

Using the previous claim, we can now argue that both $\mathsf{Sim}.\mathcal{G}_f$ and $\mathsf{Sim}.\mathcal{G}'_f$ have identical input-output behavior.

Claim 4. *For all valid inputs* $\mathsf{CT} = (c_1, c_2, \pi, vk, \sigma)$, *both* $\mathsf{Sim}.\mathcal{G}_f$ *and* $\mathsf{Sim}.\mathcal{G}'_f$ *have the same functionality.*

Proof. If both c_1 and c_2 are encryptions of the same message, then we have that $\mathsf{PKE}.\mathsf{Dec}(c_1, SK_1) = \mathsf{PKE}.\mathsf{Dec}(c_2, SK_2) = x$. Therefore both programs $\mathsf{Sim}.\mathcal{G}_f$ and $\mathsf{Sim}.\mathcal{G}'_f$ output $f(x; r)$, where $r \leftarrow \mathsf{Eval}(K, c_1\|c_2\|vk) = \mathsf{Eval}(K', c_1\|c_2\|vk)$. If $c_1\|c_2\|vk = c_1^*\|c_2^*\|vk^*$, then both $\mathsf{Sim}.\mathcal{G}_f$ and $\mathsf{Sim}.\mathcal{G}'_f$ output y^*, where y^* is $\mathsf{KeyIdeal}$'s response to query x. Therefore, for all valid inputs, $\mathsf{Sim}.\mathcal{G}_f$ and $\mathsf{Sim}.\mathcal{G}'_f$ have identical input-output behavior.

We now describe our reduction. Assume A_2 and A_3 make a total of ℓ key queries. Consider intermediate hybrids $\mathsf{H}_{5,i}$ $0 \leq i \leq \ell$. In $\mathsf{H}_{5,i}$, we use SK_1 for the first $\ell - i$ key queries, and SK_2 for the remaining i queries. Now, suppose that there exists a PPT distinguisher $\mathcal{A}$ that can distinguish the outputs of $\mathsf{H}_{5,i}$ and $\mathsf{H}_{5,i+1}$. Then, there $\exists$ an adversary $\mathcal{B}$ that can break the security of $i\mathcal{O}$. $\mathcal{B}$ is constructed as follows:

1. $\mathcal{B}$ generates $\mathsf{MPK}, \mathsf{CT}^*$ as in H_5. It sets $st' = SK_1, SK_2, \mathsf{CT}^*$.
2. For the first $(\ell - i - 1)$ key queries by A, $\mathcal{B}$ responds as in H_5. For the last i queries, $\mathcal{B}$ responds as in H_6.
3. For the $(\ell - i)$'th key query f, $\mathcal{B}$ queries $\mathsf{KeyIdeal}$ with f and receives y. Next, it chooses a PRF Key K, computes punctured key $K' \leftarrow \mathsf{Puncture}(K, c_1^*\|c_2^*\|vk^*)$ and defines functions $\mathsf{Sim}.\mathcal{G}_f$ and $\mathsf{Sim}.\mathcal{G}'_f$. $\mathcal{B}$ sends $\mathsf{Sim}.\mathcal{G}_f$ and $\mathsf{Sim}.\mathcal{G}'_f$ to the obfuscation challenger $\mathcal{C}$, receives challenge obfuscation SK_f, which it passes on to A_2.

4. $\mathcal{B}$ runs the rest of the experiment in the same manner as in H_5 and H_6.
5. Finally, $\mathcal{B}$ sends the output of the experiment to $\mathcal{A}$ and forwards $\mathcal{A}$'s response to $\mathcal{C}$.

Now, if $\mathcal{C}$ returns obfuscation of $\mathcal{G}_f$, then $\mathcal{B}$ perfectly simulates experiment $H_{5,i}$, else it simulates experiment $H_{5,i+1}$. Thus, if $\mathcal{A}$ distinguishes the outputs with non negligible advantage, then clearly $\mathcal{B}$ breaks the security of indistinguishability obfuscation with non negligible advantage.

Lemma 8. *Assuming* (Gen, Sign, Verify) *is a strongly unforgeable one time signature scheme,* NIWI *is statistically sound and* Com *is perfectly binding, hybrid experiments* H_6 *and* H_7 *are statistically indistinguishable.*

Proof. As shown in claim 3, any valid ciphertext $CT = (c_1, c_2, \pi, vk, \sigma)$ is such that either c_1 and c_2 are encryptions of the same message or $c_1\|c_2\|vk = c_1^*\|c_2^*\|vk^*$. However, recall that for decryption queries, we only require that $CT \neq CT^*$.

If both c_1 and c_2 encrypt the same value, then clearly the use of SK_1 or SK_2 is indistinguishable. Then, lets consider the case where $c_1\|c_2\|vk = c_1^*\|c_2^*\|vk^*$, yet $CT \neq CT^*$. In this case, it must be that $\pi^*\|\sigma^* \neq \pi\|\sigma$. Now, if $\pi \neq \pi^*$, then since $vk = vk^*$ and $(c_1\|c_2\|\pi) \neq (c_1^*\|c_2^*\|\pi^*)$, we have that σ is a forgery for $(c_1\|c_2\|\pi)$. On the other hand, if $\pi = \pi^*$, then it must be that $\sigma \neq \sigma^*$. In this case, we have that σ is a strong forgery for $(c_1\|c_2\|\pi) = (c_1^*\|c_2^*\|\pi^*)$. We can therefore break the security of the strongly unforgeable one time signature scheme.

Lemma 9. *Assuming* (PKE.Setup, PKE.Enc, PKE.Dec) *is* IND-CPA *secure, hybrid experiments* H_7 *and* H_8 *are computationally indistinguishable.*

Proof. Same as proof for Lemma 6.

Lemma 10. *Assuming* (Gen, Sign, Verify) *is a strongly unforgeable one time signature scheme,* NIWI *is statistically sound and* Com *is perfectly binding, hybrid experiments* H_8 *and* H_9 *are statistically indistinguishable.*

Proof. Same as in proof of Lemma 8.

Lemma 11. *Assuming* (Gen, Sign, Verify) *is a strongly unforgeable one time signature scheme,* NIWI *is statistically sound,* $i\mathcal{O}$ *is indistinguishability obfuscator and* comm *is perfectly binding, hybrid experiments* H_9 *and* H_{10} *are computationally indistinguishable.*

Proof. Same as in proof for Lemma 7.

Lemma 12. *Assuming* (Key, Eval, Puncture) *is a puncturable family of PRFs, hybrid experiments* H_{10} *and* H_{11} *are computationally indistinguishable.*

Proof. In H_{10}, on receiving a decryption query (CT, g), we sample PRF key K and decrypt CT using $sk_g \leftarrow i\mathcal{O}(\mathcal{G}_g)$, where $CT = (c_1, c_2, \pi, vk, \sigma)$ and $\mathcal{G}_g$ uses randomness $r \leftarrow \mathsf{Eval}(K, c_1\|c_2\|vk)$ to compute the output $g(x, \mathsf{Eval}(K, c_1\|c_2\|vk))$.

On the other hand, in H_{11}, the output is computed as $g(x, r) \leftarrow \mathsf{DecryptIdeal}(x, g)$ where $r \xleftarrow{\$} \mathcal{R}_\kappa$.

If there exists an efficient adversary that can distinguish between the outputs of H_{10} and H_{11} with non negligible probability, then there exists an efficient adversary that can distinguish between the output of Eval from a truly random string with non negligible probability, thereby breaking the security of a pseudorandom function.

Acknowledgements. We thank Gil Segev for helpful comments on our security definitions. We also thank Ran Canetti, Shafi Goldwasser, Brent Waters and Xiang Xe for useful discussions.

References

1. Alwen, J., Barbosa, M., Farshim, P., Gennaro, R., Gordon, S.D., Tessaro, S., Wilson, D.A.: On the relationship between functional encryption, obfuscation, and fully homomorphic encryption. In: Stam, M. (ed.) IMACC 2013. LNCS, vol. 8308, pp. 65–84. Springer, Heidelberg (2013)
2. Barak, B., Goldreich, O., Impagliazzo, R., Rudich, S., Sahai, A., Vadhan, S., Yang, K.: On the (Im)possibility of obfuscating programs. In: Kilian, J. (ed.) CRYPTO 2001. LNCS, vol. 2139, pp. 1–18. Springer, Heidelberg (2001)
3. Bellare, M., O'Neill, A.: Semantically-secure functional encryption: Possibility results, impossibility results and the quest for a general definition (2013)
4. Boneh, D., Sahai, A., Waters, B.: Functional encryption: Definitions and challenges. In: Ishai, Y. (ed.) TCC 2011. LNCS, vol. 6597, pp. 253–273. Springer, Heidelberg (2011)
5. Boneh, D., Waters, B.: Conjunctive, subset, and range queries on encrypted data. In: Vadhan, S.P. (ed.) TCC 2007. LNCS, vol. 4392, pp. 535–554. Springer, Heidelberg (2007)
6. Boneh, D., Waters, B.: Constrained pseudorandom functions and their applications. In: Sako, K., Sarkar, P. (eds.) ASIACRYPT 2013, Part II. LNCS, vol. 8270, pp. 280–300. Springer, Heidelberg (2013)
7. Boyle, E., Goldwasser, S., Ivan, I.: Functional signatures and pseudorandom functions. In: Krawczyk, H. (ed.) PKC 2014. LNCS, vol. 8383, pp. 501–519. Springer, Heidelberg (2014)
8. De Caro, A., Iovino, V., Jain, A., O'Neill, A., Paneth, O., Persiano, G.: On the achievability of simulation-based security for functional encryption. In: Canetti, R., Garay, J.A. (eds.) CRYPTO 2013, Part II. LNCS, vol. 8043, pp. 519–535. Springer, Heidelberg (2013)
9. Dolev, D., Dwork, C., Naor, M.: Non-malleable cryptography (extended abstract). In: STOC, pp. 542–552 (1991)
10. Dwork, C., Naor, M., Reingold, O., Rothblum, G.N., Vadhan, S.P.: On the complexity of differentially private data release: efficient algorithms and hardness results. In: STOC, pp. 381–390 (2009)
11. Garg, S., Gentry, C., Halevi, S., Raykova, M., Sahai, A., Waters, B.: Candidate indistinguishability obfuscation and functional encryption for all circuits. In: FOCS (2013)

12. Garg, S., Gentry, C., Halevi, S., Zhandry, M.: Fully secure functional encryption without obfuscation. IACR Cryptology ePrint Archive 2014, 666 (2014), http://eprint.iacr.org/2014/666

13. Gentry, C., Lewko, A.B., Sahai, A., Waters, B.: Indistinguishability obfuscation from the multilinear subgroup elimination assumption. IACR Cryptology ePrint Archive 2014, 309 (2014), http://eprint.iacr.org/2014/309

14. Goldreich, O., Goldwasser, S., Micali, S.: How to construct random functions. J. ACM 33(4), 792–807 (1986), http://doi.acm.org/10.1145/6490.6503

15. Goldwasser, S., Kalai, Y.T., Popa, R.A., Vaikuntanathan, V., Zeldovich, N.: Reusable garbled circuits and succinct functional encryption. In: STOC (2013)

16. Gorbunov, S., Vaikuntanathan, V., Wee, H.: Functional encryption with bounded collusions via multi-party computation. In: Safavi-Naini, R., Canetti, R. (eds.) CRYPTO 2012. LNCS, vol. 7417, pp. 162–179. Springer, Heidelberg (2012)

17. Goyal, V., Jain, A., Koppula, V., Sahai, A.: Functional encryption for randomized functionalities. IACR Cryptology ePrint Archive 2013, 729 (2013)

18. Goyal, V., Pandey, O., Sahai, A., Waters, B.: Attribute-based encryption for fine-grained access control of encrypted data. In: ACM Conference on Computer and Communications Security (2006)

19. Katz, J., Sahai, A., Waters, B.: Predicate encryption supporting disjunctions, polynomial equations, and inner products. In: Smart, N.P. (ed.) EUROCRYPT 2008. LNCS, vol. 4965, pp. 146–162. Springer, Heidelberg (2008)

20. Kiayias, A., Papadopoulos, S., Triandopoulos, N., Zacharias, T.: Delegatable pseudorandom functions and applications. In: ACM CCS (2013)

21. O'Neill, A.: Definitional issues in functional encryption. IACR Cryptology ePrint Archive 2010 (2010)

22. Pass, R., Seth, K., Telang, S.: Indistinguishability obfuscation from semantically-secure multilinear encodings. In: Garay, J.A., Gennaro, R. (eds.) CRYPTO 2014, Part I. LNCS, vol. 8616, pp. 500–517. Springer, Heidelberg (2014)

23. Sahai, A.: Non-malleable non-interactive zero knowledge and adaptive chosen-ciphertext security. In: FOCS, pp. 543–553 (1999)

24. Sahai, A., Seyalioglu, H.: Worry-free encryption: functional encryption with public keys. In: ACM Conference on Computer and Communications Security, pp. 463–472 (2010)

25. Sahai, A., Waters, B.: Fuzzy identity-based encryption. In: Cramer, R. (ed.) EUROCRYPT 2005. LNCS, vol. 3494, pp. 457–473. Springer, Heidelberg (2005)

26. Sahai, A., Waters, B.: How to use indistinguishability obfuscation: Deniable encryption, and more. In: STOC (2014)

27. Waters, B.: A punctured programming approach to adaptively secure functional encryption. IACR Cryptology ePrint Archive 2014, 588 (2014), http://eprint.iacr.org/2014/588

A Correctness of $\mathcal{FE}$

Theorem 3. *If* (Key, Puncture, Eval) *is a PRF, then the proposed scheme* $\mathcal{FE}$ *satisfies correctness.*

Proof. We first prove this theorem for a single key. Fix any $f \in \mathcal{F}_\kappa, x \in \mathcal{X}_\kappa^n$. Consider the distribution $Real_1: \{\text{rFE.Dec}(\text{CT}_i, \text{SK}_f)\}_{i=1}^n$, where $(\text{MPK}, \text{MSK}) \leftarrow$ rFE.Setup(1^κ), $\text{CT}_i = (c_{i,1}, c_{i,2}, \pi_i, vk_i, \sigma_i) \leftarrow$ rFE.Enc(x_i, MPK) for $i \in [n]$

and $K_f \leftarrow$ rFE.Keygen(f, MSK). Similarly, consider the $Ideal_1$ distribution $\{f(x_i, r_i)\}_{i=1}^n$, where $r_i \leftarrow \mathcal{R}_\kappa$.

Claim 5. *Assuming* Eval($\cdot, \cdot$) *is a PRF,* $Real_1$ *and* $Ideal_1$ *distributions are computationally indistinguishable.*

Proof. Note that rFE.Dec($\mathsf{CT}_i, \mathsf{SK}_f$) $= f(x_i, \mathsf{Eval}(K, c_{i,1} \| c_{i,2} \| vk_i))$. Therefore, the $Real_1$ distribution is $\{f(x_i, \mathsf{Eval}(K, c_{i,1} \| c_{i,2} \| vk_i))\}_{i=1}^n$. Suppose there exists an adversary $\mathcal{A}$ that can distinguish between the distributions $Real_1$ and $Ideal_1$ with non-negligible advantage. Then there exists an adversary $\mathcal{B}$ that can break the PRF security of Eval($\cdot, \cdot$). The reduction is as follows :

1. PRF challenger $\mathcal{C}$ chooses a bit $b \leftarrow \{0, 1\}$.
2. For $i = 1$ to n
   (a) $\mathcal{B}$ sends $(c_{i,1} \| c_{i,2} \| vk_i)$ to $\mathcal{C}$, and receives r. If $b = 0$, $r = \mathsf{Eval}(K, c_{i,1} \| c_{i,2} \| vk_i)$, else $r \leftarrow \mathcal{R}_\kappa$.
   (b) $\mathcal{B}$ computes $y_i = f(x_i, r)$.
3. $\mathcal{B}$ sends $\boldsymbol{y}$ to $\mathcal{A}$, and depending on $\mathcal{A}$'s guess, $\mathcal{B}$ outputs 0 or 1.

Clearly, if $\mathcal{A}$ distinguishes between the distributions $Real_1$ and $Ideal_1$ with non-negligible advantage, then $\mathcal{B}$ breaks the PRF security with non-negligible advantage.

This lemma can be extended, via a standard hybrid argument, to prove that the Real and Ideal distributions are computationally indistinguishable.

B   SIM Security Implies IND$_{\mathsf{pre}}$ and IND$_{\mathsf{post}}$ Security

We first prove that 1-SIM security implies one-message IND$_{\mathsf{pre}}$ security. We actually prove the stronger statement that 1-$\widetilde{\mathsf{SIM}}$ security implies one-message IND$_{\mathsf{pre}}$ security where in $\widetilde{\mathsf{SIM}}$ security, the adversary is restricted to making all of the key queries before receiving the public key. Let $x_0, x_1 \in \mathcal{X}_\kappa$ be any two messages. Let $\mathsf{REAL}^0(1^\kappa)$ correspond to real world experiment in Definition 3 where the challenge ciphertext corresponds to the encryption of x_0. From 1-$\widetilde{\mathsf{SIM}}$ security, we have that $\mathsf{REAL}^0(1^\kappa)$ is computationally indistinguishable to $\widetilde{\mathsf{IDEAL}}^0(1^\kappa)$ where $\widetilde{\mathsf{IDEAL}}^0(1^\kappa)$ is the corresponding ideal world in Definition 2. (In particular, in $\widetilde{\mathsf{IDEAL}}^1(1^\kappa)$, the simulator receives the output of every key query f on message x_0.) Now, since Definition 3 requires the promise that $(z, \{f(x_0)\})$ and $(z, \{f(x_1)\})$ are computationally indistinguishable, we have that $\widetilde{\mathsf{IDEAL}}^0(1^\kappa)$ is computationally indistinguishable from $\widetilde{\mathsf{IDEAL}}^1(1^\kappa)$, where $\widetilde{\mathsf{IDEAL}}^1(1^\kappa)$ is defined analogously to $\mathsf{IDEAL}^0(1^\kappa)$.[7] Now, finally, we can invoke 1-$\widetilde{SIM}$-security

---

[7] One may note that since the simulator in our definition performs the key generation in the ideal world, we actually require (MPK, MSK, $z, \{f(x_0)\}$) and (MPK, MSK, $z, \{f(x_1)\}$) to be computationally indistinguishable. This, however, follows immediately since the key queries $\{f\}$ are independent of the public key MPK.

once again to argue that $\widetilde{\mathsf{IDEAL}}^1(1^\kappa)$ and $\mathsf{REAL}^1(1^\kappa)$ are computationally indistinguishable. Combining the above, we have that $\mathsf{REAL}^0(1^\kappa)$ and $\mathsf{REAL}^1(1^\kappa)$ are computationally indistinguishable, as required.

The proof that 1-SIM security implies one-message $\mathsf{IND}_{\mathsf{post}}$ security follows in a similar manner as above. In particular, note that in this case, we have the promise from Definition 4 that $(z, \{f(x_0)\})$ and $(z, \{f(x_1)\})$ are *statistically* indistinguishable. This immediately implies that $(\mathsf{MPK}, \mathsf{MSK}, z, \{f(x_0)\})$ and $(\mathsf{MPK}, \mathsf{MSK}, z, \{f(x_1)\})$ are computationally indistinguishable. The rest of the steps of the proof follow similarly as above.

Functional Encryption for Randomized Functionalities in the Private-Key Setting from Minimal Assumptions

Ilan Komargodski[1,*], Gil Segev[2,**], and Eylon Yogev[1,*]

[1] Weizmann Institute of Science, Rehovot 76100, Israel
{ilan.komargodski,eylon.yogev}@weizmann.ac.il
[2] Hebrew University of Jerusalem, Jerusalem 91904, Israel
segev@cs.huji.ac.il

Abstract. We present a construction of a private-key functional encryption scheme for any family of randomized functionalities based on *any such scheme for deterministic functionalities* that is sufficiently expressive. Instantiating our construction with existing schemes for deterministic functionalities, we obtain schemes for any family of randomized functionalities based on a variety of assumptions (including the LWE assumption, simple assumptions on multilinear maps, and even the existence of any one-way function) offering various trade-offs between security and efficiency.

Previously, Goyal, Jain, Koppula and Sahai [TCC, 2015] constructed a public-key functional encryption scheme for any family of randomized functionalities based on indistinguishability obfuscation.

One of the key insights underlying our work is that, in the private-key setting, a sufficiently expressive functional encryption scheme may be appropriately utilized for implementing proof techniques that were so far implemented based on obfuscation assumptions (such as the punctured programming technique of Sahai and Waters [STOC, 2014]). We view this as a contribution of independent interest that may be found useful in other settings as well.

1 Introduction

The cryptographic community's vision of functional encryption [28,11,27] is rapidly evolving. Whereas traditional encryption schemes offer an all-or-nothing guarantee when accessing encrypted data, functional encryption schemes offer tremendous flexibility. Specifically, such schemes support restricted decryption

* Research supported in part by a grant from the Israel Science Foundation, the I-CORE Program of the Planning and Budgeting Committee, BSF and the Israeli Ministry of Science and Technology.
** Supported by the European Union's Seventh Framework Programme (FP7) via a Marie Curie Career Integration Grant, by the Israel Science Foundation (Grant No. 483/13), and by the Israeli Centers of Research Excellence (I-CORE) Program (Center No. 4/11).

keys that allow users to learn specific functions of the encrypted data and nothing else.

Motivated by the early examples of functional encryption schemes for specific functionalities (such as identity-based encryption [30,8,16]), extensive research has recently been devoted to the construction of functional encryption schemes for rich and expressive families of functions (see, for example, [28,11,27,23,2,7,13,17,18,22,32,19,15,5] and the references therein).

Until very recently, research on functional encryption has focused on the case of *deterministic* functions. More specifically, in a functional encryption scheme for a family $\mathcal{F}$ of deterministic functions, a trusted authority holds a master secret key msk that enables to generate a functional key sk_f for any function $f \in \mathcal{F}$. Now, anyone holding the functional key sk_f and an encryption of some value x, can compute $f(x)$ but cannot learn any additional information about x. In many scenarios, however, dealing only with deterministic functions may be insufficient, and a more general framework allowing *randomized* functions is required.

Functional Encryption for Randomized Functionalities. Motivated by various real-world scenarios, Goyal et al. [24] have recently put forward a generalization of functional encryption to randomized functionalities. In this setting, given a functional key sk_f for a randomized function f and given an encryption of a value x, one should be able to obtain a sample from the distribution $f(x)$. As Goyal et al. pointed out, the case of randomized functions presents new challenges for functional encryption. These challenge arise already when formalizing the security of functional encryption for randomized functions[1], and then become even more noticeable when designing such schemes.

Goyal et al. [24] presented a realistic framework for modeling the security of functional encryption schemes for randomized functionalities. Even more importantly, within their framework they constructed a public-key functional encryption scheme supporting the set of all randomized functionalities (that are computable by bounded-size circuits). Their construction builds upon the elegant approach of punctured programming due to Sahai and Waters [29], and they prove the security of their construction based on indistinguishability obfuscation [6,18].

Identifying the Minimal Assumptions for Functional Encryption. The work of Goyal et al. [24] naturally gives rise to the intriguing question of whether functional encryption for randomized functionalities can be based on assumptions that are seemingly weaker than indistinguishability obfuscation. On one hand, it may be the case that functional encryption for randomized functionalities is indeed a significantly more challenging primitive than functional encryption for deterministic functionalities. In this case, it would be conceivable to use

---

[1] For example, an adversary holding a functional key sk_f and an encryption of a value x, should not be able to tamper with the randomness that is used for sampling from distribution $f(x)$. This is extremely well motivated by the examples provided by Goyal et al. in the contexts of auditing an encrypted database via *randomized* sampling, and of performing differentially-private analysis on an encrypted database via *randomized* perturbations. We refer the reader to [24] for more details.

the full power of indistinguishability obfuscation for constructing such schemes. On the other hand, however, it may be possible that a functional encryption scheme for randomized functions can be constructed in a direct black-box manner from any such scheme for deterministic functions.

This question is especially interesting since various functional encryption schemes for (general) deterministic functionalities are already known to exist based on assumptions that seem significantly weaker than indistinguishability obfuscation (such as Learning with Errors assumption or even the existence of any one-way function) offering various trade-offs between security and efficiency (see Section 2.2 for more details on the existing schemes).

1.1   Our Contributions

In this work we consider functional encryption in the private-key setting, where the master secret key is used both for generating functional keys and for encryption. In this setting we provide an answer to the above question: we present a construction of a private-key functional encryption scheme for any family $\mathcal{F}$ of *randomized* functions based on *any* private-key functional encryption scheme for *deterministic* functions that is sufficiently expressive[2]. Inspired by the work of Goyal et al. [24] in the public-key setting, we prove the security of our construction within a similarly well-motivated framework for capturing the security of private-key functional encryption for randomized functions.

Instantiations. Our resulting scheme inherits the flavor of security guaranteed by the underlying scheme (e.g., full vs. selective security, and one-key vs. many-keys security), and can be instantiated by a variety of existing functional encryption schemes. Specifically, our scheme can be based either on the Learning with Errors assumption, on obfuscation assumptions, on multilinear-maps assumptions, or even on the existence of any one-way function (offering various trade-offs between security and efficiency – we refer the reader to Section 2.2 for more details on the possible instantiations).

Applicable Scenarios. Following-up on the motivating applications given by Goyal et al. [24] in the contexts of auditing an encrypted database via *randomized* sampling, and of performing differentially-private analysis on an encrypted database via *randomized* perturbations, we observe that these two examples are clearly valid in the private-key setting as well. Specifically, in both applications, the party that provides functional keys is more than likely the same one who encrypts the data.

Obfuscation-Based Techniques via Function Privacy. One of the key insights underlying our work is that in the private-key setting, where encryption is performed honestly by the owner of the master secret key, the power of indistinguishability obfuscation may not be needed. Specifically, we observe that in some

---

[2] Our only assumption on the underlying scheme is that it supports the family $\mathcal{F}$ (when viewed as a family of single-input deterministic functions), supports the evaluation procedure of a pseudorandom function family, and supports a few additional basic operations (such as conditional statements).

cases one can instead rely on the weaker notion of *function privacy* [31,9,1,15]. Intuitively, a functional encryption scheme is function private if a functional key sk_f for a function f reveals no "unnecessary" information on f. For functional encryption in the private-key setting, this essentially means that encryptions of messages $m_1, \ldots, m_T$ together with functional keys corresponding to functions $f_1, \ldots, f_T$ reveal essentially no information other than the values $\{f_i(m_j)\}_{i,j \in [T]}$. Brakerski and Segev [15] recently showed that a function-private scheme can be obtained from *any* private-key functional encryption scheme.

Building upon the notion of function privacy, we show that any *private-key* functional encryption scheme may be appropriately utilized for implementing some of the proof techniques that were so far implemented based on indistinguishability obfuscation. These include, in particular, a variant of the punctured programming approach of Sahai and Waters [29]. We view this as a contribution of independent interest that may be found useful in other settings as well.

1.2 Additional Related Work

A related generalization of functional encryption is that of functional encryption for *multiple-input* functions due to Goldwasser et al. [21]. A multiple-input functional encryption scheme for a function family $\mathcal{F}$ allows generating a functional key sk_f for any function $f \in \mathcal{F}$, and this enables to compute $f(x, y)$ given an encryption of x and an encryption of y, while not learning any additional information. Although capturing the security guarantees that can be provided by such schemes is quite challenging, multiple-input functional encryption might be useful for dealing with single-input randomized functionalities: One can view a randomized function $f(x; r)$ as a two-input function, where its first input is the actual input x, and its second input is the randomness r (that is possibly derived by a PRF key). However, the construction of Goldwasser et al. is based on indistinguishability obfuscation, and our goal is to rely on weaker assumptions. In addition, it is not clear that the notion of security of Goldwasser et al. suffices for capturing our notion of "best-possible" message privacy which allows for an a-priori non-negligible advantage in distinguishing the output distributions of two randomized functions (see Sections 1.3 and 3 for our notion of privacy).

Our construction relies on the notion of function privacy for functional encryption schemes, first introduced by Boneh et al. [9,10] in the public-key setting, and then studied by Agrawal et al. [1] and by Brakerski and Segev [15] in the private-key setting (generalizing the work on predicate privacy in the private-key setting by Shen et al. [31]). As discussed in Section 1.1, for functional encryption in the private-key setting, function privacy essentially means that encryptions of messages $m_1, \ldots, m_T$ together with functional keys corresponding to functions $f_1, \ldots, f_T$ reveal essentially no information other than the values $\{f_i(m_j)\}_{i,j \in [T]}$. In terms of underlying assumptions, we rely on the fact that Brakerski and Segev [15] showed that a function-private scheme can be obtained from any private-key functional encryption scheme.

Lastly, Alwen et al. [3] studied the relationship between functional encryption and fully homomorphic encryption. In their work, they define the notion of a public-key multi-input functional encryption scheme for randomized functionalities, and construct such a scheme assuming a public-key multi-input function encryption scheme for deterministic functionalities. This result is somewhat incomparable to ours since *public-key multi-input* functional encryption seems significantly stronger than the assumptions underlying our approach (e.g., it is known to imply indistinguishability obfuscation [21]).

1.3   Overview of Our Approach

A private-key functional encryption scheme for a family $\mathcal{F}$ of randomized functions consists of four probabilistic polynomial-time algorithms (Setup, KG, Enc, Dec). The syntax is identical to that of functional encryption for deterministic functions (see Section 2.2), but the correctness and security requirements are more subtle. In this section we begin with a brief overview of our notions of correctness and security. Then, we provide a high-level overview of our new construction, and the main ideas and challenges underlying its proof of security.

Correctness and Independence of Decrypted Values. Our notion of correctness follows that of Goyal et al. [24] by adapting it to the private-key setting. Specifically, we ask that for any sequence of messages $x_1, \ldots, x_T$ and for any sequence of functions $f_1, \ldots, f_T \in \mathcal{F}$, it holds that the distribution obtained by encrypting $x_1, \ldots, x_T$ and then decrypting the resulting ciphertexts with functional keys corresponding to $f_1, \ldots, f_T$ is computationally indistinguishable from the distribution $\{f_j(x_i; r_{i,j})\}_{i,j \in [T]}$ where the $r_{i,j}$'s are sampled independently and uniformly at random. As noted by Goyal et al. [24], unlike in the case of deterministic functions where is suffices to define correctness for a single ciphertext and a single key, here it is essential to define correctness for multiple (possibly correlated) ciphertexts and keys. We refer the reader to Section 3.1 for our formal definition.

"Best-Possible" Message Privacy. As in functional encryption for deterministic functions, we consider adversaries whose goal is to distinguish between encryptions of two challenge messages, x_0^* and x_1^*, when given access to an encryption oracle (as required in private-key encryption) and to functional keys of various functions. Recall that in the case of deterministic functions, the adversary is allowed to ask for functional keys for any function f such that $f(x_0^*) = f(x_1^*)$.

When dealing with randomized functions, however, it is significantly less clear how to prevent adversaries from choosing functions f that will enable to easily distinguish between encryptions of x_0^* and x_1^*. Our notions of message privacy ask that the functional encryption scheme under consideration will not add a non-negligible advantage to the (*possibly non-negligible*) advantage that adversaries may already have in distinguishing between the distributions $f(x_0^*)$ and $f(x_1^*)$. That is, given that adversaries are able to obtain a sample from the distribution $f(x_0^*)$ or from the distribution $f(x_1^*)$ using the functional key sk_f, and may already have some advantage in distinguishing these distributions, we ask for "best-possible" message privacy in the sense that essentially *no additional advantage can be gained*.

Concretely, if the distributions $f(x_0^*)$ and $f(x_1^*)$ can be efficiently distinguished with advantage at most $\Delta = \Delta(\lambda)$ to begin with (*where Δ does not necessarily have to be negligible*), then we require that no adversary that is given a functional key for f will be able to distinguish between encryptions of x_0^* and x_1^* with advantage larger than $\Delta + \mathsf{neg}(\lambda)$, for some negligible function $\mathsf{neg}(\cdot)$. More generally, an adversary that is given functional keys for $T = T(\lambda)$ such functions (and access to an encryption oracle), should not be able to distinguish between encryptions of x_0^* and x_1^* with advantage larger than $T \cdot \Delta + \mathsf{neg}(\lambda)$. We note that our approach for realistically capturing message privacy somewhat differs from that of Goyal et al. [24], and we refer the reader to the full version [26] for a brief comparison between the two approaches[3].

We put forward two flavors of "best-possible" message privacy, a non-adaptive flavor and an adaptive flavor, depending on the flavor of indistinguishability guarantee that is satisfied by the function family under consideration. Details follow.

Out first notion addresses function families $\mathcal{F}$ such that for a randomly sampled $f \leftarrow \mathcal{F}$, no efficient adversary given f can output x_0 and x_1 and distinguish the distributions $f(x_0)$ and $f(x_1)$ with probability larger than Δ (note again that Δ does not have to be negligible). One possible example for such a function family is a function that on input x samples a public-key pk for a public-key encryption scheme, and outputs pk together with a randomized encryption of x. Our second notion addresses function families $\mathcal{F}$ such that no efficient adversary can output $f \in \mathcal{F}$ together with two inputs x_0 and x_1, and distinguish the distributions $f(x_0)$ and $f(x_1)$ with probability larger than Δ. One possible example for such a function family is that of differentially private mechanisms, as discussed by Goyal et al. [24]. We refer the reader to Section 3.2 for more information and the formal definitions.

Our Construction. Let $(\mathsf{Setup}, \mathsf{KG}, \mathsf{Enc}, \mathsf{Dec})$ be any private-key functional encryption scheme that provides message privacy and function privacy[4]. Our new scheme is quite intuitive and is described as follows:

- The setup and decryption algorithms are identical to those of the underlying scheme.
- The encryption algorithm on input a message x, samples a string s uniformly at random, and outputs an encryption $\mathsf{ct} \leftarrow \mathsf{Enc}(\mathsf{msk}, (x, \perp, s, \perp))$ of x and s together with two additional "empty slots" that will be used in the security proof.

---

[3] We emphasize that we view the main contribution of our paper as basing the security of our scheme on any underlying functional encryption scheme (and avoiding obfuscation-related assumptions), and not as offering alternative notions of message privacy.

[4] As discussed above, function privacy can be assumed without loss of generality using the transformation of Brakerski and Segev [15].

– The key-generation algorithm on input a description of a randomized function f, samples a PRF key K, and outputs a functional key for the deterministic function $\mathsf{Left}_{f,K}$ defined as follows: On input (x_L, x_R, s, z) output $f(x_L; r)$ where $r = \mathsf{PRF}_K(s)$.

The correctness and independence of our scheme follow in a straightforward manner from the correctness of the underlying scheme and the assumption that PRF is pseudorandom. In fact, it suffices that PRF is *weakly* pseudorandom (i.e., computationally indistinguishable from a truly random function when evaluated on independent and uniformly sampled inputs).

As for the message privacy of the scheme, recall that we consider adversaries that can access an encryption oracle and a key-generation oracle, and should not be able to distinguish between an encryption $\mathsf{Enc}(\mathsf{msk}, (x_0^*, \perp, s^*, \perp))$ of x_0^* and an encryption $\mathsf{Enc}(\mathsf{msk}, (x_1^*, \perp, s^*, \perp))$ of x_1^* with advantage larger than $T \cdot \Delta + \mathsf{neg}(\lambda)$ (where T is the number of functional keys given to the adversary, and Δ is the a-priori distinguishing advantage for the functions under consideration as described above).

The first step in our proof of security is to replace the challenge ciphertext with a modified challenge ciphertext $\mathsf{Enc}(\mathsf{msk}, (x_0^*, x_1^*, s^*, \perp))$ that contains information on both challenge messages (this is made possible due to the message privacy of the underlying scheme). Next, denoting the adversary's key-generation queries by $f_1, \ldots, f_T$, our goal is to replace the functional keys $\mathsf{Left}_{f_1, K_1}, \ldots, \mathsf{Left}_{f_T, K_T}$ with the functional keys $\mathsf{Right}_{f_1, K_1}, \ldots, \mathsf{Right}_{f_T, K_T}$, where the function $\mathsf{Right}_{f,K}$ is defined as follows: On input (x_L, x_R, s, z) output $f(x_R; r)$ where $r = \mathsf{PRF}_K(s)$. At this point we note that, from the adversary's point of view, when providing only Left keys the modified challenge ciphertext is indistinguishable from an encryption of x_0^*, and when providing only Right keys the modified challenge ciphertext is indistinguishable from an encryption of x_1^*.

The most challenging part of the proof is in bounding the adversary's advantage in distinguishing the sequences of Left and Right keys, based on the function privacy and the message privacy of the underlying scheme. The basic idea is to switch the functional keys from Left to Right one by one, following different proof strategies for pre-challenge keys and for post-challenge keys[5].

When dealing with a pre-challenge key sk_f, the function f is already known when producing the challenge ciphertext. Therefore, we can use the message privacy of the underlying scheme and replace the (already-modified) challenge ciphertext with $\mathsf{Enc}(\mathsf{msk}, (x_0^*, x_1^*, s^*, z^*))$, where $z^* = f(x_0^*; r^*)$ and $r^* = \mathsf{PRF}_K(s^*)$. Then, we use the function privacy of the underlying scheme, and replace the functional key $\mathsf{Left}_{f,K}$ with a functional key for the function $\mathsf{OutputZ}$ that simply outputs z whenever $s = s^*$. From this point on, we use the pseudorandomness of PRF and replace $r^* = \mathsf{PRF}_K(s^*)$ with a truly uniform r^*, and then replace $z^* \leftarrow f(x_0^*)$ with $z^* \leftarrow f(x_1^*)$. Similar steps then enable us to replace the functional key $\mathsf{OutputZ}$ with a functional key for the function $\mathsf{Right}_{f,K}$.

---

[5] We use the term *pre-challenge* keys for all functional keys that are obtained before the challenge phase, and the term *post-challenge* keys for all functional keys that are obtained after the challenge phase.

When dealing with a post-challenge key sk_f, we would like to follow the same approach of embedding the value $f(x_0^*; r^*)$ or $f(x_1^*; r^*)$. However, for post-challenge keys, the function f is not known when producing the challenge ciphertext. Instead, in this case, the challenge messages x_0^* and x_1^* are known when producing the functional key sk_f. Combining this with the function privacy of the underlying scheme enables us to embed the above values in the functional key sk_f, and once again replace the Left keys with the Right keys. We refer the reader to Section 4 for the formal description of our scheme and its proof of security.

1.4 Paper Organization

The remainder of this paper is organized as follows. In Section 2 we provide an overview of the basic notation and standard tools underlying our construction. In Section 3 we introduce our notions of security for private-key functional encryption schemes for randomized functionalities. In Section 4 we present our new scheme and prove its security. Formal proofs of the claims that are stated in Section 4 can be found in the full version [26].

2 Preliminaries

In this section we present the notation and basic definitions that are used in this work. For a distribution X we denote by $x \leftarrow X$ the process of sampling a value x from the distribution X. Similarly, for a set $\mathcal{X}$ we denote by $x \leftarrow \mathcal{X}$ the process of sampling a value x from the uniform distribution over $\mathcal{X}$. For a randomized function f and an input $x \in \mathcal{X}$, we denote by $y \leftarrow f(x)$ the process of sampling a value y from the distribution $f(x)$. For an integer $n \in \mathbb{N}$ we denote by $[n]$ the set $\{1, \ldots, n\}$. A function $\mathsf{neg} : \mathbb{N} \to \mathbb{R}$ is *negligible* if for every constant $c > 0$ there exists an integer N_c such that $\mathsf{neg}(\lambda) < \lambda^{-c}$ for all $\lambda > N_c$.

The *statistical distance* between two random variables X and Y over a finite domain Ω is defined as $\mathsf{SD}(X, Y) = \frac{1}{2} \sum_{\omega \in \Omega} |\Pr[X = \omega] - \Pr[Y = \omega]|$. Two sequences of random variables $X = \{X_\lambda\}_{\lambda \in \mathbb{N}}$ and $Y = \{Y_\lambda\}_{\lambda \in \mathbb{N}}$ are *computationally indistinguishable* if for any probabilistic polynomial-time algorithm $\mathcal{A}$ there exists a negligible function $\mathsf{neg}(\cdot)$ such that

$$\left| \Pr[\mathcal{A}(1^\lambda, X_\lambda) = 1] - \Pr[\mathcal{A}(1^\lambda, Y_\lambda) = 1] \right| \leq \mathsf{neg}(\lambda)$$

for all sufficiently large $\lambda \in \mathbb{N}$.

2.1 Pseudorandom Functions

Let $\{\mathcal{K}_\lambda, \mathcal{X}_\lambda, \mathcal{Y}_\lambda\}_{\lambda \in \mathbb{N}}$ be a sequence of sets and let $\mathsf{PRF} = (\mathsf{PRF.Gen}, \mathsf{PRF.Eval})$ be a function family with the following syntax:

- $\mathsf{PRF.Gen}$ is a probabilistic polynomial-time algorithm that takes as input the unary representation of the security parameter λ, and outputs a key $K \in \mathcal{K}_\lambda$.

– PRF.Eval is a deterministic polynomial-time algorithm that takes as input a key $K \in \mathcal{K}_\lambda$ and a value $x \in \mathcal{X}_\lambda$, and outputs a value $y \in \mathcal{Y}_\lambda$.

The sets $\mathcal{K}_\lambda$, $\mathcal{X}_\lambda$, and $\mathcal{Y}_\lambda$ are referred to as the *key space, domain,* and *range* of the function family, respectively. For easy of notation we may denote by PRF.Eval$_K(\cdot)$ or PRF$_K(\cdot)$ the function PRF.Eval$(K, \cdot)$ for $K \in \mathcal{K}_\lambda$. The following is the standard definition of a pseudorandom function family.

Definition 1 (Pseudorandomness). *A function family* PRF $=$ (PRF.Gen, PRF.Eval) *is* pseudorandom *if for every probabilistic polynomial-time algorithm $\mathcal{A}$ there exits a negligible function* neg$(\cdot)$ *such that*

$$\mathsf{Adv}_{\mathsf{PRF},\mathcal{A}}(\lambda) \overset{\text{def}}{=} \left| \Pr_{K \leftarrow \mathsf{PRF.Gen}(1^\lambda)} \left[\mathcal{A}^{\mathsf{PRF.Eval}_K(\cdot)}(1^\lambda) = 1 \right] - \Pr_{f \leftarrow F_\lambda} \left[\mathcal{A}^{f(\cdot)}(1^\lambda) = 1 \right] \right| \leq$$
$$\mathsf{neg}(\lambda),$$

for all sufficiently large $\lambda \in \mathbb{N}$, where F_λ is the set of functions that map $\mathcal{X}_\lambda$ into $\mathcal{Y}_\lambda$.

In addition to the standard notion of a pseudorandom function family, we rely on the seemingly stronger (yet existentially equivalent) notion of a *puncturable* pseudorandom function family [25,12,29,14]. In terms of syntax, this notion asks for an additional probabilistic polynomial-time algorithm, PRF.Punc, that takes as input a key $K \in \mathcal{K}_\lambda$ and a set $S \subseteq \mathcal{X}_\lambda$ and outputs a "punctured" key K_S. The properties required by such a puncturing algorithm are capture by the following definition.

Definition 2 (Puncturable PRF). *A pseudorandom function family* PRF $=$ (PRF.Gen, PRF.Eval, PRF.Punc) *is* puncturable *if the following properties are satisfied:*

1. **Functionality:** *For all sufficiently large $\lambda \in \mathbb{N}$, for every set $S \subseteq \mathcal{X}_\lambda$, and for every $x \in \mathcal{X}_\lambda \setminus S$ it holds that*

$$\Pr_{\substack{K \leftarrow \mathsf{PRF.Gen}(1^\lambda); \\ K_S \leftarrow \mathsf{PRF.Punc}(K,S)}} [\mathsf{PRF.Eval}_K(x) = \mathsf{PRF.Eval}_{K_S}(x)] = 1.$$

2. **Pseudorandomness at Punctured Points:** *Let $\mathcal{A} = (\mathcal{A}_1, \mathcal{A}_2)$ be any probabilistic polyomial-time algorithm such that $\mathcal{A}_1(1^\lambda)$ outputs a set $S \subseteq \mathcal{X}_\lambda$, a value $x \in S$, and state information* state. *Then, for any such $\mathcal{A}$ there exists a negligible function* neg$(\cdot)$ *such that*

$$\mathsf{Adv}_{\mathsf{puPRF},\mathcal{A}}(\lambda) \overset{\text{def}}{=}$$
$$|\Pr[\mathcal{A}_2(K_S, \mathsf{PRF.Eval}_K(x), \mathsf{state}) = 1] - \Pr[\mathcal{A}_2(K_S, y, \mathsf{state}) = 1]| \leq \mathsf{neg}(\lambda)$$

for all sufficiently large $\lambda \in \mathbb{N}$, where $(S, x, \mathsf{state}) \leftarrow \mathcal{A}_1(1^\lambda)$, $K \leftarrow \mathsf{PRF.Gen}(1^\lambda)$, $K_S = \mathsf{PRF.Punc}(K, S)$, and $y \leftarrow \mathcal{Y}_\lambda$.

As observed by [25,12,29,14] the GGM construction [20] of PRFs from one-way functions can be easily altered to yield a puncturable PRF.

2.2 Private-Key Functional Encryption

A private-key functional encryption scheme over a message space $\mathcal{X} = \{\mathcal{X}_\lambda\}_{\lambda \in \mathbb{N}}$ and a function space $\mathcal{F} = \{\mathcal{F}_\lambda\}_{\lambda \in \mathbb{N}}$ is a quadruple (Setup, KG, Enc, Dec) of probabilistic polynomial-time algorithms. The setup algorithm Setup takes as input the unary representation 1^λ of the security parameter $\lambda \in \mathbb{N}$ and outputs a master-secret key msk. The key-generation algorithm KG takes as input a master-secret key msk and a function $f \in \mathcal{F}_\lambda$, and outputs a functional key sk_f. The encryption algorithm Enc takes as input a master-secret key msk and a message $x \in \mathcal{X}_\lambda$, and outputs a ciphertext ct. In terms of correctness we require that for all sufficiently large $\lambda \in \mathbb{N}$, for every function $f \in \mathcal{F}_\lambda$ and message $x \in \mathcal{X}_\lambda$ it holds that $\mathsf{Dec}(\mathsf{KG}(\mathsf{msk}, f), \mathsf{Enc}(\mathsf{msk}, x)) = f(x)$ with all but a negligible probability over the internal randomness of the algorithms Setup, KG, and Enc.

In terms of security, we rely on the private-key variants existing indistinguishability based notions for message privacy (see, for example, [11,27,7]) and function privacy (see [1,15]). When formalizing these notions it would be convenient to use the following standard notion of a *left-or-right oracle*.

Definition 3 (Left-or-Right Oracle). *Let $\mathcal{O}(\cdot, \cdot)$ be a probabilistic two-input functionality. For each $b \in \{0, 1\}$ we denote by $\mathcal{O}_b$ the probabilistic three-input functionality $\mathcal{O}_b(k, z_0, z_1) \stackrel{\mathrm{def}}{=} \mathcal{O}(k, z_b)$.*

Message Privacy

A functional encryption scheme is message private if the encryptions of any two messages x_0 and x_1 are computationally indistinguishable given access to an encryption oracle (as required in private-key encryption) and to functional keys for any function f such that $f(x_0^*) = f(x_1^*)$. We consider two variants of message privacy: (*full*) message privacy in which adversaries are fully adaptive, and *selective-function* message privacy in which adversaries must issue their key-generation queries in advance.

Definition 4 (Message Privacy). *A functional encryption scheme $\mathsf{FE} = ($ Setup, KG, Enc, Dec) over a message space $\mathcal{X} = \{\mathcal{X}_\lambda\}_{\lambda \in \mathbb{N}}$ and a function space $\mathcal{F} = \{\mathcal{F}_\lambda\}_{\lambda \in \mathbb{N}}$ is message private if for any probabilistic polynomial-time adversary $\mathcal{A}$ there exists a negligible function $\mathsf{neg}(\cdot)$ such that*

$$\mathsf{Adv}_{\mathsf{FE}, \mathcal{A}, \mathcal{F}}^{\mathsf{MP}}(\lambda) \stackrel{\mathrm{def}}{=}$$
$$\left| \Pr\left[\mathcal{A}^{\mathsf{KG}(\mathsf{msk}, \cdot), \mathsf{Enc}_0(\mathsf{msk}, \cdot, \cdot)}(1^\lambda) = 1 \right] - \Pr\left[\mathcal{A}^{\mathsf{KG}(\mathsf{msk}, \cdot), \mathsf{Enc}_1(\mathsf{msk}, \cdot, \cdot)}(1^\lambda) = 1 \right] \right|$$
$$\leq \mathsf{neg}(\lambda)$$

for all sufficiently large $\lambda \in \mathbb{N}$, where for every $(x_0, x_1) \in \mathcal{X}_\lambda \times \mathcal{X}_\lambda$ and $f \in \mathcal{F}_\lambda$ with which $\mathcal{A}$ queries the oracles Enc_b and KG, respectively, it holds that $f(x_0) = f(x_1)$. Moreover, the probability is taken over the choice of $\mathsf{msk} \leftarrow \mathsf{Setup}(1^\lambda)$ and the internal randomness of $\mathcal{A}$.

Definition 5 (Selective-Function Message Privacy). *A functional encryption scheme* $\mathsf{FE} = (\mathsf{Setup}, \mathsf{KG}, \mathsf{Enc}, \mathsf{Dec})$ *over a message space* $\mathcal{X} = \{\mathcal{X}_\lambda\}_{\lambda \in \mathbb{N}}$ *and a function space* $\mathcal{F} = \{\mathcal{F}_\lambda\}_{\lambda \in \mathbb{N}}$ *is* T-*selective-function message private, where* $T = T(\lambda)$, *if for any probabilistic polynomial-time adversary* $\mathcal{A} = (\mathcal{A}_1, \mathcal{A}_2)$ *there exists a negligible function* $\mathsf{neg}(\cdot)$ *such that*

$$\mathsf{Adv}^{\mathsf{sfMP}}_{\mathsf{FE}, \mathcal{A}, \mathcal{F}, T}(\lambda) \stackrel{\mathsf{def}}{=}$$
$$\left| \Pr \left[\mathsf{Expt}^{(0)}_{\mathsf{FE}, \mathcal{A}, \mathcal{F}, T}(\lambda) = 1 \right] - \Pr \left[\mathsf{Expt}^{(1)}_{\mathsf{FE}, \mathcal{A}, \mathcal{F}, T}(\lambda) = 1 \right] \right| \leq \mathsf{neg}(\lambda)$$

for all sufficiently large $\lambda \in \mathbb{N}$, *where for each* $b \in \{0, 1\}$ *and* $\lambda \in \mathbb{N}$ *the random variable* $\mathsf{Expt}^{(b)}_{\mathsf{FE}, \mathcal{A}, \mathcal{F}, T}(\lambda)$ *is defined as follows:*

1. $\mathsf{msk} \leftarrow \mathsf{Setup}(1^\lambda)$.
2. $(f_1, \ldots, f_T, \mathsf{state}) \leftarrow \mathcal{A}_1(1^\lambda)$, *where* $f_i \in \mathcal{F}_\lambda$ *for all* $i \in [T]$.
3. $\mathsf{sk}_{f_i} \leftarrow \mathsf{KG}(\mathsf{msk}, f_i)$ *for all* $i \in [T]$.
4. $b' \leftarrow \mathcal{A}_2^{\mathsf{Enc}_b(\mathsf{msk}, \cdot, \cdot)}(\mathsf{sk}_{f_1}, \ldots, \mathsf{sk}_{f_T}, \mathsf{state})$, *where for each of* $\mathcal{A}_2$'s *queries* (x_0, x_1) $\in \mathcal{X}_\lambda \times \mathcal{X}_\lambda$ *to* $\mathsf{Enc}_b(\mathsf{msk}, \cdot, \cdot)$ *it holds that* $f_i(x_0) = f_i(x_1)$ *for all* $i \in [T]$.
5. *Output* b'.

Such a scheme is selective-function message private *if it is* T-*selective-function message private for all polynomials* $T = T(\lambda)$.

Known Constructions. Private-key functional encryption schemes that satisfy the notions presented in Definitions 4 and 5 (and support circuits of any a-priori bounded polynomial size) are known to exist based on various assumptions. The known schemes are in fact public-key schemes, which are in particular private-key ones.

Specifically, a public-key scheme that satisfies the notion of 1-selective function message privacy was constructed by Gorbunov, Vaikuntanathan and Wee [23] under the sole assumption that public-key encryption exists. In the private-key setting, their transformation can in fact rely on any private-key encryption scheme (and thus on any one-way function). By assuming, in addition, the existence of a pseudorandom generator computable by small-depth circuits (which is known to be implied by most concrete intractability assumptions), they construct a scheme that satisfies the notion of T-selective-function message privacy for any predetermined polynomial $T = T(\lambda)$. However, the length of the ciphertexts in their scheme grows linearly with T and with an upper bound on the circuit size of the functions that the scheme allows (which also has to be known ahead of time). Goldwasser et al. [22] showed that based on the Learning with Errors (LWE) assumption, T-selective-function message privacy can be achieved where the ciphertext size grows with T and with a bound on the depth of allowed functions.

In addition, schemes that satisfy the notion of (full) message privacy (Definition 4) were constructed by Boyle et al. [13] and by Ananth et al. [4] based on differing-input obfuscation, by Waters [32] based on indistinguishability obfuscation, and by Garg et al. [19] based on multilinear maps. Very recently, Ananth et al. [5] gave a generic transformation from selective-message message privacy to full message privacy. We conclude that there is a variety of constructions offering various flavors of security under various assumptions that can be used as a building block in our construction.

Function Privacy

A private-key functional-encryption scheme is function private [31,1,15] if a functional key sk_f for a function f reveals no "unnecessary" information on f. More generally, we ask that encryptions of messages $m_1, \ldots, m_T$ together with functional keys corresponding to functions $f_1, \ldots, f_T$ reveal essentially no information other than the values $\{f_i(m_j)\}_{i,j \in [T]}$. We consider two variants of function privacy: (*full*) function privacy in which adversaries are fully adaptive, and *selective-function* function privacy in which adversaries must issue their key-generation queries in advance.

Definition 6 (Function Privacy). *A functional encryption scheme* $\mathsf{FE} = ($ $\mathsf{Setup}, \mathsf{KG}, \mathsf{Enc}, \mathsf{Dec})$ *over a message space* $\mathcal{X} = \{\mathcal{X}_\lambda\}_{\lambda \in \mathbb{N}}$ *and a function space* $\mathcal{F} = \{\mathcal{F}_\lambda\}_{\lambda \in \mathbb{N}}$ *is* function private *if for any probabilistic polynomial-time adversary* $\mathcal{A}$ *there exists a negligible function* $\mathsf{neg}(\cdot)$ *such that*

$$\mathsf{Adv}^{\mathsf{FP}}_{\mathsf{FE}, \mathcal{A}, \mathcal{F}}(\lambda) \stackrel{\mathsf{def}}{=}$$
$$\left| \Pr\left[\mathcal{A}^{\mathsf{KG}_0(\mathsf{msk}, \cdot, \cdot), \mathsf{Enc}_0(\mathsf{msk}, \cdot, \cdot)}(1^\lambda) = 1 \right] - \Pr\left[\mathcal{A}^{\mathsf{KG}_1(\mathsf{msk}, \cdot, \cdot), \mathsf{Enc}_1(\mathsf{msk}, \cdot, \cdot)}(1^\lambda) = 1 \right] \right|$$
$$\leq \mathsf{neg}(\lambda)$$

for all sufficiently large $\lambda \in \mathbb{N}$, *where for every* $(f_0, f_1) \in \mathcal{F}_\lambda \times \mathcal{F}_\lambda$ *and* $(x_0, x_1) \in \mathcal{X}_\lambda \times \mathcal{X}_\lambda$ *with which* $\mathcal{A}$ *queries the oracles* KG_b *and* Enc_b, *respectively, it holds that* $f_0(x_0) = f_1(x_1)$. *Moreover, the probability is taken over the choice of* $\mathsf{msk} \leftarrow \mathsf{Setup}(1^\lambda)$ *and the internal randomness of* $\mathcal{A}$.

Definition 7 (Selective-Function Function Privacy). *A functional encryption scheme* $\mathsf{FE} = (\mathsf{Setup}, \mathsf{KG}, \mathsf{Enc}, \mathsf{Dec})$ *over a message space* $\mathcal{X} = \{\mathcal{X}_\lambda\}_{\lambda \in \mathbb{N}}$ *and a function space* $\mathcal{F} = \{\mathcal{F}_\lambda\}_{\lambda \in \mathbb{N}}$ *is said* T-selective-function function private, *where* $T = T(\lambda)$, *if for any probabilistic polynomial-time adversary* $\mathcal{A} = (\mathcal{A}_1, \mathcal{A}_2)$ *there exists a negligible function* $\mathsf{neg}(\cdot)$ *such that*

$$\mathsf{Adv}^{\mathsf{sfFP}}_{\mathsf{FE}, \mathcal{A}, \mathcal{F}, T}(\lambda) \stackrel{\mathsf{def}}{=}$$
$$\left| \Pr\left[\mathsf{Expt}^{(0)}_{\mathsf{FE}, \mathcal{A}, \mathcal{F}, T}(\lambda) = 1 \right] - \Pr\left[\mathsf{Expt}^{(1)}_{\mathsf{FE}, \mathcal{A}, \mathcal{F}, T}(\lambda) = 1 \right] \right| \leq \mathsf{neg}(\lambda),$$

for all sufficiently large $\lambda \in \mathbb{N}$, *where for each* $b \in \{0, 1\}$ *and* $\lambda \in \mathbb{N}$ *the random variable* $\mathsf{Expt}^{(b)}_{\mathsf{FE}, \mathcal{A}, \mathcal{F}, T}(\lambda)$ *is defined as follows:*

1. $\mathsf{msk} \leftarrow \mathsf{Setup}(1^\lambda)$.
2. $((f_{0,1}, \ldots, f_{0,T}), (f_{1,1}, \ldots, f_{1,T}), \mathsf{state}) \leftarrow \mathcal{A}_1(1^\lambda)$, *where* $f_{\sigma,i} \in \mathcal{F}_\lambda$ *for all* $\sigma \in \{0, 1\}$ *and* $i \in [T]$.

3. $\mathsf{sk}_i^* \leftarrow \mathsf{KG}(\mathsf{msk}, f_{b,i})$ *for all* $i \in [T]$.
4. $b' \leftarrow \mathcal{A}_2^{\mathsf{Enc}_b(\mathsf{msk},\cdot,\cdot)}(\mathsf{sk}_1^*, \ldots, \mathsf{sk}_T^*, \mathsf{state})$, *where for each query* $(x_0, x_1) \in \mathcal{X}_\lambda \times \mathcal{X}_\lambda$ *to* $\mathsf{Enc}_b(\mathsf{msk}, \cdot, \cdot)$ *it holds that* $f_{0,i}(x_0) = f_{1,i}(x_1)$ *for all* $i \in [T]$.
5. *Output* b'.

Such a scheme is selective-function function private *if it is* T-*selective-function function private for all polynomials* $T = T(\lambda)$.

Known Constructions. Brakerski and Segev [15] showed how to transform any (selective-function or fully secure) message-private functional encryption scheme into a (selective-function or fully secure, respectively) functional encryption scheme which is also function private. Thus, any instantiation of a message-private (or selective-function message private) function encryption scheme as discussed in Definition 3 can be used as a building block in our construction.

3   Private-Key Functional Encryption for Randomized Functionalities

In this section we present a framework for capturing the security of private-key functional encryption for randomized functionalities. Our framework is inspired by that of Goyal et al. [24] in the public-key setting, but takes a slightly different approach as we discuss below.

Throughout this section, we let $\mathcal{F} = \{\mathcal{F}_\lambda\}_{\lambda \in \mathbb{N}}$ be a family of randomized functionalities, where for every $\lambda \in \mathbb{N}$ the set $\mathcal{F}_\lambda$ consists of functions of the form $f : \mathcal{X}_\lambda \times \mathcal{R}_\lambda \to \mathcal{Y}_\lambda$. That is, such a function f maps $\mathcal{X}_\lambda$ into $\mathcal{Y}_\lambda$ using randomness from $\mathcal{R}_\lambda$.

A private-key functional encryption scheme for a family $\mathcal{F}$ of randomized functions consists of four probabilistic polynomial-time algorithms (Setup, KG, Enc, Dec) with the same syntax that is described in Section 2.2 for deterministic functions. Although the syntax in this setting is the same as in the deterministic setting, the correctness and security requirements are more subtle.

3.1   Correctness and Independence

In terms of correctness we rely on the definition of Goyal et al. [24] (when adapted to the private-key setting). As discussed in Section 1.3, we ask that for any sequence of messages $x_1, \ldots, x_T$ and for any sequence of functions $f_1, \ldots, f_T \in \mathcal{F}$, it holds that the distribution obtained by encrypting $x_1, \ldots, x_T$ and then decrypting the resulting ciphertexts with functional keys corresponding to $f_1, \ldots, f_T$ is computationally indistinguishable from the distribution $\{f_j(x_i; r_{i,j})\}_{i,j \in [T]}$ where the $r_{i,j}$'s are sampled independently and uniformly at random.

Definition 8 (Correctness). *A functional encryption scheme* $\Pi = $ (Setup, KG, Enc, Dec) *for a family* $\mathcal{F}$ *of randomized functions is* correct *if for all sufficiently large* $\lambda \in \mathbb{N}$, *for every polynomial* $T = T(\lambda)$, *and for every* $x_1, \ldots, x_T \in$

$\mathcal{X}_\lambda$ and $f_1, \ldots, f_T \in \mathcal{F}_\lambda$, the following two distributions are computationally indistinguishable:

- **Real**$(\lambda) \stackrel{\text{def}}{=} \big\{ \mathsf{Dec}(\mathsf{sk}_{f_j}, \mathsf{ct}_i) \big\}_{i,j \in [T]}$, where:
  - $\mathsf{msk} \leftarrow \mathsf{Setup}(1^\lambda)$,
  - $\mathsf{ct}_i \leftarrow \mathsf{Enc}(\mathsf{msk}, x_i)$ for all $i \in [T]$,
  - $\mathsf{sk}_{f_j} \leftarrow \mathsf{KG}(\mathsf{msk}, f_j)$ for all $j \in [T]$.
- **Ideal**$(\lambda) \stackrel{\text{def}}{=} \{ f_j(x_i) \}_{i,j \in [T]}$.

As noted by Goyal et al. [24], unlike in the case of deterministic functions where is suffices to define correctness for a single ciphertext and a single key, here it is essential to define correctness for multiple (possibly correlated) ciphertexts and keys. We refer the reader to [24] for more details.

3.2 "Best-Possible" Message Privacy

We consider indistinguishability-based notions for capturing message privacy in private-key functional encryption for randomized functionalities. As in the (standard) case of deterministic functions (see Section 2.2), we consider adversaries whose goal is to distinguish between encryptions of two challenge messages x_0^* and x_1^*, when given access to an encryption oracle (as required in private-key encryption) and to functional keys of various functions. Recall that in the case of deterministic functions, the adversary is allowed to ask for functional keys for any function f such that $f(x_0^*) = f(x_1^*)$.

As discussed in Section 1.3, our notions of message privacy ask that the functional encryption scheme under consideration will not add any non-negligible advantage to the (*possibly non-negligible*) advantage that adversaries holding a functional key for a function f may already have in distinguishing between the distributions $f(x_0^*)$ and $f(x_1^*)$ to begin with. That is, given that adversaries are able to obtain a sample from the distribution $f(x_0^*)$ or from the distribution $f(x_1^*)$ using the functional key sk_f, and may already have some advantage in distinguishing these distributions, we ask for "best-possible" message privacy in the sense that essentially no additional advantage can be gained.

In what follows we put forward two flavors of "best-possible" message privacy, depending on the flavor of indistinguishability guarantee that is satisfied by the function family under consideration.

Message Privacy for Non-Adaptively-Admissible Functionalities. Our first notion is that of *non-adaptively-admissible* function families. These are families $\mathcal{F}$ such that for a randomly sampled $f \leftarrow \mathcal{F}$, no efficient adversary on input f can output x_0 and x_1 and distinguish the distributions $f(x_0)$ and $f(x_1)$ with probability larger than Δ (note again that Δ does not have to be negligible). One possible example for such a function family is a function that on input x samples a public-key pk for a public-key encryption scheme, and outputs pk together with a randomized encryption of x.

For such function families we consider a corresponding notion of message privacy in which the adversary obtains functional keys only for functions that are sampled uniformly and independently from $\mathcal{F}$. This is formally captured by the following two definitions.

Definition 9 (Non-Adaptively-Admissible Function Family). *A family* $\mathcal{F} = \{\mathcal{F}_\lambda\}_{\lambda \in \mathbb{N}}$ *of efficiently-computable randomized functions is* $\Delta(\lambda)$*-non-adaptively admissible if for any probabilistic polynomial-time algorithm* $\mathcal{A} = (\mathcal{A}_1, \mathcal{A}_2)$ *it holds that*

$$\mathsf{Adv}_{\mathcal{F},\mathcal{A}}^{\mathsf{naADM}}(\lambda) \stackrel{\text{def}}{=} \left| \Pr\left[\mathsf{Expt}_{\mathcal{F},\mathcal{A}}^{\mathsf{naADM}}(\lambda) = 1 \right] - \frac{1}{2} \right| \leq \Delta(\lambda)$$

for all sufficiently large $\lambda \in \mathbb{N}$*, where the random variable* $\mathsf{Expt}_{\mathcal{F},\mathcal{A}}^{\mathsf{naADM}}(\lambda)$ *is defined via the following experiment:*

1. $b \leftarrow \{0, 1\}$, $f \leftarrow \mathcal{F}_\lambda$.
2. $(x_0, x_1, \mathsf{state}) \leftarrow \mathcal{A}_1(1^\lambda, f)$.
3. $y = f(x_b; r)$ *for* $r \leftarrow \{0, 1\}^*$.
4. $b' \leftarrow \mathcal{A}_2(y, \mathsf{state})$.
5. *If* $b' = b$ *then output* 1*, and otherwise output* 0.

Definition 10 (Message Privacy; Non-Adaptive Case). *Let* $\mathcal{F} = \{\mathcal{F}_\lambda\}_{\lambda \in \mathbb{N}}$ *be a* $\Delta(\lambda)$*-non-adaptively admissible function family. A private-key functional encryption scheme* $\Pi = (\mathsf{Setup}, \mathsf{KG}, \mathsf{Enc}, \mathsf{Dec})$ *is* message private *with respect to* $\mathcal{F}$ *if for any probabilistic polynomial-time adversary* $\mathcal{A} = (\mathcal{A}_1, \mathcal{A}_2)$ *and for any polynomial* $T = T(\lambda)$ *there exists a negligible function* $\mathsf{neg}(\lambda)$ *such that*

$$\mathsf{Adv}_{\Pi,\mathcal{F},\mathcal{A},T}^{\mathsf{naMPRF}}(\lambda) \stackrel{\text{def}}{=} \left| \Pr\left[\mathsf{Expt}_{\Pi,\mathcal{F},\mathcal{A},T}^{\mathsf{naMPRF}}(\lambda) = 1 \right] - \frac{1}{2} \right| \leq T(\lambda) \cdot \Delta(\lambda) + \mathsf{neg}(\lambda),$$

for all sufficiently large $\lambda \in \mathbb{N}$*, where the random variable* $\mathsf{Expt}_{\Pi,\mathcal{F},\mathcal{A},T}^{\mathsf{naMPRF}}(\lambda)$ *is defined via the following experiment:*

1. $b \leftarrow \{0, 1\}$, $\mathsf{msk} \leftarrow \mathsf{Setup}(1^\lambda)$, $f_1, \ldots, f_T \leftarrow \mathcal{F}_\lambda$.
2. $\mathsf{sk}_{f_i} \leftarrow \mathsf{KG}(\mathsf{msk}, f_i)$ *for all* $i \in [T]$.
3. $(x_0^*, x_1^*, \mathsf{state}) \leftarrow \mathcal{A}_1^{\mathsf{Enc}(\mathsf{msk}, \cdot)}(1^\lambda, f_1, \ldots, f_T, \mathsf{sk}_{f_1}, \ldots, \mathsf{sk}_{f_T})$.
4. $c^* = \mathsf{Enc}(\mathsf{msk}, x_b^*)$.
5. $b' \leftarrow \mathcal{A}_2^{\mathsf{Enc}(\mathsf{msk}, \cdot)}(c^*, \mathsf{state})$.
6. *If* $b' = b$ *then output* 1*, and otherwise output* 0.

Message Privacy for Adaptively-Admissible Functionalities. Our second notion is that of *adaptively-admissible* function families. These are families $\mathcal{F}$ such that no efficient adversary can output $f \in \mathcal{F}$ together with two inputs x_0 and x_1, and distinguish the distributions $f(x_0)$ and $f(x_1)$ with probability larger than Δ. One possible example for such a function family is that of differentially private mechanisms, as discussed by Goyal et al. [24]. Specifically, these are randomized functions that on any two inputs that differ on only a few of their entries, produce output distributions whose *statistical* distance is polynomially small (i.e., Δ is polynomial in $1/\lambda$)[6].

It is easy to observe that there are function families that are non-adaptively admissible but are not adaptively admissible. One possible example is functions of the form f_{pk} that are indexed by a public encryption key pk, and on input x output a randomized encryption of x under pk. Giving adversaries the possibility of adaptively choosing such functions, they can choose a function f_{pk} for which they know the corresponding decryption key sk. In this case, although for a randomly chosen pk the distributions $f_{pk}(x_0)$ and $f_{pk}(x_1)$ are computationally indistinguishable, they may be easily distinguishable given the randomness used by the adversary (from which it may be easy to compute the corresponding decryption key sk).

For adaptively-admissible function families we consider a corresponding notion of message privacy in which the adversary obtains functional keys for functions that are adaptively chosen from $\mathcal{F}$. This is formally captured by the following two definitions.

Definition 11 (Adaptively-Admissible Function Family). *A family* $\mathcal{F} = \{\mathcal{F}_\lambda\}_{\lambda \in \mathbb{N}}$ *of efficiently-computable randomized functions is* $\Delta(\lambda)$-*adaptively admissible if for any probabilistic polynomial-time algorithm* $\mathcal{A} = (\mathcal{A}_1, \mathcal{A}_2)$ *it holds that*

$$\mathsf{Adv}_{\mathcal{F},\mathcal{A}}^{\mathsf{aADM}}(\lambda) \stackrel{\mathsf{def}}{=} \left| \Pr\left[\mathsf{Expt}_{\mathcal{F},\mathcal{A}}^{\mathsf{aADM}}(\lambda) = 1 \right] - \frac{1}{2} \right| \leq \Delta(\lambda)$$

for all sufficiently large $\lambda \in \mathbb{N}$, *where the random variable* $\mathsf{Expt}_{\mathcal{F},\mathcal{A}}^{\mathsf{aADM}}(\lambda)$ *is defined via the following experiment:*

1. $b \leftarrow \{0,1\}$.
2. $(f, x_0, x_1, \mathsf{state}) \leftarrow \mathcal{A}_1(1^\lambda)$, *where* $f \in \mathcal{F}_\lambda$.
3. $y = f(x_b; r)$ *for* $r \leftarrow \{0,1\}^*$.
4. $b' \leftarrow \mathcal{A}_2(y, \mathsf{state})$.
5. *If* $b' = b$ *then output* 1, *and otherwise output* 0.

Definition 12 (Message Privacy; Adaptively-Admissible Case). *Let* $\mathcal{F} = \{\mathcal{F}_\lambda\}_{\lambda \in \mathbb{N}}$ *be a* $\Delta(\lambda)$-*adaptively admissible function family. A private-key functional encryption scheme* $\Pi = (\mathsf{Setup}, \mathsf{KG}, \mathsf{Enc}, \mathsf{Dec})$ *is message private with respect to* $\mathcal{F}$ *if for any probabilistic polynomial-time adversary* $\mathcal{A} = (\mathcal{A}_1, \mathcal{A}_2)$ *that issues at most* $T = T(\lambda)$ *key-generation queries there exists a negligible function* $\mathsf{neg}(\lambda)$ *such that*

$$\mathsf{Adv}_{\Pi,\mathcal{F},\mathcal{A}}^{\mathsf{aMPRF}}(\lambda) \stackrel{\mathsf{def}}{=} \left| \Pr\left[\mathsf{Expt}_{\Pi,\mathcal{F},\mathcal{A}}^{\mathsf{aMPRF}}(\lambda) = 1 \right] - \frac{1}{2} \right| \leq T(\lambda) \cdot \Delta(\lambda) + \mathsf{neg}(\lambda),$$

for all sufficiently large $\lambda \in \mathbb{N}$, *where the random variable* $\mathsf{Expt}_{\Pi,\mathcal{F},\mathcal{A}}^{\mathsf{aMPRF}}(\lambda)$ *is defined via the following experiment:*

---

[6] The definitions of differential privacy are in fact stronger than requiring small statistical distance.

1. $b \leftarrow \{0,1\}$, msk $\leftarrow$ Setup(1^λ).
2. $(x_0^*, x_1^*, \text{state}) \leftarrow \mathcal{A}_1^{\text{Enc}(\text{msk},\cdot), \text{KG}(\text{msk},\cdot)}(1^\lambda)$.
3. $c^* = \text{Enc}(\text{msk}, x_b^*)$.
4. $b' \leftarrow \mathcal{A}_2^{\text{Enc}(\text{msk},\cdot), \text{KG}(\text{msk},\cdot)}(c^*, \text{state})$.
5. If $b' = b$ then output 1, and otherwise output 0.

4   Our Functional Encryption Scheme

In this section we present our construction of a private-key functional encryption scheme for randomized functionalities. Let $\mathcal{F} = \{\mathcal{F}_\lambda\}_{\lambda \in \mathbb{N}}$ be a family of randomized functionalities, where for every $\lambda \in \mathbb{N}$ the set $\mathcal{F}_\lambda$ consists of functions of the form $f : \mathcal{X}_\lambda \times \mathcal{R}_\lambda \to \mathcal{Y}_\lambda$ (i.e., f maps $\mathcal{X}_\lambda$ into $\mathcal{Y}_\lambda$ using randomness from $\mathcal{R}_\lambda$). Our construction relies on the following building blocks:

1. A private-key functional encryption scheme FE = (FE.Setup, FE.KG, FE.Enc, FE.Dec).
2. A pseudorandom function family PRF = (PRF.Gen, PRF.Eval). We assume that for every $\lambda \in \mathbb{N}$ and for every key K that is produced by PRF.Gen(1^λ), it holds that PRF.Eval$(K, \cdot) : \{0,1\}^\lambda \to \mathcal{R}_\lambda$.

As discussed in Section 1.1, we assume that the scheme FE is sufficiently expressive in the sense that it supports the function family $\mathcal{F}$ (when viewed as a family of single-input deterministic functions), the evaluation procedure of the pseudorandom function family PRF, and a few additional basic operations (such as conditional statements). Our scheme $\Pi = (\text{Setup}, \text{KG}, \text{Enc}, \text{Dec})$ is defined as follows.

- **The Setup Algorithm.** On input the security parameter 1^λ the setup algorithm Setup samples FE.msk $\leftarrow$ FE.Setup(1^λ), and outputs msk = FE.msk.
- **The Key-Generation Algorithm.** On input the master secret key msk and a function $f \in \mathcal{F}_\lambda$, the key-generation algorithm KG samples $K \leftarrow$ PRF.Gen(1^λ) and outputs sk$_f \leftarrow$ FE.KG(msk, Left$_{f,K}$), where Left$_{f,K}$ is a deterministic function that is defined in Figure 1.
- **The Encryption Algorithm.** On input the master secret key msk and a message $x \in \mathcal{X}_\lambda$, the encryption algorithm Enc samples $s \leftarrow \{0,1\}^\lambda$ and outputs ct $\leftarrow$ FE.Enc(msk, $(x, \bot, s, \bot)$).
- **The Decryption Algorithm.** On input a functional key sk$_f$ and a ciphertext ct, the decryption algorithm Dec outputs FE.Dec(sk$_f$, ct).

The correctness and independence of the above scheme with respect to any family of randomized functionalities follows in a straightforward manner from the correctness of the underlying functional encryption scheme FE and the assumption that PRF is a pseudorandom function family (in fact, it suffices that PRF is a *weak* pseudorandom function family). Specifically, consider a sequence of messages $x_1, \ldots, x_T$ and a sequence of functions $f_1, \ldots, f_T$. As the encryption FE.Enc(msk, $(x_i, \bot, s_i, \bot)$) of each message x_i uses a uniformly sampled $s_i \in$

Left$_{f,K}(x_L, x_R, s, z)$:

1. Let $r = \mathsf{PRF.Eval}(K, s)$.
2. Output $f(x_L; r)$.

Right$_{f,K}(x_L, x_R, s, z)$:

1. Let $r = \mathsf{PRF.Eval}(K, s)$.
2. Output $f(x_R; r)$.

Fig. 1. The functions Left$_{f,K}$ and Right$_{f,K}$. The function Left$_{f,K}$ is used by the actual scheme, whereas the function Right$_{f,K}$ is used in the proofs of its security.

$\{0, 1\}^\lambda$, and the functional key for a function f_j contains a freshly sampled key K_j for the pseudorandom function family, the distribution $\{f_j(x_i; \mathsf{PRF.Eval}(K_j, s_i)\}$ is computationally indistinguishable from the distribution $\{f_j(x_i; r_{i,j})\}$, where the $r_{i,j}$'s are sampled independently and uniformly at random.

The following two theorems capture the security of the scheme. These theorems state that under suitable assumptions on the underlying building blocks, the scheme is message private for non-adaptively-admissible randomized functionalities and for adaptively-admissible randomized functionalities.

Theorem 1. *Assuming that* PRF *is a pseudorandom function family and that* FE *is selective-function function private, then* Π *is message private for non-adaptively-admissible randomized functionalities.*

Theorem 2. *Assuming that* PRF *is a puncturable pseudorandom function family and that* FE *is function private, then* Π *is message private for adaptively-admissible randomized functionalities.*

As discussed in Sections 2.1 and 2.2, Theorems 1 and 2 can be instantiated based on a variety of known pseudorandom function families and functional encryption schemes. In particular, Theorem 1 can be based on the minimal assumption that a selective-function message-private functional encryption scheme exists, and Theorem 2 can be based on the minimal assumption that a message-private functional encryption scheme exists.

Due to lack of space we omit the proof of Theorem 1 and include only the proof of Theorem 2. We refer to the full version of the paper [26] for the missing details.

4.1   Proof of Theorem 2

We prove that the scheme Π is message private for adaptively-admissible functionalities (see Definition 12) based on the assumptions that PRF is a puncturable pseudorandom function family and that FE is function private (see Definition 6).

Let $\mathcal{A}$ be a probabilistic polynomial-time adversary that issues at most $T_1 = T_1(\lambda)$ pre-challenge key-generation queries, at most $T_2 = T_2(\lambda)$ post-challenge key-generation queries (where $T = T_1 + T_2$), and at most $T = T(\lambda)$ encryption queries (note that T_1, T_2 and T may be any polynomials and are not fixed in advance), and let $\mathcal{F}$ be a Δ-adaptively admissible family of randomized functionalities. We denote by $f_1, \ldots, f_T$ the key-generation queries that are issued by $\mathcal{A}$.

We present a sequence of experiments and upper bound $\mathcal{A}$'s advantage in distinguishing each two consecutive experiments. Each two consecutive experiments differ either in the distribution of their challenge ciphertexts or in the distribution of the functional keys that are produced by the key-generation oracle. The first experiment is the experiment $\mathsf{Expt}_{\Pi,\mathcal{F},\mathcal{A},T}^{\mathsf{aMPRF}}(\lambda)$ (see Definition 12), and the last experiment is completely independent of the bit b. This enables us to prove that there exists a negligible function $\mathsf{neg}(\cdot)$ such that

$$\mathsf{Adv}_{\Pi,\mathcal{F},\mathcal{A},T}^{\mathsf{aMPRF}}(\lambda) \overset{\text{def}}{=} \left| \Pr\left[\mathsf{Expt}_{\Pi,\mathcal{F},\mathcal{A},T}^{\mathsf{aMPRF}}(\lambda) = 1 \right] - \frac{1}{2} \right| \leq T(\lambda) \cdot \Delta(\lambda) + \mathsf{neg}(\lambda)$$

for all sufficiently large $\lambda \in \mathbb{N}$. Throughout the proof we use, in addition to the functions $\mathsf{Left}_{f,K}$ and $\mathsf{Right}_{f,K}$ that were defined in Figure 1, the functions $\mathsf{PuncOutputY}_{f,K',y,s^*}$ and $\mathsf{PuncOutputZ}_{f,K',s^*}$ that are defined in Figure 2. In

---

PuncOutputY$_{f,K',y,s^*}(x_L, x_R, s, z)$:

1. If $s = s^*$ then output y.
2. Otherwise, let $r = \mathsf{PRF.Eval}(K', s)$ and output $f(x_L; r)$.

PuncOutputZ$_{f,K',s^*}(x_L, x_R, s, z)$:

1. If $s = s^*$ then output z.
2. Otherwise, let $r = \mathsf{PRF.Eval}(K', s)$ and output $f(x_L; r)$.

Fig. 2. The functions $\mathsf{PuncOutputY}_{f,K',y,s^*}$ and $\mathsf{PuncOutputZ}_{f,K',s^*}$.

what follows we describe the experiments. We note that in all experiments the encryption oracle is as defined by the encryption procedure of the scheme.

Experiment $\mathcal{H}^{(0)}(\lambda)$. This is the experiment $\mathsf{Expt}_{\Pi,\mathcal{F},\mathcal{A}}^{\mathsf{aMPRF}}(\lambda)$ (see Definition 12).

Experiment $\mathcal{H}^{(1)}(\lambda)$. This experiment is obtained from the experiment $\mathcal{H}^{(0)}(\lambda)$ by modifying the encryption oracle so that on the challenge input (x_0^*, x_1^*) it samples $s^* \leftarrow \{0,1\}^\lambda$ and outputs $\mathsf{ct} \leftarrow \mathsf{FE.Enc}(\mathsf{msk}, (x_b^*, \boxed{x_1^*}, s^*, \bot))$ instead of $\mathsf{ct} \leftarrow \mathsf{FE.Enc}(\mathsf{msk}, (x_b^*, \boxed{\bot}, s^*, \bot))$.

Note that for each function $f \in \{f_1, \ldots, f_T\}$ with an associated PRF key K, for the deterministic function $\mathsf{Left}_{f,K}$ and the challenge ciphertext it holds that $\mathsf{Left}_{f,K}(x_b^*, x_1^*, s^*, \bot) = \mathsf{Left}_{f,K}(x_b^*, \bot, s^*, \bot)$. Therefore, the message privacy of the underlying scheme FE (with respect to *deterministic* functions) guarantees that the adversary $\mathcal{A}$ has only a negligible advantage in distinguishing experiments $\mathcal{H}^{(0)}$ and $\mathcal{H}^{(1)}$. Specifically, let $\mathcal{F}'$ denote the family of deterministic functions $\mathsf{Left}_{f,K}$ and $\mathsf{Right}_{f,K}$ for every $f \in \mathcal{F}$ and PRF key K (as defined in Figure 1) as well as the function $\mathsf{PuncOutputY}_{f,K',y,s^*}$ and $\mathsf{PuncOutputZ}_{f,K',s^*}$ for every $f \in \mathcal{F}$, punctured PRF key K', value $y \in \mathcal{Y}_\lambda$ and string $s^* \in \{0,1\}^\lambda$ (as defined in Figure 2). In the full version (see [26]) we prove the following lemma:

Lemma 1. *There exists a probabilistic polynomial-time adversary $\mathcal{B}^{(0) \to (1)}$ such that*

$$\left| \Pr\left[\mathcal{H}^{(0)}(\lambda) = 1 \right] - \Pr\left[\mathcal{H}^{(1)}(\lambda) = 1 \right] \right| \leq \mathsf{Adv}^{\mathsf{MP}}_{\mathsf{FE}, \mathcal{F}', \mathcal{B}^{(0) \to (1)}, T}(\lambda).$$

Experiment $\mathcal{H}^{(2,i)}(\lambda)$ where $i \in [T_2 + 1]$. This experiment is obtained from the experiment $\mathcal{H}^{(1)}(\lambda)$ by modifying the post challenge key-generation oracle to generate keys as follows. The functional keys for the $f_{T_1+1}, \ldots, f_{T_1+i-1}$ are generated as $\mathsf{PuncOutputY}_{f,K',y,s^*}$ (the definition of $\mathsf{PuncOutputY}_{f,K',y,s^*}$ appears in Figure 2), where K' is generated by sampling a PRF key $K \leftarrow \mathsf{PRF.Gen}(1^\lambda)$ and then puncturing it at s^*, and where $y \leftarrow f(x_b^*)$, and the functional keys for $f_{T_1+i}, \ldots, f_{T_1+T_2} = f_T$ are generated as $\mathsf{PuncOutputY}_{f,K',y,s^*}$, where K' and s^* are as before but $y = f(x_b^*; \mathsf{PRF}_K(s^*))$.

Note that every $x \neq x_b^*$ with which the encryption oracle is queries (with probability negligibly close to 1) it holds that $s \neq s^*$, hence, using the functionality feature of the punctured PRF, for every $f \in \{f_{T_1+1}, \ldots, f_T\}$ it holds that $\mathsf{Left}_{f,K}(x, x, s, \perp) = \mathsf{PuncOutputY}_{f,K',y,s^*}(x, x, s, \perp)$. In addition, for the challenge x_b^* it holds that $\mathsf{Left}_{f,K}(x_b^*, x_1^*, s^*, \perp) = \mathsf{PuncOutputY}_{f,K',y,s^*}(x_b^*, x_1^*, s^*, \perp)$ since $\mathsf{PuncOutputY}_{f,K',y,s^*}$ simply outputs y, where $y = f(x_b^*; \mathsf{PRF}_K(s^*))$. Thus, the function-privacy of the underlying scheme FE guarantees that the adversary $\mathcal{A}$ has only a negligible advantage in distinguishing experiments $\mathcal{H}^{(1)}(\lambda)$ and $\mathcal{H}^{(2,1)}(\lambda)$. In the full version (see [26]) we prove the following lemma:

Lemma 2. *There exists a probabilistic polynomial-time adversary $\mathcal{B}^{(1) \to (2,1)}$ such that*

$$\left| \Pr\left[\mathcal{H}^{(1)}(\lambda) = 1 \right] - \Pr\left[\mathcal{H}^{(2,1)}(\lambda) = 1 \right] \right| \leq \mathsf{Adv}^{\mathsf{FP}}_{\mathsf{FE}, \mathcal{F}', \mathcal{B}^{(1) \to (2,1)}, T}(\lambda) + \mathsf{neg}(\lambda).$$

Moreover, note that the pseudorandomness of $\mathsf{PRF}_K(\cdot)$ at punctured point s^* (see Definition 2) guarantees that the adversary $\mathcal{A}$ has only a negligible advantage in distinguishing experiments $\mathcal{H}^{(2,i)}$ and $\mathcal{H}^{(2,i+1)}$. In the full version (see [26]) we prove the following lemma:

Lemma 3. *For every $i \in [T_2]$ there exists a probabilistic polynomial-time adversary $\mathcal{B}^{(2,i) \to (2,i+1)}$ such that*

$$\left| \Pr\left[\mathcal{H}^{(2,i)}(\lambda) = 1 \right] - \Pr\left[\mathcal{H}^{(2,i+1)}(\lambda) = 1 \right] \right| \leq \mathsf{Adv}_{\mathsf{puPRF}, \mathcal{B}^{(2,i) \to (2,i+1)}}(\lambda).$$

Experiment $\mathcal{H}^{(3,i)}(\lambda)$ Where $i \in [T_2 + 1]$. This experiment is obtained from the experiment $\mathcal{H}^{(2,T_2)}(\lambda)$ by modifying the post-challenge key-generation oracle as follows. The functional keys for the $f_{T_1+1}, \ldots, f_{T_1+i-1}$ are generated as $\mathsf{PuncOutputY}_{f,K',y,s^*}$, where K' is generated by sampling a PRF key $K \leftarrow \mathsf{PRF.Gen}(1^\lambda)$ and then puncturing it at s^*, and where $y \leftarrow \boxed{f(x_1^*)}$, and the functional keys for $f_{T_1+i}, \ldots, f_{T_1+T_2}$ are generated as $\mathsf{PuncOutputY}_{f,K',y,s^*}$, where K' and s^* are as before but $y \leftarrow f(x_b^*)$. We observe that $\mathcal{H}^{(2,T+1)}(\lambda) = \mathcal{H}^{(3,1)}(\lambda)$.

The adaptive admissibility of the function family $\mathcal{F}$ (see Definition 11) guarantee that the advantage of the adversary $\mathcal{A}$ in distinguishing experiments $\mathcal{H}^{(3,i)}$ and $\mathcal{H}^{(3,i+1)}$ is at most $\Delta(\lambda)$. In the full version (see [26]) we prove the following lemma:

Lemma 4. *For every $i \in [T_2]$ there exists a probabilistic polynomial-time adversary $\mathcal{B}^{(3,i) \to (3,i+1)}$ such that*

$$\left| \Pr\left[\mathcal{H}^{(3,i)}(\lambda) = 1\right] - \Pr\left[\mathcal{H}^{(3,i+1)}(\lambda) = 1\right] \right| \leq \mathsf{Adv}^{\mathsf{aADM}}_{\mathcal{F}, \mathcal{B}^{(3,i) \to (3,i+1)}} \leq \Delta(\lambda).$$

Experiment $\mathcal{H}^{(4,i)}(\lambda)$ where $i \in [T_1 + 1]$. This experiment is obtained from the experiment $\mathcal{H}^{(3,T)}(\lambda)$ by modifying the pre-challenge key-generation oracle as follows. The functional keys for $f_1, ..., f_{i-1}$ are generated as $\mathsf{sk}_f \leftarrow$ FE.KG(msk, $\boxed{\mathsf{Right}_{f,K}}$) instead of as $\mathsf{sk}_f \leftarrow$ FE.KG(msk, $\boxed{\mathsf{Left}_{f,K}}$) (where $\mathsf{Right}_{f,K}$ is defined in Figure 1), and the functional keys for $f_i, ..., f_{T_1}$ are generated as before (i.e., as $\mathsf{sk}_f \leftarrow$ FE.KG(msk, $\mathsf{Left}_{f,K}$)). We observe that $\mathcal{H}^{(3,T+1)}(\lambda) = \mathcal{H}^{(4,1)}(\lambda)$.

Experiment $\mathcal{H}^{(5,i)}(\lambda)$ where $i \in [T_1]$. This experiment is obtained from the experiment $\mathcal{H}^{(4,i)}(\lambda)$ by modifying the encryption oracle so that on the challenge input (x_0^*, x_1^*) it samples $s^* \leftarrow \{0,1\}^\lambda$ and outputs $\mathsf{ct} \leftarrow$ FE.Enc(msk, $(x_b^*, x_1^*, s^*, \boxed{z^*})$), where $z^* = f_i(x_b^*; \mathsf{PRF.Eval}(K_i, s^*))$, instead of $\mathsf{ct} \leftarrow$ FE.Enc(msk, $(x_b^*, x_1^*, s^*, \boxed{\bot})$).

Notice that both $\mathsf{Left}_{f,K}$ and $\mathsf{Right}_{f,K}$ are defined to ignore the fourth input z, hence, for the first $i - 1$ keys it holds that $\mathsf{Right}_{f,K}(x_b^*, x_1^*, s^*, \bot) = \mathsf{Right}_{f,K}(x_b^*, x_1^*, s^*, z^*)$ and for the next $T_1 - i + 1$ keys it holds that $\mathsf{Left}_{f,K}(x_b^*, x_1^*, s^*, \bot) = \mathsf{Left}_{f,K}(x_b^*, x_1^*, s^*, z^*)$. Therefore, the message privacy of the underlying scheme FE guarantees that the adversary $\mathcal{A}$ has only a negligible advantage in distinguishing experiments $\mathcal{H}^{(4,i)}$ and $\mathcal{H}^{(5,i)}$. In the full version (see [26]) we prove the following lemma:

Lemma 5. *For every $i \in [T_1]$ there exists a probabilistic polynomial-time adversary $\mathcal{B}^{(4,i) \to (5,i)}$ such that*

$$\left| \Pr\left[\mathcal{H}^{(4,i)}(\lambda) = 1\right] - \Pr\left[\mathcal{H}^{(5,i)}(\lambda) = 1\right] \right| \leq \mathsf{Adv}^{\mathsf{MP}}_{\mathsf{FE}, \mathcal{F}', \mathcal{B}^{(4,i) \to (5,i)}, T}(\lambda).$$

Experiment $\mathcal{H}^{(6,i)}(\lambda)$ Where $i \in [T_1]$. This experiment is obtained from the experiment $\mathcal{H}^{(5,i)}(\lambda)$ by modifying the behavior of the pre-challenge key-generation oracle on the ith query f_i (without modifying its behavior on all other queries). On input the ith query f_i, the pre-challenge key-generation oracle compute $\mathsf{sk}_{f_i} \leftarrow$ FE.KG(msk, $\boxed{\mathsf{PuncOutputZ}_{f_i, K_i', s^*}}$) instead of $\mathsf{sk}_{f_i} \leftarrow$ FE.KG(msk, $\boxed{\mathsf{Left}_{f_i, K_i}}$) (where the function $\mathsf{PuncOutputZ}_{f_i, K_i', s^*}$ is defined in Figure 2).

Note that by the functionality feature of the punctured PRF (see Definition 2), for every ciphertext $(x, \bot, s, z)$ which is not the challenge ciphertext (with probability negligibly close to 1) it holds that $\mathsf{PuncOutputZ}_{f_i, K'_i, s^*}(x, \bot, s, z) = \mathsf{Left}_{f_i, K_i}(x, \bot, s, z)$ (since $s \neq s^*$ with very high probability). For the challenge ciphertext the latter also holds since $\mathsf{PuncOutputZ}_{f_i, K'_i, s^*}(x_0^*, x_1^*, s^*, z^*)$ outputs $z^* = f_i(x_b^*; \mathsf{PRF}_{K_i}(s^*))$. Thus, the function-privacy of the underlying scheme FE guarantees that the adversary $\mathcal{A}$ has only a negligible advantage in distinguishing experiments $\mathcal{H}^{(6,i)}(\lambda)$ and $\mathcal{H}^{(7,i)}(\lambda)$. In the full version (see [26]) we prove the following lemma:

Lemma 6. *For every $i \in [T_1]$ there exists a probabilistic polynomial-time adversary $\mathcal{B}^{(5,i) \to (6,i)}$ such that*

$$\left| \Pr\left[\mathcal{H}^{(5,i)}(\lambda) = 1\right] - \Pr\left[\mathcal{H}^{(6,i)}(\lambda) = 1\right] \right| \leq \mathsf{Adv}^{\mathsf{FP}}_{\mathsf{FE}, \mathcal{F}', \mathcal{B}^{(5,i) \to (6,i)}, T}(\lambda) + \mathsf{neg}(\lambda).$$

Experiment $\mathcal{H}^{(7,i)}(\lambda)$ where $i \in [T_1]$. This experiment is obtained from the experiment $\mathcal{H}^{(6,i)}(\lambda)$ by modifying the encryption oracle so that on the challenge input (x_0^*, x_1^*) it outputs $\mathsf{ct} \leftarrow \mathsf{FE.Enc}(\mathsf{msk}, (x_0^*, x_1^*, s^*, z^*))$, where $z^* = f_i(x_b^*; \boxed{r^*})$ for a fresh and uniformly sampled value r^* instead of $z^* = f_i(x_b^*; \boxed{\mathsf{PRF.Eval}(K_i, s^*)})$.

The pseudorandomness at punctured point s^* of $\mathsf{PRF.Eval}(K_i, \cdot)$ guarantees that the adversary $\mathcal{A}$ has only a negligible advantage in distinguishing experiments $\mathcal{H}^{(6,i)}$ and $\mathcal{H}^{(7,i)}$. In the full version (see [26]) we prove the following lemma:

Lemma 7. *For every $i \in [T_1]$ there exists a probabilistic polynomial-time adversary $\mathcal{B}^{(6,i) \to (7,i)}$ such that*

$$\left| \Pr\left[\mathcal{H}^{(6,i)}(\lambda) = 1\right] - \Pr\left[\mathcal{H}^{(7,i)}(\lambda) = 1\right] \right| \leq \mathsf{Adv}_{\mathsf{puPRF}, \mathcal{B}^{(6,i) \to (7,i)}}(\lambda).$$

Experiment $\mathcal{H}^{(8,i)}(\lambda)$ Where $i \in [T_1]$. This experiment is obtained from the experiment $\mathcal{H}^{(7,i)}(\lambda)$ by modifying the encryption oracle so that on the challenge input (x_0^*, x_1^*) it outputs $\mathsf{ct} \leftarrow \mathsf{FE.Enc}(\mathsf{msk}, (x_0^*, x_1^*, s^*, z^*))$, where $z^* = f_i(\boxed{x_1^*}; r^*)$ instead of $z^* = f_i(\boxed{x_b^*}; r^*)$ (both with fresh and uniform r^*).

The adaptive admissibility of the function family $\mathcal{F}$ (see Definition 11) guarantees that the advantage of the adversary $\mathcal{A}$ in distinguishing experiments $\mathcal{H}^{(7,i)}$ and $\mathcal{H}^{(8,i)}$ is at most $\Delta(\lambda)$. In the full version (see [26]) we prove the following lemma:

Lemma 8. *For every $i \in [T_1]$ there exists a probabilistic polynomial-time adversary $\mathcal{B}^{(7,i) \to (8,i)}$ such that*

$$\left| \Pr\left[\mathcal{H}^{(7,i)}(\lambda) = 1\right] - \Pr\left[\mathcal{H}^{(8,i)}(\lambda) = 1\right] \right| \leq \mathsf{Adv}^{\mathsf{aADM}}_{\mathcal{F}, \mathcal{B}^{(7,i) \to (8,i)}} \leq \Delta(\lambda).$$

Experiment $\mathcal{H}^{(9,i)}(\lambda)$ Where $i \in [T_1]$. This experiment is obtained from the experiment $\mathcal{H}^{(8,i)}(\lambda)$ by modifying the encryption oracle so that on the challenge input (x_0^*, x_1^*) it outputs $\mathsf{ct} \leftarrow \mathsf{FE.Enc}(\mathsf{msk}, (x_b^*, x_1^*, s^*, z^*))$, where $z^* = f_i(x_1^*; \boxed{\mathsf{PRF.Eval}(K_i, s^*)})$ instead of $z^* = f_i(x_1^*; \boxed{r^*})$ for a fresh and uniformly sampled value r^*.

The pseudorandomness at punctured point s^* of $\mathsf{PRF.Eval}(K_i, \cdot)$ guarantees that the adversary $\mathcal{A}$ has only a negligible advantage in distinguishing experiments $\mathcal{H}^{(9,i)}$ and $\mathcal{H}^{(10,i)}$. The proof of the following lemma is essentially identical to the proof of Lemma 7 (see [26]):

Lemma 9. *For every $i \in [T_1]$ there exists a probabilistic polynomial-time adversary $\mathcal{B}^{(8,i)\rightarrow(9,i)}$ such that*

$$\left| \Pr\left[\mathcal{H}^{(8,i)}(\lambda) = 1 \right] - \Pr\left[\mathcal{H}^{(9,i)}(\lambda) = 1 \right] \right| \leq \mathsf{Adv}_{\mathsf{puPRF}, \mathcal{B}^{(8,i)\rightarrow(9,i)}}(\lambda).$$

Experiment $\mathcal{H}^{(10,i)}(\lambda)$ Where $i \in [T_1]$. This experiment is obtained from the experiment $\mathcal{H}^{(9,i)}(\lambda)$ by modifying the behavior of the pre-challenge key-generation oracle on the ith query f_i (without modifying its behavior on all other queries). On input the ith query f_i, the key-generation oracle compute $\mathsf{sk}_{f_i} \leftarrow \mathsf{FE.KG}(\mathsf{msk}, \boxed{\mathsf{Right}_{f_i, K_i}})$ instead of $\mathsf{sk}_{f_i} \leftarrow \mathsf{FE.KG}(\mathsf{msk}, \boxed{\mathsf{PuncOutputZ}_{f_i, K_i', s^*}})$.

As in the proof of Lemma 6, the function privacy of the underlying scheme FE (with respect to *deterministic* functions) guarantees that the adversary $\mathcal{A}$ has only a negligible advantage in distinguishing experiments $\mathcal{H}^{(9,i)}$ and $\mathcal{H}^{(10,i)}$. The proof of the following lemma is essentially identical to the proof of Lemma 6 (see [26]):

Lemma 10. *For every $i \in [T_1]$ there exists a probabilistic polynomial-time adversary $\mathcal{B}^{(9,i)\rightarrow(10,i)}$ such that*

$$\left| \Pr\left[\mathcal{H}^{(9,i)}(\lambda) = 1 \right] - \Pr\left[\mathcal{H}^{(10,i)}(\lambda) = 1 \right] \right| \leq \mathsf{Adv}_{\mathsf{FE}, \mathcal{F}', \mathcal{B}^{(9,i)\rightarrow(10,i)}, T}^{\mathsf{FP}}(\lambda) + \mathsf{neg}(\lambda).$$

Next, we observe that experiment $\mathcal{H}^{(4,i+1)}(\lambda)$ is obtained from the experiment $\mathcal{H}^{(10,i)}(\lambda)$ by modifying the challenge ciphertext to be computed using $z^* = \bot$ instead of $z^* = f_i(x_1^*; \mathsf{PRF.Eval}(K_i, s^*))$.

Note that for each function $f \in \{f_1, \ldots, f_T\}$ with an associated PRF key K, for the deterministic functions $\mathsf{Left}_{f,K}$ and $\mathsf{Right}_{f,K}$ and the challenge ciphertext it holds that $\mathsf{Left}_{f,K}(x_b^*, x_1^*, s^*, \bot) = \mathsf{Left}_{f,K}(x_b^*, x_1^*, s^*, z^*)$ and $\mathsf{Right}_{f,K}(x_b^*, x_1^*, s^*, \bot) = \mathsf{Right}_{f,K}(x_b^*, x_1^*, s^*, z^*)$. Therefore, the selective-function message privacy of the underlying scheme FE (with respect to *deterministic* functions) guarantees that the adversary $\mathcal{A}$ has only a negligible advantage in distinguishing experiments $\mathcal{H}^{(10,i)}$ and $\mathcal{H}^{(4,i+1)}$. The proof of the following lemma is essentially identical to the proof of Lemma 5 (see [26]):

Lemma 11. *For every $i \in [T_1]$ there exists a probabilistic polynomial-time adversary $\mathcal{B}^{(10,i)\rightarrow(4,i+1)}$ such that*

$$\left| \Pr\left[\mathcal{H}^{(10,i)}(\lambda) = 1 \right] - \Pr\left[\mathcal{H}^{(4,i+1)}(\lambda) = 1 \right] \right| \leq \mathsf{Adv}_{\mathsf{FE}, \mathcal{F}', \mathcal{B}^{(10,i)\rightarrow(4,i+1)}, T}^{\mathsf{MP}}(\lambda).$$

Experiment $\mathcal{H}^{(11)}(\lambda)$. This experiment is obtained from the experiment $\mathcal{H}^{(4,T+1)}(\lambda)$ by modifying the encryption oracle so that on the challenge input (x_0^*, x_1^*) it outputs $\mathsf{ct} \leftarrow \mathsf{FE.Enc}(\mathsf{msk}, (\boxed{x_1^*}, x_1^*, s^*, \bot))$ instead of $\mathsf{ct} \leftarrow \mathsf{FE.Enc}(\mathsf{msk}, (\boxed{x_b^*}, x_1^*, s^*, \bot))$. Note that this experiment is completely independent of the bit b, and therefore $\Pr\left[\mathcal{H}^{(11)}(\lambda) = 1\right] = 1/2$.

In addition, note that for every function $f \in \{f_1, \ldots, f_{T_1}\}$ with an associated PRF key K, for the deterministic function $\mathsf{Right}_{f,K}$ it holds that $\mathsf{Right}_{f,K}(x_0^*, x_1^*, s^*, \bot) = \mathsf{Right}_{f,K}(x_1^*, x_1^*, s^*, \bot)$. Therefore, the message privacy of the underlying scheme FE (with respect to *deterministic* functions) guarantees that the adversary $\mathcal{A}$ has only a negligible advantage in distinguishing experiments $\mathcal{H}^{(4,T_1+1)}$ and $\mathcal{H}^{(11)}$. The proof of the following lemma is essentially identical to the proof of Lemma 1 (see [26]):

Lemma 12. *There exists a probabilistic polynomial-time adversary* $\mathcal{B}^{(4,T_1+1)\to(11)}$ *such that*

$$\left|\Pr\left[\mathcal{H}^{(4,T_1+1)}(\lambda) = 1\right] - \Pr\left[\mathcal{H}^{(11)}(\lambda) = 1\right]\right| \leq \mathsf{Adv}^{\mathsf{sfFP}}_{\mathsf{FE},\mathcal{F}',\mathcal{B}^{(4,T+1)\to(11)},T}(\lambda).$$

Finally, putting together Lemmas 1–12 with the facts that $\mathsf{Expt}^{\mathsf{aMPRF}}_{\Pi,\mathcal{F},\mathcal{A},T}(\lambda) = \mathcal{H}^{(0)}(\lambda)$, $\mathcal{H}^{(1)}(\lambda) = \mathcal{H}^{(2,1)}(\lambda)$, $\mathcal{H}^{(2,T+1)}(\lambda) = \mathcal{H}^{(3,1)}(\lambda)$, $\mathcal{H}^{(3,T+1)}(\lambda) = \mathcal{H}^{(4,1)}(\lambda)$ and $\Pr\left[\mathcal{H}^{(11)}(\lambda) = 1\right] = 1/2$, we observe that

$$
\begin{aligned}
\mathsf{Adv}^{\mathsf{aMPRF}}_{\Pi,\mathcal{F},\mathcal{A},T} &\stackrel{\mathsf{def}}{=} \left|\Pr\left[\mathsf{Expt}^{\mathsf{aMPRF}}_{\Pi,\mathcal{F},\mathcal{A},T}(\lambda) = 1\right] - \frac{1}{2}\right| \\
&= \left|\Pr\left[\mathcal{H}^{(0)}(\lambda) = 1\right] - \Pr\left[\mathcal{H}^{(11)}(\lambda) = 1\right]\right| \\
&\leq \left|\Pr\left[\mathcal{H}^{(0)}(\lambda) = 1\right] - \Pr\left[\mathcal{H}^{(1)}(\lambda) = 1\right]\right| \\
&\quad + \left|\Pr\left[\mathcal{H}^{(1)}(\lambda) = 1\right] - \Pr\left[\mathcal{H}^{(2,1)}(\lambda) = 1\right]\right| \\
&\quad + \sum_{j=2}^{3}\sum_{i=1}^{T_2} \left|\Pr\left[\mathcal{H}^{(j,i)}(\lambda) = 1\right] - \Pr\left[\mathcal{H}^{(j,i+1)}(\lambda) = 1\right]\right| \\
&\quad + \sum_{i=1}^{T_1}\sum_{j=4}^{9} \left|\Pr\left[\mathcal{H}^{(j,i)}(\lambda) = 1\right] - \Pr\left[\mathcal{H}^{(j+1,i)}(\lambda) = 1\right]\right| \\
&\quad + \sum_{i=1}^{T_1} \left|\Pr\left[\mathcal{H}^{(10,i)}(\lambda) = 1\right] - \Pr\left[\mathcal{H}^{(4,i+1)}(\lambda) = 1\right]\right| \\
&\quad + \left|\Pr\left[\mathcal{H}^{(4,T+1)}(\lambda) = 1\right] - \Pr\left[\mathcal{H}^{(11)}(\lambda) = 1\right]\right| \\
&\leq (T_1(\lambda) + T_2(\lambda)) \cdot \Delta(\lambda) + \mathsf{neg}(\lambda) \\
&= T(\lambda) \cdot \Delta(\lambda) + \mathsf{neg}(\lambda).
\end{aligned}
$$

∎

Acknowledgments. We thank Zvika Brakerski for various insightful discussions.

References

1. Agrawal, S., Agrawal, S., Badrinarayanan, S., Kumarasubramanian, A., Prabhakaran, M., Sahai, A.: Function private functional encryption and property preserving encryption: New definitions and positive results. Cryptology ePrint Archive, Report 2013/744 (2013)
2. Agrawal, S., Gorbunov, S., Vaikuntanathan, V., Wee, H.: Functional encryption: New perspectives and lower bounds. In: Canetti, R., Garay, J.A. (eds.) CRYPTO 2013, Part II. LNCS, vol. 8043, pp. 500–518. Springer, Heidelberg (2013)
3. Alwen, J., Barbosa, M., Farshim, P., Gennaro, R., Gordon, S.D., Tessaro, S., Wilson, D.A.: On the relationship between functional encryption, obfuscation, and fully homomorphic encryption. In: Stam, M. (ed.) IMACC 2013. LNCS, vol. 8308, pp. 65–84. Springer, Heidelberg (2013)
4. Ananth, P., Boneh, D., Garg, S., Sahai, A., Zhandry, M.: Differing-inputs obfuscation and applications. Cryptology ePrint Archive, Report 2013/689 (2013)
5. Ananth, P., Brakerski, Z., Segev, G., Vaikuntanathan, V.: The trojan method in functional encryption: From selective to adaptive security, generically. Cryptology ePrint Archive, Report 2014/917 (2014)
6. Barak, B., Goldreich, O., Impagliazzo, R., Rudich, S., Sahai, A., Vadhan, S.P., Yang, K.: On the (im)possibility of obfuscating programs. Journal of the ACM 59(2), 6 (2012)
7. Bellare, M., O'Neill, A.: Semantically-secure functional encryption: Possibility results, impossibility results and the quest for a general definition. In: Abdalla, M., Nita-Rotaru, C., Dahab, R. (eds.) CANS 2013. LNCS, vol. 8257, pp. 218–234. Springer, Heidelberg (2013)
8. Boneh, D., Franklin, M.K.: Identity-based encryption from the Weil pairing. SIAM Journal on Computing 32(3), 586–615 (2003), preliminary version in: Kilian, J. (ed.) CRYPTO 2001. LNCS, vol. 2139, pp. 213–229. Springer, Heidelberg (2001)
9. Boneh, D., Raghunathan, A., Segev, G.: Function-private identity-based encryption: Hiding the function in functional encryption. In: Canetti, R., Garay, J.A. (eds.) CRYPTO 2013, Part II. LNCS, vol. 8043, pp. 461–478. Springer, Heidelberg (2013)
10. Boneh, D., Raghunathan, A., Segev, G.: Function-private subspace-membership encryption and its applications. In: Sako, K., Sarkar, P. (eds.) ASIACRYPT 2013, Part I. LNCS, vol. 8269, pp. 255–275. Springer, Heidelberg (2013)
11. Boneh, D., Sahai, A., Waters, B.: Functional encryption: Definitions and challenges. In: Ishai, Y. (ed.) TCC 2011. LNCS, vol. 6597, pp. 253–273. Springer, Heidelberg (2011)
12. Boneh, D., Waters, B.: Constrained pseudorandom functions and their applications. In: Sako, K., Sarkar, P. (eds.) ASIACRYPT 2013, Part II. LNCS, vol. 8270, pp. 280–300. Springer, Heidelberg (2013)
13. Boyle, E., Chung, K.-M., Pass, R.: On extractability obfuscation. In: Lindell, Y. (ed.) TCC 2014. LNCS, vol. 8349, pp. 52–73. Springer, Heidelberg (2014)
14. Boyle, E., Goldwasser, S., Ivan, I.: Functional signatures and pseudorandom functions. In: Krawczyk, H. (ed.) PKC 2014. LNCS, vol. 8383, pp. 501–519. Springer, Heidelberg (2014)
15. Brakerski, Z., Segev, G.: Function-private functional encryption in the private-key setting. Cryptology ePrint Archive, Report 2014/550 (2014)
16. Cocks, C.: An identity based encryption scheme based on quadratic residues. In: Honary, B. (ed.) Cryptography and Coding 2001. LNCS, vol. 2260, pp. 360–363. Springer, Heidelberg (2001)

17. De Caro, A., Iovino, V., Jain, A., O'Neill, A., Paneth, O., Persiano, G.: On the achievability of simulation-based security for functional encryption. In: Canetti, R., Garay, J.A. (eds.) CRYPTO 2013, Part II. LNCS, vol. 8043, pp. 519–535. Springer, Heidelberg (2013)

18. Garg, S., Gentry, C., Halevi, S., Raykova, M., Sahai, A., Waters, B.: Candidate indistinguishability obfuscation and functional encryption for all circuits. In: Proceedings of the 54th Annual IEEE Symposium on Foundations of Computer Science, pp. 40–49 (2013)

19. Garg, S., Gentry, C., Halevi, S., Zhandry, M.: Fully secure functional encryption without obfuscation. Cryptology ePrint Archive, Report 2014/666 (2014)

20. Goldreich, O., Goldwasser, S., Micali, S.: How to construct random functions. Journal of the ACM 33(4), 792–807 (1986)

21. Goldwasser, S., Gordon, S.D., Goyal, V., Jain, A., Katz, J., Liu, F.-H., Sahai, A., Shi, E., Zhou, H.-S.: Multi-input functional encryption. In: Nguyen, P.Q., Oswald, E. (eds.) EUROCRYPT 2014. LNCS, vol. 8441, pp. 578–602. Springer, Heidelberg (2014)

22. Goldwasser, S., Kalai, Y., Popa, R.A., Vaikuntanathan, V., Zeldovich, N.: Reusable garbled circuits and succinct functional encryption. In: Proceedings of the 45th Annual ACM Symposium on Theory of Computing, pp. 555–564 (2013)

23. Gorbunov, S., Vaikuntanathan, V., Wee, H.: Functional encryption with bounded collusions via multi-party computation. In: Safavi-Naini, R., Canetti, R. (eds.) CRYPTO 2012. LNCS, vol. 7417, pp. 162–179. Springer, Heidelberg (2012)

24. Goyal, V., Jain, A., Koppula, V., Sahai, A.: Functional encryption for randomized functionalities. Cryptology ePrint Archive, Report 2013/729 (2013), to appear in TCC 2015

25. Kiayias, A., Papadopoulos, S., Triandopoulos, N., Zacharias, T.: Delegatable pseudorandom functions and applications. In: Proceedings of the 20th Annual ACM Conference on Computer and Communications Security, pp. 669–684 (2013)

26. Komargodski, I., Segev, G., Yogev, E.: Functional encryption for randomized functionalities in the private-key setting from minimal assumptions. Cryptology ePrint Archive, Report 2014/868 (2014)

27. O'Neill, A.: Definitional issues in functional encryption. Cryptology ePrint Archive, Report 2010/556 (2010)

28. Sahai, A., Waters, B.: Slides on functional encryption (2008), http://www.cs.utexas.edu/~bwaters/presentations/files/functional.ppt

29. Sahai, A., Waters, B.: How to use indistinguishability obfuscation: deniable encryption, and more. In: Proceedings of the 46th Annual ACM Symposium on Theory of Computing, pp. 475–484 (2014)

30. Shamir, A.: Identity-based cryptosystems and signature schemes. In: Blakely, G.R., Chaum, D. (eds.) CRYPTO 1984. LNCS, vol. 196, pp. 47–53. Springer, Heidelberg (1985)

31. Shen, E., Shi, E., Waters, B.: Predicate privacy in encryption systems. In: Reingold, O. (ed.) TCC 2009. LNCS, vol. 5444, pp. 457–473. Springer, Heidelberg (2009)

32. Waters, B.: A punctured programming approach to adaptively secure functional encryption. Cryptology ePrint Archive, Report 2014/588 (2014)

Separations in Circular Security for Arbitrary Length Key Cycles

Venkata Koppula, Kim Ramchen, and Brent Waters*

University of Texas at Austin, Austin, USA
{kvenkata,kramchen,bwaters}@cs.utexas.edu

Abstract. While standard notions of security suffice to protect any message supplied by an adversary, in some situations stronger notions of security are required. One such notion is *n-circular security*, where ciphertexts $\mathsf{Enc}(\mathsf{pk}_1, \mathsf{sk}_2), \mathsf{Enc}(\mathsf{pk}_2, \mathsf{sk}_3), \ldots, \mathsf{Enc}(\mathsf{pk}_n, \mathsf{sk}_1)$ should be indistinguishable from encryptions of zero.

In this work we prove the following results for n-circular security, based upon recent candidate constructions of indistinguishability obfuscation [18,16] and one way functions:

- For any n there exists an encryption scheme that is IND-CPA secure but not n-circular secure.
- There exists a bit encryption scheme that is IND-CPA secure, but not 1-circular secure.
- If there exists an encryption system where an attacker can distinguish a key encryption cycle from an encryption of zeroes, then in a transformed cryptosystem there exists an attacker which recovers secret keys from the encryption cycles.

The last result is generic and applies to any such cryptosystem.

1 Introduction

The classical notion of secure encryption, due to Goldwasser and Micali [20] demands that random encryptions of two messages submitted by the adversary should be indistinguishable. However this security notion makes no guarantees about the security of encrypting messages which the adversary is unable to generate - indeed this was observed by Goldwasser and Micali. Of particular interest is when an adversary can receive encryptions of messages which depend upon the *secret key*. The resulting notion of security against *key dependent message* attacks was first studied by Black et al [8].

* Supported by NSF CNS-0915361 and CNS-0952692, CNS-1228599 DARPA through the U.S. Office of Naval Research under Contract N00014-11-1-0382, DARPA N11AP20006, Google Faculty Research award, the Alfred P. Sloan Fellowship, Microsoft Faculty Fellowship, and Packard Foundation Fellowship. Any opinions, findings, and conclusions or recommendations expressed in this material are those of the author(s) and do not necessarily reflect the views of the Department of Defense or the U.S. Government.

Y. Dodis and J.B. Nielsen (Eds.): TCC 2015, Part II, LNCS 9015, pp. 378–400, 2015.

A particularly prominent special case of KDM security, introduced by Camenisch and Lysyanskaya [14], is *n-circular security*. Let $\mathsf{pk}_1, \ldots, \mathsf{pk}_n$ be public keys. An encryption scheme is said to be n-circular secure, if an adversary is unable to distinguish $\mathsf{Enc}(\mathsf{pk}_1, \mathsf{sk}_2), \mathsf{Enc}(\mathsf{pk}_2, \mathsf{sk}_3), \ldots, \mathsf{Enc}(\mathsf{pk}_n, \mathsf{sk}_1)$ from corresponding zero encryptions. Camenisch and Lysyanskaya used circular secure encryption to build an anonymous credentials scheme with "all or nothing" sharing [14]. In fact, circular security for $n \geq 1$ arises naturally in many other applications. A common scenario is when a disk utility is used to encrypt a partition on which the secret key has been stored. Another situation is Gentry's "bootstrapping" of a somewhat homomorphic encryption to a fully homomorphic encryption [19]. In this case the decryption circuit associated with the secret key is encrypted and published in the public parameters and used to "refresh" a ciphertext periodically. Finally, circular security is used in formal methods to prove the soundness of symbolic protocols [2,22].

There have been several postive results on circular security and more generally KDM security. In the random oracle model, Black et al. [8] and independently Camenisch and Lysyanskaya [14] gave constructions for KDM secure encryption. Some time later Boneh, Hamburg, Halevi and Ostrovsky gave the first construction of circular secure encryption in the standard model [9]. Their construction provided instantiations of n-circular secure encryption for arbitrary n and in fact provided security for a broader class of key dependent messages - namely all affine functions of the secret key. Continuing in this vein, Applebaum et al [5] presented efficient constructions for affine functions under the LWE and LPN assumptions - the former for public key encryption and both for symmetric key encryption. Later works [21,11,7,12,4,23,13,3] focussed on extending the class of functions and improving efficiency of the constructions.

While there have been many positive advances for circular secure encryption and related functionalities, fewer negative results are known. *One fundamental question is whether it might be possible that circular security is implied by semantic security?* If this held, then it would have important consequences for the design of cryptographic primitives. In particular, an affirmative answer for *any* n would imply a method to construct secure fully homomorphic encryption from mildly or leveled homomorphic encryption. For small n concrete negative results are known. Indeed for $n = 1$, a folklore counterexample exists. For $n = 2$, Acar et al. [1] presented a counterexample under the SXDH assumption. Cash et al. [15] showed how to strengthen this result, with a counterexample for $n = 2$ under a weaker definition of circular security. Despite these advances, for $n > 2$ the problem has largely remained open.

A related question is whether bit-by-bit encryption might suffice for protecting the secret key, i.e. ensure 1-circular security. Again there is partial negative information in that Rothblum [25] has showed, interestingly, that if there exist l-multilinear groups of order p, with $p \leq 2^l$, in which the SXDH assumption holds, then there exists a semantically secure encryption scheme which is not 1-circular secure. Unfortunately, existing candidates for multilinear group schemes

[17,16] do not meet the SXDH requirement.[1] Consequently there are no existing candidates for the Rothblum counterexample. As Rothblum observes, if bit by bit encryption implied circular security, this would give another avenue for utilizing Gentry's bootstrapping.

1.1   Our Contribution

We present the following results:

Counterexample for n-Circular Security. We construct an encryption scheme that is IND-CPA secure but not n-circular secure.

Bit Encryption Counterexample. We construct a bit encryption scheme that is IND-CPA secure, but not circular secure.

Key Recovery from n-Circular Insecurity. Suppose there exists an IND-CPA secure encryption system where there exists an adversary that can distinguish an encryption cycle from the encryption of zeroes. We show how to transform this into an IND-CPA security cryptosystem where the adversary can recover the secret keys from the encryption cycle.

Both the constructions utilize the recent construction of *indistinguishability obfuscation* for polynomial sized circuits by Garg et al. [18] and one way functions. An indistinguishability obfuscation of a program g is a program $i\mathcal{O}(g)$ with a weaker security guarantee: if two programs g and g' have the same input-output behavior, then $i\mathcal{O}(g)$ and $i\mathcal{O}(g')$ are computationally indistinguishable. As argued by [18,27], indistinguishability obfuscation is the weakest definition of obfuscation, and unlike black box obfuscation, there are no known impossibility results for indistinguishability obfuscation.

Counterexample for n-circular security: We begin by giving intuition for our encryption scheme. Let us consider any IND-CPA secure encryption scheme $\mathcal{PKE} = (\mathsf{Keygen}, \mathsf{Encrypt}, \mathsf{Decrypt})$. We show how this encryption scheme can by modified by providing some *auxiliary information* as part of the *ciphertext*, so that the scheme is n-circular insecure, and at the same time, remains IND-CPA secure. We approach the problem in two steps. We first design an approach that works with black box obfuscation. Then we design new techniques to move our construction and proof of security to use indistinguishability obfuscation.

To construct our counterexample we begin with a standard encryption system and then modify the encryption algorithm. When encrypting a message m, in addition to the $\mathcal{PKE}$ ciphertext c, we also give out a *cycle detection program* g^m which can be used to detect whether a cycle is present or not. The program g^m has m hardwired, takes n inputs $c_1, \ldots, c_n$, and works as follows: It decrypts, if possible, c_2 using m to obtain m_2, c_3 using m_2 to obtain m_3 and so on.

---

[1] One interesting question is whether there is a simple modification of Rothblum's candidate construction and proof that can be modified to work under the current multilinear candidates. Neither we nor the author of the construction are aware of any such modification [26].

If any decryption fails, it aborts and outputs 0. If it reaches the end of cycle, it outputs 1.

Let us consider a polynomial time adversary who is given n ciphertexts $\mathsf{ct}_1, \ldots, \mathsf{ct}_n$, where each ct_i consists of a $\mathcal{PKE}$ ciphertext c_i and a program g_i. The adversary runs program g_1 with inputs $c_1, \ldots, c_n$. If these are encryptions of secret keys $\mathsf{sk}_2, \ldots, \mathsf{sk}_n, \mathsf{sk}_1$ respectively, then g_1 runs to completion outputting 1, else it outputs 0. Therefore, using this additional information, we can detect whether there is a cycle or not. However, this scheme in itself is not IND-CPA secure since g^m may leak the value m. Therefore, as part of the ciphertext, we publish a black box obfuscation of g^m: $\mathcal{O}(g^m)$. One can then argue that black box obfuscation ensures that the value m is not leaked, and hence it is IND-CPA secure.

Unfortunately, as shown by [6], it is not possible to achieve general black box obfuscation even for simple functionalities;[2] therefore, we modify our construction so as to use the weaker indistinguishability obfuscation. Our key idea is to have a set of *valid* and *invalid* public keys for each secret key such that the valid and invalid public keys are computationally indistinguishable from just the public key, *but validity is discernible given a secret key*. In our system we use such keys. In addition, at the end of the cycle detection program, we add a validity check, to ensure that pk_1 is a valid public key corresponding to m_n.

While this modification still ensures that the scheme is n-circular insecure, we need to prove IND-CPA security. Our proof of this proceeds in two hybrid steps. First, since the valid and invalid keys are indistinguishable, the real IND-CPA security game is computationally indistinguishable from one in which we substitute invalid public keys for the real ones. Next, we observe that these invalid public keys must necessarily fail the validity check at the end of the cycle detection program, and therefore the program *always outputs 0*. Therefore, instead of outputting an obfuscation of the cycle detection program, if we output the obfuscation of a program that always outputs 0, the two hybrids remain indistinguishable by the property of indistinguishability obfuscation. Finally, a program that always aborts leaks no information about m, and therefore the scheme is IND-CPA secure.

One potential view of this is as a novel and extreme application of punctured programming [27]. Once we alter the keys to be invalid, we can completely gut the obfuscated program to be one that simply outputs 0. In indepedent and concurrent work Boneh and Zhandry [10] apply a notion similar to our invalid/valid key structure (although they do not use that terminology) to building multi-party key exchange, broadcast and traitor tracing systems. An important contribution of both papers is that they demonstrate the power of altering the structure of public keys in combination with indistinguishability obfuscation.

---

[2] It is of course possible that black box obfuscation is obtainable for this particular functionality. However, we view obtaining our negative result under indistinguishability obfuscation as an important goal.

Bit encryption counterexample: We now consider the problem of bit encryption. We first observe that the aforementioned 'chasing the cycle' technique cannot be used for bit encryption. However, in this case, all encryptions use the same public key. As a result, we can now give out useful *auxiliary cryptographic material* as part of the *public key*. Here we again use the valid-invalid public keys technique. In particular, we modify the Keygen algorithm. Suppose we have a Keygen algorithm for a valid-invalid PKE system as described above that outputs pk, sk. Let pk$'$ be the part of pk used for checking whether pk is a valid public key corresponding to sk, and sk$'$ the part of sk used for decrypting ciphertexts. Now consider the program $g^{\mathsf{pk}',\mathsf{sk}'}$ that has pk$'$, sk$'$ hardwired, and takes l inputs $c_1, \ldots, c_l$. Program $g^{\mathsf{pk}',\mathsf{sk}'}$ decrypts each of the inputs using sk$'$ and checks (using pk$'$) whether pk is a valid public key corresponding to the resulting string. In our modified encryption scheme, in addition to pk, we also give out an indistinguishability obfuscation of program $g^{\mathsf{pk}',\mathsf{sk}'}$.

Clearly, this encryption scheme is not bit circular secure. To prove IND-CPA security, we use similar hybrids as before. In the first hybrid experiment, we switch from valid to invalid public keys. Since the valid and invalid public keys are computationally indistinguishable, these hybrid experiments are computationally indistinguishable. Finally, we output an obfuscation of a program that always aborts, thereby ensuring that no information about the secret key sk is leaked by the program obfuscation.

Key recovery from n-circular insecurity: One interesting question posed in the setting of circular security is what is the right definition of security. While preventing against cycle detection is seemingly the strongest notion, in many applications such as Gentry's bootstrapping it might be sufficient if the system remained semantically secure (for other messages) in the presence of a key cycle, even if the key cycle itself were detectable. Likewise, a counterexample for such a weaker notion of security would be a stronger result. Cash et al. [15] improved upon the work of Acar et al. [1] by giving a such a stronger counterexample which allowed for an attacker to completely recover private keys for the case of key cycles of length two.

The key-recovery from cycles technique of Cash et. al. was tailored specifically to the case of bilinear maps. In this work, we show that if for any n there exists an encryption system where an attacker can distinguish a key encryption cycle from a encryption of zeroes, then we can create a transformed cryptosystem where there exists an attacker which recovers secret keys from the encryption cycles. Thus, for obtaining a strong key recovery counterexample, one only needs to work to obtain a cycle detection counterexample.

Our methods here are in spirit similar to Rothblum's result in [25] for the bit encryption case. When encrypting a message, we also publish a *hint* for each bit of the message, indicating whether the bit is 0 or 1. To determine the bit, we use the cycle detection algorithm. As a consequence, this hint works if and only if we have a cycle of secret keys, therefore ensuring both IND-CPA security and key recovery.

Relation to [24] On October 2013, we initially posted on eprint a paper that contained our three main results: (i) a construction of a public key encryption scheme that is IND-CPA secure but not n-circular secure, (ii) the construction of a bit encryption scheme that is IND-CPA secure but not 1-circular secure and (iii) a transformation of an encryption scheme in which key cycles can be distinguished from encryption of zeroes into one which secret keys can be recovered from encryption cycles. The first two results are based upon indistinguishability obfuscation [18], the last result is completely generic.

Very shortly thereafter, Marcedone and Orlandi [24] showed how to construct a public key encryption scheme that is IND-CPA secure but not n-circular secure using the virtual black box [6] notion of obfuscation. This result was similar to our result (i); however, they used virtual black box obfuscation instead of indistinguishability obfuscation. Four months later, the authors added a result showing an n-circular security counterexample using indistinguishability obfuscation, thus matching one of the results contained in this work. (There are no analogues of the other two results in their paper.) The ideas used by [24] [3] for the counterexample posted in February 2014 are very similar to the ones we used in our result (i), posted in October 2013. We strongly view our paper as the origination of the ideas behind this result.

2   Preliminaries

Definition 1 (Public Key Encryption). *A public key encryption scheme* $\mathcal{PKE}$ *is a set of three algorithms (*Keygen, Encrypt, Decrypt*) satisfying the following properties :*

- * ***Key Generation*** Keygen*(1^λ) is a randomized algorithm that takes as input the security parameter λ and outputs public key* pk *and secret key* sk.
- * ***Encryption*** Encrypt*(*pk, m*) is a randomized algorithm that takes as input a public key* pk, *message m and outputs a ciphertext* ct.
- * ***Decryption*** Decrypt*(*sk, ct*) is a deterministic algorithm that takes as input a secret key* sk, *a ciphertext* ct *and outputs m.*

For correctness, we require that for all m,

$$\Pr[\mathsf{Decrypt}(\mathsf{sk}, \mathsf{Encrypt}(\mathsf{pk}, m)) \neq m : (\mathsf{pk}, \mathsf{sk}) \leftarrow \mathsf{Keygen}(1^\lambda)] \leq \mathrm{negl}(\lambda).$$

A public key cryptosystem is called a bit encryption scheme *if its message space is* $\{0, 1\}$.

We define various security notions for public key cryptosystems.

---

[3] The updated draft [24] was accepted in SCN 2014.

Definition 2 (IND-CPA Security).

Let $\mathcal{PKE} = (\mathsf{Keygen}, \mathsf{Encrypt}, \mathsf{Decrypt})$ be a public key cryptosystem. Consider the following game between challenger $\mathcal{C}$ and adversary $\mathcal{A}$:

IND-CPA :

1. $\mathcal{C}$ computes $(\mathsf{pk}, \mathsf{sk}) \leftarrow \mathsf{Keygen}(1^\lambda)$ and sends pk to $\mathcal{A}$.
2. $\mathcal{A}$ sends challenge plaintext messages m_0, m_1 such that $|m_0| = |m_1|$ to $\mathcal{C}$.
3. $\mathcal{C}$ chooses a bit $b \xleftarrow{\$} \{0,1\}$, computes $ct \leftarrow \mathsf{Encrypt}(\mathsf{pk}, m_b)$ and sends ct to $\mathcal{A}$.
4. $\mathcal{A}$ outputs a bit b'.

The advantage of $\mathcal{A}$ is $Adv_{\mathcal{A}} = \Pr[b = b'] - \frac{1}{2}$.

$\mathcal{PKE}$ is said to be IND-CPA secure if for all PPT algorithms $\mathcal{A}$, $Adv_{\mathcal{A}} \leq \mathrm{negl}(\lambda)$.

2.1   Circular Security

Definition 3 (n-Circular Security [14]).

Let $\mathcal{PKE} = (\mathsf{Keygen}, \mathsf{Encrypt}, \mathsf{Decrypt})$ be a public key cryptosystem. Consider the following game between challenger $\mathcal{C}$ and adversary $\mathcal{A}$:

---

n-Circular Security :

1. $\mathcal{C}$ computes $(\mathsf{pk}_i, \mathsf{sk}_i) \leftarrow \mathsf{Keygen}(1^\lambda)$ for $1 \leq i \leq n$
2. $\mathcal{C}$ chooses a bit $b \xleftarrow{\$} \{0,1\}$.
   - If $b = 0$, $\mathcal{C}$ computes $y_i = \mathsf{Encrypt}(\mathsf{pk}_i, \mathsf{sk}_{(i \bmod n)+1})$ for $1 \leq i \leq n$
   - Else $\mathcal{C}$ computes $y_i = \mathsf{Encrypt}(\mathsf{pk}_i, 0^{|\mathsf{sk}_{(i \bmod n)+1}|})$ for $1 \leq i \leq n$
3. $\mathcal{C}$ sends $(\mathsf{pk}_1, \ldots, \mathsf{pk}_n, y_1, \ldots, y_n)$ to $\mathcal{A}$.
4. $\mathcal{A}$ outputs b'.

The advantage of $\mathcal{A}$ is $Adv_{\mathcal{A}} = \Pr[b = b'] - \frac{1}{2}$.

$\mathcal{PKE}$ is said to be n-circular secure if for all PPT algorithms $\mathcal{A}$, $Adv_{\mathcal{A}} \leq \mathrm{negl}(\lambda)$

A *weak* notion of circular security was defined in [15] as follows :

Definition 4 (n-Weak Circular Security). Let $\mathcal{PKE} = (\mathsf{Keygen}, \mathsf{Encrypt}, \mathsf{Decrypt})$ be a public key cryptosystem. Consider the following game between challenger $\mathcal{C}$ and adversary $\mathcal{A}$:

---

n-Weak Circular Security :

1. $\mathcal{C}$ computes $(\mathsf{pk}_i, \mathsf{sk}_i) \leftarrow \mathsf{Keygen}(1^\lambda)$ for $1 \le i \le n$.
   Next, it computes $y_i = \mathsf{Encrypt}(\mathsf{pk}_i, \mathsf{sk}_{(i \bmod n)+1})$ for $1 \le i \le n$.
   It sends $(\mathsf{pk}_1, \ldots, \mathsf{pk}_n, y_1, \ldots, y_n)$ to $\mathcal{A}$.
2. $\mathcal{A}$ sends challenge plaintext messages m_0, m_1 such that $|m_0| = |m_1|$ and $j \in [1, n]$ to $\mathcal{C}$
3. $\mathcal{C}$ chooses a bit $b \xleftarrow{\$} \{0, 1\}$ and sends $\mathsf{Encrypt}(\mathsf{pk}_j, m_b)$ to $\mathcal{A}$.
4. $\mathcal{A}$ outputs b'.

The advantage of $\mathcal{A}$ is $Adv_\mathcal{A} = \Pr[b = b'] - \frac{1}{2}$.
$\mathcal{PKE}$ is said to be n-weak circular secure if for all PPT algorithms $\mathcal{A}$, $Adv_\mathcal{A} \le$ negl(λ)

Definition 5 (n-Circular Security with respect to Key Recovery). Let $\mathcal{PKE} = (\mathsf{Keygen}, \mathsf{Encrypt}, \mathsf{Decrypt})$ be a public key cryptosystem. Consider the following game between challenger $\mathcal{C}$ and adversary $\mathcal{A}$:

---

n-Circular Security with respect to Key Recovery :

1. $\mathcal{C}$ computes $(\mathsf{pk}_i, \mathsf{sk}_i) \leftarrow \mathsf{Keygen}(1^\lambda)$ for $1 \le i \le n$.
   Next, it computes $y_i = \mathsf{Encrypt}(\mathsf{pk}_i, \mathsf{sk}_{(i \bmod n)+1})$ for $1 \le i \le n$.
   It sends $(\mathsf{pk}_1, \ldots, \mathsf{pk}_n, y_1, \ldots, y_n)$ to $\mathcal{A}$.
2. $\mathcal{A}$ outputs sk'_1.

The advantage of $\mathcal{A}$ is $Adv_\mathcal{A} = \Pr[sk_1 = sk'_1]$.
$\mathcal{PKE}$ is said to be n-circular secure with respect to key recovery if for all PPT algorithms $\mathcal{A}$, $Adv_\mathcal{A} \le$ negl(λ)

Remark. If a public key encryption scheme is n-circular secure, then it is also n-weak circular secure. Similarly, if a scheme is n-weak circular secure, then it is also n-circular secure with respect to key recovery.

The notion of circular security can be extended to bit encryption schemes. The following definition is actually equivalent to Definition 3 in the case that $n = 1$, but will be slightly more convenient to work with.

Definition 6 (1-Circular Security of Bit-by-bit Encryption). Let $\mathcal{PKE} = (\mathsf{Keygen}, \mathsf{Encrypt}, \mathsf{Decrypt})$ be a bit encryption scheme. Consider the following game between challenger $\mathcal{C}$ and adversary $\mathcal{A}$:

1-*Circular Security of Bit-by-bit Encryption* :

1. $\mathcal{C}$ chooses $b \xleftarrow{\$} \{0, 1\}$. $\mathcal{C}$ generates the public key and secret key $(\mathsf{pk}, \mathsf{sk}) \leftarrow \mathsf{Keygen}(1^\lambda)$ and sends pk to $\mathcal{A}$.
2. For a polynomial number of queries
   (a) $\mathcal{A}$ queries for encryption of j_i^{th} bit of sk.
   (b) If $b = 1$, $\mathcal{C}$ sends $\mathsf{ct} \leftarrow \mathsf{Encrypt}(\mathsf{pk}, \mathsf{sk}_{j_i})$. Else $\mathcal{C}$ sends $\mathsf{ct} \leftarrow \mathsf{Encrypt}(\mathsf{pk}, 0)$.
3. $\mathcal{A}$ outputs b'

The advantage of $\mathcal{A}$ is $Adv_{\mathcal{A}} = \Pr[b = b'] - \frac{1}{2}$.
$\mathcal{PKE}$ is said to be bit circular secure *if for all PPT algorithms $\mathcal{A}$, $Adv_{\mathcal{A}} \leq \mathrm{negl}(\lambda)$*

Rothblum in [25] showed that this notion of bit circular security, which he called circular security with respect to indistinguishability of oracles, is equivalent to the seemingly stronger notion where the adversary must extract the entire secret key, given encryptions of the secret key bits. Therefore, it suffices to restrict our attention to this notion of bit circular security.

2.2 Indistinguishability Obfuscation

Next, we recall the definition of indistinguishability obfuscation from [27].

Definition 7. *(Indistinguishability Obfuscation) A uniform PPT machine $i\mathcal{O}$ is called an indistinguishability obfuscator for a circuit class $\{\mathcal{C}_\lambda\}$ if it satisfies the following conditions:*

- *(Preserving Functionality) For all security parameters $\lambda \in \mathbb{N}$, for all $C \in \mathcal{C}_\lambda$, for all inputs x, we have that $C'(x) = C(x)$ where $C' \leftarrow i\mathcal{O}(\lambda, C)$.*
- *(Indistinguishability of Obfuscation) For any (not necessarily uniform) PPT distinguisher $(Samp, D)$, there exists a negligible function $\mathrm{negl}(\cdot)$ such that the following holds: if for all security parameters $\lambda \in \mathbb{N}, \Pr[\forall x, C_0(x) = C_1(x) : (C_0; C_1; \sigma) \leftarrow Samp(1^\lambda)] > 1 - \mathrm{negl}(\lambda)$, then*

$$| \Pr[D(\sigma, i\mathcal{O}(\lambda, C_0)) = 1 : (C_0; C_1; \sigma) \leftarrow Samp(1^\lambda)] -$$
$$\Pr[D(\sigma, i\mathcal{O}(\lambda, C_1)) = 1 : (C_0; C_1; \sigma) \leftarrow Samp(1^\lambda)]| \leq \mathrm{negl}(\lambda)$$

In a recent work, [18] showed a candidate indistinguishability obfuscator for the circuit class *P/poly*.

3   Counter Example for n-Circular Security

In this section, we describe how to build for any n, a cryptosystem $\mathcal{PKE}$ that is IND-CPA secure, but not n-circular secure.

Let $\mathcal{PKE}_A = (\mathsf{Keygen}_A, \mathsf{Encrypt}_A, \mathsf{Decrypt}_A)$ be a public key encryption scheme with message space $\mathcal{M}_A = \{0, 1\}^{2l}$, key space $\mathcal{K}_A \subseteq \{0, 1\}^l$ and ciphertext space $\mathcal{C}_A$. Let $G : \{0, 1\}^l \to \{0, 1\}^{2l}$ be a PRG family. We construct cryptosystem $\mathcal{PKE} = (\mathsf{Keygen}, \mathsf{Encrypt}, \mathsf{Decrypt})$ as follows:

- Keygen(1^λ): Let $(\mathsf{sk}_A, \mathsf{pk}_A) \leftarrow \mathsf{Keygen}_A(1^\lambda)$. Let $r \xleftarrow{\$} \{0,1\}^l$ and $t = G(r)$.
  Set $\mathsf{sk} = (\mathsf{sk}_A, r)$. Set $\mathsf{pk} = (\mathsf{pk}_A, t)$.
- Encrypt(pk, m, r): Parse $\mathsf{pk} = (\mathsf{pk}_A, t)$. Let $C \leftarrow \mathsf{Encrypt}_A(\mathsf{pk}_A, m)$.
  Let CycleFind be a circuit defined as follows :

CycleFind :

Inputs : $C_1, \ldots, C_n \in \mathcal{C}_A$

Constants : m, t.

1. Parse $m = (\mathsf{sk}_2, r)$.
2. For i=2 to n
   (a) Let $(\mathsf{sk}_{(i \bmod n)+1}, r_{(i \bmod n)+1}) = \mathsf{Decrypt}_A(\mathsf{sk}_i, C_i)$ or output $\perp$ if $\mathsf{Decrypt}_A$ fails.
3. If $G(r_1) = t$ output 1, else output $\perp$.

The circuit CycleFind takes as input n ciphertexts $C_1, \ldots, C_n$, and has constants m, t hardwired, where the circuit is appropriately padded to be of the same size as the corresponding ones in the security proof.
Compute obfuscation of circuit CycleFind$_{m,t}$ as $O \leftarrow i\mathcal{O}(\lambda, \mathsf{CycleFind}_{m,t})$.
The ciphertext $ct = (C, O)$.

- Decrypt(sk, ct): Parse $\mathsf{sk} = (\mathsf{sk}_A, r)$ and $ct = (C, O)$. Output $\mathsf{Decrypt}_A(\mathsf{sk}_A, C)$.
  String O is ignored.

Correctness follows immediately from the correctness of the original scheme $\mathcal{PKE}_A$.

3.1 The Attack

Proposition 1. *The above construction is n-circular insecure.*

Proof. We construct a polynomial time adversary $\mathcal{A}$ that breaks the n-circular security of the above construction as follows. $\mathcal{A}$ receives $(\mathsf{pk}_1, \ldots, \mathsf{pk}_n, y_1, \ldots, y_n)$ from the challenger. $\mathcal{A}$ parses y_i as $(C_i, \mathcal{O}_i)$ where $\mathcal{O}_i$ is a circuit. $\mathcal{A}$ outputs the value $b \leftarrow \mathcal{O}_1(C_1, \ldots, C_n)$. By construction this is 1 iff $(y_1, \ldots, y_n)$ is an encryption cycle with respect to $\mathcal{PKE}$.

3.2 IND-CPA Security

In order to show that our construction is IND-CPA secure, we construct a series of hybrid experiments as follows.

Game 0: IND-CPA *Game*

1. Choose $r \xleftarrow{\$} \{0,1\}^l$ and set $t = G(r)$.
2. Let $(\mathsf{sk}_A, \mathsf{pk}_A) \leftarrow \mathsf{Keygen}_A(1^\lambda)$.
3. Let $\mathsf{sk} = (\mathsf{sk}_A, r)$ and $\mathsf{pk} = (\mathsf{pk}_A, t)$.
4. Suppose $\mathcal{A}$ sends $m_0, m_1 : |m_0| = |m_1|$.
5. Choose $b \xleftarrow{\$} \{0,1\}$.

6. Let $C = \mathsf{Encrypt}_A(\mathsf{pk}_A, m_b)$.
7. Let $O = i\mathcal{O}(\lambda, \mathsf{CycleFind})$ where $\mathsf{CycleFind}$ is the circuit described above.
8. Let $ct_b = (C, O)$. Send ct_b to $\mathcal{A}$.
9. Let $b' \leftarrow \mathcal{A}_2(\delta, ct_b)$.

$\mathcal{A}$ wins if $b = b'$ and has advantage $Adv_\mathcal{A} = \Pr[b = b'] - 1/2$.

Game 1: This game proceeds identically as the IND-CPA game, except we modify Step 1 as follows.

1. Choose $r \xleftarrow{\$} \{0,1\}^l$ and choose $t \xleftarrow{\$} \{0,1\}^{2l}$. Note that r is information theoretically hidden in this experiment.

Game 2: This game proceeds identically as **Game 1**, except we modify Step 7 as follows.
Let $\mathsf{CycleReject}$ be the following circuit:

CycleReject :
Inputs : $C_1, \ldots, C_n \in \mathcal{C}_A$
Constants : $0^{w'}$

1. Output $\perp$

The circuit $\mathsf{CycleReject}$ takes as input n ciphertexts $C_1, \ldots, C_n$, has zero padding of length w'. The constant w in circuit $\mathsf{CycleFind}$ and w' in circuit $\mathsf{CycleReject}$ are chosen such that the size of circuits $\mathsf{CycleFind}$ and $\mathsf{CycleReject}$ are equal.
Let $O = i\mathcal{O}(\lambda, \mathsf{CycleReject})$.

Proposition 2. *Suppose that there exists a polynomial time adversary $\mathcal{A}$ such that $Game_0 Adv_\mathcal{A} - Game_1 Adv_\mathcal{A} = \epsilon$. Then there exists a polynomial time adversary $\mathcal{B}$ who distinguishes the output of G from random with advantage $\epsilon_{PRG} = \epsilon$.*

Proof. The only modification is that t is computed as random $2l$-bit string rather than the output of G. The algorithm $\mathcal{B}$ is defined as follows :

1. $\mathcal{B}$ receives $t \in \{0,1\}^{2l}$ from PRG Challenger $\mathcal{C}$, where t is either a pseudo-random string generated by G or a truly random string.
2. $\mathcal{B}$ computes $(\mathsf{sk}_A, \mathsf{pk}_A) \leftarrow \mathsf{Keygen}_A(1^\lambda)$. It sets $\mathsf{pk} = (\mathsf{pk}_A, t)$ and sends it to $\mathcal{A}$.
3. $\mathcal{A}$ sends challenge messages m_0, m_1.
4. $\mathcal{B}$ chooses $b \xleftarrow{\$} \{0,1\}$. It sets $C = \mathsf{Encrypt}_A(\mathsf{pk}_A, m_b)$. Next, it defines circuit $\mathsf{CycleFind}$, which has m_b and t hard-wired. Therefore, $\mathcal{B}$ can define $\mathsf{CycleFind}$, and hence compute $O \leftarrow i\mathcal{O}(\lambda, \mathsf{CycleFind})$. Hence it sets $ct = (C, O)$ and sends it to $\mathcal{A}$.
5. $\mathcal{A}$ outputs a bit b'. If $(b = b')$ $\mathcal{B}$ outputs that the string was pseudorandom. Else $\mathcal{B}$ outputs the string was random.

If $\mathcal{C}$ sends an output of G, then this experiment corresponds to Game 0. If $\mathcal{C}$ sends a truly random string t, then this corresponds to Game 1. Therefore, if $\mathcal{A}$ can distinguish between Game 0 and Game 1 with advantage ϵ, then $\mathcal{B}$ distinguishes a pseudorandom string form a truly random string with advantage ϵ.

Proposition 3. *Suppose that there exists a polynomial time adversary $\mathcal{A}$ such that $Game_1 Adv_\mathcal{A} - Game_2 Adv_\mathcal{A} = \epsilon$. Then there exists a polynomial time adversary $\mathcal{B}$ who breaks the indistinguishability obfuscation with advantage $\epsilon_{iO} = \epsilon$.*

Proof. Recall that $\mathcal{B}$ should comprise a pair of adversaries $(Samp, D)$ as in Definition 2.2. We construct these adversaries as follows.
$Samp(1^\lambda)$:

1. Choose $r \xleftarrow{\$} \{0,1\}^l$ and $t \xleftarrow{\$} \{0,1\}^{2l}$.
2. Let $(\mathsf{sk}_A, \mathsf{pk}_A) \leftarrow \mathsf{Keygen}_A(1^\lambda)$.
3. Let $\mathsf{sk} = (\mathsf{sk}_A, r)$ and $\mathsf{pk} = (\mathsf{pk}_A, t)$.
4. Let $(m_0, m_1) \leftarrow \mathcal{A}(\mathsf{pk}) : |m_0| = |m_1|$.
5. Choose $b \xleftarrow{\$} \{0,1\}$.
6. Let CycleFind be the circuit described in our construction with constants $(m_b, t, 0^w)$ hardwired.
   Let CycleReject be the circuit described in Game 2 with constant $0^{w'}$ hardwired.
7. Output $(g_0 = \mathsf{CycleFind}, g_1 = \mathsf{CycleReject})$.
8. Set $\sigma = (b, m_0, m_1, \mathsf{pk})$.
$D(\sigma, i\mathcal{O}(\lambda, g_z))$:
1. Let $C = \mathsf{Encrypt}_A(\mathsf{pk}_A, m_b)$, let $O = i\mathcal{O}(\lambda, g_z)$.
2. Let $ct = (C, O)$.
3. Let $b' \leftarrow \mathcal{A}(ct, \mathsf{pk})$.
4. D guesses 1 if $b = b'$.

We first prove that $\mathcal{B}$ produces circuits g_0, g_1 which are equivalent on all inputs, with overwhelming probability. Observe that with overwhelming probability t is not in the range of G since $t \xleftarrow{\$} \{0,1\}^{2l}$ and hence $\mathsf{CycleFind}(x)$ outputs $\bot$ for all x. Thus $Samp$ produces circuits CycleReject and CycleFind which are equivalent on all inputs with overwhelming probability, by the random choice of t.

All that remains is to show $Adv_\mathcal{B} = \epsilon$. Let $p_z = \Pr[D(\sigma, i\mathcal{O}(\lambda, g_z)) = 1]$ for $z = 0, 1$. Note that $g_0 = \mathsf{CycleFind}$, hence when $z = 0$ the event $b = b'$ occurs iff $\mathcal{A}$ wins **Game 1**. Similarly $g_1 = \mathsf{CycleReject}$, hence when $z = 1$, the event $b = b'$ occurs iff $\mathcal{A}$ wins **Game 2**. Then $p_0 = 1/2 + Game_1 Adv_\mathcal{A}$, while $p_1 = 1/2 + Game_2 Adv_\mathcal{A}$. Thus $Adv_\mathcal{B} = p_0 - p_1 = Game_1 Adv_\mathcal{A} - Game_2 Adv_\mathcal{A} = \epsilon$.

Finally, we need to show that any polynomial time adversary has only negligible advantage in Game 2. This follows from the fact that $\mathcal{PKE}_A$ is IND-CPA secure.

Proposition 4. *If there exists a polynomial time adversary $\mathcal{A}$ with non negligible advantage ϵ in Game 2, then there exists a polynomial time algorithm $\mathcal{B}$ that can break the IND-CPA security of $\mathcal{PKE}_A$ with advantage $\epsilon_A = \epsilon$.*

Proof. Suppose $\mathcal{A}$ has advantage ϵ in Game 2. We define $\mathcal{B}$ as follows :

1. $\mathcal{B}$ receives pk_A from the IND-CPA Challenger $\mathcal{C}$. It chooses $t \xleftarrow{\$} \{0,1\}^{2l}$ and sends public key $\mathsf{pk} = (\mathsf{pk}_A, t)$ to $\mathcal{A}$.
2. $\mathcal{A}$ sends challenge messages m_0, m_1, which are passed on to $\mathcal{C}$, and receives ciphertext C.
3. $\mathcal{B}$ computes $O \leftarrow i\mathcal{O}(\lambda, \mathsf{CycleReject})$ and sends ciphertext $ct = (C, O)$ to $\mathcal{A}$.
4. $\mathcal{A}$ sends bit b', which $\mathcal{B}$ passes on to $\mathcal{C}$.

Note that if $\mathcal{A}$ wins Game 2, then $\mathcal{B}$ wins the IND-CPA game. Hence the result follows.

The advantage of any polynomial time IND-CPA adversary against $\mathcal{PKE}$ is at most $\epsilon_{PRG} + \epsilon_{iO} + \epsilon_A$. Therefore we have the following theorem.

Theorem 1. *Assuming that G is a secure PRG family, $i\mathcal{O}$ is an indistinguishability obfuscator and $\mathcal{PKE}_A$ is an IND-CPA secure encryption scheme, $\mathcal{PKE}$ is IND-CPA secure but not n-circular secure.*

4   Counter Example for 1-Circular Security of Bit-by-bit Encryption

In this section, we describe a bit encryption scheme that is IND-CPA secure, but is not 1-circular secure.

Let $\mathcal{PKE}_A = (\mathsf{Keygen}_A, \mathsf{Encrypt}_A, \mathsf{Decrypt}_A)$ be a bit encryption cryptosystem with key space $\mathcal{K}_A \subseteq \{0,1\}^l$ and ciphertext space $\mathcal{C}_A$. Let $G : \{0,1\}^l \to \{0,1\}^{2l}$ be a PRG. We construct a bit encryption cryptosystem $\mathcal{PKE} = (\mathsf{Keygen}, \mathsf{Encrypt}, \mathsf{Decrypt})$ as follows :

- $\mathsf{Keygen}(1^\lambda)$: Let $(\mathsf{pk}_A, \mathsf{sk}_A) \leftarrow \mathsf{Keygen}_A(1^\lambda)$. Choose $r \xleftarrow{\$} \{0,1\}^l$ and compute $t = G(r)$. Define a circuit BitCycleFind as follows :

BitCycleFind :
Inputs : $C_1, \ldots, C_l \in \mathcal{C}_A$
Constants : $\mathsf{sk}_A, t, 0^w$ for an appropriately chosen w
1. For $i = 1$ to l
(a) Let $x_i = \mathsf{Decrypt}_A(\mathsf{sk}_A, C_i)$ or output $\perp$ if $\mathsf{Decrypt}_A$ fails.
2. Let $x = x_1 \ldots x_l$. If $G(x) = t$ output 1, else output $\perp$.

  The circuit takes as input l ciphertexts, and has constants sk_A, t and 0^w hardwired. As in the multi-bit encryption, the zero padding is required for the security proof.
  Compute obfuscation of circuit BitCycleFind as $O \leftarrow i\mathcal{O}(\lambda, \mathsf{BitCycleFind})$. Set $\mathsf{pk} = (\mathsf{pk}_A, t, O)$ and $\mathsf{sk} = (\mathsf{sk}_A, r)$.
- $\mathsf{Encrypt}(\mathsf{pk}, m)$: Parse $\mathsf{pk} = (\mathsf{pk}_A, t, O)$. Compute ciphertext $ct \leftarrow \mathsf{Encrypt}_A(\mathsf{pk}_A, m)$.
- $\mathsf{Decrypt}(\mathsf{sk}, ct)$: Parse $\mathsf{sk} = (\mathsf{sk}_A, r)$. Output $\mathsf{Decrypt}(\mathsf{sk}_A, ct)$.

The correctness of $\mathcal{PKE}$ follows directly from the correctness of $\mathcal{PKE}_A$.

4.1   The Attack

Proposition 5. *The above construction is not bit circular secure.*

Proof. We construct a polynomial time adversary $\mathcal{A}$ that breaks the *bit circular security* of the above construction as follows. $\mathcal{A}$ receives public key $\mathsf{pk} = (\mathsf{pk}_A, t, O)$. Next, it queries for encryptions of the last l bits of the second component of the secret key, and receives $ct_1, \ldots, ct_l$. $\mathcal{A}$ outputs $b = O(ct_1, \ldots, ct_l)$. By construction, it follows that $\mathcal{A}$ outputs 1 iff the challenger outputs encryptions of the bits of the secret key sk.

4.2   IND-CPA Security

In this section, we show that our construction $\mathcal{PKE} = (\mathsf{Keygen}, \mathsf{Encrypt}, \mathsf{Decrypt})$ is IND-CPA secure.
As before we construct a sequence of hybrid experiments, and show that the outputs of the hybrid experiments are computationally indistinguishable.

Game 0: IND-CPA

1. Choose $r \xleftarrow{\$} \{0,1\}^l$ and set $t = G(r)$.
2. Let $(\mathsf{pk}_A, \mathsf{sk}_A) \leftarrow \mathsf{Keygen}_A(1^\lambda)$.
3. Let $O = i\mathcal{O}(\lambda, \mathsf{BitCycleFind})$ as described in the construction.
4. Let $\mathsf{sk} = (\mathsf{sk}_A, r)$ and $\mathsf{pk} = (\mathsf{pk}_A, t, O)$. Send pk to $\mathcal{A}$.
5. Choose $b \xleftarrow{\$} \{0,1\}$.
6. Let $ct_b \leftarrow \mathsf{Encrypt}_A(\mathsf{pk}_A, b)$. Send ct to $\mathcal{A}$.
7. Let $b' \leftarrow \mathcal{A}(ct_b)$.

$\mathcal{A}$ wins if $b = b'$ and has advantage $Adv_{\mathcal{A}} = \Pr[b = b'] - 1/2$.

Game 1: This game proceeds identically as the IND-CPA game, except we modify Step 1 as follows.

1. Choose $r \xleftarrow{\$} \{0,1\}^l$ and choose $t \xleftarrow{\$} \{0,1\}^{2l}$. Note that r is information theoretically hidden in this experiment.

Game 2: This game proceeds identically as **Game 1**, except we modify Step 3 as follows.
Let $\mathsf{BitCycleReject}$ be the following circuit:

BitCycleReject :
Inputs : $C_1, \ldots, C_l \in \mathcal{C}_A$
Constants : $0^{w'}$

1. Output $\perp$

The circuit BitCycleReject takes as input l ciphertexts $C_1, \ldots, C_l$, has zero padding of length w'. The constants w in circuit BitCycleFind and w' in circuit BitCycleReject are chosen such that $|\mathsf{BitCycleFind}| = |\mathsf{BitCycleReject}|$ Let $O = i\mathcal{O}(\lambda, \mathsf{BitCycleReject})$.

The proofs of the following indistinguishability results are similar to those of the previous section and are included in Appendix A.

Proposition 6. *Suppose that there exists a polynomial time adversary $\mathcal{A}$ such that $Game_0 Adv_{\mathcal{A}}$ - $Game_1 Adv_{\mathcal{A}} = \epsilon$. Then there exists a polynomial time adversary $\mathcal{B}$ who distinguishes the output of G from random with advantage $\epsilon_{PRG} = \epsilon$.*

Proposition 7. *Suppose that there exists a polynomial time adversary $\mathcal{A}$ such that $Game_1 Adv_{\mathcal{A}}$ - $Game_2 Adv_{\mathcal{A}} = \epsilon$. Then there exists a polynomial time adversary $\mathcal{B}$ who breaks the indistinguishability obfuscation with advantage $\epsilon_{iO} = \epsilon$.*

Proposition 8. *If there exists a polynomial time adversary $\mathcal{A}$ with non-negligible advantage ϵ in Game 2, then there exists a polynomial time algorithm $\mathcal{B}$ that can break the IND-CPA security of $\mathcal{PKE}_A$ with advantage $\epsilon_A = \epsilon$.*

Then, combining the above results, we have the following theorem.

Theorem 2. *Assuming that G is a secure PRG family, iO is an indistinguishability obfuscator and $\mathcal{PKE}_A$ is an IND-CPA secure bit encryption scheme, $\mathcal{PKE}$ is IND-CPA secure but not 1-circular secure.*

5 Key Recovery from Circular Insecurity

In this section we show how to transform any IND-CPA encryption scheme which is n-circular insecure into a new IND-CPA scheme which is n-circular insecure *with respect to key recovery*. An interesting point of comparison is a result of Cash et al. [15]. As described in the introduction their counterexample is particular to a specific construction for $n = 2$ length key cycles. We show how to generically 'leap' from any cycle detection insecure construction to one which is insecure against key recovery, but maintains IND-CPA security.

Our generic transformation proceeds in two steps. We begin with an IND-CPA encryption system that is insecure against cycle detection attacks. That is there exists a polynomial $p(\cdot)$ and an infinite set $S \subseteq \mathbb{N}$ where the advantage of the attacker is greater than $1/p(\lambda)$ for all $\lambda \in S$. We show that if such a system exists, then there exists a cryptosystem with an attacker that has advantage of $1/2 - \mathrm{negl}(\lambda)$ for all $\lambda \in S$. (i.e. the probability of winning the game is $1 - \mathrm{negl}(\lambda)$ for all $\lambda \in S$.) This effectively amplifies the probability of winning within that restricted set. Our amplification technique is just a simple repetition.

Next, we show how such an amplified cycle detection encryption system can be transformed into one where a key recovery attack is possible. Our approach is to create an encryption system where the encryption algorithm will go through the message M bit by bit and encode each 1 as a M and each 0 and a string of 0's. Then if there is a key cycle, the underlying cycle detection algorithm can recover the bits of M one by one using the cycle detection algorithm/attacker of the underlying scheme.

5.1   A Circular Key Recoverable Cryptosystem

Amplification We first state our amplification lemma which is proved in Appendix B.1.

Claim 1 *Let* $\mathcal{PKE}'_A$ *be an* IND-CPA *secure public key cryptosystem that is n-circular secure i.e. there exists a polynomial time algorithm $\mathcal{D}'$ and a polynomial $p(\cdot)$ such that for infinitely many $\lambda \in \mathbb{N}$, $Adv_{\mathcal{D}'}(\lambda) > 1/p(\lambda)$. Then there exists an* IND-CPA *secure public key cryptosystem $\mathcal{PKE}_A$, which is constructed using $\mathcal{PKE}'_A$ as a black box, for which there exists an n-circular security adversary $\mathcal{D}$ with advantage $1/2 - \mathrm{negl}(\lambda)$ (i.e. with probability $1 - \mathrm{negl}(\lambda)$) for all such $\lambda \in \mathbb{N}$.*

Our Transformation Let $\mathcal{PKE}_A$ be an IND-CPA encryption scheme for which there exists an n-circular security adversary $\mathcal{D}$ with $Adv_{\mathcal{D}}(\lambda) \geq 1/2 - \mathrm{negl}(\lambda)$ for infinitely many $\lambda \in \mathbb{N}$. Let $\mathcal{M}_A = \{0,1\}^l$ be the message space. For an l-bit message M, we will let $M[i]$ denote the i-th bit of M where $i \in [l]$. We construct an IND-CPA encryption scheme $\mathcal{PKE}$ which is n-circular insecure with respect to key recovery as follows.

IND-CPA n-Circular key recoverable $\mathcal{PKE}$:
Inputs : IND-CPA n-Circular insecure $\mathcal{PKE}_A$.

- Keygen(1^λ): Let $(sk_A, pk_A) \leftarrow$ Keygen$_A(1^\lambda)$. Let $sk = sk_A$, $pk = pk_A$. Output (sk, pk).
- Encrypt(pk, M):
    - Let $C_H = $ Encrypt$_A(pk, M)$.
    1. For $i = 1 \ldots l$
       Let $C_i = $ Encrypt$_A(pk, M)$ if $M[i] = 1$, else $C_i = $ Encrypt$_A(pk, 0^{|M|})$.
    2. Output $ct = (C_H, C_1, \ldots, C_l)$.
- Decrypt(sk, ct): Compute $M \leftarrow$ Decrypt$_A(sk, C_H)$ and $M'_i \leftarrow$ Decrypt$_A(sk, C_i)$ for $i = 1, \ldots, l$. If $\forall i \in [l]\ M'_i = M \cdot M[i]$ output M, otherwise output $\bot$.

The proof of the following claim is straightforward and is included in Appendix B.2.

Claim 2 $\mathcal{PKE}$ *is* IND-CPA *secure if* $\mathcal{PKE}_A$ *is* IND-CPA .

We now formally show that if the old cryptosystem $\mathcal{PKE}_A$ is n-circular insecure, the new cryptosystem $\mathcal{PKE}$ is n-circular insecure *with respect to key recovery*. We rely on the following result which is proved in Appendix B.3. Claim 3 states that any circular security adversary can be used to distinguish an encryption cycle from a modified encryption cycle in which a zero encryption has been substituted in the last position. The proof utilizes a hybrid argument.

Claim 3 *Let* $\mathcal{PKE}_A$ *be an* IND-CPA *public key cryptosystem. Suppose that $\mathcal{D}$ has advantage $Adv_{\mathcal{D}}(\lambda)$ in the circular security game against $\mathcal{PKE}_A$. Then $\mathcal{D}$*

distinguishes the following distributions with advantage $Adv_{\mathcal{D}}(\lambda) - negl(\lambda)$.

$\mathcal{D}_0 = [\mathsf{pk}_1, \ldots, \mathsf{pk}_n, \mathsf{Encrypt}_A(\mathsf{pk}_1, \mathsf{sk}_2), \ldots, \mathsf{Encrypt}_A(\mathsf{pk}_{n-1}, \mathsf{sk}_n), \mathsf{Encrypt}_A(\mathsf{pk}_n, \mathsf{sk}_1) :$
$(\mathsf{pk}_i, \mathsf{sk}_i) \leftarrow \mathsf{Keygen}_A(1^\lambda)]$

$\mathcal{D}_1 = [\mathsf{pk}_1, \ldots, \mathsf{pk}_n, \mathsf{Encrypt}_A(\mathsf{pk}_1, \mathsf{sk}_2), \ldots, \mathsf{Encrypt}_A(\mathsf{pk}_{n-1}, \mathsf{sk}_n), \mathsf{Encrypt}_A(\mathsf{pk}_n, 0^{|\mathsf{sk}_1|}) :$
$(\mathsf{pk}_i, \mathsf{sk}_i) \leftarrow \mathsf{Keygen}_A(1^\lambda)]$

Armed with the above claims we are now ready to prove the following lemma.

Lemma 1. *Suppose there exists an algorithm $\mathcal{D}$ with advantage $Adv_{\mathcal{D}}(\lambda) = 1/2 - negl(\lambda)$ in the n-circular security game against $\mathcal{PKE}_A$ for infinitely many $\lambda \in \mathbb{N}$. Then there exists an algorithm $\mathcal{R}$ with advantage at least $1/2 - negl(\lambda)$ in the n-circular key recovery security game against $\mathcal{PKE}$ for all such $\lambda \in \mathbb{N}$.*

Proof. Let $\mathcal{D}$ be an algorithm such with advantage in the n-circular security game against $\mathcal{PKE}_A$ at least $1/2 - negl(\lambda)$ for infinitely many $\lambda \in \mathbb{N}$. Consider the following algorithm $\mathcal{R}$ interacting with the n-circular security with respect to key recovery challenger $\mathcal{C}$:

1. $\mathcal{C}$ runs $(\mathsf{pk}_i, \mathsf{sk}_i) \leftarrow \mathsf{Keygen}(1^\lambda)$.
2. $\mathcal{C}$ computes $y_i = \mathsf{Encrypt}(\mathsf{pk}_i, \mathsf{sk}_{(i \bmod n)+1})$ for $1 \leq i \leq n$.
3. $\mathcal{C}$ sends $(\mathsf{pk}_1, \ldots, \mathsf{pk}_n, y_1, \ldots, y_n)$ to $\mathcal{R}$.
4. $\mathcal{R}$ parses $y_i = (C_{i,H}, C_{i,1}, \ldots C_{i,l})$ for $1 \leq i \leq n$.
5. $\mathcal{R}$ for $j = 1 \ldots l$.
   (a) Forms the vector $w_j = (C_{1,H}, \ldots, C_{n-1,H}, C_{n,j})$.
   (b) Lets $\mathsf{sk}_1[j] \leftarrow \mathcal{D}(\mathsf{pk}_1, \ldots, \mathsf{pk}_n, w_j)$.
6. $\mathcal{R}$ output sk_1.

Fix any such $\lambda \in \mathbb{N}$. Note that $C_{n,j}$ is either a random encryption of sk_1 or 0. Note that $\mathcal{D}$ distinguishes an n-encryption cycle from n zero encryptions with advantage at least $1/2 - negl(\lambda)$. Thus Claim 3 implies that $\mathcal{D}$ on input $(\mathsf{pk}_1, \ldots, \mathsf{pk}_n, w_j)$ distinguishes whether $C_{n,j}$ is an encryption of sk_1 or 0 with advantage at least $Adv_{\mathcal{D}'}(\lambda) - negl(\lambda) = 1/2 - negl(\lambda)$. Thus $\mathcal{D}$ fails to recover the j-th bit of sk_1 with probability at most $negl(\lambda)$. Then $\mathcal{R}$ recovers sk_1 correctly except with probability atmost $n \cdot negl(\lambda)$, which is negligible.

Combining Claim 1 and Lemma 1, we get the following theorem.

Theorem 3. *Suppose there exists an algorithm $\mathcal{D}$ with non-negligible advantage in the n-circular security game against $\mathcal{PKE}'_A$ for infinitely many $\lambda \in \mathbb{N}$. Then there exists an algorithm $\mathcal{R}$ with advantage at least $1/2 - negl(\lambda)$ in the n-circular key recovery security game against $\mathcal{PKE}$ for all such $\lambda \in \mathbb{N}$.*

Acknowledgements. The authors are grateful to Ron Rothblum for helpful comments and suggestions.

References

1. Acar, T., Belenkiy, M., Bellare, M., Cash, D.: Cryptographic agility and its relation to circular encryption. In: Gilbert, H. (ed.) EUROCRYPT 2010. LNCS, vol. 6110, pp. 403–422. Springer, Heidelberg (2010)
2. Adão, P., Bana, G., Herzog, J.C., Scedrov, A.: Soundness of formal encryption in the presence of key-cycles. In: de Capitani di Vimercati, S., Syverson, P.F., Gollmann, D. (eds.) ESORICS 2005. LNCS, vol. 3679, pp. 374–396. Springer, Heidelberg (2005)
3. Alperin-Sheriff, J., Peikert, C.: Circular and KDM security for identity-based encryption. In: Fischlin, M., Buchmann, J., Manulis, M. (eds.) PKC 2012. LNCS, vol. 7293, pp. 334–352. Springer, Heidelberg (2012), http://dx.doi.org/10.1007/978-3-642-30057-8_20
4. Applebaum, B.: Key-dependent message security: Generic amplification and completeness. In: Paterson, K.G. (ed.) EUROCRYPT 2011. LNCS, vol. 6632, pp. 527–546. Springer, Heidelberg (2011)
5. Applebaum, B., Cash, D., Peikert, C., Sahai, A.: Fast cryptographic primitives and circular-secure encryption based on hard learning problems. In: Halevi, S. (ed.) CRYPTO 2009. LNCS, vol. 5677, pp. 595–618. Springer, Heidelberg (2009), http://dx.doi.org/10.1007/978-3-642-03356-8_35
6. Barak, B., Goldreich, O., Impagliazzo, R., Rudich, S., Sahai, A., Vadhan, S., Yang, K.: On the (im)possibility of obfuscating programs. In: Kilian, J. (ed.) CRYPTO 2001. LNCS, vol. 2139, pp. 1–18. Springer, Heidelberg (2001)
7. Barak, B., Haitner, I., Hofheinz, D., Ishai, Y.: Bounded key-dependent message security. In: Gilbert, H. (ed.) EUROCRYPT 2010. LNCS, vol. 6110, pp. 423–444. Springer, Heidelberg (2010)
8. Black, J., Rogaway, P., Shrimpton, T.: Encryption-scheme security in the presence of key-dependent messages (2001) (manuscript)
9. Boneh, D., Halevi, S., Hamburg, M., Ostrovsky, R.: Circular-secure encryption from decision diffie-hellman. In: Wagner, D. (ed.) CRYPTO 2008. LNCS, vol. 5157, pp. 108–125. Springer, Heidelberg (2008)
10. Boneh, D., Zhandry, M.: Multiparty key exchange, efficient traitor tracing, and more from indistinguishability obfuscation. Cryptology ePrint Archive, Report 2013/642 (2013), http://eprint.iacr.org/
11. Brakerski, Z., Goldwasser, S.: Circular and leakage resilient public-key encryption under subgroup indistinguishability. In: Rabin, T. (ed.) CRYPTO 2010. LNCS, vol. 6223, pp. 1–20. Springer, Heidelberg (2010)
12. Brakerski, Z., Goldwasser, S., Kalai, Y.T.: Black-box circular-secure encryption beyond affine functions. In: Ishai, Y. (ed.) TCC 2011. LNCS, vol. 6597, pp. 201–218. Springer, Heidelberg (2011)
13. Brakerski, Z., Vaikuntanathan, V.: Fully homomorphic encryption from ring-LWE and security for key dependent messages. In: Rogaway, P. (ed.) CRYPTO 2011. LNCS, vol. 6841, pp. 505–524. Springer, Heidelberg (2011), http://dx.doi.org/10.1007/978-3-642-22792-9_29
14. Camenisch, J.L., Lysyanskaya, A.: An efficient system for non-transferable anonymous credentials with optional anonymity revocation. In: Pfitzmann, B. (ed.) EUROCRYPT 2001. LNCS, vol. 2045, pp. 93–118. Springer, Heidelberg (2001), http://dx.doi.org/10.1007/3-540-44987-6_7
15. Cash, D., Green, M., Hohenberger, S.: New definitions and separations for circular security. In: Fischlin, M., Buchmann, J., Manulis, M. (eds.) PKC 2012. LNCS, vol. 7293, pp. 540–557. Springer, Heidelberg (2012)

16. Coron, J.-S., Lepoint, T., Tibouchi, M.: Practical multilinear maps over the integers. In: Canetti, R., Garay, J.A. (eds.) CRYPTO 2013, Part I. LNCS, vol. 8042, pp. 476–493. Springer, Heidelberg (2013)

17. Garg, S., Gentry, C., Halevi, S.: Candidate multilinear maps from ideal lattices. In: Johansson, T., Nguyen, P.Q. (eds.) EUROCRYPT 2013. LNCS, vol. 7881, pp. 1–17. Springer, Heidelberg (2013)

18. Garg, S., Gentry, C., Halevi, S., Raykova, M., Sahai, A., Waters, B.: Candidate indistinguishability obfuscation and functional encryption for all circuits. Cryptology ePrint Archive, Report 2013/451 (2013)

19. Gentry, C.: A fully homomorphic encryption scheme. Ph.D. thesis, Stanford University (2009), http://crypto.stanford.edu/craig

20. Goldwasser, S., Micali, S.: Probabilistic Encryption. J. Comput. Syst. Sci. 28(2), 270–299 (1984)

21. Haitner, I., Holenstein, T.: On the (im)possibility of key dependent encryption. In: Reingold, O. (ed.) TCC 2009. LNCS, vol. 5444, pp. 202–219. Springer, Heidelberg (2009), http://dx.doi.org/10.1007/978-3-642-00457-5_13

22. Laud, P.: Encryption cycles and two views of cryptography. In: NORDSEC 2002 - Proceedings of the 7th Nordic Workshop on Secure IT Systems Karlstad University Studies 2002:31, pp. 85–100 (2002)

23. Malkin, T., Teranishi, I., Yung, M.: Efficient circuit-size independent public key encryption with KDM security. In: Paterson, K.G. (ed.) EUROCRYPT 2011. LNCS, vol. 6632, pp. 507–526. EUROCRYPT, Heidelberg (2011)

24. Marcedone, A., Orlandi, C.: Obfuscation ==> (ind-cpa security =/=> circular security). Cryptology ePrint Archive, Report 2013/690 (2013), http://eprint.iacr.org/

25. Rothblum, R.D.: On the circular security of bit-encryption. In: Sahai, A. (ed.) TCC 2013. LNCS, vol. 7785, pp. 579–598. Springer, Heidelberg (2013)

26. Rothblum, R.: Personal communication (2014)

27. Sahai, A., Waters, B.: How to use indistinguishability obfuscation: Deniable encryption, and more. Cryptology ePrint Archive, Report 2013/454 (2013)

A   Counter Example for 1-Circular Security of Bit-by-bit Encryption

Proposition 6. *Suppose that there exists a polynomial time adversary $\mathcal{A}$ such that $Game_0 Adv_{\mathcal{A}}$ - $Game_1 Adv_{\mathcal{A}} = \epsilon$. Then there exists a polynomial time adversary $\mathcal{B}$ who distinguishes the output of G from random with advantage $\epsilon_{PRG} = \epsilon$.*

Proof. In Game 0, t is an output of G, while in Game 1, t is a truly random $2l$-bit string. The algorithm $\mathcal{B}$ is defined as follows :

1. $\mathcal{B}$ receives $t \in \{0,1\}^{2l}$ from PRG Challenger $\mathcal{C}$, where t is either a pseudorandom string generated by G or a truly random string.
2. $\mathcal{B}$ computes $(\mathsf{pk}_A, \mathsf{sk}_A) \leftarrow \mathsf{Keygen}_A(1^\lambda)$. Next, it computes $O = i\mathcal{O}(\lambda, \mathsf{BitCycleFind})$ as described in Game 0. It sets $\mathsf{pk} = (\mathsf{pk}_A, t, O)$ and sends it to $\mathcal{A}$.
3. $\mathcal{B}$ chooses $b \xleftarrow{\$} \{0,1\}$. It sets $ct_b \leftarrow \mathsf{Encrypt}_A(\mathsf{pk}_A, b)$ and sends it to $\mathcal{A}$.
4. $\mathcal{A}$ outputs a bit b'. If $(b = b')$ $\mathcal{B}$ outputs that t was pseudorandom. Else $\mathcal{B}$ outputs that t was random.

Clearly, as shown in Proposition 2, if $\mathcal{A}$ wins the game with non negligible probability, then so does $\mathcal{B}$.

Proposition 7. *Suppose that there exists a polynomial time adversary $\mathcal{A}$ such that $Game_1Adv_\mathcal{A}$ - $Game_2Adv_\mathcal{A} = \epsilon$. Then there exists a polynomial time adversary $\mathcal{B}$ who breaks the indistinguishability obfuscation with advantage $\epsilon_{iO} = \epsilon$.*

Proof. $\mathcal{B}$ comprises a pair of adversaries $(Samp, D)$ as in Definition 7. We construct these adversaries as follows.

$Samp(1^\lambda)$:

1. Choose $r \xleftarrow{\$} \{0,1\}^l$ and $t \xleftarrow{\$} \{0,1\}^{2l}$.
2. Let $(\mathsf{sk}_A, \mathsf{pk}_A) \leftarrow \mathsf{Keygen}_A(1^\lambda)$.
3. Let BitCycleFind be the circuit described in our construction with constants $(\mathsf{sk}_A, t, 0^w)$ hardwired and BitCycleReject be the circuit described in Game 2 with constant $0^{w'}$ hardwired.
4. Output $(g_0 = \mathsf{BitCycleFind}, g_1 = \mathsf{BitCycleReject})$.
5. Set $\sigma = (\mathsf{pk}_A, t)$.

$D(\sigma, i\mathcal{O}(\lambda, g_z))$:

1. Parse $\sigma = (\mathsf{pk}_A, t)$. Set $\mathsf{pk} = (\mathsf{pk}_A, t, i\mathcal{O}(\lambda, g_z))$
2. Let $b \xleftarrow{\$} \{0,1\}$. $ct \leftarrow \mathsf{Encrypt}_A(\mathsf{pk}_A, b)$.
3. Let $b' \leftarrow \mathcal{A}(\mathsf{pk}, ct)$.
4. D guesses 1 if $b = b'$.

Note that since t is chosen uniformly at random, except with negligible probability, t is not in the range of G. Hence $\mathsf{BitCycleFind}(x)$ outputs $\perp$ for all x. Thus $Samp$ produces circuits $\mathsf{BitCycleReject}$ and $\mathsf{BitCycleFind}$ which are equivalent on all inputs with overwhelming probability, by the random choice of t.

Similar to the proof for Proposition 3, we can argue that if $\mathcal{A}$ distinguishes between the outputs of Game 1 and Game 2 with advantage ϵ, then $\mathcal{B}$ breaks the indistinguishability obfuscation with advantage ϵ.

Proposition 8. *If there exists a polynomial time adversary $\mathcal{A}$ with non-negligible advantage ϵ in Game 2, then there exists a polynomial time algorithm $\mathcal{B}$ that can break the IND-CPA security of $\mathcal{PKE}_A$ with advantage $\epsilon_A = \epsilon$.*

Proof. Suppose $\mathcal{A}$ has advantage ϵ in Game 2. We define $\mathcal{B}$ as follows :

1. $\mathcal{B}$ receives pk_A, ct from the IND-CPA Challenger $\mathcal{C}$. It chooses $t \xleftarrow{\$} \{0,1\}^{2l}$ and computes $O \leftarrow i\mathcal{O}(\lambda, \mathsf{BitCycleReject})$. It sends public key $\mathsf{pk} = (\mathsf{pk}_A, t, O)$ and ciphertext ct to $\mathcal{A}$.
2. $\mathcal{A}$ sends bit b', which $\mathcal{B}$ passes on to $\mathcal{C}$.

Note that if $\mathcal{A}$ wins Game 2, then $\mathcal{B}$ wins the IND-CPA game. Hence the result follows.

B Key Recovery From Circular Insecurity

B.1

Claim 1 *Let $\mathcal{PKE}'_A$ be an IND-CPA secure public key cryptosystem that is n-circular secure i.e. there exists a polynomial time algorithm $\mathcal{D}'$ and a polynomial*

$p(\cdot)$ such that for infinitely many $\lambda \in \mathbb{N}$, $Adv_{\mathcal{D}'}(\lambda) > 1/p(\lambda)$. Then there exists an IND-CPA secure public key cryptosystem $\mathcal{PKE}_A$, which is constructed using $\mathcal{PKE}'_A$ as a black box, for which there exists an n-circular security adversary $\mathcal{D}$ with advantage $1/2 - negl(\lambda)$ (i.e. with probability $1 - negl(\lambda)$) for all such $\lambda \in \mathbb{N}$.

Proof. Let $\mathcal{PKE}'_A = (\mathsf{Keygen}'_A, \mathsf{Encrypt}'_A, \mathsf{Decrypt}'_A)$. Let $t(\lambda) = \lambda \cdot p(\lambda)^2$ be the amplification factor. We now define $\mathcal{PKE}_A = (\mathsf{Keygen}_A, \mathsf{Encrypt}_A, \mathsf{Decrypt}_A)$ as follows.

- $\mathsf{Keygen}_A(1^\lambda)$: Compute t public key, secret key pairs. $(\mathsf{pk}_i, \mathsf{sk}_i) \xleftarrow{\$}$ $\mathsf{Keygen}'_A(1^\lambda)$ for $1 \le i \le t$. The public key $\mathsf{pk} = (\mathsf{pk}_1, \ldots, \mathsf{pk}_t)$ and the secret key is $(\mathsf{sk}_1, \ldots, \mathsf{sk}_t)$.
- $\mathsf{Encrypt}_A(\mathsf{pk}, m)$: Parse $\mathsf{pk} = (\mathsf{pk}_1, \ldots, \mathsf{pk}_t)$ and $m = (m_1, \ldots, m_t)$ such that $|m_i| = |m_j|$ for all i, j. Compute t ciphertexts $ct_1, \ldots, ct_t$, where $ct_i \xleftarrow{\$}$ $\mathsf{Encrypt}'_A(\mathsf{pk}_i, m_i)$. The ciphertext $ct = (ct_1, \ldots, ct_t)$.
- $\mathsf{Decrypt}_A(\mathsf{sk}, ct)$: Parse $\mathsf{sk} = (\mathsf{sk}_1, \ldots, \mathsf{sk}_t)$ and $ct = (ct_1, \ldots, ct_t)$. Output $\mathsf{Decrypt}'_A(\mathsf{sk}_1, ct_1)$.

IND-CPA security of $\mathcal{PKE}_A$ follows from hybrid argument. We need to show that there exists an algorithm $\mathcal{D}$ such that for infinitely many λ, $Adv_{\mathcal{D}}(\lambda) > 1/2 - negl(\lambda)$ in the n-circular security game. Note that each ciphertext ct_i consists of t ciphertexts $(ct_{i1}, \ldots, ct_{it})$, and for all $1 \le j \le t$, either $(ct_{1j}, \ldots, ct_{nj})$ is an encryption cycle or an encryption of zeroes. By construction, it follows that each of these cycles is independent, since we have t independent invocations of Keygen'_A during Keygen_A.

$\mathcal{D}$ is defined as follows :

$\mathcal{D}'$:

1. For $1 \le i \le t$, compute $d_i \xleftarrow{\$} \mathcal{D}'(ct_{1i}, \ldots, ct_{ni})$
2. Output majority of $\{d_1, \ldots, d_t\}$.

If we have an encryption cycle, then, for each $1 \le j \le t$, we have $\Pr[\mathcal{D}'(ct_{1i}, \ldots, ct_{ni}) = 1] > 1/2 + 1/p(\lambda)$. Since we have $t = \lambda \cdot p(\lambda)^2$ invocations, using Chernoff bounds, it follows that $\Pr[\mathcal{D}(ct_1, \ldots, ct_n) = 1] > 1 - negl(\lambda)$.

Similarly, if we have encryptions of zeroes, then for each $1 \le j \le t$, $\Pr[\mathcal{D}'(ct_{1i}, \ldots, ct_{ni}) = 1] < 1/2 - 1/p(\lambda)$. Using Chernoff bounds, we get that $\Pr[\mathcal{D}(ct_1, \ldots, ct_n) = 1] < negl(\lambda)$.

B.2

Claim 2 $\mathcal{PKE}$ is IND-CPA secure if $\mathcal{PKE}_A$ is IND-CPA .

Proof. To prove this claim it will be convenient to define $C_0 =: C_H$ and $M[0] =: 1$. Suppose that adversary $\mathcal{A}$ has advantage $\epsilon(\lambda)$ in the IND-CPA game against $\mathcal{PKE}$. We construct an adversary $\mathcal{B}$ which has advantage $\epsilon(\lambda)/(l+1)$ in the IND-CPA game against $\mathcal{PKE}_A$.

1. $\mathcal{B}$ receives pk_A from the challenger and forwards it to $\mathcal{A}$.
2. $\mathcal{A}$ makes some ciphertext queries to Encrypt which are answered using Encrypt$_A$.
3. $\mathcal{B}$ receives two l-bit message M_0, M_1 from $\mathcal{A}$.
4. $\mathcal{B}$ chooses $i^* \xleftarrow{\$} \{0,\ldots,l\}$ and forms $M_0' = M_0 \cdot M_0[i^*]$ and $M_1' = M_1 \cdot M_1[i^*]$. If $M_0' = M_1'$ it aborts, otherwise it sends M_0' and M_1' to the challenger.
5. $\mathcal{B}$ receives $ct_b' = \text{Encrypt}_A(M_b')$ from the challenger.
6. $\mathcal{B}$ forms the ciphertext $ct = (C_0, \ldots C_l)$ where

$$C_i = \begin{cases} \text{Encrypt}_A(pk, M_0 \cdot M_0[i]) : i < i^* \\ ct_b' : i = i^* \\ \text{Encrypt}_A(pk, M_1 \cdot M_1[i]) : i > i^* \end{cases}$$

and forwards ct to $\mathcal{A}$.
7. $\mathcal{B}$ receives bit z from $\mathcal{A}$.
8. Step 2 may be repeated.
9. $\mathcal{B}$ sends guess $b' = z$ to the challenger.

Define for $i = 0 \ldots l$, $p_i = \Pr[b' = 0 | i^* = i, b = 0]$ and $q_i = \Pr[b' = 0 | i^* = i, b = 1]$. Since $\mathcal{A}$ has advantage ϵ in the IND-CPA game against $\mathcal{PKE}$, we have $\epsilon = 1/2 \cdot (p_l - q_0)$. By inspection $p_{i-1} = q_i$ hence $\epsilon = 1/2 \cdot (\sum_{i=0}^l (p_i - q_i))$. Then $\epsilon = 1/2 \cdot (\sum_{i=0}^l p_i - \sum_{i=0}^l q_i) = 1/2 \cdot (\Pr[b' = 0 | b = 0] - \Pr[b' = 0 | b = 1]) \cdot (l+1) = Adv_{\mathcal{B}} \cdot (l+1)$. Thus $\mathcal{B}$ has advantage $\epsilon/(l+1)$ which is non-negligible if ϵ is non-negligible.

B.3

Claim 3 *Let $\mathcal{PKE}_A$ be an IND-CPA public key cryptosystem. Suppose that $\mathcal{D}$ has advantage $Adv_{\mathcal{D}}(\lambda)$ in the circular security game against $\mathcal{PKE}_A$. Then $\mathcal{D}$ distinguishes the following distributions with advantage $Adv_{\mathcal{D}}(\lambda) - \text{negl}(\lambda)$.*

$[\mathsf{pk}_1, \ldots, \mathsf{pk}_n, \text{Encrypt}_A(\mathsf{pk}_1, \mathsf{sk}_2), \ldots, \text{Encrypt}_A(\mathsf{pk}_{n-1}, \mathsf{sk}_n), \text{Encrypt}_A(\mathsf{pk}_n, \mathsf{sk}_1) : (\mathsf{pk}_i, \mathsf{sk}_i) \leftarrow \text{Keygen}_A(1^\lambda)]$

$[\mathsf{pk}_1, \ldots, \mathsf{pk}_n, \text{Encrypt}_A(\mathsf{pk}_1, \mathsf{sk}_2), \ldots, \text{Encrypt}_A(\mathsf{pk}_{n-1}, \mathsf{sk}_n), \text{Encrypt}_A(\mathsf{pk}_n, 0^{|\mathsf{sk}_1|}) : (\mathsf{pk}_i, \mathsf{sk}_i) \leftarrow \text{Keygen}_A(1^\lambda)]$

Proof. In order to prove this result, we define n intermediate hybrid experiments $H_j : 1 \leq j \leq n$, and show that $\mathcal{D}$ has overwhelming advantage in each of the hybrids. Hybrid H_j is defined as follows :

H_j :

1. $\mathcal{C}$ computes $(\mathsf{pk}_i, \mathsf{sk}_i) \leftarrow \mathsf{Keygen}_\mathsf{A}(1^\lambda)$ for $1 \leq i \leq n$
2. $\mathcal{C}$ chooses a bit $b \xleftarrow{\$} \{0, 1\}$.
   - If $b = 0$, $\mathcal{C}$ computes $y_i = \mathsf{Encrypt}_\mathsf{A}(\mathsf{pk}_i, \mathsf{sk}_{(i \bmod n)+1})$ for $1 \leq i \leq n$
   - Else $\mathcal{C}$ computes $y_i = \mathsf{Encrypt}_\mathsf{A}(\mathsf{pk}_i, \mathsf{sk}_{(i \bmod n)+1})$ for $i < j$ and $y_i = \mathsf{Encrypt}_\mathsf{A}(\mathsf{pk}_i, 0^{|\mathsf{sk}_{(i \bmod n)+1}|})$ for $i \geq j$
3. $\mathcal{C}$ sends $(\mathsf{pk}_1, \ldots, \mathsf{pk}_n, y_1, \ldots, y_n)$ to $\mathcal{D}$.
4. $\mathcal{D}$ outputs b'.

H_1 corresponds to the n-circular security game, while H_n corresponds to the case where an encryption cycle might be modified by substituting a zero encryption in the last position. Let $Adv_\mathcal{D}(H_j)$ denote the advantage of $\mathcal{D}$ in hybrid experiment H_j. Suppose $Adv_\mathcal{D}(H_j) - Adv_\mathcal{D}(H_{j+1})$ is non-negligible. Then there exists a polynomial time adversary $\mathcal{A}$ that can break the IND-CPA security of $\mathcal{PKE}$ using $\mathcal{D}$.

1. $\mathcal{A}$ receives public key pk from the IND-CPA challenger $\mathcal{C}$.
2. $\mathcal{A}$ generates $n - 1$ public key, secret key pairs $(\mathsf{pk}_i, \mathsf{sk}_i) \xleftarrow{\$} \mathsf{Keygen}(1^\lambda)$ for $2 \leq i \leq n$.
3. $\mathcal{A}$ sends $\mathsf{sk}_{j+1}, 0^{|sk_{j+1}|}$ as challenge messages to $\mathcal{C}$ and receives ct as the ciphertext.
4. $\mathcal{A}$ computes the remaining $n - 1$ ciphertexts $(ct_1, \ldots, ct_{j-1}, ct_{j+1}, \ldots, ct_n)$ as in the hybrids, and then runs $\mathcal{D}$ on this input.
5. Depending on the output of $\mathcal{D}$, $\mathcal{A}$ sends its guess to $\mathcal{C}$.

Note that the advantage of $\mathcal{A}$ is equal to $Adv_\mathcal{D}(H_j) - Adv_\mathcal{D}(H_{j+1})$. We have $Adv_\mathcal{D}(H_1) = \epsilon$. Therefore, the advantages of $\mathcal{D}$ in each of the successive hybrids is $\epsilon - \mathsf{negl}(\lambda)$, and in particular, its advantage in H_n is $\epsilon - \mathsf{negl}(\lambda)$.

ZAPs and Non-Interactive Witness Indistinguishability from Indistinguishability Obfuscation

Nir Bitansky[1,*] and Omer Paneth[2,**]

[1] MIT, USA
[2] Boston University, USA

Abstract. We present new constructions of two-message and one-message witness-indistinguishable proofs (ZAPs and NIWIs). This includes:

- ZAPs (or, equivalently, non-interactive zero-knowledge in the common random string model) from indistinguishability obfuscation and one-way functions.
- NIWIs from indistinguishability obfuscation and one-way permutations.

The previous construction of ZAPs [Dwork and Naor, FOCS 00] was based on trapdoor permutations. The two previous NIWI constructions were based either on ZAPs and a derandomization-type complexity assumption [Barak, Ong, and Vadhan CRYPTO 03], or on a specific number theoretic assumption in bilinear groups [Groth, Sahai, and Ostrovsky, CRYPTO 06].

1 Introduction

Zero-knowledge proofs [GMR89] and their feasibility for **NP** [GMW91] are fundamental to modern cryptography, allowing to prove any **NP** statement while guaranteeing total privacy of the witness. One of the main pillars on which this exquisite guarantee relies is interaction. Minimizing interaction from zero-knowledge protocols has drawn significant efforts. Since even two-message zero-knowledge, without any setup assumptions, is impossible [GO94], these efforts have either focused on non-interactive zero-knowledge (NIZK) in the trusted *common random string* (or common reference string) model [BFM88], or on achieving non-interactive systems with weaker security guarantees. One notable relaxation is that of *witness indistinguishability* (WI), guaranteeing that the proof does not reveal which one of many witnesses is used.

Following this direction, Dwork and Naor [DN07] show that, unlike zero-knowledge, two-message WI, or as they term *ZAPs*, can be achieved without any

---

* Part of this work was done while at Tel Aviv university and supported by an IBM Ph.D. Fellowship and the Check Point Institute for Information Security.
** Supported by the Simons award for graduate students in theoretical computer science and an NSF Algorithmic foundations grant 1218461.

Y. Dodis and J.B. Nielsen (Eds.): TCC 2015, Part II, LNCS 9015, pp. 401–427, 2015.

setup. Specifically, they show that assuming one-way functions, the existence of ZAPs in the plain model is equivalent to that of NIZKs in the common random string model. The latter were constructed by Feige, Lapidot, and Shamir from trapdoor permutations [FLS99]. Furthermore, in ZAPs, the verifier's message is a random string that can be used for multiple proofs, giving rise to a completely non-interactive WI system where the first message is fixed non-uniformly. This provided evidence that diverging from the strong notion of zero-knowledge may very well allow to remove interaction altogether.

Barak, Ong, and Vadhan [BOV07] then constructed completely non-interactive WI (NIWI), without any non-uniformity, by derandomizing the ZAP verifier under the assumption that $\mathbf{Dtime}(2^{O(n)})$ has a problem of non-deterministic circuit complexity $2^{\Omega(n)}$. Groth, Ostrovsky, and Sahai [GOS12] subsequently constructed NIWIs under the decision linear assumption on bilinear groups.

ZAPs and NIWIs have found a great number of applications in cryptography, but only few candidate constructions, from specific assumptions, are known. In particular, both [DN07] and [BOV07] rely on trapdoor permutations, for which there are very few candidates, all based on factoring-related assumptions. Alternatively, the [GOS12] construction is based on a specific assumption on bilinear groups. Different constructions from different computational assumptions are still sought after. In this work, we provide new constructions of ZAPs and NIWIs under a rather different type of assumption: *indistinguishability obfuscation*.

Indistinguishability Obfuscation. The goal of obfuscation is to make code unintelligible while preserving its functionality. It has been long considered to be a holy grail of cryptography, with diverse and far reaching applications. Up until recently, there were no candidates obfuscators, except for very restricted classes of programs, and in fact, some classes were shown to be unobfuscatable under the natural virtual black-box notion [BGI+01]. This dramatically changed with the work by Garg et al. [GGH+13b] who proposed a candidate construction of general-purpose obfuscators, based on graded multilinear encodings [GGH13a], and conjectured that it satisfies the seemingly weak notion of indistinguishability obfuscation (iO) [BGI+01], for which no impossibility results are known. This notion only requires that it is hard to distinguish an obfuscation of C_0 from an obfuscation of C_1, for any two circuits C_0 and C_1 of the same size that compute the exact same function.

Perhaps surprisingly, iO has been shown to have variety of powerful applications, such as functional encryption, deniable encryption, two-message multiparty computation [GGH+13b, SW14, GGHR14], and many more. Still, some basic primitives have so far evaded the long arms of iO, including collision-resistant hashing, fully-homomorphic encryption, and also trapdoor permutations, which as said above, are essential in generic ZAP and NIWI constructions.

1.1   Results

We provide new constructions of ZAPs and NIWIs based on iO.

Our first result is a construction of ZAPs (or NIZKs):

Theorem 1.1 (informal). *Assuming indistinguishability obfuscation for a certain family of polysize circuits and one-way functions, there exist ZAPs in the plain model and NIZKs in the common random string model, for every language in* **NP**.[1]

The new ZAP can, in particular, be plugged into the result of Barak, Ong, and Vadhan [BOV07] to obtain a NIWI for all of **NP**, assuming in addition the existence of a language in **Dtime**$(2^{O(n)})$ with circuit complexity $2^{\Omega(n)}$. We give a new construction of NIWIs based on indistinguishability obfuscation and one-way permutations.

Theorem 1.2 (informal). *Assuming indistinguishability obfuscation for a certain family of polysize circuits and one-way permutations, there exist NIWI proofs, for every language in* **NP**.

As explained below, in our construction of NIWI, one-way permutations are used to construct a *dense* non-interactive commitment scheme. We show that such commitments are somewhat inherent (see more details below).

Comparison to Previous Constructions. Sahai and Waters [SW14] constructed, from iO, non-adaptive NIZK arguments in the common reference string model; these are insufficient to obtain ZAPs (let alone, NIWIs).

The assumptions that we rely on for either NIWIs, or NIZKs in the random string model (or equivalently, ZAPs) are incomparable to previously used assumptions. Our main assumption is iO, which is not formally known to be either weaker or stronger. While perhaps not weaker, iO does seem to be of different nature than the previous assumptions. Indeed, previous constructions are based on primitives with an *exact* combinatorial or algebraic structure, such as trapdoor permutations [DN07, BOV07], or bilinear maps in appropriate groups [GOS12]. Finding candidates adhering to such exact structures has proven to be challenging, and such candidates remain scarce. In contrast, iO has candidates based on *noisy* graded encodings [GGH+13b], which by now already have several proposed instantiations [GGH13a, CLT13, GGH14]. Future constructions of iO may be based on primitives with even less algebraic structure.

While our construction of NIZKs relies solely on iO and one-way functions, the NIWI construction also requires (certifiable) one-way permutations, which are already a rather structured object, with only few more candidates than trapdoor permutations (based on the hardness of discrete logs). We find that the main appeal of the suggested NIWI construction, compared to previous ones, is its relative simplicity.

1.2 Techniques

We now overview the techniques behind our constructions. We start with the construction of ZAPs, and then move on to the NIWI construction.

---

[1] The assumption of one-way functions can be replaced with assuming **NP** $\neq$ **coRP** [KMN+14].

ZAPs. Our main technical contribution towards obtaining ZAPs is a construction of *invariant signatures in the common random string model*, a concept presented by Goldwasser and Ostrovsky [GO92]. We then apply a series of generic transformations from the literature: in the common random string model, invariant signatures imply NIZKs [GO92], which imply ZAPs [DN07] (in the plain model). As a secondary contribution, we also give a full description and proof of the first transformation, previously sketched in [GO92]. Details follow.

Invariant Signatures. Invariant signatures, introduced by Goldwasser and Ostrovsky [GO92], are digital signatures where all valid signatures of any message are either identical, or share a common property. Concretely, we say that a signature scheme is invariant if there is some efficiently computable property P of signatures such that for any message m^* and any verification key vk there is a unique value $P_{\mathsf{vk}}(m^*)$ such that $P(\sigma) = P_{\mathsf{vk}}(m^*)$ for any valid signature σ with respect to vk. Furthermore, it is required that for every message m^*, for an honestly generated verification key (sampled independently of m^*), the property value $P_{\mathsf{vk}}(m^*)$ is pseudo-random, even given the verification key and a signature oracle on messages $m \neq m^*$. Like in [GO92], we consider a relaxed notion of invariant signatures in the common random string model (CRS). Here we require that the property value P of valid signatures is unique for every verification key vk, with overwhelming probability over the choice of the CRS. Pseudo-randomness of P_{vk} should hold even given the CRS.[2]

Before explaining how we construct invariant signatures, let us first motivate them by recalling how they are utilized in the construction of NIZKs.

NIZKs from Invariant Signatures. Goldwasser and Ostrovsky gave a transformation from invariant signatures to NIZKs [GO92]. Their transformation is based on the construction of Feige, Lapidot and Shamir [FLS99] of NIZKs in the *hidden-bits model*. In this model, a random hidden string is available to the prover but is hidden from the verifier. The prover can reveal to the verifier specific bits of the hidden string in the locations of its choice, but it cannot change the value of these bits. [FLS99] also show how to compile a NIZK in the hidden-bits model into a NIZK in the CRS model assuming trapdoor permutations. [GO92] gave a different compilation technique based on invariant signatures. Next, we describe such a compilation following the same high-level idea as [GO92, BGRV09].

We interpret the CRS (available to both prover and verifier) as containing a CRS for the invariant signature, as well as a sequence of messages $\{m_i\}$ and one-time pad bits $\{s_i\}$ where every (m_i, s_i) will be used to obtain a single hidden bit b_i. The prover will sample keys $(\mathsf{sk}, \mathsf{vk})$ for the invariant signature and send the verification key vk to the verifier as part of the proof. The hidden bit b_i is then defined as the bit $P_{\mathsf{vk}}(m_i)$, the property value of the message m_i, XORed with

---

[2] In the original definition of [GO92], pseudo-randomness is also required for messages m^* sampled adaptively after the verification key. While we do not achieve such *adaptive* pseudo-randomness, the above *selective* pseudo-randomness will suffice for our purpose.

the the one-time pad bit s_i. To reveal the bit b_i, the prover sends to the verifier a signature σ_i on m_i. The verifier can compute b_i by computing $P(\sigma_i) = P_{\mathsf{vk}}(m_i)$.

The uniqueness of the signature guarantees that the prover cannot affect the value of the hidden bits. Note that the prover can always affect the value of b_i by choosing the verification key vk. However, since the length of vk is bounded, this issue can be overcome via soundness amplification [FLS99, GO92]). Another problem is that the prover might affect the distribution of the hidden bits by choosing a verification key such that $P_{\mathsf{vk}}(\cdot)$ is unbalanced. In [GO92, BGRV09], this is addressed by certifying the fact that $P_{\mathsf{vk}}(\cdot)$ is balanced, using a similar approach to [BY96]. In our construction, we guarantee that the hidden bits are uniformly distributed by simply XORing $P_{\mathsf{vk}}(m_i)$ with the random one-time pad bit s_i. Finally, the pseudo-randomness of P_{vk} guarantees that the bits not revealed by the prover remain hidden from the verifier.

Constructing Invariant Signatures. The starting point of our construction is the selectively-secure signature scheme of Sahai-Waters based on iO and one-way functions [SW14]. The basic idea behind the construction is as follows. The secret signing key is simply a key K for a pseudo-random function PRF_K, and a signature σ on message m is simply $\sigma = \mathsf{PRF}_K(m)$. The public verification key is an obfuscation $\tilde{C} \leftarrow i\mathcal{O}(C_K)$ of a circuit C_K that given any m returns $y_m = f(\mathsf{PRF}_K(m))$ for some one-way function f. Verification of any σ for m is simply done by computing $f(\sigma)$ and comparing to the value y_m output by $\tilde{C}_K$.

Sahai and Waters show, based on the indistinguishability obfuscation guarantee, that their scheme is selectively-secure; namely, it is impossible to forge a signature for any preselected message m^*, even given a signature oracle on other messages. The idea is to consider an alternative to the the the circuit C_K that computes the same function, but while only "knowing" y_{m^*}, and without actually "knowing" the preimage $\mathsf{PRF}_K(m^*)$. This is achieved using their elegant *puncturing technique*. Specifically, instead of using any PRF family, they use a *puncturable PRF*. In such PRFs, it is possible to puncture a given key K at an arbitrary point m^* in the domain of the function. The punctured function $\mathsf{PRF}_{K_{m^*}}$, with punctured key K_{m^*}, preserves functionality at any other point, but hides any information on the point $\mathsf{PRF}_K(m^*)$; namely, the value $\mathsf{PRF}_K(m^*)$ is pseudo-random, even given (m^*, K_{m^*}). Such puncturable PRFs follow from the GGM [GGM86] construction [BW13, BGI14, KPTZ13].

Using a puncturable PRF in the implementation of C_K, it can be shown that if a forger succeeds in finding a preimage of $y_{m^*} = f(\mathsf{PRF}_K(m^*))$, it would also succeed had we provided it with an obfuscation of the alternative circuit $C_{K_{m^*}, y_{m^*}}$. The circuit $C_{K_{m^*}, y_{m^*}}$ computes the same function as C_K, but in a different way: it only has the punctured key K_{m^*}, and has the value y_{m^*} directly hardwired into it, so that it does not have to evaluate the PRF in order to compute it. Thus, the fact that the forger still succeeds follows by the guarantee of indistinguishability obfuscation. However, now by the pseudo-randomness guarantee at the punctured point m^*, we know that $\mathsf{PRF}_K(m^*)$ is pseudo random, even given the circuit $C_{K_{m^*}, y_{m^*}}$, and thus the forger can be used to invert the one-way function f.

We next observe that the Sahai-Waters signature scheme can be made invariant as follows. To get uniqueness, we can use an injective one-way function f instead of an arbitrary one-way function. Indeed, this guarantees that for any, even malicious, verification key $\tilde{C}$ and any message m, the value $y^* = \tilde{C}(m)$ has a unique preimage under f that will be accepted in verification. To get (selective) pseudo-randomness, rather than just (selective) unforgeability, we can define the property P to extract a Goldreich-Levin hardcore bit [GL89] from the unique signature with respect to a fixed seed put in the verification key; this preserves uniqueness.

The above solution requires the extra assumption that injective one-way functions exist. We show that a more significant modification of the SW scheme allows to rely on any one-way function. Unlike the solution above that did not rely on a CRS, the new solution will (as explained before, this is still sufficient for our purposes). The basic idea is the following. Imagine we had at our expense a non-interactive perfectly-binding commitment scheme Com. We could then augment the circuit C_K to output, instead of a one-way function $f(\mathsf{PRF}_K(m^*))$, a commitment $c_{m^*} = \mathsf{Com}(b; r)$, to plaintext $b = \mathsf{PRF}_K(m^*)$, where the randomness r is derived, say, by applying another $\mathsf{PRF}_{K'}$ to m^* (or simply setting $(b, r) = \mathsf{PRF}_K(m^*)$). A signature would then include the plaintext underlying the commitment $\mathsf{PRF}_K(m^*)$ and the randomness $r = \mathsf{PRF}_{K'}(m^*)$. The unique property P will simply be the plaintext b.

Indeed, uniqueness will now follow by the perfect binding of the commitment, and pseudo-randomness of b will follow using a similar puncturing argument to the one above, coupled with the hiding of the commitment Com. However, non-interactive perfectly-binding commitments are only known based on injective one-way functions [Blu81], which may take us back to square one. Here the CRS comes to our aid. We can use Naor's [Nao91] two-message statistically-binding commitment scheme, where the first receiver message is simply a random string that can be put in the CRS; indeed, this commitment can be based on any one-way function.

NIWIs. The first stepping stone in our NIWI construction is a natural idea suggested by Niu et al. [NLLT14], where it is described using the terminology of witness encryption [GGSW13]. In witness encryption, anyone can encrypt a message m under a public candidate instance x for some **NP** language $\mathcal{L}$ (x is thought of as the public key); if $x \in \mathcal{L}$, anyone holding a corresponding witness w can decrypt the encrypted $\mathsf{Enc}_x(m)$; however, if $x \notin \mathcal{L}$, the encryption is semantically secure; namely, $\mathsf{Enc}_x(m)$ is computationally indistinguishable from $\mathsf{Enc}_x(m')$ for any two messages m, m'. Such a scheme can be easily constructed from any indistinguishability obfuscator (as we shall soon see).

Given a witness encryption scheme, Niu et al. suggest the following candidate for a NIWI. Given $(x, w) \in \mathcal{R}_{\mathcal{L}}$, a proof that $x \in \mathcal{L}$ is simply an indistinguishability obfuscation $\tilde{D} \leftarrow i\mathcal{O}(D_{x,w})$ of the witness decryption circuit $D_{x,w}$ that given a witness encryption $\mathsf{Enc}_x(m)$, decrypts it with the witness w and outputs m. Verification is done by running the circuit $\tilde{D}$ on an encryption $\mathsf{Enc}_x(m)$ of a random $m \leftarrow \{0,1\}^n$, and testing whether it successfully decrypts m. Indeed, if $x \notin \mathcal{L}$, $\tilde{D}$ fails with overwhelming probability due to semantic security.

What about witness indistinguishability? at first it seems that regardless of which witness w is used by $D_{x,w}$, it has the same functionality, since any witness can be used for decryption. Thus, WI should follow by the iO guarantee. However, this argument is flawed—while, for valid (honestly generated) witness encryptions $\mathsf{Enc}_x(m)$, $D_{x,w}$ behaves the same regardless of the witness, *this might not be true for maliciously generated encryptions*.

To illustrate this consider a witness encryption scheme implemented using indistinguishability obfuscation, where $\mathsf{Enc}_x(m)$ consists of an obfuscation $\tilde{E} \leftarrow i\mathcal{O}(E_x^m)$ for the circuit E_x^m that given as input a proper witness $w \in \mathcal{R}_\mathcal{L}(x)$ outputs m and otherwise $\bot$. For $x \notin \mathcal{L}$, such a circuit always returns $\bot$, regardless of m, and thus semantic security follows from iO. However, if we instantiate the above candidate NIWI with this witness encryption scheme, the result will be completely insecure. A malicious verifier may obfuscate an arbitrary circuit, instead of a proper circuit E_x^m, and distinguish between different witnesses. Taken to the extreme, it could just obfuscate the identity, and recover from $\tilde{D}$ the entire witness w.

Fixing the NIWI Using ZAPs. the above problem can be resolved by requiring that the malicious verifier proves to $\tilde{D}$ that its witness encryption is indeed a proper encryption of some plaintext with some randomness. However, to maintain soundness, this should be done while keeping the one-wayness of m. To achieve this, we rely on ZAPs, and the Feige-Shamir trapdoor paradigm [FS89]: the prover will hard-code into $\tilde{D}$ a random first message for a ZAP, and the verifier will prove to $\tilde{D}$ that either $\tilde{E}$ was generated properly or that some "trapdoor" statement is true. In order to assure that the verifier's encryption is proper, the trapdoor statement is usually chosen such that it is true, but it is hard for the verifier to find a witness for it, for example, stating that a random sting is in the image of a one-way permutation. However, in our setting, since the ZAP is not a proof of knowledge, such a trapdoor statement is insufficient.

In our protocol, we do not rely on the fact that the trapdoor statement is hard to prove, but rather we aim to design a trapdoor statement such that, if true, certifies the validity of the witness encryption $\tilde{E}$. The problem is that such certification cannot use the encryption's randomness or the plaintext m as a witness, otherwise we cannot argue that the ZAP hides m. The key observation is that it is enough to certify that the encryption $\tilde{E}$ behaves in the same way on any two potential witnesses. To implement this idea, the trapdoor statement will include a pair of perfectly binding commitments c_1, c_2 chosen by the prover (the honest prover commits to all-zero strings). The statement asserts that there is a pair of candidate witnesses w_1, w_2 such that c_1, c_2 are commitments to w_1, w_2 and the verifier's encryption $\tilde{E}$ decrypts to the same value when decrypted with either w_1 or w_2.

Let us describe the intuition behind the proof of security. To prove soundness, we rely on the fact that c_1, c_2 are are computed using a *dense* commitment scheme, where every string has some valid decommitment. Assume there exists an accepting proof $(\tilde{D}, c_1, c_2)$ for a false statement $x \notin \mathcal{L}$, meaning that $\tilde{D}$ manages to invert a witness encryption $\tilde{E}$ for a random message, given also the ZAP described above. We show that, $\tilde{D}$ must break the semantic security of the witness encryption. Indeed, letting w_1, w_2 be the plaintexts underlying c_1, c_2, we note that, since $x \notin \mathcal{L}$, decrypting with either one results in the same value $\bot$. Therefore, the trapdoor statement is true, and we could have used it to compute a ZAP proof π, without compromising the semantic security of $\tilde{E}$. Since $\tilde{D}$ cannot tell the difference between the two ZAP proofs, it would still invert $\tilde{E}$, and thus violate semantic security. We note that for the above argument to go through, we rely on the fact that the ZAP guarantees witness-indistinguishability against *non-uniform verifiers*; indeed, the reduction described above gets a non-uniform advice: the decommitment information for the commitments c_1, c_2.

To show that the proof is WI, consider any instance $x \in \mathcal{L}$ with two valid witnesses w_1, w_2. We go through several hybrid experiments. We start by using the hiding property of the commitment to replace c_1 and c_2 with commitments to w_1 and w_2, instead of all-zero strings. Now if the verifier generates an encryption $\tilde{E}$ with a valid ZAP proof π it follows from the the soundness of the ZAP and the binding of the commitment that either $\tilde{E}$ is a valid witness encryption, or $\tilde{E}$ decrypts to the same value when decrypted with either w_1 or w_2. In any case, the witness decryption circuits D_{x,w_1} and D_{x,w_2} agree on the input $(\tilde{E}, \pi)$. By the iO guarantee, the obfuscated decryption circuits are thus indistinguishable, and we can replace one with the other.

A Note on Statistical Soundness. At first glance, it may seem that our reliance on computational primitives such as witness encryption implies that the resulting system is only computationally sound; we stress, however, that soundness is statistical.[3] To cheat in our protocol, the (unbounded) prover must produce a proof consisting of a *small* circuit (allegedly an obfuscated witness description circuit). The soundness of the system is based on the fact that this computationally-bounded circuit cannot break the security of the underling primitives. Indeed, the computational assumptions imply that such a circuit simply does not exist.

Additionally, we note that the soundness we get is statistical and not *perfect* as in [BOV07, GOS12]. In the language of [BOV07], we get **MA** proofs rather than **NP** proofs (for all languages in **NP**).

On the Necessity of Dense Commitments. The NIWI construction described above can be based on a non-interactive commitment scheme satisfying the following properties. First, it is computationally hiding. Second, it is statistically binding, but only against *honest* committers; namely, honestly generated commitments can only be opened to a single value. Finally, the commitment is *dense*; that is, every string in the range of the commitment, can be opened to at least one value (commitments that are not generated honestly may potentially

---

[3] In fact, any single message argument system that is sound against non-uniform provers must be statistically sound, as accepting proofs for false statements may be hardwired to the prover.

be opened to more than one value). We observe that such dense commitments are somewhat necessary. Specifically, using NIWIs, we can transform any non-interactive statistically binding commitment into a commitment satisfying the above three requirements (note that statistically-binding commitments can be constructed from any injective one-way function [Blu81].)

The basic idea is to commit twice to the same value and add a NIWI proof that one of the two commitments were honestly generated. A valid opening of this commitment would consist of an opening of any one of the two underlying commitments. If the NIWI is not accepting, the committed value is set arbitrarily to zero.[4] The hiding of the new commitment follows from that of the original commitment, together with the witness-indistinguishability of the NIWI. Binding, for honestly generated commitments (where the two underlying plaintexts are identical), follows from the binding property of the original commitment. Finally, the fact that the new commitment is dense follows directly from the soundness of the NIWI.

We note that it may still be possible that NIWIs, and in particular, dense commitments as above, can be based on iO and any one-way function.

Organization. In Section 2, we present the basic definitions used in the paper. In Section 3, we define and construct invariant signatures. In Section 4, we describe the Goldwasser-Ostrovsky transformation from invariant signatures to NIZKs. Section 5 describes the NIWI construction.

2   Definitions

2.1   Non-Interactive Zero-Knowledge

Definition 2.1. *A pair of PPT algorithms* $(\mathcal{P}, \mathcal{V})$ *is a NIZK proof in the CRS model if they satisfy the following properties:*

1. *Completeness: there exists a polynomial r denoting the length of the common random string such that for every $(x, w) \in \mathcal{R}_\mathcal{L}$ we have that:*

$$\Pr_{\mathcal{P}, \mathsf{crs} \leftarrow \{0,1\}^{r(|x|)}} [\mathcal{V}(x, \mathsf{crs}, \pi) = 1 : \pi \leftarrow \mathcal{P}(x, w, \mathsf{crs})] = 1 .$$

2. *Soundness: for every $x \notin \mathcal{L}$ we have that:*

$$\Pr_{\mathsf{crs} \leftarrow \{0,1\}^{r(|x|)}} [\exists \pi : \mathcal{V}(x, \mathsf{crs}, \pi) = 1] < 2^{-|x|} .$$

3. *Zero-Knowledge: there exists a PPT algorithm $\mathcal{S}$ such that:*

$$\left\{ (\mathsf{crs}, \mathcal{P}(x, w, \mathsf{crs})) : \mathsf{crs} \leftarrow U_{r(|x|)} \right\}_{(x,w) \in \mathcal{R}_\mathcal{L}} \approx_c \left\{ \mathcal{S}(x) \right\}_{(x,w) \in \mathcal{R}_\mathcal{L}}$$

---

[4] Here we assume NIWI with perfect soundness. In particular we assume that the verification procedure of the NIWI is deterministic. Dense commitments satisfying a slightly more involved definition can be constructed from NIWI with only statistical soundness.

Remark 2.1. Definition 2.1 considers only non-adaptive soundness and zero-knowledge. Additionally, zero-knowledge is not guaranteed when multiple statement are proven with respect to the same CRS. We note that any NIZK proof system for **NP** can be transformed into a system that does not have these disadvantages assuming only OWFs [FLS99].

2.2   ZAPs

ZAPs [DN07] are two-message public-coin witness-indistinguishable proofs, defined as follows.

Definition 2.2. *A pair of algorithms* $(\mathcal{P}, \mathcal{V})$, *where* $\mathcal{P}$ *is PPT and* $\mathcal{V}$ *is (deterministic) polytime, is a ZAP for an* **NP** *relation* $\mathcal{R}_\mathcal{L}$ *if it satisfies:*

1. *Completeness: there exists a polynomial r such that for every $(x, w) \in \mathcal{R}_\mathcal{L}$,*

$$\Pr_{\mathcal{P}, \mathsf{r} \leftarrow \{0,1\}^{r(|x|)}} [\mathcal{V}(x, \pi, \mathsf{r}) = 1 : \pi \leftarrow \mathcal{P}(x, w, \mathsf{r})] = 1 \ .$$

2. *Adaptive soundness: for every malicious prover $\mathcal{P}^*$ and every $n \in \mathbb{N}$:*

$$\Pr_{\mathsf{r} \leftarrow \{0,1\}^{r(n)}} \left[\exists \begin{matrix} x \in \{0,1\}^n \setminus \mathcal{L} \\ \pi \in \{0,1\}^* \end{matrix} : \mathcal{V}(x, \pi, \mathsf{r}) = 1 \right] \le 2^{-n} \ .$$

3. *Witness indistinguishability: for   any   sequence $\mathcal{I} =$ $\{(x, w_1, w_2) : w_1, w_2 \in \mathcal{R}_\mathcal{L}(x)\}$ and any first-message sequence $\mathcal{R} =$ $\{\mathsf{r}_{x, w_1, w_2} \in \{0,1\}^{r(|x|)} : (x, w_1, w_2) \in \mathcal{I}\}$:*

$$\{\pi_1 \leftarrow \mathcal{P}(x, w_1, \mathsf{r}_{x, w_1, w_2})\}_{(x, w_1, w_2) \in \mathcal{I}} \approx_c \{\pi_2 \leftarrow \mathcal{P}(x, w_2, \mathsf{r}_{x, w_1, w_2})\}_{(x, w_1, w_2) \in \mathcal{I}} \ .$$

2.3   NIWIs

NIWIs [BOV07] are completely non-interactive witness-indistinguishable proofs.

Definition 2.3. *A pair of PPT algorithms* $(\mathcal{P}, \mathcal{V})$ *is a NIWI for an* **NP** *relation* $\mathcal{R}_\mathcal{L}$ *if it satisfies:*

1. *Completeness: for every $(x, w) \in \mathcal{R}_\mathcal{L}$,*

$$\Pr_{\mathcal{P}} [\mathcal{V}(x, \pi) = 1 : \pi \leftarrow \mathcal{P}(x, w)] = 1 \ .$$

2. *Soundness: there exists a negligible function μ, such that for every $x \notin \mathcal{L}$ and $\pi \in \{0,1\}^*$:*

$$\Pr_{\mathcal{V}} [\mathcal{V}(x, \pi) = 1] \le \mu(|x|) \ .$$

3. *Witness indistinguishability: for   any   sequence $\mathcal{I} =$ $\{(x, w_1, w_2) : w_1, w_2 \in \mathcal{R}_\mathcal{L}(x)\}$:*

$$\{\pi_1 : \pi_1 \leftarrow \mathcal{P}(x, w_1)\}_{(x, w_1, w_2) \in \mathcal{I}} \approx_c \{\pi_2 : \pi_2 \leftarrow \mathcal{P}(x, w_2)\}_{(x, w_1, w_2) \in \mathcal{I}} \ .$$

2.4   Indistinguishability Obfuscation

Indistinguishability obfuscation (iO) was introduced in [BGI+01] and given a candidate construction in [GGH+13b], and subsequently in [BR13, BGK+13].

Definition 2.4 (Indistinguishability Obfuscation [BGI+01]). *A PPT algorithm iO is said to be an* indistinguishability obfuscator *for a collection of polysize circuits* $\mathcal{C} = \bigcup_{n \in \mathbb{N}} \mathcal{C}_n$, *if it satisfies:*

1. *Functionality: For any* $C \in \mathcal{C}$,

$$\Pr_{iO} [\forall x : iO(C)(x) = C(x)] = 1 \; .$$

2. *Indistinguishability: For any polysize distinguisher* $\mathcal{D}$ *there negligible function* μ, *such that for any* $n \in \mathbb{N}$ *and* $C_1, C_2 \in \mathcal{C}_n$ *of the same size and functionality*

$$\left| \Pr_{iO} [\mathcal{D}(iO(C_1)) = 1] - \Pr_{iO} [\mathcal{D}(iO(C_2)) = 1] \right| \leq \mu(n) \; .$$

3   Invariant Signatures from Indistinguishability Obfuscation

In this section, we recall the definition of invariant signatures [GO92] in the common random string (CRS) model and construct them based on iO.

Roughly, invariant signatures are digital signatures where valid signatures of any message are either identical, or share a common property. More accurately, there is an efficiently computable property P of signatures such that for any message m^* and any verification key vk there is a unique value $P_{vk}(m^*)$ such that $P(\sigma) = P_{vk}(m^*)$ for any valid signature σ with respect to vk. Furthermore, it is required that for every message m^*, for an honestly generated verification key (sampled independently of m^*), the property value $P_{vk}(m^*)$ is pseudo-random, even given the verification key and a signature oracle on messages $m \neq m^*$. Like in [GO92], we consider a relaxed notion of invariant signatures in the common random string model (CRS). Here the property value P is unique for every verification key vk, with overwhelming probability over the choice of the CRS, and pseudo-randomness of P_{vk} should hold even given the CRS. (In the original definition of [GO92], pseudo-randomness is also required for messages m^* sampled adaptively after the verification key. While we do not achieve such *adaptive* pseudo-randomness, the above *selective* pseudo-randomness will suffice for our purpose.)

Definition 3.1 (Invariant Signatures in the CRS Model). *A triple of poly-time algorithms* (Gen, Sign, Ver), *where* Gen *is randomized, is a digital signature scheme with invariant signatures and selective security in the CRS model if it satisfies the following properties:*

1. *Syntax and completeness: There exists a polynomial r such that for every security parameter $n \in \mathbb{N}$, and for every message $m \in \{0,1\}^*$ we have that:*

$$\Pr_{crs \leftarrow U_{r(n)}} [\text{Ver}_{vk}(crs, m, \sigma) = 1 : \sigma \leftarrow \text{Sign}_{sk}(m), (sk, vk) \leftarrow \text{Gen}(crs)] = 1 \; .$$

2. *Uniqueness:* There exists a deterministic, efficiently computable, predicate $P : \{0,1\}^* \to \{0,1\}$, and a negligible function μ such that:

$$\Pr_{\mathsf{crs} \leftarrow U_{r(n)}} [\exists m, \mathsf{vk}, \sigma_1, \sigma_2 : P(\sigma_1) \neq P(\sigma_2) \wedge \mathsf{Ver}_{\mathsf{vk}}(\mathsf{crs}, m, \sigma_1) = \mathsf{Ver}_{\mathsf{vk}}(\mathsf{crs}, m, \sigma_2) = 1] \leq \mu(n) \ .$$

3. *Pseudo-randomness:* For every poly-size adversary A, there exists a negligible function μ such that for every security parameter $n \in \mathbb{N}$, and for every message $m \in \{0,1\}^n$:

$$\left| \begin{array}{l} \Pr_{\mathsf{A},\mathsf{crs} \leftarrow U_{r(n)}} [\mathsf{A}^{\mathsf{Sign}^*_{\mathsf{sk},m}}(\mathsf{crs}, \mathsf{vk}, m, P(\mathsf{Sign}_{\mathsf{sk}}(m))) = 1 : (\mathsf{sk}, \mathsf{vk}) \leftarrow \mathsf{Gen}(\mathsf{crs})] - \\ \Pr_{\mathsf{A},\mathsf{crs} \leftarrow U_{r(n)}, b \leftarrow U_1} [\mathsf{A}^{\mathsf{Sign}^*_{\mathsf{sk},m}}(\mathsf{crs}, \mathsf{vk}, m, b) = 1 : (\mathsf{sk}, \mathsf{vk}) \leftarrow \mathsf{Gen}(\mathsf{crs})] \end{array} \right| \leq \mu(n) \ ,$$

where $\mathsf{Sign}^*_{\mathsf{sk},m}$ is an oracle that is identical to $\mathsf{Sign}_{\mathsf{sk}}$ excpet that on input m it outputs $\perp$.

Remark 3.1 (Unforgeability). We do not explicitly require that the signature scheme is unforgeable against selective attackers. Unforgeability is, in fact, implied by uniqueness and the pseudo-randomness properties. In particular, if an adversary can forge a signature σ on a message m, it can compute $P(\sigma)$ and break pseudo-randomness.

Sahai and Waters construct digital signature scheme with based on iO and one-way functions [SW14]. As outline in the introduction, we observe that a modification of their construction is also invariant assuming also *injective* one-way functions.

Theorem 3.1 (follows from [SW14]). *Assuming indistinguishability obfuscation and injective OWFs, there exists a selectively secure invariant signature scheme.*

We show that, in the CRS model, we can in fact construct selectively secure invariant signatures based on iO and *any* one-way function.

Theorem 3.2. *Assuming indistinguishability obfuscation and one-way functions, there exists a selectively-secure invariant signature scheme in the CRS model.*

Like the Shahi-Waters construction, the construction here relies on their punctured program paradigm. We next define puncturable pseudo-random functions, a central tool in our construction, and then move to describe the construction.

3.1 Puncturable PRFs

We consider a simple case of the puncturable PRFs where any PRF might be punctured at a single point. The definition is formulated as in [SW14].

Definition 3.2 (Puncturable PRFs). *Let ℓ, m be polynomially bounded length functions. An efficiently computable family of functions*

$$\mathcal{PRF} = \left\{ \mathsf{PRF}_K : \{0,1\}^{m(n)} \to \{0,1\}^{\ell(n)} \,\middle|\, K \in \{0,1\}^n, n \in \mathbb{N} \right\} ,$$

associated with an efficient (probabilistic) key sampler $\mathcal{K}_{\mathcal{PRF}}$, is a puncturable PRF if there exists a puncturing algorithm Punc that takes as input a key $K \in \{0,1\}^n$, and a point x^*, and outputs a punctured key K_{x^*}, so that the following conditions are satisfied:

1. *Functionality is preserved under puncturing: For every $x^* \in \{0,1\}^{\ell(n)}$,*

$$\Pr_{K \leftarrow \mathcal{K}_{\mathcal{PRF}}(1^n)} \left[\forall x \neq x^* : \mathsf{PRF}_K(x) = \mathsf{PRF}_{K_{x^*}}(x) \mid K_{x^*} = \mathsf{Punc}(K, x^*) \right] = 1 .$$

2. *Indistinguishability at punctured points: The following ensembles are computationally indistinguishable:*
   - $\left\{ x^*, K_{x^*}, \mathsf{PRF}_K(x^*) \mid K \leftarrow \mathcal{K}_{\mathcal{PRF}}(1^n), K_{x^*} = \mathsf{Punc}(K, x^*) \right\}_{x^* \in \{0,1\}^{m(n)}, n \in \mathbb{N}}$
   - $\left\{ x^*, K_{x^*}, u \mid K \leftarrow \mathcal{K}_{\mathcal{PRF}}(1^n), K_{x^*} = \mathsf{Punc}(K, x^*), u \leftarrow \{0,1\}^{\ell(n)} \right\}_{x^* \in \{0,1\}^{m(n)}, n \in \mathbb{N}}$

To be explicit, we include x^* in the distribution; throughout, we shall assume for simplicity that a punctured key K_{x^*} includes x^* in the clear. As shown in [BGI14, BW13, KPTZ13], the GGM [GGM86] PRF yield puncturable PRFs as defined above.

3.2 Invariant Signatures Construction

We now present the details of our construction, an overview is given in the introduction. We shall rely on the following primitives:

- A two-message statistically binding commitment Com with a random first message based on any one-way function [Nao91]. We denote by C, S the polynomials such that $\mathsf{Com}_s(b; r) \in \{0,1\}^{C(n)}$ is a commitment to a bit b, and where the first commitment message is $s \in \{0,1\}^{S(n)}$, and the randomness is $r \in \{0,1\}^n$.
- A family of puncturable PRFs $\mathcal{PRF} = \{\mathsf{PRF}_K\}$ from $\{0,1\}^n$ to $\{0,1\}^{n+1}$ associated with a key sampler $\mathcal{K}_{\mathcal{PRF}}$ and a puncturing algorithm Punc.
- An indistinguishability obfuscator $i\mathcal{O}$.

Construction 3.3 (A selectively-secure invariant signature)

The CRS. *The CRS consists of a random first message s for Com. Throughout the construction, one may identify the notation crs with that of s.*

The algorithm Gen. *Given the CRS s, Gen samples a PRF key $K \leftarrow \mathcal{K}_{\mathcal{PRF}}(1^n)$. Gen sets $\mathsf{sk} = K$ and sets $\mathsf{vk} = i\mathcal{O}([C_{s,K}]_\ell)$ where $[C_K]_\ell$ is a "commitment to pseudo-random property" circuit $C_{s,K}$, given by Figure 1, padded up to length the maximum size ℓ of the circuits given in Figures 1,2.*

Hardwired: CRS containing a first message $s \in \{0,1\}^{S(n)}$ for Com and a PRF key $K \leftarrow \mathcal{K}_{\mathcal{PRF}}(1^n)$.
Input: Message $m \in \{0,1\}^n$.
Output: Obtain $(b', r') = \mathsf{PRF}_K(m)$ and output $\mathsf{Com}_s(b', r')$.

Fig. 1. The "commitment to pseudo-random property" circuit $C_{s,K}$

Hardwired:
  1. CRS containing a first message $s \in \{0,1\}^{S(n)}$ for Com.
  2. Punctured PRF key $K_{m^*} = \mathsf{Punc}(K, m^*)$ where $K \leftarrow \mathcal{K}_{\mathcal{PRF}}(1^n)$.
  3. Commitment $\mathsf{c}^* \in \{0,1\}^{C(n)}$.
Input: Message $m \in \{0,1\}^n$.
Output:
  1. If $m = m^*$, output c^*.
  2. Else, obtain $(b', r') = \mathsf{PRF}_{K_{m^*}}(m)$ and output $\mathsf{Com}_s(b', r')$.

Fig. 2. The circuit $C_{s,K_{m^*},\mathsf{c}^*}$

The algorithm Sign. *Given the secret key K and a message m output $(b', r') = \mathsf{PRF}_K(m)$.*

The algorithm Ver. *Given the obfuscated circuit* vk, *the CRS s, a message m and a signature $\sigma = (b, r)$, obtain $\mathsf{c} = \mathsf{vk}(m)$. Output 1 if $\mathsf{c} = \mathsf{Com}_s(b; r)$. Otherwise, output 0.*

Proposition 3.1. *Construction 3.3 is a selectively-secure invariant signature scheme in the CRS model.*

Proof. It is straightforward to verify the completeness of the construction. Next we prove the uniqueness and pseudo-randomness properties.

Uniqueness. For a signature $\sigma = (b, r)$ let $P(\sigma)$ be the a predicate that outputs b. Let G be the event, over the choice of the CRS crs that there exist a message $m \in \{0,1\}^n$, a verification key vk and a pair of signatures $\sigma_1 = (b_1, r_1), \sigma_2 = (b_2, r_2)$ such that:

$$P(\sigma_1) \neq P(\sigma_2) \wedge \mathsf{Ver}_{\mathsf{vk}}(\mathsf{crs}, m, \sigma_1) = \mathsf{Ver}_{\mathsf{vk}}(\mathsf{crs}, m, \sigma_2) = 1 \ .$$

Equivalently, $b_1 \neq b_2$ and $\mathsf{Com}_s(b_1; r_1) = \mathsf{Com}_s(b_2; r_2) = \mathsf{vk}(m)$. Since with overwhelming probability Com_s is perfectly binding property of Com, it holds that $\Pr_{\mathsf{crs} \leftarrow U_{r(n)}}[G] \leq \mathsf{negl}(n)$, as required.

Pseudo-Randomness. Fix any polysize adversary A, and for every message $m \in \{0,1\}^n$, let $p_0(m)$ denote the probability that it outputs 1 given the unique property b of any signature (b, r) on m:

$$p_0(m) = \Pr \left[A^{\mathsf{Sign}^*_{\mathsf{sk},m}(\cdot)}(\mathsf{crs}, \mathsf{vk}, m, P(\mathsf{Sign}_{\mathsf{sk}}(m))) = 1 : \begin{array}{c} \mathsf{crs} \leftarrow U_{r(n)} \\ (\mathsf{sk}, \mathsf{vk}) \leftarrow \mathsf{Gen}(\mathsf{crs}) \end{array} \right]$$

$$= \Pr \left[A^{\mathsf{PRF}^*_{K,m}(\cdot)}(s, \mathsf{vk}, m, b) = 1 : \begin{array}{c} s \leftarrow U_{r(n)} \\ K \leftarrow \mathcal{K}_{\mathcal{PRF}}(1^n) \\ (b, r) = \mathsf{PRF}_K(m) \\ \mathsf{vk} \leftarrow i\mathcal{O}([C_{s,K}]_\ell) \end{array} \right] ,$$

where $\mathsf{Sign}^*_{\mathsf{sk},m}(\cdot) \equiv \mathsf{PRF}^*_{K,m}(\cdot)$ is an oracle that is identical to $\mathsf{Sign}_{\mathsf{sk}}(\cdot) \equiv \mathsf{PRF}_K(\cdot)$, except that on input m it outputs $\perp$.

Consider an alternative experiment where vk is chosen to be an obfuscation of the circuit C_{s,K_m,c^*}, rather than $C_{s,K}$, where $c^* = \mathsf{Com}_s(b, r)$, and (b, r) are computed as before. Let $p_1(m)$ denote the probability that A outputs 1 in this augmented experiment:

$$p_1(m) = \Pr \left[A^{\mathsf{PRF}^*_{K,m}(\cdot)}(s, \mathsf{vk}, m, b) = 1 : \begin{array}{c} s \leftarrow U_{r(n)} \\ K \leftarrow \mathcal{K}_{\mathcal{PRF}}(1^n) \\ K_m \leftarrow \mathsf{Punc}(K, m) \\ (b, r) = \mathsf{PRF}_K(m) \\ \boxed{c^* = \mathsf{Com}_s(b, r)} \\ \boxed{\mathsf{vk} \leftarrow i\mathcal{O}([C_{s,K_m,c^*}]_\ell)} \end{array} \right] .$$

Since the circuits $C_{s,K}$ and C_{s,K_m,c^*} are equivalent, it follows from the security of $i\mathcal{O}$ that the circuits $i\mathcal{O}([C_{s,K}]_\ell)$ and $i\mathcal{O}([C_{s,K_m,c^*}]_\ell)$ are computationally indistinguishable, and therefore

$$|p_0(m) - p_1(m)| < \mathsf{negl}(n) .$$

Next, consider another experiment where the signature (b, r) is chosen uniformly at random, instead of being set to $\mathsf{PRF}_K(m)$. We denote by $p_2(m)$ be the probability that A outputs 1 in this experiment:

$$p_2(m) = \Pr \left[A^{\mathsf{PRF}^*_{K,m}(\cdot)}(s, \mathsf{vk}, m, b) = 1 : \begin{array}{c} s \leftarrow U_{r(n)} \\ K \leftarrow \mathcal{K}_{\mathcal{PRF}}(1^n) \\ K_m \leftarrow \mathsf{Punc}(K, m) \\ \boxed{(b, r) = U_{n+1}} \\ c^* = \mathsf{Com}_s(b, r) \\ \mathsf{vk} \leftarrow i\mathcal{O}([C_{s,K_m,c^*}]_\ell) \end{array} \right] .$$

By the indistinguishability at punctured points property of $\mathcal{PRF}$:

$$|p_1(m) - p_2(m)| \leq \mathsf{negl}(n) ;$$

Indeed, to distinguish between $K_m, \mathsf{PRF}_K(m)$ and $K_m, U_{|m|+1})$, a distinguisher can perfectly emulate A, by answering its oracle queries $m' \neq m$ using the punctured key K_m.

Consider yet another experiment where, instead of giving A the bit b, we replace it with a random independent bit. We denote by $p_3(m)$ the probability that the adversary outputs 1 in this experiment:

$$
p_3(m) = \Pr \left[\mathsf{A}^{\mathsf{PRF}^*_{K,m}(\cdot)}(s, \mathsf{vk}, m, \boxed{b'}) = 1 : \begin{array}{c} \boxed{b' \leftarrow U_1} \\ s \leftarrow U_{r(n)} \\ K \leftarrow \mathcal{K}_{\mathcal{PRF}}(1^n) \\ K_m \leftarrow \mathsf{Punc}(K, m) \\ (b, r) = U_{n+1} \\ \mathsf{c}^* = \mathsf{Com}_s(b, r) \\ \mathsf{vk} \leftarrow i\mathcal{O}([C_{s,K_m,\mathsf{c}^*}]_\ell) \end{array} \right] .
$$

Then, by the computational hiding property of Com:

$$|p_2(m) - p_3(m)| \leq \mathrm{negl}(n) .$$

We define the probabilities p_4, p_5 in the same way we defined p_1, p_0 respectively, except that in these experiments, A gets a random independent bit b'; that is,

$$
p_4(m) = \Pr \left[\mathsf{A}^{\mathsf{PRF}^*_{K,m}(\cdot)}(s, \mathsf{vk}, m, b') = 1 : \begin{array}{c} b' \leftarrow U_1 \\ s \leftarrow U_{r(n)} \\ K \leftarrow \mathcal{K}_{\mathcal{PRF}}(1^n) \\ K_m \leftarrow \mathsf{Punc}(K, m) \\ \boxed{(b, r) = \mathsf{PRF}_K(m)} \\ \mathsf{c}^* = \mathsf{Com}_s(b, r) \\ \mathsf{vk} \leftarrow i\mathcal{O}([C_{s,K_m,\mathsf{c}^*}]_\ell) \end{array} \right] ,
$$

$$
p_5(m) = \Pr \left[\mathsf{A}^{\mathsf{PRF}^*_{K,m}(\cdot)}(s, \mathsf{vk}, m, b') = 1 : \begin{array}{c} b' \leftarrow U_1 \\ s \leftarrow U_{r(n)} \\ K \leftarrow \mathcal{K}_{\mathcal{PRF}}(1^n) \\ (b, r) = \mathsf{PRF}_K(m) \\ \boxed{\mathsf{vk} \leftarrow i\mathcal{O}([C_{s,K}]_\ell)} \end{array} \right] ,
$$

Following the same arguments as before:

$$|p_3(m) - p_4(m)| \leq \mathrm{negl}(n) \quad , \quad |p_4(m) - p_5(m)| \leq \mathrm{negl}(n) ,$$

and overall:

$$|p_0(m) - p_5(m)| \leq \mathrm{negl}(n) .$$

Thus, we have shown as required that for every $m \in \{0, 1\}^n$:

$$
\left| \Pr \left[\mathsf{A}^{\mathsf{PRF}^*_{K,m}(\cdot)}(s, \mathsf{vk}, m, b) = 1 : \begin{array}{c} s \leftarrow U_{r(n)} \\ K \leftarrow \mathcal{K}_{\mathcal{PRF}}(1^n) \\ (b, r) = \mathsf{PRF}_K(m) \\ \mathsf{vk} \leftarrow i\mathcal{O}([C_{s,K}]_\ell) \end{array} \right] \right.
$$
$$
\left. - \Pr \left[\mathsf{A}^{\mathsf{PRF}^*_{K,m}(\cdot)}(s, \mathsf{vk}, m, b') = 1 : \begin{array}{c} b' \leftarrow U_1 \\ s \leftarrow U_{r(n)} \\ K \leftarrow \mathcal{K}_{\mathcal{PRF}}(1^n) \\ (b, r) = \mathsf{PRF}_K(m) \\ \mathsf{vk} \leftarrow i\mathcal{O}([C_{s,K}]_\ell) \end{array} \right] \right| \leq \mathrm{negl}(n) .
$$

4   NIZKs and ZAPs from Invariant Signatures

In this section, we show how to construct NIZKs in the CRS model based on invariant signatures. A construction of ZAPs from NIZKs is given in [DN07]. Feige, Lapidot and Shamir constructed a NIZK proof system that is unconditionally secure in the *hidden-bits model.* They also showed how to transform NIZK in the hidden-bits model to NIZK in the CRS model. Goldwasser and Ostrovsky give a different transformation based on invariant signatures. We present a transformation that follows [GO92], in most parts, and provide a full proof of security.

We start by formally defining NIZK in the hidden-bits model. In this model, a random string crs is sampled as a trusted setup. The prover can read all the bits of crs and reveal a subset of these bits to the verifier, corresponding to indices $\mathcal{I}$. The prover cannot change the bits of crs, and the verifier gets no information about the bits of crs that where not revealed by the prover.

Definition 4.1 (NIZK Proof in the Hidden-Bits Model). *A pair of PPT algorithms* $(\mathcal{P}, \mathcal{V})$ *is a NIZK proof in the hidden-bits model if it satisfies the following properties:*

1. *Completeness: there exists a polynomial r denoting the length of the hidden random string, such that for every* $(x, w) \in \mathcal{R}_{\mathcal{L}}$ *we have that:*

$$\Pr_{\mathcal{P}, \mathsf{crs} \leftarrow \{0,1\}^{r(|x|)}} [\mathcal{V}(x, \mathsf{crs}|_{\mathcal{I}}, \pi) = 1 : (\pi, \mathcal{I}) \leftarrow \mathcal{P}(x, w, \mathsf{crs})] = 1 \ ,$$

   where $\mathcal{I} \subseteq [r(|x|)]$ *and* $\mathsf{crs}|_{\mathcal{I}} = \{(i, \mathsf{crs}[i]) : i \in \mathcal{I}\}$.
2. *Soundness: for every* $x \notin \mathcal{L}$ *we have that:*

$$\Pr_{\mathsf{crs} \leftarrow \{0,1\}^{r(|x|)}} [\exists \pi, \mathcal{I} : \mathcal{V}(x, \mathsf{crs}|_{\mathcal{I}}, \pi) = 1] < 2^{-n} \ .$$

3. *Zero-Knowledge: there exists a PPT algorithm* $\mathcal{S}$ *such that:*

$$\left\{ (\mathsf{crs}|_{\mathcal{I}}, \pi) : \mathsf{crs} \leftarrow U_{r(|x|)}, (\pi, \mathcal{I}) \leftarrow \mathcal{P}(x, w, \mathsf{crs}) \right\}_{(x,w) \in \mathcal{R}_{\mathcal{L}}} \approx_c \left\{ \mathcal{S}(x) \right\}_{(x,w) \in \mathcal{R}_{\mathcal{L}}} \ .$$

Next we construct a NIZK proof in the CRS model.

Construction 4.1 (NIZK in the CRS Model). *We make use of the following primitives:*

- *A selectively secure invariant signature scheme* (Gen, Sign, Ver) *with an invariant predicate P. For security parameter n, let* $r_\sigma = r_\sigma(n)$ *be the length of the CRS, and let* $k = k(n)$ *be the length of the verification key.*
- *A NIZK proof system* $(\mathcal{P}_{\mathsf{hb}}, \mathcal{V}_{\mathsf{hb}})$ *in the hidden-bits model with hidden random string of length* $r = r(n)$.

The NIZK system $(\mathcal{P}, \mathcal{V})$ *in the CRS model is defined as follows:*

The CRS. *The common random string is of length* $r_\sigma + k \cdot r \cdot (n+1)$. *The first* r_σ *bits of the CRS are interpreted as a CRS for the signature scheme* crs_σ. *We think of the rest of the CRS as divided into* $k \cdot r$ *blocks, each of length* $n+1$. *For every* $i \in [k], j \in [r]$, *we think of the* (i,j)-th *block as divided into a message* $m_{i,j} \in \{0,1\}^n$ *and a one-time pad bit* $s_{i,j} \in \{0,1\}$.

The prover $\mathcal{P}$. *Given* $(x,w) \in \mathcal{R}_\mathcal{L}$, *and the CRS* $(\mathsf{crs}_\sigma, \{m_{i,j}, s_{i,j}\})$ $\mathcal{P}$

1. *samples a pair of keys* $(\mathsf{sk}, \mathsf{vk}) \leftarrow \mathsf{Gen}(1^n)$,
2. *computes the strings* $\{\widetilde{\mathsf{crs}}_i \in \{0,1\}^r : i \in [k]\}$ *such that* $\widetilde{\mathsf{crs}}_i[j] = P(\sigma_{i,j}) \oplus s_{i,j}$ *and* $\sigma_{i,j} = \mathsf{Sign}_{\mathsf{sk}}(m_{i,j})$,
3. *for* $i \in [k]$, *emulates* $\mathcal{P}_{\mathsf{hb}}(x, w, \widetilde{\mathsf{crs}}_i)$ *and obtains a proof string* π_i *and a set of indices* $\mathcal{I}_i$,
4. *outputs a proof that contains the verification key* vk *and the hidden-bits proofs* $\{\pi_i, \Sigma_i : i \in [k]\}$, *where* $\Sigma_i = \{(j, \sigma_{i,j}) : j \in \mathcal{I}_i\}$.

The verifier $\mathcal{V}$. *Given* x, *the CRS* $(\mathsf{crs}_\sigma, \{m_{i,j}, s_{i,j}\})$, *and a proof* $(\mathsf{vk}, \{\pi_i, \Sigma_i\})$, $\mathcal{V}$

1. *for every* $i \in [k], j \in [r]$ *such that* $(j, \sigma_{i,j}) \in \Sigma_i$ *verifies that* $\mathsf{Ver}_{\mathsf{vk}}(\mathsf{crs}_\sigma, m_{i,j}, \sigma_{i,j}) = 1$; *otherwise,* $\mathcal{V}$ *rejects,*
2. *for* $i \in [k]$, *computes the set* $\widetilde{\mathsf{crs}}_i(\Sigma_i) = \{(j, P(\sigma_{i,j}) \oplus s_{i,j}) : (j, \sigma_{i,j}) \in \Sigma_i\}$,
3. *for each* $i \in [k]$, *emulates* $\mathcal{V}_{\mathsf{hb}}(x, \mathsf{crs}_i(\Sigma_i), \pi_i)$,
4. *accepts iff the emulation of* $\mathcal{V}_{\mathsf{hb}}$ *accepts for every* i.

Proposition 4.1. *The protocol given by Construction 4.1 is a NIZK proof in the CRS model.*

Proof. The completeness property of $(\mathcal{P}, \mathcal{V})$ follows from the completeness of $(\mathcal{P}_{\mathsf{hb}}, \mathcal{V}_{\mathsf{hb}})$ by construction. Next we prove the soundness and zero-knowledge properties.

Soundness. Fix some $x \in \{0,1\}^n \setminus \mathcal{L}$. Let $(\mathsf{crs}_\sigma, \{m_{i,j}, s_{i,j} : i \in [k], j \in [r]\})$ be uniform random variables describing the content of the CRS. Let $\{\widetilde{\mathsf{crs}}_i : i \in [k]\}$ be the set of hidden random strings for the protocol $(\mathcal{P}_{\mathsf{hb}}, \mathcal{V}_{\mathsf{hb}})$ computed by the honest prover $\mathcal{P}$ from $\{m_{i,j}, s_{i,j}\}$.

We prove that with overwhelming probability over the CRS, there is no proof $(\mathsf{vk}', \{\pi_i', \Sigma_i'\})$ that will make $\mathcal{V}$ accept. The uniqueness property of the signature holds with overwhelming probability over the crs_σ; from hereon, we condition on this event. Fix some $i \in [k]$ and a verification key vk'. Recall that for every $i \in [k], j \in [r]$, we have that $\widetilde{\mathsf{crs}}_i[j] = P(\sigma_{i,j}) \oplus s_{i,j}$. By the uniqueness property of the signature, the value of $P(\sigma_{i,j})$ is determined by the CRS of the signature crs_σ, the verification key vk and the messages $\{m_{i,j}\}$ and is independent of the pad bits $\{s_{i,j}\}$. It follows that $\widetilde{\mathsf{crs}}_i$ is uniformly distributed.

Let $\mathcal{I}_i' \subseteq [r]$ be a set of indices such that Σ_i' is of the form $\Sigma_i' = \{(j, \sigma_{i,j}') : j \in \mathcal{I}_i'\}$ for some signatures $\{\sigma_{i,j}'\}$. By the uniqueness property of the signature we have that if Σ_i' contains an element $(j, \sigma_{i,j}')$ such that $\widetilde{\mathsf{crs}}_i[j] \neq P(\sigma_{i,j}) \oplus s_{i,j}$ the verifier $\mathcal{V}$ rejects the proof. Therefore, if $\mathcal{V}$ accepts, it must be that $\widetilde{\mathsf{crs}}_i(\Sigma_i') = \widetilde{\mathsf{crs}}_i|_{\mathcal{I}_i'}$, where:

$$\widetilde{\mathsf{crs}}_i(\Sigma_i') = \left\{ (j, \sigma_{i,j}' \oplus s_{i,j}) : (j, \sigma_{i,j}') \in \Sigma_i' \right\} \ ,$$

$$\widetilde{\mathsf{crs}}_i'|_{\mathcal{I}_i'} = \left\{ (j, \widetilde{\mathsf{crs}}_i'[j]) : j \in \mathcal{I}_i' \right\} \ .$$

It follows that:

$$\Pr_{m_{i,j}, s_{i,j}} [\exists \pi_i', \Sigma_i' : \mathcal{V}_{\mathsf{hb}}(x, \widetilde{\mathsf{crs}}_i(\Sigma_i'), \pi_i') = 1] = \Pr_{\widetilde{\mathsf{crs}}_i \leftarrow U_r} [\exists \pi_i', \mathcal{I}_i' : \mathcal{V}_{\mathsf{hb}}(x, \widetilde{\mathsf{crs}}_i|_{\mathcal{I}_i'}, \pi_i') = 1] \ .$$

By the soundness of $(\mathcal{P}_{\mathsf{hb}}, \mathcal{V}_{\mathsf{hb}})$ we have that the above probability is at most 2^{-n}. Since this is true independently for every i and since $\mathcal{V}$ accepts iff all k executions of $\mathcal{V}_{\mathsf{hb}}$ accept, we have that:

$$\Pr_{m_{i,j}, s_{i,j}} [\exists \{\pi_i', \Sigma_i'\} : \mathcal{V}_{\mathsf{hb}}(x, \{m_{i,j}, s_{i,j}\}, (\mathsf{vk}', \{\pi_i', \Sigma_i'\})) = 1] \leq 2^{-n \cdot k} \ .$$

Since there are at most 2^k verification keys, by a union bound:

$$\Pr_{m_{i,j}, s_{i,j}} [\exists \mathsf{vk}', \{\pi_i', \Sigma_i'\} : \mathcal{V}_{\mathsf{hb}}(x, \{m_{i,j}, s_{i,j}\}, (\mathsf{vk}', \{\pi_i', \Sigma_i'\})) = 1] \leq 2^{-n} \ ,$$

as required.

Zero-Knowledge. We start be describing the simulator $\mathcal{S}$.

1. $\mathcal{S}$ is given as input a statement $x \in \mathcal{L}$ of length n.
2. For every $i \in [k]$, execute the simulator $\mathcal{S}_{\mathsf{hb}}$ of the protocol $(\mathcal{P}_{\mathsf{hb}}, \mathcal{V}_{\mathsf{hb}})$ and obtain:

$$(B_i, \pi_i) \leftarrow \mathcal{S}(x) \ ,$$

   where $B_i = \{(j, b_{i,j}) : j \in \mathcal{I}_i\}$ for some set $\mathcal{I}_i \subseteq [r]$ and bits $\{b_{i,j}\}$.
3. Sample $\mathsf{crs}_\sigma \leftarrow U_{r_\sigma(n)}$ and $(\mathsf{sk}, \mathsf{vk}) \leftarrow \mathsf{Gen}(\mathsf{crs}_\sigma)$.
4. For every $i \in [k]$, $j \in [r]$ sample $m_{i,j} \leftarrow U_n$
5. For every $i \in [k]$, $j \in [r]$ if $j \notin \mathcal{I}_i$ sample $s_{i,j} \leftarrow U_1$, otherwise set:

$$s_{i,j} = P(\mathsf{Sign}_{\mathsf{sk}}(m_{i,j})) \oplus b_i \ .$$

6. Output the CRS $(\mathsf{crs}_\sigma, \{(m_{i,j}, s_{i,j}) : i \in [k], j \in [r]\})$. Output a simulated proof containing the verification key vk and the simulated hidden-bits proofs $\{\pi_i, \Sigma_i : i \in [k]\}$ where

$$\Sigma_i = \{(j, \mathsf{Sign}_{\mathsf{sk}}(m_{i,j})) : j \in \mathcal{I}_i\} \ .$$

Next we prove that the output of the simulator is indistinguishable from an honestly generated proof. For $0 \leq i \leq k$ consider the experiment H_i where for every $i' \leq i$ the messages and pad bits $\{(m_{i',j}, s_{i',j}) : j \in [r]\}$ are chosen uniformly, and the hidden-bits proof $(\pi_{i'}, \Sigma_{i'})$ is computed following the honest prover strategy, and for every $i < i'$ they are computed according to the simulated strategy as above. For every $(x, w) \in \mathcal{R}_\mathcal{L}$ we have that:

$$H_0(\dot{x}, w) \approx \mathcal{S}(x) \ ,$$

$$H_{k(|x|)}(x, w) \approx \left\{ (\mathsf{crs}, P(x, w, \mathsf{crs})) : m_{i,j} \leftarrow U_{|x|}, s_{i,j} \leftarrow U_1 \right\} \ .$$

Therefore, the correctness of the simulation follows from the next claim.

Claim. For every polysize distinguisher D, there exists a negligible function μ such that for every $(x, w) \in \mathcal{R}_\mathcal{L}$ and for every $i \in [k(|x|)]$

$$|\Pr[D(H_{i-1}(x, w)) = 1] - \Pr[D(H_i(x, w)) = 1]| \leq \mu(|x|) .$$

Proof. For $i \in [k(|x|)]$, consider the experiment H_i' that is defined just like H_i except that instead of sampling $(B_i, \pi_i) \leftarrow \mathcal{S}(x)$, we do the following:

1. Sample a random string $\widetilde{\mathsf{crs}}_i \leftarrow U_{r(|x|)}$.
2. Emulate $\mathcal{P}_{\mathsf{hb}}(x, w, \widetilde{\mathsf{crs}}_i)$ and obtain the proof string π_i and the set of indices $\mathcal{I}_i$.
3. Set $B_i = \widetilde{\mathsf{crs}}_i|_{\mathcal{I}_i} = \{(j, \widetilde{\mathsf{crs}}_i[j]) : j \in \mathcal{I}_i\}$.

By the zero-knowledge property of $(\mathcal{P}_{\mathsf{hb}}, \mathcal{V}_{\mathsf{hb}})$:

$$\left\{ (\widetilde{\mathsf{crs}}_i|_{\mathcal{I}_i}, \pi_i) : \widetilde{\mathsf{crs}}_i \leftarrow U_{r(|x|)}, (\pi_i, \mathcal{I}_i) \leftarrow \mathcal{P}_{\mathsf{hb}}(x, w, \widetilde{\mathsf{crs}}_i) \right\}_{(x,w) \in \mathcal{R}_\mathcal{L}} \approx_c \{\mathcal{S}(x)\}_{(x,w) \in \mathcal{R}_\mathcal{L}} ,$$

and therefore, for every $(x, w) \in \mathcal{R}_\mathcal{L}$ and for every $i \in [k(|x|)]$:

$$|\Pr[D(H_i'(x, w)) = 1] - \Pr[D(H_i(x, w)) = 1]| \leq \mathsf{negl}(|x|) . \tag{1}$$

For every $0 \leq j \leq r(|x|)$ consider the experiment $H_{i,j}'$ that is defined just like H_i' except that for all $j' \leq j$ we set:

$$s_{i,j'} = P(\mathsf{Sign}_{\mathsf{sk}}(m_{i,j'})) \oplus \widetilde{\mathsf{crs}}_i[j'] . \tag{2}$$

For $j' > j$ choose $s_{i,j'}$ as in the experiment H_i'. That is, if $j' \in \mathcal{I}_i$ we set $s_{i,j'}$ as in 2 and if $j' \notin \mathcal{I}_i$ we sample $s_{i,j'}$ uniformly.

Note that the output distribution of the experiments $H_{i,j-1}'$ and $H_{i,j}'$ may differ when $j \notin \mathcal{I}_i$. This is due to the fact that conditioned on $j \notin \mathcal{I}_i$, the bit $\widetilde{\mathsf{crs}}_i[j]$ may no longer be uniform. However, based on the pseudo-randomness property of the signature we will show that the experiments are computationally indistinguishable.

Claim. For every polysize distinguisher D, there exists a negligible function μ such that for every $(x, w) \in \mathcal{R}_\mathcal{L}$ and for every $i \in [k(|x|)], j \in [r(|x|)]$:

$$\left|\Pr[D(H_{i,j-1}'(x, w)) = 1] - \Pr[D(H_{i,j}'(x, w)) = 1]\right| < \mu(|x|) .$$

Proof. Assume towards contradiction that there is a distinguisher D and a polynomial p such that for infinitely many $(x, w) \in \mathcal{R}_\mathcal{L}$ there exist $i \in [k(|x|)], j \in [r(|x|)]$ such that:

$$\left|\Pr[D(H_{i,j-1}'(x, w)) = 1] - \Pr[D(H_{i,j}'(x, w)) = 1]\right| > \frac{1}{p(|x|)} . \tag{3}$$

We construct a distinguisher $\tilde{D}$ that breaks the pseudo-randomness property of the signature. That is for infinity many values of n:

$$\left| \begin{matrix} \Pr\left[\begin{matrix} m \leftarrow U_n, (\mathsf{sk}, \mathsf{vk}) \leftarrow \mathsf{Gen}(\mathsf{crs}_\sigma), \mathsf{crs}_\sigma \leftarrow U_{r_\sigma(n)}, b = P(\mathsf{Sign}_{\mathsf{sk}}(m)) : \\ \tilde{D}^{\mathsf{Sign}_{\mathsf{sk}}(\cdot)}(\mathsf{crs}_\sigma, \mathsf{vk}, m, b) = 1 \end{matrix} \right] - \\ \Pr\left[\begin{matrix} m \leftarrow U_n, (\mathsf{sk}, \mathsf{vk}) \leftarrow \mathsf{Gen}(\mathsf{crs}_\sigma), \mathsf{crs}_\sigma \leftarrow U_{r_\sigma(n)}, b \leftarrow U_1 : \\ \tilde{D}^{\mathsf{Sign}_{\mathsf{sk}}(\cdot)}(\mathsf{crs}_\sigma, \mathsf{vk}, m, b) = 1 \end{matrix} \right] \end{matrix} \right| > \frac{1}{p(|x|)} ,$$
$$\tag{4}$$

where $\tilde{D}$ never queries its oracle on m. $\tilde{D}$ will have hardcoded $(x, w) \in \mathcal{R}_{\mathcal{L}}$ and i, j for which (3) holds. Then $\tilde{D}(\mathsf{crs}_\sigma, \mathsf{vk}, m, b)$ emulates $H'_{i,j}(x, w)$ with the following modifications:

1. When the experiment $H'_{i,j}(x, w)$ samples crs_σ and vk, D uses its input crs_σ and vk instead.
2. Every time a the emulation needs to sign a message $\tilde{D}$ forwards the message to the signing oracle (note that the experiment $H'_{i,j}(x, w)$ does not use the secret key sk except for signing messages).
3. If $j \notin \mathcal{I}_i$ set $s_{i,j} = b \oplus \widetilde{\mathsf{crs}}_i[j]$.

We have that:

$$\Pr[D(H'_{i,j}(x, w)) = 1] = \Pr\left[\begin{array}{l} m \leftarrow U_n, (\mathsf{sk}, \mathsf{vk}) \leftarrow \mathsf{Gen}(\mathsf{crs}_\sigma), \mathsf{crs}_\sigma \leftarrow U_{r_\sigma(n)}, b = P(\mathsf{Sign}_{\mathsf{sk}}(m)) : \\ \tilde{D}^{\mathsf{Sign}_{\mathsf{sk}}(\cdot)}(\mathsf{crs}_\sigma, \mathsf{vk}, m, b) = 1 \end{array}\right] ,$$

$$\Pr[D(H'_{i,j-1}(x, w)) = 1] = \Pr\left[\begin{array}{l} m \leftarrow U_n, (\mathsf{sk}, \mathsf{vk}) \leftarrow \mathsf{Gen}(\mathsf{crs}_\sigma), \mathsf{crs}_\sigma \leftarrow U_{r_\sigma(n)}, b \leftarrow U_1 : \\ \tilde{D}^{\mathsf{Sign}_{\mathsf{sk}}(\cdot)}(\mathsf{crs}_\sigma, \mathsf{vk}, m, b) = 1 \end{array}\right] ,$$

and therefore, (4) follows from (3) and we get a contradiction to the pseudo-randomness property of the signature.

The experiment $H'_{i,0}$ is identical to the experiment H'_i by definition. It follows from Claim 4 that for every $(x, w) \in \mathcal{R}_{\mathcal{L}}$ and for every $i \in [k(|x|)]$:

$$\left|\Pr[D(H'_i(x, w)) = 1] - \Pr[D(H'_{i,r(|x|)}(x, w)) = 1]\right| \leq \mathsf{negl}(|x|) . \tag{5}$$

Note that for $i \in [k]$, the experiment H_{i-1} is identical to the experiment $H'_{i,r(|x|)}$ except for the order in which the the pad bits $\{s_{i,j}\}$ and the random hidden string $\widetilde{\mathsf{crs}}_i$ are sampled. Specifically, in the experiment H_{i-1}:

1. First sample $m_{i,j} \leftarrow U_{|x|}, s_{i,j} \leftarrow U_1$, for every $j \in [r(|x|)]$.
2. Then compute $\widetilde{\mathsf{crs}}_i$ where $\widetilde{\mathsf{crs}}_i[j] = P(\mathsf{Sign}_{\mathsf{sk}}(m_{i,j})) \oplus s_{i,j}$.

Since in both experiments H_{i-1} and $H'_{i,r(|x|)}$ we have that $\widetilde{\mathsf{crs}}_i$ is uniform and $m_{i,j}, s_{i,j}$ are uniform conditioned on the fact that:

$$\widetilde{\mathsf{crs}}_i[j] = P(\mathsf{Sign}_{\mathsf{sk}}(m_{i,j})) \oplus s_{i,j} ,$$

we have that the experiments H_{i-1} and $H'_{i,r(|x|)}$ are identical. Combining this with (1) and (5) we get that for every $(x, w) \in \mathcal{R}_{\mathcal{L}}$ and for every $i \in [k(|x|)]$:

$$|\Pr[D(H_{i-1}(x, w)) = 1] - \Pr[D(H_i(x, w)) = 1]| \leq \mathsf{negl}(|x|) ,$$

as required.

5   Non-Interactive Witness-Indistinguishability

In this section, we construct a NIWI proof system based on indistinguishability obfuscation and one-way permutations.

Theorem 5.1. *Assuming iO for* **P/poly** *and one-way permutations, there exist NIWI proof for every language in* **NP**.[5]

We now describe the NIWI system yielding the theorem. A high-level overview of the construction and the main ideas behind it are provided in the introduction.

Primitives and Notation. The construction relies on an indistinguishability obfuscator iO, a ZAP system (that can be constructed from iO and OWFs as in Section 4), and a non-interactive (one message) statistically binding commitment Com. We require that Com is dense, in the sense that every string of appropriate length is a valid commitment to some message. Such a commitment can be constructed from one-way permutations [Blu81].

Let $\mathcal{L}$ be any **NP** language. For every candidate instance $x \in \{0,1\}^n$ and message $m \in \{0,1\}^n$, denote by E_x^m the canonical "witness-encryption" circuit that given any $w \in \mathcal{R}_{\mathcal{L}}(x)$ outputs m and otherwise outputs $\bot$. Let $\mathcal{T}$ be the **NP** language containing instances of the form $(x, c_1, c_2, \tilde{E})$ where x is candidate instance for $\mathcal{L}$, c_1, c_2 are commitments, and $\tilde{E}$ is an obfuscation such that at least one of the following conditions holds:

1. $\tilde{E}$ is a valid obfuscation of a witness-encryption circuit. That is, there exist randomness r and a message m such that $\tilde{E} = iO(E_x^m; r)$.
2. $\tilde{E}$ has the same output on the plaintexts underlying the commitments c_1, c_2. That is, there exist decommitments (w_1, r_1) and (w_2, r_2) such that:

$$c_1 = \mathsf{Com}(w_1; r_1) \quad \wedge \quad c_2 = \mathsf{Com}(w_2; r_2) \quad \wedge \quad \tilde{E}(w_1) = \tilde{E}(w_2) .$$

Finally, let $D_{x,w}^{s,c_1,c_2}$ be a "witness-decryption" circuit as described in Figure 3.

Hardwired:
  1. Instance and witness $(x, w) \in \mathcal{R}_{\mathcal{L}}$,
  2. first ZAP message s,
  3. commitments c_1, c_2.
Input:
  1. A circuit $\tilde{E}$,
  2. second ZAP message π.
Output:
  1. Verify that π is a valid proof for the statement $(x, c_1, c_2, \tilde{E}) \in \mathcal{T}$ with respect to the first message s. If not, output $\bot$.
  2. Output $\tilde{E}(w)$.

Fig. 3. The "witness-decryption" circuit $D_{x,w}^{s,c_1,c_2}$

---

[5] We assume iO for all circuits for simplicity of exposition; naturally, it suffices to have iO for a certain restricted class of circuits that we use in our construction and analysis.

Construction 5.2 (NIWI Proof). *The NIWI system* $(\mathcal{P}, \mathcal{V})$ *is defined as follows:*

The prover $\mathcal{P}$ *given* $x \in \{0,1\}^n \cap \mathcal{L}$ *and* $w \in \mathcal{R}_{\mathcal{L}}(x)$:
  1. *Sample a first ZAP message* $s \in \{0,1\}^{\mathrm{poly}(n)}$,
  2. *compute a pair of commitments to the all zero string* $c_1, c_2 \leftarrow \mathsf{Com}(0^{|w|})$,
  3. *compute the obfuscation* $\tilde{D} \leftarrow i\mathcal{O}(D_{x,w}^{s,c_1,c_2})$,
  4. *output* $(s, c_1, c_2, \tilde{D})$ *as the proof.*

The verifier $\mathcal{V}$ *given* x *and the proof* $(s, c_1, c_2, \tilde{D})$:
  1. *Sample a message* $m \leftarrow \{0,1\}^n$,
  2. *compute the obfuscation* $\tilde{E} \leftarrow i\mathcal{O}(E_x^m)$,
  3. *compute a proof* π *for the statement* $(c_1, c_2, \tilde{E}) \in \mathcal{T}$ *with respect to the first message* s. *Use* m *and the randomness used to compute* $\tilde{E}$ *as a witness for the fact that* $\tilde{E}$ *is a valid obfuscation of a witness-encryption circuit,*
  4. *accept if* $m = \tilde{D}(\tilde{E}, \pi)$ *accept, otherwise reject.*

Proposition 5.1. *The protocol given by Construction 5.2 is a NIWI proof.*

Proof. The completeness of the system follows readily from the completeness of the ZAP and the functionality of $i\mathcal{O}$. We focus on proving soundness and then witness-indistinguishability.

Soundness. Assume towards contradiction that there exist a polynomial p such that for infinitely many $x \notin \mathcal{L}$ there exists a proof $(s, c_1, c_2, \tilde{D})$ such that:

$$\Pr[\mathcal{V}(x, (s, c_1, c_2, \tilde{D})) = 1] \geq \frac{1}{p(|x|)} .$$

Let m be the random message sampled by $\mathcal{V}$ in a random execution, and let $(\tilde{E}, \pi)$ be the obfuscation and proof computed by $\mathcal{V}$. By our assumption:

$$\Pr[\tilde{D}(\tilde{E}, \pi) = m] \geq \frac{1}{p(|x|)} .$$

Let (w_1, r_1) and (w_2, r_2) be decommitments of c_1, c_2 respectively (such decommitments exist since Com is dense). Since $x \notin \mathcal{L}$, and by the (perfect) functionality of $i\mathcal{O}$, the circuit $\tilde{E}$ outputs $\perp$ on all inputs. Therefore, the decommitments $(w_1, r_1), (w_2, r_2)$ can be used as a witness for the statement $(c_1, c_2, \tilde{E}) \in \mathcal{T}$. Let π' be a proof for the statement $(c_1, c_2, \tilde{E}) \in \mathcal{T}$ with respect to the first message s computed using the witness $(w_1, r_1), (w_2, r_2)$. By the witness indistinguishability of the ZAP: $\pi \approx_c \pi'$. Therefore,

$$\Pr[\tilde{D}(\tilde{E}, \pi') = m] \geq \frac{1}{p(|x|)} - \mathrm{negl}(|x|) .$$

Let $\tilde{E}'' = i\mathcal{O}(E_x^{0^n})$. Since the circuits $\tilde{E}$ and $\tilde{E}''$ are of the same size, and since both output $\perp$ on all inputs, it follows from the security of $i\mathcal{O}$ that $\tilde{E} \approx_c \tilde{E}''$. Let π'' be a proof with respect to $\tilde{E}''$ rather than for $\tilde{E}$. Then $(\tilde{E}, \pi') \approx_c$

$(\tilde{E}'', \pi'')$. Indeed, a distinguisher between $(\tilde{E}, \pi'), (\tilde{E}'', \pi'')$ can be reduced to a distinguisher between $\tilde{E}, \tilde{E}''$, since computing π' and π'' does not require the randomness underlying $\tilde{E}_1, \tilde{E}_2$). Hence, it also holds that:

$$\Pr[\tilde{D}(\tilde{E}'', \pi'') = m] \geq \frac{1}{p(|x|)} - \mathrm{negl}(|x|) \ .$$

Since m is uniform in $\{0,1\}^{|x|}$ and in the above experiment the view of $\tilde{D}$ is independent of m, we get a contradiction.

Witness Indistinguishability. Let $\mathcal{I} = \{(x, w_1, w_2) : w_1, w_2 \in \mathcal{R}_{\mathcal{L}}(x)\}$, be a sequence of instances $x \in \mathcal{L}$, with two corresponding witnesses w_1, w_2. We show that

$$\left\{ (s, c_1, c_2, \tilde{D}) \leftarrow \mathcal{P}(x, w_1)) \right\}_{(x, w_1, w_2) \in \mathcal{I}} \approx_c \left\{ (s, c_1, c_2, \tilde{D}) \leftarrow \mathcal{P}(x, w_2)) \right\}_{(x, w_1, w_2) \in \mathcal{I}} ,$$

by considering a sequence of hybrid distributions.

Hyb_1: Here $(c_1, c_2, \tilde{D}) \leftarrow \mathcal{P}(x, w_1)$ corresponds to a proof using the first witness w_1.

Hyb_2: Here for each $b \in \{0, 1\}$, $c_b \leftarrow \mathsf{Com}(w_b)$ is a commitment to the corresponding witness rather than to the all-zero string. By the computational-hiding of Com, $\mathsf{Hyb}_2 \approx_c \mathsf{Hyb}_1$.

Hyb_3: Here the first ZAP message s is sampled conditioned the on the ZAP being absolutely sound; that is, there exists no accepting proof, with respect to s, for any false statement. By the soundness of the ZAP, this holds with overwhelming probability and thus $\mathsf{Hyb}_3 \approx_s \mathsf{Hyb}_2$.

Hyb_4: Here instead of sampling $\tilde{D} \leftarrow i\mathcal{O}(D^{s, c_1, c_2}_{x, w_1})$ using w_1, it is sampled using w_2, i.e., $\tilde{D} \leftarrow i\mathcal{O}(D^{s, c_1, c_2}_{x, w_2})$. To show that $\mathsf{Hyb}_4 \approx_c \mathsf{Hyb}_5$, we show that for any realization of s, c_1, c_2 (which have the same distribution in $\mathsf{Hyb}_4, \mathsf{Hyb}_5$), the two circuits $D^{s, c_1, c_2}_{x, w_1}, D^{s, c_1, c_2}_{x, w_2}$ have the exact same functionality and thus, by the iO guarantee, $i\mathcal{O}(D^{s, c_1, c_2}_{x, w_1}) \approx_c i\mathcal{O}(D^{s, c_1, c_2}_{x, w_2})$. Indeed, for any input $(\tilde{E}, \pi)$ for D^{s, c_1, c_2}_{x, w_b}, there are two options:

1. π is not a valid proof for the statement $(x, c_1, c_2, \tilde{E}) \in \mathcal{T}$ with respect to the first message s. In this case, by the definition of D^{s, c_1, c_2}_{x, w_b}, it holds that $D^{s, c_1, c_2}_{x, w_1}(\tilde{E}, \pi) = D^{s, c_1, c_2}_{x, w_2}(\tilde{E}, \pi) = \perp$.
2. π is a valid proof. In this case, by the soundness of the ZAP, $(x, c_1, c_2, \tilde{E}) \in \mathcal{T}$. This in turn implies one of two cases
   (a) $\tilde{E}$ is a valid obfuscation $i\mathcal{O}(E^m_x)$, in which case by the definition of E^m_x, and the functionality of $i\mathcal{O}$, $\tilde{E}(w_1) = \tilde{E}(w_2)$.
   (b) c_1, c_2 can be opened to $\tilde{w}_1, \tilde{w}_2$, such that $\tilde{E}(\tilde{w}_1) = \tilde{E}(\tilde{w}_2)$, in which case by the binding of Com, for both $b \in \{0, 1\}$, $w_b = \tilde{w}_b$, and thus also $\tilde{E}(w_1) = \tilde{E}(w_2)$.

So in either case $D^{s, c_1, c_2}_{x, w_1}(\tilde{E}, \pi) = D^{s, c_1, c_2}_{x, w_2}(\tilde{E}, \pi)$, as required.

Hyb_5: Here we remove the requirement that s is sampled conditioned on absolute soundness. Like before, it holds that $\mathsf{Hyb}_5 \approx_s \mathsf{Hyb}_4$ by the soundness of the ZAP.

Hyb_6: Here $(\mathsf{c}_1, \mathsf{c}_2, \tilde{D}) \leftarrow \mathcal{P}(x, w_2)$ corresponds to a proof using the first witness w_2. This hybrid differs from Hyb_5 only in that $\mathsf{c}_1, \mathsf{c}_2$ are commitments to all-zero strings rather than to w_1, w_2. Like before, it holds that $\mathsf{Hyb}_6 \approx_c \mathsf{Hyb}_5$ by the computational hiding of the commitment Com.

Remark 5.1 (Relying on relaxed dense commitments). The non-interactive commitment scheme used in the NIWI construction can be somewhat relaxed. Indeed, it suffices to require a non-interactive commitment scheme that is statistically binding, but only against *honest* committers; namely, honestly generated commitments can only be opened to a single value. The commitment should still be *dense* in the sense that every string in the range of the commitment, can be opened to at least one value (commitments that are not generated honestly may potentially be opened to more than one value).

Remark 5.2 (Using witness-encryption generically). We note that we refrain from explicitly defining witness encryption, and in the above construction, directly implement witness encryption using iO (which we anyhow rely on). While we find that thinking about witness encryption in terms of obfuscation is helpful in this context, it is possible to state the construction in terms of generic witness encryption.

Acknowledgements. We thank Ran Canetti for discussions and valuable advice. We thank Rafail Ostrovsky for discussions on the the way that unbalanced properties are dealt with in [GO92], and for referring us to [BGRV09]. We also thank Sanjam Garg for discussing NIZKs based on graded encodings.

References

[BFM88]   Blum, M., Feldman, P., Micali, S.: Non-interactive zero-knowledge and its applications (extended abstract). In: STOC, pp. 103–112 (1988)

[BGI+01]   Barak, B., Goldreich, O., Impagliazzo, R., Rudich, S., Sahai, A., Vadhan, S.P., Yang, K.: On the (Im)possibility of obfuscating programs. In: Kilian, J. (ed.) CRYPTO 2001. LNCS, vol. 2139, pp. 1–18. CRYPTO, Heidelberg (2001)

[BGI14]   Boyle, E., Goldwasser, S., Ivan, I.: Functional signatures and pseudorandom functions. In: Krawczyk, H. (ed.) PKC 2014. LNCS, vol. 8383, pp. 501–519. Springer, Heidelberg (2014)

[BGK+13]   Barak, B., Garg, S., Kalai, Y.T., Paneth, O., Sahai, A.: Protecting obfuscation against algebraic attacks. Cryptology ePrint Archive, Report 2013/631 (2013), http://eprint.iacr.org/

[BGRV09]   Brakerski, Z., Goldwasser, S., Rothblum, G.N., Vaikuntanathan, V.: Weak verifiable random functions. In: Reingold, O. (ed.) TCC 2009. LNCS, vol. 5444, pp. 558–576. Springer, Heidelberg (2009)

[Blu81]   Blum, M.: Coin flipping by telephone. In: Proceedings of the 18th Annual International Cryptology Conference, pp. 11–15 (1981)

[BOV07]   Barak, B., Ong, S.J., Vadhan, S.P.: Derandomization in cryptography. SIAM J. Comput. 37(2), 380–400 (2007)

[BR13]   Brakerski, Z., Rothblum, G.N.: Virtual black-box obfuscation for all circuits via generic graded encoding. Cryptology ePrint Archive, Report 2013/563 (2013), http://eprint.iacr.org/

[BW13]   Boneh, D., Waters, B.: Constrained pseudorandom functions and their applications. In: Sako, K., Sarkar, P. (eds.) ASIACRYPT 2013, Part II. LNCS, vol. 8270, pp. 280–300. Springer, Heidelberg (2013)

[BY96]   Bellare, M., Yung, M.: Certifying permutations: Noninteractive zero-knowledge based on any trapdoor permutation. J. Cryptology 9(3), 149–166 (1996)

[CLT13]   Coron, J.-S., Lepoint, T., Tibouchi, M.: Practical multilinear maps over the integers. In: Canetti, R., Garay, J.A. (eds.) CRYPTO 2013, Part I. LNCS, vol. 8042, pp. 476–493. Springer, Heidelberg (2013)

[DN07]   Dwork, C., Naor, M.: Zaps and their applications. SIAM J. Comput. 36(6), 1513–1543 (2007)

[FLS99]   Feige, U., Lapidot, D., Shamir, A.: Multiple noninteractive zero knowledge proofs under general assumptions. SIAM J. Comput. 29(1), 1–28 (1999)

[FS89]   Feige, U., Shamir, A.: Zero knowledge proofs of knowledge in two rounds. In: Brassard, G. (ed.) CRYPTO 1989. LNCS, vol. 435, pp. 526–544. Springer, Heidelberg (1990)

[GGH13a]   Garg, S., Gentry, C., Halevi, S.: Candidate multilinear maps from ideal lattices. In: Johansson, T., Nguyen, P.Q. (eds.) EUROCRYPT 2013. LNCS, vol. 7881, pp. 1–17. Springer, Heidelberg (2013)

[GGH$^+$13b] Garg, S., Gentry, C., Halevi, S., Raykova, M., Sahai, A., Waters, B.: Candidate indistinguishability obfuscation and functional encryption for all circuits. In: FOCS (2013)

[GGH14]   Gentry, C., Gorbunov, S., Halevi, S.: Graded multilinear maps from lattices. Cryptology ePrint Archive, Report 2014/645 (2014), http://eprint.iacr.org/

[GGHR14]   Garg, S., Gentry, C., Halevi, S., Raykova, M.: Two-round secure MPC from indistinguishability obfuscation. In: Lindell, Y. (ed.) TCC 2014. LNCS, vol. 8349, pp. 74–94. Springer, Heidelberg (2014)

[GGM86]   Goldreich, O., Goldwasser, S., Micali, S.: How to construct random functions. J. ACM 33(4), 792–807 (1986)

[GGSW13]   Garg, S., Gentry, C., Sahai, A., Waters, B.: Witness encryption and its applications. In: STOC, pp. 467–476 (2013)

[GL89]   Goldreich, O., Levin, L.A.: A hard-core predicate for all one-way functions. In: STOC 1989: Proceedings of the Twenty-first Annual ACM Symposium on Theory of Computing, pp. 25–32. ACM, New York (1989)

[GMR89]   Goldwasser, S., Micali, S., Rackoff, C.: The knowledge complexity of interactive proof systems. SIAM J. Comput. 18(1), 186–208 (1989)

[GMW91]   Goldreich, O., Micali, S., Wigderson, A.: Proofs that yield nothing but their validity for all languages in np have zero-knowledge proof systems. J. ACM 38(3), 691–729 (1991)

[GO92]   Goldwasser, S., Ostrovsky, R.: Invariant signatures and non-interactive zero-knowledge proofs are equivalent. In: Brickell, E.F. (ed.) CRYPTO 1992. LNCS, vol. 740, pp. 228–245. Springer, Heidelberg (1993)

[GO94]   Goldreich, O., Oren, Y.: Definitions and properties of zero-knowledge proof systems. Journal of Cryptology 7(1), 1–32 (1994)

[GOS12]   Groth, J., Ostrovsky, R., Sahai, A.: New techniques for noninteractive zero-knowledge. J. ACM 59(3), 11 (2012)

[KMN⁺14]   Komargodski, I., Moran, T., Naor, M., Pass, R., Rosen, A., Yogev, E.: One-way functions and (im)perfect obfuscation. IACR Cryptology ePrint Archive, 2014:347 (2014)

[KPTZ13]   Kiayias, A., Papadopoulos, S., Triandopoulos, N., Zacharias, T.: Delegatable pseudorandom functions and applications. In: ACM Conference on Computer and Communications Security, pp. 669–684 (2013)

[Nao91]   Naor, M.: Bit commitment using pseudorandomness. J. Cryptology 4(2), 151–158 (1991)

[NLLT14]   Niu, Q., Li, H., Liang, B., Tang, F.: One-round witness indistinguishability from indistinguishability obfuscation. IACR Cryptology ePrint Archive, 2014:176 (2014)

[SW14]   Sahai, A., Waters, B.: How to use indistinguishability obfuscation: Deniable encryption, and more. In: STOC (2014)

Random-Oracle Uninstantiability from Indistinguishability Obfuscation

Christina Brzuska[1], Pooya Farshim[2], and Arno Mittelbach[3]

[1] Microsoft Research Cambridge, UK
[2] Queen's University Belfast, Northern Ireland, UK
[3] Darmstadt University of Technology, Germany
{christina.brzuska,pooya.farshim}@gmail.com
mail@arno-mittelbach.de

Abstract. Assuming the existence of indistinguishability obfuscation (iO), we show that a number of prominent transformations in the random-oracle model are uninstantiable in the standard model. We start by showing that the Encrypt-with-Hash transform of Bellare, Boldyreva and O'Neill (CRYPTO 2007) for converting randomized public-key encryption schemes to deterministic ones is not instantiable in the standard model. To this end, we build on the recent work of Brzuska, Farshim and Mittelbach (CRYPTO 2014) and rely on the existence of iO for Turing machines or for circuits to derive two flavors of uninstantiability. The techniques that we use to establish this result are flexible and lend themselves to a number of other transformations such as the classical Fujisaki–Okamoto transform (CRYPTO 1998) and transformations akin to those by Bellare and Keelveedhi (CRYPTO 2011) and Douceur et al. (ICDCS 2002) for obtaining KDM-secure encryption and de-duplication schemes respectively. Our results call for a re-assessment of scheme design in the random-oracle model and highlight the need for new transforms that do not suffer from iO-based attacks.

Keywords: Random oracle, uninstantiability, indistinguishability obfuscation, deterministic encryption, UCE, Fujisaki–Okamoto transform, KDM security, message-locked encryption.

1 Introduction

1.1 Background

The random-oracle model (ROM) [18] is an idealized model of computation where all parties, honest or otherwise, have oracle access to a uniformly chosen random function. Random oracles model ideal hash functions and have found a plethora of applications in cryptography. They have enabled the security proofs of a wide range of practical cryptosystems which include, amongst others, digital signature schemes, CCA-secure encryption, key-exchange protocols, identity-based encryption, cryptosystems that are resilient to related-key and key-dependent-message attacks, as well as more advanced security goals such

Y. Dodis and J.B. Nielsen (Eds.): TCC 2015, Part II, LNCS 9015, pp. 428–455, 2015.

as deterministic encryption of high-entropy messages, de-duplication schemes, and point-function obfuscators. After designing and analyzing the scheme in the random-oracle model, one then instantiates the oracle via a concrete, possibly keyed, hash function. In this paper we revisit this methodology and show that a number of prominent ROM cryptosystems cannot be securely instantiated in the standard model.

1.2 Uninstantiability

The power and practicality of random oracles drew early attention to their standard-model instantiations. Canetti, Goldreich and Halevi (CGH) [33] demonstrated a general negative result by constructing digital signature and encryption schemes which are secure in the random-oracle model but become insecure as soon as the oracle is instantiated with *any* concrete hash function. Such *uninstantiable* schemes rely on the existence of a compact description for concrete hash functions and lack of one for truly random functions. Roughly speaking, the idea is to take a secure ROM scheme and tweak it slightly so that it behaves securely unless it is run on messages that match the code of the hash function used in the instantiation, in which case it does something "obviously insecure" (e.g., returns the signing key or the message).

A number of other works have further studied uninstantiability problems associated with random oracles. In a follow-up work [34], CGH extend their result to signature schemes which only support short messages. Bellare, Boldyreva and Palacio [8] show that no instantiation of the hashed ElGamal key-encapsulation mechanism composes well with symmetric schemes, even though it enjoys this property in the ROM. Goldwasser and Kalai [44] study the Fiat–Shamir heuristic and establish uninstantiability results for it. Nielsen [51] gives an uninstantiable cryptographic task, namely that of non-interactive, non-committing encryption, which although achievable in the ROM, is infeasible in the standard model. CGH-type uninstantiability has been adapted to other models of computations such as the ideal-cipher model [21] and the generic-group model [36].

A number of recent works have looked into ROM (un)instantiability in light of the recently proposed candidate for indistinguishability obfuscation (iO) [39]. A secure indistinguishability obfuscator guarantees that the obfuscations of any two functionally equivalent programs (modeled as circuits or Turing machines) are computationally indistinguishable. On the positive side, Hohenberger, Sahai and Waters [46] show how to instantiate the hash function in full-domain hash (FDH) signatures using iO. Bellare, Stepanovs and Tessaro [19] show the first standard-model construction for polynomially many hardcore bits for *any* one-way function. Recently, Brzuska and Mittelbach [31] have shown how to use iO to instantiate certain forms of Universal Computational Extractors (UCEs). UCE is a novel framework of security notions introduced by Bellare, Hoang and Keelveedhi [12] and can be used to generically instantiate random oracles in many protocols.

On the negative side, Brzuska, Farshim and Mittelbach [27] show that under the existence of iO, several security notions in the UCE framework are uninstantiable in the standard model, and proposed fixes to salvage many of the applications. Brzuska and Mittelbach [30] show that assuming iO, multi-bit output point-function obfuscation secure in the presence of auxiliary information cannot be realized. Both results can be interpreted as conditional uninstantiability results as ROM constructions for both UCEs [12,50] and strong multi-output bit point obfuscation [48] exist. Bitansky et al. [20] show that indistinguishability obfuscation rules out the existence of certain types of extractable one-way function families which can be constructed in the random-oracle model [32].

1.3 Our Results

Our work continues the study of uninstantiability of random oracles and shows that a number of well-known and widely deployed ROM transforms are provably uninstantiable if indistinguishability obfuscators exist. More specifically, we are interested in ROM transformations T^{RO} that take as input *any* standard-model scheme S which is guaranteed to satisfy a mild form of security, and convert S into a new scheme $T^{RO}[S]$ in the random-oracle model that meets a stronger level of security. A fundamental question for such transforms is their instantiability, that is, whether or not there exists an efficient hash function H such that $T^H[S]$ is strongly secure for *any* mildly secure S. We show a number of negative results in this direction, which take the form: there is a mildly secure scheme S^* such that no matter which hash function H is picked, scheme $T^H[S^*]$ is provably insecure.

Our results come in two flavors depending on the class of programs that the indistinguishability obfuscator supports. Assuming iO for circuits of a priori bounded size b, we show there is a ROM cryptosystem which is uninstantiable with respect to keyed hash functions of description size at most b. This means that there exists a scheme S_b such that for any hash function H of description size at most b the scheme $T^H[S_b]$ is insecure. This, in particular, yields an uninstantiability result for any fixed and finite set of hash functions. This result, however, does not rule out instantiating the oracle with hash functions which have larger description size and are in some sense "more complex" than the base scheme. By assuming the existence of iO for *Turing machines* we are able to further strengthen this result to one which rules out instantiations with respect to *any*, possibly scheme-dependent, hash function.

Overview of BFM. We build on techniques of Brzuska, Farshim and Mittelbach (BFM) [27] to construct our uninstantiable schemes and briefly recall their technique here. BFM utilize the power of indistinguishability obfuscation to show that a recent notion of security for hash functions known as UCE1 is uninstantiable in the standard model.[1] To this end, BFM construct an adversary which outputs an indistinguishability obfuscation of the Boolean circuit

$$C[x, y](\mathsf{hk}) := (\mathsf{H}(\mathsf{hk}, x) = y) ,$$

where x is a random domain point and y is the corresponding hash value which could be real or ideal. That is, the circuit $C[x, y]$ has x and y hard-coded into it and gets as input a hash key hk, computes $\mathsf{H}(\mathsf{hk}, x)$ and outputs 1 if and only if this value is equal to y.

BFM need to argue that an indistinguishability obfuscation of this circuit hides x whenever y is truly random (and not computed by applying the hash-function to x). They prove this by a counting argument that establishes that, under appropriate restrictions on the lengths of y and the length of the key hk, the above circuit implements the constant zero circuit with overwhelming probability. They then employ the security of the obfuscator to conclude as the zero circuit is independent of x. The restriction that they require, is that the number of hash keys hk is much smaller than the size of the range $2^{|y|}$, which means that y (with overwhelming probability) is outside the image of the function $\mathsf{H}(\cdot, x)$ that has a fixed x and maps hash-keys hk to $\mathsf{H}(\mathsf{hk}, x)$. On the other hand, the above circuit returns 1 when the hash value y is computed as $\mathsf{H}(\mathsf{hk}, x)$ and hk as the correct hash key is plugged into $C[x, y]$.

Techniques. In our uninstantiability results for encryption, we will embed an obfuscated program into the ciphertext.[2] We now describe this program which is a *universal* variant of the BFM circuit. This program takes as input the full description of a hash function H_{hk}, including its key hk if there is one, and returns the result of running the BFM circuit on the input hash-function description. It performs the latter in the standard way by using a universal evaluator UEval, which could be a universal Turing machine or a universal circuit evaluator, depending on the considered model of computation.

$$P[x, y](\mathsf{H}_{\mathsf{hk}}) := (\mathsf{UEval}(\mathsf{H}_{\mathsf{hk}}, x) = y) .$$

So, the program $P[x, y]$ has x and y hard-coded and takes as input a description of H_{hk}, computes $\mathsf{H}_{\mathsf{hk}}(x)$ and checks whether this value is equal to y. In other words, we no longer consider a fixed keyed hash function, but instead (potentially) look at the set of *all* hash functions on a given range and domain.[3] (Similar ideas have

---

[1] In UCE1 (later renamed to UCE[$\mathcal{S}^{\mathrm{cup}}$]) security a two-stage adversary needs to distinguish a hash function from a random oracle. The first-stage adversary is given oracle access to either the hash function under a random key or the random oracle. It does not get to see the hash key but can leak a message to the second-stage adversary on termination, which additionally gets the hash key and outputs a bit. The second-stage adversary can no longer call the oracle. UCE1 security requires that the leaked message should be such that it does not computationally reveal any of the oracle queries when the oracle is a random function.

[2] We speak of programs which can be modeled either as circuits or as Turing machines.

[3] Alternatively, we are looking at the universal hash function.

been used by Brzuska and Mittelbach [30] to study the feasibility of multi-bit output point function obfuscation in the presence of auxiliary inputs under the iO assumption.) Note that $P[x, y]$ is either a circuit or a Turing machine depending on the underlying universal evaluator UEval. In adopting this approach, a number of technicalities need to be addressed, which we discuss next.

Our ultimate goal is to derive a strong result which rules out instantiations (of a transformation) by arbitrary hash functions. This means that program P above should accept inputs of arbitrary length. This, however, lies beyond the powers of the circuit model of computation which current indistinguishability obfuscators support. We address this problem in two incomparable ways. First, we weaken target uninstantiability and under iO for circuits rule out instantiations by a priori bounded-size hash functions. Second, in order to strengthen this result to full uninstantiability, we consider a stronger form of iO which supports *Turing machines*. For our purposes, the crucial difference between iO for circuits and iO fro Turing machines is that an obfuscated Turing machine is still a Turing machine which can process inputs of *arbitrary* length. (Note that the actual Turing machine that we need to obfuscate is a universal Turing machine and has an a priori fixed size.) Our theorem statements will therefore contain two parts to reflect this trade off between the strength of assumptions and the reach of the uninstantiability result obtained.

A second problem arises from the fact that the number of possible hash function descriptions might be greater than $2^{|y|}$ so that we cannot directly apply BFM's counting argument. We overcome this obstacle by composing both sides of the equality in P with a pseudorandom generator (PRG) and look at

$$P[x, y](H_{hk}) := (PRG(UEval(H_{hk}, x)) = PRG(y)) .$$

This does not affect the success probability of the attack and allows us to argue that x remains hidden as follows: First note that the right-hand side $PRG(y)$ is a constant that does not depend on the program input and can thus be hard-coded into the program. Now, in a first step we can replace the right hand-side value with a truly random value by the security of the PRG. Note that in this step we *do not* rely on the security of the obfuscator and merely use the indistinguishability of program descriptions. Indeed, the two programs might implement significantly different functionalities. Next, we use the fact that a truly random value is, with overwhelming probability, outside the range of a PRG with sufficiently long stretch. Hence, the obfuscations of the above program are computationally indistinguishable from those of the zero program. We note that our usage of the PRG is somewhat similar to that by Sahai and Waters in their construction of a CCA-secure PKE scheme from iO [55], the range extension of Matsuda and Hanaoka [49] of a multi-bit point function to obtain shorter point values, the range-extension of a UCE1-secure hash function by Bellare, Hoang and Keelveedhi [14], and the negative result of Brzuska and Mittelbach [30] on multi-bit point-function obfuscation with auxiliary inputs.

Assumptions. Garg et al. [39] construct an indistinguishability obfuscator for $\mathcal{NC}^1$ circuits based on intractability assumptions related to multi-linear maps, and show how to bootstrap it to support all polynomial-time circuits via a fully homomorphic encryption scheme with a decryption circuit in $\mathcal{NC}^1$. The authors validate their multi-linear intractability assumption in a generic model of computation. Recent results show how to improve the assumptions used in constructing indistinguishability obfuscators [52,26,4,3,42], further supporting their plausibility.

Indistinguishability obfuscation for Turing machines has been constructed in the works of Boyle, Chung and Pass [25] and Ananth et al. [2]. The authors study a stronger primitive called *extractability* or *differing-inputs* obfuscation (diO) which extends iO to circuits (and Turing machines) that are not necessarily functionally equivalent. The requirement is that any adversary that can break the indistinguishability property can be converted to an extractor that can output a point on which the two circuits differ. Boyle, Chung and Pass [25] and Ananth et al. [2] show how to build iO for *Turing machines* assuming diO for circuits. The plausibility of differing-inputs obfuscation, however, has become somewhat controversial due to a recent result of Garg et al. [40]. These authors show that the existence of a special-purpose obfuscator for a signature scheme implies that diO with arbitrary auxiliary input cannot exist. Although we currently do not know how to build this special-purpose obfuscator, its existence appears to be a milder assumption than diO, one can consider its existence to be more likely. It is therefore important to seek alternative instantiations of iO for Turing machines from assumptions that are weaker than diO for circuits. Indeed, very recently and shortly after the appearance of this work, Koppula, Lewko and Waters [47] have succeeded in constructing iO for Turing machines without relying on diO, and using iO for circuits, one-way functions and injective pseudorandom generators.

Deterministic encryption. Our first result establishes the uninstantiability of the Encrypt-with-Hash (EwH) transform of Bellare, Boldyreva and O'Neill [7], whereby one converts a randomized IND-CPA public-key encryption scheme into a deterministic public-key encryption (D-PKE) scheme D-PKE by extracting the randomness needed for encryption via hashing the message and the public key, that is, the encryption algorithm D-PKE.Enc$^{\mathcal{RO}(\cdot,\cdot)}(m, (\mathsf{hk}, pk))$ first computes random coins $r \leftarrow \mathcal{RO}(\mathsf{hk}, pk\|m)$ and then invokes the base encryption algorithm on message m, public key pk and random coins r to generate a ciphertext. This simple transformation meets the strongest notion of security that has been proposed for deterministic encryption (that is, PRIV security) in the ROM if the underlying encryption scheme is IND-CPA secure. Standard-model constructions, on the other hand, achieve weaker levels of security, e.g., security against block sources [10,22] or q-bounded adversaries [38,29]. To this end, we ask if any hash function can be used to instantiate the random oracle within the EwH transform. Assuming iO for circuits/Turing machines, we build an IND-CPA secure encryption scheme such that when the EwH transform is applied to this specially devised encryption scheme together with some (b-bounded) hash-function, the resulting scheme is not PRIV-secure, not even for block-sources or 1-bounded PRIV-security.

Starting with an arbitrary scheme PKE we consider a new scheme PKE* which includes an indistinguishability obfuscation of the following program as part of its ciphertexts.

$$P[pk, m, r](H_{hk}) := \text{if } (PRG(UEval(H_{hk}, pk\|m)) = PRG(r))$$

$$\text{return } m$$

$$\text{else return } 0$$

This program performs a check similar to that of the universal BFM circuit, but instead of returning a Boolean value returns the encrypted messages when the check passes. That is, in $P[pk, m, r]$, the public-key pk, the message m and the randomness r are parameters, and the program takes as input a hash-function H_{hk} (potentially with some hard-coded key hk), evaluates H_{hk} on $pk\|m$ to get some value y. Then, it applies PRG to y and checks whether $PRG(y)$ is equal to $PRG(r)$. If this is the case, it returns the message m. Else, it returns 0.

We can use an obfuscation of this program to attack the security of $EwH^H[PKE^*]$. The second stage of the adversary runs this program on the description H_{hk} of the hash function that is used in the instantiation (with hard-coded hk) to obtain the encrypted message. A corollary of this result is that under iO, no security assumption (single or multi-staged, falsifiable or not) is strong enough to build D-PKEs via EwH. In particular, a new UCE assumption used to instantiate EwH [15] is uninstantiable assuming iO for Turing machines (and b-bounded uninstantiable assuming iO for circuits). We remark that our results are incomparable to those of Wichs [57] who shows an unconditional unprovability result for D-PKEs using *arbitrary* techniques from single-stage assumptions. (Our results are conditional and show uninstantiability of EwH regardless of the assumptions used.) This result naturally extends to the Randomized-Encrypt-with-Hash [9] transform for building hedged PKEs.

The Fujisaki–Okamoto transform. The above result generalizes to a wider class of (possibly randomized) *admissible* transformations that use their underlying PKE schemes in a structured way and admit recovery algorithms that satisfy certain correctness properties. (We leave the details to the main body.) Somewhat surprisingly, the Fujisaki–Okamoto (FO) transform for converting CPA into CCA security is admissible and thus suffers from uninstantiability. The FO transform, which dates back to the 1990s, is a simple and flexible technique to boost security of various schemes and has been widely used in identity-based encryption [24], its hierarchical and fuzzy variants [43,56], forward-secure encryption [35], and certificateless and certificate-based encryption [1,41] to mention a few. Our results, once again, come in two flavors depending on the strength of the underlying obfuscator. Our techniques can be further tweaked to show that one cannot instantiate the oracle used within the asymmetric component of the FO transform. This means that the POWHF-encryption assumption of Boldyreva and Fischlin [23] used for partial instantiation of the oracles in FO is also uninstantiable if iO/iO for Turing machines exists.

Other constructs. The uninstantiability problems arising from the existence of indistinguishability obfuscators are not limited to deterministic encryptions and its generalizations. We revisit the work of Bellare and Keelveedhi (BK) [16] on authenticated and misuse-resistant encryption of key-dependent data and show that it too suffers from uninstantiability problems. Roughly speaking, BK give a transformation called RHtE to convert authenticated encryption into one which resists key-dependent-message (KDM) attacks. This is done by hashing the key with a random nonce to derive the actual key used in encryption: one encrypts m as $(N, \mathsf{Enc}(\mathsf{H}(\mathsf{hk}, N\|k), m))$ for a random nonce N. Our iO-based uninstantiability result describes an IND-CPA and INT-CTXT-secure authenticated encryption (AE) scheme whose BK transformation is not KDM secure.

Interestingly, BK require the base scheme to meet a stronger security level than IND-CPA: ciphertexts should be indistinguishable from random strings. BK do not consider this difference to be of major importance; in the abstract of their paper they write that they *present a RO-transform RHtE that endows any AE-scheme with this security.* Our result brings this stronger requirement to light, and shows that assuming that ciphertexts are pseudorandom might be a way to circumvent uninstantiability as the current state-of-the-art obfuscators produce programs that are structured and do not look random. Conversely, if an indistinguishability obfuscator can produce obfuscations of the zero circuit that look random,[4] then reverting to the stronger security notion would no longer be of any help.

As a final example we show that the Convergent-Encryption transform of Douceur et al. [37] formalized by Bellare, Keelveedhi and Ristenpart (BKR) [17] for building message-locked encryption is also uninstantiable. Once again, BKR formally rely on pseudorandomness of ciphertexts but similar observation to those given above for BK apply here too.

Comparison with CGH. Recall that Canetti, Goldreich and Halevi (CGH) [33] show the uninstantiability of certain ROM digital signature and encryption schemes without relying on iO. Their technique is to give a (contrived) scheme that is secure in the random oracle model but behaves anomalously on certain inputs that are related to a compact description of the hash function. Our uninstantiability results share these features, that is, neither their nor our uninstantiability results apply to "natural" schemes. For instance, it is not known if Encrypt-with-Hash when used with ElGamal is uninstantiable or not. On the other hand, our results apply to natural transformations.

It is natural to ask if CGH-like techniques can be directly applied here so as to obtain uninstantiability results that do not rely on the iO machinery. For uninstantiability with respect to *unkeyed* hash functions, one can indeed construct anomalous PKE schemes which follow the CGH paradigm and give the desired

---

[4] Note that generally, obfuscations of circuits cannot look random, because obfuscation maintains functionality and thus, the obfuscations of the zero circuit would be distinguishable from those of the constant one circuit. This trivial attack, however, does not apply here if we require pseudorandomness only for the zero circuit.

uninstantiability result for Encrypt-with-Hash. For keyed hash functions, on the other hand, there seems to be an inherent limitation to CGH-like techniques. For instance, the security model for D-PKEs do *not* allow message distributions to depend on the hash key as this value is included in the public key and the latter is denied to the first-stage adversary. Consequently there is no way to generate messages which contain the full description of the hash function used, *including its key*, which seems to be necessary when applying CGH-like techniques. It might appear that this issue can be easily resolved by noting that the encryption routine *does* have access to the hash key, and a full description of the hash function can be formed at this point. The caveat, however, is that such an uninstantiable scheme no longer falls under the umbrella of schemes arising from the Encrypt-with-Hash transform. More precisely, although we can freely modify the base PKE to prove uninstantiability, the transformation is *fixed* and it only allows black-box access to the hash function and denies encryption access to the hash key.[5] This observation applies to other transformations as well. For instance, in the FO transformation the message that is asymmetrically encrypted is chosen uniformly at random and thus cannot be set to the description of the hash function. To summarize, although the description of the hash function will be eventually made public, the adversarial scheme never gets to the hash function in full and needs to coordinate the attack with the actual adversary, who sees the hash key, to be successful. Indistinguishability obfuscation allows this distributed attack to be carried out.

Concurrent work. In concurrent and independent work, Green et al. [45] use iO and techniques similar to ours to demonstrate the uninstantiability of random-oracle schemes. Like us, they embed an obfuscated program into schemes in order to make them uninstantiable. Our results, however, rule out the instantiability of (existing) random-oracle *transformations* whereas Green at al. construct uninstantiable *schemes* for primitives which cannot be targeted with CGH-like techniques. For instance bit encryption falls outside the reach of CGH as its input space is too short and cannot be made to behave anomalously on special long inputs. Green et al. show that indistinguishability obfuscation can be used to extended CGH to such constrained primitives.

Primitive design. The shortcomings of ROM primitives that we have identified call for a re-assessment of primitives whose security analyses have only been carried out in idealized models of computation. To highlight the importance of this task, we propose a new transform for building deterministic encryption that is specifically designed to bypass our attacks. In this transform one encrypts two values independently across two invocations of the underlying encryption algorithm to make sure that the information needed for the attack is not available to any of the invocations. (This transform, in particular, is not admissible.) We prove this scheme secure in the ROM, but show that the program that one

---

[5] Despite this, CGH-like techniques render Encrypt-with-Hash uninstantiable when stronger notions of security are considered [53].

would need to successfully attack the construction (assuming the availability of all needed information) can be *split* into several programs such that by feeding obfuscations of one program into the obfuscations of another an attack can be launched. We leave the characterization of the class of transformations which fall prey to extensions of the iO attack as an interesting open problem.

We believe that the structural soundness of ROM schemes should be further validated by studying if attacks similar to those given in this work can be launched against them. To provably rule out such attacks one needs to reduce security to assumptions, which although strong, are not known to be uninstantiable under existence of (d)iO. Candidate examples include UCEs against statistically and/or strongly unpredictable sources [27,31] and indeed indistinguishability obfuscation itself. We note that recently Bellare and Hoang [11] have proposed a D-PKE transform starting from lossy trapdoor function and statistical UCEs. This approach can be further combined with stronger assumptions on the base schemes (such as pseudorandomness of ciphertexts). Indeed, it would be interesting to derive positive results that circumvent iO-based uninstantiability by merely exploiting the pseudorandomness of ciphertexts, even for somewhat artificial tasks. These would strengthen our confidence in applying the random-oracle methodology despite the broad uninstantiability results presented in this paper.

2   Preliminaries

Notation. We denote the security parameter by $\lambda \in \mathbb{N}$ and assume that it is implicitly given to all algorithms in the unary representation 1^λ. We denote the set of all bit strings of length ℓ by $\{0,1\}^\ell$, the set of all bit strings of finite length by $\{0,1\}^*$, the length of $x \in \{0,1\}^*$ by $|x|$, the concatenation of two strings $x_1, x_2 \in \{0,1\}^*$ by $x_1\|x_2$, and the exclusive-or of two strings $x_1, x_2 \in \{0,1\}^*$ of the same length by $x_1 \oplus x_2$. The i-th bit of a string x is indicated by $x[i]$. We denote the empty string by ε. A vector of strings $\mathbf{x}$ is written in boldface, and $\mathbf{x}[i]$ denotes its i-th entry. The number of entries of $\mathbf{x}$ is denoted by $|\mathbf{x}|$. For a finite set X, we denote the cardinality of X by $|X|$ and the action of sampling x uniformly at random from X by $x \leftarrow_\$ X$. For a random variable X we denote the support of X by $[X]$. A real-valued function $\nu(\lambda)$ is negligible if $\nu(\lambda) \in \mathcal{O}(\lambda^{-\omega(1)})$. We denote the set of all negligible functions by negl.

An algorithm is a randomized, stateless Turing machine unless otherwise stated. We call an algorithm efficient or PPT if its runtime on any choice of inputs and random coins is at most a polynomial function of the security parameter. The action of running an algorithm $\mathcal{A}$ on input x and random coins r is denoted by $y \leftarrow \mathcal{A}(x; r)$. If $\mathcal{A}$ is randomized and no randomness is specified, then we assume that $\mathcal{A}$ is run with freshly and uniformly generated random coins and write $y \leftarrow_\$ \mathcal{A}(x)$. An adversary is a tuple of stateful PPT algorithms. We omit explicit input and output states to ease notations. When an adversary $\mathcal{A} = (\mathcal{A}_1, \mathcal{A}_2)$ consists of two stages $\mathcal{A}_1$ and $\mathcal{A}_2$, these two stages are assumed to be distinct algorithms that do *not* share any state, unless explicitly permitted to do so by a game.

Turing machines and circuits. Throughout the paper we consider two models of computation: Turing machines and circuits. Recall that a Turing machine can take inputs of arbitrary length whereas the input length to a circuit is fixed. We denote the runtime of a Turing machine M on input x by $\mathsf{time}_M(x)$ and its description size by $|M|$. We denote the size (a.k.a. runtime) of a circuit C by $|C|$. A *universal Turing machine* UM is a machine that takes two inputs (M, x), interprets M as the description of a Turing machine and returns $M(x)$. A *universal circuit* UC is defined analogously on descriptions of circuits C and inputs x for them. Note that UC only accepts inputs (C, x) of a specific total length, whereas UM can take inputs of arbitrary length. In order to simplify the presentation we use the term *program* to refer to either a Turing machine or a circuit. We may, therefore, speak of a universal program UEval, which denotes either a universal Turing machine UM or a universal circuit UC, and evaluates a program P on some input x. When defining a program, we use the notation $P[z](\cdot)$ to emphasize the fact that the value z is hard-coded into P.

Indistinguishability obfuscation. We define indistinguishability obfuscation for circuits and Turing machines under a single definition. Roughly speaking, an indistinguishability obfuscator (iO) ensures that the obfuscations of any two functionally equivalent programs (that is, circuits or Turing machines) are computationally indistinguishable. Indistinguishability obfuscation was originally proposed by Barak et al. [6,5] as a potential weakening of the virtual-black-box obfuscation property, for which wide infeasibility results are known. Here we give a game-based definition of indistinguishability obfuscation in the style of [19] with extensions to also cover obfuscation for Turing machines [2]. We only consider the setting where both the sampler and distinguisher are uniform but allow the sampler to output inequivalent programs with negligible probability. This game-based formulation is convenient for use in proofs of security.

A PPT algorithm iO is called an *indistinguishability obfuscator* for a program class $\mathcal{P} = \{\mathcal{P}_\lambda\}_{\lambda \in \mathbb{N}}$ if iO on input the security parameter 1^λ and (the description of) a program P outputs a program P' and furthermore the following conditions are satisfied:

- CORRECTNESS. For all $\lambda \in \mathbb{N}$, all $P \in \mathcal{P}_\lambda$, and all $P' \leftarrow_\$ iO(1^\lambda, P)$, the programs P and P' are functionally equivalent. That is, $P(x) = P'(x)$ for all input values x.
- SUCCINCTNESS. There is a polynomial poly such that for all $\lambda \in \mathbb{N}$, all $P \in \mathcal{P}_\lambda$ and all $P' \leftarrow_\$ iO(1^\lambda, P)$ we have that $|P'| \in \mathcal{O}(\mathsf{poly}(\lambda + |P|))$.
- INPUT-SPECIFIC RUNTIME. There is a polynomial poly such that for all $\lambda \in \mathbb{N}$, all $P \in \mathcal{P}_\lambda$ and all $P' \leftarrow_\$ iO(1^\lambda, P)$ and all input values x we have that $\mathsf{Time}_{P'}(x) \in \mathcal{O}(\mathsf{poly}(\lambda + \mathsf{Time}_P(x)))$.
- SECURITY. For any pair of PPT adversaries $(\mathcal{S}, \mathcal{D})$, where $\mathcal{S}$ is an equivalent sampler, i.e., where

$$\mathsf{Adv}_{\mathcal{S}}^{\mathsf{eq}}(\lambda) := \Pr[\exists x \text{ s.t. } P_0(x) \neq P_1(x) \vee \mathsf{Time}_{P_0}(x) \neq \mathsf{Time}_{P_1}(x) :$$
$$(P_0, P_1, aux) \leftarrow_\$ \mathcal{S}(1^\lambda)]$$

is negligible, we have that

$$\mathsf{Adv}^{\mathsf{io}}_{\mathsf{iO},\mathcal{S},\mathcal{D}}(\lambda) := 2 \cdot \Pr\left[\mathsf{IO}^{\mathcal{S},\mathcal{D}}_{\mathsf{iO}}(\lambda)\right] - 1 \in \mathsf{negl} \ ,$$

where game IO is shown in Figure 1 on the left.

When working with circuits, succinctness and runtime requirements are redundant and follow from the facts that iO is polynomial time and that the size and runtime of a circuit are identical.

Garg et al. [39] prove that under intractability assumptions related to multi-linear maps an indistinguishability obfuscator supporting all $\mathcal{NC}^1$ circuits exists. Assuming the existence of a perfectly correct, leveled fully homomorphic encryption scheme and a perfectly sound non-interactive witness-indistinguishable proof system, they also show how to extended this to support all polynomial-size circuits, i.e., the family $\mathcal{C} := \{\mathcal{C}_{\mathsf{b}(\lambda)}\}_{\lambda \in \mathbb{N}}$ where b is a polynomial and

$$\mathcal{C}_{\mathsf{b}(\lambda)} := \{\mathsf{C} : \mathsf{C} \text{ is a valid circuit of size at most } \mathsf{b}(\lambda)\} \ .$$

Several follow-up works improved the assumptions underlying indistinguishability obfuscators as well as the performance [52,26,3,4,42]. As mentioned above, circuits and obfuscations thereof admit fixed-length inputs only.

Remark. We define indistinguishability obfuscation with respect to circuit samplers that are overwhelmingly equivalent, i.e., where

$$\mathsf{Adv}^{\mathsf{eq}}_{\mathcal{S}}(\lambda) \in \mathsf{negl} \ .$$

Although we allow samplers to not always output functionally equivalent circuits, the randomized sampler only errs with negligible probability. For any bound b, existence of iO for $\mathcal{C}_{\mathsf{b}(\lambda)}$ under our definition is implied by the (non-uniform) definition of Garg et al. [39].

Ananth et al. [2] and Boyle et al. [25] give constructions of indistinguishability obfuscators for Turing machines which admit inputs of arbitrary lengths. Their constructions achieve the stronger notion of *differing-inputs* (a.k.a. extractability) obfuscation, initially also suggested in the work of Barak et al. [6,5]. This type of obfuscation can be regarded as a generalization of indistinguishability obfuscation to programs which are not necessarily functionally equivalent. We recall [2, Theorem 3] and refer the reader to the original works for details and discussion.

Theorem 1 (Ananth et al. [2]). *Under the existence of CPA-secure leveled fully homomorphic encryption, succinct non-interactive arguments of knowledge (SNARKs), differing-inputs obfuscation for all circuits in $\mathcal{P}/\mathrm{poly}$, and collision-resistant hash functions, there exists a differing-inputs obfuscator for the class of all Turing machines $\mathcal{M} := \{\mathcal{M}_\lambda\}_{\lambda \in \mathbb{N}}$, where*

$$\mathcal{M}_\lambda := \{\mathsf{M} : \mathsf{M} \text{ is a valid Turing machine of description size at most } \lambda\} \ .$$

Koppula, Lewko and Waters [47] have succeeded in constructing iO for Turing machines without relying on diO, and using iO for circuits, one-way functions and injective pseudorandom generators.

$\mathrm{IO}_{\mathrm{iO}}^{\mathcal{S},\mathcal{D}}(\lambda)$	$\mathrm{IND\text{-}CPA}_{\mathsf{PKE}}^{\mathcal{A}}(\lambda)$	$\mathrm{IND}_{\mathsf{D\text{-}PKE}}^{\mathcal{A}_1,\mathcal{A}_2}(\lambda)$
$(\mathrm{P}_0,\mathrm{P}_1,aux) \leftarrow\!\!\text{\tiny\$}\ \mathcal{S}(1^\lambda)$	$(sk,pk) \leftarrow\!\!\text{\tiny\$}\ \mathsf{PKE.Kg}(1^\lambda)$	$(\mathbf{m}_0,\mathbf{m}_1) \leftarrow\!\!\text{\tiny\$}\ \mathcal{A}_1(1^\lambda)$
$b \leftarrow\!\!\text{\tiny\$}\ \{0,1\}$	$(m_0,m_1) \leftarrow\!\!\text{\tiny\$}\ \mathcal{A}(pk)$	$(sk,pk) \leftarrow\!\!\text{\tiny\$}\ \mathsf{D\text{-}PKE.Kg}(1^\lambda)$
$\mathrm{P}' \leftarrow\!\!\text{\tiny\$}\ \mathrm{iO}(1^\lambda,\mathrm{P}_b)$	$b \leftarrow\!\!\text{\tiny\$}\ \{0,1\}$	$b \leftarrow\!\!\text{\tiny\$}\ \{0,1\}$
$b' \leftarrow\!\!\text{\tiny\$}\ \mathcal{D}(\mathrm{P}',aux)$	$c \leftarrow\!\!\text{\tiny\$}\ \mathsf{PKE.Enc}(pk,m_b)$	$\mathbf{for}\ i=1\dots|\mathbf{m}_0|\ \mathbf{do}$
$\mathbf{return}\ (b=b')$	$b' \leftarrow\!\!\text{\tiny\$}\ \mathcal{A}(c)$	$\quad\mathbf{c}[i] \leftarrow \mathsf{D\text{-}PKE.Enc}(pk,\mathbf{m}_b[i])$
	$\mathbf{return}\ (b=b')$	$b' \leftarrow\!\!\text{\tiny\$}\ \mathcal{A}_2(pk,\mathbf{c})$
		$\mathbf{return}\ (b=b')$

Fig. 1. Left: IO game defining the security of an indistinguishability obfuscator. **Middle**: The IND-CPA game for a public-key encryption scheme. **Right**: The IND security game for deterministic PKEs.

Public-key encryption. A public-key encryption scheme $\mathsf{PKE} := (\mathsf{PKE.Kg}, \mathsf{PKE.Enc}, \mathsf{PKE.Dec})$ consists of three PPT algorithms as follows. On input the security parameter, the randomized key-generation algorithm $\mathsf{PKE.Kg}(1^\lambda)$ generates a key pair (sk, pk). The randomized encryption algorithm $\mathsf{PKE.Enc}(pk, m; r)$ gets a message m, a public key pk and possibly some explicit random coins r and outputs a ciphertext c. The deterministic decryption algorithm $\mathsf{PKE.Dec}(sk, c)$ is given a ciphertext c and secret key sk and outputs a plaintext m or a special symbol $\perp$. We denote the supported message length by $\mathsf{PKE.il}(\lambda)$ and the maximum length of random strings used to encrypt a $\mathsf{PKE.il}(\lambda)$-bit message by $\mathsf{PKE.rl}(\lambda)$. We say that scheme PKE is correct if for all $\lambda \in \mathbb{N}$, all $m \in \mathsf{PKE.il}(\lambda)$, all $(sk, pk) \in [\mathsf{PKE.Kg}(1^\lambda)]$ and all $c \in [\mathsf{Enc}(pk, m)]$ we have that $\mathsf{PKE.Dec}(sk, c) = m$. We say that PKE is IND-CPA secure, if the advantage of any PPT adversary $\mathcal{A}$ in the IND-CPA game (shown in Figure 1; center) defined by

$$\mathsf{Adv}_{\mathsf{PKE},\mathcal{A}}^{\mathrm{ind\text{-}cpa}}(\lambda) := 2 \cdot \Pr\left[\mathrm{IND\text{-}CPA}_{\mathsf{PKE}}^{\mathcal{A}}(\lambda)\right] - 1$$

is negligible.

Function families. Following [19], we define a function family FF as a five tuple of PPT algorithms $(\mathsf{FF.Kg}, \mathsf{FF.Ev}, \mathsf{FF.kl}, \mathsf{FF.il}, \mathsf{FF.ol})$ where the algorithms $\mathsf{FF.kl}$, $\mathsf{FF.il}$, and $\mathsf{FF.ol}$ are deterministic and on input 1^λ specify the key, input, and output lengths, respectively. The key-generation algorithm $\mathsf{FF.Kg}$ gets the security parameter 1^λ as input and outputs a key $\mathsf{fk} \in \{0,1\}^{\mathsf{FF.kl}(\lambda)}$. The deterministic evaluation algorithm $\mathsf{FF.Ev}$ takes as input the security parameter 1^λ, a key fk, a message $x \in \{0,1\}^{\mathsf{FF.il}(\lambda)}$ and generates a hash value $\mathsf{FF.Ev}(1^\lambda, \mathsf{fk}, x) \in \{0,1\}^{\mathsf{FF.ol}(\lambda)}$. We will often refer to function families as hash functions in this work.

PRFs and PRGs. We say that a function family FF is pseudorandom if for any PPT adversary $\mathcal{A}$ we have that

$$\mathsf{Adv}_{\mathsf{FF},\mathcal{A}}^{\mathrm{prf}}(\lambda) := \Pr\left[\mathcal{A}^{\mathsf{FF.Ev}(\mathsf{fk},\cdot)}(1^\lambda) = 1\right] - \Pr\left[\mathcal{A}^{\mathcal{RO}(\cdot)}(1^\lambda) = 1\right] \in \mathsf{negl}\ .$$

In the first term above, the probability is taken over a random choice of a key $\mathsf{fk} \in \{0,1\}^{\mathsf{FF.kl}(\lambda)}$ and in the second over a random choice of $\mathcal{RO}$ with domain $\{0,1\}^{\mathsf{FF.il}}(\lambda)$ and range $\{0,1\}^{\mathsf{FF.ol}(\lambda)}$.

We say $(\mathsf{PRG}, \mathsf{PRG.il}, \mathsf{PRG.ol})$ is a secure pseudorandom generator if PRG on strings of length $\mathsf{PRG.il}(\lambda)$ outputs strings of length $\mathsf{PRG.ol}(\lambda)$ and for any PPT adversary $\mathcal{A}$ we have that

$$\mathsf{Adv}^{\mathrm{prg}}_{\mathsf{PRG},\mathcal{A}}(\lambda) := \Pr[\mathcal{A}(1^\lambda, \mathsf{PRG}(s)) = 1 : s \leftarrow_\$ \{0,1\}^{\mathsf{PRG.il}(\lambda)}]$$
$$- \Pr[\mathcal{A}(1^\lambda, y) = 1 : y \leftarrow_\$ \{0,1\}^{\mathsf{PRG.ol}(\lambda)}]$$

is negligible.

Keyed random oracles. Most random-oracle transformations and schemes in the literature are analyzed in the "unkeyed" random-oracle model, and this reflects the fact that a fixed unkeyed hash function will be used in their instantiations. Keyed hash functions, however, are more powerful when it comes to instantiating random oracles and this leaves the question of how the scheme is to be instantiated with a keyed hash function, that is, how the hash key is to be generated and who gets access to it is rather unclear. For example, if we consider a transformation of symmetric encryption schemes, the hash key could be part of the key-generation process in which case it remains hidden from the adversary, or it could be a parameter generated during set-up, in which case it would be available to the adversary. We therefore use a generalization of the standard random-oracle model whereby all parties get access to a *keyed* random function. More precisely, in the $(\mathsf{kl}, \mathsf{il}, \mathsf{ol})$-ROM, where $(\mathsf{kl}, \mathsf{il}, \mathsf{ol})$ specify various lengths as before, on security parameter λ all parties get access to a random function of the form

$$\mathcal{RO}(\cdot, \cdot) : \{0,1\}^{\mathsf{kl}(\lambda)} \times \{0,1\}^{\mathsf{il}(\lambda)} \longrightarrow \{0,1\}^{\mathsf{ol}(\lambda)} .$$

Note that we recover the standard unkeyed random-oracle model when $\mathsf{kl}(\lambda) = 0$ (there is only one key ε, the empty string). In defining the security of a cryptosystem, the underlying probability space is extended to include a random choice of a keyed function (and choices of random key as specified by the scheme). Whether or not a party gets to see the hash key depends on the specification of the scheme and its security model. For instance, if a keyed ROM scheme includes hash keys under its public keys, an honest or malicious party gets to sees the hash key whenever it gets to see the public key. As our result is a negative result, it suffices to consider weak adversaries that do not get oracle access and/or the hash key in some of their stages, because weaker adversaries correspond to a stronger negative result.

(Un)instantiability. Given a scheme in the keyed ROM, we consider its standard-model instantiations via (concrete) keyed hash functions. Formally, this entails: (1) using a hash function that has key, input and output lengths that are identical to those of the keyed random oracle, (2) running the key-generation algorithm whenever a hash key is generated in the ideal scheme, and (3) calling the evaluation routine of the hash function whenever an oracle query is placed. Given a

keyed ROM scheme and a security model for it, we say that the scheme is *instantiable* if there exists a hash function which when used to instantiate the scheme (and its security model) results in a secure scheme (with respect to the instantiated security model). Conversely, we say that a scheme is *(strongly) uninstantiable* if no hash function can securely instantiate the ideal scheme. Finally, for a polynomial bound p, we call a scheme b-*uninstantiable*, if no hash function of size at most b(λ) can securely instantiate the scheme.

3   Deterministic Encryption

We start by studying the Encrypt-with-Hash (EwH) transform of Bellare, Boldyreva and O'Neill (BBO) [7] for building deterministic encryption from standard (randomized) encryption schemes. We show that under the existence of indistinguishability obfuscation there is an IND-CPA public-key encryption scheme that cannot be safely used within EwH. We begin by formally defining the syntax and security of deterministic PKEs and the EwH transform. We then prove uninstantiability, and end with two corollaries of this result.

3.1   Definitions

Deterministic public-key encryption. Deterministic public-key encryption was first introduced by Bellare, Boldyreva and O'Neill [7]. The syntax and correctness of a deterministic public-key encryption (D-PKE) scheme D-PKE := (D-PKE.Kg, D-PKE.Enc, D-PKE.Dec) is defined similarly to a randomized PKE scheme with the difference that the encryption routine is deterministic (i.e., D-PKE.rl(λ) = 0). BBO [7] model the security of D-PKEs via a form of simulation-based notion called PRIV. In later works, Bellare et al. [10] and independently Boldyreva, Fehr and O'Neill [22] introduce an indistinguishability-based notion called IND and show that it implies is equivalent to PRIV security. The IND game is formally defined in Figure 1 on the right.[6] Roughly speaking, an IND adversary $\mathcal{A} := (\mathcal{A}_1, \mathcal{A}_2)$ consists of two stages. On input the security parameter, adversary $\mathcal{A}_1$ outputs a pair of message vectors $(\mathbf{m}_0, \mathbf{m}_1)$ of the same dimension that have distinct components and component-wise contain messages of the same length. (Adversary $\mathcal{A}_1$ does not get to see the public key.) Furthermore, each component is required to have *super-logarithmic min-entropy*. This condition is formalized by requiring that for any $x \in \{0,1\}^{\text{D-PKE.il}(\lambda)}$, any $b \in \{0,1\}$ and any $i \in [\|\mathbf{m}_b\|]$,

$$\Pr\left[x = \mathbf{m}_b[i] : (\mathbf{m}_0, \mathbf{m}_1) \leftarrow_{\$} \mathcal{A}_1(1^\lambda)\right] \in \mathsf{negl} .$$

A key pair $(pk, sk) \leftarrow_{\$} \text{D-PKE.Kg}(1^\lambda)$ is then chosen, and according to the challenge bit b, one of the two message vectors is encrypted component-wise. The second-stage adversary $\mathcal{A}_2$ is run on the resulting vector of ciphertexts and the

---

[6] Bellare et al. [10] allow an additional zeroth-stage adversary to output shared state for adversaries $\mathcal{A}_1$ and $\mathcal{A}_2$. As we prove an impossibility result we choose the weaker definition where this shared state is empty.

public key, and wins the game if it correctly guesses the hidden bit b. We define the advantage of an adversary $\mathcal{A}$ in the IND game (see Figure 1) against scheme D-PKE by

$$\mathsf{Adv}^{\mathrm{ind}}_{\mathsf{D\text{-}PKE},\mathcal{A}_1,\mathcal{A}_2}(\lambda) = 2 \cdot \Pr\left[\mathrm{IND}^{\mathcal{A}_1,\mathcal{A}_2}_{\mathsf{D\text{-}PKE}}(\lambda)\right] - 1 \ .$$

We say that scheme D-PKE is IND secure if the advantage of any PPT adversary $\mathcal{A} = (\mathcal{A}_1, \mathcal{A}_2)$ in the IND game is negligible. The 1-bounded version of this security model demands that the two vectors $(\mathbf{m}_0, \mathbf{m}_1)$ only contain a single message each.

The Encrypt-with-Hash transform. The Encrypt-with-Hash (EwH) transform constructs a deterministic public-key encryption scheme from a (randomized) public-key encryption scheme PKE in the random-oracle model [7]. We present this transform in the keyed ROM, and note that it matches the original transform for singleton key spaces. The keyed RO is assumed to have a range which matches the randomness space of the PKE scheme and a domain which consisting of all bit strings of length the maximum length of public keys plus the length of messages. The EwH transform operates as follows.

The key-generation generates a key pair using the key-generation algorithm of the base PKE scheme. It also generates a hash key $\mathsf{hk} \leftarrow_\$ \{0,1\}^{\mathsf{kl}(\lambda)}$ and returns $(sk, (\mathsf{hk}, pk))$. Algorithm $\mathsf{D\text{-}PKE.Enc}^{\mathcal{RO}(\cdot,\cdot)}(m, (\mathsf{hk}, pk))$ first computes random coins $r \leftarrow \mathcal{RO}(\mathsf{hk}, pk\|m)$ and then invokes the base encryption algorithm on m and pk and coins r to generate a ciphertext. The decryption routine is identical to that of the underlying scheme (plus a ciphertext re-computation check to ensure non-malleability). EwH results in an IND-secure D-PKE scheme in the keyed ROM when starting from an IND-CPA public-key encryption scheme.

Key access in EwH. With the formalism introduced above, both adversaries $\mathcal{A}_1$ and $\mathcal{A}_2$ get oracle access to $\mathcal{RO}(\cdot,\cdot)$. The first-stage adversary, however, does *not* get to see hk since the hash key is distributed as a component of the public keys. The second-stage adversary, on the other hand, does get to see it. A stronger model where the hash key *is* given out in the first stage can be considered. EwH meets this stronger notion of security, but since our results are negative we use the conventional (and weaker) IND model.

3.2 Uninstantiability of EwH

When the EwH transformation is instantiated with an *unkeyed* random oracle a CGH-style uninstantiability result can be directly established [33]. (This in particular shows that the use of a keyed hash function is necessary to instantiate EwH.) Given an arbitrary PKE scheme PKE, consider a tweaked variant of it PKE' which first interprets parts of the message m as the description of a hash function H (together with its single key) and checks if the provided random coins r match the hash value $\mathsf{H}(pk\|m)$. If so, it returns $0\|m$ and else it returns $1\|\mathsf{PKE.Enc}(pk, m; r)$. Scheme PKE' is still IND-CPA secure because the probability that a truly random value r matches $\mathsf{H}(pk\|m)$ is negligible. On the other hand, when the random coins are generated deterministically by applying a hash function, an IND adversary which asks for encryptions of $m_i \| \mathsf{H}$ for any two high min-entropy messages m_0 and m_1 which differ, say, on their most significant bits can easily win the game.[7] The standard IND game, however, restricts the first-stage adversary not to learn the public key, and thus, it cannot guess the (high min-entropy) hash key.

We show how to use indistinguishability obfuscation to extend the above uninstantiability to keyed hash functions. As mentioned in the introduction, our result comes in the weak and strong flavors depending on the programs that the obfuscator is assumed to support. Assuming iO for Turing machines we obtain a strong uninstantiability result: there exists an IND-CPA encryption scheme that cannot be securely used in EwH in conjunction with *any* keyed hash function. Assuming the weaker notion of iO for circuits, we get b-uninstantiability: for any polynomial bound b there exists an IND-CPA scheme that cannot be securely used in EwH for any hash function whose description size is at most b. The latter result is still quite strong as, in particular, it means that for any finite set of hash functions (e.g., those which are standardized), we can give a PKE scheme that when used within EwH yields an insecure D-PKE scheme for any choice of hash function from the set. We note that the adversarial PKE scheme that we construct depends only on an upper bound on description sizes and not on their implementation details.

Theorem 2 (Uninstantiability of EwH). *Assuming the existence of indistinguishability obfuscation for Turing machines $\mathcal{M}$ (resp. b-bounded circuits $\mathcal{C}_b$), the EwH transform is uninstantiable (resp. b-uninstantiable) with respect to IND security in the standard model.*

We start by giving a high-level description of the proof before presenting the details. We may assume, without loss of generality, that an IND-CPA-secure PKE scheme exists as otherwise uninstantiability trivially holds. This, in turn, implies that we can also assume the existence of a secure pseudorandom generator.

Now given an IND-CPA-secure PKE scheme PKE, we construct a tweaked scheme PKE^* that is also IND-CPA secure but the D-PKE scheme $\mathsf{EwH}^\mathsf{H}[\mathsf{PKE}^*]$ fails to be IND secure.

To construct the adversarial scheme PKE^* we follow a similar strategy to CGH. The fundamental difference here is that $\mathsf{PKE}^*.\mathsf{Enc}$ does not have access to the hash key. To overcome this problem, we consider the obfuscation of a program P' that implements a *universal* variant of the BFM circuit [27], i.e., it takes as input the description of a hash function $\mathsf{H}(\mathsf{hk}, \cdot)$, with a hard-wired key, runs it on two values m and pk embedded into P', and outputs m if the result matches a third hard-wired value r:

$$\mathsf{P}'[pk, m, r]\Big(\mathsf{H}(\mathsf{hk}, \cdot)\Big) := \text{if } \mathsf{H}(\mathsf{hk}, pk\|m) = r \text{ return } m \text{ else return } 0 \ .$$

---

[7] This attack generalizes to the setting where the first-stage adversary can guess the hash key with non-negligible probability and in particular, EwH is uninstantiable with respect to the stronger IND model discussed above.

The tweak that we introduce in PKE* is that the encryption operation appends obfuscations of P'[pk, m, r] to its ciphertexts, where pk, m and r are the values input to the encryption routine.

We need to argue (1) that this tweak allows an adversary to break the scheme whenever the hash function is instantiated and (2) that outputting such an indistinguishability obfuscation of P' does not hurt the IND-CPA security of PKE*.

For (1), note that given an obfuscation of P'[pk, m, r] as well as a description of H(hk, ·), an adversary can recover m by running the above circuit on H(hk, ·). Now the second stage of the IND adversary gets the public key and thus the description of the hash-function H(hk, ·). Furthermore, it also gets a ciphertext which contains an obfuscation of P'[pk, m, r]. Hence, the second-stage adversary has all the information needed to break the IND security of the deterministic encryption scheme EwH^{H}[PKE*].

Now, intuitively, this insecurity might have nothing to do with the transform because the tweaked scheme PKE* is already insecure anyway. Hence, we also need to argue that PKE*, as a randomized encryption scheme, is IND-CPA secure. Following BFM, we try to prove this by showing that the obfuscated circuit is functionally equivalent to the *zero* circuit and hence it does not leak any information about m.

We would like to argue that for a *truly random* r—such an r is used in randomized encryption—P' implements the constant zero program Z. Indeed, if r is sufficiently longer than $|pk| + |m|$ then for any fixed H(hk, ·), over a random choice of r the check performed by P' would fail with all but negligible probability. This, however, does not necessarily mean that the circuit is functionally equivalent to Z as there could *exist* a hash function H(hk, ·) which passes the check. Contrary to BFM, we cannot bound the probability of this event via the union bound as the number of hash descriptions might exceed the size of the randomness space.

To resolve this issue, we consider a further tweak to the base scheme. We consider a scheme which has a much smaller randomness space and instead uses coins that are *pseudorandomly generated*. This ensures that the randomness space used by PKE is sparse within the set of all possible coins, allowing a counting argument to go through. We adapt the program above to cater for the new tweaks:

$$\text{P}[pk, m, \text{PRG}(r)]\left(\text{H}(hk, ·)\right) := \text{if } \text{PRG}(\text{H}(hk, pk\|m)) = \text{PRG}(r)$$

$$\text{return } m$$

$$\text{else return } 0 \ .$$

At this point it might appear that no progress has been made as the above program, for reasons similar to those given above, is not functionally equivalent to Z. We note, however, that for a truly random $s \in \{0,1\}^{\text{PRG.ol}(\lambda)}$ the program P[pk, m, s] has a *description* which is indistinguishable from that of P[pk, m, PRG(r)] down to the security of PRG. Furthermore for such an s, this program *can* be shown to be functionally equivalent to the zero circuit with overwhelming probability as s will be outside the range of the PRG with overwhelming probability. These two steps allow us to prove that obfuscations of P leak no information about m, and show that scheme PKE* is IND-CPA secure.

Finally, notice that obfuscations of P (similarly to those of P') allow an IND adversary to break the resulting EwH-transformed scheme: simply run the obfuscation of P on the description of the hash function used in the instantiation (with a hard-wired key) to recover the encrypted message.

Not that formally program P will use a universal program evaluator to run its input hash-function descriptions. If the (obfuscated) program is a Turing machine, it can be run on arbitrary large descriptions and arbitrarily sized hash functions are ruled out. On the other hand, if the program is a circuit, it has an a priori *fixed* input length, and thus can only be run on hash functions that respect the input-size restrictions. We next formalize this proof intuition.

Proof (of Theorem 2). Let PKE be an IND-CPA-secure public-key encryption scheme, PRG be a pseudorandom generator of appropriate stretch and iO be an indistinguishability obfuscator supporting either Turing machines or circuits. We define a modified PKE scheme PKE* as follows. The key-generation algorithm is unchanged. The adapted encryption algorithm is defined as shown below by appending an obfuscated program $\overline{P}$ to its outputs. UEval denotes a universal program evaluator. The modified decryption algorithm ignores the $\overline{P}$ component and decrypts as in the base scheme.

ALGO. PKE*.Enc($pk, m; r\|r'$)	PROG. P[pk, m, s](H(hk, ·))
$s \leftarrow \mathsf{PRG}(r)$	$r\|r' \leftarrow \mathsf{UEval}(\mathsf{H}(\mathsf{hk}, \cdot), pk\|m)$
$c \leftarrow \mathsf{PKE.Enc}(pk, m; s)$	$s' \leftarrow \mathsf{PRG}(r)$
$\overline{P} \leftarrow \mathsf{iO}(\mathsf{P}[pk, m, s](\cdot); r')$	if $(s' = s)$ then return m
return $(c, \overline{P})$	return 0

When we consider the above construction with respect to circuits, we need to specify an extra parameter b that upper-bounds the size of the inputs to the universal circuit evaluator. This maximum size of programs that the universal circuit admits corresponds to the maximum size of the hash functions that our uninstantiability proof applies to. Note that when the construction is considered for Turing machines, the input size is arbitrary.

We show that the above tweaked scheme PKE* is IND-CPA secure via a sequence of four games that we describe next. We present the pseudocode in Figure 2.

Game$_0$: This game is identical to the IND-CPA game for the randomized base scheme PKE* and an arbitrary adversary $\mathcal{A}$.

Game$_1$: In this game the randomness s used in encryption is no longer generated via a PRG call and is sampled uniformly at random.

Game$_2$: In this game the ciphertext component $\overline{P}$ is generated as an indistinguishability obfuscation of the zero program (that is, Turing machine or circuit) Z padded to the appropriate length (and running time).

We now show that each of the above transitions negligibly changes the game's output with respect to any adversary $\mathcal{A}$.

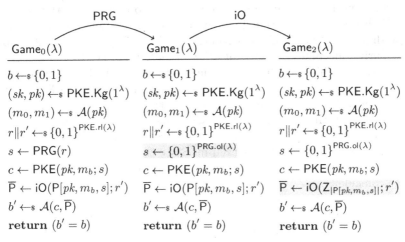

Fig. 2. Hybrids used in the proof of Theorem 2. The highlighted lines show the changes in game transitions.

Game_0 to Game_1. We bound the difference in these games by the security of PRG. Note that a PRG adversary that gets as input y, a PRG image under a uniformly random seed or a truly uniformly random value, can perfectly simulate games Game_0 and Game_1 for $\mathcal{A}$ by using y in place of s. If y is a PRG image, then Game_0 is run and if y is uniformly random the Game_1 is run:

$$\Pr[\mathsf{Game}_0(\lambda)] - \Pr[\mathsf{Game}_1(\lambda)] \leq \mathsf{Adv}^{\mathrm{prg}}_{\mathsf{PRG},\mathcal{A}}(\lambda) \ .$$

Game_1 to Game_2. We show that this hop negligibly affects the winning probability of $\mathcal{A}$ down to the security of the indistinguishability obfuscator. We let $\mathcal{S}$ to be the sampler which runs all the steps of Game_1 using the first phase of $\mathcal{A}$ up to the generation of $\overline{\mathsf{P}}$. It then sets $\mathsf{P}_0 := \mathsf{P}[pk, m_b, s]$, $\mathsf{P}_1 := \mathsf{Z}_{|\mathsf{P}_0|}$ and aux to be the ciphertext component c and the internal state of the first phase of the IND-CPA adversary. Algorithm $\mathcal{D}$ receives an obfuscation $\overline{\mathsf{P}}$ of either P_0 or P_1, and resumes the second phase of $\mathcal{A}$ on $(c, \overline{\mathsf{P}})$ using the state recovered from aux. When P_0 is obfuscated $\mathcal{A}$ is run according to the rules of Game_1 and when P_1 is obfuscated $\mathcal{A}$ is run according to the rules of Game_2. Hence,

$$\Pr[\mathsf{Game}_1(\lambda)] - \Pr[\mathsf{Game}_2(\lambda)] \leq \mathsf{Adv}^{\mathrm{io}}_{\mathsf{iO},\mathcal{S},\mathcal{D}}(\lambda) \ .$$

We must show that the sampler $\mathcal{S}$ constructed above outputs functionally equivalent circuits with overwhelming probability. Assuming that the stretch of the PRG is sufficiently large, i.e., $\mathsf{PRG.ol}(\lambda) \geq 2 \cdot \mathsf{PRG.il}(\lambda)$, by the union bound the probability over a random choice of s that there *exists* an $r \in \{0,1\}^{\mathsf{PRG.il}(\lambda)}$ such that $\mathsf{PRG}(r) = s$ is upper bounded by $2^{\mathsf{PRG.il}(\lambda) - \mathsf{PRG.ol}(\lambda)} \leq 2^{-\mathsf{PRG.il}(\lambda)}$. Hence, the probability that P_0 is functionally inequivalent to the zero circuit is upper bounded by $2^{-\mathsf{PRG.il}(\lambda)}$, that is,

$$\Pr\left[\exists x \, \mathsf{P}_0(x) \neq 0 : (\mathsf{P}_0, \mathsf{P}_1, aux) \leftarrow_{\$} \mathcal{S}(1^\lambda)\right] \leq 2^{-\mathsf{PRG.il}(\lambda)} \ .$$

When working with Turing machines, we also need to ensure that the two programs used above respect the run-time requirements of the definition of a secure indistinguishability obfuscator for Turning machines. Formally, we will implement the Turing machines P and Z *obliviously* as follows. We first consider an oblivious Turing machine which takes in the description of the hash function *and a message* as input and performs exactly the same computation that P does. We then implement P by fixing the message input of this machine to that passed to the encryption algorithm, retaining the machine's oblivious structure. The same strategy will be used in constructing the zero circuit, where the constant zero message (of correct length) is hard-wired in. Since these machines are oblivious, their runtimes depend only on the *sizes* of the message and the hash description and hence coincide.

$\mathsf{Game_2}$. We reduce the advantage of $\mathcal{A}$ in $\mathsf{Game_2}$ to the IND-CPA security of scheme PKE. The only difference between this game and the usual IND-CPA game for PKE is that an obfuscation of $\mathsf{Z}_{|\mathsf{P}[pk,m_b,s]|}$ is attached to the ciphertexts. This program has a public description and hence its obfuscations can be perfectly simulated. Hence,

$$2 \cdot \Pr[\,\mathsf{Game_2}(\lambda)\,] - 1 \leq \mathsf{Adv}^{\text{ind-cpa}}_{\mathsf{PKE^*},\mathcal{A}}(\lambda)\ .$$

THE ATTACK. To conclude the proof, we show there exists an adversary $(\mathcal{A}_1, \mathcal{A}_2)$ that breaks the IND security of $\mathsf{EwH}^{\mathsf{H}}[\mathsf{PKE^*}]$ for any function H that respects the input requirements of P (arbitrary if P is a Turing machine, and b-bounded if a circuit). Adversary $\mathcal{A}_1$ chooses two values $x_0, x_1 \leftarrow_\$ \{0,1\}^{\mathsf{PKE.il}(\lambda)-1}$ uniformly at random and outputs messages $m_0 := x_0 \| 0$ and $m_1 := x_1 \| 1$. Observe that $\mathcal{A}_1$ adheres to the entropy requirements of admissible IND adversaries. Adversary $\mathcal{A}_2$ gets as input the public key (pk, hk) and a ciphertext $(c, \overline{\mathsf{P}})$. It then evaluates $\overline{\mathsf{P}}$ on the description of hash function $\mathsf{H}(\mathsf{hk}, \cdot)$ with key hk recovered from the public key and hard-coded into the program description. (Note that if we are considering circuits, the description of this circuit must have size at most $\mathsf{b}(\lambda)$.) Adversary $\mathcal{A}_2$ returns the least significant bit of P's output. This adversary and its operation within the IND game is shown in Figure 3. By the correctness of the obfuscator, $(\mathcal{A}_1, \mathcal{A}_2)$ always win IND with probability 1 irrespectively of the message that is encrypted:

$$\mathsf{Adv}^{\text{ind}}_{\mathsf{D\text{-}PKE},\mathcal{A}_1,\mathcal{A}_2}(\lambda) = 1\ .$$

$\square$

3.3   Consequences for UCEs

We turn to Universal Computational Extractors (UCEs), a novel notion introduced by Bellare, Hoang and Keelveedhi (BHK) [12] to generically instantiate random oracles across a number of cryptographic protocols. UCEs constitute a set of assumptions that roughly speaking model the strong extractor properties

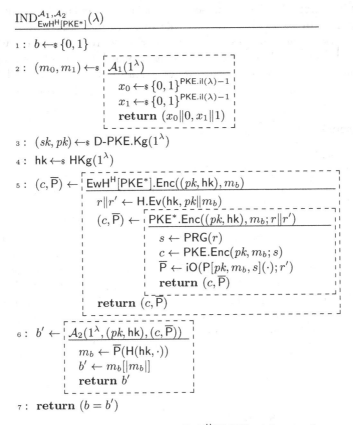

Fig. 3. The IND-security game for scheme $\mathsf{EwH}^H[\mathsf{PKE}^*]$ with our adversary $(\mathcal{A}_1, \mathcal{A}_2)$ as constructed in the proof of Theorem 2. The boxed algorithms are to be understood as subroutines.

enjoyed by (keyed) random oracles. One application of this new framework has been to the EwH transform. BHK [15] show that if a scheme PKE is IND-CPA secure and a hash function H meets what they call $\mathrm{UCE}[\mathcal{S}^{\mathrm{cup}} \cap \mathcal{S}_{\mathsf{PKE}}]$ security then $\mathsf{EwH}^H[\mathsf{PKE}]$ is IND secure. (We refer the reader to the May 2014 version of the paper for the details.) We emphasize that this security definition *depends* on the PKE scheme, because the source class $\mathcal{S}_{\mathsf{PKE}}$ is restricted to those which run the PKE scheme as a subroutine. Our negative results on EwH show that $\mathrm{UCE}[\mathcal{S}^{\mathrm{cup}} \cap \mathcal{S}_{\mathsf{PKE}}]$ security is uninstantiable.

Corollary 1 ($\mathrm{UCE}[\mathcal{S}^{\mathrm{cup}} \cap \mathcal{S}_{\mathsf{PKE}}]$ Uninstantiability). *Assuming the existence of indistinguishability obfuscation for Turing machines $\mathcal{M}$ (resp. b-bounded circuits $\mathcal{C}_{\mathsf{b}}$), $\mathrm{UCE}[\mathcal{S}^{\mathrm{cup}} \cap \mathcal{S}_{\mathsf{PKE}}]$ security for hash functions is uninstantiable (resp. b-uninstantiable) in the standard model.*

We remark that BHK based the security of EwH on other stronger UCE assumptions [12,13]. Our results also show the uninstantiability of these notions assuming indistinguishability obfuscation and in particular imply the negative results of [27]. In particular, we can rule out the instantiatiability of the so-called *bounded paralell sources*[13] by considering sources that internally run an obfuscator. (This translates to D-PKE schemes which run an obfuscator in their encryption routine as we constrcut above.) The results of BFM [27], however, rule out a wider choice of parameters for bounded paralell sources.

3.4   Extension to Hedged PKEs

Hedged public-key encryption, introduced by Bellare et al. [9] models the security of public-key encryption schemes where the random coins used in encryption might have low entropy. Indistinguishability under chosen-distribution attacks (IND-CDA) shown in Figure 4 formalizes the security of hedged PKEs. This notion is similar to IND and the only difference is that the adversary additionally to the two message vectors also outputs a randomness vector. The high min-entropy restriction is spread over the message and randomness vectors. When the length of the randomness entries is 0, one recovers the IND model for D-PKEs. A transform similar to EwH, called Randomized Encrypt-with-Hash, can be defined for hedged PKEs [9]: hash the message, public key and the randomness to obtain new coins, and use them in encryption. Our uninstantiability result can be immediately adapted to this transform as long as the message space has super-polynomial size:

PROG. $\mathsf{P}[pk, m, s](\mathsf{H}, \rho)$

$r \leftarrow \mathsf{UEval}(\mathsf{H}, pk\|m\|\rho)$
$s' \leftarrow \mathsf{PRG}(r)$
if $(s' = s)$ **then return** m
return 0

That is, the program takes an additional input ρ that allows the attacker to specify the randomness. We note that this requires the adversary to choose the randomness in a predictable way, which does not violate the min-entropy requirements as long as the min-entropy of the messages is sufficiently high. We note that if one strengthens the IND-CDA notion to require the randomness distribution to have super-logarithmic min entropy, our attacks would no longer work. This in particular is the case if the message space of the scheme is small.

3.5   Other Uninstantiability Results

In the full version of the paper [28] we show that our uninstantiability results can be further leveraged to rule out standard-model instantiations of a number of other known transformations. We generalize the iO attack to what we call *admissible transformations*, and show that the classical and widely deployed Fujisaki–Okamoto transformation [FO99] falls under it. We also show that a generic approach to building secure symmetric encryption in the presence of

$$\underline{\text{IND-CDA}_{\text{H-PKE}}^{\mathcal{A}_1, \mathcal{A}_2}(\lambda)}$$

$b \leftarrow_{\$} \{0, 1\}$

$(\mathbf{m}_0, \mathbf{m}_1, \mathbf{r}) \leftarrow_{\$} \mathcal{A}_1(1^{\lambda})$

$(sk, pk) \leftarrow_{\$} \text{H-PKE.Kg}(1^{\lambda})$

for $i = 1 \dots |\mathbf{m}_0|$ **do**

$\quad \mathbf{c}[i] \leftarrow \text{H-PKE.Enc}(pk, \mathbf{m}_b[i]; \mathbf{r}[i])$

$b' \leftarrow_{\$} \mathcal{A}_2(pk, \mathbf{c})$

return $(b' = b)$

Fig. 4. The IND-CDA security game for hedged public-key encryption without initial adversaries. Our results carry over to a setting where an initial adversary that passes state to the first and second phase of the attack is present [54].

key-dependent messages, and another one for building de-duplication schemes are uninstantiable.

In the full version, we also explore new classes of D-PKE transformations that lie beyond those captured by admissible transformations. We present a candidate transformation that is specifically designed to foil our iO attack. We first show that this transformation is structurally sound by proving it secure in the ROM. We then show how to extend our techniques to this (and potentially other classes of) transformations. Our goal is to illustrate the flexibility of our main technique and show that it can be tweaked and extended in many ways.

4   Concluding Remarks

The uninstantiability results presented in this paper (and the generalization presented in the full version [28]) demonstrate the applicability of our techniques to a more general class of transforms beyond those captured by admissible transformations. It seems an intricate task to characterize the class of transformations which are subject to our iO-based attacks. It is also an interesting and non-trivial question to propose a D-PKE transformation that is not subject to our uninstantiability result.

One promising avenue is to build schemes based on assumptions from the framework of Universal Computational Extractors (UCEs) [15]. For instance, Bellare, Hoang and Keelveedhi [15] show that message-locked encryption can be based on UCE[$\mathcal{S}^{\text{sup}}$], that is, UCEs with statistically unpredictable sources. This result, however, is not generic with respect to symmetric encryption schemes but rather fixes the base symmetric scheme. Note also that iO is not known to contradict statistical UCEs [27]. Very recently, Bellare and Hoang [11] have proposed a similar transform for D-PKE starting from lossy trapdoor functions.

Alternatively, one could switch to schemes that meet stronger notions of security. For instance, IND\$-type security notions that require the ciphertexts to be indistinguishable from random do not lend themselves to out attacks as it is unclear if obfuscation schemes can provide circuits which are indistinguishable from random strings.

Acknowledgments. Part of this work was done while Christina Brzuska was a post-doctoral researcher at Tel Aviv University and supported by the Israel Science Foundation (grant 1076/11 and 1155/11), the Israel Ministry of Science and Technology (grant 3-9094), and the German-Israeli Foundation for Scientific Research and Development (grant 1152/2011). Pooya Farshim was supported in part by EPSRC research grant EP/L018543/1. Arno Mittelbach was supported by CASED (www.cased.de) and the German Research Foundation (DFG) SPP 1736.

References

1. Al-Riyami, S.S., Paterson, K.G.: Certificateless public key cryptography. In: Laih, C.-S. (ed.) ASIACRYPT 2003. LNCS, vol. 2894, pp. 452–473. Springer, Heidelberg (2003)
2. Ananth, P., Boneh, D., Garg, S., Sahai, A., Zhandry, M.: Differing-inputs obfuscation and applications. Cryptology ePrint Archive, Report 2013/689 (2013), http://eprint.iacr.org/2013/689
3. Ananth, P.V., Gupta, D., Ishai, Y., Sahai, A.: Optimizing obfuscation: Avoiding barrington's theorem. In: Ahn, G.J., Yung, M., Li, N. (eds.) ACM CCS 2014: 21st Conference on Computer and Communications Security, November 3–7, pp. 646–658. ACM Press, Scottsdale (2014)
4. Barak, B., Garg, S., Kalai, Y.T., Paneth, O., Sahai, A.: Protecting obfuscation against algebraic attacks. In: Nguyen, P.Q., Oswald, E. (eds.) EUROCRYPT 2014. LNCS, vol. 8441, pp. 221–238. Springer, Heidelberg (2014)
5. Barak, B., Goldreich, O., Impagliazzo, R., Rudich, S., Sahai, A., Vadhan, S., Yang, K.: On the (im)possibility of obfuscating programs. J. ACM 59(2), 6:1–6:48 (2012), http://doi.acm.org/10.1145/2160158.2160159
6. Barak, B., Goldreich, O., Impagliazzo, R., Rudich, S., Sahai, A., Vadhan, S.P., Yang, K.: On the (im)possibility of obfuscating programs. In: Kilian, J. (ed.) CRYPTO 2001. LNCS, vol. 2139, pp. 1–18. Springer, Heidelberg (2001)
7. Bellare, M., Boldyreva, A., O'Neill, A.: Deterministic and efficiently searchable encryption. In: Menezes, A. (ed.) CRYPTO 2007. LNCS, vol. 4622, pp. 535–552. Springer, Heidelberg (2007)
8. Bellare, M., Boldyreva, A., Palacio, A.: An uninstantiable random-oracle-model scheme for a hybrid-encryption problem. In: Cachin, C., Camenisch, J.L. (eds.) EUROCRYPT 2004. LNCS, vol. 3027, pp. 171–188. Springer, Heidelberg (2004)
9. Bellare, M., Brakerski, Z., Naor, M., Ristenpart, T., Segev, G., Shacham, H., Yilek, S.: Hedged public-key encryption: How to protect against bad randomness. In: Matsui, M. (ed.) ASIACRYPT 2009. LNCS, vol. 5912, pp. 232–249. Springer, Heidelberg (2009)
10. Bellare, M., Fischlin, M., O'Neill, A., Ristenpart, T.: Deterministic encryption: Definitional equivalences and constructions without random oracles. In: Wagner, D. (ed.) CRYPTO 2008. LNCS, vol. 5157, pp. 360–378. Springer, Heidelberg (2008)
11. Bellare, M., Hoang, V.T.: UCE+LTDFs: Efficient, subversion-resistant PKE in the standard model. Cryptology ePrint Archive, Report 2014/876 (2014), http://eprint.iacr.org/2014/876

12. Bellare, M., Hoang, V.T., Keelveedhi, S.: Instantiating random oracles via uCEs. In: Canetti, R., Garay, J.A. (eds.) CRYPTO 2013, Part II. LNCS, vol. 8043, pp. 398–415. Springer, Heidelberg (2013)

13. Bellare, M., Hoang, V.T., Keelveedhi, S.: Instantiating random oracles via UCEs. Cryptology ePrint Archive, Report 2013/424. (September 22, 2013) (Version after initial BFM attack), http://eprint.iacr.org/2013/424/20130924:163256)

14. Bellare, M., Hoang, V.T., Keelveedhi, S.: Personal communication (September 2013)

15. Bellare, M., Hoang, V.T., Keelveedhi, S.: Instantiating random oracles via UCEs. Cryptology ePrint Archive, Report 2013/424 (May 20, 2014), (Latest version at the time of writing), http://eprint.iacr.org/2013/424

16. Bellare, M., Keelveedhi, S.: Authenticated and misuse-resistant encryption of key-dependent data. In: Rogaway, P. (ed.) CRYPTO 2011. LNCS, vol. 6841, pp. 610–629. Springer, Heidelberg (2011)

17. Bellare, M., Keelveedhi, S., Ristenpart, T.: Message-locked encryption and secure deduplication. In: Johansson, T., Nguyen, P.Q. (eds.) EUROCRYPT 2013. LNCS, vol. 7881, pp. 296–312. Springer, Heidelberg (2013)

18. Bellare, M., Rogaway, P.: Random oracles are practical: A paradigm for designing efficient protocols. In: Ashby, V. (ed.) ACM CCS 93: 1st Conference on Computer and Communications Security, November 3–5, pp. 62–73. ACM Press, Fairfax (1993)

19. Bellare, M., Stepanovs, I., Tessaro, S.: Poly-many hardcore bits for any one-way function and a framework for differing-inputs obfuscation. In: Sarkar, P., Iwata, T. (eds.) ASIACRYPT 2014, Part II. LNCS, vol. 8874, pp. 102–121. Springer, Heidelberg (2014)

20. Bitansky, N., Canetti, R., Paneth, O., Rosen, A.: On the existence of extractable one-way functions. In: Shmoys, D.B. (ed.) 46th Annual ACM Symposium on Theory of Computing, May 31–June 3, pp. 505–514. ACM Press, New York (2014)

21. Black, J.: The ideal-cipher model, revisited: An uninstantiable blockcipher-based hash function. In: Robshaw, M. (ed.) FSE 2006. LNCS, vol. 4047, pp. 328–340. Springer, Heidelberg (2006)

22. Boldyreva, A., Fehr, S., O'Neill, A.: On notions of security for deterministic encryption, and efficient constructions without random oracles. In: Wagner, D. (ed.) CRYPTO 2008. LNCS, vol. 5157, pp. 335–359. Springer, Heidelberg (2008)

23. Boldyreva, A., Fischlin, M.: Analysis of random oracle instantiation scenarios for OAEP and other practical schemes. In: Shoup, V. (ed.) CRYPTO 2005. LNCS, vol. 3621, pp. 412–429. Springer, Heidelberg (2005)

24. Boneh, D., Franklin, M.: Identity-based encryption from the weil pairing. In: Kilian, J. (ed.) CRYPTO 2001. LNCS, vol. 2139, pp. 213–229. Springer, Heidelberg (2001)

25. Boyle, E., Chung, K.M., Pass, R.: On extractability obfuscation. In: Lindell, Y. (ed.) TCC 2014. LNCS, vol. 8349, pp. 52–73. Springer, Heidelberg (2014)

26. Brakerski, Z., Rothblum, G.N.: Virtual black-box obfuscation for all circuits via generic graded encoding. In: Lindell, Y. (ed.) TCC 2014. LNCS, vol. 8349, pp. 1–25. Springer, Heidelberg (2014)

27. Brzuska, C., Farshim, P., Mittelbach, A.: Indistinguishability obfuscation and uCEs: The case of computationally unpredictable sources. In: Garay, J.A., Gennaro, R. (eds.) CRYPTO 2014, Part I. LNCS, vol. 8616, pp. 188–205. Springer, Heidelberg (2014)

28. Brzuska, C., Farshim, P., Mittelbach, A.: Random oracle uninstantiability from indistinguishability obfuscation. Cryptology ePrint Archive, Report 2014/867 (2014), http://eprint.iacr.org/2014/867

29. Brzuska, C., Mittelbach, A.: Deterministic public-key encryption from indistinguishability obfuscation and point obfuscation (September 2014)

30. Brzuska, C., Mittelbach, A.: Indistinguishability obfuscation versus multi-bit point obfuscation with auxiliary input. In: Sarkar, P., Iwata, T. (eds.) ASIACRYPT 2014, Part II. LNCS, vol. 8874, pp. 142–161. Springer, Heidelberg (2014)

31. Brzuska, C., Mittelbach, A.: Using indistinguishability obfuscation via UCEs. In: Sarkar, P., Iwata, T. (eds.) ASIACRYPT 2014, Part II. LNCS, vol. 8874, pp. 122–141. Springer, Heidelberg (2014)

32. Canetti, R., Dakdouk, R.R.: Extractable perfectly one-way functions. In: Aceto, L., Damgård, I., Goldberg, L.A., Halldórsson, M.M., Ingólfsdóttir, A., Walukiewicz, I. (eds.) ICALP 2008, Part II. LNCS, vol. 5126, pp. 449–460. Springer, Heidelberg (2008)

33. Canetti, R., Goldreich, O., Halevi, S.: The random oracle methodology, revisited (preliminary version). In: 30th Annual ACM Symposium on Theory of Computing, May 23–26, pp. 209–218. ACM Press, Dallas (1998)

34. Canetti, R., Goldreich, O., Halevi, S.: On the random-oracle methodology as applied to length-restricted signature schemes. Cryptology ePrint Archive, Report 2003/150 (2003), http://eprint.iacr.org/2003/150

35. Canetti, R., Halevi, S., Katz, J.: A forward-secure public-key encryption scheme. In: Biham, E. (ed.) EUROCRYPT 2003. LNCS, vol. 2656, pp. 255–271. Springer, Heidelberg (2003)

36. Dent, A.W.: Adapting the weaknesses of the random oracle model to the generic group model. In: Zheng, Y. (ed.) ASIACRYPT 2002. LNCS, vol. 2501, pp. 100–109. Springer, Heidelberg (2002)

37. Douceur, J.R., Adya, A., Bolosky, W.J., Simon, D., Theimer, M.: Reclaiming space from duplicate files in a serverless distributed file system. In: International Conference on Distributed Computing Systems, pp. 617–624 (2002)

38. Fuller, B., O'Neill, A., Reyzin, L.: A unified approach to deterministic encryption: New constructions and a connection to computational entropy. In: Cramer, R. (ed.) TCC 2012. LNCS, vol. 7194, pp. 582–599. Springer, Heidelberg (2012)

39. Garg, S., Gentry, C., Halevi, S., Raykova, M., Sahai, A., Waters, B.: Candidate indistinguishability obfuscation and functional encryption for all circuits. In: 54th Annual Symposium on Foundations of Computer Science, October 26–29, pp. 40–49. IEEE Computer Society Press, Berkeley (2013)

40. Garg, S., Gentry, C., Halevi, S., Wichs, D.: On the implausibility of differing-inputs obfuscation and extractable witness encryption with auxiliary input. In: Garay, J.A., Gennaro, R. (eds.) CRYPTO 2014, Part I. LNCS, vol. 8616, pp. 518–535. Springer, Heidelberg (2014)

41. Gentry, C.: Certificate-based encryption and the certificate revocation problem. In: Biham, E. (ed.) EUROCRYPT 2003. LNCS, vol. 2656, pp. 272–293. Springer, Heidelberg (2003)

42. Gentry, C., Lewko, A., Sahai, A., Waters, B.: Indistinguishability obfuscation from the multilinear subgroup elimination assumption. Cryptology ePrint Archive, Report 2014/309 (2014), http://eprint.iacr.org/2014/309

43. Gentry, C., Silverberg, A.: Hierarchical ID-based cryptography. In: Zheng, Y. (ed.) ASIACRYPT 2002. LNCS, vol. 2501, pp. 548–566. Springer, Heidelberg (2002)

44. Goldwasser, S., Kalai, Y.T.: On the (in)security of the Fiat-Shamir paradigm. In: 44th Annual Symposium on Foundations of Computer Science, October 11-14, pp. 102–115. IEEE Computer Society Press, Cambridge (2003)

45. Green, M.D., Katz, J., Malozemoff, A.J., Zhou, H.S.: A unified approach to idealized model separations via indistinguishability obfuscation. Cryptology ePrint Archive, Report 2014/863 (2014), http://eprint.iacr.org/2014/863

46. Hohenberger, S., Sahai, A., Waters, B.: Replacing a random oracle: Full domain hash from indistinguishability obfuscation. In: Nguyen, P.Q., Oswald, E. (eds.) EUROCRYPT 2014. LNCS, vol. 8441, pp. 201–220. Springer, Heidelberg (2014)

47. Koppula, V., Lewko, A.B., Waters, B.: Indistinguishability obfuscation for turing machines with unbounded memory. Cryptology ePrint Archive, Report 2014/925 (2014), http://eprint.iacr.org/2014/925

48. Lynn, B., Prabhakaran, M., Sahai, A.: Positive results and techniques for obfuscation. In: Cachin, C., Camenisch, J.L. (eds.) EUROCRYPT 2004. LNCS, vol. 3027, pp. 20–39. Springer, Heidelberg (2004)

49. Matsuda, T., Hanaoka, G.: Chosen ciphertext security via point obfuscation. In: Lindell, Y. (ed.) TCC 2014. LNCS, vol. 8349, pp. 95–120. Springer, Heidelberg (2014)

50. Mittelbach, A.: Salvaging indifferentiability in a multi-stage setting. In: Nguyen, P.Q., Oswald, E. (eds.) EUROCRYPT 2014. LNCS, vol. 8441, pp. 603–621. Springer, Heidelberg (2014)

51. Nielsen, J.B.: Separating random oracle proofs from complexity theoretic proofs: The non-committing encryption case. In: Yung, M. (ed.) CRYPTO 2002. LNCS, vol. 2442, pp. 111–126. Springer, Heidelberg (2002)

52. Pass, R., Seth, K., Telang, S.: Indistinguishability obfuscation from semantically-secure multilinear encodings. In: Garay, J.A., Gennaro, R. (eds.) CRYPTO 2014, Part I. LNCS, vol. 8616, pp. 500–517. Springer, Heidelberg (2014)

53. Raghunathan, A., Segev, G., Vadhan, S.P.: Deterministic public-key encryption for adaptively chosen plaintext distributions. In: Johansson, T., Nguyen, P.Q. (eds.) EUROCRYPT 2013. LNCS, vol. 7881, pp. 93–110. Springer, Heidelberg (2013)

54. Ristenpart, T., Shacham, H., Shrimpton, T.: Careful with composition: Limitations of the indifferentiability framework. In: Paterson, K.G. (ed.) EUROCRYPT 2011. LNCS, vol. 6632, pp. 487–506. Springer, Heidelberg (2011)

55. Sahai, A., Waters, B.: How to use indistinguishability obfuscation: deniable encryption, and more. In: Shmoys, D.B. (ed.) 46th Annual ACM Symposium on Theory of Computing, May 31–June 3, pp. 475–484. ACM Press, New York (2014)

56. Sahai, A., Waters, B.R.: Fuzzy identity-based encryption. In: Cramer, R. (ed.) EUROCRYPT 2005. LNCS, vol. 3494, pp. 457–473. Springer, Heidelberg (2005)

57. Wichs, D.: Barriers in cryptography with weak, correlated and leaky sources. In: Kleinberg, R.D. (ed.) ITCS 2013: 4th Innovations in Theoretical Computer Science, January 9–12, pp. 111–126. Association for Computing Machinery, Berkeley (2013)

On Obfuscation with Random Oracles

Ran Canetti[1,2,*], Yael Tauman Kalai[3], and Omer Paneth[1,**]

[1] Boston University, USA
[2] Tel Aviv University, Israel
[3] Microsoft Research, USA

Abstract. Assuming trapdoor permutations, we show that there exist function families that cannot be VBB-obfuscated even if both the obfuscator and the obfuscated program have access to a random oracle. Specifically, these families are the robust unobfuscatable families of [Bitansky-Paneth, STOC 13].

Our result stands in contrast to the general VBB obfuscation algorithms in more structured idealized models where the oracle preserves certain algebraic homomorphisms [Canetti-Vaikuntanathan, ePrint 13; Brakerski-Rothblum, TCC 14; Barak et al., Eurocrypt 14].

1 Introduction

Program obfuscators, namely efficient compilers that transform an arbitrary program into one that has the same functionality but is otherwise "impenetrable", are an intriguing concept. The widely applicable interpretation of "impenetrable," called virtual black-box (VBB) [BGI+01], requires that the obfuscated version of a program helps learn any predicate of the program no more than does oracle access to the program's input-output functionality.

While a number of program families of interest are known to be VBB obfuscatable (under some strong hardness assumptions), e.g. [Can97, Wee05, BCKP14], no general-purpose VBB-obfuscators of all programs can exist. Indeed [BGI+01] show that, assuming one way functions, there exist *unobfuscatable functions*. These are functions that have a succinct description that cannot be effectively learned when having only oracle access to the function. At the same time, however, this succinct description can be extracted from *any* program that computes the function. Clearly, no program that computes such a function can possibly be VBB-obfuscated.

The construction of [BGI+01] makes crucial use of the fact that programs can be represented as strings and in particular can be executed with their own specification as input. In contrast, in some abstract models where programs do not necessarily have succinct representations as strings VBB obfuscation is in fact obtainable. One example is "hardware assisted" obfuscation, where some

---

* Supported by the Check Point Institute for Information Security, ISF grant 1523/14, the NSF MACS Frontier project, and NSF Algorithmic Foundations grant 1218461.
** Supported by the Simons award for graduate students in theoretical computer science and an NSF Algorithmic foundations grant 1218461.

Y. Dodis and J.B. Nielsen (Eds.): TCC 2015, Part II, LNCS 9015, pp. 456–467, 2015.

part of the computation is modeled as a black-box representing impenetrable secure hardware [GIS+10, BCG+11].

Another example is motivated by the recent candidate construction of obfuscation for all circuits of Garg et. al. [GGH+13b], that is based on an algebraic primitive called graded encodings [GGH13a]. The works of [BR14, BGK+14] prove that close variants of the proposed candidate are VBB secure in a model where the graded encodings are implemented by an ideal oracle. [CV13] study a different construction based on ideal pseudofree groups. Here, idealized models serve as an intermediate steps on the way to full-fledged obfuscation, namely as a model for developing potentially viable obfuscation algorithms and for understanding their security properties, as well as the computational assumptions on which their security might be based.

This raises natural questions: What are the simplest and minimally-structured abstract models that allow for general-purpose VBB obfuscation? For instance, do general-purpose VBB obfuscators exist in the random-oracle model? Do they exist in the generic group model [Sho97, BS84]? In fact, is there *any* non-trivial abstract model of computation where general-purpose VBB obfuscation is impossible?

Answers to the above question may shed light on what algebraic structure (if any) is inherent for secure obfuscation — even in the plain model, and even when attempting to obtain only weaker notions of obfuscation such as indistinguishability obfuscation.

We note that Barak et al. show that their impossiblity holds even when all entities, namely the program to be obfuscated, the obfuscator and the obfuscated program have access to a random oracle. [1] Goldwasser and Rothblum [GR14] extend this to show that even the considerably weaker notion of *indistinguishability obfuscation* is unobtainable in general in this setting. However, these results do not answer the above questions. Specifically, they leave open the possibility of obfuscating fully specified programs that do not access the random oracle. Indeed, Lynn et al. ask whether general purpose obfuscation is possible in that setting [LPS04].

1.1 This Work

We consider obfuscation in the setting of Lynn et al. [LPS04], where both the obfuscator and the obfuscated program have access to a random oracle, and where the obfuscator is only required to operate on fully specified programs that do not have access to the random oracle. Furthermore, we give the adversary access to the same oracle. Here we show:

Theorem 1.1 (Main Theorem, informal). *Assume trapdoor permutations exist. Then there exist function families that cannot be VBB obfuscated, even in a model where the obfuscator and the obfuscated function have access to a random oracle.*

---

[1] In fact, [BGI+01] prove that their negative result holds in the more general settings of *bounded relativization*.

Our impossibility extends to the case where the obfuscator and obfuscated program have access to an *invertible random permutation* rather than a random function. That is, the oracle represents a random permutation, and can be asked both to evaluate the function and to invert it. It also extends to the case of approximate obfuscation, where the obfuscated program is only required to agree with the original program on a significant fraction of the inputs.

1.2   Techniques

The starting point of our proof is the existence of *robust unobfuscatable functions (RUFs)* which are a strengthening of the *unobfuscatable functions* of [BGI$^+$01]. Essentially, RUFs have a succinct description that cannot be effectively learned having only oracle access to the function. At the same time, this description can be extracted from any program that *approximates* the function, namely agrees with the function on some large fraction of the inputs, say 90%. Bitansky and Paneth [BP13] construct RUFs from any trapdoor permutation.

Our proof now proceeds by transforming any obfuscator in the RO model into an obfuscator in the plain model, namely one where the RO is not used. The transformation loses in correctness: the resulting plain-model obfuscator generates a program that computes the function correctly only on some fraction of the inputs. Still, impossibility is demonstrated by applying the transformation to an obfuscator for a family of RUFs.

We describe in more detail the transformation from obfuscation in the RO model to obfuscation in the plain model. Let $\mathcal{O}^R$ be an obfuscator in the RO model. Our goal is to transform $\mathcal{O}^R$ into an obfuscator $\mathcal{O}$ in the plain model. We start by describing a simple warm-up. Let $\mathcal{O}$ be the following plain-model obfuscator: given a description of a program C, the obfuscator $\mathcal{O}$ emulates an execution of the RO obfuscator $\mathcal{O}^R(C)$, answering every oracle query of $\mathcal{O}^R$ randomly and independently (repeated queries are answered consistently), and obtains a RO obfuscation $\tilde{C}^R$ of C. Let $\mathcal{R}_C$ be the set of RO query-answer pairs that occurred during the emulation of $\mathcal{O}^R(C)$. The obfuscator $\mathcal{O}$ then outputs an obfuscated program $\tilde{C}$ that has hard-coded to it the description of the RO obfuscation $\tilde{C}^R$ and the set $\mathcal{R}_C$. Given an input x, the obfuscation $\tilde{C}$ emulates the RO obfuscation $\tilde{C}^R(x)$. $\tilde{C}$ answers any RO query made by $\tilde{C}^R$ as follows: if the query appears in the set $\mathcal{R}_C$ it is answered consistently with $\mathcal{R}_C$, otherwise, a random answer is given.[2]

The correctness of $\mathcal{O}$ follows directly from the correctness of $\mathcal{O}^R$ in the RO model since, when $\tilde{C}$ emulates the program $\tilde{C}^R$, all the RO queries made by $\tilde{C}^R$ are answered randomly and consistently with the answers given to the obfuscator $\mathcal{O}^R(C)$ that generated $\tilde{C}^R$. However, even if $\mathcal{O}^R$ is a VBB obfuscator in the RO model, the obfuscator $\mathcal{O}$ may be completely insecure, since the obfuscation $\tilde{C}$ includes the set $\mathcal{R}_C$ in the clear. This may reveal information about the program C.

---

[2] This results in a *randomized* obfuscated program. In the full construction we make the obfuscated program deterministic by including in the description of the obfuscated program a list of random oracle answers that are reused in every evaluation.

In our actual transformation, the obfuscation $\tilde{C}$ will include a different set of RO query-answer pairs $\mathcal{R}_X$ that on the one hand, will give no information about the program C, but on the other hand, will result in a obfuscation that is only approximately correct.

The actual plain-model obfuscator $\mathcal{O}$ starts by emulating the random oracle obfuscator $\mathcal{O}^R(C)$ and obtains the RO obfuscation $\tilde{C}^R$ and the set $\mathcal{R}_C$ as before. Next, $\mathcal{O}$ "tests" the RO obfuscation $\tilde{C}^R$ to learn which oracle queries are often made by $\tilde{C}^R$ when executed on a random input. Specifically, $\mathcal{O}$ samples random inputs $x_1, \ldots, x_\ell$ used to test the program $\tilde{C}^R$. The set $\mathcal{R}_X$ is initially empty. For every $i \in [\ell]$, $\mathcal{O}$ emulates the RO obfuscation $\tilde{C}^R(x_i)$ and answers any RO query made by $\tilde{C}^R$ as follows: if the query appears in the set $\mathcal{R}_C$ or in the set $\mathcal{R}_X$ it is answered consistently, otherwise, a random answer is given. In both cases, the query-answer pair is added to the set $\mathcal{R}_X$. Note that the final set $\mathcal{R}_X$ may not contain all the queries in $\mathcal{R}_C$ and it may also contain queries outside $\mathcal{R}_C$.

Finally, the obfuscator $\mathcal{O}$ outputs an obfuscated program $\tilde{C}$ that has hard-coded to it the description of the RO obfuscation $\tilde{C}^R$ and the set $\mathcal{R}_X$. As before, the obfuscation $\tilde{C}$ on an input x emulates the RO obfuscation $\tilde{C}^R(x)$ and answers any RO query made by $\tilde{C}^R$ as follows: if the query appears in the set $\mathcal{R}_X$ it is answered consistently with $\mathcal{R}_X$, otherwise, a random answer is given.

We argue that the new set $\mathcal{R}_X$ gives no information about the program C: Consider the following alternative way to sample the set $\mathcal{R}_X$. Let $\mathcal{R}$ be a random function that is consistent with the query-answer pairs in $\mathcal{R}_C$. Now execute the RO obfuscation $\tilde{C}^R$ on random inputs $x_1, \ldots, x_\ell$ and given oracle access to $\mathcal{R}$. The set $\mathcal{R}_X$ simply contains all the query-answer pairs that occur in these executions. Intuitively, since $\mathcal{R}_X$ can be sampled given the RO obfuscation $\tilde{C}^R$ and oracle access to $\mathcal{R}$, it follows from the VBB security of the RO obfuscator $\mathcal{O}^R$ that $\mathcal{R}_X$ reveals no information about the program C.

To argue that $\tilde{C}$ is approximately correct, consider an evaluation of $\tilde{C}$ on a random input x. $\tilde{C}$ emulates the RO obfuscation $\tilde{C}^R(x)$ and answers any RO query made by $\tilde{C}^R$ randomly and consistently with the set $\mathcal{R}_X$. As discussed in the warm-up, if all of the queries made by $\tilde{C}^R$ were answered consistently with the set $\mathcal{R}_C$, perfect correctness would have followed from the correctness of $\mathcal{O}^R$ in the RO model. However, the emulation of $\tilde{C}^R(x)$ may make a query that is in the set $\mathcal{R}_C$ but not in the set $\mathcal{R}_X$. Such a query will be answered randomly in a way that may not be consistent with the answer in $\mathcal{R}_C$ and correctness may be lost. We can therefore bound the probability that $\tilde{C}(x)$ disagrees with $C(x)$ by the probability that $\tilde{C}^R(x)$ makes a query $q \in \mathcal{R}_C \setminus \mathcal{R}_X$. Such a query q must not have been asked by any of the test executions of $\tilde{C}^R$ on the random inputs $x_1, \ldots, x_\ell$, otherwise it would have been added to the set $\mathcal{R}_X$. The probability that a query in $\mathcal{R}_C$ is asked by $\tilde{C}(x)$ but is not asked by $\tilde{C}(x_i)$ for any $i \in [\ell]$ is inversely proportional to ℓ. Therefore, by making ℓ large enough, we can make the correctness error sufficiently small (recall that any constant correctness error that is bounded away from 1 is sufficient for the negative result of [BP13] to hold).

Connection to [IR89]. Our proof follows the same outline as the proof of Impagli-azzo and Rudich [IR89] separating key-agreement protocols from one-way func-tions (as well as many subsequent works). In essence, Impagliazzo and Rudich rule out existence of key-agreement protocols secure gainst unbounded adver-saries in the RO model. They do so in two steps: first they transform any key-agreement protocol in the RO model into a key-agreement protocol in the plain model. Next they rely on the impossibility for information-theoretically secure key-agreement. We follow the same two steps: first we transform any general (possibly approximate) obfuscator in the RO model to a general approximate obfuscator in the plain model. Next we rely on the impossibility of the latter. Note that in our case the impossibility in the plain model is stronger in the sense that it rules out existence of a primitive that provides only computational security.

2   Impossibility of Obfuscation in the RO Model

In this section we prove an impossibility result for general purpose obfuscation in the RO model. We start by defining approximate obfuscation and state the known impossibility result for obfuscation with approximate correctness.

2.1   Approximate Obfuscation

We define approximate obfuscation, both in the RO model and in the plain model.

Let $\mathcal{F} = \{F_k\}_{k \in \{0,1\}^*}$ be a family of functions such that F_k has a domain $D_{|k|}$.

Definition 2.1 (Approximate Obfuscation). *For a function $\epsilon : \mathbf{N} \to [0,1]$, a PPT algorithm $\mathcal{O}$ is a secure ϵ-approximate obfuscator for $\mathcal{F}$ if it satisfies the following requirements:*

- *Approximate Functionality: for all $n \in \mathbb{N}, k \in \{0,1\}^n$:*

$$\Pr_{x \leftarrow D_n} [\mathcal{O}(k)(x) \neq F_k(x)] \leq \epsilon(n) \ ,$$

  where the probability is also over the coins of the obfuscator $\mathcal{O}$.
- *Virtual Black-Box: for every poly-size adversary $\mathcal{A}$ there exists a poly-size simulator $\mathcal{S}$ and a negligible function μ such that for every $k \in \{0,1\}^*$:*

$$\left| \Pr[\mathcal{A}(\mathcal{O}(k)) = 1] - \Pr[\mathcal{S}^{F_k}(1^{|k|}) = 1] \right| \leq \mu(|k|) \ ,$$

  where the probabilities are over the coins of the obfuscator $\mathcal{O}$, the adversary $\mathcal{A}$ and the simulator $\mathcal{S}$.

Definition 2.2 (Approximate Obfuscation in the RO model). *For a function $\epsilon : \mathbf{N} \to [0,1]$, a PPT algorithm $\mathcal{O}$ is a secure ϵ-approximate obfus-cator for $\mathcal{F}$ in the RO model if it satisfies the following requirements:*

- *Approximate Functionality:* for all $n \in \mathbb{N}, k \in \{0,1\}^n$:

$$\Pr_{x \leftarrow D_n} [\mathcal{O}^{\mathcal{R}}(k)(x) \neq F_k(x)] \leq \epsilon(n) \ ,$$

  where $\mathcal{R} : \{0,1\}^* \to \{0,1\}^*$ is a random function, and the probability is also over $\mathcal{R}$ and the coins of the obfuscator $\mathcal{O}$.
- *Virtual Black-Box:* for every poly-size adversary $\mathcal{A}$ there exist a poly-size simulator $\mathcal{S}$ and a negligible function μ such that for every $k \in \{0,1\}^*$:

$$\left| \Pr[\mathcal{A}^{\mathcal{R}}(\mathcal{O}^{\mathcal{R}}(k)) = 1] - \Pr[\mathcal{S}^{F_k}(1^{|k|}) = 1] \right| \leq \mu(|k|) \ ,$$

  where the probabilities are over $\mathcal{R}$, the coins of the obfuscator $\mathcal{O}$, the adversary $\mathcal{A}$, and the simulator $\mathcal{S}$.

Next we formally state the known impossibility results for approximate obfuscation in the plain model. The following is a direct corollary of [BP13, Theorem 3.1, Theorem 4.1, Lemma 4.1].

Corollary 2.1 ([BP13]). *Assuming trapdoor permutations, there exists a family of functions $\mathcal{F}$ such that an $\left(\frac{1}{2} - \epsilon\right)$-approximate obfuscator for $\mathcal{F}$ does not exist for every noticeable function ϵ.*

Remark 2.1 (More on the impossibility of approximate obfuscation). The work of [BP13] constructs a family of *error-robust* unobfuscatable functions. These are families $\{F_k\}_{k \in \{0,1\}^*}$ such that given oracle access to F_k for a random key k, the key remains completely hidden. However, given the code of any function that agrees with F_k on $\frac{1}{2} + \epsilon$ of the inputs, it is possible to fully recover the key k. This implies the following strong impossibility for approximate obfuscation: For any $\left(\frac{1}{2} - \epsilon\right)$-approximate obfuscator for $\{F_k\}$, with probability at least $\frac{\epsilon}{2}$ over the coins the the obfuscation, the obfuscated function agrees with the original function with probability at least $\frac{1+\epsilon}{2}$. Therefore, with noticeable probability over the coins the the obfuscation, it is always possible to reconstruct the entire key from the obfuscated program.

2.2 The Impossibility

We start by describing a transformation from any (possibly approximate) obfuscation in the RO model to an approximate obfuscation in the plain model. The approximation error of the resulting obfuscation will be slightly larger then that of the original obfuscation.

Theorem 2.1. *If a family of functions $\mathcal{F}$ has a secure ϵ-approximate obfuscator in the RO model then it has a secure $(\epsilon + \delta)$-approximate obfuscator in the plain model for every noticeable function δ.*

Then, we combine the transformation in Theorem 2.1 with the known impossibility result for approximate obfuscation (Corollary 2.1) to derive the following impossibility for obfuscation in the RO model:

Corollary 2.2. *Assuming trapdoor permutations, there exists a family of functions $\mathcal{F}$ such that an $\left(\frac{1}{2} - \epsilon\right)$-approximate obfuscator for $\mathcal{F}$ in the RO model does not exist for every noticeable function ϵ.*

Next we prove Theorem 2.1. See Section 1.2 for a high-level overview of the proof.

Proof. Let $\mathcal{O}$ be a secure ϵ-approximate obfuscator for $\mathcal{F}$ in the RO model, making at most $\ell = \ell(|k|)$ queries to the oracle. We construct a secure $(\epsilon + \delta)$-approximate obfuscator $\mathcal{O}'$ for $\mathcal{F}$ in the plain model.

The obfuscator $\mathcal{O}'$:

1. On input k, emulate $\mathcal{O}(k)$ as follows. Run $\mathcal{O}$ on input k, answer every oracle query made by $\mathcal{O}(k)$ randomly (assume w.l.o.g that $\mathcal{O}$ never makes the same query twice), and obtain an obfuscated oracle circuit C. Set $\mathcal{R}_k$ to be all the queries made by $\mathcal{O}(k)$ and their answers.
2. Set $\mathcal{R}_C$ to be the empty set.
3. For $i = 1$ to $\left\lceil \frac{|C| \cdot \ell}{\delta} \right\rceil$:
   (a) Sample $x_i \leftarrow D_{|k|}$.
   (b) Execute $C(x_i)$. For every oracle query made by $C(x_i)$, if it is in $\mathcal{R}_C \cup \mathcal{R}_k$ then answer consistently, otherwise answer randomly (assume w.l.o.g that C never makes the same query twice). Add all new pairs of queries made by $C(x_i)$ and their answers to $\mathcal{R}_C$.
4. Sample $|C|$ random oracle answers $r_1, \ldots, r_{|C|}$.
5. Output the description of a circuit C' as follows:
   (a) The circuit C' has the description of C, the set $\mathcal{R}_C$ and the answers $\{r_i\}$ hardcoded into it.
   (b) On input x, C' emulates $C(x)$. Let q_i be the i-th oracle query made by $C(x)$. If q_i is in $\mathcal{R}_C$, C' answers consistently, otherwise, C' answers with r_i.
   (c) C' outputs the same as $C(x)$.

Next we show that $\mathcal{O}'$ is a secure $(\epsilon + \delta)$-approximate obfuscator. That is, $\mathcal{O}'$ satisfies the approximate functionality and the virtual black-box requirements.

Approximate functionality. Fix a key $k \in \{0,1\}^n$, let $\epsilon = \epsilon(n), \delta = \delta(n)$, and let x be a random input sampled from D_n. By the approximate functionality of $\mathcal{O}$, the circuit C produced by $\mathcal{O}(k)$ satisfies:

$$\Pr_x[C^{\mathcal{R}}(x) \neq F_k(x)] \leq \epsilon . \tag{1}$$

Let C' be the obfuscated circuit generated by the plain-model obfuscator $\mathcal{O}'(k)$. Recall that $C'(x)$ emulates the execution of $C(x)$ and the answers the oracle queries made by C. Queries that are in $\mathcal{R}_C$ are answered consistently with $\mathcal{R}$, and queries outside $\mathcal{R}_C$ are answered from the set of random answers $\{r_i\}$. Since every distinct query made by $C(x)$ is answered randomly and independently, we can consider an identical experiment where C' answers all of C's queries using a random oracle $\mathcal{R}'$ which agrees with $\mathcal{R}$ on all the queries in $\mathcal{R}_C$. Additionally, all the answers of $\mathcal{R}$ and $\mathcal{R}'$ outside the set $\mathcal{R}_k \cup \mathcal{R}_C$ are random independent of C. Let $G(x)$ be the event that the execution of $C^{\mathcal{R}'}(x)$ does not query $\mathcal{R}'$ on any query in the set $\mathcal{R}_k \setminus \mathcal{R}_C$. We have that conditioned on $G(x)$, the output of $C^{\mathcal{R}'}(x)$ and of $C^{\mathcal{R}}(x)$ are identically distributed, and specifically:

$$\Pr_x[(C^{\mathcal{R}'}(x) \neq F_k(x)) \wedge G(x)] = \Pr_x[(C^{\mathcal{R}}(x) \neq F_k(x)) \wedge G(x)] \leq \epsilon \ .$$

Therefore, we can bound the probability of error on x by bounding the probability of the event $\neg G(x)$ as follows:

$$\Pr_x[C^{\mathcal{R}'}(x) \neq F_k(x)] \leq \Pr_x[(C^{\mathcal{R}'}(x) \neq F_k(x)) \wedge G(x)] + \Pr_x[\neg G(x)] \leq \epsilon + \Pr_x[\neg G(x)] \ .$$

Thus, to prove approximate functionality it suffices to prove the following claim, bounding the probability of the event $\neg G(x)$.

Claim 2.2

$$\Pr_x[\neg G(x)] \leq \delta.$$

Proof (Proof of Claim 2.2.) We start by giving a high-level overview of the proof. For a random input x, the execution of $C(x)$ makes at most $|C|$ oracle queries. To bound the probability of the event $\neg G(x)$ we bound the probability that the i'th query of $C^{\mathcal{R}'}(x)$ is the first query to fall in the set $\mathcal{R}_k \setminus \mathcal{R}_C$, for every $i \in |C|$. To this end, we argue that the for every query $q \in \mathcal{R}_k$, the probability that the i'th query of $C^{\mathcal{R}'}(x)$ is indeed q, but q was never queried during the "testing phase" of $\mathcal{O}'$ (Step 3) is small. (if q is queried queried during testing phase then $q \in \mathcal{R}_C$.)

Recall that in the testing phase of $\mathcal{O}'$ we execute $C^{\mathcal{R}}$ on many random inputs. Since we are only bounding the probability that the i'th query of $C^{\mathcal{R}'}(x)$ is the *first* query to fall outside the set $\mathcal{R}_k \setminus \mathcal{R}_C$, we can condition on the event that all previous queries do not fall in the set $\mathcal{R}_k \setminus \mathcal{R}_C$. Conditioned on this event, by the definition of the oracles $\mathcal{R}$ and $\mathcal{R}'$, the i-th query of $C^{\mathcal{R}'}$ and of $C^{\mathcal{R}}$ are identically distributed. Therefore, the probability that the i'th query of $C^{\mathcal{R}'}(x)$ is q, but q was never queried in any of the test executions is bounded by the inverse of the number of test executions. Since the number of different queries $q \in \mathcal{R}_k$ is bounded by ℓ we get the required bound on probability that the i'th query of $C^{\mathcal{R}'}(x)$ falls in the set $\mathcal{R}_k \setminus \mathcal{R}_C$, and therefore also on the probability of the event $\neg G(x)$.

We continue with the formal proof of the claim. Let:

$$I = \left\lceil \frac{|C| \cdot \ell}{\delta} \right\rceil \ ,$$

be the number if iterations of the loop in Step 3 of $\mathcal{O}'$. Let q_j be the j-th query $C(x)$ makes. Let $q_{i,j}$ be the j-th query made by the emulation of C on the random input x_i in the i-th iteration of the loop in Step 3. For every $j \in [\ell]$, let $G_j(x)$ the event that $q_j \notin R_k \setminus \{q_{i,j}\}_{i \in [I]}$. Note that

$$G_j(x) \Rightarrow q_j \notin R_k \setminus R_C \ ,$$

and therefore,

$$\forall_j G_j(x) \Rightarrow G(x) \ .$$

Thus we can bound the probability of the event $\neg G(x)$ as follows:

$$\Pr_x[\neg G(x)] \leq \Pr_x[\neg G_1(x) \vee \cdots \vee \neg G_{|C|}(x)]$$
$$= \sum_{j \in |C|} \Pr_x [G_1(x) \wedge \cdots \wedge G_{j-1}(x) \wedge \neg G_j(x)] \ .$$

It is therefore sufficient to show that for every $j \in [|C|]$,

$$\Pr_x[G_1(x) \wedge \cdots \wedge G_{j-1}(x) \wedge \neg G_j(x)] \leq \frac{\delta}{|C|} \ .$$

To this end, fix $j \in [|C|]$ and fix the oracles R and R'. Let $\tilde{G}_{j-1}(x)$ denote the event:

$$G_1(x) \wedge \cdots \wedge G_{j-1}(x) \ .$$

Note that:

$$\Pr_x[\tilde{G}_{j-1}(x) \wedge \neg G_j(x)] \leq \Pr_x[\neg G_j(x)|\tilde{G}_{j-1}(x)]$$

and therefore, it suffices to prove that:

$$\Pr_x[\neg G_j(x)|\tilde{G}_{j-1}(x)] \leq \frac{\delta}{|C|} \ .$$

For every query q denote by

$$p_q \triangleq \Pr_x[q_j = q|\tilde{G}_{j-1}(x)] \ .$$

Since for every $i \in [I]$, x and x_i are both uniform in D_n and since the oracles R and R' only differ on queries in the set $R_k \cap R_C$ we have that conditioned on $\tilde{G}_{j-1}(x)$ the view of the two executions:

$$C^{R'}(x) \quad \text{and} \quad C^R(x_i)$$

up until the j-th query, are identically distributed. Therefore, for every $i \in [I]$:

$$p_q = \Pr_x[q_j = q|\tilde{G}_{j-1}(x)] = \Pr_x[q_{i,j} = q|\tilde{G}_{j-1}(x)] \ .$$

Thus, as desired,

$$\Pr_x[\neg G_j(x)|\tilde{G}_{j-1}(x)] \leq$$

$$\sum_{q \in \mathcal{R}_k} \Pr_x[(q_j = q) \wedge (\forall i, q_{i,j} \neq q) \,|\tilde{G}_{j-1}(x)] \leq$$

$$\sum_{q \in \mathcal{R}_k} p_q(1-p_q)^{\frac{|C| \cdot \ell}{\delta}} \leq \tag{2}$$

$$\sum_{q \in \mathcal{R}_k} \frac{\delta}{|C| \cdot \ell} \leq \frac{\delta}{|C|} \,,$$

where (2) follows from the fact that the expression $p_q(1-p_q)^e$ is maximized by $p_q = \frac{1}{e+1}$. This completes the proof of Claim 2.2.

Virtual Black-Box. Fix a key $k \in \{0,1\}^n$ and let $\mathcal{A}'$ be an adversary that tries to learn some information from the obfuscation $\mathcal{O}'(k)$. We show how to use the code of $\mathcal{A}'$ to construct an adversary $\mathcal{A}$ that learns the same information from the obfuscation $\mathcal{O}(k)$ where both $\mathcal{A}$ and $\mathcal{O}$ have access to the same random oracle. That is, we will show that:

$$\Pr[\mathcal{A}^{\mathcal{R}}(\mathcal{O}^{\mathcal{R}}(k)) = 1] = \Pr[\mathcal{A}'(\mathcal{O}'(k)) = 1] \,, \tag{3}$$

where the probabilities are over $\mathcal{R}$, the coins of the obfuscators $\mathcal{O}$ and $\mathcal{O}'$, and the coins of the adversaries $\mathcal{A}$ and $\mathcal{A}'$. By the security of $\mathcal{O}$, there exist a simulator $\mathcal{S}$ and a negligible function μ such that:

$$\left|\Pr[\mathcal{A}^{\mathcal{R}}(\mathcal{O}^{\mathcal{R}}(k)) = 1] - \Pr[\mathcal{S}^{F_k}(1^n) = 1]\right| \leq \mu(n) \,. \tag{4}$$

It follows from Equations (3) and (4) that $\mathcal{S}$ is a good simulator for $\mathcal{A}'$ as well. It is left to show how to construct an adversary $\mathcal{A}$ that satisfies Equation (3). Loosely speaking, given an obfuscation $\mathcal{O}(k)$, $\mathcal{A}$ will use the same strategy of the obfuscator $\mathcal{O}'$ to transform the obfuscation $\mathcal{O}(k)$ into an obfuscation $\mathcal{O}'(k)$ and then execute $\mathcal{A}'$ on $\mathcal{O}'(k)$. $\mathcal{A}$ will use its random oracle to answer queries made by $\mathcal{O}(k)$. Formally, $\mathcal{A}$ is defined as follows:

1. Given an obfuscated input circuit C and given access to oracle $\mathcal{R}$, repeat the following for $i = 1$ to $\left\lceil \frac{|C| \cdot \ell}{\delta} \right\rceil$:
   (a) Sample $x_i \leftarrow D_n$.
   (b) Execute $C(x_i)$ and forward its oracle queries to $\mathcal{R}$.
2. Sample $|C|$ random oracle answers $r_1, \ldots, r_{|C|}$.
3. Set $\mathcal{R}_C$ to be the set of queries made by C in Step 1 and their answers. Construct a circuit C' from C, $\mathcal{R}_C$ and $\{r_i\}$ as in Step 5 of the obfuscator $\mathcal{O}'$.
4. Output the same as $\mathcal{A}'(C')$.

By construction, the circuit C' used by $\mathcal{A}$ in Step 4 is distributed identically to the output of $\mathcal{O}'(k)$ and therefore Equation (3) holds.

References

[BCG+11]   Bitansky, N., Canetti, R., Goldwasser, S., Halevi, S., Kalai, Y.T., Rothblum, G.N.: Program obfuscation with leaky hardware. In: Lee, D.H., Wang, X. (eds.) ASIACRYPT 2011. LNCS, vol. 7073, pp. 722–739. Springer, Heidelberg (2011)

[BCKP14]   Bitansky, N., Canetti, R., Kalai, Y.T., Paneth, O.: On virtual grey box obfuscation for general circuits. In: Garay, J.A., Gennaro, R. (eds.) CRYPTO 2014, Part II. LNCS, vol. 8617, pp. 108–125. Springer, Heidelberg (2014)

[BGI+01]   Barak, B., Goldreich, O., Impagliazzo, R., Rudich, S., Sahai, A., Vadhan, S.P., Yang, K.: On the (im)possibility of obfuscating programs. In: Kilian, J. (ed.) CRYPTO 2001. LNCS, vol. 2139, pp. 1–18. Springer, Heidelberg (2001)

[BGK+14]   Barak, B., Garg, S., Kalai, Y.T., Paneth, O., Sahai, A.: Protecting obfuscation against algebraic attacks. In: Nguyen, P.Q., Oswald, E. (eds.) EUROCRYPT 2014. LNCS, vol. 8441, pp. 221–238. Springer, Heidelberg (2014)

[BP13]   Bitansky, N., Paneth, O.: On the impossibility of approximate obfuscation and applications to resettable cryptography. In: STOC, pp. 241–250 (2013)

[BR14]   Brakerski, Z., Rothblum, G.N.: Virtual black-box obfuscation for all circuits via generic graded encoding. In: Lindell, Y. (ed.) TCC 2014. LNCS, vol. 8349, pp. 1–25. Springer, Heidelberg (2014)

[BS84]   Babai, L., Szemerédi, E.: On the complexity of matrix group problems I. In: 25th Annual Symposium on Foundations of Computer Science, West Palm Beach, Florida, USA, October 24-26, pp. 229–240 (1984)

[Can97]   Canetti, R.: Towards realizing random oracles: Hash functions that hide all partial information. In: Kaliski Jr., B.S. (ed.) CRYPTO 1997. LNCS, vol. 1294, pp. 455–469. Springer, Heidelberg (1997)

[CV13]   Canetti, R., Vaikuntanathan, V.: Obfuscating branching programs using black-box pseudo-free groups. IACR Cryptology ePrint Archive, 2013:500 (2013)

[GGH13a]   Garg, S., Gentry, C., Halevi, S.: Candidate multilinear maps from ideal lattices. In: Johansson, T., Nguyen, P.Q. (eds.) EUROCRYPT 2013. LNCS, vol. 7881, pp. 1–17. Springer, Heidelberg (2013)

[GGH+13b] Garg, S., Gentry, C., Halevi, S., Raykova, M., Sahai, A., Waters, B.: Candidate indistinguishability obfuscation and functional encryption for all circuits. In: FOCS (2013)

[GIS+10]   Goyal, V., Ishai, Y., Sahai, A., Venkatesan, R., Wadia, A.: Founding cryptography on tamper-proof hardware tokens. In: Micciancio, D. (ed.) TCC 2010. LNCS, vol. 5978, pp. 308–326. Springer, Heidelberg (2010)

[GR14]   Goldwasser, S., Rothblum, G.N.: On best-possible obfuscation. J. Cryptology 27(3), 480–505 (2014)

[IR89]   Impagliazzo, R., Rudich, S.: Limits on the provable consequences of one-way permutations. In: Proceedings of the 21st Annual ACM Symposium on Theory of Computing, Seattle, Washigton, USA, May 14-17, pp. 44–61 (1989)

[LPS04]   Lynn, B., Prabhakaran, M., Sahai, A.: Positive results and techniques for obfuscation. In: Cachin, C., Camenisch, J.L. (eds.) EUROCRYPT 2004. LNCS, vol. 3027, pp. 20–39. Springer, Heidelberg (2004)

[Sho97]   Shoup, V.: Lower bounds for discrete logarithms and related problems. In: Fumy, W. (ed.) EUROCRYPT 1997. LNCS, vol. 1233, pp. 256–266. Springer, Heidelberg (1997)

[Wee05]   Wee, H.: On obfuscating point functions. IACR Cryptology ePrint Archive, 2005:1 (2005)

Obfuscation of Probabilistic Circuits and Applications

Ran Canetti[1], Huijia Lin[2], Stefano Tessaro[2], and Vinod Vaikuntanathan[3]

[1] Boston University, USA and Tel Aviv University, Israel
[2] UC Santa Barbara, USA
[3] MIT CSAIL, USA

Abstract. This paper studies the question of how to define, construct, and use obfuscators for *probabilistic programs*. Such obfuscators compile a possibly randomized program into a *deterministic* one, which achieves computationally indistinguishable behavior from the original program as long as it is run on each input at most once. For obfuscation, we propose a notion that extends *indistinguishability obfuscation* to probabilistic circuits: It should be hard to distinguish between the obfuscations of any two circuits whose output distributions at each input are computationally indistinguishable, possibly in presence of some auxiliary input. We call the resulting notion *probabilistic indistinguishability obfuscation (pIO)*.

We define several variants of pIO, and study relations among them. Moreover, we give a construction of one of our variants, called *X*-pIO, from sub-exponentially hard indistinguishability obfuscation (for deterministic circuits) and one-way functions.

We then move on to show a number of applications of pIO. In particular, we first give a general and natural methodology to achieve fully homomorphic encryption (FHE) from variants of pIO and of semantically secure encryption schemes. In particular, one instantiation leads to FHE from any *X*-pIO obfuscator and any re-randomizable encryption scheme that's slightly super-polynomially secure.

We note that this constitutes the first construction of full-fledged FHE that does not rely on encryption with circular security.

Moreover, assuming sub-exponentially secure puncturable PRFs computable in $\mathbf{NC}^1$, sub-exponentially-secure indistinguishability obfuscation for (deterministic) $\mathbf{NC}^1$ circuits can be bootstrapped to obtain indistinguishability obfuscation for arbitrary (deterministic) poly-size circuits (previously such bootstrapping was known only assuming FHE with $\mathbf{NC}^1$ decryption algorithm).

1 Introduction

Program obfuscation, namely the algorithmic task of turning input programs into "unintelligible" ones while preserving their functionality, has been a focal point for cryptography for over a decade. However, while the concept is intuitively attractive and useful, the actual applicability of obfuscation has been limited. Indeed, the main notion to be considered has been virtual black box (VBB) [7]

Y. Dodis and J.B. Nielsen (Eds.): TCC 2015, Part II, LNCS 9015, pp. 468–497, 2015.

which, while natural and intuitively appealing, is very strong, hard to satisfy, and also not easy to use. In fact, for many program classes of interest, VBB obfuscation is unattainable [7,26,10].

All this changed with the recent breakthrough results of [21,37]. Their contribution is twofold: First they demonstrate a candidate general obfuscation algorithm for all circuits, thus reviving the hope in the possibility of making good of the initial intuitive appeal of program obfuscation as an important and useful cryptographic primitive. Second, they demonstrate how to make use of a considerably weaker notion of secure obfuscation than VBB, namely indistinguishability obfuscation (IO), initially defined in [7]. Indeed, following [21,37] there has been a gush of works demonstrating how to apply IO to a plethora of situations and applications, and even resolving long standing open problems.

Obfuscating Probabilistic Programs. Still, exiting notions of obfuscation, VBB and IO included, predominantly address the task of obfuscating *deterministic* programs. That is, the program to be obfuscated is a sequence of deterministic operations. This leaves open the question of obfuscating *probabilistic programs,* namely programs that make random choices as part of their specification, and whose output, on each input, is a random variable that depends on the internal random choices.

A priori it may not be clear what one wants to obtain when obfuscating such programs, or why is the problem different than that of obfuscating deterministic programs. Indeed, why not just obfuscate the deterministic program that takes both "main" and "random" input, and leave it to the evaluator to choose some of the input at random, if she so desires?

The main drawback of this approach is that it does not allow the obfuscation mechanism to hide the random choices of the program from the evaluator. Consider for instance the task of creating a program that allows generating elements of the form g^r, h^r for a random r, where g, h are two generators of a large group, and where r should remain hidden even from the evaluator of the program. Alternatively, consider the task of obfuscation-based re-encryption: Here we wish to "obfuscate" the program that decrypts a ciphertext using an internal decryption key, and then re-encrypts the obtained plaintext under a different key, using fresh randomness — all this while keeping the plaintext hidden from the evaluator.

Indeed, in both these examples, the goal is to create an obfuscation mechanism with two additional properties, stated very informally as follows: (a) the internal random choices of the obfuscated program should "remain hidden" from the evaluator, up to what is learnable from the output, and (b) the random choices of the program should remain "random", or "unskewed", as much as possible.

How can we define these properties in a sensible way? Barak et al. [7] take a first stab by defining the concept of *obfuscators for sampling algorithms,* namely algorithms that take only random input and at each execution output a sample from a distribution. Essentially, their definition requires that the (one bit) output of any adversary that has access to an obfuscated version of such a sampling algorithm be simulatable given only poly-many random samples from the distribution. However, while this definition does capture much of the essence of the problem, it is subject to essentially the same unattainability results that apply to VBB obfuscation.

Probabilistic IO. We propose an alternative definition for what it means to obfuscate probabilistic circuits. Our starting point is IO, rather than VBB, and hence we refer to the resulting general notion as *indistinguishability obfuscation for probabilistic circuits*, or pIO for short. This both reduces the susceptibility to unattainability results and allows making stronger distributional requirements on the outputs.

Consider a randomized circuit, namely a circuit C that takes an input x and a uniformly chosen random input r, and returns the random variable $C(x, r)$. The basic idea is to compile such a circuit into a *deterministic* obfuscated circuit $\Lambda = \mathcal{O}(C)$ that has essentially the same output distribution as the original circuit — with the one caveat that if Λ is run multiple times on the same input then it will give the same output.

The security requirements from a pIO obfuscator $\mathcal{O}$ for a family of circuits $\mathcal{C}$ are thus three: First, polynomial slowdown should hold as usual. Second, functionality should be preserved in the sense that for any $C \in \mathcal{C}$ and for any input x it should hold that $C(x) \approx_c \Lambda(x)$. Note that in $C(x)$ the probability is taken over the random choices of C (i.e., the sampling of r), whereas Λ is a deterministic circuit and the probability is taken *only over the random choices of $\mathcal{O}$*. In fact, we make the stronger requirement that no efficient adversary can distinguish whether it is given oracle access to the randomized oracle $C(\cdot)$ or the deterministic oracle $\Lambda(\cdot)$, as long as it does not submit repetitive queries to the oracles.)

Third, obfuscation should hold in the sense that $\mathcal{O}(C_1) \approx_c \mathcal{O}(C_2)$ for any two circuits C_1 and C_2 where the output distributions of $C_1(x)$ and $C_2(x)$ are "similar" for all inputs x, where similar means in general computationally indistinguishable. This property is trickiest to define, and to stress this even further, we note that the indistinguishability of $\mathcal{O}(C_1)$ and $\mathcal{O}(C_2)$ does *not* follow from IO even if the distributions of $C_1(x)$ and $C_2(x)$ are *identical*. Another important aspect is that we often need to consider programs that are parameterized by some additional system parameters, such as a public key of a cryptosystem. We thus extend the definition to consider also *families* of circuits with auxiliary input.

Concretely, we consider four variants of the above intuitive notion, depending on the specific notion of indistinguishability of probabilistic circuits assumed on the distribution. The four variants we consider differ in the level of adaptivity in choosing the inputs on which the programs are run in the experiment that determines whether programs are indistinguishable. Our formalization follows the approach of [9,2], capturing the strength of an IO algorithm $\mathcal{O}$ in terms of the distributions on triples (C_1, C_2, z) on which it succeeds in making $\mathcal{O}(C_1)$ and $\mathcal{O}(C_2)$ indistinguishable (given z).

A Construction for X-pIO. As our first main result, we show how to construct a general X-pIO scheme, where X-pIO is one of our four variants (see definition within), from any subexponentially secure IO scheme and one way function. The scheme is natural: X-$pi\mathcal{O}(C)$ is the result of applying an indistinguishability obfuscator to the following circuit. First apply the puncturable PRF to the input x to obtain a pseudorandom value r, using a hard-coded PRF secret key. Next, we run the circuit C on input x and random input r.

We show by reduction that if the underlying IO and puncturable PRF are subexponentially secure then the scheme is X-pIO. Furthermore, one can consider the same natural construction as a candidate implementation of any of the other variants of pIO.

Applications: FHE and Bootstrapping. To demonstrate the usefulness of pIO we present two natural applications of the notion, which are arguably of independent interest.

Our first application (see Section 2.2) is to constructing fully homomorphic encryption schemes. Here we provide a natural construction of fully homomorphic encryption from pIO (or, in turn from sub-exponentially secure IO and puncturable PRFs.) In fact, we provide the *first* full-fledged FHE scheme that does not rely on circular security assumptions for encryption.

We proceed in two steps. First we show how to obtain leveled homomorphic encryption (LHE), where only a prespecified number of homomorphic operations can be made securely. The basic idea is to use pIO to transform an underlying encryption scheme with some mild structural properties (such as rerandomizability) into an LHE. We give a number of different instantiations of the general scheme, where each instantiation uses a different variant of pIO and a different type of encryption scheme as a starting point.

The second step transforms the resulting LHE into a full-fledged FHE, again assuming IO and puncturable PRFs. (All primitives from LHE to IO to PRFs are required to be slightly super-polynomially secure.) While this transformation works in general for any LHE with a-priori fixed polynomial decryption depth, it is particularly suitable for LHEs that result from our pIO based construction in that it uses the same underlying primitives and assumptions. These constructions use in an inherent way the concept of obfuscation of randomized circuits, and in particular probabilistic IO.

As a second application, discussed in Section 2.3, we consider variants of bootstrapping, transforming IO obfuscation (both probabilistic and not probabilistic) for weak classes (such as low-depth circuits) into obfuscation for arbitrary polynomial-size circuits.

Organization. Section 2 gives a detailed high-level and self-contained overview of the contributions of this paper, both at the definitional level, as well as in terms of applications.

Further down, Section 3 presents our definitions of pIO and studies relations among them. Moreover, it presents the construction of X-pIO from IO and puncturable PRFs. Section 4 present the application to FHE, whereas the application to bootstrapping IO is deferred to the full version for lack of space.

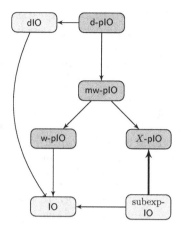

Fig. 1. Notions of obfuscation for probabilistic circuits: Arrows indicate implication, whereas lack of arrows among azure boxes implies a formal separation. The thicker line indicates that the implication holds under the assumption of subexponentially-hard puncturable PRFs.

2   Overview

We provide an overview of the definitions and results in this work.

2.1   Our Definitional Framework: IO for Probabilistic Circuits

The first contribution of this paper, found in Section 3, is the definition and study of IO notions for *probabilistic* circuits, or pIO. For our purposes, a probabilistic obfuscator $pi\mathcal{O}$ transforms a (usually probabilistic, i.e. randomized) circuit C into a *deterministic* circuit $\Lambda := pi\mathcal{O}(C)$ with the property that $\Lambda(x)$ is computationally indistinguishable from $C(x)$ the *first* time it is invoked, even when the circuits are invoked as oracles multiple times on *distinct* inputs. (Across multiple calls with the *same* input, $\Lambda(x)$ returns the same value over and over, whereas $C(x)$ returns a fresh random output.)

As for security, we want to ensure indistinguishability of $pi\mathcal{O}(C_0)$ and $pi\mathcal{O}(C_1)$ whenever $C_0(x)$ and $C_1(x)$ are computationally indistinguishable for every input x, rather than identical as in IO. However, formalizing this concept is challenging, due to the exponential number of inputs and the fact that C_0, C_1 are usually chosen from some distribution.

Four pIO Notions. Following the approach of [9,2], we capture different pIO notions via classes of *samplers*, where such a sampler is a distributions D (parametrized by the security parameter) outputting triples (C_0, C_1, z), such that C_0, C_1 are circuits, and z is some (generally correlated) auxiliary input. Different pIO notions result from different requirement in terms of the class of samplers for which a certain obfuscator $pi\mathcal{O}$ guarantees indistinguishability of the obfuscations of C_0 and C_1 (given the auxiliary input z), in addition to the above correctness requirement. Concretely, we consider four different notions matching different approaches to formalizing the above computational indistinguishability requirement on all inputs:

X-Ind pIO (X-pIO). In the simplest notion, we require that for every *statically* chosen input x, the distributions of $C_0(x)$ and $C_1(x)$ are indistinguishable, given z, where the randomness is over the sampled (C_0, C_1, z). While this results in an unachievable notion, we additionally require the distinguishing advantage to be very small, smaller than $\mathsf{negl} \cdot X^{-1}$, for some negligible function, where X is the number of inputs of the circuits. The requirement on the small distinguishing gap seems stringent and leads to a weak notion, but we show that it is necessary.

Dynamic-input pIO (d-pIO). A d-pIO obfuscator is required to work on samplers D such that any PPT attacker, given a triple (C_0, C_1, z) sampled from D, cannot find (adaptively) an input x for which, when given additionally $C_b(x)$ for a random b, it can guess the value of b with noticeable advantage over random guessing.

Worst-case-input pIO (w-pIO and mw-pIO). A w-pIO obfuscator weakens the above notion by only working on samplers for which the above indistinguishability requirement holds for (much) stronger attackers where the choice of x after sampling (C_0, C_1, z) is made without any computational restrictions, whereas the final guess, *after* learning $C_b(x)$, is restricted to be polynomial-time. This captures the fact that the choice of the input x is *worst case* as to maximize the guessing probability in the second stage. The (stronger) notion where we enlarge our sampler class to only require indistinguishability for attackers not passing such state is referred to as *memory-less worst-case-input pIO* (or mw-pIO for short).

We prove that d-pIO implies mw-pIO, and mw-pIO implies both w-pIO and X-pIO, but the latter two notions do not imply each other. These relations are summarized in Figure 1 below. The fact that mw-pIO implies X-pIO is surprising at first, as on one hand we are *restricting* the power of the attacker, but on the other hand we are simplifying its task by choosing our barrier at negl/X advantage, and it is not clear what prevails.

The notion of d-pIO is a natural generalization of the notion of *differing inputs obfuscation* [7,13,2], and therefore directly suffers from recent implausibility results [22] in its most general form. In contrast, achievability of mw-pIO and the even weaker notion of w-pIO is not put in question by similar results, and the original IO notion is recovered from both w-pIO and mw-pIO when restricting them to deterministic circuits only. We in fact feel comfortable in conjecturing that w-pIO is achieved by a construction first transforming a randomized circuit C into a deterministic one $D^k(x) = C(x; \mathsf{PRF}(k, x))$ for a PRF key k, then applying an existing obfuscator $\mathcal{O}$ to D^k, such as those from [21,6,17].

X-Ind pIO **from Sub-Exponential IO.** The main technical result of this part is a proof that for X-pIO, the above approach indeed *provably* yields a secure obfuscator if the PRF is puncturable and if the obfuscator $\mathcal{O} = i\mathcal{O}$ is an IO, as long as additionally PRF and $i\mathcal{O}$ are *subexponentially secure*. In this context, subexponential means that no PPT attacker can achieve better than sub-exponential advantage, an assumption which we believe to be reasonable.

2.2   Application 1: Fully-Homomorphic Encryption

The first testbed for our pIO notions, discussed in Section 4, is a generic construction of leveled homomorphic-encryption (or LHE, for short) from a regular encryption scheme. We are then going to boost this to achieve fully-homomorphic encryption (FHE) without any circular security assumptions using a technique of independent interest.

The LHE Construction. When trying to build a LHE scheme using ofuscation, the following natural and straightforward idea came up immediately. Starting from a CPA-secure encryption, we generate public-key and secret-key pairs for all levels $(\mathsf{pk}_0, \mathsf{sk}_0), \ldots, (\mathsf{pk}_L, \mathsf{sk}_L)$, and then, as part of the evaluation key, add for every level $i \in \{1, \ldots, L\}$, the pIO obfuscation of the circuit $\mathsf{Prog}^{(\mathsf{sk}_{i-1}, \mathsf{pk}_i)}$ which takes two ciphertexts $\alpha = \mathsf{Enc}(\mathsf{pk}_{i-1}, a)$ and $\beta = \mathsf{Enc}(\mathsf{pk}_{i-1}, b)$ (where a and b are bits), decrypts them using sk_{i-1}, and then outputs a fresh encryption $c = \mathsf{Enc}(\mathsf{pk}_i, a \text{ NAND } b)$. The outputs of this circuit, given sk_{i-1} and pk_i (but not sk_i) are computationally indistinguishable from those of a "trapdoor" circuit $\mathsf{tProg}^{(\mathsf{pk}_i)}$ which instead ignores its inputs, and simply outputs a fresh encryption $c = \mathsf{Enc}(\mathsf{pk}_i, 0)$ of 0. Note that this circuit is independent of sk_{i-1}. We therefore hope that by relying on some pIO notion for the sampler $D^{\mathsf{sk}_{i-1}}$ that outputs $(\mathsf{Prog}^{(\mathsf{sk}_{i-1}, \mathsf{pk}_i)}, \mathsf{tProg}^{(\mathsf{pk}_i)}, \mathsf{pk}_i)$ (and through a careful hybrid argument), one might transform the honest evaluation key to one that contains only obfuscations of the "trapdoor" circuits; in the latter case, since the evaluation key depends only on *public* keys, the semantic security of the LHE scheme reduces down to that of the underlying CPA scheme. The nice feature of this approach is that it builds on top of any already existing encryption scheme (say ElGamal), and that for all levels, ciphertexts are of the same type and size. A similar generic approach was for example abstracted in the work of Alwen et al. [1], and proved secure under ad-hoc obfuscation assumptions.

Unfortunately, it turns out that the above approach generically works for every CPA-secure scheme only when using d-pIO, which, as we discussed above, is somewhat brittle. Indeed, the above sampler $D^{\mathsf{sk}_{i-1}}$ is *not* contained in the classes associated with X-pIO and w-pIO. With respect to X-pIO there is no guarantee that encryptions (of values (a NAND b) or 0) are negl/X close to each other (note that here the domain size X corresponds to the length $|\alpha| + |\beta|$ of the *two* input ciphertexts). It seems that to fix the problem, one could simply re-encrypt under an encryption scheme which is negl/X secure (which exists assuming sub-exponentially secure CPA encryption), but this results in a longer output ciphertext of size $\mathsf{poly}(\log X)$ (i.e., $\mathsf{poly}(|\alpha| + |\beta|)$), leading to exponentially growing ciphertext with the depth.

With respect to w-pIO (and to mw-pIO also), the main challenge with the above sampler is that given the two circuits, the adversarial first stage is computationally unbounded and can (for example) find a secret key corresponding to the public key, and pass it on to the second stage, which proceeds in distinguishing encryptions (of values (a NAND b) and 0 again) using the secret key efficiently.

LHE via trapdoor encryption. We get around the above conundrum by using a generalization of CPA encryption—called trapdoor encryption: The idea here is that the encryption scheme can generate a *special* trapdoor key which is indistinguishable from a real public-key, but it does not guarantee decryption any more. In this way, we expect to be able to guarantee stronger ciphertext indistinguishability (even statistical) under a trapdoor key which cannot be satisfied by normal encryption scheme as long as correctness needs to be guaranteed. In particular, we modify the proof in the above approach as follows: In the hybrids, the obfuscations in the evaluation key are changed one by one in the reverse order; to change the obfuscation of circuit $\mathsf{Prog}^{(\mathsf{sk}_{i-1}, \mathsf{pk}_i)}$, first replace the public key with a trapdoor key tpk_i, and then move to an obfuscation of a modified trapdoor circuit $\mathsf{tProg}^{(\mathsf{tpk}_i)}$ with the trapdoor key built in. Now thanks to the stronger ciphertext indistinguishability under the trapdoor key, it suffices to use weak notions of pIO. In this paper, we provide the following instantiations of this paradigm:

- **Lossy encryption + w-pIO.** In order to instantiate the construction from w-pIO, we consider encryption schemes which are statistically secure under a trapdoor key, so-called *lossy* encryption schemes [8]. Such schemes can be built using techniques from a variety of works [31,34,8,35], and admit instantiations from most cryptographic assumptions. This gives an LHE construction from w-pIO and any lossy encryption schemes.[1]
- **Re-randomizable encryption + sub-exponential IO.** Existing constructions of lossy encryption unfortunately do not allow a distinguishing gap of negl/X without having the ciphertext size growing polynomially in $\log X$. Instead, we construct a trapdoor encryption scheme with such a tiny distinguishing gap under the trapdoor key, from any re-randomizable (secret or public-key) encryption scheme: The (honest) public key of the trapdoor encryption scheme consists of two encryptions (c_0, c_1) of 0 and 1 of the underlying re-randomizable encryption scheme, and to encrypt a bit b, one simply re-randomizes c_b; the trapdoor key, on the other hand, simply consists of two encryptions (c_0, c_0') of both 0. By the semantic security of the underlying scheme, the honest and trapdoor keys are indistinguishable. Furthermore, if the re-randomizability of the underlying scheme guarantees that re-randomization of one ciphertext or another of the same plaintext yields identical distributions, then encryptions under the trapdoor keys are perfectly hiding. Many encryption schemes such as ElGamal, Goldwasser-Micali [27], Paillier [33], Damgård-Jurik [20], satisfy

---

[1] In fact, this instantiation only requires an even weaker w-pIO notion where sampler indistinguishability must hold against computationally unbounded adversaries in both stages.

the perfect re-randomizability. Therefore, when relying on such a scheme, the corresponding samplers is negl/X-indistinguishable, for any X; hence X-pIO suffices. Combined with the aforementioned construction of X-pIO, this also gives us leveled LHE from any re-randomizable encryption scheme and sub-exponentially hard IO and one-way functions.

We also note that the instantiation from d-pIO mentioned above from any CPA-secure encryption scheme is also a (trivial) application of the above general result.

From LHE to FHE. As a final contribution of independent interest, we use IO to turn an LHE scheme info an FHE scheme via techniques inspired by the recent works of Bitansky, Garg, and Telang [11], and of Lin and Pass [32].

The basic idea is to instantiate the above LHE construction on *super-polynomially* many levels, but to represent these keys *succinctly*. This is done by considering a circuit Γ that on input i genarates the i-th level evaluation key, i.e., the pIO obfuscation of $\mathsf{Prog}^{(\mathsf{sk}_{i-1},\mathsf{pk}_i)}$ (in the evaluation key for super-polynomially many levels), where the key pairs $(\mathsf{pk}_{i-1},\mathsf{sk}_{i-1})$ and $(\mathsf{pk}_i,\mathsf{sk}_i)$ are generated using pseudo-random coins $\mathsf{PRF}(k, i-1)$ and $\mathsf{PRF}(k, i)$ computed using a puncturable PRF on a hard-coded seed k; (the pIO obfuscations also use pseudo-random coins as well). The new succinct evaluation key is the *IO-obfuscation of this circuit* Γ, while the public key is pk_0 (generated using coins $\mathsf{PRF}(k, 0)$) and the secret key is the PRF seed k. In order for this approach to be secure, we need the IO obfuscation to be slightly super-polynomially secure (not necessarily sub-exponentially secure), in order to accommodate for a number of hybrids in the proof which accounts to the (virtual) super-polynomial number of levels implicitly embedded in the succinct representation. In particular, we get this step almost for free (in terms of assumptions) when starting with our LHE constructions, either because we assume sub-exponential IO in the first place, or assuming just a slightly stronger form of w-pIO and d-pIO than what necessary above.

We also observe that this is a special case of a more general paradigm of using IO to turn any LHE with a fixed decryption depth (independent of the maximum evaluation level) into an FHE, which applies to almost all known LHE schemes (e.g. [23,18,16,15,24]). We believe that this general transformation is of independent interest, especially because it does not rely on any encryption scheme with circular security.

2.3   Application 2: Bootstrapping IO

Our second contribution is to use the notion of pIO to provide a simple way of bootstrapping (standard, deterministic) IO for weak circuit classes, such as $\mathbf{NC}^1$, into ones for all polynomial-size circuits. In the very first candidate construction of IO for $\mathbf{P}/poly$, Garg et al. [21] show how to obtain full fledged IO assuming the existence of indistinguishability obfuscation for a weak circuit class **WEAK**, as well as a fully homomorphic encryption scheme whose decryption can be computed in **WEAK** (given the known FHE schemes, one can think of **WEAK** as $\mathbf{NC}^1$). The natural question that remained is: Can we achieve bootstrapping without the FHE assumption?

We show a new way to bootstrap indistinguishability obfuscation, without assuming that FHE schemes exist. Instead, our assumption is the existence of sub-exponentially hard indistinguishability obfuscation for a complexity class **WEAK** and a sub-exponentially secure puncturable PRF computable in **WEAK**. Our technique is inspired by the recent work of Applebaum [3] that shows how to bootstrap VBB obfuscations from **WEAK** to $\mathbf{P}/poly$ using randomized encodings; however his transformation strongly relies on the fact that the starting point is a VBB obfuscation.

The idea is to apply the "randomized encodings" paradigm which was originally proposed in the context of multiparty computation [29,4] and has found many further uses ever since. A randomized encoding RE for a circuit family $\mathcal{C}$ is a probabilistic algorithm that takes as input a circuit $C \in \mathcal{C}$ and an input x, and outputs its randomized encoding $(\hat{C}, \hat{x})$. The key properties of RE are that: (1) given $\hat{C}$ and $\hat{x}$, one can efficiently recover $C(x)$; (2) given $C(x)$, one can efficiently simulate the pair $(\hat{C}, \hat{x})$, implying that the randomized encoding reveals no information beyond the output $C(x)$; and (3) computing RE is very fast in parallel. In particular, the work of Applebaum, Ishai and Kushilevitz [5], building on Yao's garbled circuits, showed a way to perform randomized encoding of any circuit in $\mathbf{P}/poly$ using a circuit $\mathsf{RE} \in \mathbf{NC}^0$, assuming a PRG in $\oplus \mathbf{L}/poly$ (which is implied by most cryptographic assumptions). The typical use of randomized encodings is to reduce computing a circuit C to the easier task of computing its randomized encoding $\mathsf{RE}(C, \cdot)$.

Therefore, to obfuscate a circuit $C \in \mathbf{P}/poly$, the natural idea is obfuscating its randomized encoding $\mathsf{RE}(C, x; r)$ using an appropriate pIO scheme for $\mathbf{NC}^0$. (Here, pIO comes into play naturally, since RE is a randomized circuit.) We show that, in fact, X-pIO suffices for this purpose: Assuming that randomized encoding is sub-exponentially secure, then for any two functionally equivalent circuits C_1 and C_2, their randomized encoding $\mathsf{RE}(C_1, x; r)$ and $\mathsf{RE}(C_2, x; r)$ have indistinguishable outputs for every input x, where the distinguishing gap is as small as $\mathsf{negl}(\lambda)2^{-|x|^2}$. Therefore, obfuscating $\mathsf{RE}(C_1, \cdot)$ and $\mathsf{RE}(C_2, \cdot)$ using an X-pIO scheme $pi\mathcal{O}$ yields indistinguishable obfuscated programs, and hence $i\mathcal{O}(C) = pi\mathcal{O}(\mathsf{RE}(C, \cdot))$ is an indistinguishable obfuscator for all $\mathbf{P}/poly$. Since, our construction of X-pIO from sub-exponentially indistinguishable IO preserves the class of circuits modulo the complexity of the sub-exponentially indistinguishable puncturable PRF. Put together, we are able to bootstrap sub-exponentially indistinguishable IO for a weak class, say $\mathbf{NC}^1$, to IO for all of $\mathbf{P}/poly$, assuming a sub-exponentially indistinguishable PRF computable in the weak class of circuits.

Bootstrapping pIO. The same technique above can be applied to bootstrap worst-case-input pIO from $\mathbf{NC}^0$ to $\mathbf{P}/poly$, assuming the existence of a PRG in $\oplus \mathbf{L}/poly$. The key observation here is that since pIO handles directly randomized

---

[2] This can be done by using a sufficiently large security parameter when generating the randomized encoding.

circuits, it can be used to obfuscate the randomized encoding $\mathsf{RE}(C, \cdot)$ (without relying on pseudorandom functions). Furthermore, the security of the randomized encoding holds for any input and auxiliary information (even ones that are not efficiently computable). Then, given any two circuits $C_1(x; r), C_2(x; r)$ whose outputs are indistinguishable even for dynamically chosen worst-case inputs, their randomized encoding $C_1'(x; r, r') = \mathsf{RE}(C_1, (x, r); r')$ and $C_2'(x; r, r') = \mathsf{RE}(C_2, (x, r); r')$ are also indistinguishable on dynamically chosen worst case inputs. This is because, over the random choice of r and r', the distributions of $C_1'(x; r, r')$ and $C_2'(x; r, r')$ can be simulated using only $C_1(x; r)$ and $C_2(x; r)$, which are indistinguishable. Therefore a worst-case-input pIO scheme for $\mathbf{NC}^0$ suffices for obfuscating the circuit $C'(x; r, r') = \mathsf{RE}(C, (x, r); r)$, leading to a worst-case-input pIO scheme for all $\mathbf{P}/poly$. Following the same approach, we can bootstrap dynamic-input pIO for $\mathbf{NC}^0$ to dynamic-input pIO for $\mathbf{P}/poly$ assuming a PRG in $\oplus\mathbf{L}/poly$. Similarly, we can also bootstrap X-pIO for $\mathbf{NC}^0$ to X-pIO for $\mathbf{P}/poly$, but relying on the sub-exponential security of the PRG. The stronger security of PRG is needed so that the randomized encoding can be made $\mathsf{negl}(\lambda)/X(\lambda)$ indistinguishable.

3   IO for Probabilistic Circuits

3.1   IO for General Samplers over Probabilistic Circuits

We start with the notion of indistinguishability obfuscation for general classes of samplers over *potentially probabilistic* circuits, called pIO for samplers in class **S**. Here, a sampler is a distribution ensemble over pairs of potentially randomized circuits, together with an auxiliary input. Below, we define various notions of obfuscation for probabilistic circuits instantiating the general definition with classes of samplers that produce pairs of probabilistic circuits satisfying different variants of our point-wise indistinguishability requirement.

More formally, let $\mathcal{C} = \{\mathcal{C}_\lambda\}_{\lambda \in \mathbb{N}}$ be a family of sets of (randomized) circuits, where $\mathcal{C}_\lambda$ contains circuits of size $\mathsf{poly}(\lambda)$. Extending the notation of [9], a *circuit sampler* for $\mathcal{C}$ is a distribution ensemble $D = \{D_\lambda\}_{\lambda \in \mathbb{N}}$, where the distribution D_λ ranges over triples (C_0, C_1, z) with $C_0, C_1 \in \mathcal{C}_\lambda$ such that C_0, C_1 take inputs of the same length, and $z \in \{0, 1\}^{\mathsf{poly}(\lambda)}$. Moreover, a *class* **S** of samplers for $\mathcal{C}$ is a set of circuit samplers for $\mathcal{C}$.

The following definition captures the notion of pIO for a class of samplers.

Definition 1 (pIO for a Class of Samplers). *A uniform PPT machine $pi\mathcal{O}$ is an* indistinguishability obfuscator *for a class of samplers* **S** *over the (potentially randomized) circuit family $\mathcal{C} = \{\mathcal{C}_\lambda\}_{\lambda \in \mathbb{N}}$ if the following two conditions hold:*

Correctness: *$pi\mathcal{O}$ on input a (potentially probabilistic) circuit $C \in \mathcal{C}_\lambda$ and the security parameter $\lambda \in \mathbb{N}$ (in unary), outputs a deterministic circuit Λ of size $\mathsf{poly}(|C|, \lambda)$.*

Furthermore, for every non-uniform PPT distinguisher $\mathcal{D}$, every (potentially probabilistic) circuit $C \in \mathcal{C}_\lambda$, and string z, we define the following two experiments:

– $\mathsf{Exp}^1_{\mathcal{D}}(1^\lambda, C, z)$: $\mathcal{D}$ on input $1^\lambda, C, z$, participates in an unbounded number of iterations of his choice. In iteration i, it chooses an input x_i; if x_i is the same as any of the previously chosen input x_j for $j < i$, then abort; otherwise, $\mathcal{D}$ receives $C(x_i; r_i)$ using fresh random coins r_i ($r_i = $ null if C is deterministic). At the end of all iterations, $\mathcal{D}$ outputs a bit b. (Note that $\mathcal{D}$ can keep state across iterations.)

– $\mathsf{Exp}^2_{\mathcal{D}}(1^\lambda, C, z)$: Obfuscate circuit C to obtain $\Lambda \xleftarrow{\$} pi\mathcal{O}(1^\lambda, C; r)$ using fresh random coins r. Run $\mathcal{D}$ as described above, except that in each iteration, feed $\mathcal{D}$ with $\Lambda(x_i)$ instead.

Overload the notation $\mathsf{Exp}^i_{\mathcal{D}}(1^\lambda, C, z)$ as the output of $\mathcal{D}$ in experiment $\mathsf{Exp}^i_{\mathcal{D}}$. We require that for every non-uniform PPT distinguisher $\mathcal{D}$, there is a negligible function μ, such that, for every $\lambda \in \mathbb{N}$, every $C \in \mathcal{C}_\lambda$, and every auxiliary input $z \in \{0,1\}^{\text{poly}(\lambda)}$,

$$\mathsf{Adv}_{\mathcal{D}}(1^\lambda, C, z) = |\Pr[\mathsf{Exp}^1_{\mathcal{D}}(1^\lambda, C, z)] - \Pr[\mathsf{Exp}^2_{\mathcal{D}}(1^\lambda, C, z)]| = \mu(\lambda) .$$

Security with Respect to S: For every sampler $D = \{D_\lambda\}_{\lambda \in \mathbb{N}} \in \mathbf{S}$, and for every non-uniform PPT machine $\mathcal{A}$, there exists a negligible function μ such that

$$\big| \Pr[(C_1, C_2, z) \xleftarrow{\$} D_\lambda : \mathcal{A}(C_1, C_2, pi\mathcal{O}(1^\lambda, C_1), z) = 1] -$$
$$- \Pr[(C_1, C_2, z) \xleftarrow{\$} D_\lambda : \mathcal{A}(C_1, C_2, pi\mathcal{O}(1^\lambda, C_2), z) = 1] \big| = \mu(\lambda) .$$

where μ is called the **distinguishing gap**.

Furthermore, we say that $pi\mathcal{O}$ is δ-**indistinguishable** if the distinguishing gap μ bounded by δ. Especially, $pi\mathcal{O}$ is **sub-exponentially indistinguishable** if $\mu(\lambda)$ is bounded by $2^{-\lambda^\epsilon}$ for a constant ϵ.

Note that the sub-exponential indistinguishability defined above is weaker than usual sub-exponential hardness assumptions in that the distinguishing gap only needs to be small for PPT distinguishers, rather than sub-exponential ones.

An obvious (but important) remark is that an obfuscator $pi\mathcal{O}$ for the class $\mathbf{S}$ is also an obfuscator for any class $\mathbf{S}' \subseteq \mathbf{S}$, whereas conversely, if no obfuscator exists for $\mathbf{S}'$ (or its existence is implausible), then the same is true for $\mathbf{S} \supseteq \mathbf{S}'$.

3.2 Static-input pIO for Circuits

Arguably, the simplest way to formulate the property that two circuits are indistinguishable on every input is to require that this true for every statically chosen input, i.e., chosen independently of the random choice of the sampler. This results in the following definition, which we state for completeness, but that we will have to further restrict below to by-pass impossibility:

Definition 2 (*Static-input* Indistinguishable Samplers). The class $\mathbf{S}^{\text{s-Ind}}$ of static-input indistinguishable samplers for a circuit family $\mathcal{C}$ contains all circuit samplers $D = \{D_\lambda\}_{\lambda \in \mathbb{N}}$ for $\mathcal{C}$ with the following property: For all non-uniform PPT $\mathcal{A} = (\mathcal{A}_1, \mathcal{A}_2)$, the advantage of $\mathcal{A}$ in the following experiment is negligible.

Experiment *static-input-*$\mathsf{IND}_{\mathcal{A}}^D(1^\lambda)$:

1. $(x, st) \xleftarrow{\$} \mathcal{A}_1(1^\lambda)$ // $\mathcal{A}_1$ *chooses challenge input* x *statically.*
2. $(C_0, C_1, z) \xleftarrow{\$} D_\lambda$
3. $y \xleftarrow{\$} C_b(x)$, *where* $b \xleftarrow{\$} \{0, 1\}$.
4. $b' \xleftarrow{\$} \mathcal{A}_2(st, C_0, C_1, z, x, y)$

The advantage of $\mathcal{A}$ *is* $\Pr[b' = b] - 1/2$.

Unfortunately, we now show that pIO for general static-input indistinguishable samplers is (unconditionally) impossible, but we will see below that a further restriction of the class $\mathbf{S}^{\text{s-Ind}}$ will bypass this impossibility.

Proposition 1. *There exists a static-input indistinguishable sampler* D^* *over deterministic circuits, such that, there is no* pIO *for* D^*.

Proof. Consider the following sampler D^*: D_λ^* samples (C_0, C_1, z) where C_0 is an all zero circuit, C_1 computes a point function that outputs 1 at a single point s chosen uniformly randomly, and z is set to s. Clearly, D^* is a static-input indistinguishable sampler. Indeed, for any fixed input x, with overwhelming probability D^* samples (C_0, C_1, s), with a differing input $s \neq x$. Thus, the outputs $C_0(x) = C_1(x) = 0$ cannot be distinguished.

However, any $pi\mathcal{O}$ achieving correctness cannot be secure for this sampler D^*: An adversary can easily tell apart $(C_0, C_1, s, \Lambda_0 = pi\mathcal{O}(C_0))$ from $(C_0, C_1, s, \Lambda_1 = pi\mathcal{O}(C_1))$ by simply evaluating Λ_0 and Λ_1 on input s.   □   □

X**-Ind pIO.** To circumvent impossibility, we consider a smaller class of static-input indistinguishable samplers, $\mathbf{S}^{X\text{-Ind}} \subset \mathbf{S}^{\text{s-Ind}}$. In fact, in Section 3.6 we give a construction for such a pIO assuming sub-exponentially indistinguishable IO.

The samplers D we consider satisfy that the distinguishing gap of any PPT adversary in the above static-input-IND experiment is bounded by $\mathsf{negl} \cdot X^{-1}$, where X is the number of "differing inputs" that circuits C_0, C_1 sampled from D have, and negl is some negligible function. More precisely:

Definition 3 ((Static-input) X-Ind-Samplers). *Let* $X(\lambda)$ *be a function bounded by* 2^λ. *The class* $\mathbf{S}^{X\text{-Ind}}$ *of (static-input)* X-Ind-*samplers for a circuit family* $\mathcal{C}$ *contains all circuit samplers* $D = \{D_\lambda\}_{\lambda \in \mathbb{N}}$ *for* $\mathcal{C}$ *with the following property: For every* $\lambda \in \mathbb{N}$, *there is a set* $\mathcal{X} = \mathcal{X}_\lambda \subseteq \{0, 1\}^*$ *of size at most* $X(\lambda)$ *(called the differing domain), such that,*

X **differing inputs:** *With overwhelming probability over the choice of* $(C_0, C_1, z) \xleftarrow{\$} D_\lambda$, *for every input outside the differing domain,* $x \notin \mathcal{X}$, *it holds that* $C_0(x'; r) = C_1(x'; r)$ *for every random string* r.
X**-indistinguishability:** *For all non-uniform PPT* $\mathcal{A} = (\mathcal{A}_1, \mathcal{A}_2)$, *the advantage of* $\mathcal{A}$ *in the experiment static-input-*$\mathsf{IND}_{\mathcal{A}}^D(1^\lambda)$ *defined in Definition 2 is* $\mathsf{negl} X^{-1}$.

Definition 4 (X-Ind pIO for Randomized Circuits). *Let X be any function bounded by 2^λ. A uniform PPT machine X-$pi\mathcal{O}$ is an X-pIO for randomized circuits, if it is a pIO for the class of X-Ind samplers $\mathbf{S}^{X\text{-Ind}}$ over $\mathcal{C}$ that includes all randomized circuits of size at most λ.*

We note that the notion of a differing set is added for flexibility purposes, as our constructions below will allow for it. We stress that its definition is not allowed to depend on the circuits which are actually sampled, and must be fixed a-priori. Also, note that the notion encompasses the setting where $C_0(x)$ and $C_1(x)$ are identically distributed, or are statistically very close.

The notion of X-Ind pIO is the "best-possible" achievable with respect of static input. Indeed, one can modify the distribution D^* constructed in Proposition 1 to have $C_1(s)$ output 1 with probability $\frac{1}{p(\lambda)}$ for a polynomial p. The differing domain there is the whole domain, i.e., $\mathcal{X}_\lambda = \{0,1\}^\lambda$ (since the circuit *may* differ at any point.) This makes the sampler *exactly* $X \cdot p^{-1}$ indistinguishable for static adversaries, as $C_1(x) \neq C_0(x)$ with probability $X \cdot p^{-1}$ over the choice of (C_0, C_1, z). Yet, pIO for this sampler is impossible, as again, the circuits differ on input $z = s$ with probability $\frac{1}{p(\lambda)}$. This impossibility cannot be pushed any further, and indeed general X-pIO is possible, as shown in Section 3.6.

3.3 Dynamic-input pIO for Circuits

The above notion, while achievable, makes an unnaturally strong indistinguishability requirement. We explore alternative notions where the distinguishing gap is not required to be as small. We start with a natural sampler notion asking for indistinguishability on every input x *adaptively* chosen by a (PPT) adversary $\mathcal{A}_1$ on input (C_0, C_1, z).

Definition 5 (*Dynamic-input* Indistinguishable Samplers). *The class $\mathbf{S}^{d\text{-Ind}}$ of dynamic-input indistinguishable samplers for a circuit family $\mathcal{C}$ contains all circuit samplers $D = \{D_\lambda\}_{\lambda \in \mathbb{N}}$ for $\mathcal{C}$ with the following property: For all non-uniform PPT $\mathcal{A} = (\mathcal{A}_1, \mathcal{A}_2)$, the advantage of $\mathcal{A}$ in the following experiment is negligible.*

Experiment. *dynamic-input-*$\mathsf{IND}^D_\mathcal{A}(1^\lambda)$:

1. $(C_0, C_1, z) \xleftarrow{\$} D_\lambda$
2. $(x, st) \xleftarrow{\$} \mathcal{A}_1(C_0, C_1, z)$
3. $y \xleftarrow{\$} C_b(x)$, where $b \xleftarrow{\$} \{0,1\}$
4. $b' \xleftarrow{\$} \mathcal{A}_2(st, C_0, C_1, z, x, y)$

The advantage of $\mathcal{A}$ is $\Pr[b' = b] - 1/2$.

We note that the restriction to requiring indistinguishability on a single input is without loss of generality, as it follows from a standard hybrid argument that the advantage of any efficiency adversary is still negligible even if it receives samples from $C_b(x)$ for an unbounded number of *adaptively* chosen inputs.

We can now use the above sampler class to directly obtain the notion of Dynamic-input pIO for randomized circuits.

Definition 6 (Dynamic-input pIO for Randomized Circuits). *A uniform PPT machine* d-$pi\mathcal{O}$ *is a* dynamic-input pIO (or d-pIO) *for randomized circuits, if it is a pIO for the class of dynamic-input indistinguishable samplers* $\mathbf{S}^{\text{d-Ind}}$ *over $\mathcal{C}$ that includes all randomized circuits of size at most λ.*

Differing-Input Indistinguishability Obfuscation. It is not hard to see that we can recover the notion of differing-inputs indistinguishability obfuscation (dIO) for circuits [7,13,2], by just restricting the above definition of d-pIO to the class $\mathcal{C}' = \{\mathcal{C}'_\lambda\}_{\lambda \in \mathbb{N}}$ of *deterministic* circuits.

This means that the notion of dynamic-input pIO generalizes dIO to randomized circuits. In a recent work by Garg et al. [22], it was shown that assuming strong obfuscation for a specific sampler of circuits and auxiliary inputs, it is impossible to construct differing-input IO for general differing-input samplers over circuits. Since dynamic-input pIO implies differing-input IO, a construction of d-pIO for general dynamic-input indistinguishable samplers is also implausible. However, a construction of d-pIO for specific dynamic-input indistinguishable samplers remains possible, as in the case of dIO.

3.4 Worst-case-input pIO for Circuits

In light of the implausibility of general dynamic-input pIO, we seek for a weaker notion, which is possibly achievable. The resulting notion is what we consider the most natural formalization of IO in the probabilistic setting, but in contrast to X-pIO above, we are only able to *conjecture* the existence of suitable obfuscators.

We first introduce the following class of samplers:

Definition 7 (*Worst-case-input* Indistinguishable Samplers). *The class* $\mathbf{S}^{\text{w-Ind}}$ *of worst-case-input indistinguishable samplers for a circuit family $\mathcal{C}$ contains all circuit samplers $D = \{D_\lambda\}_{\lambda \in \mathbb{N}}$ for $\mathcal{C}$ with the following property: For all adversary $\mathcal{A} = (\mathcal{A}_1, \mathcal{A}_2)$ where $\mathcal{A}_1$ is an unbounded non-uniform machine and $\mathcal{A}_2$ is PPT, the advantage of $\mathcal{A}$ in the following experiment is negligible.*

Experiment *worst-case-input*-$\mathsf{IND}^D_{\mathcal{A}}(1^\lambda)$:

1. $(C_0, C_1, z) \xleftarrow{\$} D_\lambda$
2. $(x, st) = \mathcal{A}_1(C_0, C_1, z)$ // $\mathcal{A}_1$ is unbounded.
3. $y \xleftarrow{\$} C_b(x)$, where $b \xleftarrow{\$} \{0,1\}$
4. $b' \xleftarrow{\$} \mathcal{A}_2(st, C_0, C_1, z, x, y)$ // $\mathcal{A}_2$ is PPT.

The advantage of $\mathcal{A}$ is $\Pr[b' = b] - 1/2$.

This directly yields the notion of worst-case-input pIO:

Definition 8 (Worst-case-input pIO for Randomized Circuits). *A uniform PPT machine* w-$pi\mathcal{O}$ *is a* worst-case-input pIO *(or* w-pIO*) for randomized circuits, if it is a pIO for the class of worst-case-input indistinguishable samplers* $\mathbf{S}^{\text{w-Ind}}$ *over* $\mathcal{C}$ *that includes all randomized circuits of size at most* λ.

Note that in the above definition, since $\mathcal{A}_1$ is computationally unbounded, its best strategy on input (C_0, C_1, z) is to choose (x^*, st^*) that maximizes the guessing advantage of $\mathcal{A}_2$ and hence worst-case-input indistinguishable samplers can be seen as producing pairs of probabilistic circuits satisfying that no efficient adversary ($\mathcal{A}_2$) can distinguish their output $C_0(x)$ or $C_1(x)$ on *any* input x.

The above definition implies a limited form of multi-input indistinguishability: By a hybrid argument, for a worst-case-input sampler the advantage of any adversary ($\mathcal{A}_1, \mathcal{A}_2$) in the above experiment is negligible even if $\mathcal{A}_1$ can choose a polynomial number of inputs $(x_1, \cdots, x_\ell, st)$ at once and $\mathcal{A}_2$ receives output samples $y_i \xleftarrow{\$} C_b(x_i)$ for all these inputs, i.e., it is given $(st, C_0, C_1, z, \{x_i\}, \{y_i\})$. However, we cannot prove an adaptive form of multi-input indistinguishability, due to the asymmetric computational powers of $\mathcal{A}_1$ and $\mathcal{A}_2$.

Memory-less worst-case-input pIO: *Forbidding state-passing.* Passing state between $\mathcal{A}_1$ and $\mathcal{A}_2$ in the above definition of worst-case-input indistinguishable samplers appears somewhat unavoidable for any "meaningful" way of defining $\mathbf{S}^{\text{w-Ind}}$. Indeed, if we have a sampler $D \in \mathbf{S}^{\text{w-Ind}}$, then for any length function ℓ, we also would like any sampler D' constructed as follows to be also in $\mathbf{S}^{\text{w-Ind}}$: D'_λ samples (C_0, C_1, z) from the same distribution as D_λ, but instead returns a triple (C'_0, C'_1, z) where C'_b is a circuit such that $C'_b(x; x') = C_b(x)$ for any $x' \in \{0, 1\}^{\ell(\lambda)}$, i.e., the last $\ell(\lambda)$ bits of the input are ignored. For such a pair, the adversary $\mathcal{A}_1$ can always use the last $\ell(\lambda)$ bits of the input (which are ignored by the circuit) to pass on some helpful, not efficiently computable, information to $\mathcal{A}_2$ that would help distinguish.

Explicitly forbidding state passing will however be useful when establishing the landscape of relationships among notions below. In particular, we define $\mathbf{S}^{\text{mw-Ind}}$ as the class of memory-less worst-case-input indistinguishable samplers, which consists of all samplers D for which the advantage in worst-case-input-$\text{IND}_{\mathcal{A}}^{D}(1^\lambda)$ is negligible for any $\mathcal{A} = (\mathcal{A}_1, \mathcal{A}_2)$ such that $\mathcal{A}_1$ is unbounded and outputs $st = \bot$, whereas $\mathcal{A}_2$ is PPT. Note that clearly $\mathbf{S}^{\text{w-Ind}} \subseteq \mathbf{S}^{\text{mw-Ind}}$. This then is used in the following definition.

Definition 9 (Memory-less worst-case-input pIO for Randomized Circuits). *A uniform PPT machine* mw-$pi\mathcal{O}$ *is a* memory-less worst-case-input pIO *(or* mw-pIO*) for randomized circuits, if it is a pIO for the class of memory-less worst-case-input indistinguishable samplers* $\mathbf{S}^{\text{mw-Ind}}$ *over* $\mathcal{C}$ *that includes all randomized circuits of size at most* λ.

Indistinguishability Obfuscation. The notion of worst-case-input pIO is a direct generalization of IO to the case of randomized circuits. We can recover the original notion of indistinguishability obfuscation (IO) for circuits [7,21] by restricting Definition 8 of worst-case-input pIO to the class $\mathcal{C}' = \{\mathcal{C}'_\lambda\}_{\lambda \in \mathbb{N}}$ of *deterministic* circuits. Also, note that the classes $\mathbf{S}^{\mathsf{mw\text{-}Ind}}$ and $\mathbf{S}^{\mathsf{w\text{-}Ind}}$ are the same (and thus the notion of memory-less worst-case-input and worst-case-input pIO) when restricted to deterministic circuits.

3.5   Relations

In the full version, we prove a number of relations among notions, some of which are quite non-trivial to establish. They are summarized by the following theorem.

Theorem 1 (Relations Among pIO notions).

- *A dynamic-input* pIO *obfuscator is also a memory-less worst-case-input* pIO *obfuscator for randomized circuits.*
- *A memory-less worst-case-input* pIO *obfuscator is also a worst-case-input* pIO *obfuscator.*
- *A memory-less worst-case-input* pIO *obfuscator is also an* X-Ind pIO *obfuscator.*

Moreover, all of these implications are strict, i.e., their converses are not true, assuming subexponentially-secure (trapdoor) one-way permutations exist.

3.6   Construction of X-Ind pIO from Sub-exp Indistinguishable IO

In this section, we prove the existence of a construction of an X-Ind pIO obfuscator (as in Definition 4) from sub-exponentially hard IO. It relies on sub-exponentially secure puncturable PRFs, which we now recall

Definition 10 (Puncturable PRFs). *A puncturable family of PRFs is given by a triple of uniform PPT machines* Key, Puncture, *and* PRF, *and a pair of computable functions* $n(\cdot)$ *and* $m(\cdot)$, *satisfying the following conditions:*

Correctness. *For all outputs K of* Key(1^λ), *all points $i \in \{0,1\}^{n(\lambda)}$, and $K_{-i} =$* Puncture(K, i), *we have that* PRF$(K_{-i}, x) =$ PRF(K, x) *for all $x \neq i$.*
Pseudorandom at Punctured Point. *For every PPT adversary $(\mathcal{A}_1, \mathcal{A}_2)$, there is a neligible function μ, such that in an experiment where $\mathcal{A}_1(1^\lambda)$ outputs a point $i \in \{0,1\}^{n(\lambda)}$ and a state σ, $K \xleftarrow{\$}$ Key(1^λ) and $K_{-i} =$ Puncture(K, i), the following holds*

$$\left| \Pr[\mathcal{A}_2(\sigma, K_{-i}, i, \mathsf{PRF}(K, i)) = 1] - \Pr[\mathcal{A}_2(\sigma, K_{-i}, i, U_{m(\lambda)}) = 1] \right| \leq \mu(\lambda)$$

As observed by [12,14,30], the GGM tree-based construction of PRFs [25] from PRGs yields puncturable PRFs. Furthermore, if the PRG underlying the GGM construction is sub-exponentially hard (and this can in turn be built from sub-exponentially hard OWFs), then the resulting puncturable PRF is sub-exponentially pseudo-random.

We are now ready to move to our theorem. Its formal proof is deferred to the full version, but we give a detailed description of the main ideas below.

Theorem 2 (Existence of X-Ind pIO.). *Assume the existence of a sub-exponentially indistinguishable indistinguishability obfuscator $i\mathcal{O}$ for circuits and a sub-exponentially secure puncturable PRF* (Key, Puncture, PRF). *Then, there exists a X-Ind pIO obfuscator X-$pi\mathcal{O}$ for randomized circuits.*

We first describe our construction of X-Ind pIO, denoted as X-$pi\mathcal{O}$. Recall that by our assumption, both $i\mathcal{O}$ and the puncturable PRF (Key, Puncture, PRF) have a $2^{-\lambda^\epsilon}$ distinguishing gap for some constant $\epsilon \in (0,1)$ and any non-uniform PPT distinguisher. Also, in the following, we implicitly identify strings with integers (via their binary encoding) and vice versa.

Construction X-$pi\mathcal{O}$: On input 1^λ and a probabilistic circuit C of size at most λ, proceed as follows:

1. Let $\lambda' = \lambda'(\lambda) = (\lambda \log^2(\lambda))^{1/\epsilon}$. Sample a key of the PRF function $K \leftarrow \mathsf{Key}(1^{\lambda'})$.
2. Construct deterministic circuit $E^{(C,K)}$ which outputs $C(x\ ;\ \mathsf{PRF}(K,x))$. By construction the size of $E^{(C,K)}$ is bounded by a polynomial $p(\lambda') \geq \lambda'$ in λ'.
3. Let $\lambda'' = p(\lambda') \geq \lambda'$. Obfuscate $E^{(C,K)}$ using $i\mathcal{O}$, $\Lambda \xleftarrow{\$} i\mathcal{O}(1^{\lambda''}, E^{(C,K)})$.
4. Output Λ.

To see why the construction works, consider two circuits C_1, C_2 sampled satisfying the indistinguishability requirement imposed by X-Ind pIO, their obfuscation are the IO obfuscated programs Λ_1, Λ_2 of the two derandomized circuits D_1^k, D_2^k. The challenge lies in how to apply the security guaranetees of IO on two circuits D_1^k, D_2^k that have completely different functionality. Our hope is to leverage the fact that the original circuits C_1, C_2 are strongly indistinguishable together with the sub-exponential pseudo-randomness of PRF; indeed, when the PRF key is sufficiently long, it holds that for every x, the output pair $D_1^k(x)$ and $D_2^k(x)$ is $\frac{1}{X 2^{\omega(\log(\lambda))}}$-indistinguishable. Thus by a simple union bound over all X inputs, even the entire truth tables $\left\{ D_1^{k_1}(x) \right\}, \left\{ D_2^{k_2}(x) \right\}$ are indistinguishable. However, even given such strong guarantees, it is still not clear how to apply IO.

We overcome the challenge by considering a sequence of $X + 1$ hybrids $\{H_i\}$, in which we obfuscate a sequence of "hybrid circuits" $\left\{ E_i^k(x) \right\}$ that "morph" gradually from D_1^k to D_2^k. More specifically, circuit E_i^k evaluates the first i inputs using D_2^k, and the rest using D_1^k. In any two subsequent hybrids, the circuits E_{i-1}^k and E_i^k only differ at whether the i'th input is evaluated using D_1^k or D_2^k. Consider additionally two auxiliary hybrids H_{i-1}^+, H_i^+ where two circuits $F_{i-1}^{k-i,y}, F_i^{k-i,y'}$ modified from E_{i-1}^k, E_i^k are obfuscated; they proceed the same as E_{i-1}^k, E_i^k respectively, except that they use internally a PRF key k_{-i} punctured at point i, and output directly y and y' for input i respectively. Then, when y and y' are programmed to be exactly $y = E_{i-1}^k(i) = D_1^k(i)$ and $y' = E_i^k(i) = D_2^k(i)$, the two circuits compute exactly the same functionality as E_{i-1}^k, E_i^k. By IO, these auxilary hybrids are indistinguishable from hybrids H_{i-1} and H_i respectively.

Then, by the fact that $y = E^k_{i-1}(i) = D^k_1(i)$ and $y' = E^k_i(i) = D^k_2(i)$ are indistinguishable (which in turn relies on the pseudo-randomness of the PRF function), the two auxiliary hybrids H^+_{i-1}, H^+_i are indistinguishable, and thus so are H_{i-1} and H_i. Furthermore, since the distinguishing gap of IO and PRF are bounded by $\frac{1}{X 2^{\omega(\log \lambda)}}$, it follows from a hybrid argument that H_0 and H_X, which contain the IO obfuscations of $D^k_1(x)$ and $D^k_2(x)$, respectively, are $\frac{1}{2^{\omega(\log \lambda)}}$-indistinguishable.

Other notions. While we cannot prove this statement in any meaningful model, we also conjecture that the same construction is w-pIO obfuscator for randomized circuits.

4   Application 1: Fully Homomorphic Encryption

We now describe how to construct leveled and fully homomorphic encryption schemes using different notions of pIO. (See the Introduction for an overview of the constructions.)

4.1   Trapdoor Encryption Schemes

Trapdoor encryption schemes have two modes: In the honest mode, an honest public key is sampled and the encryption and decryption algorithms work as in a normal CPA-secure encryption scheme with semantic security and correctness; additionally, there is a "trapdoor mode", in which a indistinguishable "trapdoor public key" is sampled and the encryption algorithm produces ciphertexts that may have stronger indistinguishability properties than these in the honest mode, at the price of losing correctness. More precisely,

Definition 11 (Trapdoor Encryption Scheme). *We say that $\Pi =$ (KeyGen, Enc, Dec, tKeyGen) is a trapdoor encryption scheme, if (KeyGen, Enc, Dec) is a CPA-secure encryption scheme and the trapdoor key generation algorithm tKeyGen satisfies the following additionally properties:*

Trapdoor Public Keys: *The following two ensembles are indistinguishable:*

$$\left\{ (\mathsf{pk}, \mathsf{sk}) \xleftarrow{\$} \mathsf{KeyGen}(1^\lambda) \; : \; \mathsf{pk} \right\}_\lambda \approx \left\{ \mathsf{tpk} \xleftarrow{\$} \mathsf{tKeyGen}(1^\lambda) \; : \; \mathsf{tpk} \right\}_\lambda$$

Computational Hiding: *The following ensembles are indistinguishable.*

$$\left\{ \mathsf{tpk} \xleftarrow{\$} \mathsf{tKeyGen}(1^\lambda) \; : \; \mathsf{Enc}_{\mathsf{tpk}}(0) \right\}_\lambda \approx \left\{ \mathsf{tpk} \xleftarrow{\$} \mathsf{tKeyGen}(1^\lambda) \; : \; \mathsf{Enc}_{\mathsf{tpk}}(1) \right\}_\lambda$$

The basic definition of trapdoor encryption scheme only requires encryption of different bits under a freshly generated trapdoor public key to be computationally indistinguishable. As discussed before, this definition is a generalization of CPA encryption in the following sense,

Lemma 1. *Let $\Pi' =$ (KeyGen, Enc, Dec) be a CPA-encryption scheme. Then $\Pi =$ (KeyGen, Enc, Dec, tKeyGen = KeyGen) is a trapdoor encryption scheme.*

The basic trapdoor encryption scheme does not provide any advantage in the trapdoor mode than the honest mode. Below, we consider two stronger security properties in the trapdoor mode.

Definition 12 (Statistical Trapdoor Encryption Scheme). *We say that trapdoor encryption scheme* $\Pi = (\mathsf{KeyGen}, \mathsf{Enc}, \mathsf{Dec}, \mathsf{tKeyGen})$ *is a statistical trapdoor encryption scheme, if the computational hiding property in Definition 11 is replaced by the following.*

Statistical hiding: *The following ensembles are statistically close.*

$$\left\{ \mathsf{tpk} \xleftarrow{\$} \mathsf{tKeyGen}(1^\lambda) \; : \; \mathsf{Enc}_{\mathsf{tpk}}(0) \right\}_\lambda \approx_s \left\{ \mathsf{tpk} \xleftarrow{\$} \mathsf{tKeyGen}(1^\lambda) \; : \; \mathsf{Enc}_{\mathsf{tpk}}(1) \right\}_\lambda$$

We note that any lossy encryption scheme as defined by Bellare, Hofheinz and Yilek [8] implies a statistical trapdoor encryption scheme. A lossy encryption scheme has a key generation algorithm KeyGen that takes as input the security parameter 1^λ and additionally a variable $m \in \{\mathsf{injective}, \mathsf{lossy}\}$ indicating whether to generate a key in the injective mode or in the lossy mode. A key generated in the injective mode ensures decryption correctness and semantic security, whereas a key generated in the lossy mode statistically loses information of the plaintexts, that is, encryption of different bits are statistically close. Therefore, we have:

Lemma 2. *Let* $\Pi' = (\mathsf{Gen}', \mathsf{Enc}, \mathsf{Dec})$ *be a lossy encryption scheme. Then* $\Pi = (\mathsf{KeyGen}, \mathsf{Enc}, \mathsf{Dec}, \mathsf{tKeyGen})$ *where* $\mathsf{KeyGen}(1^\lambda) = \mathsf{Gen}'(1^\lambda, \mathsf{injective})$ *and* $\mathsf{KeyGen}(1^\lambda) = \mathsf{Gen}'(1^\lambda, \mathsf{lossy})$, *is a statistical trapdoor encryption scheme.*

Definition 13 (μ-Hiding Trapdoor Encryption Scheme). *Let* μ *be any function We say that trapdoor encryption scheme* $\Pi = (\mathsf{KeyGen}, \mathsf{Enc}, \mathsf{Dec}, \mathsf{tKeyGen})$ *is a* μ*-Lossy trapdoor encryption scheme, if the computational hiding property in Definition 11 is replaced by the following.*

μ**-hiding:** *For any non-uniform PPT adversary* $\mathcal{A}$, *the following holds:*

$$\left| \Pr[\mathsf{tpk} \xleftarrow{\$} \mathsf{tKeyGen}(1^\lambda) \; : \; \mathcal{A}(\mathsf{Enc}_{\mathsf{tpk}}(0)) = 1] \right.$$
$$\left. - \Pr[\mathsf{tpk} \xleftarrow{\$} \mathsf{tKeyGen}(1^\lambda) \; : \; \mathcal{A}(\mathsf{Enc}_{\mathsf{tpk}}(1)) = 1] \right| \le \mu(\lambda)$$

where μ *is called the distinguishing gap.*

One of the instantiations of our general transformation for obtaining FHE relies on sub-exponentially indistinguishable IO and a μ-hiding trapdoor encryption scheme where μ is bounded by $\mathsf{negl}(\lambda)2^{-2l(\lambda)}$ and $l(\lambda)$ is an upper bound on the length of the ciphertext. In other words, the distinguishing gap is much smaller than the inverse exponentiation of the ciphertext length. We construct such a μ-hiding trapdoor encryption scheme using a μ-rerandomizable encryption. In fact, our construction achieves the stronger property of perfect hiding, that is, $\mu = 0$.

Definition 14 (μ-Rerandomizable Encryption Scheme). *We say that a quadruple of uniform PPT algorithms $\Pi = (\mathsf{Gen}, \mathsf{Enc}, \mathsf{Dec}, \mathsf{reRand})$ is a μ-rerandomizable encryption scheme, if $(\mathsf{Gen}, \mathsf{Enc}, \mathsf{Dec})$ is a CPA-secure encryption scheme, and additionally the algorithm reRand satisfies the following property:*

μ-**Rerandomizability:** *For every non-uniform PPT adversary $\mathcal{A}$, the following holds for every $\lambda \in \mathbb{N}$.*

$$\Big| \Pr[\mathcal{A}(\mathsf{pk}, c_0, c_1, \mathsf{reRand}_{\mathsf{pk}}(c_0)) = 1]$$
$$- \Pr[\mathcal{A}(\mathsf{pk}, c_0, c_1, \mathsf{reRand}_{\mathsf{pk}}(c_1)) = 1]\Big| \leq \mu(\lambda)$$

where $(\mathsf{pk}, \mathsf{sk}) \xleftarrow{\$} \mathsf{Gen}(1^\lambda)$, $c_0 \xleftarrow{\$} \mathsf{Enc}_{\mathsf{pk}}(b)$ and $c_1 \xleftarrow{\$} \mathsf{Enc}_{\mathsf{pk}}(b)$.

We way that Π is perfectly re-randomizable, if the distinguishing gap μ above is zero.

Many encryption scheme such as ElGamal, Goldwasser-Micali [27], Paillier [33], Damgård-Jurik [20], are in fact perfectly rerandomizable as per [36,28] and satisfy our definition. Furthermore, we show that

Lemma 3. *Let μ be a negligible function. Every μ-rerandomizable CPA encryption scheme can be transformed into a μ-hiding trapdoor encryption scheme.*

An overview of the construction was provided in the Introduction (See "LHE via trapdoor encryption"). We defer the formal construction and proof to the full version [19].

4.2   From Trapdoor Encryption to Leveled Homomorphic Encryption

In this section, we present our general transformation from a trapdoor encyrption scheme $\Pi = (\mathsf{KeyGen}, \mathsf{Enc}, \mathsf{Dec}, \mathsf{tKeyGen})$ to a leveled fully homomorphic encryption scheme LHE, relying on a pIO scheme $pi\mathcal{O}$ for a specific class $\mathbf{S}^\Pi$ of samplers defined by Π as described in Figure 3; (more explanation on the class is provided in the proof of semantic security).

Proposition 2. *Let Π be any trapdoor encryption scheme. Assume the existence of pIO for the class of samplers $\mathbf{S}^\Pi$ defined by Π as in Figure 3. Then, Π can be transformed into a leveled homomorphic encryption scheme.*

Below we first describe our construction and then prove its correctness and semantic security in Lemma 4 and 5. Without loss of generality, we assume that the public, secret keys and ciphertexts of Π have lengths bounded by $l(\lambda)$. Below we first describe our construction.

Construction of LHE: Let $L = L(\lambda)$ be the depth of the circuits that we want to evaluate. The four algorithms of the scheme proceed as follows:

- **Key generation:** LHE.Keygen($1^\lambda, 1^L$) does the following for every level i from 0 to L.
  - samples a pair of keys $(\mathsf{pk}_i, \mathsf{sk}_i) \xleftarrow{\$} \mathsf{KeyGen}(1^\lambda)$ of Π;
  - for $i \geq 1$, obfuscate the circuit $P_i = \mathsf{Prog}^{(\mathsf{sk}_{i-1}, \mathsf{pk}_i)}$ as described in Figure 2, that is, sample $\Lambda_i \xleftarrow{\$} pi\mathcal{O}(1^s, P_i)$ where the security parameter $s = s(\lambda)$ for obfuscation is an upper-bound on the size of all P_i's. [3]

  Finally outputs $\mathsf{pk} = \mathsf{pk}_0$, $\mathsf{sk} = \mathsf{sk}_L$, $\mathsf{evk} = \{P_i\}_{0 \leq i \leq L}$.
- **Encryption:** LHE.Enc$_{\mathsf{pk}}(m)$ outputs a fresh encryption of m under $\mathsf{pk} = \mathsf{pk}_0$ using Π, $c \xleftarrow{\$} \mathsf{Enc}_{\mathsf{pk}_0}(m)$.
- **Decryption:** LHE.Dec$_{\mathsf{sk}}(c)$ decrypts c using the secret key $\mathsf{sk} = \mathsf{sk}_L$ to obtain $m = \mathsf{Dec}_{\mathsf{sk}_L}(c)$.
- **Homomorphic evaluation:** LHE.Eval$_{\mathsf{evk}}(C, c_1, \ldots, c_\ell)$ on input a layered circuit C (consisting of only NAND gates) of depth at most L, evaluate C layer by layer; in iteration i, layer $i \in [L]$ is evaluated (the first layer is connected with the input wires): At the onset of this iteration, the values of the input wires of layer i has been homomorphically evaluated in the previous iteration and encrypted under key pk_{i-1} (in the first iteration, these encryptions are simply $c_1, \cdots, c_\ell$); for each NAND gate g in this layer i, let $\alpha(g), \beta(g)$ be encryption of the values of its input wires; evaluate g homomorhpically by computing $\gamma(g) = \Lambda_i(\alpha(g), \beta(g))$ to obtain an encryption of the value of g's output wire under public key pk_i. At the end, output the encryptions generated in the last iteration L.

---

sk, pk, tpk, α, and β are strings of length $l(\lambda)$.

Circuit Prog$^{(\mathsf{sk},\mathsf{pk})}(\alpha, \beta)$: Decrypt α and β to obtain $a = \mathsf{Dec}_{\mathsf{sk}}(\alpha)$ and $b = \mathsf{Dec}_{\mathsf{sk}}(\beta)$; output $\gamma \xleftarrow{\$} \mathsf{Enc}_{\mathsf{pk}}(a \text{ NAND } b)$.

Circuit tProg$^{(\mathsf{tpk})}(\alpha, \beta)$: Output $\gamma \xleftarrow{\$} \mathsf{Enc}_{\mathsf{tpk}}(0)$.

Both circuits are padded to their maximum size. Let $s(\lambda)$ be an upper bound on their sizes.

Fig. 2. Circuits used in the construction of LHE and its analysis

It follows from the correctness of pIO and Π that the scheme LHE is correct; we refer the reader to the full version [19] for a formal proof.

Lemma 4. *If* pIO *and* Π *are correct, then* LHE *has homomorphism.*

---

[3] This is because the obfuscator $pi\mathcal{O}(1^\lambda, C)$ works with classes of circuits $\mathcal{C}_\lambda$ of size at most λ.

Proof of Semantic Security of LHE. Towards establishing the semantic security of LHE, we rely on the security property of pIO for the class of samplers $\mathbf{S}^\Pi$ defined by the trapdoor encryption scheme Π used in LHE. Roughly speaking, samplers in $\mathbf{S}^\Pi$ samples pairs of circuits where one of them is identical the "honest" program used for generating the evaluation key in LHE, except that a trapdoor public key tpk (instead of an honest public key) is hardwired in (that is, $\mathsf{Prog}^{(\mathsf{sk},\mathsf{tpk})}$), and the other one is a "trapdoor" program $\mathsf{tProg}^{(\mathsf{tpk})}$ as described in Figure 2 that always generates a ciphertext of 0 under the "trapdoor" public key hardwired inside. More precisely, we describe the class of samplers in Figure 3.

$\Pi = (\mathsf{KeyGen}, \mathsf{Enc}, \mathsf{Dec}, \mathsf{tKeyGen})$ is a trapdoor encryption scheme, $SK = \{sk_\lambda\}$ is a sequence of strings of length $l(\lambda)$, and $s(\lambda)$ is an upper bound on the sizes of programs $\mathsf{Prog}^{(\mathsf{sk},\mathsf{tpk})}$ and $\mathsf{tProg}^{(\mathsf{tpk})}$.

The Sampler D^{SK}: The distribution D_s^{SK} samples a trapdoor public key $\mathsf{tpk} \xleftarrow{\$} \mathsf{tKeyGen}(1^\lambda)$, and outputs $C_0 = \mathsf{Prog}^{(\mathsf{sk},\mathsf{tpk})}$, $C_1 = \mathsf{tProg}^{(\mathsf{tpk})}$ and $z = \mathsf{tpk}$, where $sk = \mathsf{sk}_\lambda$.

The Class $\mathbf{S}^\Pi$: Let $\mathbf{S}^\Pi$ be the class of samplers that include distribution ensembles D^{SK} for all sequence of strings SK of length $l(\lambda)$.

Fig. 3. The class of samplers for proving the semantic security of LHE

Next we show that LHE is semantic secure. We note that for the proof to go through, we only rely on the fact that $pi\mathcal{O}$ is a pIO for the above described class $\mathbf{S}^\Pi$ and the fact that trapdoor public keys of the trapdoor encryption scheme Π are indistinguishable from honest public keys. The proof actually does not depend on any hiding property in the trapdoor mode, which will only play a role later when instantiating pIO for $\mathbf{S}^\Pi$.

Lemma 5. *Assume that Π is a trapdoor encryption scheme and $pi\mathcal{O}$ is a pIO for the class of samplers $\mathbf{S}^\Pi$ in Figure 3. Then, LHE is semantically secure.*

Proof. Fix any polynomial time adversary $\mathcal{A}$. We want to show that for every $\lambda \in \mathbb{N}$, it holds that, the advantage of the adverary $\mathsf{Adv}_{\mathrm{CPA}}[\mathcal{A}]$ is negligible.

$$|\Pr[\mathcal{A}(\mathsf{pk}, \mathsf{evk}, \mathsf{LHE}.\mathsf{Enc}_{\mathsf{pk}}(0)) = 1] - \Pr[\mathcal{A}(\mathsf{pk}, \mathsf{evk}, \mathsf{LHE}.\mathsf{Enc}_{\mathsf{pk}}(1)) = 1]| < \mathsf{negl}(\lambda),$$

where $(\mathsf{pk}, \mathsf{evk}, \mathsf{sk}) \leftarrow \mathsf{LHE}.\mathsf{Keygen}(1^\lambda)$.

Towards this, we consider two sequences of hybrids $H_0^b, \cdots, H_L^b$ for $b \in \{0,1\}$. H_0^b is exactly an honest CPA game with the adversary $\mathcal{A}$ where it receives a challenge ciphertext that is an encryption of b; in intermediate hybrids, the adversary $\mathcal{A}$ participates in a modified game. We show that for every two subsequent hybrids H_i^b, H_{i+1}^b, as well as H_L^0, H_L^1, the view of $\mathcal{A}$ is indistinguishable. Below we formally describe all the hybrids.

Hybrid H_0^b: Hybrid H_0^b is an honest CPA game with $\mathcal{A}$, where $\mathcal{A}$ receives $(\mathsf{pk}, \mathsf{evk}, c^* = \mathsf{LHE.Enc}_{\mathsf{pk}}(b))$ for freshly sampled $(\mathsf{pk}, \mathsf{evk}, \mathsf{sk}) \xleftarrow{\$} \mathsf{LHE.Keygen}(1^\lambda)$. By construction of LHE, the view of $\mathcal{A}$ is,

$$\mathsf{view}[\mathcal{A}]_0^b = \left(\mathsf{pk} = \mathsf{pk}_0, \mathsf{evk} = (\Lambda_1, \cdots, \Lambda_L), c_b = \mathsf{Enc}_{\mathsf{pk}_0}(b)\right)$$

Hybrid H_i^b for $i > 0$: Hybrid H_i^b proceeds identically to H_0^b except that the evaluation key evk is sampled in a different way. Recall that in H_0^b, evk consists of the obfuscated circuits $\Lambda_1, \cdots, \Lambda_L$ of circuits $P_1, \cdots, P_L$, where $P_j = \mathsf{Prog}^{(sk_{j-1}, pk_j)}$. In H_i^b, the last i circuits $P_{L-i+1}, \cdots, P_L$ are replaced with $tP_{L-i+1}, \cdots, tP_L$, where $tP_j = \mathsf{tProg}^{(tpk_j)}$ (see Figure 2) hardwired with a freshly sampled "trapdoor" public key $tpk_j \xleftarrow{\$} \mathsf{tKeyGen}(1^\lambda)$. Let $t\Lambda_{L-i+1}, \cdots, t\Lambda_L$ be the obfuscated circuits of $tP_{L-i+1}, \cdots, tP_L$. Then evk_i in H_i^b consists of $\mathsf{evk}_i = \Lambda_1, \cdots, \Lambda_{L-i}, t\Lambda_{L-i+1}, \cdots, t\Lambda_L$. The view of A in H_i^b is

$$\mathsf{view}[\mathcal{A}]_i^b = \left(\mathsf{pk}_0, \mathsf{evk}_i = (\Lambda_1, \cdots, \Lambda_{L-i}, t\Lambda_{L-i+1}, \cdots, t\Lambda_L), c_b = \mathsf{Enc}_{\mathsf{pk}_0}(b)\right)$$

To show that the $\mathcal{A}$ cannot distinguish the two CPA games, it is equivalent to show that $\mathcal{A}$ cannot distinguish hybrids H_0^0 and H_0^1. Towards this, it suffices to prove that $\mathcal{A}$ cannot distinguish any of the neighboring hybrids, that is,

- The views of $\mathcal{A}$ in H_L^0 and H_L^1 are indistinguishable,
- For every b and $0 \leq i \leq L$, the views of $\mathcal{A}$ in H_i^b and H_{i+1}^b are indistinguishable,

Towards showing the first indistinguishability, we observe that in H_L^0 and H_L^1, the evaluation key evk_L consists of only obfuscation of the "trapdoor" programs $\{t\Lambda_i \xleftarrow{\$} pi\mathcal{O}(1^s, \mathsf{tProg}^{(tpk_j)})\}$ which does not depend on any secret key sk_j. Thus by the semantic security of Π, encryption $\mathsf{Enc}_{\mathsf{pk}_0}(0)$ and $\mathsf{Enc}_{\mathsf{pk}_0}(1)$ are indistinguishable, and hence so are the views of $\mathcal{A}$ in H_L^0 and H_L^1.

Towards showing the second indistinguishability, we observe that the only difference between H_i^b and H_{i+1}^b lies in whether the evaluation key contains an obfuscation Λ_{L-i} of the honest program $\mathsf{Prog}^{(sk_{L-i-1}, pk_{L-i})}$ for layer $L-i$, or an obfuscation $t\Lambda_{L-i}$ of the trapdoor program $\mathsf{tProg}^{(tpk_{L-i})}$. Furthermore, in both H_i^b and H_{i+1}^b the generation of the evaluation key does not depend on sk_{L-i}, and hence neither do the views of $\mathcal{A}$. Thus to show the indistinguishability of the views of $\mathcal{A}$ it suffices to show the indistinguishability of the following ensembles, from which the views of A in H_i^b and H_{i+1}^b can be reconstructed.

$$\left\{\Lambda_{L-i}, \mathsf{pk}_{L-i}, \mathsf{pk}_{L-i-1},\right)\right\}_\lambda \approx \left\{t\Lambda_{L-i}, \mathsf{tpk}_{L-i}, \mathsf{pk}_{L-i-1}\right\}_\lambda$$

where in the above distributions $(\mathsf{pk}_{L-i}, \mathsf{sk}_{L-i})$ and $(\mathsf{pk}_{L-i-1}, \mathsf{sk}_{L-i-1})$ are all randomly sampled honest keys of Π, tpk_{L-i} is a randomly sampled trapdoor public key, and Λ_{L-i} and $t\Lambda_{L-i}$ are obfuscations of the honest program or the trapdoor program as in H_i^b and H_{i+1}^b. We argue why the views of $\mathcal{A}$ in H_i^b and H_{i+1}^b can be reconstructed from the left and right random variables respectively: This is because Λ_{L-i} and $t\Lambda_{L-i}$ correspond respectively to the $(L-i)$'th obfuscation in the evaluation key in H_i^b and H_{i+1}^b, and the other obfuscated programs $\Lambda_1, \cdots \Lambda_{L-i-1}, t\Lambda_{L-i+1}, \cdots, t\Lambda_L$ in the evaluation key can be sampled efficiently given pk_{L-i-1} together with pk_{L-i} or tpk_{L-i}; finally, encryption of b under pk_0 can be sampled independently.

We show the above indistinguishability in two steps, via an intermediate hybrid where an obfuscation Λ'_{L-i} of the hybrid program $\mathsf{Prog}^{(\mathsf{sk}_{L-i-1}, \mathsf{tpk}_{L-i})}$ is sampled; the hybrid program is the same as the honest program except that a trapdoor public key tpk_{L-i} is hardwired.

$$\left\{ \boxed{\Lambda_{L-i}, \mathsf{pk}_{L-i}}, \mathsf{pk}_{L-i-1},) \right\}_\lambda \approx \left\{ \boxed{\Lambda'_{L-i}, \mathsf{tpk}_{L-i}}, \mathsf{pk}_{L-i-1},) \right\}_\lambda \tag{1}$$

$$\left\{ \boxed{\Lambda'_{L-i}}, \mathsf{pk}_{L-i}, \mathsf{pk}_{L-i-1},) \right\}_\lambda \approx \left\{ \boxed{t\Lambda_{L-i}}, \mathsf{tpk}_{L-i}, \mathsf{pk}_{L-i-1}) \right\}_\lambda \tag{2}$$

Equation (1) follows directly from the fact that a randomly sampled trapdoor public key is indistinguishable from an honest public key.

Equation (2) holds following the pIO security for the class of samplers $\mathbf{S}^\Pi$. More specifically, to show the equation, it suffices to show that it holds for every fixed sequence of pairs $S = \left\{ (\mathsf{pk}_{L-i-1}, \mathsf{sk}_{L-i-1}) \right\}$ of length $l(\lambda)$ each. Fix such a sequence S and let $SK = \{ \mathsf{sk}_{L-i-1} \}$ be the sequence of secret keys only. Notice that the sampler D^{SK} described in Figure 3 produces exactly the hybrid and trapdoor programs as above, that is,

$$(C_0 = \mathsf{Prog}^{(\mathsf{sk}_{L-i-1}, \mathsf{tpk}_{L-i})}, C_1 = \mathsf{tProg}^{(\mathsf{tpk}_{L-i})}, z = \mathsf{tpk}_{L-i}) \xleftarrow{\$} D_s^{SK}$$

Thus for the fixed sequence S, Equation (2) is equivalent to the following:

$$\left\{ (C_0, C_1, z) \xleftarrow{\$} D_s^{\mathsf{sk}_{L-i-1}} : (C_0, C_1, pi\mathcal{O}(1^s, C_0), z) \right\}_\lambda$$
$$\approx \left\{ (C_0, C_1, z) \xleftarrow{\$} D_s^{\mathsf{sk}_{L-i-1}} : (C_0, C_1, pi\mathcal{O}(1^s, C_1), z) \right\}_\lambda$$

This indistinguishability follows directly from the premise that $pi\mathcal{O}$ is a pIO for the sampler D^{SK}. Thus the views of $\mathcal{A}$ in H_i^b and H_{i+1}^b are indistinguishable. $\square$

Instantiation of LHE. We show how to instantiate our general transformation from any trapdoor encryption scheme to a LHE scheme, more precisely, how to realize the premise of Proposition 2.

Instantiation 1: Rerandomizable Encryption + Sub-exponential IO. The first instantiation uses a ν-hiding trapdoor encryption scheme Π and a X-Ind pIO for appropriate functions ν and X. Let us specify the functions: First, set $\nu(\lambda) = \mathsf{negl}(\lambda)2^{-2l(\lambda)}$, where $l(\lambda)$ is an upper bound on the lengths of the ciphertexts of Π. Second, to set the function X, recall that every sampler D_s^{SK} [4] in the class $\mathbf{S}^\Pi$ produces circuits $C_0 = \mathsf{Prog}^{(\mathsf{sk}, \mathsf{tpk})}$ and $C_1 = \mathsf{tProg}^{(\mathsf{tpk})}$ of size $s(\lambda)$ and input

---

[4] We remind the reader that all variables related with the encryption scheme Π, such as $\mathsf{pk}, \mathsf{sk}, \mathsf{tpk}$, are generated using security parameter λ, while the pIO scheme $pi\mathcal{O}$ and the related samplers all use security parameter $s = s(\lambda)$.

length $2l(\lambda)$; by setting $X(s(\lambda)) = 2^{2l(\lambda)}$, we have that the two sampled circuits C_0, C_1 differ at most $X(s)$ inputs and the output distributions of C_0 and C_1 are $\mathsf{negl}(\lambda)X(\lambda)^{-1}$-indistinguishable following from the ν-hiding property of Π. Therefore D^{SK} is an X-Ind sampler.

Therefore, any X-Ind pIO scheme is a pIO scheme matching the ν-hiding trapdoor encryption scheme Π. Furthermore, by Lemma 3, the existence of a ν-rerandomizable encryption scheme (in particular, a perfectly rerandomizable one) implies that of a ν-hiding trapdoor encryption scheme. By Theorem 2, X-Ind pIO can be constructed from any sub-exponentially indistinguishable IO and sub-exponentially secure OWFs. Therefore, following Proposition 2, we have

Corollary 1 (LHE from Rerandomizable Encryption and Sub-exp Secure IO and OWF.). *Let Π be a perfectly rerandomizable encryption scheme. Assume the existence of sub-exponentially indistinguishable IO for circuits and sub-exponentially secure one-way functions. Π can be turned into a leveled homomorphic encryption scheme.*

Instantiation 2: Lossy Encryption + worst-case-input pIO. The second instantiation combines a lossy encryption scheme, which by Lemma 2 directly implies a statistical trapdoor encryption scheme Π, with a worst-case-input pIO. By the statistical hiding property of the trapdoor mode of Π, every sampler D^{SK} in the class $\mathbf{S}^{\Pi}$ corresponding to Π samples circuits $C_0 = \mathsf{Prog}^{(\mathsf{sk},\mathsf{tpk})}$ and $C_1 = \mathsf{tProg}^{(\mathsf{tpk})}$ with statistically close output distributions for every input. Therefore, D^{SK} is a worst-case-input indistinguishable sampler. In other words, any worst-case-input pIO is a pIO for the class $\mathbf{S}^{\Pi}$. Following Proposition 2,

Corollary 2 (LHE from Lossy Encryption and worst-case-input pIO). *Let Pi be a lossy encryption scheme. Assume the existence of a worst-case-input* pIO *scheme $pi\mathcal{O}$. Then, Π can be transformed into a leveled homomorphic encryption scheme.*

Instantiation 3: CPA Encryption + Dynamic-input pIO *for Specific Class.* Finally, we observe that any CPA encryption Π can be turned into a LHE, if there exits a strong notion of pIO, namely dynamic-input pIO for $\mathbf{S}^{\Pi}$. As observed in Lemma 1, any CPA encryption scheme $\Pi = (\mathsf{Gen}, \mathsf{Enc}, \mathsf{Dec})$ directly implies a trapdoor encryption scheme $\Pi' = (\mathsf{Gen}, \mathsf{Enc}, \mathsf{Dec}, \mathsf{tKeyGen} = \mathsf{Gen})$ with a computationally hiding trapdoor mode. This implies that every sampler D^{SK} in the matching class $\mathbf{S}^{\Pi}$ is a dynamic-input indistinguishable sampler. Therefore,

Corollary 3. *Let Π be any CPA encryption scheme and Π' the corresponding trapdoor encryption scheme. Assume the existence of a dynamic-input* pIO *scheme $pi\mathcal{O}$ for $\mathbf{S}^{\Pi'}$. Then, Π can be transformed into a leveled homomorphic encryption scheme.*

We note that although general pIO for all dynamic-input indistinguishable samplers is implausible by [22], pIO for the specific class of samplers $\mathbf{S}^{\Pi'}$ circumvents the implausibility result. This is because the implausibility of [22]

applies only to a specific class of samplers that produce (C_0, C_1, z) where z is an obfuscated program that essentially distinguishes circuits with the same functionality as C_0 from ones with the same functionality as C_1 using only their I/O interfaces. However, samplers in $\mathbf{S}^{\Pi'}$ produce auxiliary input that is a public key pk of Π, which cannot be used to tell apart circuits of functionalities identical to $\mathsf{Prog}^{(sk,pk)}$ or $\mathsf{tProg}^{(pk)}$ through only their I/O interfaces, due to the semantic security of Π. Therefore, dynamic-input pIO for $\mathbf{S}^{\Pi'}$ circumvents the implausibility. We consider the same construction of X-Ind pIO as a potential candidate construction of dynamic-input pIO for $\mathbf{S}^{\Pi'}$.

4.3 From LHE to FHE

In this section, we show how to transform any leveled homomorphic encryption scheme LHE with a fixed decryption depth into a fully homomorphic one, *without relying on circular security*. More specifically,

- we say that a LHE scheme LHE = (LHE.Keygen, LHE.Enc, LHE.Dec, LHE.Eval) has a *fixed decryption depth* $D_{\mathsf{LHE.Dec}}(\cdot)$, if for every polynomial depth L, every (pk, sk, evk) in the support of $\mathsf{LHE.Keygen}(1^\lambda, 1^{L(\lambda)})$, every freshly generated or homomorphically evaluated ciphertext c^* in the support of $\mathsf{LHE.Enc(pk, \cdot)}$ or $\mathsf{LHE.Eval(pk, (C, \cdots))}$ with a depth $L(\lambda)$ circuit C, the decryption algorithm $\mathsf{LHE.Dec_{sk}}(c^*)$ has depth bounded by $D_{\mathsf{LHE.Dec}}(\lambda)$.

We now sketch a general transformation that turns any LHE scheme with a fixed decryption depth into a FHE. The transformation proceeds in two steps.

A "imaginary" FHE with a non-succinct evaluation key: In a first step, imagine a FHE scheme with an evaluation key evk that consists of a super-polynomial number $\mathcal{L}(\lambda)$ of layer evaluation keys each of size $\mathrm{poly}(\lambda)$. Each layer $\ell \in [\mathcal{L}]$ is associated with a key tuple $(\mathsf{pk}_\ell, \mathsf{sk}_\ell, \mathsf{evk}_\ell)$ of LHE that supports evaluating circuits of depth $D' = D_{\mathsf{LHE.Dec}} + 1$; moreover, for each layer, an encryption of the secret key $\mathsf{sk}_{\ell-1}$ under the public key pk_ℓ is released, that is, $\Lambda_\ell = (\mathsf{pk}_\ell, \mathsf{evk}_\ell, c_\ell)$ where $(\mathsf{pk}_\ell, \mathsf{sk}_\ell, \mathsf{evk}_\ell) \xleftarrow{\$} \mathsf{LHE.Keygen}(1^\lambda, 1^{D'})$ for $D' = D_{\mathsf{LHE.Dec}} + 1$ and $c_\ell = \mathsf{LHE.Enc_{pk_\ell}}(\mathsf{sk}_{\ell-1})$. Each Λ_ℓ is a layer evaluation key: Given two ciphertexts α, β of bits a and b under $\mathsf{pk}_{\ell-1}$, we can obtain an encryption γ of a NAND b under pk_ℓ, by evaluating homomorphically over c_ℓ the function $f_{\alpha,\beta}(\mathsf{sk}_{\ell-1})$ that decrypts α, β using $\mathsf{sk}_{\ell-1}$ and computes NAND of the decrypted bits. Since $f_{\alpha,\beta}$ has depth exactly $D_{\mathsf{LHE.Dec}} + 1$, the homomorphic computation yields a ciphertext γ of a NAND b under pk_ℓ correctly.

Therefore by publishing a super-polynomially number $\mathcal{L}$ of layer evaluation keys $\mathsf{evk} = (\Lambda_1, \cdots \Lambda_{\mathcal{L}})$, the scheme supports homomorphic evaluation of any polynomial depth circuits.

"Compress" the size of the evaluation key: The next step is to "compress" the size of the super-polynomially long evaluation key to obtain a FHE with succinct evaluation key. This step relies on an IO for circuits and a puncturable PRF. The idea is to obfuscate a master circuit Γ that on input

$\ell \in [\mathcal{L}]$ computes the ℓ'th layer evaluation key Λ_ℓ produced using pseudo-random coins generated with a puncturable PRF and hardwired PRF keys k, k'. That is,

$$\Lambda_\ell = \Gamma^{(k,k')}(\ell), \quad \text{where } (\mathsf{pk}_\ell, \mathsf{sk}_\ell, \mathsf{evk}_\ell) = \mathsf{LHE.Keygen}(1^\lambda, 1^{D'}; \mathsf{PRF}(k, \ell)),$$
$$(\mathsf{pk}_{\ell-1}, \mathsf{sk}_{\ell-1}, \mathsf{evk}_{\ell-1}) = \mathsf{LHE.Keygen}(1^\lambda, 1^{D'}; \mathsf{PRF}(k, \ell-1)),$$
$$c_\ell = \mathsf{LHE.Enc}_{\mathsf{pk}_\ell}(\mathsf{sk}_{\ell-1}; \mathsf{PRF}(k', \ell))$$
$$\Lambda_\ell = (\mathsf{pk}_\ell, \mathsf{evk}_\ell, c_\ell)$$

Since the size of the master program $\Gamma^{(k,k')}$ is a fixed polynomial in λ, the new evaluation key $\mathsf{evk} \xleftarrow{\$} i\mathcal{O}(1^s, \Gamma^{k,k'})$ is succinct, of a fixed polynomial size in λ (where s an upperbound on the size of Γ and k, k' are randomly sampled PRF keys). It follows from a careful hybrid argument over the virtual super-polynomial number of levels (similar to that in [11,32]) that the semantic security of LHE remains even when the new evaluation key is additionally released, provided that all primitives from LHE, to $i\mathcal{O}$ to PRF all have a slightly inverse super-polynomially small distinguishing gap $\mu(\lambda) = \mathsf{negl}(\lambda)\mathcal{L}(\lambda)^{-1}$.

Finally, we note that any LHE scheme with decryption in $\mathbf{NC}^1$ have a fixed decryption depth (in particular, the depth is bounded by λ). Many known constructions, for example [23,18,16,15,24] satisfy this property. Thus, if these constructions are slightly super-polynomially secure, by assuming slightly stronger underlying assumptions (for instance the LHE scheme of [18] can be made slightly super-polynomially secure if assuming that the underlying learning with error assumption is slightly super-polynomially secure), they can be directly transformed into a FHE assuming slightly super-polynomially secure IO and OWFs (without assuming circular security).

We also note that our LHE scheme constructed in Section 4.2 has a fixed decryption depth, since its decryption algorithm is identical to that of the underlying trapdoor encryption scheme. It can also be transformed into a FHE using the above general transformation. In the full version [19], we provide a formal description and security proof of the FHE transformed from our LHE.

Acknowledgements. Ran Canetti's research is supported by the Check Point Institute for Information Security, ISF grant 1523/14, the NSF MACS Frontier project, and NSF Algorithmic Foundations grant 1218461. Huijia Lin's research is partially supported by a gift from the Gareatis Foundation. Stefano Tessaro's research was partially supported by NSF Grant CNS-1423566 and a gift from the Gareatis Foundation. Vinod Vaikuntanathan's research was supported by DARPA Grant number FA8750-11-2-0225, an Alfred P. Sloan Research Fellowship, an NSF CAREER Award CNS-1350619, NSF Frontier Grant CNS-1414119, a Microsoft Faculty Fellowship, and a Steven and Renee Finn Career Development Chair from MIT.

References

1. Alwen, J., Barbosa, M., Farshim, P., Gennaro, R., Dov Gordon, S., Tessaro, S., Wilson, D.A.: On the relationship between functional encryption, obfuscation, and fully homomorphic encryption. In: Stam, M. (ed.) IMACC 2013. LNCS, vol. 8308, pp. 65–84. Springer, Heidelberg (2013)

2. Ananth, P., Boneh, D., Garg, S., Sahai, A., Zhandry, M.: Differing-inputs obfuscation and applications. Cryptology ePrint Archive, Report 2013/689 (2013), http://eprint.iacr.org/

3. Applebaum, B.: Bootstrapping obfuscators via fast pseudorandom functions. IACR Cryptology ePrint Archive, 2013:699 (2013)

4. Applebaum, B., Ishai, Y., Kushilevitz, E.: Cryptography in NC^0. In: FOCS, pp. 166–175. IEEE Computer Society (2004)

5. Applebaum, B., Ishai, Y., Kushilevitz, E.: Computationally private randomizing polynomials and their applications. Computational Complexity 15(2), 115–162 (2006)

6. Barak, B., Garg, S., Kalai, Y.T., Paneth, O., Sahai, A.: Protecting obfuscation against algebraic attacks. In: Nguyen, P.Q., Oswald, E. (eds.) EUROCRYPT 2014. LNCS, vol. 8441, pp. 221–238. Springer, Heidelberg (2014)

7. Barak, B., Goldreich, O., Impagliazzo, R., Rudich, S., Sahai, A., Vadhan, S.P., Yang, K.: On the (im)possibility of obfuscating programs. J. ACM 59(2), 6 (2012)

8. Bellare, M., Hofheinz, D., Yilek, S.: Possibility and impossibility results for encryption and commitment secure under selective opening. In: Joux, A. (ed.) EUROCRYPT 2009. LNCS, vol. 5479, pp. 1–35. Springer, Heidelberg (2009)

9. Bellare, M., Stepanovs, I., Tessaro, S.: Poly-many hardcore bits for any one-way function and a framework for differing-inputs obfuscation. In: Sarkar, P., Iwata, T. (eds.) ASIACRYPT 2014, Part II. LNCS, vol. 8874, pp. 102–121. Springer, Heidelberg (2014)

10. Bitansky, N., Canetti, R., Cohn, H., Goldwasser, S., Kalai, Y.T., Paneth, O., Rosen, A.: The impossibility of obfuscation with auxiliary input or a universal simulator. In: Garay, J.A., Gennaro, R. (eds.) CRYPTO 2014, Part II. LNCS, vol. 8617, pp. 71–89. Springer, Heidelberg (2014)

11. Bitansky, N., Garg, S., Telang, S.: Succinct randomized encodings and their applications. Cryptology ePrint Archive, Report 2014/771 (2014), http://eprint.iacr.org/

12. Boneh, D., Waters, B.: Constrained pseudorandom functions and their applications. In: Sako, K., Sarkar, P. (eds.) ASIACRYPT 2013, Part II. LNCS, vol. 8270, pp. 280–300. Springer, Heidelberg (2013)

13. Boyle, E., Chung, K.-M., Pass, R.: On extractability obfuscation. In: Lindell, Y. (ed.) TCC 2014. LNCS, vol. 8349, pp. 52–73. Springer, Heidelberg (2014)

14. Boyle, E., Goldwasser, S., Ivan, I.: Functional signatures and pseudorandom functions. In: Krawczyk, H. (ed.) PKC 2014. LNCS, vol. 8383, pp. 501–519. Springer, Heidelberg (2014)

15. Brakerski, Z.: Fully homomorphic encryption without modulus switching from classical gapSVP. In: Safavi-Naini, R., Canetti, R. (eds.) CRYPTO 2012. LNCS, vol. 7417, pp. 868–886. Springer, Heidelberg (2012)

16. Brakerski, Z., Gentry, C., Vaikuntanathan, V.: (Leveled) fully homomorphic encryption without bootstrapping. In: ITCS, pp. 309–325 (2012)

17. Brakerski, Z., Rothblum, G.N.: Obfuscating conjunctions. In: Canetti, R., Garay, J.A. (eds.) CRYPTO 2013, Part II. LNCS, vol. 8043, pp. 416–434. Springer, Heidelberg (2013)

18. Brakerski, Z., Vaikuntanathan, V.: Efficient fully homomorphic encryption from (standard) LWE. In: FOCS, pp. 97–106 (2011), References are to full version: http://eprint.iacr.org/2011/344
19. Canetti, R., Lin, H., Tessaro, S., Vaikuntanathan, V.: Obfuscation of probabilistic circuits and applications. Cryptology ePrint Archive, Report 2014/882 (2014), http://eprint.iacr.org/
20. Damgård, I., Jurik, M.: A generalisation, a simplification and some applications of paillier's probabilistic public-key system. In: Kim, K.-C. (ed.) PKC 2001. LNCS, vol. 1992, pp. 119–136. Springer, Heidelberg (2001)
21. Garg, S., Gentry, C., Halevi, S., Sahai, A., Raikova, M., Waters, B.: Candidate Indistinguishability Obfuscation and Functional Encryption for all circuits. In: FOCS (2013)
22. Garg, S., Gentry, C., Halevi, S., Wichs, D.: On the implausibility of differing-inputs obfuscation and extractable witness encryption with auxiliary input. In: Garay, J.A., Gennaro, R. (eds.) CRYPTO 2014, Part I. LNCS, vol. 8616, pp. 518–535. Springer, Heidelberg (2014)
23. Gentry, C.: Fully homomorphic encryption using ideal lattices. In: STOC, pp. 169–178 (2009)
24. Gentry, C., Sahai, A., Waters, B.: Homomorphic encryption from learning with errors: Conceptually-simpler, asymptotically-faster, attribute-based. In: Canetti, R., Garay, J.A. (eds.) CRYPTO 2013, Part I. LNCS, vol. 8042, pp. 75–92. Springer, Heidelberg (2013)
25. Goldreich, O., Goldwasser, S., Micali, S.: How to construct random functions. J. ACM 33(4), 792–807 (1986)
26. Goldwasser, S., Kalai, Y.T.: On the impossibility of obfuscation with auxiliary input. In: FOCS, pp. 553–562. IEEE Computer Society (2005)
27. Goldwasser, S., Micali, S.: Probabilistic encryption. J. Comput. Syst. Sci. 28(2), 270–299 (1984)
28. Hemenway, B., Libert, B., Ostrovsky, R., Vergnaud, D.: Lossy encryption: Constructions from general assumptions and efficient selective opening chosen ciphertext security. In: Lee, D.H., Wang, X. (eds.) ASIACRYPT 2011. LNCS, vol. 7073, pp. 70–88. Springer, Heidelberg (2011)
29. Ishai, Y., Kushilevitz, E.: Randomizing polynomials: A new representation with applications to round-efficient secure computation. In: FOCS, pp. 294–304. IEEE Computer Society (2000)
30. Kiayias, A., Papadopoulos, S., Triandopoulos, N., Zacharias, T.: Delegatable pseudorandom functions and applications. In: Sadeghi, A.R., Gligor, V.D., Yung, M. (eds.) CCS, pp. 669–684. ACM (2013)
31. Kol, G., Naor, M.: Games for exchanging information. In: STOC, pp. 423–432 (2008)
32. Lin, H., Pass, R.: Succinct garbling schemes and applications. Cryptology ePrint Archive, Report 2014/766 (2014), http://eprint.iacr.org/
33. Paillier, P.: Public-key cryptosystems based on composite degree residuosity classes. In: Stern, J. (ed.) EUROCRYPT 1999. LNCS, vol. 1592, pp. 223–238. Springer, Heidelberg (1999)
34. Peikert, C., Vaikuntanathan, V., Waters, B.: A framework for efficient and composable oblivious transfer. In: Wagner, D. (ed.) CRYPTO 2008. LNCS, vol. 5157, pp. 554–571. Springer, Heidelberg (2008)
35. Peikert, C., Waters, B.: Lossy trapdoor functions and their applications. SIAM J. Comput. 40(6), 1803–1844 (2011)
36. Prabhakaran, M., Rosulek, M.: Rerandomizable RCCA encryption. In: Menezes, A. (ed.) CRYPTO 2007. LNCS, vol. 4622, pp. 517–534. Springer, Heidelberg (2007)
37. Sahai, A., Waters, B.: How to use indistinguishability obfuscation: deniable encryption, and more. In: STOC, pp. 475–484 (2014)

Graph-Induced Multilinear Maps from Lattices

Craig Gentry[1], Sergey Gorbunov[2], and Shai Halevi[1]

[1] IBM Research, Yorktown, NY, USA
cbgentry@us.ibm.com, shaih@alum.mit.edu
[2] MIT, Cambridge, MA, USA
sergeyg@mit.edu

Abstract. Graded multilinear encodings have found extensive applications in cryptography ranging from non-interactive key exchange protocols, to broadcast and attribute-based encryption, and even to software obfuscation. Despite seemingly unlimited applicability, essentially only two candidate constructions are known (GGH and CLT). In this work, we describe a new graph-induced multilinear encoding scheme from lattices. In a graph-induced multilinear encoding scheme the arithmetic operations that are allowed are restricted through an explicitly defined directed graph (somewhat similar to the "asymmetric variant" of previous schemes). Our construction encodes Learning With Errors (LWE) samples in short square matrices of higher dimensions. Addition and multiplication of the encodings corresponds naturally to addition and multiplication of the LWE secrets. Security of the new scheme is not known to follow from LWE hardness (or any other "nice" assumption), at present it requires making new hardness assumptions.

Keywords: Multilinear maps, Lattices, LWE.

1 Introduction

Cryptographic multilinear maps are an amazingly powerful tool: like homomorphic encryption schemes, they let us encode data in a manner that simultaneously hides it and permits processing on it. But they go even further and let us recover some limited information (such as equality) on the processed data without needing any secret key. Even in their simple bi-linear form (that only supports quadratic processing) they already give us pairing-based cryptography [28,39,5], enabling powerful applications such as identity- and attribute-based encryption [6,40,26], broadcast encryption [8] and many others. In their general form, cryptographic multilinear maps are so useful that we had a body of work examining their applications even before we knew of any candidate constructions to realize them [9,37,35,38]. Formally, a non-degenerate map between order-q algebraic groups, $e : G^d \to G_T$, is $d-$multilinear if for all $a_1, \ldots, a_d \in \mathbb{Z}_q$ and $g \in G$,

$$e(g^{a_1}, \ldots, g^{a_d}) = e(g, \ldots, g)^{a_1 \cdot \ldots \cdot a_d}.$$

We say that the map e is "cryptographic" if we can evaluate it efficiently and at least the discrete-logarithm in the groups G, G_T is hard.

Y. Dodis and J.B. Nielsen (Eds.): TCC 2015, Part II, LNCS 9015, pp. 498–527, 2015.

In a recent breakthrough, Garg, Gentry and Halevi [19] gave the first candidate construction that "approximate" multilinear maps from ideal lattices, followed by a second construction by Coron, Lepoint and Tibouchi [16] over the integers. (Some optimizations to the GGH scheme were proposed in [30]). In these constructions there are no explicit algebraic groups, and the transformation $a \mapsto g^a$ is replaced by some (randomized) encoding function. These constructions are called *graded encoding schemes*, where the "graded" adjective refers to the ability to carry out intermediate computations. One way to think of these intermediate computations is as a sequence of levels (or groups) $G_1, \ldots, G_d$ and a set of maps e_{ij} such that for all $g_i^a \in G_i, g_j^b \in G_j$ (satisfying $i + j \leqslant d$), $e_{ij}(g_i^a, g_j^b) = g_{i+j}^{ab}$. Asymmetric variant of graded encoding schemes provides additional structure on how these encodings can be combined. Each encoding is assigned with a set of levels $S \subseteq [N]$. Given two encodings $g_S^a, g_{S'}^b$, the map allows to compute $g_{S \cup S'}^{ab}$ only if $S \cap S' = \varnothing$.

Both [19] and [16] constructions begin from some variant of homomorphic encryption and use public-key encryption as the encoding method. The main new ingredient, however, is that they also publish a defective version of the secret key, which cannot be used for decryption but can be used to test if a ciphertext encrypts a zero. (This defective key is called the "zero-test parameter".) Over the last two years, the applications of (graded) multilinear maps have expanded much further, supporting applications such as witness encryption, general-purpose obfuscation, functional encryption, and many more [21,20,18,7,11].

1.1 Our Results

We present a new "graph-induced" variant of multilinear maps. In this variant, the multilinear map is defined with respect to a directed acyclic graph. Namely, encoded value are associated with paths in the graph, and it is only possible to add encoding relative to the same paths, or to multiply encodings relative to "connected paths" (i.e., one ends where the other begins) Our candidate construction of graph-induced multilinear maps does not rely on ideal lattices or hard-to-factor integers. Rather, *we use standard random lattices* such as those used in LWE-based cryptography. We follow a similar outline to the previous constructions, except our instance generation algorithm takes as input a description of a graph. Furthermore, our zero-tester *does not* include any secrets about the relevant lattices. Rather, in our case the zero-tester is just a random matrix, similar to a *public key* in common LWE-based cryptosystems.

Giving up the algebraic structure of ideal lattices and integers could contribute to a better understanding of the candidate itself, reducing the risk of unforeseen algebraic crypt-analytical attacks. On the flip side, using our construction is sometimes harder than previous construction, exactly because we give up some algebraic structure. In terms of security, we were not able so far to reduce any of our new construction to "nice" hardness assumptions, currently they are all just candidate constructions, that withstood our repeated cryptanalytic attempts at breaking them. Still we believe that our new construction is a well needed addition to our cryptographic toolbox, providing yet another avenue for implementing multilinear maps. This is particularly important in light of the new techniques for attacking these schemes [15]. For more discussion see Section 4.2.

Our Techniques. Our starting point is the new homomorphic encryption (HE) scheme of Gentry, Sahai and Waters [25]. The secret key in that scheme is a vector $\mathbf{a} \in \mathbb{Z}_q^m$, and a ciphertext encrypting $\mu \in \mathbb{Z}_q$ is a matrix $\mathbf{C} \in \mathbb{Z}_q^{m \times m}$ with small entries such that $\mathbf{C} \cdot \mathbf{a} = \mu \cdot \mathbf{a} + \mathbf{e}$ for some small error vector $\mathbf{e}$. In other words, valid ciphertexts all have the secret key $\mathbf{a}$ as an "approximate eigenvector", and the eigenvalue is the message. Given the secret eigenvector $\mathbf{a}$, decoding arbitrary μ's becomes easy.

This HE scheme supports addition and multiplication, but we also need a *public* equivalent of the approximate eigenvector for zero-testing. The key idea is to replace the "approximate eigenvector" with an "approximate eigenspace" by increasing the dimensions. Instead of having a single approximate eigenvectors, our "approximate eigenspace" is described by n vectors $\mathbf{A} \in \mathbb{Z}_q^{m \times n}$. The approximate eigenvalues will not merely be elements of $\mathbb{Z}_q$, but rather matrices $\mathbf{S} \in \mathbb{Z}_q^{n \times n}$ with small entries. An encoding of $\mathbf{S}$ is a matrix $\mathbf{C} \in \mathbb{Z}^{m \times m}$ with small entries such that

$$\mathbf{C} \cdot \mathbf{A} = \mathbf{A} \cdot \mathbf{S} + \mathbf{E}$$

for small noise matrix $\mathbf{E} \in \mathbb{Z}_q^{m \times n}$. In other words, $\mathbf{C}$ is a matrix that maps any column vector in $\mathbf{A}$ to a vector that is very close to the span of $\mathbf{A}$. In that sense, $\mathbf{A}$ is an approximate eigenspace. In the HE scheme, $\mathbf{a}$ was a secret key that allowed us to easily recover μ. However, for the eigenspace setting, assuming $\mathbf{A}$ is just a uniformly random matrix and $\mathbf{S}$ is a random small matrix, $\mathbf{A} \cdot \mathbf{S} + \mathbf{E}$ is an LWE instance that looks uniform even when given $\mathbf{A}$.

Overview of Our Construction. Our construction is parametrized by a directed acyclic graph $G = (V, E)$. For each node $v \in V$, we assign a random matrix $\mathbf{A}_v \in \mathbb{Z}_q^{m \times n}$. Any path $u \rightsquigarrow v$ (which can be a single edge) can be assigned with an encoding $\mathbf{D} \in \mathbb{Z}_q^{m \times m}$ of some plaintext secret $\mathbf{S} \in \mathbb{Z}_q^{n \times n}$ satisfying

$$\mathbf{D} \cdot \mathbf{A}_u = \mathbf{A}_v \cdot \mathbf{S} + \mathbf{E} \tag{1}$$

for some small error $\mathbf{E} \in (\chi)^{m \times n}$.

Adding and multiplying encodings corresponds to addition and multiplication of matrices. Addition of encodings can only be performed relative to the same path $u \rightsquigarrow v$. For example, given encodings $\mathbf{D}_1, \mathbf{D}_2$ at path $u \rightsquigarrow v$, we have that:

$$(\mathbf{D}_1 + \mathbf{D}_2) \cdot \mathbf{A}_u \approx \mathbf{A}_v \cdot \mathbf{S}_1 + \mathbf{A}_v \cdot \mathbf{S}_2 = \mathbf{A}_v \cdot (\mathbf{S}_1 + \mathbf{S}_2).$$

Multiplication of encodings can only be performed when they form a complete path. That is, given encodings $\mathbf{D}_1$ and $\mathbf{D}_2$ relative to paths $u \rightsquigarrow v$ and $v \rightsquigarrow w$ respectively, we have:

$$
\begin{aligned}
\mathbf{D}_2 \cdot \mathbf{D}_1 \cdot \mathbf{A}_u &= \mathbf{D}_2 \cdot (\mathbf{A}_v \cdot \mathbf{S}_1 + \mathbf{E}_1) \\
&= (\mathbf{A}_w \cdot \mathbf{S}_2 + \mathbf{E}_2) \cdot \mathbf{S}_1 + \mathbf{D}_2 \cdot \mathbf{E}_1 \\
&= \mathbf{A}_w \cdot \mathbf{S}_2 \cdot \mathbf{S}_1 + \underbrace{\mathbf{E}_2 \cdot \mathbf{S}_1 + \mathbf{D}_2 \cdot \mathbf{E}_1}_{\mathbf{E}'}
\end{aligned}
\tag{2}
$$

where $\mathbf{E}'$ is small since the errors and matrices $\mathbf{S}_1, \mathbf{D}_2$ have small entries. Furthermore, it is possible to compare two encodings with the same sink node. That is, given $\mathbf{D}_1$ and $\mathbf{D}_2$ relative to paths $u \rightsquigarrow v$ and $w \rightsquigarrow v$, it is sufficient to check if $\mathbf{D}_1 \cdot \mathbf{A}_u - \mathbf{D}_2 \cdot \mathbf{A}_w$ is small since if $\mathbf{S}_1 = \mathbf{S}_2$, then we have

$$\mathbf{D}_1 \cdot \mathbf{A}_u - \mathbf{D}_2 \cdot \mathbf{A}_w = (\mathbf{A}_v \cdot \mathbf{S}_1 + \mathbf{E}_1) - (\mathbf{A}_v \cdot \mathbf{S}_2 + \mathbf{E}_2) = \mathbf{E}_1 - \mathbf{E}_2 \quad (3)$$

Hence, the random matrices $\mathbf{A}_u, \mathbf{A}_w \in \mathbb{Z}_q$, which are commonly available in the public parameters, is sufficient for comparison and zero-testing.

As we explain in Section 3, generating the encoding matrices requires knowing a trapdoor for the matrices $\mathbf{A}_i$. But for the public-sampling setting, it is possible to generate encodings of many random matrices during setup, and later anyone can take a random linear combinations of them to get "fresh" random encodings.

We remark that since $\mathbf{S}$ needs to be small in Eqn. (2), our scheme only supports encoding of *small plaintext elements*, as opposed to arbitrary plaintext elements as in previous schemes.[1] Another difference is that in the basic construction our plaintext space is a non-commutative ring (i.e. square matrices). We extend to the commutative setting in Section 3.2.

Variations and Parameters. One standard way of improving parameters is to switch to a ring-LWE setting, where scalars are taken from a large polynomial ring (rather than being just integers), and the dimension of vectors and matrices is reduced proportionally. In our context, we can also use the same approach to move to a commutative plaintext space, see Section 3.2.

1.2 Applications

Our new constructions support many of the known cryptographic uses of graded encoding. Here we briefly sketch two of them.

Non-interactive Multipartite Key-Exchange. Consider k-partite key-exchange. We design a graph in a star topology with k-branches each of length $k+1$ nodes, where each player is associated with one of these branches. All branches meet at the common sink node $\mathbf{A}_0$. For each branch, we associate encodings of small LWE secrets $t_1, \ldots, \ldots, t_k$ in a specific order, where the same values are used in all the branches, but in different order. The public parameters consists of the encoding of many such plaintext values. Each player then takes random linear combinations of these encodings so as to obtain the encoding of the same plaintext value relative to one edge on each branch. The player stores the encoding along its own branch as its secret key and broadcasts the rest of to other players. Assume some canonical ordering of the players. Each player computes the $k-1$ product

---

[1] The only exception is that the leftmost plaintext matrix $\mathbf{S}$ in a product could encode a large element, as Eqn. (2) is not affected by the size of $\mathbf{S}_1$. Similarly the rightmost encoding matrix $\mathbf{D}$ in a product need not be small. We do not use these exceptions in the current paper, however.

of the other players' encodings along its own branch and multiplied also by its secret encoding. This yields an encoding $\mathbf{D}$ of $\mathbf{T}^* = \prod_{i \in [k]} s_i$, satisfying

$$\mathbf{D} \cdot \mathbf{A}_{j,1} = \mathbf{A}_0 \cdot \prod_{i \in [k]} s_i + \text{noise}$$

And the players obtain the shared secret key by applying a randomness extractor on the most significant bits.

Branching-Program Obfuscation. Perhaps the "poster application" of cryptographic graded encodings is to obtain general-purpose obfuscation [20], [12,4,36,23], with the crucial step being the use of graded encoding to obfuscate branching programs . These branching programs are represented as a sequence of pairs of encoded matrices, and the user just picks one matrix from each pair and then multiply them all in order.

This usage pattern of graded encoding fits very well into our graph-induced scheme since these matrices are given in a pre-arranged order. We describe a candidate obfuscation construction from our multilinear map based on a path graph. Informally, to obfuscate a length-L matrix branching program $\{\mathbf{B}_{i,b}\}$, we first perform Kilian's randomization and then encode values $\mathbf{R}_{i-1}^{-1}\mathbf{B}_{i,0}\mathbf{R}_i$ and $\mathbf{R}_{i-1}^{-1}\mathbf{B}_{i,1}\mathbf{R}_i$ relative to the edge i. The user can then compute an encoding of a product of matrices corresponding to its input. If the product $\prod_{i \in [L]} \mathbf{B}_{i,x_{\text{var}i}} = \mathbf{I}$, then the user obtains an encoding $\mathbf{D}$ satisfying:

$$\mathbf{D} \cdot \mathbf{A}_0 = \mathbf{A}_L \cdot \mathbf{I} + \text{noise}$$

Given $\mathbf{A}_L \cdot \mathbf{I} + \text{noise}'$ in the public parameters (or its encoding), the user can then learn the result of the computation by a simple comparison. We note that our actual candidate construction is more involved as we deploy additional safeguards from the literature (See Section 5.2).

1.3   Organization

In Section 2, we provide some background and present the syntax of graph-induced multilinear maps. In Section 3, we describe our basic construction in the non-commutative variant. In Subsection 3.2 we show how to extend our basic construction to commutative variant. In Section 4, we analyze the security of our construction. In Section 5 we present applications of our construction to key-exchange and obfuscation.

2   Preliminaries

Notation. For any integer $q \geqslant 2$, we let $\mathbb{Z}_q$ denote the ring of integers modulo q and we represent $\mathbb{Z}_q$ as integers in $(-q/2, q/2]$. We let $\mathbb{Z}_q^{n \times m}$ denote the set of $n \times m$ matrices with entries in $\mathbb{Z}_q$. We use bold capital letters (e.g. $\mathbf{A}$) to denote matrices, bold lowercase letters (e.g. $\mathbf{x}$) to denote vectors.

If $\mathbf{A}_1$ is an $n \times m$ matrix and $\mathbf{A}_2$ is an $n \times m'$ matrix, then $[\mathbf{A}_1 | \mathbf{A}_2]$ denotes the $n \times (m + m')$ matrix formed by concatenating $\mathbf{A}_1$ and $\mathbf{A}_2$. Similarly, if $\mathbf{A}_1, \mathbf{A}_2$ have dimensions $n \times m$ and $\mathbf{A}_2$ is an $n' \times m$, respectively, then we denote by $(\mathbf{A}_1/\mathbf{A}_2)$ the $(n + n') \times m$ matrix formed by putting $\mathbf{A}_1$ on top of $\mathbf{A}_2$. Similar notations apply to vectors. When doing matrix-vector multiplication we usually view vectors as column vectors.

A function $f(n)$ is *negligible* if it is $o(n^{-c})$ for all $c > 0$, and we use negl(n) to denote a negligible function of n. We say that $f(n)$ is *polynomial* if it is $O(n^c)$ for some $c > 0$, and we use poly(n) to denote a polynomial function of n. An event occurs with *overwhelming probability* if its probability is $1 -$ negl(n). The notation $\lfloor x \rceil$ denotes the nearest integer to x, rounding toward 0 for half-integers.

The ℓ_∞ norm of a vector is denoted by $\|\mathbf{x}\| = \max_i |x_i|$. We identify polynomials with their representation in some standard basis (e.g., the standard coefficient representation), and the norm of a polynomial is the norm of the representation vector. The norm of a matrix, $\|\mathbf{A}\|$, is the norm of its largest column.

Extractors. An efficient (n, m, ℓ, ϵ)-strong extractor is a poly-time algorithm Extract : $\{0,1\}^n \rightarrow \{0,1\}^\ell$ such that for any random variable W over $\{0,1\}^n$ with min-entropy m, it holds that the statistical distance between $(\mathsf{Extract}_\alpha(W), \alpha)$ and (U_ℓ, α) is at most ϵ. Here, α denotes the random bits used by the extractor. Universal hash functions [14,41] can extract $\ell = m - 2 \log \frac{1}{\epsilon} + 2$ nearly random bits, as given by the leftover hash lemma [27]. This will be sufficient for our applications.

2.1 Lattice Preliminaries

Gaussian Distributions. For a real parameter $\sigma > 0$, define the spherical Gaussian function on $\mathbb{R}^n$ with parameter σ as $\rho_\sigma(\mathbf{x}) = \exp(-\pi \|\mathbf{x}\|^n / \sigma^2)$ for all $\mathbf{x} \in \mathbb{R}^n$. This generalizes to ellipsoid Gaussians, where we replace the parameter $\sigma \in \mathbb{R}$ by the (square root of the) covariance matrix $\Sigma \in \mathbb{R}^{n \times n}$: For a rank-$n$ matrix $\mathbf{S} \in \mathbb{R}^{m \times n}$, the ellipsoid Gaussian function on $\mathbf{R}^n$ with parameter $\mathbf{S}$ is defined by $\rho_S(\mathbf{x}) = \exp(-\pi \mathbf{x}^T (\mathbf{S}^T \mathbf{S})^{-1} \mathbf{x})$ for all $\mathbf{x} \in \mathbb{R}^n$. The ellipsoid discrete Gaussian distribution with parameter $\mathbf{S}$ over a set $L \subset \mathbb{R}^n$ is $D_{L,\mathbf{S}}(\mathbf{x}) = \rho_S(\mathbf{x})/\rho_S(L)$, where $\rho_S(L)$ denotes $\sum_{\mathbf{x} \in L} \rho_S(\mathbf{x})$ and serves as just a normalization factor. The same notations also apply the to spherical case, $D_{L,\sigma}(\cdot)$, and in particular $D_{\mathbb{Z}^n, r}$ denotes the n-dimensional discrete Gaussian distribution.

It follows from [33] that when L is a lattice and σ is large enough relative to its "smoothing parameter" (alternatively its λ_n or the Gram-Schmidt norm of one of its bases), then for every point $\mathbf{c} \in \mathbb{R}^n$ we have

$$\Pr\left[\|\mathbf{x} - \mathbf{c}\| > \sigma \sqrt{n} : \mathbf{x} \xleftarrow{\text{R}} D_{L,\sigma,\mathbf{c}} \right] \leqslant \text{negl}(n).$$

Also under the same conditions, the probability for a random sample from $D_{\mathbb{Z}^m, \sigma}$ to be $\mathbf{0}$ is negligible.

Trapdoors for Lattices

Lemma 1 (Lattice Trapdoors [3,24,32]). *There is an efficient randomized algorithm* TrapSamp$(1^n, 1^m, q)$ *that, given any integers* $n \geq 1$, $q \geq 2$, *and sufficiently large* $m = \Omega(n \log q)$, *outputs a parity check matrix* $\mathbf{A} \in \mathbb{Z}_q^{m \times n}$ *and some 'trapdoor information'* τ *that enables sampling small solutions to* $\mathbf{rA} = \mathbf{u}$ (mod q).

Specifically, there is an efficient randomize algorithm PreSample *such that for large enough* $s = \Omega(\sqrt{n \log q})$ *and with overwhelming probability over* $(\mathbf{A}, \tau) \leftarrow$ TrapSamp$(1^n, 1^m, q)$, *the following two distributions are within* negl(n) *statistical distance:*

- $\mathcal{D}_1[\mathbf{A}, \tau]$ *chooses a uniform* $\mathbf{u} \in \mathbb{Z}_q^n$ *and uses* τ *to solve for* $\mathbf{rA} = \mathbf{u}$ (mod q),

$$\mathcal{D}_1[\mathbf{A}, \tau] \stackrel{\text{def}}{=} \{(\mathbf{u}, \mathbf{r}) \ : \ \mathbf{u} \leftarrow \mathbb{Z}_q^n; \ \mathbf{r} \leftarrow \mathsf{PreSample}(\mathbf{A}, \tau, \mathbf{u}, s)\}.$$

- $\mathcal{D}_2[\mathbf{A}]$ *chooses a Gaussian* $\mathbf{r} \leftarrow D_{\mathbb{Z}^m, s}$ *and sets* $\mathbf{u} := \mathbf{rA} \bmod q$,

$$\mathcal{D}_2[\mathbf{A}] \stackrel{\text{def}}{=} \{(\mathbf{u}, \mathbf{r}) \ : \ \mathbf{r} \leftarrow D_{\mathbb{Z}^m, s}; \ \mathbf{u} := \mathbf{rA} \bmod q\}.$$

We can extend PreSample from vectors to matrices by running it k times on k different vectors $\mathbf{u}$ and concatenating the results, hence we write $\mathbf{R} \leftarrow$ PreSample$(\mathbf{A}, \tau, \mathbf{U}, s)$.

We also note that any small-enough full rank matrix $\mathbf{T}$ (over the integers) such that $\mathbf{TA} = 0$ (mod q) can be used as the trapdoor τ above. This is relevant to our scheme because in many cases an "encoding of zero" can be turned into such a trapdoor (see Section 4).

2.2   Graded Multilinear Encodings

The notion of graded encoding scheme that we relaize is similar (but not exactly identical) to the GGH notion from [19]. Very roughly, a graded encoding scheme for an algebraic "plaintext ring R" provides methods for encoding ring elements and manipulating these encodings. Namely we can sample random plaintext elements together with their encoding, can add and multiply encoded elements, can test if a given encoding encodes zero, and can also extract a "canonical representation" of a plaintext element from an encoding of that element. k

Syntax of Graph-Induced Graded Encoding Schemes. There are several variations of graded-encoding systems in the literature, such as public/secret encoding, with/without re-randomization, symmetric/asymmetric, etc. Below we define the syntax for our scheme, which is still somewhat different than all of the above. The main differences are that our encodings are defined relative to edges of a directed graph (as opposed to levels/sets/vectors as in previous schemes), and that we only encode "small elements" from the plaintext space. Below we provide the relevant definitions, modifying the ones from [19].

Definition 1 (Graph-Induced Encoding Scheme). *A graph-based graded encoding scheme with secret sampling consists of the following (polynomial-time) procedures,* $\mathcal{G}_{es}$ = (PrmGen, InstGen, Sample, Enc, add, neg, mult, ZeroTest, Extract):

- PrmGen($1^\lambda, G, \mathcal{C}$): *The parameter-generation procedure takes the security parameter* λ, *underlying directed graph* $G = (V, E)$, *and the class* $\mathcal{C}$ *of supported circuits. It outputs some global parameters of the system* gp, *which includes in particular the graph* G, *a specification of the plaintext ring* R *and also a distribution* χ *over* R.
  For example, in our case the global parameters consists of the dimension n *of matrices, the modulus* q *and the Gaussian parameter* σ.
- InstGen(gp): *The randomized instance-generation procedure takes the global parameters* gp, *and outputs the public and secret parameters* sp, pp.
- Sample(pp): *The sampling procedure samples an element in the the plaintext space, according to the distribution* χ.
- Enc(sp, p, α): *The encoding procedure takes the secret parameters* pp, *a path* $p = u \rightsquigarrow v$ *in the graph, and an element* $\alpha \in R$ *from the support of the* Sample *procedure, and outputs an encoding* u_p *of* α *relative to* p. [2]
- neg(pp, u), add(pp, u, u'), mult(pp, u, u'). *The arithmetic procedures are deterministic, and they all take as input the public parameters and use them to manipulate encodings.*
  Negation takes an encoding of $\alpha \in R$ *relative to some path* $p = u \rightsquigarrow v$ *and outputs encoding of* $-\alpha$ *relative to the same path. Addition takes* u, u' *that encode* $\alpha, \alpha' \in R$ *relative to the same path* p, *and outputs an encoding of* $\alpha + \alpha$ *relative to* p. *Multiplication takes* u, u' *that encode* $\alpha, \alpha' \in R$ *relative to consecutive paths* $p = u \rightsquigarrow v$ *and* $p' = v \rightsquigarrow w$, *respectively. It outputs an encoding of* $\alpha \cdot \alpha'$ *relative to the combined path* $u \rightsquigarrow w$.
- ZeroTest(pp, u): *Zero testing is a deterministic procedure that takes the public parameters* pp *and an encoding* u *that is tagged by its path* p. *It outputs 1 if* u *is an encoding of zero and 0 if it is an of a non-zero element.*
- Extract(pp, u): *The extraction procedure takes as input the public parameters* pp *and an encoding* u *that is tagged by its path* p. *It outputs a* λ*-bit string that serves as a "random canonical representation" of the underlying plaintext element* α *(see below).*

Correctness. The graph G, in conjunction with the procedures for sampling, encoding, and arithmetic operations, and the class of supported circuits, implicitly define the set S_G of "valid encodings" and its partition into sets $S_G^{(\alpha)}$ of "valid encoding of α".

Namely, we consider arithmetic circuits whose wires are labeled by paths in G in a way that respects the permitted operations of the scheme (i.e., negation and addition have all the same labels, and multiplication has consecutive input paths and the output is labeled by their concatenation). Then S_G consists of all the

---

[2] See the description below for the meaning of "u_p is an encoding of α relative to p", formally u_p is just a bit string, which is tagged with its path p.

encoding that can be generated by running the sampling/encoding procedures to sample plaintext elements and compute their encoding, then computing the operations of the scheme according to Π, and finally collecting the encoding at the output of Π. An encoding $u \in S_G$ belongs to $S_G^{(\alpha)}$ if there exists such circuit Π and inputs for which Π outputs α when evaluated on plaintext elements. Of course, to be useful we require that the sets $S_G^{(\alpha)}$ form a partition of S_G.

We can also sub-divide each $S_G^{(\alpha)}$ into $S_{G,p}^{(\alpha)}$ for different paths p in the graph, depending on the label of the output wire of Π (but here it is not important that these sets are disjoint), and define $S_{G,p} = \bigcup_{\alpha \in R} S_{G,p}^{(\alpha)}$.

Note that the sets $S_{G,p}^{(\alpha)}$ can be empty, for example in our construction the sampling procedure only outputs "small" plaintext values α, so a "large" β would have $S_{G,p}^{(\beta)} = \varnothing$. Below we denote the set of α's with non-empty encoding sets (relative to path p) by $\mathsf{SMALL}_{G,p} \overset{\text{def}}{=} \{\alpha \in R : S_{G,p}^{(\alpha)} \neq \varnothing\}$, and similarly $\mathsf{SMALL}_G \overset{\text{def}}{=} \{\alpha \in R : S_G^{(\alpha)} \neq \varnothing\}$.

We assume for simplicity that the sets SMALL *depend only on the global parameters* gp *and not the specific parameters* sp, pp. (This assumption holds for our construction and it simplifies the syntax below.)

We can now state the correctness conditions for zero-testing and extraction. For zero-testing we require that $\mathsf{ZeroTest}(\mathsf{pp}, u) = 1$ for every $u \in S^{(0)}$ (with probability one), and for every $\alpha \in \mathsf{SMALL}_G$, $\alpha \neq 0$ it holds with overwhelming probability over instance-generation that $\mathsf{ZeroTest}(\mathsf{pp}, u) = 0$ for every encoding $u \in S_G^{(\alpha)}$.

For extraction, we roughly require that $\mathsf{Extract}$ outputs the same string on all the encodings of the same α, different strings on encodings of different α's, and random strings on encodings of "random α's." Formally, we require the following for any global parameters gp output by PrmGen:

- For any plaintext element $\alpha \in \mathsf{SMALL}_G$ and path p in G, with overwhelming probability over the parameters $(\mathsf{sp}, \mathsf{pp}) \leftarrow \mathsf{InstGen}(\mathsf{gp})$, there exists a single value $x \in \{0,1\}^{\lambda}$ such that $\mathsf{Extract}(\mathsf{pp}, u) = x$ holds for all $u \in S_{G,p}^{(\alpha)}$.
- For any $\alpha \neq \alpha' \in \mathsf{SMALL}_G$ and path p in G, it holds with overwhelming probability over the parameters $(\mathsf{sp}, \mathsf{pp}) \leftarrow \mathsf{InstGen}(\mathsf{gp})$ that for any $u \in S_{G,p}^{(\alpha)}$ and $u' \in S_{G,p}^{(\alpha')}$, $\mathsf{Extract}(\mathsf{pp}, u) \neq \mathsf{Extract}(\mathsf{pp}, u')$.
- For any path p in G and distribution $\mathcal{D}$ over $\mathsf{SMALL}_{G,p}$ with min-entropy 3λ or more, it holds with overwhelming probability over the parameters $(\mathsf{sp}, \mathsf{pp}) \leftarrow \mathsf{InstGen}(\mathsf{gp})$ that the induced distribution $\{\mathsf{Extract}(\mathsf{pp}, u) : \alpha \leftarrow \mathcal{D}, u \in S_d^{(\alpha)}\}$ is nearly uniform over $\{0,1\}^{\lambda}$.

In some applications these conditions can be weakened. For example we often only need them to hold for some paths in G rather than all of them (e.g., we only care about source-to-sink paths).

Variations

No Re-randomization. In some applications one may want to re-randomize a given encoding, obtaining a "fresh" encoding of the same value. The common way of obtaining this functionality is by providing encoding of zeros in the public parameter and adding a subset-sum of them. For our construction this turns out to be insecure, see Section 4. Hence this construction does not support re-randomization.

Public Sampling of Encoded Elements. Another useful variation allows a public sampling procedure that takes as input pp rather than sp and outputs both a plaintext α and its encoding u_p relative to some path p. The common way of implementing it is to add to the public parameters many plaintext-encoding pairs (α_i, u_i) (e.g., wrt the edges in G). Then a public sampling procedure can just use a subset sum of these tuples as a new sample. (One can make the distribution of these new samples "nice", e.g., by using the leftover-hash over Gaussians from [2,1].) In our construction this is sometimes insecure, specifically when applied to our commutative variant, see Section 4. We do not know of attacks on public sampling in the non-commutative case, but the fact that the commutative case is insecure seems worrisome.

3 Our Graph-Induced Multilinear Maps

The plaintext space in our basic scheme is the non-commutative ring of matrices $R = \mathbb{Z}_q^{n \times n}$, later in Section 3.2 we describe a commutative variant. In this section we only deal with correctness of these schemes, their security is discussed in Section 4.

As sketched in the introduction, for the basic scheme we have an underlying directed acyclic graph $G = (V, E)$, we identify a random matrix $\mathbf{A}_v \in \mathbb{Z}_q^{m \times n}$ with each node $v \in V$, and encodings in the scheme are defined relative to paths. A small plaintext matrix $\mathbf{S} \in R$ is encoded wrt to the path $u \rightsquigarrow v$ via another small matrix $\mathbf{D} \in \mathbb{Z}_q^{m \times m}$ such that $\mathbf{D} \cdot \mathbf{A}_u \approx \mathbf{A}_v \cdot \mathbf{S}$. In more detail, we have the following graded encoding scheme $\mathcal{G}_{es} = ($PrmGen, InstGen, Sample, Enc, add, neg, mult, ZeroTest, Extract$)$:

- PrmGen($1^\lambda, G, \mathcal{C}$): On input the security parameter λ, an underlying DAG $G = (V, E)$, and class $\mathcal{C}$ of supported circuits, we compute:
    1. LWE parameters n, m, q and error distribution $\chi = D_{\mathbb{Z},s}$.
    2. A Gaussian parameters σ for PreSample.
    3. Another parameter t for the number of most significant bits used for zero-test and extraction.

    The constraints that dictate these parameters are described in Appendix A. The resulting parameters for a DAG of diameter d are $n = \Theta(d\lambda \log(d\lambda))$, $q = (d\lambda)^{\Theta(d)}$, $m = \Theta(nd \log q)$, $s = \sqrt{n}$, $\sigma = \sqrt{n(d+1)\log q}$, and $t = \lfloor (\log q)/4 \rfloor - 1$. These global parameters gp (including the graph G) are given to all the procedures below.

- InstGen(gp): Given the global parameters, instance-generation proceeds as follows:
  1. Use trapdoor-sampling to generate $|V|$ matrices with trapdoors, one for each node.
  $$\forall v \in V, \quad (\mathbf{A}_v, \tau_v) \leftarrow \mathsf{TrapSamp}(1^n, 1^m, q)$$
  2. Choose the randomness-extractor seed β from a pairwise-independent function family, and a uniform "shift matrix" $\mathbf{\Delta} \in \mathbb{Z}_q^{m \times n}$.
  The public parameters are $\mathsf{pp} := (\{\mathbf{A}_v : v \in V\}, \beta, \mathbf{\Delta})$ and the secret parameters include also the trapdoors $\{\tau_v : v \in V\}$.
- Sample(pp): This procedure just samples an LWE secret $\mathbf{S} \leftarrow (\chi)^{n \times n}$ as the plaintext.
- Enc(sp, p, $\mathbf{S}$): On input the matrices $\mathbf{A}_u, \mathbf{A}_v$, the trapdoor τ_u, and the small matrix $\mathbf{S}$, sample an LWE error matrix $\mathbf{E}_i \leftarrow (\chi)^{m \times n}$, set $\mathbf{V} = \mathbf{A}_v \cdot \mathbf{S} + \mathbf{E} \in \mathbb{Z}_q^{m \times n}$, and then use the trapdoor τ_u to compute the encoding $\mathbf{D}_p$ s.t. $\mathbf{D}_p \cdot \mathbf{A}_u = \mathbf{V}$, $\mathbf{D}_p \leftarrow \mathsf{PreSample}(\mathbf{A}_u, \tau_u, \mathbf{V}, \sigma)$. The output is the plaintext S and encoding $\mathbf{D}_p$.
- The arithmetic operations are just matrix operations in $\mathbb{Z}_q^{m \times m}$:
$$\mathsf{neg}(\mathsf{pp}, \mathbf{D}) := -\mathbf{D}, \quad \mathsf{add}(\mathsf{pp}, \mathbf{D}, \mathbf{D}') := \mathbf{D} + \mathbf{D}', \quad \text{and} \quad \mathsf{mult}(\mathsf{pp}, \mathbf{D}, \mathbf{D}') := \mathbf{D} \cdot \mathbf{D}'.$$

To see that negation and addition maintain the right structure, let $\mathbf{D}, \mathbf{D}' \in \mathbb{Z}_q^{m \times m}$ be two encodings reltive to the same path $u \rightsquigarrow v$. Namely $\mathbf{D} \cdot \mathbf{A}_u = \mathbf{A}_v \cdot \mathbf{S} + \mathbf{E}$ and $\mathbf{D}' \cdot \mathbf{A}_u = \mathbf{A}_v \cdot \mathbf{S}' + \mathbf{E}'$, with the matrices $\mathbf{D}, \mathbf{D}', \mathbf{E}, \mathbf{E}', \mathbf{S}, \mathbf{S}'$ all small. Then we have

$$-\mathbf{D} \cdot \mathbf{A}_u = \mathbf{A}_v \cdot (-\mathbf{S}) + (-\mathbf{E}),$$

and $(\mathbf{D} + \mathbf{D}') \cdot \mathbf{A}_u = (\mathbf{A}_v \cdot \mathbf{S} + \mathbf{E}) + (\mathbf{A}_v \cdot \mathbf{S}' + \mathbf{E}') = \mathbf{A}_v \cdot (\mathbf{S} + \mathbf{S}') + (\mathbf{E} + \mathbf{E}')$,

and all the matrices $-\mathbf{D}, -\mathbf{S}, -\mathbf{E}, \ \mathbf{D} + \mathbf{D}', \ \mathbf{S} + \mathbf{S}', \ \mathbf{E} + \mathbf{E}'$ are still small. For multiplication, consider encodings $\mathbf{D}, \mathbf{D}'$ relative to paths $v \rightsquigarrow w$ and $u \rightsquigarrow v$, respectively, then we have

$$(\mathbf{D} \cdot \mathbf{D}') \cdot \mathbf{A}_u = \mathbf{D} \cdot (\mathbf{A}_v \cdot \mathbf{S}' + \mathbf{E}')$$
$$= (\mathbf{A}_w \cdot \mathbf{S} + \mathbf{E}) \cdot \mathbf{S}' + \mathbf{D} \cdot \mathbf{E}' = \mathbf{A}_w \cdot (\mathbf{S} \cdot \mathbf{S}') + \underbrace{(\mathbf{E} \cdot \mathbf{S}' + \mathbf{D} \cdot \mathbf{E}')}_{\mathbf{E}''},$$

and the matrices $\mathbf{D} \cdot \mathbf{D}'$, $\mathbf{S} \cdot \mathbf{S}'$, and $\mathbf{E}''$ are still small.
Of course, the matrices $\mathbf{D}, \mathbf{S}, \mathbf{E}$ all grow with arithmetic operations, but our parameter-choice enures that for any encoding relative to any path in the graph $u \rightsquigarrow v$ (of length $\leq d$) we have $\mathbf{D} \cdot \mathbf{A}_u = \mathbf{A}_v \cdot \mathbf{S} + \mathbf{E}$ where $\mathbf{E}$ is still small, specifically $\|\mathbf{E}\| < q^{3/4} \leq q/2^{t+1}$.
- ZeroTest(pp, $\mathbf{D}$). Given an encoding $\mathbf{D}$ relative to path $u \rightsquigarrow v$ and the matrix $\mathbf{A}_u$, our zero-test procedure outputs 1 if and only if $\|\mathbf{D} \cdot \mathbf{A}_u\| < q/2^{t+1}$.
- Extract(pp, $\mathbf{D}$): Given an encoding $\mathbf{D}$ relative to path $u \rightsquigarrow v$, the matrix $\mathbf{A}_u$ and shift-matrix $\mathbf{\Delta}$, and the extrator seed β, we compute $\mathbf{D} \cdot \mathbf{A}_0 + \mathbf{\Delta}$, collect the t most-significant bits from each entry (when mapped to the interval $[0, q-1]$), and apply the randomness extractor, outputting

$$w := \mathsf{RandExt}_\beta \left(\mathsf{msb}_t (\mathbf{D} \cdot \mathbf{A}_u + \mathbf{\Delta}) \right)$$

3.1   Correctness

Correctness of the scheme follows from our invariant, which says that encoding of some plaintext matrix $\mathbf{S}$ relative to any path $u \rightsquigarrow v$ of legnth $\leq d$ satisfies $\mathbf{D} \cdot \mathbf{A}_u = \mathbf{A}_v \cdot \mathbf{S} + \mathbf{E}$ for $\|\mathbf{E}\| < q/2^{t+1}$.

Correctness of Zero-Test. An encoding of zero satisfies $\mathbf{D} \cdot \mathbf{A}_u = \mathbf{E}$, hence $\|\mathbf{D} \cdot \mathbf{A}_u\| < q/2^{t+1}$. On the other hand, since $\mathbf{A}_v$ is uniform then for any nonzero $\mathbf{S}$ we only get $\|\mathbf{A}_v \cdot \mathbf{S}\| \leq q/2^t$ with exponentially small probability, and since $\|\mathbf{E}\| < q/2^{t+1}$ then

$$\|\mathbf{D} \cdot \mathbf{A}_u\| \geq \|\mathbf{A}_v \cdot \mathbf{S}\| - \|\mathbf{E}\| > q/2^t - q/2^{t+1} \geq q/2^{t+1}.$$

Hence with overwhelming probability over the choise of $\mathbf{A}_v$, our zero-test will output 0 on *all* the encoding of $\mathbf{S}$.

Correctness of Extraction. We begin by proving that for any plaintext matrix $\mathbf{S}$ and any encoding $\mathbf{D}$ of $\mathbf{S}$ (relative to $u \rightsquigarrow v$), with overwhelming probability over the parameters we have that $\mathsf{msb}_t(\mathbf{D} \cdot \mathbf{A}_u + \mathbf{\Delta}) = \mathsf{msb}_t(\mathbf{A}_v \cdot \mathbf{S} + \mathbf{\Delta})$.

Since the two matrices $\mathbf{M} = \mathbf{A}_v \cdot \mathbf{S} + \mathbf{\Delta}$ and $\mathbf{M}' = \mathbf{D} \cdot \mathbf{A}_u + \mathbf{\Delta}$ differ in each entry by at most $q/2^{t+1}$ modulo q, they can only differ in their top t bits due to the mod-q reduction, i.e., if for some entry we have $[\mathbf{M}]_{k,\ell} \approx 0$ but $[\mathbf{M}']_{k,\ell} \approx q$ or the other way around. (Recall that here we reduce mod-q into the interval $[0, q-1]$.) Clearly, this only happens when $\mathbf{M} \approx \mathbf{M}' \approx 0 \pmod{q}$, in particular we need

$$- < q/2^{t+1} < [\mathbf{A}_v \mathbf{S} + \mathbf{\Delta}]_{k,\ell} < q/2^{t+1}.$$

For any $\mathbf{S}$ and $\mathbf{A}_v$, the last condition occurs only with exponentially small probability over the choise of $\mathbf{\Delta}$. We conclude that if all the entries of $|\mathbf{A}_v \cdot \mathbf{S} + \mathbf{\Delta}|$ are larger than $q/2^{t+1}$ (modulo q), which happens with overwhelming probability, then for all level-i encodings $\mathbf{D}$ of $\mathbf{S}$, the top t bits of $\mathbf{D} \cdot \mathbf{A}_u$ agree with the top t bits of $\mathbf{A}_v \cdot \mathbf{S}$. We call a plaintext matrix $\mathbf{S}$ "v-good" if the above happens, and denote their set by GOOD_v. With this notation, the arguments above say that for any fixed $\mathbf{S}, v$, we have $\mathbf{S} \in \mathsf{GOOD}_v$ with overwhelming probability over the instance-generation randomness.

Same Input Implies Same Extracted Value. For any plaintext matrix $\mathbf{S} \in \mathsf{GOOD}_v$, clearly all its encodings relative to $u \rightsquigarrow v$ agree on the top t bits of $\mathbf{D} \cdot \mathbf{A}_u$ (since they all agree with $\mathbf{A}_v \cdot \mathbf{S}$). Hence they all have the same extracted value.

Different Inputs Imply Different Extracted Values. If $\mathbf{D}, \mathbf{D}'$ encode different plaintext matrices then $\mathbf{D} - \mathbf{D}'$ is an encoding of non-zero, hence $\|(\mathbf{D} - \mathbf{D}') \cdot \mathbf{A}_u\| \gg q/2^t$ except with negligible probability, $\mathbf{D} \cdot \mathbf{A}_u + \mathbf{\Delta}$ and $\mathbf{D}' \cdot \mathbf{A}_u + \mathbf{\Delta}$ must differ somewhere in their top t bits. Since we use universal hashing for our randomness extractor, then with high probability (over the hash function β) we get $\mathsf{RandExt}_\beta(\mathsf{msb}_t(\mathbf{D} \cdot \mathbf{A}_u + \mathbf{\Delta})) \neq \mathsf{RandExt}_\beta(\mathsf{msb}_t(\mathbf{D}' \cdot \mathbf{A}_u + \mathbf{\Delta}))$.

Random Input Implies Random Extracted Value. Fix some high-entropy distribution $\mathcal{D}$ over inputs $\mathbf{S}$. Since for every $\mathbf{S}$ we have $\Pr[\mathbf{S} \in \mathsf{GOOD}_v] = 1 - \mathsf{negl}(\lambda)$ then also with overwheling probability over the parameters we have $\Pr_{\mathbf{S}\leftarrow\mathcal{D}}[\mathbf{S} \in \mathsf{GOOD}_v] = 1 - \mathsf{negl}(\lambda)$. It is therefore enough to show that $\mathsf{RandExt}_\beta(\mathsf{msb}_t(\mathbf{A}_v \cdot \mathbf{S} + \boldsymbol{\Delta}))$ is nearly uniform on $\mathbf{S} \leftarrow \mathcal{D}$.
We observe that the function $H(\mathbf{S}) = \mathbf{A}_v \cdot \mathbf{S} + \boldsymbol{\Delta}$ is itself pairwise independent on each column of the output separately, and therefore so is the function $H'(\mathbf{S}) = \mathsf{msb}_t(H(\mathbf{S}))$. [3] We note that H' has very low collision probability, its range has many more than 6λ bits in every column, so for every $\mathbf{S} \neq \mathbf{S}'$ we get $\Pr_{H'}[H'(\mathbf{S}) = H'(\mathbf{S}')] \ll 2^{-6\lambda}$. Therefore H' is a good condenser, i.e., if the min-entropy of $\mathcal{D}$ is above 3λ, then with overwhelming probability over the choise of H, the min-entropy of $H'(\mathcal{D})$ is above $3\lambda - 1$ (say). By the extraction properties of $\mathsf{RandExt}$, this implies that $\mathsf{RandExt}_\beta(H'(\mathcal{D}))$ is close to uniform (whp over β).

3.2   A Commutative Variant

In some applications it may be convenient or even necessary to work with a commutative plaintext space. Of course, simply switching to a commutative sub-ring of the ring of matrices (such as $s \cdot I$ for a scalar s and the identity I) would be insecure, but we can make it work by moving to a larger ring.

Cyclotomic Rings. We switch from working over the ring of integers to working over polynomial rings, $R = \mathbb{Z}[x]/(F(X))$ and $R_q = R/qR$ for some degree n irreducible integer polynomial $F(X) \in \mathbb{Z}[X]$ and an integer $q \in \mathbb{Z}$. Elements of this ring correspond to degree-$(n-1)$ polynomials, and hence they can be represented by n-vectors of integers in some convenient basis. The norm of a ring element is the norm of its coefficient vector, and this can be extended as usual for norm of vectors and matrices over R. Addition and multiplication are just polynomial addition and multiplication modulo $F(X)$ (and also modulo q when talking about R_q).

As usual, we need a ring where the norm of a product is not much larger than the product of the norms, and this can be achieved for example by using $F = \Phi_M(X)$, the M'th cyclotomic polynomial (of degree $n = \phi(M)$). All the required operations and lemmas that we need (such as trapdoor and pre-image sampling etc.) can be extended also to this setting, see e.g. [31].

The construction remains nearly identical, except all operations are now performed over the rings R and R_q and the dimensions are changed to match. We now have the "matrices" $\mathbf{A}_v \in R_q^{m \times 1}$ with only one column (and similarly the error matrices are $\mathbf{E} \in R_q^{m \times 1}$), and the plaintext space is R_q itself. An encoding of plaintext element $s \in R_q$ relative to path $u \rightsquigarrow v$ is a small matrix $\mathbf{D} \in R_q^{m \times m}$ such that

$$\mathbf{D} \cdot \mathbf{A}_u = \mathbf{A}_v \cdot s + \mathbf{E}$$

---

[3] If q is not a power of two then H' does not produce uniformly random t-bit strings. But still its outputs on any two $\mathbf{S}' \neq \mathbf{S}$ are independent, and each has almost full (min-)entropy, which suffices for our purposes.

where $\mathbf{E}'$ is some small error term. As before, we only encode small plaintext elements, i.e., the sampling procedure draws s from a Gaussian distribution with small parameter. The operations all remain the same as in the basic scheme.

We emphasize that it is the plaintext space that is commutative, not the space of encoding. Indeed, if we have $\mathbf{D}, \mathbf{D}'$ that encode s, s' relative to paths $v \rightsquigarrow w$ and $u \rightsquigarrow v$, respectively, we can only multiply them in the order $\mathbf{D} \cdot \mathbf{D}'$. Multiplying in the other order is inconsistent with the graph G and hence is unlikely to yield a meaningful result. What makes the commutative scheme useful is the ability to multiply the *plaintext elements* in arbitrary order. For example for $\mathbf{D}, \mathbf{D}'$ that encode s, s' relative to paths $u \rightsquigarrow w$ and $v \rightsquigarrow w$, we can compute either $\mathbf{D} \cdot \mathbf{A}_u \cdot s'$ or $\mathbf{D}' \cdot \mathbf{A}_v \cdot s$ and the results will both be close $\mathbf{A}_v \cdot ss'$ (and hence also close to each other).

3.3   Public Sampling and Some Other Variations

As mentioned in Section 2.2, for the non-commutative version we can provide a public sampling procedure relative to any desired path $p = u \rightsquigarrow v$ by publishing with the public parameters a collection of pairs generated by the secret sampling procedure above, $\{(\mathbf{S}_k, \mathbf{D}_k) : k = 1, \ldots, \ell\}$ (for some large enough ℓ). The public sampling procedure then takes a random linear combination of these pairs as a new sample, namely it chooses $\mathbf{r} \leftarrow D_{\mathbb{Z}^\ell, \sigma'}$ and compute the encoding pair as:

$$(\mathbf{S}, \mathbf{D}) := \left(\sum\nolimits_{i \in [\ell]} \mathbf{r}_i \mathbf{S}_i , \ \sum\nolimits_{i \in [\ell]} \mathbf{r}_i \mathbf{D}_i \right).$$ It is easy to see that the resulting $\mathbf{D}$

encodes $\mathbf{S}$ relative to the edge e. Also by the leftover-hash over Gaussians [2,1], the plaintext matrix $\mathbf{S}$ is distributed according to a Gaussian distribution whp. We again caution that adding these encodings is insecure in the commutative case, as noted in Section 4.

Some Safeguards. Since our schemes are graph-based, and hence the order of products is known in advance, we can often provide additional safeguards using Kilian-type randomization [29] "on the encoding side". Namely, for each internal node v in the graph we choose a random invertible $m \times m$ matrix modulo q $\mathbf{R}_v$, and for the sinks and sources we set $\mathbf{R}_v = I$. Then we replace each encoding $\mathbf{C}$ relative to the path $u \rightsquigarrow v$ by the masked encoding $\mathbf{C}' := \mathbf{R}_v^{-1} \cdot \mathbf{C} \cdot \mathbf{R}_u$.

Clearly, this randomization step does not affect the product on any source-to-sink path in the graph, but the masked encodings relative to any other path no longer consist of small entries, and this makes it harder to mount the attacks from Section 4.

Other safeguards of this type includes the observations that encoding matrices relative to paths that end at a sink node need not have small entries since the size of the last matrix on a path does not contribute to the size of the final error matrix. Similarly plaintext elements encoded on paths that begin at source nodes need not be small, for the same reason.

We remark that applying the safeguards from above comes with a price tag: namely the encoding matrices no longer consist of small entries, hence it takes more bits to represent them.

Finally, we observe that sometimes we do not need to give explicitly the matrices $\mathbf{A}_u$ corresponding to source nodes, and can instead "fold them" into the encoding matrices. That is, instead of providing both $\mathbf{A}$ and $\mathbf{C}$ such that $\mathbf{B} = \mathbf{D} \cdot \mathbf{A} \approx \mathbf{A}' \cdot \mathbf{S}$, we can publish only the matrix $\mathbf{B}$ and keep $\mathbf{A}, \mathbf{D}$ hidden. This essentially amounts to shortening the path by one, starting it at the matrix $\mathbf{B}$. (Of course, trying to repeat this process and further process the path will lead to exponential growth in the number of matrices that we need to publish.)

3.4   Hardness Assumptions

One can verify that the hardness of some simple tasks related to the construction above follows from the hardness of standard LWE. For example, when the graph has just a single edge $A_1 \to A_0$ then given A_0, A_1 and C it is hard to determine whether C is a valid encoding relative to this edge: The reduction begins with (A_0, B) (where either $B = A_0 S + E$ or B is random), then chooses A_1 with a trapdoor and samples C as a small solution to $CA_1 = B$. For the same reason, given A_0, A_1 and a valid encoding C on the edge $A_1 \to A_0$, it is hard to recover the plaintext S which is encoded by C (assuming the hardness of search-LWE).

However, such simple hardness assumptions do not seem too useful, since in most applications we presumably need an underlying set of plaintext matrices and some expression that evaluates to zero in them (so that we can use the zero-test). Hence it appears that useful hardness assumptions would have to argue about the hardness of a collection of LWE instances with related secrets. Moreover, in settings where we have multiple secrets encoded on each edge, we do not know how to generate these encodings without knowing trapdoors for all the non-sink matrices, which makes it impossible to reduce hardness to LWE instances involving these matrices.

Below we describe one type of "useful hardness assumptions" for our scheme. (This is similar to what we need in our key-exchange protocol for $n = 2$, but without the commutativity.) Although this assumption is neither necessary nor sufficient for any application that we know of, it may still be interesting to study as an instrument for gaining better understanding of the security properties of this scheme.

The underlying graph has three chains that end at a common sink, two of length 2 and one of length 1, and we have two "target" random plaintext matrices S_1, S_2 and a few "auxiliary" random plaintext matrices T_1, T_2, T_2. The adversary gets the source matrices (A, A', B'' in the picture below) and multiple encoding of the plaintext matrices, relative to edges as depicted below:

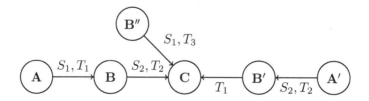

The adversary also gets a matrix $\mathbf{U} \in \mathbb{Z}_q^{m \times n}$, and it needs to distinguish the case $\mathbf{U} = \mathbf{C} \times (\mathbf{S}_1 \times \mathbf{S}_2) + \mathbf{E}$ (for a small Gaussian $\mathbf{E}$) from the case where $\mathbf{U}$ is random. It is easy to see that given these encodings anyone can compute $\mathbf{U}' \approx \mathbf{C} \times (\mathbf{S}_2 \times \mathbf{S}_1)$, but this is the non-commutative case so we do not immediately get an approximation of $\mathbf{C} \times (\mathbf{S}_1 \times \mathbf{S}_2)$. Note that the encoded auxiliary $\mathbf{T}_i$'s do not play much of a role here, but nontheless their presence seems to hinder reduction to LWE. Also the choice of what plaintext matrix to encode relative to what edge was made so as to avoid the attacks from Section 4 below.

4   Cryptanalysis

Below we describe several attacks and "near attacks" on some variations of our scheme, these attacks guided our choices in designing these scheme.

4.1   Encoding of Zero is a Weak Trapdoor

The main observation in this section is that an encoding of zero relative to a path $u \rightsquigarrow v$ can sometimes be used as a weak form of trapdoor for the matrix $\mathbf{A}_u$. Recall from [3,24] that a full-rank $m \times m$ matrix $\mathbf{T}$ with small entries satisfying $\mathbf{TA} = 0 \pmod{q}$ can be used as a trapdoor for the matrix $\mathbf{A}$ as per Lemma 1. An encoding of zero relative the path $u \rightsquigarrow v$ is a matrix $\mathbf{C}$ such that $\mathbf{CA}_u = \mathbf{E} \pmod{q}$ for a small matrix $\mathbf{E}$. This is not quite a trapdoor, but it appears close and indeed we show that it can often be used as if it was a real trapdoor.

Let us denote by $\mathbf{A}_u' = (\mathbf{A}_u/I)$ the $(m + n) \times n$ matrix whose first m rows are those of $\mathbf{A}_u$ and whose last n rows are the $n \times n$ identity matrix. Given the matrices $\mathbf{A}_u$ and $\mathbf{C}$ as above, we can compute the small matrix $\mathbf{E} = \mathbf{CA}_u \bmod q$, then set $\mathbf{C}' = [\mathbf{C}|(-\mathbf{E})]$ to be the $m \times (m + n)$ matrix whose first m columns are the columns of $\mathbf{C}$ and whose last n columns are the negation of the columns of $\mathbf{E}$. Clearly $\mathbf{C}'$ is a small matrix satisfying $\mathbf{C}'\mathbf{A}_u' = 0 \pmod{q}$, but it is not a trapdoor yet because it has rank m rather than $m + n$.

However, assume that we have two encodings of zero, relative to two (possibly different) paths that begin at the same node u. Then we can apply the procedure above to get two such matrices $\mathbf{C}_1'$ and $\mathbf{C}_2'$, and now we have $2m$ rows that are all orthogonal to $\mathbf{A}_u' \bmod q$, and it is very likely that we can find $m + n$ among them that are linearly independent. This gives a full working trapdoor $\mathbf{T}_u'$ for the matrix $\mathbf{A}_u'$, what can we do with this trapdoor?

Assume now that the application gives us, in addition to the zero encodings for path that begin with u, also an encoding of a plaintext elements $\mathbf{S} \neq 0$ relative to some path that ends at u, say $w \rightsquigarrow u$. This is a matrix $\mathbf{D}$ such that $\mathbf{DA}_w = \mathbf{A}_u\mathbf{S} + \mathbf{E}$, namely $\mathbf{B} = \mathbf{DA}_w \bmod q$ is an LWE instance relative to public matrix $\mathbf{A}_u$, secret $\mathbf{S}$, and error term $\mathbf{E}$. Recalling that the plaintext $\mathbf{S}$ in our scheme must be small, it is easy to convert $\mathbf{B}$ into an LWE instance relative to matrix $\mathbf{A}_u' = (\mathbf{A}_u/I)$, for which we have a trapdoor: Simply add n zero rows at the bottom, thus getting $\mathbf{B}' = (\mathbf{B}/0)$, and we have $\mathbf{B}' = \mathbf{A}_u'\mathbf{S} + \mathbf{E}'$, with $\mathbf{E}' = (\mathbf{E}/(-\mathbf{S}))$ a small matrix.[4] Given $\mathbf{B}'$ and $\mathbf{A}'_u$, in conjunction with the trapdoor $\mathbf{T}'_u$, we can now recover the plaintext $\mathbf{S}$.

We note that a consequence of this attack is that in our scheme it is unsafe for the application to allow computation of zero-encoding, except perhaps relative to source-nodes in the graph. As we show in Section 5, it is possible to design applications that get around this problem.

Extensions. The attacks from above can be extended even to some cases where we are not given encodings of zero. Suppose that instead we are given pairs $\{(\mathbf{C}_i, \mathbf{C}'_i)\}_i$, where the two encodings in each pair encode *the same plaintext* $\mathbf{S}_i$ relative to two paths with a common end point, $u \rightsquigarrow v$ and $u' \rightsquigarrow v$. In this case we can use the same techniques to find a "weak trapdoor" for the concatenated matrix $\mathbf{A}' = (\mathbf{A}_u/\mathbf{A}_{u'})$ of dimension $2m \times n$, using the fact that $[\mathbf{C}_i | (-\mathbf{C}'_i)] \cdot \mathbf{A}' = (\mathbf{A}_v \mathbf{S}_i + \mathbf{E}_i) - (\mathbf{A}_v \mathbf{S}_i + \mathbf{E}'_i) = \mathbf{E}_i - \mathbf{E}'_i$.

If we are also given a pair $(\mathbf{D}, \mathbf{D}')$ that encodes the same element $\mathbf{S}$ relative to two paths that end at u, u', respectively, then we can use these approximate trapdoors to find $\mathbf{S}$, since $(\mathbf{D}, \mathbf{D}')$ (together with the start points of these paths) yield an LWE instance relative to public matrix $\mathbf{A}'$ and the secret $\mathbf{S}$.

Corollary 1: No Re-randomization. A consequence of the attacks above is that in our scheme we usually cannot provide encoding-of-zero in the public parameters. Hence the re-randomization technique by adding encodings of zero usually cannot be used in our case.

Corollary 2: No Commutative plaintext/encoding pairs. Another consequence of the attacks above is that at least in the commutative case it is not safe to provide many pairs (s_i, C_i) s.t. C_i is an encoding of the scalar s_i along a path $u \rightsquigarrow v$. The reason is that given two such pairs $(s_1, C_1), (s_2, C_2)$ we can compute an encoding of zero along the path $u \rightsquigarrow v$ as $s_1 C_2 - s_2 C_1$.

4.2   The Cheon et al. Attacks

Very recently, Cheon, Han, Lee, Ryu, and Stehlé described in [15] a serious attack on the CLT encoding scheme, and their techniques were shown to extend also to other settings [10,22,17]. This attack leverages a large number of expressions that evaluate to zero — multiplied by the zero-test parameter — in order to setup a system of euqations in the secret parameters of the scheme. That system of equations is over the integers without any modular reduction, and it is often possible to use linear-algebra tools to extract from it useful information and break the scheme.

In principle, the techniques used in these attacks may be applicable also to our new scheme, since here too we obtain small values after multiplying by the zero-test parameter (and hence get a system of equations without modular reduction).

---

[4] $\mathbf{B}'$ does not have the right distribution for an LWE instance, but using the trapdoor we can solve the worst-case BDD, not just the average-case LWE, so the attack still stands.

But so far we were not able to find any actual case where this line of attacks is applicable to our scheme. The main reason is that the attacks from above are often more powerful: in many cases where the Cheon et al. attacks could be applied we get an easier break using the encoding-of-zero-as-trapdoor technique from above. Another reason why it may be harder to apply these attacks is that the quantities of interest here are matrices rather than single elements (at least in the non-commutative case), which could make solving the equations harder.

4.3   Recovering Hidden $\mathbf{A}_v$'s.

As we noted earlier, in many applications we only need to know the matrices $\mathbf{A}_u$ for source nodes u and there is no need to publish the matrices $\mathbf{A}_v$ for internal nodes. This raises the possibility that we might get better security by withholding the $\mathbf{A}_v$'s of internal nodes.

Trying to investigate this possibility, we show below two "near attacks" for recovering the public matrices of internal nodes from those of source nodes in the graph. The first attack applies to the commutative setting, and is able to recover an approximate version of the internal matrices (with the approximation deteriorating as we move deeper into the graph). The second attack can recover the internal matrices exactly, but it requires a full trapdoor for the matrices of the source nodes (and we were not able to extend it to work with the "approximate trapdoors" that one gets from an encoding of zero).

The conclusion from these "near attacks" is uncertain. Although is still possible that withholding the internal-node matrices helps security, it seems prudent to examine the security of candidate applications that use our scheme in a setting where the $\mathbf{A}_v$'s are all public.

Recovering the $\mathbf{A}_v$'s in the commutative setting. For this attack we are given a matrix $\mathbf{A}_u$, and many encodings relative to the path $u \rightsquigarrow v$, *together with the corresponding plaintext elements* (e.g., as needed for the public-encoding variant). Namely, we have $\mathbf{A}_u$, small matrices $\mathbf{C}_1, \ldots, \mathbf{C}_t$ (for $t > 1$) and small ring elements $s_1, \ldots, s_t$ such that $\mathbf{C}_j \cdot \mathbf{A}_u = \mathbf{A}_v \cdot s_j + \mathbf{E}_j$ holds for all j, with small $\mathbf{E}_j$'s. Our goal is to find $\mathbf{A}_v$.

We note that the matrix $\mathbf{A}_v$ and the error vectors $\mathbf{E}_j$ are only defined upto small additive factors, since adding 1 to any entry in $\mathbf{A}_v$ can be offset by subtracting the s_j's from the corresponding entry in the $\mathbf{E}_j$'s. Hence the best we can hope for is to solve for $\mathbf{A}_v$ upto a small additive factor (resp. for the $\mathbf{E}_j$'s upto a small additive multiple of the s_j's). Denoting $\mathbf{B}_j := \mathbf{C}_j \cdot \mathbf{A}_u = \mathbf{A}_v \cdot s_j + \mathbf{E}_j$, we compute for $j = 1, \ldots, t-1$,

$$\mathbf{F}_j := \mathbf{B}_j \cdot s_{j+1} - \mathbf{B}_{j+1} \cdot s_j$$
$$= (\mathbf{A}_v \cdot s_j + \mathbf{E}_j) \cdot s_{j+1} - (\mathbf{A}_v \cdot s_{j+1} + \mathbf{E}_{j+1}) \cdot s_j = \mathbf{E}_j \cdot s_{j+1} - \mathbf{E}_{j+1} \cdot s_j.$$

This gives us a non-homogeneous linear system of equations (with the s_j's and $\mathbf{F}_j$'s as coefficients), which we want to solve for the small solution $\mathbf{E}_j$'s. Writing this system explicitly we have

$$
\begin{pmatrix}
[s_2] & [-s_1] & & \\
 & [s_3] & [-s_2] & \\
 & & \ddots & \ddots \\
 & & & [s_t] & [-s_{t-1}]
\end{pmatrix}
\begin{pmatrix}
\mathbf{X}_1 \\
\mathbf{X}_2 \\
\vdots \\
\mathbf{X}_{t-1} \\
\mathbf{X}_t
\end{pmatrix}
=
\begin{pmatrix}
\mathbf{F}_1 \\
\mathbf{F}_2 \\
\vdots \\
\mathbf{F}_{t-1}
\end{pmatrix},
$$

where $[s]$ denotes the $m \times m$ matrix $I_{m \times m} \cdot s$. Clearly this system is partitioned into m independent systems, each of the form

$$
\begin{pmatrix}
s_2 & -s_1 & & \\
 & s_3 & -s_2 & \\
 & & \ddots & \ddots \\
 & & & s_t & -s_{t-1}
\end{pmatrix}
\begin{pmatrix}
x_{1,\ell} \\
x_{2,\ell} \\
\vdots \\
x_{t-1,\ell} \\
x_{t,\ell}
\end{pmatrix}
=
\begin{pmatrix}
f_{1,\ell} \\
f_{2,\ell} \\
\vdots \\
f_{t,\ell}
\end{pmatrix},
$$

with $x_{j,\ell}$, $f_{j,\ell}$ being the ℓ'th entries of the vectors $\mathbf{X}_j, \mathbf{F}_j$, respectively. These systems are under-defined, and to get the $\mathbf{E}_i$'s we need to find *small solutions* for them. Suppressing the index ℓ, we denote these systems in matrix form by $\mathbf{M}\boldsymbol{x} = \boldsymbol{f}$, and show how to find small solutions for them.

At first glance this seems like a SIS problem so one might expect it to be hard, but here we already know a small solution for the corresponding homogeneous system, namely the solution $x_j = s_j$ for all j. Below we assume that the s_j do not all share a prime factor (i.e., that $GCD(s_1, s_2, \ldots, s_t) = 1$), and also that at least one of them has a small inverse in the field of fractions of R. (These two conditions hold with good probability, see discussion in [19, Sec 4.1].)

To find a small solution for the inhomogeneous system, we begin by computing an arbitrary solution for it over the ring R (not modulo q). We note that a solution exists (in particular the E_j's solve this system over R without mod-q reduction), and we can use Gaussian elimination in the field of fractions of R to find it. Denote that solution that was found by $\boldsymbol{g} \in R$, namely we have $\mathbf{M}\boldsymbol{g} = \boldsymbol{f}$.[5] Since over R this is a $(t-1) \times t$ system then its solution space is one-dimensional. Hence every solution to this system (and in particular the small solution that we seek) is of the form $\boldsymbol{e} = \boldsymbol{g} + \boldsymbol{s} \cdot k$ for some $k \in R$.[6]

Choosing one index j such that the element $1/s_j$ in the field of fractions is small, we compute a candidate for the scalar k simply by rounding, $k' := -\lfloor g_j/s_j \rceil$, where division happens in the field of fractions. We next prove that indeed the vector $\boldsymbol{e}' = \boldsymbol{g} + \boldsymbol{s} \cdot k'$ is a small vector over R. Clearly $\boldsymbol{e}' \in R^t$ since

---

[5] Using Gaussian elimination may yield a fractional solution $\boldsymbol{g}'$, but we can "round it" to an integral solution by solving for k' the equation $\boldsymbol{g}' + \boldsymbol{s} \cdot k' = 0 \pmod 1$, then setting $\boldsymbol{g} = \boldsymbol{g}' + \boldsymbol{s} \cdot k'$.

[6] In general the scalar k may be fractional, but if $GCD(s_1, s_2, \ldots, s_t) = 1$ then k must be integral.

$k' \in R$ and $\boldsymbol{g}, \boldsymbol{s} \in R^t$, we next prove that it must be small by showing that "the right scalar k" must be close to the scalar k' that we computed. First, observe that $e'_j = g_j + s_j \cdot k'$ must be small, since

$$e'_j = g_j + s_j \cdot k' = g_j - \lfloor g_j/s_j \rceil \cdot s_j = g_j - (g_j/s_j + \epsilon_j) \cdot s_j = -\epsilon_j \cdot s_j,$$

with ϵ_j the rounding error. Since both ϵ_j and s_j are small, then so is e'_j.

Now consider the "real value" e_j, it too is small and is obtained as $g_j + s_j \cdot k$ for some $k \in R$. It follows that $e_j - e'_j = s_j \cdot (k - k')$ is small, and since we know that $1/s_j$ is also small then it follows that so is $k - k' = (e_j - e'_j)/s_j$. We thus conclude that $e' = g + k' \cdot s = e + (k - k') \cdot e$ is also small.

Repeating the same procedure for all the m independent systems, we get a small solution $\{\mathbf{E}'_j, j = 1, \ldots, t\}$ to the system $\mathbf{B}_j = \mathbf{A}_v \cdot s_j + \mathbf{E}'_j$. Subtracting the $\mathbf{E}'_j$'s from the $\mathbf{B}_j$'s and dividing by the s_j's give us (an approximation of) $\mathbf{A}_v$.

Recovering the $\mathbf{A}_v$'s using trapdoors. Suppose that we are given $\mathbf{A}_u$, encodings $\mathbf{C}_j$ and the corresponding plaintext matrices $\mathbf{S}_j$, s.t. $\mathbf{B}_j := \mathbf{C}_j \cdot \mathbf{A}_u = \mathbf{A}_v \cdot \mathbf{S}_j + \mathbf{E}_j$ (mod q) for small errors $\mathbf{E}_j$. Suppose that in addition we are also given a *full working trapdoor* for the matrix $\mathbf{A}_v$, say, in the form of a small full-rank matrix $\mathbf{T}$ over R s.t. $\mathbf{T} \cdot \mathbf{A}_v = 0$ (mod q). We can then use $\mathbf{T}$ to recover the errors $\mathbf{E}_j$ from the LWE instances $\mathbf{B}_j$, which can be done without knowing $\mathbf{A}_v$: Let $\mathbf{T}^{-1}$ be the inverse of $\mathbf{T}$ over R, we compute $\mathbf{E}_j \leftarrow \mathbf{T}^{-1} \cdot (\mathbf{T} \cdot \mathbf{B}_j \bmod q)$. Once we have the error matrices $\mathbf{E}_j$ we can subtract them and get the set of equations $\mathbf{B}_j - \mathbf{E}_j = \mathbf{A}_v \cdot \mathbf{S}_j$ (mod q), where the entries of $\mathbf{A}_v$ are the unknowns. With sufficiently many of these equations, we can then solve for $\mathbf{A}_v$.

We note that so far we were unable to extend this attack to using the "weak trapdoor" that one gets from an encoding of zero wrt paths of the form $v \rightsquigarrow w$. Indeed the procedure from Section 4.1 for recovering a stronger trapdoor from the weak one relies on knowing $\mathbf{A}_v$.

5   Applications

5.1   Multipartite Key-Agreement

For our first application, we describe a candidate construction for a non-interactive multipartite key-agreement protocol using the commutative variant of our graph-based encoding scheme. As is usual with multipartite key-agreement from multilinear maps, each party i is contributing an encoding of some secret s_i and the shared secret is derived from an encoding of the product $s = \prod_i s_i$. However in our case we need to use extra caution to protect against the "weak trapdoor attacks" from Section 4.1.

To that end, we design our graph to ensure that the adversary is never given encodings of the same element on two paths with a common end-point, and also is not given an encoding and the corresponding plaintext on any edge. For an k-partite protocol we use a graph topology of k directed chains that meet at a common point, where the contribution of any given party appears at different edges on different chains (i.e. the first edge on one chain, the second edge on another, the third edge on a third chain, etc.)

That is, each player i has a directed path of matrices, $\mathbf{A}_{i,1}, \ldots, \mathbf{A}_{i,k+1}$, all sharing the same end-point, i.e., $\mathbf{A}_{i,k+1} = \mathbf{A}_0$ for all i. Note that every chain has k edges, and for the chain "belonging" to party i we will broadcast on its edges encodings of all the secrets s_j, $j \neq i$, but not an encoding of s_i, that last encoding will only be known to party i. Party i will multiply these encodings (the one that only it knows, and all the ones that are publicly available) to get an encoding of $\prod_i s_j$ relative to the path $\mathbf{A}_{i,1} \rightsquigarrow \mathbf{A}_0$. Namely, a matrix $\mathbf{D}_i$ such that $\mathbf{D}_i \cdot \mathbf{A}_{i,1} \approx \mathbf{A}_0 \cdot \prod_i s_j$. The shared secret is then obtained by applying the extraction procedure to this $\mathbf{D}_i$.

The assignment of which secret is encoded on what edge of what chain is done in a "round robin" fashion. Specifically, the i'th secret s_i is encoded on the j'th edge of the chain belonging to party $i' = j - i + 1$. In other words, the secret that we encode on the edge $\mathbf{A}_{i,j} \to \mathbf{A}_{i,j+1}$ in the graph is s_{j-i+1}, with index arithmetic modulo k. An example of the assignment of secrets to edges for a 4-partite protocol is depicted in Figure 1.

Of course, we must publish encodings that would allow the parties to choose their secrets and provide encodings for them. This means that together with the public parameters we also publish encodings of many plaintext elements $\{t_{i,\ell} : i = 1, \ldots, k, \ \ell = 1, \ldots, N\}$ (for a sufficiently large N), for each $t_{i,\ell}$ we publish encoding of it relative to all the edges $\mathbf{A}_{i',i+i'-1} \to \mathbf{A}_{i',i+i'}$ for all i, i' (index arithmetic modulo $k+1$). Party i then chooses random small coefficients $r_{i,\ell}$ and computes its encoding relative to each edge $\mathbf{A}_{i',i+i'-1} \to \mathbf{A}_{i',i+i'}$ as the linear combination of the encodings on that edge with the coefficient $r_{i,\ell}$. We are now ready to describe our scheme $\mathcal{NMKE} = (\mathsf{KE.Setup}, \mathsf{KE.Publish}, \mathsf{KE.Keygen})$.

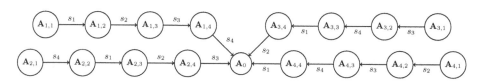

Fig. 1. Graph for a 4-partite key-agreement protocol

- $\mathsf{KE.Setup}(1^\lambda, k)$: The setup algorithm takes as input the security parameter 1^λ and the total number of players k.
  1. Run the parameter-generation and instance-generation of our graph-based encoding scheme for the graph consisting of k chains with a common end-point, each of length k edges. Let $e_{i,j}$ denote the j'th edge on the i'th chain.
  2. Using the secret parameters, run the sampling procedure of the encoding scheme to choose random plaintext elements $t_{i,\ell}$ for $i = 1, \ldots, k$ and $\ell = 1, \ldots, N$, and for each $t_{i,\ell}$ compute also an encoding of it relative to

all the edges $e_{i',j}$ for $j = i + i'$ (mod k). Denote the encoding of $t_{i,\ell}$ on chain i' (relative to edge $e_{i',i+i' \bmod k}$) by $\mathbf{C}_{i,\ell,i'}$.

The public parameters of the key-agreement protocol include the public parameters of the encoding scheme (i.e., the matrices for all the source nodes $\mathbf{A}_{i,1}$), and also the encoding matrices $\{\mathbf{C}_{i,\ell,i'} : i, i' = 1, \ldots, k, \ \ell = 1, \ldots, N\}$.

- KE.Publish(pp, i) : The i'th party chooses random small plaintext elements $r_{i,\ell} \leftarrow \chi$ for $\ell = 1, \ldots, N$ and then sets $\mathbf{D}_{i,i'} \leftarrow \sum_\ell \mathbf{C}_{i,\ell,i'} \cdot r_{i,\ell}$ for all i'. It keeps $\mathbf{D}_{i,i}$ as its secret and broadcast all the other $\mathbf{D}_{i,i'}$'s.

- KE.Keygen(pp, i, sk_i, $\{\mathsf{pub}_j\}_{j \neq i}$) : Party i collects all the matrices $\mathbf{D}_{i',i}$ (encoding the secrets $s_{i'}$ relative to "its chain" i) and orders them according to $j = i + i'$. Namely, it sets $\mathbf{F}_{j,i} = \mathbf{D}_{i+j \bmod k,i}$ for $j = 1, \ldots k$, then computes the product $\mathbf{F}_i^* = (\prod_{j=1}^k F_{j,i}) \cdot \mathbf{A}_{i,1}$. Finally, party i applies the extraction procedure of the encoding scheme to obtain the secret key, setting $\mathsf{ssk} = \mathsf{Extract}(\mathbf{F}_i^*)$.

Security. Unfortunately, we were not able to reduce the security of this candidate scheme to any "nicer" assumption. As such, at present the only evidence of security that we can offer is the failure of our attempts to cryptanalyze it.

The basic attack from Section 4.1 does not seem to apply here since the public parameters do not provide any encoding of zero (not even relative to $\mathbf{A}_0$). Similarly, the extended attacks do not seem to apply since the only common endpoint in the graph is $\mathbf{A}_0$, and no two paths that end at $\mathbf{A}_0$ include an encoding of the same element.

We note that the attacker can use the public parameters to compute an approximate trapdoors for concatenated matrices of the form $(\mathbf{A}_0 \cdot t_{i,\ell,i'}/(-\mathbf{A}_0))$ (or similar), but the broadcast messages of the parties do not provide LWE instances relative to these matrices.

Finally, we note that as for any other application of this encoding scheme, it seems that security would be enhanced by applying the additional safeguards that were discussed at the end of Section 3. That is, we can use Kilian-style randomization on the encoding side, by choosing k invertible matrices for each chain, $\mathbf{R}_{i,1}, \ldots, \mathbf{R}_{i,k}$, where the first and last are set to the identity and the others are chosen at random. Then we can replacing each encoding matrix $\mathbf{C}$ in the public parameters by $\mathbf{C}' := \mathbf{R}^{-1} \cdot \mathbf{C} \cdot \mathbf{R}'$ using the randomizer matrices $\mathbf{R}, \mathbf{R}'$ belonging to the two adjacent nodes. We can also choose the first encoding matrix in each chain to have large entries.

This has no effect on the product of all the encoding matrices along the i'-th chain, but the new matrices $\mathbf{C}'$ no longer have small entries, which seems to aid security. On the down side, this increases the size of the encodings roughly by a $\log q / \log n$ factor.

5.2   Candidate Branching-Program Obfuscation

We next describe how to adapt the branching-program obfuscation constructions from previous work [18,13,4,36] to use our encoding schemes. We remark that on some level this is the simplest type of constructions to adapt to our setting, since we essentially need only a single chain and there almost no issues of providing zero-encoding in the public parameters (or encodings of the same plaintext relative to different nodes in the graph).

Roughly speaking, previous works all followed a similar strategy for obfuscating branching programs. Starting from a given oblivious branching program, encoded as permutation matrices, they all applied Kilian's randomization strategy to randomized these matrices, then added some extra randomization steps (mostly multiplication by random scalars) to protect against partial-evaluation and mixed-input attacks, and finally encoded the resulting matrices relative to specially-designed sets/levels. The specific details of the extra randomization steps are somewhat different between the previous schemes, but all these techniques have their counterparts in our setting. Below we explain how to adapt the randomization techniques from previous work to our setting, and then describe one specific BP-obfuscation candidate.

Matrices vs. individual elements. Our scheme natively encodes matrices, rather than individual elements. This has some advantages, for example we need not worry about attacks that mix and match encoded elements from different matrices. At the same time it also poses some challenges, in particular some of the prior schemes worked by comparing to zero one element of the resulting matrix at the end of evaluation, an operation which is not available in our case.

To be able to examine sub-matrices (or single elements), we adopt the "bookend encoding" trick from [18]. That is, we add to our chain a new source u^* and a new sink v^*, with edges from u^* to the old source and from the old sink to v^*. On the edge from u^* we encode a matrix $\mathbf{T}$ which is only nonzero in the columns that we want to examine, and on the edge to v^* we encode a matrix $\mathbf{S}$ which is only nonzero in the rows that we wish to examine. This way, we should have the matrix $\mathbf{T} \cdot \mathbf{U} \cdot \mathbf{S}$ encoded relative to a path $u^* \rightsquigarrow v^*$, and that matrix is only nonzero in the sub-matrix of interest. In the candidate below we somewhat improve on this by folding the source matrix $\mathbf{A}_{u^*}$ into the encoding of $\mathbf{T}$, publishing instead the matrix $\mathbf{A}_{u^*} \cdot \mathbf{T}$ (and in fact making $\mathbf{T}$ a single column vector).

Only small plaintexts. In our scheme we can only encode "small plaintext elements", not every plaintext element. This is particularly relevant for Kilian randomization technique, since it requires that we multiply by both $\mathbf{R}$ and $\mathbf{R}^{-1}$ for each randomizer matrix $\mathbf{R}$. One way to get randomizer matrices with both $\mathbf{R}$ and $\mathbf{R}^{-1}$ small matrices with four quadrants consisting of $\mathbf{I}, \mathbf{R}, \mathbf{0}, \mathbf{I}$ (in any order that yields determinant 1). Multiplying a sequence of these types of matrices above yields a high-entropy distribution of randomizer matrices with the desired property, and seemingly without obvious algebraic structure. Another family of matrices where both the matrix and its inverse are small are permutation matrices (and of course we can mix and match these families). Concretely, we speculate that a randomizer of the form

$$\mathbf{R} = \Pi_1 \cdot \begin{pmatrix} \mathbf{0} & \mathbf{I} \\ \mathbf{I} & \mathbf{R}_1 \end{pmatrix} \cdot \Pi_2 \cdot \begin{pmatrix} \mathbf{I} & \mathbf{0} \\ \mathbf{R}_2 & \mathbf{I} \end{pmatrix} \cdot \Pi_3 \cdot \begin{pmatrix} \mathbf{R}_3 & \mathbf{I} \\ \mathbf{I} & \mathbf{0} \end{pmatrix} \cdot \Pi_4 \cdot \begin{pmatrix} \mathbf{I} & \mathbf{R}_4 \\ \mathbf{0} & \mathbf{I} \end{pmatrix} \cdot \Pi_5 \quad (4)$$

(with the Π_i's random permutations and the $\mathbf{R}_i$'s random small matrices) has sufficient entropy and lack of algebraic structure to serve as randomizers for our scheme.

We note that although these randomizers are far from uniform, there may still be hope of using some of the tools developed in [13,4,36] (where the analysis includes a reduction to Kilian's information-theoretic argument). This is because the matrices before randomization are permutation matrices, and hence the random permutations Π_i can be used to perfectly randomize them. In this way, one can view the $\mathbf{R}_i$'s are merely "safeguards" to protect against possible weaknesses in the encoding scheme, and the Π_i's are "ideal model randomizers" than can be used in an ideal-model analysis. So far we did not attempt such analysis, however.

Another way to introduce Kilian-type rerandomization in our setting is the aforementioned option of applying it "on the encoding side," i.e., choosing random $m \times m$ invertible matrices $\mathbf{P}$ modulo q and set $\mathbf{C}' \leftarrow \mathbf{P}^{-1} \cdot \mathbf{C} \cdot \mathbf{P}'$.

Multiplicative binding and sraddling sets. Another difference between our setting and that of GGH or CLT is that the previous schemes support encoding relative to arbitrary subsets of a universe set, so there are exponentially many potential sets to use. In our scheme the encoding is relative to edges of a given graph, and there can be only polynomial many of them. This difference seems particularly critical in the design of sraddling sets [4,36].

On a second look, however, this issue is more a question of modeling, rather than a real difference. The different encoding sets in the "asymmetric variants" of [19,16] are obtained just by multiplying by different random secret constants (e.g., the z_i's from GGH), and we can similarly multiply our encoding matrices by such random constants mod q (especially when working over a large polynomial ring). We use that option in the candidate scheme that we describe below.

We note that similar effects can be obtained by the multiplicative binding technique of [18]. Roughly speaking, the main difference between multiplicative binding and sraddling sets is that the former multiplies by constants "on the plaintext side" while the latter multiplies "on the encoding side." In our setting we can do both, and indeed it seems prudent to do so.

A Concrete BP-Obfuscation Candidate.

For our concrete candidate below we work over a large polynomial ring of dimension k, and we will use small-dimension matrices over this ring (roughly as high as the dimension of the underlying branching program).

Let $\mathsf{Sym}(w)$ be the set of $w \times w$ permutation matrices and consider a length-n branching program over ℓ bit inputs:

$$\mathsf{BP} = \{(\mathsf{ind}(i), \mathbf{B}_{i,0}, \mathbf{B}_{i,1} : i \in [n], \mathsf{ind}(i) \in [\ell], \mathbf{B}_{i,b} \in \mathsf{Sym}(w)\}$$

For a bit position $j \in [\ell]$, let I_j be the steps in the branching program that examines j'th input bit: $I_j = \{i \in [n] : \mathsf{ind}(i) = j\}$. We obfuscate BP as follows:

- Following the original construction of [20] we embed the $\mathbf{B}_{i,\sigma}$'s inside higher-dimension matrices with random elements on the diagonal, but in our case it is sufficient to have only two such random entries (so the dimension only grows form w to $w + 2$). Denote the higher-dimension matrices by $\mathbf{B}'_{i,\sigma}$.
  We also follow the original construction of [20] by applying the same transformation to a "dummy program" DP of the same length that consists of only the identity matrices, let $\mathbf{D}'_{i,\sigma}$ be the higher-dimension dummy matrices.
- We proceed to randomize these branching programs a-la-Kilian "on the plaintext side," by choosing randomizing matrices $\mathbf{R}_i$'s as per the form of Eqn. (4) such that both $\mathbf{R}_i$ and $\mathbf{R}_i^{-1}$ are small, and setting $\mathbf{B}''_{i,\sigma} = \mathbf{R}_{i-1}\mathbf{B}'_{i,\sigma}\mathbf{R}_i^{-1}$. The dummy program is randomized similarly.
- We then prepare $(w + 2) \times (w + 2)$ "bookend matrices" $\mathbf{S}, \mathbf{S}'$, and "bookend column vectors" $\mathbf{t}, \mathbf{t}'$. $\mathbf{S}$ is random and small except the first row which is zero, $\mathbf{t}$ is random and small except the second entry which is zero, and similarly for $\mathbf{S}'$ and $\mathbf{t}'$, subject to $\mathbf{S}' \cdot \mathbf{t}' = \mathbf{S} \cdot \mathbf{t}$. Then we set $\tilde{\mathbf{S}} = \mathbf{SR}_0^{-1}$ and $\tilde{\mathbf{t}} = \mathbf{R}_n\mathbf{t}$, and similarly $\tilde{\mathbf{S}}' = \mathbf{S}'\mathbf{R}_0^{-1}$ and $\tilde{\mathbf{t}}' = \mathbf{R}_n\mathbf{t}'$.
- Next we use our encoding scheme to encode these matrices relative to a graph with two chains with a common end-point, each of length $n + 2$. Namely we have $\mathbf{A}_1 \to \ldots \to \mathbf{A}_{n+2}$ and $\mathbf{A}'_1 \to \ldots \to \mathbf{A}'_{n+1} \to \mathbf{A}_{n+2}$.
  For each $i \in [n]$, we encode the two matrices $\mathbf{B}''_{n-i+1,b}$ relative to the edge $\mathbf{A}_i \to \mathbf{A}_{i+1}$, i.e., we have

$$\mathbf{C}_{n-i+1,b} \cdot \mathbf{A}_i = \mathbf{A}_{i+1} \cdot \mathbf{B}''_{n-i+1,b} + \mathbf{E}_{i,b}$$

  for some small error $\mathbf{E}_{i,b}$. Similarly we encode the dummy program with the two matrices $\mathbf{D}''_{n-i+1,b}$ encoded relative to the edge $\mathbf{A}'_i \to \mathbf{A}'_{i+1}$, i.e.,

$$\mathbf{C}'_{n-i+1,b} \cdot \mathbf{A}'_i = \mathbf{A}'_{i+1} \cdot \mathbf{D}''_{n-i+1,b} + \mathbf{E}'_{i,b}$$

- Encode $\tilde{\mathbf{S}}, \tilde{\mathbf{S}}'$ relative to the edges leading to the common sink, i.e. compute the encoding matrices $\mathbf{C}_S, \mathbf{C}'_{S'}$ such that

$$\mathbf{C}_S \cdot \mathbf{A}_{n+1} = \mathbf{A}_{n+2} \cdot \tilde{\mathbf{S}} + \mathbf{E}_S \text{ and } \mathbf{C}'_{S'} \cdot \mathbf{A}'_{n+1} = \mathbf{A}_{n+2} \cdot \tilde{\mathbf{S}}' + \mathbf{E}'_{S'}$$

- Compute the encoded bookend vectors, folded into the two sources $\mathbf{A}_1$ and $\mathbf{A}'_1$, namely $\mathbf{a} = \mathbf{A}_1 \cdot \tilde{\mathbf{t}} + \mathbf{e}_t$ and $\mathbf{a}' = \mathbf{A}'_1 \cdot \tilde{\mathbf{t}}' + \mathbf{e}'_t$.
- We next apply both the multiplicative bundling and the the Kilian-style randomization *also on the encoding side*. Namely we sample random full-rank matrices $\mathbf{P}_0, \ldots, \mathbf{P}_n$ and $\mathbf{P}'_0, \ldots, \mathbf{P}'_n$, and also random scalars modulo q $\{\beta_{i,0}, \beta_{i,1}, \beta'_{i,0}, \beta'_{i,1} : i \in [n]\}$, subject to constraints $\prod_{i \in I_j} \beta_{i,0} = \prod_{i \in I_j} \beta'_{i,0} = \prod_{i \in I_j} \beta_{i,1} = \prod_{i \in I_j} \beta'_{i,1} = 1$.
  We then set $\hat{\mathbf{C}}_{i,\sigma} = \mathbf{P}_{i-1}^{-1} \cdot \mathbf{C}_{i,\sigma} \cdot \mathbf{P}_i \cdot \beta_{i,\sigma}$ and $\hat{\mathbf{C}}'_{i,\sigma} = \mathbf{P}'^{-1}_{i-1} \cdot \mathbf{C}'_{i,\sigma} \cdot \mathbf{P}'_i \cdot \beta'_{i,\sigma}$, and also $\hat{\mathbf{C}}_S = \mathbf{C}_S \cdot \mathbf{P}_0$ and $\hat{\mathbf{C}}'_{S'} = \mathbf{C}'_{S'} \cdot \mathbf{P}'_0$ and $\hat{\mathbf{a}} = \mathbf{P}_n^{-1}\mathbf{a}$ and $\hat{\mathbf{a}}' = \mathbf{P}'^{-1}_n\mathbf{a}'$.
- The obfuscation consists of all the matrices and vectors above, namely

$$\mathcal{O}(\mathsf{BP}) = \left(\left\{ \hat{\mathbf{C}}_S, \{\hat{\mathbf{C}}_{i,\sigma} : i \in [n], \sigma \in \{0,1\}\}, \hat{\mathbf{a}} \right\}, \right.$$
$$\left. \left\{ \hat{\mathbf{C}}'_{S'}, \{\hat{\mathbf{C}}'_{i,\sigma} : i \in [n], \sigma \in \{0,1\}\}, \hat{\mathbf{a}}' \right\} \right)$$

Evaluation. On input $\mathbf{x} \in \{0,1\}^{\ell}$ the user choose the appropriate encoding matrices $\hat{\mathbf{C}}_{i,0}$ or $\hat{\mathbf{C}}_{i,1}$ depending on the relevant input bit (and the same for $\hat{\mathbf{C}}'_{i,0}$ or $\hat{\mathbf{C}}'_{i,1}$) and then multiply in order setting

$$\mathbf{y} = \hat{\mathbf{C}}_S \cdot (\prod_{i=1}^{n} \hat{\mathbf{C}}_{i,x[\mathsf{ind}(i)]}) \cdot \mathbf{a} = \mathbf{A}_{n+2} \cdot (\mathbf{S} \cdot (\prod_{i=1}^{n} \mathbf{B}''_{i,x[\mathsf{ind}(i)]}) \cdot \mathbf{t}) + \mathbf{e}$$

and

$$\mathbf{y}' = \hat{\mathbf{C}}'_{S'} \cdot (\prod_{i=1}^{n} \hat{\mathbf{C}}'_{i,x[\mathsf{ind}(i)]}) \cdot \mathbf{a}' = \mathbf{A}_{n+2} \cdot (\mathbf{S}' \cdot (\prod_{i=1}^{n} \mathbf{D}''_{i,x[\mathsf{ind}(i)]}) \cdot \mathbf{t}') + \mathbf{e}',$$

The output is 1 if $\|\mathbf{y} - \mathbf{y}'\| < q^{3/4}$ and 0 otherwise. Note that indeed if $\prod_{i=1}^{n} \mathbf{D}_{i,x[\mathsf{ind}(i)]} = \mathbf{I}$ then both $\mathbf{y}$ and $\mathbf{y}'$ are roughly equal to $\mathbf{A}_{n+2} \cdot \mathbf{S} \cdot \mathbf{t} \cdot (\prod_{i=1}^{n} \alpha_{i,x[\mathsf{ind}(i)]})$, as needed.

Security. As before, this is merely a candidate and we do not know how to reduce its security to any "nice" assumption. However the type of attacks that we know against these scheme do not seem to apply to this candidate.

Acknowledgments. We thank Zvika Brakerski for pointing out to us vulnerabilities in earlier versions of this work. We also thank Vinod Vaikuntanathan for insightful discussions.

References

1. Aggarwal, D., Regev, O.: A note on discrete gaussian combinations of lattice vectors. CoRR abs/1308.2405 (2013)
2. Agrawal, S., Gentry, C., Halevi, S., Sahai, A.: Discrete gaussian leftover hash lemma over infinite domains. In: Sako, K., Sarkar, P. (eds.) ASIACRYPT 2013, Part I. LNCS, vol. 8269, pp. 97–116. Springer, Heidelberg (2013)
3. Ajtai, M.: Generating hard instances of the short basis problem. In: Wiedermann, J., Van Emde Boas, P., Nielsen, M. (eds.) ICALP 1999. LNCS, vol. 1644, pp. 1–9. Springer, Heidelberg (1999)
4. Barak, B., Garg, S., Kalai, Y.T., Paneth, O., Sahai, A.: Protecting obfuscation against algebraic attacks. In: Nguyen and Oswald [34], pp. 221–238
5. Boneh, D., Franklin, M.: Identity-based encryption from the Weil pairing. SIAM J. of Computing 32(3), 586–615 (2003); extended abstract in Kilian, J. (ed.) CRYPTO 2001. LNCS, vol. 2139, pp. 213–615. Springer, Heidelberg (2001)
6. Boneh, D., Franklin, M.: Identity-based encryption from the weil pairing. In: Kilian, J. (ed.) CRYPTO 2001. LNCS, vol. 2139, pp. 213–229. Springer, Heidelberg (2001)
7. Boneh, D., Gentry, C., Gorbunov, S., Halevi, S., Nikolaenko, V., Segev, G., Vaikuntanathan, V., Vinayagamurthy, D.: Fully key-homomorphic encryption, arithmetic circuit ABE and compact garbled circuits. In: Nguyen, P.Q., Oswald, E. (eds.) EUROCRYPT 2014. LNCS, vol. 8441, pp. 533–556. Springer, Heidelberg (2014), http://dx.doi.org/10.1007/978-3-642-55220-5_30

8. Boneh, D., Gentry, C., Waters, B.: Collusion resistant broadcast encryption with short ciphertexts and private keys. In: Shoup, V. (ed.) CRYPTO 2005. LNCS, vol. 3621, pp. 258–275. Springer, Heidelberg (2005), http://dx.doi.org/10.1007/11535218_16

9. Boneh, D., Silverberg, A.: Applications of multilinear forms to cryptography. Contemporary Mathematics 324, 71–90 (2003)

10. Boneh, D., Wu, D.J., Zimmerman, J.: Immunizing multilinear maps against zeroizing attacks. Cryptology ePrint Archive, Report 2014/930 (2014), http://eprint.iacr.org/

11. Boneh, D., Zhandry, M.: Multiparty key exchange, efficient traitor tracing, and more from indistinguishability obfuscation. In: Garay, J.A., Gennaro, R. (eds.) CRYPTO 2014, Part I. LNCS, vol. 8616, pp. 480–499. Springer, Heidelberg (2014), http://dx.doi.org/10.1007/978-3-662-44371-2_27

12. Brakerski, Z., Rothblum, G.N.: Virtual black-box obfuscation for all circuits via generic graded encoding. In: Lindell, Y. (ed.) TCC 2014. LNCS, vol. 8349, pp. 1–25. Springer, Heidelberg (2014)

13. Brakerski, Z., Rothblum, G.: Virtual black-box obfuscation for all circuits via generic graded encoding. In: Lindell, Y. (ed.) TCC 2014. LNCS, vol. 8349, pp. 1–25. Springer, Heidelberg (2014), http://dx.doi.org/10.1007/978-3-642-54242-8_1

14. Carter, J., Wegman, M.N.: Universal classes of hash functions. Journal of Computer and System Sciences 18(2), 143–154 (1979)

15. Cheon, J.H., Han, K., Lee, C., Ryu, H., Stehlé, D.: Cryptanalysis of the multilinear map over the integers. Cryptology ePrint Archive, Report 2014/906 (2014), http://eprint.iacr.org/

16. Coron, J.-S., Lepoint, T., Tibouchi, M.: Practical multilinear maps over the integers. In: Canetti, R., Garay, J.A. (eds.) CRYPTO 2013, Part I. LNCS, vol. 8042, pp. 476–493. Springer, Heidelberg (2013), http://dx.doi.org/10.1007/978-3-642-40041-4_26

17. Coron, J.S., Lepoint, T., Tibouchi, M.: Cryptanalysis of two candidate fixes of multilinear maps over the integers. Cryptology ePrint Archive, Report 2014/975 (2014), http://eprint.iacr.org/

18. Garg, S., Gentry, C., Halevi, S., Raykova, M., Sahai, A., Waters, B.: Candidate indistinguishability obfuscation and functional encryption for all circuits. In: 2013 IEEE 54th Annual Symposium on Foundations of Computer Science (FOCS), pp. 40–49 (October 2013)

19. Garg, S., Gentry, C., Halevi, S.: Candidate multilinear maps from ideal lattices. In: Johansson, T., Nguyen, P.Q. (eds.) EUROCRYPT 2013. LNCS, vol. 7881, pp. 1–17. Springer, Heidelberg (2013), http://dx.doi.org/10.1007/978-3-642-38348-9_1, Full version at http://eprint.iacr.org/2013/451

20. Garg, S., Gentry, C., Halevi, S., Sahai, A., Waters, B.: Attribute-based encryption for circuits from multilinear maps. Cryptology ePrint Archive, Report 2013/128 (2013)

21. Garg, S., Gentry, C., Sahai, A., Waters, B.: Witness encryption and its applications. In: Proceedings of the Forty-fifth Annual ACM Symposium on Theory of Computing, STOC 2013, pp. 467–476. ACM, New York (2013), http://doi.acm.org/10.1145/2488608.2488667

22. Gentry, C., Halevi, S., Maji, H.K., Sahai, A.: Zeroizing without zeroes: Cryptanalyzing multilinear maps without encodings of zero. Cryptology ePrint Archive, Report 2014/929 (2014), http://eprint.iacr.org/

23. Gentry, C., Lewko, A., Sahai, A., Waters, B.: Indistinguishability obfuscation from the multilinear subgroup elimination assumption. Cryptology ePrint Archive, Report 2014/309 (2014), http://eprint.iacr.org/

24. Gentry, C., Peikert, C., Vaikuntanathan, V.: Trapdoors for hard lattices and new cryptographic constructions. In: Proceedings of the 40th Annual ACM Symposium on Theory of Computing, STOC 2008, pp. 197–206. ACM (2008)

25. Gentry, C., Sahai, A., Waters, B.: Homomorphic encryption from learning with errors: Conceptually-simpler, asymptotically-faster, attribute-based. In: Canetti, R., Garay, J.A. (eds.) CRYPTO 2013, Part I. LNCS, vol. 8042, pp. 75–92. Springer, Heidelberg (2013), http://dx.doi.org/10.1007/978-3-642-40041-4_5

26. Goyal, V., Pandey, O., Sahai, A., Waters, B.: Attribute-based encryption for fine-grained access control of encrypted data. In: ACM CCS, pp. 89–98 (2006)

27. Hastad, J., Impagliazzo, R., Levin, L.A., Luby, M.: A pseudorandom generator from any one-way function. SIAM J. Comput. 28, 1364–1396 (1999), http://dx.doi.org/10.1137/S0097539793244708

28. Joux, A.: A one round protocol for tripartite diffie hellman. Journal of Cryptology 17(4), 263–276 (2004), http://dx.doi.org/10.1007/s00145-004-0312-y

29. Kilian, J.: Founding crytpography on oblivious transfer. In: Proceedings of the Twentieth Annual ACM Symposium on Theory of Computing, STOC 1988, pp. 20–31. ACM, New York (1988), http://doi.acm.org/10.1145/62212.62215

30. Langlois, A., Stehlé, D., Steinfeld, R.: Gghlite: More efficient multilinear maps from ideal lattices. In: Nguyen and Oswald [34], pp. 239–256

31. Lyubashevsky, V., Peikert, C., Regev, O.: A toolkit for ring-LWE cryptography. In: Johansson, T., Nguyen, P.Q. (eds.) EUROCRYPT 2013. LNCS, vol. 7881, pp. 35–54. Springer, Heidelberg (2013)

32. Micciancio, D., Peikert, C.: Trapdoors for lattices: Simpler, tighter, faster, smaller. In: Pointcheval, D., Johansson, T. (eds.) EUROCRYPT 2012. LNCS, vol. 7237, pp. 700–718. Springer, Heidelberg (2012)

33. Micciancio, D., Regev, O.: Worst-case to average-case reductions based on gaussian measures. SIAM J. Comput. 37(1), 267–302 (2007)

34. Nguyen, P.Q., Oswald, E. (eds.): EUROCRYPT 2014. LNCS, vol. 8441. Springer, Heidelberg (2014)

35. Papamanthou, C., Tamassia, R., Triandopoulos, N.: Optimal authenticated data structures with multilinear forms. In: Joye, M., Miyaji, A., Otsuka, A. (eds.) Pairing 2010. LNCS, vol. 6487, pp. 246–264. Springer, Heidelberg (2010), http://dx.doi.org/10.1007/978-3-642-17455-1_16

36. Pass, R., Seth, K., Telang, S.: Indistinguishability obfuscation from semantically-secure multilinear encodings. In: Garay, J.A., Gennaro, R. (eds.) CRYPTO 2014, Part I. LNCS, vol. 8616, pp. 500–517. Springer, Heidelberg (2014)

37. Rückert, M., Schröder, D.: Aggregate and verifiably encrypted signatures from multilinear maps without random oracles. In: Park, J.H., Chen, H.-H., Atiquzzaman, M., Lee, C., Kim, T.-H., Yeo, S.-S. (eds.) ISA 2009. LNCS, vol. 5576, pp. 750–759. Springer, Heidelberg (2009), http://dx.doi.org/10.1007/978-3-642-02617-1_76

38. Rothblum, R.: On the circular security of bit-encryption. In: Sahai, A. (ed.) TCC 2013. LNCS, vol. 7785, pp. 579–598. Springer, Heidelberg (2013), http://dx.doi.org/10.1007/978-3-642-36594-2_32

39. Sakai, R., Ohgishi, K., Kasahara, M.: Cryptosystems based on pairing. In: SCIS 2000, Okinawa, Japan (January 2000)

40. Waters, B.: Efficient identity-based encryption without random oracles. In: Cramer, R. (ed.) EUROCRYPT 2005. LNCS, vol. 3494, pp. 114–127. Springer, Heidelberg (2005)
41. Wegman, M.N., Carter, J.: New hash functions and their use in authentication and set equality. Journal of Computer and System Sciences 22(3), 265–279 (1981)

A   Parameter Selection

We now describe the parameter-generation procedure PrmGen, showing how to choose the parameters for our scheme. The procedure takes as input the security parameter λ, a DAG G with diameter d, and the class $\mathcal{C}$ of supported circuits. It outputs n, m, q and the Gaussian parameters s, σ. The constraints that these parameters need to satisfy are the following:

- It should be possible to efficiently sample from the input/error distribution $\chi = D_{\mathbb{Z},s}$, and the LWE problem with parameters n, m, q, χ should be hard. This means that we need (say) $s = \sqrt{n}$ and $q/s < 2^{n/\lambda}$.
- It should possible to generate trapdoor for the $\mathbf{A}_v$'s, that enables sampling from PreSample with parameter σ. By Lemma 1, this means that we need $m = \Omega(n \log q)$ and $\sigma = \Omega(\sqrt{n \log q})$.
- For any supported circuit, the size of the error $\mathbf{E}$ at the output of the circuit must remain below $q^{3/4}$. Namely if the output is an encoding $\mathbf{D}$ of the plaintext matrix $\mathbf{S}$ relative to path $u \rightsquigarrow v$, then we need $\|[\mathbf{DA}_u - \mathbf{A}_b\mathbf{S}]_q\| < q^{3/4}$.

Let us now analyze the error size in the system. We assume here that we use truncated Gaussian distributions, i.e. we condition $D_{\mathbb{Z},s}$ on the output being smaller than $b \overset{\text{def}}{=} s\sqrt{\lambda}$ (which only affect the distribution negligibly.) We similarly condition PreSample on the output being shorter than $B \overset{\text{def}}{=} \sigma\sqrt{\lambda}$. With our settings, we get $b \leqslant n$ and $B \leqslant n\sqrt{\log q}$. Hence the sample procedure always outputs $(\mathbf{S}, \mathbf{C}, \mathbf{D})$ with the plaintext satisfying $\|\mathbf{S}\| < b$ and the encoding matrices satisfying $\|\mathbf{C}\|, \|\mathbf{D}\| < B$.

To analyze the noise development, recall that when multiplying $\mathbf{A} \in \mathbb{Z}^{u \times v}$ by $\mathbf{B} \in \mathbb{Z}^{v \times w}$ we have $\|\mathbf{AB}\| \leqslant \|\mathbf{A}\| \cdot \|\mathbf{B}\| \cdot v$. This means in particular that multiplying i encoding matrices we get an encoding matrix $\mathbf{D} \in \mathbb{Z}_q^{m \times m}$ with $\|\mathbf{D}\| < B^i m^{i-1}$ and similarly multiplying i plaintext matrices we get a plaintext matrix $\mathbf{S} \in \mathbb{Z}_q^{n \times n}$ with $\|\mathbf{S}\| < b^i n^{i-1}$. Regarding the error, one can show by induction that the product of i encoding matrices has an error $\mathbf{E} \in \mathbb{Z}_q^{m \times n}$ with

$$\|\mathbf{E}\| < b \cdot \sum_{j=0}^{i-1} B^j m^j b^{d-1-j} n^{d-1-j} < b \cdot i \cdot B^{i-1} m^{i-1}.$$

Given a class $\mathcal{C}$ of permitted circuits, we consider the canonical representation of the polynomials in this class as sums of monomials. Let D be a bound on the degree of these polynomials, R be a bound on the size of the coefficients, and N

be a bound on the number of monomials. Note that in our setting, the degree-bound D cannot be larger than the diameter of the graph G (since G is acyclic and hence cannot have directed paths longer than d). The size of the error in this case could grow as much as $N \cdot R \cdot b \cdot D \cdot B^{D-1} m^{D-1}$. With $b \leqslant n$ and $B \leqslant n\sqrt{\log q}$, we thus have the constraint

$$q^{3/4} > N \cdot R \cdot n \cdot D \cdot \left(n\sqrt{\log q}\right)^{D-1} m^{D-1}$$
$$= N \cdot R \cdot n^D \cdot m^{D-1} \cdot D \cdot \left(\log q\right)^{(D-1)/2}. \tag{5}$$

Substituting $m = \Theta(n \log q)$, and $q = 2^{n/\lambda}$, we can use Eqn. (5) to solve for n in terms of λ, N, R and D. With $D \leqslant d$ and assuming the (typical) case of $R = \text{poly}(\lambda)$ and $N < d^d$, it can be verified that this bound is satisfied using $n = \Theta(d\lambda \log(d\lambda))$. This setting yields $q = 2^{n/\lambda} = 2^{\Theta(d \log(d\lambda))} = (d\lambda)^{\Theta(d)}$ and $m = \Theta(n \log q) = \Theta(d^2\lambda^2 \log^2(d\lambda))$.

Note that with this setting, each matrix $\mathbf{A}_v \in \mathbb{Z}_q^{m \times n}$ is of size $mn \log q = \Theta(d^4\lambda^2 \log^4(d\lambda))$ bits. The public parameters typically contain just one or a handful of such matrices (corresponding to the source nodes in G), but the secret parameters contain all of them. Hence the secret parameters are of size $\Theta(|V| \times d^4\lambda^2 \log^4(d\lambda)) = \Omega(d^5\lambda^2 \log^4(d\lambda))$ bits. (We have $|V| > d$ since the diameter of G is d.) The encoding matrices are of dimension $m \times m$, but their entries are small, so they can be represented by roughly $m^2 \log n = \Theta(d^4\lambda^2 \log^5(d\lambda))$ bits.

Working over a larger ring. As usual, we can get better parameters by working over larger rings, and let n denote the extension degree of the ring. In this case the matrices $\mathbf{A}$ are $m \times 1$ column vectors over the larger ring, and we can find trapdoors for these matrices already for $m = \Theta(\log q)$, and also the plaintext elements are now scalars (or constant-degree matrices).

This only affects Eqn. (5) or the solution $n = \Theta(d\lambda \log(d\lambda))$ by a constant factor, and hence shaves only a constant factor from the number of bits in $q = 2^{\Theta(d \log(d\lambda))}$, but now we have $m = \Theta(\log q) = \Theta(d \log(d\lambda))$. With each scalar in R_q represented by $n \log q$ bits, it takes $mn \log q = \Theta(d^3\lambda \log^3(d\lambda))$ bits to represent each matrix $\mathbf{A}_v \in R_q^{m \times 1}$, and $\Theta(m^2 \log n) = \Theta(d^3\lambda \log^4(d\lambda))$ bits to represent each encoding matrix with small entries.

Obfuscating Circuits via Composite-Order Graded Encoding

Benny Applebaum[1,*] and Zvika Brakerski[2,**]

[1] School of Electrical Engineering, Tel-Aviv University, Israel
benny.applebaum@gmail.com
[2] Weizmann Institute of Science, Israel
zvika.brakerski@weizmann.ac.il

Abstract. We present a candidate obfuscator based on composite-order Graded Encoding Schemes (GES), which are a generalization of multilinear maps. Our obfuscator operates on circuits directly without converting them into formulas or branching programs as was done in previous solutions. As a result, the time and size complexity of the obfuscated program, measured by the number of GES elements, is directly proportional to the circuit complexity of the program being obfuscated. This improves upon previous constructions whose complexity was related to the formula or branching program size. Known instantiations of Graded Encoding Schemes allow us to obfuscate circuit classes of polynomial degree, which include for example families of circuits of logarithmic depth.

We prove that our obfuscator is secure against a class of generic algebraic attacks, formulated by a generic graded encoding model. We further consider a more robust model which provides more power to the adversary and extend our results to this setting as well.

As a secondary contribution, we define a new simple notion of *algebraic security* (which was implicit in previous works) and show that it captures standard security relative to an ideal GES oracle.

1 Introduction

General-purpose program obfuscation allows us to transform an arbitrary computer program into an "unintelligible" form while preserving its functionality. Syntactically, an obfuscator for a function family $\mathcal{C} = \{C_K\}$ is a randomized algorithm that maps a function $C_K \in \mathcal{C}$ (represented by an identifier K) into a "program" $\hat{C} \in \{0,1\}^*$. The obfuscated program should preserve the same functionality as C_K while hiding all other information about C_K. The first property is formalized via the existence of an efficient universal evaluation algorithm Eval

---

* Supported by the European Unions Horizon 2020 Programme (ERC-StG-2014-2020) under grant agreement no. 639813 ERC-CLC, ISF grant 1155/11, Israel Ministry of Science and Technology (grant 3-9094), GIF grant 1152/2011, and the Check Point Institute for Information Security.
** Supported by ISF grant 468/14 and by an Alon Young Faculty Fellowship.

Y. Dodis and J.B. Nielsen (Eds.): TCC 2015, Part II, LNCS 9015, pp. 528–556, 2015.

which, given an input x and an obfuscated program $\hat{C}$, outputs $C_K(x)$. The second property has several different formulations, most notably *Virtual Black-Box* (VBB) and *Indistinguishability Obfuscation* (iO) [5].

The first candidate general-purpose obfuscator has been introduced by Garg et al. [11]. Their work and follow-ups such as [8,4] relied on *Graded Encoding Schemes* (GES) [10,9] which generalize the more traditional notion of multilinear maps. All these works share a similar two-step outline. First it is shown how to use the GES to obfuscate function families from a weak complexity class such as $\mathcal{NC}^1$ (the class of polynomial-size circuits with logarithmic depth and bounded fan-in gates), and then the weak obfuscator is bootstrapped into a general-purpose obfuscator for arbitrary polynomial-size circuits based on low-complexity fully homomorphic encryption [11,8,4] or on low-complexity pseudorandom functions [14,2]. Following [1], we refer to the first step as the "core obfuscator".

Somewhat mysteriously, all known core obfuscators are applied to the branching program representation of the function. Hence, in order to obfuscate some family of circuits (say in $\mathcal{NC}^1$) one has to first convert the given circuit into a branching program, and only then use the core obfuscator. This is both unnatural and inefficient. Indeed, the complexity of existing core obfuscators (in terms of computation, program size, and number of multilinear levels) grow with the *formula size* or *branching program* size of the obfuscated program, which are polynomially larger than the *circuit size*. From a more principal point of view, one may wonder whether the use of branching programs is inherent or is just a limitation of our current techniques. In this paper, we study the existence of "direct circuit obfuscators". Specifically, we ask:

> Is it possible to obfuscate a function family $\mathcal{C}$ with complexity which is linear in the *circuit complexity* of $\mathcal{C}$?

Following [1], we assume that the family $\mathcal{C} = \{C_K\}$ is represented by some universal evaluator $\mathcal{U}$ which given an index K of a function $C_K \in \mathcal{C}$ and an input x outputs the value $C_K(x)$. The (circuit) complexity of $\mathcal{C}$ is measured by the (circuit) complexity of $\mathcal{U}$ and the size of $\mathcal{C}$ is measured by the bit-length of the identifiers K. (These are natural complexity measures which lower-bound the time/size complexity of the any obfuscator for $\mathcal{C}$. See Remark 1.2.)

1.1   Our Results

We take a step towards answering the above question in the affirmative: We construct new core obfuscators for any circuit family $\mathcal{C}$, where the size of an obfuscated program, measured by the number of GES elements, is proportional to the size of $\mathcal{C}$, and its time-complexity, measured by the number of GES operations, is proportional to the circuit complexity of $\mathcal{C}$. This falls short of a full answer to the above question since in current GES candidates, the element size depends on the total evaluated *degree*, a property which is inherited by our constructions. Our constructions are based on composite order Graded Encoding Scheme [13], and are proved to achieve indistinguishability against adversaries which are limited to algebraic attacks allowed in a generic GES model. In fact, we study two different variants of the generic model (one is weaker than the other) and provide corresponding constructions in each of these models. Before stating our results, a few words about generic models are in order.

Generic Graded Encoding Schemes. A *Graded Encoding Scheme* is parameterized by a ring $\mathcal{R}$ and a top level multiset $\mathbf{v}_{zt}$ over the universe $[\tau]$. Intuitively, the GES defines (exponentially) many encodings of the ring where each encoding is indexed by a multiset $\mathbf{v} \subseteq \mathbf{v}_{zt}$. (A multiset is represented by an integer vector $\mathbb{N}^\tau$.) In our first (and weaker) generic model $\mathcal{MRG}$ (for *Multiple-encoding Random GES*), the encoding of a ring element $g \in \mathcal{R}$ under an index $\mathbf{v}$, denoted by $[g]_\mathbf{v}$, is distributed uniformly over an exponentially large set of random strings.[1] As a result, the adversary can manipulate encodings only via the use of a GES oracle that supports some restricted set of "legal" operations. In particular, the adversary is allowed to: (1) compute addition/subtraction for encodings that share the same index $[g]_\mathbf{v} \pm [g']_\mathbf{v} = [g \pm g']_\mathbf{v}$; (2) multiply elements with distinct indices as long as the union of their indices is still a subset of $\mathbf{v}_{zt}$, i.e., $[g]_\mathbf{v} \times [g']_{\mathbf{v}'} = [g \cdot g']_{\mathbf{v} \bigcup \mathbf{v}'}$; and (3) zero-test a top-level encoding, i.e., $\mathsf{isZero}([g]_{\mathbf{v}_{zt}}) = 1 \Leftrightarrow g = 0$. In this model we prove the following theorem:

Theorem 1.1. *There exists an indistinguishability obfuscator* SimpleObf *with respect to* $\mathcal{MRG}$ *for any circuit family* $\mathcal{C}$ *in* $\mathcal{NC}^1$. *Moreover, for a function family* $\mathcal{C} = \{C_K\}_{K \in \{0,1\}^m}$ *which is indexed by m-bit strings, operates on n-bit inputs, and can be universally computed by a t-size circuit of depth d, the following hold:*

- *(Size) The obfuscated program contains* $4n + 2m + 2$ *ring elements.*
- *(Evaluation complexity) The complexity of the evaluation algorithm is* $O(t)$, *and it can be represented as an* $O(t)$*-size arithmetic circuit with oracle gates to the GES oracle.*
- *(GES parameters) The underlying ring is a composite order ring* $\mathbb{Z}_{p_1} \times \mathbb{Z}_{p_2}$ *where* p_1 *and* p_2 *are large co-primes (whose bit length is polynomial in the security parameter), and the* L_1 *norm of the zero-testing level* $\mathbf{v}_{zt}$ *is upper-bounded by* 2^d.[2]

---

[1] An alternative way to model a generic attack is to assume that the adversaries is given abstract handles to the encodings. We prefer the current model due to its simplicity and for the sake of consistency with previous works. We futrher note that security in the current (random string) model immediately implies security in the "handle model". Since our model only gives more power to the adversary and not to the honest user (since correctness should hold even for non random GES instantiations).

[2] The L_1 norm of $\mathbf{v}_{zt}$ essentially corresponds to the (maximal) total degree of polynomials that can be evaluated on the GES elements. The upper-bound of 2^d can be replaced with the more refined bound of the total degree of the universal evaluation algorithm. All known instantiations of graded encoding schemes require that $\|\mathbf{v}_{zt}\|_1$ is polynomial in the security parameter, hence the importance of bounding this parameter.

Remark 1.2. The above parameters are essentially optimal (up to the dependency in the element size of the GES and the security parameter). Indeed, by the correctness property, any obfuscation scheme for $\mathcal{C}$ with size M and evaluation complexity T naturally defines an M-bit length indexing for $\mathcal{C}$ and a corresponding T-time universal evaluation algorithm. Hence, the size/time-complexity of the obfuscated program cannot beat the size/complexity of $\mathcal{C}$.

While the above scheme has a fairly low overhead, it relies on a strong generic model. Specifically, the scheme becomes insecure if the adversary manages to zero-test element whose encoding $[g]_\mathbf{v}$ lies in a low-level $\mathbf{v} \subsetneq \mathbf{v}_{zt}$. This vulnerability puts a strong restriction on the class of potential implementations. For example, GES in which each ring element has a unique encoding (in each level) cannot be used as it trivially permits low-level zero-testing.[3] Note that known candidates for *bilinear* maps have exactly this property. Currently, only few candidates for GES are known [10,9], and based on our current understanding, it is possible to tweak these candidates into forbidding low-level zero-testing.[4] Still, it is desirable to obtain results in a more robust model which allows the adversary to zero-test low-level encodings. Formally, we define a different ideal GES oracle $\mathcal{URG}$ (for *Unique-encoding Random GES*) which, for every level $\mathbf{v}$, assigns for every ring element $g \in \mathcal{R}$ a *unique* (randomly chosen) encoding. In this model, we prove the following result.[5]

Theorem 1.3. *There exists an indistinguishability obfuscator* RobustObf *with respect to* $\mathcal{URG}$ *for any circuit family* $\mathcal{C}$ *in* $\mathcal{NC}^1$*, where the* $\mathcal{URG}$ *oracle is defined over an* $(n+2)$*-composite order ring* $\mathcal{R} = \mathbb{Z}_{p_1} \times \cdots \times \mathbb{Z}_{p_{n+2}}$*. Furthermore, the size, evaluation complexity and the* L_1 *norm of the zero-testing level are exactly as in Theorem 1.1.*

The use of $O(n)$-composite order rings introduces an indirect efficiency overhead. Indeed, the description length of ring elements and the computational cost of oracle operations now grow with the input length n (and the security parameter). It is important to note that this overhead is independent of the size/complexity

---

[3] To test if $[g]_\mathbf{v}$ is an encoding of zero simply check if the string $[g]_\mathbf{v} + [g]_\mathbf{v}$ equals to $[g]_\mathbf{v}$.

[4] It may be surprising that such tweaking could exist, since one can always increase the level from $\mathbf{v}$ to $\mathbf{v}_{zt}$ by multiplying with a non-zero element of level $(\mathbf{v}_{zt} - \mathbf{v})$. However, in known instantiations, it is possible generate public parameters that cannot be used for generating new encodings, thus restricting the adversary to only have access to those levels provided in the obfuscated program. Those, in turn, are designed so there is no way to obtain an encoding at level $(\mathbf{v}_{zt} - \mathbf{v})$ for "dangerous" values of $\mathbf{v}$. (This is somewhat similar to the "straddling sets" technique [4].)

[5] It is important to emphasize that the honest parties (the obfuscator and evaluator) do not exploit the fact that elements have unique encodings, as they are required to work with respect to any GES implementation. (See Section 3). Therefore, the $\mathcal{URG}$ model is strictly better than the $\mathcal{MRG}$ model in the sense that an obfuscation which is secure relatively to $\mathcal{URG}$ is also secure relatively to $\mathcal{MRG}$. The reverse direction does not necessarily hold as demonstrated by SimpleObf from Theorem 1.1.

of $\mathcal{C}$, and so, for sufficiently large/complicated circuit families, we may still get an asymptotic advantage over alternative branching-program based approaches.

Definitional contribution. As already mentioned, we prove security relative to an ideal GES oracle (either $\mathcal{URG}$ or $\mathcal{MRG}$). Towards this goal, we propose a new algebraic abstraction of *canonical GES-based obfuscators* and define a corresponding notion of *algebraic security* (which was explicit in [8,4,1]). Roughly speaking, a GES-based obfuscator is in canonical form, if given a program identifier K, it samples a tuple $a = (a_1, \ldots, a_\ell)$ from the underlying ring $\mathcal{R}$, and then outputs a GES-encoding of these elements under the labels $(\mathbf{v}_1, \ldots, \mathbf{v}_\ell)$ which depend only on the length parameters but otherwise are independent of the program identifer. Hence, an obfuscator is essentially a mapping from a program identifier K to distribution D_K over $\mathcal{R}^\ell$.

For algebraic security, an adversary A is a polynomial over ℓ variables (described by an arithmetic circuit) which is evaluated over the tuple $a = (a_1, \ldots, a_\ell)$ sampled from D_K. The outcome of the attack is the bit $\mathsf{isZero}(A(a))$. Security requires the existence of a simulator S that given an oracle access to C_K can predict $\mathsf{isZero}(A(D_K))$ with all but negligible probability.

Abstracting previous works, we show that algebraic security implies (standard) security relative to an ideal GES oracles.[6] Note that algebraic security considers only a static attack (a single query to D_K) whereas standard security (VBB or iO) corresponds to an adaptive game with possibly many queries. Correspondingly, algebraic security is much easier to work with. (Indeed, a notable part of the proofs of [8,4] is devoted to essentially reducing standard security to algebraic security.) More importantly, algebraic security *does not* depend on the GES oracle at all. As such, it crystalizes the information-theoretic properties that are required in order to achieve security in an ideal GES model. We hope that this abstraction will be valuable for future constructions, and that our general lemma (algebraic security $\Rightarrow$ standard security) will allow to work directly with algebraic security.

1.2   Our Techniques

To illustrate our techniques let us consider the following simplified physical model. The obfuscator is allowed to put ring elements in locked boxes (marked by multisets) and everyone can add/multiply boxes at most T times. After performing T operations the box is opened only if its content is equal to zero.

A naive way to obfuscate a function C_K in this model is to put the identifier's bits $K_1, \ldots, K_m$ in m separate boxes (labeled by $\mathbf{v}_1, \ldots, \mathbf{v}_m$), and prepare for each input x_i a pair of boxes (labeled by $\mathbf{v}_{i,0}, \mathbf{v}_{i,1}$) with the values 0 and 1. Given these boxes and an input x, the evaluator can choose the boxes $\mathbf{v}_{i,x_i}$, propagate

---

[6] The difference between $\mathcal{URG}$ and $\mathcal{MRG}$ will arise by putting different syntactic restrictions on the class of "legal" polynomials A. Furthermore, an efficient simulator corresponds to VBB security while inefficient simulator corresponds to indistinguishability obfuscation.

the values according to the (arithmetization of the) universal circuit $\hat{\mathcal{U}}(\cdot, \cdot)$ and obtain a box which holds the value $\hat{\mathcal{U}}(x, K)$. Assuming that the circuit's size is T, the resulting box will be opened if and only if $C_K(x) = 0$.

This construction is insecure for several reasons. First and foremost, one can ignore the structure of the universal circuit $\hat{\mathcal{U}}$ and compute any other T-size circuit $F(x, K)$. As a result one can easily extract the identifier K and completely learn the function. We resolve this problem via a novel use of *authenticators*. Assume that the each box has two slots. We keep the second slot untouched as in the previous obfuscator, and fill the first slot with $n+m$ random authenticators $y_1, \ldots, y_{n+m}$ where the zero-box and the one-box that correspond to the same input variable x_i share the same authenticator y_i. In addition, let us add another box (labeled by $\mathbf{v}_0$) that contains the pair $(0, y_0)$ where $y_0 = \hat{\mathcal{U}}(y_1, \ldots, y_{n+m})$, and let us increase the bound on the number of operations to $T + 1$. Given an input x, we can apply $\hat{\mathcal{U}}$ to the boxes $\mathbf{v}_{i,x_i}$, obtain a box with the pair $(\hat{\mathcal{U}}(x, K), \hat{\mathcal{U}}(y))$, subtract the result from the last $\mathbf{v}_0$ box and check for zero.

In terms of security, we are now protected from attacks that respect the input x. Specifically, if the adversary apply some $(n + m + 1)$-variate polynomial
$$F(V_0, (V_1, \ldots, V_m), (V_1', \ldots, V_n')) \neq \left(\hat{\mathcal{U}}((V_1', \ldots, V_n'), (V_1, \ldots, V_m)) - V_0 \right) \text{ to the}$$
boxes $\mathbf{v}_0, (\mathbf{v}_1, \ldots, \mathbf{v}_m), (\mathbf{v}_{1,x_1}, \ldots, \mathbf{v}_{n,x_n})$ for some $x \in \{0, 1\}^n$, then the result is almost surely non-zero. To see this, let us focus on the first slot of $F(\mathbf{v}_0, (\mathbf{v}_1, \ldots, \mathbf{v}_m), (\mathbf{v}_{1,x_1}, \ldots, \mathbf{v}_{n,x_n}))$ and substitute $y_0 = \hat{\mathcal{U}}(y_1, \ldots, y_{n+m})$. Then the polynomial F simplifies to a non-trivial low-degree polynomial over the random values $y_1, \ldots, y_{n+m}$ and therefore (by Schwartz-Zippel) vanishes with negligible probability.[7]

We adopt the above outline to the GES setting where the two-slot boxes are emulated via the use of a composite-order ring $\mathcal{R} = \mathbb{Z}_{p_1} \times \mathbb{Z}_{p_2}$ where p_1, p_2 are two (large) co-prime integers. Note that there are still several technicalities. First, unlike the simplified boxes model, in the GES setting, addition can be applied only over encodings from the same level. We solve this problem by representing each value $w \in \mathcal{R}$ by a pair $([r]_\mathbf{v}, [r \cdot w]_\mathbf{v})$ where $r \overset{R}{\leftarrow} \mathcal{R}$ is a (unique) randomizer. This "El-Gamal" encoding (which was also used in prior works, cf. [6]) naturally supports addition and multiplication. Namely, if two ring elements w, w' are in El-Gamal form $([r]_\mathbf{v}, [rw]_\mathbf{v})$ and $([r']_{\mathbf{v}'}, [r'w']_{\mathbf{v}'})$ then their sum can be computed by "cross-multiplication" and their product can be computed by computing component-wise product.

In addition, the resulting construction is vulnerable to "input-mixing" attacks in which the adversary uses two boxes $\mathbf{v}_{i,0}, \mathbf{v}_{i,1}$ which correspond to the same input variable. We solve this issue via the use of *straddling sets* similarly to [4]. Roughly speaking, straddling sets force consistency by making sure that input-mixing attacks cannot reach to the top (zero-testing) level. These modifications eventually lead to our basic obfuscator SimpleObf.

---

[7] The above argument is somewhat inaccurate as one has to take into account the case where F is a multiple of $(\hat{\mathcal{U}}(\cdots) - V_0)$. A formal proof appears in Section 5.

Interestingly, SimpleObf is completely broken if low-level zero-testing is allowed. Recall that y_i is shared among the zero and one encoding of the i-th input. Therefore, one can zero-out y_i in a low-level encoding by subtracting their El-Gamal encodings, thus obtaining an element that has zero in one of the slot. At this point one can zero-out all authenticators as well, and fully recover the string K using low-level zero-testing.

Solving this issue is the main technical challenge addressed by our more robust obfuscator RobustObf. As a first step, we add more slots to the encoding (using $(n+2)$ subrings $\mathcal{R} = \mathbb{Z}_{p_1} \times \ldots \cdots \mathbb{Z}_{p_{n+2}}$) and make sure that the pair of encodings which share the same y_i, hold distinct (random) values on all other slots. In order to preserve the functionality we must allow the honest evaluator to zero-out the additional slots (while preventing the adversary from doing so in a low-level). To this end, we publish some auxiliary elements $\hat{w}_i$ whose i-th slot is zero. We publish two copies of each $\hat{w}_i$, each in a different level $\hat{\mathbf{v}}_{i,0}, \hat{\mathbf{v}}_{i,1}$, and an appropriate straddling set structure that guarantees that the $\hat{\mathbf{v}}_{i,0}$ copy can only interact with $\mathbf{v}_{i,0}$ and vice versa. Now, if the previous attack is sought, the attacker will attempt to subtract $\mathbf{v}_{i,0}$ from $\mathbf{v}_{i,1}$, but then it will need to multiply by one of $\hat{\mathbf{v}}_{i,0}$ or $\hat{\mathbf{v}}_{i,1}$. Since both operations are forbidden by the straddling sets, the attack seems to be prevented.

Alas, we recall that functionality needs to hold as well. The element w_0, which generalizes the y_0 that we had before, now must have 0 in all of the new slots, since after the honest evaluator finishes multiplying with the $\hat{w}$ values, it needs to compare against w_0. This leaves us vulnerable to an attacker that will use w_0 instead of the $\hat{w}$ to zero out coordinates ahead of time. To solve this last problem, we present our final trick, the *shifted El-Gamal encoding*. Instead of encoding $([r]_\mathbf{v}, [rw]_\mathbf{v})$, we will now use $([r]_\mathbf{v}, [rw]_{\mathbf{v}+\mathbf{v}^*})$, where $\mathbf{v}^*$ is a special vector used by all of the encodings. The result of this change is that now, if addition/subtraction is performed, the $\mathbf{v}^*$ part of the result is the same as of the operands, but if multiplication is performed, the $\mathbf{v}^*$ part is the sum of the $\mathbf{v}^*$'s of the operands. Therefore the $\mathbf{v}^*$ part keeps track of the multiplicative degree of the evaluation process. Finally, the element w_0 will be encoded as $([r]_{\mathbf{v}_0}, [rw_0]_{\mathbf{v}_0+D\mathbf{v}^*})$, where D is the total multiplicative degree of our evaluation process. This means that one can only add/subtract with w_0, and never multiply (otherwise the $\mathbf{v}^*$ multiple goes beyond D and we set the zero-test level to not allow this). This prevents misuse of w_0 and completes the description of RobustObf.

See Section 4 for the construction and Section 5 for the proof of security.

Remark 1.4 (The degree restriction). Due to noise issues, current instantiations of GES only support $poly(\lambda)$-multiplicative degree. In particular, the representation length of each element is proportional to the degree. In our context, this restriction translates to a degree restriction on the universal circuit $\mathcal{U}$.[8] For the (typical) case of balanced circuits, this results in a logarithmic-depth restriction.

---

[8] For our purposes, the degree of a boolean circuit is its formula size.

1.3 Related Works

Ananth et al. [1] explored the efficiency of obfuscating formulae. They considered two settings. One where the formula is represented as a sequence of variables and gates, and another more similar to our formulation where there is a universal evaluator (in the form of a formula in their case), and the specific function is specified as a key to this evaluator. In the latter case, which is more relevant for the sake of comparison, they show how to obfuscate classes with formula size s with obfuscated program size and complexity almost as low as $O(s)$. This is in comparison to previous methods that used Barrington's theorem and achieved $O(s^2)$ for balanced formulae or $O(s^{3.64})$ for unbalanced. Still, their complexity measure remained the *formula size* of the function family, whereas in this work we show that one can obfuscate relative to the *circuit size* of the family which may be smaller. On the flip side, we use composite order graded encoding schemes that are even newer and less substantiated (and possibly less efficient) than standard prime-order graded encoding schemes.

Composite order graded encoding schemes have been used by Gentry, Lewko and Waters [13] and by Gentry, Lewko, Sahai and Waters [12] to introduce improved *security reductions* for witness encryption and for obfuscation (respectively). In particular they showed that in this setting one can construct a witness encryption scheme or an obfuscator, and prove security in the standard model based on exponential hardness assumptions.

Concurrent and Independent Work. In a very recent concurrent and independent work, Zimmerman [18] presented an obfuscator which is almost identical to our simpler obfuscator SimpleObf. Zimmerman also presents applications for this new obfuscation method for circuits. Security is proven in a generic model where zero testing below the last level is impossible, similar to our $\mathcal{MRG}$ oracle. Both the obfuscator from [18] and our SimpleObf are completely broken in a more challenging model where it is possible to test for zero at low levels. Our second obfuscator RobustObf addresses this issue and provides security in the more challenging setting represented by the GES oracle $\mathcal{URG}$, at the expense of being less efficient. On the other hand, we only prove that our obfuscators are secure indistinguishability obfuscator in the generic model, whereas [18] proves the more stringent notion of virtual black box security.

Road map. Section 2 defines Graded Encoding over Composite Order Groups and ideal GES oracles. Section 3 defines GES-based obfuscation, suggests two alternative security definitions (standard oracle-based definition and algebraic security) and shows that one implies the other. Section 4 describes our new constructions and Section 5 is devoted to the proof of their security. In this extended abstract, most of the proofs are omitted. Full proofs appear in [3].

2   Graded Encoding over Composite Order Groups

2.1   General Notation

Partial Order of Natural Valued Vectors. For an integer $\tau \in \mathbb{N}$, we view vectors in $\mathbb{N}^\tau$ as multisets over the universe $[\tau]$. Correspondingly, we define a partial ordering on vectors $\mathbb{N}^\tau$ which corresponds to inclusion. In particular, we say that $\mathbf{v} \leq \mathbf{w}$ if for all $i \in [\tau]$ it holds that $\mathbf{v}[i] \leq \mathbf{w}[i]$. If there exists a coordinate i for which the above does not hold, we say that $\mathbf{v} \not\leq \mathbf{w}$. We note that since our vectors are defined over the naturals, this relation is monotonous: If $\mathbf{v} \leq \mathbf{w}$ then for all $\mathbf{w}' \in \mathbb{N}^\tau$ it also holds that $\mathbf{v} \leq (\mathbf{w} + \mathbf{w}')$, and dually if $\mathbf{v} \not\leq \mathbf{w}$ then for all $\mathbf{v}' \in \mathbb{N}^\tau$ it holds that $(\mathbf{v} + \mathbf{v}') \not\leq \mathbf{w}$.

CRT representation. Let $\sigma \in \mathbb{N}$, let $p_1, \ldots, p_\sigma$ be distinct coprime numbers and let $P = \prod_{i=1}^{\sigma} p_i$. Considering the ring $\mathbb{Z}_P$, the Chinese Remainder Theorem (CRT) asserts that there is an isomorphism $\mathbb{Z}_P \cong \mathbb{Z}_{p_1} \times \cdots \times \mathbb{Z}_{p_\sigma}$ such that if $a \cong (a_1, \ldots, a_\sigma)$ and $b \cong (b_1, \ldots, b_\sigma)$, then $a + b \cong (a_1 + b_1, \ldots, a_\sigma + b_\sigma)$ and $a \cdot b \cong (a_1 \cdot b_1, \ldots, a_\sigma \cdot b_\sigma)$. For a given isomorphism, we will denote by $a[\![i]\!]$ the component $a_i = a \pmod{p_i}$.

2.2   Syntax

We begin with the definition of a graded encoding scheme in composite order groups. The definition is adapted from [10] and follow-up works, but our notation deviates somewhat from that of some previous work.

Definition 2.1 (Graded Encoding Scheme). *Let $\mathcal{R}$ be a ring, and let $\mathbf{v}_{zt} \in \mathbb{N}^\tau$ be an integer vector of dimension $\tau \in \mathbb{N}$. A graded encoding scheme for $\mathcal{R}$, $\mathbf{v}_{zt}$ is a collection of sets $\{[\alpha]_\mathbf{v} \subset \{0,1\}^* : \mathbf{v} \in \mathbb{N}^\tau, \mathbf{v} \leq \mathbf{v}_{zt}, \alpha \in \mathcal{R}\}$ with the following properties:*

1. *For every index $\mathbf{v} \leq \mathbf{v}_{zt}$, the sets $\{[\alpha]_\mathbf{v} : \alpha \in \mathcal{R}\}$ are disjoint, and so they are a partition of the indexed set $[\mathcal{R}]_\mathbf{v} = \bigcup_{\alpha \in \mathcal{R}} [\alpha]_\mathbf{v}$. We slightly abuse notation and often denote $a = [\alpha]_\mathbf{v}$ instead of $a \in [\alpha]_\mathbf{v}$.*
2. *There are binary operations "$+$" and "$-$" such that for all $\mathbf{v} \in \{0,1\}^\tau, \alpha_1, \alpha_2 \in \mathcal{R}$ and for all $u_1 = [\alpha_1]_\mathbf{v}, u_2 = [\alpha_2]_\mathbf{v}$:*

$$u_1 + u_2 = [\alpha_1 + \alpha_2]_\mathbf{v} \quad and \quad u_1 - u_2 = [\alpha_1 - \alpha_2]_\mathbf{v} \, ,$$

   where $\alpha_1 + \alpha_2$ and $\alpha_1 - \alpha_2$ are addition and subtraction in $\mathcal{R}$.
3. *There is an associative binary operation "$\times$" such that for all $\mathbf{v}_1, \mathbf{v}_2 \in \mathbb{N}^\tau$ such that $\mathbf{v}_1 + \mathbf{v}_2 \leq \mathbf{v}_{zt}$, for all $\alpha_1, \alpha_2 \in \mathcal{R}$ and for all $u_1 = [\alpha_1]_{\mathbf{v}_1}, u_2 = [\alpha_2]_{\mathbf{v}_2}$, it holds that*

$$u_1 \times u_2 = [\alpha_1 \cdot \alpha_2]_{\mathbf{v}_1 + \mathbf{v}_2},$$

   where $\alpha_1 \cdot \alpha_2$ is multiplication in $\mathcal{R}$.

The above definition does not touch upon the computational aspects of graded encoding schemes, which are described below. We note that there is a difference between the definition below and the definitions for the prime order definitions.

Definition 2.2 (Efficient Procedures for Graded Encoding Scheme). *We consider a graded encoding schemes (see above) where the following procedures are efficiently computable.*

- *Composite-Order Instance Generation:* InstGen$(1^\lambda, 1^\sigma, \mathbf{v}_{zt}, 1^{\|\mathbf{v}_{zt}\|_1})$ *outputs the set of parameters params, a description of a Graded Encoding Scheme relative to $\mathbf{v}_{zt}$ and relative to a ring $\mathcal{R}$ such that $\mathcal{R} \cong \mathbb{Z}_{p_1} \times \cdots \times \mathbb{Z}_{p_\sigma}$, where all p_i are pairwise coprime numbers, i.e. $\mathcal{R} \cong \mathbb{Z}_N$ for $N = \prod p_i$.*[9]
  In addition, the procedure outputs a subset evparams $\subset$ params that is sufficient for computing addition, multiplication and zero testing, but may be insufficient for sampling, encoding or for randomization.
  We note that for known GES candidates, the running time of the setup procedure (and all other procedures) scales with $\|\mathbf{v}_{zt}\|_1$, and hence we require that this value is provided in unary representation in addition to $\mathbf{v}_{zt}$ itself. It is conceivable that more efficient instantiations that do not require this additional input will be discovered in the future.
- *Ring Sampler:* samp$(params)$ *outputs a "level zero encoding" $A \in [a]_0$ for a nearly uniform $a \xleftarrow{R} \mathcal{R}$.*
- *Sub-Ring Sampler:* subsamp$(params, i^*)$, *where $i^* \in [\sigma]$ outputs a "level zero encoding" in a CRT sub-ring of $\mathcal{R}$. Namely, it outputs $A \in [a]_0$ for an element $a \cong (a_1, \ldots, a_\sigma)$, such that a_{i^*} is nearly uniform in p_{i^*}, and for all $i \neq i^*$ it holds that $a_i = 0$. We stress it is very important for the security of our constructions that evparams does not enable such functionality.*
- *Encode and Re-Randomize:* encRand$(params, i, a)$ *takes as input an index $\mathbf{v} \leq \mathbf{v}_{zt}$ and $A = [a]_0$, and outputs an encoding $B = [a]_\mathbf{v}$, where the distribution of B is (statistically close to being) only dependent on a and not otherwise dependent on A.*
- *Addition and Negation:* add$(evparams, A_1, A_2)$ *takes $A_1 = [a_1]_\mathbf{v}, A_2 = [a_2]_\mathbf{v}$, and outputs $B = [a_1 + a_2]_\mathbf{v}$. (If the two operands are not in the same indexed set, then add returns $\bot$). We often use the notation $u_1 + u_2$ to denote this operation when evparams is clear from the context. Similarly, negate$(evparams, A_1) = [-a_1]_\mathbf{v}$.*
- *Multiplication:* mult$(evparams, A_1, A_2)$ *takes $A_1 = [a_1]_{\mathbf{v}_1}, A_2 = [a_2]_{\mathbf{v}_2}$. If $\mathbf{v}_1 + \mathbf{v}_2 \leq \mathbf{v}_{zt}$, then mult outputs $B = [a_1 \cdot a_2]_{\mathbf{v}_1+\mathbf{v}_2}$. Otherwise, mult outputs $\bot$. We often use the notation $A_1 \times A_2$ to denote this operation when evparams is clear from the context.*
- *Zero Test:* isZero$(evparams, A)$ *outputs 1 if $A = [0]_{\mathbf{v}_{zt}}$, and 0 otherwise.*

Noisy encodings. In known candidate constructions, encodings are *noisy* and the noise level increases with addition and multiplication operations, so one has to be

---

[9] In our security model, we will require that each prime factor of N is chosen from a distribution with roughly $(\|\mathbf{v}_{zt}\|_1 + \omega(\log \lambda))$ bits of entropy. See Section 5.

careful not to go over a specified noise bound. However, the parameters can be set so as to support $O(\|\mathbf{v}_{zt}\|_1)$ operations, so long as InstGen is allowed to run in $poly(\|\mathbf{v}_{zt}\|_1)$ time, as our function interface compels. This will be sufficient for our purposes and we therefore ignore noise management throughout this manuscript.

Remark 2.3. Given *params*, we can use subsamp to efficiently generate level-0 encodings of related elements, so long as each of their CRT components can be expressed as a polynomial size arithmetic circuit applied to a set of uniformly distributed variables. These variables may not be shared across CRT components, but they can be shared between elements. E.g. in a 2-composite GES, one can generate $[((a_1 + a_2) \cdot a_3, b_1)]_0$, $[(a_3 + a_4, b_2)]_0$, $[(a_1 \cdot a_2, b_1 + b_2)]_0$ (but cannot generate in addition $[(b_1, a_1)]_0$. (Note that the product of level zero-encoding results in a level zero encoding.) Combining the above with access to encRand allows, given *params* to encode the aforementioned elements to arbitrary indices $\mathbf{v} \leq \mathbf{v}_{zt}$.

Remark 2.4. For our application we require that it is intractable to execute subsamp using only *evparams* and without access to *params*. Our application involves an adversary that is given a set of encodings and *evparams*. If the adversary is able to perform sub-ring sampling or to modify the level of an encoded element, then our obfuscator will be insecure.

Further, in our first construction, the adversary should not be able to apply zero-testing to encodings in level $\mathbf{v} < \mathbf{v}_{zt}$, and these encodings need to appear the same as encodings of non-zero elements. This in turn means that we must forbid the adversary to run samp, encRand as well, since these will allow to "lift" an encoding from level $\mathbf{v}$ to level $\mathbf{v}_{zt}$ and run isZero. While this may seem like a severe limitation, known candidates appear to have this property.

Concrete instantiations. The candidate constructions of [10,9] do not support the above functionality out of the box. Specifically, [10] only allows $\mathcal{R}$ of prime order, whereas [9] does natively support composite order groups, but its security features are unclear if sub-ring sampling is allowed. This issue has been extensively addressed in [13, Appendix B of full version]. In particular the authors there present a variant of [9] that appears to overcome the aforementioned security issues. This variant supports a σ-product ring $\mathcal{R} \cong (\mathbb{Z}_{p_1} \times \cdots \times \mathbb{Z}_{p_\sigma})$ where the p_i's are composite numbers with large prime divisors. Note that this is compatible with our requirements which allow the p_i's to be non-primes. Furthermore, this variant adheres to the constraints we need to impose as per Remark 2.4. Overall, to the best of our knowledge, this candidate is consistent with the requirements of our obfuscator (although we prove security only in a generic model and not under explicit assumptions).

2.3   Ideal GES Oracles

We would like to prove the security of our construction against *generic adversaries*. To this end, we will use the *generic graded encoding scheme* model, adapted from [6,7,8,4], which is analogous to the *generic group model* (see Shoup [17] and

Maurer [15]). Intuitively, we would like to guarantee that the encoding of a ring element is independent of the element itself, and so the adversary can manipulate elements only via the GES oracle. One way to formulate this restriction is to prove security relative to an oracle that implements a truly random GES. We focus on two particular (inefficient) GES oracles: the unique random generic encoding scheme oracle $\mathcal{URG}$ and the multiple-encoding random GES oracle $\mathcal{MRG}$. Both variants will be defined with respect to some probability distribution ensemble $\{\mathcal{R}_{\lambda,\sigma,\mathbf{v}_{zt}}\}$ over rings.

The $\mathcal{URG}$ Oracle. Upon initialization of $\mathsf{InstGen}(1^\lambda, 1^\sigma, \mathbf{v}_{zt}, 1^{\|\mathbf{v}_{zt}\|_1})$, the oracle $\mathcal{URG}$ samples a ring $\mathcal{R} \overset{R}{\leftarrow} \mathcal{R}_{\lambda,\sigma,\mathbf{v}_{zt}}$ and encodes each element $a \in \mathcal{R}$ in level $\mathbf{v} \leq \mathbf{v}_{zt}$ by a string $(\mathbf{v}, \rho)$ where ρ is random string of length $t = (\log |\mathcal{R}| \cdot \lambda)$. The oracle also releases random private/public parameters $evparams, secparams \in \{0,1\}^\lambda$ which are associated with this encoding. From now on, the oracle supports all the GES-operations with resect to the above encoding. It is not hard to see that the only way that $\mathcal{A}$ can obtain valid encodings is by calls to the oracle $\mathcal{URG}$ (except with negligible probability).

The oracle $\mathcal{URG}$ is practically identical to the random GES oracle of [8], and similarly to that work we will also consider an online variant of $\mathcal{URG}$, or rather a variant that approximates $\mathcal{URG}$ to within negligible statistical distance. This is done by an *online polynomial time process*, which samples the representations on-the-fly. Specifically, the oracle will maintain a table of entries of the form $(\mathbf{v}, a, \mathsf{label}_{\mathbf{v},a})$, where $\mathsf{label}_{\mathbf{v},a} \in \{0,1\}^t$ is the representation of $[a]_{\mathbf{v}}$ in $\mathcal{URG}$. The table is initially empty. Every time $\mathcal{URG}$ is called for some functionality, it checks that its operands indeed correspond to an entry in the table, in which case it can retrieve the appropriate $(\mathbf{v}, a)$ to perform the operation. If the operands are not in the table, $\mathcal{URG}$ returns $\perp$. Whenever $\mathcal{URG}$ needs to return a value $[a]_{\mathbf{v}}$, it checks whether $(\mathbf{v}, a)$ is already in the table, and if so returns the appropriate $\mathsf{label}_{\mathbf{v},a}$. Otherwise it samples a new uniform label, and inserts a new entry into the table.

When interacting with an adversary that only makes a polynomial number of calls, the online version of $\mathcal{URG}$ is within negligible statistical distance of the offline version (in fact, the statistical distance is exponentially small in λ). This is because the only case when the online oracle implementation differs from the offline one is when when the adversary guesses a valid label that it has not seen (in the offline setting). This can only occur with exponentially small probability due to the sparsity of the labels. The running time of the online oracle is polynomial in the number of oracle calls.

Defining Multiple-encoding random GES. We would like to define a similar random oracle which assigns exponentially many possible encodings for each element in each level. The interface to this oracle has to be defined carefully. Consider, for example, the case where we have three labels A, B, C where $A = [a]_{\mathbf{v}}, B = [b]_{\mathbf{v}}$, $C = [c]_{\mathbf{v}}$ and we compute the term $(A + B) \times C$ and the term $A \times C + B \times C$. We have to specify whether the resulting label will be identical or not. We choose the more conservative approach and assume that in such a case the label will be indeed identical. In contrast, the labels of $A + B$ and $A' + B$ should disagree when

A, A' are two *independent* labels of a (e.g., both A and A' were generated using two different calls to encRand on some label $A_0 = [a]_0$). To formalize these requirements we define an online version of the Multiple-encoding Random GES ($\mathcal{MRG}$) oracle.

The (online) $\mathcal{MRG}$ Oracle. The oracle $\mathcal{MRG}$ is initialized similarly to the $\mathcal{URG}$ oracle, except that each ring element $a \in \mathcal{R}$ in level $\mathbf{v} \leq \mathbf{v}_{zt}$ is encoded by 2^λ strings of the form $(\mathbf{v}, \rho_i)$ where ρ_i is random string of length $t = (\log |\mathcal{R}| \cdot \lambda^2)$. Whenever a sampling query is made, $\mathcal{MRG}$ generates an element a from $\mathcal{R}$ or the appropriate sub-ring, a uniform length t label, but it also generates a new formal variable X_i, it then stores the tuple $(\mathbf{0}, a, X_i, \mathsf{label}_{0,a,X_i})$ in its table. Whenever an encRand query is made, again a random label and a new formal variable $X_{i'}$ are chosen, and the tuple $(\mathbf{v}, a, X_{i'}, \mathsf{label}_{\mathbf{v},a,X_{i'}})$ is stored. Whenever an "arithmetic" query is made, $\mathcal{MRG}$ looks up the input labels and finds the appropriate labels in its table, and adds or multiplies the respective formal variables (which will now become formal polynomial). Thus, the table will now contain tuples of the form $(\mathbf{v}, a, poly(\vec{X}), \mathsf{label}_{\mathbf{v},a,poly(\vec{X})})$, and labels will be unique if the respective formal polynomials are distinct. Finally, for zero-test queries, $\mathcal{MRG}$ will test whether the actual value is the zero value in $\mathcal{R}$ and respond accordingly.

Both oracles support the standard GES operations with respect to the resulting encodings. We note that $\mathcal{URG}$ (which essentially corresponds to the traditional notion of multilinear maps) is more robust than $\mathcal{MRG}$ as it gives more power to the adversary (for example it can easily detect if it has two encodings of the same element). Specifically, it is not hard to show that if a construction is secure with respect to $\mathcal{URG}$ then it is also secure with respect to $\mathcal{MRG}$. (Formally, the $\mathcal{MRG}$ oracle can be efficiently emulated using a $\mathcal{URG}$ oracle.)

3   GES-Based Obfuscators

In this section we define the notion of GES-based obfuscators. Our definitions somewhat deviate from the more traditional definitions formulated in [5]. Specifically, to allow a more fine-grained notions of efficiency, we distinguish between the description-length and the time complexity of the obfuscated program. Furthermore, we adopt the definition to the GES setting and distinguish between correctness, which should hold for any syntactically valid (possibly trivial) GES, and security, which should hold with respect to some "ideal" GES oracle. Finally, we show (Section 3.2) that for natural GES-based obfuscators, security with respect to ideal oracles boils down to certain algebraic properties of the obfuscator's output (referred to as *algebraic security*). This abstraction (which was implicit in previous works) allows us to decouple the computational properties of the GES from the information-theoretic properties of the obfuscator. Indeed, the security of our obfuscator will be established using the algebraic definition.

3.1 Main Definitions

We begin by recalling the notion of efficient function families.

Function family. Let $\mathcal{C} = \{\mathcal{C}_K\}_{K \in \{0,1\}^*}$ be a family of efficiently computable functions, where for every $K \in \{0,1\}^{m(n)}$ the function C_K operates on inputs of length n. We will assume that $\mathcal{C}$ is represented by a uniform family of polynomial-size *universal evaluation* circuits $\mathcal{U} = \{\mathcal{U}_n\}_{n \in \mathbb{N}}$, where $\mathcal{U}_n$ maps an identifier $K \in \{0,1\}^{m(n)}$ and input $x \in \{0,1\}^n$ to the output $C_K(x)$. The *computational complexity* of $\mathcal{C}$ (with respect to the representation $\mathcal{U}$) is the circuit size of $\mathcal{U}$ and the *representation size* of $\mathcal{C}$ is $m(n)$. We say that $\mathcal{C}$ is in $\mathcal{NC}^1$ if $\mathcal{U}$ is computed by polynomial-size circuits of logarithmic depth.[10]

GES-based Obfuscators: Syntax. A GES-based obfuscation scheme for a family of efficiently computable functions $\mathcal{C}$ consists of a pair of PPT algorithms: an obfuscator Obf and an evaluator Eval, which have oracle access to a GES. The input to the obfuscator is an identifier $K \in \{0,1\}^{m(n)}$ of a function $C_K \in \mathcal{C}$, an unary representation of the security parameter 1^λ, and an unary representation 1^n of the input length of C_K. The obfuscator outputs an obfuscated program $\hat{C} \in \{0,1\}^*$. The evaluation algorithm Eval maps an obfuscated program $\hat{C}$, an input $x \in \{0,1\}^n$, and an unary representations of the security parameter 1^λ to a string y. We note that the efficiency requirement on the obfuscator Obf implicitly puts a polynomial restriction on the size of the obfuscated program $\hat{C}$.

Correctness should hold with respect to an arbitrary GES implementation.

Definition 3.1 (Preserving Functionality). *A GES-based obfuscation scheme* (Obf, Eval) *for $\mathcal{C}$ is* functionality preserving *if for every instantiation of GES $\mathcal{G}$, every $n \in \mathbb{N}$, every $C_K \in \mathcal{C}$ where $K \in \{0,1\}^{m(n)}$, and every $x \in \{0,1\}^n$, with all but $negl(\lambda)$ probability over the coins of Obf, Eval and the GES oracle $\mathcal{G}$ it holds that:*

$$\mathsf{Eval}^{\mathcal{G}}(1^n, 1^\lambda, \hat{C}, x) = C_K(x), \qquad where\ \hat{C} \xleftarrow{R} \mathsf{Obf}^{\mathcal{G}}(1^n, 1^\lambda, K).$$

We define Indistinguishability Obfuscator with respect to some (possibly inefficient) GES instantiation. Our definition is formulated in terms of unbounded simulation which is equivalent to the more standard indistinguishability-based definition (cf. [8]).

Definition 3.2 (Indistinguishability Security [5]). *A GES-based obfuscation scheme* (Obf, Eval) *for $\mathcal{C}$ is called an* Indistinguishability Obfuscator *(iO) with respect to some GES instantiation $\mathcal{G}$ if for every (non-uniform) polynomial size adversary $\mathcal{A}$, there exists a (computationally unbounded) simulator $\mathcal{S}$, such that for every $n \in \mathbb{N}$ and for every $C_K \in \mathcal{C}$ where $K \in \{0,1\}^{m(n)}$:*

$$\left| \Pr[\mathcal{A}^{\mathcal{G}}(1^\lambda, \hat{C}) = 1] - \Pr[\mathcal{S}^{C_K}(1^{|K|}, 1^n, 1^\lambda) = 1] \right| = negl(\lambda),$$

---

[10] We note that the family of all depth-d size-s circuits for some $s(n) \in poly(n)$ and $d(n) \in O(\log n)$ admit a universal evaluation circuit in $\mathcal{NC}^1$ of size $s(n) \cdot 2^{d(n)}$.

where $\hat{C} \xleftarrow{R} \mathsf{Obf}^{\mathcal{G}}(1^n, 1^\lambda, K)$. If the simulator can be implemented by (non-uniform) polynomial size circuits than the obfuscator is Virtually Black-Box *(VBB) secure.*

We will instantiate the above definition with the ideal oracles $\mathcal{URG}$ and $\mathcal{MRG}$ defined in Section 2.3.

3.2 Algebraic Security

In this section we present a notion of security that will be easier to work with, and prove its equivalence to the random GES model above. This model and the equivalence are implicit in previous works. As before, we let $\mathcal{R}_{\lambda,\sigma,\mathbf{v}_{zt}}$ be some ensemble of probability distributions over rings.

Definition 3.3 (Obfuscator in Canonical Form). *An obfuscator is in* canonical form *if it can be presented as follows. (Recall that the obfuscator is given a security parameter 1^λ, an input length 1^n, and a program identifier $K \in \{0,1\}^{m(n)}$.)*

1. *Based on n, the obfuscator deterministically generates $\ell = \ell(n)$ integer-valued vectors $\mathbf{v}_1, \ldots, \mathbf{v}_\ell$, a zero-testing vector $\mathbf{v}_{zt}$ and a ring arity $\sigma \in \mathbb{N}$.*
2. *Based on λ, K, n, the obfuscator defines a joint distribution $\mathcal{D}_\lambda(n, K)$ over ℓ (generic) ring elements $(a_1, \ldots, a_\ell)$.[11]*
3. *Then, the obfuscator initializes the GES which samples $\mathcal{R} \xleftarrow{R} \mathcal{R}_{\lambda,\sigma,\mathbf{v}_{zt}}$ the obfuscator samples the tuple $(a_1, \ldots, a_\ell)$ from $\mathcal{R}$ according to the distribution $\mathcal{D}_\lambda(n, K)$, and outputs the vector of encodings $([a_1]_{\mathbf{v}_1}, \ldots, [a_\ell]_{\mathbf{v}_\ell})$ together with the evaluation parameters evparams.*

Overall, such a canonical obfuscator can be defined by the length function $\ell = \ell(n)$, the ring arity $\sigma(n)$, the vectors $V_n = (\mathbf{v}_1, \ldots, \mathbf{v}_\ell, \mathbf{v}_{zt})$, the distribution $\mathcal{D}_\lambda(n, K)$, and the ring distribution $\mathcal{R}_{\lambda,\sigma,\mathbf{v}_{zt}}$.

Intuitively, an adversary who gets an obfuscated program $([a_1]_{\mathbf{v}_1}, \ldots, [a_\ell]_{\mathbf{v}_\ell})$ can choose some polynomial and check if it is evaluated to zero on the ring elements $(a_1, \ldots, a_\ell)$. Security should guarantee that such an attack gives no information on the program K beyond what follows from an oracle access to C_K. That is, we would like to have a simulator that given an oracle access to C_K can tell whether a given an adversary A (i.e., some arithmetic circuits) evaluates to zero on $\mathcal{D}_\lambda(n, K)$.

We will formalize this notion of security in Definition 3.6, but before that we define a family of ring-independent adversaries. We will focus on the class of purely arithmetic circuits with arbitrary fan-out. These circuits will not have any constants and will contain only input, addition and multiplication gates. Since it contains no constants, it is not ring-specific and one can consider the evaluation of the same circuit over various rings. Formally, each such circuit naturally defines a polynomial with integer coefficients.

---

[11] More precisely, the distribution is defined by randomized arithmetic circuits with GES oracle-gates to the underlying ring, as explained in Remark 2.3.

Definition 3.4. *A purely arithmetic circuit* A *is a circuit which contains input gates (no fan-in, fan out > 0), an output gate (fan-in 1, fan out 0) and operator gates for addition (+), subtraction ($-$) and multiplication ($\times$) with fan-in 2 and fanout > 0. The size of A is the number of gates in A. Given a purely arithmetic circuit A with ℓ input gates and a ring $\mathcal{R}$, we let $\mathcal{P}_{A,\mathcal{R}} \in \mathcal{R}[X_1,\dots,X_\ell]$ denote the ℓ-variate polynomial defined by the circuit A by associating a formal variable X_i with each input gate. When the subscript $\mathcal{R}$ is omitted we view $\mathcal{P}_A$ as a polynomial over the integers.*

We will consider adversaries A that respect the GES-indexing, namely, addition and multiplication can be applied only according to the algebra induced by the GES indexing.

Definition 3.5 (V-Compatible Circuits). *A purely arithmetic circuit A is evaluated over the integer-valued vectors $(\mathbf{v}_1,\dots,\mathbf{v}_\ell)$ via the following recursive process. The i-th input gate takes the value $\mathbf{v}_i$, a multiplication gate with inputs $\mathbf{v}, \mathbf{v}'$ takes the value $\mathbf{v} + \mathbf{v}'$, and an addition (or subtraction) gate with identical inputs $\mathbf{v} = \mathbf{v}'$ takes the value $\mathbf{v}$. If there exists an addition (subtraction) gate with non-identical inputs $\mathbf{v} \neq \mathbf{v}'$ then the circuit is defined to be* syntactically-illegal*. We say that A is compatible with $V = ((\mathbf{v}_1,\dots,\mathbf{v}_\ell), \mathbf{v}_{zt})$ if the computation $A(\mathbf{v}_1,\dots,\mathbf{v}_\ell)$ is syntactically legal and the level $\mathbf{v}$ of the output gate is lower or equal to the zero-test level $\mathbf{v}_{zt}$, i.e., $\mathbf{v} \leq \mathbf{v}_{zt}$. When $\mathbf{v} = \mathbf{v}_{zt}$ we say that A is strongly compatible with V.*

We can now define a simulation-based definition of security which is parameterized by some family of arithmetic circuits $\mathcal{A}_\lambda$.

Definition 3.6 (Algebraic Security). *Let $\mathcal{A} = \{\mathcal{A}_{\lambda,n}\}$ be some class of purely arithmetic circuits where every circuit $A \in \mathcal{A}_{\lambda,n}$ has $\ell(n)$ inputs. We say that a canonical obfuscator $(\ell, \sigma, V, \mathcal{D}, \mathcal{R}_{\lambda,\sigma,\mathbf{v}_{zt}})$ is secure against $\mathcal{A}$ if there exists a (possibly unbounded) randomized algorithm $\mathcal{S}$ (simulator) such that for every input length n, function identifier $K \in \{0,1\}^{m(n)}$, and adversary $A \in \mathcal{A}_{\lambda,n}$ we have*

$$\left| \Pr_{\mathcal{R} \xleftarrow{R} \mathcal{R}_{\lambda,\sigma,\mathbf{v}_{zt}}} [\mathcal{P}_{A,\mathcal{R}}(\mathcal{D}_\lambda(n,K)) = 0] - \Pr[\mathcal{S}^{C_K}(1^\lambda, 1^n, A) = 0] \right| \leq negl(\lambda).$$

By default, we consider security against the class of all $poly(\lambda, n)$-size purely arithmetic circuits $\mathcal{A} = \{\mathcal{A}_{\lambda,n}\}$ which are V_n-compatible (resp., strongly V_n-compatible) and refer to this notion as algebraic security (resp., strong algebraic security).

We note that the case of efficient simulator $\mathcal{S}$ corresponds to VBB security and the inefficient case to the notion of iO. Also, different choices of adversaries $\mathcal{A}$ may be considered in order to capture the operations accessible for the adversary in other generic models. A larger class provides stronger security. Note that the class of V_n-compatible adversaries is strictly larger than the class of *strongly V_n-*compatible, and so security against the former strictly implies security against the latter.

The following lemma, which is implicit in previous works (cf. [8]), shows that security in the algebraic model implies security in the generic model.

Lemma 3.7. *If a canonical GES-based obfuscator $(\ell, \sigma, V, \mathcal{D}, \mathcal{R})$ is algebraically secure (resp., strong algebraically secure) then it is a secure indistinguishably obfuscator relative to the GES oracle $\mathcal{URG}$ (resp., $\mathcal{MRG}$) over the ring distribution $\mathcal{R}$. Furthermore, if the above holds with efficient simulation then the conclusion is strengthened to VBB security in the corresponding model.*

The proof of the lemma (which is implicit in previous works) is differed to the full version.

4   Description of the Obfuscator and Correctness

4.1   Setting and Definitions

Let $\mathcal{C} = \{\mathcal{C}_K\}_{K \in \{0,1\}^*}$ be a family of efficiently computable functions with n-bit inputs, representation size $m = m(n)$ and universal evaluator $\mathcal{U}$. Let $\hat{\mathcal{U}}$ be the arithmetized version of $\mathcal{U}$. Namely an arithmetic circuit with $\{+, -, \times\}$ gates such that for any ring $\mathcal{R}$, if $(x, K) \in \{0,1\}^{n+m} \subseteq \mathcal{R}^{n+m}$, then $\hat{\mathcal{U}}(x, K) = \mathcal{C}_K(x)$. We let $D_{\hat{\mathcal{U}}}$ denote the degree of the polynomial computed by $\hat{\mathcal{U}}$.

Consider an enumeration of the wires of $\hat{\mathcal{U}}$ in topological order, such that the first $n + m$ wires refer to the wires of the x, K inputs. For each wire i, we define a vector $\mathbf{s}_i \in \mathbb{Z}^{n+m+1}$ as follows. If $i \leq n+m$, then $\mathbf{s}_i = \mathbf{e}_i$ (the ith indicator vector). For a wire i which is the output wire of a gate whose input wires are j_1, j_2, we define $\mathbf{s}_i = \mathbf{s}_{j_1} + \mathbf{s}_{j_2}$. We define the *multiplicity* of input wire i to be $M_i = \mathbf{s}_{\text{out}}[i]$, where "out" is the output wire of $\hat{\mathcal{U}}$. (Note that we only used the first $(n+m)$ coordinates of the vectors. The last coordinate will be utilized in the actual construction for the purpose of checking the consistency of the computation.)

4.2   The Obfuscator SimpleObf

For all $i \in [n]$, $b \in \{0,1\}$, we define $\mathbf{v}_{i,b} \in \mathbb{Z}^{(n+m+1)\times 3}$ as $\mathbf{v}_{i,b} = \mathbf{e}_i \otimes [b, 1, 1 - b]$. We further define $\hat{\mathbf{v}}_{i,b} = \mathbf{e}_i \otimes [(1 - b) \cdot M[i], 0, b \cdot M[i]]$.

For all $i \in \{n + 1, \ldots, n + m\}$ we define $\mathbf{v}_i = \mathbf{e}_i \otimes [1, 1, 1]$ and similarly $\mathbf{v}_0 = \mathbf{e}_{n+m+1} \otimes [1, 1, 1]$. Lastly, we define $\mathbf{v}_{\text{zt}} = (\mathbf{s}_{\text{out}} + \mathbf{e}_{n+m+1}) \otimes [1, 1, 1] \in \mathbb{Z}^{(n+m+1)\times 3}$. We note that for all $x \in \{0,1\}^n$ it holds that $\mathbf{v}_{\text{zt}} = \mathbf{v}_0 + \sum_{i=1}^{n}(M[i] \cdot \mathbf{v}_{i,x_i} + \hat{\mathbf{v}}_{i,x_i}) + \sum_{i=n+1}^{n+m} M[i] \cdot \mathbf{v}_i$.

We illustrate the various level vectors in Figure 1.

$$\mathbf{v}_{i,0} = \begin{bmatrix} 0 \cdots 0 \cdots 0 | 0 \\ 0 \cdots 1 \cdots 0 | 0 \\ 0 \cdots 1 \cdots 0 | 0 \end{bmatrix}, \quad \mathbf{v}_{i,1} = \begin{bmatrix} 0 \cdots 1 \cdots 0 | 0 \\ 0 \cdots 1 \cdots 0 | 0 \\ 0 \cdots 0 \cdots 0 | 0 \end{bmatrix}, \quad \mathbf{v}_i = \begin{bmatrix} 0 \cdots 1 \cdots 0 | 0 \\ 0 \cdots 1 \cdots 0 | 0 \\ 0 \cdots 1 \cdots 0 | 0 \end{bmatrix}$$

$$\hat{\mathbf{v}}_{i,0} = \begin{bmatrix} 0 \cdots M[i] \cdots 0 | 0 \\ 0 \cdots 0 \cdots 0 | 0 \\ 0 \cdots 0 \cdots 0 | 0 \end{bmatrix}, \quad \hat{\mathbf{v}}_{i,1} = \begin{bmatrix} 0 \cdots 0 \cdots 0 | 0 \\ 0 \cdots 0 \cdots 0 | 0 \\ 0 \cdots M[i] \cdots 0 | 0 \end{bmatrix}$$

$$\mathbf{v}_0 = \begin{bmatrix} 0 \cdots 0 | 1 \\ 0 \cdots 0 | 1 \\ 0 \cdots 0 | 1 \end{bmatrix}, \quad \mathbf{v}_{zt} = \begin{bmatrix} M[1] \cdots M[n+m] | 1 \\ M[1] \cdots M[n+m] | 1 \\ M[1] \cdots M[n+m] | 1 \end{bmatrix}$$

Fig. 1. The level vectors for obfuscator SimpleObf

Obfuscator SimpleObf:

- **Input:** Circuit identifier $K \in \{0,1\}^m$ where $C_K \in \mathcal{C}$.
- **Output:** Obfuscated program with the same functionality as C_K.
- **Algorithm:**

1. Instantiate a 2-composite graded encoding scheme

$$(params, evparams) = \mathsf{InstGen}(1^\lambda, 1^2, \mathbf{v}_{zt}, 1^{\|\mathbf{v}_{zt}\|_1}) .$$

2. For all $i \in [n]$, $b \in \{0,1\}$, compute random encodings $R_{i,b} = [r_{i,b}]_{\mathbf{v}_{i,b}}$ as well as encodings of $Z_{i,b} = [r_{i,b} \cdot w_{i,b}]_{\mathbf{v}_{i,b}}$, where $w_{i,b} = (y_i, b)$ and y_i is uniform.

3. For all $i \in [n]$, $b \in \{0,1\}$, compute random encodings $\hat{R}_{i,b} = [\hat{r}_{i,b}]_{\hat{\mathbf{v}}_{i,b}}$ as well as encodings of $\hat{Z}_{i,b} = [\hat{r}_{i,b} \cdot \hat{w}_i]_{\hat{\mathbf{v}}_{i,b}}$, where $\hat{w}_i = (\hat{y}_i, \hat{\beta}_i)$ are uniform.

4. For all $i \in \{n+1, \ldots, n+m\}$, compute random encodings $R_i = [r_i]_{\mathbf{v}_i}$ as well as encodings of $Z_i = [r_i \cdot w_i]_{\mathbf{v}_i}$, where $w_i = (y_i, K_{i-n})$, where K_i is the ith bit of the circuit description and y_i is uniform.

5. Compute random encoding $R_0 = [r_0]_{\mathbf{v}_0}$ and $Z_0 = [r_0 w_0]_{\mathbf{v}_0}$, where $w_0 = \left(\prod_{i \in [n]} \hat{w}_i \right) \cdot (y_0, 1)$ and $y_0 = \hat{\mathcal{U}}(y_1, \ldots, y_{n+m})$.

6. The obfuscated program will contain the following:
    - The evaluation parameters $evparams$.
    - For all $i \in [n]$, $b \in \{0,1\}$ the elements $R_{i,b}, Z_{i,b}, \hat{R}_{i,b}, \hat{Z}_{i,b}$.
    - For all $i \in \{n+1, \ldots, n+m\}$ the elements R_i, Z_i.
    - The elements R_0, Z_0.

We note that all of the required encodings can be efficiently generated using $params$, as explained in Remark 2.3.

An important feature of our obfuscator that will be used in the proof is that all of the information that depends on the circuit C_K resides in the second element of the CRT representation, and the distribution of the first element is completely independent of C_K.

4.3   The Obfuscator **RobustObf**

For simplicity of presentation, we assume w.l.o.g that $\hat{\mathcal{U}}$ is such that the inputs to every multiplication gate have the same degree (as polynomials in the input variables and program description). This is straightforward to achieve by adding the constant 1 as one of the elements of the program description, and multiplying by this variable (raised to the proper degree) to balance the input degrees.

For all $i \in [n]$, $b \in \{0, 1\}$, we define $\mathbf{v}_{i,b} \in \mathbb{Z}^{(n+m+1)\times 4}$ as $\mathbf{v}_{i,b} = \mathbf{e}_i \otimes [b, 1, 1 - b, 0]$. We further define $\hat{\mathbf{v}}_{i,b} = \mathbf{e}_i \otimes [(1 - b) \cdot M[i], 0, b \cdot M[i], 1]$.

For all $i \in \{n + 1, \ldots, n + m\}$ we define $\mathbf{v}_i = \mathbf{e}_i \otimes [1, 1, 1, 1]$. We define $\mathbf{v}_0 = \mathbf{e}_{n+m+1} \otimes [1, 1, 1, 0]$ and $\mathbf{v}^* = \mathbf{e}_{n+m+1} \otimes [0, 0, 0, 1]$. Lastly, we define $\mathbf{v}_{zt} = (\mathbf{s}_{out} + \mathbf{e}_{n+m+1}) \otimes [1, 1, 1, 0] + (\sum_{i=1}^{n+m} \mathbf{e}_i) \otimes [0, 0, 0, 1] + D \cdot \mathbf{v}^* \in \mathbb{Z}^{(n+m+1)\times 4}$, where $D = D_{\hat{\mathcal{U}}} + n$ (and $D_{\hat{\mathcal{U}}}$, as defined above, is the degree of the polynomial computed by $\hat{\mathcal{U}}$). We note that for all $x \in \{0, 1\}^n$ it holds that $\mathbf{v}_{zt} = \mathbf{v}_0 + \sum_{i=1}^{n}(M[i] \cdot \mathbf{v}_{i,x_i} + \hat{\mathbf{v}}_{i,x_i}) + \sum_{i=n+1}^{n+m} M[i] \cdot \mathbf{v}_i + D \cdot \mathbf{v}^*$.

We illustrate the various level vectors in Figure 2.

Obfuscator RobustObf:

- **Input:** Circuit identifier $K \in \{0, 1\}^m$ where $C_K \in \mathcal{C}$.
- **Output:** Obfuscated program with the same functionality as C_K.
- **Algorithm:**

1. Instantiate a $(n + 2)$-composite graded encoding scheme

$$(params, evparams) = \mathsf{InstGen}(1^\lambda, 1^{n+2}, \mathbf{v}_{zt}, 1^{\|\mathbf{v}_{zt}\|_1}).$$

2. For all $i \in [n]$, $b \in \{0, 1\}$, compute random encodings $R_{i,b} = [r_{i,b}]_{\mathbf{v}_{i,b}}$ as well as encodings of $Z_{i,b} = [r_{i,b} \cdot w_{i,b}]_{\mathbf{v}_{i,b}+\mathbf{v}^*}$, where $w_{i,b} = (y_i, b, \rho_{i,b,1}, \ldots, \rho_{i,b,n})$ and $y_i, \rho_{i,b,j}$ are uniform.

3. For all $i \in [n]$, $b \in \{0, 1\}$, compute random encodings $\hat{R}_{i,b} = [\hat{r}_{i,b}]_{\hat{\mathbf{v}}_{i,b}}$ as well as encodings of $\hat{Z}_{i,b} = [\hat{r}_{i,b} \cdot \hat{w}_i]_{\hat{\mathbf{v}}_{i,b}+\mathbf{v}^*}$, where $\hat{w}_i = (\hat{y}_i, \hat{\beta}_i, \hat{\rho}_{i,1}, \ldots, \hat{\rho}_{i,n})$, where $\hat{y}_i, \hat{\beta}_i, \{\hat{\rho}_{i,j}\}_{j\neq i}$ are all uniform, but $\hat{\rho}_{i,i} = 0$.

4. For all $i \in \{n + 1, \ldots, n + m\}$, compute random encodings $R_i = [r_i]_{\mathbf{v}_i}$ as well as encodings of $Z_i = [r_i \cdot w_i]_{\mathbf{v}_i+\mathbf{v}^*}$, where $w_i = (y_i, K_{i-n}, \rho_{i,1}, \ldots, \rho_{i,n})$, where K_i is the ith bit of the circuit description and $y_i, \rho_{i,j}$ are uniform.

5. Compute random encoding $R_0 = [r_0]_{\mathbf{v}_0}$ and $Z_0 = [r_0 w_0]_{\mathbf{v}_0+D\mathbf{v}^*}$, where $w_0 = \left(\prod_{i\in[n]} \hat{w}_i\right) \cdot (y_0, 1, 0, \ldots, 0)$ and $y_0 = \hat{\mathcal{U}}(y_1, \ldots, y_{n+m})$.

6. The obfuscated program will contain the following:
   - The evaluation parameters $evparams$.

$$\mathbf{v}_{i,0} = \begin{bmatrix} 0 \cdots & 0 & \cdots 0|0 \\ 0 \cdots & 1 & \cdots 0|0 \\ 0 \cdots & 1 & \cdots 0|0 \\ 0 \cdots & 0 & \cdots 0|0 \end{bmatrix}, \quad \mathbf{v}_{i,1} = \begin{bmatrix} 0 \cdots 1 \cdots 0|0 \\ 0 \cdots 1 \cdots 0|0 \\ 0 \cdots 0 \cdots 0|0 \\ 0 \cdots 0 \cdots 0|0 \end{bmatrix}, \quad \mathbf{v}_i = \begin{bmatrix} 0 \cdots 1 \cdots 0|0 \\ 0 \cdots 1 \cdots 0|0 \\ 0 \cdots 1 \cdots 0|0 \\ 0 \cdots 0 \cdots 0|0 \end{bmatrix}$$

$$\hat{\mathbf{v}}_{i,0} = \begin{bmatrix} 0 \cdots & M[i] & \cdots 0|0 \\ 0 \cdots & 0 & \cdots 0|0 \\ 0 \cdots & 0 & \cdots 0|0 \\ 0 \cdots & 1 & \cdots 0|0 \end{bmatrix}, \quad \hat{\mathbf{v}}_{i,1} = \begin{bmatrix} 0 \cdots & 0 & \cdots 0|0 \\ 0 \cdots & 0 & \cdots 0|0 \\ 0 \cdots & M[i] & \cdots 0|0 \\ 0 \cdots & 1 & \cdots 0|0 \end{bmatrix}$$

$$\mathbf{v}_0 = \begin{bmatrix} 0 \cdots 0|1 \\ 0 \cdots 0|1 \\ 0 \cdots 0|1 \\ 0 \cdots 0|0 \end{bmatrix}, \quad \mathbf{v}^* = \begin{bmatrix} 0 \cdots 0|0 \\ 0 \cdots 0|0 \\ 0 \cdots 0|0 \\ 0 \cdots 0|1 \end{bmatrix}$$

$$\mathbf{v}_{zt} = \begin{bmatrix} M[1] \cdots M[n] & M[n+1] \cdots M[n+m] & 1 \\ M[1] \cdots M[n] & M[n+1] \cdots M[n+m] & 1 \\ M[1] \cdots M[n] & M[n+1] \cdots M[n+m] & 1 \\ 1 \cdots 1 & 0 \cdots 0 & D \end{bmatrix}$$

Fig. 2. The level vectors for obfuscator RobustObf.

- For all $i \in [n]$, $b \in \{0,1\}$ the elements $R_{i,b}, Z_{i,b}, \hat{R}_{i,b}, \hat{Z}_{i,b}$.
- For all $i \in \{n+1, \ldots, n+m\}$ the elements R_i, Z_i.
- The elements R_0, Z_0.

As in our previous obfuscator, all of the required encodings can be efficiently generated using *params*, as explained in Remark 2.3.

Note that again all of the information that depends on C_K appears in the second component of $\mathcal{R}$, and the distributions in all other components are independent of K.

4.4 Evaluating an Obfuscated Program

We will now describe the evaluator for our obfuscators SimpleObf and RobustObf. Due to their very similar structure, we are able to present a single evaluator that works for both obfuscators. In the context of SimpleObf we will define $\mathbf{v}^* = \mathbf{0}$ and ignore the last n sub-rings of the ring $\mathcal{R}$.

As can be seen in the description of our obfuscator above, the obfuscated circuit is encoded in the w variables, and each w variable in turn is encoded relative to an r variable. We first show that these pairs of encodings of r and $r \cdot w$ can be manipulated algebraically while keeping the invariant that each value is encoded relative to an r. This is demonstrated by the following procedure.

Procedure PairOp:

- **Input:** GES evaluation parameters $evparams$, pairs of encodings $(R_1 = [r_1]_{\mathbf{v}_1}, Z_1 = [r_1 w_1]_{\mathbf{v}_1 + k\mathbf{v}^*})$, $(R_2 = [r_2]_{\mathbf{v}_2}, Z_1 = [r_2 w_2]_{\mathbf{v}_2 + k\mathbf{v}^*})$, operation op $\in \{\times, +, -\}$.
- **Output:** Pair of encodings $(R^* = [r_1 r_2]_{\mathbf{v}_1 + \mathbf{v}_2}, Z = [r_1 r_2 \cdot (w_1 \text{ op } w_2)]_{\mathbf{v}_1 + \mathbf{v}_2 + tk \cdot \mathbf{v}^*})$, where $t = 1$ for op $\in \{+, -\}$ and $t = 2$ for op $\in \{\times\}$. If $(\mathbf{v}_1 + \mathbf{v}_2 + tk \cdot \mathbf{v}^*) > \mathbf{v}_{zt}$, the procedure outputs $\bot$.
- **Algorithm:**

1. Compute $R^* = R_1 \times R_2$.
2. If op $= \times$ compute $Z^* = Z_1 \times Z_2$.
3. If op $= +$ compute $Z^* = Z_1 \times R_2 + R_1 \times Z_2$.
4. If op $= -$ compute $Z^* = Z_1 \times R_2 - R_1 \times Z_2$.

We note that PairOp can be applied iteratively to evaluate any arithmetic circuit on pairs of encodings. The multiplicity of $\mathbf{v}^*$ will be exactly the multiplicative degree of the evaluated circuit. We can now describe our evaluator for obfuscated programs.

Procedure Eval:

- **Input:** Obfuscated program as produced by $\mathsf{SimpleObf}(K)$ for some identifier K:

$$\mathcal{O} = \left(evparams, \{R_{i,b}, Z_{i,b}, \hat{R}_{i,b}, \hat{Z}_{i,b}\}_{\substack{i \in [n], \\ b \in \{0,1\}}}, \{R_i, Z_i\}_{i=n+1}^{n+m}, \{R_0, Z_0\} \right),$$

  input $x \in \{0,1\}^n$.
- **Output:** Value $\mathcal{O}(x) \in \{0,1\}$.
- **Algorithm:**

1. We consider the pairs of elements (R_{i,x_i}, Z_{i,x_i}) for $i \in [n]$, and R_i, Z_i for $i = n+1, \ldots, n+m$. We apply the circuit $\hat{\mathcal{U}}$ on these pairs of encodings as described above, to obtain a pair:

$$R_{\mathcal{U}} = [r_{\mathcal{U}}]_{\mathbf{v}_{\mathcal{U}}}, \quad Z_{\mathcal{U}} = [r_{\mathcal{U}} \cdot w_{\mathcal{U}}]_{\mathbf{v}_{\mathcal{U}} + D_{\hat{u}}},$$

where $\mathbf{v}_{\mathcal{U}} = \sum_{i=1}^{n} M[i] \cdot \mathbf{v}_{i,x_i} + \sum_{i=n+1}^{n+m} M[i] \cdot \mathbf{v}_i$ and

$$\begin{aligned} w_{\mathcal{U}} &= \hat{\mathcal{U}}(w_{1,x_1}, \ldots, w_{n,x_n}, w_{n+1}, \ldots, w_{n+m}) \\ &= (\hat{\mathcal{U}}(y_1, \ldots, y_n, y_{n+1}, \ldots, y_{n+m}), \hat{\mathcal{U}}(x, K), -) \\ &= (\hat{\mathcal{U}}(\vec{y}), C_K(x), -) \end{aligned}$$

where the values denoted by "—" will not matter for correctness so we will not explicitly mention them to avoid cluttering (recall that the simpler obfuscator SimpleObf does not need these values at all).

2. We take the product of the pair of elements $(R_\mathcal{U}, Z_\mathcal{U})$ with the pairs $(\hat{R}_{i,x_i}, \hat{Z}_{i,x_i})$ to obtain

$$\hat{R}_\mathcal{U} = [\hat{r}_\mathcal{U}]_{\hat{\mathbf{v}}_\mathcal{U}}, \quad \hat{Z}_\mathcal{U} = [\hat{r}_\mathcal{U} \cdot \hat{w}_\mathcal{U}]_{\hat{\mathbf{v}}_\mathcal{U}+D\mathbf{v}^*},$$

where $\hat{w}_\mathcal{U} = \prod_{i=1}^n \hat{w}_i \cdot w_\mathcal{U}$, and

$$\hat{\mathbf{v}}_\mathcal{U} = \sum_{i=1}^n M[i] \cdot \mathbf{v}_{i,x_i} + \sum_{i=n+1}^{n+m} M[i] \cdot \mathbf{v}_i + \sum_{i=1}^n \hat{\mathbf{v}}_{i,x_i} = \mathbf{v}_{zt} - \mathbf{v}_0.$$

3. We subtract the pair $(\hat{R}_\mathcal{U}, \hat{Z}_\mathcal{U})$ from the pair (R_0, Z_0), to obtain

$$R'' = [r'']_{\hat{\mathbf{v}}_\mathcal{U}+\mathbf{v}_0}, \quad Z'' = [r'' \cdot w'']_{\hat{\mathbf{v}}_\mathcal{U}+D\mathbf{v}^*+\mathbf{v}_0},$$

and we notice that indeed $(\hat{\mathbf{v}}_\mathcal{U} + D\mathbf{v}^* + \mathbf{v}_0) = \mathbf{v}_{zt}$ and

$$w'' = w_0 - \prod_{i=1}^n \hat{w}_i \cdot (\hat{\mathcal{U}}(\vec{y}), C_K(x), -) = \prod_{i=1}^n \hat{w}_i \cdot (\hat{\mathcal{U}}(\vec{y}) - \hat{\mathcal{U}}(\vec{y}), 1 - C_K(x), -).$$

Recalling that $\prod_{i=1}^n \hat{w}_i = (\alpha, \beta, 0, \ldots, 0)$, for some values α, β, we have that

$$w'' = (0, \beta(1 - C_K(x)), 0, \ldots, 0).$$

4. Finally, zero testing is applied to Z''. If isZero$(Z'') = 1$ then output 1, otherwise output 0.

---

5 Generic Security of Our Construction

5.1 Useful Algebraic Tools

We will use the following corollary of the Schwartz-Zippel lemma [16,19].

Fact 5.1. *Let $\sigma \in \mathbb{N}$, let $p_1, \ldots, p_\sigma$ be distinct primes and let $P = \prod_{i=1}^\sigma p_i$. Then a multivariate polynomial of total degree d has at most d^σ roots over $\mathbb{Z}_P$.*

The proof is deferred to the full version.

For a univariate polynomial P, defined over a field, it holds that $(x - a)|P(x)$ if and only if $P(a) = 0$. The following lemma generalizes this fact to the case of multivariate polynomials over the integers.

Fact 5.2. *Let $P(x_1, \ldots, x_n) \in \mathbb{Z}[x_1, \ldots, x_n]$ and let $A(x_2, \ldots, x_n) \in \mathbb{Z}[x_1, \ldots, x_n]$ (however x_1 does not appear in A). Then*

$$((x_1 - A(x_2, \ldots, x_n)) | P(x_1, \ldots, x_n)) \leftrightarrow P(A(x_2, \ldots, x_n), x_2, \ldots, x_n) \equiv 0 .$$

The proof is deferred to the full version.

Next we present a bound on the size of the coefficients of a polynomial computed by a purely arithmetic circuit of bounded size and bounded degree.

Fact 5.3. *Let C be a purely arithmetic circuit (as per Definition 3.4) of size s and degree d. Then the polynomial $\|\mathcal{P}_C\|_1 \leq 2^{sd}$ (where the norm refers to the ℓ_1 norm of the coefficient vector of $\mathcal{P}_C$).*

The proof is deferred to the full version. A polynomial is *free* of some variable or monomial if this variable/monomial does not appear in its expansion. A formal definition follows.

Definition 5.4. *Let $P(X_1, \ldots, X_n)$ be a polynomial. We say that P is X_i-free if all monomials that contain X_i take zero value in P's coefficient vector. We extend this notation to monomials and say that P is $(\prod X_i^{d_i})$-free if all monomials that are divisible by $(\prod X_i^{d_i})$ take zero value in P's coefficient vector. For a set of monomials $\{M_1, \ldots, M_k\}$ we say that P is $\{M_1, \ldots, M_k\}$-free if it is M_j-free for all $j = 1, \ldots, k$.*

Obfuscators in El-Gamal form. Recall that a canonical obfuscator is defined by length function $\ell = \ell(n)$, ring arity $\sigma(n)$, integer-valued vectors $(\mathbf{v}_1, \ldots, \mathbf{v}_\ell, \mathbf{v}_{zt})$, and a distribution $\mathcal{D}_\lambda(n, K)$ over $\ell(n)$ ring elements $(a_1, \ldots, a_\ell)$. A canonical form obfuscator is in *El-Gamal form* (EG in short) if the ring elements $(a_i)_{i \in [\ell]}$ can be partitioned to pairs $(r_i, z_i)_{\ell/2}$ where the vector $(r_1, \ldots, r_{\ell/2})$ is a vector of uniformly and independently chosen ring elements, and $z_i = r_i \cdot w_i$ for some $w_i \in \mathcal{R}$. (The same w_i may appear twice and may not be uniformly distributed, and furthermore r_i, z_i may not be encoded in the same level.)

Note that both of our obfuscators are in El-Gamal form.

5.2   Admissible Distributions on Composites and Rings

We define the notion of admissible distributions over composite numbers (and by extension over rings). Intuitively, a probability distribution $\mathcal{N}_k$ is k-admissible if it samples a $poly(k)$-bit integer with the property that the min-entropy of every prime factor of $\mathcal{N}_k$ is at least $\Omega(k)$. A formal definition follows.

Definition 5.5. *An ensemble of probability distributions $\{\mathcal{N}_k\}$ is k-admissible if $\mathcal{N}_k$ samples a $poly(k)$-bit integer with the property that the min-entropy of every prime factor of $\mathcal{N}_k$ is at least $\Omega(k)$. An ensemble of probability distributions over rings $\{\mathcal{R}_k\}$ is k-admissible if $\mathcal{R}_k \cong \mathbb{Z}_N$ and the random variable N is k-admissible.*

It is not hard to see that every small fixed integer x is likely to be co-prime to $y \xleftarrow{R} \mathcal{N}_k$.

Lemma 5.6. *Let $\mathcal{N}_k$ be some k-admissible distribution. Then for all $x \in \mathbb{Z} \setminus \{0\}$, it holds that*

$$\Pr_{y \xleftarrow{R} \mathcal{N}_{t,k}} [\gcd(|x|, y) > 1] \leq \log |x| \cdot poly(k) \cdot 2^{-\Omega(k)} \leq \log |x| \cdot 2^{-\Omega(k)} .$$

The proof is deferred to the full version.

It follows for an admissible ring distribution, any fixed (short) list of (small) integers is unlikely to hit non-invertible ring element.

Corollary 5.7. *Let $L \in \mathbb{N}$ and let $\mathcal{L} \subseteq \mathbb{Z} \setminus \{0\}$ be a list of L integers such that all $x \in \mathcal{L}$, $|x| \leq 2^{poly(\lambda)}$. Let $\mathcal{R} \cong \mathbb{Z}_N$ be a ring where N is chosen from some $(\log L + \omega(\log \lambda))$-admissible distribution. Then, the probability that there exists $x \in \mathcal{L}$ which is not a unit in $\mathcal{R}$ is $negl(\lambda)$.*

5.3   Proof Outline

In this section we describe the common general outline of the proof that will be applied both to SimpleObf (Section 5.4) and to RobustObf (Section 5.5).

Since our constructions are in canonical form (in fact, in El-Gamal form) it suffices, by Lemma 3.7, to prove algebraic security according to Definition 3.6.

Fix some function identifier K and polynomial P. We note that P is associated with a purely arithmetic circuit of polynomial size and polynomial degree. The latter is since the degree cannot go above $\|\mathbf{v}_{zt}\|_1$. Since our obfuscator is in EG form, we can re-write P as a sum of terms of the form

$$M(r) \cdot Q(w) ,$$

where M is a monomial and Q is a polynomial. It suffices to show that given an oracle access to C_K, we can determine if the above product equals to zero with more than negligible probability (where the probability is taken over $\mathcal{D}_\lambda(K)$).

In the simulation we use the min-entropy of the orders p_i of the sub-rings of $\mathcal{R}$ as follows. We present a simulator that needs not know any information about the order of $\mathcal{R}$ or its sub-rings. However, this simulator succeeds only as long as a list of non-zero integers $\mathcal{L}$ generated during the simulation does not contain any element that is not a unit in $\mathcal{R}$. The length of the list will be bounded by $2^{\|\mathbf{v}_{zt}\|_1} \cdot poly(\lambda)$, and the absolute value of each of these numbers will be at most $2^{poly(\lambda)}$. Fact 5.3 and Corollary 5.7 thus guarantee that the simulation fails only with negligible probability.

Formally, our simulator (which is oblivious to the order of the ring) is going to have the following properties:

1. The simulator will generate, as a by product, a list $\mathcal{L}$ of $L = 2^{\|\mathbf{v}_{zt}\|_1} \cdot poly(\lambda)$ integers of absolute value at most $2^{poly(\lambda)}$. In particular, the list is a subset of the coefficients of the polynomial P. Since P is computable by a purely arithmetic circuit of size $poly(\lambda)$ and degree at most $\|\mathbf{v}_{zt}\|_1$, the bounds will follow from Fact 5.3.

2. We will prove that as long as all of the elements of $\mathcal{L}$ (cast into $\mathcal{R}$) are units in $\mathcal{R}$, the simulation is successful.
3. The distribution of $\mathcal{R}$ as described in our model will guarantee, by Corollary 5.7, that the event of simulation failure due to $\mathcal{L}$ containing a non-unit is negligible.

The specifics of applying this outline will vary between the specific obfuscators.

5.4   Algebraic Security Proof for SimpleObf

Theorem 5.8. *The obfuscator* SimpleObf *is secure relative to the GES oracle* $\mathcal{MRG}$ *defined over any* $(\|\mathbf{v}_{zt}\|_1 + \omega(\log(\lambda)))$-*admissible ring distribution.*

We follow the outline from Section 5.3. We start with structural claims on Q, viewed as a polynomial over the integers.

Lemma 5.9. *There exists* $x = (x_1, \ldots, x_n) \in \{0,1\}^n$ *such that* Q *is* $\{w_{i,1-x_i}\}_{i \in [n]}$-*free.*

The proof is deferred to the full version.

Lemma 5.10. *The polynomial* Q *is* $\left\{ w_{i,b}^{(M[i]+1)}, w_i^{(M[i]+1)}, \hat{w}_i^2, \hat{w}_0^2 \right\}_{i \in [n], b \in \{0,1\}}$-*free.*

The proof is deferred to the full version.

Lemma 5.11. *The polynomial* $P(r, w)$ *can be written as a sum of at most* $T = 2^{\|\mathbf{v}_{zt}\|_1}$ *terms of the form* $M(r)Q(w)$, *where* M *is a monomial.*

The proof is deferred to the full version.
    We can now distinguish between two cases.

1. It holds that
$$\left(w_0 - \left(\prod_{i \in [n]} \hat{w}_i \right) \cdot \hat{\mathcal{U}}(w_{1,x_1}, \ldots, w_{n,x_n}, w_{n+1}, \ldots, w_{n+m}) \right) \nmid Q$$

   for any $x = (x_1, \ldots, x_n) \in \{0,1\}^n$.
2. There exists $x = (x_1, \ldots, x_n) \in \{0,1\}^n$ such that
$$\left(w_0 - \left(\prod_{i \in [n]} \hat{w}_i \right) \cdot \hat{\mathcal{U}}(w_{1,x_1}, \ldots, w_{n,x_n}, w_{n+1}, \ldots, w_{n+m}) \right) \mid Q .$$

Lemma 5.12 (Case 1). *Let* $x = (x_1, \ldots, x_n) \in \{0,1\}^n$ *be the value guaranteed to exist in Lemma 5.9. If it holds that*
$$\left(w_0 - \left(\prod_{i \in [n]} \hat{w}_i \right) \cdot \hat{\mathcal{U}}(w_{1,x_1}, \ldots, w_{n,x_n}, w_{n+1}, \ldots, w_{n+m}) \right) \nmid Q , \tag{1}$$

then
$$\Pr_{\mathcal{D}_\lambda(K)} [Q = 0] = negl(\lambda) .$$

The proof is deferred to the full version.

Lemma 5.13. *Let $x = (x_1, \ldots, x_n) \in \{0,1\}^n$ be the value guaranteed to exist in Lemma 5.9. If it holds that*

$$\left(w_0 - \big(\prod_{i \in [n]} \hat{w}_i \big) \cdot \hat{\mathcal{U}}(w_{1,x_1}, \ldots, w_{n,x_n}, w_{n+1}, \ldots, w_{n+m}) \right) \Big| Q(w),$$

then there exists a constant a' such that

$$Q(w) = a' \cdot \left(w_0 - \big(\prod_{i \in [n]} \hat{w}_i \big) \cdot \hat{\mathcal{U}}(w_{1,x_1}, \ldots, w_{n,x_n}, w_{n+1}, \ldots, w_{n+m}) \right).$$

The proof is deferred to the full version.

Lemma 5.14 (Case 2). *If there exists $x = (x_1, \ldots, x_n) \in \{0,1\}^n$ such that*

$$Q(w) = a \cdot \left(w_0 - \left(\prod_{i \in [n]} \hat{w}_i \right) \cdot \hat{\mathcal{U}}(w_{1,x_1}, \ldots, w_{n,x_n}, w_{n+1}, \ldots, w_{n+m}) \right),$$

Then if $C_K(x) = 0$ then $\Pr_{\mathcal{D}_\lambda(K)}[Q = 0] = negl(\lambda)$, and if $C_K(x) = 1$ then $\Pr_{\mathcal{D}_\lambda(K)}[Q = 0] = 1$.

The proof is deferred to the full version.

Overall, the simulator can determine whether Q evaluates to zero. In the first case, it will simply say that Q does not evaluates to zero, and in the second case it will test if $C_K(x) = 1$ and, only if this test passes, it will output Yes (meaning that Q evaluates to zero). By lemmas 5.12 and 5.14 the simulator errs with no more than negligible probability.

5.5 Algebraic Security Proof for RobustObf

We will prove the following theorem.

Theorem 5.15. *The obfuscator RobustObf is secure relative to the GES oracle $\mathcal{U}\mathcal{R}\mathcal{G}$ defined over any $(\|\mathbf{v}_{zt}\|_1 + \omega(\log(\lambda)))$-admissible ring distribution.*

Since we aim for security relative to $\mathcal{M}\mathcal{R}\mathcal{G}$ (using Lemma 3.7), we should take into account the possibility that the adversary P is of level smaller than $\mathbf{v}_{zt}$. We follow the outline from Section 5.3 by viewing P as a sum of terms of the form $M(r) \cdot Q(w)$. We will analyze the probability that $Q(\mathcal{D}_\lambda(K))$ evaluates to zero, starting with a few structural claims on Q.

Lemma 5.16. *There exists a constant a and a w_0-free polynomial $Q'(w)$ such that*

$$Q(w) = a \cdot w_0 - Q'(w).$$

The proof is deferred to the full version.

Lemma 5.17. *For all $i \in [n]$, the polynomial Q (and therefore also Q' from Lemma 5.16) is $\hat{w}_i^2$-free.*

The proof is deferred to the full version.

We can now distinguish between three main cases.

1. It holds that $\left(\prod_{i \in [n]} \hat{w}_i \right) \nmid Q'(w)$.
2. It holds that $\left(\prod_{i \in [n]} \hat{w}_i \right) | Q'(w)$, namely there exists a polynomial $Q''(w)$ which is $\{w_0, \hat{w}_1, \ldots, \hat{w}_n\}$-free and

$$Q(w) = aw_0 - \left(\prod_{i \in [n]} \hat{w}_i \right) \cdot Q''(w) .$$

However,

$$Q''(w) \neq a \cdot \hat{\mathcal{U}}(w_{1,x_1}, \ldots, w_{n,x_n}, w_{n+1}, \ldots, w_{n+m})$$

for any $x = (x_1, \ldots, x_n) \in \{0,1\}^n$.
3. There exists $x = (x_1, \ldots, x_n) \in \{0,1\}^n$ such that

$$Q(w) = a \cdot \left(w_0 - \left(\prod_{i \in [n]} \hat{w}_i \right) \cdot \hat{\mathcal{U}}(w_{1,x_1}, \ldots, w_{n,x_n}, w_{n+1}, \ldots, w_{n+m}) \right) . \quad (2)$$

Lemma 5.18 (Case 1). *If $\exists i. \hat{w}_i \nmid Q'(w)$, then*

$$\Pr_{w \xleftarrow{R} \mathcal{D}_\lambda(K)} [Q(w) = 0] = negl(\lambda) .$$

The proof is deferred to the full version.

Lemma 5.19. *If $Q'(w) = \left(\prod_{i \in [n]} \hat{w}_i \right) \cdot Q''$ for some Q'', then there exists $x = (x_1, \ldots, x_n) \in \{0,1\}^n$ such that Q'' is $\{w_{i,1-x_i}\}_{i \in [n]}$-free.*

We recall that by Lemma 5.17, $Q''(w)$ is also $\{\hat{w}_i\}_{i \in [n]}$-free. The proof of Lemma 5.19 is deferred to the full version.

Lemma 5.20 (Case 2). *If $Q' = \left(\prod_{i \in [n]} \hat{w}_i \right) \cdot Q''$ and for the x from Lemma 5.19 it holds that $Q'' \neq a \cdot \hat{\mathcal{U}}(w_{1,x_1}, \ldots, w_{n,x_n}, w_{n+1}, \ldots, w_{n+m})$, then $\Pr_{\mathcal{D}_\lambda(K)}[Q = 0] = negl(\lambda)$.*

The proof is deferred to the full version.

Lemma 5.21 (Case 3). *If there exists $x = (x_1, \ldots, x_n) \in \{0,1\}^n$ such that*

$$Q = a \cdot \left(w_0 - \left(\prod_{i \in [n]} \hat{w}_i \right) \cdot \hat{\mathcal{U}}(w_{1,x_1}, \ldots, w_{n,x_n}, w_{n+1}, \ldots, w_{n+m}) \right) ,$$

Then if $C_K(x) = 0$ then $\Pr_{\mathcal{D}_\lambda(K)}[Q = 0] = negl(\lambda)$, and if $C_K(x) = 1$ then $\Pr_{\mathcal{D}_\lambda(K)}[Q = 0] = 1$.

The proof is deferred to the full version.

Overall, the simulator can determine whether Q evaluates to zero. In the first and second cases, it will simply say that Q does not evaluates to zero, and in the third case it will test if $C_K(x) = 1$ and, only if this test passes, it will output Yes (meaning that Q evaluates to zero). By lemmas 5.18, 5.20 and 5.21 the simulator errs with no more than negligible probability.

References

1. Ananth, P., Gupta, D., Ishai, Y., Sahai, A.: Optimizing obfuscation: Avoiding barrington's theorem. Cryptology ePrint Archive, Report 2014/222 (2014), http://eprint.iacr.org/

2. Applebaum, B.: Bootstrapping obfuscators via fast pseudorandom functions. In: Sarkar, P., Iwata, T. (eds.) ASIACRYPT 2014, Part II. LNCS, vol. 8874, pp. 162–172. Springer, Heidelberg (2014)

3. Applebaum, B., Brakerski, Z.: Obfuscating circuits via composite-order graded encoding (2014); Full version of this paper. Available at the authors homepage

4. Barak, B., Garg, S., Kalai, Y.T., Paneth, O., Sahai, A.: Protecting obfuscation against algebraic attacks. In: Nguyen, P.Q., Oswald, E. (eds.) EUROCRYPT 2014. LNCS, vol. 8441, pp. 221–238. Springer, Heidelberg (2014)

5. Barak, B., Goldreich, O., Impagliazzo, R., Rudich, S., Sahai, A., Vadhan, S.P., Yang, K.: On the (im)possibility of obfuscating programs. J. ACM 59(2), 6 (2012); Preliminary version in Kilian, J. (ed.) CRYPTO 2001. LNCS, vol. 2139, pp. 1–18. Springer, Heidelberg (2001)

6. Brakerski, Z., Rothblum, G.N.: Obfuscating conjunctions. In: Canetti, R., Garay, J.A. (eds.) CRYPTO 2013, Part II. LNCS, vol. 8043, pp. 416–434. Springer, Heidelberg (2013)

7. Brakerski, Z., Rothblum, G.N.: Black-box obfuscation for d-cnfs. In: Innovations in Theoretical Computer Science, ITCS 2014, Princeton, NJ, USA, January 12-14, pp. 235–250. ACM (2014)

8. Brakerski, Z., Rothblum, G.N.: Virtual black-box obfuscation for all circuits via generic graded encoding. In: Lindell, Y. (ed.) TCC 2014. LNCS, vol. 8349, pp. 1–25. Springer, Heidelberg (2014)

9. Coron, J.-S., Lepoint, T., Tibouchi, M.: Practical multilinear maps over the integers. In: Canetti, R., Garay, J.A. (eds.) CRYPTO 2013, Part I. LNCS, vol. 8042, pp. 476–493. Springer, Heidelberg (2013)

10. Garg, S., Gentry, C., Halevi, S.: Candidate multilinear maps from ideal lattices. In: Johansson, T., Nguyen, P.Q. (eds.) EUROCRYPT 2013. LNCS, vol. 7881, pp. 1–17. Springer, Heidelberg (2013)

11. Garg, S., Gentry, C., Halevi, S., Raykova, M., Sahai, A., Waters, B.: Candidate indistinguishability obfuscation and functional encryption for all circuits. In: 54th Annual IEEE Symposium on Foundations of Computer Science, FOCS 2013, October 26-29, pp. 40–49. IEEE Computer Society, Berkeley (2013)

12. Gentry, C., Lewko, A.B., Sahai, A., Waters, B.: Indistinguishability obfuscation from the multilinear subgroup elimination assumption. IACR Cryptology ePrint Archive, 2014:309 (2014)

13. Gentry, C., Lewko, A.B., Waters, B.: Witness encryption from instance independent assumptions. In: Garay, J.A., Gennaro, R. (eds.) CRYPTO 2014, Part I. LNCS, vol. 8616, pp. 426–443. Springer, Heidelberg (2014)

14. Goyal, V., Ishai, Y., Sahai, A., Venkatesan, R., Wadia, A.: Founding cryptography on tamper-proof hardware tokens. In: Micciancio, D. (ed.) TCC 2010. LNCS, vol. 5978, pp. 308–326. Springer, Heidelberg (2010)

15. Maurer, U.: Abstract models of computation in cryptography. In: Smart, N.P. (ed.) Cryptography and Coding 2005. LNCS, vol. 3796, pp. 1–12. Springer, Heidelberg (2005)

16. Schwartz, J.: Fast probabilistic algorithms for verification of polynomial identities. J. ACM 27(2), 701–717 (1980)

17. Shoup, V.: Lower bounds for discrete logarithms and related problems. In: Fumy, W. (ed.) EUROCRYPT 1997. LNCS, vol. 1233, pp. 256–266. Springer, Heidelberg (1997)

18. Zimmerman, J.: How to obfuscate programs directly. Cryptology ePrint Archive, Report 2014/776 (2014), http://eprint.iacr.org/

19. Zippel, R.: Probabilistic algorithms for sparse polynomials. In: Ng, E.W. (ed.) EUROSAM 1979 and ISSAC 1979. LNCS, vol. 72, pp. 216–226. Springer, Heidelberg (1979)

Adaptively Secure Two-Party Computation from Indistinguishability Obfuscation[*]

Ran Canetti[1,3,**], Shafi Goldwasser[2], and Oxana Poburinnaya[3]

[1] Tel-Aviv University, Israel
`canetti@bu.edu`
[2] Weizmann, Israel and MIT, USA
`shafi@csail.mit.edu`
[3] Boston University, USA
`oxanapob@bu.edu`

Abstract. We present the first two-round, two-party general function evaluation protocol that is secure against honest-but-curious adaptive corruption of both parties. In addition, the protocol is incoercible for one of the parties, and fully leakage tolerant. It requires a *global* (non-programmable) reference string and is based on one way functions and general-purpose indistinguishability obfuscation with sub-exponential security, as well as augmented non-committing encryption.

A Byzantine version of the protocol, obtained by applying the Canetti et al. [STOC 02] compiler, achieves UC security with comparable efficiency parameters, but is no longer incoercible.[1]

1 Introduction

Obtaining adaptive security, namely guaranteeing security against adversaries that decide who to corrupt in an adaptive way depending on their view of the computation so far, has been a major challenge in secure computation since its inception. Indeed, adaptive security provides a more realistic modeling of adversarial behavior and party infection in modern communication networks. Furthermore, when combined with an additional property called *corruption oblivious simulation,* adaptive security implies a strong variant of leakage tolerance [BCH12], namely resilience to side channel attacks on the participating computational devices.

Guaranteeing adaptive security turns out to be considerably more challenging than guaranteeing security in the static setting where the set of corrupted parties is fixed in advance. As in the static setting, the security guarantees become stronger when the adversary is allowed to corrupt more parties. Furthermore, while in the static case the situation where all the parties are corrupted is trivial,

---

[*] Research Supported by the NSF MACS Frontier project.

[**] Supported in addition by the Check Point Institute for Information Security, ISF grant 1523/14, and NSF Algorithmic Foundations grant 1218461.

[1] ©IACR 2015. This article is the final version submitted by the authors to the IACR and to Springer-Verlag on January 13. The version published by Springer-Verlag is available at the beginning of March 2015.

Y. Dodis and J.B. Nielsen (Eds.): TCC 2015, Part II, LNCS 9015, pp. 557–585, 2015.
© International Association for Cryptologic Research 2015

in the adaptive case protecting against adversaries that can eventually corrupt all parties is by far the hardest case. Note that withstanding corruption of all parties is crucial for guaranteeing meaningful security of a protocol within a larger system or context. Also, the transformation from adaptive security to leakage tolerance is most meaningful in this case (namely, leakage from all parties). In particular:

- The best round complexity of a fully adaptively secure protocol (namely a protocol that does not rely on secure erasure of information and that withstands adaptive corruption of all parties) is $\tilde{\Omega}(d)$, where d is the depth of the circuit being evaluated [BGW88, CFGN96, CLOS02]. (The works of [IPS08], [GS12] obtain constant number of rounds; however they cannot support corruption of all parties.) Furthermore, this is the best known round complexity even in the case of two party computation, even for the honest but curious setting, and even in the common reference string model.
- No fully leakage-tolerant (hence also no non-erasing oblivious simulation adaptively secure) general function evaluation protocol is known, with any number of rounds. Again, this holds even for honest-but-curious corruptions and even for two party protocols. (The protocol of [BDL14] obtains leakage tolerance in a setting with an initial, leakage free interactive set-up state.)

Our results. We present a *two-message*, two party secure function evaluation protocol that is secure against adaptive honest-but-curious corruption of all parties — thereby resolving a long standing open problem in the theory of secure computation. Furthermore, the protocol has non-erasing oblivious simulation, implying leakage tolerance. Security is based on subexponentially secure indistinguishability obfuscation for all circuits and one way functions, as well as augmened non-committing encryption as in [DN00, CLOS02].

The protocol requires a global, non-programmable reference string. Specifically, the string contains an obfuscated program to be run by parties. We call this mild version of the reference string model the *factory model*, since it is reminiscent of a setting where the obfuscated program is generated by a "trusted factory".

The protocol is also *incoercible* [CG96] for one of the parties. That is, it provides one of the parties with a mechanism to present "convincing evidence" that explains its outgoing messages as resulting from any arbitrary input value (that may be different than the input value actually used). This holds even when the "coercer" expects to see the full internal state of the party. That is, we show:

Theorem 1. *Assume existence of sub-exponentially secure indistinguishability obfuscators for all circuits and one way functions, as well as augmented non-committing encryption. Then there exists a two-message, two party protocol,*

in the factory model, for evaluating any function with UC security in the presence of adaptive, honest-but-curious corruption of both parties. Furthermore:
(a) The protocol is leakage tolerant as in [BCH12].
(b) The protocol is incoercible with respect to one of the parties.

In fact we show that the protocol satisfies a stronger variant of the [CG96] definition, that avoids a weakness in the original definition and is also universally composable. Furthermore, new definition may be of interest independently of the present protocol; in particular, it applies also to multi-party protocols and general (Byzantine) corruptions.

Compiling this protocol via the [GMW87, CLOS02] compiler, we obtain a constant-round, adaptively secure UC protocol for Byzantine adversaries in the standard CRS model. While the protocol remains leakage resilient, it is no longer incoercible.

The protocol and techniques. Before presenting the protocol, let us recall the definition of adaptive security. Security requires existence of a simulator that has access only to the trusted party for the function, and still emulates for the adversary (or, rather, the environment) an execution with the actual protocol. Since we are in the honest but curious model, we can assume without loss of generality that the adversary first waits to see the entire communication of the protocol to the end, and then corrupts all parties. The simulator should first create a simulated public transcript of the computation; then, when a party is corrupted and the simulator learns the input and the output of that party, the simulator should present the adversary with the appropriate random choices of the party that are consistent with the party's input and messages sent.

Our starting point is Yao's garbled circuit two party protocol, together with a two-message oblivious transfer. Recall that the first message in the protocol is the first OT message from the evaluator to the garbler. The second message, from the garbler to the evaluator, consists of the second OT message together with the garbled circuit. The evaluator then outputs the result of the computation. (If both parties wish to learn the output then they run another copy of the protocol in parallel, with reverse roles; or the evaluator can send the result to the garbler, but this adds one more round.)

When the OT is adaptively secure (as in, say, [CLOS02]) and the garbler's message is encrypted using non-commiting encryption, the protocol becomes adaptively secure with respect to the corruption of the evaluator. That is, the simulator can indeed create the transcript of the communication ahead of time (this is just ciphertexts of non-committing encryption) and when the evaluator is corrupted, provide the receiver message for the adaptively secure OT protocol. Note however that here the simulator has to commit to the garbled circuit, without knowing the garbler's input.

Now, simulating the corruption of the garbler gets stuck: Here the environment expects to see the internal randomness of the garbler, including the random choices used for the generation of the garbled circuit. This we do not know how to do efficiently. In fact, in some cases such valid opening simply does not exist.

Our approach to get around this apparently inherent difficulty is to provide the garbler with an *obfuscated* version of his program. That is, let the common reference string contain an obfuscated version of the garbler's program. The garbler will then run the obfuscated program on its input and random input and send the resulting message. The hope is that this will hide the internal randomness of the garbling, even when the environment sees the random input of the party.

However, this naive attempt does not work by itself, since the randomness for the protocol may well be correlated with the internal randomness that's not supposed to be leaked. We address this issue by applying a pseudorandom function to the random and real inputs, and using the result as randomness to the protocol. In addition, to make the simulation go through with only indistinguishability obfuscation we follow the lead of Sahai and Waters [SW14] and use puncturable PRFs and an "explain" algorithm that allows the simulator to generate randomness that "explains" any given outgoing message.

As simple as the protocol is, the proof of security is rather delicate. One subtle point that deserves highlighting is the treatment of adaptivity in the choice of inputs. We first prove security in a model where the inputs are "selective": the environment determines the inputs to the computation before it sees the reference string (namely the obfuscated programs). This is a rather weak security model. We then extend the analysis to the setting where the environment chooses the inputs adaptively. Here is where we use the sub-exponential security of the indistinguishability obfuscator: the analysis here requires as many hybrids as the number of potential inputs to the computation. This number can be exponential. We note, however, that since the parameters of the obfuscation can be chosen to be larger than the size of the inputs to the computation, this requires only *sub*-exponential security of the $i\mathcal{O}$ in use.

Finally we remark that the trust requirements from the reference string are relatively mild. First, it is non-programmable, in the sense that the simulator need not know any secret information related to the string. This means that the same instance of the reference string (namely, the same obfuscated program) can be used by multiple protocols and instances thereof without compromising security [CDPW07]. Second, static security holds even if the secrets associated with the reference string, namely the secrets of the obfuscation and the secret keys, are exposed.

Concurrent work. Concurrently to and independently from this work, two other works develop fully adaptively secure protocols using indistinguishability obfuscation. Both of these works appear in these proceedings [GP14, DKR14]. We give account of these works. Like ours, both works describe protocols for evaluating general functions, not only adaptively well formed ones as in [CLOS02]. Furthermore, all works obtain resilience against adaptive corruption for *all* parties. Finally, all works use the CRS model, where the CRS contains indistinguishability-obfuscated programs.

Dachman-Soled, Katz and Rao [DKR14] describe a general mechanism to transfrom programs into deniable ones, and use this mechanism to construct a four-round, multiparty, adaptively secure protocol against honest-but-curious corruptions. They then compile their protocol using the [CLOS02] compiler to handle Byzantine corruptions. Their analysis assumes only indistinguishability obfuscator and one way functions that are secure against polysize adversaries. Garg and Polychroniadou [GP14] directly describe a multi-party, two round, adaptively secure protocol against Byzantine corruptions. Similarly to this work, their analysis assumes sub-exponentially secure indistinguishability obfuscation and one way functions. Both protocols need a *programmable* (i.e., non-global) CRS, and neither protocol is incoercible.

Organization. Section 2 sketches the models of computation and recalls the main results of this work. Section 3 provides an overview of the construction. Section 4 provides a detailed presentation and analysis of the main protocol.

2   The Models of Computation

We consider the standard UC model of computation (as in [Can01]) with adaptive, honest-but-curious party corruptions. The parties and the environment have access to a *global, public* common reference string functionality. That is, the functionality first draws the reference string from a predefined distribution; next, all parties, including the adversary and the environment, obtain that string.

In our protocol the reference string is a description of programs run by parties; these programs are obfuscated and contain secret keys which shouldn't be known to the parties. We refer to such a global reference string model as "the factory model", since it is intended to represent a situation where all parties obtain the program from a trusted "factory".

Leakage tolerance. We will show that our protocol is leakage-tolerant. The leakage tolerance model we consider is the one in [BCH12], which is aimed as capturing protocols that are tolerant to arbitrary amount of leakage, and where the security loss grows gradually with the amount of leakage. More specifically, in that model a protocol π computes a function f *in a leakage tolerant way* if no adversarial environment can tell whether it is interacting with the parties running π, while obtaining some ℓ-bit leakage function of the individual internal states of the participants, or alternatively with a simulator and an ideal process for evaluating f, in which the simulator obtains some arbitrary ℓ-bit function of each of the inputs of the parties.

It is shown there that if a protocol is shown to be adaptively secure with a special type of simulator, called *corruption oblivious simulator* (defined below), then the protocol is leakage tolerant.

A simulator is *corruption oblivious* if the information it gathers upon corruption of a party, namely the secret input (and potentially also the secret output) of that party, is used *only* to generate a simulated view of the local state of that party. This information is not used anywhere else in the simulation. Formally, the simulator creates a special subroutine for simulating the internal state of that party. The newly learned input of the corrupted party does not leave the confines of this subroutine. It is shown in [BCH12] that if a protocol is adaptively secure with a corruption oblivious simulator then it is also leakage tolerant.

Incoercibility. Incoercibility aims to protect the protocol participants from external authoritative (or otherwise coercive) entities that try to entice a party to reveal its state voluntarily. The idea is to provide parties with a "faking" algorithm that takes any desired fake value of the secret input, and exhibits "fake randomness" that is consistent with both the newly decided fake value and all the past messages sent by the party so far. Incoercible computation was defined in [CG96], where a generic construction from any deniable encryption scheme [CDNO97, SW14] is given. However, the construction there has a large number of rounds and works only when strictly less than half of the parties are either coerced or corrupted.

We revisit the definition of coercion-free computation, providing a new definition that is significantly stronger than the one in [CG96]. Specifically, the security guarantees provided by the new definition are preserved under universal composition. The definition also overcomes another weakness in the [CG96], as explained below. The definition here fleshes out ideas from [Can01, P. 59].

Informally, the definition captures incoercibility by asking that the protocol in question emulates an ideal functionality that employs the following "ideally incoercible" corruption process. Whenever the ideal functionality is asked by the ideal-model adversary to provide the internal information of some participant P in the protocol, the ideal functionality first asks the environment (representing the entity that invoked party P to participate in the protocol) whether to reveal the real input that P contributed to the computation, or alternatively whether to report some fake input. If the environment instructs to reveal the real input, then the functionality returns the real input of P to the adversary. If the environment provides a fake value x, then the functionality returns x to the adversary. Crucially, the adversary does not learn whether the value provided was fake or real.[2]

Now, consider a protocol π that realizes such an ideal functionality $\mathcal{F}$, and consider a party P that runs π. Now, upon receiving a corruption message from the adversary, π must instruct P to first ask the environment (which, again, represents the entity that invoked party P to participate in the protocol) whether to report the real internal state or to provide a fake one. If the response is to reveal the real input, then we require that P reveals its true internal state. If

---

[2] We remark that the definition in [CG96] reveals to the ideal-model adversary whether the value provided by the functionality is real or fake. This renders that definition weak. For instance, consider a protocol with a faking algorithm that outputs the empty string as "fake randomness". While this protocol should clearly not be considered as "incoercible", it could be accepted by a simulation based definition — as long as the simulator knows which parties are coerced and which ones are corrupted, since there is no problem for the simualtor to output an empty string upon coercion request.

the response is fake input value x, then P follows the instructions of π for such a case.

We argue that if π emulates $\mathcal{F}$ in the usual (UC) sense then this means that π is incoercible. Indeed, $\mathcal{F}$ provides "ideal incoercibility" in the sense that there the ideal adversary learns nothing about whether a party revealed the real or the fake input - beyond what is revealed by the legitimate outputs of the corrupted parties. Thus, the same must hold also with respect to the real adversary that interacts with π - or else the environment could tell the difference between the two interactions. Note however that this arument hinges on two facts: (a) in the real world the corrupted party must reveal its real input upon corruption, when instructed so by the environment, and (b) that the ideal adversary is not being told whether the input value it received upon corruption is real or fake.

More formally, we define incoercible protocols in two steps. First we define what it means for an ideal functionality to be incoercible. Next, we define what it means for a protocol to be *corruption-compliant*. A protocol will be incoercible if it is corruption compliant and in addition it UC-emulates an incoercible ideal functionality.

We consider ideal functionalities $\mathcal{F}$ where each input to $\mathcal{F}$ is associated with two party identities: the first, P represents the identity of the protocol participant that holds this input, and the second, C_P is the identity of the "calling party", namely the party that provided the input value(s) to P, and will obtain the output value(s) from the computation. Such an ideal functionality $\mathcal{F}$ is incoercible if it behaves as follows upon receiving a **corrupt** P message from the adversary. $\mathcal{F}$ first outputs to C_P a **corrupted** value. Next, if C_P responds with **do not fake** then $\mathcal{F}$ returns to the adversary all the inputs received from C_P and all the output passed to C_P in this interaction so far. If C_P responds with **fake to** x then $\mathcal{F}$ interprets x as a list of inputs and outputs and hands this list to the adversary instead of the real one.

We consider protocols π that attempt to UC-emulate an incoercible ideal functionality. We say that such a protocol π is *corruption compliant* if, after having received a **corrupt** P message from the adversary, followed by a **do not fake** input from its caller, C_P, P forwards its entire internal state to the adversary. (Note that we do not restrict what π instructs to do in case that C_P responds by **fake to** x. Indeed, this is the essence of the "faking procedure" that should be specified in π.)

In general, a protocol π is incoercible if it is corruption-compliant and it UC-emulates an incoercible ideal functionality. We also provide a definition of incoercible distributed function evaluation. Let $f : (\{0,1\}^*)^n \to (\{0,1\}^*)^n$ be an n-party function, and let $\mathcal{F}_f$ be the incoercible ideal functionality that computes f, say with respect to some fixed set of party identities $P_1, ..., P_n$. That is, upon receiving inputs from the calling parties of $P_1, ..., P_n$, $\mathcal{F}_f$ evaluates f on these inputs and provides the caller of each P_i with its corresponding output value $f(x_1, ..., x_n)_i$. Party corruptions are handled in an incoercible way as described above.

Definition 1. *Protocol π evaluates an n-party function f : $(\{0,1\}^*)^n \rightarrow (\{0,1\}^*)^n$ if it is corruption-compliant and it UC-emulates $\mathcal{F}_f$.*

Note that the above definition of incoercibility did not specify whether the corruptions are honest-but-curious or Byzantine. Indeed, this definition is meaningful in both cases.

3   Protocol Overview

Let's first recall how the original Yao protocol looks like. Let's say parties P_0 and P_1 have inputs x_0 and x_1 and they want to evaluate $y = C(x_0, x_1)$ for some circuit C. P_0 generates a garbled circuit: that is, for every wire of C P_0 creates two random labels l_0, l_1, and a garbled circuit consists of 4 encryptions of output label under input labels as keys, and the result table, which lists 0 and 1 labels for output gates.

P_0 sends to P_1 the garbled circuit together with the labels corresponding to P_0's input. Then for every P_0's input bit P_0 and P_1 run OT protocol, after which P_1 learns the keys corresponding to his input. At this point P_1 has all information he needs to evaluate the circuit: it has all input labels, and it keeps evaluating the circuit gate by gate, until finally it learns output labels. Then it uses result table to learn the output.

As shown in [LP09], the original Yao protocol is statically secure, given augmented non-committing encryption [DN00, CLOS02]. In particular, when P_1 is corrupted, Simulator learns x_1 and y and shows a fake garbled circuit which always evaluates to y and is indistinguishable from the real garbled circuit. (It cannot show the real garbled circuit since it doesn't know x_0.) Also the simulator shows labels corresponding to P_0's and P_1's inputs. Here it is crucial that an adversary sees only one label per each input bit and therefore cannot distinguish between a fake circuit and a real one.

The same simulation works in adaptive case with erasures: P_0 should erase its internal state before sending the second message. However, in the adaptive case without erasures this simulation fails: an adversary could corrupt P_0 after corrupting P_1 and learning a fake garbled circuit. For every P_1's input bit, a simulator has to show *both* labels since these labels were P_0's input in OT protocol. Now the adversary sees one label for each one of P_0's input bits and both labels for P_1's input bit. This allows the adversary to detect that the garbled circuit is not valid.

Indeed, consider a circuit that consists of just one AND gate. The simulator corrupts P_1 and learns its input $x_1 = 0$ and $y = 0$. At this point the simulator still doesn't know P_0's input, but it has to show the garbled circuit, therefore it shows fake circuit where all four ciphertexts encrypt the same key l_0, and it shows the result table where l_0 is decrypted to 0. Now the simulator corrupts P_0 and learns $x_0 = 1$. It has to show keys corresponding to both $x_1 = 0$ and $x_1 = 1$. This means that the adversary knows the keys for $x_0 = 1$, $x_1 = 0$ and $x_1 = 1$ and can evaluate the circuit on inputs $(1, 1)$ and $(1, 0)$. Since the circuit is just an AND gate, the result should be different. However, since our garbled circuit contained the same key in all four encryptions, an adversary trying to evaluate the circuit will get 0 in both cases and will detect cheating.

The problem is that an adversary learns too much at the moment of corruption: learning both keys for P_1's input allows him to evaluate the circuit on many inputs and to check that the circuit is a fake. To avoid this problem, we change the protocol such that P_0 himself doesn't know the keys for P_1's input. In order to achieve this, we "glue together" the garbled circuit generation, the input labels generation and the OT into one program P which outputs the next message function for the Yao protocol. This program will be obfuscated by the factory. Now, P_0 will run this program on his input and local randomness and send its output to P_1.

Naively one may hope that, since the program is obfuscated, P_0 himself doesn't know more than just inputs he used and output it sent to P_1 (in particular, it doesn't know the keys for P_1's input). However, this is not enough: it might be the case that the input itself reveals the keys (say, if the keys are just set to be some substring of the random input). To deal with this problem, we don't use the random input directy in the protocol. Instead, we first apply a pseudorandom function to the input and random input, and then use the output of the pseudorandom function as the random input to the protocol.

The next set of challenges deals with making the above plan to work with an obfuscation mechanism that only guarantees indistinguishability obfuscation. Here we follow the lead of Sahai and Waters [SW14] and use similar constructs and techniques as there. Specifically, we use the technique of embedding "hidden triggers" in the random input to the program P. If the program recognizes a hidden trigger then it just outputs the value encrypted in that trigger. Else, the program used the randomness as in the Yao protocol. We publish P together with a "faking" algorithm Explain that allows anyone to generate hidden triggers of one's choice. This addition has a twofold effect: For one it provides for incoercibility for the garbler. In addition it also simplifies the proof of security.

Throughout, and following [SW14], we employ *constrained,* or *puncturable* pseudorandom functions [GGM86, BGI13, BW13], which enables applying indistinguishability obfuscation to pseudorandom function in a meaningful way.

We describe and analyze the scheme in a simple setting where the parties have secure communication channels, and with only honest but curious corruptions. Once we have such a protocol, we can implement secure channels using non-committing encryption. We can also deal with Byzantine corruptions by forcing semi-honest behavior.

We also assume without loss of generality that only the evaluator learns the output. If both parties need to obtain outputs from the computation then they can run the same protocol twice, on the same inputs but with reverse roles. (Alternatively, at the cost of adding a message to the protocol, the evaluator can send the function value to the garbler.)

Implementing secure channels. As we will see later, only the second message in our protocol should be sent over a secure channel. This means that P_1 can send EK_{NCE} in the first message, and the protocol still remains two-round after implementing secure channels.

Corruption obliviousness and leakage resilience. The naive protocol, described above, does not naturally lend to corruption-oblivious simulation. Indeed, to simulate the corruption of the garbler, the simulator needs to come up with a second message, namely a garbled circuit, that outputs the correct output of the computation. This needs to be done without knowing the input or output of the evaluator, and only using the input of the garbler. Furthermore, when the evaluator is corrupted, the simulator needs to come up with the *same garbled circuit*, without knowing the input of the garbler. This is not known to be possible in general. We get around this issue by making a simple modification to the protocol: Instead of evaluating $f(x_0, x_1)$, the parties will use the above protocol to evaluate $f'(x_0, (x_1, z)) = f(x_0, x_1) \oplus z$. The evaluator, P_1, will choose z at random, and after obtaining the output value y, it will set its output to be $y \oplus z$.

With this modification in place, the simulator can set the output of the garbled circuit to be a random value fixed in advance and then deal with the corruption of the parties in an oblivious way.

Incoercibility. We provide incoercibility for the garbler. This is done in a straighforward way: Since the explain procedure is public, a coerced garbler can demonstrate random input that explains any input value of its choice, in the same way as in [SW14].

Handling Byzantine corruptions. Here we use the generic transformation of [CLOS02] (based on [GMW87]) that transforms a protocol that is secure against adaptive honest but curious corruptions into a protocol that is secure against adaptive Byzantine corruptions.

4   Detailed Description and Analysis

Preliminaries.   In our construction we use the following primitives. The reader is referred to the papers cited for detailed definitions.

1. Indistinguishability obfuscation $i\mathcal{O}$ for polynomial-size circuits, as defined, constructed and used in [BGI+01, GR14, GGH+13, SW14].
2. Augmented non-committing encryption scheme Enc ([DN00, CLOS02]). We denote its generation, oblivious generation and inverting algorithms as $Enc.Gen$, $Enc.oGen$ and $Enc.Inv$.
3. Puncturable PRFs which are additionally extracting or injective [BGI13, BW13, SW14].
4. The garbled circuit generation algorithm Gen together with an algorithm SimGen for generating fake garbled circuit from [LP09]. These programs use a special encryption scheme which they call a public key encryption with elusive efficiently verifiable range.

Deterministic single-party-output functionalities. First, we recall that it suffices to be able to compute deterministic functionalities: indeed, there exists a standard reduction of any randomized functionality to a deterministic one, given by $f_{det}((x_0, r_0), (x_1, r_1)) = f_{rand}(x_0, x_1; r_0 \oplus r_1)$. Second, it is enough to compute functionalities where only one party gets the output (and the other party gets nothing): parties can run in parallel two instances of the protocol with the same input, where in the first execution only the first party generates output and in the second execution only the second party generates output.

In our protocol P_0 is the garbler and P_1 is the evaluator for the Yao protocol. The natural thing to do would be to create a garbled circuit for the functionality they want to compute $(-; f(x_0, x_1))$. However, in this case the simulation is not corruption-oblivious.[3] We therefore slightly modify a protocol: P_1 first generates random z, and P_0 generates a garbled circuit for the function $f'(x_0, (x_1, z)) = f(x_0, x_1) \oplus z$. As we'll see, this will suffice for making the simulation corruption-oblivious.

Oblivious transfer. We use the following one out of two OT protocol, based on [EGL85]: assume P_0 has k_0, k_1 and P_1 has a bit b; we want P_1 to learn k_b. First, P_1 generates keys (EK_b, DK_b) and EK_{1-b} without corresponding decryption key (this encryption scheme, in addition to normal key generation, should have oblivious key generation algorithm which outputs encryption keys without corresponding decryption keys, in such a way that this encryption keys are indistinguishable from normal encryption keys. For this we use augmented non-committing encryption). P_1 sends EK_0, EK_1 to P_0. P_0 sends back encryptions $c_0 = Enc(EK_0; k_0)$ and $c_1 = Enc(EK_1; k_1)$. Since P_1 has DK_b, he can decrypt $k_b = Dec(DK_b; c_b)$. However, since there is no DK_{1-b} generated, the second value k_{1-b} remains unknown to P_1. Following [CLOS02], we make the OT adaptively secure by using non-committing encryption for the encryption scheme.

With this implementation of OT, the Yao protocol consists of the following two messages:

1. First, P_1 generates two sets of encryption keys $\overline{PK_0}, \overline{PK_1}$ and one set of decryption keys $\overline{SK_{x_1}}$ (such that for every input bit x_1^i P_1 only knows $DK_{x_i}^i$). P_1 sends $\overline{PK_0}, \overline{PK_1}$ to P_0.
2. P_0 generates a garbled circuit GC and sends to P_1 GC, keys for P_0's input bits, and keys for all possible P_1's input bits encrypted under $\overline{PK_0}, \overline{PK_1}$ (we will call this a *Yao message*). P_1 decrypts the keys corresponding to its input, and, since it has GC and all input labels, it evaluates the circuit gate by gate.

Protocol description. We have parties P_0, P_1 with inputs x_0, x_1 respectively. The protocol for allowing P_1 to learn the value $f(x_0, x_1)$ for some function f is described

---

[3] Indeed, for the simulation to be corruption-oblivious, the subroutine for generating P_1's internal state should be able to create a fake garbled circuit without knowing x_0. At the same time, the subroutine for creating P_0 internal state should be able to create (the same) fake garbled circuit without knowing the output y. It is not clear how to do that for the above "natural" garbling method.

in Figure 1. The referece string consists of programs P and Explain, described in Figures 2 and 3. The circuit C that prorgam P evaluates will be the circuit that computes the function $f'(x_0, (x_1, z)) = f(x_0, x_1) \oplus z$. (The value z will be chosen by P_1 at random as part of the protocol.)

The protocol consists of two rounds. In round one, P_1 (the evaluator) chooses randomness s and z and sets $x'_1 = (x_1, z)$ to be its new input. It samples secret and public keys for oblivious transfer using s (public keys which do not correspond to P_1's input are sampled obliviously). P_1 sends all public keys to P_0. In the second round P_0 chooses its randomness r and runs a program P on its input x_0, randomness r and a set of public keys from P_1. The program P internally generates new randomness u and runs the underlying subroutine Gen to generate a Yao message, which becomes the program output. P_0 sends this message to P_1. P_1 gets the labels for x_0, decrypts the labels for x_1 and evaluates the circuit, obtaining $f(x_0, x_1) \oplus z$. Then P_1 xor's the result with z and gets the output $f(x_0, x_1)$.

The program Explain is not used in the protocol directly. However, it is used in the case when parties want to deny their inputs, as well as in the proof.

The Protocol:

1. P_1 chooses random z and sets $x'_1 \leftarrow (x_1, z)$. Then it chooses random s and generates $\overline{PK_{x'_1}}, \overline{SK_{x'_1}} \leftarrow Enc.Gen(s[0])$ and $PK_{1-x'_1} \leftarrow Enc.oGen(s[1])$. It sets $\alpha^* \leftarrow \overline{PK_0}, \overline{PK_1}$ and sends α^* to P_0.
2. P_0 chooses random r^*, runs $\beta^* \leftarrow P(x_0, \alpha^*; r^*)$ and sends β^*
3. P_1 evaluates the garbled circuit taken from β^*, using the labels and output table from β^*, and outputs the result xor'ed with z.

Fig. 1. Protocol description

We show:

Theorem 2. *Let:*

- *SEnc be CPA-secure symmetric key encryption scheme with an elusive efficiently verifiable range ([LP09])*
- *Enc be an augmented non-committing encryption scheme*
- *$E = \{E_{k_E}\}$ be an extracting puncturable PRF family*
- *$I = \{I_{k_I}\}$ be an injective puncturable PRF family*
- *$F = \{F_k\}$ be a puncturable PRF family*
- *PRG be an input-doubling PRG*
- *iO be indistinguishability obfuscator*

then the protocol is adaptively secure with oblivious simulation in the factory model in the presence of semi-honest adversaries given secure channels.

Program P

inputs: P_0's input x, P_1's 1-round message α, randomness $r = r[1]r[2]$
$P(x, \alpha; r)$:

1. check if r has encoded value inside:
   (a) $M' \leftarrow F_k(r[2]) \oplus r[1]$; if $I_{k_I}(M') \neq r[2]$ then goto 2;
   (b) parse M' as $\beta', x', \alpha', \hat{\rho}'$. If $(x', \alpha') \neq (x, \alpha)$ then goto 2;
   (c) output β'
2. else run Gen:
   (a) $u \leftarrow E_{k_E}(x, \alpha, r)$
   (b) output $Gen(x, \alpha; u)$

Program Gen.

Constants: circuit C with m wires and s output wires; let's assume that first $2n$ wires are input wires and last s wires are output wires
Input: P_0's input x_0; P_1's two sets of public keys $\overline{PK_0}, \overline{PK_1}$;
randomness $u = u_1 u_2 u_3 u_4$
$Gen(x_0, \overline{PK}; u)$:

1. Create labels for wires: $(k_1^0, k_1^1), \ldots, (k_m^0, k_m^1) \leftarrow u_1$
2. Create encryptions of labels:
   (a) Partition u_2 into $u_{21}, \ldots, u_{2m}$, and each u_{2t} into $u_{2t1}, \ldots, u_{2t4}$
   (b) Partition u_3 into $u_{31}, \ldots, u_{3m}$, and each u_{3t} into $u_{3t1}, \ldots, u_{3t4}$
   (c) For every gate t in C create 4 encryptions:
       − if t is an AND gate:
       $$GC_t[0,0] \leftarrow SEnc_{k_i^0}(SEnc_{k_j^0}(k_l^0; u_{2t1}); u_{3t1})$$
       $$GC_t[0,1] \leftarrow SEnc_{k_i^0}(SEnc_{k_j^1}(k_l^0; u_{2t2}); u_{3t2})$$
       $$GC_t[1,0] \leftarrow SEnc_{k_i^1}(SEnc_{k_j^0}(k_l^0; u_{2t3}); u_{3t3})$$
       $$GC_t[1,1] \leftarrow SEnc_{k_i^1}(SEnc_{k_j^1}(k_l^1; u_{2t4}); u_{3t4})$$

       − if t is an OR gate:
       $$GC_t[0,0] \leftarrow SEnc_{k_i^0}(SEnc_{k_j^0}(k_l^0; u_{2t1}); u_{3t1})$$
       $$GC_t[0,1] \leftarrow SEnc_{k_i^0}(SEnc_{k_j^1}(k_l^1; u_{2t2}); u_{3t2})$$
       $$GC_t[1,0] \leftarrow SEnc_{k_i^1}(SEnc_{k_j^0}(k_l^1; u_{2t3}); u_{3t3})$$
       $$GC_t[1,1] \leftarrow SEnc_{k_i^1}(SEnc_{k_j^1}(k_l^1; u_{2t4}); u_{3t4})$$

   (d) shuffle $GC_t[0,0], GC_t[1,0], GC_t[0,1], GC_t[1,1]$
3. Create encryptions of labels for P_1's input:
   (a) Partition u_4 into $u_{401}, \ldots, u_{40n}, u_{411}, \ldots, u_{41n}$
   (b) For all $i = 1, \ldots, n$ $c_i^0 \leftarrow Enc_{PK_0^i}(k_{n+i}^0; u_{40i}), c_i^1 \leftarrow Enc_{PK_1^i}(k_{n+i}^1; u_{41i})$
4. output:
   (a) $GC_i[0,0], GC_i[0,1], GC_i[1,0], GC_i[1,1]$ for $i = 1..m$ (garbled circuit)
   (b) $(0 : k_{m-s+1}^0; 1 : k_{m-s+1}^1), \ldots, (0 : k_m^0; 1 : k_m^1)$ (the result table)
   (c) $k_1^{x_0^1}, \ldots, k_n^{x_0^n}$ (labels for P_0's input)
   (d) $(c_1^0, c_1^1) \ldots (c_n^0, c_n^1)$ (encrypted labels for P_1's input)

Fig. 2. Program P is used by P_0 to generated the second protocol message. It calls Gen as a subroutine; Gen is a program which outputs a Yao message: that is, a garbled circuit, labels for P_0's input and encrypted labels for all possible P_1's inputs.

Program Explain

inputs: message m which should be encoded; randomness ρ
$P(m; \rho)$:

1. $M \leftarrow m, prg(\rho)$
2. $r[2] \leftarrow I_{k_I}(M), \quad r[1] \leftarrow F_k(r[2]) \oplus M$
3. output $r = r[1]r[2]$

Fig. 3. Program Explain

The choice of parameters. Since we use different types of PRFs (in particular, extracting PRFs and injective PRFs) in the construction, we must ensure that the lengths of all values fit the requirements for these PRFs. Indeed, as shown in [SW14], there exist:

- injective puncturable PRFs which map $n(\lambda)$ bits to $m(\lambda)$ bits where injectivity holds with probability $1 - 2^{-e(\lambda)}$ (over the choice of a key), as long as $m(\lambda) \geq 2n(\lambda) + e(\lambda)$;
- extracting puncturable PRFs which map $n(\lambda)$ bits to $m(\lambda)$ bits for distribution X with min-entropy $k(\lambda)$ with statistical distance between $(k, F_k(X))$ and (k, U_m) at most $2^{-e(\lambda)}$, as long as $n(\lambda) \geq k(\lambda) \geq m(\lambda) + 2e(\lambda) + 2$.

Let's recall how we use these PRFs in the computation. Let's denote the lengths of a Yao message β and randomness used to create it u as $|\beta|$ and $|u|$; also we denote the length of M (the hidden value prepared by a simulator and encoded inside randomness) as $|M|$. All these lengths are polynomial in security parameter as well as a circuit size and inputs length. We have to choose randomness length to guarantee that both injective and extracting PRFs exist. Recall that randomness r (denoted as er in simulated case) consists of two parts $r[1]$ and $r[2]$. Note that the way $er[1]$, the first part of randomness, is generated $(er[1] \leftarrow F_k(er_2) \oplus M)$ implies that its length is exactly $|M|$.

1. I_{k_I} should be an injective PRF with negligible failure. It takes as input M and outputs $er[2]$. Thus, it should be the case that $|er[2]| \geq 2|M| + \lambda$.
2. E_{k_E} should be an extracting PRF with negligible distance. It takes as input $(x_0, \overline{PK}, r[1]r[2])$ and outputs u. We are going to use extracting property when $r = r[1]r[2]$ is chosen at random, and min-entropy of input is at least $|r| = |r[1]| + |r[2]|$. Thus, it should be the case that $|x_0| + |\overline{PK}| + |r[1]| + |r[2]| \geq |r[1]| + |r[2]| \geq |u| + 2\lambda + 2$.

Once a security parameter and a circuit are fixed, all values above are also fixed except $|r[2]|$. Note that by choosing $|r[2]|$ large enough (but still polynomial in the security parameter), we can satisfy both inequalities.

Proof. The outline of the proof is the following. First, we give a description of our simulator. Then we prove that no environment can distinguish between a real execution and a simulation. We do this in two steps. In step one we deal with the case of non-adaptively chosen inputs; that is, the environment first chooses parties' inputs and only then sees a CRS. In order to show indistinguishability in non-adaptive case, we consider an intermidiate middle hybrid where all protocol messages are generated as in a real execution, but the randomness is explained. In two lemmas we prove that this middle hybrid is indistinguishable from both real execution and simulation. In step two we consider the case of adaptive inputs choice, thus proving the theorem statement.

Simplifying assumptions. In our honest-but-curious setting we can assume that corruptions happen after the protocol execution and that both parties are corrupted. Since our simulator, as we see later, is corruption-oblivious (information learned in one party corruption is not used in the other party corruption), we don't need to think about different order of corrupting parties. Also we assume secure channels, therefore our simulator has to show the protocol transcript only after one of the parties is corrupted.

In our proofs of lemmas instead of having an interactive game with the adversary we just run an experiment and show to the adversary the resulting distribution, asking it to guess which hybrid it sees. Indeed, by itself the security definition is interactive: an environment first sees a CRS and then outputs inputs; after this, it sees protocol messages. Then it can send corruption requests and get back parties' internal states. Given this information, the adversary chooses which hybrid it sees. However, in the case of non-adaptively chosen inputs, we can use a non-interactive security definition: the inputs are fixed in advance, therefore we can send a CRS later with other values the adversary should see. Next, we assumed that all parties are corrupted, and therefore the adversary doesn't need to send corruption requests; the simulator will send it all parties' internal states itself. Therefore, instead of playing an interactive game with the adversary, in our security definitions the simulator generates all protocol information (protocol messages, parties' internal states) and sends it to the adversary, who should distinguish between hybrids.

Description of the simulator. Our simulator is described in Figure 4. It gets a CRS, generates randomness needed (s_{PKE} to create P_1's keys for encryption scheme, s_{GC} to create a fake garbled circuit, and s_y, a random value which is the result of $z \oplus y$ in a real execution), and sets its state to be $s = (CRS, s_{PKE}, s_{GC}, s_y)$.

Since we assume secure channels, the simulator doesn't need to show a transcript before corruptions. Upon corruption of a party P_i, the simulator calls its subroutine $Sim_{P_i}(CRS, s_{PKE}, s_{GC}, s_y)$ to simulate P_i's internal state. Each subroutine has to show randomness used by a party and the communication it sees. Sim_{P_i} first generates secret and public keys for P_1 and sets α^* to be P_1's public keys (note that since all three programs (Sim, Sim_{P_0} and Sim_{P_1}) use the same state to generate values, they get the same result - public keys and

garbled circuit). Then it generates a fake garbled circuit and encryptions for
OT $\beta^* \leftarrow SimGen(s_y, \alpha^*; s_{GC})$. The next step depends on the party. A simulator for P_0 computes explained randomness $er^* \leftarrow Explain((\beta^*; x_0, \overline{PK}; \rho^*)$ for
randomly chosen ρ^* and shows er^* (internal state) and α^* (communication). A
simulator for P_1 sets its randomness z to be consistent with the garbled circuit
output and the protocol output (that is, $z = y \oplus s_y$) and then, using an in-
vertion algorithm, creates randomness es^*, which produces obliviously sampled
keys $\overline{PK_{1-x_1}}$. The simulator shows es^* and z as P_1's internal state and β^* as
the communication seen.

Note that to simulate a party during corruption, the simulator doesn't use
internal information of the other party; only this party's input/output is used,
together with randomness s which acts as a state of the simulator. Therefore
this simulator is corruption oblivious.

The simulation:

1. Obtain the public programs $CRS = P, Explain$
2. Choose randomness for simulation (s_{PKE}, s_{GC}, s_y). Set the state to be $s = (CRS, s_{PKE}, s_{GC}, s_y)$
3. upon corruption of P_0: output $Sim_{P_0}(s)$
4. upon corruption of P_1: output $Sim_{P_1}(s)$

$Sim_{P_0}(CRS, s_{PKE}, s_{GC}, s_y)$

1. learn x_0
2. generate $\overline{PK_0}, \overline{SK_0}, \overline{PK_1}, \overline{SK_1} \leftarrow Enc.Gen(s_{PKE})$; set $\alpha^* \leftarrow \overline{PK_0}, \overline{PK_1}$
3. set $\beta^* \leftarrow SimGen(s_y, \overline{PK}; s_{GC})$
4. choose random ρ^* and set $er^* \leftarrow Explain(\beta^*; x_0, \overline{PK}; \rho^*)$
5. output (er^*, α^*)

$Sim_{P_1}(CRS, s_{PKE}, s_{GC}, s_y)$

1. learn x_1, y
2. generate $\overline{PK_0}, \overline{SK_0}, \overline{PK_1}, \overline{SK_1} \leftarrow Enc.Gen(s_{PKE})$
3. set $\beta^* \leftarrow SimGen(s_y, \overline{PK}; s_{GC})$
4. set $z \leftarrow s_y \oplus y$, $x_1' \leftarrow (x_1, z)$
5. set $es^* \leftarrow Enc.Inv(s, x_1')$
6. output $(es^*, z; \beta^*)$

Fig. 4. Simulation

Step one - non-adaptive inputs case.   In the following two lemmas, we prove
that real and simulated experiments are indistinguishable. To achieve this we
consider a middle hybrid where all protocol messages are generated honestly
like in a real execution, but the randomness shown to the adversary is obtained
using Explain algorithm. In the first lemma we show that this middle hybrid is
indistinguishable from the simulation; indistinguishability between the middle

---

Program SimGen

Constants: circuit C with m wires and s output wires; let's assume that first $2n$ wires are input wires and last s wires are output wires

Input: the result of the computation y; P_1's two sets of public keys $\overline{PK_0}, \overline{PK_1}$; randomness $u = u_1 u_2 u_3 u_4$

$Gen(y, \overline{PK}; u)$:

1. Create labels for wires: $(k_1^0, k_1^1), \ldots, (k_m^0, k_m^1) \leftarrow u_1$
2. Create encryptions of labels:
   (a) Partition u_2 into $u_{21}, \ldots, u_{2m}$, and each u_{2t} into $u_{2t1}, \ldots, u_{2t4}$
   (b) Partition u_3 into $u_{31}, \ldots, u_{3m}$, and each u_{3t} into $u_{3t1}, \ldots, u_{3t4}$
   (c) For every gate t in C create 4 encryptions (all 4 encryptions encrypt the same label):
   $$GC_t[0,0] \leftarrow SEnc_{k_i^0}(SEnc_{k_j^0}(k_l^0; u_{2t1}); u_{3t1})$$
   $$GC_t[0,1] \leftarrow SEnc_{k_i^0}(SEnc_{k_j^1}(k_l^0; u_{2t2}); u_{3t2})$$
   $$GC_t[1,0] \leftarrow SEnc_{k_i^1}(SEnc_{k_j^0}(k_l^0; u_{2t3}); u_{3t3})$$
   $$GC_t[1,1] \leftarrow SEnc_{k_i^1}(SEnc_{k_j^1}(k_l^0; u_{2t4}); u_{3t4})$$

   (d) shuffle $GC_t[0,0], GC_t[1,0], GC_t[0,1], GC_t[1,1]$
3. Create encryptions of labels for P_1's input:
   (a) Partition u_4 into $u_{401}, \ldots, u_{40n}, u_{411}, \ldots, u_{41n}$
   (b) For all $i = 1, \ldots, n$ $c_i^0 \leftarrow Enc_{PK_0^i}(k_{n+i}^0; u_{40i}), c_i^1 \leftarrow Enc_{PK_1^i}(k_{n+i}^1; u_{41i})$
4. output:
   (a) $GC_i[0,0], GC_i[0,1], GC_i[1,0], GC_i[1,1]$ for $i = 1..m$ (garbled circuit)
   (b) $(y_1 : k_{m-s+1}^0; 1 - y_1 : k_{m-s+1}^1), \ldots, (y_s : k_m^0; 1 - y_s : k_m^1)$ (the result table)
   (c) $k_1^0, \ldots, k_n^0$ (labels for P_0's input)
   (d) $(c_1^0, c_1^1) \ldots, (c_n^0, c_n^1)$ (encrypted labels for P_1's input)

Fig. 5. Program SimGen, used by a simulator to create a fake garbled circuit

hybrid and a real execution is shown in lemma 2. In both proofs we first give an overview of hybrids, and then present a detailed description with reductions.

Our notation. To denote the first and the second part of randomness, we write $r[1]$ and $r[2]$. By $\overline{PK}$ we denote a set of public keys for each possible input bit of P_1's input; $\overline{PK_0}$ and $\overline{PK_1}$ mean sets of public keys for input bits 0 and input bit 1. By PK_{x_1} we mean the set of public keys corresponding to P_1's input, that is, $PK_{x_1} = (PK_{x_1^1}^1, \ldots, PK_{x_1^n}^n)$. By PK_{1-x_1} we mean the opposite set of public keys.

We mark the values obtained in the experiment with a star to distinguish these values from variables in programs. We denote the first round message (P_1's public keys) as α^* and the second round message (a garbled circuit, an output table, labels for P_0's input, encrypted labels for all possible P_1's inputs) as β^*.

Lemma 1. *The results of the following two experiments are indistinguishable:*

Experiment Simulation:

1. choose randomness $s_{PKE}, s_{CRS}, s_{GC}, s_y$. Set $z = y \oplus s_y$. Set $x'_1 \leftarrow (x_1, z)$
2. generate a CRS: prf keys k_E, k_I, k, internal keys for Gen, and choose randomness for obfuscation x_P, x_{Expl} using s_{CRS}. Create obfuscated programs $P \leftarrow O(P_{k_E, k_I, k}; Gen; x_P), Explain \leftarrow O(Explain_{k_I, k}; x_{Expl})$.
3. sample P_0's keys $\overline{PK_0}, \overline{PK_1}, \overline{SK_0}, \overline{SK_1} \leftarrow PKE.Gen(s_{PKE})$. Set $\alpha^* \leftarrow \overline{PK_0}, \overline{PK_1}$
4. run $\beta^* \leftarrow SimGen(s_y, \alpha^*; s_{GC})$
5. choose ρ^* at random
6. $er^* \leftarrow Explain(\beta^*; x_0, \alpha^*; \rho^*), es^* \leftarrow Enc.Inv(s_{PKE}, x'_1)$

An adversary sees protocol transcript (α^*, β^*), internal states er^* and (es^*, z), programs $(P, Explain)$.
    and

Experiment Middle:

1. choose randomness $s_{PKE}, s_{CRS}, s_{GC}, s_y$. Choose random z. Set $x'_1 \leftarrow (x_1, z)$
2. generate a CRS: prf keys k_E, k_I, k, Gen internal keys and choose randomness for obfuscation x_P, x_{Expl} using s_{CRS}. Create obfuscated programs $P \leftarrow O(P_{k_E, k_I, k}; Gen; x_P), Explain \leftarrow O(Explain_{k_I, k}; x_{Expl})$.
3. sample P_0's keys $\overline{PK_{x'_1}}, \overline{SK_{x'_1}} \leftarrow PKE.Gen(s_{PKE}[0])$, $\overline{PK_{1-x'_1}} \leftarrow PKE.oGen(s_{PKE}[1])$. Set $\alpha^* \leftarrow \overline{PK_0}, \overline{PK_1}$
4. choose random r^*
5. run $\beta^* \leftarrow P(x_0, \alpha^*; r^*)$
6. choose ρ^* at random
7. $er^* \leftarrow Explain(\beta^*; x_0, \alpha^*; \rho^*)$

An adversary sees protocol transcript (α^*, β^*), internal states er^* and (s_{PKE}, z), programs $(P, Explain)$.

Proof. We show indistinguishability using several hybrids as described below:

1. H0 = Simulation
2. H1: like a simulation, but OT public keys $\overline{PK}_{1-x_1}$ (which do not correspond to P_1's input) are sampled obliviously
3. H2: like H1, but β^* is chosen as a result of $Gen(x_0; \alpha^*; u^*)$ for some random u^*; previously β^* was the result of $SimGen$. Based on indistinguishability between a fake and a real garbled circuit.
4. H3: Like H2, but u^* is chosen as $E_{k_E}(x_0, \alpha^*, r^*)$ for random r^*; previously it was chosen at random. Based on extracting property of E_{k_E}
5. H4 = Middle: Like H3, but $\beta^* \leftarrow P(x_0, \alpha^*; r^*)$ (which means that now first check 1 is performed on randomness r^* before generating the output). Based on the fact that r^* is random and for a random value this check passes with negligible probability.

H1.

1. choose randomness $s_{PKE}, s_{CRS}, s_{GC}, s_y$. Set $z = y \oplus s_y$. Set $x'_1 \leftarrow (x_1, z)$
2. generate a CRS: prf keys k_E, k_I, k, Gen internal keys and choose randomness for obfuscation x_P, x_{Expl} using s_{CRS}. Create obfuscated programs $P \leftarrow O(P_{k_E, k_I, k}; Gen; x), Explain \leftarrow O(Explain_{k_I, k}; x_{Expl})$.
3. sample P_0's keys $\overline{PK_{x'_1}}, \overline{SK_{x'_1}} \leftarrow PKE.Gen(s_{PKE}[0])$, $\overline{PK_{1-x'_1}} \leftarrow PKE.oGen(s_{PKE}[1])$. Set $\alpha^* \leftarrow \overline{PK_0}, \overline{PK_1}$
4. run $\beta^* \leftarrow SimGen(s_y, \alpha^*; s_{GC})$
5. choose ρ^* at random
6. $er^* \leftarrow Explain(\beta^*; x_0, \alpha^*; \rho^*)$

An adversary sees protocol transcript (α^*, β^*), internal states er^* and (s_{PKE}, z), programs $(P, Explain)$.

In this hybrid we generate public keys for OT which do not correspond to P_1' input obliviously and show to the adversary the real randomness s_{PKE} which was used to generate these keys. Indistinguishability holds because of the property of augmented non-committing encryption: no adversary can distinguish between a real randomness used for oblivious key generation and a randomness obtained as a result of inverting algorithm.

H2.

1. choose randomness $s_{PKE}, s_{CRS}, s_{GC}, s_y$. Choose random z. Set $x'_1 \leftarrow (x_1, z)$
2. generate a CRS: prf keys k_E, k_I, k, Gen internal keys and choose randomness for obfuscation x_P, x_{Expl} using s_{CRS}. Create obfuscated programs $P \leftarrow O(P_{k_E, k_I, k}; Gen; x_P), Explain \leftarrow O(Explain_{k_I, k}; x_{Expl})$.
3. sample P_0's keys $\overline{PK_{x'_1}}, \overline{SK_{x'_1}} \leftarrow PKE.Gen(s_{PKE}[0])$, $\overline{PK_{1-x'_1}} \leftarrow PKE.oGen(s_{PKE}[1])$. Set $\alpha^* \leftarrow \overline{PK_0}, \overline{PK_1}$
4. choose random u^*
5. run $\beta^* \leftarrow Gen(x_0, \alpha^*; u^*)$
6. choose ρ^* at random
7. $er^* \leftarrow Explain(\beta^*; x_0, \alpha^*; \rho^*)$

An adversary sees protocol transcript (α^*, β^*), internal states er^* and (s_{PKE}, z), programs $(P, Explain)$.

In this hybrid we changed the way β^* is generated. Previously it contained a fake garbled circuit which always evaluates to s_y, now it contains a real garbled circuit. Indistinguishability is based on indistinguishability between a fake garbled circuit and a real one, as shown in [LP09].

H3.

1. choose randomness $s_{PKE}, s_{CRS}, s_{GC}, s_y$. Choose random z. Set $x'_1 \leftarrow (x_1, z)$
2. generate a CRS: prf keys k_E, k_I, k, Gen internal keys and choose randomness for obfuscation x_P, x_{Expl} using s_{CRS}. Create obfuscated programs $P \leftarrow O(P_{k_E, k_I, k}; Gen; x_P), Explain \leftarrow O(Explain_{k_I, k}; x_{Expl})$.
3. sample P_0's keys $\overline{PK_{x'_1}}, \overline{SK_{x'_1}} \leftarrow PKE.Gen(s_{PKE}[0])$, $\overline{PK_{1-x'_1}} \leftarrow PKE.oGen(s_{PKE}[1])$. Set $\alpha^* \leftarrow \overline{PK_0}, \overline{PK_1}$

4. choose random r^*. Set $u^* \leftarrow E_{k_E}(x_0, \alpha^*, r^*)$
5. run $\beta^* \leftarrow Gen(x_0, \alpha^*; u^*)$
6. choose ρ^* at random
7. $er^* \leftarrow Explain(\beta^*; x_0, \alpha^*; \rho^*)$

An adversary sees protocol transcript (α^*, β^*), internal states er^* and (s_{PKE}, z), programs $(P, Explain)$.

In this hybrid we choose u^* as $u^* \leftarrow E_{k_E}(x_0, \alpha^*, r^*)$, instead of choosing it at random. Indistinguishability holds because of extracting property of E_{k_E}. Indeed, since min-entropy of the PRF input is at least $|r^*|$, then by our choice of parameters the output of this PRF is indistinguishable from random. We can reduce these hybrids to an extracting prf game as follows: given k_E and random w or $w = E_{k_E}(x_0, \alpha^*, r^*)$ for random r^*, we choose other keys and obfuscate programs, and then compute other variables using $u^* = w$. Depending on whether w is random or not, we are either in H2 or in H3.

H4 (Middle).

1. choose randomness $s_{PKE}, s_{CRS}, s_{GC}, s_y$. Choose random z. Set $x_1' \leftarrow (x_1, z)$
2. generate a CRS: prf keys k_E, k_I, k, Gen internal keys and choose randomness for obfuscation x_P, x_{Expl} using s_{CRS}. Create obfuscated programs $P \leftarrow O(P_{k_E, k_I, k}; Gen; x_P), Explain \leftarrow O(Explain_{k_I, k}; x_{Expl})$.
3. sample P_0's keys $\overline{PK_{x_1'}}, \overline{SK_{x_1'}} \leftarrow PKE.Gen(s_{PKE}[0])$, $\overline{PK_{1-x_1'}} \leftarrow PKE.oGen(s_{PKE}[1])$. Set $\alpha^* \leftarrow \overline{PK_0}, \overline{PK_1}$
4. choose random r^*.
5. run $\beta^* \leftarrow P(x_0, \alpha^*; r^*)$
6. choose ρ^* at random
7. $er^* \leftarrow Explain(\beta^*; x_0, \alpha^*; \rho^*)$

An adversary sees protocol transcript (α^*, β^*), internal states er^* and (s_{PKE}, z), programs $(P, Explain)$.

In this hybrid we generate β^* as a result of a program P. In other words, before computing u^* we perform check 1 in P. Since for randomly chosen r^* this check passes with negligible probability, hybrids are statistically close to each other.

Thus lemma 1 is proved.

Lemma 2. *No PPT adversary can distinguish between the following two distributions:*

Experiment Middle:

1. choose randomness s_{PKE}, s_{CRS}, s_y. Choose random z. Set $x_1' \leftarrow (x_1, z)$
2. generate a CRS: prf keys k_E, k_I, k, Gen internal keys and choose randomness for obfuscation x_P, x_{Expl} using s_{CRS}. Create obfuscated programs $P \leftarrow O(P_{k_E, k_I, k}; Gen; x_P), Explain \leftarrow O(Explain_{k_I, k}; x_{Expl})$.
3. sample P_0's keys $\overline{PK_{x_1'}}, \overline{SK_{x_1'}} \leftarrow PKE.Gen(s_{PKE}[0])$, $\overline{PK_{1-x_1'}} \leftarrow PKE.oGen(s_{PKE}[1])$. Set $\alpha^* \leftarrow \overline{PK_0}, \overline{PK_1}$
4. choose random r^*

5. run $\beta^* \leftarrow P(x_0, \alpha^*; r^*)$
6. choose ρ^* at random
7. $er^* \leftarrow Explain(\beta^*; x_0, \alpha^*; \rho^*)$

An adversary sees $(\alpha^*, \beta^*, er^*, s_{PKE}, z)$, programs $(P, Explain)$.

Experiment Real:

1. choose randomness s_{PKE}, s_{CRS}, s_y. Choose random z. Set $x_1' \leftarrow (x_1, z)$
2. generate a CRS: prf keys k_E, k_I, k, Gen internal keys and choose randomness for obfuscation x_P, x_{Expl} using s_{CRS}. Create obfuscated programs $P \leftarrow O(P_{k_E, k_I, k}; Gen; x_P), Explain \leftarrow O(Explain_{k_I, k}; x_{Expl})$.
3. sample P_0's keys $\overline{PK}_{x_1'}, \overline{SK}_{x_1'} \leftarrow PKE.Gen(s_{PKE}[0])$, $\overline{PK}_{1-x_1'} \leftarrow PKE.oGen(s_{PKE}[1])$. Set $\alpha^* \leftarrow \overline{PK}_0, \overline{PK}_1$
4. choose random r^*
5. run $\beta^* \leftarrow P(x_0, \alpha^*; r^*)$

An adversary sees $(\alpha^*, \beta^*, r^*, s_{PKE}, z)$, programs $(P, Explain)$.

Proof. The lemma states that the view of an adversary in the real execution is indistinguishable from its view in the experiment when instead of real randomness, explained randomness is shown (which we called a middle experiment). To prove the lemma statement, we consider a sequence of hybrids $Real = H_0^0 \sim \ldots \sim H_6^0 \sim H_6^1 \sim \ldots \sim H_0^1 = Middle$. For $b = 0, 1$ we will show that H_0^b is indistinguishable from H_6^b. After this, we show that H_6^0 and H_6^1 are indistinguishable as well. This proves that a middle hybrid and a real execution are indistinguishable.

Hybrids overview:

1. In $H1^b$ we skip check 1 in the program P and directly compute $u^* \leftarrow E_{k_E}(x_0, \alpha^*; r^*), \beta^* \leftarrow Gen(x_0, \alpha^*; u^*)$. Since r^* is random, the check passes with negligible probability.
2. In $H2^b$, instead of computing $\hat{\rho}^* \leftarrow prg(\rho^*)$ (and then evaluating er^* using this $\hat{\rho}^*$), we choose $\hat{\rho}^*$ at random. Indistinguishability is based on security of a PRG.
3. In $H3^b$ we show punctured programs $P : 1$ and $Explain : 1$ instead of original ones. We prove that new programs have the same functionality and rely the indistinguishability on the security of iO.
4. In $H4^b$ we choose u^* at random instead of $E_{k_E}(x_0^*, \alpha^*; r^*)$. Based on punctured PRF E_{k_E}.
5. In $H5^b$ we choose $er^*[2]$ at random instead of $I_{k_I}(\beta^*; x_0, \alpha^*; \hat{\rho}^*)$. Based on punctured PRF I_{k_I}.
6. In $H6^b$ we choose $er^*[1]$ at random instead of $F_k(er^*[2]) \oplus (\beta^*; x_0, \alpha^*; \hat{\rho}^*)$. Based on punctured PRF F_k.

$H0^b$

1. choose randomness s_{PKE}, s_{CRS}, s_y. Choose random z. Set $x_1' \leftarrow (x_1, z)$
2. generate a CRS: prf keys k_E, k_I, k, Gen internal keys and choose randomness for obfuscation x_P, x_{Expl} using s_{CRS}. Create obfuscated programs $P \leftarrow O(P_{k_E, k_I, k}; Gen; x_P), Explain \leftarrow O(Explain_{k_I, k}; x_{Expl})$.
3. sample P_0's keys $\overline{PK_{x_1'}}, \overline{SK_{x_1'}} \leftarrow PKE.Gen(s_{PKE}[0])$, $\overline{PK_{1-x_1'}} \leftarrow PKE.oGen(s_{PKE}[1])$. Set $\alpha^* \leftarrow \overline{PK_0}, \overline{PK_1}$
4. choose random r^*
5. run $\beta^* \leftarrow P(x_0, \alpha^*; r^*)$
6. choose ρ^* at random
7. $er^* \leftarrow Explain(\beta^*; x_0, \alpha^*; \rho^*)$

If $b = 0$, an adversary sees $(\alpha^*, \beta^*, r^*, s_{PKE}, z)$, programs $(P, Explain)$.
If $b = 1$, an adversary sees $(\alpha^*, \beta^*, er^*, s_{PKE}, z)$, programs $(P, Explain)$.

$H1^b$

1. choose randomness s_{PKE}, s_{CRS}, s_y. Choose random z. Set $x_1' \leftarrow (x_1, z)$
2. generate a CRS: prf keys k_E, k_I, k, Gen internal keys and choose randomness for obfuscation x_P, x_{Expl} using s_{CRS}. Create obfuscated programs $P \leftarrow O(P_{k_E, k_I, k}; Gen; x_P), Explain \leftarrow O(Explain_{k_I, k}; x_{Expl})$.
3. sample P_0's keys $\overline{PK_{x_1'}}, \overline{SK_{x_1'}} \leftarrow PKE.Gen(s_{PKE}[0])$, $\overline{PK_{1-x_1'}} \leftarrow PKE.oGen(s_{PKE}[1])$. Set $\alpha^* \leftarrow \overline{PK_0}, \overline{PK_1}$
4. choose random r^*, $u^* \leftarrow E_{k_E}(x_0, \alpha^*; r^*)$,
5. $\beta^* \leftarrow Gen(x_0, \alpha^*; u^*)$.
6. choose ρ^* at random
7. $er^* \leftarrow Explain(\beta^*; x_0, \alpha^*; \rho^*)$

If $b = 0$, an adversary sees $(\alpha^*, \beta^*, r^*, s_{PKE}, z)$, programs $(P, Explain)$.
If $b = 1$, an adversary sees $(\alpha^*, \beta^*, er^*, s_{PKE}, z)$, programs $(P, Explain)$.

In this hybrid we omit check 1 in the program P while computing β^*. Since for randomly chosen r^* the check passes with negligible probability, hybrids are statistically close.

$H2^b$

1. choose randomness s_{PKE}, s_{CRS}, s_y. Choose random z. Set $x_1' \leftarrow (x_1, z)$
2. generate a CRS: prf keys k_E, k_I, k, Gen internal keys and choose randomness for obfuscation x_P, x_{Expl} using s_{CRS}. Create obfuscated programs $P \leftarrow O(P_{k_E, k_I, k}; Gen; x_P), Explain \leftarrow O(Explain_{k_I, k}; x_{Expl})$.
3. sample P_0's keys $\overline{PK_{x_1'}}, \overline{SK_{x_1'}} \leftarrow PKE.Gen(s_{PKE}[0])$, $\overline{PK_{1-x_1'}} \leftarrow PKE.oGen(s_{PKE}[1])$. Set $\alpha^* \leftarrow \overline{PK_0}, \overline{PK_1}$
4. choose random r^*, $u^* \leftarrow E_{k_E}(x_0, \alpha^*; r^*)$,
5. $\beta^* \leftarrow Gen(x_0, \alpha^*; u^*)$.
6. choose $\hat{\rho}^*$ at random
7. set $M^* \leftarrow \beta^*; x_0, \alpha^*; \hat{\rho}^*$
8. $er^*[2] \leftarrow I_{k_I}(M^*)$
9. $er^*[1] \leftarrow F_k(er^*[2]) \oplus M^*$

If $b = 0$, an adversary sees $(\alpha^*, \beta^*, r^*, s_{PKE}, z)$, programs $(P, Explain)$.

If $b = 1$, an adversary sees $(\alpha^*, \beta^*, er^*, s_{PKE}, z)$, programs $(P, Explain)$.

In this hybrid we use randomly chosen $\hat{\rho}^*$ instead of the result of applying a PRG to ρ^* while generating er^*. Indistinguishability of hybrids immediately follows from the security of a PRG.

$H3^b$

1. choose randomness s_{PKE}, s_{CRS}, s_y. Choose random z. Set $x_1' \leftarrow (x_1, z)$
2. generate a CRS: prf keys k_E, k_I, k, Gen internal keys and choose randomness for obfuscation x_P, x_{Expl} using s_{CRS}. Create obfuscated programs $P \leftarrow O(P_{k_E,k_I,k}; Gen; x_P)$, $Explain \leftarrow O(Explain_{k_I,k}; x_{Expl})$.
3. sample P_0's keys $\overline{PK_{x_1'}}, \overline{SK_{x_1'}} \leftarrow PKE.Gen(s_{PKE}[0])$, $\overline{PK_{1-x_1'}} \leftarrow PKE.oGen(s_{PKE}[1])$. Set $\alpha^* \leftarrow \overline{PK_0}, \overline{PK_1}$
4. choose random r^*, $u^* \leftarrow E_{k_E}(x_0, \alpha^*; r^*)$,
5. $\beta^* \leftarrow Gen(x_0, \alpha^*; u^*)$.
6. choose $\hat{\rho}^*$ at random
7. set $M^* \leftarrow \beta^*; x_0, \alpha^*; \hat{\rho}^*$
8. $er^*[2] \leftarrow I_{k_I}(M^*)$
9. $er^*[1] \leftarrow F_k(er^*[2]) \oplus M^*$

If $b = 0$, an adversary sees $(\alpha^*, \beta^*, r^*, s_{PKE}, z)$, programs $(P : 1, Explain : 1)$.

If $b = 1$, an adversary sees $(\alpha^*, \beta^*, er^*, s_{PKE}, z)$, programs $(P : 1, Explain : 1)$.

In this hybrid we show punctured programs $P : 1$ (Fig. 6) and $Explain : 1$ (Fig. 7) instead of their normal versions. We rely the indistinguishability on $i\mathcal{O}$ security: modified programs have the same functionality as original ones, as proven in [SW14] in their proof for deniable encryption scheme (with a natural modification of the input from their input m, r to our input $(x_0, \overline{PK}, r)$). However, for the sake of self-containment we briefly sketch it here:

Program P:

1. we add a line "if $(x, \alpha, r) = (x_0, \alpha^*, r^*)$ or $(x, \alpha, r) = (x_0, \alpha^*, er^*)$ then output β^*", this is exactly what the original program outputs on these inputs.
2. add "f $r[2] = r^*[2]$ or $r[2] = er^*[2]$ then goto 2". If $r[2] = r^*[2]$, then the check in step one will not pass since a random $r^*[2]$ with high probability is outside the image of I_{k_I}, so we can go to step 2. If $r[2] = er^*[2]$, then either the check doesn't pass and we can go to step 2, or, if it passes, then the encoded message $M' = M^*$ (due to injectivity of I_{k_I}), and therefore $r[1] = er^*[1]$, $(x', \alpha') = (x_0, \alpha^*)$, which would be detected in the first added line in P:1.
3. now F_k is never called on $r^*[2]$ or $er^*[2]$, therefore we can safely puncture at these points.
4. add "if $M' = M^*$ then goto 2". If $M' = M^*$ and the check passes, then $r[2] = er^*[2]$, $r[1] = er^*[1]$, and this would be detected in the first line in P:1.

5. now I_{k_I} will not be called on M^*, and we can puncture at this point.
6. we can puncture $F_{k_1\{(x_0,\alpha^*,r^*),(x_0,\alpha^*,er^*)\}}$, since these inputs are treated in the first line of P:1.

Program Explain:

1. we puncture k_I at M^*, since $\hat{\rho}^*$ (which is a part of M^*) is generated at random (instead of $prg(\rho^*)$) and with high probability is outside the image of a PRG; therefore no input results in $M = M^*$ in Explain.
2. we puncture k at both points $r^*[2]$ and $er^*[2]$. Since $r^*[2]$ is randomly chosen, with high probability it is outside the image of a PRF I_{k_I}, therefore no input for Explain results in $r[2] = r^*[2]$ and therefore F_k is never called on $r^*[2]$. Furthermore, as we said no input for Explain results in $M = M^*$, and due to I_{k_I} injectivity no input for Explain results in $er^*[2] = I_{k_I}(M^*)$, which means that F_k is not called on $er^*[2]$ as well.

Program P:1

constants: $\alpha^*, r^*, er^*, \beta^*, M^*, x_0$.
inputs: protocol input x, 1-round message α, randomness $r = r[1]r[2]$
$P(x, \alpha; r)$:

1. check if r has encoded value inside:
   (a) if $(x, \alpha, r) = (x_0, \alpha^*, r^*)$ or $(x, \alpha, r) = (x_0, \alpha^*, er^*)$ then output β^*
   (b) if $r[2] = r^*[2]$ or $r[2] = er^*[2]$ then goto 2
   (c) $M' \leftarrow F_{k\{r^*[2],er^*[2]\}}(r[2]) \oplus r[1]$;
   (d) if $M' = M^*$ then goto 2;
   (e) if $I_{k_I\{M^*\}}(M') \neq r[2]$ then goto 2;
   (f) parse M' as $\beta', x', \alpha', \hat{\rho}'$. If $(x', \alpha') \neq (x, \alpha)$ then goto 2;
   (g) output β'
2. else run Gen:
   (a) $u \leftarrow E_{k_E\{(x_0,\alpha^*,r^*),(x_0,\alpha^*,er^*)\}}(x, \alpha, r)$
   (b) output $Gen(x, \alpha; u)$

Fig. 6. Program P:1

$H4^b$

1. choose randomness s_{PKE}, s_{CRS}, s_y. Choose random z. Set $x_1' \leftarrow (x_1, z)$
2. generate a CRS: prf keys k_E, k_I, k, Gen internal keys and choose randomness for obfuscation x_P, x_{Expl} using s_{CRS}. Create obfuscated programs $P : 1 \leftarrow O(P : 1_{k_E, k_I, k}; Gen; x_P), Explain : 1 \leftarrow O(Explain : 1_{k_I, k}; x_{Expl})$.
3. sample P_0's keys $\overline{PK_{x_1'}}, \overline{SK_{x_1'}} \leftarrow PKE.Gen(s_{PKE}[0])$,
   $\overline{PK_{1-x_1'}} \leftarrow PKE.oGen(s_{PKE}[1])$. Set $\alpha^* \leftarrow \overline{PK_0}, \overline{PK_1}$
4. choose random u^*, r^*

Program Explain:1

constants: M^*, r^*, er^*
inputs: message m which should be encoded; randomness ρ
$P(m; \rho):$

1. $M \leftarrow m, prg(\rho)$
2. $r[2] \leftarrow I_{k_I\{M^*\}}(M), \quad r[1] \leftarrow F_{k\{r^*[2], er^*[2]\}}(r[2]) \oplus M$
3. output $r = r[1]r[2]$

Fig. 7. Program Explain:1

5. $\beta^* \leftarrow Gen(x_0, \alpha^*; u^*)$.
6. choose $\hat{\rho}^*$ at random
7. set $M^* \leftarrow \beta^*; x_0, \alpha^*; \hat{\rho}^*$
8. $er^*[2] \leftarrow I_{k_I}(M^*)$
9. $er^*[1] \leftarrow F_k(er^*[2]) \oplus M^*$

If $b = 0$, an adversary sees $(\alpha^*, \beta^*, r^*, s_{PKE}, z)$, programs $(P : 1, Explain : 1)$.
If $b = 1$, an adversary sees $(\alpha^*, \beta^*, er^*, s_{PKE}, z)$, programs $(P : 1, Explain : 1)$.

In this hybrid we choose u^* at random instead of choosing it as $E_{k_E}(x_k, \alpha^*_{1-k}, r^*)$. Security follows from pseudorandomness of a puncturable PRF. Indeed, given a punctured key $k_E\{(x_k, \alpha^*_{1-k}, r^*)\}$ and w, which is random or $E_{k_E}(x_k, \alpha^*_{1-k}, r^*)$, we choose other keys ourselves and create programs. Then we evaluate variables in the experiment setting $u^* = w$ and showing the resulting distribution to the adversary. If w was random, then the adversary sees H_4^b, otherwise H_3^b.

$\quad H5^b$

1. choose randomness s_{PKE}, s_{CRS}, s_y. Choose random z. Set $x_1' \leftarrow (x_1, z)$
2. generate a CRS: prf keys k_E, k_I, k, Gen internal keys and choose randomness for obfuscation x_P, x_{Expl} using s_{CRS}. Create obfuscated programs $P : 1 \leftarrow O(P : 1_{k_E, k_I, k}; Gen; x_P), Explain : 1 \leftarrow O(Explain : 1_{k_I, k}; x_{Expl})$.
3. sample P_0's keys $\overline{PK_{x_1'}}, \overline{SK_{x_1'}} \leftarrow PKE.Gen(s_{PKE}[0])$,
   $\overline{PK_{1-x_1'}} \leftarrow PKE.oGen(s_{PKE}[1])$. Set $\alpha^* \leftarrow \overline{PK_0}, \overline{PK_1}$
4. choose random u^*, r^*
5. $\beta^* \leftarrow Gen(x_0, \alpha^*; u^*)$.
6. choose $\hat{\rho}^*$ at random
7. set $M^* \leftarrow \beta^*; x_0, \alpha^*; \hat{\rho}^*$
8. choose random $er^*[2]$
9. $er^*[1] \leftarrow F_k(er^*[2]) \oplus M^*$

If $b = 0$, an adversary sees $(\alpha^*, \beta^*, r^*, s_{PKE}, z)$, programs $(P : 1, Explain : 1)$.
If $b = 1$, an adversary sees $(\alpha^*, \beta^*, er^*, s_{PKE}, z)$, programs $(P : 1, Explain : 1)$.

In this hybrid we choose $er^*[2]$ at random instead of choosing it as $I_{k_I}(M^*)$. Security follows from pseudorandomness of a puncturable PRF. Indeed, given a punctured key $k_I\{M^*\}$ and w, which is random or $I_{k_I}(M^*)$, we choose other keys ourselves and create programs. Then we evaluate variables in the experiment setting $er^*[2] = w$ and showing the resulting distribution to the adversary. If w was random, then the adversary sees H_5^b, otherwise H_4^b.

$H6^b$

1. choose randomness s_{PKE}, s_{CRS}, s_y. Choose random z. Set $x_1' \leftarrow (x_1, z)$
2. generate a CRS: prf keys k_E, k_I, k, Gen internal keys and choose randomness for obfuscation x_P, x_{Expl} using s_{CRS}. Create obfuscated programs $P \leftarrow O(P_{k_E,k_I,k}; Gen; x_P)$, $Explain \leftarrow O(Explain_{k_I,k}; x_{Expl})$.
3. sample P_0's keys $\overline{PK}_{x_1'}, \overline{SK}_{x_1'} \leftarrow PKE.Gen(s_{PKE}[0])$, $\overline{PK}_{1-x_1'} \leftarrow PKE.oGen(s_{PKE}[1])$. Set $\alpha^* \leftarrow \overline{PK}_0, \overline{PK}_1$
4. choose random u^*, r^*
5. $\beta^* \leftarrow Gen(x_0, \alpha^*; u^*)$.
6. choose $\hat{\rho}^*$ at random
7. set $M^* \leftarrow \beta^*; x_0, \alpha^*; \hat{\rho}^*$
8. choose random $er^*[2]$
9. choose random $er^*[1]$

If $b = 0$, an adversary sees $(\alpha^*, \beta^*, r^*, s_{PKE}, z)$, programs $(P : 1, Explain : 1)$.
If $b = 1$, an adversary sees $(\alpha^*, \beta^*, er^*, s_{PKE}, z)$, programs $(P : 1, Explain : 1)$.

In this hybrid we choose $er^*[1]$ at random instead of choosing it as $F_k(er^*[2]) \oplus M$. Security follows from pseudorandomness of a puncturable PRF. Indeed, given a punctured key $k\{er^*[2]\}$ and w, which is random or $F_k(M^*)$, we choose other keys ourselves and create programs. Then we evaluate variables in the experiment setting $er^*[2] = w$ and showing the resulting distribution to the adversary. If w was random, then the adversary sees H_6^b, otherwise H_5^b.

Finally we notice that distributions H_6^0 and H_6^1 are the same, since both programs and the experiment treat r^* and er^* in the same manner (i.e. both r^* and er^* are chosen at random and are not connected to other variables in the protocol). Therefore no adversary can distinguish between these two hybrids, and lemma statement is proved.

Step two - dealing with adaptive inputs.   In this part we show how to deal with the case of adaptive inputs. In order to do this, for all possible pairs of inputs $(x_0^*, x_1^*) = (0^n, 0^n), \ldots, (x_0^*, x_1^*) = (1^n, 1^n)$, sorted lexicographically, we consider a hybrid $M_{x_0^*, x_1^*}$. In this hybrid we use x_0^*, x_1^* as a guess for inputs which an adversary will choose. We create a CRS and show it to the adversary. If it chooses (lexicographically) smaller pair of inputs (x_0', x_1'), then we run a simulation experiment with new inputs x_0', x_1'; otherwise we run a real execution experiment with new inputs x_0', x_1' (it is crucial that in both a real execution and a simulation, a CRS has the same distribution; this allows us to choose which experiment to run *after* we show a CRS). Note that $M_{0^n, 0^n}$ is always a real execution and $M_{1^n, 1^n}$ is a real execution only if an adversary chooses $(1^n, 1^n)$.

Indistinguishability between M_k and M_{k+1} (and also between $M_{1^n, 1^n}$ and a simulation) follows from selective security of the protocol proven in part one. If an adversary which sees a CRS chooses an input which is smaller than k, then in both cases it sees the same distribution (real). If it chooses an input greater or equal than $k+1$, then it again sees the same distribution (a simulation). Finally, if an adversary chooses an input k, then it sees a real execution in M_k and a simulation in M_{k+1}. As we proved in part one, for any fixed input these distributions are indistinguishable. This implies that for every $k = 0^{2n}, \ldots, 1^{2n}$ M_k and M_{k+1} are indistinguishable (where $M_{1^{2n}+1}$ is a simulation), and therefore a real execution and a simulation are indistinguishable even in the case of adaptively chosen inputs.

It should be noted that we have as many hybrids as the number of potential inputs to the protocol, thus the security loss is also linear in the number of possible inputs to the computation. Consequently, the parameters of the underlying primitives (especially, the obfuscation and the puncturable PRFs) need to be set accordingly.

4.1   Obtaining Incoercibility

Recall that, to be incoercible, the protocol should be augmented by *faking algorithms* for the two parties. The faking algorithm for a party takes as input a value x', representing a fake input value for the party, as well as the party's local state and the messages sent by that party so far, and outputs a "fake random input" r' for the party, such that running the party's program on input x' and random input r' results in the messages sent by the party so far, and furthermore r' "looks random" given the rest of the view of the adversary. More precisely, the protocol together with the faking algorithm should be simulatable as in the definition of incoercible computation presented in Section 2.

To show incoercibility for the garbler, we demonstrate a faking algorithm: Having received message α, sent message β, and given the fake input value x', simply run the Explain algorithm with input message $m = \beta, x', \alpha$ and some fresh randomness. Then output the output of Explain.

It is straightforward to see that the same simulation actually demonstrates incoercibility for the garbler. Indeed, the simulator exhibits the same information for coercion and corruption attacks.

References

BCH12.   Bitansky, N., Canetti, R., Halevi, S.: Leakage-tolerant interactive protocols. In: Cramer, R. (ed.) TCC 2012. LNCS, vol. 7194, pp. 266–284. Springer, Heidelberg (2012)

BDL14.   Bitansky, N., Dachman-Soled, D., Lin, H.: Leakage-tolerant computation with input-independent preprocessing. In: Garay, J.A., Gennaro, R. (eds.) CRYPTO 2014, Part II. LNCS, vol. 8617, pp. 146–163. Springer, Heidelberg (2014)

BGI+01. Barak, B., Goldreich, O., Impagliazzo, R., Rudich, S., Sahai, A., Vadhan, S.P., Yang, K.: On the (im)possibility of obfuscating programs. Electronic Colloquium on Computational Complexity (ECCC) 8(057) (2001)

BGI13.   Boyle, E., Goldwasser, S., Ivan, I.: Functional signatures and pseudorandom functions. IACR Cryptology ePrint Archive, 2013:401 (2013)

BGW88.   Ben-Or, M., Goldwasser, S., Wigderson, A.: Completeness theorems for non-cryptographic fault-tolerant distributed computation (extended abstract). In: Proceedings of the 20th Annual ACM Symposium on Theory of Computing, Chicago, Illinois, USA, May 2-4, pp. 1–10 (1988)

BW13.   Boneh, D., Waters, B.: Constrained pseudorandom functions and their applications. In: Sako, K., Sarkar, P. (eds.) ASIACRYPT 2013, Part II. LNCS, vol. 8270, pp. 280–300. Springer, Heidelberg (2013)

Can01.   Canetti, R.: Universally composable security: A new paradigm for cryptographic protocols. In: FOCS, pp. 136–145 (2001), Full version in IACR Eprint Archive, record 2000/067 (2013 revision)

CDNO97.   Canetti, R., Dwork, C., Naor, M., Ostrovsky, R.: Deniable encryption. In: Kaliski Jr., B.S. (ed.) CRYPTO 1997. LNCS, vol. 1294, pp. 90–104. Springer, Heidelberg (1997)

CDPW07.   Canetti, R., Dodis, Y., Pass, R., Walfish, S.: Universally composable security with global setup. In: Vadhan, S.P. (ed.) TCC 2007. LNCS, vol. 4392, pp. 61–85. Springer, Heidelberg (2007)

CFGN96.   Canetti, R., Feige, U., Goldreich, O., Naor, M.: Adaptively secure multiparty computation. In: Proceedings of the Twenty-Eighth Annual ACM Symposium on the Theory of Computing, Philadelphia, Pennsylvania, USA, May 22-24, pp. 639–648 (1996)

CG96.   Canetti, R., Gennaro, R.: Incoercible multiparty computation (extended abstract). In: 37th Annual Symposium on Foundations of Computer Science, FOCS 1996, Burlington, Vermont, USA, October 14-16, pp. 504–513 (1996)

CLOS02.   Canetti, R., Lindell, Y., Ostrovsky, R., Sahai, A.: Universally composable two-party and multi-party secure computation. In: Proceedings on 34th Annual ACM Symposium on Theory of Computing, Montréal, Québec, Canada, May 19-21, pp. 494–503 (2002)

DKR14.   Dachman-Soled, D., Katz, J., Rao, V.: Adaptively secure, universally composable, multi-party computation in constant rounds. IACR Cryptology ePrint Archive, 2014:858 (2014)

DN00.   Damgård, I., Nielsen, J.B.: Improved non-committing encryption schemes based on a general complexity assumption. In: Bellare, M. (ed.) CRYPTO 2000. LNCS, vol. 1880, pp. 432–450. Springer, Heidelberg (2000)

EGL85.   Even, S., Goldreich, O., Lempel, A.: A randomized protocol for signing contracts. Commun. ACM 28(6), 637–647 (1985)

GGH+13.   Garg, S., Gentry, C., Halevi, S., Raykova, M., Sahai, A., Waters, B.: Candidate indistinguishability obfuscation and functional encryption for all circuits. In: 54th Annual IEEE Symposium on Foundations of Computer Science, FOCS 2013, Berkeley, CA, USA, October 26-29, pp. 40–49 (2013)

GGM86.   Goldreich, O., Goldwasser, S., Micali, S.: How to construct random functions. J. ACM 33(4), 792–807 (1986)

GMW87.   Goldreich, O., Micali, S., Wigderson, A.: How to play any mental game or A completeness theorem for protocols with honest majority. In: Proceedings of the 19th Annual ACM Symposium on Theory of Computing, New York, USA, pp. 218–229 (1987)

GP14.   Garg, S., Polychroniadou, A.: Two-round adaptively secure MPC from indistinguishability obfuscation. IACR Cryptology ePrint Archive, 2014:844 (2014)

GR14.   Goldwasser, S., Rothblum, G.N.: On best-possible obfuscation. J. Cryptology 27(3), 480–505 (2014)

GS12.   Garg, S., Sahai, A.: Adaptively secure multi-party computation with dishonest majority. In: Safavi-Naini, R., Canetti, R. (eds.) CRYPTO 2012. LNCS, vol. 7417, pp. 105–123. Springer, Heidelberg (2012)

IPS08.   Ishai, Y., Prabhakaran, M., Sahai, A.: Founding cryptography on oblivious transfer – efficiently. In: Wagner, D. (ed.) CRYPTO 2008. LNCS, vol. 5157, pp. 572–591. Springer, Heidelberg (2008)

LP09.   Lindell, Y., Pinkas, B.: A proof of security of Yao's protocol for two-party computation. J. Cryptology 22(2), 161–188 (2009)

SW14.   Sahai, A., Waters, B.: How to use indistinguishability obfuscation: deniable encryption, and more. In: Symposium on Theory of Computing, STOC 2014, New York, NY, USA, 2014, May 31-June 03, pp. 475–484 (2014)

Adaptively Secure, Universally Composable, Multiparty Computation in Constant Rounds

Dana Dachman-Soled[1], Jonathan Katz[1,*], and Vanishree Rao[2,**]

[1] University of Maryland, USA
danadach@ece.umd.edu, jkatz@cs.umd.edu
[2] University of California at Los Angeles, USA
vanishri@cs.ucla.edu

Abstract. Cryptographic protocols with *adaptive security* ensure that security holds against an adversary who can dynamically determine which parties to corrupt as the protocol progresses—or even after the protocol is finished. In the setting where all parties may potentially be corrupted, and secure erasure is not assumed, it has been a long-standing open question to design secure-computation protocols with adaptive security running in *constant* rounds.

Here, we show a constant-round, universally composable protocol for computing any functionality, tolerating a malicious, adaptive adversary corrupting any number of parties. Interestingly, our protocol can compute *all* functionalities, not just adaptively well-formed ones. The protocol relies on indistinguishability obfuscation, and assumes a common reference string.

1 Introduction

When designing and analyzing protocols for secure computation, there are several different adversarial models one can consider. The original definitions of security assume a *static* adversary who decides which parties to corrupt before execution of the protocol begins. Subsequently [3,11], researchers began to consider the more challenging setting in which the adversary may *adaptively* decide which parties to corrupt as the protocol progresses—or even after the protocol ends. It is easy to come up with examples of protocols that are secure in a static-corruption model, but that are trivially insecure in the adaptive setting.

Even in a setting where adaptive corruptions are considered, there are different assumptions one can make. Initial work on adaptive security [3] made the assumption that honest parties can securely *erase* local data (e.g., randomness or other internal state) when no longer needed. Later work, led by Canetti et al. [11], sought to avoid this assumption, arguing that it is unwise to rely on other parties to erase data (since there is no way such erasure can be verified) and that it is generally difficult—even for an honest party who intends

---

* Work supported in part by NSF awards #1111599 and #1223623.
** Work done while visiting the University of Maryland.

Y. Dodis and J.B. Nielsen (Eds.): TCC 2015, Part II, LNCS 9015, pp. 586–613, 2015.

to erase data—to ensure that all traces of data are gone. Whether or not erasure is assumed has a significant impact on the complexity of adaptively secure protocols; for example, adaptively secure public-key encryption is fairly simple and efficient [3] if erasure is assumed, but much more complicated (and much less efficient) [11,2,18,16] without this assumption. Similarly, adaptively secure two-party computation is much easier with the assumption of secure erasure [30] than without [14].

Designing protocols without the assumption of secure erasure is difficult, in part, due to the need to deal with *post-execution corruption* (PEC), whereby an adversary can corrupt parties (and hence obtain the randomness they used) even after execution of the protocol has concluded. Handling PEC is inherent to the setting of universal composability (UC) [9], and is important for ensuring sequential composition even in the stand-alone setting [8]. If secure erasure is assumed, the definition of adaptive security does not change whether or not PEC is allowed [10], but without erasure the requirement of dealing with PEC adds significant additional complications.

Prior Work. We are interested in adaptive security, with PEC, in a model where secure erasure is not assumed. Some prior protocols for secure computation in this setting (e.g., [11,2]) assume a majority of the parties remain uncorrupted. Other work [28,27,22,25], including concurrent work of [19], allows *all but one* of the parties to be corrupted. While it may seem strange to worry about corruption of *all* parties, consideration of this case is important when a protocol Π_{outer} invokes some protocol Π_{inner} (not involving all parties running Π_{outer}) as a subroutine. In this case, all parties running Π_{inner} may eventually be corrupted, and security of Π_{outer} should still be guaranteed.

To the best of our knowledge, all prior work giving adaptively secure protocols for general functionalities (without erasure), and tolerating an arbitrary number of corruptions, are based on the Goldreich-Micali-Wigderson [23] paradigm for semi-honest computation, and thus have round complexity linear in the depth of the circuit being computed. These include protocols in the common reference string model [14], the "sunspots" model [15], the key-registration model [1], and, more generally, based on adaptively secure UC puzzles [17]. In addition, all prior work in this setting handles only "adaptively well-formed functionalities" (see [14] for a definition).

1.1 Our Result

We show a constant-round, universally composable protocol for multiparty computation of arbitrary functionalities, with security against a malicious, adaptive adversary corrupting any number of parties. We highlight that our protocol can be used to securely compute *all* functionalities, not just adaptively well-formed ones. Our protocol relies on indistinguishability obfuscation, and assumes a common reference string.

Overview of Our Techniques. The main difficulty in our setting is to construct a constant-round protocol with security against a *semi-honest*, adaptive adversary corrupting any number of parties. Given any such protocol, we can compile it as in [14] to obtain a universally composable protocol with security against a *malicious*, adaptive adversary, and still running in constant rounds. We may also assume secure channels, which can be implemented using adaptively secure encryption.

Our protocol in the semi-honest setting relies on a common reference string (CRS). While it would be more elegant to avoid this assumption, a CRS—or some other form of setup—is anyway needed [12] in order to obtain universally composable computation in the presence of *malicious* adversaries corrupting half or more of the parties, even in a static-corruption model. Thus, as far as our final result (i.e., our protocol with security in the malicious setting) is concerned, some form of setup is unavoidable. Moreover, results of Garg and Sahai [22] indicate that a CRS (or some other form of setup) is needed to obtain constant-round, universally composable, multiparty protocols with adaptive security even in the semi-honest case; see further discussion below.

At its core, our protocol relies on the ability to make arbitrary algorithms *explainable*, an idea we explain in more detail now. Fix some randomized algorithm Alg. Informally, an *explainable* version of Alg is an algorithm $\widetilde{\mathsf{Alg}}$ along with an associated *explain* algorithm $\mathsf{Explain}$ such that, for any input, (1) the distributions over the outputs of $\mathsf{Alg}(\mathsf{input})$ and $\widetilde{\mathsf{Alg}}(\mathsf{input})$ are statistically close, and (2) choosing random coins r, computing $\mathsf{output} := \widetilde{\mathsf{Alg}}(\mathsf{input}; r)$, and outputting (output, r) is computationally indistinguishable from choosing random coins r, computing $\mathsf{output} := \widetilde{\mathsf{Alg}}(\mathsf{input}; r)$, and then outputting $(\mathsf{output}, \mathsf{Explain}(\mathsf{input}, \mathsf{output}))$. That is, the $\mathsf{Explain}$ algorithm provides the ability to sample random coins for $\widetilde{\mathsf{Alg}}$ that "explain" any given input/output pair. (A related notion was considered by Ishai et al. [26], though without any construction being given.)

Sahai and Waters [31] introduced the notion of explainability for the specific case of public-key encryption schemes, in the context of constructing a deniable encryption scheme. We observe that their techniques can be suitably generalized to give an explainable version of *arbitrary* algorithms based on indistinguishability obfuscation for general circuits (and one-way functions). We refer the reader to Section 3 for a formal statement of this result.

Let C be a circuit taking n-bit inputs.[1] Consider the following functionality $\mathsf{NextMsg}$ that (essentially) computes the next-message function for a two-round secure-computation protocol for C based on garbled circuits: $\mathsf{NextMsg}$ takes as input a sequence of first-round messages $\mathsf{OT}_{1,1}, \ldots, \mathsf{OT}_{1,n}$ for a two-round, adaptively secure, oblivious-transfer (OT) protocol (e.g., the protocol of [14]); it then (1) computes a garbled circuit GC corresponding to C, along with input-wire labels $\{(y_{i,0}, y_{i,1})\}_{i=1}^{n}$, and (2) computes a sequence of OT responses $\mathsf{OT}_{2,1}, \ldots, \mathsf{OT}_{2,n}$. (These responses allow the party that generated $\mathsf{OT}_{1,i}$ using input bit b to recover $y_{i,b}$ while learning nothing about $y_{i,1-b}$.) The output of $\mathsf{NextMsg}$ is $(\mathsf{GC}, \mathsf{OT}_{2,1}, \ldots, \mathsf{OT}_{2,n})$. The CRS for our protocol will be $\widetilde{\mathsf{NextMsg}}$,

---

[1] We assume for simplicity here that C is deterministic. Randomized functionalities are handled in Section 4.

an explainable version of NextMsg.[2] We note that, in contrast to [31], in the real-world execution no parties have access to the Explain algorithm corresponding to NextMsg.

Our multiparty protocol computing C can now be described quite simply. The protocol proceeds in four rounds. Say we have n parties $P_1, \ldots, P_n$ holding inputs $x_1, \ldots, x_n$, respectively. These parties generate first-round OT messages $OT_{1,1}, \ldots, OT_{1,n}$ (with the party who is supposed to provide the ith input generating $OT_{1,i}$), and send these to P_n. Party P_n then runs $NextMsg(OT_{1,1}, \ldots, OT_{1,n})$ to obtain $GC, OT_{2,1}, \ldots, OT_{2,n}$, and sends $OT_{2,i}$ to the corresponding party (which might be itself). Each party P_i then locally recovers y_i, the label for the ith input wire of the garbled circuit, and sends y_i to P_n. Finally, P_n evaluates the garbled circuit GC using the provided input-wire labels to obtain the output z, and sends z to all the other parties.[3] Only the third- and fourth-round messages need to be sent via a secure channel.

We now describe the simulator informally. Our simulator begins by generating NextMsg along with its associated Explain algorithm, and letting NextMsg be the CRS. It simulates $OT_{1,1}, \ldots, OT_{1,n}$ and $OT_{2,1}, \ldots, OT_{2,n}$ using the simulator for the OT protocol (recall the OT protocol is adaptively secure), and uses these for the first two rounds of the protocol. Upon corruption of party P_i, the simulator corrupts that party in the ideal world and learns its input x_i and the output z. Then:

- If this is the first corruption, the simulator generates a simulated garbled circuit GC consistent with output z, along with n input-wire labels $y_1, \ldots, y_n$. It also uses the Explain algorithm to generate random coins r^* consistent with running NextMsg on input $OT_{1,1}, \ldots, OT_{1,n}$ and obtaining output $GC, OT_{2,1}, \ldots, OT_{2,n}$.
- The simulator uses the simulator for the OT protocol to generate internal state for P_i consistent with input x_i and output y_i, and returns this to the adversary. In addition, if $P = P_n$ then it returns r^* to the adversary.

Notably, our simulator is "corruption oblivious" [4]. Roughly, this means the behavior of the simulator upon corruption of a party is independent of the ideal state learned previously. By [4, Theorem 1.2], this means our protocol is also the first *leakage-tolerant* protocol with arbitrary leakage for general functionalities under semi-honest corruption. Moreover, by [5, Theorem 1], we also obtain the first construction of a *2-component OCL compiler* (see [5] for a definition).

Impossibility Results? We briefly mention two impossibility results regarding (constant-round) adaptively secure computation, and explain why they do not apply in our setting.

---

[2] As described, the CRS depends on the circuit C. However, by taking C to be a universal circuit, the CRS can be fixed independently of the "actual" function f the parties wish to compute (other than the size of a circuit for f).

[3] As described, all parties learn the output of the computation. Standard techniques can be used to handle the general case in which each party learns a possibly different function of the inputs.

First, our protocol can compute *arbitrary* randomized functionalities, not just *adaptively well-formed* ones. (We refer to [14] for a definition of this term.) This may seem somewhat surprising in light of an impossibility result of Ishai et al. [26] showing that adaptively secure computation of functionalities that are not adaptively well-formed is impossible. A closer examination of their result, however, reveals that it does *not* hold in the CRS model.

Second, Garg and Sahai [22] show that no constant-round, adaptively secure, multiparty protocol can be proven secure using black-box techniques; although they only claim this result for protocols with security against malicious adversaries, their proof appears to extend to the case of semi-honest adversaries as well. Again, however, their impossibility result only applies to the "plain" model with no setup, whereas we assume a CRS.

Concurrent Work. Independent of our work, two other groups of researchers have also studied the problem of constant-round adaptively secure computation. Canetti et al. [13] give a protocol that is similar in spirit to ours, but works only for the *two-party* case and requires sub-exponentially hard indistinguishability obfuscation. Garg and Polychroniadou [21], though also relying on indistinguishability obfuscation, follow a different approach. They give a *round-optimal*, adaptively secure protocol for the multiparty setting. We remark that both these other works only consider adaptively well-formed functionalities.

1.2   Organization of the Paper

We review some standard cryptographic background and primitives in Section 2. In Section 3, we introduce the notion of an *explainable* algorithm, and show how the Sahai-Waters compiler [31] can be used to make any algorithm explainable. Finally, in Section 4 we present a constant-round multiparty computation protocol tolerating a semi-honest, adaptive adversary corrupting any number of parties. Applying the compiler of Canetti et al. [14] yields a constant-round protocol tolerating a *malicious*, adaptive adversary corrupting any number of parties.

2   Preliminaries

We let λ denote the security parameter. We refer to previous work [8,10,30] for definitions of secure computation in the adaptive-corruption setting (with PEC).

2.1   Garbled Circuits

We rely on the standard notion of garbled circuits [32]. However, we use slightly non-standard notation that we introduce here. Let C be a randomized circuit taking n-bit inputs and using λ bits of randomness. We abstract the construction/evaluation of a garbled circuit for C via algorithms GenGC, EvalGC with the following properties. GenGC is a randomized algorithm that takes as input 1^λ and C, and outputs a garbled circuit GC along with $2n$ input-wire labels

$y_{1,0}, y_{1,1}, \ldots, y_{n,0}, y_{n,1} \in \{0,1\}^\lambda$ and 2λ random-wire labels $w_{1,0}, w_{1,1}, \ldots, w_{\lambda,0}$, $w_{\lambda,1} \in \{0,1\}^\lambda$. Deterministic algorithm EvalGC takes as input GC and $n + \lambda$ labels $y_1, \ldots, y_n, w_1, \ldots, w_\lambda$, and outputs a value z.

Correctness requires that for any GC, $(\{y_{i,0}, y_{i,1}\}_{i=1}^n, \{w_{i,0}, w_{i,1}\}_{i=1}^\lambda)$ output by GenGC$(1^\lambda, C)$, any $x \in \{0,1\}^n$ and any $r \in \{0,1\}^\lambda$, we have

$$\mathsf{EvalGC}\left(\mathsf{GC}, \{y_{i,x_i}\}_{i=1}^n, \{w_{i,r_i}\}_{i=1}^\lambda\right) = C(x;r).$$

Security requires an efficient simulator SimGC such that for all x, r, the distribution

$$\left\{\left(\mathsf{GC}, \{(y_{i,0}, y_{i,1})\}_{i=1}^n, \{(w_{i,0}, w_{i,1})\}_{i=1}^\lambda\right) \leftarrow \mathsf{GenGC}(1^\lambda, C) : \right.$$
$$\left. \left(\mathsf{GC}, \{y_{i,x_i}\}_{i=1}^n, \{w_{i,r_i}\}_{i=1}^\lambda\right)\right\}$$

is computationally indistinguishable from the output of $\mathsf{SimGC}(1^\lambda, C, C(x;r))$.

2.2 Adaptively Secure Oblivious Transfer

Our protocol uses a two-round, semi-honest, adaptively secure OT protocol as a building block. A suitable construction can be found in [14].

A two-round OT protocol Π_{OT} comprises three algorithms: a receiver algorithm R_{OT}, a sender algorithm S_{OT}, and an evaluation algorithm E_{OT}. Algorithm R_{OT} takes as input a bit b and random coins r_R, and outputs initial message OT_1. Algorithm S_{OT} takes as input an initial message OT_1, a pair of λ-bit strings (y_0, y_1), and randomness r_S, and outputs message OT_2. The evaluation algorithm E_{OT} takes as input b, r_R, and OT_2 and outputs the λ-bit string y_b.

For our purposes we require the following property that is implied by semi-honest, adaptive security of Π_{OT}. There is exist an efficient simulator $\mathsf{SimOT} = (\mathsf{SimOT}_1, \mathsf{SimOT}_2)$, where SimOT_2 is deterministic, such that (1) SimOT_1 outputs a transcript $(\mathsf{OT}_1, \mathsf{OT}_2)$ along with state st and (2) SimOT_2, given as input b, y, and st, outputs coins r_R for the receiver consistent with $(\mathsf{OT}_1, \mathsf{OT}_2)$ and the receiver holding input b and obtaining output y; for any b, y_0, y_1, the distribution

$$\left\{r_R, r_S \leftarrow \{0,1\}^*; \mathsf{OT}_1 := R_{\mathsf{OT}}(b; r_R) : \left(r_R, \mathsf{OT}_1, S_{\mathsf{OT}}(\mathsf{OT}_1, y_0, y_1; r_S)\right)\right\}$$

is computationally indistinguishable from

$$\left\{\begin{array}{l}(\mathsf{OT}_1, \mathsf{OT}_2, \mathsf{st}) \leftarrow \mathsf{SimOT}_1(1^\lambda); \\ r_R := \mathsf{SimOT}_2(1^\lambda, b, y_b, \mathsf{st})\end{array} : (r_R, \mathsf{OT}_1, \mathsf{OT}_2)\right\}.$$

That is, we only require "one-sided security" [25] for adaptive corruption of the receiver.

If we define algorithm $\mathsf{SimOT}_1'(1^\lambda)$ to run $\mathsf{SimOT}_1(1^\lambda)$ and output only $(\mathsf{OT}_1, \mathsf{st})$, and define the algorithm $\mathsf{SimOT}_2'(1^\lambda, b, \mathsf{st})$ to simply run $\mathsf{SimOT}_2(1^\lambda, b, 0^\lambda, \mathsf{st})$, then for any b the distribution $\left\{r_R \leftarrow \{0,1\}^* : \left(r_R, R_{\mathsf{OT}}(b; r_R)\right)\right\}$ is computationally indistinguishable from

$$\left\{\begin{array}{l}(\mathsf{OT}_1, \mathsf{st}) \leftarrow \mathsf{SimOT}_1'(1^\lambda); \\ r_R := \mathsf{SimOT}_2'(1^\lambda, b, \mathsf{st})\end{array} : (r_R, \mathsf{OT}_1)\right\}.$$

2.3   Indistinguishability Obfuscation

We use an indistinguishability obfuscator as a building block. A PPT machine $i\mathcal{O}$ is an *indistinguishability obfuscator* for a circuit class $\{\mathcal{C}_\lambda\}$ if the following conditions are satisfied:

Correctness. For all λ, and all $C \in \mathcal{C}_\lambda$, it holds that C and $i\mathcal{O}(1^\lambda, C)$ compute the same function.

Polynomial slowdown. There is a polynomial $p(\cdot)$ such that $|i\mathcal{O}(1^\lambda, C)| \leq p(\lambda) \cdot |C|$ for all $C \in \mathcal{C}_\lambda$.

Indistinguishability. For any sequence $\{(C_{\lambda,0}, C_{\lambda,1}, \mathsf{aux}_\lambda)\}_\lambda$ where $C_{\lambda,0}, C_{\lambda,1} \in \mathcal{C}_\lambda$, $C_{\lambda,0} \equiv C_{\lambda,1}$, and $|C_{\lambda,0}| = |C_{\lambda,1}|$, and any PPT distinguisher D, there is a negligible function negl such that:

$$\left| \Pr[D(i\mathcal{O}(1^\lambda, C_{\lambda,0}), \mathsf{aux}_\lambda) = 1] - \Pr[D(i\mathcal{O}(1^\lambda, C_{\lambda,1}), \mathsf{aux}_\lambda) = 1] \right| \leq \mathsf{negl}(\lambda).$$

When clear from the context, we will often omit the security parameter 1^λ as an input to $i\mathcal{O}$ and as a subscript for C.

$i\mathcal{O}$ is an *indistinguishability obfuscator* for P/poly if there is a polynomial p such that $i\mathcal{O}$ is an indistinguishability obfuscator for $\{\mathcal{C}_\lambda\}$, where $\mathcal{C}_\lambda$ contains all circuits of size at most $p(\lambda)$. Garg et al. [20] have shown the first candidate construction of indistinguishability obfuscators for P/poly.

3   Explainability Compilers

Sahai and Waters [31] define a notion of *explainability* for public-key encryption, and show a compiler that transforms any public-key encryption scheme into an explainable version. Here, we generalize the notion of explainability for an *arbitrary* algorithm Alg, and show that the Sahai-Waters compiler can be used to transform any algorithm Alg into an explainable version $\widetilde{\mathsf{Alg}}$.

At a high level, an explainability compiler takes as input (a description of) a randomized algorithm Alg, and outputs two algorithms $\widetilde{\mathsf{Alg}}$, Explain. The first of these is a randomized algorithm computing the same functionality as Alg. The second algorithm, roughly speaking, takes an input/output pair input, output and produces random coins r consistent with running $\widetilde{\mathsf{Alg}}(\mathsf{input})$ and obtaining the result output. That is, the algorithm "explains" the input/output pair input, output. We now give a formal definition.

Definition 1. *A PPT algorithm* Comp *is an* explainability compiler *if for every efficient, randomized circuit* Alg, *the following hold:*

Polynomial slowdown. *There is a polynomial $p(\cdot)$ such that, for any $(\widetilde{\mathsf{Alg}},$ Explain) output by $\mathsf{Comp}(1^\lambda, \mathsf{Alg})$ it holds that $|\widetilde{\mathsf{Alg}}| \leq p(\lambda) \cdot |\mathsf{Alg}|$.*

Statistical functional equivalence. *With overwhelming probability over choice of $(\widetilde{\mathsf{Alg}}, \star)$ as output by $\mathsf{Comp}(1^\lambda, \mathsf{Alg})$, the distribution of $\widetilde{\mathsf{Alg}}(\mathsf{input})$ is statistically close to the distribution of $\mathsf{Alg}(\mathsf{input})$ for all* input.

$$\overline{\text{Alg}}$$

Hardwired constants: Keys K_1, K_2, and K_3.
Input: Input input and randomness $u = (u[1], u[2])$.

1. Let $\text{input}', \text{output}', r') := F_3(K_3, u[1]) \oplus u[2]$. If it is the case that $\text{input} = \text{input}'$ and $u[1] = F_2(K_2, (\text{input}', \text{output}', r'))$, then output $\text{output} := \text{output}'$ and end.
2. Else let $x := F_1(K_1, (\text{input}, u))$ and output $\text{output} := \text{Alg}(\text{input}; x)$.

Fig. 1. Program $\overline{\text{Alg}}$

Explainability. *The success probability of every non-uniform, polynomial-time adversary $\mathcal{A}$ in the following experiment is negligibly close to $1/2$:*

1. *$\mathcal{A}(1^\lambda)$ outputs input^* of its choice.*
2. *$\text{Comp}(1^\lambda, \text{Alg})$ is run to obtain $(\widetilde{\text{Alg}}, \text{Explain})$.*
3. *Choose uniform coins $r_0 \in \{0, 1\}^*$ and compute $\text{output}^* := \widetilde{\text{Alg}}(\text{input}^*; r_0)$.*
4. *Compute $r_1 \leftarrow \text{Explain}(\text{input}^*, \text{output}^*)$.*
5. *Choose a uniform bit b and give $\widetilde{\text{Alg}}, \text{output}^*, r_b$ to $\mathcal{A}$.*
6. *$\mathcal{A}$ outputs a bit b', and succeeds if $b' = b$.*

We highlight one key difference between our definition and the corresponding one from [31]: in our case input^* is an arbitrary length value (depending on the domain of Alg) chosen by the adversary, whereas in [31] the input to the explainable algorithm is a *single bit* chosen uniformly (and given to the adversary). Because of this, and due to the way the explainability compiler is constructed, we require the adversary to choose input^* "non-adaptively," i.e., before being given $\widetilde{\text{Alg}}$. This definition of explainability suffices for our eventual protocol.

3.1 Constructing an Explainability Compiler

Following [31], we now show how to construct an explainability compiler. As in [31], we rely on an indistinguishability obfuscator, $i\mathcal{O}$, for P/poly and three different pseudorandom function (PRF) variants (cf. Appendix A):

- A *puncturable, extracting* PRF $F_1(K_1, \cdot)$ that accepts inputs of length $\ell_1 + \ell_2 + \ell_{\text{in}}$, and outputs strings of length ℓ_r. It is extracting when the input min-entropy is greater than $\ell_r + 2\lambda + 4$, with statistical closeness less than $2^{-(\lambda+1)}$. Observe that $\ell_{\text{in}} + \ell_1 + \ell_2 \geq \ell_r + 2\lambda + 4$, and thus if one-way functions exist then such a PRF exists by Theorem 4.
- A *puncturable, statistically injective* PRF $F_2(K_2, \cdot)$ that accepts inputs of length $2\lambda + \ell_{\text{in}} + \ell_{\text{out}}$, and outputs strings of length ℓ_1. Observe that $\ell_1 \geq 2 \cdot (2\lambda + \ell_{\text{in}} + \ell_{\text{out}}) + \lambda$, and thus if one-way functions exist then such a PRF exists by Theorem 3.
- A *puncturable* PRF $F_3(K_3, \cdot)$ that accepts inputs of length ℓ_1 and outputs strings of length ℓ_2. If one-way functions exist, then such a PRF exists by Theorem 2.

We define $\mathsf{Comp}(1^\lambda, \mathsf{Alg})$ as follows. Let $\mathsf{Alg} : \{0,1\}^{\ell_\text{in}} \times \{0,1\}^{\ell_r} \to \{0,1\}^{\ell_\text{out}}$ be an algorithm with domain $\{0,1\}^{\ell_\text{in}}$, range $\{0,1\}^{\ell_\text{out}}$, and randomness length ℓ_r. Our compiled program $\widetilde{\mathsf{Alg}}$ will take input $\mathsf{input} \in \{0,1\}^{\ell_\text{in}}$ and randomness $u = (u[1], u[2])$ of length $\ell_1 + \ell_2$, where $|u[1]| = \ell_1 = 5\lambda + 2(\ell_\text{in} + \ell_\text{out}) + \ell_r$ and $|u[2]| = \ell_2 = 2\lambda + \ell_\text{in} + \ell_\text{out}$. Our compiler first samples keys K_1, K_2, and K_3 for PRFs F_1, F_2, and F_3, respectively. It then defines algorithms $\overline{\mathsf{Alg}}$ and $\overline{\mathsf{Explain}}$ as in Figures 1 and 2, respectively. Finally, it computes $\widetilde{\mathsf{Alg}} \leftarrow i\mathcal{O}(\overline{\mathsf{Alg}})$ and $\mathsf{Explain} \leftarrow i\mathcal{O}(\overline{\mathsf{Explain}})$, and outputs $(\widetilde{\mathsf{Alg}}, \mathsf{Explain})$.

The proofs of security for our compiler, given for completeness in Appendix B, follow closely along the lines of the analogous proofs in [31]. Specifically, the proof of statistical functional equivalence closely follows the proof used by Sahai and Waters to establish IND-CPA security of their deniable encryption scheme, and the proof of explainability follows the Sahai-Waters proof establishing explainability of their deniable encryption scheme. We highlight, however, that in our proof of explainability a difference arises because in our case the input input^* is an arbitrary length value (depending on the domain of Alg), whereas in [31] the input is just a single bit. We are able to adapt the proof to this case because we do not allow input^* to depend on $\widetilde{\mathsf{Alg}}$.

4 A Semi-Honest, Adaptively Secure Protocol

We describe here a protocol for secure computation of a randomized circuit C by a set of parties $P_1, \ldots, P_n$. We assume for simplicity that all parties learn the output of C; using standard techniques, we can handle the general case in which each party learns a possibly different function of the inputs. For ease of notation, we assume that the domain of C is $\{0,1\}^n$ with party P_i providing the ith input $x_i \in \{0,1\}$. (One can easily verify that our protocol and proof generalize to the case of arbitrary-length inputs.) We also assume without loss of generality that C uses λ random bits.

The CRS of our protocol is an "explainable" version $\widetilde{\mathsf{NextMsg}}$ of the algorithm $\mathsf{NextMsg}$ defined in Figure 3. That is, the CRS is generated by computing $(\widetilde{\mathsf{NextMsg}}, \mathsf{Explain}) \leftarrow \mathsf{Comp}(1^\lambda, \mathsf{NextMsg})$ and letting the CRS be $\widetilde{\mathsf{NextMsg}}$. As described, the CRS depends on C (since $\mathsf{NextMsg}$ does); however, by letting C

$\overline{\mathsf{Explain}}$

Hardwired constants: Keys K_2 and K_3.
Input: input, output, and randomness $r \in \{0,1\}^\lambda$.

   1. Set $\alpha := F_2(K_2, (\mathsf{input}, \mathsf{output}, \mathsf{PRG}(r)))$ and let $\beta := F_3(K_3, \alpha) \oplus (\mathsf{input}, \mathsf{output}, \mathsf{PRG}(r))$.
     Output (α, β).

Fig. 2. Program $\overline{\mathsf{Explain}}$

NextMsg

Inputs: $OT_{1,1}, \ldots, OT_{1,n}$;   randomness $r_1, \ldots, r_\lambda \in \{0,1\}$ and
$r_{GC}, r_{S,1}, \ldots, r_{S,n} \in \{0,1\}^*$.

1. Run $\mathsf{GenGC}(1^\lambda, C; r_{GC})$ to produce the garbled circuit GC along with n
   pairs of input-wire labels $\{(y_{i,0}, y_{i,1})\}_{i=1}^n$ and λ pairs of random-wire
   labels $\{(w_{i,0}, w_{i,1})\}_{i=1}^\lambda$.
2. For $i \in [n]$, run S_{OT} on input $OT_{1,i}$ and $(y_{i,0}, y_{i,1})$ using randomness $r_{S,i}$, to obtain $OT_{2,i}$.
3. Output GC, OT messages $\{OT_{2,i}\}_{i=1}^n$, and random-wire labels
   $w_{1,r_1}, \ldots, w_{\lambda,r_\lambda}$.

Fig. 3. Algorithm NextMsg. The security parameter 1^λ and circuit C are hardwired.

be a universal circuit the CRS can be fixed independently of the "actual" function the parties wish to compute. We note that we allow the environment $\mathcal{Z}$ to
choose the parties' inputs depending on the CRS.

Let $\Pi_{OT} = (R_{OT}, S_{OT}, E_{OT})$ be a two-round, semi-honest, adaptively secure
OT protocol (cf. Section 2.2). Our secure-computation protocol Π is defined
in Figure 4. We describe the protocol assuming the existence of secure channels; these can be instantiated using any adaptively secure public-key encryption
scheme.

Theorem 1. *Assume* Comp *is an explainability compiler, and* GenGC *and* Π_{OT}
satisfy the definitions from Sections 2.1 and 2.2, respectively. Then protocol Π *in
Figure 4 UC-realizes functionality* C *in the presence of a semi-honest, adaptive
adversary corrupting any number of parties.*

Proof. Let SimGC, SimOT denote appropriate simulators as defined in Section 2.
Fix an environment $\mathcal{Z}$ and a dummy adversary $\mathcal{A}$ attacking protocol Π. Recall
that we allow the environment $\mathcal{Z}$ to adaptively choose the inputs of all parties
after seeing the common reference string. Without loss of generality, we assume
$\mathcal{Z}$ first observes the entire protocol transcript (which, since we use secure channels in rounds 3 and 4, consists only of the messages sent in the first two rounds)
before corrupting any parties. Our simulator Sim for this adversary proceeds as
follows:

1. Compute $(\widetilde{\mathsf{NextMsg}}, \mathsf{Explain}) \leftarrow \mathsf{Comp}(1^\lambda, \mathsf{NextMsg})$, and give $\widetilde{\mathsf{NextMsg}}$ to $\mathcal{Z}$
   as the CRS.
2. Run $\mathsf{SimOT}_1(1^\lambda)$ a total of n times to obtain $\{(OT_{1,i}, OT_{2,i}, \mathsf{st}_i)\}_{i=1}^n$. Give
   $OT_{1,1}, \ldots, OT_{1,n-1}$ to $\mathcal{Z}$ as the first-round message, and $OT_{2,1}, \ldots, OT_{2,n-1}$
   to $\mathcal{Z}$ as the second-round message.
3. When $\mathcal{Z}$ requests to corrupt party P_i, corrupt P_i in the ideal world to learn
   its input x_i and the output z. Then:
   - If this is the first party to be corrupted, compute $(GC, \{y_i\}_{i=1}^n, \{w_i\}_{i=1}^\lambda)$
     $\leftarrow \mathsf{SimGC}(1^\lambda, C, z)$ and $r_n^* \leftarrow \mathsf{Explain}((OT_{1,1}, \ldots, OT_{1,n}), (GC, OT_{2,1}, \ldots,$

Semi-Honest, Adaptively Secure Multiparty Computation

Common input:

 - $CRS = \widetilde{\mathsf{NextMsg}}$.
 - Description of a randomized circuit C.

Private inputs: Every party P_i has private input $x_i \in \{0, 1\}$.

Each P_i: *Compute first-round OT messages*:

 - Sample random coins $r_{R,i} \leftarrow \{0,1\}^*$ of appropriate length.
 - Compute $\mathsf{OT}_{1,i} := R_{\mathsf{OT}}(x_i; r_{R,i})$ and, for $i \in [n-1]$, send $\mathsf{OT}_{1,i}$ to P_n.

P_n: *Compute garbled circuit and second-round OT messages*:

 - Sample random coins $r_n \leftarrow \{0,1\}^*$ of appropriate length.
 - Compute

   $$(\mathsf{GC}, \mathsf{OT}_{2,1}, \ldots, \mathsf{OT}_{2,n}, w_1, \ldots, w_\lambda) := \widetilde{\mathsf{NextMsg}}(\mathsf{OT}_{1,1}, \ldots, \mathsf{OT}_{1,n}; r_n).$$

 - For $i \in [n-1]$, send $\mathsf{OT}_{2,i}$ to P_i.

Each P_i: *Recover OT output*:

 - Compute $y_i := E_{\mathsf{OT}}(x_i, r_{R,i}, \mathsf{OT}_{2,i})$ and, for $i \in [n-1]$, send y_i to P_n over a secure channel.

P_n: *Evaluate garbled circuit and broadcast output*:

 - Compute $z := \mathsf{EvalGC}(\mathsf{GC}, \{y_i\}_{i=1}^n, \{w_i\}_{i=1}^\lambda)$.
 - For $i \in [n-1]$, send z to P_i over a secure channel.

Output: Each party P_i outputs z.

Fig. 4. Protocol Π for computing randomized circuit C.

$\mathsf{OT}_{2,n}, w_1, \ldots, w_n))$. Store these values to be used, as needed, in the rest of the simulation.
 - In any case, compute $r_{R,i} := \mathsf{SimOT}_2(1^\lambda, x_i, y_i, \mathsf{st}_i)$ and give x_i, z, y_i, and $r_{R,i}$ to $\mathcal{Z}$. In addition, if $i = n$ give $\{y_i\}_{i=1}^{n-1}$ and r_n^* to $\mathcal{Z}$.
4. Output whatever $\mathcal{Z}$ outputs.

We prove that the output of $\mathcal{Z}$ when interacting with $\mathcal{A}$ and parties in a real-world execution of protocol Π is indistinguishable from the output of $\mathcal{Z}$ when interacting with Sim and the functionality C in an ideal-world execution of the protocol. We do so by considering a sequence of hybrid experiments, beginning with the real-world execution and ending with the ideal-world execution, and showing that each experiment is computationally indistinguishable from the preceding one.

Hybrid 0. This corresponds to the real-world execution of the protocol. We write the experiment in a format convenient for the proof. This experiment proceeds via the following steps:

1. Compute $(\widetilde{\mathsf{NextMsg}}, \mathsf{Explain}) \leftarrow \mathsf{Comp}(1^\lambda, \mathsf{NextMsg})$, and give $\widetilde{\mathsf{NextMsg}}$ to $\mathcal{Z}$ as the CRS. $\mathcal{Z}$ chooses inputs $x_1, \ldots, x_n$.
2. For $i \in [n]$, sample coins $r_{R,i}$ and compute $\mathsf{OT}_{1,i} := R_{\mathsf{OT}}(x_i; r_{R,i})$. Give the sequence of values $\mathsf{OT}_{1,1}, \ldots, \mathsf{OT}_{1,n-1}$ to $\mathcal{Z}$ as the first-round message.
3. Sample coins r_n and compute

$$(\mathsf{GC}, \mathsf{OT}_{2,1}, \ldots, \mathsf{OT}_{2,n}, w_1, \ldots, w_\lambda) := \widetilde{\mathsf{NextMsg}}(\mathsf{OT}_{1,1}, \ldots, \mathsf{OT}_{1,n}; r_n).$$

Give $\mathsf{OT}_{2,1}, \ldots, \mathsf{OT}_{2,n-1}$ to $\mathcal{Z}$ as the second-round message.
4. When $\mathcal{Z}$ requests to corrupt party P_i, compute $y_i := E_{\mathsf{OT}}(x_i, r_{R,i}, \mathsf{OT}_{2,i})$ and give x_i, z, y_i, and $r_{R,i}$ to $\mathcal{Z}$. In addition, if $i = n$ then compute $y_i := E_{\mathsf{OT}}(x_i, r_{R,i}, \mathsf{OT}_{2,i})$ for $i \in [n-1]$ and give $\{y_i\}_{i=1}^{n-1}$ and r_n to $\mathcal{Z}$.

Hybrid 1. This experiment is similar to the previous one, except that the OT_1 messages and the random coins $\{r_{R,i}\}$ are generated by the simulator for the OT protocol (cf. Section 2.2). That is, the experiment proceeds via the following steps:

1. Compute $(\widetilde{\mathsf{NextMsg}}, \mathsf{Explain}) \leftarrow \mathsf{Comp}(1^\lambda, \mathsf{NextMsg})$, and give $\widetilde{\mathsf{NextMsg}}$ to $\mathcal{Z}$ as the CRS. $\mathcal{Z}$ chooses inputs $x_1, \ldots, x_n$.
2. Run $\mathsf{SimOT}'_1(1^\lambda)$ a total of n times to obtain $\{(\mathsf{OT}_{1,i}, \mathsf{st}_i)\}_{i=1}^n$. Give the sequence of values $\mathsf{OT}_{1,1}, \ldots, \mathsf{OT}_{1,n-1}$ to $\mathcal{Z}$ as the first-round message.
3. Sample coins r_n and compute

$$(\mathsf{GC}, \mathsf{OT}_{2,1}, \ldots, \mathsf{OT}_{2,n}, w_1, \ldots, w_\lambda) := \widetilde{\mathsf{NextMsg}}(\mathsf{OT}_{1,1}, \ldots, \mathsf{OT}_{1,n}; r_n).$$

Give $\mathsf{OT}_{2,1}, \ldots, \mathsf{OT}_{2,n-1}$ to $\mathcal{Z}$ as the second-round message.
4. When $\mathcal{Z}$ corrupts party P_i, compute $r_{R,i} := \mathsf{SimOT}'_2(1^\lambda, x_i, \mathsf{st}_i)$ and $y_i := E_{\mathsf{OT}}(x_i, r_{R,i}, \mathsf{OT}_{2,i})$, and give x_i, z, y_i, and $r_{R,i}$ to $\mathcal{Z}$. In addition, if $i = n$ then for $i \in [n-1]$ compute $r_{R,i} := \mathsf{SimOT}'_2(1^\lambda, x_i, \mathsf{st}_i)$ and $y_i := E_{\mathsf{OT}}(x_i, r_{R,i}, \mathsf{OT}_{2,i})$, and give $\{y_i\}_{i=1}^{n-1}$ and r_n to $\mathcal{Z}$.

It follows immediately by security of the OT protocol (and a straightforward hybrid argument) that this experiment is computationally indistinguishable from the previous one.

Hybrid 2. This experiment is similar to the previous one, except that we now use the Explain algorithm to generate the random coins r_n. That is, the experiment proceeds as follow:

1. Compute $(\widetilde{\mathsf{NextMsg}}, \mathsf{Explain}) \leftarrow \mathsf{Comp}(1^\lambda, \mathsf{NextMsg})$, and give $\widetilde{\mathsf{NextMsg}}$ to $\mathcal{Z}$ as the CRS. $\mathcal{Z}$ chooses inputs $x_1, \ldots, x_n$.
2. Run $\mathsf{SimOT}'_1(1^\lambda)$ a total of n times to obtain $\{(\mathsf{OT}_{1,i}, \mathsf{st}_i)\}_{i=1}^n$. Give the sequence of values $\mathsf{OT}_{1,1}, \ldots, \mathsf{OT}_{1,n-1}$ to $\mathcal{Z}$ as the first-round message.
3. Sample coins r_n and compute

$$(\mathsf{GC}, \mathsf{OT}_{2,1}, \ldots, \mathsf{OT}_{2,n}, w_1, \ldots, w_\lambda) := \widetilde{\mathsf{NextMsg}}(\mathsf{OT}_{1,1}, \ldots, \mathsf{OT}_{1,n}; r_n).$$

In addition, let $\mathsf{input}^* = (\mathsf{OT}_{1,1}, \ldots, \mathsf{OT}_{1,n})$ and $\mathsf{output}^* = (\mathsf{GC}, \mathsf{OT}_{2,1}, \ldots, \mathsf{OT}_{2,n}, w_1, \ldots, w_\lambda)$, and compute $r_n^* \leftarrow \mathsf{Explain}(\mathsf{input}^*, \mathsf{output}^*)$. Give $\mathsf{OT}_{2,1}, \ldots, \mathsf{OT}_{2,n-1}$ to $\mathcal{Z}$ as the second-round message.

4. When $\mathcal{Z}$ corrupts party P_i, compute $r_{R,i} := \mathsf{SimOT}_2'(1^\lambda, x_i, \mathsf{st}_i)$ and $y_i := E_{\mathsf{OT}}(x_i, r_{R,i}, \mathsf{OT}_{2,i})$, and give x_i, z, y_i, and $r_{R,i}$ to $\mathcal{Z}$. In addition, if $i = n$ then for $i \in [n-1]$ compute $r_{R,i} := \mathsf{SimOT}_2'(1^\lambda, x_i, \mathsf{st}_i)$ and $y_i := E_{\mathsf{OT}}(x_i, r_{R,i}, \mathsf{OT}_{2,i})$, and give $\{y_i\}_{i=1}^{n-1}$ and r_n^* to $\mathcal{Z}$.

Computationally indistinguishability of this experiment from the previous one follows from the definition of explainability (cf. Definition 1), and the fact that Comp is an explainability compiler. To see this, say there is an efficient adversary $\mathcal{Z}$ and a non-uniform, polynomial-time distinguisher D that distinguishes the outcome of Hybrid 1 from that of Hybrid 2. We show how to use this to construct an attacker $\mathcal{A}'$ violating explainability. $\mathcal{A}'$ works as follows: it runs $\mathsf{SimOT}_1'(1^\lambda)$ a total of n times to obtain $\{(\mathsf{OT}_{1,i}, \mathsf{st}_i)\}_{i=1}^n$, and outputs the value $\mathsf{input}^* = (\mathsf{OT}_{1,1}, \ldots, \mathsf{OT}_{1,n})$. Given $\widetilde{\mathsf{NextMsg}}, \mathsf{output}^*, r$ in response, where $\mathsf{output}^* = (\mathsf{GC}, \mathsf{OT}_{2,1}, \ldots, \mathsf{OT}_{2,n}, w_1, \ldots, w_\lambda)$, it then does:

1. Give $\widetilde{\mathsf{NextMsg}}$ to $\mathcal{Z}$ as the CRS. $\mathcal{Z}$ chooses inputs $x_1, \ldots, x_n$.
2. Give $\mathsf{OT}_{1,1}, \ldots, \mathsf{OT}_{1,n-1}$ to $\mathcal{Z}$ as the first-round message, and $\mathsf{OT}_{2,1}, \ldots, \mathsf{OT}_{2,n-1}$ to $\mathcal{Z}$ as the second-round message.
3. When $\mathcal{Z}$ corrupts party P_i, compute $r_{R,i} := \mathsf{SimOT}_2'(1^\lambda, x_i, \mathsf{st}_i)$ and $y_i := E_{\mathsf{OT}}(x_i, r_{R,i}, \mathsf{OT}_{2,i})$, and give x_i, z, y_i, and $r_{R,i}$ to $\mathcal{Z}$. In addition, if $i = n$ then for $i \in [n-1]$ compute $r_{R,i} := \mathsf{SimOT}_2'(1^\lambda, x_i, \mathsf{st}_i)$ and $y_i := E_{\mathsf{OT}}(x_i, r_{R,i}, \mathsf{OT}_{2,i})$, and give $\{y_i\}_{i=1}^{n-1}$ and r to $\mathcal{Z}$.

Finally, run D on the output of $\mathcal{Z}$ and output the result. It is easy to see that if the coins r are those used to run $\widetilde{\mathsf{NextMsg}}$, then the view of $\mathcal{Z}$ when run as a subroutine by $\mathcal{A}'$ corresponds to Hybrid 1, whereas if the coins r are those output by Explain, then the view of $\mathcal{Z}$ when run as a subroutine by $\mathcal{A}'$ corresponds to Hybrid 2. Indistinguishability of the two experiments follows.

Hybrid 3. This is similar to the previous experiment, except that $\widetilde{\mathsf{NextMsg}}$ and Explain are used in place of $\widetilde{\mathsf{NextMsg}}$. That is, the experiment proceeds as follows:

1. Compute $(\widetilde{\mathsf{NextMsg}}, \mathsf{Explain}) \leftarrow \mathsf{Comp}(1^\lambda, \mathsf{NextMsg})$, and give $\widetilde{\mathsf{NextMsg}}$ to $\mathcal{Z}$ as the CRS. $\mathcal{Z}$ chooses inputs $x_1, \ldots, x_n$.
2. Run $\mathsf{SimOT}_1'(1^\lambda)$ a total of n times to obtain $\{(\mathsf{OT}_{1,i}, \mathsf{st}_i)\}_{i=1}^n$. Give the sequence of values $\mathsf{OT}_{1,1}, \ldots, \mathsf{OT}_{1,n-1}$ to $\mathcal{Z}$ as the first-round message.
3. Compute

$$(\mathsf{GC}, \mathsf{OT}_{2,1}, \ldots, \mathsf{OT}_{2,n}, w_1, \ldots, w_\lambda) \leftarrow \mathsf{NextMsg}(\mathsf{OT}_{1,1}, \ldots, \mathsf{OT}_{1,n}).$$

In addition, let $\mathsf{input}^* = (\mathsf{OT}_{1,1}, \ldots, \mathsf{OT}_{1,n})$ and $\mathsf{output}^* = (\mathsf{GC}, \mathsf{OT}_{2,1}, \ldots, \mathsf{OT}_{2,n}, w_1, \ldots, w_\lambda)$, and compute $r_n^* \leftarrow \mathsf{Explain}(\mathsf{input}^*, \mathsf{output}^*)$. Give $\mathsf{OT}_{2,1}, \ldots, \mathsf{OT}_{2,n-1}$ to $\mathcal{Z}$ as the second-round message.
4. When $\mathcal{Z}$ corrupts party P_i, compute $r_{R,i} := \mathsf{SimOT}_2'(1^\lambda, x_i, \mathsf{st}_i)$ and $y_i := E_{\mathsf{OT}}(x_i, r_{R,i}, \mathsf{OT}_{2,i})$, and give x_i, z, y_i, and $r_{R,i}$ to $\mathcal{Z}$. In addition, if $i = n$ then for $i \in [n-1]$ compute $r_{R,i} := \mathsf{SimOT}_2'(1^\lambda, x_i, \mathsf{st}_i)$ and $y_i := E_{\mathsf{OT}}(x_i, r_{R,i}, \mathsf{OT}_{2,i})$, and give $\{y_i\}_{i=1}^{n-1}$ and r_n^* to $\mathcal{Z}$.

Indistinguishability of this experiment from the previous one follows by statistical equivalence of NextMsg and $\widetilde{\mathsf{NextMsg}}$.

Hybrid 4. In this experiment, we first make explicit the steps of NextMsg. (This is just a syntactic rewriting, and does not affect the experiment.) In addition, we now set $y_i = y_{i,x_i}$ instead of computing y_i using the OT-evaluation algorithm E_{OT}. This experiment proceeds as follows:

1. Compute $(\widetilde{\mathsf{NextMsg}}, \mathsf{Explain}) \leftarrow \mathsf{Comp}(1^\lambda, \mathsf{NextMsg})$, and give $\widetilde{\mathsf{NextMsg}}$ to $\mathcal{Z}$ as the CRS. $\mathcal{Z}$ chooses inputs $x_1, \ldots, x_n$.
2. Run $\mathsf{SimOT}'_1(1^\lambda)$ a total of n times to obtain $\{(\mathsf{OT}_{1,i}, \mathsf{st}_i)\}_{i=1}^n$. Give the sequence of values $\mathsf{OT}_{1,1}, \ldots, \mathsf{OT}_{1,n-1}$ to $\mathcal{Z}$ as the first-round message.
3. Compute $(\mathsf{GC}, \{(y_{i,0}, y_{i,1})\}_{i=1}^n, \{(w_{i,0}, w_{i,1})\}_{i=1}^\lambda) \leftarrow \mathsf{GenGC}(1^\lambda, C)$ and set $y_i = y_{i,x_i}$ for all i. For $i \in [n]$, run $\mathsf{OT}_{2,i} \leftarrow \mathcal{S}_{\mathsf{OT}}(\mathsf{OT}_1, y_{i,0}, y_{i,1})$. Choose uniform $r_1, \ldots, r_\lambda \in \{0,1\}$, and let $\mathsf{input}^* = (\mathsf{OT}_{1,1}, \ldots, \mathsf{OT}_{1,n})$ and $\mathsf{output}^* = (\mathsf{GC}, \mathsf{OT}_{2,1}, \ldots, \mathsf{OT}_{2,n}, w_{r_1}, \ldots, w_{r_\lambda})$. Compute $r_n^* \leftarrow \mathsf{Explain}(\mathsf{input}^*, \mathsf{output}^*)$. Give $\mathsf{OT}_{2,1}, \ldots, \mathsf{OT}_{2,n-1}$ to $\mathcal{Z}$ as the second-round message.
4. When $\mathcal{Z}$ corrupts party P_i, compute $r_{R,i} := \mathsf{SimOT}'_2(1^\lambda, x_i, \mathsf{st}_i)$. Give x_i, z, y_i, and $r_{R,i}$ to $\mathcal{Z}$. In addition, if $i = n$ then give $\{y_i\}_{i=1}^{n-1}$ and r_n^* to $\mathcal{Z}$.

Computational indistinguishability of this experiment from the previous one follows from security of the OT protocol.

Hybrid 5. In the previous experiment the OT_2 messages were generated honestly as part of NextMsg. Here, we have the OT simulator output them instead. That is, we now do:

1. Compute $(\widetilde{\mathsf{NextMsg}}, \mathsf{Explain}) \leftarrow \mathsf{Comp}(1^\lambda, \mathsf{NextMsg})$, and give $\widetilde{\mathsf{NextMsg}}$ to $\mathcal{Z}$ as the CRS. $\mathcal{Z}$ chooses inputs $x_1, \ldots, x_n$.
2. Run $\mathsf{SimOT}_1(1^\lambda)$ a total of n times to obtain $\{(\mathsf{OT}_{1,i}, \mathsf{OT}_{2,i}, \mathsf{st}_i)\}_{i=1}^n$. Give the sequence of values $\mathsf{OT}_{1,1}, \ldots, \mathsf{OT}_{1,n-1}$ to $\mathcal{Z}$ as the first-round message, and give $\mathsf{OT}_{2,1}, \ldots, \mathsf{OT}_{2,n-1}$ to $\mathcal{Z}$ as the second-round message.
3. Compute $(\mathsf{GC}, \{(y_{i,0}, y_{i,1})\}_{i=1}^n, \{(w_{i,0}, w_{i,1})\}_{i=1}^\lambda) \leftarrow \mathsf{GenGC}(1^\lambda, C)$ and set $y_i = y_{i,x_i}$ for all i. Choose uniform values $r_1, \ldots, r_\lambda \in \{0,1\}$, and let $\mathsf{input}^* = (\mathsf{OT}_{1,1}, \ldots, \mathsf{OT}_{1,n})$ and $\mathsf{output}^* = (\mathsf{GC}, \mathsf{OT}_{2,1}, \ldots, \mathsf{OT}_{2,n}, w_{r_1}, \ldots, w_{r_\lambda})$. Compute $r_n^* \leftarrow \mathsf{Explain}(\mathsf{input}^*, \mathsf{output}^*)$.
4. When $\mathcal{Z}$ corrupts party P_i, compute $r_{R,i} := \mathsf{SimOT}_2(1^\lambda, x_i, y_i, \mathsf{st}_i)$. Give x_i, z, y_i, and $r_{R,i}$ to $\mathcal{Z}$. In addition, if $i = n$ then give $\{y_i\}_{i=1}^{n-1}$ and r_n^* to $\mathcal{Z}$.

Again, computational indistinguishability between this experiment and the previous one follows by security of the OT protocol.

Hybrid 6. Here we use the garbled-circuit simulator (cf. Section 2.1) instead of the garbled-circuit generation algorithm. Thus, the experiment now proceeds as follows:

1. Compute $(\widetilde{\mathsf{NextMsg}}, \mathsf{Explain}) \leftarrow \mathsf{Comp}(1^\lambda, \mathsf{NextMsg})$, and give $\widetilde{\mathsf{NextMsg}}$ to $\mathcal{Z}$ as the CRS. $\mathcal{Z}$ chooses inputs $x_1, \ldots, x_n$.

2. Run $\mathsf{SimOT}_1(1^\lambda)$ a total of n times to obtain $\{(\mathsf{OT}_{1,i}, \mathsf{OT}_{2,i}, \mathsf{st}_i)\}_{i=1}^n$. Give $\mathsf{OT}_{1,1}, \ldots, \mathsf{OT}_{1,n-1}$ to $\mathcal{Z}$ as the first-round message, and $\mathsf{OT}_{2,1}, \ldots, \mathsf{OT}_{2,n-1}$ to $\mathcal{Z}$ as the second-round message.

3. Compute $(\mathsf{GC}, \{y_i\}_{i=1}^n, \{w_i\}_{i=1}^\lambda) \leftarrow \mathsf{SimGC}(1^\lambda, C, z)$. Let $\mathsf{input}^* = (\mathsf{OT}_{1,1}, \ldots, \mathsf{OT}_{1,n})$ and $\mathsf{output}^* = (\mathsf{GC}, \mathsf{OT}_{2,1}, \ldots, \mathsf{OT}_{2,n}, w_{r_1}, \ldots, w_{r_\lambda})$. Compute $r_n^* \leftarrow \mathsf{Explain}(\mathsf{input}^*, \mathsf{output}^*)$.

4. When $\mathcal{Z}$ corrupts party P_i, compute $r_{R,i} := \mathsf{SimOT}_2(1^\lambda, x_i, y_i, \mathsf{st}_i)$. Give x_i, z, y_i, and $r_{R,i}$ to $\mathcal{Z}$. In addition, if $i = n$ then for $i \in [n-1]$ give $\{y_i\}_{i=1}^{n-1}$ and r_n^* to $\mathcal{Z}$.

Computational indistinguishability between this experiment and the previous one follows from security of garbled circuits.

We conclude the proof by noting that Hybrid 6 is simply a syntactic rewriting of the ideal-world execution involving the simulator originally defined.

5   Conclusions and Open Questions

In this work we have shown the first constant-round, universally composable protocol tolerating a malicious, adaptive adversary that can corrupt any number of parties, in a setting where secure erasure is not assumed. In addition, we have shown the first adaptively secure protocol, regardless of round complexity, that can compute arbitrary functionalities (and not only adaptively well-formed ones) in the presence of any number of corruptions and without erasures.

Several interesting open questions remain. Although a CRS (or some other form of setup) is necessary if we wish to obtain a universally composable protocol with security against malicious adversaries corrupting an arbitrary number of parties, it is still possible that the CRS can be avoided in the semi-honest case, or in the stand-alone setting. Moreover, our protocol assumes that the CRS depends on the circuit C being computed or, if we let C be a universal circuit (cf. footnote 2), an a priori bound on the size of the circuit being computed. It would be interesting to see if this can be avoided.

References

1. Barak, B., Canetti, R., Nielsen, J.B., Pass, R.: Universally composable protocols with relaxed set-up assumptions. In: 45th Annual Symposium on Foundations of Computer Science (FOCS), pp. 186–195. IEEE (2004)
2. Beaver, D.: Plug and play encryption. In: Kaliski Jr., B.S. (ed.) CRYPTO 1997. LNCS, vol. 1294, pp. 75–89. Springer, Heidelberg (1997)
3. Beaver, D., Haber, S.: Cryptographic protocols provably secure against dynamic adversaries. In: Rueppel, R.A. (ed.) EUROCRYPT 1992. LNCS, vol. 658, pp. 307–323. Springer, Heidelberg (1993)
4. Bitansky, N., Canetti, R., Halevi, S.: Leakage-tolerant interactive protocols. In: Cramer, R. (ed.) TCC 2012. LNCS, vol. 7194, pp. 266–284. Springer, Heidelberg (2012)

5. Bitansky, N., Dachman-Soled, D., Lin, H.: Leakage-tolerant computation with input-independent preprocessing. In: Garay, J.A., Gennaro, R. (eds.) CRYPTO 2014, Part II. LNCS, vol. 8617, pp. 146–163. Springer, Heidelberg (2014)
6. Boneh, D., Waters, B.: Constrained pseudorandom functions and their applications. In: Sako, K., Sarkar, P. (eds.) ASIACRYPT 2013, Part II. LNCS, vol. 8270, pp. 280–300. Springer, Heidelberg (2013)
7. Boyle, E., Goldwasser, S., Ivan, I.: Functional signatures and pseudorandom functions. In: Krawczyk, H. (ed.) PKC 2014. LNCS, vol. 8383, pp. 501–519. Springer, Heidelberg (2014)
8. Canetti, R.: Security and composition of multiparty cryptographic protocols. Journal of Cryptology 13(1), 143–202 (2000)
9. Canetti, R.: Universally composable security: A new paradigm for cryptographic protocols. In: 42nd Annual Symposium on Foundations of Computer Science (FOCS), pp. 136–145. IEEE (2001), Full version at http://eprint.iacr.org/2000/067/
10. Canetti, R., Damgård, I., Dziembowski, S., Ishai, Y., Malkin, T.: Adaptive versus non-adaptive security of multi-party protocols. J. Crypto. 17(3), 153–207 (2004)
11. Canetti, R., Feige, U., Goldreich, O., Naor, M.: Adaptively secure multiparty computation. In: 28th Annual ACM Symposium on Theory of Computing (STOC), pp. 639–648. ACM Press (1996)
12. Canetti, R., Fischlin, M.: Universally composable commitments. In: Kilian, J. (ed.) CRYPTO 2001. LNCS, vol. 2139, pp. 19–40. Springer, Heidelberg (2001)
13. Canetti, R., Goldwasser, S., Poburinnaya, O.: Adaptively secure two-party computation from indistinguishability obfuscation. Cryptology ePrint Archive, Report 2014/845 (2014)
14. Canetti, R., Lindell, Y., Ostrovsky, R., Sahai, A.: Universally composable two-party and multi-party secure computation. In: 34th Annual ACM Symposium on Theory of Computing (STOC), pp. 494–503. ACM Press (2002), Full version available at http://eprint.iacr.org/2002/140
15. Canetti, R., Pass, R., Shelat, A.: Cryptography from sunspots: How to use an imperfect reference string. In: 48th Annual Symposium on Foundations of Computer Science (FOCS), pp. 249–259. IEEE (2007)
16. Choi, S.G., Dachman-Soled, D., Malkin, T., Wee, H.: Improved non-committing encryption with applications to adaptively secure protocols. In: Matsui, M. (ed.) ASIACRYPT 2009. LNCS, vol. 5912, pp. 287–302. Springer, Heidelberg (2009)
17. Dachman-Soled, D., Malkin, T., Raykova, M., Venkitasubramaniam, M.: Adaptive and concurrent secure computation from new adaptive, non-malleable commitments. In: Sako, K., Sarkar, P. (eds.) ASIACRYPT 2013, Part I. LNCS, vol. 8269, pp. 316–336. Springer, Heidelberg (2013)
18. Damgård, I.B., Nielsen, J.B.: Improved non-committing encryption schemes based on a general complexity assumption. In: Bellare, M. (ed.) CRYPTO 2000. LNCS, vol. 1880, pp. 432–450. Springer, Heidelberg (2000)
19. Damgård, I., Polychroniadou, A., Rao, V.: Secure UC constant round multi-party computation. Cryptology ePrint Archive, Report 2014/830 (2014)
20. Garg, S., Gentry, C., Halevi, S., Raykova, M., Sahai, A., Waters, B.: Candidate indistinguishability obfuscation and functional encryption for all circuits. In: 54th Annual Symposium on Foundations of Computer Science (FOCS), pp. 40–49. IEEE (2013)
21. Garg, S., Polychroniadou, A.: Two-round adaptively secure mpc from indistinguishability obfuscation. Cryptology ePrint Archive, Report 2014/844 (2014)

22. Garg, S., Sahai, A.: Adaptively secure multi-party computation with dishonest majority. In: Safavi-Naini, R., Canetti, R. (eds.) CRYPTO 2012. LNCS, vol. 7417, pp. 105–123. Springer, Heidelberg (2012)

23. Goldreich, O., Micali, S., Wigderson, A.: How to play any mental game, or a completeness theorem for protocols with honest majority. In: 19th Annual ACM Symposium on Theory of Computing (STOC), pp. 218–229. ACM Press (1987)

24. Goldreich, O., Goldwasser, S., Micali, S.: On the cryptographic applications of random functions. In: Blakely, G.R., Chaum, D. (eds.) CRYPTO 1984. LNCS, vol. 196, pp. 276–288. Springer, Heidelberg (1985)

25. Hazay, C., Patra, A.: One-sided adaptively secure two-party computation. In: Lindell, Y. (ed.) TCC 2014. LNCS, vol. 8349, pp. 368–393. Springer, Heidelberg (2014)

26. Ishai, Y., Kumarasubramanian, A., Orlandi, C., Sahai, A.: On invertible sampling and adaptive security. In: Abe, M. (ed.) ASIACRYPT 2010. LNCS, vol. 6477, pp. 466–482. Springer, Heidelberg (2010)

27. Ishai, Y., Prabhakaran, M., Sahai, A.: Founding cryptography on oblivious transfer – efficiently. In: Wagner, D. (ed.) CRYPTO 2008. LNCS, vol. 5157, pp. 572–591. Springer, Heidelberg (2008)

28. Katz, J., Ostrovsky, R.: Round-optimal secure two-party computation. In: Franklin, M. (ed.) CRYPTO 2004. LNCS, vol. 3152, pp. 335–354. Springer, Heidelberg (2004)

29. Kiayias, A., Papadopoulos, S., Triandopoulos, N., Zacharias, T.: Delegatable pseudorandom functions and applications. In: 20th ACM Conf. on Computer and Communications Security (CCS), pp. 669–684. ACM Press (2013)

30. Lindell, A.Y.: Adaptively secure two-party computation with erasures. In: Fischlin, M. (ed.) CT-RSA 2009. LNCS, vol. 5473, pp. 117–132. Springer, Heidelberg (2009)

31. Sahai, A., Waters, B.: How to use indistinguishability obfuscation: Deniable encryption, and more. In: 46th Annual ACM Symposium on Theory of Com- puting (STOC), pp. 475–484. ACM Press (2014)

32. Yao, A.C.-C.: How to generate and exchange secrets. In: 27th Annual Sympo- sium on Foundations of Computer Science (FOCS), pp. 162–167. IEEE (1986)

A   Puncturable PRFs

Puncturable PRFs are a type of constrained PRF [6,7,29] whereby it is possible to generate a key that defines the function everywhere except on some polynomial-size set of inputs:

Definition 2. *A* puncturable family of PRFs *is defined by polynomials* $n(\cdot)$ *and* $m(\cdot)$ *and a triple of Turing machines* Key_F, $\mathsf{Puncture}_F$, *and* Eval_F *satisfying the following conditions:*

Functionality preserved under puncturing. *For all polynomial-size sets* $S \subseteq \{0,1\}^{n(\lambda)}$ *and all* $x \in \{0,1\}^{n(\lambda)} \setminus S$, *we have:*

$$\Pr\left[K \leftarrow \mathsf{Key}_F(1^\lambda); K_S = \mathsf{Puncture}_F(K,S) : \mathsf{Eval}_F(K,x) = \mathsf{Eval}_F(K_S,x)\right] = 1.$$

Pseudorandom at punctured points. *For every* PPT *adversary* (A_1, A_2) *such that* $A_1(1^\lambda)$ *outputs a set* $S \subseteq \{0,1\}^{n(\lambda)}$ *and state* σ, *consider an experiment* *where $K \leftarrow \mathsf{Key}_F(1^\lambda)$ and $K_S = \mathsf{Puncture}_F(K, S)$. Then we have*

$$\left| \Pr\left[A_2(\sigma, K_S, S, \mathsf{Eval}_F(K, S)) = 1\right] - \Pr\left[A_2(\sigma, K_S, S, U_{m(\lambda)\cdot|S|}) = 1\right] \right|$$
$$\leq \mathsf{negl}(\lambda)$$

where $\mathsf{Eval}_F(K, S)$ is the concatenation of $\mathsf{Eval}_F(K, x_1), \ldots, \mathsf{Eval}_F(K, x_k)$, and $S = \{x_1, \ldots, x_k\}$ denoted the elements of S in lexicographic order.

For ease of notation, we write $F(K, x)$ to represent $\mathsf{Eval}_F(K, x)$. We also represent the punctured key $\mathsf{Puncture}_F(K, S)$ by $K(S)$.

As observed by [6,7,29], the GGM construction [24] of PRFs from one-way functions yields puncturable PRFs. Thus:

Theorem 2. *[6,7,29] Assuming one-way functions exist, for all polynomials $n(\lambda), m(\lambda)$ there is a puncturable PRF family that maps $n(\lambda)$ bits to $m(\lambda)$ bits.*

We follow [31] for the following definitions of puncturable PRFs with enhanced properties:

Definition 3. *A statistically injective (puncturable) PRF family with failure probability $\epsilon(\cdot)$ is a family of (puncturable) PRFs F such that with probability $1 - \epsilon(\lambda)$ over the random choice of key $K \leftarrow \mathsf{Key}_F(1^\lambda)$, we have that $F(K, \cdot)$ is injective.*

Definition 4. *An extracting (puncturable) PRF family with error $\epsilon(\cdot)$ for min-entropy $k(\cdot)$ is a family of (puncturable) PRFs F mapping $n(\lambda)$ bits to $m(\lambda)$ bits such that for all λ, if X is any distribution over $n(\lambda)$ bits with min-entropy greater than $k(\lambda)$, then the statistical distance between $(K \leftarrow \mathsf{Key}_F(1^\lambda), F(K, X))$ and $(K \leftarrow \mathsf{Key}_F(1^\lambda), U_{m(\lambda)})$ is at most $\epsilon(\lambda)$.*

The following results were proved in [31]:

Theorem 3 ([31]). *If one-way functions exist, then for all efficiently computable functions $n(\lambda)$, $m(\lambda)$, and $e(\lambda)$ such that $m(\lambda) \geq 2n(\lambda) + e(\lambda)$, there exists a puncturable statistically injective PRF family with failure probability $2^{-e(\lambda)}$ that maps $n(\lambda)$ bits to $m(\lambda)$ bits.*

Theorem 4. *If one-way functions exist, then for all efficiently computable functions $n(\lambda)$, $m(\lambda)$, $k(\lambda)$, and $e(\lambda)$ such that $n(\lambda) \geq k(\lambda) \geq m(\lambda) + 2e(\lambda) + 2$, there exists an extracting puncturable PRF family that maps $n(\lambda)$ bits to $m(\lambda)$ bits with error $2^{-e(\lambda)}$ for min-entropy $k(\lambda)$.*

B   Proof of Security for Our Explainability Compiler

In this section we prove security of our explainability compiler. We must show two properties: statistical functional equivalence and explainability. (Polynomial slowdown is obvious.) The proof of statistical functional equivalence is largely identical to the analogous proof in [31], and is omitted. Instead, we focus on explainability.

We first state the following lemma, whose proof is the same as in [31].

Lemma 1. *Except with negligible probability over the choice of key K_2, the following hold:*

1. *For any fixed $u[1] = \alpha$, there exists at most one pair (input, β) such that the input* input *with randomness $u = (\alpha, \beta)$ will cause the Step 1 check of $\overline{\text{Alg}}$ to be satisfied.*
2. *There are at most $2^{2\lambda + \ell_{in} + \ell_{out}}$ values for the randomness u that can cause the Step 1 check of $\overline{\text{Alg}}$ to be satisfied.*

Given the above, we prove:

Theorem 5. *If F_1, F_2, F_3 are PRFs that satisfy the properties specified in Section 3.1, and $i\mathcal{O}$ is an indistinguishability obfuscator for P/poly, then our construction $\text{Comp}(\cdot, \cdot)$ satisfies explainability.*

Proof. Recall the explainability experiment from Definition 1:

1. $\mathcal{A}(1^\lambda)$ outputs input* of its choice.
2. $\text{Comp}(1^\lambda, \text{Alg})$ is run to obtain $(\widetilde{\text{Alg}}, \widetilde{\text{Explain}})$.
3. Choose random coins $r_0 \leftarrow \{0,1\}^*$, and compute output* $\leftarrow \widetilde{\text{Alg}}(\text{input}^*; r_0)$.
4. Compute $r_1 \leftarrow \widetilde{\text{Explain}}(\text{input}^*, \text{output}^*)$.
5. Choose a uniform bit b and give $\widetilde{\text{Alg}}, \text{output}^*, r_b$ to $\mathcal{A}$.
6. $\mathcal{A}$ outputs a bit b', and succeeds if $b' = b$.

Let $\text{Expl}_{\text{Alg}, \mathcal{A}}$ be a random variable set to 1 if $\mathcal{A}$ succeeds in outputting $b' = b$ in the above experiment. Security of $\text{Comp}(1^\lambda, \text{Alg})$ requires that for every PPT $\mathcal{A}$ and for every efficient algorithm Alg, we have $\Pr[\text{Expl}_{\text{Alg}, \mathcal{A}} = 1] \leq 1/2 + \text{negl}(\lambda)$.

Assume towards a contradiction that there is some PPT adversary $\mathcal{A}$ and some efficient algorithm Alg such that $\Pr[\text{Expl}_{\text{Alg}, \mathcal{A}} = 1] \geq 1/2 + \varepsilon(\lambda)$, for non-negligible $\varepsilon(\cdot)$. We derive a contradiction via a sequence of hybrid experiments. The change between each experiment and the previous one will be denoted by underlined text.

Original Experiment. We consider the probability that $b' = b$ in the following experiment:

1. $b \leftarrow \{0,1\}$.
2. input* $\leftarrow \mathcal{A}(1^\lambda)$.
3. Choose K_1, K_2, K_3 at random.
4. Select u^* and r^* at random.
5. If $F_3(K_3, u[1]) \oplus u[2] = (\text{input}', \text{output}', r')$ for (proper length) strings output', r', input', and input' $=$ input*, and $u[1] = F_2(K_2, (\text{input}', \text{output}', r'))$, then let output* $=$ output' and jump to the next step. Otherwise, let $x^* = F_1(K_1, (\text{input}^*, u^*))$ and output* $= \text{Alg}(\text{input}^*; x^*)$.
6. Set $\alpha^* = F_2(K_2, (\text{input}^*, \text{output}^*, \text{PRG}(r^*)))$. Let $\beta^* = F_3(K_3, \alpha^*) \oplus (\text{input}^*, \text{output}^*, \text{PRG}(r^*))$, and set $e^* = (\alpha^*, \beta^*)$.
7. <u>Let $\widetilde{\text{Alg}} \leftarrow i\mathcal{O}(\overline{\text{Alg}})$ for $\overline{\text{Alg}}$ as in Figure 1. Let $\widetilde{\text{Explain}} \leftarrow i\mathcal{O}(\overline{\text{Explain}})$ for $\overline{\text{Explain}}$ as in Figure 2.</u>

$$\overline{\mathsf{Alg}}$$

Constants: Keys K_1, K_2, and K_3.
Input: Input input, randomness $u = (u[1], u[2])$.

1. If $F_3(K_3, u[1]) \oplus u[2] = (\mathsf{input}', \mathsf{output}', r')$ for (proper length) strings $\mathsf{output}', r', \mathsf{input}'$, and $\mathsf{input}' = \mathsf{input}$, and $u[1] = F_2(K_2, (\mathsf{input}', \mathsf{output}', r'))$, then output $\mathsf{output} = \mathsf{output}'$ and end.
2. Else let $x = F_1(K_1, (\mathsf{input}, u))$. Output $\mathsf{output} = \mathsf{Alg}(\mathsf{input}; x)$.

Fig. 5. Program $\overline{\mathsf{Alg}}$

8. If $b = 0$, set $b' \leftarrow \mathcal{A}(\widetilde{\mathsf{Alg}}, \mathsf{output}^*, u^*)$. If $b = 1$, set $b' \leftarrow \mathcal{A}(\widetilde{\mathsf{Alg}}, \mathsf{output}^*, e^*)$.

Hybrid 0. Next, we eliminate the check in Step 1 from the $\overline{\mathsf{Alg}}$ program when preparing the outputs of the fixed challenge input^*. Hybrid 0 is statistically close to the original experiment by Lemma 1. Consider the probability that $b' = b$ in the following experiment:

1. $b \leftarrow \{0, 1\}$.
2. $\mathsf{input}^* \leftarrow \mathcal{A}(1^\lambda)$.
3. Choose K_1, K_2, K_3 at random.
4. Select u^* and r^* at random.
5. If $F_3(K_3, u[1]) \oplus u[2] = (\mathsf{input}', \mathsf{output}', r')$ for (proper length) strings output', r', input', and $\mathsf{input}' = \mathsf{input}^*$, and $u[1] = F_2(K_2, (\mathsf{input}', \mathsf{output}', r'))$, then let $\mathsf{output}^* = \mathsf{output}'$ and jump to the next step. Otherwise, let $x^* = F_1(K_1, (\mathsf{input}^*, u^*))$ and $\mathsf{output}^* = \mathsf{Alg}(\mathsf{input}^*; x^*)$.
6. Set $\alpha^* = F_2(K_2, (\mathsf{input}^*, \mathsf{output}^*, \mathsf{PRG}(r^*)))$. Let $\beta^* = F_3(K_3, \alpha^*) \oplus (\mathsf{input}^*, \mathsf{output}^*, \mathsf{PRG}(r^*))$, and set $e^* = (\alpha^*, \beta^*)$.
7. Let $\widetilde{\mathsf{Alg}} \leftarrow i\mathcal{O}(\overline{\mathsf{Alg}})$ for $\overline{\mathsf{Alg}}$ as in Figure 5. Let $\mathsf{Explain} \leftarrow i\mathcal{O}(\overline{\mathsf{Explain}})$ for $\overline{\mathsf{Explain}}$ as in Figure 6.
8. If $b = 0$, set $b' \leftarrow \mathcal{A}(\widetilde{\mathsf{Alg}}, \mathsf{output}^*, u^*)$. If $b = 1$, set $b' \leftarrow \mathcal{A}(\widetilde{\mathsf{Alg}}, \mathsf{output}^*, e^*)$.

$$\overline{\mathsf{Explain}}$$

Constants: Keys K_2 and K_3.
Input: Input input, output output, randomness $r \in \{0, 1\}^\lambda$.

1. Set $\alpha = F_2(K_2, (\mathsf{input}, \mathsf{output}, \mathsf{PRG}(r)))$. Let $\beta = F_3(K_3, \alpha) \oplus (\mathsf{input}, \mathsf{output}, \mathsf{PRG}(r))$. Output $e = (\alpha, \beta)$.

Fig. 6. Program $\overline{\mathsf{Explain}}$

Hybrid 1. Here, we modify the $\overline{\mathsf{Alg}}$ program as follows: First, we add constants $\mathsf{input}^*, \mathsf{output}^*, u^*, e^*$ to the program. Then, we add an "if" statement at the

---

$$\overline{\mathsf{Alg}}$$

Constants: $\mathsf{input}^*, \mathsf{output}^*, u^*, e^*$ and   PRF   keys
$K_1((\mathsf{input}^*, u^*), (\mathsf{input}^*, e^*)), K_2, \text{ and } K_3.$
Input: Input input, randomness $u = (u[1], u[2]).$

1. If $(\mathsf{input}, u) = (\mathsf{input}^*, u^*)$ or (input^*, e^*), output output^* and stop.
2. If $F_3(K_3, u[1]) \oplus u[2] = (\mathsf{input}', \mathsf{output}', r')$ for (proper length) strings $\mathsf{output}', r', \mathsf{input}',$ and $\mathsf{input}' = \mathsf{input},$ and $u[1] = F_2(K_2, (\mathsf{input}', \mathsf{output}', r')),$ then output $\mathsf{output} = \mathsf{output}'$ and end.
3. Else let $x = F_1(K_1, (\mathsf{input}, u))$. Output $\mathsf{output} = \mathsf{Alg}(\mathsf{input}; x)$.

Fig. 7. Program $\overline{\mathsf{Alg}}$

start that outputs output^* if the input is either (input^*, u^*) or (input^*, e^*), as this is exactly what the original $\overline{\mathsf{Alg}}$ program would do by our choice of u^*, e^*. Because this "if" statement is in place, we know that F_1 cannot be evaluated at either (input^*, u^*) or (input^*, e^*) within the program, and therefore we can safely puncture K_1 at those two positions.

By construction, the new $\overline{\mathsf{Alg}}$ program is functionally equivalent to the original $\overline{\mathsf{Alg}}$ program. Therefore, indistinguishability of Hybrid 0 and Hybrid 1 follows by the security of $i\mathcal{O}$. Thus, the difference in the probabilities that $\mathcal{A}$ outputs $b' = b$ in Hybrid 0 and Hybrid 1 is negligible.

1. $b \leftarrow \{0, 1\}.$
2. $\mathsf{input}^* \leftarrow \mathcal{A}(1^\lambda).$
3. Choose K_1, K_2, K_3 at random.
4. Select u^* and r^* at random.
5. Let $x^* = F_1(K_1, (\mathsf{input}^*, u^*))$ and let $\mathsf{output}^* = \mathsf{Alg}(\mathsf{input}^*; x^*).$
6. Set $\alpha^* = F_2(K_2, (\mathsf{input}^*, \mathsf{output}^*, \mathsf{PRG}(r^*)))$. Let $\beta^* = F_3(K_3, \alpha^*) \oplus (\mathsf{input}^*, \mathsf{output}^*, \mathsf{PRG}(r^*)),$ and set $e^* = (\alpha^*, \beta^*).$
7. Let $\widetilde{\mathsf{Alg}} \leftarrow i\mathcal{O}(\overline{\mathsf{Alg}})$ for $\overline{\mathsf{Alg}}$ as in Figure 7. Let $\widetilde{\mathsf{Explain}} \leftarrow i\mathcal{O}(\overline{\mathsf{Explain}})$ for $\overline{\mathsf{Explain}}$ as in Figure 8.
8. If $b = 0$, set $b' \leftarrow \mathcal{A}(\widetilde{\mathsf{Alg}}, \mathsf{output}^*, u^*)$. If $b = 1$, set $b' \leftarrow \mathcal{A}(\widetilde{\mathsf{Alg}}, \mathsf{output}^*, e^*).$

$$\overline{\mathsf{Explain}}$$

Constants: PRF keys $K_2,$ and $K_3.$
Input: Input input, output output, randomness $r.$

1. Set $\alpha = F_2(K_2, (\mathsf{input}, \mathsf{output}, \mathsf{PRG}(r)))$. Let $\beta = F_3(K_3, \alpha) \oplus (\mathsf{input}, \mathsf{output}, \mathsf{PRG}(r))$. Output $e = (\alpha, \beta).$

Fig. 8. Program $\overline{\mathsf{Explain}}$

Hybrid 2. Here, the value x^* is chosen uniformly instead of as the output of $F_1(K_1, (\text{input}^*, u^*))$. Pseudorandomness of F_1 thus implies that the difference in the probabilities that $\mathcal{A}$ outputs $b' = b$ in Hybrid 1 and Hybrid 2 is negligible.

1. $b \leftarrow \{0, 1\}$.
2. $\text{input}^* \leftarrow \mathcal{A}(1^\lambda)$.
3. Choose K_1, K_2, K_3 at random.
4. Select u^* and r^* at random.
5. Choose uniform x^* and let $\text{output}^* = \text{Alg}(\text{input}^*; x^*)$.
6. Set $\alpha^* = F_2(K_2, (\text{input}^*, \text{output}^*, \text{PRG}(r^*)))$. Let $\beta^* = F_3(K_3, \alpha^*) \oplus (\text{input}^*, \text{output}^*, \text{PRG}(r^*))$, and set $e^* = (\alpha^*, \beta^*)$.
7. Let $\widetilde{\text{Alg}} \leftarrow i\mathcal{O}(\overline{\text{Alg}})$ for $\overline{\text{Alg}}$ as in Figure 9. Let $\widetilde{\text{Explain}} \leftarrow i\mathcal{O}(\overline{\text{Explain}})$ for $\overline{\text{Explain}}$ as in Figure 10.
8. If $b = 0$, set $b' \leftarrow \mathcal{A}(\widetilde{\text{Alg}}, \text{output}^*, u^*)$. If $b = 1$, set $b' \leftarrow \mathcal{A}(\widetilde{\text{Alg}}, \text{output}^*, e^*)$.

$$\overline{\text{Alg}}$$

Constants: $\text{input}^*, \text{output}^*, u^*, e^*$ and   PRF   keys $K_1((\text{input}^*, u^*), (\text{input}^*, e^*)), K_2$, and K_3.
Input: Input input, randomness $u = (u[1], u[2])$.

1. If $(\text{input}, u) = (\text{input}^*, u^*)$ or (input^*, e^*), output output^* and stop.
2. If $F_3(K_3, u[1]) \oplus u[2] = (\text{input}', \text{output}', r')$ for (proper length) strings $\text{output}', r', \text{input}'$, and $\text{input}' = \text{input}$, and $u[1] = F_2(K_2, (\text{input}', \text{output}', r'))$, then output $\text{output} = \text{output}'$ and end.
3. Else let $x = F_1(K_1, (\text{input}, u))$. Output $\text{output} = \text{Alg}(\text{input}; x)$.

Fig. 9. Program $\overline{\text{Alg}}$

$$\overline{\text{Explain}}$$

Constants: PRF keys K_2, and K_3.
Input: Input input, output output, randomness r.

1. Set $\alpha = F_2(K_2, (\text{input}, \text{output}, \text{PRG}(r)))$. Let $\beta = F_3(K_3, \alpha) \oplus (\text{input}, \text{output}, \text{PRG}(r))$. Output $e = (\alpha, \beta)$.

Fig. 10. Program $\overline{\text{Explain}}$

Hybrid 3. Here, instead of choosing uniform r^* and applying a PRG to it, a value $\tilde{r}$ is chosen uniformly from the range of the PRG. Security of the PRG implies that the difference in the probabilities that $\mathcal{A}$ outputs $b' = b$ in Hybrid 2 and Hybrid 3 is negligible.

1. $b \leftarrow \{0,1\}$.
2. $\mathsf{input}^* \leftarrow \mathcal{A}(1^\lambda)$.
3. Choose K_1, K_2, K_3 at random.
4. Select u^* <u>and $\tilde{r}$</u> at random.
5. Choose uniform x^* and let $\mathsf{output}^* = \mathsf{Alg}(\mathsf{input}^*; x^*)$.
6. Set $\alpha^* = F_2(K_2, (\mathsf{input}^*, \mathsf{output}^*, \underline{\tilde{r}}))$. Let $\beta^* = F_3(K_3, \alpha^*) \oplus (\mathsf{input}^*, \mathsf{output}^*, \underline{\tilde{r}})$, and set $e^* = (\alpha^*, \beta^*)$.
7. Let $\widetilde{\mathsf{Alg}} \leftarrow i\mathcal{O}(\overline{\mathsf{Alg}})$ for $\overline{\mathsf{Alg}}$ as in Figure 11. Let $\widetilde{\mathsf{Explain}} \leftarrow i\mathcal{O}(\overline{\mathsf{Explain}})$ for $\overline{\mathsf{Explain}}$ as in Figure 12.
8. If $b = 0$, set $b' \leftarrow \mathcal{A}(\widetilde{\mathsf{Alg}}, \mathsf{output}^*, u^*)$. If $b = 1$, set $b' \leftarrow \mathcal{A}(\widetilde{\mathsf{Alg}}, \mathsf{output}^*, e^*)$.

$$\overline{\mathsf{Alg}}$$

Constants: $\mathsf{input}^*, \mathsf{output}^*, u^*, e^*$ and   PRF   keys $K_1((\mathsf{input}^*, u^*), (\mathsf{input}^*, e^*))$, K_2, and K_3.
Input: Input input, randomness $u = (u[1], u[2])$.

1. If $(\mathsf{input}, u) = (\mathsf{input}^*, u^*)$ or (input^*, e^*), output output^* and stop.
2. If $F_3(K_3, u[1]) \oplus u[2] = (\mathsf{input}', \mathsf{output}', r')$ for (proper length) strings $\mathsf{output}', r', \mathsf{input}'$, and $\mathsf{input}' = \mathsf{input}$, and $u[1] = F_2(K_2, (\mathsf{input}', \mathsf{output}', r'))$, then output $\mathsf{output} = \mathsf{output}'$ and end.
3. Else let $x = F_1(K_1, (\mathsf{input}, u))$. Output $\mathsf{output} = \mathsf{Alg}(\mathsf{input}; x)$.

Fig. 11. Program $\overline{\mathsf{Alg}}$

Hybrid 4. Here, the $\overline{\mathsf{Alg}}$ and $\overline{\mathsf{Explain}}$ programs are modified as shown below. In Lemma 2, (proven below), we argue that except with negligible probability over choice of constants, these modifications do not alter the functionality of either program. Thus, the $i\mathcal{O}$ security property implies that the difference in the probabilities that $\mathcal{A}$ outputs $b' = b$ in Hybrid 3 and Hybrid 4 is negligible.

1. $b \leftarrow \{0,1\}$.
2. $\mathsf{input}^* \leftarrow \mathcal{A}(1^\lambda)$.
3. Choose K_1, K_2, K_3 at random.
4. Select u^* and $\tilde{r}$ at random.
5. Choose uniform x^* and let $\mathsf{output}^* = \mathsf{Alg}(\mathsf{input}^*; x^*)$.
6. Set $\alpha^* = F_2(K_2, (\mathsf{input}^*, \mathsf{output}^*, \tilde{r}))$. Let $\beta^* = F_3(K_3, \alpha^*) \oplus (\mathsf{input}^*, \mathsf{output}^*, \tilde{r})$, and set $e^* = (\alpha^*, \beta^*)$.
7. Let $\widetilde{\mathsf{Alg}} \leftarrow i\mathcal{O}(\overline{\mathsf{Alg}})$ for $\overline{\mathsf{Alg}}$ as in Figure 13. Let $\widetilde{\mathsf{Explain}} \leftarrow i\mathcal{O}(\overline{\mathsf{Explain}})$ for $\overline{\mathsf{Explain}}$ as in Figure 14.

Explain

Constants: PRF keys K_2, and K_3.
Input: Input input, output output, randomness r.

1. Set $\alpha = F_2(K_2, (\text{input}, \text{output}, \text{PRG}(r)))$. Let $\beta = F_3(K_3, \alpha) \oplus (\text{input}, \text{output}, \text{PRG}(r))$. Output $e = (\alpha, \beta)$.

Fig. 12. Program $\overline{\text{Explain}}$

Alg

Constants: input*, output*, u^*, e^* and   PRF   keys
$K_1((\text{input}^*, u^*), (\text{input}^*, e^*))$, K_2, and $K_3(u^*[1], e^*[1])$.
Input: Input input, randomness $u = (u[1], u[2])$.

1. If $(\text{input}, u) = (\text{input}^*, u^*)$ or (input^*, e^*), output output* and stop.
2. If $u[1] = e^*[1]$ or $u[1] = u^*[1]$, then skip this step. If $F_3(K_3, u[1]) \oplus \overline{u[2]} = (\text{input}', \text{output}', r')$ for (proper length) strings output$'$, r', input$'$, and input$' = $ input, and $u[1] = F_2(K_2, (\text{input}', \text{output}', r'))$, then output output $=$ output$'$ and end.
3. Else let $x = F_1(K_1, (\text{input}, u))$. Output output $=$ Alg(input; x).

Fig. 13. Program $\overline{\text{Alg}}$

Explain

Constants: PRF keys K_2, and $K_3(u^*[1], e^*[1])$.
Input: Input input, output output, randomness r.

1. Set $\alpha = F_2(K_2, (\text{input}, \text{output}, \text{PRG}(r)))$. Let $\beta = F_3(K_3, \alpha) \oplus (\text{input}, \text{output}, \text{PRG}(r))$. Output $e = (\alpha, \beta)$.

Fig. 14. Program $\overline{\text{Explain}}$

8. If $b = 0$, set $b' \leftarrow \mathcal{A}(\widetilde{\text{Alg}}, \text{output}^*, u^*)$. If $b = 1$, set $b' \leftarrow \mathcal{A}(\widetilde{\text{Alg}}, \text{output}^*, e^*)$.

Hybrid 5. Here, the value $e^*[2]$, denoted β^*, is chosen at random instead of being chosen as $\beta^* = F_3(K_3, \alpha^*) \oplus (\text{input}^*, \text{output}^*, \tilde{r})$. Pseudorandomness of F_3 thus implies that the difference in the probabilities that $\mathcal{A}$ outputs $b' = b$ in Hybrid 4 and Hybrid 5 is negligible.

1. $b \leftarrow \{0, 1\}$.
2. input$^* \leftarrow \mathcal{A}(1^\lambda)$.
3. Choose K_1, K_2, K_3 at random.
4. Select u^* and $\tilde{r}$ at random.
5. Choose uniform x^* and let output$^* = $ Alg(input*; x^*).
6. Set $\alpha^* = F_2(K_2, (\text{input}^*, \text{output}^*, \tilde{r}))$. Choose uniform β^*, and set $e^* = (\alpha^*, \beta^*)$.

7. Let $\widetilde{\mathsf{Alg}} \leftarrow i\mathcal{O}(\overline{\mathsf{Alg}})$ for $\overline{\mathsf{Alg}}$ as in Figure 15. Let $\widetilde{\mathsf{Explain}} \leftarrow i\mathcal{O}(\overline{\mathsf{Explain}})$ for $\overline{\mathsf{Explain}}$ as in Figure 16.
8. If $b = 0$, set $b' \leftarrow \mathcal{A}(\widetilde{\mathsf{Alg}}, \mathsf{output}^*, u^*)$. If $b = 1$, set $b' \leftarrow \mathcal{A}(\widetilde{\mathsf{Alg}}, \mathsf{output}^*, e^*)$.

$$\overline{\mathsf{Alg}}$$

Constants: $\mathsf{input}^*, \mathsf{output}^*, u^*, e^*$ and   PRF   keys
$K_1((\mathsf{input}^*, u^*), (\mathsf{input}^*, e^*))$, K_2, and $K_3(u^*[1], e^*[1])$.
Input: Input input, randomness $u = (u[1], u[2])$.

1. If $(\mathsf{input}, u) = (\mathsf{input}^*, u^*)$ or (input^*, e^*), output output^* and stop.
2. If $u[1] = e^*[1]$ or $u[1] = u^*[1]$, then skip this step. If $F_3(K_3, u[1]) \oplus u[2] = (\mathsf{input}', \mathsf{output}', r')$ for (proper length) strings $\mathsf{output}', r', \mathsf{input}'$, and $\mathsf{input}' = \mathsf{input}$, and $u[1] = F_2(K_2, (\mathsf{input}', \mathsf{output}', r'))$, then output $\mathsf{output} = \mathsf{output}'$ and end.
3. Else let $x = F_1(K_1, (\mathsf{input}, u))$. Output $\mathsf{output} = \mathsf{Alg}(\mathsf{input}; x)$.

Fig. 15. Program $\overline{\mathsf{Alg}}$

Hybrid 6. First we modify the $\overline{\mathsf{Alg}}$ program to add a condition to the check in Step 2 to determine if $(\mathsf{input}', \mathsf{output}', r') = (\mathsf{input}^*, \mathsf{output}^*, \tilde{r})$ and, if so, to skip this check. This does not change the functionality of the program, because $e^*[1] = F_2(K_2, (\mathsf{input}^*, \mathsf{output}^*, \tilde{r}))$, and therefore the check cannot be satisfied if $(\mathsf{input}', \mathsf{output}', r') = (\mathsf{input}^*, \mathsf{output}^*, \tilde{r})$, since Step 2 is skipped entirely if $u[1] = e^*[1]$. Furthermore, both the $\overline{\mathsf{Alg}}$ and $\overline{\mathsf{Explain}}$ programs are modified to have K_2 punctured at the points $(\mathsf{input}^*, \mathsf{output}^*, \tilde{r})$. This puncturing does not change the functionality of the $\overline{\mathsf{Alg}}$ program because of the new "if" condition just added. With overwhelming probability, $\tilde{r}$ is not in the image of the PRG and therefore this puncturing also does not change the functionality of the $\overline{\mathsf{Explain}}$ program. Thus, the difference in the probabilities that $\mathcal{A}$ outputs $b' = b$ in Hybrid 5 and Hybrid 6 is negligible.

$$\overline{\mathsf{Explain}}$$

Constants: PRF keys K_2, and $K_3(u^*[1], e^*[1])$.
Input: Input input, output output, randomness r.

1. Set $\alpha = F_2(K_2, (\mathsf{input}, \mathsf{output}, \mathsf{PRG}(r)))$. Let $\beta = F_3(K_3, \alpha) \oplus (\mathsf{input}, \mathsf{output}, \mathsf{PRG}(r))$. Output $e = (\alpha, \beta)$.

Fig. 16. Program $\overline{\mathsf{Explain}}$

1. $b \leftarrow \{0,1\}$.
2. $\mathsf{input}^* \leftarrow \mathcal{A}(1^\lambda)$.
3. Choose K_1, K_2, K_3 at random.
4. Select u^* and $\tilde{r}$ at random.
5. Choose uniform x^* and let $\mathsf{output}^* = \mathsf{Alg}(\mathsf{input}^*; x^*)$.
6. Set $\alpha^* = F_2(K_2, (\mathsf{input}^*, \mathsf{output}^*, \tilde{r}))$. Choose uniform β^*, and set $e^* = (\alpha^*, \beta^*)$.
7. Let $\widetilde{\mathsf{Alg}} \leftarrow i\mathcal{O}(\overline{\mathsf{Alg}})$ for $\overline{\mathsf{Alg}}$ as in Figure 17. Let $\widetilde{\mathsf{Explain}} \leftarrow i\mathcal{O}(\overline{\mathsf{Explain}})$ for $\overline{\mathsf{Explain}}$ as in Figure 18.
8. If $b = 0$, set $b' \leftarrow \mathcal{A}(\widetilde{\mathsf{Alg}}, \mathsf{output}^*, u^*)$. If $b = 1$, set $b' \leftarrow \mathcal{A}(\widetilde{\mathsf{Alg}}, \mathsf{output}^*, e^*)$.

$\overline{\mathsf{Alg}}$

Constants: $\mathsf{input}^*, \mathsf{output}^*, u^*, e^*, \tilde{r}$ and PRF keys $K_1((\mathsf{input}^*, u^*), (\mathsf{input}^*, e^*))$, $K_2((\mathsf{input}^*, \mathsf{output}^*, \tilde{r}))$, and $K_3(u^*[1], e^*[1])$.
Input: Input input, randomness $u = (u[1], u[2])$.

1. If $(\mathsf{input}, u) = (\mathsf{input}^*, u^*)$ or (input^*, e^*), output output^* and stop.
2. If $u[1] = e^*[1]$ or $u[1] = u^*[1]$, then skip this step. If $F_3(K_3, u[1]) \oplus u[2] = (\mathsf{input}', \mathsf{output}', r')$ for (proper length) strings $\mathsf{output}', r', \mathsf{input}'$, and $\mathsf{input}' = \mathsf{input}$, and $(\mathsf{input}', \mathsf{output}', r') \neq (\mathsf{input}^*, \mathsf{output}^*, \tilde{r})$, then also check if $u[1] = F_2(K_2, (\mathsf{input}', \mathsf{output}', r'))$, then output $\mathsf{output} = \mathsf{output}'$ and end.
3. Else let $x = F_1(K_1, (\mathsf{input}, u))$. Output $\mathsf{output} = \mathsf{Alg}(\mathsf{input}; x)$.

Fig. 17. Program $\overline{\mathsf{Alg}}$

$\overline{\mathsf{Explain}}$

Constants: PRF keys $K_2((\mathsf{input}^*, \mathsf{output}^*, \tilde{r}))$, and $K_3(u^*[1], e^*[1])$.
Input: Input input, output output, randomness r.

1. Set $\alpha = F_2(K_2, (\mathsf{input}, \mathsf{output}, \mathsf{PRG}(r)))$. Let $\beta = F_3(K_3, \alpha) \oplus (\mathsf{input}, \mathsf{output}, \mathsf{PRG}(r))$. Output $e = (\alpha, \beta)$.

Fig. 18. Program $\overline{\mathsf{Explain}}$

Hybrid 7. Finally, we modify $e^*[1]$, denoted α^*, to be uniform, instead of being computed as $\alpha^* = F_2(K_2, (\mathsf{input}^*, \mathsf{output}^*, \tilde{r}))$. Pseudorandomness of F_2 implies that the difference in the probabilities that $\mathcal{A}$ outputs $b' = b$ in Hybrid 6 and Hybrid 7 is negligible.

1. $b \leftarrow \{0,1\}$.
2. $\mathsf{input}^* \leftarrow \mathcal{A}(1^\lambda)$.
3. Choose K_1, K_2, K_3 at random.
4. Select u^* and $\tilde{r}$ at random.
5. Choose uniform x^* and let $\mathsf{output}^* = \mathsf{Alg}(\mathsf{input}^*; x^*)$.
6. <u>Choose uniform α^*</u> and β^*, and set $e^* = (\alpha^*, \beta^*)$.
7. Let $\widetilde{\mathsf{Alg}} \leftarrow i\mathcal{O}(\overline{\mathsf{Alg}})$ for $\overline{\mathsf{Alg}}$ as in Figure 19. Let $\widetilde{\mathsf{Explain}} \leftarrow i\mathcal{O}(\overline{\mathsf{Explain}})$ for $\overline{\mathsf{Explain}}$ as in Figure 20.
8. If $b = 0$, set $b' \leftarrow \mathcal{A}(\widetilde{\mathsf{Alg}}, \mathsf{output}^*, u^*)$. If $b = 1$, set $b' \leftarrow \mathcal{A}(\widetilde{\mathsf{Alg}}, \mathsf{output}^*, e^*)$.

$$\overline{\mathsf{Alg}}$$

Constants: $\mathsf{input}^*, \mathsf{output}^*, u^*, e^*, \tilde{r}$ and   PRF   keys $K_1((\mathsf{input}^*, u^*), (\mathsf{input}^*, e^*))$, $K_2((\mathsf{input}^*, \mathsf{output}^*, \tilde{r}))$, and $K_3(u^*[1], e^*[1])$.
Input: Input input, randomness $u = (u[1], u[2])$.

1. If $(\mathsf{input}, u) = (\mathsf{input}^*, u^*)$ or (input^*, e^*), output output^* and stop.
2. If $u[1] = e^*[1]$ or $u[1] = u^*[1]$, then skip this step. If $F_3(K_3, u[1]) \oplus u[2] = (\mathsf{input}', \mathsf{output}', r')$ for (proper length) strings $\mathsf{output}', r', \mathsf{input}'$, and $\mathsf{input}' = \mathsf{input}$, and, $(\mathsf{input}', \mathsf{output}', r') \neq (\mathsf{input}^*, \mathsf{output}^*, \tilde{r})$, then also check if $u[1] = F_2(K_2, (\mathsf{input}', \mathsf{output}', r'))$, then output $\mathsf{output} = \mathsf{output}'$ and end.
3. Else let $x = F_1(K_1, (\mathsf{input}, u))$. Output $\mathsf{output} = \mathsf{Alg}(\mathsf{input}; x)$.

Fig. 19. Program $\overline{\mathsf{Alg}}$

In Hybrid 7 we observe that the variables e^*, u^* are now uniform and independent. Thus, the inputs to $\mathcal{A}$ are distributed identically regardless of whether $b = 0$ or $b = 1$ are identical, and so $b = b'$ with probability exactly $1/2$. The lemma below thus concludes the proof.

$$\overline{\mathsf{Explain}}$$

Constants: PRF keys $K_2((\mathsf{input}^*, \mathsf{output}^*, \tilde{r}))$, and $K_3(u^*[1], e^*[1])$.
Input: Input input, output output, randomness r.

1. Set $\alpha = F_2(K_2, (\mathsf{input}, \mathsf{output}, \mathsf{PRG}(r)))$. Let $\beta = F_3(K_3, \alpha) \oplus (\mathsf{input}, \mathsf{output}, \mathsf{PRG}(r))$. Output $e = (\alpha, \beta)$.

Fig. 20. Program $\overline{\mathsf{Explain}}$

The proof above relies on the following lemma showing that the programs obfuscated in Hybrid 3 are equivalent to the corresponding programs in Hybrid 4.

Lemma 2. *Except with negligible probability over the choice of $u^*[1]$ and $e^*[1]$, the $\overline{\text{Alg}}$ and $\overline{\text{Explain}}$ programs in Hybrid 4 are equivalent to the $\overline{\text{Alg}}$ and $\overline{\text{Explain}}$ programs in Hybrid 3.*

Proof. We consider below each change to the programs.

First, an "if" statement is added to Step 2 of the $\overline{\text{Alg}}$ program, to skip the check in Step 2 if either $u[1] = e^*[1]$ or $u[1] = u^*[1]$. To see why this change does not affect the functionality of the program, let us consider each case in turn. By Lemma 1, if $u[1] = e^*[1]$, then the only way the Step 2 check can be satisfied is if input $=$ input* and $u[2] = e^*[2]$. But this case is already handled in Step 1; therefore, skipping Step 2 if $u[1] = e^*[1]$ does not affect the functionality of the program. On the other hand, recall that $u^*[1]$ is chosen at random, and therefore the probability that $u^*[1]$ is in the image of $F_2(K_2, \cdot)$ is negligible. Thus, with overwhelming probability over the choice of constants $u^*[1]$, the check in Step 2 cannot be satisfied if $u[1] = u^*[1]$. Therefore, the addition of this "if" statement does not alter the functionality of the $\overline{\text{Alg}}$ program.

Also, the key K_3 is punctured at $u^*[1], e^*[1]$ in both the $\overline{\text{Alg}}$ and $\overline{\text{Explain}}$ programs. The new "if" statement above ensures that $F_3(K_3, \cdot)$ is never called at these values in the $\overline{\text{Alg}}$ program. Recall that the $\overline{\text{Explain}}$ program only calls $F_3(K_3, \cdot)$ on values computed as $F_2(K_2, (\text{input}, \text{output}, \text{PRG}(r)))$ for some bit input and strings output and r. Furthermore, F_2 is statistically injective with a very sparse image set, by our choice of parameters. Since every $u^*[1]$ is randomly chosen, it is very unlikely to be in the image of $F_2(K_2, \cdot)$. Since every $e^*[1]$ is chosen based on a random $\tilde{r}$ value instead of a PRG output, it is very unlikely to correspond to $F_2(K_2, (\text{input}, \text{output}, \text{PRG}(r)))$ for any $(\text{input}, \text{output}, r)$. Thus, these values are not called by the $\overline{\text{Explain}}$ program, except with negligible probability over the choice of these constants $u^*[1]$ and $e^*[1]$.

Two-Round Adaptively Secure MPC from Indistinguishability Obfuscation

Sanjam Garg[1] and Antigoni Polychroniadou[2,*]

[1] University of California, Berkeley, USA
sanjamg@berkeley.edu
[2] Aarhus University, Denmark
antigoni@cs.au.dk

Abstract. Adaptively secure Multi-Party Computation (MPC) first studied by Canetti, Feige, Goldreich, and Naor in 1996, is a fundamental notion in cryptography. Adaptive security is particularly hard to achieve in settings where arbitrary number of parties can be corrupted and honest parties are not trusted to properly erase their internal state. We did not know how to realize constant round protocols for this task even if we were to restrict ourselves to semi-honest adversaries and to the simpler two-party setting. Specifically the round complexity of known protocols grows with the depth of the circuit the parties are trying to compute.

In this work, using indistinguishability obfuscation, we construct a UC two-round Multi-Party computation protocol secure against any active, adaptive adversary corrupting an arbitrary number of parties.

1 Introduction

The notion of *secure computation* is central in cryptography. Introduced in the seminal work of [41,30] secure multiparty computation (MPC) allows several mutually distrustful parties $P_1, \ldots, P_n$ to compute a joint function f on their private inputs $(x_1, \ldots, x_n)$, in a way that ensures that honest parties obtain the correct outputs and no group of colluding malicious parties learns anything beyond their own inputs and the prescribed output. For this problem, we are interested in the natural setting where the attacker can on-the-fly decide on which parties to corrupt. This model of *adaptive* corruption was first studied by Canetti et al. [9]. In this paper we consider adaptive adversaries that are allowed to corrupt *arbitrary number* of honest parties. Additionally we only consider *non-erasure* protocols, specifically the protocols whose security does not depend on having honest parties erase any of their internal state. We refer the reader to [9, Section 1] for discussion on the importance of considering adaptive adversaries.

---

* Research supported by the Danish National Research Foundation and the National Science Foundation of China (under the grant 61061130540) for the Sino-Danish Center for the Theory of Interactive Computation and from the Center for Research in Foundations of Electronic Markets (CFEM), supported by the Danish Strategic Research Council. Also supported by ISF grant no. 1255/12.

Y. Dodis and J.B. Nielsen (Eds.): TCC 2015, Part II, LNCS 9015, pp. 614–637, 2015.

One fundamental complexity measure of an MPC protocol is its *round complexity*. For the static setting, Yao's original two-party secure computation protocol [41] was already round-optimal. Analogous results for the multi-party setting were obtain recently [1,22].

However achieving similar results in the adaptive setting has remained open. In the case where all but one of the parties can be corrupted, [36,34] and [35,27] including the concurrent work of [18], propose constant round two-party and multi-party protocols, respectively. On the other hand, round complexity of all know adaptively secure protocols secure against an arbitrary number of corruptions grows (see, e.g. [14,27,16]) linearly in the depth of the circuit that the parties are trying to compute. We stress that for this problem, this limitation holds for essentially every special case of interest — namely, even if we were to restrict to semi-honest/passive adversaries or to the special case of two-party protocols. In this work we ask the following fundamental question:

> *Is it possible to construct a constant round protocol secure against adaptive corruption of arbitrary number of parties?*

1.1 Our Result

We answer the above question in the affirmative and show how to obtain a two-round adaptively secure MPC protocol. Specifically:

Theorem 1 (Informal). *Assuming sub-exponentially secure indistinguishability obfuscation and other standard assumptions, we show that arbitrary functions can be UC-securely [8] computed in the presence of adaptive, active corruption of arbitrary number of parties with just two rounds of broadcast messages.*

We stress that in the above claim we are in the setting where security holds against an adversary corrupting any arbitrary number of parties. Furthermore, honest parties in our case are not required to erase anything. Also note that our results are for the strongest notion of security, the UC security. This means that our protocol remains secure even when multiple instances of our protocol are executed simultaneously. Since it is impossible to achieve UC security for dishonest majority without assuming trusted setup assumptions [10,12,37], we base our construction in the common reference string model.

In our results we consider an asynchronous multi-party network[1] where the communication is open (i.e. all the communication between the parties is seen by the adversary) and delivery of messages is not guaranteed. For simplicity, we assume that the delivered messages are authenticated. This can be achieved using standard methods.

1.2 Independent Work

In two very recent concurrent and independent works, [15,11] construct constant round protocols with security against a semi-honest, adaptive adversary

---

[1] The fact that the network is asynchronous means that the messages are not necessarily delivered in the order which they are sent.

corrupting any number of parties. Both works can obtain a constant round malicious version of their protocols by applying the [14] compiler.

In our paper we construct a two-round multi-party protocol with security against a malicious, adaptive adversary corrupting any number of parties. In contract, the protocols of [15] and [11] require more rounds. Furthermore, the construction of [11] solves the problem only for the special case of two parties. Note that the reduction in our result and the result of [11] involves a sub-exponential loss of security.

Last but not least, our protocol and the protocol of [11] are also leakage tolerant. The semi-honest version of [11] is also incoercible with respect to one of the parties.

1.3   Technical Difficulties and New Ideas

The key technical tool that we use in our construction is obfuscation so let us start by recalling it briefly.

Obfuscation. Obfuscation was first rigorously defined and studied by Barak et al. [4]. Most famously, they defined the notion of *virtual black box (VBB)* obfuscation, and proved that this notion is impossible to realize in general — i.e., there exist functions, though a bit unnatural, that are VBB unobfuscatable.

Barak et al. also defined a weaker notion of *indistinguishability obfuscation (iO)*, which avoids their impossibility results. Indistinguishability obfuscation requires that for any two circuits C_0, C_1 of similar size that *compute the same function*, it is hard to distinguish an obfuscation of C_0 from an obfuscation of C_1. In a recent result, Garg et al. [23] proposed a construction of iO for all circuits, basing security on assumptions related to multilinear maps [21].

Starting point — Garg et al. [22] construction. In a recent work, Garg et al. [22] constructed a two-round multiparty computation protocol secure against static adversaries. Though our goal is to realize a protocol secure in the adaptive setting it would be illustrative to see how Garg et al.'s construction works.

With the goal of explaining intuition [22] better we will describe the ideas assuming we have access to VBB obfuscation, rather than just indistinguishability obfuscation. We start by noting that two rounds of interaction are essential for realizing multiparty secure computation. This is because a 1-round protocol is inherently susceptible to the "residual function" attack in which a corrupted party could repeatedly evaluate the "residual function" with the inputs of the honest parties fixed on many different inputs of its own (e.g., see [33]). This attack can be circumvented by having two rounds of interaction — where in the first round the parties commit to their inputs and the output is generated only in the second round. The first round commitments help guarantee that the "residual function" attack can not be performed in this setting.

The key idea of the Garg et al. construction is to have every party commit to its input along with its randomness in the first round. The second round of the Garg et al. protocol is actually a simple compiler: it takes any (possibly highly interactive) underlying MPC protocol, and has each party obfuscate their "next-message" function in that protocol, providing one obfuscation for each round. This enables each party to independently evaluate the obfuscations one by one, generating messages of the underlying MPC protocol and finally obtain the output. Party i's next-message circuit for round j in the underlying MPC protocol depends on its input x_i and randomness r_i (which are hard-coded in the obfuscation). This circuit takes as input the transcript through round $j-1$, and it produces as output the next broadcast message.

However, there is another complication. Unlike the initial interactive protocol being compiled, the obfuscations are susceptible to a "reset" attack – i.e., they can be evaluated on multiple inputs. To prevent such an attack, we need to limit the obfuscations to be used for evaluation only on a unique set of values – namely, values consistent with the inputs and randomness that the parties committed to in the first round, and the current transcript of the underlying MPC protocol. Note that this would implicitly fix the transcript to a unique value. To ensure this consistency, Garg et al. [22] use non-interactive zero-knowledge (NIZK) proofs. Since the NIZKs apply not only to the committed values of the first round, but also to the transcript as it develops in the second round, the obfuscations themselves must also generate these NIZKs "on the fly". In other words, the obfuscations are now augmented to perform not only the next-message function, but also to prove that their output is consistent with their input, randomness and transcript so far. Also, obfuscations in round j of the underlying MPC protocol verify NIZKs associated to obfuscations in previous rounds before providing any output.

Garg et al. show that this construction can be adapted so that security can be based on indistinguishability obfuscation alone but we will not delve into that. Instead we will argue that this approach is fundamentally problematic for achieving the task at hand.

Our approach – starting afresh. Note that the above intuitive description uses multiple obfuscations that are generated by honest parties. This however only works in the static setting and our goal is adaptive security. The challenge in proving adaptive security is that, a simulator would have a hard time explaining these obfuscations as being honestly generated. Towards solving this problem we first would like to limit the use of obfuscation in our construction; specifically not requiring honest parties to generate any obfuscations.

Still assuming we have access to VBB obfuscation, we need a fresh direction to solve the above problem. Here is our first stab at the problem: assume the parties had access to a trusted third party. In this case each party could encrypt its input and deliver it to the trusted party. The trusted party could then decrypts these values to obtain the inputs of all the parties, compute the function on the inputs and then deliver the output back to the parties. Our idea is to have an obfuscated program given out as part of the CRS implement this trusted party. Just like the Garg et al. construction, in order to make this construction secure against "residual function" attack we will need to consider a setting with two rounds. In the first round, we will have all parties commit to their inputs and then in the second round we will have them provide encryptions of the openings previously committed.

Making this construction adaptively secure seems more amenable — specifically, by using adaptive commitments for the first round and a deniable encryption scheme for the second. We actually need the first round commitments to be simulation extractable. This allows our simulator to extract the values committed to by the adversary on behalf of corrupted parties, even as it equivocates on its own commitments. Once the simulator has access to the inputs of the corrupted outputs it can force the output by including it in its own second round encryption.

Basing it on Indistinguishability Obfuscation. The protocol described so far relies on VBB and we would like to instantiate our construction based on $i\mathcal{O}$. The CRS of the scheme includes an obfuscation that takes as input encryptions of inputs of all the parties and computes the desired functionality on their decryptions. A reader might have observed that this bears resemblance with functional encryption or even multi-input functional encryption [31]. One might wonder if the use of "two key trick" can help us realize this construction using just indistinguishability obfuscation — in a way similar to the functional encryption construction of Garg et al. [23]. In particular the idea would be that each party encrypts its input along with the opening twice under two different keys and attach along with them a NIZK proof proving that they indeed encrypt the same value.

Unfortunately, this solution is fundamentally problematic as we are in the adaptive setting. Even if we were to use an adaptively secure NIZK the problem is that NIZKs given on deniable encryptions are useless. This is because the encryption scheme is deniable. The deniability of the encryption scheme allows the adversary to encrypt two different plaintexts under the two public keys but still succeed in explaining them as encrypting the same message. This also allows the attacker to successfully prove that the two ciphertexts indeed encrypt the same message.

In summary, what we really need is a system with two ciphertexts and a proof proving that the two ciphertexts encrypt the same message with the property that only valid proofs exists. Additionally we need the property that both the ciphertexts and the proof can be denied upon in the proof of security. These requirements indeed seem to be in conflict with each other. For example, simultaneously achieving perfect soundness for NIZK and the ability to explain the simulated proofs as though they were honestly generated seems like a bottleneck.

Our solution to this seemingly paradoxical problem is to first construct a core two key encryption scheme which comes attached with a NIZK and then build deniability on top of it. In particular, the underlying core encryption scheme consists of two copies of a perfectly correct encryption scheme along with a NIZK proving that the two ciphertexts encrypt the same message and it is combined with a language which also binds it with the commitments of the first round. The NIZK we use will have statistical soundness. This underlying encryption scheme is then made *deniable* using the Sahai and Waters [40] Universal

Deniable Encryption (UDE) transformation. Briefly, UDE takes *any* encryption scheme and converts it to *deniable* so that ciphertexts are still indistinguishable from the usual ciphertexts of the underlying core encryption. Hence, our resulting encryption is deniable in a very strong sense — specifically, it allows the encryptor to deny not just on the two ciphertexts but also on the NIZK directly. However, interestingly proofs for invalid statements do not exist.

Finally various other technical challenges arise in the security proof. For example, in the proof of security the simulator needs to hardcode the output that the adversary gets as part of its ciphertext in a way that remains indistinguishable from an honest execution. In order to force the output, the core encryption scheme which is plugged into the UDE transformation is combined with the language which implicitly includes a trapdoor mode. In its trapdoor mode, the simulator can in particular plant the output of the function inside the encryptions it generates on behalf of honest parties. Then the obfuscation checks for such a trapdoor and acts accordingly. We refer the reader to the full construction and proof for details on how we resolve this and other issues.

1.4 Application to Leakage Tolerant Protocols

As another application of our techniques, we observe that our adaptively secure protocol is also leakage tolerant in a way that previous protocols failed to be. The study of leakage tolerant protocols was initiated by Bitansky et al. [5] and Garg et al. [25]. Very roughly, leakage tolerant protocols preserve security even when the adversary can obtain arbitrary leakage on the entire internal state of honest parties, however only up to the leaked information.

One limitation of known leakage tolerant secure computation protocols [7] (see also [17]) from the literature is that the leakage in the ideal world queries needs to depend on the inputs of all honest parties rather than just on the input of the party being leaked upon. Our adaptively secure protocol also turns out to be leakage resilient further avoiding this limitation. Another advantage of our protocol is that it is much simpler than the Boyle et al. [7] construction.

In a recent result, Garg et al. [24] show an alternative way of avoiding this limitation, without using indistinguishability obfuscation. However their result is restricted to a setting where at least one of the parties is never leaked on. We do not make such an assumption.

2 Preliminaries

In this section we recall preliminary notions needed in this work. We will start by recalling notions of indistinguishability obfuscation and non-interactive zero-knowledge. Next we recall the notion of publicly deniable encryption scheme that we adapt from [40].

2.1 Notation

Throuhgout the paper $\lambda \in \mathbb{N}$ will denote the security parameter. We say that a function $f : \mathbb{N} \to \mathbb{R}$ is negligible if $\forall c \ \exists \ n_c$ such that if $n > n_c$ then $f(n) < n^{-c}$.

We will use $\mathsf{negl}(\cdot)$ to denote an unspecified negligible function. We often use $[n]$ to denote the set $\{1, ..., n\}$. The concatenation of a with b is denoted by $a\|b$. Moreover, we use $d \leftarrow \mathcal{D}$ to denote the process of sampling d from the distribution $\mathcal{D}$ or, if $\mathcal{D}$ is a set, a uniform choice from it. If $\mathcal{D}_1$ and $\mathcal{D}_2$ are two distributions, then we denote that they are statistically close by $\mathcal{D}_1 \approx_s \mathcal{D}_2$; we denote that they are computationally indistinguishable by $\mathcal{D}_1 \approx_c \mathcal{D}_2$; and we denote that they are identical by $\mathcal{D}_1 \equiv \mathcal{D}_2$.

2.2 Indistinguishability Obfuscators

We will start by recalling the notion of indistinguishability obfuscation ($i\mathcal{O}$) recently realized in [23] using candidate multilinear maps [21].

Definition 1 (Indistinguishability Obfuscator ($i\mathcal{O}$)). *A uniform PPT machine $i\mathcal{O}$ is called an* indistinguishability obfuscator *for a circuit class $\{\mathcal{C}_\lambda\}$ if the following conditions are satisfied:*

- *For all security parameters $\lambda \in \mathbb{N}$, for all $C \in \mathcal{C}_\lambda$, for all inputs x, we have that*

$$\Pr[C'(x) = C(x) : C' \leftarrow i\mathcal{O}(\lambda, C)] = 1$$

- *For any (not necessarily uniform) PPT distinguisher D, there exists a negligible function α such that the following holds: For all security parameters $\lambda \in \mathbb{N}$, for all pairs of circuits $C_0, C_1 \in \mathcal{C}_\lambda$, we have that if $C_0(x) = C_1(x)$ for all inputs x, then*

$$\left| \Pr\left[D(i\mathcal{O}(\lambda, C_0)) = 1 \right] - \Pr\left[D(i\mathcal{O}(\lambda, C_1)) = 1 \right] \right| \leq \mathsf{negl}(\lambda)$$

2.3 Non-Interactive Zero-Knowledge Proofs

Let $\mathcal{R}$ be an NP-relation. For pairs $(x, w) \in \mathcal{R}$ we call x the statement and w the witness. Let $\mathcal{L}$ be the language consisting of statements in $\mathcal{R}$. A Non-Interactive Zero Knowledge (NIZK) Proof system [6,19] consists of three PPT algorithms (K, P, V), a common reference string generation algorithm K, a prover P and a verifier V.

- $K(1^\lambda)$ expects as input the unary representation of the security parameter λ and outputs a common reference string σ of length $\Omega(\lambda)$.
- $P(\sigma, x, w)$ takes as input a common reference string σ, a statement x together with a witness w such that $\mathcal{R}(x, w)$ and produces a proof π.
- $V(\sigma, x, \pi)$ takes as input a common reference string σ, a statement x, a proof π and outputs 1 if the proof is accepting and 0 otherwise.

We call (K, P, V) a non-interactive proof system for $\mathcal{R}$ if it satisfies the following properties.

PERFECT COMPLETENESS. A proof system is complete if an honest prover with a valid witness can convince an honest verifier. Formally, $\forall x \in \mathcal{L}$, $\forall w$ witness of x

$$\Pr\left[\sigma \leftarrow K(1^\lambda); \pi \leftarrow P(\sigma, x, w) : V(\sigma, x, \pi) = 1\right] = 1.$$

STATISTICAL SOUNDNESS. A proof system is sound if it is infeasible to convince an honest verifier when the statement is false. Formally, we have

$$\Pr\left[\sigma \leftarrow K(1^\lambda); \exists (x, \pi) : x \notin \mathcal{L} \wedge V(\sigma, x, \pi) = 1\right] < \mathsf{negl}(\lambda).$$

COMPUTATIONAL ZERO-KNOWLEDGE. We say a non-interactive proof (K, P, V) is computational zero-knowledge if there exists a PPT simulator $S = (S_1, S_2)$, where S_1 returns a simulated common reference string σ together with a simulation trapdoor τ that enables S_2 to simulate proofs without having access to the witness. For all non-uniform PPT adversaries $\mathcal{A} = (\mathcal{A}_1, \mathcal{A}_2)$ the following quantity is upper bounded by a negligible function:

$$\left| \Pr\left[\sigma \leftarrow K(1^\lambda); (x, state) \leftarrow \mathcal{A}_1(\sigma); \pi \leftarrow P(\sigma, x, w) : \mathcal{A}_2(x, \sigma, \pi, state) = 1\right] \right. -$$

$$\left. \Pr\left[(\sigma, \tau) \leftarrow S_1(1^\lambda); (x, state) \leftarrow \mathcal{A}_1(\sigma); \pi \leftarrow S_2(\sigma, \tau, x) : \mathcal{A}_2(x, \sigma, \pi, state) = 1\right] \right|.$$

2.4 Double Key Encryption and Its Deniable Variant

Our protocol will use a special publicly deniable encryption scheme that we construct by first describing a core public-key encryption scheme that we then transform to its deniable variant using the Universal Deniable Encryption (UDE) transformation of [40].

Let (Setup, Enc, Dec) be a perfectly correct IND-CPA secure public-key encryption scheme and let (K, P, V) be a NIZK proof system with statistical soundness and computational zero-knowledge. The core encryption scheme we consider is very similar to the Naor-Yung CCA [39] secure encryption scheme. Recall that in the Naor-Yung construction a ciphertext consists of encryptions of a message under two different public keys and a NIZK proof certifying that the two ciphertexts indeed encrypt the same message. In our encryption scheme a ciphertext will also consist of two ciphertexts under the two public keys but the NIZK proof will be used to certify a more sophisticated requirement. More formally:

Definition 2 (Double Key Encryption Scheme). *Let* (Setup, Enc, Dec) *be a IND-CPA secure encryption scheme with perfect correctness. Let* (K, P, V) *be a NIZK proof system for an NP-Language $\mathcal{L}$. A Double Key encryption scheme, parametrised by a language $\mathcal{L}$, consists of three algorithms* $\mathsf{DK}_{\mathcal{L}} = (\mathsf{Setup}_{\mathsf{DK}}, \mathsf{Enc}_{\mathsf{DK}}, \mathsf{Dec}_{\mathsf{DK}})$.

- $\mathsf{Setup}_{\mathsf{DK}}(1^\lambda, 1^\ell)$ *is a polynomial time procedure that takes as input the unary representation of the security parameter λ, the description of length of messages encrypted 1^ℓ. It computes $(pk_0, sk_0), (pk_1, sk_1) \leftarrow \mathsf{Setup}(1^\lambda)$ and the common reference string $\sigma \leftarrow K(1^\lambda)$ for the NIZK proof. It outputs the public key $PK = (pk_0, pk_1, \sigma)$ and the secret key $SK = (sk_0, sk_1)$.*
- $\mathsf{Enc}_{\mathsf{DK}}(PK, m_0, m_1, \mathsf{aux}, w; r)$: *This polynomial time procedure takes as input public key $PK = (pk_0, pk_1, \sigma)$, messages $m_0, m_1 \in \{0,1\}^\ell$, auxiliary information aux and some w which will be used as part of the witness for the language $\mathcal{L}$. It generates $c = \mathsf{Enc}(pk_0, m_0; s_0)$ and $c' = \mathsf{Enc}(pk_1, m_1; s_1)$ and outputs (c, c', π), where $\pi \leftarrow P(\sigma, (c, c', \mathsf{aux}), (m_0, m_1, s_0, s_1, w))$ for the language $\mathcal{L}$.*
- $\mathsf{Dec}_{\mathsf{DK}}(PK, SK, (c, c', \pi), \mathsf{aux}))$: *is a polynomial time procedure that takes as input $PK = (pk_0, pk_1, \sigma)$, $SK = (sk_0, sk_1)$, ciphertext (c, c', π) and auxiliary information aux. Outputs $\perp$, in case that $V(\sigma, (c, c', \mathsf{aux}), \pi) = 0$ else output $(\mathsf{Dec}(sk_0, c), \mathsf{Dec}(sk_1, c'))$.*

Double Key Deniable Encryption Scheme. Next we want to transform the above core public key encryption into its deniable variant using the UDE transformation of Sahai and Waters [40, Section 4.2]. In particular, once we plug the above $\mathsf{DK}_\mathcal{L}$ double key encryption scheme in the UDE transformation, we obtain a double key *deniable* encryption scheme $\mathsf{DDK}_\mathcal{L} = (\mathsf{Setup}_{\mathsf{DDK}}, \mathsf{Enc}_{\mathsf{DDK}}, \mathsf{Dec}_{\mathsf{DDK}}, \mathsf{DenEnc}_{\mathsf{DDK}}, \mathsf{Explain}_{\mathsf{DDK}})$ parametrized by the language $\mathcal{L}$ with associate relation $\mathcal{R}_\mathcal{L}$ where the procedures $\mathsf{Enc}_{\mathsf{DDK}}$ and $\mathsf{Dec}_{\mathsf{DDK}}$ are same as $\mathsf{Enc}_{\mathsf{DK}}$ and $\mathsf{Dec}_{\mathsf{DK}}$. Here $\mathsf{Setup}_{\mathsf{DDK}}$ is obtained by augmenting the procedure $\mathsf{Setup}_{\mathsf{DK}}$ to additionally output a public denying key DK generated using $\mathsf{UniversalSetup}(PK)$ as defined in [40, Section 4.2] which is going to be included in PK. Further the scheme is augmented with the following two procedures where $PK = (\sigma, pk_0, pk_1, DK)$.

- $\mathsf{DenEnc}_{\mathsf{DDK}}(PK, m_0, m_1, \mathsf{aux}, w; r)$ is a polynomial time procedure that takes as input PK which includes the public denying key DK, $m_0, m_1 \in \{0,1\}^\ell$, auxiliary information aux and witness w and uses random coins r. It then outputs (c, c', π).
- $\mathsf{Explain}_{\mathsf{DDK}}(PK, (c, c', \pi), (m_0, m_1, \mathsf{aux}, w); u)$: This polynomial time procedure takes as input public key PK which includes the public denying key DK, messages $m_0, m_1 \in \{0,1\}^\ell$, auxiliary information aux and witness w. It also takes as input a value (c, c', π) and outputs a string e, that is the same size as the randomness r taken by $\mathsf{DenEnc}_{\mathsf{DDK}}$ above.

This new scheme has the following two additional properties.

Indistinguishability of source of ciphertext. We say that the scheme has indistinguishability of source of ciphertext if for any λ and any PPT adversary $\mathcal{A} = (\mathcal{A}_1, \mathcal{A}_2)$ the following quantity can be upper bounded by a negligible function:

$$\left| \Pr \left[\begin{array}{l} (PK, SK) \leftarrow \mathsf{Setup}_{\mathsf{DDK}}(1^\lambda, 1^\ell), \\ (m_0, m_1, \mathsf{aux}, w) \leftarrow \mathcal{A}_1(PK), \\ ct = \mathsf{Enc}_{\mathsf{DDK}}(PK, m_0, m_1, \mathsf{aux}, w; r) \\ \mathcal{A}_2(PK, ct) = 1 \end{array} \right] - \Pr \left[\begin{array}{l} (PK, SK) \leftarrow \mathsf{Setup}_{\mathsf{DDK}}(1^\lambda, 1^\ell), \\ (m_0, m_1, \mathsf{aux}, w) \leftarrow \mathcal{A}_1(PK), \\ ct = \mathsf{DenEnc}_{\mathsf{DDK}}(PK, m_0, m_1, \mathsf{aux}, w; r) \\ \mathcal{A}_2(PK, ct) = 1 \end{array} \right] \right|$$

Indistinguishability of explanation. We say that the scheme has indistinguishability of explanation if for any λ and any PPT adversary $\mathcal{A} = (\mathcal{A}_1, \mathcal{A}_2)$ the following quantity can be upper bounded by a negligible function:

$$\left| \Pr \left[\begin{array}{l} (PK, SK) \leftarrow \mathsf{Setup}_{\mathsf{DDK}}(1^\lambda, 1^\ell), \\ (m_0, m_1, \mathsf{aux}, w) \leftarrow \mathcal{A}_1(PK), \\ ct = \mathsf{DenEnc}_{\mathsf{DDK}}(PK, m_0, m_1, \mathsf{aux}, w; r) \\ \\ \mathcal{A}_2(PK, ct, r) = 1 \end{array} \right] - \Pr \left[\begin{array}{l} (PK, SK) \leftarrow \mathsf{Setup}_{\mathsf{DDK}}(1^\lambda, 1^\ell), \\ (m_0, m_1, \mathsf{aux}, w) \leftarrow \mathcal{A}_1(PK), \\ ct = \mathsf{DenEnc}_{\mathsf{DDK}}(PK, m_0, m_1, \mathsf{aux}, w; r) \\ e = \mathsf{Explain}_{\mathsf{DDK}}(PK, ct, (m_0, m_1, \mathsf{aux}, w); u) \\ \mathcal{A}_2(PK, ct, e) = 1 \end{array} \right] \right|$$

We remark that the [40] deniable encryption scheme immediately implies a deniable encryption scheme for multi-bit messages of any polynomial length k bits by creating a ciphertext for a k-bit message as a sequence of k single bit encryptions. Our construction cannot support the above bit-by-bit encryption since every single encryption takes longer messages. However the Sahai-Waters construction is selectivly secure and the security can be amplified to the adaptive setting (as defined above) at the cost of a sub-exponential loss in the security. In other words we can realize the above definition assuming sub-exponential hardness on the assumptions made by Sahai-Waters.

2.5   Equivocal and Extractable Commitments

An Equivocal and Extractable Commitment scheme COM consists of a tuple of PPT algorithms ($\mathsf{Setup}^{\mathsf{bind}}_{\mathsf{Com}}, \mathsf{Setup}^{\mathsf{equiv}}_{\mathsf{Com}}, \mathsf{Com}, \mathsf{Extr}, \mathsf{Equiv}$). We will describe our definitions for the setting of bit commitment and note that they extend to the setting of strings in a natural way.

- $\mathsf{Setup}^{\mathsf{bind}}_{\mathsf{Com}}(1^\lambda)$ expects as input the unary representation of the security parameter λ and outputs a public parameter CK together with a trapdoor μ (used for extraction).
- $\mathsf{Setup}^{\mathsf{equiv}}_{\mathsf{Com}}(1^\lambda)$ expects as input the unary representation of the security parameter λ and outputs a public parameter CK together with trapdoors μ and ν (used for extraction and equivocation).
- $\mathsf{Com}(CK, b; r)$ takes as input CK, a bit $b \in \{0, 1\}$ and randomness $r \in \{0, 1\}^\lambda$ and outputs a commitment β.

Let us define the following language (the extraction procedure Extr is defined below):

$$\mathcal{L}_{\mathsf{Com}} = \{(\beta, b) \mid \exists t : \beta = \mathsf{Com}(CK, b; t) \vee b = \mathsf{Extr}(CK, t, \beta)\}.$$

We note that the language naturally extends to a setting where commitments are defined over strings instead of just bits. Also we defined associated relation $\mathcal{R}_{\text{Com}}$. The above commitment scheme should satisfy the following properties.

INDISTINGUISHABILITY OF PUBLIC PARAMETERS. We require that:

$$\left| \Pr\left[(CK, \mu) \leftarrow \text{Setup}_{\text{Com}}^{\text{bind}}(1^\lambda) : \mathcal{A}(CK, \mu) = 1 \right] - \right.$$
$$\left. \Pr\left[(CK, \mu, \nu) \leftarrow \text{Setup}_{\text{Com}}^{\text{equiv}}(1^\lambda) : \mathcal{A}(CK, \mu) = 1 \right] \right| < \text{negl}(\lambda).$$

COMPUTATIONAL HIDING. Hiding means that no computationally bounded adversary can distinguish as to which bit is locked in the commitment. Let $\mathcal{A}$ be any non-uniform adversary running in time $poly(\lambda)$. We say that the commitment scheme is computationally hiding if:

$$\Pr\left[b = b' \left| \begin{array}{l} b \leftarrow \{0,1\}; (CK, \mu) \leftarrow \text{Setup}_{\text{Com}}^{\text{bind}}(1^\lambda); \\ \beta = \text{Com}(CK, b; r); b' \leftarrow \mathcal{A}(\beta) \end{array} \right. \right] = \frac{1}{2} + \text{negl}(\lambda) .$$

The same applies to the setup algorithm $\text{Setup}_{\text{Com}}^{\text{equiv}}$.

PERFECTLY BINDING. Intuitively speaking, binding requires that no (even unbounded) adversary can open the commitment in two different ways. Here, we define the strongest variant known as *perfectly binding*. Formally we require that for all $(CK, \mu) \leftarrow \text{Setup}_{\text{Com}}^{\text{bind}}(1^\lambda)$ there exists no values (r_0, r_1) such that $\text{Com}(CK, 0; r_0) = \text{Com}(CK, 1; r_1)$. For perfectly binding we require that either $(c, 0) \in \mathcal{L}_{\text{Com}}$ or $(c, 1) \in \mathcal{L}_{\text{Com}}$, but not both.

POLYNOMIAL EQUIVOCALITY. The setup algorithm $\text{Setup}_{\text{Com}}^{\text{equiv}}$ generates public parameters together with trapdoors μ and ν such that Equiv using ν is able to produce polynomially many fake commitments, using the same CK, which can then be explained to either 0 and 1. More formally, Equiv can be viewed as a pair of PPT algorithms $(\text{Equiv}_1, \text{Equiv}_2)$ such that the following holds. Let $(CK, \mu, \nu) \leftarrow \text{Setup}_{\text{Com}}^{\text{equiv}}(1^\lambda)$ then $(\beta, state) \leftarrow \text{Equiv}_1(CK, \nu)$ and $r_b \leftarrow \text{Equiv}_2(CK, \nu, \beta, state, b)$ such that $\text{Com}(CK, b; r_b) = \beta$. Furthermore we require that for $b \in \{0, 1\}$ the distribution of $\{(CK, \beta, r_b)\}$ generated in this way is computationally indistinguishable from the distribution $\{(CK, \beta, r_b)\}$ where $\beta = \text{Com}(CK, b; r_b)$.

SIMULATION EXTRACTABILITY. We require that the commitment remains binding for any adversary $\mathcal{A}$, even after $\mathcal{A}$ obtains polynomially many equivocal commitments generated by Equiv along with their openings. More formally, the following quantity is negligible:

$$\Pr\left[b \neq b' \left| \begin{array}{l} (CK, \mu, \nu) \leftarrow \text{Setup}_{\text{Com}}^{\text{equiv}}(1^\lambda); (\beta, b, r) \leftarrow \mathcal{A}^{\text{Equiv}^*(CK,\nu)}(CK); \\ \text{Com}((CK, b, r) = \beta \wedge \text{Extr}(CK, \mu, \beta) = b' \end{array} \right. \right]$$

where Equiv^* is either invoked as Equiv_1 without revealing the *state*, or as Equiv_2 which only expects as input fake commitments generated by previous invocations of Equiv_1.

In this paper, we use the non-interactive equivocal and extractable commitment scheme of [14] (CLOS commitment) which assumes the existence of enhanced trapdoor permutations. At the heart of their commitment scheme is the Feige-Shamir trapdoor commitment scheme [20], which they transform to obtain a UC Commitment scheme secure against adaptive adversaries.

3   Our Protocol

In this section we will present our adaptively secure two-round MPC protocol, described in Figure 1. For simplicity, we assume that the delivered messages are authenticated. Also for simplicity of exposition, in the sequel, we will assume that random coins are an implicit input to the commitment and encryption functions, unless specified explicitly.

Theorem 2. *Let f be any deterministic poly-time function with n inputs and single output. Assume the existence of an Indistinguishability Obfuscator $i\mathcal{O}$, a Double Key Deniable encryption scheme $\mathsf{DDK}_{\mathcal{L}} = (\mathsf{Setup}_{\mathsf{DDK}}, \mathsf{Enc}_{\mathsf{DDK}}, \mathsf{Dec}_{\mathsf{DDK}}, \mathsf{DenEnc}_{\mathsf{DDK}}, \mathsf{Explain}_{\mathsf{DDK}})$ and an adaptively secure Commitment scheme $\mathsf{COM} = (\mathsf{Setup}_{\mathsf{Com}}^{\mathsf{bind}}, \mathsf{Setup}_{\mathsf{Com}}^{\mathsf{equiv}}, \mathsf{Com}, \mathsf{Extr}, \mathsf{Equiv})$. Then the protocol Π presented in Figure 1 UC-securely realizes the ideal functionality $\mathcal{F}_f$ in the $\mathcal{F}_{CRS}$-hybrid model with computational security against any adaptive, active adversary corrupting an arbitrary number of parties in two rounds of broadcast.*

Corollary 1. *Assume the existence of a sub-exponentially secure indistinguishability obfuscation and doubly enhanced trapdoor permutation then any ideal functionality $\mathcal{F}_f$ can be UC-securely realized in the $\mathcal{F}_{CRS}$- model against any adaptive, active adversary corrupting an arbitrary number of parties. Furthermore this protocol involves only two rounds of broadcast.*

We start by noting that the protocol is correct. Observe that if all the parties behave honestly then the protocol ends us executing the circuit f on the inputs of all parties, leading to the correct output. Security is proved via a simulator provided in Section 4 and indistinguishability is argued in Section 5.

3.1   Extensions

Now we give some natural extensions of our protocol and remove assumptions that were made to simplify exposition.

Protocol Π

Protocol Π uses an Indistinguishability Obfuscator $i\mathcal{O}$, a Double Key Deniable encryption scheme $\mathsf{DDK}_{\mathcal{L}} = (\mathsf{Setup_{DDK}}, \mathsf{Enc_{DDK}}, \mathsf{Dec_{DDK}}, \mathsf{DenEnc_{DDK}}, \mathsf{Explain_{DDK}})$ based on the scheme $(\mathsf{Setup}, \mathsf{Enc}, \mathsf{Dec})$ with perfect correctness, where the relation $\mathcal{L}$ is defined below, and an adaptively secure Commitment scheme $\mathsf{COM} = (\mathsf{Setup_{Com}^{bind}}, \mathsf{Com})$.[a] Let $f : (\{0,1\}^{\ell_{in}})^n \rightarrow \{0,1\}^{\ell_{out}}$ be the circuit parties want to evaluate on their private inputs.

Private Inputs: Party P_i for $i \in [n]$, receives its input x_i.

CRS: Output $(PK, CK, \mathsf{o}P)$ as the common reference string generated as follows:

- Generate $(PK, SK) \leftarrow \mathsf{Setup_{DDK}}(1^\lambda, 1^{\ell_{in}+\ell_{out}})$ where $PK = (\sigma, pk_0, pk_1, DK)$ and $SK = (sk_0, sk_1)$
- Generate $(CK, \mu) \leftarrow \mathsf{Setup_{Com}^{bind}}(1^\lambda)$.
- Let $\mathsf{o}P = i\mathcal{O}_{\mathsf{Prog}_{sk_0, PK, CK, f}}$ be the obfuscation of the program $\mathsf{Prog}_{sk_0, PK, CK, f}$, described in Figure 2.

Round 1: Each party P_i generates $\beta_i = \mathsf{Com}(CK, x_i; \omega_i)$ and broadcasts it to all parties.

Round 2: Each party P_i generates $(c_i, c_i', \pi_i) = \mathsf{DenEnc_{DDK}}(PK, x_i||\phi^{\ell_{out}}, x_i||\phi^{\ell_{out}}, (i, \{\beta_j\}_{j\in[n]}), (0^{n \cdot \ell_{in}}, 0^{\ell_{out}}, \{t_j\}_{j\in[n]}); r_i)$ where ϕ is a special fixed symbol and $t_i = \omega_i$ and $t_j = 0^*$ for all $j \in [n]$ such that $j \neq i$. It then broadcasts (c_i, c_i', π_i) to all parties.

Output phase: Each party P_i outputs $\mathsf{o}P(\{\beta_j\}_{j\in[n]}, \{c_j, c_j', \pi_j\}_{j\in[n]})$.

Language $\mathcal{L}$ for the Double Key deniable encryption scheme $\mathsf{DDK}_{\mathcal{L}}$:
Recall $\mathcal{L}_{\mathsf{Com}}$ as the language defined in Section 2.5, and let $\mathcal{R}_{\mathsf{Com}}$ be the associated relation. We have that $(c, c', (i, \{\beta_j\}_{j\in[n]})) \in \mathcal{L}$ if $(c, c', (i, \{\beta_j\}_{j\in[n]})) \in \mathcal{L}_1 \vee \mathcal{L}_2$ defined as follows:[b]

$$
\mathcal{L}_1 = \left\{ (c, c', (i, \{\beta_j\}_{j\in[n]})) \middle| \begin{array}{l} \exists\, (m_0, m_1, s_0, s_1, (\{x_j\}_{j\in[n]}, out, \{t_j\}_{j\in[n]})) \text{ s.t.} \\ c = \mathsf{Enc}(pk_0, m_0; s_0) \wedge c' = \mathsf{Enc}(pk_1, m_1; s_1) \\ \wedge\, m_0 = m_1 = x_i||\phi^{\ell_{out}} \\ \wedge\, \mathcal{R}_{\mathsf{Com}}((\beta_i, x_i), t_i) \end{array} \right\}
$$

$$
\mathcal{L}_2 = \left\{ (c, c', (i, \{\beta_j\}_{j\in[n]})) \middle| \begin{array}{l} \exists\, (m_0, m_1, s_0, s_1, (\{x_j\}_{j\in[n]}, out, \{t_j\}_{j\in[n]})) \text{ s.t.} \\ c = \mathsf{Enc}(pk_0, m_0; s_0) \wedge c' = \mathsf{Enc}(pk_1, m_1; s_1) \\ \wedge\, m_0 = x_i||\phi^{\ell_{out}} \wedge m_1 = \phi^{\ell_{in}}||out \\ \wedge\, \forall j \in [n], \mathcal{R}_{\mathsf{Com}}((\beta_j, x_j), t_j) \\ \wedge\, out = f(\{x_j\}_{j\in[n]}) \end{array} \right\}
$$

---

[a] We note that COM provides more procedures but we note that they only affect the proof. Hence for simplicity of exposition we skip mentioning them here.

[b] Changes in $\mathcal{L}_2$ from $\mathcal{L}_1$ are highlighted in red.

Fig. 1. The Π Protocol

Program $\mathsf{Prog}_{SK_b, PK, CK, f}$

Input: $(\{\beta_j\}_{j \in [n]}, \{c_j, c'_j, \pi_j\}_{j \in [n]})$.
Description:

1. If there exists $j \in [n]$ such that $\mathsf{Dec}_{DDK}(PK, SK_b, (c_j, c'_j, \pi_j), \{\beta_j\}_{j \in [n]}) = \perp$ then output $\perp$.
2. Parse c_j as $d_{j,0} \| e_{j,0}$ where $d_{j,0}$ is the encryption of the first ℓ_{in} bits and $e_{j,0}$ is the encryption of the rest of the bits. Similarly parse c'_j as $d_{j,1} \| e_{j,1}$. If $\exists j \in [n]$ such that $\mathsf{Dec}(sk_b, e_{j,b}) \neq \phi^{\ell_{out}}$, then let i be the first such j. If this is the case then output $\mathsf{Dec}(sk_b, e_{i,b})$.
3. Otherwise for each $j \in [n]$, let $x_j = \mathsf{Dec}(sk_b, d_{j,b})$ and output $f(\{x_j\}_{j \in [n]})$.

Fig. 2. The Program $\mathsf{Prog}_{SK_b, PK, CK, f}$

General Functionality. Our basic MPC protocol as described in Figure 1, only considers deterministic functionalities where all the parties receive the same output. We would like to generalize it to handle randomized functionalities and individual outputs (just as in [2]). First, the standard transformation from a randomized functionality to a deterministic one (See [29, Section 7.3]) works for this case as well. In this transformation, instead of computing some randomized function $g(x_1, \ldots x_n; r)$, the parties compute the deterministic function $f((r_1, x_1), \ldots, (r_n, x_n)) \overset{def}{=} g(x_1, \ldots, x_n; \oplus_{i=1}^n r_i)$. We note that this computation does not add any additional rounds. We note that since we are in the setting of adaptive security we can only realize *adaptively well-formed* [14] functionalities, which reveals its randomness if all the parties are corrupted.

Next, we move to individual outputs. Again, we use a standard transformation (See [38], for example). Given a function $g(x_1, \ldots, x_n) \to (y_1, \ldots, y_n)$, the parties can evaluate the following function which has a single output:

$$f((k_1, x_1), \ldots, (k_n; x_n)) = (g_1(x_1, \ldots, x_n) \oplus k_1 \| \ldots \| g_n(x_1, \ldots, x_n) \oplus k_n)$$

where g_i indicates the ith output of g, and k_i is randomly chosen by the ith party. Then, the parties can evaluate f, which is a single output functionality, instead of g. Subsequently every party P_i uses its secret input k_i to recover its own output. The only difference is that f has one additional exclusive-or gate for every circuit-output wire. Again, this transformation does not add any additional rounds of interaction.

Making CRS independent of the circuit being computed. Note that in our construction the obfuscation oP that is given as part of the CRS depends on the circuit f parties are trying to compute on their joint inputs. We can remove this dependence by using a universal circuit. Then the parties can feed in the universal circuit the actual circuit that they want along with their private inputs. However, the CRS will still depend on the size of the circuit. This is also the case for the protocols in [11,15]. We can avoid this by setting a priori bound on the size of the circuit being computed. It would be interesting to remove the dependence of the CRS on the size of the circuit.

4   Description of Our Simulator

Let $\mathcal{A}$ be an active, adaptive adversary that interacts with parties running the protocol Π from Figure 1 in the $\mathcal{F}_{CRS}$-hybrid model. We construct a simulator $\mathcal{S}$ (the ideal world adversary) with access to the ideal functionality $\mathcal{F}_f$, which simulates a real execution of Π with $\mathcal{A}$ such that no environment $\mathcal{Z}$ can distinguish the ideal world experiment with $\mathcal{S}$ and $\mathcal{F}_f$ from a real execution of Π with $\mathcal{A}$.

Recall that $\mathcal{S}$ interacts with the ideal functionality $\mathcal{F}_f$ and with the environment $\mathcal{Z}$. The ideal adversary $\mathcal{S}$ starts by invoking a copy of $\mathcal{A}$ and running a simulated interaction of $\mathcal{A}$ with the environment $\mathcal{Z}$ and the parties running the protocol. Our simulator $\mathcal{S}$ proceeds as follows:

Simulated CRS: The common reference string is chosen by $\mathcal{S}$ in the following manner (recall that $\mathcal{S}$ chooses the CRS for the simulated $\mathcal{A}$ as we are in the $\mathcal{F}_{CRS}$-hybrid model):

1. $\mathcal{S}$ runs the setup algorithm $\mathsf{Setup}_{\mathsf{DDK}}(1^\lambda, 1^{\ell_{in}+\ell_{out}})$ of the Double Key deniable encryption scheme, but replaces its internal call to the algorithm K with $S = (S_1, S_2)$ of the NIZK proof system. More specifically, $\mathcal{S}$ generates $(pk_0, sk_0), (pk_1, sk_1) \leftarrow \mathsf{Setup}(1^\lambda)$, $(\sigma, \tau) \leftarrow S_1(1^\lambda)$, along with the public denying key DK. It sets the public key $PK = (pk_0, pk_1, \sigma, DK)$.
2. $\mathcal{S}$ runs the algorithm $\mathsf{Setup}_{\mathsf{Com}}^{\mathsf{equiv}}(1^\lambda)$ of the adaptively secure commitment scheme COM and obtains (CK, μ, ν).
3. $\mathcal{S}$ computes $\mathsf{o}P = i\mathcal{O}_{\mathsf{Prog}_{sk_1, PK, CK, f}}$ where the latter is the obfuscation of the program Prog, as described in Figure 2, parameterized with the key sk_1.

$\mathcal{S}$ sets the common reference string equal to $(PK, CK, \mathsf{o}P)$ and locally stores (SK, τ, μ, ν).

Looking ahead, the trapdoor μ will be used to extract the inputs of the corrupted parties and ν to equivocate on the commitment $\mathcal{S}$ provides on behalf of honest parties. The trapdoor τ for the simulated σ will be used to generate simulated proofs.

Simulating the communication with $\mathcal{Z}$: Every input value that $\mathcal{S}$ receives from $\mathcal{Z}$ is written on $\mathcal{A}$'s input tape. Similarly, every output value written by $\mathcal{A}$ on its own output tape is directly copied to the output tape of $\mathcal{S}$.

Simulating actual protocol messages in Π: Note that there might be multiple sessions executing concurrently. Let sid be the session identifier for one specific session. We will specify the simulation strategy corresponding to this specific session. The simulator strategy for all other sessions will be the same. Let $\mathcal{P} = \{P_1, \ldots, P_n\}$ be the set of parties participating in the execution of Π corresponding to the session identified by the session identifier sid. Also let $\mathcal{P}^{\mathcal{A}} \subseteq \mathcal{P}$ be the set of parties corrupted by the adversary $\mathcal{A}$ at any time. Recall that we are in the setting of adaptive corruption so more parties could be added to this set as the protocol proceeds. At any point $\mathcal{S}$ only generates messages on behalf of parties $\mathcal{P}\backslash\mathcal{P}^{\mathcal{A}}$. In the following, if at the end of some round all parties are corrupted then $\mathcal{S}$ does not need to go to do anything else.

Round 1 Messages $\mathcal{S} \to \mathcal{A}$: In the first round $\mathcal{S}$ must generate messages on behalf of the honest parties, i.e. parties in the set $\mathcal{P}\backslash\mathcal{P}^{\mathcal{A}}$. For each party $P_i \in \mathcal{P}\backslash\mathcal{P}^{\mathcal{A}}$ our simulator proceeds as:

1. Generate a fake commitment $(\beta_i, state_i) \leftarrow \mathsf{Equiv}_1(CK, \nu)$.

It then sends β_i to $\mathcal{A}$ on behalf of party P_i and it internally saves $state_i$.

Round 1 Messages $\mathcal{A} \to \mathcal{S}$: Also in the first round the adversary $\mathcal{A}$ generates the messages on behalf of corrupted parties in $\mathcal{P}^{\mathcal{A}}$. For each party $P_i \in \mathcal{P}^{\mathcal{A}}$ our simulator proceeds as follows:

1. Extracting inputs of corrupted parties: Let β_i be the commitment that $\mathcal{A}$ sends on behalf of P_i. Our simulator $\mathcal{S}$ runs the extraction algorithm $\mathsf{Extr}(CK, \mu, \beta_i)$ in order to obtain x_i.
   Note that it is possible that $\mathcal{A}$ sends a commitment β_i on behalf of P_i such that it is not well-formed, or in other words extraction using the function Extr fails. In this case $\mathcal{S}$ sets $x_i = \bot$ and proceeds to the next step. (Looking ahead, we note that in this case the adversary will not be able to generate a valid second round message.)
2. Next $\mathcal{S}$ sends $(\mathsf{input}, \mathsf{sid}, \mathcal{P}, P_i, x_i)$ to $\mathcal{F}_f$ on behalf of the corrupted party P_i.

Simulating corruption of parties in Round 1: When $\mathcal{A}$ corrupts a real world party P_i, then $\mathcal{S}$ first corrupts the corresponding ideal world party P_i and obtains its input x_i. Next $\mathcal{S}$ prepares the internal state on behalf of P_i such that it will be consistent with the commitment value β_i that it had provided to $\mathcal{A}$ earlier. Specifically $\mathcal{S}$ computes $\mathsf{Equiv}_2(CK, \nu, \beta_i, state_i, x_i)$ in order to obtain randomness ω_i such that $\beta_i = \mathsf{Com}(CK, \beta_i; \omega_i)$. $\mathcal{S}$ provides ω_i as the randomness of party P_i to $\mathcal{A}$. Note that $\mathcal{S}$ can do this at any point during 1st round.

Completion of Round 1: After $\mathcal{S}$ has submitted the inputs of all the corrupted parties to $\mathcal{F}_f$ then it responds by sending back the message $(\mathsf{output}, \mathsf{sid}, \mathcal{P}, out)$ where $out = f(\{x_j\}_{j\in[n]})$. Note that in case $\mathcal{S}$ had failed to extract an input for some player P_i then it would have sent $\bot$ to $\mathcal{F}_f$ and would have received $\bot$ as the output from the ideal functionality.

Round 2 Messages $\mathcal{S} \to \mathcal{A}$: In the second round $\mathcal{S}$ generates messages on behalf of the honest parties, i.e. parties in the set $\mathcal{P}\backslash\mathcal{P}^{\mathcal{A}}$ as follows:

1. For each party $P_i \in \mathcal{P}\backslash\mathcal{P}^{\mathcal{A}}$, $\mathcal{S}$ generates $c_i = \mathsf{Enc}(pk_0, \phi^{\ell_{in}}\|out)$, $c_i' = \mathsf{Enc}(pk_1, \phi^{\ell_{in}}\|out)$ and generates π_i as a simulated proof for the statement $(c_i, c_i', (i, \{\beta_j\}_{j\in[n]}))$. More specifically it generates $\pi_i \leftarrow S_2(\sigma, \tau, (c_i, c_i', (i, \{\beta_j\}_{j\in[n]})))$.

$\mathcal{S}$ sends (c_i, c_i', π_i) to $\mathcal{A}$ on behalf of P_i.

Round 2 Messages $\mathcal{A} \to \mathcal{S}$: In the second round the adversary $\mathcal{A}$ generates the messages on behalf of corrupted parties in $\mathcal{P}^{\mathcal{A}}$. For each party $P_i \in \mathcal{P}^{\mathcal{A}}$ our simulator proceeds as:

1. Let (c_i, c_i', π_i) be the message that $\mathcal{A}$ sends on behalf of party P_i. $\mathcal{S}$ checks to see if $V(\sigma, (c_i, c_i', (i, \{\beta_j\}_{j\in[n]})), \pi_i) = 1$ for each $P_i \in \mathcal{P}^{\mathcal{A}}$.

If all the proofs verify then $\mathcal{S}$ sends the message (generateOutput, sid, $\mathcal{P}$) to the ideal functionality $\mathcal{F}_f$.

Simulating corruption of parties during/at the end of Round 2: When $\mathcal{A}$ corrupts a party P_i in the real word, then $\mathcal{S}$ corrupts the corresponding party P_i in the ideal world and obtains its input x_i. Next $\mathcal{S}$ prepares the internal state on behalf of P_i such that it will be consistent with messages it had sent on behalf of P_i. As explained before, $\mathcal{S}$ generates randomness ω_i that explains the commitment β_i to the value x_i running the algorithm $\omega_i = \mathsf{Equiv}_2(CK, \nu, \beta_i, state_i, x_i)$. Next $\mathcal{S}$ needs to explain the second round message (c_i, c_i', π_i). $\mathcal{S}$ has to explain the message (c_i, c_i', π_i) by computing the randomness as $\psi_i = \mathsf{Explain}_{\mathsf{DDK}}(PK, (c_i, c_i', \pi_i), (x_i||\phi^{\ell_{out}}, x_i||\phi^{\ell_{out}}, (i, \{\beta_j\}_{j\in[n]}), (0^{n\cdot\ell_{in}}, 0^{\ell_{out}}, \{t_j\}_{j\in[n]}); u)$ where $t_i = \omega_i$ and $t_j = 0^*$ for all $j \in [n]$ such that $j \neq i$. $\mathcal{S}$ provides $\omega_i||\psi_i$ as the randomness of party P_i to $\mathcal{A}$. Note that $\mathcal{S}$ can do this at any point during or after the round 2 of the protocol.

This completes the description of the simulator.

5 Proof of Security

In this section, via a sequence of hybrids, we will prove that no environment $\mathcal{Z}$ can distinguish the ideal world experiment with $\mathcal{S}$ and $\mathcal{F}_f$ (as defined above) from a real execution of Π with $\mathcal{A}$. We will start with the real world execution in which the adversary $\mathcal{A}$ interacts directly with the honest parties holding their inputs and step-by-step make changes till we finally reach the simulator as described in Section 4. At each step we will argue that the environment cannot distinguish the change except with negligible probability.

Hybrid 0. This hybrid corresponds to the $\mathcal{Z}$ interacting with the real world adversary $\mathcal{A}$ and honest parties that hold their private inputs.

We can restate the above experiment with the simulator as follows. We replace the real world adversary $\mathcal{A}$ with the ideal world adversary $\mathcal{S}$. The ideal adversary $\mathcal{S}$ starts by invoking a copy of $\mathcal{A}$ and running a simulated interaction of $\mathcal{A}$ with the environment $\mathcal{Z}$ and the honest parties. $\mathcal{S}$ forwards the messages that $\mathcal{A}$ generates for it environment directly to $\mathcal{Z}$ and vice versa (as explained in the description of the simulator $\mathcal{S}$). In this hybrid the simulator $\mathcal{S}$ holds the private inputs of the honest parties and generates messages on their behalf using the honest party strategies as specified by Π.

Hybrid 1. In this hybrid, we change how the internal randomness of the corrupted party is explained to $\mathcal{A}$ on being adaptively corrupted. Specifically we change the randomness that is used to explain the ciphertext S generates on behalf of parties in round 2 of protocol Π.

Recall that in the second round S on behalf of an honest party P_i generates the second message as $(c_i, c_i', \pi_i) = \mathsf{DenEnc_{DDK}}(PK, x_i||\phi^{\ell_{out}}, x_i||\phi^{\ell_{out}}, (i, \{\beta_j\}_{j\in[n]}), (0^{n\cdot\ell_{in}}, 0^{\ell_{out}}, \{t_j\}_{j\in[n]}); r_i)$ where t_i is the randomness used in generating commitment β_i and $t_j = 0^*$ for all $j \in [n]$ such that $j \neq i$. So if $\mathcal{A}$ corrupts P_i then the randomness r_i would be reveal to $\mathcal{A}$. In Hybrid 1, instead we provide $\psi_i = \mathsf{Explain_{DDK}}(PK, (c_i, c_i', \pi_i), (x_i||\phi^{\ell_{out}}, x_i||\phi^{\ell_{out}}, (i, \{\beta_j\}_{j\in[n]}), (0^{n\cdot\ell_{in}}, 0^{\ell_{out}}, \{t_j\}_{j\in[n]}); u)$ where t_j values are as before.

Lemma 1. $\mathrm{Hybrid}_0 \approx_c \mathrm{Hybrid}_1$.

Proof. The indistinguishability of Hybrid_1 from Hybrid_0 follows from the indistinguishability of explanation property of the Double Key deniable encryption scheme.

Hybrid 2. In this hybrid we change the way S generates the message (c_i, c_i', π) on behalf of the honest parties.

Recall that in the second round in Hybrid 1, S on behalf of an honest party P_i generates the second message as $(c_i, c_i', \pi_i) = \mathsf{DenEnc_{DDK}}(PK, x_i||\phi^{\ell_{out}}, x_i||\phi^{\ell_{out}}, (i, \{\beta_j\}_{j\in[n]}), (0^{n\cdot\ell_{in}}, 0^{\ell_{out}}, \{t_j\}_{j\in[n]}); r_i)$ where t_i is the randomness used in generating commitment β_i and $t_j = 0^*$ for all $j \in [n]$ such that $j \neq i$. We will change this by generating the ciphertexts directly using procedures Enc and the prover P.

Specifically, $c_i = \mathsf{Enc}(pk_0, x_i||\phi^{\ell_{out}}; s_0^i)$ and $c_i' = \mathsf{Enc}(pk_1, x_i||\phi^{\ell_{out}}; s_1^i)$ and outputs (c_i, c_i', π_i), where $\pi_i \leftarrow P(\sigma, (c_i, c_i', \{i, \{\beta\}_{j\in[n]}\}), (x_i||\phi^{\ell_{out}}, x_i||\phi^{\ell_{out}}, s_0^i, s_1^i, (0^{n\cdot\ell_{in}}, 0^{\ell_{out}}, \{t_j\}_{j\in[n]})))$ where t_i is the randomness used in generating commitment β_i and $t_j = 0^*$ for all $j \in [n]$ such that $j \neq i$.

Lemma 2. $\mathrm{Hybrid}_1 \approx_c \mathrm{Hybrid}_2$.

Proof. The indistinguishability of Hybrid_2 from Hybrid_1 follows immediately from the indistinguishability of *source of ciphertext* property of the Double Key deniable encryption scheme.

Hybrid 3. In this hybrid, we change how σ, which is a part of PK, and the proofs π_i for every $P_i \in \mathcal{P}\backslash\mathcal{P}^{\mathcal{A}}$ are generated.

More specifically, S runs the setup algorithm $\mathsf{Setup_{DDK}}(1^\lambda, 1^{\ell_{in}+\ell_{out}})$ of the Double Key deniable encryption scheme, but replaces its internal call to the algorithm K with $S = (S_1, S_2)$ of the NIZK proof system. More specifically, S generates $(pk_0, sk_0), (pk_1, sk_1) \leftarrow \mathsf{Setup}(1^\lambda), (\sigma, \tau) \leftarrow S_1(1^\lambda)$, along with the public denying key DK. It sets the public key $PK = (\sigma, pk_0, pk_1, DK)$. We also generate fake proofs π_i using trapdoor τ. Specifically we generate $\pi_i \leftarrow S_2(\sigma, \tau, (c_i, c_i', (i, \{\beta_j\}_{j\in[n]})))$.

Lemma 3. $\mathrm{Hybrid}_2 \approx_c \mathrm{Hybrid}_3$.

Proof. The indistinguishability of Hybrid_3 from Hybrid_2 follows immediately from the computational zero-knowledge property of the NIZK proof system.

Hybrid 4. We don't change anything in the output of the hybrid itself. We just use knowledge of μ to extract the inputs $\mathcal{A}$ commits to in the 1st round messages that it sends on behalf of the corrupted parties.
More specifically, $\mathcal{S}$ for every $P_i \in \mathcal{P}^{\mathcal{A}}$ obtains $x_i = \text{Extr}(CK, \mu, \beta_i)$. If extraction fails then it sets $x_i = \bot$.

Hybrid 5. In this hybrid, we change how the simulator $\mathcal{S}$ generates c_i' in the second round message (c_i, c_i', π_i) on behalf of honest parties $P_i \in \mathcal{P}\backslash\mathcal{P}^{\mathcal{A}}$. In particular, $\mathcal{S}$ instead of computing the ciphertext $c_i' = \text{Enc}(pk_1, x_i||\phi^{\ell_{out}}; s_1^i)$, generates $c_i' = \text{Enc}(pk_1, \phi^{\ell_{in}}||out; s_1^i)$, where out is the output computed as $f(\{x_j\}_{j \in [n]})$ using the inputs x_i of the honest parties, that the simulator has access to, and extracted inputs of the malicious parties.

Lemma 4. $\text{Hybrid}_4 \approx_c \text{Hybrid}_5$.

Proof. We base the indistinguishability between hybrids Hybrid_4 and Hybrid_5 on the semantic security of the encryption scheme (Setup, Enc, Dec).

Hybrid 6. In this hybrid we essentially reverse the change that was made in going from Hybrid 2 to Hybrid 3. In particular we change the σ so that it is sampled from the honest distribution and generate the proof honestly. Note that since now we have changed the ciphertext c_i' the proof will have to be generated with respect to language $\mathcal{L}_2$.
More specifically, $\mathcal{S}$ uses K to generate σ instead of S_1. Also for every $P_i \in \mathcal{P}\backslash\mathcal{P}^{\mathcal{A}}$, $\mathcal{S}$ generates $\pi_i \leftarrow P(\sigma, (c_i, c_i', (i, \{\beta_j\}_{j \in [n]})), (x_i||\phi^{\ell_{out}}, \phi^{\ell_{in}}||out, s_0^i, s_1^i, (\{x_i\}_{i \in [n]}, out, \{t_j\}_{j \in [n]})))$ where t_j is the witness that $\beta_j \in \mathcal{L}_{\text{Com}}$.

Lemma 5. $\text{Hybrid}_5 \approx_c \text{Hybrid}_6$.

Proof. The indistinguishability of Hybrid_5 from Hybrid_6 follows immediately from the computational zero-knowledge property of the NIZK proof system.

Hybrid 7. In this hybrid we change how oP, the obfuscated program in the CRS is generated. More specifically, oP is generated as an obfuscation of $\text{Prog}_{sk_1, PK, CK, f}$ instead of $\text{Prog}_{sk_0, PK, CK, f}$.
In the following we show that the program Prog is equivalent under sk_0 and sk_1 with overwhelming probability. This allows us to conclude that the Hybrid 6 and Hybrid 7 are indistinguishable based on indistinguishability obfuscation.

Lemma 6. $\text{Prog}_{sk_0, PK, CK, f} \equiv \text{Prog}_{sk_1, PK, CK, f}$.

Proof. Recall that the underlying language $\mathcal{L}$ of the Double Key deniable encryption scheme consists of two languages, namely $\mathcal{L}_1$ and $\mathcal{L}_2$. Note that since the NIZK has statistical soundness with overwhelming probability over the choices of σ we have that all ciphertexts with an accepting proof must be from one of the two languages. We refer to the two types of ciphertexts corresponding to the language $\mathcal{L}_1$ and $\mathcal{L}_2$, as Type-1 and Type-2 ciphertext, respectively.

Recall that the program Prog takes $\{\beta_j\}_{j\in[n]}$ and $\{c_j, c'_j, \pi_j\}_{i\in[n]}$ as input. Recall from Figure 2 that in Step 1, Prog checks to see that all the proofs π_i are accepting and otherwise it outputs $\perp$. This means that for the program to do anything interesting all the proofs must be valid. Next we will show that in such cases the output of the program is identical regardless of whether sk_0 or sk_1 is used.

All ciphertexts are of Type-1: In this case, c_j and c'_j for $j \in [n]$ encrypted under pk_0 and pk_1 respectively, encrypt the same value. Hence, regardless of whether sk_0 is used or sk_1 is used the program outputs the exact same value $f(\{x_j\}_{j\in[n]})$.

There is at least one Type-2 ciphertext: Note that, in case sk_0 is used then we have that Step 2 of Prog is never invoked. On the other hand in case sk_1 is used then we have that Step 2 of Prog is necessarily invoked.

In other words if sk_0 is used then the x_j values are decrypted and output is calculated. On the other hand if sk_1 is used then a hard-coded *out* value is generated. We will argue that in both cases the output generated by Prog is identical. We argue this by showing that the only acceptable value for the hard-coded value *out* is $f(\{x_j\}_{j\in[n]})$, where x_j are the inputs parties commit to in the first round. Recall that the commitment scheme is perfectly binding, meaning that for every commitment β_i there is exactly one x_i such that $(\beta_i, x_i) \in \mathcal{L}_{\mathsf{COM}}$. This proves our claim. $\square$

Hybrid 8. In this hybrid we do the same change that was made in going from Hybrid 2 to Hybrid 3. In this hybrid, we change how σ, which is a part of PK, and the proofs π_i for every $P_i \in \mathcal{P}\backslash\mathcal{P}^{\mathcal{A}}$ are generated.

More specifically, $\mathcal{S}$ runs the setup algorithm $\mathsf{Setup}_{\mathsf{DDK}}(1^\lambda, 1^{\ell_{in}+\ell_{out}})$ of the Double Key deniable encryption scheme, but replaces its internal call to the algorithm K with $S = (S_1, S_2)$ of the NIZK proof system. More specifically, $\mathcal{S}$ generates $(pk_0, sk_0), (pk_1, sk_1) \leftarrow \mathsf{Setup}(1^\lambda)$, $(\sigma, \tau) \leftarrow S_1(1^\lambda)$, along with the public denying key DK. It sets the public key $PK = (\sigma, pk_0, pk_1, DK)$. We also generate fake proofs π_i using trapdoor τ. Specifically, it generates $\pi_i \leftarrow S_2(\sigma, \tau, (c_i, c'_i, (i, \{\beta_j\}_{j\in[n]})))$.

Lemma 7. $\mathrm{Hybrid}_7 \approx_c \mathrm{Hybrid}_8$.

Proof. The indistinguishability of Hybrid_7 from Hybrid_8 follows immediately from the computational zero-knowledge of the NIZK proof system.

Hybrid 9. In this hybrid, we change how the simulator $\mathcal{S}$ generates c_j in the second round message (c_j, c'_j, π_j) on behalf of honest parties $P_j \in \mathcal{P}\backslash\mathcal{P}^{\mathcal{A}}$. More specifically, $\mathcal{S}$ instead of computing $c_j = \mathsf{Enc}(pk_0, x_i || \phi^{\ell_{out}})$, it computes $c_j = \mathsf{Enc}(pk_0, \phi^{\ell_{in}} || out)$ where $out = f(\{x_j\}_{j\in[n]})$.

Lemma 8. $\text{Hybrid}_8 \approx_c \text{Hybrid}_9$.

Proof. We base the indistinguishability between hybrids Hybrid_8 and Hybrid_9 on the semantic security of the encryption scheme, (Setup, Enc, Dec).

Hybrid 10. In this hybrid we change the way the public parameters of the commitment scheme COM are generated. In particular, S runs the setup algorithm $\text{Setup}_{\text{Com}}^{\text{equiv}}(1^\lambda)$ (instead of $\text{Setup}_{\text{Com}}^{\text{bind}}(1^\lambda)$) of the adaptively secure commitment scheme COM and obtains (CK, μ, ν) where the trapdoor μ is still being used for extraction of adversary's inputs.

Lemma 9. $\text{Hybrid}_9 \approx_c \text{Hybrid}_{10}$.

Proof. Indistinguishability between hybrids Hybrid_9 and Hybrid_{10} follows from the indistinguishability of the public parameters of the commitment scheme COM.

Hybrid 11. In this hybrid we change the way S generates the commitments on behalf of the honest parties. In particular we will remove the inputs and make these commitments equivocal. More specifically, for every party $P_i \in \mathcal{P} \backslash \mathcal{P}^{\mathcal{A}}$ the first round message is computed by S running $(\beta_i, state_i) \leftarrow \text{Equiv}_1(CK, \nu)$. If the party later gets corrupted then S will produce randomness ω_i to equivocate the commitment β_i to the prescribed input x_i. To this end, S will run $\omega_i = \text{Equiv}_2(CK, \nu, \beta_i, state_i, x_i)$.

Lemma 10. $\text{Hybrid}_{10} \approx_c \text{Hybrid}_{11}$.

Proof. We base the indistinguishability between hybrids Hybrid_{10} and Hybrid_{11} on the polynomial equivocality of the adaptively secure commitment scheme COM.

Note that Hybrid_{11} is identical to the simulation strategy described in Section 4. This concludes the proof.

6 Extending to Leakage Tolerant Secure Computation

The adaptively secure protocol presented in this paper also turns out to be leakage tolerant. The model of leakage can be found in the full version [26].

Lemma 11. *Assume the existence of indistinguishability obfuscation and doubly enhanced trapdoor permutation then any ideal functionality $\mathcal{F}_f$ can be UC-securely realized in the $\mathcal{F}_{CRS}$- model against any adaptive, active adversary corrupting an arbitrary number of parties and allowed with arbitrary leakage. Furthermore this protocol involves only two rounds of broadcast.*

This lemma follows immediately from our construction and proof except for some syntactic differences. We explain this next. We will only describe how our simulator for adaptive security (from Section 4) can be converted into a simulator for the setting of leakage tolerance. The proof of indistinguishability for the adaptive simulator was already provided in Section 5.

Recall that that the simulator for arguing adaptive security, on corruption of an honest party, uses the honest party's input alone in order to explain the messages it had previously sent on behalf of the honest party. In the setting of leakage, we note that this method of explanation can directly be expressed by a circuit that on input the input of the honest party outputs the internal secret state of that party. Furthermore note that the way in which the simulator explains its first round messages of honest parties remains the same even after it has sent the second round messages.

Using this explanation procedure as a translation method, allows us to immediately conclude that any leakage query of the real-world adversary can be reduced directly to a leakage query in the ideal-world.

Acknowledgments. We would like to thank Oxana Poburinnaya and Ran Canetti for pointing out that sub-exponential iO is needed to instantiate our double key deniable encryption scheme.

References

1. Asharov, G., Jain, A., López-Alt, A., Tromer, E., Vaikuntanathan, V., Wichs, D.: Multiparty computation with low communication, computation and interaction via threshold FHE. In: Pointcheval, D., Johansson, T. (eds.) EUROCRYPT 2012. LNCS, vol. 7237, pp. 483–501. Springer, Heidelberg (2012)
2. Asharov, G., Jain, A., Wichs, D.: Multiparty computation with low communication, computation and interaction via threshold FHE. Cryptology ePrint Archive, Report 2011/613 (2011), http://eprint.iacr.org/2011/613
3. Barak, B., Canetti, R., Lindell, Y., Pass, R., Rabin, T.: Secure computation without authentication. In: Shoup, V. (ed.) CRYPTO 2005. LNCS, vol. 3621, pp. 361–377. Springer, Heidelberg (2005)
4. Barak, B., Goldreich, O., Impagliazzo, R., Rudich, S., Sahai, A., Vadhan, S.P., Yang, K.: On the (im)possibility of obfuscating programs. J. ACM 59(2), 6 (2012)
5. Bitansky, N., Canetti, R., Halevi, S.: Leakage-tolerant interactive protocols. In: Cramer, R. (ed.) TCC 2012. LNCS, vol. 7194, pp. 266–284. Springer, Heidelberg (2012)
6. Blum, M., Feldman, P., Micali, S.: Proving security against chosen cyphertext attacks. In: Goldwasser, S. (ed.) CRYPTO 1988. LNCS, vol. 403, pp. 256–268. Springer, Heidelberg (1990)
7. Boyle, E., Garg, S., Jain, A., Kalai, Y.T., Sahai, A.: Secure computation against adaptive auxiliary information. In: Canetti, R., Garay, J.A. (eds.) CRYPTO 2013, Part I. LNCS, vol. 8042, pp. 316–334. Springer, Heidelberg (2013)
8. Canetti, R.: Universally composable security: A new paradigm for cryptographic protocols. In: 42nd Annual Symposium on Foundations of Computer Science, Las Vegas, Nevada, USA, October 14–17, pp. 136–145. IEEE Computer Society Press (2001)
9. Canetti, R., Feige, U., Goldreich, O., Naor, M.: Adaptively secure multi-party computation. In: 28th Annual ACM Symposium on Theory of Computing, Philadephia, Pennsylvania, USA, May 22–24, pp. 639–648. ACM Press (1996)

10. Canetti, R., Fischlin, M.: Universally composable commitments. In: Kilian, J. (ed.) CRYPTO 2001. LNCS, vol. 2139, pp. 19–40. Springer, Heidelberg (2001)

11. Canetti, R., Goldwasser, S., Poburinnaya, O.: Adaptively secure two-party computation from indistinguishability obfuscation. IACR Cryptology ePrint Archive, 2014:845 (2014)

12. Canetti, R., Kushilevitz, E., Lindell, Y.: On the limitations of universally composable two-party computation without set-up assumptions. In: Biham, E. (ed.) EUROCRYPT 2003. LNCS, vol. 2656, pp. 68–86. Springer, Heidelberg (2003)

13. Canetti, R., Lin, H., Pass, R.: Adaptive hardness and composable security in the plain model from standard assumptions. In: 51st Annual Symposium on Foundations of Computer Science, Las Vegas, Nevada, USA, October 23–26, pp. 541–550. IEEE Computer Society Press (2010)

14. Canetti, R., Lindell, Y., Ostrovsky, R., Sahai, A.: Universally composable two-party and multi-party secure computation. In: 34th Annual ACM Symposium on Theory of Computing, Montréal, Québec, Canada, May 19–21, pp. 494–503. ACM Press (2002)

15. Dachman-Soled, D., Katz, J., Rao, V.: Adaptively secure, universally composable, multi-party computation in constant rounds. IACR Cryptology ePrint Archive, 2014:858 (2014)

16. Dachman-Soled, D., Malkin, T., Raykova, M., Venkitasubramaniam, M.: Adaptive and Concurrent Secure Computation from New Adaptive, Non-malleable Commitments. In: Sako, K., Sarkar, P. (eds.) ASIACRYPT 2013, Part I. LNCS, vol. 8269, pp. 316–336. Springer, Heidelberg (2013)

17. Damgård, I., Hazay, C., Patra, A.: Leakage resilient secure two-party computation. IACR Cryptology ePrint Archive, 2011:256 (2011)

18. Damgård, I., Polychroniadou, A., Rao, V.: Adaptively secure UC constant round multi-party computation protocols. IACR Cryptology ePrint Archive, 2014:830 (2014)

19. Feige, U., Lapidot, D., Shamir, A.: Multiple non-interactive zero knowledge proofs based on a single random string (extended abstract). In: 31st Annual Symposium on Foundations of Computer Science, St. Louis, Missouri, October 22–24, pp. 308–317. IEEE Computer Society Press (1990)

20. Feige, U., Shamir, A.: Zero Knowledge Proofs of Knowledge in Two Rounds. In: Brassard, G. (ed.) CRYPTO 1989. LNCS, vol. 435, pp. 526–544. Springer, Heidelberg (1990)

21. Garg, S., Gentry, C., Halevi, S.: Candidate multilinear maps from ideal lattices. In: Johansson, T., Nguyen, P.Q. (eds.) EUROCRYPT 2013. LNCS, vol. 7881, pp. 1–17. Springer, Heidelberg (2013)

22. Garg, S., Gentry, C., Halevi, S., Raykova, M.: Two-round secure MPC from indistinguishability obfuscation. In: Lindell, Y. (ed.) TCC 2014. LNCS, vol. 8349, pp. 74–94. Springer, Heidelberg (2014)

23. Garg, S., Gentry, C., Halevi, S., Raykova, M., Sahai, A., Waters, B.: Candidate indistinguishability obfuscation and functional encryption for all circuits. In: 54th Annual Symposium on Foundations of Computer Science, October 26–29, pp. 40–49. IEEE Computer Society Press (2013)

24. Garg, S., Gupta, D., Khurana, D., Sahai, A.: All-but-one leakage resilient multiparty computation and incoercible multiparty computation. Personal Communication (2014)

25. Garg, S., Jain, A., Sahai, A.: Leakage-Resilient Zero Knowledge. In: Rogaway, P. (ed.) CRYPTO 2011. LNCS, vol. 6841, pp. 297–315. Springer, Heidelberg (2011)

26. Garg, S., Polychroniadou, A.: Two-round adaptively secure MPC from indistinguishability obfuscation. IACR Cryptology ePrint Archive, 2014:844 (2014)

27. Garg, S., Sahai, A.: Adaptively secure multi-party computation with dishonest majority. In: Safavi-Naini, R., Canetti, R. (eds.) CRYPTO 2012. LNCS, vol. 7417, pp. 105–123. Springer, Heidelberg (2012)

28. Goldreich, O.: Foundations of Cryptography: Basic Tools, vol. 1. Cambridge University Press, Cambridge (2001)

29. Goldreich, O.: Foundations of Cryptography: Basic Applications, vol. 2. Cambridge University Press, Cambridge (2004)

30. Goldreich, O., Micali, S., Wigderson, A.: How to play any mental game or A completeness theorem for protocols with honest majority. In: Aho, A. (ed.) 19th Annual ACM Symposium on Theory of Computing, May 25–27, pp. 218–229. ACM Press, New York (1987)

31. Goldwasser, S., Gordon, S.D., Goyal, V., Jain, A., Katz, J., Liu, F.-H., Sahai, A., Shi, E., Zhou, H.-S.: Multi-input functional encryption. In: Nguyen, P.Q., Oswald, E. (eds.) EUROCRYPT 2014. LNCS, vol. 8441, pp. 578–602. Springer, Heidelberg (2014)

32. Goldwasser, S., Micali, S., Rackoff, C.: The knowledge complexity of interactive proof systems. SIAM Journal on Computing 18(1), 186–208 (1989)

33. Halevi, S., Lindell, Y., Pinkas, B.: Secure Computation on the Web: Computing without Simultaneous Interaction. In: Rogaway, P. (ed.) CRYPTO 2011. LNCS, vol. 6841, pp. 132–150. Springer, Heidelberg (2011)

34. Hazay, C., Patra, A.: One-sided adaptively secure two-party computation. In: Lindell, Y. (ed.) TCC 2014. LNCS, vol. 8349, pp. 368–393. Springer, Heidelberg (2014)

35. Ishai, Y., Prabhakaran, M., Sahai, A.: Founding cryptography on oblivious transfer - efficiently. In: Wagner, D. (ed.) CRYPTO 2008. LNCS, vol. 5157, pp. 572–591. Springer, Heidelberg (2008)

36. Katz, J., Ostrovsky, R.: Round-optimal secure two-party computation. In: Franklin, M. (ed.) CRYPTO 2004. LNCS, vol. 3152, pp. 335–354. Springer, Heidelberg (2004)

37. Lindell, Y.: Bounded-concurrent secure two-party computation without setup assumptions. In: 35th Annual ACM Symposium on Theory of Computing, San Diego, California, USA, June 9–11, pp. 683–692. ACM Press,

38. Lindell, Y., Pinkas, B.: A proof of security of Yao's protocol for two-party computation. Journal of Cryptology 22(2), 161–188 (2009)

39. Naor, M., Yung, M.: Public-key cryptosystems provably secure against chosen ciphertext attacks. In: 22nd Annual ACM Symposium on Theory of Computing, Baltimore, Maryland, USA, May 14–16, pp. 427–437. ACM Press (1990)

40. Sahai, A., Waters, B.: How to use indistinguishability obfuscation: deniable encryption, and more. In: Shmoys, D.B. (ed.) 46th Annual ACM Symposium on Theory of Computing, May 31–June 3, pp. 475–484. ACM Press, New York (2014)

41. Yao, A.C.-C.: Protocols for secure computations (extended abstract). In: 23rd Annual Symposium on Foundations of Computer Science, Chicago, Illinois, November 3–5, pp. 160–164. IEEE Computer Society Press (1982)

Obfuscation-Based Non-black-box Simulation and Four Message Concurrent Zero Knowledge for NP

Omkant Pandey[1,2,*,**], Manoj Prabhakaran[1,*], and Amit Sahai[2,**]

[1] University of Illinois at Urbana Champaign, USA
{omkant,mmp}@uiuc.edu
[2] UCLA and Center for Encrypted Functionalities, USA
sahai@cs.ucla.edu

Abstract. We show the following result: Assuming the existence of *public-coin differing-input obfuscation* (pc-diO) for the class of all polynomial time Turing machines, then there exists a four message, fully concurrent zero-knowledge proof system for all languages in **NP** with negligible soundness error. This result is constructive: given pc-diO, our reduction yields an explicit protocol along with an *explicit* simulator that is "straight line" and runs in strict polynomial time. The obfuscation security property is used only to prove soundness.

Public-coin differing-inputs obfuscation is a notion of obfuscation closely related to indistinguishability obfuscation. Most importantly for our result, pc-diO does not suffer from any known impossibility results: recent negative results on standard differing-inputs obfuscation do not apply to pc-diO. Furthermore, candidate constructions for pc-diO for the class of all polynomial-time Turing Machines are known.

Our reduction relies on a new non-black-box simulation technique which does not use the PCP theorem. We view the development of this new non-black-box simulation technique as the main contribution of our work. In addition to assuming pc-diO, our reduction also assumes (standard and polynomial time) cryptographic assumptions such as collision-resistant hash functions.

1 Introduction

Zero-Knowledge and Program Obfuscation. Zero-knowledge proofs, introduced by Goldwasser, Micali and Rackoff [GMR85] are the classical example of the

* Research supported in part by the NSF Grant 1228856.
** Research supported in part from a DARPA/ONR PROCEED award, NSF Frontier Award 1413955, NSF grants 1228984, 1136174, 1118096, and 1065276, a Xerox Faculty Research Award, a Google Faculty Research Award, an equipment grant from Intel, and an Okawa Foundation Research Grant. This material is based upon work supported by the Defense Advanced Research Projects Agency through the U.S. Office of Naval Research under Contract N00014-11- 1-0389. The views expressed are those of the author and do not reflect the official policy or position of the Department of Defense, the National Science Foundation, or the U.S. Government.

Y. Dodis and J.B. Nielsen (Eds.): TCC 2015, Part II, LNCS 9015, pp. 638–667, 2015.

simulation paradigm. They allow a *prover* to convince a *verifier* that a mathematical statement $x \in \mathbf{L}$ is true while giving *no additional knowledge* to the verifier. Prior to 2001, all known zero-knowledge simulators used the (cheating) verifier V^* as a *black-box* to produce their output (called the simulated view). Barak [Bar01] demonstrated how to take advantage of verifier's program to build more powerful *non-black-box* simulation techniques.

Constructing and analyzing non-black-box simulators can be a challenging task.The reason why taking advantage of verifier's code is difficult is because of the intriguing possibility of *program obfuscation.* Roughly speaking, program obfuscation is a method to transform a computer program (say described as a Boolean circuit) into a form that is executable but otherwise completely "unintelligible." In its strongest form, an obfuscated program leaks no information about the program beyond its "functionality" or the "input-output behavior". Therefore, access to the obfuscated program is no better than having black box access to it. This property, as formalized by Barak, Goldreich, Impagliazzo, Rudich, Sahai, Vadhan, and Yang [BGI$^+$01], is called the *virtual black box* (VBB) security. It was shown in [BGI$^+$01] that VBB-secure obfuscation is impossible in general. In hindsight, this negative result shows why non-black-box (NBB) simulation is possible, despite the possibility that program obfuscation could hide nearly every useful aspect of the verifier's code.

Zero-knowledge, in particular non-black-box simulation, is intimately connected to program obfuscation. This connection has been explicitly studied in the works of Hada [Had00], and Bitansky and Paneth [BP12b, BP12a, BP13a], and alluded to in several other works, e.g., [HT99, Bar01]). In this work, we explore this line of research further, particularly in light of recent work showing the first plausible constructions of general-purpose obfuscation schemes [GGH$^+$13]. In particular, for the first time, we show that program obfuscation can be useful for designing new non-black-box simulation strategies that yield constant-round concurrent zero knowledge protocols.

General-Purpose Obfuscation. In 2013, Garg, Gentry, Halevi, Raykova, Sahai, and Waters [GGH$^+$13] presented the first candidate construction for general-purpose obfuscation. Several formalizations for obfuscation have been proposed as alternatives to the impossible-to-achieve notion of VBB obfuscation. A basic definition, called *indistinguishability obfuscation* (iO) [BGI$^+$01], roughly speaking, guarantees that if two (same-size) programs C_0, C_1 are functionally equivalent, then their obfuscations are computationally indistinguishable. A closely related notion is that of *differing input obfuscation* (diO) [BGI$^+$01] which, roughly speaking, guarantees that the obfuscations of C_0 and C_1 are computationally indistinguishable *provided that* it is hard to find an input x such that $C_0(x) \neq C_1(x)$. Unfortunately, recently evidence was shown [BP13b, GGHW14] that the notion of diO is impossible to achieve in general, due to the existence of problematic contrived auxiliary inputs. However, very recently, Ishai, Pandey, and Sahai [IPS15] formulated the notion of *public-coin differing-inputs obfuscation* (pc-diO) in which no auxiliary input is allowed except for the *random coins of the sampler.* This modification avoids the negative results of [BP13b,

GGHW14], and indeed all previous negative results on obfuscation using auxiliary input [GK05, GK13], as all previous negative results using auxiliary input critically relied on the possibility of a secret being embedded within the auxiliary input. Because in pc-diO the auxiliary input is only allowed to be public randomness, this possibility is eliminated (please see [IPS15] for further details). Furthermore, [IPS15], building on [ABG+13, BCP14], present candidate constructions of pc-diO for the class of all polynomial-time Turing machines which can accept inputs of unbounded polynomial length.

Our Results. In this work we show how to use program obfuscation to build a new non-black-box simulation strategy that works for fully *concurrent* zero-knowledge. More specifically, we show that:

- If public-coin differing-input obfuscation (pc-diO) exists for the class of all polynomial time Turing machines with unbounded inputs, then there exists a constant round, fully concurrent zero knowledge protocol for **NP** with negligible soundness error. The protocol has an *explicit* simulator;[1] the simulator is "straight line" and runs in strict polynomial time. The security of the obfuscation is used only prove the soundness of our protocol.
- We also show how to implement the core ideas of the above protocol in only *four* rounds. That is, our new protocol requires sending only four messages between the prover and the verifier.

Our protocol can be instantiated using the constructions of [BCP14, ABG+13, IPS15] which obfuscates polynomial time Turing machines that can accept inputs of variable length (at most polynomial in the security parameter). We stress that we are able to obtain an *explicit* simulator for our protocol *irrespective of the computational assumptions* underlying the constructions of differing-inputs obfuscation. This is because we use the security—i.e., the public-coin differing-inputs security property—of obfuscation only in proving the soundness of our protocol. The simulator only depends on the correctness or the *functionality* of the obfuscated program, and hence can be described explicitly.

Other than pc-diO, our reduction only assumes standard (polynomial time hardness) assumptions, namely injective one-way functions and collision-resistant hash functions. Interestingly, our reduction does not *explicitly* depend on CS-proofs or universal-arguments [Kil92, Mic94, Kil95, BG02]; in particular, if we instantiate the constructions of [IPS15, ABG+13] using the SNARKs of Bitansky et al. [BCCT13] based on bilinear maps (which do not rely on the PCP theorem), we obtain an instantiation of our protocol that also does not rely on the PCP theorem.

The round complexity of our final protocol also sheds new light on the *exact* (as opposed to asymptotic) round-complexity of concurrent zero-knowledge.

---

[1] In some protocols, specifically those based on knowledge-type assumptions [HT99], by virtue of the assumption that there exists an "extractor," it is only possible to obtain an *existential* result that a simulator exists; however, the actual program of the simulator is not explicitly given in the security proofs.

Even in the simpler case of *stand alone* zero knowledge, the best known constructions require at least four rounds [FS89], and historically, concurrent zero-knowledge has always required more rounds than stand-alone zero-knowledge.[2] Our four-round protocol, for the first time, closes the gap between the best known upper bounds on round complexities of concurrent versus standalone zero-knowledge protocols (whose simulators can be explicitly described).

In retrospect, the fact that obfuscation actually *helps* non-black-box simulation can be perplexing. Indeed, in all prior works along this line [Had00, BP12b, BP13a], the core ideas for simulation are of *opposite* nature: it is the *inability* to obfuscate the "unobfuscatable functions" that helps the simulator. In our case, similar to [BP12a], it is the *ability* to obfuscate programs that allows polynomial time simulation. We believe that our method can be useful in other settings as well where non-black-box simulation seems essential such as constant-round leakage-resilient zero-knowledge [Pan14, GJS11] or CCA secure commitments in sub-logarithmic rounds [CLP10, GLP+15].

Paper Organization. We start by discussing how to use program obfuscation to avoid the use of universal arguments in Barak's protocol in Section 1.1. This results in a stand alone ZK protocol with a "straight line" simulator. In Section 4, we discuss why the simulator of this protocol fails in the concurrent setting, and present a (substantially) different constant-round protocol which is concurrent ZK along with main proof ideas. In Section 5, we present an overview of our four-round concurrent-ZK protocol. The full details can be found in the full version of this work [PPS15].

1.1 Technical Overview: Non-black-box Simulation via Program Obfuscation

Let us start by considering the simplest approach to zero-knowledge from (the possibility of) program obfuscation. For now, let us restrict ourselves to the case of *stand alone* zero-knowledge for **NP**-languages. Let $x \in \mathbf{L}$ be the statement and R be the witness-relation.

One simple approach is to have the verifier send an obfuscation of the following program $M_{x,s}$ which contains a secret string $s \in \{0,1\}^n$: $M_{x,s}(a) = s$ if and only $R(x,a) = 1$ and $M_{x,s}(a) = 0^n$ otherwise. Let $\widetilde{M}_{x,s}$ denote the iO-secure obfuscation of $M_{x,s}$. The real prover can recover s by using a witness w to x. Further, if x is false, $M_{x,s}$ is identical to $M_{x,0^n}$ and therefore must hide s, ensuring the soundness.[3] This gives us a two-message, *honest verifier* ZK proof. However, this idea does not help the simulation against malicious verifiers.

To fix this, let us try to use Barak's preamble (called GenStat [Bar01]) which has the following three rounds: first, the verifier sends a collision-resistant hash

---

[2] Barak's (bounded-concurrent ZK) protocol [Bar01] and recent construction of Chung, Lin, and Pass [CLP13b] require at least six rounds even after optimizations; the recent protocol of Gupta and Sahai [GS12b] requires five rounds and does not have an explicit simulator.

[3] By security of iO, $\widetilde{M}_{x,s} \overset{c}{\approx} \widetilde{M}_{x,0^n}$ and $\widetilde{M}_{x,0^n}$ has no information about s.

function $h : \{0,1\}^* \rightarrow \{0,1\}^n$, then the prover sends a commitment c to 0^n (using a perfectly binding scheme Com), and then the verifier sends a string $r \in \{0,1\}^n$. The transcript defines a "fake statement" $\lambda = \langle h, c, r \rangle$. A "fake witness" ω for the statement λ consists of a pair (Π, u) such that $c = \mathsf{Com}(h(\Pi) \; ; \; u)$ and Π is a program of length $\mathsf{poly}(n)$ which outputs the string r on input the string c (say, in $n^{\log \log n}$ steps). If h is a good collision-resistant hash function, then it was shown in [Bar01, BG02], no efficient prover P^* can output a satisfying witness ω to the statement λ (sampled in an interaction with the honest verifier). However, a simulator can commit to $h(V^*)$ (instead of 0^n) so that it will have a valid witness to the resulting transcript λ.

Coming back to our protocol, we use this idea as follows. We modify our first idea, and require the verifier to send a the obfuscation of a new program $M_{\lambda,s}$ (instead of $M_{x,s}$) where $\lambda = \langle h, c, r \rangle$ is the transcript of GenStat. The new program $M_{\lambda,s}$ outputs s if and only if it receives a valid witness ω to the statement λ (as described earlier) and 0^n on all other inputs. To prove the statement x will be proven by proving the knowledge of either a witness w to x or the secret s (using an ordinary *witness-indistinguishable proof-of-knowledge* (WIPOK)). A simulator can "succeed" in the simulation as before: it commits to verifier's program in c to obtain (an indistinguishable statement) λ, then uses the fake witness ω (which it now has) to execute the program $\widetilde{M}_{\lambda,s}(\omega)$ and learn s and complete the WIPOK using s.

We now draw attention to some important points arising due to the use of λ in the obfuscation (instead of x). First, the length of the fake witness ω that the simulator has depends on the length of the program of V^*. Since the protocol needs to take into account V^* of *every* polynomial length, the obfuscated program $\widetilde{M}_{\lambda,s}$ must accept inputs ω of arbitrary, a-priori unknown, polynomial length. In other words, the obfuscated program $\widetilde{M}_{\lambda,s}$ must be a *Turing machine* which accepts inputs of arbitrary, a-priori unknown, (polynomial) length. Therefore, we will have to use program obfuscation for Turing machines.

Second, the statement $\lambda = \langle h, c, r \rangle$ is not a "false" statement since an all powerful prover can always find collisions in h and obtain a satisfying input to $M_{\lambda,s}$. The only guarantee we have is that if λ is sampled as above, then it would be *hard* for any efficient prover—even those with a valid witness to x—to find a satisfying input for $M_{\lambda,s}$. Therefore, unlike before (when x was used instead of λ), obfuscations $\widetilde{M}_{\lambda,s}$ and $\widetilde{M}_{\lambda,0^n}$ are not guaranteed to be indistinguishable if we use an iO-secure obfuscation; this is because the Turing machines $M_{\lambda,s}$ and $M_{\lambda,0^n}$ are not functionally equivalent. Therefore, we will have to use diO-secure obfuscation (since finding a differing input is still hard for these programs). The security of diO is a subtle issue and we discuss it shortly.

By putting these ideas together, we actually a get a standalone ZK protocol for **NP** (summarized below). The protocol needs to use some kind of reference to s other than the obfuscated program. This is done by using a $f(s)$ where f is a one-way function. This protocol has a "straight line" simulator. Further, unlike Barak's protocol, this protocol does not use universal arguments (and hence the PCP theorem).

Standalone Zero-Knowledge using Obfuscation. The protocol has three stages.

1. Stage-1 is the 3 round preamble GenStat: V sends a CRHF h, P sends a commitment $c = \mathsf{Com}(0^n; u)$ and V sends a random $r \leftarrow \{0,1\}^n$.
2. In stage 2, V sends $(f, \widetilde{s}, \widetilde{M}_{\lambda,s})$ where f is a one-way function, $\widetilde{s} = f(s)$, and $\widetilde{M}_{\lambda,s}$ is the obfuscation of *Turing machine* $M_{\lambda,s}$ described earlier and $\lambda = \langle h, c, r \rangle$ is the transcript of stage-1. V also proves that $(f, \widetilde{s}, \widetilde{M}_{\lambda,s})$ are correctly constructed (using a standard ZK proof).
3. In stage-3 P provess, using a standard WIPOK, the knowledge of "either a witness w to x or secret s such that $\widetilde{s} = f(s)$."

Standalone ZK of this protocol can be proven by following Barak's simulator which commits to the code of V^* and therefore has an ω for simulated statement λ such that $\widetilde{M}_{\lambda,s}(\omega) = s$ within a polynomial number of steps; the simulator computes s and uses it in the WIPOK. The soundness of the protocol relies on the diO-security of obfuscation. Indeed, following [Bar01], for a properly sampled λ, it is hard to find ω such that $M_{\lambda,s}(\omega) \neq M_{\lambda,0^n}(\omega)$, and therefore it is hard to distinguish $\widetilde{M}_{\lambda,s}$ from $\widetilde{M}_{\lambda,0^n}$ by diO-security of obfuscation. Now, soundness is argued using three hybrid experiments: first use the simulator of the ZK protocol in stage 2, then replace $\widetilde{M}_{\lambda,s}$ from $\widetilde{M}_{\lambda,0^n}$, and finally extract s from the WIPOK in stage 3 and violate the hardness of one-way function f (since x is false, extraction must yield s).

In Section 4, we will discuss why the simulator of this protocol fails in the concurrent setting, and new ideas will be needed to obtain a concurrent ZK protocol. In particular, we will make use of the DGS-oracle idea [DGS09].

Security of diO *and the issue of auxiliary information.* Continuing from our discussion above, it is clear that for our approach to work obfuscations $\widetilde{M}_{\lambda,s}$ and $\widetilde{M}_{\lambda,0^n}$ should be indistinguishable to a cheating P^*. However, this is not all: in addition to one of these programs, P^* also has access to the statement λ, which is *auxiliary information* about the two programs. Therefore, our approach works if we have diO secure w.r.t. auxiliary information (distributed according to λ).

Recent implausibility results of Garg et al. [GGHW14] cast serious doubts about the existence of diO w.r.t. arbitrary auxiliary information. While their result does not rule out the possibility of diO w.r.t. specific distributions, we should be extra careful to not rely on auxiliary information which keeps some "secret" such as an obfuscated code [GGHW14].

In our approach, the distribution of λ does not have to keep any secrets. In the language of [IPS15], this is public-coin auxiliary information. We show that our approach indeed works by only assuming that the obfuscation is *public-coin differing-input* secure.

1.2   Related Work

Concurrent zero-knowledge. From early on, it was understood and explicitly proven in [FS90, GK96], that zero-knowledge is not preserved under parallel repetition where multiple sessions of the protocol run at the same time. The more complex notion of concurrent zero-knowledge (cZK) was introduced and achieved by Dwork, Naor, and Sahai [DNS98] (assuming "timing constraints" on the underlying network). A large body of research on cZK studied the round-complexity of black-box concurrent ZK with improving lower bounds on the same [KPR98, Ros00, CKPR03]. The state of art is the lower-bound is by Canetti, Kilian, Petrank, and Rosen [CKPR03] who prove that black-box cZK requires at least $O(\log n / \log \log n)$ rounds where n is the length of the statements being proven. Prabhakaran, Rosen, and Sahai [PRS02], building upon the prior works of Richardson and Kilian [RK99] and Kilian and Petrank [KP01], presented a cZK protocol for **NP** which has $\widetilde{O}(\log n)$ rounds, matching the lower bound of [CKPR03].

The central open question in this area is to construct a constant round cZK protocol for **NP** languages based on standard (or at least reasonable) assumptions. Barak [Bar01] showed that in the *bounded concurrent* setting where there is an a-priori upper bound on the number of sessions, there exists a constant round non-black-box cZK protocol for **NP**; the protocol is based on the existence of collision-resistant hash functions [Bar01] and uses universal arguments [Kil92, Mic94, Kil95, BG02]. The communication complexity of Barak's protocol depends on the a-priori bound on the sessions.

It has proven difficult to extend Barak's NBB techniques to the setting of fully concurrent ZK (i.e., to unbounded polynomially many sessions) in $o(\log n)$ rounds. Nevertheless, NBB techniques have enjoyed great success resulting in the construction of resettable protocols [BGGL01, DL07, DGS09, GM11], non-malleable protocols [Bar02, PR05b, PR05a], leakage-resilient ZK [Pan14], bounded-concurrent secure computation [PR03, Pas04], adaptive security [GS12a], and so on. Bitanksy and Paneth [BP12a] showed that it is possible to perform non-black-box simulation using oblivious transfer (instead of collision-resistant hash functions and universal arguments). This eventually led to the construction of resettablly-sound ZK under one-way functions [BP13a, CPS13, COPV13]. Goyal [Goy13] presents a non-black-box simulation technique in the fully concurrent setting and achieves the first public-coin cZK protocol in the plain model.[4]

An alternative approach to construct round-efficient zero-knowledge proofs is to use "knowledge assumptions" [Dam91, HT99, BP04]. The recent work of Gupta and Sahai [GS12b] shows that such assumptions also yield a constant round concurrent ZK protocol for **NP**. However, all known ZK protocols based on knowledge-type assumptions do not yield an *explicit* simulator. This is because the knowledge-type assumptions assume the existence of a special "extractor" machine (which is not explicitly known); this extractor is used by the simulator of ZK protocols and only provides an "existential" result.

Chung, Lin, and Pass [CLP13b] recently presented the first construction of a constant-round fully concurrent ZK protocol which has an explicit simulator.

---

[4] The protocol requires $\mathsf{poly}(n)$ rounds. Canetti et al. [CLP13a] obtain a similar result, albeit in the "global hash" model where a global hash function—which the simulator cannot program—is known to all parties.

Their result is based on a new complexity-theoretic assumption, namely the existence of so called "strong **P**-certificates."

Another alternative proposed in the literature is to assume some kind of a setup such as timing constraints, (untrusted) public-key infrastructure, and so on [DNS98, DS98, CGGM00, Dam00, Gol02, PTV10, GJO$^+$13] or switch to super-polynomial time simulation [Pas03, PV08]. We will not consider such models further in this work.

Program obfuscation. After the strong impossibility results of [BGI$^+$01], research in program obfuscation proceeded in two main directions. The first line of research focussed on constructing obfuscation for specific functionalities such as point functions and their variants, proxy re-encryption, encrypted signatures, hyperplanes, conjunctions, and so on[Wee05, LPS04, HRSV07, Had10, CRV10, BR13a]. The other line of research focussed on finding weaker definitions and alternative models. Goldwasser and Rothblum [GR07] considered the notion of *best possible obfuscation* (and is equivalent to iO when the obfusactor is polynomial time); and Bitansky and Canetti [BC10] considered *virtual grey box* security. Alternative models for obfuscation such as the hardware model were considered in [GIS$^+$10, BCG$^+$11].

After [GGH$^+$13], an improved construction of iO was presented by Barak et. al. [BGK$^+$13]. Further, in an idealized "generic encodings" model it is shown that VBB-obfuscation for all circuits can be achieved [CV13, BR13b, BGK$^+$13]. These results often involve a "bootstrapping step"; Applebaum [App13] presents an improved technique for bootstrapping obfuscation. Further complexity-theoretic results appear in recent works of Moran and Rosen [MR13], and Barak et. al. [BBC$^+$14].

Sahai and Waters [SW13] show that indistinguishability obfuscation is a powerful tool and use it to successfully construct several (old and new) cryptographic primitives; further applications of iO appear in [HSW13, BZ13, BCP14, KRW13, MO13]

Differing input obfuscation was studied by Ananth et. al. [ABG$^+$13], who present a candidate construction of diO for the class of polynomial time Turing machines and demonstrate new applications. Another variant of their construction allows the Turing machines to accept variable length inputs. Concurrent work of Boyle, Chung, and Pass [BCP14] introduces a related notion of *extractability obfuscation* and shows this notion (and diO) are implied by iO when the programs differ only on polynomially many inputs. In addition, it also presents obfuscation for the class of polynomial time Turing machines, building upon the work of Brakerski and Rothblum [BR13a].Very recently, in concurrent and independent works, construction for bounded-space RAM programs were presented by relying on iO and OWFs [BGT14] and other additional assumptions [CHJV14, LP14].

The issue of *auxiliary information* in program obfuscation was first considered by Goldwasser and Kalai [GK05], and further explored in [GK13, BCPR13, BP13b, GGHW14, IPS15]. The work of Bitansky, Canetti, Paneth, and Rosen [BCPR13] shows that if iO exists then "extractability primitives" such as knowledge-types assumptions and extractable one-way functions [CD09] cannot exist in the presence of *arbitrary* auxiliary information. Boyle and Pass [BP13b] strengthen this result further by showing a pair of (universal) distributions $\mathcal{Z}, \mathcal{Z}'$ on auxiliary information such that either extractable OWF w.r.t. $\mathcal{Z}$ do not exist or extractability-obfuscations w.r.t. $\mathcal{Z}'$ do not exist. The work of Garg, Gentry, Halevi, and Wichs [GGHW14] shows that diO w.r.t. arbitrary auxiliary information cannot exist if certain specific obfuscation assumption is true. Ishai, Pandey, and Sahai [IPS15] formulate the notion of public-coin diO in which the auxiliary is restricted to be merely the random coins of the sampler, and recover much of the existing applications of diO under this new notion.

2   Preliminaries

We use standard notations which are recalled here. This section can be skipped without affecting readability.

Notation. For a randomized algorithm A we write $A(x; r)$ the process of evaluating A on input x with random coins r. We write $A(x)$ the process of sampling a uniform r and then evaluating $A(x; r)$. We define $A(x, y; r)$ and $A(x, y)$ analogously. We denote by $\mathbb{N}$ and $\mathbb{R}$ the set of natural and real numbers respectively. The concatenation of two string a and b is denoted by $a \parallel b$.

We assume familiarity with interactive Turing machines (ITMs). For two randomized ITMs A and B, we denote by $[A(x, y) \leftrightarrow B(x, z)]$ the interactive computation between A and B, with A's inputs (x, y) and B's inputs (x, z), and uniform randomness; and $[A(x, y; r_A) \leftrightarrow B(x, z; r_B)]$ when we wish to specify randomness. We denote by $\mathsf{VIEW}_P[A(x, y) \leftrightarrow B(x, z)]$ and $\mathsf{OUT}_P[A(x, y) \leftrightarrow B(x, z)]$ the view and output of machine $P \in \{A, B\}$ in this computation. Finally, $\mathsf{TRANS}[A(x, y) \leftrightarrow B(x, z)]$ denotes the *transcript* of the interaction $[A(x, y) \leftrightarrow B(x, z)]$ which consists of all messages exchanged in the computation.

We also assume familiarity with *oracle* Turing machines, which are ordinary TMs with an extra tape called the *oracle communication tape*. An oracle TMs A will be written as $A^{\langle \cdot \rangle}$ to insist that it is an oracle TM; in addition, we write $A^{\mathcal{I}}$ when A's oracle is fixed to $\mathcal{I}$. Recall that each query to $\mathcal{I}$ counts as one step towards the running time of $A^{\mathcal{I}}$.

Unless specified otherwise, all algorithms receive a parameter $n \in \mathbb{N}$, called the security parameter, as their first input. Often, the security parameter will not be mentioned explicitly and dropped from the notation. With some exceptions, all algorithms run in $\mathsf{poly}(n)$ steps and all inputs have $\mathsf{poly}(n)$ length. A function $\mathsf{negl} : \mathbb{N} \to \mathbb{R}$ is negligible if it approaches zero faster than every polynomial.

Two ensembles $\{\mathcal{X}_\setminus\}_{n \in \mathbb{N}}$ and $\{\mathcal{Y}_\setminus\}_{n \in \mathbb{N}}$ are said to be *computationally indistinguishable*, denoted $\{\mathcal{X}_\setminus\} \stackrel{c}{\approx} \{\mathcal{Y}_\setminus\}$, if for all non-uniform probabilistic polynomial time (PPT) distinguishers D, sufficiently large n, and every advice string z_n: $|\Pr_{x \leftarrow \mathcal{X}_n}[D_n(x) = 1] - \Pr_{y \leftarrow \mathcal{Y}_n}[D_n(y) = 1]| \leq \mathsf{negl}(n)$, where we write $D_n(a)$ to denoted $D(n, z_n, a)$, and negl is a negligible function. The statistical distance between two probability distributions $\mathcal{X}$ and $\mathcal{Y}$ over the same support S is denoted by $\Delta(X, Y) = \frac{1}{2} \sum_{a \in S} |\Pr[X = a] - \Pr[Y = a]|$. We say that en-sembles $\{\mathcal{X}_n\}_{n \in \mathbb{N}}$ and $\{\mathcal{Y}_n\}_{n \in \mathbb{N}}$ are *statistically indistinguishable* (or statistically close), denoted $\{\mathcal{X}_n\} \stackrel{s}{\approx} \{\mathcal{Y}_n\}$, if there exists a negligible function negl such that $\Delta(\mathcal{X}_n, \mathcal{Y}_n) \leq \mathsf{negl}(n)$ for all sufficiently large n.

Standard primitives. In this work, we will be using a family of *injective* one-way functions. In addition, unless specified otherwise, we assume that all functions $f \in \mathcal{F}_n$ in the family have an *efficiently testable range membership*: i.e., there exists a polynomial time algorithm to test that $y \in \mathsf{Range}(f)$ where $\mathsf{Range}(f)$ denotes the range of f.

We will also be using a family of *collision resistant hash functions* (CRHF) $\{\mathcal{H}_n\}$ where $h : \{0,1\}^* \to \{0,1\}^{\mathsf{poly}(n)}$ for $h \in \mathcal{H}_n$; recall that $\{\mathcal{H}_n\}$ is a CRHF family if there exists a negligible function negl such that for every non-uniform PPT machines A, every sufficiently large n, and every advice string z_n:
$\Pr_{h \leftarrow \mathcal{H}_n}[h(x) = h(y) : (x, y) \leftarrow A(z_n, h)] \leq \mathsf{negl}(n)$.

Finally, we will also be using a non-interactive, perfectly binding *commitment scheme* for committing strings of polynomial length. A commitment to a string m using randomness u will be denoted by $c = \mathsf{Com}(m; u)$. Without loss of generality, we assume that the message m committed to in c can be recovered given the randomness u and the string c. We assume perfectly binding schemes purely for the simplicity of exposition. One can replace Com by the 2-round statistically-binding commitment scheme of Naor [Nao89] without affecting our results.

2.1 Interactive Proofs, Proofs of Knowledge, and Witness Indistinguishability

We recall the standard definitions of interactive proofs [GMR85], witness indistinguishability [FS90], and proofs of knowledge [GMR85, TW87, FFS88, FS90, BG92, PR05b].

Definition 1 (Interactive Proofs). *A pair of probabilistic polynomial time interactive Turing machines $\langle P, V \rangle$ is called an interactive argument system for a language $\mathbf{L} \in \mathbf{NP}$ with witness relation $\mathbf{R}$ if there exists a negligible function $\mathsf{negl} : \mathbb{N} \to \mathbb{R}$ such that the following two conditions hold:*

- Completeness: *for every $x \in \mathbf{L}$, and every witness w such that $\mathbf{R}(x, w) = 1$, it holds that*
$$\Pr[\mathsf{OUT}_V[P(x, w) \leftrightarrow V(x)] = 1] = 1.$$

- Soundness: *for every $x \notin \mathbf{L}$, every interactive Turing machine P^* running in time at most $\mathsf{poly}(|x|)$, and every $y \in \{0,1\}^*$,*
$$\Pr[\mathsf{OUT}_V[P^*(x, y) \leftrightarrow V(x)] = 1] \leq \mathsf{negl}(|x|).$$

If the soundness condition holds for every (not necessarily PPT) machine P^* then $\langle P, V \rangle$ is called an interactive proof *system.* □

The probability in the soundness condition is called the *soundness error* of the system, and we say that the system has *negligible* soundness error since this probability is at most $\mathsf{negl}(|x|)$. Although, traditionally soundness error is defined in terms of the statement length $|x|$, in cryptographic contexts, it is convenient to define it in terms of the security parameter n, and write $\mathsf{negl}(n)$. This is without loss of generality, since in our setting since $|x| = \mathsf{poly}(n)$. Also, in this work, we will use words "argument" and "proof" interchangeably throughout the paper.

Definition 2 (Proof of Knowledge). *Let $\langle P, V \rangle$ be an interactive proof system for a language $\mathbf{L} \in \mathbf{NP}$ with witness relation $\mathbf{R}$. We say that $\langle P, V \rangle$ is a proof of knowledge (POK) for relation $\mathbf{R}$ if there exists a polynomial p and a probabilistic oracle machine E (called the* extractor*) such that for every PPT ITM P^*, there exists a negligible function negl such that for every $x \in \mathbf{L}$, and every $(y, r) \in \{0, 1\}^*$ such that $q_{x,y,r} := \Pr[\mathsf{OUT}_V[P^*_{x,y,r} \leftrightarrow V(x)] = 1] > 0$ where $P^*_{x,y,r}$ denotes the machine P^* whose common input, auxiliary input, and randomness are fixed to x, y and r respectively and the probability is taken over the randomness of V, the following conditions holds:*

- *the expected number of steps taken by $E^{P^*_{x,y,r}}$ is bounded by $\frac{p(|x|)}{q_{x,y,r}}$, where $E^{P^*_{x,y,r}}$ is machine E with oracle access to $P^*_{x,y,r}$;*
- *except with negligible probability, $E^{P^*_{x,y,r}}$ outputs w^* such that $\mathbf{R}(x, w^*) = 1$.*

□

Definition 3 (Witness Indistinguishable Proofs). *Let $\langle P, V \rangle$ be an interactive proof system for a language $\mathbf{L} \in \mathbf{NP}$ with witness relation $\mathbf{R}$. We say that $\langle P, V \rangle$ is witness indistinguishable (WI) for relation $\mathbf{R}$ if for every PPT ITM V^*, every statement $x \in \mathbf{L}$, every pair of witnesses (w_1, w_2) such that $\mathbf{R}(x, w_i) = 1$ for every $i \in \{1, 2\}$, and every (advice) string $z \in \{0, 1\}^*$, it holds that $\{\mathsf{VIEW}^{(1)}_{|x|}\} \overset{c}{\approx} \{\mathsf{VIEW}^{(2)}_{|x|}\}$ where $\{\mathsf{VIEW}^{(i)}_{|x|}\} := \mathsf{VIEW}_{V^*}[P(x, w_i) \leftrightarrow V^*(x, z)]$.*

□

As before, w.l.o.g., we can replace $|x|$ by the security parameter n in all definitions above. We remark that there exists a WIPOK with *strict* polynomial time extraction in *constant rounds* using non-black-box techniques [BL04] and in $\omega(1)$ rounds using black-box techniques [GMR85, Blu87].

Three round, public-coin WIPOK *and* ZAP*s.* The classical protocols of [GMR85, Blu87], based on the existence of non-interactive perfectly binding commitment schemes, are 3-round *witness indistinguishable, proof of knowledge* (WIPOK) protocols (for every language in $\mathbf{NP}$). We will use Blum's protocol [Blu87] as a building block and denote its three messages by $\langle \alpha, \beta, \gamma \rangle$, where β is random string of sufficient length.[5]

---

[5] We remark that this protocol has a black-box extractor whose *expected* running time is proportional to the inverse of a cheating prover's success probability. However, there also exist WIPOK with *strict* polynomial time extraction in *constant rounds* using non-black-box techniques [BL04] and in $\omega(1)$ rounds using black-box techniques [GMR85, Blu87].

A ZAP for a language $\mathbf{L}$, introduced by Dwork and Naor [DN00], is a *two round witness indistinguishable* interactive proof for $\mathbf{L}$. ZAPs can be constructed from a variety of assumptions such as non-interactive zero-knowledge proofs [BFM88, BSMP91] (which in turn can be based on trapdoor permutations [FLS99]) and verifiable random functions [MRV99]. In fact, even *non-interarctive* (i.e., one round) constructions for ZAPs for all of $\mathbf{NP}$ exist based on bilinear pairings [GOS06] and derandomization techniques [BOV03].

We will use the two round construction of [DN00] based on NIZK as a building block and denote its two messages by $\langle \sigma, \pi \rangle$ where σ is a randomly string of sufficient length. An important property of this construction is *adaptive soundness*: the statement to be proven can be chosen *after* the string σ has been sent by the verifier. We will rely on this property in our security proofs.

2.2 Concurrent Zero Knowledge

We now recall the notion of concurrent zero-knowledge [DNS98] in which one considers a "concurrent adversary" V^* who interacts in many copies of P, proving adaptively chosen, possibly correlated, polynomially many statements. We follow conventions established in [DNS98, PRS02, Ros04].

Concurrent attack. The *concurrent attack* on an interactive proof systems $\langle P, V \rangle$ for language $\mathbf{L} \in \mathbf{NP}$ with witness relation $\mathbf{R}$ considers an arbitrary interactive TM V^* which opens at most $m = m(n)$ sessions for an arbitrary polynomial m with arbitrary auxiliary input $z \in \{0,1\}^*$. Let $\boldsymbol{x} := \{x_i\} \in \mathbf{L}^m$ be set of statements in $\mathbf{L}$ of length at most $\mathsf{poly}(n)$, and $\boldsymbol{w} := \{w_i\}_{i \in [m]}$ be such that $\mathbf{R}(x_i, w_i) = 1$. The attack proceeds by uniformly fixing the random coins of V^* and initiating its execution on input the security parameter $n \in \mathbb{N}$ and auxiliary input z. At each step, V^* either initiates a new *session*—in which case a new prover instance $P(x_i, w_i)$ with fresh randomness is fixed who interacts with V^* in session i; or V^* schedules the delivery of a message of an existing session in which the corresponding prover instance responds with corresponding message. There is no restriction on how V^* schedules the messages of various sessions. We say that V^* *launches m-concurrent attack* on $\langle P, V \rangle$. The output of the attack consists of the view of V^*, denoted $\mathsf{VIEW}_{V^*}^{\langle P, V \rangle}(n, m, \boldsymbol{x}, \boldsymbol{w}, z)$.

Definition 4 (Concurrent Zero Knowledge). *We say that an interactive proof system $\langle P, V \rangle$ for a language $\mathbf{L} \in \mathbf{NP}$ (with witness relation $\mathbf{R}$) is concurrent zero knowledge if for every polynomial $m : \mathbb{N} \to \mathbb{N}$, every PPT ITM V^* launching a m-concurrent attack, there exists a PPT machine S_{V^*} such that for every set $\boldsymbol{x} := \{x_i\} \in \mathbf{L}^m$ of statements of length at most $\mathsf{poly}(n)$, every $\boldsymbol{w} := \{w_i\}_{i \in [m]}$ such that $\mathbf{R}(x_i, w_i) = 1$, and every auxiliary input $z \in \{0,1\}^*$ it holds that*

$$\left\{ S_{V^*}(n, \boldsymbol{x}, z) \right\}_{n \in \mathbb{N}} \overset{c}{\approx} \left\{ \mathsf{VIEW}_{V^*}^{\langle P, V \rangle}(n, m, \boldsymbol{x}, \boldsymbol{w}, z) \right\}_{n \in \mathbb{N}}.$$

Machine S_{V^*} is called the simulator. □

In what follows, we will sometimes abuse the notation and write V^* to also mean the *description* of the Turing machine V^*. However, when we want to be explicit about the description of a Turing machine M (including V^*), we will actually write $\mathsf{desc}(M)$. For the simulator, we may sometimes write $S_{V^*}(\cdot) := S(V^*, \cdot)$ to insist that the program of V^* is given as an explicit input to the simulator (and drop n from the notation). Further, we will assume a (unique) session identifier for each session represented by a string of length n; this session identifier can be chosen by V^* so long as it is unique for every session. W.l.o.g. we assume that the all-ones string 1^n (not to be confused with the unary representation of the security parameter) is never used as a session identifier and denotes a special symbol.

3   Differing Input Obfuscation for Turing Machines

In this section, we recall the notion of public-coin differing input obfuscation (pc-diO) for Turing machines [IPS15]. Let $\mathsf{Steps}(M, x)$ denote the number of steps taken by a TM M on input x; we use the convention that if M does not halt on x then $\mathsf{Steps}(M, x)$ is defined to be the *special symbol* ∞. Let $\mathcal{M} = \{\mathcal{M}_n\}$ denote a parameterized collection of polynomial size and polynomial time TMs, i.e., there exists a global polynomial a such that for every $n \in \mathbb{N}$, every $M \in \mathcal{M}_n$, $|M| \leq a(n)$ and $\mathsf{Steps}(M, x) \leq a(|x|)$ where x can be of arbitrary polynomial length.

We say that a pair of TMs (M_0, M_1) in the class $\mathcal{M}_n$ (for any n) is compatible if they have the same size and for every x, $\mathsf{Steps}(M_0, x) = \mathsf{Steps}(M_1, x)$.

Definition 5 (Compatible TMs). *A pair of Turing machines* $(M_0, M_1) \in \mathcal{M}_n \times \mathcal{M}_n$ *for* $n \in \mathbb{N}$ *is said to be* compatible *if* $|M_0| = |M_1|$ *and for every string* $x \in \{0, 1\}^*$ *it holds that* $\mathsf{Steps}(M_0, x) = \mathsf{Steps}(M_1, x)$.

We remark that the notion of compatible TMs allows the obfuscation to leak the running time of the obfuscated TMs. This is standard requirement; we can also use the convention of [ABG+13, IPS15] where the TMs also output their running time in addition to the "official" output.

We now recall the notion of public-coin differing inputs sampler [IPS15]. Roughly speaking, Samp is public-coin differing-inputs sampler if, on input the random coins z, it output a pair of compatible TMs (M_0, M_1) such that no PPT adversary A having access to z can produce an x such that: $M_0(x) \neq M_1(x)$. We use a slightly different notation form [IPS15], and require that in addition to outputting M_0, M_1, Samp also outputs its random coins. The randomness z of Samp will then not be mentioned as an explicit input. The definition follows.

Definition 6 (Public-coin Differing-Inputs Sampler for TMs). *We say that a (possibly non-uniform)* PPT *Turing machine* Samp *is a* public-coin differing-inputs sampler for Turing machines *if the following conditions hold:*

1. *the output of* $\mathsf{Samp}(1^n)$ *is a triplet* (z, M_0, M_1) *such that* z *is the randomness of* Samp *and* $(M_0, M_1) \in \mathcal{M}_n \times \mathcal{M}_n$ *is always a pair of compatible TMs;*

2. *for every (possibly non-uniform)* PPT TM *A there exists a negligible function* negl *such that for all* $n \in \mathbb{N}$:

$$\Pr \left[M_0(x) \neq M_1(x) : (z, M_0, M_1) \leftarrow \mathsf{Samp}(1^n; z) \; ; \; A(z, M_0, M_1) = x \right] \leq \mathsf{negl}(n).$$

For convenience, a public-coin differing-input sampler will also be referred to as a **nice sampler**.   □

Public-coin differing-input obfuscator. We now present the definition of a *public-coin differing input obfuscator* for Turing machines. Roughly speaking, the notion states that a machine $\mathcal{O}$ is a pc-diO if the following holds: if there exists a PPT distinguisher D who distinguishes $\mathcal{O}(M_0)$ from $\mathcal{O}(M_1)$ when given as auxiliary input the random coins z of the sampler who samples (M_0, M_1), then it is easy to find an x (given z) such that $M_0(x) \neq M_1(x)$. In other words, if it is hard to find the "differing input" x then the two obfuscations are indistinguishable.

Definition 7 (Public-coin Differing-Inputs Obfuscator for Turing Machines, [IPS15]). *A uniform* PPT *machine $\mathcal{O}$ is called a* public-coin differing input obfuscator (pc-diO) *for a class of Turing machines $\{\mathcal{M}_n\}$ if the following conditions are satisfied:*

1. Polynomial slowdown and functionality: *there exists a polynomial a_{dio} such that for every $n \in \mathbb{N}$, every $M \in \mathcal{M}_n$, every input x, and every $\widetilde{M} \leftarrow \mathcal{O}(n, M)$, the following conditions hold:*
   - $\mathsf{Steps}(\widetilde{M}, x) \leq a_{\mathrm{dio}}\left(n, \; \mathsf{Steps}(M, x) \right)$
   - $\widetilde{M}(x) = M(x)$
   Polynomial a_{dio} is called the slowdown polynomial *of $\mathcal{O}$.*
2. Indistinguishability: *for every public-coin differing-input (a.k.a. nice) sampler* Samp *(i.e., satisfying definition 6), for every (possibly non-uniform)* PPT *distinguisher D, there exists a negligible function* negl *such that for all n:*

$$\left| \Pr \left[D\left(z, \mathcal{O}(n, M_0)\right) = 1 : (z, M_0, M_1) \leftarrow \mathsf{Samp}\left(1^n; z\right) \right] \right.$$
$$\left. - \Pr \left[D\left(z, \mathcal{O}(n, M_1)\right) = 1 : (z, M_0, M_1) \leftarrow \mathsf{Samp}\left(1^n; z\right) \right] \right| \leq \mathsf{negl}(n).$$

where the probability is taken over the randomness of both Samp *and* $\mathcal{O}$.   □

Candidate constructions. In [IPS15], a candidate construction of pc-diO for all polynomial time TMs with variable length input of polynomial size is presented. The functionality of the construction allows the TMs to accept inputs of any length, even larger than polynomial. The security states that if a PPT machine distinguishes the obfuscation of the given TMs, there will exist an input of *polynomial* size, which can be extracted, such that the two machines will differ on that input. The assumptions underlying this construction are: pc-diO for NC^1 circuits, fully homomorphic encryption, and SNARKs for **NP**. The construction of [GGH+13] is seen as a plausible candidate for pc-diO for NC^1 and existing implausibility results [GGHW14] are unlikely to have a consequence to this assumption. Construction of diO with stronger forms of auxiliary input—worst case in which the security must hold for all auxiliary strings μ, and distributional where it holds for specific distributions of μ—were presented in [ABG+13, BCP14].

4   Constant Round Concurrent Zero-knowledge

The simplest way to see why the protocol of previous section does not work in the concurrent setting is to consider its execution in a recursively interleaved schedule (described by Dwork, Naor, and Sahai [DNS98]). In the context of our protocol, this schedule will have n sessions interleaved recursively as follows: session n does not "contain" any messages of any other session, and all messages of session i are contained between messages c_{i-1} and r_{i-1} of session $i-1$ for every i, starting from $i = n$. A pictorial representation of this scheduling is given in the full version [PPS15]. The double-headed arrows marked by π_i represent the rest of the messages of the i-th session. Roughly speaking, the simulation fails because of the following: in order to simulate session i, the simulator needs to extract the secret s_i by running the program $\widetilde{M}_{\lambda_i, s_i}$; however, the execution of $\widetilde{M}_{\lambda_i, s_i}$ contains an execution of $\widetilde{M}_{\lambda_{i+1}, s_{i+1}}$ and due to this recursion, simulator's total running time in session 1 is exponential in n.

Formally, let $t_3 \geq 1$ be the time taken by the verifier in computing r_3 on input the string c_3. Then clearly, the time taken by the simulator in running the obfuscated machine $\widetilde{M}_{\lambda_3, s_3}$ is $T_3 \geq t_3$. Then, if t_2 denotes the time taken by the simulator to obtain string r_2, we have that $t_2 \geq t_3 + T_3 \geq 2t_3$. Clearly, the time taken by the simulator to extract s_2 by running the program $\widetilde{M}_{\lambda_2, s_2}$ will be at least $T_2 \geq t_2 \geq 2t_3$. By repeating this argument for session 1, we have that $T_1 \geq t_1 \geq t_2 + T_2 \geq 2t_2 \geq 2^2 t_3$. Repeating this argument for n sessions in the DNS schedule, the total time taken by the simulator will be $\geq 2^{n-1}$.

Avoiding recursive computation via DGS-oracle. It is clear that the reason our stand-alone simulator runs in exponential time is because in order to compute s_i for session i, the simulator runs (the obfuscation of) a program which recursively runs such a program for every interleaved session between c_i and r_i. That is, the program $\widetilde{M}_{\lambda_i, s_i}$ ends up *recomputing* all of the secrets of the interleaved sessions even though they have already been computed.

We can avoid this recomputation as follows. Let $\mathcal{I}$ be an oracle which takes as input queries of the form $(f, \widetilde{s})$—where f is an *injective* one-way function and $\widetilde{s}$ is in the range of f—and returns the unique value s such that $f(s) = \widetilde{s}$.[6] Now consider an arbitrary program $\Pi^{\mathcal{I}}$ which has access to the inversion oracle $\mathcal{I}$. Clearly, if r is chosen randomly, then for any (fixed) program $\Pi^{\mathcal{I}}$ and any fixed input a, the probability that $\Pi^{\mathcal{I}}(a) = r$ is at most 2^{-n}. This is because

---

[6] We assume that it is easy to test that f is injective and that $\widetilde{s}$ is in the range of f. These requirements are only for simplicity and the protocol works even if it is not easy to test these properties.

once the description of the oracle program $\Pi^{\langle \cdot \rangle}$ is fixed, the output of $\Pi^{\mathcal{I}}(a)$ is deterministically fixed (for any fixed input a chosen prior to seeing r) and r hits this value with probability at most 2^{-n}.

Our main point here is that it is hard to come up with a satisfying "fake witness" ω to the transcripts $\lambda = \langle h, c, r \rangle$ *even if* the program committed in c is given access to the inversion oracle $\mathcal{I}$. On the other hand, the simulator can still predict r as before. However, more importantly, by means of the oracle $\mathcal{I}$ we can avoid the recursive re-computation of the secrets in the concurrent setting as follows.

Consider an *alternative* simulator $S^{\langle \cdot \rangle}$ which will be given access to the oracle $\mathcal{I}$. This simulator will have access to both, the program of the verifier V^* as well as *its own program*, given as explicit inputs, collectively denoted as $\Pi^{\langle \cdot \rangle}_{S, V^*}$. The simulator, on input a session index i, will work by initiating an execution of V^*. It will commit to program $\Pi^{\langle \cdot \rangle}_{S, V^*}(j)$ in session j (ignoring for the moment the fact that simulator needs fresh randomness); finally, this simulator *does not run any obfuscated program* to compute the secrets. Instead it queries the oracle $\mathcal{I}$ on "well formed" $(f_j, \tilde{s}_j)$ for every session $j \neq i$; when $j = i$ it simply returns the string r_i. Then, if all goes well, observe that program $\Pi^{\langle \cdot \rangle}(i)$ predicts string r_i in polynomial time (given $\mathcal{I}$) and this holds for every session i. In particular, there is no recursive recomputation of the secrets since they can be fed to the program directly once they have been computed. We note that such an oracle was first used by Deng, Goyal, and Sahai [DGS09] to construct the first *resettably-sound resettable zero-knowlege* protocol for **NP**.

It should be clear that the actual simulation will be performed by a "main" simulator S_{main} which will *not have access to any* inversion oracle, and run in (strict) polynomial time. The main simulator will run in the same manner as the alternative simulator $S^{\langle \cdot \rangle}$ except that instead of using $\mathcal{I}$, it will run the obfuscated programs (only once for each session) to recover the secrets. To ensure efficient simulation, once a session secret has been recovered, it will be stored in a global table $\mathcal{T}$ (which will be used to simulate answers of $\mathcal{I}$). Therefore the "fake witness" will now have the form $\omega = \langle u, \Pi^{\langle \cdot \rangle}, \mathcal{T} \rangle$, but the statements will still have the same form $\lambda = \langle h, c, r \rangle$; and we require that $\Pi^{\mathcal{T}}$ outputs r within finite steps (see details below).

Relation $\mathbf{R}_{\mathrm{sim}}$ *and the simple variant of our protocol.* To formally capture the above mentioned requirement for the transcripts λ, we define a relation $\mathbf{R}_{\mathrm{sim}}$ in figure 1. The family of injective one-way functions is denoted by $\{\mathcal{F}_n\}_{n \in \mathbb{N}}$ and that of collision-resistant hash functions by $\{\mathcal{H}_n\}_{n \in \mathbb{N}}$. An important observation regrading $\mathbf{R}_{\mathrm{sim}}$ is that since table $\mathcal{T}$ is not a part of the commitment c (and it should not be), we must enforce that $\Pi^{\langle \cdot \rangle}$ *never makes any invalid queries to* $\mathcal{T}$. This is because after seeing r, it is easy to design a "bad" table $\mathcal{T}$ which will encode r by means of "bad" entries and "satisy" λ.

Relation $\mathbf{R}_{\mathrm{sim}}$ allows us to prove that no efficient prover can compute ω such that $\mathbf{R}_{\mathrm{sim}}(\lambda, \omega) = 1$ with noticeable probability where λ is the transcript of GenStat with an honest verifier. We prove this claim formally in lemma 1 under the collision-intractability of $\{H_n\}$. We note that $\mathbf{R}_{\mathrm{sim}}$ is not decidable in polynomial time in general, but this will not be an issue for our reductions since we will ensure that it is checked only on "good" instances (which can be verified in polynomial time).

Instance: A tuple $\langle h, c, r \rangle \in \mathcal{H}_n \times \{0,1\}^{\mathsf{poly}(n)} \times \{0,1\}^n$ where $h : \{0,1\}^* \to \{0,1\}^n$.

Witness: A tuple $\langle u, \Pi^{\langle \cdot \rangle}, \mathcal{T} \rangle \in \{0,1\}^{\mathsf{poly}(n)} \times \{0,1\}^* \times \{0,1\}^*$ where $\Pi^{\langle \cdot \rangle}$ is an *oracle* Turing machine, and $\mathcal{T}$ is a table containing entries of the form $(f, \widetilde{s}, s)$ such that when queried on $(f, \widetilde{s})$, $\mathcal{T}$ returns s, denoted $\mathcal{T}(f, \widetilde{s}) = s$.

Relation: $\mathbf{R}_{\mathrm{sim}}\big(\langle h, c, r \rangle,\ \langle u, \Pi^{\langle \cdot \rangle}, \mathcal{T} \rangle\big) = 1$ if and only if all of the following conditions hold:

1. $c = \mathsf{Com}\left(h\left(\Pi^{\langle \cdot \rangle} \right)\ ;\ u \right)$
2. $\forall\ (f, \widetilde{s}, s) \in \mathcal{T}$ it holds that $f \in \mathcal{F}_n$ is an *injective* function and $f(s) = \widetilde{s}$
3. Program $\Pi^{\mathcal{T}}$, takes no input, outputs the string r, and halts within 2^n steps.
4. Program $\Pi^{\mathcal{T}}$ makes oracle queries of the form $(f, \widehat{s})$ such that:

$$\forall \text{ queries } (f, \widehat{s})\ \exists\ s \text{ s.t. } (f, \widehat{s}, s) \in \mathcal{T}$$

Fig. 1. Relation $\mathbf{R}_{\mathrm{sim}}$ based on a perfectly binding commitment Com.

We are now ready to describe the simpler variant of our protocol which is constant round and concurrent zero-knowledge. The protocol has three stages: in first stage λ is sampled, in second stage the verifier sends $(f, \widetilde{s})$ and *Turing machine* $\widetilde{M}_{\lambda,s}$ (and also proves that is is a correctly generated), in stage 3, P proves the knowledge of either a witness to x (the statement) or s s.t. $f(s) = \widetilde{s}$. The formal description of our protocol, named Simple-c$\mathcal{ZK}$, appears in figure 2. In the protocol description, we have renamed the machine $M_{\lambda,s}$ (which was only informally stated earlier) to $\mathsf{SimLock}_{\lambda,s}(\cdot) := \mathsf{SimLock}(\lambda, \cdot, s)$[7] where, formally:

$$\underline{\mathsf{SimLock}(\lambda, \omega, s):}$$
- Test if $\mathbf{R}_{\mathrm{sim}}(\lambda, \omega) = 1$, and if so output s; else, output 0^n.

Relation $\mathbf{R}^a_{\mathrm{sim}}$, *language* $\mathbf{L}^a_{\mathrm{sim}}$. Relation $\mathbf{R}_{\mathrm{sim}}$ is undecidable in polynomial in general. We define a polynomial time decidable version of $\mathbf{R}_{\mathrm{sim}}$. For a polynomial $a : \mathbb{N} \to \mathbb{N}$, relation $\mathbf{R}^a_{\mathrm{sim}}$ is defined as follows: $\mathbf{R}^a_{\mathrm{sim}}$ is identical to $\mathbf{R}_{\mathrm{sim}}$ except that the witness $(u, \Pi^{\langle \cdot \rangle}, \mathcal{T})$ satisfies following additional constraints: (1) $|\mathcal{T}| \le a(n)$, and (2) $\Pi^{\mathcal{T}}$ halts in at most $a(n)$ steps.

---

[7] SimLock stands for "simulator's lock," i.e., only the simulator will be able to "unlock" the secret s from this program.

Inputs.   The common input to P and V is a statement $x \in \mathbf{L}$ where language $\mathbf{L} \in \mathbf{NP}$. The prover's auxiliary input is a witness w such that $\mathbf{R}(x, w) = 1$. The security parameter n is an implicit input to both parties.

Protocol.   The protocol proceeds in three stages.

    Stage 1: P and V execute the GenStat protocol in which V sends the first message $h \leftarrow \mathcal{H}_n$, P sends the second message $c = \mathsf{Com}(0^n; u)$ for a random u, and V sends the final message $r \leftarrow \{0, 1\}^n$. Let $\lambda = \langle h, c, r \rangle$ be the transcript.

    Stage 2: V samples an injective one-way functions $f \leftarrow \mathcal{F}_n$, a random input $s \in \{0, 1\}^n$, and a sufficiently long random tape $\zeta \in \{0, 1\}^{\mathsf{poly}(n)}$ and computes:

$$\widetilde{s} = f(s), \qquad \widetilde{M}_{\lambda, s} \leftarrow \mathcal{O}(\ \mathsf{SimLock}_{\lambda, s} \ ; \ \zeta) \qquad (1)$$

    V sends $(f, \widetilde{s}, \widetilde{M}_{\lambda, s})$, and proves using a constant round ZK protocol (say Π_{ZK}) that there exist (s, ζ) satisfying equation (1) above.

    Stage 3: P proves to V, using a 3-round WIPOK (say Π_{WIPOK}) the knowledge of either:
      – w such that $\mathbf{R}(x, w) = 1$; OR
      – s such that $f(s) = \widetilde{s}$.

    Verifier's output: V accepts if the proof in stage 3 succeeds; otherwise, it rejects.

Fig. 2. The simpler variant of our protocol: Simple-c$\mathcal{ZK}$.

We define $\mathbf{L}_{\mathrm{sim}}$ and $\mathbf{L}_{\mathrm{sim}}^a$ to be the languages corresponding to $\mathbf{R}_{\mathrm{sim}}$ and $\mathbf{R}_{\mathrm{sim}}^a$ respectively. Note that for every polynomial a, it holds that $\mathbf{L}_{\mathrm{sim}}^a \in \mathbf{NP}$. We say that $\mathcal{Z} = \{\mathcal{Z}_n\}$ is a *hard distribution* over the statements of $\mathbf{L}_{\mathrm{sim}}^a$ if there exists a negligible function negl such that for every non-uniform PPT algorithm A^* and every sufficiently large n it holds that $\Pr[\lambda \leftarrow \mathcal{Z}_n; \omega \leftarrow A^*(1^n, \lambda); \mathbf{R}_{\mathrm{sim}}^a(\lambda, \omega) = 1] \leq \mathsf{negl}(n)$.

The main result of this section is the following theorem.

Theorem 1. *Assume the existence of collision-resistant hash functions and injective one-way functions. Further,* public-coin differing-inputs obfuscation (pc-diO) *for the class of all polynomial-size Turing machines that halt in a polynomial number of steps.*[8] *Then, there exists a constant round, fully concurrent zero-knowledge protocol with negligible soundness, for all languages in* **NP**.

We prove the above theorem by proving that protocol Simple-c$\mathcal{ZK}$ is a fully concurrent zero-knowledge protocol with negligible soundness error. It is clear that the protocol has constant rounds and perfect completeness. We have already

---

[8] We note that we actually do not need obfuscation for the class of all PPT Turing machines. Instead, we only need obfuscation for those Turing machines of the form $\mathsf{SimLock}^a$ where a is a polynomial and $\mathsf{SimLock}^a$ is the same as $\mathsf{SimLock}$ except that it runs for at most $a(|x|)$ steps on input x.

discussed briefly the main ideas for proving the soundness and concurrent ZK of this protocol. We discuss a few more points here and provide the full proofs in [PPS15].

To prove the soundness, we start by proving some claims about the obfuscation of Turing machine $\mathsf{SimLock}_{\lambda,s}$. In Section 6 we prove that it is hard for any (non-uniform) prover P^* to write a "fake witness" ω to the statements λ sampled using the preamble $\mathsf{GenStat}$ (see lemma 1). Using this lemma, we show that a sampling algorithm that outputs the pair of machines $(\mathsf{SimLock}_{\lambda,s}, \mathsf{SimLock}_{\lambda,0^n})$ is a *nice* sampler for Turing machines—which, roughly speaking, means that it is hard to produce an input y such that $\mathsf{SimLock}_{\lambda,s}(y) \neq \mathsf{SimLock}_{\lambda,0^n}(y)$. Therefore, by security of pc-diO, the obfuscation of $\mathsf{SimLock}_{\lambda,s}$ will be indistinguishable from that of $\mathsf{SimLock}_{\lambda,0^n}$. This however requires some care since we have to ensure that λ can indeed be correctly sampled using public-coins. But λ consists of (h, r) which are completely random strings, and c is the output of P^* which is a publicly known deterministic TM. Therefore we can actually sample λ and still ensure hardness of finding a differing-input given (h, r). Now, the soundness follows by considering three hybrids as before and violating the hardness of one-way functions (similar to the soundness of standalone ZK in section 1.1). The full proof appears in [PPS15].

To prove concurrent ZK we consider two simulators. The first one is called the internal simulator which requires access to an inversion oracle $\mathcal{I}$ for (injective) one-way functions. The second is the main simulator which essentially runs exactly as the internal simulator (by committing its description in c) and extracting the secrets using the obfuscated programs. The full descriptions of both the simulators as well as the full proof of indistinguishability of simulation are given in [PPS15].

An important issue that we did not discuss relates to the randomness used in the simulation. For the simulation to work, it is essential the internal simulator and the main simulator must use identical randomness in computing the messages that are "fed" to the verifier. This creates a circularity in the security proof: how can the commitments sent by the main simulator be "secure" when the message in the commitment (i.e., the program of the internal simulator) is correlated to the randomness used to create the commitment. We address this issue as follows: we do not include the randomness as part of the internal simulator's description in the "plain;" instead, we include it in the "committed" form using a perfectly binding commitment *which can be recovered using the inversion oracle $\mathcal{I}$*—e.g., using commitments based on hard-core bits [GL89].

5   The Four Round Construction

In the previous section, we presented a reduction from *constant round, concurrent zero-knowledge* to diO based on standard cryptographic assumptions. In this section, we present a similar reduction for four message concurrent zero-knowledge.

Let us start by optimizing the number of rounds in our constant round protocol of previous section. The standalone ZK protocol used in stage 2 has at least four rounds.[9] Since the last message of this ZK protocol must come from the verifier, our resulting protocol will have at least five rounds even after optimizations.

We consider two approaches to obtain a four round protocol. First, we can use a two-round ZK protocol with *super polynomial time simulation*[Pas03]. This approach gives us a reduction where the soundness of the resulting protocol must assume sub-exponential hardness assumptions. The second approach is to use a WI protocol to prove the correctness of the obfuscated program. However, in typical applications of WI, to get any useful security we must somehow ensure that the statement being proven has at least two witnesses.

The standard approach in such cases is to consider two independently sampled statements, in this case, two obfuscated programs $\widetilde{M}_{\lambda,s}$ and $\widetilde{M}_{\lambda,s'}$; and prove that at least one of them is correctly constructed using a WI proof. However, this approach actually fails for a very interesting reason. Although it does hide one of the secrets s, s', it actually breaks the simulation. Indeed, the internal simulator committed to in the preamble, will have no efficient way of knowing which of these two programs is actually correctly prepared. In particular, it will have to ask for the inversion of *two* challenges per session but the main simulator might be able to return only one of them (since one of the obfuscated programs could have been maliciously prepared). Attempting to overcome this subtle issue actually breaks the hardness of $\mathbf{R}_{\mathsf{sim}}$.

We therefore use a different approach; we set up an "intermediate statement" which is selected by the prover, and require the prover to provide a WIPOK of its correctness. The verifier then proves that either this intermediate statement is true or the obfuscated program is correctly prepared. The intermediate statement is prepared in such a way that it is possible to make it false and succeed (using the real witness for x) without the verifier noticing. This allows us to ensure that the obfuscated program must be correctly prepared and simulation still continues to go through. For the soundness, roughly speaking, we can extract the witness corresponding to the "intermediate statement" by using the extractor of WIPOK; we then use it to simulate the WI proof that comes from verifier's side. This allows us to again enforce the ideas we developed to prove the soundness of the Simple-c$\mathcal{ZK}$ protocol.

To setup the "intermediate statement" we use perfectly binding commitments to specially prepared strings. In the final proof, we will need to actually extract the secret s to violate the hardness of one-way functions. We get around this difficulty by using a combination of the WIPOK used by the prover and a ZAP proof. We now present a sketch of our four round protocol below. The formal description of the protocol appears [PPS15].

Four round protocol for concurrent zero-knowledge. The protocol has four components whose messages will be sent in parallel:

---

[9] To keep our *reduction* from concurrent ZK to obfuscation free from "knowledge assumptions," we cannot use 3-round ZK protocols based on such assumptions.

1. The first component is the GenStat protocol, producing statements of the form $\lambda = \langle h, c, r \rangle$.

2. The second component is a three round WIPOK given by the *prover to the verfier*. The prover prepares two commitments, namely $\widetilde{t}_1 = \mathsf{Com}(0 \parallel t_1; v_1)$ and $\widetilde{t}_2 = \mathsf{Com}(0 \parallel t_2; v_2)$ and proves that either $(\widetilde{t}_1, \widetilde{t}_2)$ are correctly prepared or x is true. The 3 messages of this WIPOK will be denoted by $\langle \alpha, \beta, \gamma \rangle$.

3. The final component is a ZAP for a specially prepared statement, which will let us extract either a witness to x or the secret s in the proof of soundness. The special statement is prepared as follows.

   The prover creates two commitments τ_1, τ_2 such that τ_1 uses string t_1 (defined above in item 2) as its randomness; likewise τ_2 uses t_2. Further, the value committed to in one of them is the witness w for statement x. The prover then proves, using a ZAP, that there exists $i \in \{1, 2\}$ such that τ_i is a commitment to w using t_i. The two messages of this ZAP are denoted by $\langle \sigma', \pi' \rangle$.

To get four rounds, the messages of these components are piggy backed with each other. The main result of this section is the following theorem.

Theorem 2. *Assume the existence of collision-resistant hash functions and trapdoor one-way permutations (alternatively, injective one-way functions and ZAP proofs for* **NP***). Further, for every polynomial* $a : \mathbb{N} \to \mathbb{N}$*, and every hard distribution* $\mathcal{Z}$ *over the statements of* $\mathbf{L}_{\mathrm{sim}}^a$*, assume the existence of* $\mathcal{Z}$*-auxiliary differing-input obfuscation (*diO*) for the class of all polynomial-size Turing machines that halt in a polynomial number of steps. Then, there exists a four message, fully concurrent zero-knowledge protocol with negligible soundness, for all languages in* **NP***.*

We have already discussed main ideas behind the proof of soundness and concurrent zero-knowledge of this protocol. The full details are given in [PPS15].

References

[ABG+13] Wee, H.: On obfuscating point functions. In: STOC, pp. 523–532 (2005)

[App13]   Applebaum, B.: Bootstrapping obfuscators via fast pseudorandom functions. Cryptology ePrint Archive, Report 2013/699 (2013), http://eprint.iacr.org/2013/699.pdf

[Bar01]   Barak, B.: How to go beyond the black-box simulation barrier. In: FOCS, pp. 106–115 (2001)

[Bar02]   Barak, B.: Constant-round coin-tossing with a man in the middle or realizing the shared random string model. In: FOCS (2002)

[BBC+14] Barak, B., Bitansky, N., Canetti, R., Kalai, Y.T., Paneth, O., Sahai, A.: Obfuscation for evasive functions. In: Lindell, Y. (ed.) TCC 2014. LNCS, vol. 8349, pp. 26–51. Springer, Heidelberg (2014)

[BC10]   Bitansky, N., Canetti, R.: On strong simulation and composable point obfuscation. In: Rabin, T. (ed.) CRYPTO 2010. LNCS, vol. 6223, pp. 520–537. Springer, Heidelberg (2010)

[BCCT13] Bitansky, N., Canetti, R., Chiesa, A., Tromer, E.: Recursive composition and bootstrapping for snarks and proof-carrying data. In: STOC, pp. 111–120 (2013)

[BCG+11] Bitansky, N., Canetti, R., Goldwasser, S., Halevi, S., Kalai, Y.T., Rothblum, G.N.: Program obfuscation with leaky hardware. In: Lee, D.H., Wang, X. (eds.) ASIACRYPT 2011. LNCS, vol. 7073, pp. 722–739. Springer, Heidelberg (2011)

[BCP14] Boyle, E., Chung, K.-M., Pass, R.: On extractability obfuscation. In: Lindell, Y. (ed.) TCC 2014. LNCS, vol. 8349, pp. 52–73. Springer, Heidelberg (2014), Preliminary version on Eprint 2013: http://eprint.iacr.org/2013/650.pdf

[BCPR13] Bitansky, N., Canetti, R., Paneth, O., Rosen, A.: More on the impossibility of virtual-black-box obfuscation with auxiliary input. Cryptology ePrint Archive, Report 2013/701 (2013), http://eprint.iacr.org/2013/701.pdf

[BFM88] Blum, M., Feldman, P., Micali, S.: Non-interactive zero-knowledge and its applications (extended abstract). In: STOC, pp. 103–112 (1988)

[BG92] Bellare, M., Goldreich, O.: On defining proofs of knowledge. In: Brickell, E.F. (ed.) CRYPTO 1992. LNCS, vol. 740, pp. 390–420. Springer, Heidelberg (1993)

[BG02] Barak, B., Goldreich, O.: Universal arguments and their applications. In: Annual IEEE Conference on Computational Complexity (CCC), vol. 17 (2002), Preliminary full version available as Cryptology ePrint Archive, Report 2001/105.

[BGGL01] Barak, B., Goldreich, O., Goldwasser, S., Lindell, Y.: Resettably-sound zero-knowledge and its applications. In: FOCS 2001, pp. 116–125 (2001)

[BGI+01] Barak, B., Goldreich, O., Impagliazzo, R., Rudich, S., Sahai, A., Vadhan, S.P., Yang, K.: On the (im)possibility of obfuscating programs. In: Kilian, J. (ed.) CRYPTO 2001. LNCS, vol. 2139, pp. 1–18. Springer, Heidelberg (2001)

[BGK+13] Barak, B., Garg, S., Kalai, Y.T., Paneth, O., Sahai, A.: Protecting obfuscation against algebraic attacks. IACR Cryptology ePrint Archive, 2013:631 (2013)

[BGT14] Bitansky, N., Garg, S., Telang, S.: Succinct randomized encodings and their applications. Cryptology ePrint Archive, Report 2014/771 (2014), http://eprint.iacr.org/

[BL04] Barak, B., Lindell, Y.: Strict polynomial-time in simulation and extraction. SIAM Journal on Computing 33(4), 783–818 (2004), Extended abstract appeared in STOC 2002

[Blu87] Blum, M.: How to prove a theorem so no one else can claim it. In: Proceedings of the International Congress of Mathematicians, pp. 1444–1451 (1987)

[BOV03] Barak, B., Ong, S.J., Vadhan, S.P.: Derandomization in cryptography. In: Boneh, D. (ed.) CRYPTO 2003. LNCS, vol. 2729, pp. 299–315. Springer, Heidelberg (2003)

[BP04] Bellare, M., Palacio, A.: The knowledge-of-exponent assumptions and 3-round zero-knowledge protocols. In: Franklin, M. (ed.) CRYPTO 2004. LNCS, vol. 3152, pp. 273–289. Springer, Heidelberg (2004)

[BP12a] Bitansky, N., Paneth, O.: From the impossibility of obfuscation to a new non-black-box simulation technique. In: FOCS, pp. 223–232 (2012)

[BP12b]   Bitansky, N., Paneth, O.: Point obfuscation and 3-round zero-knowledge. In: Cramer, R. (ed.) TCC 2012. LNCS, vol. 7194, pp. 190–208. Springer, Heidelberg (2012)

[BP13a]   Bitansky, N., Paneth, O.: On the impossibility of approximate obfuscation and applications to resettable cryptography. In: STOC, pp. 241–250 (2013)

[BP13b]   Boyle, E., Pass, R.: Limits of extractability assumptions with distributional auxiliary input. Cryptology ePrint Archive, Report 2013/703 (2013), http://eprint.iacr.org/2013/703.pdf

[BR13a]   Brakerski, Z., Rothblum, G.N.: Obfuscating conjunctions. In: Canetti, R., Garay, J.A. (eds.) CRYPTO 2013, Part II. LNCS, vol. 8043, pp. 416–434. Springer, Heidelberg (2013)

[BR13b]   Brakerski, Z., Rothblum, G.N.: Virtual black-box obfuscation for all circuits via generic graded encoding. Cryptology ePrint Archive, Report 2013/563 (2013), http://eprint.iacr.org/2013/563.pdf

[BSMP91]   Blum, M., De Santis, A., Micali, S., Persiano, G.: Noninteractive zero-knowledge. SIAM J. Comput. 20(6), 1084–1118 (1991)

[BZ13]   Boneh, D., Zhandry, M.: Multiparty key exchange, efficient traitor tracing, and more from indistinguishability obfuscation. Cryptology ePrint Archive, Report 2013/642 (2013), http://eprint.iacr.org/

[CD09]   Canetti, R., Dakdouk, R.R.: Towards a theory of extractable functions. In: Reingold, O. (ed.) TCC 2009. LNCS, vol. 5444, pp. 595–613. Springer, Heidelberg (2009)

[CGGM00]   Canetti, R., Goldreich, O., Goldwasser, S., Micali, S.: Resettable zero-knowledge. In: Proc. 32th STOC, pp. 235–244 (2000)

[CHJV14]   Canetti, R., Holmgren, J., Jain, A., Vaikuntanathan, V.: Indistinguishability obfuscation of iterated circuits and ram programs. Cryptology ePrint Archive, Report 2014/769 (2014), http://eprint.iacr.org/

[CKPR03]   Canetti, R., Kilian, J., Petrank, E., Rosen, A.: Black-box concurrent zero-knowledge requires (almost) logarithmically many rounds. SIAM Journal on Computing 32(1), 1–47 (2003), Preliminary version in STOC 2001

[CLP10]   Canetti, R., Lin, H., Pass, R.: Adaptive hardness and composable security in the plain model from standard assumptions. In: FOCS, pp. 541–550 (2010), Full version: http://www.cs.cornell.edu/~rafael/papers/ccacommit.pdf

[CLP13a]   Canetti, R., Lin, H., Paneth, O.: Public-coin concurrent zero-knowledge in the global hash model. In: Sahai, A. (ed.) TCC 2013. LNCS, vol. 7785, pp. 80–99. Springer, Heidelberg (2013)

[CLP13b]   Chung, K.-M., Lin, H., Pass, R.: Constant-round concurrent zero knowledge from p-certificates. In: FOCS (2013)

[COPV13]   Chung, K.-M., Ostrovsky, R., Pass, R., Visconti, I.: Simultaneous resettability from one-way functions. In: FOCS, pp. 231–240 (2013)

[CPS13]   Chung, K.-M., Pass, R., Seth, K.: Non-black-box simulation from one-way functions and applications to resettable security. In: STOC, pp. 231–240 (2013)

[CRV10]   Canetti, R., Rothblum, G.N., Varia, M.: Obfuscation of hyperplane membership. In: Micciancio, D. (ed.) TCC 2010. LNCS, vol. 5978, pp. 72–89. Springer, Heidelberg (2010)

[CV13]   Canetti, R., Vaikuntanathan, V.: Obfuscating branching programs using black-box pseudo-free groups. IACR Cryptology ePrint Archive, 2013:500 (2013)

[Dam91]   Damgård, I.B.: Towards practical public key systems secure against cho-
            sen ciphertext attacks. In: Feigenbaum, J. (ed.) CRYPTO 1991. LNCS,
            vol. 576, pp. 445–456. Springer, Heidelberg (1992)

[Dam00]   Damgård, I.B.: Efficient concurrent zero-knowledge in the auxiliary string
            model. In: Preneel, B. (ed.) EUROCRYPT 2000. LNCS, vol. 1807, pp.
            418–430. Springer, Heidelberg (2000)

[DGS09]   Deng, Y., Goyal, V., Sahai, A.: Resolving the simultaneous resettability
            conjecture and a new non-black-box simulation strategy. In: FOCS (2009)

[DL07]   Deng, Y., Lin, D.: Instance-dependent verifiable random functions and
            their application to simultaneous resettability. In: Naor, M. (ed.) EURO-
            CRYPT 2007. LNCS, vol. 4515, pp. 148–168. Springer, Heidelberg (2007)

[DN00]   Dwork, C., Naor, M.: Zaps and their applications. In: Proc. 41st FOCS,
            pp. 283–293 (2000)

[DNS98]   Dwork, C., Naor, M., Sahai, A.: Concurrent zero knowledge. In: Proc. 30th
            STOC, pp. 409–418 (1998)

[DS98]   Dwork, C., Sahai, A.: Concurrent zero-knowledge: Reducing the need
            for timing constraints. In: Krawczyk, H. (ed.) CRYPTO 1998. LNCS,
            vol. 1462, pp. 442–457. Springer, Heidelberg (1998)

[FFS88]   Feige, U., Fiat, A., Shamir, A.: Zero-knowledge proofs of identity. Journal
            of Cryptology 1(2), 77–94 (1987), Preliminary version in STOC 1987

[FLS99]   Feige, Lapidot, Shamir.: Multiple noninteractive zero knowledge proofs
            under general assumptions. SIAM Journal on Computing 29 (1999)

[FS89]   Feige, U., Shamir, A.: Zero knowledge proofs of knowledge in two rounds.
            In: Brassard, G. (ed.) CRYPTO 1989. LNCS, vol. 435, pp. 526–544.
            Springer, Heidelberg (1990)

[FS90]   Feige, U., Shamir, A.: Witness indistinguishable and witness hiding pro-
            tocols. In: Proc. 22nd STOC, pp. 416–426 (1990)

[GGH+13]   Garg, S., Gentry, C., Halevi, S., Raykova, M., Sahai, A., Waters, B.: Can-
            didate indistinguishability obfuscation and functional encryption for all
            circuits. In: FOCS (2013)

[GGHW14]   Garg, S., Gentry, C., Halevi, S., Wichs, D.: On the implausibility of
            differing-inputs obfuscation and extractable witness encryption with aux-
            iliary input. In: Garay, J.A., Gennaro, R. (eds.) CRYPTO 2014, Part I.
            LNCS, vol. 8616, pp. 518–535. Springer, Heidelberg (2014)

[GIS+10]   Goyal, V., Ishai, Y., Sahai, A., Venkatesan, R., Wadia, A.: Founding cryp-
            tography on tamper-proof hardware tokens. In: Micciancio, D. (ed.) TCC
            2010. LNCS, vol. 5978, pp. 308–326. Springer, Heidelberg (2010)

[GJO+13]   Goyal, V., Jain, A., Ostrovsky, R., Richelson, S., Visconti, I.: Concurrent
            zero knowledge in the bounded player model. In: Sahai, A. (ed.) TCC
            2013. LNCS, vol. 7785, pp. 60–79. Springer, Heidelberg (2013)

[GJS11]   Garg, S., Jain, A., Sahai, A.: Leakage-resilient zero knowl-
            edge. In: Rogaway, P. (ed.) CRYPTO 2011. LNCS, vol. 6841,
            pp. 297–315. Springer, Heidelberg (2011), Full version at:
            http://www.cs.ucla.edu/~abhishek/papers/lrzk.pdf

[GK96]   Goldreich, O., Krawczyk, H.: On the composition of zero-knowledge proof
            systems. SIAM Journal on Computing 25(1), 169–192 (1996), Preliminary
            version appeared in Paterson, M. (ed.): ICALP 1990. LNCS, vol. 443, pp.
            268–282. Springer, Heidelberg (1990)

[GK05]   Goldwasser, S., Kalai, Y.T.: On the impossibility of obfuscation with aux-
            iliary input. In: FOCS, pp. 553–562 (2005)

[GK13]   Goldwasser, S., Kalai, Y.T.: A note on the impossibility of obfuscation with auxiliary input. Cryptology ePrint Archive, Report 2013/665 (2013), http://eprint.iacr.org/2013/665.pdf

[GL89]   Goldreich, O., Levin, L.A.: A hard-core predicate for all one-way functions. In: Proc. 21st STOC, pp. 25–32 (1989)

[GLP+15]   Goyal, V., Lin, H., Pandey, O., Pass, R., Sahai, A.: Round-efficient concurrently composable secure computation via a robust extraction lemma. In: TCC (2015), Full version of this work available as IACR Eprint Report 2012/65

[GM11]   Goyal, V., Maji, H.K.: Stateless cryptographic protocols. In: FOCS, pp. 678–687 (2011)

[GMR85]   Goldwasser, S., Micali, S., Rackoff, C.: The knowledge complexity of interactive proof-systems. In: Proc. 17th STOC, pp. 291–304. ACM (1985)

[Gol02]   Goldreich, O.: Concurrent zero-knowledge with timing, revisited. In: Proc. 34th STOC, pp. 332–340 (2002)

[GOS06]   Groth, J., Ostrovsky, R., Sahai, A.: Non-interactive zaps and new techniques for NIZK. In: Dwork, C. (ed.) CRYPTO 2006. LNCS, vol. 4117, pp. 97–111. Springer, Heidelberg (2006)

[Goy13]   Goyal, V.: Non-black-box simulation in the fully concurrent setting. In: STOC, pp. 221–230 (2013)

[GR07]   Goldwasser, S., Rothblum, G.N.: On best-possible obfuscation. In: Vadhan, S.P. (ed.) TCC 2007. LNCS, vol. 4392, pp. 194–213. Springer, Heidelberg (2007)

[GS12a]   Garg, S., Sahai, A.: Adaptively secure multi-party computation with dishonest majority. In: Safavi-Naini, R., Canetti, R. (eds.) CRYPTO 2012. LNCS, vol. 7417, pp. 105–123. Springer, Heidelberg (2012)

[GS12b]   Gupta, D., Sahai, A.: On constant-round concurrent zero-knowledge from a knowledge assumption. CoRR, abs/1210.3719 (2012)

[Had00]   Hada, S.: Zero-knowledge and code obfuscation. In: Okamoto, T. (ed.) ASIACRYPT 2000. LNCS, vol. 1976, pp. 443–457. Springer, Heidelberg (2000)

[Had10]   Hada, S.: Secure obfuscation for encrypted signatures. In: Gilbert, H. (ed.) EUROCRYPT 2010. LNCS, vol. 6110, pp. 92–112. Springer, Heidelberg (2010)

[HRSV07]   Hohenberger, S., Rothblum, G.N., Shelat, A., Vaikuntanathan, V.: Securely obfuscating re-encryption. In: Vadhan, S.P. (ed.) TCC 2007. LNCS, vol. 4392, pp. 233–252. Springer, Heidelberg (2007)

[HSW13]   Hohenberger, S., Sahai, A., Waters, B.: Replacing a random oracle: Full domain hash from indistinguishability obfuscation. In: Cryptology ePrint Archive, Report 2013/509 (2013), http://eprint.iacr.org/2013/509.pdf

[HT99]   Hada, S., Tanaka, T.: On the existence of 3-round zero-knowledge protocols. Cryptology ePrint Archive, Report 1999/009 (1999), http://eprint.iacr.org/

[IPS15]   Ishai, Y., Pandey, O., Sahai, A.: Public Coin Differing-Inputs Obfuscation. In: TCC (2015), Cryptology Eprint Archive Report 2014/942

[Kil92]   Kilian, J.: A note on efficient zero-knowledge proofs and arguments (extended abstract). In: Proc. 24th STOC, pp. 723–732 (1992)

[Kil95]   Kilian, J.: Improved efficient arguments. In: Coppersmith, D. (ed.) CRYPTO 1995. LNCS, vol. 963, pp. 311–324. Springer, Heidelberg (1995)

[KP01]   Kilian, J., Petrank, E.: Concurrent and resettable zero-knowledge in poly-
          logarithm rounds. In: STOC, pp. 560–569 (2001)

[KPR98]   Kilian, J., Petrank, E., Rackoff, C.: Lower bounds for zero knowledge on
          the Internet. In: Proc. 39th FOCS, pp. 484–492 (1998)

[KRW13]   Koppula, V., Ramchen, K., Waters, B.: Separations in circular security for
          arbitrary length key cycles. Cryptology ePrint Archive, Report 2013/683
          (2013), http://eprint.iacr.org/2013/683.pdf

[LP14]   Lin, H., Pass, R.: Succinct garbling schemes and applications. Cryptology
          ePrint Archive, Report 2014/766 (2014), http://eprint.iacr.org/

[LPS04]   Lynn, B., Prabhakaran, M., Sahai, A.: Positive results and techniques for
          obfuscation. In: Cachin, C., Camenisch, J.L. (eds.) EUROCRYPT 2004.
          LNCS, vol. 3027, pp. 20–39. Springer, Heidelberg (2004)

[Mic94]   Micali, S.: CS proofs. In: Proc. 35th FOCS, pp. 436–453 (1994)

[MO13]   Marcedone, A., Orlandi, C.: Obfuscation == (ind-cpa security =/=
          circular security). Cryptology ePrint Archive, Report 2013/690 (2013),
          http://eprint.iacr.org/2013/690.pdf

[MR13]   Moran, T., Rosen, A.: There is no indistinguishability obfuscation
          in pessiland. Cryptology ePrint Archive, Report 2013/643 (2013),
          http://eprint.iacr.org/2013/643.pdf

[MRV99]   Micali, S., Rabin, M.O., Vadhan, S.P.: Verifiable random functions. In:
          FOCS, pp. 120–130 (1999)

[Nao89]   Naor, M.: Bit commitment using pseudo-randomness (extended abstract).
          In: Brassard, G. (ed.) CRYPTO 1989. LNCS, vol. 435, pp. 128–136.
          Springer, Heidelberg (1990)

[Pan14]   Pandey, O.: Achieving constant round leakage-resilient zero-knowledge.
          In: Lindell, Y. (ed.) TCC 2014. LNCS, vol. 8349, pp. 146–166.
          Springer, Heidelberg (2014), Preliminary version on Eprint 2012:
          http://eprint.iacr.org/2012/362.pdf

[Pas03]   Pass, R.: Simulation in quasi-polynomial time, and its application to proto-
          col composition. In: Biham, E. (ed.) EUROCRYPT 2003. LNCS, vol. 2656,
          pp. 160–176. Springer, Heidelberg (2003)

[Pas04]   Pass, R.: Bounded-concurrent secure multi-party computation with a dis-
          honest majority. In: Proc. 36th STOC, pp. 232–241 (2004)

[PPS15]   Pandey, O., Prabhakaran, M., Sahai, A.: Obfuscation-based non-black-
          box simulation and four message concurrent zero knowledge for np. In:
          TCC 2015 (2015), Full version of this work available as Cryptology ePrint
          Archive Report 2013/754

[PR03]   Pass, R., Rosen, A.: Bounded-concurrent secure two-party computation in
          a constant number of rounds. In: Proc. 44th FOCS (2003)

[PR05a]   Pass, R., Rosen, A.: Concurrent non-malleable commitments. In: FOCS
          (2005)

[PR05b]   Pass, R., Rosen, A.: New and improved constructions of non-malleable
          cryptographic protocols. In: STOC (2005)

[PRS02]   Prabhakaran, M., Rosen, A., Sahai, A.: Concurrent zero knowledge with
          logarithmic round-complexity. In: FOCS (2002)

[PTV10]   Pass, R., Tseng, W.-L.D., Venkitasubramaniam, M.: Eye for an eye: Ef-
          ficient concurrent zero-knowledge in the timing model. In: Micciancio, D.
          (ed.) TCC 2010. LNCS, vol. 5978, pp. 518–534. Springer, Heidelberg (2010)

[PV08]   Pass, R., Venkitasubramaniam, M.: On constant-round concurrent zero-
          knowledge. In: Canetti, R. (ed.) TCC 2008. LNCS, vol. 4948, pp. 553–570.
          Springer, Heidelberg (2008)

[RK99]   Richardson, R., Kilian, J.: On the concurrent composition of zero-knowledge proofs. In: Stern, J. (ed.) EUROCRYPT 1999. LNCS, vol. 1592, pp. 415–432. Springer, Heidelberg (1999)

[Ros00]   Rosen, A.: A note on the round-complexity of concurrent zero-knowledge. In: Bellare, M. (ed.) CRYPTO 2000. LNCS, vol. 1880, pp. 451–468. Springer, Heidelberg (2000)

[Ros04]   Rosen, A.: The Round-Complexity of Black-Box Concurrent Zero-Knowledge. PhD thesis, Department of Computer Science and Applied Mathematics, Weizmann Institute of Science, Rehovot, Israel (2004)

[SW13]   Sahai, A., Waters, B.: How to use indistinguishability obfuscation: Deniable encryption, and more. IACR Cryptology ePrint Archive, 2013:454 (2013)

[TW87]   Tompa, M., Woll, H.: Random self-reducibility and zero-knowledge interactive proofs of possession of information. In: Proc. 28th FOCS, pp. 472–482 (1987)

[Wee05]   Wee, H.: On obfuscating point functions. In: STOC, pp. 523–532 (2005)

6   Hardness of **GenStat** and a Nice Sampler

In this section, we prove that a randomly sampled transcript of GenStat is a hard distribution over the statements of $\mathbf{L}_{\text{sim}}^a$ for every polynomial a. Recall that $\mathcal{Z} = \{\mathcal{Z}_n\}$ is a *hard distribution* over the statements of $\mathbf{L}_{\text{sim}}^a$ if there exists a negligible function negl such that for every non-uniform PPT algorithm A^* and every sufficiently large n it holds that $\Pr[\lambda \leftarrow \mathcal{Z}_n; \omega \leftarrow A^*(1^n, \lambda); \mathbf{R}_{\text{sim}}^a(\lambda, \omega) = 1] \leq \text{negl}(n)$. The preamble GenStat, is recalled below. For convenience, we use a non-interactive perfectly binding commitment scheme; the two-round statistically-binding commitment scheme of [Nao89] also works.

6.1   Preamble **GenStat**

Statement generation protocol. Let $\{\mathcal{H}_n\}$ be a family of collision-resistant hash functions (CRHF) $h \in \mathcal{H}_n$ such that $h : \{0,1\}^* \to \{0,1\}^n$ and Com be a non-interactive perfectly-binding commitment scheme for $\{0,1\}^n$. The statement generation protocol GenStat $:= \langle P_1, V_1 \rangle$ is a three round protocol between P_1 and V_1 which proceeds as follows:

> **Protocol GenStat** $:= \langle P_1, V_1 \rangle$:
> 1. V_1 sends a random $h \leftarrow \mathcal{H}_n$
> 2. P_1 sends a commitment $c = \text{Com}(0^n; u)$ where u is a randomly chosen
> 3. V_1 sends a random string $r \leftarrow \{0,1\}^n$
> The transcript of the protocol is $\lambda := \langle h, c, r \rangle$.   □

6.2   Hardness of **GenStat** with respect to relation $\mathbf{R}_{\text{sim}}$

We defined $\mathbf{R}_{\text{sim}}$ in figure 1.   Recall that $\mathbf{R}_{\text{sim}}$ is undecidable in polynomial time in general. But our analysis will ensure that $\mathbf{R}_{\text{sim}}$ is tested only on inputs on which $\Pi^{\mathcal{T}}$ does halt (and in fact halts in a polynomial number of steps). To capture this, we defined a bounded variant of this relation, namely $\mathbf{R}_{\text{sim}}^a$ for every polynomial $a : \mathbb{N} \to \mathbb{N}$.

Relation $\mathbf{R}^a_{\mathrm{sim}}$:   Let $a : \mathbb{N} \to \mathbb{N}$ be a polynomial; relation $\mathbf{R}^a_{\mathrm{sim}}$ is identical to $\mathbf{R}_{\mathrm{sim}}$ except that the witness $(u, \Pi^{\langle \cdot \rangle}, \mathcal{T})$ satisfies following additional constraints:

1. $|\mathcal{T}| \leq a(n)$,
2. $\Pi^{\mathcal{T}}$ halts in at most $a(n)$ steps.

Note that $\mathbf{R}^a_{\mathrm{sim}}$ can be tested in time $\mathsf{poly}(a(n)) = \mathsf{poly}(n)$. $\mathbf{L}_{\mathrm{sim}}$ (resp. $\mathbf{L}^a_{\mathrm{sim}}$) is the language corresponding to relation $\mathbf{R}_{\mathrm{sim}}$ (resp., $\mathbf{R}^a_{\mathrm{sim}}$) and $\mathbf{L}^a_{\mathrm{sim}} \in \mathbf{NP}$.

The following lemma states that it is hard for any PPT machine P_1^* to compute a witness ω to statements λ when λ is the transcript of GenStat between P_1^* and an honest V_1. The proof follows [Bar01].

Lemma 1 (Hardness of GenStat). *Assume that $\{\mathcal{H}_n\}$ is a family of collision-resistant hash functions against (non-uniform) PPT algorithms. There exists a negligible function negl such that for every (non-uniform) PPT Turing machine P_1^* , the probability that P_1^*, after interacting with an honest V_1 in protocol GenStat, writes a string ω on its (private) output tape such that $\mathbf{R}_{\mathrm{sim}}(\lambda, \omega) = 1$ is at most $\mathsf{negl}(n)$, where λ is the transcript of interaction between P_1^* and V_1, and the probability is taken over the randomness of both P_1^* and V_1.*

Remark. We note that since P_1^* is polynomial time, it can only write a ω of polynomial length. However, since we have to consider all polynomial time P_1^*, it is not known in advance how large ω will be even though it will be of polynomial size.

Proof of lemma 1. Assume, on the contrary, that there exist polynomials p, q and a prover P_1^* such that P_1^* takes at most $p(n)$ steps and writes a string ω on its private output tape such that for infinitely many values of n, $\delta(n) \geq 1/q(n)$ where $\delta(n)$ is the probability that $\mathbf{R}_{\mathrm{sim}}(\lambda, \omega) = 1$ (where λ is sampled as defined in the lemma). Now consider the machine P_1^* in an execution of GenStat and let (h, c) be the first two messages in this interaction. Let the machine $P_{1,h,c}^*$ denote the machine P_1^* whose state has been frozen up to the point where c is sent in this execution. By a standard averaging argument, it follows that with probability at least $\delta/2$ over the sampling of (h, c) in this interaction, the probability that $P_{1,h,c}^*$ writes a valid witness ω at the end of the interaction is at least $\delta/2$. We call such (h, c) "good."

The following procedure finds collisions in h provided (h, c) are good: the procedure chooses two random strings r_1, r_2 each of length n, feeds P_1^* with r_1 and then with r_2 separately; let $\omega_i = (u_i, \Pi_i^{\langle \cdot, \cdot \rangle}, \mathcal{T}_i)$ be the contents of the private output tape of $P_{1,h,c}^*$ when fed with string r_i for $i \in \{1, 2\}$. The procedure outputs (Π_1, Π_2) as the potential collision on h.

We claim that the procedure finds collisions in h with noticeable probability as follows. Note that since (h, c) is good, with probability $\delta^2/4$, it holds that $\mathbf{R}_{\mathrm{sim}}(\lambda_i, \omega_i) = 1$ where $\lambda_i = (h, c, r_i)$. Hence, $\Pi_i^{\mathcal{T}_i} = r_i$ and $h(\Pi_1) = h(\Pi_2)$ w.h.p. since c is perfectly binding.

Now, define $\mathcal{I}$ to be an inversion oracle which on input a query of the form $(f, \tilde{s})$ for $f \in \mathcal{F}_n$ and $\tilde{s} \in \mathrm{Range}(f)$ outputs $s = f^{-1}(\tilde{s})$. Then, by definition of

$\mathbf{R}_{\text{sim}}$ (in particular, due to condition 4 in figure 1), we have that the output of $\Pi_i^{T_i}$ is the same as that of $\Pi_i^{\mathcal{I}}$. I.e., $\Pi_i^{\mathcal{I}}$ outputs r_i. Since $\Pi_i^{\mathcal{I}}$ is a *deterministic* computation, it holds that Π_1 and Π_2 are different programs whenever $r_1 \neq r_2$ (which happens with prob. $1 - 2^{-n}$). Further, since P_1^* runs in time at most $p(n)$, programs Π_1, Π_2 are of size at most $p(n)$. Therefore, Π_1 and Π_2 are collisions in h, found with probability at least $\frac{\delta^2}{4} \cdot (1 - 2^{-n}) \geq \delta^2/8$.

It follows that collisions can be found for a noticeable (specifically, at least $\delta/2$) fraction of functions in $\{\mathcal{H}_n\}$ with noticeable probability (specifically, $\delta^2/8$). This concludes the proof. ∎

6.3   A Nice Sampler for TM

Protocol GenStat allows us to build a (non uniform) sampling algorithm Samp which will be nice according to definition 6. Samp uses the following simple TM, which was defined earlier:

> SimLock(λ, ω, s):
>> Test if $\mathbf{R}_{\text{sim}}(\lambda, \omega) = 1$, and if so output s;
>> Else, output the empty string 0^n.

Also, for a fixed (λ, s), define $\text{SimLock}_{\lambda,s}(\cdot) := \text{SimLock}(\lambda, \cdot, s)$. Machine $\text{SimLock}_{\lambda,s}$ essentially tests whether the input is a valid witness to λ, and if so outputs the fixed value s, and nothing otherwise. Note that it is possible that SimLock takes 2^n steps on some inputs. However, no such inputs will be returned by any PPT adversary who uses (an obfuscation of) SimLock. Also, w.l.o.g., we assume that $\text{Steps}(\text{SimLock}_{\lambda,s_1}, \omega) = \text{Steps}(\text{SimLock}_{\lambda,s_2}, \omega)$ for every λ, ω and $(s_1, s_2) \in \{0,1\}^n \times \{0,1\}^n$.

The sampler. The sampling algorithm, $\text{Samp}_{P_1^*}$ is defined with respect to an arbitrary PPT interactive TM P_1^*. ITM P_1^* follows the instructions of GenStat protocol and interacts with algorithm V_1.

> $\text{Samp}_{P_1^*}(1^n; z)$.
> − z is of the form $(h, r, s) \in \mathcal{H}_n \times \{0,1\}^n \times \{0,1\}^n$.
> − Sample a random transcript λ of GenStat by interacting with P_1^* honestly by sending h as the first message and r as the third message. Let c be the output of P_1^* so that $\lambda = (h, c, r)$.
> − Output $\big(z, \text{SimLock}_{\lambda,s}, \text{SimLock}_{\lambda,0^n}\big)$

When the third component of z is fixed to a specific s, we will denote the sampler by Samp_{s,P_1^*} to emphasize a fixed s. The following lemma is essentially a corollary of lemma 1. It proves a stronger claim by directly about Samp_{s,P_1^*}; it is easy to see that the claim will trivially follow for a random s since it follows for each one of them.

Lemma 2. *For every non-uniform PPT TM P_1^*, and every $s \in \{0,1\}^n$, Samp_{s,P_1^*} is a nice sampler for Turing machines (according to definition 6).*

Proof. Observe that the pair $(\mathsf{SimLock}_{\lambda,s}, \mathsf{SimLock}_{\lambda,0^n})$ is *always* a pair of compatible TMs, by definition of SimLock. Now suppose that the second property of definition 6 is not satisfied. Then there exists an A, running in time $a(n)$ for some polynomial a, which outputs an x with noticeable probability such that $\mathsf{SimLock}_{\lambda,s}(x) \neq \mathsf{SimLock}_{\lambda,0^n}(x)$, and $|x| \leq a(n)$; here the probability is taken over the sampling of λ. It follows, from the definition of $\mathsf{SimLock}_{\lambda,s}$, that x must be a witness to λ and therefore A is a PPT machine which finds witnesses to statements $\lambda \in \mathbf{L}_{\mathrm{sim}}^a$ with noticeable probability. We can use A to violate lemma 1 as follows.

Consider the machine $B_{1,s}^*$ which incorporates P_1^* and A. It then samples λ by routing messages between P_1^* and an external (honest) V_1, and returns the output of $A(z, \mathsf{SimLock}_{\lambda,s}, \mathsf{SimLock}_{\lambda,0^n})$. It is straightforward to see that $B_{1,s}^*$ violates lemma 1 (for every fixed s).

Public-Coin Differing-Inputs Obfuscation and Its Applications

Yuval Ishai[1,*], Omkant Pandey[2,3,**], and Amit Sahai[3,**]

[1] Technioin, Israel
yuvali@cs.technion.ac.il
[2] University of Illinois at Urbana Champaign, USA
omkant@uiuc.edu
[3] UCLA and Center for Encrypted Functionalities, USA
sahai@cs.ucla.edu

Abstract. *Differing inputs obfuscation* (diO) is a strengthening of indistinguishability obfuscation (iO) that has recently found applications to improving the efficiency and generality of obfuscation, functional encryption, and related primitives. Roughly speaking, a diO scheme ensures that the obfuscations of two efficiently generated programs are indistinguishable not only if the two programs are equivalent, but also if it is hard to find an input on which their outputs differ. The above "indistinguishability" and "hardness" conditions should hold even in the presence of an auxiliary input that is generated together with the programs.

The recent works of Boyle and Pass (ePrint 2013) and Garg et al. (Crypto 2014) cast serious doubt on the plausibility of general-purpose diO with respect to general auxiliary inputs. This leaves open the existence of a variant of diO that is plausible, simple, and useful for applications.

We suggest such a diO variant that we call *public-coin* diO. A public-coin diO restricts the original definition of diO by requiring the auxiliary input to be a public random string which is given as input to all relevant algorithms. In contrast to standard diO, we argue that it remains very plausible that current candidate constructions of iO for circuits satisfy the public-coin diO requirement.

We demonstrate the usefulness of the new notion by showing that several applications of diO can be obtained by relying on the public-coin

---

* Research supported by the European Union's Tenth Framework Programme (FP10/2010-2016) under grant agreement no. 259426 ERC-CaC, ISF grant 1361/10, and BSF grant 2012378. Research done in part while visiting UCLA and the Center for Encrypted Functionalities.
** Research supported in part from a DARPA/ONR PROCEED award, NSF Frontier Award 1413955, NSF grants 1228984, 1136174, 1118096, and 1065276, a Xerox Faculty Research Award, a Google Faculty Research Award, an equipment grant from Intel, and an Okawa Foundation Research Grant. This material is based upon work supported by the Defense Advanced Research Projects Agency through the U.S. Office of Naval Research under Contract N00014-11- 1-0389. The views expressed are those of the author and do not reflect the official policy or position of the Department of Defense, the National Science Foundation, or the U.S. Government.

Y. Dodis and J.B. Nielsen (Eds.): TCC 2015, Part II, LNCS 9015, pp. 668–697, 2015.

variant instead. These include constructions of *succinct* obfuscation and functional encryption schemes for Turing Machines, where the size of the obfuscated code or keys is essentially independent of the input-length, running time and space.

1 Introduction

General-purpose obfuscation refers to the concept of transforming an arbitrary program so that its functionality is preserved, but otherwise rendering the program "unintelligible." This concept has intrigued cryptographers for decades, and led to multiple attempts at formalization (most notably [BGI+12]). A critical goal in obfuscation research has been to identify the strongest notions of obfuscation that are *plausible* and have wide applicability. General-purpose obfuscation, however, has proven to be perched precariously between possibility and impossibility.

On the one extreme, *virtual black-box obfuscation* (VBB) is an ideal form of obfuscation that captures the intuitive notion of obfuscation and often can be directly used in applications. Unfortunately, this notion is impossible in the sense that it provably cannot be realized for certain contrived classes of programs [BGI+12], or for quite large classes of programs under contrived auxiliary inputs [GK05].

On the other extreme, the most liberal notion of general-purpose obfuscation is *indistinguishability obfuscation* (iO) [BGI+12, GR07]. An iO scheme for a class of "programs" is an efficient randomized algorithm that maps any program P into a functionally equivalent obfuscated program P' such that if P_1 and P_2 compute the same function then their obfuscations P_1' and P_2' are computationally indistinguishable.

The first plausible construction of a general-purpose iO scheme was given in 2013 by Garg et al. [GGH+13b]. This construction and similar ones from [BR14, BGK+14] render the existence of an iO scheme a *plausible assumption*, since there are currently no attacks or other evidence suggesting that these constructions fail to meet the iO requirements. In particular, no theoretical impossibility results are known for iO schemes even for contrived classes of programs and auxiliary inputs.

On the downside, the security guarantee of iO appears to be too weak for most natural applications of obfuscation. A recent line of work, originating from [GGH+13b, SW14], has made impressive progress on applying iO towards a wide array of cryptographic applications. However, these applications are still not as broad as one might expect, and the corresponding constructions and their analysis are significantly more complex than those that could be obtained from an ideal obfuscation primitive. Indeed, this may be the case because the definition of iO seems to capture only a quite minimal property of obfuscation.

In search of the "strongest plausible assumption." The above limitations of iO motivate the search for stronger notions of obfuscation that support more applications and give rise to simpler constructions and security proofs. Such a stronger notion should be *plausible*, in the sense that current candidate obfuscation constructions can be conjectured to satisfy the stronger requirements without contradicting other theoretical or empirical evidence. Another important feature is *succinct description*, ruling out contrived notions whose security requirements refer separately to each application. This leads to the following question, which is at the center of our work:

Is there a plausible, useful, *and* succinctly described *notion of obfuscation that captures stronger security requirements than indistinguishability obfuscation?*

Differing Inputs Obfuscation. A seemingly promising positive answer to our question was given by the notion of *differing inputs obfuscation* (diO). First proposed in [BGI+12] and recently revisited in [ABG+13, BCP14], diO has found a diverse array of applications that do not seem to follow from traditional iO (see below for more details). Roughly speaking, a diO scheme ensures that the obfuscations of two efficiently generated programs P_1 and P_2 are indistinguishable not only if they compute the same function, but also if it is hard for an adversary to find an input x on which the two programs differ, namely x such that $P_1(x) \neq P_2(x)$. The above "indistinguishability" and "hardness" conditions should hold even in the presence of an auxiliary input aux that is generated together with the programs and is given as input both to the adversary trying to find an input x as above and to the distinguisher who tries to distinguish between the two obfuscated programs. Indeed, different applications give rise to different distributions of aux.

However, the recent works of [BP13, GGHW14] cast serious doubts on the plausibility of general-purpose diO with respect to general auxiliary inputs. In particular, [GGHW14] showed that the existence of diO with respect to arbitrary auxiliary inputs contradicts a certain "special-purpose obfuscation" conjecture. At a high level, the impossibility result of [GGHW14] proceeds as follows: Consider a pair of programs P_1 and P_2 that produce different one-bit outputs only on inputs $x = (m, \sigma)$ that consist of valid message-signature pairs with respect to a fixed unforgeable signature scheme verification key. Now we consider another program D which takes a program P as input, and then hashes P to compute $m = h(P)$ together with a signature σ on m. It then feeds $x = (m, \sigma)$ as input to P, and outputs the first bit of $P(x)$. Now, the auxiliary input given to the adversary will be a "special-purpose obfuscation" of this program D. The special-purpose obfuscation conjecture of [GGHW14] is that even given this auxiliary input, it is still hard for the adversary to obtain any valid message-signature pair. This assumption seems quite plausible, for instance if D is obfuscated using the obfuscators of [GGH+13b, BR14, BGK+14]. Now, it is evident that the adversary can distinguish between any obfuscations of P_1 and P_2 using the auxiliary input, and yet by the special-purpose assumption, the adversary cannot compute any valid message-signature pair, and therefore cannot find a differing input.

What Causes Impossibility for diO? If we would like to consider general notions of obfuscation that capture security requirements beyond indistinguishability obfuscation, it is imperative that we understand the roots of impossibility for diO. Indeed, it is not difficult to evade the impossibility results of [BP13, GGHW14] by simply assuming that diO only holds with respect to *specific* auxiliary input distributions, as has been suggested in [ABG+13, BCP14, BP13]. However, this approach would yield disparate special-purpose variants of the diO assumption for each potential application scenario, with little clarity on why any particular such assumption should be valid. This would defeat our goal of obtaining a general and *succinctly described* assumption. Therefore, we seek to understand the essential flaw that the works of [BP13, GGHW14], and others like it, can exploit using auxiliary inputs.

Our starting point is the suggestion, made in several previous works [BCP14, BP13, BCCT13, BCPR14], to specifically consider an auxiliary input that is *uniformly random*, since at least some applications of diO and other suspect primitives seem to work with just uniformly random auxiliary inputs. This certainly seems a great deal safer, and does avoid the specific impossibility results known. However, our starting observation is that even a uniformly random auxiliary input could potentially be problematic in that the auxiliary input could be chosen so that it is the output of a one-way permutation – thus there would still be a secret about the auxiliary input that is hidden from the adversary. Although we don't currently see a way to exploit this to obtain an impossibility result, could this eventually lead to trouble?

Indeed, in the negative results of [BP13, GGHW14], and similarly in other impossibility results using auxiliary inputs (e.g. [GK05]), it is critical that the auxiliary input can contain *trapdoor information*. In other words, secrets can be used to choose the auxiliary input, and these secrets are not themselves revealed to the adversary. (In the case of [GGHW14], this trapdoor information includes the secret signing key of the signature scheme, and the randomness used to obtain the obfuscation of the program D.) Our objective, then, is to formulate a notion of diO that avoids this possibility altogether.

Public-Coin Differing Inputs Obfuscation. Building upon the observations above, we introduce the notion of *public-coin* diO. A public-coin diO restricts the original definition of diO by requiring the auxiliary input aux to be the *actual random coins* that were given to the program that sampled P_1 and P_2. Thus, in public-coin diO, the auxiliary input is not chosen by a sampling algorithm, but rather the auxiliary input is simply a set of public coins that are made available to all algorithms. In particular, this means that it must be hard to find an input x such that $P_1(x) \neq P_2(x)$ even given all information about how P_1 and P_2 were generated. This rules out the possibility of planting a trapdoor in the auxiliary input, an option that was critical for proving the negative evidence against diO [BP13, GGHW14].

Indeed, we know of no evidence of impossibility for public-coin diO. The public coin restriction appears to cut a wide path around the impossibility result of [GGHW14]. Intuitively, public-coin diO requires that even "nature" –

which is computationally bounded but all-seeing – cannot find any inputs on which the two programs P_1 and P_2 will differ. This is important because not only [BP13, GGHW14], but also all previous impossibility results on VBB obfuscation (e.g [BGI+12, GK05]) used the input/output behavior of the program to plant hidden inputs on which the output of the program is too revealing. But in public-coin diO, the existence of such planted inputs would automatically rule out any security guarantee from diO, since given knowledge of these planted inputs it is easy to find a differing input. Thus, intuitively speaking, this suggests that an impossibility result for public-coin diO would need to find actual weaknesses in the obfuscation mechanism itself – some way to distinguish obfuscations that does not use the input/output behavior of the underlying programs in any way. Existing security proofs in generic[1] models [BR14, BGK+14] offer strong evidence that such an impossibility result is unlikely to exist.

We also view our public coin restriction as being a natural limitation to place on diO. Indeed, while our notion is novel in the context of obfuscation, it is reminiscent of (but also quite different from) other scenarios in cryptography where the public-coin restriction was introduced in order to prevent the existence of trapdoor information. For example, in the context of trapdoor permutations, it was observed that allowing the input sampler to use general auxiliary information can lead to problematic constructions technically satisfying the definition of a trapdoor permutation but rendering applications of trapdoor permutations insecure [GR13]. To prevent this, the notion of enhanced trapdoor permutations limits the input samplers to be given only public coins as input. Separately, in the context of collision-resistant hash functions, the distinction between secret-coin and public-coin collision-resistant hash families was studied in [HR04], where it was noted that some applications of collision-resistant hashing require public coins, since secret coins may enable the party picking the key to know impermissible trapdoor information. While these other public-coin primitives are quite different in nature from ours, we view our notion of public-coin diO to be as natural a variant as enhanced trapdoor permutations or public-coin collision resistant hash functions.

Bellare, Stepanovs, and Tessaro [BST14] presented a definitional framework for diO where security of obfuscation is parameterized by a class of samplers (instead of applying for all circuits). This allows one to define and study restricted forms of diO by considering different types of samplers. The central object in this framework is then to identify appropriate types of samplers (which, for example, do not suffer from the negative results of [BP13, GGHW14]).

Our notion of public-coin diO can be cast in the framework of [BST14] by considering samplers that are *public-coin*. We put forward the case of public-coin samplers as an important notion worthy of further study. Our work demonstrates that the public-coin case is of general interest, evades the implausibility results of [BP13, GGHW14] at a fundamental level, and yields several applications which we discuss shortly.

---

[1] The idealized adversary model considered in [BR14, BGK+14] is a generic model for multilinear maps [GGH13a, CLT13].

On Non-uniformity. Often, because auxiliary input can also capture non-uniformity, the issues of auxiliary input and non-uniformity are treated jointly in definitions. However, since we are introducing nontrivial constraints on auxiliary inputs, we deal with non-uniformity separately. We formulate our definitions to separate out the contributions of auxiliary input (which is a public coin input to the potentially non-uniform sampler), and non-uniform advice. Specifically, we take care to ensure that no secrets can be embedded in the non-uniform advice provided to the sampler, by allowing the non-uniform advice given to the differing-input finding algorithm to depend arbitrarily on the non-uniform advice given to the sampler. Thus, in particular, the non-uniform advice given to the differing-input finding algorithm can contain all secrets present in the non-uniform advice given to the sampler.

Applications of Public-Coin diO. While the public-coin limitation allows us to give a general definition that avoids all known techniques for proving impossibility, one may wonder whether this limitation also rules out known applications of diO. Indeed, at first glance, the situation may seem quite problematic, since an auxiliary input is typically used to capture the partial progress of a security reduction, which will almost always contain various secrets that must be kept from the adversary. Indeed, existing security proofs for applications of diO [ABG+13, BCP14] proceed along these lines, and therefore do not carry over for public-coin diO.

In order to make use of public-coin diO, we need to ensure that a stronger property is true in the application scenario and throughout the hybrids in the security proof where the diO property is being used: We need to ensure that whenever the diO property is used, the two programs P_1 and P_2 being considered have the property that it is infeasible to find a differing input between P_1 and P_2 even given all the randomness used in the entire description of the hybrid experiment (except for the random coins of the obfuscation itself). This is sufficient: When using the diO property across two hybrids, we need that the obfuscations are indistinguishable to an all-knowing adversary that is privy to all randomness in these hybrids (except for the random coins of the obfuscation itself). But if the obfuscations are indistinguishable to an all-knowing adversary, then they are also indistinguishable to a more ignorant adversary. Thus, even if some secrets need to be hidden from the adversary in other hybrid experiments, the proof of security can go through.

Despite the flexibility of the above approach, there are still applications of diO in the literature where we do not know how to use public-coin diO in a similar way, because of the nature of the programs being obfuscated. For example, in [BCP14], diO is used to obtain full security for functional encryption by obfuscating programs that deal explicitly with signatures, where a secret verification key of a signature scheme is hidden within obfuscated programs, and given the signing key it is possible to discover differing inputs. Since trapdoors are crucial in this approach, we do not know how to apply public-coin diO. The fact that public-coin diO does not generically replace diO for all applications illustrates the nontrivial nature of our restriction.

Nevertheless, we can use public-coin diO to obtain several interesting applications, as we now detail. Separate from the applications below, building on our work, public-coin diO has been used to replace the need for diO to achieve constant-round concurrent zero knowledge based on obfuscation [PPS15].

Obfuscating Turing Machines / RAMs with Unbounded Inputs. Generally, obfuscation has been studied in the context of circuits [BGI+12, GGH+13b]. Of course, given a bound on both (1) the input length and (2) the running time, any Turing Machine or RAM program can be converted to an equivalent circuit. However, if either or both of these variables can be unbounded, then obfuscating Turing Machines presents new feasibility questions beyond obfuscating circuits.

Moreover, note that transforming the TM into an equivalent circuit results in a circuit whose size is proportional to the *worst case* running time of the TM. This leads to severe inefficiency since one would have to evaluate a rather large circuit for every input. Indeed, motivated by this issue, Goldwasser et al. [GKP+13a, GKP+13b] introduced and studied the notion of *input-specific* run time in the context of several cryptographic primitives such as fully homomorphic encryption [Gen09], functional encryption [SW05, BSW11], and attribute-based encryption [SW05, GPSW06].

Using indistinguishability obfuscation alone, there has recently been exciting progress towards obfuscating Turing Machines directly (i.e., without first transforming it into a circuit). The recent works of Lin and Pass [LP14], Canetti, Holmgren, Jain, and Vaikuntanathan [CHJV14], and Bitansky, Garg, and Telang [BGT14], show how to obfuscate Turing machines or RAM programs directly when both the *input-length* and the overall *space* of the computation is a-priori bounded. More specifically, [LP14, CHJV14, BGT14] first construct garbling schemes for Turing machines (with bounded input-length and space) and use them to obtain obfuscation for Turing machines under same constraints. The size of the obfuscated program in these constructions only depends on the maximum input length *and* space used by the computation (as opposed to worst case running time of the original TM). However, obtaining obfuscation of TMs from garbling schemes introduces its own subtleties due to which current constructions additionally require cryptographic assumptions of sub-exponential hardness.

The recent work of Koppula, Lewko, and Waters [KLW14] presents a novel construction of indistinguishability obfuscation for Turing Machines with *bounded input length* (and unbounded space), based only on iO for circuits and standard assumptions. In other words, the size of the obfuscated TM in the [KLW14] construction is polynomial in the maximum *input length* to be accepted by the obfuscated TM, the description-size of the TM M to be obfuscated, and the security parameter. (Note that, in particular, it is independent of the maximum space of the computation.) While this is a remarkable result, the dependence upon the maximum input length is still a drawback of this work – a drawback that our work does not encounter. In applications of iO for TMs such as non-black-box simulation [PPS15], it is crucial that there is no a-priori polynomial upper bound on the input length of the obfuscated TM. Furthermore, we note that our construction is significantly simpler than the iO-based construction of [KLW14] and relies only on *polynomial* hardness assumptions; in contrast [KLW14] (as well as [LP14, CHJV14, BGT14]) require sub-exponential hardness assumptions.

In [BCP14, ABG+13], diO for circuits, together with SNARKs [BCCT12, BCCT13, BCC+14], was shown to imply diO for Turing Machines with *unbounded* inputs, running time, and space complexity (we will refer to this setting as the setting of *general* Turing Machines). However, given the evidence of impossibility for diO, prior to our work, there was no method known to bootstrap from obfuscation for circuits to obfuscation for general Turing Machines based on a plausible and general obfuscation definition. We show that the construction and proofs of [BCP14, ABG+13] can be adapted to the setting of public-coin diO: Specifically, we show that public-coin diO for NC^1, together with fully homomorphic encryption with decryption in NC^1, and public-coin SNARKs, imply diO for general Turing Machines. We note that our formulation of public-coin SNARK also avoids all known impossibility results for SNARKs and other extractability assumptions [BCPR14].

Functional Encryption for Turing Machines with Unbounded Inputs.
We next tackle the problem of (selectively secure) functional encryption [SW05, BSW11] for Turing Machines with unbounded inputs. Here, we are able to show that public-coin diO for general Turing Machines together with standard cryptographic assumptions, implies selectively secure functional encryption for general Turing Machines. As mentioned above, the approach given in [BCP14] achieves full security for functional encryption, but does not adapt to the setting of public-coin diO. The starting point for our scheme is the functional encryption construction given by [ABG+13], however in the case of functional encryption, we must make several adjustments to both the construction and the proof of security in order to make use of public-coin diO, and avoid the need for security with respect to general auxiliary inputs. We note that for the case of single-key functional encryption [SS10], the problem of supporting Turing machines and achieving input-specific runtimes was previously introduced and resolved by Goldwasser et al. [GKP+13a] under cryptographic assumptions that are incomparable to our work, but nevertheless, subject to the same criticism as the existence of diO.

Functional encryption is strict strengthening of many cryptographic notions including garbling schemes [Yao82, FKN94, BHR12] (also known as randomized encoding of functions [IK00, AIK06]). Thus, our results for functional encryption imply results for garbling schemes (as well as other notions that are implied by functional encryption) under the public-coin diO assumptions of only *polynomial* hardness. In particular, this applies to several applications of garbling schemes discussed in recent works of [LP14, CHJV14, BGT14] (under incomparable assumptions). We refer the reader to [App11] for a survey of applications of garbling schemes in different areas of cryptography.

OTHER RELATED WORKS. Another general and plausible notion of obfuscation that strengthens iO is *virtual gray box* (VGB) obfuscation [BC14, BCKP14]. While conceptually appealing, this notion does not seem as useful as diO for natural applications.

As briefly discussed above, current iO obfuscation candidates can be backed by security proofs in a generic multilinear model [BR14, BGK+14]. One can draw an analogy between the broad challenge addressed in the present work and earlier works on instantiating random oracles. Similarly to the practical use of the random oracle model [BR93], provable constructions in the generic multilinear model can give rise to heuristic real-world constructions by plugging in multilinear map candidates such as those from [GGH13a, CLT13]. This may be a reasonable heuristic leap of faith in the context of concrete natural applications. However, similarly to the negative results on instantiating random oracles [CGH04], this methodology is provably not sound in general. Thus, one is left with the challenge of formulating a *succinct* and *plausible* assumption that can be satisfied by an explicit random oracle instantiation and suffices for a wide array of applications. Despite some partial progress (e.g., [Can97, BHK13]), this challenge is still quite far from being fully met.

2   Our Definitions

Notation. We denote by $\mathbb{N}$ the set of all natural numbers, and use $n \in \mathbb{N}$ to denote the security parameter. An efficient non-uniform algorithm A is denoted by a family of circuits $A = \{A_n\}_{n \in \mathbb{N}}$ and an associated polynomial s such the size of A_n is at most $s(n)$ for all $n \in \mathbb{N}$.

We denote by $\mathcal{C} = \{\mathcal{C}_n\}_{n \in \mathbb{N}}$ a parameterized *collection* of circuits such that $\mathcal{C}_n$ is the set of all circuits of size at most n. Likewise, we denote by $\mathcal{M} = \{\mathcal{M}_n\}_{n \in \mathbb{N}}$ a parameterized collection of *Turing machines* (TM) such that $\mathcal{M}_n$ is the set of all TMs of size at most n which halt within polynomial number steps on all inputs. For $x \in \{0, 1\}^*$, if M halts on input x, we denote by $\mathsf{steps}(M, x)$ the number of steps M takes to output $M(x)$. Following [ABG+13], we adopt the convention that the output $M(x)$ includes the number of steps M takes on x, in addition to the "official" output. When clear from the context, we drop $n \in \mathbb{N}$ from the notation.

2.1   Circuits

We first define the notion of a public-coin differing-inputs sampler.

Definition 1 (Public-Coin Differing-Inputs Sampler for Circuits). *An efficient non-uniform sampling algorithm* $\mathsf{Sam} = \{\mathsf{Sam}_n\}$ *is called a* public-coin differing-inputs sampler *for the parameterized collection of circuits* $\mathcal{C} = \{\mathcal{C}_n\}$ *if the output of* Sam_n *is distributed over* $\mathcal{C}_n \times \mathcal{C}_n$ *and for every efficient non-uniform algorithm* $A = \{A_n\}$ *there exists a negligible function* ε *such that for all* $n \in \mathbb{N}$:

$$\Pr_r\left[C_0(x) \neq C_1(x) : (C_0, C_1) \leftarrow \mathsf{Sam}_n(r), x \leftarrow A_n(r)\right] \leq \varepsilon(n). \quad \square$$

The definition insists that the sampler and the attacker circuits both receive the same random coins as input. Therefore, Sam cannot keep any "secret" from A. We now define the notion of public-coin differing-inputs obfuscator. The crucial change from existing diO definitions is that the distinguisher now gets the actual coins of the sampler as the auxiliary input.

Definition 2 (Public-Coin Differing-Inputs Obfuscator for Circuits). *A uniform PPT algorithm $\mathcal{O}$ is a* public-coin differing-inputs obfuscator *for the parameterized collection of circuits $\mathcal{C} = \{\mathcal{C}_n\}$ if the following requirements hold:*

- **Correctness:** $\forall n, \forall C \in \mathcal{C}_n, \forall x$ *we have that*
  $\Pr[C'(x) = C(x) : C' \leftarrow \mathcal{O}(1^n, C)] = 1.$
- **Security:** *for every public-coin differing-inputs samplers* $\mathsf{Sam} = \{\mathsf{Sam}_n\}$ *for the collection $\mathcal{C}$, every efficient non-uniform (distinguishing) algorithm $\mathcal{D} = \{\mathcal{D}_n\}$, there exists a negligible function ε s.t. for all n:*

$$| \Pr[\mathcal{D}_n(r, C') = 1 : (C_0, C_1) \leftarrow \mathsf{Sam}_n(r), C' \leftarrow \mathcal{O}(1^n, C_0)] - \\ \Pr[\mathcal{D}_n(r, C') = 1 : (C_0, C_1) \leftarrow \mathsf{Sam}_n(r), C' \leftarrow \mathcal{O}(1^n, C_1)] | \leq \varepsilon(n)$$

where the probability is taken over r and the coins of $\mathcal{O}$. $\square$

2.2 Turing Machines

We now present our definitions for the case of Turing machines.

Definition 3 (Public-Coin Differing-Inputs Sampler for TMs). *An efficient non-uniform sampling algorithm* $\mathsf{Sam} = \{\mathsf{Sam}_n\}$ *is called a* public-coin differing-inputs sampler *for the parameterized collection of TMs $\mathcal{M} = \{\mathcal{M}_n\}$ if the output of Sam_n is always a pair of Turing machines $(M_0, M_1) \in \mathcal{M}_n \times \mathcal{M}_n$ such that $|M_0| = |M_1|$ and for all efficient non-uniform (attacker) algorithms $A = \{A_n\}$ there exists a negligible function ε such that for all $n \in \mathbb{N}$:*

$$\Pr_r \left[\begin{array}{c} M_0(x) \neq M_1(x) \ \wedge \\ \mathsf{steps}(M_0, x) = \mathsf{steps}(M_1, x) = t \end{array} : \begin{array}{c} (M_0, M_1) \leftarrow \mathsf{Sam}_n(r), \\ (x, 1^t) \leftarrow A_n(r) \end{array} \right] \leq \varepsilon(n). \quad \square$$

Remark. By requiring A_n to output 1^t, we rule out all inputs x for which M_0, M_1 may take more than polynomial steps.

Definition 4 (Public-Coin Differing-Inputs Obfuscator for TMs). *A uniform PPT algorithm $\mathcal{O}$ is a* public-coin differing-inputs obfuscator *for the parameterized collection of TMs $\mathcal{M} = \{\mathcal{M}_n\}$ if the following requirements hold:*

- **Correctness:** $\forall n, \forall M \in \mathcal{M}_n, \forall x \in \{0,1\}^*$ *we have that*
  $\Pr[M'(x) = M(x) : M' \leftarrow \mathcal{O}(1^n, M)] = 1.$
- **Security:** *for every public-coin differing-inputs samplers* $\mathsf{Sam} = \{\mathsf{Sam}_n\}$ *for the collection $\mathcal{M}$, for every efficient non-uniform (distinguishing) algorithm $\mathcal{D} = \{\mathcal{D}_n\}$, there exists a negligible function ε s.t. for all n:*

$$| \Pr[\mathcal{D}_n(r, M') = 1 : (M_0, M_1) \leftarrow \mathsf{Sam}_n(r), M' \leftarrow \mathcal{O}(1^n, M_0)] - \\ \Pr[\mathcal{D}_n(r, M') = 1 : (M_0, M_1) \leftarrow \mathsf{Sam}_n(r), M' \leftarrow \mathcal{O}(1^n, M_1)] | \leq \varepsilon(n)$$

where the probability is taken over r and the coins of $\mathcal{O}$.

- **Succinctness and input-specific running time:** *there exists a (global) polynomial s' such that for all n, for all $M \in \mathcal{M}_n$, for all $M' \leftarrow \mathcal{O}(1^n, M)$, and for all $x \in \{0,1\}^*$, $\mathsf{steps}(M', x) \leq s'(n, \mathsf{steps}(M, x))$.* $\qquad \square$

Remark. The size of the obfuscated machine M' is always bounded by the running time of $\mathcal{O}$ which is polynomial in n. More importantly, the size of M' is independent of the running time of M. This holds even if we consider TMs which always run in polynomial time. This is because the polynomial bounding the running time of $\mathcal{O}$ is independent of the collection $\mathcal{M}$ being obfuscated.

It is easy to obtain a uniform formulation from our current definitions.

3   Preliminaries

Succinct Non-Interactive Arguments. The universal relation [BG02] is defined to be the set $\mathcal{R}_\mathcal{U}$ of instance-witness pairs (y, w) such that y is of the form (M, x, t), $|w| \leq t$, and M is a TM which accepts (x, w) within t steps where t is an arbitrary number in $\mathbb{N}$. For constant $c \in \mathbb{N}$, we define $\mathcal{R}_c$ to the subset of $\mathcal{R}_\mathcal{U}$ consisting of those pairs $\{(y, w) = ((M, x, t), w)\}$ for which $t \leq |x|^c$. The language corresponding to a relation $\mathcal{R} \subseteq \mathcal{R}_\mathcal{U}$ will be denoted by $L_\mathcal{R}$.

We recall the definitions of *succinct non-interactive arguments* (SNARG) and succinct non-interactive arguments *of knowledge* (SNARK) below. We require that these systems be *publicly verifiable* and work in the common *random* string model where any uniformly random string of sufficient length can act as the CRS. Our definition follows the standard formulations [BCCT12, BCP14].

Definition 5 (SNARG). *A pair of algorithms (P, V) is a (publicly verifiable) SNARG for a relation $\mathcal{R} \subseteq \mathcal{R}_\mathcal{U}$ in the common* random *string model if there exist polynomials p, q, ℓ (independent of $\mathcal{R}$) such that the following conditions are satisfied:*

- **Completeness:** $\forall (y, w) \in \mathcal{R}$, *it holds that*
  $\Pr\left[V(\mathsf{crs}, y, \pi) = 1 : \mathsf{crs} \leftarrow \{0,1\}^{\mathsf{poly}(n)}, \pi \leftarrow P(\mathsf{crs}, y, w)\right] = 1$, *and for every* crs, $P(\mathsf{crs}, y, w)$ *halts within* $p(n, |y|, t)$ *where* $y = (M, x, t)$.
- **Succinctness:** *for every* $(\mathsf{crs}, y, w) \in \{0,1\}^{\mathsf{poly}(n)} \times \mathcal{R}$ *the size of* $\pi \leftarrow P(\mathsf{crs}, y, w)$ *is bounded by* $\ell(n, \log t)$ *and the running time of* $V(\mathsf{crs}, y, \pi)$ *is bounded by* $q(n + |y|) = q(n + |M| + |x| + \log t)$.
- **Adaptive soundness:** *for every polynomial-size prover* $P^* = \{P_n^*\}$, *there exists a negligible function* ε *such that for all* n:

$$\Pr\left[V(\mathsf{crs}, y, \pi) = 1 \wedge y \notin L_\mathcal{R} : \mathsf{crs} \leftarrow \{0,1\}^{\mathsf{poly}(n)}, (y, \pi) \leftarrow P_n^*(\mathsf{crs})\right] \leq \varepsilon(n). \quad \square$$

Observe that the soundness condition is not required to hold with respect to *common auxiliary input* of any kind. This notion suffices for the restricted cases where obfuscation size grows with the maximum supported input length (a.k.a. bounded-input case). To deal with inputs of unbounded polynomial length, we need the following stronger notion.

Definition 6 (SNARK). *A pair of algorithms (P, V) is a (publicly verifiable) SNARK for the relation $\mathcal{R} \subseteq \mathcal{R}_\mathcal{U}$ in the common* random *string model if it satisfies the completeness and succinctness conditions of definition 5 (above) and the following argument-of-knowledge property:*

- **Adaptive Argument of Knowledge:** *for every polynomial-size prover $P^* = \{P_n^*\}$, there exists a polynomial-size extractor $\mathcal{E}_{P^*} = \{\mathcal{E}_n\}$ and a negligible function ε such that for all n:*

$$\Pr\left[\begin{array}{l} (V(\mathsf{crs}, y, \pi) = 1) \\ \\ \wedge \left((y, w) \notin \mathcal{R}\right) : \left\{ \begin{array}{l} \mathsf{crs} \leftarrow \{0,1\}^{\mathsf{poly}(n)}, \quad z \leftarrow \{0,1\}^{\mathsf{poly}(n)}, \\ (y, \pi) \leftarrow P_n^*(\mathsf{crs}, z), \ (y, w) \leftarrow \mathcal{E}_n(\mathsf{crs}, z) \end{array} \right\} \end{array} \right] \qquad (3.1)$$

$$\leq \varepsilon(n). \ \square$$

Observe that in this definition a uniformly distributed auxiliary input z is allowed. As noted in [BCCT12], none of the existing implausibility results regarding the existence of SNARKs or extractable one-way/collision-resistant-hash functions apply to the case where auxiliary input is a uniformly random string. A candidate construction (and perhaps the only one at this time) for such SNARKs is Micali's CS proof system [Mic94].

We remark that the above definition requires the extraction to succeed with probability almost 1. Our results do not require this strong form of extraction, and work with a weaker notion as well where extraction probability is only required to be negligibly close to the success probability of the cheating prover.

We shall also use fully homomorphic encryption and non-interactive strong witness indistinguishable proofs, e.g., [FLS99]. We discuss them in appendix A.

4   Bootstrapping Obfuscation from NC^1 to Turing Machines

In this section, we show that given a public-coin differing-inputs obfuscator for the class NC^1, we can construct a public-coin differing-inputs obfuscator for the parameterized collection $\mathcal{M}_n$ of all polynomial-time TMs. The construction is a slightly simplified version of [ABG$^+$13] where we get rid of the hash functions. We shall prove the following theorem.

Theorem 1. *If there exists a public-coin differing-inputs obfuscator for circuits in the class NC^1, a fully homomorphic encryption scheme with decryption in NC^1, and a public-coin SNARK for $\mathcal{R}_\mathcal{U}$ in the common* random *string model, there exists a public-coin differing-inputs obfuscator for the class of all polynomial-time Turing machines accepting inputs of unbounded polynomial length.*

We first present the construction, and then prove the theorem. Let $\mathcal{M} = \{\mathcal{M}_n\}_{n \in \mathbb{N}}$ be a parameterized collection of polynomial-time TMs that accepts inputs of unbounded polynomial length, i.e., there exists a constant $c \in \mathbb{N}$ such that every $M \in \mathcal{M}_n$ is of size n, takes inputs of length at most n^c, and halts within n^c steps. We adopt the convention that c is included in the description of M. Let $\mathsf{FHE} = (\mathsf{Gen}, \mathsf{Enc}, \mathsf{Dec}, \mathsf{Eval})$ be a fully homomorphic encryption scheme with decryption in NC^1 and $\Pi = (P, V)$ be a SNARK for the relation $\mathcal{R}_\mathcal{U}$ defined earlier. The description of our obfuscator for $\mathcal{M}$, and its evaluation algorithm, are as follows.

Obfuscator $\mathcal{O}(1^n, M \in \mathcal{M}_n)$: By convention, description of M includes a constant c bounding the running time of M on all inputs by n^c. Let U_n be an *oblivious* universal TM which on input the description of a TM B, and a string x executes B on x for no more than n^c steps. The obfuscator proceeds in the following steps:

1. Generate two FHE public-keys $(pk_1, sk_1) \leftarrow \mathsf{Gen}(1^n; u_1)$ and $(pk_2, sk_2) \leftarrow \mathsf{Gen}(1^n; u_2)$.
2. Encrypt M under both FHE public-keys: $g_1 \leftarrow \mathsf{Enc}_{pk_1}(M; v_1)$ and $g_2 \leftarrow \mathsf{Enc}_{pk_2}(M; v_2)$. Here M is assumed to be encoded as a bit string of length n for use by the universal TM U_n.
3. Uniformly sample $\mathsf{crs} \leftarrow \{0,1\}^{\mathsf{poly}(n)}$ of sufficient length (for the SNARK Π).
4. Generate an obfuscation of the NC^1-program $P1_{sk_1, g_1, g_2}^{\mathsf{crs}}$ given in figure 1:

$$P' \leftarrow \mathcal{O}_{\mathrm{NC}^1}\left(1^n, P1_{sk_1, g_1, g_2}^{\mathsf{crs}}\right).$$

5. Output $M' = (P', \mathsf{crs}, pk_1, pk_2, g_1, g_2)$.

Evaluation of M': Evaluate $M' = (P', \mathsf{crs}, pk_1, pk_2, g_1, g_2)$ on input x as follows:

1. Compute $(e_1, e_2) = M_{\mathsf{Eval}}(x)$. This takes at most n^{2c} steps. See fig. 1 for M_{Eval}.
2. Compute a SNARK proof π using x as the witness and $t = n^{4c}$ as the time-bound:

$$\pi \leftarrow P\left(\mathsf{crs}, \; (\widetilde{M}_{\mathsf{Eval}}, (e_1, e_2), t), \; x\right)$$

3. Compute a low-depth proof ϕ for the **NP**-statement $1 = V(\mathsf{crs}, (\widetilde{M}_{\mathsf{Eval}}, (e_1, e_2), t), \pi)$. This can be done by providing the entire computation of V on these inputs.
4. Execute $P'(e_1, e_2, t, \pi, \phi)$ and output the result.

The construction is now analyzed in the proof below. We denote by $a\|b$ the concatenation of two bit strings a and b.

Program $P1^{\mathsf{crs}}_{sk_1,g_1,g_2}$:

- **Input:** a tuple (e_1, e_2, t, π, ϕ),   **Constants:** $\mathsf{crs}, sk_1, g_1, g_2, pk_1, pk_2$.
- Check that $t \leq 2^n$ and ϕ is a valid low-depth proof for the **NP**-statement:
$$1 = V\left(\mathsf{crs},\ (\widetilde{M}_{\mathsf{Eval}},\ (e_1, e_2),\ t),\ \pi\right)$$
  where $\widetilde{M}_{\mathsf{Eval}}$ is defined as follows. Let M_{Eval} be the computation that takes x as input, has (pk_1, pk_2, g_1, g_2) hardcoded, and homomorphically evaluates $U_n(\cdot, x)$ on g_1 and g_2 to produce e_1 and e_2 respectively. I.e.,
$$e_1 = \mathsf{Eval}_{pk_1}(U_n(\cdot, x), g_1) \text{ and } e_2 = \mathsf{Eval}_{pk_2}(U_n(\cdot, x), g_2).$$
  $\widetilde{M}_{\mathsf{Eval}}$ takes as input an instance of the form (e_1, e_2) and a witness x; it accepts if and only if $M_{\mathsf{Eval}}(x)$ outputs (e_1, e_2) within 2^n steps.
- If the check fails, output $\perp$; otherwise output $\mathsf{Dec}_{sk_1}(e_1)$.

Program $P2^{\mathsf{crs}}_{sk_2,g_1,g_2}$:

- Same as $P1^{\mathsf{crs}}_{sk_1,g_1,g_2}$ except that if the check is successful, it outputs $\mathsf{Dec}_{sk_2}(e_2)$.

Fig. 1. The programs $P1$ and $P2$

Proof of theorem 1. The correctness and succinctness of this construction are relatively straightforward to verify, and in particular, closely follow the analyses in [ABG+13, BCP14]. We analyze its security.

Security. Fix any public-coin differing-inputs sampler $\mathsf{Sam} = \{\mathsf{Sam}_n\}$ for the family $\mathcal{M}$ and any efficient distinguisher $\mathcal{D} = \{\mathcal{D}_n\}$. For a bit b, let $\mathcal{X}_n(b)$ denote the output of the following experiment over a random choice of r and the coins of $\mathcal{O}$:

$$\mathcal{X}_n(b) := \{(M_0, M_1) \leftarrow \mathsf{Sam}_n(r), M' \leftarrow \mathcal{O}(1^n, M_b), \text{ output } \mathcal{D}_n(r, M')\}$$

We need to show that $\mathcal{X}_n(0) \approx_c \mathcal{X}_n(1)$. Consider the following sequence of hybrid experiments.

- H_0: This hybrid corresponds to an honest sampling of $\mathcal{X}_n(0)$. In this case, M' creates two FHE encryptions of M_0, namely g_1 and g_2 (where M_0 is the first output of Sam_n).
- H_1: Same as H_0 except that the second FHE ciphertext is now generated as an encryption of M_1, i.e., $g_2 = \mathsf{Enc}_{pk_2}(M_1)$ (where M_1 is the second output of Sam_n).
- H_2: Same as H_1 except that the obfuscated program P' is now generated as an obfuscation of $P2^{\mathsf{crs}}_{sk_2,g_1,g_2}$ which decrypts the *second* ciphertext using sk_2, i.e., $P' \leftarrow \mathcal{O}_{\mathrm{NC}^1}(1^n, P2^{\mathsf{crs}}_{sk_2,g_1,g_2})$.

- H_3: Same as H_2 except that the *first* FHE ciphertext g_1 is now also generated as an encryption of M_1, i.e., $g_1 \leftarrow \mathsf{Enc}_{pk_1}(M_1)$.
- H_4: Same as H_3 except that the obfuscated program P' is once again generated as an obfuscation of $P1^{\mathsf{crs}}_{sk_1,g_1,g_2}$, i.e., $P' \leftarrow \mathcal{O}_{\mathsf{NC}^1}(1^n, P1^{\mathsf{crs}}_{sk_1,g_1,g_2})$. Note that H_4 is identical to $\mathcal{X}_n(1)$.

We now prove that each hybrid in this sequence is indistinguishable from the previous one.

Step 1: $H_0 \approx_c H_1$. This follows from the IND-CPA security of FHE. Formally, consider an adversary A_{FHE}, who receives a challenge public-key pk, then samples $(M_0, M_1) \leftarrow \mathsf{Sam}_n(r)$ for a random r, and receives an honestly generated ciphertext g to either M_0 or M_1 under pk. A_{FHE} then generates an obfuscation of M_0 following the instruction of $\mathcal{O}$ except that it sets $pk_2 = pk$ and $g_2 = g$. Note that all instructions of $\mathcal{O}$ can indeed be performed efficiently knowing only (pk_2, g_2). Let M' denote the resulting obfuscation which includes an NC^1-obfuscation P' of program $P1^{\mathsf{crs}}_{sk_1,g_1,g_2=g}$. A_{FHE} outputs whatever $\mathcal{D}_n(r, M')$ outputs. The output of A_{FHE} is distributed identically to that of H_b when g is an encryption of M_b where $b \in \{0, 1\}$. Because Sam_n and $\mathcal{D}_n$ are of polynomial-size, it follows that A_{FHE} is a polynomial-size circuit violating IND-CPA security of FHE unless $H_0 \approx_c H_1$.

Step 2: $H_1 \approx_c H_2$. We use the soundness of SNARK and diO-security of $\mathcal{O}_{\mathsf{NC}^1}$ to argue that $H_1 \approx_c H_2$. Suppose that H_1 and H_2 are not computationally indistinguishable. We use Sam_n and $\mathcal{D}_n$ to construct a public-coin differing-inputs sampler $\mathsf{Sam}_n^{\mathsf{NC}^1}$ along with a distinguisher $\mathcal{D}_n^{\mathsf{NC}^1}$ such that $\mathsf{Sam}_n^{\mathsf{NC}^1}$ outputs circuits in NC^1 and $\mathcal{D}_n^{\mathsf{NC}^1}$ violates the security of $\mathcal{O}_{\mathsf{NC}^1}$ w.r.t. $\mathsf{Sam}_n^{\mathsf{NC}^1}$. We start by constructing $\mathsf{Sam}_n^{\mathsf{NC}^1}$.

Sampler $\mathsf{Sam}_n^{\mathsf{NC}^1}(\rho)$.
1. Parse ρ as $\rho = (r, \rho_1, , u_1, u_2, v_1, v_2)$.
2. Sample $(M_0, M_1) \leftarrow \mathsf{Sam}_n(r)$.   // comment: this is the given sampler.
3. Set $\mathsf{crs} = \rho_1$, $(pk_1, sk_1) \leftarrow \mathsf{Gen}(1^n; u_1)$, $(pk_2, sk_2) \leftarrow \mathsf{Gen}(1^n; u_2)$.
4. Set $g_1 \leftarrow \mathsf{Enc}_{pk_1}(M_0; v_1)$ and $g_2 \leftarrow \mathsf{Enc}_{pk_2}(M_1; v_2)$.
5. Output (C_0, C_1) corresponding to the programs $\left(P1^{\mathsf{crs}}_{sk_1,g_1,g_2}, P2^{\mathsf{crs}}_{sk_2,g_1,g_2} \right)$.

Note that input to the circuits C_0, C_1 above are of the form $m = (e_1, e_2, t, \pi, \phi)$.

Claim. $\forall n \in \mathbb{N}$ $\mathsf{Sam}_n^{\mathsf{NC}^1}$ is a public-coin differing-inputs sampler for a family $\mathcal{C} \in \mathsf{NC}^1$.

Proof. We have to show that every non-uniform PPT attacker $\{A_n^{\mathsf{NC}^1}\}$ fails to find a differing-input for circuits sampled by $\mathsf{Sam}_n^{\mathsf{NC}^1}$. Given an attacker $A_n^{\mathsf{NC}^1}$ which succeeds against our sampler, we construct an attacker A_n which will succeed against the given sampler Sam_n We shall rely on the soundness of SNARK to prove this.

Formally, suppose that the claim is false, and there exists a polynomial-size attacker family $\{A_n^{\mathrm{NC}^1}\}$, a polynomial p, and infinitely many n s.t.

$$\Pr_\rho \left[C_0(x) \neq C_1(x) : (C_0, C_1) \leftarrow \mathsf{Sam}_n^{\mathrm{NC}^1}(\rho), x \leftarrow A_n^{\mathrm{NC}^1}(\rho) \right] \geq 1/p(n).$$

We start by defining a prover family which receives a uniformly random auxiliary input, denoted by z, and then use it to define an attacker $\widetilde{A}_n$ which also receives a uniform auxiliary input. Later on, this auxiliary input will be completely removed from $\widetilde{A}_n$.

Prover $P_n^*(\mathsf{crs}, z)$: String z is of the form (r, u_1, u_2, v_1, v_2) and the circuit has adversary $A_n^{\mathrm{NC}^1}$ hardcoded in it. The circuit proceeds as follows:
1. Define $\rho := (r, \mathsf{crs}, u_1, u_2, v_1, v_2)$ using z and crs.
2. Compute $m \leftarrow A_n^{\mathrm{NC}^1}(\rho)$ which is of the form $m = (e_1, e_2, t, \pi, \phi)$. Recall that $A_n^{\mathrm{NC}^1}(\rho)$ defines a TM $\widetilde{M}_{\mathsf{Eval}}$ and π is a SNARK proof for the statement $y := (\widetilde{M}_{\mathsf{Eval}}, (e_1, e_2), t)$.
3. Output (y, π).

Let $\{\mathcal{E}_n^*\}$ be a family of extractor circuits w.r.t. the prover family $\{P_n^*\}$ defined above. Now we define the following attacker circuit $\widetilde{A}_n$ which receives a uniformly random auxiliary input z^* and outputs a differing input for the given sampler Sam_n. Later we will choose an appropriate z^* and hardcode it as part of the circuit description to achieve an attacker circuit without auxiliary input.

Circuit $\widetilde{A}_n(r, z^*)$: String z^* is of the form $(\rho_1, u_1, u_2, v_1, v_2)$ and extractor circuit $\mathcal{E}_n^*$ is hardcoded in this circuit. The circuit computes as follows:
1. Define $\mathsf{crs} = \rho_1$ and $z = (r, u_1, u_2, v_1, v_2)$ using r and z^*.
2. Compute $(y, w) \leftarrow \mathcal{E}_n^*(\mathsf{crs}, z)$ where y is of the form $y := (\widetilde{M}_{\mathsf{Eval}}, (e_1, e_2), t)$.
3. Output $x = w$ as the differing input.

For any given r, z^*, the concatenation $\rho = r \| z^*$ is of the form $(r, \rho_1, u_1, u_2, v_1, v_2)$, and defines a valid random string for $A_n^{\mathrm{NC}^1}$. We say that a fixed string z^* is good if, the success probability of $A_n^{\mathrm{NC}^1}(\rho)$ is at least $\frac{1}{2p}$ over the choice of r. Formally, a string z^* is good if for a randomly chosen r, defining the tape $\rho = r \| z^*$, the probability that $A_n^{\mathrm{NC}^1}(\rho)$ outputs m such that $C_0(m) \neq C_1(m)$ where $(C_0, C_1) \leftarrow \mathsf{Sam}_n^{\mathrm{NC}^1}(\rho)$ is at least $1/2p$. By simple averaging, at least $\frac{1}{2p}$ fraction of z^* are good.

Now let us define sound strings. Roughly speaking, we say that z^* is sound if the probability that the output of $A_n^{\mathrm{NC}^1}(\rho)$ contains a valid proof π but the output of the extractor (in step 2 above) is not a valid witness, is less than $1/4p$.

Formally, we say that a fixed string $z^* = (\rho_1, u_1, u_2, v_1, v_2)$ is sound if for a randomly chosen r, defining the tape $\rho = r \| z^*$, the probability of the following event, taken over r, is at most $1/4p$: the output of $\mathcal{E}^*(\mathsf{crs}, z)$ (in step 2 of $\widetilde{A}_n(r, z^*)$) is (y, w) and output of $A_n^{\mathrm{NC}^1}(\rho)$ is $m = (e_1, e_2, t, \pi, \phi)$ such that V accepts the proof π for the statement y but w is not a valid witness, i.e. $(y, w) \notin \mathcal{R}_{2c}$.

A randomly chosen z^* contains a uniformly distributed crs string; therefore, it follows that at least $1 - \varepsilon'$ fraction of z^* are sound where ε' is the soundness error of SNARK.

Therefore, at least $\frac{1}{2p} - \varepsilon' \geq \frac{1}{4p}$ fraction of z^* are both good and sound. Fix such a z^*. By definition of good it follows that w.r.t. this z^*, at least $1/2p$ fraction of inputs r are such that $A_n^{\mathrm{NC}^1}(r\|z^*)$ outputs a differing-input for C_0, C_1. Further, by definition of sound, at most a $1/4p$ fraction of such inputs r are such that the extractor $\mathcal{E}_n^*$ (in step 2 above) will not output a valid witness w. Therefore, at least $\frac{1}{2p} - \frac{1}{4p} \geq \frac{1}{4p}$ of inputs r result in a differing-input where the extractor's output is a valid witness. We call such inputs nice.

By construction, for nice r, we have that:

$$C_0(m) \neq C_1(m) \implies P1_{sk_1,g_1,g_2}^{\mathrm{crs}=\rho_1}(m) \neq P2_{sk_2,g_1,g_2}^{\mathrm{crs}=\rho_1}(m) \implies M_0(x) \neq M_1(x)$$

where $x = w$ is the differing-input output by $\widetilde{A}_n$, and the last implication follows because x is a valid witness, i.e. if m contains ciphertexts e_1, e_2 then the values in these ciphertexts will indeed be $M_0(x)$ and $M_1(x)$ respectively. Here M_0, M_1 are the TMs sampled in (step 2 of) the execution of $\mathsf{Sam}_n^{\mathrm{NC}^1}(r\|z^*)$. We now observe that, by construction, (M_0, M_1) are also the output of $\mathsf{Sam}_n(r)$. Therefore, the output of $\widetilde{A}_n$ outputs a differing input x for the outputs of Sam_n whenever z^* is good and sound and r is nice.

To have a deterministic attacker A_n which on input r outputs a differing-input x, we choose a z^* that is good and sound, and hardcode it in the description of $\widetilde{A}_n$. It follows that, since the fraction of nice strings r is at least $1/4p$, A_n violates the public-coin differing-input property of Sam_n with noticeable property. The proof is completed by observing that circuits output by $\mathsf{Sam}_n^{\mathrm{NC}^1}$ are indeed in the complexity class NC^1.   □

We now present distinguisher $\mathcal{D}_n^{\mathrm{NC}^1}$ which violates the security of $\mathcal{O}_{\mathrm{NC}^1}$ w.r.t. sampler $\mathsf{Sam}_n^{\mathrm{NC}^1}$.

Distinguisher $\mathcal{D}_n^{\mathrm{NC}^1}(\rho, C')$. The input consists a string ρ and an obfuscated circuit C'. C' is an obfuscation of either C_0 or C_1 which are output by $\mathsf{Sam}_n^{\mathrm{NC}^1}(\rho)$. The distinguisher attempts to create a valid obfuscation M' of the TM implicitly present in C'. Since entire string ρ is available, it can be efficiently done as follows.

1. Parse ρ as $\rho = (r, \rho_1, u_1, u_2, v_1, v_2)$, and set $\mathsf{crs} = \rho_1$, $(pk_1, sk_1) \leftarrow \mathsf{Gen}(1^n; u_1)$, $(pk_2, sk_2) \leftarrow \mathsf{Gen}(1^n; u_2)$, $g_1 \leftarrow \mathsf{Enc}_{pk_1}(M_0; v_1)$ and $g_2 \leftarrow \mathsf{Enc}_{pk_2}(M_1; v_2)$, where $(M_0, M_1) \leftarrow \mathsf{Sam}(1^n; r)$.
2. Define $M' = (C', \mathsf{crs}, pk_1, pk_2, g_1, g_2)$, and output whatever $\mathcal{D}_n(r, M')$ outputs. (Recall that $\mathcal{D}_n$ is the given distinguisher).

By construction of $\mathsf{Sam}_n^{\mathrm{NC}^1}$, we can see that if C' is a correctly generated obfuscation of C_b, then M' is distributed as in hybrid H_{b+1}. It follows that if outputs of H_1 and H_2 are distinguishable then $\mathcal{D}_n^{\mathrm{NC}^1}$ is a good distinguisher against $\mathcal{O}_{\mathrm{NC}^1}$ w.r.t. $\mathsf{Sam}_n^{\mathrm{NC}^1}$.

Final step: $H_2 \approx_c H_3$ and $H_3 \approx_c H_4$. Proof for the claim $H_2 \approx_c H_3$ is nearly identical to step 1. The proof for $H_3 \approx_c H_4$ is nearly identical to step 2. We omit the details. ∎

5 Functional Encryption for Turing Machines

In this section, we shall construct a functional encryption scheme. The scheme can encrypt messages of arbitrary polynomial length. The secret key SK_M is given corresponding to a TM $\mathcal{M}$ of polynomial size which can accept inputs of arbitrary polynomial length and halts in polynomial time. The holder of SK_M can learn the value of $M(x)$ given an encryption of x. The size of the public-parameters of our scheme is polynomial in the security parameter, and the size of secret-keys, say SK_M, is polynomial in the security parameter, $|M|$, and $\log t$ where t is an arbitrary polynomial bounding the worst case running time of M.

We assume familiarity with the definition of functional encryption (FE) schemes. Our scheme will satisfy indistinguishability based notion of security in the selective model of security which we recall here. In this model, we consider the following experiment Expt between an attacker A and a challenger. The experiment takes a bit b as input, and proceeds as follows:

Init A sends two messages x_0^*, x_1^* such that $|x_0^*| = |x_1^*|$.

Phase 1 The challenger samples $(\mathsf{pp}, \mathsf{msk}) \leftarrow \mathsf{F.Setup}(1^n)$ and sends pp to A.

Phase 2 A adaptively asks polynomially secret-key queries where in each query it sends the description of a TM $M \in \mathcal{M}$ such that $M(x_1^*) = M(x_2^*)$.[2] The challenger responds with $SK_M \leftarrow \mathsf{F.KeyGen}(\mathsf{pp}, \mathsf{msk}, M)$.

Challenge The challenger sends an encryption $e = \mathsf{F.Enc}(\mathsf{pp}, x_b^*)$.

Phase 3 Phase 2 is repeated.

Output The output A is the output of the experiment, which is a bit without loss of generality.

The scheme is said to be selectively secure if Adv_A is negligible in n where we define $\mathsf{Adv}_A := |\Pr[\mathsf{Expt}(0) = 1] - \Pr[\mathsf{Expt}(1) = 1]|$.

Our Construction. Let $\mathcal{O}$ be a public-coin differing-inputs obfuscation for the class of polynomial-size and polynomial-time TMs taking inputs of arbitrary polynomial length. Let $\Pi = (\mathsf{CRSGen}, P, V)$ be a statistically sound, non-interactive, *strong* witness-indistinguishable proof system for **NP** where CRSGen simply outputs its own random coins—therefore, we are in the common *random* string model where any random string of sufficient length can act as the CRS. Let $\mathcal{H} = \{\mathcal{H}_n\}$ be a family of collision-resistant hash functions such that every $h \in \mathcal{H}_n$ maps strings from $\{0,1\}^*$ to $\{0,1\}^n$.

---

[2] Recall that according to our convention, the output of M on an input x includes its running time as well.

Let PKE $=$ (Gen, Enc, Dec) be an ordinary, semantically secure public-key encryption scheme, and com be a statistically binding commitment scheme.[3] We assume that PKE (resp., com) encrypts (resp., commits) to a string of unbounded polynomial length by individually encrypting (resp., committing) to each bit. We assume w.l.o.g. that PKE (resp., com) uses randomness of length n to encrypt (resp., commit) to a single bit (and therefore sn random bits for a string of length s will be needed).

The algorithms of our functional encryption scheme are as follows. Recall that $a\|b$ denotes the concatenation of two bit strings a and b.

- F.Setup(1^n): Generate $(pk_1, sk_1) \leftarrow$ Gen(1^n), $(pk_2, sk_2) \leftarrow$ Gen(1^n), and $(pk_3, sk_3) \leftarrow$ Gen(1^n). Generate two commitments $\alpha_1 = $ com($0^n; u_1$) and $\alpha_2 = $ com($0^n; u_2$). Sample crs $\leftarrow$ CRSGen(1^n) and $h \leftarrow \mathcal{H}_n$. Output (pp, msk) where:
$$\text{pp} := (pk_1, pk_2, pk_3, \alpha_1, \alpha_2, \text{crs}, h), \qquad \text{msk} := sk_1.$$

- F.Enc(pp, x): On input a message $x \in \{0,1\}^*$ of arbitrary polynomial length, generate two ciphertexts $c_1 = $ Enc$_{pk_1}(x; r_1)$ and $c_2 = $ Enc$_{pk_2}(x; r_2)$. Define string $a := x\|r_1\|0^{n^2}\|x\|r_2\|0^{n^2}$ and encrypt it under the third public-key to get ciphertext $c_3 = $ Enc$_{pk_3}(a; r_3)$. Finally, compute a proof π for the statement that $y \in L_{\text{FE}}$ using $w = (a, r_3)$ as the witness where $y = (c_1, c_2, c_3, pk_1, pk_2, pk_3, \alpha_1, \alpha_2)$: i.e., $\pi \leftarrow P(\text{crs}, y, w)$.[4] Here L_{FE} is the language corresponding to the relation R_{FE} defined below.

Relaton R_{FE}:
Instance: $y' = (c_1', c_2', c_3', pk_1', pk_2', pk_3', \alpha_1', \alpha_2')$
Witness: $w' = (a', r_3')$ where $a' = x_1'\|r_1'\|u_1'\|x_2'\|r_2'\|u_2'$

$R_{\text{FE}}(y', w') = 1$ if and only if the following condition holds:

1. $c_3' = $ Enc$_{pk_3}(a'; r_3')$; AND

2. The OR of the following two statements is true:

   (a) c_1', c_2' encrypt the same message which is one of x_1' or x_2', i.e.:
   $(c_1' = $ Enc$_{pk_1'}(x_1'; r_1')$ AND $c_2' = $ Enc$_{pk_2'}(x_1'; r_2')$); OR
   $(c_1' = $ Enc$_{pk_1'}(x_2'; r_1')$ AND $c_2' = $ Enc$_{pk_2'}(x_2'; r_2')$);
   (b) c_1', c_2' encrypt x_1', x_2' respectively, which may be different but then the hash of one them is committed in α_1', α_2'; i.e.,
   i. $(c_1' = $ Enc$_{pk_1'}(x_1'; r_1')$ AND $c_2' = $ Enc$_{pk_2'}(x_2'; r_2')$); AND
   ii. $(\alpha_1' = $ com($h(x_1'); u_1'$) OR $\alpha_1' = $ com($h(x_2'); u_1'$)); AND
   iii. $(\alpha_2' = $ com($h(x_1'); u_2'$) OR $\alpha_2' = $ com($h(x_2'); u_2'$)).

Proof π is computed for the AND of statements 1 and 2(a) of R_{FE}. The algorithm outputs $e = (c_1, c_2, c_3, \pi)$ as the ciphertext.

---

[3] We view com as a non-interactive algorithm; we can also use two-round schemes where the first message is fixed as part of the public-parameters by the setup algorithm and then com is viewed w.r.t. such a fixed message.

[4] Observe that here no a-priori bound is known on $|x|$; any multi-theorem proof system such as [FLS99] is capable of proving statements of unbounded polynomial length.

- F.KeyGen(pp, msk, M): The secret-key SK_M corresponding to a TM M is an obfuscation of the program $\mathsf{Prog}_{M,\mathsf{msk}}$, i.e., $SK_M \leftarrow \mathcal{O}\left(1^n, \mathsf{Prog}_{M,\mathsf{msk}}\right)$, where $\mathsf{Prog}_{M,\mathsf{msk}}$ is the following program.

  Program $\mathsf{Prog}_{M,\mathsf{msk}}$:
  - Input: a ciphertext e of the form $e = (c_1, c_2, c_3, \pi)$.
  - Constants: $\mathsf{msk} = sk_1$ and $\mathsf{pp} = (pk_1, pk_2, pk_3, \alpha_1, \alpha_2, \mathsf{crs}, h)$.
  - The program checks that $1 = V(\mathsf{crs}, y, \pi)$ where $y = (c_1, c_2, c_3, pk_1, pk_2, pk_3, \alpha_1, \alpha_2)$.
  - If the check fails, output $\bot$; otherwise output $M\left(\mathsf{Dec}_{sk_1}(c_1)\right)$.

- F.Dec(SK_M, e): Evaluate the program SK_M on input e and output whatever it outputs.

Theorem 2. *Let $\mathcal{M}$ be the class of all polynomial-time Turing machines accepting inputs of unbounded polynomial length. If there exists a public-coin differing-inputs obfuscator for the class $\mathcal{M}$, a non-interactive zero-knowledge proof system (i.e., with statistical soundness) for $\mathbf{NP}$ in the common random string model, a public-key encryption scheme, a non-interactive perfectly-binding commitment scheme, and a family of collision-resistant hash functions with publicly samplable index, then there exists a selectively-secure functional encryption scheme with indistinguishability-based security for Turing machines in the class $\mathcal{M}$.*

Proof of Theorem 2. The correctness and succinctness of our scheme is easy to verify, and in particular, is similar to analyses in [GGH+13b, ABG+13]. We shall provide this analysis in the full version. We now analyze the security of this construction. We prove that the scheme satisfies indistinguishability based security for FE in the selective security model. We prove this by considering the following sequence of hybrid experiments:

- Hybrid H_0 : This hybrid is identical to experiment Expt(0). The public-parameters pp in Phase 1 are of the form $\mathsf{pp} := (pk_1, pk_2, pk_3, \alpha_1, \alpha_2, \mathsf{crs}, h)$ where $\alpha_1 = \mathsf{com}(0^n; u_1)$ and $\alpha_2 = \mathsf{com}(0^n; u_2)$.
- Hybrid H_1 : This hybrid is identical to H_0 except that α_1 and α_2 are computed as commitments to $h(x_0^*)$ and $h(x_1^*)$ respectively: $\alpha_1 = \mathsf{com}(h(x_0^*); u_1)$ and $\alpha_2 = \mathsf{com}(h(x_1^*); u_2)$. We recall that the challenge ciphertext is of the form (c_1, c_2, c_3, π) where both c_1, c_2 encrypt x_0^*, c_3 encrypts $a_0 = x_0^* \| r_1 \| 0^{n^2} \| x_0^* \| r_2 \| 0^{n^2}$ using randomness r_3, and π is computed using $w = (a_0, r_3)$ as the witness. This is identical to how these values were computed in the previous hybrid.
- Hybrid H_2 : Identical to H_2 except that string a_0 is now changed to $a^* = x_0^* \| r_1 \| u_1 \| x_1^* \| r_2 \| u_2$. Consequently, ciphertext c_3 is an encryption of a^* which we denote by c_3^*. Since a^* has changed, the witness used in computing the proof π has also changed, and we shall denote the new proof by π^*. The challenge ciphertext is therefore (c_1, c_2, c_3^*, π^*) where both c_1, c_2 still encrypt x_0^*.
- Hybrid H_3 : Same as H_2 except that c_2 now encrypts x_1^*. Furthermore, π^* is computed w.r.t. the AND of statements 1 and 2(b) (see the description of R_{FE}). That is, the witness corresponding to condition 2(b.i) will now be

(x_0^*, r_1, x_1^*, r_2), and for 2(b.ii) and 2(b.iii) it will be (x_0^*, u_1) and (x_1^*, u_2) respectively. Note that a^*, c_3^* and everything else remains the same.

- Hybrid H_4 : Same as H_3 except that the keys are generated differently. The challenger sets $\mathsf{msk} = sk_2$ and answers the secret-key queries corresponding to a TM M by obfuscating the following program $\mathsf{Prog}_{M,\mathsf{msk}}^*$:

    Program $\mathsf{Prog}_{M,\mathsf{msk}}^*$: The program is identical to $\mathsf{Prog}_{M,\mathsf{msk}}$ except that it decrypts the second ciphertext using $\mathsf{msk} = sk_2$ if the check succeeds. That is, the input to the program is a ciphertext $e = (c_1, c_2, c_3, \pi)$, the constants are $(\mathsf{msk}, \mathsf{pp})$. The program outputs $\perp$ if π is not a valid proof; otherwise it outputs $M(\mathsf{Dec}_{sk_2}(c_2))$.

- Hybrid H_5 Same as H_4 except that c_1 is now changed to encrypt x_1^*. Furthermore, π^* is computed by using the witness corresponding to condition 2(a), i.e., using (x_1^*, r_1, r_2).
- Hybrid H_6 : Same as H_5 except that all secret-key queries are now switched back to using $\mathsf{msk} = sk_1$ and the key for TM M is an obfuscation of the program $\mathsf{Prog}_{M,\mathsf{msk}}$.
- Hybrid H_7: Same as H_6 except that a^* is changed to string $a_1 = x_1^* \| r_1 \| 0^{n^2} \| x_1^* \| r_2 \| 0^{n^2}$. The witness corresponding to 2(a) does not change, but corresponding to statement in 1 changes (see R_{FE}). Therefore, proof π also changes.
- Hybrid H_8: Same as H_7 except that α_1, α_2 are switched back to the commitments of 0^n. Observe that H_8 is identical to the experiment $\mathsf{Expt}(1)$.

We now prove the indistinguishability of every two consecutive hybrids in this experiment.

Step 1: $H_0 \approx_c H_1$. This follows from computational hiding of the commitment scheme. Formally, we consider the following adversary A_{com}, which internally executes the hybrid H_0 except that it does not generate commitments (α_1, α_2) on its own. Instead, after receiving values (x_1^*, x_2^*) during **Init** phase from A, it sends two sets of strings, namely $(0^n, 0^n)$ and $(h(x_1^*), h(x_2^*))$, to the outside challenger and receives in return two commitments (α_1, α_2) corresponding to either the first or the second set of strings. It is clear that A_{com} is a polynomial time machine, and violates the hiding of com unless $H_0 \approx_c H_1$.

Step 2: $H_1 \approx_c H_2$. The proof of this claim relies on the semantic security of PKE and the *strong* witness indistinguishability of the proof system Π for polynomially many statements.[5] Recall that strong WI asserts the following: let $\mathcal{D}_0$ and $\mathcal{D}_1$ be distributions which output an instance-witness pair for an **NP**-relation $\mathbf{R}$ and suppose that the first components of these distributions are computationally indistinguishable, i.e., $\{y : (y, w) \leftarrow \mathcal{D}_0(1^n)\} \approx_c \{y : (y, w) \leftarrow \mathcal{D}_1(1^n)\}$; then $\mathcal{X}_0 \approx_c \mathcal{X}_1$ where $\mathcal{X}_b : \{(\mathsf{crs}, y, \pi) : \mathsf{crs} \leftarrow \mathsf{CRSGen}(1^n); (y, w) \leftarrow \mathcal{D}_0(1^n); \pi \leftarrow$

---

[5] Strictly speaking, we only need strong WI w.r.t. a single statement y of *unbounded polynomial* length. Any multi-theorem NIZK proof system such as [FLS99] generically yields a strong WI proof system for unbounded polynomial length statements.

$P(\mathsf{crs}, y, w)\}$ for $b \in \{0, 1\}$. Strong WI for polynomially many statements is implied by any multi-theorem NIZK proof such as [FLS99].

Suppose that H_1 and H_2 can be distinguished with noticeable advantage δ. Observe that both distribution internally sample the following values in an identical manner: $z := (h, pk_1, pk_2, pk_3, c_1, c_2, \alpha_1, \alpha_2)$ which is all but crs, c_3 and π. By simple averaging, there are at least $\delta/2$ fraction of string st s.t. the two hybrids can be distinguished with advantage at least $\delta/2$ when $z = st$. Call such a z good. Fix one such z, and denote the resulting hybrids by $H_1^{(z)}, H_2^{(z)}$. Note that the hybrids have inbuilt into them all other values used to sample z namely: (x_0^*, x_1^*) received from A, randomness u_1, u_2, r_1, r_2 for $(\alpha_1, \alpha_2, c_1, c_2)$ respectively, and $\mathsf{msk} = sk_1$.

Define distribution $\mathcal{D}_0^{(z)}$ as follows: set $a_0 = (x_0^* \| r_1 \| 0^{n^2} \| x_0^* \| r_2 \| 0^{n^2})$, compute $c_3 = \mathsf{Enc}_{pk_3}(a_0; r_3)$, and let statement $y = (c_1, c_2, c_3, pk_1, pk_2, pk_3, \alpha_1, \alpha_2)$, witness $w_0 = (a_0, r_3)$; output (y, w_0). Note that y has identical to z except that h has been removed and c_3 has been added. Now define a second distribution $\mathcal{D}_1^{(z)}$ which is identical to $\mathcal{D}_0^{(z)}$ except that instead of string a_0, it uses string $a^* = (x_0^* \| r_1 \| u_1 \| x_0^* \| r_2 \| u_2)$, sets $c_3 = \mathsf{Enc}_{pk_3}(a^*; r_3)$, and $w = (a^*, r_3)$. It follows from the security of the encryption scheme that the distribution of y sampled by $\mathcal{D}_0^{(z)}$ is computationally indistinguishable from when it is sampled by $\mathcal{D}_1^{(z)}$. Therefore, we must have that $\mathcal{X}_0 \approx \mathcal{X}_1$ w.r.t. these distributions. We show that this is not the case unless $H_1^{(z)} \approx_c H_2^{(z)}$.

Consider an adversary for strong WI who incorporates A and z (along with sk_1 and all values for computing z described above), and receives a challenge (crs, y, π) distributed according to either $D_0^{(z)}$ or $D_1^{(z)}$; here y has a component c_3 (and all other parts of y are identical to the respective parts of $z \setminus \{h\}$). The adversary uses crs to completely define pp and feeds it to A; it uses sk_1 to complete phase 2 and 4, and (c_3, π) to define the challenge ciphertext $e = (c_1, c_2, c_3)$. The adversary outputs whatever A outputs. We observe that the output of this adversary is distributed according to $H_1^{(z)}$ (resp., $H_2^{(z)}$) when it receives a tuple from distribution $\mathcal{X}_0$ (resp., $\mathcal{X}_1$). A randomly sampled z is good with probability at least $\delta/2$, and therefore it follows that with probability at least $\delta^2/4$ the strong WI property will be violated unless δ is negligible.

Step 3: $H_2 \approx_c H_3$. The proof of this part follows exactly the same ideas as in step 2, and relies on the semantic security of encryption and strong WI property of Π. Roughly speaking, changing c_2 to encrypt x_1^* results in a computationally indistinguishable distribution over the statement (to be proven by proof π). Due to this, although the resulting proof π will use a different witness, strong WI guarantees that the joint distribution of statement and proof (present in the challenge ciphertext) remains computationally indistinguishable in these two hybrids. The details are omitted.

Step 4: $H_3 \approx_c H_4$. This is the key part of our proof where we shall rely on the indistinguishability security of public-coin diO. Suppose that the claim is false and A's output in H_3 is noticeably different from its output in H_4. Suppose that A's running time is bounded by a polynomial t so that there are at most t secret-key queries it can make in phase 2 and 3 combined. We consider a sequence of t hybrid experiments between H_3 and H_3 such that hybrid H_3^i, for $i \in [t]$ is as follows.

Hybrid H_3^i is identical to H_3 except that it answers the secret-key queries as follows. For $j \in [t]$, if $j \leq i$, the secret-key corresponding to the j-th query, denoted M_j, is an obfuscation of program Prog_{M_j,sk_1}; otherwise, for $j > i$, it is an obfuscation of program $\mathsf{Prog}^*_{M_j,sk_2}$. We define H_3^0 to be H_3 and observe that H_3^t is the same as H_4.

By simple calculation, we see that if A's advantage in distinguishing H_3 and H_4 is δ, then there exists an $i \in [t]$ such that A distinguishes between H_3^{i-1} and H_3^i with advantage at least δ/t. We show that if δ is not negligible than we can us A to violate the indistinguishability of $\mathcal{O}$. To do so, we define a sampling algorithm Sam_A^i and a distinguishing algorithm $\mathcal{D}_A^i$ and prove that Sam_A^i is a public-coin differing-input sampler outputting a pair of differing-input TMs yet $\mathcal{D}_A^i$ can distinguish an obfuscation of left TM from that of right output by Sam_A^i. The description of these two algorithms is as follows:

Sampler $\mathsf{Sam}_A^i(\rho)$:
1. Receive (x_0^*, x_1^*) from A.
2. Parse ρ as (crs, h, τ).
3. Proceed identically to H_4 using τ as randomness for all tasks except for sampling the hash function which is set to h, and the CRS, which is set to crs. This involves the following steps:
   (a) Parse $\tau = (\tau_1, \tau_2, \tau_3, r_1, r_2, r_3, u_1, u_3)$.
   (b) Use τ_i as randomness to generate $(pk_i, sk_i) \leftarrow \mathsf{Gen}(1^n; \tau_i)$ for $i \in [3]$, r_1, r_2 to generate $c_1 = \mathsf{Enc}_{pk_1}(x_0^*; r_1)$, $c_2 = \mathsf{Enc}_{pk_1}(x_1^*; r_2)$, and u_1, u_2 to generate $\alpha_1 = \mathsf{com}(h(x_0^*); u_1)$, $\alpha_2 = \mathsf{com}(h(x_1^*); u_2)$.
   (c) Use $a^* = x_0^* \| r_1 \| u_1 \| x_1^* \| r_2 \| u_2$ and r_3 to compute $c_3^* = \mathsf{Enc}_{pk_3}(a^*; r_3)$, and then use $w = (a^*, r_3)$ to compute proof π^* corresponding to conditions 1 and 2(b) in R_{FE}.
4. Define $\mathsf{pp} = (pk_1, pk_2, pk_3, \alpha_1, \alpha_2, \mathsf{crs}, h)$ and challenge $e = (c_1, c_2, c_3^*, \pi^*)$.
5. Send pp to A and answer its secret-key queries as follows. For all queries M_j until $j < i$, send an obfuscation of Prog_{M,sk_1}.
6. If i-th secret-key query comes in phase 2, send ciphertext e in the challenge phase.
7. Upon receiving the i-th secret-key query M_i, output (M_0, M_1) and halt where:
$$M_0 := \mathsf{Prog}_{M_i,sk_1}, \quad M_1 := \mathsf{Prog}^*_{M_i,sk_2}.$$

Distinguisher $\mathcal{D}_A^i(\rho, M')$: on input a random tape ρ and an obfuscated TM M', the distinguisher simply executes all steps of the sampler $\mathsf{Sam}_A^i(\rho)$, answering secret-keys for all $j < i$, as described above. The distinguisher,

however, does not halt when i-th query is sent, and continues the execution of A answering secret-key queries for M_j as follows:
- if $j = i$: send M' (which is an obfuscation of either M_0 or M_1)
- if $j > i$: send an obfuscation of $\mathsf{Prog}^*_{M_j,sk_2}$

The distinguisher outputs whatever A outputs.

It is straightforward to see that if M' is an obfuscation of M_1, the output of $\mathcal{D}^i_A$ is identical to A's output in H^{i-1}_3; and if M' is an obfuscation of M_0, it is identical to A's output in H^i_3. We have that $\mathcal{D}^i_A$ distinguishes H^{i-1}_3 and H^i_3 with at least δ/t advantage. All that remains to prove now is that Sam^i_A is a public-coin differing-inputs sampler.

Claim. Sam^i_A is a public-coin differing-inputs sampler.

Proof. We show that if there exists an adversary $\mathcal{B}$ who can find differing-inputs to the pair of TMs sampled by Sam^i_A with noticeable probability, say μ, we can use $\mathcal{B}$ and Sam^i_A to construct an efficient algorithm $\mathsf{CollFinder}_{\mathcal{B},\mathsf{Sam}^i_A}$ which finds collisions in h with probability $\mu - \mathsf{negl}(n)$. The algorithm works as follows:

$\mathsf{CollFinder}_{\mathcal{B},\mathsf{Sam}^i_A}(h)$:

The algorithm incorporates $\mathcal{B}, \mathsf{Sam}^i_A$. On input a random hash function $h \leftarrow \mathcal{H}_n$, the algorithm works as follows:
- sample uniformly random strings (crs, τ) to define a random tape $\rho := (\mathsf{crs}, h, \tau)$.
- sample $(M_0, M_1) \leftarrow \mathsf{Sam}^i_A(\rho)$ and $e \leftarrow \mathcal{B}(\rho)$.
- recall that e is of the form (c_1, c_2, c_3, π) where c_3 is an encryption under pk_3 where (pk_3, sk_3) are sampled using parts of the randomness τ.
- let $x^*_0 \neq x^*_1$ be the strings output by A during the **Init** phase in the execution of Sam^i_A.
- if π is a valid proof, compute $a = \mathsf{Dec}_{sk_3}(c_3)$ and let $a = x'_1 \| r'_1 \| u'_1 \| x'_2 \| r'_2 \| u'_2$.
- if $h(x^*_0) = h(x^*_1)$, output (x^*_0, x^*_1) as the collisions; otherwise, if $\exists\, m_1 \in \{x'_1, x'_2\}$ and $\exists\, m_2 \in \{x^*_0, x^*_1\}$ s.t. $m_1 \neq m_2$ and $h(m_1) = h(m_2)$ output (m_1, m_2) as collisions; if none of the two conditions hold, output $\perp$.

Let us now analyze the success probability of this algorithm. Since h is uniformly sampled, ρ is a uniform random tape, and therefore with probability μ, $\mathcal{B}$ outputs an e such that $M_0(e) \neq M_1(e)$. Recall that $M_0 = \mathsf{Prog}_{M_i,sk_1}$ and $M_2 = \mathsf{Prog}^*_{M_i,sk_2}$ for some TM M_i such that $M_i(x^*_0) = M^*_i(x^*_1)$. Furthermore, both of these programs output $\perp$ if proof π is not valid. Since the output these two programs differ on e, it must be that π is a valid proof so that $M_0(e) = M_i(\mathsf{Dec}_{sk_1}(c_1))$ and $M_1(e) = M_i(\mathsf{Dec}_{sk_2}(c_2))$. By construction, since π is a statistically sound proof, except with negligible probability it holds that $x'_1 = \mathsf{Dec}_{sk_1}(c_1))$ and $x'_2 = dec_{sk_2}(c_2)$ where x'_1, x'_2 are part of the string a obtained by the collision finding algorithm by decrypting c_3 above. Therefore, we have that $M_0(e) = M_i(x'_1)$ and $M_1(e) = M_i(x'_2)$. However, we also have that

$M_0(e) \neq M_1(e) \implies M_i(x_1') \neq M_i(x_2') \implies x_1' \neq x_2'$. Since $M_i(x_0^*) = M_i(x_1^*)$ it holds that the sets $\{x_1', x_2'\} \neq \{x_0^*, x_1^*\}$.

Since π is valid, and c_1, c_2 are encryptions of (unequal strings) x_1', x_2', from the statistical soundness of π statements 2(b.ii) and 2(b.iii) must be true. That is, α_1 (likewise α_2) must be a commitment to one of $h(x_1')$ or $h(x_2')$. But α_1 is a commitment to $h(x_0^*)$ and α_2 is a commitment to $h(x_1^*)$ and commitment is statistically binding. Since at least one of x_1', x_2' is not equal to any of x_0^*, x_1^* the collision must occur on one of the four possible pairs of these strings. □

Step 5: Indistinguishability of H_4–H_8. Hybrids H_4 to H_8 are applying changes very similar to the first four hybrids except in the reverse order. The proof of their indistinguishability can be obtained by following previous proofs in a near identical fashion. In particular we can prove $H_4 \approx_c H_5$ by relying on the security of encryption and strong WI (following the proof in step 2 or 3), $H_5 \approx_c H_6$ following the proof in step 4, $H_6 \approx_c H_7$ following the proof in step 2, and $H_7 \approx_c H_8$ following the proof in step 1.

This completes the proof of security of our functional encryption scheme. ■

References

[ABG+13] Ananth, P., Boneh, D., Garg, S., Sahai, A., Zhandry, M.: Differing-inputs obfuscation and applications. IACR Cryptology ePrint Archive, 2013 (2013)

[AIK06] Applebaum, B., Ishai, Y., Kushilevitz, E.: Cryptography in nc^0. SIAM J. Comput. 36(4), 845–888 (2006)

[App11] Applebaum, B.: Randomly encoding functions: A new cryptographic paradigm. In: Fehr, S. (ed.) ICITS 2011. LNCS, vol. 6673, pp. 25–31. Springer, Heidelberg (2011)

[BC14] Bitansky, N., Canetti, R.: On strong simulation and composable point obfuscation. J. Cryptology 27(2), 317–357 (2014)

[BCC+14] Bitansky, N., Canetti, R., Chiesa, A., Goldwasser, S., Lin, H., Rubinstein, A., Tromer, E.: The hunting of the SNARK. IACR Cryptology ePrint Archive, 2014:580 (2014)

[BCCT12] Bitansky, N., Canetti, R., Chiesa, A., Tromer, E.: From extractable collision resistance to succinct non-interactive arguments of knowledge, and back again. In: Innovations in Theoretical Computer Science 2012, Cambridge, MA, USA, January 8-10, pp. 326–349 (2012)

[BCCT13] Bitansky, N., Canetti, R., Chiesa, A., Tromer, E.: Recursive composition and bootstrapping for SNARKS and proof-carrying data. In: Symposium on Theory of Computing Conference, STOC 2013, Palo Alto, CA, USA, June 1-4, pp. 111–120 (2013)

[BCKP14] Bitansky, N., Canetti, R., Kalai, Y.T., Paneth, O.: On virtual grey box obfuscation for general circuits. In: Garay, J.A., Gennaro, R. (eds.) CRYPTO 2014, Part II. LNCS, vol. 8617, pp. 108–125. Springer, Heidelberg (2014)

[BCP14] Boyle, E., Chung, K.-M., Pass, R.: On extractability obfuscation. In: Lindell, Y. (ed.) TCC 2014. LNCS, vol. 8349, pp. 52–73. Springer, Heidelberg (2014), Preliminary version on Eprint 2013:
http://eprint.iacr.org/2013/650.pdf

[BCPR14] Bitansky, N., Canetti, R., Paneth, O., Rosen, A.: On the existence of extractable one-way functions. In: Symposium on Theory of Computing, STOC, New York, NY, USA, May 31 - June 03, pp. 505–514 (2014)

[BG02]   Barak, B., Goldreich, O.: Universal arguments and their applications. In: Annual IEEE Conference on Computational Complexity (CCC), vol. 17 (2002), Preliminary full version available as Cryptology ePrint Archive, Report 2001/105

[BGI+12] Barak, B., Goldreich, O., Impagliazzo, R., Rudich, S., Sahai, A., Vadhan, S.P., Yang, K.: On the (im)possibility of obfuscating programs. J. ACM 59(2), 6 (2012)

[BGK+14] Barak, B., Garg, S., Kalai, Y.T., Paneth, O., Sahai, A.: Protecting obfuscation against algebraic attacks. In: Nguyen, P.Q., Oswald, E. (eds.) EUROCRYPT 2014. LNCS, vol. 8441, pp. 221–238. Springer, Heidelberg (2014)

[BGT14]   Bitansky, N., Garg, S., Telang, S.: Succinct randomized encodings and their applications. Cryptology ePrint Archive, Report 2014/771 (2014), http://eprint.iacr.org/

[BHK13]   Bellare, M., Hoang, V.T., Keelveedhi, S.: Instantiating random oracles via uces. In: Canetti, R., Garay, J.A. (eds.) CRYPTO 2013, Part II. LNCS, vol. 8043, pp. 398–415. Springer, Heidelberg (2013)

[BHR12]   Bellare, M., Hoang, V.T., Rogaway, P.: Foundations of garbled circuits. In: The ACM Conference on Computer and Communications Security, CCS 2012, Raleigh, NC, USA, October 16-18, pp. 784–796 (2012), Cryptoglogy Eprint Archive Report 2012/265

[BP13]   Boyle, E., Pass, R.: Limits of extractability assumptions with distributional auxiliary input. Cryptology ePrint Archive, Report 2013/703 (2013), http://eprint.iacr.org/2013/703.pdf

[BR93]   Bellare, M., Rogaway, P.: Random oracles are practical: A paradigm for designing efficient protocols. In: ACM Conference on Computer and Communications Security, pp. 62–73 (1993)

[BR14]   Brakerski, Z., Rothblum, G.N.: Virtual black-box obfuscation for all circuits via generic graded encoding. In: Lindell, Y. (ed.) TCC 2014. LNCS, vol. 8349, pp. 1–25. Springer, Heidelberg (2014), Preliminary version on Eprint at http://eprint.iacr.org/2013/563.pdf

[BST14]   Bellare, M., Stepanovs, I., Tessaro, S.: Poly-many hardcore bits for any one-way function and a framework for differing-inputs obfuscation. In: Sarkar, P., Iwata, T. (eds.) ASIACRYPT 2014, Part II. LNCS, vol. 8874, pp. 102–121. Springer, Heidelberg (2014), Earlier version: IACR Cryptology ePrint Archive 2013:873 (December 2013)

[BSW11]   Boyle, E., Segev, G., Wichs, D.: Fully leakage-resilient signatures. In: Paterson, K.G. (ed.) EUROCRYPT 2011. LNCS, vol. 6632, pp. 89–108. Springer, Heidelberg (2011)

[Can97]   Canetti, R.: Towards realizing random oracles: Hash functions that hide all partial information. In: Kaliski Jr., B.S. (ed.) CRYPTO 1997. LNCS, vol. 1294, pp. 455–469. Springer, Heidelberg (1997)

[CGH04]   Canetti, R., Goldreich, O., Halevi, S.: The random oracle methodology, revisited. J. ACM 51(4), 557–594 (2004)

[CHJV14] Canetti, R., Holmgren, J., Jain, A., Vaikuntanathan, V.: Indistinguishability obfuscation of iterated circuits and ram programs. Cryptology ePrint Archive, Report 2014/769 (2014), http://eprint.iacr.org/

[CLT13]   Coron, J.-S., Lepoint, T., Tibouchi, M.: Practical multilinear maps over the integers. In: Canetti, R., Garay, J.A. (eds.) CRYPTO 2013, Part I. LNCS, vol. 8042, pp. 476–493. Springer, Heidelberg (2013)

[FKN94]   Feige, U., Kilian, J., Naor, M.: A minimal model for secure computation. In: STOC, pp. 554–563 (1994)

[FLS99]   Feige, Lapidot, Shamir.: Multiple noninteractive zero knowledge proofs under general assumptions. SIAM Journal on Computing 29 (1999)

[Gen09]   Gentry, C.: Fully homomorphic encryption using ideal lattices. In: STOC, pp. 169–178 (2009)

[GGH13a]   Garg, S., Gentry, C., Halevi, S.: Candidate multilinear maps from ideal lattices. In: Johansson, T., Nguyen, P.Q. (eds.) EUROCRYPT 2013. LNCS, vol. 7881, pp. 1–17. Springer, Heidelberg (2013)

[GGH+13b] Garg, S., Gentry, C., Halevi, S., Raykova, M., Sahai, A., Waters, B.: Candidate indistinguishability obfuscation and functional encryption for all circuits. In: FOCS, pp. 40–49 (2013)

[GGHW14]   Garg, S., Gentry, C., Halevi, S., Wichs, D.: On the implausibility of differing-inputs obfuscation and extractable witness encryption with auxiliary input. In: Garay, J.A., Gennaro, R. (eds.) CRYPTO 2014, Part I. LNCS, vol. 8616, pp. 518–535. Springer, Heidelberg (2014)

[GK05]   Goldwasser, S., Kalai, Y.T.: On the impossibility of obfuscation with auxiliary input. In: FOCS, pp. 553–562 (2005)

[GKP+13a] Goldwasser, S., Kalai, Y.T., Popa, R.A., Vaikuntanathan, V., Zeldovich, N.: How to run turing machines on encrypted data. In: Canetti, R., Garay, J.A. (eds.) CRYPTO 2013, Part II. LNCS, vol. 8043, pp. 536–553. Springer, Heidelberg (2013)

[GKP+13b] Goldwasser, S., Kalai, Y., Popa, R.A., Vaikuntanathan, V., Zeldovich, N.: Succinct functional encryption and applications: Reusable garbled circuits and beyond. In: STOC (2013)

[Gol01]   Goldreich, O.: Foundations of Cryptography: Basic Tools. Cambridge University Press (2001), Earlier version available on http://www.wisdom.weizmann.ac.il/~oded/frag.html

[GPSW06]   Goyal, V., Pandey, O., Sahai, A., Waters, B.: Attribute-based encryption for fine-grained access control of encrypted data. In: CCS (2006)

[GR07]   Goldwasser, S., Rothblum, G.N.: On best-possible obfuscation. In: Vadhan, S.P. (ed.) TCC 2007. LNCS, vol. 4392, pp. 194–213. Springer, Heidelberg (2007)

[GR13]   Goldreich, O., Rothblum, R.D.: Enhancements of trapdoor permutations. J. Cryptology 26(3), 484–512 (2013)

[HR04]   Hsiao, C.-Y., Reyzin, L.: Finding collisions on a public road, or do secure hash functions need secret coins? In: Franklin, M. (ed.) CRYPTO 2004. LNCS, vol. 3152, pp. 92–105. Springer, Heidelberg (2004)

[IK00]   Ishai, Y., Kushilevitz, E.: Randomizing polynomials: A new representation with applications to round-efficient secure computation. In: 41st Annual Symposium on Foundations of Computer Science, FOCS 2000, Redondo Beach, California, USA, November 12-14, pp. 294–304 (2000)

[KLW14]   Koppula, V., Lewko, A.B., Waters, B.: Indistinguishability obfuscation for turing machines with unbounded memory. Cryptology ePrint Archive, Report 2014/925 (2014)

[LP14]   Lin, H., Pass, R.: Succinct garbling schemes and applications. Cryptology ePrint Archive, Report 2014/766 (2014), http://eprint.iacr.org/

[Mic94]   Micali, S.: CS proofs. In: FOCS, pp. 436–453 (1994)

[PPS15]   Pandey, O., Prabhakaran, M., Sahai, A.: Obfuscation-based non-black-box simulation and four message concurrent zero knowledge for NP. In: TCC (2015), Earlier version: IACR Cryptology ePrint Archive 2013:754

[SS10]   Sahai, A., Seyalioglu, H.: Worry-free encryption: functional encryption with public keys. In: CCS, pp. 463–472 (2010)

[SW05]   Sahai, A., Waters, B.: Fuzzy identity-based encryption. In: Cramer, R. (ed.) EUROCRYPT 2005. LNCS, vol. 3494, pp. 457–473. Springer, Heidelberg (2005)

[SW14]   Sahai, A., Waters, B.: How to use indistinguishability obfuscation: deniable encryption, and more. In: STOC, pp. 475–484 (2014)

[Yao82]   Yao, A.C.: Theory and applications of trapdoor functions. In: Proc. 23rd FOCS, pp. 80–91 (1982)

A   Other Primitives

Fully Homomorphic Encryption with Decryption in NC^1. A fully homomorphic encryption (FHE) scheme is a public-key encryption scheme with an additional evaluation algorithm Eval. Formally, given a public-key pk, ciphertexts $c_1, \ldots, c_m$ corresponding to the bits $b_1, \ldots, b_m$ (under pk), and a circuit $f : \{0,1\}^m \to \{0,1\}$, algorithm Eval outputs a ciphertext c' such that except with negligible probability over the randomness of all algorithms, the decryption of c' is the bit $f(b_1, \ldots, b_m)$ where $m = m(n)$ is an arbitrary polynomial.

The encryption of a long message $x \in \{0,1\}^n$ under pk consists of encrypting each bit of x under pk, and will be denoted by $c = \text{Enc}_{pk}(x)$. Given c, the homomorphic evaluation of an *oblivious Turing machine* M with known running time t consists of t homomorphic evaluations of the the circuit corresponding to the transition function of M where in the i-th iteration the transition function is applied on the contents of the encrypted input/work tape (containing x at the start) and an encrypted state; it results in a new encrypted state as well as new encrypted contents on the work tape.

A FHE scheme has decryption in NC^1 if there exists a constant $c \in \mathbb{N}$ such that for all $n \in \mathbb{N}$ the depth of the circuit corresponding to the decryption function $\text{Dec}(1^n, pk, \cdot)$ is at most $c \log n$.

Strong Non-Interactive Witness Indistinguishable Proofs for NP. As a tool for our functional encryption application, we need non-interactive proofs for **NP** in the common random string model. We require that the proof system be capable of proving statements of unbounded polynomial length. In terms of soundness, we require the system to be a *proof* system where the soundness guarantee is statistical: i.e., even unbounded provers cannot prove a false statement with noticeable probability. In terms of prover security, we only require the proof system to satisfy *strong* witness indistinguishability [Gol01] which is implied by the zero-knowledge property. The NIZK proof system of Feige, Lapidot, and Shamir [FLS99] satisfies all of these requirements.

B   Bounded-Input Case

In this section we consider the ℓ-*bounded-input* case, in which we consider the class of TMs whose input is bounded by a polynomial ℓ, and the size of the obfuscation is allowed to depend on ℓ; however it does not depend on the running time of TMs in the class, which could be much larger than ℓ. To emphasize that a class is a bounded-input TM class, we will explicitly include ℓ in the notation.

Definition 7 (Public-Coin Differing-Inputs Obfuscator for ℓ-*Bounded-Input* TMs). *For every polynomial $\ell : \mathbb{N} \to \mathbb{N}$, let $\mathcal{M} = \{\mathcal{M}_n^\ell\}_{n \in \mathbb{N}}$ denote the class of all TMs such that every $M \in \mathcal{M}_n^\ell$ is of size n, accepts inputs of length at most $\ell(n)$, and halts within a polynomial, say $t(n)$, number of steps on all inputs. A uniform PPT algorithm $\mathcal{O}$ is a public-coin differing-inputs obfuscator for the class of all bounded-input TMs if it satisfies the* correctness *and* security *conditions of definition 4 and the following modified succinctness condition: there exists a (global) polynomial $s' : \mathbb{N} \to \mathbb{N}$ such that for all n, for all $M \in \mathcal{M}_n$, and for all $M' \leftarrow \mathcal{O}(1^n, M)$, the size of M' is bounded by $s'(n, \ell(n))$ and its running time is bounded by $s'(n, t(n))$ for all $x \in \{0, 1\}^{\leq \ell(n)}$.*

We show that given a public-coin differing-inputs obfuscator for the class NC^1, we can construct a public-coin differing-inputs obfuscator for *bounded-input* Turing machines' class.

Theorem 3. *Suppose that there exists a public-coin differing-inputs obfuscator for circuits in the class NC^1. Then, assuming the existence of a fully homomorphic encryption scheme with decryption in NC^1 and a (publicly verifiable) SNARG system for $\mathbf{P}$ (alternatively, a $\mathbf{P}$-certificate) in the common random string model, there exists a public-coin differing-inputs obfuscator for bounded-input Turing machines as defined in 7.*

We first present the construction, and then prove the theorem. Let ℓ and t be polynomials, and let $\mathcal{M} = \{\mathcal{M}_n^\ell\}_{n \in \mathbb{N}}$ be the family of bounded-input TMs where every $M \in \mathcal{M}_n^\ell$ is of size n, accepts inputs of length at most $\ell(n)$ and halts within $t(n)$ steps on every x. Let $\mathsf{FHE} = (\mathsf{Gen}, \mathsf{Enc}, \mathsf{Dec}, \mathsf{Eval})$ be a fully homomorphic encryption scheme with decryption in NC^1, and $\Pi = (P, V)$ be a SNARG for the relation $\mathcal{R}_c$ defined earlier where c is a constant such that $t(n) \leq n^c$ for all n. The description of our obfuscator and its evaluation algorithm, are as follows.

Obfuscator $\mathcal{O}\left(1^n, M \in \mathcal{M}_n^\ell\right)$: By convention, description of M includes the bounds t and ℓ. Let U_n be an *oblivious* universal TM which on input the description of a TM B, and a string $x \in \{0, 1\}^{\leq \ell(n)}$ executes B on x for no more than $t(n)$ steps. The obfuscator proceeds in the following steps:

1. Generate two FHE public-keys $(pk_1, sk_1) \leftarrow \mathsf{Gen}(1^n; u_1)$ and $(pk_2, sk_2) \leftarrow \mathsf{Gen}(1^n; u_2)$.
2. Encrypt M under both FHE public-keys: $g_1 \leftarrow \mathsf{Enc}_{pk_1}(M; v_1)$ and $g_2 \leftarrow \mathsf{Enc}_{pk_2}(M; v_2)$. Here M is assumed to be encoded as a bit string of length n for use by the universal TM U_n.

3. Uniformly sample $\mathsf{crs} \leftarrow \{0,1\}^{\mathsf{poly}(n)}$ of sufficient length for SNARG Π.
4. Generate an obfuscation of the $\mathbf{NC}^1$-program $P1^{\mathsf{crs}}_{sk_1,g_1,g_2}$ given in figure 2:

$$P' \leftarrow \mathcal{O}_{\mathbf{NC}^1}\left(1^n, P1^{\mathsf{crs}}_{sk_1,g_1,g_2}\right).$$

5. Output $M' = (P', \mathsf{crs}, pk_1, pk_2, g_1, g_2)$.

Program $P1^{\mathsf{crs}}_{sk_1,g_1,g_2}$:

- **Input:** a tuple (x, e_1, e_2, π, ϕ),   **Constants:** $\mathsf{crs}, sk_1, g_1, g_2, pk_1, pk_2$.
- Check that ϕ is a valid low-depth proof for the **NP**-statement:

$$1 = V\left(\mathsf{crs}, (\widetilde{M}_{\mathsf{Eval}}, \langle x, e_1, e_2 \rangle, t^2), \pi\right)$$

  where $\widetilde{M}_{\mathsf{Eval}}$ simply checks that computation $M_{\mathsf{Eval}}(x)$ outputs (e_1, e_2) in $\leq 2t \log t$ steps. $M_{\mathsf{Eval}}(x)$ is defined as follows: it has (pk_1, pk_2, g_1, g_2) hardcoded, and homomorphically evaluates $U_n(\cdot, x)$ on g_1 and g_2 to produce e_1 and e_2 respectively.
  I.e., $e_1 = \mathsf{Eval}_{pk_1}(U_n(\cdot, x), g_1)$ and $e_2 = \mathsf{Eval}_{pk_2}(U_n(\cdot, x), g_2)$.
- If the check fails, output $\bot$; otherwise output $\mathsf{Dec}_{sk_1}(e_1)$.

Program $P2^{\mathsf{crs}}_{sk_2,g_1,g_2}$:

- Same as $P1^{\mathsf{crs}}_{sk_1,g_1,g_2}$ except that if the check is successful, it outputs $\mathsf{Dec}_{sk_2}(e_2)$.   •

Fig. 2. The programs $P1$ and $P2$

Evaluation of M'. Evaluate $M' = (P', \mathsf{crs}, pk_1, pk_2, g_1, g_2)$ on input x as follows:

1. Let $(e_1, e_2) = M_{\mathsf{Eval}}(x)$. Recall that (fig. 1): $e_1 = \mathsf{Eval}_{pk_1}(U_n(\cdot, x), g_1)$, $e_2 = \mathsf{Eval}_{pk_2}(U_n(\cdot, x), g_2)$
2. W.l.o.g, the running time of $M_{\mathsf{Eval}}(x)$ is at most $2t \log t \leq t^2$. Compute the proof π:[6]

$$\pi \leftarrow P\left(\mathsf{crs}, (\widetilde{M}_{\mathsf{Eval}}, \langle x, e_1, e_2 \rangle, t^2), \bot\right)$$

3. Compute a low-depth proof ϕ that π is a valid SNARG, i.e., $V(\mathsf{crs}, (\widetilde{M}_{\mathsf{Eval}}, \langle \sigma, e_1, e_2 \rangle, t^2), \pi) = 1$. This can be done by providing the entire computation of V on these inputs.
4. Execute $P'(x, e_1, e_2, \pi, \phi)$ and output the result.

The analysis of this construction is similar to the case of unbounded input. It can be found in the Eprint version of this work (Report 2014/942).

---

[6] No witness is necessary as this is computation in **P**.

Author Index